Analysis and Design of Planar Microwave Components

IEEE PRESS
445 Hoes Lane, PO Box 1331
Piscataway, NJ 08855-1331

Analysis and Design of Planar Microwave Components

Edited by

K. C. Gupta

University of Colorado at Boulder

M. D. Abouzahra

Massachusetts Institute of Technology
Lincoln Laboratory

IEEE PRESS

A Selected Reprint Volume
IEEE Microwave Theory and Techniques Society, *Sponsor*

The Institute of Electrical and Electronics Engineers, Inc., New York

This book may be purchased at a discount from the publisher
when ordered in bulk quantities. For more information contact:

IEEE PRESS Marketing
Attn: Special Sales
PO Box 1331
445 Hoes Lane
Piscataway, NJ 08855-1331
Fax: (908) 981-8062

Printed in the United States of America

10 9 8 7 6 5 4 3 2 1

ISBN 0-7803-0437-3

IEEE Order Number: PC0334-3

Library of Congress Cataloging-in-Publication Data

Analysis and design of planar microwave components / edited by K.C.
Gupta, M.D. Abouzahra.
p. cm.
"A selected reprint volume."
"IEEE Microwave Theory and Techniques Society, sponsor."
"IEEE order number: PC0334-3"—T.p. verso.
Includes bibliographical references and indexes.
ISBN 0-7803-0437-3
1. Microwave integrated circuits. 2. Planar components.
3. Electric circuit analysis. I. Gupta, K.C. II. Abouzahra,
Mohamed Deeb. III. IEEE Microwave Theory and Techniques Society.
TK7876.A5 1994
621.381'32—dc20 94-2477
CIP

Contents

Preface

IN recent years, microwave and millimeter-wave integrated circuits have undergone impressive technological advancements. GaAs-based monolithic integrated circuits are now replacing ceramic-based hybrid microwave integrated circuits in several system applications. Both the hybrid and monolithic microwave integrated circuits make extensive use of what has come to be popularly known as "planar components." These components are derived from planar transmission structures such as microstrip lines and coplanar waveguides. Planar components make use of two-dimensional metallization patterns on the surface of a dielectric substrate, or at an interface between two dielectric layers. A number of analysis and design methods for these planar components have been reported in the literature. However, because of the intensive interest in this field (motivated by applications in microwave IC design), research and development efforts are still vigorously active.

This book is an attempt to present an up-to-date overview of analysis and design of planar microwave components. Since the editors have been working only in a subset of topics of interest in this field, the hybrid format has been selected. This book includes 4 chapters of original material written by the editors. The remaining 7 chapters present a collection of various outstanding papers published in the literature. This assemblage is intended to provide the reader with an up-to-date overview of the material. It must be noted that there is a large number of good and useful papers on every topic discussed in this book. However, only a small fraction of these papers could be included here to keep the volume at a manageable size. To partially compensate for the papers not included in this book, additional references are listed in the introduction to each of the chapters which includes reprinted papers.

The analytical techniques reviewed in this book can be divided into two broad classes: two-dimensional field analysis techniques based on the planar waveguide model of microstrip structures and full-wave three-dimensional numerical analysis techniques. The first approach (two-dimensional analysis) is addressed in papers 1.1 and 1.2 in Chapter 1, as well as Chapters 2–4. Full-wave analysis techniques are discussed in papers 1.3 and 1.4 of Chapter 1, sec. 3.3 of Chapter 3, and Chapters 5–6. Chapters 7–11 describe applications of these techniques to the analysis and design of various types of planar components.

Chapter 1 is a collection of four overview papers. Two of these papers describe the two-dimensional analysis approach, while the more recent papers, from 1992, review time-domain and frequency domain full-wave analysis methods.

Chapter 2 is an exposition of the Green's function approach as applied to the two-dimensional analysis of planar components of regular shapes. This approach yields impedance matrix characterization of these components with reference to the number of external ports that exist at selected locations on the edges of, or inside, the components.

Chapter 3 describes multiport network techniques known as segmentation and desegmentation methods. These techniques have been developed specifically for two-dimensional analysis of microwave planar components, and microstrip patch antennas. Application of these methods to full-wave analysis techniques (of three-dimensional fields), which are based on mode-matching approach is discussed in Sec. 3.3.

Chapter 4 is a collection of six reprinted papers which describe applications of the boundary element method to planar components of arbitrary shape. This method is more popularly known as the contour-integral approach. It allows analysis of generalized geometrical shapes unlike the methods discussed in Chapter 2—for regular shapes, and Chapter 3—for composites of regular shapes.

Chapter 5 contains seven reprinted papers on the integral equation approach for three-dimensional field analysis of planar components. This approach employs moment-method solution for current distribution on the metallization of planar components. Characteristics of the components are derived from this current distribution. Both space-domain and spectral-domain formulations are discussed. This method is used frequently, and is the basis for several commercially available electromagnetic simulation software packages.

Chapter 6 emphasizes difference equation–based analyses for planar components. It is a collection of eight papers on finite-difference, finite-element, and transmission line matrix (TLM) methods. The finite-difference and the TLM approaches have been used extensively for time-domain analysis of planar circuits. All these approaches are more generalized formulations, as compared to those in Chapter 5, for incorporating vertically oriented currents in planar circuits. However, larger computer memory allocations and run times are often needed.

Chapter 7 contains reprints on nonreciprocal planar components such as circulators, isolators and so on. The papers selected span the research which has been reported over almost three decades (1964–1992). The nonreciprocal nature of the material used in these components necessitates modifications of the analytical methods discussed in earlier chapters. The contour-integral approach and the finite-element analysis are discussed in some of the included papers.

Chapter 8 deals with characterization of discontinuities present in transmission line structures (like microstrips and coplanar waveguides) when used in microwave integrated circuits. Accurate characterization of these discontinuities and junctions is needed for first-pass success in the design of monolithic MICs. Of the various analytical techniques that may be used for this purpose are quasi-static methods, two-dimensional approach, and full-wave methods. These methods are discussed in the first six chapters of this book.

Chapter 9 addresses another important class of passive components used in microwave integrated circuits. Filters and stubs are present in almost any functional circuit. The seven papers reprinted in this chapter provide a representative sample of the analysis techniques and design used. Two-dimensional analysis is frequently used. Application of the coupled finite–boundary-element method has been reported recently.

Chapter 10 is a description of microwave hybrids, power dividers, and power combiner circuits written specifically for this book. Planar hybrids and *n*-way power dividers/combiners are designed by making extensive use of the two-dimensional analysis techniques discussed in earlier chapters.

Chapter 11 is a review of microstrip patch antennas. These antennas make use of printed circuit technology, and, have thus made it possible to integrate the circuit and antenna functions in a single module. This approach has encouraged the development of a new class of components known as active antennas. This chapter discusses application of two-dimensional and three-dimensional field analysis methods for the design of microstrip patch antennas.

Motivation for this book developed from the research work by the two editors and several of their colleagues. Chapters 2, 3, 10 and 11 incorporate many of these research results, and, are therefore, presented in a detailed and descriptive manner. Other chapters are samples of results from other colleagues, which presents a much broader picture of the field, and allows us to recognize how narrow individual research efforts have become. We are thankful to all of the authors whose research results are included in this volume.

Acknowledgments

PREPARATION of this book has been possible only with the positive contribution of several colleagues. At the University of Colorado, K. C. Gupta acknowledges contributions of several of his former students whose work is reviewed here. Mr. Donald Hasting has provided superb secretarial assistance needed for preparation of this volume. At MIT Lincoln Laboratory, M. D. Abouzahra acknowledges the Department of the Air Force, the Aerospace and the Radar Measurement Divisions, for their support, and Ms. Denise Kolek for editing part of the manuscript. Many thanks are also due to Dr. J. A. Weiss for his suggestions and review of Chapter 7. We are grateful to various publishers and authors for their kind permission to include their papers. Finally, the support of Dr. Kris Agarwal of the IEEE Microwave Theory and Techniques Society and Mr. Dudley Kay of IEEE Press during the evolution of this hybrid book project is thankfully acknowledged.

Chapter 1

Overview Papers

THIS chapter is intended to serve as a general introduction or tutorial on some of the analytical and numerical methods that have been used to analyze planar microwave and millimeter-wave circuits. Four papers are included in this chapter. The first paper (Paper 1.1) is a classic by Okoshi and Miyoshi. It introduces the planar circuit approach as an effective technique for analyzing two-dimensional circuits of arbitrary shape. It also describes how the wave equation, and the associated boundary conditions governing the planar circuit under consideration, can be converted into a contour (or boundary) integral equation. This integral equation is then discretized, resulting in a matrix equation that relates the terminal voltages and currents at the coupling ports. This approach is validated by applying it to triplate-type planar circuits having regular as well as arbitrary shapes. A thorough examination of this approach will be seen in Chapter 4 of this book. In addition to the contour-integral approach, this paper also discusses the Green's function approach, which is appropriate for planar circuits with regular shapes, as it will be seen in Chapter 2 where a more detailed discussion is presented. It should be noted here that Eqs. (21) and (22) in this paper are not complete. For the correct versions, the reader is referred to Eqs. (44) and (166) on pages 227 and 313 of [1.5].

Paper 1.2 is an invited review paper that appeared in a special issue of the *IEEE Transactions on Microwave Theory and Techniques* on the subject of numerical methods [1.1]. This paper illustrates the main features, and also presents a rather good review, of the planar circuit approach and its theoretical basis. The paper starts with Maxwell's equation and then derives a terminal description for the planar circuit in terms of Green's function. It then briefly describes the segmentation and desegmentation techniques as effective network methods for characterizing planar circuits with more complicated planar shapes. An in-depth description of the segmentation-desegmentation techniques is presented later in Chapter 3 of this book. Additional issues such as fringe field effects and radiation loss are also briefly discussed in Paper 1.2.

Recently, there has been increasing interest in applying time-domain electromagnetic simulation techniques to the analysis of planar circuits. The most prominent of these techniques are the finite-difference time-domain (FD-TD) method, the transmission line matrix (TLM) method, and the finite-element method. Paper 1.3 discusses the basic properties and the principal advantages of the various time domain simulators. It highlights the important developments in time-domain field modeling and discusses the practical implementation of these techniques in the form of computer-aided design tools for designing microwave circuits. This paper also gives a critical assessment of the TLM method, from the standpoint of computational speed and accuracy. Various measures to accelerate the simulation speed by several orders of magnitude and hence reduce the computation time are discussed. Finally, trends in the future development of these techniques are also discussed.

In addition to the planar circuit approach and the time-domain methods, there has been a strong trend in applying integral equation techniques to the analysis of planar circuits. The last paper (Paper 1.4) in this chapter is also an invited paper that appeared in a recent special issue of the *International Journal of Microwave and Millimeter-Wave Computer-Aided Engineering* [1.2] on the subject of electromagnetic simulation of planar microwave and millimeter-wave circuits. This paper reviews the theoretical basis of the full-wave spectral-domain integral equation approach as applied to open microstrip discontinuities of arbitrary shapes. It outlines how the appropriate current expansion functions can be chosen and then used to solve the resulting integral equation by means of the method of moments approach. After finding the current distribution on the planar component, the scattering parameters are determined. The flexibility and accuracy as well as the utility of this approach is demonstrated by using it to investigate several practical planar structures.

Additional tutorial material on the numerical and analytical techniques that have been shown to be suitable for the analysis of planar circuits can be found in books [1.3–1.5], in workshops [1.6], and in regular as well as special issues [1.1,1.2,1.7–1.13] of refereed journals.

References

[1.1] "Special issue on numerical methods." In James W. Mink and Felix K. Schwering, eds., *IEEE Trans. Microwave Theory Tech.*, Vol. MTT-33, October 1985.

[1.2] "Special issue on electromagnetic simulation of planar and millimeter-wave circuits." In Fred Gardiol, ed., *Int'l. J. Microwave and Millimeter-wave Computer-Aided Engineering*, Vol. 2, October 1992.

[1.3] Gupta, K. C. et al. *Computer-Aided Design of Microwave Circuits.* Dedham, MA: Artech House, 1981.

[1.4] Okoshi, T. *Planar Circuits for Microwaves and Lightwaves.* New York: Springer-Verlag, 1985.

[1.5] Itoh, T., ed. *Numerical Techniques for Microwave and Millimeter-Wave Passive Structures.* New York: John Wiley & Sons, 1988.

[1.6] Workshop on "Simulation of hybrid and monolithic microwave and millimeter-wave components using full wave approaches." In *1992 IEEE MTT-S Int'l Microwave Symp.*, Albuquerque, New Mexico, June 1–5, 1992.

[1.7] "Focus on computer oriented design techniques for microwave circuits." In V. A. Monaco, ed., Special Issue of *Alta Frequ.*, Pt. I, Vol. LVII, June 1988.

[1.8] Sorrentino, R., ed. *Numerical Methods for Passive Microwave and Millimeter Wave Structures*, Reprint Volume. Piscataway, NJ: IEEE Press, 1989.

[1.9] Itoh, T. "An overview on numerical techniques for modeling miniaturized passive components." *Annales des Telecommunications*, Vol. 41, pp. 449–462, September–October 1986.

[1.10] Gupta, K. C. "Two-dimensional analysis of microstrip circuits and antennae." *J. Inst. Electron. Telecommun. Eng.* (New Delhi), Vol. 28, pp. 346–364, July 1982.

[1.11] "Special issue on computer-aided design." In K. C. Gupta and T. Itoh, eds., *IEEE Trans. Microwave Theory Tech.*, Vol. MTT-36, February 1992.

[1.12] "Special issue on process-oriented microwave CAD and modeling." In J. W. Bandler and R. H. Jansen, eds., *IEEE Trans. Microwave Theory Tech.*, Vol. MTT-40, July 1992.

[1.13] "Special issue on the workshop on discrete time domain modelling of electromagnetic fields and networks," Parts 1 and 2. *Int'l. J. Numerical Modelling*, Vol. 5, No. 3, August 1992, and Vol. 6, No. 1, February 1993.

Paper 1.1

Planar Circuits, Waveguide Models, and Segmentation Method

ROBERTO SORRENTINO, SENIOR MEMBER, IEEE

(Invited Paper)

Abstract —**The planar-circuit approach to the analysis and design of microwave integrated circuits (MIC's), with specific reference to microstrip circuits, is reviewed. The planar approach overcomes the limitations inherent to the more conventional transmission-line approach. As the operating frequency is increased and/or low-impedence levels are required, in fact, the transverse dimensions of the circuit elements become comparable with the wavelength and/or the longitudinal dimensions. In such cases, one-dimensional analyses give inaccurate or even erroneous results.**

The analysis of planar elements is formulated in terms of an *N*-port circuit and results in a generalized impedance-matrix description. Analysis techniques for simple geometries, such as the resonant mode expansion, and for more complicated planar configurations, such as the segmentation method, are discussed along with planar models for accounting for fringing fields effects and radiation loss.

I. Introduction

As MICROWAVE TECHNOLOGY evolves toward the use of higher frequencies and more sophisticated circuits and components, a considerable theoretical effort is required in order to improve the characterization and modeling of microwave structures. This is the basis for reliable computer-aided design (CAD) techniques.

In the setup of CAD techniques, one has to compromise between accuracy and simplicity. Exact analyses are often impractical because of the exceedingly high computer time required. From this viewpoint, the planar-circuit approach is a very powerful technique, which has been basically developed for the analysis of microstrip circuits, but can be extended to other microwave circuit configurations, such as reduced-height waveguide, stripline, suspended microstrip, etc.

Though the planar circuit is an approximate model of microstrip components, it constitutes a substantial improvement over conventional transmission-line models, providing accurate descriptions of their performances. On the other hand, planar-circuit models are simple enough to keep computer analyses reasonably inexpensive.

It is the scope of this paper to review the theoretical basis of the planar-circuit approach and to stress its suitability to the characterization, modeling, and design of two-dimensional microwave structures, with specific reference to microstrip circuits. This paper is not intended to provide details on planar-circuit analysis and design, which can be found in the referenced papers and overview books [1]–[3], but to illustrate the main features of the planar approach in contrast with the more conventional transmission-line approach.

The concept and definition of planar circuits are introduced in the next section, and the advantages of such an approach are briefly described. Starting from Maxwell's equations, the theoretical bases for the analysis of planar microwave components in terms of a two-dimensional circuit model are assessed in Section III. The terminal description of planar circuit is derived in Section IV; this is the basis for a brief discussion on the filtering properties and lumped-element equivalent circuits of planar elements. Once the terminal description of a single planar element has been obtained, the techniques mentioned in Section V, such as the segmentation method, can be applied to the analysis of more complicated planar configurations. The techniques for modeling a microstrip component such as a planar circuit, so as to account for effects of fringe fields and radiation loss, are discussed in Section VI. Finally, in order to describe the effects of planarity in microstrip circuits, a simple stub structure is taken as an example and its behavior illustrated in some detail in Section VII.

II. The Planar Circuit

The concept of a planar circuit was introduced by Okoshi and Miyoshi [4] as an approach to the analysis of microwave integrated circuits (MIC's). Depending on the number of dimensions which are comparable with the operating wavelength, conventional circuit elements can be classified into three categories: zero-dimensional (lumped), one-dimensional (uniform transmission lines), and three-dimensional (waveguides). The fourth category is represented by two-dimensional or planar circuits (Fig. 1). A planar circuit is defined as an electrical circuit having two dimensions comparable with the wavelength, while the third dimension is a negligible fraction of the wavelength. Strictly speaking, a distinction should be made between a microwave planar element and a planar circuit, the latter being the mathematical model, phrased in terms of voltage and current, of the former; in some instances throughout this paper, however, the two terms can be used indistinctly.

As will be shown in the next section, a two-dimensional circuit theory can be developed for planar components by extending to the two-dimensional case the concepts of

Manuscript received February 5, 1985; revised June 3, 1985.
The author is with the Department of Electronics, University of Rome La Sapienza, Rome, Italy.

Reprinted from *IEEE Trans. Microwave Theory Tech.*, vol. MTT-33, no. 10, pp. 1057–1066, Oct. 1985.

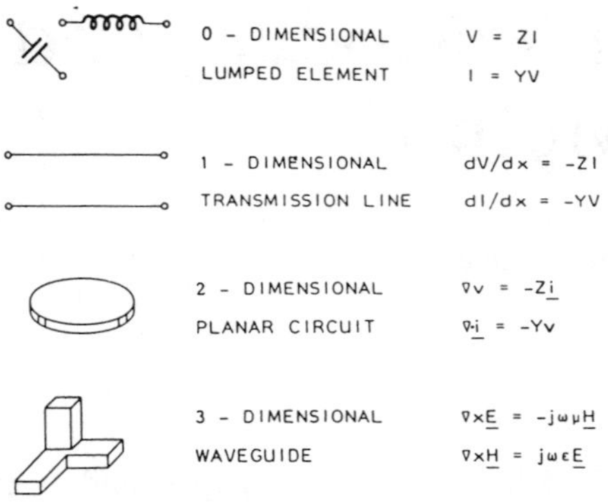

Fig. 1. Classification of electrical components.

voltage and current usually defined in transmission-line theory.

The planar approach can be used to characterize a number of MIC components, basically in stripline or microstrip configuration, which typically have one dimension, the substrate thickness, much smaller than the operating wavelength. Our attention will be focused on microstrip circuits, which presently play a major role in the area of MIC's.

With reference to a microstrip component, it should be observed that it can be only approximately considered as a planar circuit, as the electromagnetic (EM) field is not entirely confined to the substrate region but, particularly near the edges of the metallization, extends into air outside the dielectric substrate. In other words, the presence of stray fields makes the planar-circuit concept not rigorously applicable to microstrip components. Nonetheless, as discussed in Section VI, provided suitable modifications in terms of effective parameters are made, planar models provide accurate enough characterizations of microstrip circuits and components.

The planar-circuit model is intermediate between transmission-line and full-wave three-dimensional models. In some respects, it combines advantages of both approaches. On the one hand, with respect to the usual transmission-line description of microstrip circuits, the planar description is far more accurate, while, on the other hand, it is much more simple and computationally affordable than a full-wave description.

The advantages associated with the planar-circuit approach can be summarized as follows.

1) The planar-circuit approach provides accurate descriptions of microstrip components and discontinuities. As the operating frequency is increased and low-impedance values are required, the performance of microstrip circuits designed on a transmission-line basis deteriorates because of unwanted reactances associated with discontinuities. The EM field cannot any longer be assumed to have a uniform distribution in the transverse direction so that a planar approach is required to obtain accurate characterizations of the circuit performances.

2) New classes of components can be analyzed and designed using the planar-circuit approach. The wider degree of freedom of planar elements can be used to obtain specific performances and to overcome the limitation inherent to the one-dimensional approach. Several new components have been designed which utilize the planar concept, such as 3-dB hybrid circuits [5], bias filter elements [6], coupled-mode filters [7], in-phase 3-dB power dividers [8], etc.; circular polarization in microstrip antennas is obtained exciting two degenerate orthogonal modes in a planar structure [9].

3) Planar circuits are simpler to analyze than three-dimensional circuits. Although a three-dimensional full-wave analysis is the only rigorous approach to characterize microstrip circuits and components, it is too laborious and computer time consuming for most practical purposes. Planar-circuit analyses, on the contrary, require reasonably short computer times, while providing descriptions which are generally accurate enough for the needs of the microstrip circuit designer.

III. Planar-Circuit Analysis

The basic equations for the analysis of N-port planar circuits are derived in this section. The case of magnetic wall boundaries is considered, as is usually assumed for representing a microstrip or stripline component. A terminal description in terms of an impedance matrix can be derived for this type of planar circuit. The case of electrically conducting boundaries, which is representative for reduced-height waveguide, can be treated in a similar manner and described in terms of an admittance matrix [11].

Fig. 2 shows a schematic of a N-port planar element. The EM field is confined by two parallel perfectly conducting plates (top and bottom) bounded by the contour $\mathscr{C}$ and, laterally, by a cylindrical magnetic-wall surface. The excitation of the EM field inside this structure may take place either through some apertures produced at the lateral wall to couple the planar element to the external circuit (edge-fed microstrip) or by some internal current sources J_z. The latter case is normally encountered only in antenna applications, while the former is the only one usually considered in MIC applications. Both cases, however, can be formally treated in the same way; it can be easily demonstrated, in fact, that the coupling aperture produced in the magnetic wall is equivalent to an electric-current density flowing on the aperture surface.

Because of planarity (thus $\partial/\partial z = 0$) and open-circuit boundary conditions, Maxwell's equations reduce to

$$\nabla_t E_z = -j\omega\mu\hat{z} \times \boldsymbol{H}_t \tag{1}$$

$$\nabla_t \times \boldsymbol{H}_t = (j\omega\epsilon E_z + J_z)\hat{z} \tag{2}$$

where ∇_t is the two-dimensional nabla operator, $\hat{z}$ is the unit vector normal to the plane of the circuit, μ and ϵ are the permeability and permittivity of the filling substrate material. The E-field has only the z-component, while the H-field lies in the xy plane.

A two-dimensional form of telegraphists' equations can be obtained from (1) and (2) defining at each point $\boldsymbol{r}$ of the planar circuit a voltage v and a surface current density $\boldsymbol{J}_s$

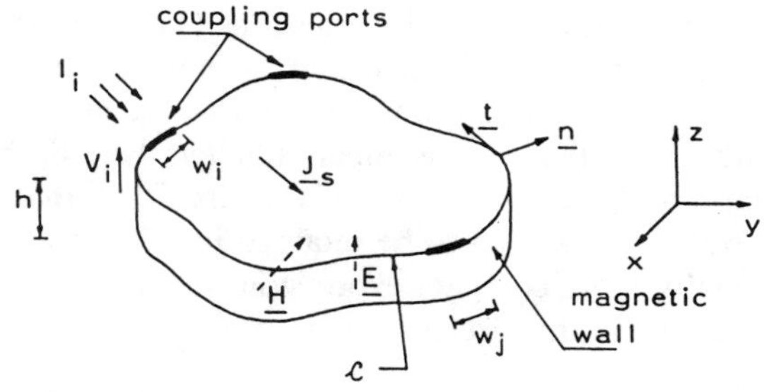

Fig. 2. Geometry of the planar circuit.

flowing on the top conductor as

$$v(\boldsymbol{r}) = -hE_z \qquad V \tag{3}$$

$$\boldsymbol{J}_s(\boldsymbol{r}) = -\hat{z} \times \boldsymbol{H}_t \qquad A/m \tag{4}$$

where h is the substrate thickness.

Note that the Poynting vector is given by

$$\boldsymbol{P} = \frac{1}{2}\boldsymbol{E} \times \boldsymbol{H}^* = \frac{1}{2h} v \boldsymbol{J}_s^* \qquad W/m^2$$

so that the quantity

$$\boldsymbol{P} = \frac{1}{2} v \boldsymbol{J}_s^* \qquad W/m$$

represents the linear power-density vector flowing on the planar circuit.

Inserting (3) and (4) into (1) and (2), we get

$$\nabla_t v = -j\omega\mu h \boldsymbol{J}_s \tag{5}$$

$$\nabla_t \cdot \boldsymbol{J}_s = -j\frac{\omega\epsilon}{h} v + J_z, \tag{6}$$

These equations represent a two-dimensional form of inhomogeneous telegraphists' equations, involving the voltage v and surface current density on the top metallic plate. The voltage wave equation is obtained taking the divergence of (5) and substituting into (6)

$$\nabla_t^2 v + k^2 v = -j\omega\mu h J_z \tag{7}$$

where

$$k^2 = \omega^2 \mu\epsilon.$$

The boundary condition associated with (7) is

$$\frac{\partial v}{\partial n} = \begin{cases} -j\omega\mu h \boldsymbol{J}_s \cdot \boldsymbol{n}, & \text{on } w_i,\ i = 1, 2, \cdots N \\ 0, & \text{elsewhere on } \mathscr{C} \end{cases} \tag{8}$$

w_i being the ith port of the planar circuit and $\boldsymbol{n}$ the outward directed normal to the periphery $\mathscr{C}$.

It can be demonstrated easily that the inhomogeneous boundary condition (8) can be replaced by a homogeneous boundary condition along the whole periphery $\mathscr{C}$, provided an additional equivalent current density

$$J_z = -\boldsymbol{J}_s \cdot \boldsymbol{n} \delta(\boldsymbol{r} - \boldsymbol{r}') \tag{9}$$

is assumed in (7), $\boldsymbol{r}'$ being the source location at w_i on the periphery $\mathscr{C}$.

A formal solution of (7) and (8) for the voltage v can be obtained using either the resonant-mode expansion technique [10], [11] or the Green's function approach [4]. The two approaches are substantially equivalent, since the Green's function is normally not known in closed form but is itself expressed in terms of a resonant mode expansion. In some cases, however, as discussed in Section VII, the Green's function approach can provide more compact analytical expressions leading to more efficient computer analyses. On the other hand, however, the resonant-mode technique provides a deeper physical insight, as it lends itself to a physical interpretation of the filtering properties of planar elements and is the basis for the modeling in terms of equivalent circuits.

The resonant-mode technique for planar structures with magnetic-wall boundaries can be obtained by suitably modifying the theory on field expansion in resonant cavities [12]. Let $\phi_\nu (\nu = 1, 2, \cdots)$ be the orthonormalized eigenfunctions of the following eigenvalue problem:

$$\nabla_t^2 \phi_\nu + k_\nu^2 \phi_\nu = 0, \qquad \text{in } S$$

$$\frac{\partial \phi_\nu}{\partial n} = 0, \qquad \text{on } \mathscr{C} \tag{10}$$

where S is the planar region bounded by the contour $\mathscr{C}$. The lowest eigenvalue of (10) is $k_0^2 = 0$, corresponding to the electrostatic mode.

Once the set of eigenfunctions ϕ_ν is known, the solution of (7) can be expressed as

$$v(\boldsymbol{r}) = \sum_{\nu=0}^{\infty} A_\nu \phi_\nu \tag{11}$$

where

$$A_\nu = \frac{-j\omega\mu h}{k^2 - k_\nu^2} \int_S \phi_\nu J_z \, dS. \tag{12}$$

When the planar element is edge fed and no volume current sources are present inside it, because of (9), the integral over the planar surface S reduces to the integral over the ports at the periphery $\mathscr{C}$, so that (12) reduces to

$$A_\nu = \frac{-j\omega\mu h}{k^2 - k_\nu^2} \sum_{i=1}^{N} \int_{w_i} \boldsymbol{J}_s \cdot (-\boldsymbol{n}) \, dl. \tag{13}$$

The solution of (7) for a unit current density pulse δ located at $\boldsymbol{r}'$ gives the Green's function in terms of the set of eigenfunctions

$$G(\boldsymbol{r}, \boldsymbol{r}') = j\omega\mu h \sum_{\nu=0}^{\infty} \frac{\phi_\nu(\boldsymbol{r})\phi_\nu(\boldsymbol{r}')}{k_\nu^2 - k^2} \tag{14}.$$

Using (14), the expression (11) for v can be replaced by

$$v(\boldsymbol{r}) = \int_S G(\boldsymbol{r}, \boldsymbol{r}') J_z(\boldsymbol{r}') \, dS \tag{15}$$

or, when (9) applies

$$v(\boldsymbol{r}) = \sum_{i=1}^{N} \int_{w_i} G(\boldsymbol{r}, \boldsymbol{r}')(-\boldsymbol{n} \cdot \boldsymbol{J}_s) \, dl. \tag{16}$$

It is worth noting that the Green's function is a frequency-dependent function, while eigenfunctions ϕ_ν are not. As a consequence, the frequency dependence of v is not apparent in (15) while it results in the form of a partial fraction expansion in (11) and (12).

IV. Terminal Description

From now on we shall assume the planar element to be edge fed, as in usual microstrip circuit applications.

Voltage and surface currents on the planar circuit are excited by a linear current density $-\boldsymbol{n}\cdot\boldsymbol{J}_s$ injected through the various ports. At the ith port, these quantities can be expressed by their Fourier expansions

$$v=\sum_{m=0}^{\infty} V_i^{(m)}\sqrt{\delta_m}\cos\frac{m\pi l}{w_i} \tag{17}$$

$$-\boldsymbol{n}\cdot\boldsymbol{J}_s=\sum_{n=0}^{\infty} J_i^{(n)}\sqrt{\delta_n}\cos\frac{n\pi l}{w_i} \tag{18}$$

where l is the coordinate along the ith port ($0\leqslant l\leqslant w_i$), w_i the port width, δ_m the Neumann delta ($\delta_m=1$ for $m=0$, $\delta_m=2$ for $m\neq 0$). If the port is terminated by a transmission line having the same width w_i, $V_i^{(n)}$ and $J_i^{(n)}$ represent voltage and longitudinal current density amplitudes of the mode of nth order, $n=0$ being the dominant TEM mode. The voltage v and the current density entering the ith port, $-\boldsymbol{n}\cdot\boldsymbol{J}_s$, can be thus represented by their Fourier expansion coefficients

$$V_i^{(m)}=\frac{\sqrt{\delta_m}}{w_i}\int_{w_i} v\cos\frac{m\pi l}{w_i}\,dl \tag{19}$$

$$J_i^{(n)}=\frac{\sqrt{\delta_n}}{w_i}\int_{w_i}(-\boldsymbol{n}\cdot\boldsymbol{J}_s)\cos\frac{n\pi l}{w_i}\,dl. \tag{20}$$

The above definitions are such that the complex power entering the circuit through the ith port is given by

$$P_i=\frac{1}{2}\int_{w_i}\boldsymbol{E}\times\boldsymbol{H}^*\cdot(-\boldsymbol{n})\,dl=\frac{1}{2}\sum_{m=0}^{\infty}V_i^{(m)}I_i^{(m)*} \tag{21}$$

where we have defined

$$I_i^{(m)}=w_iJ_i^{(m)} \tag{22}$$

as the current entering the ith port associated with the mth order mode.

Inserting (11) and (13) into (19) and using (18) and (22), the relationship between voltage and currents is obtained in terms of the generalized impedance matrix of the planar circuit

$$V_i^{(m)}=\sum_{n=0}^{\infty}\sum_{j=1}^{N}Z_{ij}^{(mn)}I_j^{(n)} \tag{23}$$

where

$$Z_{ij}^{(mn)}=\frac{j\omega\mu h\sqrt{\delta_m\delta_n}}{w_iw_j}\sum_{\nu=0}^{\infty}\frac{g_{\nu i}^{(m)}g_{\nu j}^{(n)}}{k_\nu^2-k^2} \tag{24}$$

$$g_{\nu i}^{(m)}=\int_{w_i}\phi_\nu\cos\frac{m\pi l}{w_i}\,dl \tag{25}$$

$Z_{ij}^{(mn)}$ gives the mth order voltage at the ith port due to a unit nth order current injected at the jth port, all other currents being zero. With this procedure, each physical port corresponds to an infinite number of electrical ports, relative to the spatial harmonics of voltage and current. In practice, only a few terms of expansions (17) and (18) are required to represent with good approximation the voltage and current distributions along the ports. In most cases, the width w_i of the port is much smaller than both the wavelength and the dimensions of the circuit, so that only the 0th order terms need to be retained in (17) and (18).

A formally more compact expression for the Z-elements is obtained using the Green's function

$$Z_{ij}^{(mn)}=\frac{\sqrt{\delta_m\delta_n}}{w_iw_j}\int_{w_i}dl\int_{w_j}G(\boldsymbol{r},\boldsymbol{r}')\cos\frac{m\pi l}{w_i}\cos\frac{n\pi l'}{w_j}\,dl'. \tag{26}$$

Expression (24), or (26), forms the basis for the description of the microwave planar element as an electrical circuit. Descriptions in terms of a generalized scattering matrix can be easily obtained through known formulas.

It is worth noting that the frequency dependence of the Z-parameters is not apparent in (26), as it is incorporated in the Green's function. Since coefficients (25) are frequency independent, on the contrary, the partial fraction expansion (24) explicitly shows the frequency dependence of the impedance parameters. Expression (24) is therefore useful to interpret the filtering properties of planar circuits and lends itself to the evaluation of equivalent-lumped circuits. These aspects are briefly discussed herein in the case of two-port ($N=2$) planar elements.

As mentioned above, a notable simplification can be made when, as in many practical circuits, the ports are very narrow with respect to both the wavelength and the dimensions of the planar element. In such cases, the contribution of higher order modes ($n\geqslant 1$) at the ports can be neglected, so that voltages and current densities are assumed to be constant along the ports. For two-port elements, the Z-matrix reduces to a 2×2 matrix relating the voltages and currents of the dominant (0th order) modes at the two ports. Omitting for simplicity the indexes $m=n=0$, (24) and (25) become

$$Z_{ij}=\frac{j\omega h}{w_iw_j\epsilon}\sum_{\nu=0}^{\infty}\frac{g_{\nu i}g_{\nu j}}{\omega_\nu^2-\omega^2} \tag{27}$$

$$g_{\nu i}=\int_{w_i}\phi_\nu\,dl \tag{28}$$

where we have put $k_\nu^2=\omega_\nu^2\mu\epsilon$ and $k^2=\omega^2\mu\epsilon$. The above expressions can be used to obtain a general equivalent-lumped circuit in the form of a series connection, through ideal transformers, of anti-resonant LC cells, each cell corresponding to a resonant mode of the structure. If the planar element is symmetrical, then $|g_{\nu 1}|=|g_{\nu 2}|$, and the equivalent circuit can be put in the form of a symmetrical lattice network without any use of transformers [13].

For practical applications, only a finite number of cells are to be included in the equivalent circuit; such a number depends on the frequency range of interest and on the approximation required. In a low-frequency approximation, only the first two resonant modes can be taken into account, i.e., the static mode resonating at zero frequency and the first higher mode. When the latter is an odd mode, it can be easily shown that the equivalent circuit has the same structure as a third-order elliptic filter [14]. On this basis, low-pass filters with elliptic function responses have been designed by cascading microstrip rectangular elements [15].

The physical nature of the transmission zeros occurring in planar circuits can be easily interpreted on the basis of (27) [10]. It is well known that transmission zeros ($s_{21}=0$) in two-port networks occur in two cases: a) for $Z_{21}=0$ and b) for Z_{21} finite and Z_{11} or Z_{22} infinite. For planar elements, the latter case can occur at some specific resonant frequencies of the structure. This type of transmission zero has been called a modal transmission zero and occurs when a resonant mode 'ν' can be excited from one port ($g_{\nu 1} \neq 0$), but is uncoupled to the other port ($g_{\nu 2}=0$). The former type of transmission zero ($Z_{21}=0$), on the contrary, is due to the destructive interaction between resonant modes. While the frequency location of a modal transmission zero is determined only by the shape and dimensions of the planar element, the frequency location of an interaction transmission zero can be controlled by varying the position of the ports [6].

The same distinction between transmission zeros applies to waveguide circuits and has been used to improve selectivity in cylindrical filters [16]. A liquid-crystal field mapping technique for MIC's [17], [18] can been used to give perceptible evidence to these results [19].

V. Segmentation of Planar Elements

Numerical techniques [1], [4], [21], [22], [47], should be used for the analysis of planar elements with completely arbitrary shapes. Eigenfunctions ϕ_ν and Green's function G, in fact, are known for a limited number of simple geometrical shapes, such as rectangles, circles, circular sectors, rings, and annular sectors, etc. A number of Green's functions for such and other simple shapes are listed in [2] and [20].

The analysis of planar elements, however, can be easily extended to those geometries which result from the connection of elementary shapes. This is illustrated in a simple example. The structure of Fig. 3 can be decomposed into the cascade of two subelements (Fig. 3(b)) for which the impedance matrices $[Z_a]$ and $[Z_b]$ can be computed as discussed in the previous section. By grouping together voltages and currents at the connected ports, $[Z_a]$ and $[Z_b]$ can be put in the form

$$[Z_a]=\begin{bmatrix}[Z_{a11}] & [Z_{a12}]\\ [Z_{a21}] & [Z_{a22}]\end{bmatrix} \quad [Z_b]=\begin{bmatrix}[Z_{b11}] & [Z_{b12}]\\ [Z_{b21}] & [Z_{b22}]\end{bmatrix}. \tag{29}$$

Using the conditions imposed by the cascade connection, i.e.,

$$[V_{a2}]=[V_{b1}] \quad [I_{a2}]=-[I_{b1}] \tag{30}$$

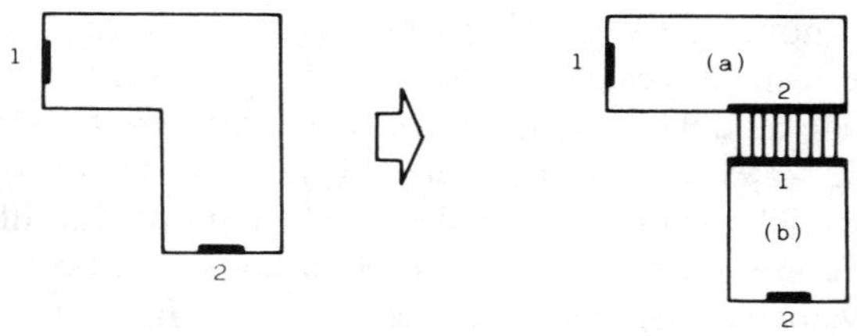

Fig. 3. Segmentation of planar elements.

the Z-matrix of the resulting network is expressed by

$$[Z]=\begin{bmatrix}[Z_{11}] & [Z_{12}]\\ [Z_{21}] & [Z_{22}]\end{bmatrix}$$

$$\begin{aligned}[Z_{11}]&=[Z_{a11}]-[Z_{a12}][Y_{ab}][Z_{a21}]\\ [Z_{12}]&=[Z_{a12}][Y_{ab}][Z_{b12}]\\ [Z_{21}]&=[Z_{b21}][Y_{ab}][Z_{a21}]\\ [Z_{22}]&=[Z_{b22}]-[Z_{b21}][Y_{ab}][Z_{b12}]\end{aligned} \tag{31}$$

where

$$[Y_{ab}]=([Z_{a22}]+[Z_{b11}])^{-1}. \tag{32}$$

Note that, according to the technique described in the previous section, voltage and current distributions at the interconnection between the planar elements are expressed in terms of their Fourier expansions, so that the same physical port corresponds to different (infinite, in theory) electrical ports. In practice, voltage and currents are approximated by truncated Fourier expansions, corresponding to a finite number of electrical ports.

An alternative technique is the segmentation method, in which, on the contrary, the interconnection is discretized into a finite number of ports. Voltage and current are assumed to be constant along each port, which thus corresponds to one electrical port. Voltage and current distributions along the interconnection are so approximated by stepped functions. The segmentation method has been originally formulated in terms of scattering matrix [23], but can be also implemented with higher computational efficiency in terms of impedance matrices [24].

The analysis of complicated geometries can be further extended by the desegmentation method [25], [26]. This technique can be applied to those geometries which, after addition of a simple element, can be analyzed by either the elementary or segmentation methods. In other words, the desegmentation method applies to geometries resulting from the subtraction of a simple shape from another geometry which can be analyzed by segmentation.

VI. Planar Models of Microstrip Circuits

As previously mentioned, a microstrip circuit cannot be considered as a planar circuit but only in an approximate way. Substantial discrepancies may arise due to the fields at the edges of the metallization. Nevertheless, an equivalent planar circuit can be used to model the microstrip circuit. This section will discuss how to define such an equivalent planar model.

As is known, the dynamic properties of a uniform microstripline can be calculated using an equivalent planar waveguide model [27], [28]. This is a waveguide with lateral magnetic walls, having the same height h as the substrate thickness. The width w_e and the permittivity of the filling dielectric are determined by the conditions that both the phase velocity and the characteristic impedance be the same as for the microstripline. As the dominant mode of the planar waveguide is a TEM mode, the equality of the phase velocities imposes that the filling dielectric has the same effective permittivity ϵ_e of the quasi-TEM mode of the microstripline, while the condition on the characteristic impedance imposes that

$$Z_0 = \eta_o h / \left(w_e \sqrt{\epsilon_e} \right)$$

where Z_0 is the characteristic impedance defined for the microstripline, $\eta_0 = \sqrt{\mu_0/\epsilon_0}$ is the free-space impedance. A frequency dependence of ϵ_e and w_e can be introduced to account for dispersion on the phase velocity and characteristic impedance [28].

The planar waveguide model has been found to provide a good approximation of the cutoff frequencies of higher order modes of the microstripline. The usefulness of this model relies on the reduction of an open structure into a closed one; it therefore substantially simplifies the calculation of microstrip discontinuities. In some sense, the planar waveguide is a two-dimensional model, as it takes into account the transverse variations of the EM field. This model has found a number of applications from the analysis of the frequency-dependent properties of microstrip discontinuities [29], [30] to the design of stepped microstrip components [31], microstrip power dividers [32], and computer-aided design of microstrip filters [33].

It seems reasonable to extend the planar waveguide model to the case of two-dimensional microstrip circuits using a planar circuit with effective dimensions and an effective permittivity. The case of two-dimensional elements, however, is more complicated, as the EM field is allowed to vary along two directions. To illustrate this point with an example, let us consider the case of a two-port circular microstrip. Fig. 4(a) shows the experimental frequency behavior of the scattering parameter $|s_{21}|$. The theoretical behavior of Fig. 4(b) has been obtained by applying the mode expansion technique and by suitably choosing the effective parameters of the circular microstrip so as to optimize the agreement with Fig. 4(a).

A good agreement between theory and experiment is observed up to ~12 GHz. Above this frequency, the theory appears to be inconsistent with the experiment. Such an inconsistency can be explained by the following argument.

The effective permittivity is used to account for the electric-field lines being more or less confined to the substrate material and therefore depends on the electric-field distribution along the edge of the planar element. Considering the EM field as the superposition of the resonant modes of the structure, it is evident that a different effective permittivity should be ascribed to resonant modes having a different field distribution along the periphery of the circuit.

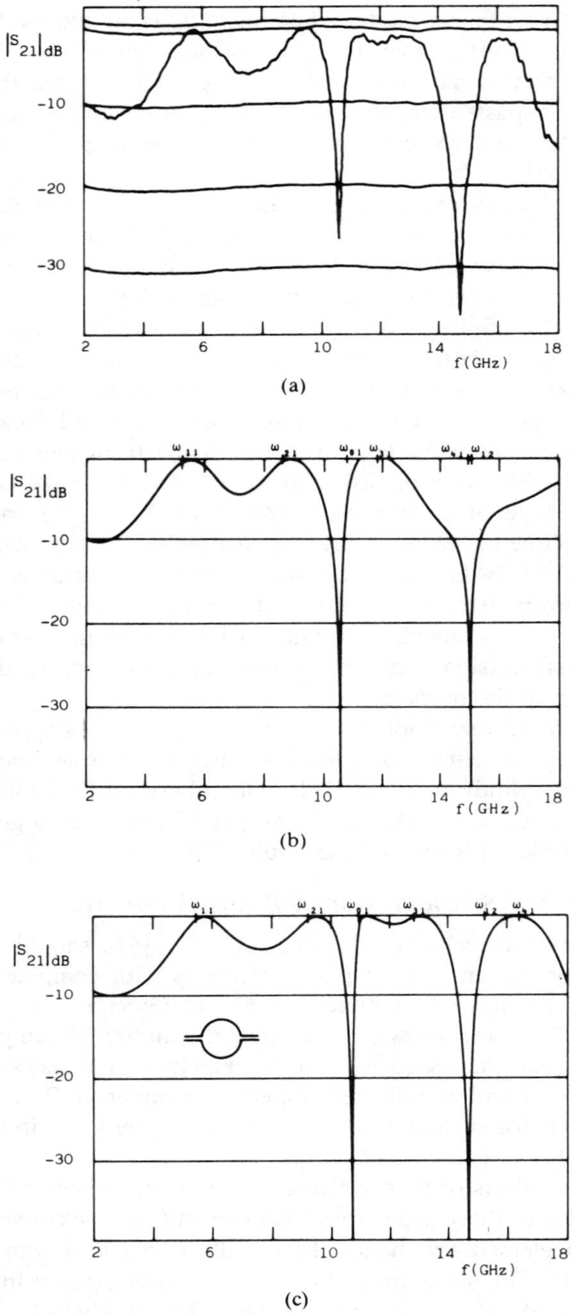

Fig. 4. Scattering parameter $|s_{21}|$ of a circular microstrip. (a) Experiment (after [10]). (b) Theoretical analysis using a single effective model. (c) Theoretical analysis using a different effective model for each resonant mode (after [10]).

Wolff and Knoppik have developed a theory [34] for computing the resonant frequencies of circular and rectangular microstrip resonators using a planar model. The theory can be extended to ring resonators [35]–[37]. This model is characterized by effective dimensions and effective permittivities which depend on the resonant mode; it can be used in conjunction with the resonant-mode technique to get the results of Fig. 4(c) [7]. The agreement with the experiment of Fig. 4(a) is quite good up to 18 GHz. The comparison between Fig. 4(b) and (c) shows that each resonant frequency undergoes a different shift; in particular, the resonant frequencies ω_{14} and ω_{21} are interchanged.

This example demonstrates that an accurate characterization of a two-dimensional microstrip circuit can be achieved through an effective planar-circuit model in conjunction with the resonant-mode technique, provided a suitable effective permittivity is used for each resonant mode. The effective permittivity accounts for the reactive energy associated with the fringing field of the corresponding mode and may produce modal inversions. Since, as previously demonstrated, the response of a planar element is tightly related to the sequence of resonant modes, the alteration of that sequence produces strong alterations of the frequency behavior of the circuit.

An additional source of discrepancy may be due to radiation loss. Differently from dielectric and conductor losses, which are associated with the field distribution inside the microstrip element, radiation loss depends on the field distribution along the periphery of the circuit. Conductor and dielectric losses do not alter substantially the circuit performance, but usually only introduce some degradations. On the contrary, radiation loss may produce effects substantially different from those predicted on the basis of a lossless theory.

A typical example is represented by nonsymmetrical structures. For a lossless reciprocal two-port network, $|s_{11}| = |s_{22}|$ at any frequency; this equality is no longer valid in the presence of losses. It has been experimentally observed in nonsymmetrical rectangular microstrips, in fact, that at some particular frequencies $|s_{11}| \cong 0$ while $|s_{22}| \cong 1$ [41]. This phenomenon occurs when a resonant mode, which strongly radiates, can be excited from the first port, but is uncoupled with the second one.

An exact theory for calculating radiation loss from microstrip structures has not been developed. The resonant-mode technique applied to planar circuits, however, can be extended to include radiation loss in an approximate way. This technique, in fact, has been extensively applied to the analysis of microstrip antennas [9], [38]–[40]. Basically, one has to account for radiation by assuming that the tangential magnetic field at the periphery of the circuit is different from zero, so that the field expansion coefficients must be modified accordingly. The general formulation presented in [41] is of impractical use; neglecting the coupling between resonant modes, which arises from the inhomogeneous boundary conditions, a simplified theory can be derived which reduces the problem to the evaluation of the complex power radiated by each unperturbed resonant mode. In spite of the approximations involved, which include neglecting surface waves, this theory was shown to accurately predict in a quantitative way the frequency behavior of the scattering parameters of rectangular and circular microstrip structures.

VII. Planar Analysis of Stub Structure

The limitations of the transmission-line (one-dimensional) approach to the analysis and design of MIC's are illustrated and discussed in this section, through the simple but typical and significant example of a stub structure.

The stub is used to provide a zero-impedance level, thus a transmission zero, at the frequency corresponding to a quarter of wavelength. Low-characteristic impedances of the stub are required for broad-band applications, so that, in microstrip circuits, this function is more conveniently realized as the parallel of two stubs. We are therefore led to the consideration of the double-stub structure of Fig. 5.

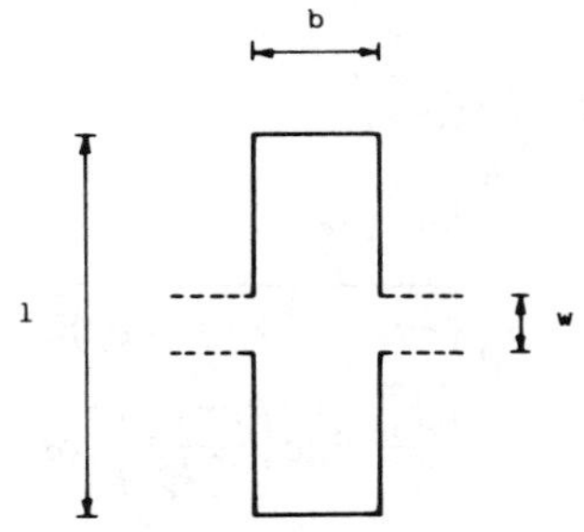

Fig. 5. Geometry of a double stub.

From a planar point of view, this is a rectangular element $b \times l$ symmetrically connected to a main line of width w. With reference to the discussion of Section VI on fringe field effects, a different effective model should be used for each resonant mode of the structure. In order to simplify the discussion, which is aimed to point out two-dimensional effects arising even in a simplified model independently of fringe field effects, this structure will be characterized by a unique effective model, i.e., by effective dimensions w_e, b_e, l_e and effective dielectric constant ϵ_e. (This is a planar waveguide model, which is strictly valid as long as $l_e > b_e$ and higher order modes excited at the connection with the main line are rapidly decaying toward the open ends of the stub structure.)

Depending on the dimensions and the frequency range, the rectangular structure behaves as: a) a shunt capacitor $C = \epsilon_e l_e b_e / h$ in the limit of very low frequencies, so that $w_e, b_e, l_e \ll \lambda$; b) a shunt stub of transmission line with characteristic impedance $Z_0/2 = (1/2) h\eta_0 / (b_e\sqrt{\epsilon_e})$ and length $l_e/2$ as long as $w_e, b_e \ll \lambda$ and $w_e \ll l_e$; c) a planar circuit in all other cases.

The general case c), which includes a) and b) as special cases, can be treated as in Section III by evaluating the generalized impedance matrix, which accounts also for reflected and transmitted higher order modes on the main line. If these modes are evanescent and the line is long enough at both ends so that the discontinuity represented by the stub does not interact with other possible discontinuities, higher order modes have a negligible effect. In such a case, the Z-matrix computation can be reduced to the only terms relative to the dominant modes ($m = n = 0$ in (24)).

Using the resonant-mode technique, the Z-matrix is expressed in the form of a double series over indexes r and s corresponding to the resonances along l and b, respectively. More specifically, one obtains

$$Z_{11} = Z_{22} = j\omega c Z_0 \sum_{r=0}^{\infty} \sum_{s=0}^{\infty} \frac{g_{rs}^2}{\omega_{rs}^2 - \omega^2}$$

$$Z_{12} = Z_{21} = j\omega c Z_0 \sum_{r=0}^{\infty} \sum_{s=0}^{\infty} \frac{(-1)^s g_{rs}^2}{\omega_{rs}^2 - \omega^2} \qquad (33)$$

with

$$c=1/\sqrt{\mu_0\epsilon_0\epsilon_e}$$

$$Z_0=\frac{h}{b_e}\eta_0/\sqrt{\epsilon_e}$$

$$\omega_{rs}=c\sqrt{(r\pi/l_e)^2+(s\pi/b_e)^2}$$

$$g_{rs}=\sqrt{\frac{\delta_r\delta_s}{l_e}}\operatorname{sinc}\left(\frac{r\pi w_e}{2l_e}\right)\cos\frac{r\pi}{2}. \quad (34)$$

In the numerical computation of the Z-parameters, it would be convenient to evaluate analytically the series in (33). Actually, it could be shown that either the series over r or that over s can be expressed in a closed form. This can be done by regarding the rectangular structure as a section of planar waveguide with its longitudinal axis directed along either the length l or the width b, respectively. In both cases, the Green's function is obtained as a single series involving the modes of the planar waveguide. It is found, in particular, that (33)–(34) can be replaced by

$$Z_{11}=Z_{22}=j\sum_{s=0}^{\infty}X_s \quad Z_{12}=Z_{21}=j\sum_{s=0}^{\infty}(-1)^sX_s \quad (35)$$

with

$$X_s=-Z_0\cdot\frac{\beta_0}{\beta_s}\delta_s\left[\frac{1}{2}\operatorname{cotan}\left(\frac{\beta_sl_e}{2}\right)\operatorname{sinc}^2\left(\frac{\beta_sw_e}{2}\right)+\frac{\beta_sw_e-\sin(\beta_sw_e)}{(\beta_sw_e)^2}\right] \quad (36)$$

$$\beta_s^2=\omega^2\mu\epsilon_0\epsilon_e-(s\pi/b_e)^2.$$

Trigonometric functions in (36) reduce to corresponding hyperbolic functions when the frequency and the index s are such that $\beta_s^2<0$. This corresponds to the sth order mode being below cutoff. Clearly, (35) and (36) are computationally much more efficient than (33) and (34). Alternative expressions can be obtained by evaluating analytically the series over s in (33) [42], [43]. In such a case, the rectangular structure is viewed as a longitudinally symmetric cascade of two step discontinuities.

Expressions (35) and (36) permit one to point out the differences between the planar approach and the transmission-line approach; they reduce to the usual expressions for the parallel of two shunt open stubs: a) retaining only the 0th order terms in (35) and b) in the limit for $w_e/\lambda\rightarrow 0$

$$\lim X_0=\frac{Z_0}{2}\operatorname{cotan}\left(\frac{\beta_0l_e}{2}\right).$$

It is worth noting that discrepancies between planar and transmission-line models arise not only because of the excitation of higher order modes ($s>0$) in the stub structure, but also because of the finite width w_e of the ports. Even if the stub has a high characteristic impedance, and higher order modes can therefore be neglected, in fact, the finite width of the ports produces both a shift of the zero impedance frequency f_0 (because of the additive term appearing in (36)), and a different impedance slope (because of the coefficient of the cotangent).

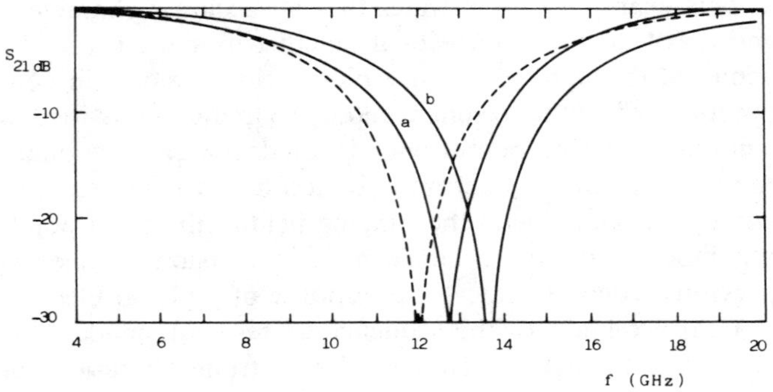

(a)

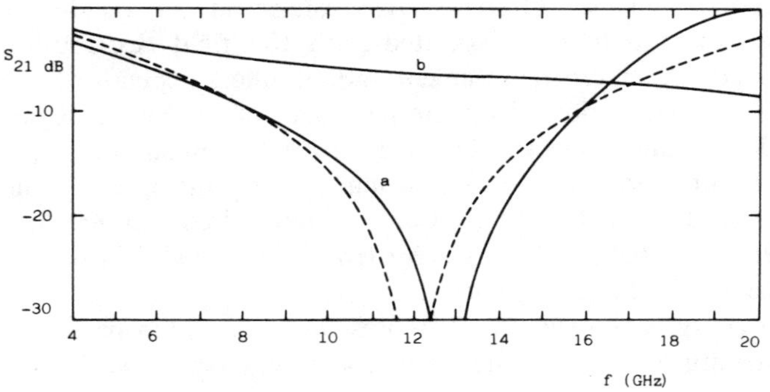

(b)

Fig. 6. (a) Scattering parameter $|s_{21}|$ of a double stub with $Z_0=90\ \Omega$, $f_0=12$ GHz: ideal response (dashed line); a planar analysis neglecting higher mode and b with higher modes.(b) Same as Fig. 6(a), but with $Z_0=30\ \Omega$.

These effects are illustrated in Fig. 6(a) and (b). Fig. 6(a) shows the frequency behavior of the scattering parameter $|s_{21}|$ of a double stub with $Z_0=90\ \Omega$ inserted on a 50-Ω line. The length of the stub has been chosen so that $(l_e-w_e)=\lambda/2$ at the frequency $f_0=12$ GHz. The dotted curve represents the response obtained by an ideal transmission-line model. Curve a has been computed including in (35) only the 0th order terms. A notable shift of the transmission zero frequency is observed because of the finite width of the 50-Ω line. The inclusion of higher order terms in (35), curve (b), gives rise to a further shift of the frequency f_0, which is about 13.7 GHz instead of 12 GHz.

These effects become even more marked if the stub impedance is reduced. Because of the excitation of higher order modes, the transmission zero may eventually disappear, as shown in Fig. 6(b), where the double stub impedance has been chosen as 30 Ω.

The difficulties of designing microstrip stubs with low-characteristic impedances have suggested the use of alternative structures, such as radial line stubs [44]–[46]. Linear stubs, however, can still be used, provided a planar approach is used in the design. This is demonstrated in Fig. 7, where the response of the planar structure designed is compared with that of an ideal transmission-line single stub of 15 Ω. Although the rectangular structure exhibits a somewhat more selective behavior, nevertheless the response appears to be satisfactory for practical applications. This simple example shows the wider design possibilities of the planar approach, which permits one to overcome the limitations inherent to the one-dimensional approach.

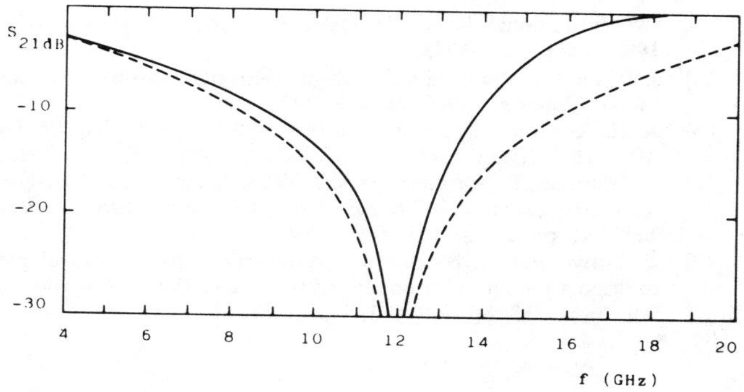

Fig. 7. Behavior of a planar stub designed to have a transmission zero at $f_0 = 12$ GHz, compared with the ideal response of a 15-Ω stub (----.)

VIII. Conclusion

An attempt has been made to review the planar-circuit concept and its theoretical basis for analysis and design of microwave planar components.

It has been stressed that the planar approach to MIC's is an approximate technique and therefore cannot be expected to provide extremely accurate results in all actual problems: to this scope, hybrid-mode full-wave techniques should be used. In conjunction with the planar models discussed in Section VI, however, accurate characterizations are obtained in most practical cases, including radiation loss, in such a way as to overcome the limitations inherent to the conventional transmission-line approach in the analysis and design of MIC's.

Acknowledgment

Since 1976, most of the work the author has done in the area of MIC's has been promoted and encouraged by Prof. F. Giannini. Prof. M. Salerno is gratefully acknowledged for helpful discussions.

References

[1] K. C. Gupta *et al.*, *Microstriplines and Slotlines*, Dedham, MA: Artech House, 1979.

[2] K. C. Gupta *et al.*, *Computer-Aided Design of Microwave Circuits*, Dedham, MA: Artech House, 1981.

[3] T. C. Edwards, *Foundations for Microstrip Circuit Design*, New York: Wiley, 1981.

[4] T. Okoshi and T. Miyoshi, "The planar circuit—An approach to microwave integrated circuitry," *IEEE Trans. Microwave Theory Tech.*, vol. MTT-20, pp. 245–252, 1972.

[5] T. Okoshi, T. Takeuchi, and J. P. Hsu, "Planar 3-dB hybrid circuit," *Electronics Comm. Japan*, vol. 58-B, pp. 80–90, 1975.

[6] G. D'Inzeo, F. Giannini, and R. Sorrentino, "Design of circular planar networks for bias filter elements in microwave integrated circuits," *Alta Freq.*, vol. 48, pp. 251e–257e, July 1979.

[7] T. Miyoshi and T. Okoshi, "Analysis of microwave planar circuits," *Electronics Comm. Japan*, vol. 55-B, pp. 24–31, 1972.

[8] K. C. Gupta, R. Chadha, and P. C. Sharma, "Two-dimensional analysis for stripline microstrip circuits," *IEEE MTT-S Int. Microwave Symp. Dig.* (Los Angeles), 1981, pp. 504–506.

[9] K. R. Carver and J. W. Mink, "Microstrip antenna technology," *IEEE Trans. Antennas Propagat.*, vol. AP-29, pp. 2–24, Jan. 1981.

[10] G. D'Inzeo, F. Giannini, C. M. Sodi, and R. Sorrentino, "Method of analysis and filtering properties of microwave planar networks," *IEEE Trans. Microwave Theory Tech.*, vol. MTT-26, pp. 462–471, July 1978.

[11] P. P. Civalleri and S. Ridella, "Impedance and admittance matrices of distributed three-layer *N*-ports," *IEEE Trans. Circuit Theory*, vol. CT-17, pp. 392–398, Aug. 1970.

[12] K. Kurokawa, *An Introduction to the Theory of Microwave Circuits.* New York: Academic Press, 1967, ch. 4.

[13] G. D'Inzeo, F. Giannini, and R. Sorrentino, "Wide-band equivalent circuits of microwave planar networks," *IEEE Trans. Microwave Theory Tech.*, vol. MTT-24, pp. 1107–1113, Oct. 1980.

[14] G. D'Inzeo, F. Giannini, and R. Sorrentino, "Novel integrated low-pass filter," *Electron. Lett.*, vol. 15, pp. 258–260, Apr. 1979.

[15] F. Giannini, M. Salerno, and R. Sorrentino, "Design of low-pass elliptic filters by means of cascaded microstrip rectangular elements," *IEEE Trans. Microwave Theory Tech.*, vol. MTT-30, pp. 1348–1353, Sept. 1982.

[16] D. E. Kreinheder and T. D. Lingren, "Improved selectivity in cylindrical TE_{011} filters by TE_{211}/TE_{311} mode control," *IEEE Trans. Microwave Theory Tech.*, vol. MTT-30, pp. 1383–1387, Sept. 1982.

[17] F. Giannini, P. Maltese, and R. Sorrentino, "Liquid crystal technique for field detection in microwave integrated circuitry," *Alta Freq.*, vol. 46, pp. 80e–88E, Apr. 1977.

[18] F. Giannini, P. Maltese, and R. Sorrentino, "Liquid crystal improved technique for thermal field measurements," *Appl. Opt.*, vol. 18, no. 17, pp. 3048–3052, Sept. 1979.

[19] G. D'Inzeo, F. Giannini, P. Maltese, and R. Sorrentino, "On the double nature of transmission zeros in microstrip structures," *Proc. IEEE*, vol. 66, pp. 800–802, July 1978.

[20] K. C. Gupta, "Two-dimensional analysis of microstrip circuits and antennae," *J. Inst. Electronics Telecomm. Engrs.*, vol. 28, pp. 346-364, July 1982.

[21] P. Silvester, "Finite element analysis of planar microwave circuits," *IEEE Trans. Microwave Theory Tech.*, vol. MTT-21, pp. 104–108, Feb. 1973.

[22] G. D'Inzeo, F. Giannini, and R. Sorrentino, "Theoretical and experimental analysis of non-uniform microstrip lines in the frequency range 2–18 GHz," in *Proc. 6th Euro. Microwave Conf.*, 1976, pp. 627–631.

[23] T. Okoshi, Y. Uehara, and T. Takeuchi, "The segmentation method —An approach to the analysis of microwave planar circuits," *IEEE Trans. Microwave Theory Tech.*, vol. MTT-24, pp. 662–668, Oct. 1976.

[24] R. Chadha and K. C. Gupta, "Segmentation method using impedance matrices for analysis of planar microwave circuits," *IEEE Trans. Microwave Theory Tech.*, vol. MTT-29, pp. 71–74, Jan. 1981.

[25] P. C. Sharma and K. C. Gupta, "Desegmentation method for analysis of two-dimensional microwave circuits," *IEEE Trans. Microwave Theory Tech.*, vol. MTT-29, pp. 1094–1097, 1981.

[26] P. C. Sharma and K. C. Gupta, "An alternative procedure for implementing the desegmentation method," *IEEE Trans. Microwave Theory Tech.*, vol. MTT-32, pp. 1–4, Jan. 1984.

[27] I. Wolff, G. Kompa, and R. Mehran, "Calculation method for microstrip discontinuities and T-junctions," *Electron. Lett.*, vol. 8, pp. 177–179, Apr. 1972.

[28] G. Kompa and R. Mehran, "Planar waveguide model for calculating microstrip components," *Electron. Lett.*, vol. 11, pp. 459–460, Sept. 1975.

[29] G. Kompa, "*S*-matrix computation of microstrip discontinuities with a planar waveguide model," *Arch. Elek. Übertragung.*, vol. 30, pp. 58–64, Feb. 1976.

[30] W. Menzel and I. Wolff, "A method for calculating the frequency-dependent properties of microstrip discontinuities," *IEEE Trans. Microwave Theory Tech.*, vol. MTT-25, pp. 107–112, Feb. 1977.

[31] G. Kompa, "Design of stepped microstrip components," *Radio Electron. Eng.*, vol. 48, pp. 53–63, Jan./Feb. 1978.

[32] W. Menzel, "Design of microstrip power dividers with simple geometry," *Electron. Lett.*, vol. 12, no. 24, pp. 639–640, Nov. 1976.

[33] R. Mehran, "Computer-aided design of microstrip filters considering dispersion, loss and discontinuity effects," *IEEE Trans. Microwave Theory Tech.*, vol. MTT-27, pp. 239–245, Mar. 1978.

[34] I. Wolff and N. Knoppik, "Rectangular and circular microstrip disk capacitors and resonators," *IEEE Trans. Microwave Theory Tech.*, vol. MTT-22, pp. 857–864, Oct. 1974.

[35] G. D'Inzeo, F. Giannini, R. Sorrentino, and J. Vrba, "Microwave planar networks: The annular structure," *Electron. Lett.*, vol. 14, no. 16, pp. 526–528, Aug. 1978.

[36] J. Vrba, "Dynamic permittivity of microstrip ring resonator," *Electron. Lett.*, vol. 15, no. 16, pp. 504–505, Aug. 1979.

[37] I. Wolff and V. K. Tripathi, "The microstrip open-ring resonator," *IEEE Trans. Microwave Theory Tech.*, vol. MTT-32, pp. 102–107, Jan. 1984.

[38] K. R. Carver, "A modal expansion theory for the microstrip antenna," in *AP-S Int. Symp. Dig.*, vol. I, June 1979, pp. 101–104.

[39] A. G. Derneryd and A. G. Lind, "Cavity model of the rectangular microstrip antenna," *IEEE Trans. Antennas Propagat.*, vol. AP-27, pp. 12-1/12-11, Oct. 1979.
[40] Y. T. Lo, D. Solomon, and W. F. Richards, "Theory and experiment on microstrip antennas," *IEEE Trans. Antennas Propagat.*, vol. AP-27, pp. 137–145, Mar. 1979.
[41] R. Sorrentino and S. Pileri, "Method of analysis of planar networks including radiation loss," *IEEE Trans. Microwave Theory Tech.*, vol. MTT-29, pp. 942–948, Sept. 1981.
[42] B. Bianco and S. Ridella, "Nonconventional transmission zeros in distributed rectangular structures," *IEEE Trans. Microwave Theory Tech.*, vol. MTT-20, pp. 297–303, May 1972.
[43] B. Bianco, M. Granara, and S. Ridella, "Filtering properties of two-dimensional lines' discontinuities," *Alta Freq.*, vol. 42, pp. 140E–148E, June 1973.
[44] J. P. Vinding, "Radial line stubs as elements in strip line circuits," in *NEREM Rec.*, 1967, pp. 108–109.
[45] A. H. Atwater, "Microstrip reactive circuit elements," *IEEE Trans. Microwave Theory Tech.*, vol. MTT-31, pp. 488–491, June 1983.
[46] F. Giannini, R. Sorrentino, and J. Vrba, "Planar circuit analysis of microstrip radial stub," *IEEE Trans. Microwave Theory Tech.*, vol. MTT-32, pp. 1652–1655, Dec. 1984.
[47] E. Tonye and H. Baudrand, "Multimode *S*-parameters of planar multiport junctions by boundary element method," *Electron. Lett.*, vol. 20, no. 19, pp. 799–802, Sept. 1984.

Paper 1.2

The Planar Circuit—An Approach to Microwave Integrated Circuitry

TAKANORI OKOSHI, MEMBER, IEEE, AND TANROKU MIYOSHI, STUDENT MEMBER, IEEE

Abstract—Three principal categories have been known in electrical circuitry so far. They are the lumped-constant (0-dimensional) circuit, distributed-constant (1-dimensional) circuit, and waveguide (3-dimensional) circuit. The planar circuit to be discussed in general in this paper is a circuit category that should be positioned as a 2-dimensional circuit. It is defined as an "electrical circuit having dimensions comparable to the wavelength in two directions, but much less thickness in one direction."

The main subject of this paper is the computer analysis of an arbitrarily shaped, triplate planar circuit. It is shown that a computer analysis based upon a contour-integral solution of the wave equation offers an accurate and efficient tool in the design of the planar circuit. Results of some computer calculations are described.

It is also shown that the circuit parameters can be derived directly from Green's function of the wave equation when the shape of the circuit is relatively simple. Examples of this sort of analysis are also shown for comparison with the computer analysis.

I. Introduction

THREE PRINCIPAL categories have been known in electrical circuitry so far. They are the lumped-constant (0-dimensional) circuit, distributed-constant (1-dimensional) circuit, and waveguide (3-dimensional) circuit. The planar circuit to be discussed in general in this paper is a circuit category that should be positioned as a 2-dimensional circuit. It is defined as an "electrical circuit having dimensions comparable to the wavelength in two directions, but much less thickness in one direction."

Then three types of the planar circuit are possible. They are the triplate type, the open type, and the cavity type, as shown in Fig. 1. However, in this paper mainly the triplate-type planar circuit will be dealt with to avoid confusion.

There are three reasons that the planar circuit should be investigated in general at present [1], [2].

1) The planar circuit has wider freedom in the circuit design than the stripline circuit does. In other words, the former includes the latter as a special case. Therefore, if the design technique for the planar circuit is established in future, it will offer an exact and efficient tool in the design of microwave integrated circuits.

2) The planar circuit can offer a lower impedance level than the stripline circuit does. The recently developed microwave semiconductor devices, such as Gunn, IMPATT, or Schottky-barrier diodes, usually require a low-impedance circuitry.

Manuscript received March 10, 1971; revised September 7, 1971.
The authors are with the Department of Electronic Engineering, University of Tokyo, Tokyo, Japan.

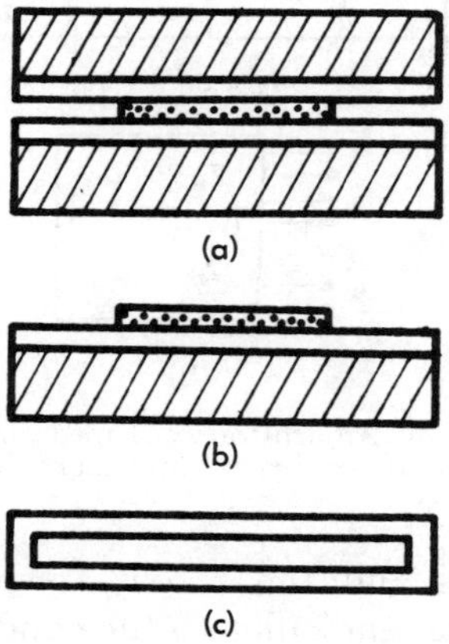

Fig. 1. Three types of the planar circuit. (a) Triplate type. (b) Open type. (c) Cavity type.

3) The planar circuit is easier to analyze and design than the waveguide circuit. By virtue of the recent progress in the computer, the analysis of an arbitrarily shaped planar circuit is within our reach if we rely on the computer.

We should note that the planar circuit is not an entirely new concept. A special case of this circuitry, the disk-shaped resonator, has been used in the stripline circulator or even as a filter [3]–[5]. The so-called "radial line" is also a special case of the planar circuit. However, to the authors' knowledge, general treatment of the planar circuit, or, in other words, the analysis of an arbitrarily shaped planar circuit, has never been presented.

The main subject of this paper is the analysis of an arbitrarily shaped, triplate planar circuit. The term "analysis" denotes here the determination of the circuit parameters of the equivalent multiport as shown in Fig. 2.

II. Basic Equations

A symmetrically excited, triplate planar circuit as shown in Fig. 2(a) will be considered throughout this paper. The model to be considered is as follows.

An arbitarily shaped, thin conductor plate is sandwiched between two ground conductors, with a spacing d from each of them. The circuit is assumed to be excited symmetrically with respect to the upper and lower ground conductors. There are several coupling ports, and their widths are denoted by W_i, W_j, $\cdots$. The rest of the periphery is assumed to be open circuited. The

Reprinted from *IEEE Trans. Microwave Theory Tech.*, vol. MTT-20, no. 4, pp. 245–252, April 1972.

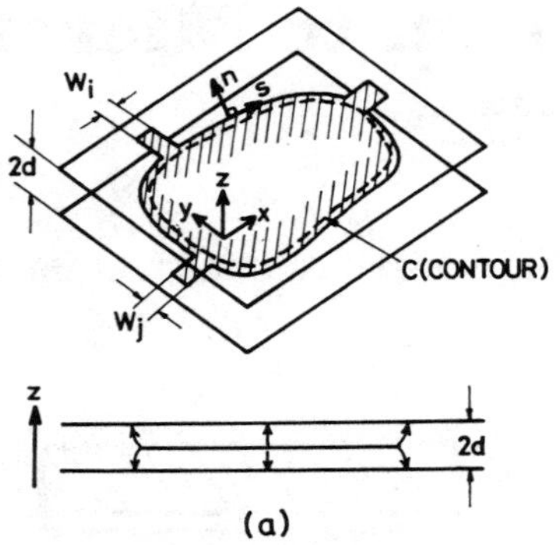

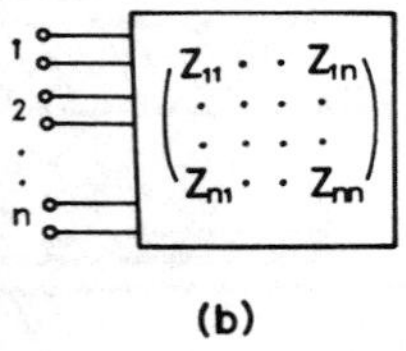

Fig. 2. (a) An arbitrarily shaped planar circuit. (b) Its equivalent multiport circuit.

xy coordinates and the z axis, respectively, are set parallel and perpendicular to the conductors.

When the spacing d is much smaller than the wavelength and the spacing material is homogeneous and isotropic, it is deduced directly from Maxwell's equation that a two-dimensional Helmholtz equation dominates the electromagnetic field in the planar circuit:

$$(\nabla_T^2 + k^2)V = 0, \qquad k^2 = \omega^2\epsilon\mu, \quad \nabla_T^2 = \frac{\partial^2}{\partial x^2} + \frac{\partial^2}{\partial y^2} \tag{1}$$

where V denotes the RF voltage of the center conductor with respect to the ground conductors; ω, ϵ, and μ are the angular frequency, permittivity, and permeability of the spacing material, respectively.[1] The network characteristics can be determined by solving (1) under given boundary conditions.

At most of the periphery where the coupling ports are absent, no current flows at the edge of the center conductor in the direction normal to the edge, because the circuit is excited symmetrically with respect to the upper and lower ground conductors.[2] Hence, the following boundary condition must hold:

$$\partial V/\partial n = 0 \tag{2}$$

where n is normal.

[1] In most of the discussions in this paper the circuit is assumed to be lossless. When we consider a small circuit dissipation, k is given, approximately, as

$$k = k' - jk'', \qquad k' \gg k'' \tag{F1}$$

where

$$k' = \omega\sqrt{\epsilon\mu}$$

$$k'' = \omega\sqrt{\epsilon\mu}\,(\tan\delta + r/d)/2 \tag{F2}$$

δ is loss angle of the spacing material, and r is skin depth.

[2] This is equivalent to assuming that the periphery is a perfect magnetic wall. Actually, however, a fringing field [see Fig. 2(a)] is always present. A simple but reasonable correction for it is to extend the periphery outwards by $2d(\log_e 2)/\pi$ to simulate the *static* fringing capacitance.

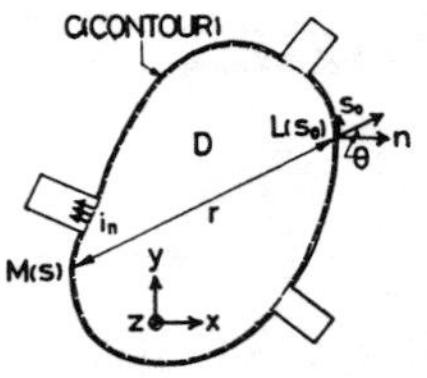

Fig. 3. Symbols used in the integral equation.

At a coupling port, (2) is no longer valid. Let the width of the port and the surface current density normal to the periphery C be denoted by W and i_n, respectively. If an admittance Y is connected to this port,

$$Y \doteqdot \frac{2\int_W i_n ds}{\int_W V ds/W} = \frac{-2jW\int_W \left(\frac{\partial V}{\partial n}\right) ds}{\omega\mu d\int_W V ds} \tag{3}$$

holds. The factor 2 expresses the fact that the current flows on both the upper and lower surfaces of the center conductor.

III. Computer Analysis

A. Integral Equation

The main feature of the planar circuit, as compared with waveguide circuit, is that we can analyze an arbitrarily shaped planar circuit within a reasonable computer time.

We consider an arbitrarily shaped, triplate planar circuit with several coupling ports, as shown in Fig. 3. Solving the wave equation over the entire area inside the contour C will require a long computer time. However, when we are concerned only with the RF voltage along the periphery, such a computation is not necessary. Using Weber's solution for cylindrical waves [6], the potential at a point upon the periphery is found to satisfy the following equation (refer to the Appendix for the detail of the derivation):

$$2jV(s) = \oint_c \{k\cos\theta H_1^{(2)}(kr)V(s_0) - j\omega\mu d\, i_n(s_0)H_0^{(2)}(kr)\}ds_0. \tag{4}$$

In this equation $H_0^{(2)}$ and $H_1^{(2)}$ are the zeroth-order and first-order Hankel functions of the second kind, respectively, i_n denotes the current density flowing outwards along the periphery, s and s_0 denote the distance along contour C. The variable r denotes distance between points M and L represented by s and s_0, respectively, and θ denotes the angle made by the straight line from point M to point L and the normal at point L, as shown in Fig. 3. When i_n is given, (4) is a second-kind Fredholm equation in terms of the RF voltage.

B. Circuit Parameters of an Equivalent N-Port [7]

For numerical calculation we divide the periphery into N incremental sections numbered as 1, 2, $\cdots$, N,

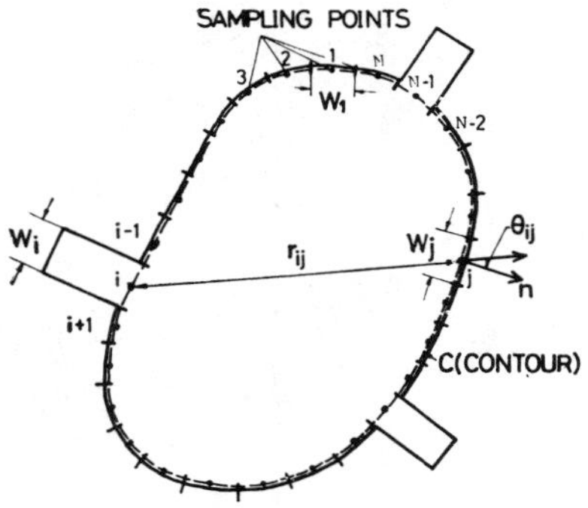

Fig. 4. Symbols used in the computer analysis.

having widths $W_1, W_2, \cdots, W_N$, respectively, as illustrated in Fig. 4. Coupling ports are assumed to occupy each one of those sections. Further, we set N sampling points at the center of each section.

When we assume that the magnetic and electric field intensities are constant over each width of those sections, the above integral equation results in a matrix equation:

$$2jV_i = \sum_{j=1}^{N} \{kV_jG_{ij} + j\omega\mu dI_jF_{ij}\} \tag{5}$$

where

$$F_{ij} = \begin{cases} \dfrac{1}{W_j}\displaystyle\int_{W_j} H_0^{(2)}(kr)ds, & (i \neq j) \\ 1 - \dfrac{2j}{\pi}\left(\log\dfrac{kW_i}{4} - 1 + \gamma\right), & (i = j) \end{cases}$$

$$G_{ij} = \begin{cases} \displaystyle\int_{W_j} \cos\theta H_1^{(2)}(kr)ds, & (i \neq j) \\ 0, & (i = j) \end{cases} \tag{6}$$

$\gamma = 0.5772 \cdots$ is Euler's constant, and $I_j = -i_nW_j$ represents the total current flowing into the jth port. The formulas for F_{ii} and G_{ii} in (6) have been derived assuming that the ith section is straight.

We can temporarily consider that all the N sections upon the periphery are coupling ports and that the planar circuit is represented by an N-port equivalent circuit. Then, from the above relations, the impedance matrix of the equivalent N-port circuit is obtained as

$$Z = U^{-1}H \tag{7}$$

where U and H denote N-by-N matrices determined by the shape of the circuit, whose components are given as

$$\begin{cases} u_{ij} = -kG_{ij}, & (i \neq j) \\ u_{ii} = 2j \end{cases} \qquad h_{ij} = j\frac{\omega\mu d}{2}F_{ij} \tag{8}$$

and U^{-1} denotes the inverse matrix to U.

In practice, most of the N ports described above are open circuited. When external admittances are connected to several of them and the rest of the ports are left open circuited, the reduced impedance matrix can be derived without difficulty.

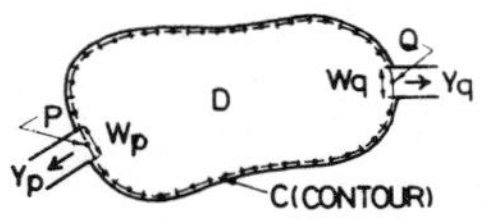

Fig. 5. Center conductor of a two-port planar circuit.

C. Transfer Parameters of a Two-Port Circuit [8]

In the case of a two-port circuit, the transfer parameters A, B, C, and D of the equivalent two-port can be given more simply as follows.

Suppose P and Q denote the driving terminal and load terminal, respectively, as shown in Fig. 5. Admittances Y_p and Y_q are connected to those terminals:

$$\begin{aligned} Y_p &= 2i_n(P)W_p/V_p \\ Y_q &= 2i_n(Q)W_q/V_q. \end{aligned} \tag{9}$$

Then Y_p has a negative conductance component. Equation (5) can be applied to all the N sampling points. Thus the RF voltage at each point can be given by the following matrix equation:

$$[U + Y_pV + Y_qW]\begin{bmatrix} V_1 \\ \vdots \\ V_N \end{bmatrix} = 0 \tag{10}$$

where V and W are again matrices determined by the shape of the circuit:

$$V = \begin{bmatrix} 0 & \cdot & \overset{p}{v_{1p}} & \cdot & 0 \\ \cdot & \cdot & \cdot & \cdot & \cdot \\ \cdot & \cdot & \cdot & \cdot & \cdot \\ 0 & \cdot & v_{Np} & \cdot & 0 \end{bmatrix}, \qquad v_{ip} = h_{ip}$$

$$W = \begin{bmatrix} 0 & \cdot & \overset{q}{w_{1q}} & \cdot & 0 \\ \cdot & \cdot & \cdot & \cdot & \cdot \\ \cdot & \cdot & \cdot & \cdot & \cdot \\ 0 & \cdot & w_{Nq} & \cdot & 0 \end{bmatrix}, \qquad w_{iq} = h_{iq}.$$

In order that a steady field exists in the circuit, from the nontrivial condition,

$$\det[U + Y_pV + Y_qW] = 0 \tag{11}$$

must hold. This equation directly gives a bilinear relation between $-Y_p$, the driving point admittance, and Y_q, the load admittance, as

$$-Y_p = \frac{C' + D'Y_q}{A' + B'Y_q} \tag{12}$$

where A', B', C', and D' are given as the following determinants:

$$A' = \begin{vmatrix} u_{11} & \cdot & \overset{p}{v_{1p}} & \cdot & u_{1N} \\ \cdot & \cdot & \cdot & \cdot & \cdot \\ \cdot & \cdot & \cdot & \cdot & \cdot \\ u_{N1} & \cdot & v_{Np} & \cdot & u_{NN} \end{vmatrix}$$

$$B' = \begin{vmatrix} u_{11} & \cdot & \overset{p}{v_{1p}} & \cdot & \overset{q}{w_{1q}} & \cdot & u_{1N} \\ \cdot & \cdot & \cdot & \cdot & \cdot & \cdot & \cdot \\ \cdot & \cdot & \cdot & \cdot & \cdot & \cdot & \cdot \\ u_{N1} & \cdot & v_{Np} & \cdot & w_{Nq} & \cdot & u_{NN} \end{vmatrix}$$

$$C' = \det [U]$$

$$D' = \begin{vmatrix} u_{11} & \cdot & \overset{q}{w_{1q}} & \cdot & u_{1N} \\ \cdot & \cdot & \cdot & \cdot & \cdot \\ \cdot & \cdot & \cdot & \cdot & \cdot \\ u_{N1} & \cdot & w_{Nq} & \cdot & u_{NN} \end{vmatrix}.$$

Equation (12) shows that A', B', C', and D' are quantities proportional to the so-called transfer parameters A, B, C, and D of the equivalent two-port circuit. In order that the reciprocity condition $(\sqrt{AD-BC}=1)$ holds, we should divide A', B', C', and D' by $\sqrt{A'D'-B'C'}$ to get A, B, C, and D, respectively, as

$$\begin{pmatrix} A & B \\ C & D \end{pmatrix} = \frac{1}{\sqrt{A'D' - B'C'}} \begin{pmatrix} A' & B' \\ C' & D' \end{pmatrix}. \tag{13}$$

When the circuit is a one-port circuit, the input admittance is given simply as (C'/A').

When the circuit has no coupling port and no circuit loss, $C'=0$ gives the proper frequency; that is, the resonant frequency of the circuit. In this situation the planar circuit is the Babinet dual of a metal wall TE-mode waveguide at its cutoff frequency.

D. Examples of Computer Analysis

In computing G_{ij} and F_{ij}, the integrals in (6) can be subdivided into as many subsections as necessary to assure the desired accuracy. However, in the following calculations the simplest approximation is used:

$$G_{ij} = \cos \theta_{ij} H_1^{(2)}(kr_{ij}) W_j \tag{14}$$

$$F_{ij} = H_0^{(2)}(kr_{ij}). \tag{15}$$

1) One-Port Disk-Shaped Circuit: As an example of the computer analysis, the input admittance of a one-port disk-shaped circuit with $\epsilon_r = 2.62$, $a = 1.841$ [m], $d = 0.628$ [m] was computed first. (These values are not realistic ones; $a = 1.841$ [m] is used so that the fundamental resonant mode is given by $k = 1$ [m^{-1}].) The result is shown in Fig. 6. This figure shows the variation of the input admittance, given by (C'/A'), around the fundamental resonant frequency $f_0 = 1.841/2\pi a\sqrt{\epsilon\mu}$ where 1.841 is the first root of $J_1'(x)$. The parameter N denotes the number of the sampling points along the periphery.

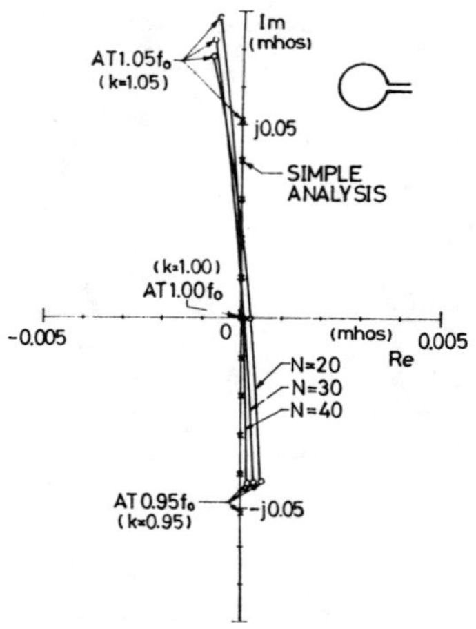

Fig. 6. Input admittance of an one-port disk-shaped circuit.

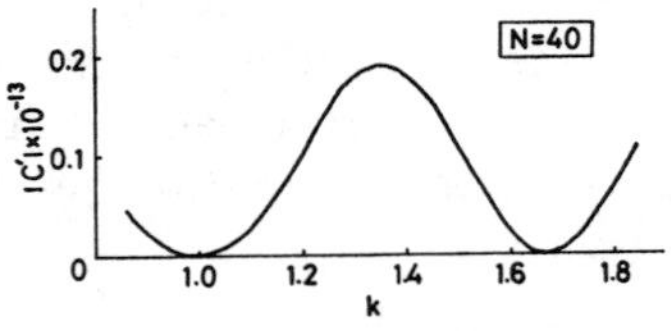

Fig. 7. The variation of $|C'|$ as a functions of k of a disk-shaped circuit.

TABLE I

COMPUTED FIRST EIGENVALUE k CORRESPONDING TO DIPOLE MODE OF A DISK-SHAPED CIRCUIT FOR VARIOUS N

Number of Sections N	Eigenvalue k
20	1.00013
30	1.00008
40	1.00007

As N increases, the real frequency locus approaches the values obtained by the simple theories as described in Section IV, shown as the small crosses along the ordinate in Fig. 6. Note that the abscissa is expanded by a factor of ten to exaggerate the computation error.

The values of k giving $C' = \det [U] = 0$ corresponds to the resonant frequencies of the circuit. From the simple analyses to be described in Section IV, they should satisfy $J_m'(ka) = 0$. For $a = 1.841$ [m], k should then be 1.000 [m^{-1}], 1.659 [m^{-1}], and so forth. This fact gives a good check of the computation accuracy.

Since C' is complex due to the computation error and $C' = 0$ is never realized for real k, we define k which gives the minimum of $|C'|$ as the eigenvalue. The variation of $|C'|$ is shown as a function of k in Fig. 7, which shows the first ($k = 1.00$) and the second ($k = 1.66$) minima. The former corresponds to the fundamental dipole mode (the first root of $J_1'(ka) = 0$) and the latter to the quadrupole mode (the first root of $J_2'(ka) = 0$). Table I shows the former k obtained for various N. As N increases k tends toward unity.

2) Two-Port Disk-Shaped Circuit: Next the transfer parameters A, B, C, and D of a disk-shaped circuit

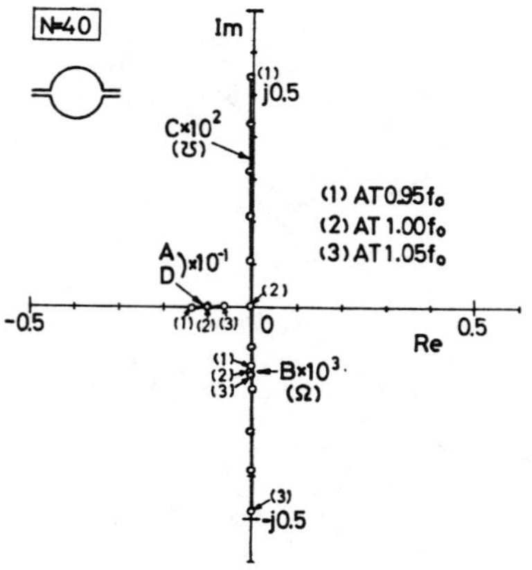

Fig. 8. Transfer parameters of a two-port disk-shaped resonator.

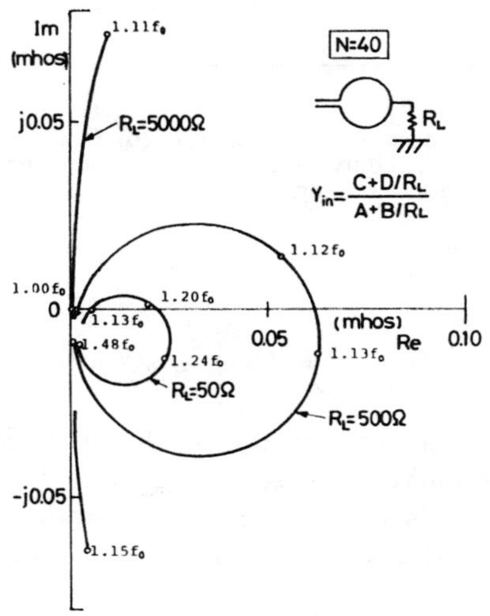

Fig. 9. Input admittance of a two-port disk-shaped resona tor loaded by various load resistances R_L.

having two-ports on its opposite sides were computed, and the result is shown in Fig. 8. In this figure the abscissa gives the real part and the ordinate the imaginary part of the transfer parameters obtained in the case $N=40$. Parameters A and D are equal to each other, as the circuit is symmetrical, and they take -1.0 at the resonant frequency.

By using the obtained transfer parameters and the relation $Y_{in}=(C+D/R_L)/(A+B/R_L)$, the input admittance of the disk-shaped circuit loaded by a pure resistance R_L was computed as shown in Fig. 9. The curves show the computer calculations for load resistances of 50, 500, and 5000 Ω. The loci in this figure cover a frequency range from the dipole mode of resonance to the quadrupole mode of resonance. In between these two low-admittance, parallel resonant points we find the frequency where the input admittance is very high, that is, a series resonant point, on the right-hand sides of loci. Note that such frequencies can never be found except by computer analysis.

Fig. 10 shows the RF voltage distribution along the

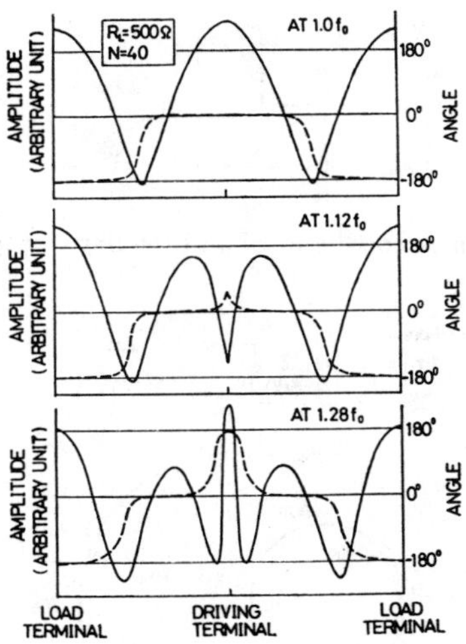

Fig. 10. The RF voltage distribution, magnitude (solid curve), and phase (broken curve), along the periphery of a disk-shaped circuit for $R_L=500$ Ω and $N=40$, for various frequencies.

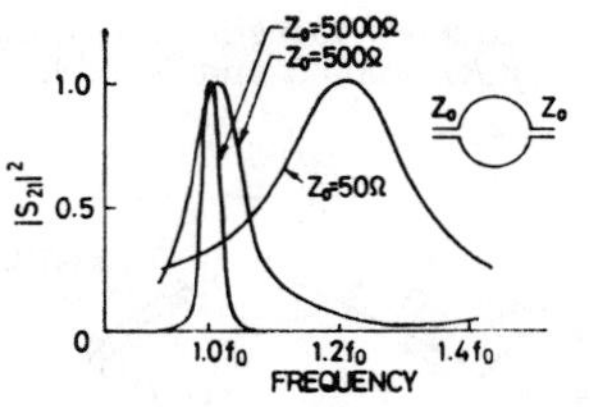

Fig. 11. The power transmission of a disk-shaped circuit calculated numerically for various characteristic impedances.

periphery for $R_L=500$ Ω and $N=40$, for various frequencies. In this figure, both ends and the center of the abscissa correspond to the load terminal and the driving terminal, respectively. The solid and broken curves show the magnitude (arbitrary scale) and phase of the RF voltage along the periphery, respectively. It is found that as the frequency increases, the distribution of the RF voltage changes from a dipole mode to a quadrupole mode. At the frequency of $1.12f_0$, the RF voltage at the input port is minimized; this corresponds to the series resonance of the circuit.

Fig. 11 shows the power transmission calculated numerically by using the relation $S_{21}=2Z_0/(AZ_0+B+CZ_0^2+DZ_0)$ for the case when the characteristic impedances Z_0 of the input and output lines are equal to each other and are pure resistance. It is found that both the frequency giving the maximum power transmission and the transmission bandwidth increase as the line impedance Z_0 is lowered.

3) Arbitrarily Shaped Circuit: As an example of more irregularly shaped circuit, a planar circuit, as shown in Fig. 12, was studied. Fig. 13 shows the frequency loci of the input admittance for $R_L=500$ Ω and $N=32$. In this figure f_0 denotes the resonant frequency of the quadrupole mode of a regular square circuit having dimensions $2a$ by $2a$. In this figure parallel resonances are found at $0.54f_0$ and $0.86f_0$, and a series resonance at $0.64f_0$.

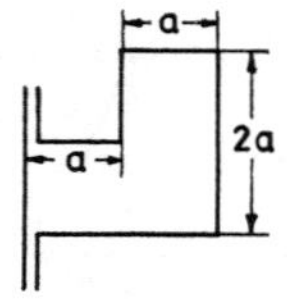

Fig. 12. Center conductor of an irregularly shaped circuit.

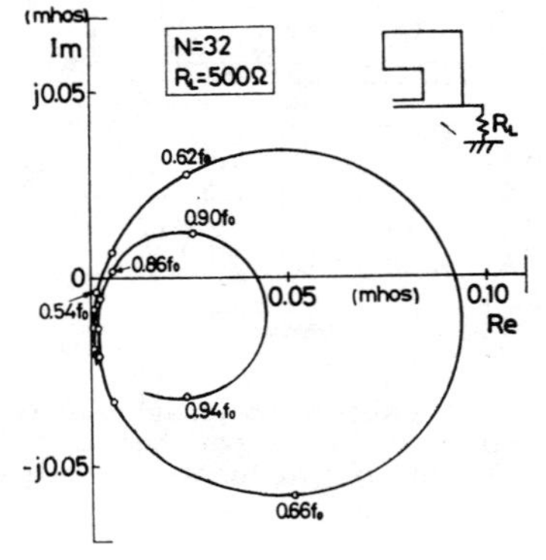

Fig. 13. The frequency locus of the input admittance for $R_L = 500\ \Omega$ and $N = 30$.

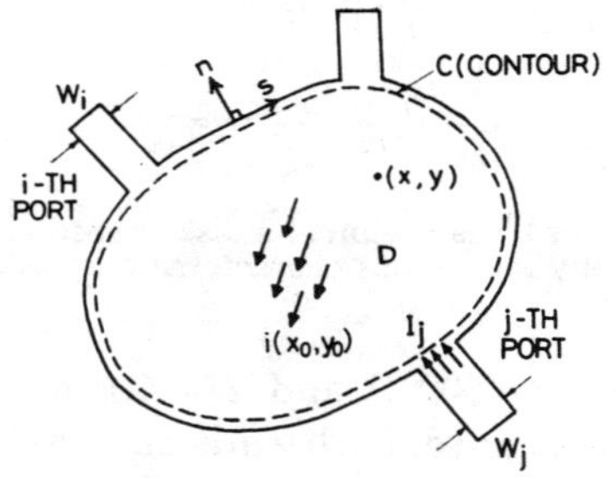

Fig. 14. Symbols used in the Green's function analysis.

IV. Analysis Based Upon Green's Function

When the shape of the circuit is relatively simple (a disk, for example) and we can get the Green's function of the wave equation analytically, the equivalent circuit parameters can be derived directly from the Green's function as follows.

We introduce the Green's function G of the second kind, having a dimension of impedance which satisfies

$$V(x, y) = \iint_D G(x, y \mid x_0, y_0) i(x_0, y_0) dx_0 dy_0 \tag{16}$$

inside the contour C shown in Fig. 14, and an open boundary condition

$$\partial G / \partial n = 0 \tag{17}$$

along C. In (16), $i(x_0, y_0)$ denotes an assumed (fictitious) RF current density injected normally into the circuit (see Fig. 14).

In a real planar circuit, current is injected from the periphery where a coupling port is present. Hence the RF voltage at a point upon the periphery is given by a line integral

$$V(s) = -\oint_c G(s \mid s_0) i_n(s_0) ds_0 \tag{18}$$

where s and s_0 are used to denote distance along C, and i_n is the line current density normal to C at coupling ports. Since i_n is present only at coupling ports, the RF voltage at the ith port is given approximately as

$$V_i \doteqdot \sum_j I_j \frac{1}{2W_i W_j} \int_{W_i} \int_{W_j} G(s \mid s_0) ds_0 ds \tag{19}$$

where $I_j = -2\int_{W_j} i_n(s_0) ds_0$ represents the current flowing into the jth port on both the upper and lower surfaces. Hence, the elements of the impedance matrix of the equivalent N-port circuit are

$$z_{ij} = \frac{1}{2W_i W_j} \int_{W_i} \int_{W_j} G(s \mid s_0) ds_0 ds. \tag{20}$$

As examples of the analysis based upon Green's function, the input impedances of one port disk and square circuits as shown in Fig. 15 are calculated.

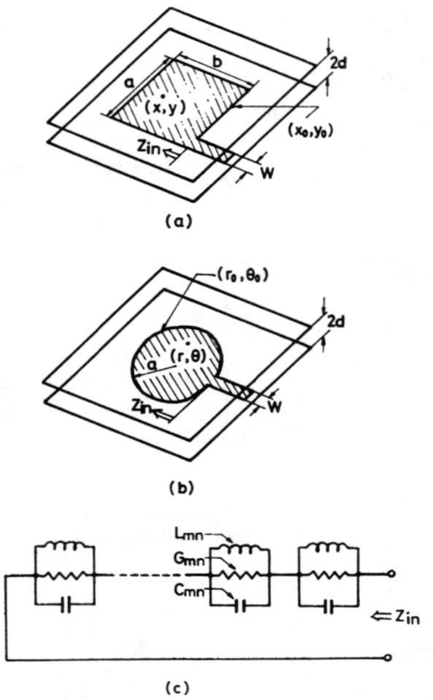

Fig. 15. (a) One-port square resonator. (b) One-port disk-shaped resonator. (c) The equivalent circuit describing the input admittance of one-port resonator.

A. Square Circuit

For a square circuit pattern [see Fig. 15(a)] having $a \times b$, the Green's function is given as [9]

$$G(x, y \mid x_0, y_0) = j\omega\mu d \frac{4}{ab} \sum_n \sum_m \frac{\cos(k_x x_0) \cos(k_y y_0)}{k_x^2 + k_y^2 - k^2} \cdot \cos(k_x x) \cos(k_y y) \tag{21}$$

where $k_x = m\pi/a$ and $k_y = n\pi/b$.

We compute the input impedance [Z_{in} shown in Fig. 15(a)] of a one-port square circuit having the

TABLE II

THE PROPER FUNCTION AND EQUIVALENT CIRCUIT PARAMETERS OF THE TRIPLATE-TYPE, SQUARE, AND DISK CIRCUITS

Planar Resonator	Square Resonator [Fig. 15(a)]	Disk Resonator [Fig. 15(b)]
Proper function	$\cos(k_x x)\cos(k_y y)$	$J_m(k_{mn}r)\cos(m\theta)$
Resonant frequency f_{mn}	$\frac{\sqrt{(m/a)^2+(n/b)^2}}{2\sqrt{\epsilon\mu}}$	$\frac{k_{mn}}{2\pi\sqrt{\epsilon\mu}}$
C_{mn}	$\frac{\epsilon ab}{2d}\frac{1}{F}$	$\epsilon\frac{\pi a^2}{d}\{1-m^2/(ak_{mn})^2\}\frac{1}{F}$
L_{mn}	$\frac{2\mu d}{ab\{(m\pi/a)^2+(n\pi/b)^2\}}F$	$\frac{\mu d}{\pi}\frac{1}{(ak_{mn})^2-m^2}F$
G_{mn}	$2\pi f_{mn}C_{mn}/Q_0$	$2\pi f_{mn}C_{mn}/Q_0$
F	$\left(\frac{\sin(k_x W)}{k_x W}\right)^2$	$\left(\frac{\sin(mW/a)}{mW/a}\right)^2$
Q_0	$Q_0^{-1}=Q_d^{-1}+Q_c^{-1}$; $Q_d=1/\tan\delta$ (δ is the loss angle of the dielectrics); $Q_c=d/r$ (r is the skin depth of the conductor)	

coupling port at one of the corners as shown in Fig. 15(a). Equations (20) and (21) directly give

$$Z_{\text{in}} = \sum_n \sum_m \frac{j\omega\mu d(\sin(k_x W)/k_x W)^2}{ab(k_x^2 + k_y^2 - k^2)}. \tag{22}$$

When we use (F1) and (F2) to consider the circuit loss, we obtain, after some computations,

$$Z_{\text{in}} = \sum_n \sum_m \frac{1}{\left(j\omega C_{mn} - j\dfrac{1}{\omega L_{mn}} + G_{mn}\right)} \tag{23}$$

where C_{mn}, L_{mn}, and G_{mn} are the equivalent circuit parameters corresponding to each mode, and are tabulated in Table II.[3] Equation (23) suggests that the equivalent circuit describing the input impedance is given [see Fig. 15(c)] as a series connection of many parallel resonating circuits representing each resonance.

B. Disk Circuit

The disk circuit is shown in Fig. 15(b). The Green's function is given as

$$G(r, \theta \mid r_0, \theta_0) = \sum_n \sum_m \frac{2j\omega\mu d J_m(k_{mn}a)J_m(k_{mn}r)\cos(m(\theta - \theta_0))}{(k_{mn}^2 - k^2)a^2(1 - m^2/a^2k_{mn}^2)J_m^2(k_{mn}a)} \tag{24}$$

where k_{mn} satisfies

$$\frac{\partial}{\partial r} J_m(k_{mn}r)\Big|_{r=a} = 0 \quad (n\text{th root}). \tag{25}$$

[3] When we are concerned only with the circuit performance near a single resonant frequency, we can also derive the equivalent circuit parameters from the resonant frequency, stored energy, and the unloaded Q factor. The parameters thus obtained agree with those shown in Table II, except for the factor F describing the effect of the width of the terminal. This sort of analysis is fairly common in microwave circuit analyses. For example, one of the reviewers of this paper called the authors' attention to [10].

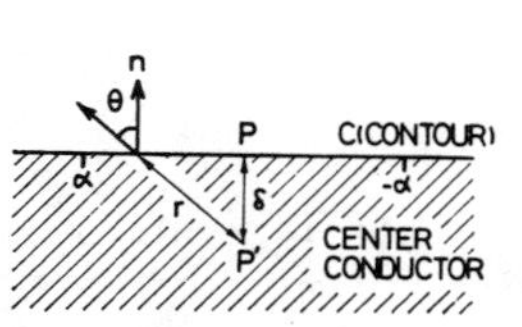

Fig. 16. Symbols used in the derivation of (4).

The equivalent circuit parameters of a one-port disk-shaped circuit can be computed by using (20), and are tabulated in Table II.

C. Multiport Disk and Square Circuits

The Green's function analysis can be applied to the circuit of this sort, which is useful in practical integrated circuitry as filters or hybrids. However, those examples are omitted for space limitations and will be reported elsewhere.

V. CONCLUSION

What is emphasized is that we can analyze an arbitrarily shaped planar circuit within a reasonable computer time. The *design* of a planar circuit, based upon the high-speed computer analysis and the trial-and-error principle, will also be possible within several years.[4]

Among possible applications of the planar circuit, the applications in Gunn and IMPATT oscillators seem to be promising. Since they are oscillation devices having relatively low impedances, the oscillator performance can be improved by using the planar circuit instead of the stripline circuitry.

APPENDIX

DERIVATION OF EQUATION (4)

The RF voltage at a piont P' *inside* the periphery satisfies the following Weber's solution for cylindrical waves [6],

$$4jV(P') = \oint_c \left\{ H_0^{(2)}(kr) \frac{\partial V(Q)}{\partial n} - V(Q) \frac{\partial H_0^{(2)}(kr)}{\partial n} \right\} ds. \tag{A1}$$

To obtain the RF voltage of the point P just *upon* the periphery, a little algebra is required. We first define a point P' just inside the point P as shown in Fig. 16, where we assume that $\delta \ll \alpha \ll k^{-1}$. By using the following approximations of the Hankel function near the origin

$$H_0^{(2)}(kr) \doteqdot -\frac{2j}{\pi}\log\frac{k\sqrt{s^2+\delta^2}}{2}$$

$$\frac{\partial H_0^{(2)}(kr)}{\partial n} \doteqdot -\frac{2j}{\pi}\frac{\delta}{s^2+\delta^2}$$

[4] For example, the time required to obtain the entire data in Fig. 8 is about 100 s using a typical Japanese high-speed computer HITAC-5020E. The improvement in the speed by a factor of (1/10) may be needed for the design.

we can rewrite

$$4jV(P')$$

$$= \int_{-\alpha}^{\alpha} \left\{ -\frac{2j}{\pi} \log \frac{k\sqrt{s^2+\delta^2}}{2} \frac{\partial V}{\partial n} + \frac{2j}{\pi} \frac{\delta}{s^2+\delta^2} V \right\} ds$$

$$+ \int_{\Gamma} \left\{ H_0^{(2)}(kr) \frac{\partial V}{\partial n} - V \frac{\partial H_0^{(2)}(kr)}{\partial n} \right\} ds \qquad \text{(A2)}$$

where Γ denotes the contour excluding the section $-\alpha \sim +\alpha$. If V and $\partial V/\partial n$ vary slowly in the minute section between $-\alpha$ and α, the integrals in (A2) become

$$-\frac{2j}{\pi} \int_{-\alpha}^{\alpha} \log \frac{k\sqrt{s^2+\delta^2}}{2} \frac{\partial V}{\partial n} ds$$

$$= -\frac{4j}{\pi} \frac{\partial V}{\partial n} \left\{ \alpha \left(\log \frac{k\sqrt{\alpha^2+\delta^2}}{2} - 1 \right) - \frac{k\delta}{2} \left(\text{arc cosec} \frac{\sqrt{\alpha^2+\delta^2}}{\delta} - \frac{\pi}{2} \right) \right\} \qquad \text{(A3)}$$

$$\frac{2j}{\pi} \int_{-\alpha}^{\alpha} \frac{s}{s^2+\delta^2} V ds = \frac{2j}{\pi} V \tan^{-1} \frac{\alpha}{\delta} \cdot \qquad \text{(A4)}$$

When P' approaches P, and hence δ tends to zero, (A2), (A3), (A4) give

$$4jV(P') = -\frac{4j}{\pi} \frac{\partial V}{\partial n} \left\{ \alpha \left(\log \frac{k\alpha}{2} - 1 \right) \right\} + 2jV(P)$$

$$+ \int_{\Gamma} \left\{ H_0^{(2)}(kr) \frac{\partial V}{\partial n} - V \frac{\partial H_0^{(2)}(kr)}{\partial n} \right\} ds. \qquad \text{(A5)}$$

Next, as α tends to zero, the first term in the right-hand side of (A5) vanishes, and hence

$$2jV(P) = \oint_{c} \left\{ H_0^{(2)}(kr) \frac{\partial V}{\partial n} - V \frac{\partial H_0^{(2)}(kr)}{\partial n} \right\} ds. \qquad \text{(A6)}$$

This equation and the relations

$$\frac{\partial V}{\partial n} = -j\omega\mu d\, i_n$$

$$\frac{\partial H_0^{(2)}(kr)}{\partial n} = -k \cos\theta H_1^{(2)}(kr)$$

give (4) in the text.

Acknowledgment

The authors wish to thank M. Hashimoto of Osaka University, Osaka, Japan, for stimulating discussions.

References

[1] T. Okoshi, "The planar circuit," in *Rec. of Professional Groups, IECEJ*, Paper SSD68-37/CT68-47, Feb. 17, 1969.

[2] ——, "The planar circuit," *J. IECEJ*, vol. 52, no. 11, pp. 1430–1433, Nov. 1969.

[3] S. Mao, S. Jones, and G. D. Vendelin, "Millimeter-wave integrated circuits," *IEEE Trans. Microwave Theory Tech.* (*Special Issue on Microwave Integrated Circuits*), vol. MTT-16, pp. 455–461, July 1968.

[4] Y. Tajima and I. Kuru, "An integrated Gunn oscillator," in *Rec. of Professional Groups, IECEJ*, Paper MW70-9, June 26, 1970.

[5] H. Bosma, "On stripline *Y*-circulation at UHF," *IEEE Trans. Microwave Theory Tech.* (*1963 Symposium Issue*), vol. MTT-12, pp. 61–72, Jan. 1964.

[6] J. A. Stratton, *Electromagnetic Theory*. New York: McGraw-Hill, 1941, p. 460.

[7] T. Okoshi and T. Miyoshi, "The planar circuit—An approach to microwave IC," in *Proc. 1971 European Microwave Conf.*, Paper C4, Aug. 1971.

[8] —— "The planar circuit—A novel approach to microwave circuitry," in *Proc. Kyoto Int. Conf. on Circuit and System Theory*, Paper B-5-1, Sept. 1970.

[9] P. M. Morse and H. Feshbach, *Method of Theoretical Physics*, pt. II. New York: McGraw-Hill, 1953, p. 1360.

[10] S. B. Cohn, P. M. Sherk, J. K. Shimizu, and E. M. T. Jones, "Final report on strip transmission lines and components," Stanford Res. Institute, Contract DA36-0393SC-63232, Final Rep., pp. 79–162.

Paper 1.3

Time Domain Electromagnetic Simulation for Microwave CAD Applications

Wolfgang J. R. Hoefer, *Fellow, IEEE*

Abstract—**Rapid progress in time domain modeling and computer technology have brought practical time domain simulators within reach. The next decade will see the emergence of sophisticated time domain simulation tools linking geometry, layout, physical and processing parameters of a microwave or high speed digital circuit with its system specifications and the desired time and frequency performance, including electromagnetic susceptibility and emissions. These CAD systems will most likely employ dedicated parallel processors configured specifically for modeling three-dimensional field problems. Furthermore, the nature of discrete time domain algorithms allows the designer to employ optimization and synthesis procedures which differ from those employed in traditional frequency domain CAD tools. In this paper, recent progress in time domain modeling will be highlighted, and the possible impact of these development on future CAD procedures and systems will be discussed.**

I. Introduction

COMPUTER AIDED DESIGN (CAD) of microwave circuits, including nonlinear analysis, is almost exclusively performed in the frequency domain. The principal reason lies in the historical development of electromagnetic field analysis in communication and broadcasting engineering. Time domain analysis was, at least in the first half of the twentieth century, mostly performed in high voltage engineering in order to deal with transients due to switching, loading changes, breakdown and lightning. Later, the development of digital techniques for communications, measurement (time domain reflectometry) and high speed logic have generated greater interest in time domain analysis, both at the circuit and the field levels.

In order to reduce the computational complexity of a field problem it is, of course, both reasonable and desirable to reduce the number of independent variables by using Fourier transform techniques. In this way, the time dimension can be eliminated by assuming sinusoidal time dependence. Furthermore, if a structure is uniform in one or even two space dimensions, the problem can be broken down into simpler steady-state periodic solutions. The spectral domain approach is a good example of this procedure. In more general terms, traditional field analysis

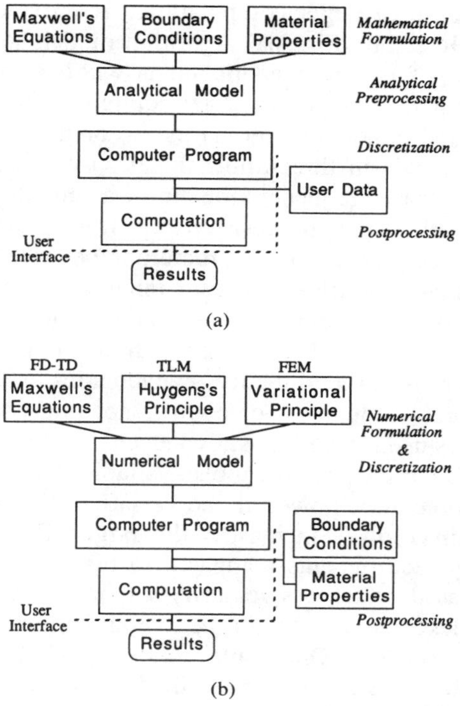

Fig. 1. Traditional (a) and alternative (b) approach to electromagnetic modeling of microwave structures.

transforms the physical situation into a mathematical model by projecting or mapping the space-time problem into one or several abstract domains where the solution procedure is simplified. This traditional approach to field analysis is summarized in Fig. 1(a). The important point here is that the formulation of the problem includes the specific boundary topology and material composition, as reflected, for example, by the choice of basis functions in a spectral domain solution, or the choice of eigenmodes in a mode-matching solution. Typical examples are spectral domain programs for multi-layer planar circuit structures, or mode-matching programs for waveguide filters. This extensive analytical pre-processing leads to computationally efficient algorithms which, when implemented on powerful computers, result in efficient field solving tools.

Naturally, these tools are well suited for specific types of circuit topologies associated with well defined manufacturing technologies. However, the user of such tools can only control the geometrical and electrical parameters

Manuscript received September 4, 1991; revised February 17, 1992.
The author is with the Department of Electrical and Computer Engineering, University of Victoria, P.O. Box 1700, Victoria, BC, Canada V8W 2Y2.
IEEE Log Number 9108315.

Reprinted from *IEEE Trans. Microwave Theory Tech.*, vol. 40, no. 7, pp. 1517–1527, July 1992.

of the structure under study, but not its essential configuration. It is also very difficult to obtain with such methods the response of a circuit to arbitrary waveform excitation, particularly in the presence of dispersion and nonlinearities. However, this is usually acceptable in return for computational efficiency, provided that the circuit technology is well defined.

In recent years, however, alternative approaches have emerged which provide greater flexibility to the user at the expense of larger computational requirements. (Fig. 1(b)). Here, the principal algorithm models the intrinsic behavior of electromagnetic fields without reference to specific boundary and material configurations; it can be formulated either for time-harmonic or general time-dependent fields in three-dimensional space, thus forming the basis for a frequency domain or a time domain electromagnetic simulator, respectively. It is up to the user to specify every detail of the structure to be modeled. While it demands a considerable user input, such an approach extremely general and flexible; it allows the user to tackle previously unsolved structures without the need for special expertise in field theory, mathematics and numerical techniques. This does, of course, not imply that circuit designers need to be less knowledgeable when using such a tool, but rather that the tools available to circuit designers become more powerful and versatile. As the available computing power increases, the larger CPU time and memory requirements compared to that of a specialized code based on extensive analytical pre-processing, becomes less relevant within the overall context of electromagnetic design. One could almost speak of a paradigmatic shift in electromagnetic field modeling. Among the factors that are driving this evolution, the following are the most obvious:

a) The rapid increase in speed and memory of digital computers, as well as the availability of vector computers and parallel processors.
b) The increasing complexity, density and operating frequency of electromagnetic structures, as well as the need to study their response to non-sinusoidal waveforms. Typical examples include monolithic integrated microwave circuits, high speed digital electronics, and EMI/EMC problems.
c) The requirements of potential users of numerical field modeling tools. The most important are the ability to solve realistic problems of arbitrary geometry, reliability, accuracy and numerical stability, time and frequency domain capability, user friendliness, extraction of relevant parameters, compatibility with already existing CAD tools, and the possibility of field visualization and animation.

One of the principal virtues of powerful processors is that they make it possible to model electromagnetic processes directly in space and time, the natural dimensions in which dynamic physical events are experienced. For maximum efficiency of digital processing, problems are formulated directly in numerical, algorithmic rather than in analytical form, and executed following a time-stepping procedure. This has the advantage that boundary conditions, electrical composition and excitation of a structure which are characteristic of a particular problem, do not restrict the basic algorithm, but are input by the user, preferably through a graphics interface (See Fig. 1(b)).

The evolution of microwave CAD closely follows that of electromagnetic analysis while depending strongly on available computer performance. Thus, earlier tools for microwave design were based on equivalent circuit representation and closed-form expressions, but modern versions employ increasingly some form of frequency domain field analysis and computer-generated multidimensional lookup tables. Design is performed by repeated analysis combined with appropriate optimization strategies.

Another important development is the combination of electromagnetic modeling with interactive computer graphics. Through field visualization and animation, electromagnetic field behavior can be observed directly, enabling the designer to relate the properties of a circuit to the behaviour of the fields and thus give physical meaning to abstract formalism.

However, modeling electromagnetic fields in the time domain requires a number of new concepts and procedures which, when implemented efficiently in the form of a time domain electromagnetic simulator, are opening new horizons in microwave CAD. In this paper, important developments in time domain field modeling will be highlighted, and their practical implementation in the form of CAD tools for process-oriented microwave circuit design will be demonstrated. Finally, future development trends will be discussed.

II. Basic Properties of Time Domain Simulators

A. *Time-Stepping Algorithms*

Discrete time domain models of electromagnetic fields can be obtained by discretizing time domain differential or integral formulations, by discrete spatial network representation of fields, or by finite element approximation. The large majority of time-domain simulators, however, is based either on a discretization of Maxwell's equations in differential form (Finite Difference—Time Domain, or FD-TD method) or on a discrete spatial network model (Transmission Line Matrix, or TLM method).

While FD-TD is formulated in terms of electric and magnetic field components, TLM exploits the analogy between field propagation in space and voltage/current propagation in a spatial transmission line network, which is formulated as a multiple scattering process following Huygens' principle. As a general rule, both methods lead to practically identical results; in fact, for each TLM scheme there exists an equivalent FD-TD formulation, and *vice versa*, which means that one can be derived from the

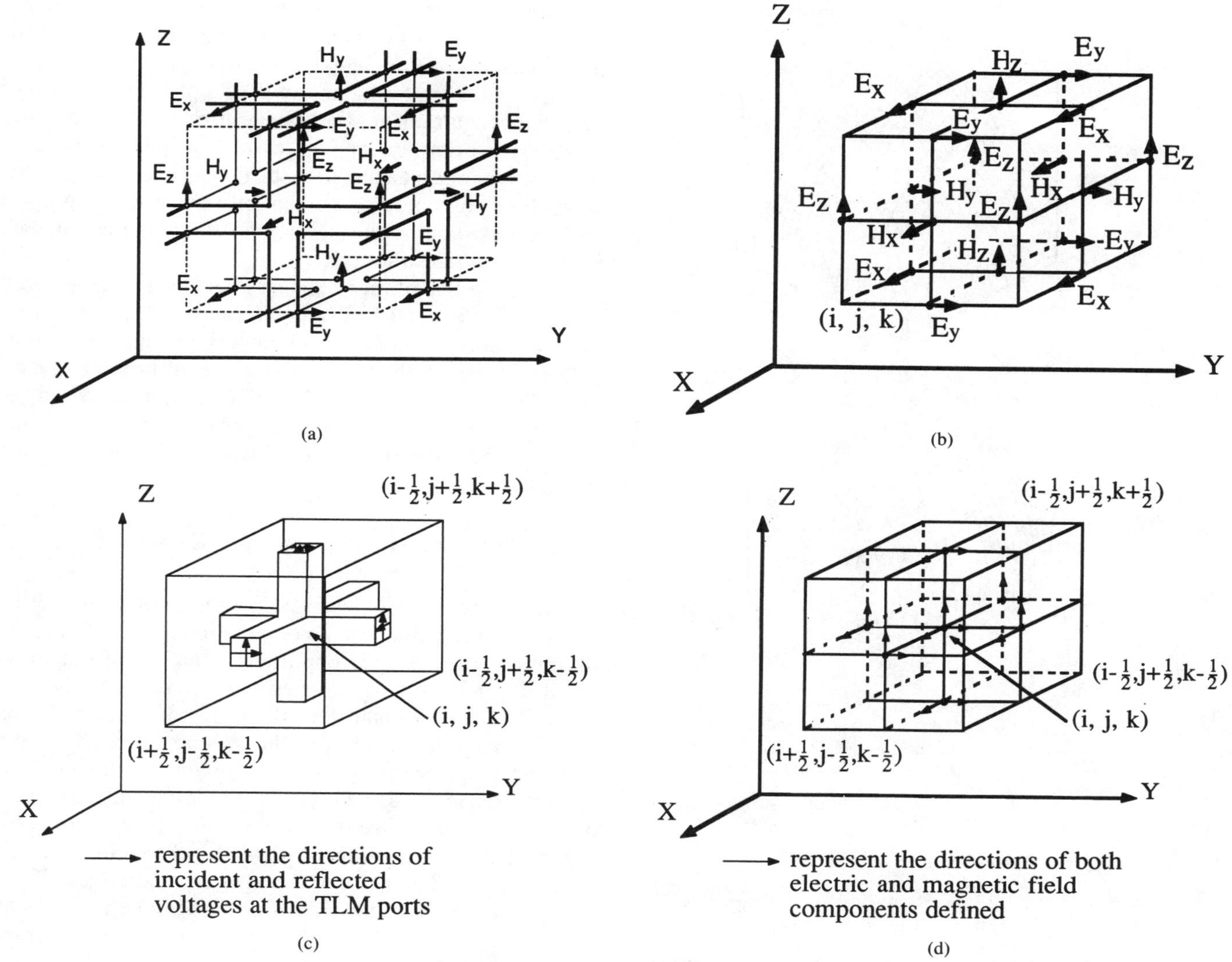

Fig. 2. Various schemes for time-space discretization of electromagnetic field problems. (a) Johns' distributed 3-D TLM node [1]; (b) Yee's Finite Difference—Time Domain grid [2]; (c) Johns' condensed 3-D TLM node [3]; (d) FD-TD grid proposed by Chen *et al.* [4].

other. Fig. 2 shows two such pairs of equivalent schemes. Fig. 2(a) and (b) show Johns' distributed TLM node [1] and Yee's unit FD-TD cell [2]. Fig. 2(c) and (d) compare Johns' condensed TLM node [3] and the equivalent FD-TD scheme derived by Chen *et al.* [4]. One basis for comparing discrete time domain field models is their dispersion of the propagation vector. It is well known that the dispersion in a discrete system is equal to that in its generic continuous system only in the infinitesimal limit of the discretization step ($(\Delta l/\lambda) \rightarrow 0$), and it deviates from it the more $\Delta l/\lambda$ increases. Fig. 3 shows the dispersion characteristics of the discretization schemes in Fig. 2 as derived by Nielsen and Hoefer [5]. These dispersion surfaces are the loci for the propagation vector in the three-dimensional discretized propagation space. For low frequencies ($(\Delta l/\lambda << 1$) the dispersion surfaces form a unit sphere in all cases, which means that propagation is isotropic and nondispersive. However, at higher frequencies, the dispersion of the condensed TLM node and Chen's FD-TD scheme is considerably smaller than that of the other two schemes in Fig. 2(a) and (b).

Clearly, equivalent TLM and FD-TD schemes possess identical dispersion characteristics. Furthermore, optimized codes for equivalent schemes require a similar computer memory and execution time. Nevertheless, they have their respective advantages and disadvantages when implementing boundaries, dispersive constitutive parameters, and nonlinear devices. In the final analysis, the choice between TLM or FD-TD is mostly based more on personal preferences and familiarity with one or the other method. In the following, the salient features of time domain simulators will thus be described in terms of TLM formalism with the understanding that there exists, or can be derived, an equivalent FD-TD formulation.

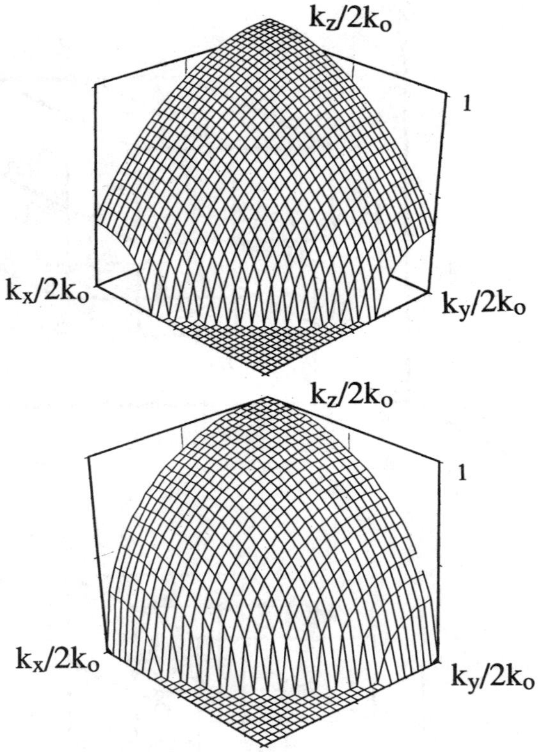

Fig. 3. Plots of the dispersion surfaces for the schemes shown in Fig. 2(a) Expanded TLM node and Yee's FD-TD scheme; (b) Condensed TLM node and Chen's FD-TD scheme. (Normalized frequency $2\pi\Delta l/\lambda = 0.7$. Stability factor for the FD-TD schemes $s = 0.5$. The surfaces are perfect unit spheres when $2\pi\Delta l/\lambda = 0$).

B. Advantages of Time Domain Analysis

The principal advantages of modeling electromagnetic fields in the time domain are the following:

1) Non-sinusoidal waveforms and transient phenomena can be studied directly.
2) Nonlinear and frequency dispersive behaviour can be modeled more physically in the time domain than in the frequency domain where the representation of these properties is mostly phenomenological.
3) Properties over a wide frequency band can be obtained with a single impulsive analysis.
4) The geometry and the electromagnetic material properties of a structure can be varied during a simulation through modeling of moving boundaries and time varying constitutive parameters.
5) Direct numerical synthesis is possible through reversal of electromagnetic processes in time.
6) Due to the localized character of time domain algorithms, they are particularly well suited for implementation on parallel or vector processors.

However, in order to implement these capabilities in a practical simulator, dispersive and nonlinear properties, moving boundaries, and sophisticated signal processing procedures must be implemented, which include forward and inverse Fourier transforms, numerical convolution, and wideband absorbing boundaries. The latter are required for the extraction of scattering parameters from the impulse response of dispersive guiding structures, and for the truncation of the computational domain when modeling free-space scattering and radiation problems. Another important requirement is a graphic user interface for 3-D geometry editing, parameter extraction and display, as well as dynamic visualization of fields, charges and currents. Last, but not least, compatibility with existing CAD tools is important, requiring the transfer of simulation data in the appropriate formats.

The feasibility of these features has been demonstrated both in TLM and FD-TD environments [6]–[8]. However, the computational requirements for modeling complex structures with such methods are still extremely severe. Therefore, research efforts are being focused on the development of techniques to improve computational efficiency, the most important of which will be discussed next.

III. Techniques for Improving Computational Efficiency

One of the main objectives in time domain modeling research is the reduction of computational expenditure. In the following, the most important techniques for achieving this objective will be briefly described. The first exploits the localised nature of the time domain algorithms through parallel processing, the second is based on the numerical processing of the time domain output signal using Prony's Method, and the third involves the reduction of the so-called coarseness error due to insufficient resolution of highly nonuniform fields by the finite discretization in the vicinity of sharp corners and edges. Even though these techniques will be described below for TLM modeling, they can be implemented in FD-TD schemes as well.

A. Parallel Processing

The principle of causality ensures that any local change in the discretized field space affects only its immediate neighbourhood at the next computational step. The highly localised nature of time domain algorithms is therefore perfectly suited for parallel processing. The type of machine most suitable for the implementation of such algorithms is the SIMD system (Single Instruction Multiple Data) such as the "Connection Machine" of Thinking Machines Inc., which consists of a large number of processors (up to 16 384) that are networked together. The processors have their own memory banks and operate on instructions broadcast to them by a host computer.

This allows implementation in a form quite different from the program on a serial machine. Since each processor has its own memory, it is practical to assign, in the TLM case, to each of them in an impulse scattering matrix and a set of boundary conditions. The impulse scattering matrix incorporates the local properties of the computational space such as permittivity, permeability, conductivity, and mesh size in the three co-ordinate directions. The boundary conditions specify whether there are boundaries between a node and its neighbours, or whether the nodes are connected together. This parallel implementation greatly facilitates variable mesh grading, conformal boundary modeling, and the simulation of highly inhomogeneous materials and complicated geometries.

In order to assess the impact of parallel processing on the computing speed we have implemented the 3-D TLM condensed node on the Connection Machine of the INRIA Computing Centre at Sophia Antipolis, France. The gain in performance is impressive. Fig. 4 compares, on a logarithmic scale, the improvements made over the last year in computing speed using various programming techniques [9] and parallelisation. The original matrix formulation by Johns [3] requires 144 multiplications and 126 additions/subtractions per scattering per node. Through manipulation of the highly symmetrical impulse scattering matrix, Tong and Fujino [9] have reduced the scattering to six multiplications, 66 additions/subtractions and 12 divisions by four, increasing computing speed over six times. Programming in Assembler rather than C++ accelerates the process again four times. Finally, parallel processing increases speed by more than two orders of magnitude over the fastest serial version implemented on a 386 computer in C++ language. The combined measures effectively reduce computation times from hours to seconds.

This comparison strongly suggests that future implementations of time domain simulators for CAD purposes will be based on dedicated parallel processors or supercomputers that emulate parallel processing.

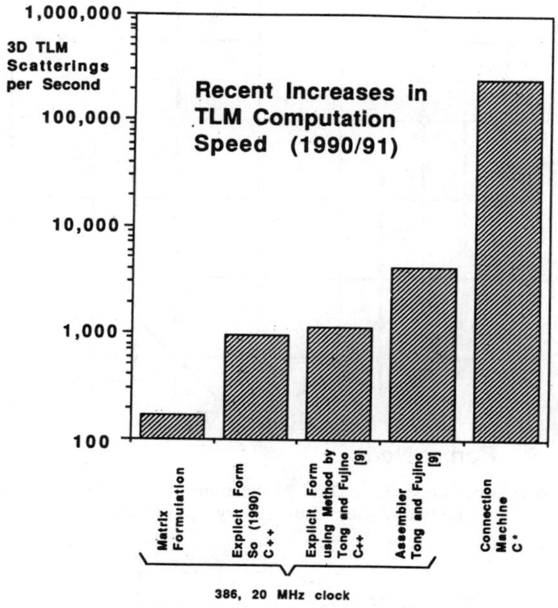

Fig. 4. Recent increases in the speed of TLM computations.

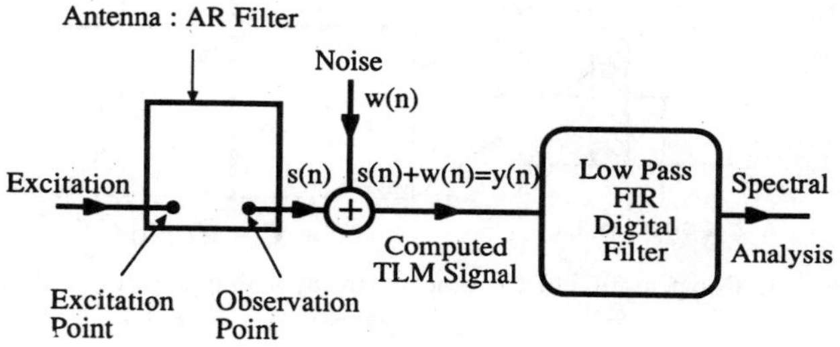

Fig. 5. Model of the Prony–Pisarenko method for TLM signal processing (after Dubard *et al.* [10]).

B. Signal Processing

The fast Fourier Transform (FFT) is the most frequently used signal processing method for extracting the spectral characteristics of a structure from a time domain simulation. For a realistic structure such as a planar antenna, typically 60 000 iterations (time samples) at 3,000 space points requiring 23 Mbytes of memory storage are needed to get the radiation pattern, but even this is insufficient for computing the input impedance of the antenna. It is therefore of prime importance to reduce the number of time samples required to extract a meaningful frequency response. To achieve this goal, processing of the time response using Prony's method has been applied successfully [10], [11].

In this approach the discrete time domain output signal is treated as a deterministic signal drowned in noise. (Fig. 5). The signal is then approximated by a superposition of damped exponential functions (Prony's method), and the noise is minimized using Pisarenko's model. This signal processing technique reduces the number of required time samples by typically one order of magnitude.

C. Reduction of Coarseness Error

One of the principal sources of error in the TLM analysis of structures with sharp edges and corners is the so-called coarseness error. It is due to the insufficient resolution of the edge field by the discrete TLM network. The error is particularly severe when boundaries and their corners are placed halfway between nodes as shown in Fig. 6. It is clearly seen that the nodes situated diagonally in front of an edge are not interacting directly with the boundary but receive information about its presence only across their neighbours who have one branch connected to it. The network is thus not sufficiently "stiff" at the edge, i.e., the edge field is not sufficiently coupled to the edge current, and results obtained in these cases are always shifted towards lower frequencies. The classical remedy for this problem is to use a finer mesh in the vicinity of the edge, but this introduces additional complications and computational requirements. On the other hand, the dispersion characteristics of the condensed 3-D TLM node (see Fig. 3(b)) are so good that the velocity error is practically negligible even for rather coarse meshes. A much better and more efficient way is thus to modify the corner node such that it can interact directly with the boundary corner through an additional stub as shown in Fig. 7 for the 2-D case [12]. Since this stub is longer than the other branches by a factor $\sqrt{2}$ it is simply assumed to have a correspondingly larger propagation ve-

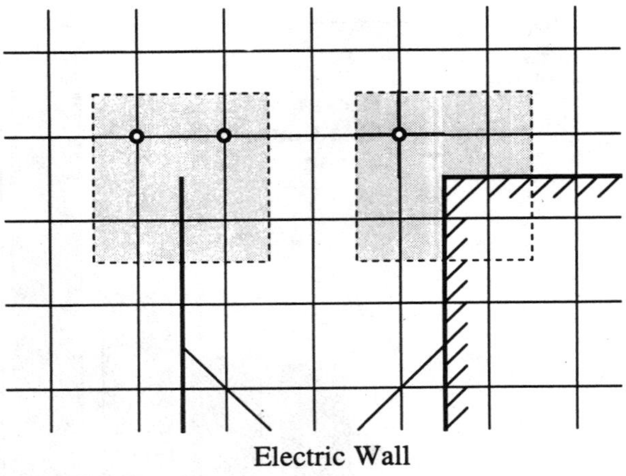

Fig. 6. Corner nodes in 2-D TLM mesh are not interacting directly with the boundaries, causing large coarseness error.

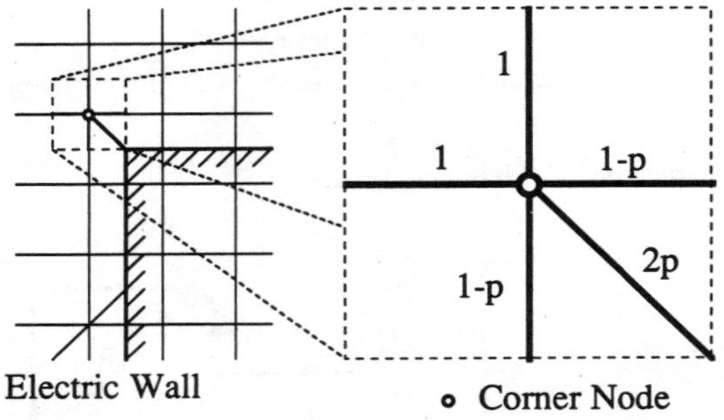

Fig. 7. Compensation of coarseness error by adding a fifth branch to the corner node.

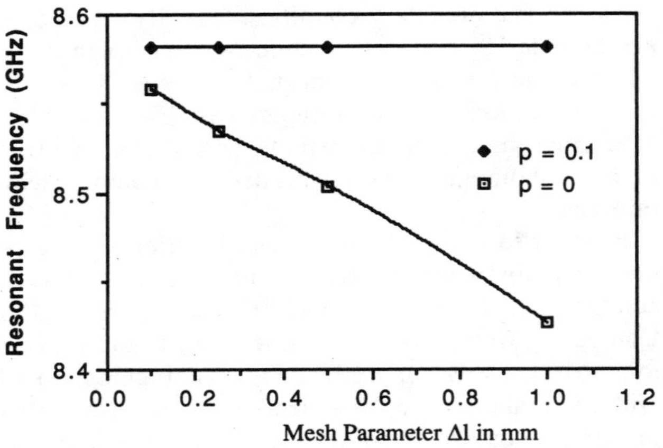

Fig. 8. Effect of the fifth branch of a corner node on the accuracy of TLM simulations of structures with sharp edges or corners.

locity. The effect of this corner correction is demonstrated in Fig. 8 which shows typical results for the first resonant frequency of a cavity containing a sharp edge as a function of the mesh parameter Δl (see [12]). The parameter p is proportional to the fraction of power carried by the fifth branch of the corner node and is equal to half the characteristic admittance of the corner branch when normalized to the link line admittance (see Fig. 7). For $p = 0$ (no corner correction) the coarseness error increases almost linearly with increasing Δl, while for $p = 0.1$ the frequency remains accurate even for a very coarse mesh. The simplest and most accurate method for finding this optimum value of p is to determine it empirically in such a way that the resonant frequency becomes independent of the mesh size as demonstrated in Fig. 8. Numerous numerical experiments using different geometries and frequencies have confirmed that the optimal $p-$ value is insensitive to these factors. For more detail the reader is referred to [12].

IV. Arbitrary Positioning of Boundaries

A. Accurate Dimensioning and Curved Boundaries

The accurate modeling of waveguide components, discontinuities and junctions requires a precision in the positioning of boundaries that is identical to, or better than, the manufacturing tolerances. If boundaries could only be introduced either across nodes or halfway between nodes, then the mesh parameter Δl would have to be very small indeed, leading to unacceptable computational requirements. Similar considerations apply when curved boundaries with relatively small radii of curvature must be modeled. It is therefore important to provide for arbitrary positioning of boundaries. The basis for this feature has been described already in 1973 by Johns [13] who, at the time, thought that the advantage of this procedure over stepped contour modeling was too small to warrant the additional complexity of the algorithm. However, this is not true when analyzing narrowband waveguide components such as filters. Furthermore, as mentioned in Section III-A, parallel implementation facilitates the inclusion of such features without penalty in computation speed.

Fig. 9 shows the concept of arbitrary wall positioning in two-dimensional TLM. The boundary branch which has a length different from $\Delta l/2$ is simply replaced by an equivalent branch of length $\Delta l/2$ having the same input admittance. This ensures synchronism, but requires a different characteristic admittance for the boundary branch and hence, a modification of the impulse scattering matrix of the boundary node (see [13]). The effect of such boundary tuning is shown in Fig. 10 which indicates that the length of the boundary branch can be continuously tuned over a range of more than one mesh parameter Δl without appreciable error. This important technique definitely removes the restriction that dimensions of TLM models can only be integer multiples of the mesh parameter.

An alternative technique which can easily be applied to the 3-D case as well consists of replacing the excess length of an irregular boundary branch beyond the $\Delta l/2$ position by an equivalent inductance or capacitance (which will be independent of frequency as long as $\Delta l/\lambda << 1$) and to discretize the differential equation governing the relation between voltage and current in this element [14]. This results in the following general recursive formula:

$$ {}_kV^i = \rho \frac{\kappa - 1}{\kappa + 1} {}_kV^r + \frac{\kappa}{\kappa + 1} (\rho_{k-1}V^r + {}_{k-1}V^i) \quad (1) $$

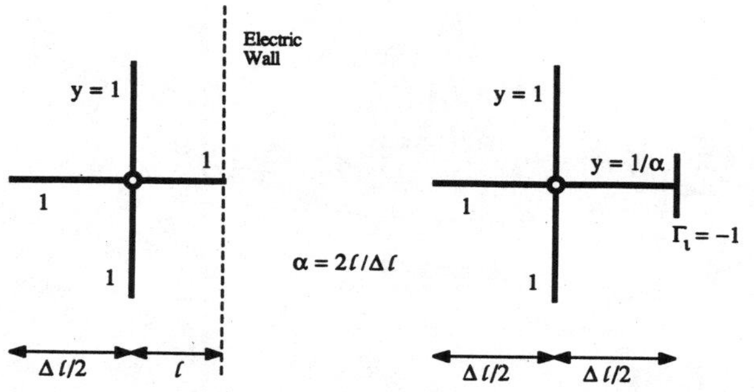

Fig. 9. Modification of boundary node for arbitrary position of boundary.

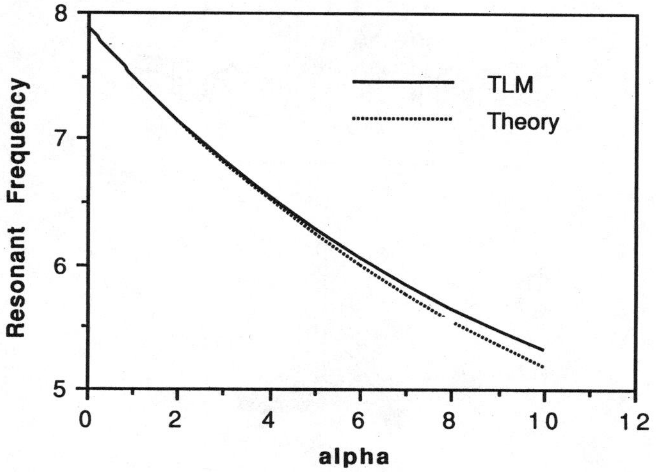

Fig. 10. Resonant frequency of a quarter-wave resonator terminated by a tunable electric wall as a function of the relative position $\alpha = 2l/\Delta l$.

where $\rho = +1$ for a magnetic wall and $\rho = -1$ for an electric wall. κ is equal to $2l/\Delta l$ in the 3-D TLM case, and $\sqrt{2l/\Delta l}$ in the 2D TLM case. Equation (1) indicates that the present impulse reflected from the boundary in the reference plane at $\Delta l/2$ depends on the present incident impulse as well as on the previous incident and reflected impulses, which need to be stored. This recursive algorithm amounts to a numerical procedure for integrating the differential equation describing the behavior of the reactive stub in the time domain.

These techniques effectively free the modeler from the restrictions of the "Manhattan-style" or staircase approximation of curved boundaries and provides the necessary flexibility for fine adjustments of structure dimensions without resorting to graded mesh techniques.

B. Implementation of Moving Boundaries

Since the position parameter $\alpha = 2l/\Delta l$, or the factor κ in (1), can be continuously varied during a simulation, this feature can be used to model moving boundaries. This means that the position of a boundary can be changed by an arbitrarily small amount between computational steps. This opens possibilities not available in frequency domain simulators, such as the modeling of expansion or contraction of objects during a simulation (temperature effects), the influence of wall vibrations on the electromagnetic behaviour of structures, Doppler effect, etc. It also allows the modeler to adjust or tune circuit dimensions during a simulation without the need for exciting the structure anew after every modification. This is because the field solution before the modification can be regarded as a slightly perturbed state of the field in the modified structure, and the transition between the two solutions is faster than the buildup of a new solution from the zero field state. It becomes thus feasible to tune, for example, a resonator in the time domain by exciting it only once, and when the field is built up, by moving (tuning) one or several of its walls and observing the change in its resonant frequencies. This tuning process may be combined with an optimization strategy, or be executed directly by the operator using visual feedback, thus realistically simulating the tweaking of the dimensions of a physical circuit.

V. Numerical Synthesis through Time Reversal

It has been shown recently by Sorrentino *et al.* [15] that the impulsive excitation of a linear lossless TLM network can be exactly reconstructed from the solution it produces, by inverting the TLM process. This is a direct consequence of the properties of the TLM impulse scattering matrix which is always equal to its inverse. In order to relate this fact to the numerical synthesis of microwave structures, consider a perfectly conducting body in space or in a waveguide. Finding the topology of such a scatterer amounts to reconstructing from the radiated fields the position of the current sources induced on its surface. In a forward simulation of a scattering process, the field function ϕ_{tot} in space is a superposition of the incident and the scattered fields:

$$\phi_{\text{tot}} = \phi_{\text{inc.}} + \phi_{\text{scatt.}} \tag{2}$$

Hence, in order to obtain the field $\phi_{\text{scatt.}}$ due to the induced sources alone, the incident field $\phi_{\text{inc.}}$ (or homogeneous solution) must be computed separately and then subtracted at each node from the total solution ϕ_{tot}. From this difference, the induced source and hence, the topology of the scatterer, can then be reconstructed by inverse TLM simulation.

To demonstrate this process for a very simple case, consider the situation shown in Fig. 11. It shows two identical sections of parallel plate waveguide which are terminated at each end with wideband absorbing boundaries. The upper guide is empty, while the lower guide contains a discontinuity in the form of a thin, perfectly conducting septum. A Gaussian pulse is now injected from the left into both structures (Fig. 11(a)). It travels unchanged through the upper homogeneous section, but is scattered by the septum in the lower section (Fig. 11(b) and (c)). The output nodes situated at both extremities capture the response of the upper section (incident or homogeneous solution), and that of the lower section (total field solution), shown here after 30, 70, and 100 iterations (a, b, and c, respectively). The difference of the responses of the empty and the loaded section is then computed and

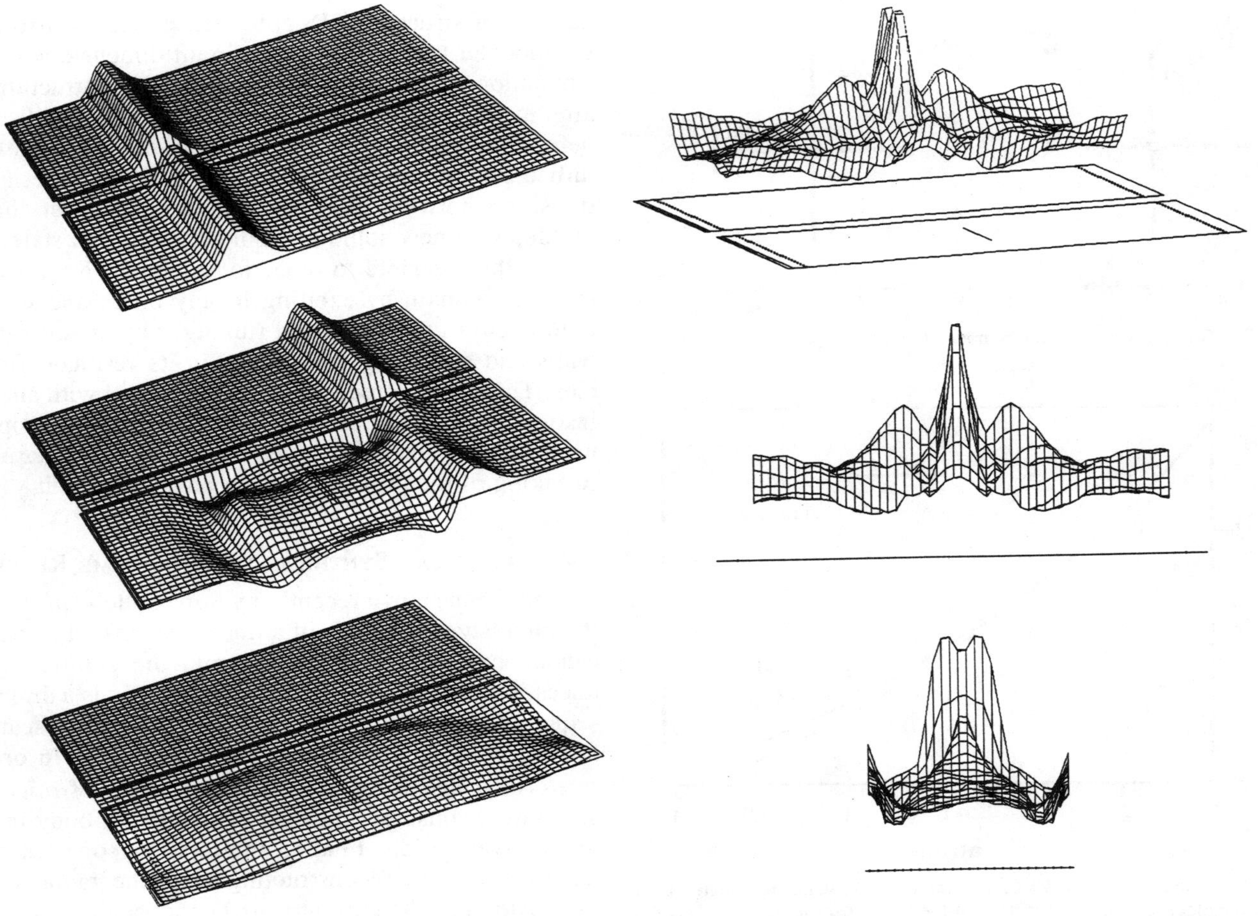

(a) (b) (c) (d) (e) (f)

Fig. 11. Two identical matched parallel plate waveguide sections are shown. The upper section is empty, and the lower contains a conducting septum. (a) to (c) show the fields due to injection of a Gaussian pulse from the left after 30, 70, and 100 iterations. The difference betweern the two responses is re-injected into the empty section in reverse time sequence. (d) to (f) show the maximum field after re-injection, reconstructing the position of the septum within the rather coarse resolution of the pulse.

re-injected into the empty section through the same rows of nodes in reversed time sequence, and the maximum field value occurring at each node during the entire reverse simulation is recorded. Fig. 11(d) to (f) show the result of this inverse process in perspective, longitudinal and transversal view. It can be seen that this procedure yields the exact position of the septum as expected. Obviously, the spatial resolution of the reconstruction in the transverse dimension is inversely proportional to the spatial width of the exciting Gaussian pulse, which is rather coarse in this case, and directly proportional to the mesh parameter Δl. However, the exact dimensions can be extracted by further processing (see [16]).

In a practical CAD problem, however, the specifications are not available as a time domain response, but more likely as a frequency response over a restricted frequency range. This information is insufficient for the reconstruction of the circuit topology yielding this response, and the designer must therefore generate the missing information somehow. It will be shown below how this is done in the case of the septum in Fig. 11. (see also [16]). Obviously, this septum acts as a shunt inductance in the operating band of the dominant mode. It is well known that the frequency response of a shunt inductance can be produced by many types of obstacles other than a septum (posts of rectangular, elliptical or irregular cross-section, irises, etc.) in various lateral positions. Therefore, the designer must first select an appropriate type of obstacle based on other considerations, such as available technology or ease of manufacturing. Then, an obstacle with approximately the right dimensions can be selected as a starting guess using available formulas for the shunt susceptance in terms of the dimensions.

For this starting obstacle the impulse response is then generated with a forward TLM analysis. The dominant mode content of this first response $\Phi^t_{\text{left}}(i, k)$ and $\Phi^t_{\text{right}}(i, k)$ is extracted at both extremities of the waveguide section, Fourier transformed and replaced by the desired (specified) dominant mode content in the frequency domain. The modified total response is then transformed back into the time domain and, reduced by the homogeneous response of the empty wáveguide, reinjected into the computational domain in the inverse time sequence. This procedure will now be described in more detail.

Let the impulse response of the approximate obstacle be $f_1(t)$ with its Fourier transform $F_1(\omega)$. The latter is most likely different from the desired (specified) frequency response $F_2(\omega)$ which is usually defined over a restricted frequency range and for the dominant mode only. We can thus modify $F_1(\omega)$ so that it is identical to $F_2(\omega)$ in the bandwidth of interest, and for the dominant mode or propagation. Now the modified response $F_1'(\omega)$ must be converted to a time domain signal for reinjection into the empty waveguide.

Since $F_1(\omega) - F_1'(\omega)$ is limited to the frequency band in which only the dominant mode can propagate, the transverse field distribution of the corresponding time function $f_1(t) - f_1'(t)$ is known. The time domain signal corresponding to the modified frequency response is thus

$$f_1'(t) = f_1(t) - \underbrace{\mathfrak{F}^{-1}[F_1(\omega) - F_2(\omega)]}_{\text{known transversal distribution}}. \tag{3}$$

This new time domain impulse response $f_1'(t)$ approximates the wanted impulse response since it contains, for the bandwidth of interest, the dominant mode response of the obstacle to be synthesized.

Finally, the difference between $f_1'(t)$ and the homogeneous response is injected into the empty structure in reversed time sequence, yielding an image of the synthesized obstacle as described above. This represents an improvement over the initial guess and in many cases, is already acceptable as a final result. However, if a new forward analysis still yields an unsatisfactory answer, the same sequence is repeated until the analysis satisfies specifications. In the case of single obstacles this process converges rather quickly, and typically two or three forward-and-backward simulation cycles yield the final geometry. Synthesis of larger circuit configurations is currently under study.

VI. User Interface and Typical Simulation Results

Since the user of a time domain electromagnetic simulator must enter every physical and electrical detail of a structure under test, the graphical interface must be very friendly and well conceived. It consists typically of a 3-D geometrical editor window which is similar to a 3-D drafting environment, a source window for specifying the excitation waveforms, an analyzer window for displaying time domain and frequency domain output as well as S-parameters, and an animator window for visualizing propagating fields. Figs. 12 to 15 show some typical examples (screen dumps) generated with a 3-D geometry-based TLM-Simulator prototype developed at the University of Ottawa.

Fig. 12 reproduces the geometrical editor window containing a microstrip meander line and a straight microstrip section side by side, shown from three co-ordinate directions and in perspective. The structures are enclosed by absorbing boundaries simulating open space. Fig. 13 visualizes the propagation of a Gaussian pulse through both the meander and reference lines. Obviously, this representation does not fully reproduce the effect of the dynamic display which shows the motion of the fields. (To give some idea of the time needed to run this simulation, each iteration requires about one second on a DEC 3100 RISC workstation. This includes the time for graphics. More powerful workstations with visualization hardware will require only a small fraction of a second for one iteration). Fig. 14 compares three time domain signals picked up at the extremities of the lines. They clearly show the effects of dispersion on the travelling waveform, the delay in the meander line, and the effects of surface wave

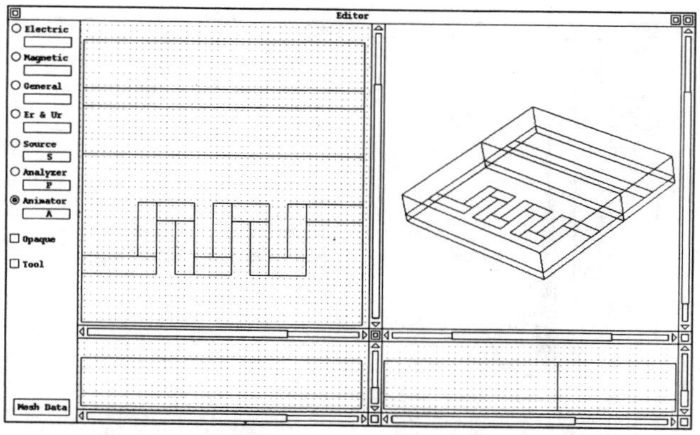

Fig. 12. Geometrical editor window of 3-D TLM electromagnetic simulator. It shows a microstrip meander line and a reference section in a box with absorbing top and side walls.

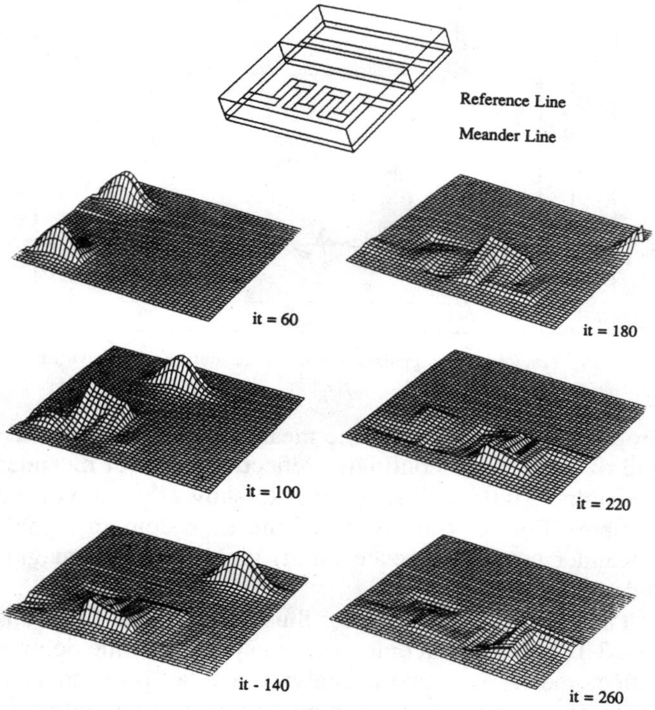

Fig. 13. Propagation of a Gaussian pulse through the meander line and the straight reference line.

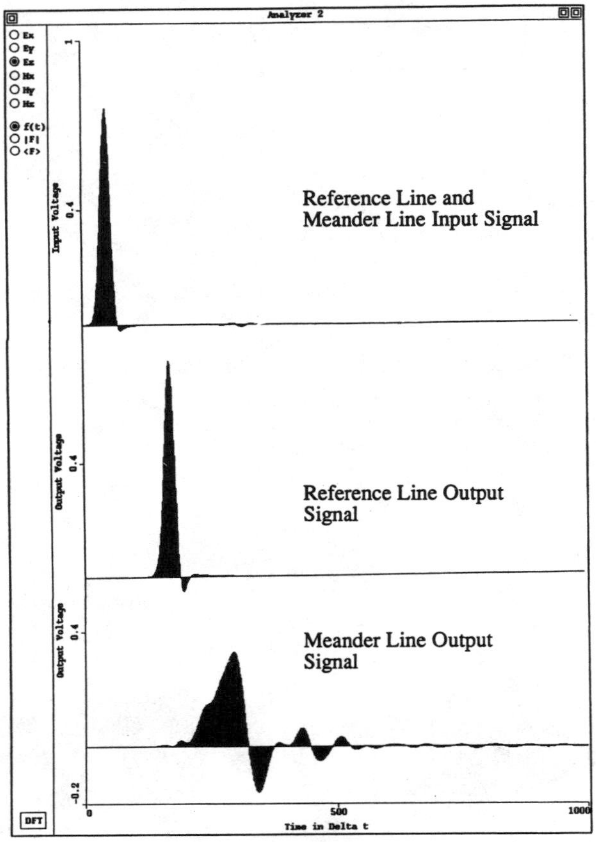

Fig. 14. Time domain signals at the input and output of the two lines.

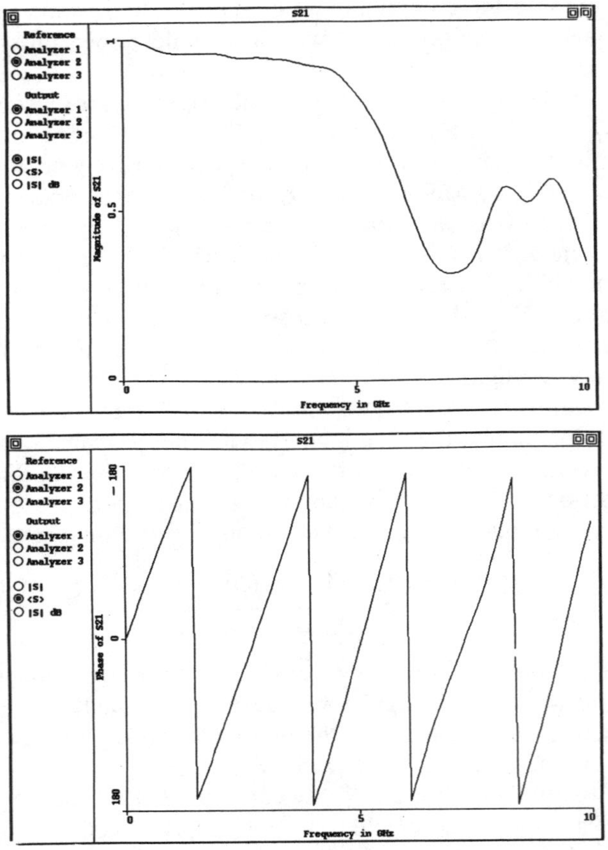

Fig. 15. Magnitude and phase of S_{21} extracted from the time domain signals in Fig. 14 by discrete Fourier transform.

propagation (precursor in the meander line output signal) and of multiple discontinuity reflections (tail of meander line output signal). Finally, Fig. 15 shows the ratio of the complex Fourier transforms of the time domain signals (meander output/reference input) yielding S_{21} in magnitude and phase.

The time domain simulator thus emulates the functions of a 3-D drafting machine, a video system, a time domain reflectometer, a network analyser and a spectrum analyzer, in other words a complete microwave laboratory in a single computer tool.

VII. Discussion and Conclusion

While most of the basic time domain modeling procedures have now been demonstrated and implemented in various computer programs, the considerable computer time and memory required to model realistic electromagnetic structures is still an obstacle when it comes to realizing professional CAD tools based on these techniques. Therefore, considerable research efforts are concentrating on measures to reduce the computation count to manageable levels. In this paper three different ways to achieve this have been described, namely parallel processing to accelerate the TLM process itself, Prony–Pisarenko signal processing to reduce the required number of computation steps, and coarseness error compensation at sharp corners and edges. All these methods can be combined to accelerate TLM simulations by several orders of magnitude. Since the computation count for TLM analyses increases faster than the fourth power of the linear mesh density, these accelerating features enhance our ability to model complex structures to a much greater extent than the mere increase of memory size and speed of computers. Procedures for fine tuning of wall positions have also been described. The ability to position boundaries at arbitrary distances from TLM mesh nodes not only provides higher modeling accuracy and resolution but also minimizes required computer resources. Since a time domain simulator explicitly stores all geometrical and electrical characteristics of a structure under test, it can directly be linked with manufacturing and processing facilities. Furthermore, it can produce time domain and frequency domain data in a format suitable for input into other CAD tools.

Future time domain CAD systems will most likely employ dedicated parallel processors configured in a 3-D array. Furthermore, the nature of discrete time domain algorithms gives rise to simulation procedures which differ considerably from those employed in traditional frequency domain CAD tools. These include the implementation of moving boundaries for geometrical tuning during a simulation as well as numerical synthesis through reversal of the TLM process in time. Considerable research is still required to bring all these procedures to full maturity, but it is conceivable that at the present rate of progress in time domain modeling, these techniques will equal, and in many aspects, surpass the capabilities of frequency domain CAD tools in the next decade.

References

[1] S. Akhtarzad and P. B. Johns, "Solution of Maxwell's equations in three space dimensions and time by the T.L.M. method of analysis," *Proc. Inst. Elec. Eng.*, vol. 122, no. 12, pp. 1344–1348, Dec. 1975.

[2] K. S. Yee, "Numerical solution of initial boundary value problems involving Maxwell's equations, in isotropic media," *IEEE Trans. Antennas Propagat.*, vol. AP-14, no. 5, pp. 302–307, May 1966.

[3] P. B. Johns, "A Symmetrical condensed node for the TLM method," *IEEE Trans. Microwave Theory Tech.*, vol. MTT-35, no. 4, pp. 370–377, Apr. 1987.

[4] Z. Chen, W. J. R. Hoefer, and M. Ney, "A new finite-difference time-domain formulation and its equivalence with the TLM symmetrical condensed Node," *IEEE Trans. Microwave Theory Tech.*, vol. 39, no. 12, pp. 2160–2169, Dec. 1991.

[5] J. Nielsen and W. J. R. Hoefer, "A complete dispersion analysis of the condensed node TLM mesh," in *4th Biennial IEEE Conf. Electromagnetic Field Computation Dig.*, Toronto, ON., Oct. 22–24, 1990.

[6] P. P. M. So and W. J. R. Hoefer, "3D-TLM time domain electromagnetic wave simulator for microwave circuit modeling," in *1991 IEEE MTT-S Int. Microwave Symp. Dig.*, Boston, June 11–13, 1991, pp. 631–634.

[7] P. P. M. So, Eswarappa, and W. J. R. Hoefer, "A two-dimensional TLM microwave field simulator using new concepts and procedures," *IEEE Trans. Microwave Theory Tech.*, vol. 37, no. 12, pp. 1877–1884, Dec. 1989.

[8] M. A. Morgan, Ed., "Finite element and finite difference methods in electromagnetic scattering," in *PIER 2 Progress in Electromagnetics Research*, New York: Elsevier, 1990.

[9] C. T. Tong and Y. Fujino, "An efficient algorithm for transmission line matrix analysis of electromagnetic problems using the symmetrical condensed node," *IEEE Trans. Microwave Theory Tech.*, vol. 39, no. 8, pp. 1420–1424, Aug. 1991.

[10] J. L. Dubard, D. Pompei, J. Le Roux, and A. Papiernik, "Characterization of microstrip antennas using the TLM simulation associated with a Prony–Pisarenko method," *Int. J. Numerical Modelling*, vol. 3, no. 4, pp. 269–285, Dec. 1990.

[11] W. L. Ko and R. Mittra, "A combination of FD-TD and Prony's method for analyzing microwave integrated circuits," *IEEE Trans. Microwave Theory Tech.*, vol. 39, no. 12, pp. 2176–2181, Dec. 1991.

[12] U. Müller, P. P. M. So, and W. J. R. Hoefer, "The Compensation of Coarseness Error in 2D TLM Modeling of Microwave Structures," in *1992 IEEE MTT-S Int. Microwave Symp. Dig.*, Albuquerque, NM, June 1–5, 1992.

[13] P. B. Johns, "Transient analysis of waveguides with curved boundaries," *Electron. Lett.*, vol. 9, no. 21, Oct. 18, 1973.

[14] U. Müller, A. Beyer, and W. J. R. Hoefer, "The implementation of smoothly moving boundaries in 2D and 3D TLM simulations," in *1922 IEEE MTT-S Int. Microwave Symp. Dig.*, Albuquerque, NM, June 1–5, 1992.

[15] R. Sorrentino, P. P. M. So, and W. J. R. Hoefer, "Numerical microwave synthesis by inversion of the TLM Process," in *21st European Microwave Conf. Dig.*, Stuttgart, Germany, Sept. 9–12, 1991.

[16] M. Forest and W. J. R. Hoefer, "TLM synthesis of microwave structures using time reversal," in *1992 IEEE MTT-S Int. Microwave Symp. Dig.*, Albuquerque, NM, June 1–5, 1992.

Full-Wave Spectral-Domain Analysis for Open Microstrip Discontinuities of Arbitrary Shape Including Radiation and Surface-Wave Losses (Invited Paper)

Tzyy-Sheng Horng, Nicolaos G. Alexopoulos, Shih-Chang Wu, and Hung-Yu Yang

Electrical Engineering Department, University of California, Los Angeles, Los Angeles, California 90024

Received January 9, 1992; revised April 13, 1992.

ABSTRACT

This article is a survey of the theoretical background for full-wave spectral-domain analysis of open microstrip discontinuities of arbitrary shape. The spectral-domain dyadic Green's function, which takes into account all the physical effects, such as radiation and surface waves, is used to formulate an electric field integral equation. The method of moments is then employed to find the current distribution on the microstrips, and subsequently, the scattering parameters of the junctions. Since all field components can be expressed in terms of the dyadic Green's function and the current distribution, the losses due to both radiation and surface waves are further determined through a rigorous Poynting vector analysis. To model the discontinuities of arbitrary shape, both rectangular and triangular subdomain functions are used as the current expansion functions in the moment method procedure. In addition, the semi-infinite traveling wave functions are applied to simulate the feeding structure and isolate individual junction effects. Several examples are demonstrated to illustrate the utility of different techniques in this analysis. Comparison of some numerical results with available experimental data shows excellent agreement. Finally, this approach is most natural for the characterization of 3-D integrated circuits and the design of printed antennas including excitation circuit effects. **© 1992 John Wiley & Sons, Inc.**

1. INTRODUCTION

In the past, planar waveguide models [1–8] as well as quasistatic methods [8–21] were used to characterize open microstrip discontinuities. In these techniques, a complicated microstrip problem was approximated either by adding equivalent magnetic side walls and substituting an equivalent dielectric constant or by solving Poisson's equation instead of Maxwell's equations. Both models are restricted to either low frequencies or electrically thin substrates. In general, their application is of limited value, since it does not rigorously account for losses due to radiation and surface waves.

As the operating frequency of integrated circuits and printed circuit antennas moves into the millimeter-wave range, radiation and surface wave excitation both become significant. From a circuit design point of view, a thin substrate may reduce radiation and surface-wave losses; however, as frequency increases, the substrate thickness cannot be thinned indefinitely. Thus, both types of losses may not be too small to be ignored in the modeling. An enclosed housing may provide another way to prevent both losses, but it limits the dimension of circuits and the applications to antennas. Moreover, as the technology advances, antenna elements may be integrated

Reprinted with permission from "Full-Wave Spectral-Domain Analysis for Open Microstrip Discontinuities of Arbitrary Shape Including Radiation and Surface-Wave Losses," T.-S. Horng, N. G. Alexopoulos, S.-C. Wu, and H.-Y. Yang, *Int. J. of MIMICAE*, vol. 2, no. 4, pp. 224–240, © 1992. Reprinted by permission of John Wiley & Sons, Inc.

with MMICs. To accurately characterize antenna gain and efficiency, a full-wave analysis, which takes into account radiation as well as surface waves and the mutual coupling between circuit and antenna elements in a 3-D integrated circuit structure, is highly desirable.

Full-wave analysis using the finite-difference time-domain (FDTD) method has been reported in studying microstrip discontinuities [22–25]. The FDTD method is formulated by discretizing Maxwell's equations over a finite volume and approximating the derivatives with a centered difference approximation. This method has solved shielded microstrip problems successfully. Because of the limitation to a finite volume, when it is applied to an open structure, artificial absorbing boundaries must be imposed to simulate the radiation condition. These artificial absorbing boundaries usually adopt only the first-order approximation, which assumes a plane-wave incidence.

The most general and rigorous treatment for open microstrip discontinuities is governed by the well-known electric field integral equation (EFIE), which can be formulated in both the space [26–36] and spectral domains [37–48]. In the space domain, the dyadic kernel in the EFIE is the Green's function for the electric field which can be obtained from a Sommerfeld-type integral. Since the kernel is highly singular, the evaluation of the reaction integrals in the method of moments is difficult when the observation point is within the integration range. The mixed potential integral equation (MPIE), which is a modification of the EFIE, is usually solved in the space domain. In comparison with the EFIE in the space domain, the kernel in the MPIE is less singular, which makes the evaluation of reaction integrals more numerically simple and stable for two nearby elements [48–54].

The spectral-domain analysis performs an integral transformation, usually Fourier or Hankel transforms, to transform a partial differential equation into an ordinary differential equation. After satisfying the boundary conditions at the interfaces of multilayer stratified dielectric media, this approach can lead to a closed form expression for the so-called spectral-domain dyadic Green's function. The space-domain electromagnetic fields can be further expressed by taking the inverse Fourier transform of the vector product of this spectral-domain dyadic Green's function and Fourier transform of the microstrip currents. Such an EFIE formulation in the spectral domain has several advantages over the other two space-domain analyses. One advantage is that the integration path may be chosen to avoid singularities, thus yielding a smoothly varying integrand. A second advantage is the opportunity to independently determine both radiation and surface-wave losses [55]. One can compute the impedance matrix in the method of moments very accurately using the spectral-domain analysis such that all the circuit parameters in a junction can be rigorously determined. The disadvantage is that it may take more computation time to calculate the impedance matrix compared to the space domain approaches. An efficient method to improve computation time is to employ a space-domain technique for the asymptotic solution of the spectral-domain reaction integral [56–58].

A review of past work using full-wave spectral-domain analysis for open microstrip discontinuities reveals that the moment method using rectangular subdomain functions is limited to a multiport junction whose shape can be divided into a number of rectangles, such as printed dipole and rectangular patch antennas [37–40], or open-end, step, gap, stub, and T- and cross-junction discontinuities [41–46]. To model the discontinuities of arbitrary shape, in addition to rectangular subdomain functions, vector-valued triangular subdomain functions are also applied as both expansion and testing functions in the moment method procedure. These triangular subdomain functions were apparently first employed by Rao et al. [59]. They are suitable for modeling electric currents on arbitrary 3-D PEC surfaces. However, the present application will restrict the surface patches to lie in the plane of microstrip discontinuities.

Semi-infinite traveling wave functions have been successfully employed in the characterization of some microstrip discontinuities [41–46]. Since they simulate the current distribution on a microstrip line so well, the subdomain functions are required only in the vicinity of discontinuities. Semi-infinite traveling wave functions represent the currents on transmission lines far away from the discontinuities. Thus, in this scheme, the number of subdomain expansion functions used for a discontinuity problem can be dramatically reduced. In addition, individual microstrip junctions can be isolated by these assumed semi-infinite transmission lines and then characterized by scattering parameters without involving complicated excitation or impedance-matching problems. In this study, several open microstrip discontinuities are investigated using the different kinds of expansion functions introduced above to illustrate their flexiblity and utility.

2. THEORY

2.1 Electric Field Integral Equation

The current distribution over all the microstrip discontinuities is usually treated in terms of a number of infinitesimal dipoles continuously distributed on each interface of the dielectric layers. From linear superposition, the tangential electric field on each interface can be expressed by a two-dimensional spatial convolution of the dyadic Green's function with the current distribution. Each component of this two-dimensional spatial convolution corresponds to multiplication of the two-dimensional Fourier transforms in the spectral domain. Therefore, the space-domain tangential electric field can be expressed as an inverse Fourier transform of the vector product of spectral-domain dyadic Green's function and the Fourier transform of the microstrip currents. An EFIE can be obtained by imposing the boundary condition that the total tangential field is zero on the conductors.

$$\vec{E}^{\mathrm{inc}}(x, y) + \frac{1}{4\pi^2}\int_{-\infty}^{\infty}\int_{-\infty}^{\infty}\sum_j \overline{\overline{G}}_{ij}(k_x, k_y) \times \vec{J}_j(k_x, k_y)e^{-jk_x x}e^{-jk_y y}dk_x dk_y = 0 \qquad (1)$$

for x, y on microstrips

where $\overline{\overline{G}}_{ij}$ and $\vec{J}_j$ are the spectral-domain dyadic Green's functions and the current distribution, respectively, for a multilayer structure. Substrips i, j identify different interfaces. The closed-form expression of the spectral-domain dyadic Green's function [44,60–63] takes into account all the physical phenomena including radiation and surface waves. $\vec{E}^{\mathrm{inc}}$ is an impressed electric field used to excite a microstrip line attached to a microstrip discontinuity.

2.2 Plane-Wave Spectrum

Figure 1 shows an arbitrarily shaped microstrip junction residing in a multilayered medium. Assume a dominant microstripline mode is incident upon port #1. All types of electromagnetic waves which satisfy the boundary conditions are generated, and the electromagnetic fields near the junction are quite complicated. If the observation point is chosen far away from the junction, however, the field expressions can be greatly simplified. These far fields can be classified according to the propagation properties into several types of waves, i.e., radiated space waves, surface waves, and dominant as well as higher-order microstripline modes. The radiated space waves propagate upward from the top of the dielectric layers into free space, while surface waves propagate along the planar direction of the dielectric layers and decay exponentially toward free space. Both types of waves contribute to power loss at the junction. The higher-order microstrip line modes in an electrically narrow microstrip line are highly attenuated [64], and they are confined nearby the microstrip junction. Therefore, the plane-wave spectrum implied in eq. (1) for the far fields can be approximated by a combination of the spectrum for radiated space waves, surface waves, and dominant microstripline modes. An example of the spectrum for the configuration of Figure 1 is shown in Figure 2. The plane-wave spectrum for the radiated space waves satisfies $k_x^2 + k_y^2 < k_0^2$, which corresponds to a range within a circle with radius k_0. For the surface waves, the plane-wave spectrum corresponds to circles with radii equal to the surface-wave poles, λ_p. For a dominant mode propagating in the $x(y)$ direction,

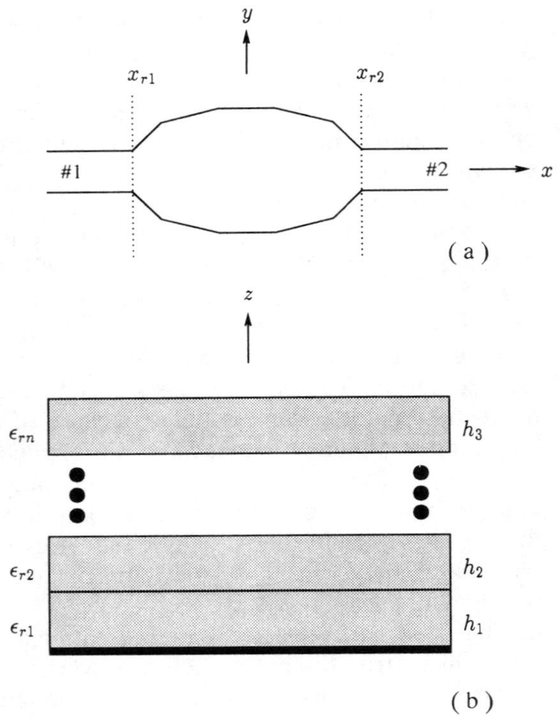

Figure 1. (a) Top view. (b) Side view of an arbitrarily shaped microstrip junction residing in a multilayered medium.

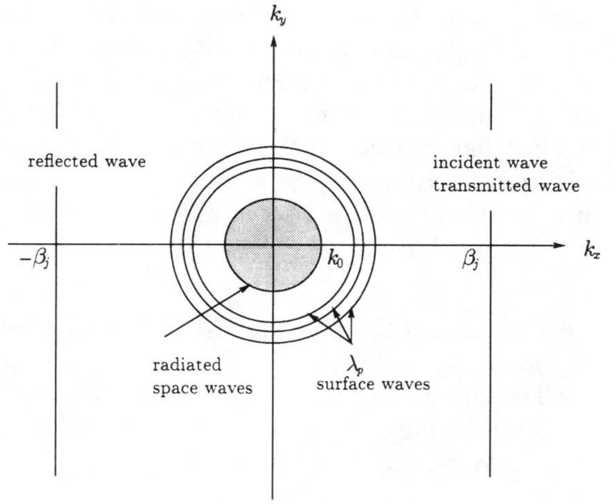

Figure 2. Plane wave spectrum for the far fields.

the plane-wave spectrum corresponds to the line perpendicular to the $k_x(k_y)$ axis with intersection equal to the propagation constant β_j. An inequality relating k_0, λ_p, and β_j is

$$k_0 \leq \lambda_p \leq \beta_j, \quad \text{for all } p \text{ and each } j \tag{2}$$

More detailed mathematical expressions, and physical interpretation for each traveling wave, can be found elsewhere [55]. Due to power conservation, the incident power will be equal to the summation of the powers in individual propagating waves on microstrip lines plus the losses due to radiation and surface waves.

2.3 The Choice of Expansion Functions

The solution of the electric field integral equation given in eq. (1) is obtained by the method of moments [65]. The first step of this procedure is to represent the microstrip currents approximately by a set of expansion functions with unknown coefficients. The accuracy of the moment method solution depends, in part, on how closely these expansion functions can represent the exact current distribution. In this analysis, three types of current expansion functions are utilized. They are rectangular and triangular subdomain functions along with semi-infinite traveling wave (entire domain) functions.

2.3.1. Rectangular Subdomain Functions ($r\vec{e}c$). The mathematical expression of a rectangular subdomain function (shown in Fig. 3) is multiplication of a piecewise sinusoidal function (pws) in the longitudinal direction (direction of current flow), and a pulse function (p) in the transverse direction. In general, y-directed rectangular subdomains may be used. The Fourier transform pairs of a rectangular subdomain function are given as

$$r\vec{e}c_n(x, y) = \begin{cases} pws_n(x)p_n(y)\hat{x}, & \text{for } x\text{-directed currents} \\ pws_n(y)p_n(x)\hat{y}, & \text{for } y\text{-directed currents} \end{cases} \tag{3}$$

and

$$R\vec{E}C_n(k_x, k_y) = \begin{cases} PWS_n(k_x)P_n(k_y)\hat{x}, & \text{for } x\text{-directed currents} \\ PWS_n(k_y)P_n(k_x)\hat{y}, & \text{for } y\text{-directed currents} \end{cases} \tag{4}$$

where

$$pws_n(\zeta) = \begin{cases} \dfrac{\sin k_e(d - |\zeta - \zeta_n|)}{\sin k_e d_1}; & \text{for } |\zeta - \zeta_n| \leq d_1 \\ 0; & \text{otherwise} \end{cases} \tag{5}$$

$$PWS_n(k_\zeta) = \frac{2k_e}{\sin k_e d_1} \frac{\cos k_e d_1 - \cos k_\zeta d_1}{k_\zeta^2 - k_e^2} e^{jk_\zeta \zeta_n} \tag{6}$$

$$p_n(\zeta) = \begin{cases} \dfrac{1}{d_2}; & \text{for } |\zeta - \zeta_n| \leq \dfrac{d_2}{2} \\ 0; & \text{otherwise} \end{cases} \tag{7}$$

$$p_n(k_\zeta) = \sin\left(\frac{k_\zeta d_2}{2}\right) \bigg/ \left(\frac{k_\zeta d_2}{2}\right) e^{jk_\zeta \zeta_n} \tag{8}$$

where $\zeta = x$ or y, and subscript n identifies the nth rectangular subdomain expansion function whose center is at (x_n, y_n). It is noted that k_e is a constant to describe the shape of the piecewise

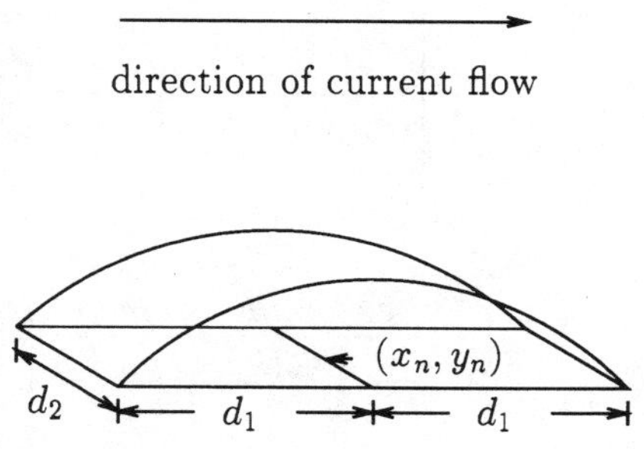

Figure 3. Rectangular subdomain function.

sinusoidal function and must be less than $\pi/2d_1$ to avoid a dip at the center. One can recognize that the rectangular subdomain function approaches zero at the edges perpendicular to the current flow direction. Thus, this expansion function satisfies the normal boundary edge condition of any irregular discontinuity with 90° corners very well.

2.3.2. Triangular Subdomain Functions ($t\vec{r}i$).

The triangular subdomain functions, as shown in Figure 4, are vector-valued functions defined as

$$t\vec{r}i_n(x, y) = \begin{cases} \dfrac{l_n}{2A_n^+}[(x - x_3)\hat{x} + (y - y_3)\hat{y}], & x, y \text{ in } T_n^+ \\ \dfrac{l_n}{2A_n^-}[(x_4 - x)\hat{x} + (y_4 - y)\hat{y}], & x, y \text{ in } T_n^- \\ 0, & \text{otherwise} \end{cases} \tag{9}$$

where T_n^+ and T_n^- denote the faces of two triangles with areas A_n^+ and A_n^-, respectively. The nth expansion function is uniquely associated with the nth interior edge, whose length is l_n. This basis function has two salient features which can make it uniquely suited to approximate electric surface currents on microstrip discontinuities of arbitrary shape [59]:

- The current has no component normal to the boundary edge.
- The component of current normal to the nth interior edge is constant and continuous across the edge.

This expansion function consists of a pair of linearly varying functions with triangular support whose Fourier transform was derived elsewhere [66]. After arrangement, the Fourier transform of triangular subdomain function is expressed as:

$$\begin{aligned} T\vec{R}I_n(k_x, k_y) = \; & \frac{l_n}{2A_n^+}\left\{ \frac{e^{-jk_x x_1}e^{-jk_y y_1}(m_{31} - m_{12})}{(k_x + m_{12}k_y)(k_x + m_{31}k_y)} \left[\left(x_1 - x_3 - \frac{2k_x + (m_{12} + m_{31})k_y}{j(k_x + m_{12}k_y)(k_x + m_{31}k_y)} \right)\hat{x} + \left(y_1 - y_3 - \frac{2k_y m_{12}m_{31} + (m_{12} + m_{31})k_x}{j(k_x + m_{12}k_y)(k_x + m_{31}k_y)} \right)\hat{y} \right] \right. \\ & + \frac{e^{-jk_x x_2}e^{-jk_y y_2}(m_{12} - m_{23})}{(k_x + m_{23}k_y)(k_x + m_{12}k_y)} \left[\left(x_2 - x_3 - \frac{2k_x + (m_{23} + m_{12})k_y}{j(k_x + m_{23}k_y)(k_x + m_{12}k_y)} \right)\hat{x} + \left(y_2 - y_3 - \frac{2k_y m_{23}m_{12} + (m_{23} + m_{12})k_x}{j(k_x + m_{23}k_y)(k_x + m_{12}k_y)} \right)\hat{y} \right] \\ & \left. + \frac{e^{-jk_x x_3}e^{-jk_y y_3}(m_{23} - m_{31})}{(k_x + m_{31}k_y)(k_x + m_{23}k_y)} \left[\left(- \frac{2k_x + (m_{31} + m_{23})k_y}{j(k_x + m_{31}k_y)(k_x + m_{23}k_y)} \right)\hat{x} + \left(- \frac{2k_y m_{31}m_{23} + (m_{31} + m_{23})k_x}{j(k_x + m_{31}k_y)(k_x + m_{23}k_y)} \right)\hat{y} \right] \right\} \\ & - \frac{l_n}{2A_n^-}\left\{ \frac{e^{-jk_x x_1}e^{-jk_y y_1}(m_{41} - m_{12})}{(k_x + m_{12}k_y)(k_x + m_{41}k_y)} \left[\left(x_1 - x_4 - \frac{2k_x + (m_{12} + m_{41})k_y}{j(k_x + m_{12}k_y)(k_x + m_{41}k_y)} \right)\hat{x} + \left(y_1 - y_4 - \frac{2k_y m_{12}m_{41} + (m_{12} + m_{41})k_x}{j(k_x + m_{12}k_y)(k_x + m_{41}k_y)} \right)\hat{y} \right] \right. \\ & \left. + \frac{e^{-jk_x x_2}e^{-jk_y y_2}(m_{12} - m_{24})}{(k_x + m_{24}k_y)(k_x + m_{12}k_y)} \left[\left(x_2 - x_4 - \frac{2k_x + (m_{24} + m_{12})k_y}{j(k_x + m_{24}k_y)(k_x + m_{12}k_y)} \right)\hat{x} + \left(y_2 - y_4 - \frac{2k_y m_{24}m_{12} + (m_{24} + m_{12})k_x}{j(k_x + m_{24}k_y)(k_x + m_{12}k_y)} \right)\hat{y} \right] \right\} \end{aligned} \tag{10}$$

triangular subdomain functions are superior to the rectangular subdomain functions for modeling discontinuities of arbitrary shape. However, the disadvantages in comparison with the rectangular subdomain functions include:

- More complicated Fourier transform expressions.
- Fewer redundancies in the database of reaction elements.
- More expansion functions are necessary to accurately model a discontinuity.

2.3.3. Semi-Infinite Traveling Wave Functions ($s\vec{e}m$).

The semi-infinite traveling wave functions, as shown in Figure 5, are used to simulate the current distribution along microstrip lines attached to a microstrip discontinuity. These expansion functions consist of a pulse function in the transverse direction and a traveling wave function terminated at a reference plane in the longitudinal direction (direction of current flow). Mathematically, they are defined as

$$s\vec{e}m_{mn}(x, y, \zeta_r, \beta) = \begin{cases} s_m(x, x_r, \beta)p_n(y)\hat{x}, & \text{for } x\text{-directed currents} \\ s_m(y, y_r, \beta)p_n(x)\hat{y}, & \text{for } y\text{-directed currents} \end{cases} \quad (11)$$

where β is the propagation constant of the dominant mode of an infinitely long microstrip line and $\zeta = \zeta_r$ is the reference plane to terminate the traveling wave function. Subscript $m = 1, 2, 3$ represents incident type, reflected type, and transmitted type semi-infinite traveling wave functions, respectively. Each type of function has a general form of

$$s_m(\zeta, \zeta_r, \beta) = e^{\pm j\beta(\zeta-\zeta_r)}U(\pm(\zeta - \zeta_r)). \quad (12)$$

$U(\tau)$ herein is a unit-step function. In eq. (12), the cosine term of the traveling wave function is

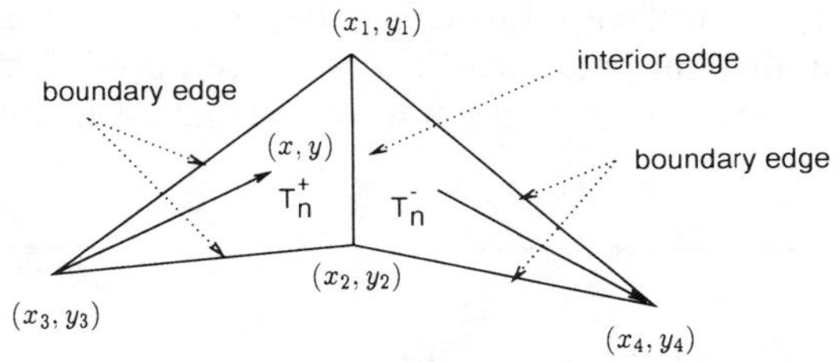

Figure 4. Triangular subdomain function.

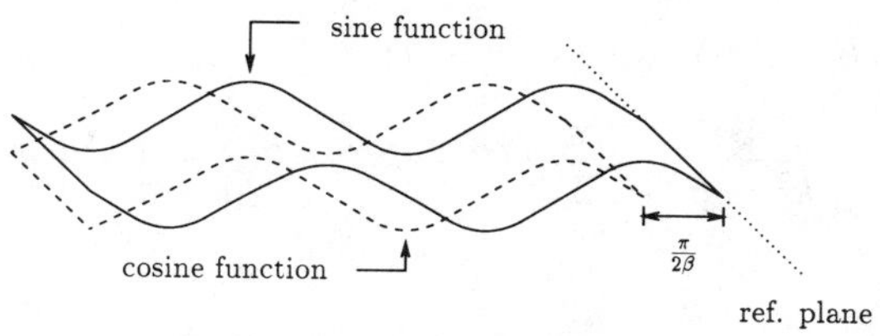

Figure 5. Semi-infinite traveling wave function.

not zero at the reference plane $\zeta = \zeta_r$, which is, in general, not appropriate for modeling a discontinuity with an open-end right at this reference plane. We may slightly modify these expansion functions by terminating the cosine functions at a quarter-guided wavelength away from the reference plane, while the sine functions are still terminated at the reference discontinuity with an open-end right at this reference plane. We may slightly modify these expansion functions by terminating the cosine functions at a quarter-guided wavelength away from the reference plane, while the sine functions are still terminated at the reference plane [41]. As a result, these modified functions are given as

$$s_1(\zeta, \zeta_r, \beta) = \cos(\beta(\zeta - \zeta_r))U\left(\zeta_r - \frac{\pi}{2\beta} - \zeta\right) - j\sin(\beta(\zeta - \zeta_r))U(\zeta_r - \zeta) \quad (13)$$

$$s_2(\zeta, \zeta_r, \beta) = \cos(\beta(\zeta - \zeta_r))U\left(\zeta_r - \frac{\partial}{2\beta} - \zeta\right) + j\sin(\beta(\zeta - \zeta_r))U(\zeta_r - \zeta) \quad (14)$$

$$s_3(\zeta, \zeta_r, \beta) = \cos(\beta(\zeta - \zeta_r))U\left(\zeta - \zeta_r + \frac{\pi}{2\beta}\right) - j\sin(\beta(\zeta - \zeta_r))U(\zeta - \zeta_r) \quad (15)$$

The Fourier transforms of eqs. (13–15) are further expressed as follows.

$$S_1(k_\zeta, \zeta_r, \beta) = (e^{-jk_\zeta\pi/2\beta} - j) \times \left[\frac{\beta}{k_\zeta^2 - \beta^2} + \frac{j\pi}{2}[\delta(\beta - k_\zeta) - \delta(\beta + k_\zeta)]\right] e^{jk_\zeta\zeta_r} \quad (16)$$

$$S_2(k_\zeta, \zeta_r, \beta) = (e^{-jk_\zeta\pi/2\beta} + j) \times \left[\frac{\beta}{k_\zeta^2 - \beta^2} + \frac{j\pi}{2}[\delta(\beta - k_\zeta) - \delta(\beta + k_\zeta)]\right] e^{jk_\zeta\zeta_r} \quad (17)$$

$$S_3(k_\zeta, \zeta_r, \beta) = (e^{jk_\zeta \pi/2\beta} + j) \times \left[\frac{\beta}{k_\zeta^2 - \beta^2} - \frac{j\pi}{2} [\delta(\beta - k_\zeta) - \delta(\beta + k_\zeta)] \right] e^{jk_\zeta \zeta_r} \tag{18}$$

where $\delta(\tau)$ is the delta function.

With the help of these three types of expansion functions provided above, the current distribution on any microstrip discontinuities of arbitrary shape can be expanded effectively. In general, both rectangular and triangular subdomains are used in the discontinuity region, and rectangular subdomains along with the semi-infinite traveling wave functions are used on the microstrip lines. As an example, the current distribution on the discontinuity shown in Figure 1 is approximated as

$$\begin{aligned} \vec{J}(x, y) \simeq & \sum_{n=1}^{N} A_n r\vec{e}c_n(x, y) \\ & + \sum_{m=1}^{N} B_m t\vec{r}i_m(x, y) \\ & + \sum_{p=1}^{P} C_p[s\vec{e}m_{1p}(x, y, x_{r1}, \beta) \\ & + S_{11} s\vec{e}m_{2p}(x, y, x_{r1}, \beta) \\ & + S_{21} s\vec{e}m_{3p}(x, y, x_{r2}, \beta)] \end{aligned} \tag{19}$$

where A_n, B_m, and S_{ij} are the unknown coefficients and scattering parameters to be obtained through the method of moments. C_p represents the transverse current dependence of the dominant mode found from a spectral-domain analysis of an infinitely long microstrip line [60–62].

2.4. The Method of Moments and Matrix Formulation

After substituting the expanded current expression (19) into (1), the method of moments can be applied to transform this integral equation into a matrix equation. The testing functions in the moment method procedure are chosen to be subdomain functions only. As a result, the matrix equation for the microstrip discontinuity shown in Figure 1 is formulated as

Each submatrix in (20) represents a set of mutual impedances between expansion functions and testing functions. For instance, the matrix elements in submatrix [ZRR] are mathematically expressed as

$$Z_{RR}^{mn} = \frac{1}{4\pi^2} \int_{-\infty}^{\infty} \int_{-\infty}^{\infty} [\overline{\overline{G}}(k_x, k_y) \cdot R\vec{E}C_n(k_x, k_y)] \cdot R\vec{E}C_M^*(k_x, k_y) dk_x dk_y. \tag{21}$$

After performing a matrix inversion for (20), one can obtain the coefficients A_n, B_m, and scattering parameters S_{ij} simultaneously.

2.5 Numerical Integration for Matrix Elements

The double infinite integration in each matrix element is carried out numerically after transforming into polar coordinates. This procedure reduces the integration to a finite integral (0 to 2π) with respect to an angular variable $\theta = \tan^{-1}(k_y/k_x)$ and an infinite integral (0 to ∞) with respect to a radial variable, $\lambda = \sqrt{k_x^2 + k_y^2}$. For the finite integral, it may be broken into several smaller subregions, and a lower-order (8–16-point) Gaussian quadrature formula is used in each subregion. For the infinite integral, special numerical methods are required. One can break the infinite integration range into two parts, $(0, A)$ and (A, ∞), such that in the first section the integrand contains all the singularities, namely the surface-wave poles, while for the second section the integrand is well behaved but slowly convergent. The choice of A is quite flexible, but it should exceed all surface losses due to radiation and surface waves, and the second integral contributes mainly to the circuit reactances.

A pole extraction technique [27–30] can be employed in the first section to include both residue and Cauchy principal value associated with each surface-wave pole. Another method of performing the integration from 0 to A is to deform the contour off the real axis so that the integrand is well behaved [39]. This method is particularly useful in a multilayered structure, because knowledge of individual pole locations is not required. This deformed integration contour is depicted in Figure 6.

$$\begin{bmatrix} [Z_{RR}]_{(N+2)\times N} & [Z_{RT}]_{(N+2)\times M} & [Z_{RS_2}]_{N+2} & [Z_{RS_3}]_{N+2} \\ [Z_{TR}]_{M\times N} & [Z_{TT}]_{M\times M} & [Z_{TS_2}]_M & [Z_{TS_3}]_M \end{bmatrix} \begin{bmatrix} [A]_N \\ [B]_M \\ S_{11} \\ S_{21} \end{bmatrix} = - \begin{bmatrix} [Z_{RS_1}]_{N+2} \\ [Z_{TS_1}]_M \end{bmatrix}. \tag{20}$$

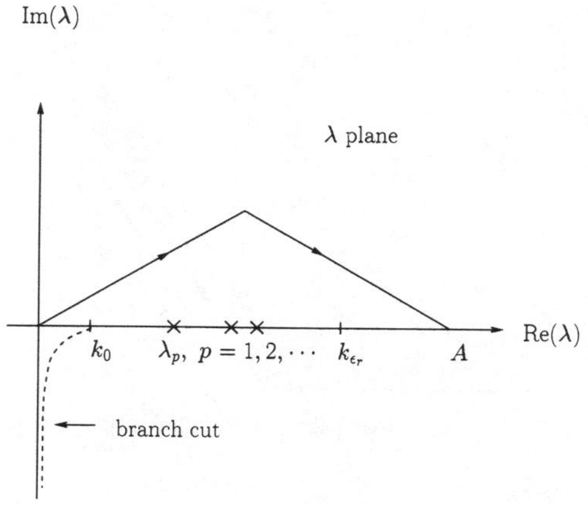

Figure 6. Contour of integration deformed around surface-wave poles.

The second integration, from A to ∞, is the so-called tail integration. This integration converges slowly when the testing and observation points are on the same plane (no separation in z). Also, as λ gets larger, the integrand becomes a highly oscillating function. Although Filon's integration method can be applied, difficulty still exists, especially for small-size basis function [63]. A more efficient and accurate method is to use an asymptotic extraction technique [44, 57]. This technique requires additional computations of the self and mutual reactions between any two expansion functions in a homogeneous medium with dielectric constant, ϵ_{eff}, equal to the average of the dielectric constants immediately above and below the source point. With this technique, the convergence of the tail integral can be improved by the order of λ^2.

2.6 Excitation Structure

Two types of excitation, an incident traveling wave and a delta-gap voltage excitation, are discussed in this analysis. The former type of excitation uses a semi-infinite incident traveling wave function, $\vec{sem}_{1p}$, as a pseudo-feed for the circuit. One advantage of using this excitation is that the microstrip junction may be isolated from the real, physical feed. Therefore, one can characterize the junction accurately without considering the coupling from the feed. The latter type uses an ideal delta-gap voltage source impressed near the end of a finite length of microstrip line. This type of excitation does not require any semi-infinite lines. One terminates the ports with an open-end instead of a matched load. In this scheme, the entire junction has a finite dimension where only subdomain expansion functions are required to expand the current distribution on the junction. Once the current distribution on the entire junction is obtained through the method of moments, the scattering parameters can be further extracted from knowledge of this current distribution and the propagation constant of the dominant mode. The advantage using a gap excitation is that one can avoid the complicated computation for the reaction integrals associated with semi-infinite traveling wave functions; however, the disadvantages include more unknowns in the moment method procedure to expand the currents, and less accuracy in computing scattering parameters. The detailed mathematical formulations for both excitations are described as follows.

2.6.1. Incident Traveling Wave Excitation. In eq. (20), the right-hand side matrix represents an incident traveling wave excitation for the microstrip junction. The submatrices, $[Z_{RS_1}]$ and $[Z_{TS_1}]$, identify the reactions of the semi-infinite incident traveling wave function, with rectangular and triangular subdomain testing functions, respectively. Mathematically, the matrix elements in each submatrix are expressed as

$$Z_{RS_1}^{m} = \frac{1}{4\pi^2} \sum \int_{-\infty}^{\infty} \int_{-\infty}^{\infty} \times \left[\overline{\overline{G}}(k_x, k_y) \cdot \sum_p C_p \vec{SEM}_{1p}(k_x, k_y, x_{r1}, \beta) \right] \cdot \vec{REC}_m^*(k_x, k_y) dk_x dk_y, \quad (22)$$

$$Z_{TS_1}^{m} = \frac{1}{4\pi^2} \int_{-\infty}^{\infty} \int_{-\infty}^{\infty} \times \left[\overline{\overline{G}}(k_x, k_y) \cdot \sum_p C_p \vec{SEM}_{1p}(k_x, k_y, x_{r1}, \beta) \right] \cdot \vec{TRI}_m^*(k_x, k_y) dk_x dk_y, \quad (23)$$

It is noted that, in (20), excitation by a semi-infinite incident traveling wave function ($\vec{sem}_{1p}$) is usually accompanied by two other semi-infinite traveling wave functions, the reflected type ($\vec{sem}_{2p}$) and the transmitted type ($\vec{sem}_{3p}$), which theoretically simulate matched loads for the input and output ports, respectively. All the scattering parameters of the junction in this scheme are solutions of the method of moments.

2.6.2. *Delta-Gap Voltage Excitation.* For a delta-gap voltage excitation, the matrix equation in (20) needs to be modified. On the junction as well as the finite lengths of microstrip lines, only the left four submatrices in the left-hand side of (20), which are the mutual impedance between two subdomain functions, remain. For the excitation matrix in the right-hand side, since a delta-function electric field distribution is assumed across the gap on the perfectly conducting surface, integrating over the testing functions will yield nonzero matrix elements. Mathematically, we express this source as

$$\vec{E}^{\text{inc}}(x, y) = V_x \delta(x - x_s)\hat{x} + V_y \delta(y - y_s)\hat{y} \tag{24}$$

where V_x is nonzero for a gap parallel to the y axis and V_y is nonzero for a gap parallel to the x axis. The excitation matrix elements representing the reaction between the source and the rectangular and triangular subdomain testing functions can be further given as

$$Z_{RV}^{m} = V_x \int_{-\infty}^{\infty} r\vec{e}c_m(x_s, y) \cdot \hat{x}dy + V_y \int_{-\infty}^{\infty} r\vec{e}c_m(x, y_s) \cdot \hat{y}dx, \tag{25}$$

$$Z_{TV}^{m} = V_x \int_{-\infty}^{\infty} t\vec{r}i_m(x_s, y) \cdot \hat{x}dy + V_y \int_{-\infty}^{\infty} t\vec{r}i_m(x, y_s) \cdot \hat{y}dx, \tag{26}$$

In eqs. (25) and (26), a constant value is obtained for the matrix elements whose subdomain testing functions straddle the source point. A zero value appears on the other matrix elements.

2.7. Evaluation of Radiation and Surface-Wave Losses

Radiation and surface waves are unavoidable physical effects for open microstrip discontinuities. In recent years, a full-wave analysis, which includes these physical effects, has been developed for various microstrip discontinuities. Although both radiation and surface waves are included [31,35,41,45], the analysis can provide only the total normalized losses by subtracting the sum of the square of the absolute value of scattering coefficients from unity. A generalized method to independently calculate the power components associated with radiation and surface waves by integrating the corresponding Poynting vectors over the appropriate regions of space was reported by Davidovitz and Lo [67] for a microstrip patch antenna. However, this method assumes that a magnetic current is flowing on the outer wall of a cavity model. A more rigorous method to determine both power components can be developed in the spectral domain [55]. It investigates the radiation loss or radiated power by performing the surface integral of the Poynting vector associated with the radiated space waves in the direction ($\hat{z}$) normal to the substrate. The integration plane is chosen as an infinite plane, Σ (shown in Fig. 7) in free space above the microstrip discontinuities. The surface integral for radiation loss is given as

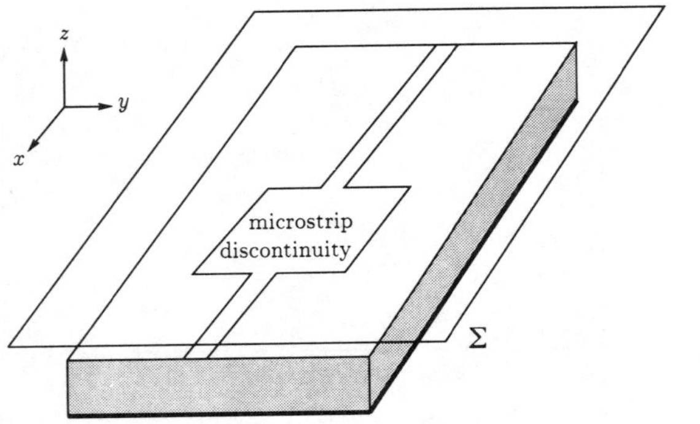

Figure 7. Integration plane for calculating radiation loss.

$$P_{\text{rad}} = \frac{1}{2} R_e \iint_{\Sigma} (\vec{E}_r(\vec{r}) \times \vec{H}_r^*)(\vec{r}) \cdot \hat{z}dxdy \tag{27}$$

where $\vec{E}_r$ and $\vec{H}_r$ denote the electric and magnetic field in the free space, respectively. With the specific characteristics that surface waves propagate along the surface, the surfface-wave loss can be found by integrating the Poynting vector over a cylinder (shown in Fig. 8) of large radius, ρ, with

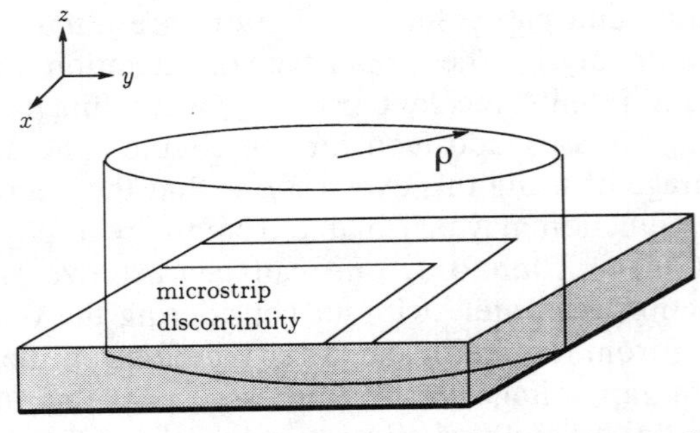

Figure 8. Integration cylinder for calculating surface-wave loss.

height extending from the ground plane to infinity. The surface integral for surface-wave loss is given as

$$P_{\text{sur}} = \frac{1}{2} R_e \int_0^{\infty} \int_0^{2\pi} \sum_i (\vec{E}_{si}(\vec{r}) \times \vec{H}_{si}^*(\vec{r})) \cdot \hat{\rho}\rho d\phi dz, \quad (28)$$

where subscript i identifies a different dielectric layer and $\vec{E}_{si}$, $\vec{H}_{si}$ denote the surface wave electric and magnetic field, respectively. It is noted that the integration in (27) over the x–y plane can be converted into an integration over the $k_x k_y$ plane such that the spectral-domain expressions for all the field components can be utilized. Similarly, the integration in (28) with respect to spatial angular variable, ϕ, can also be performed in the spectral, domain while the integration with respect to z can be evaluated in closed form. The advantage of using spectral-domain field expressions is that the spectra for both radiated space waves and surface waves are finite (shown in Fig. 2). Hence, the spectral-domain surface integrals for both losses can be accurately calculated.

3. EXAMPLES

3.1. An Open-End Discontinuity

To demonstrate the numerical accuracy of this analysis, the phase term of S_{11} calculated from a microstrip open-end discontinuity (shown in Fig. 9) has been compared with the measurements in Gronau and Wolff [68]. The theoretical results include two different expansion-excitation mechanisms, i.e., rectangular subdomain expansion functions with an incident traveling-wave excitation (denoted by REC-SEM) [46] and triangular subdomain expansion functions with a delta-gap voltage excitation (denoted by TRI-GAP) [69]. The layout to depict both mechanisms is also shown in Figure 9. It is noted that the voltage source for the TRI-GAP mechanism is considered as a concentrated slot field where the slot or gap is cut along the edges of the triangular patch model [70]. Figure 10 shows the comparisons, and it is seen that the difference among three curves is less than 1° over the frequency range of the measurement. Figure 11 shows the percentage of radiation loss, surface-wave loss, and total loss as a function of frequency for this open-end discontinuity. It is observed that the losses due to both radiation and surface waves increase with frequency. At low frequencies, the losses are mainly due to radiation. When the frequency increases, surface-wave loss increases faster than the radiation loss. Above a certain frequency, the surface-wave loss is more significant than the radiation loss.

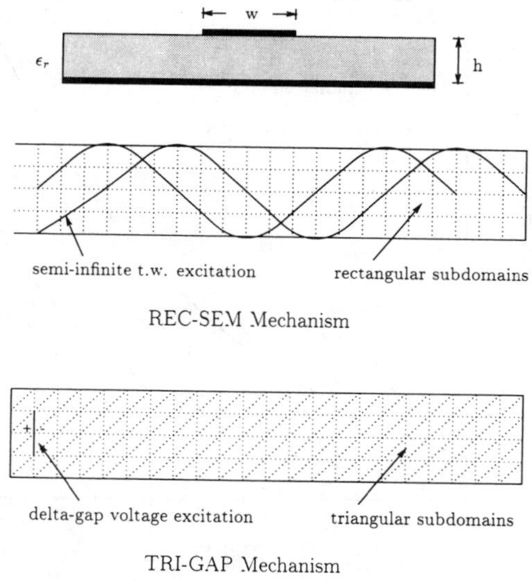

Figure 9. Expansion-excitation mechanism on a microstrip open-end discontinuity.

3.2. Right-Angle and Mitered-Bend Discontinuities

In the design of microwave and millimeter-wave circuits, compensation of microstrip discontinuities is widely used to reduce the effects of the

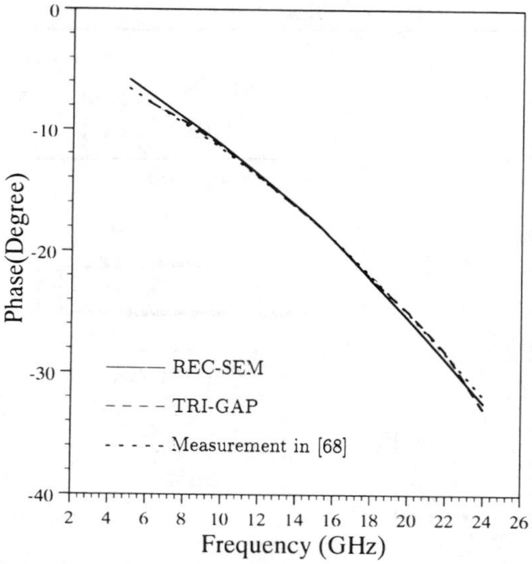

Figure 10. Phase of S_{11} for a microstrip open-end discontinuity (ϵ_r = 9.9, w = 24 mil, h = 25 mil).

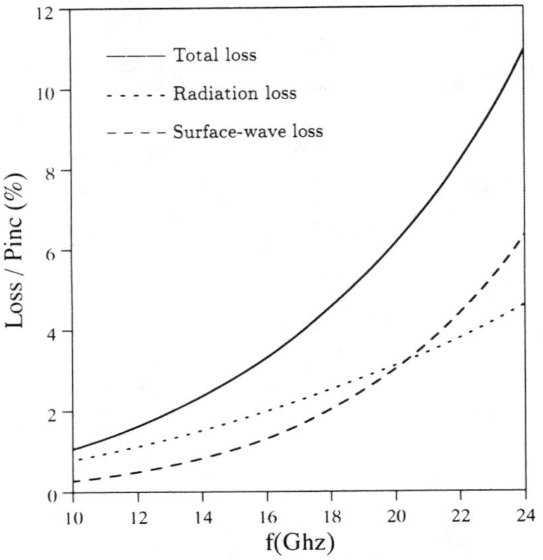

Figure 11. Power losses vs. frequency for a microstrip open-end discontinuity (ϵ_r = 9.9, w = 24 mil, h = 25 mil).

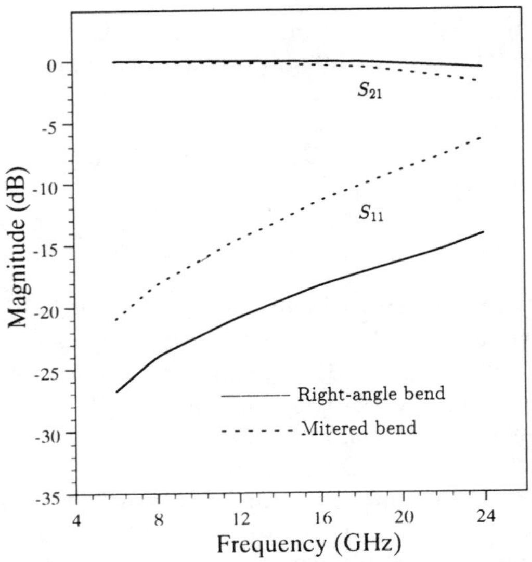

Figure 13. Magnitude of S-parameters of right-angle bend and mitered-bend discontinuities (ϵ_r = 9.9, w = 24 mil, h = 25 mil).

discontinuity reactances. As an example, the improvements provided by geometrical modification to the outer portion of the right-angle bend with a 45° miter are investigated. For modeling discontinuities with miters as well as 90° corners, a TRI-GAP mechanism is employed in the moment method procedure. The layout of this mechanism for both right-angle and mitered-bend discontinuities is shown in Figure 12. Figure 13 shows the numerical results for the magnitude of scattering parameters as a function of frequency. As expected, the mitered bend has a smaller reflection coefficient than the right-angle one over a wide frequency range of interest.

3.3 Basic and Shaped T-Junctions

Both basic and shaped T-junctions [shown in Fig. 14(a) and (b), respectively] are analyzed for comparison using a REC-SEM mechanism [46]. The problem for a basic T-junction is that the power transmission distributed in each port is usually re-

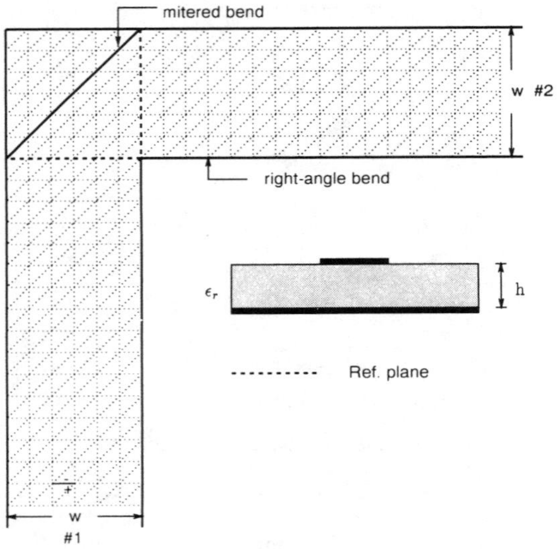

Figure 12. The geometry of right-angle and mitered-bend discontinuities.

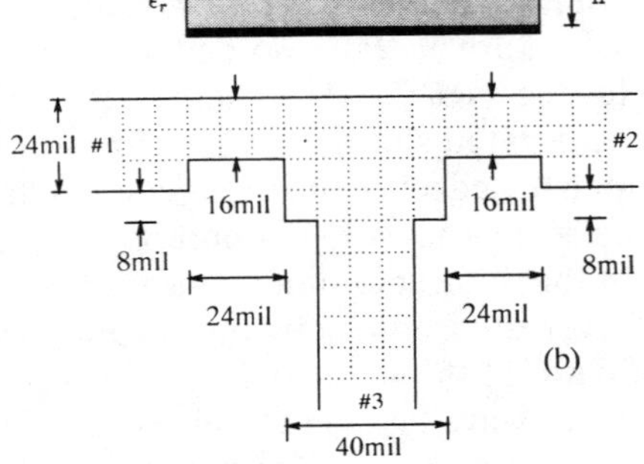

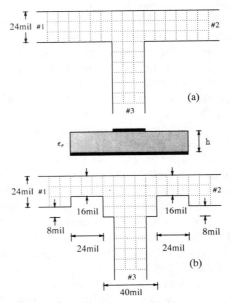

Figure 14. (a) Basic T-junction. (b) Shaped T-junction (ϵ_r = 9.9, h = 25 mil).

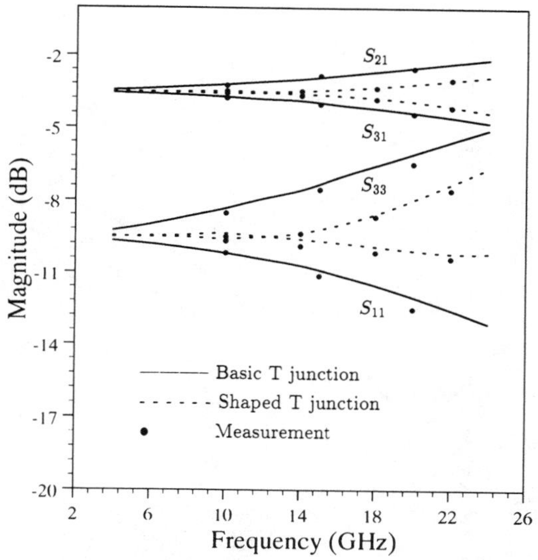

Figure 15. Magnitude of S-parameters of basic and shaped T-junctions ($\epsilon_r = 9.9$, $h = 25$ mil).

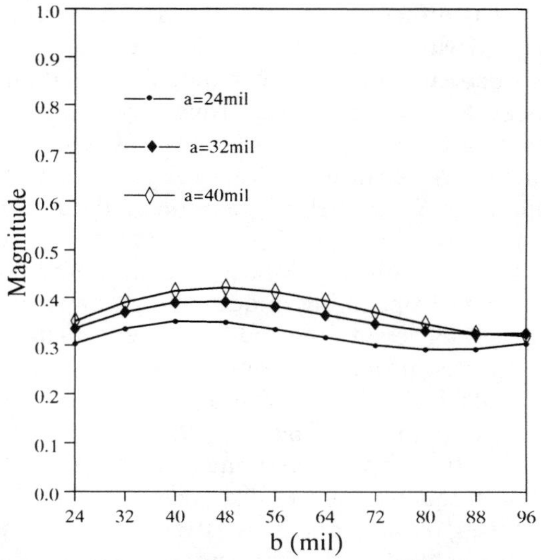

Figure 17. Magnitude of S_{11} vs. b for a 2-way power divider ($\epsilon_r = 9.9$, $w = 24$ mil, $h = 25$ mil).

stricted. For example, in a basic T-junction, there is always more power transmitted in port 2 than in port 3, and the reflection coefficient increases as the frequency increases. In order to make the circuit design flexible, a shaped T-junction, which is another example of circuit compensation, is designed as a broad-band equal power divider. Figure 15 shows the comparisons of these two junctions. It is seen that, for the shaped T-junction, the equal power transmission is very broad band (from dc up to 16 GHz), while, for the basic T-junction, equal power transmission is valid only up to 6 GHz. Furthermore, S_{33} of the shaped T-junction is usually a few decibels lower than that of the basic T-junction. The measured data of both junctions are also shown in Figure 15. The difference in the magnitudes of the S-parameters between theory and measurement is less than 0.4 dB in the frequency range 4–25 GHz.

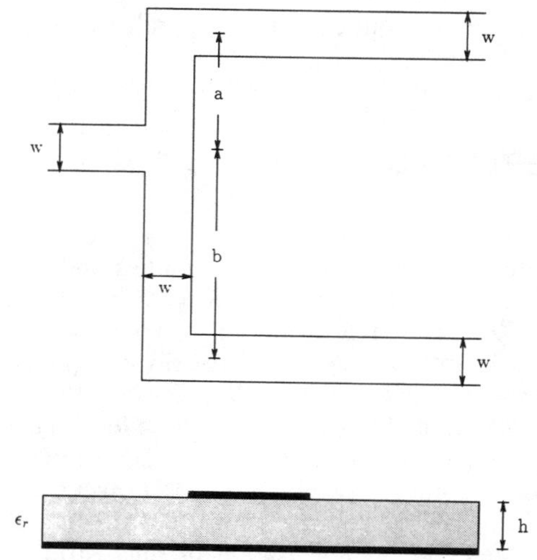

Figure 16. The geometry of a 2-way power divider.

3.4. A 2-Way Power Divider

Figure 16 shows a 2-way power divider which consists of a T-junction and two right-angle bend discontinuities. In the output ports, since there are two dominant modes (even-mode and odd-mode) propagating on the two parallel microstrip lines with different propagation characteristics (β and

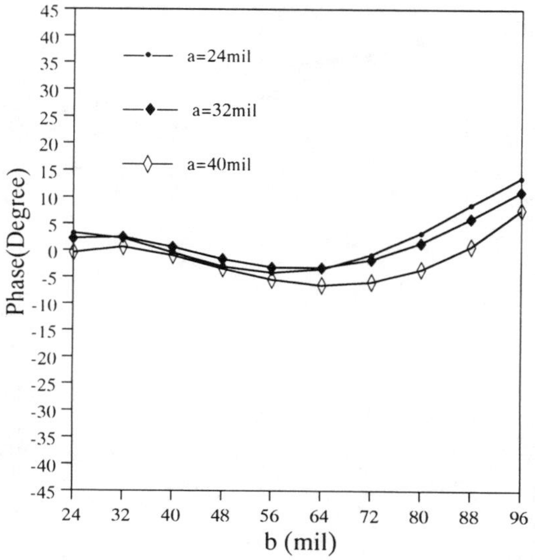

Figure 18. Phase of S_{11} vs. b for a 2-way power divider ($\epsilon_r = 9.9$, $w = 24$ mil, $h = 25$ mil).

C_p), it is difficult to extract the S-parameters associated with these two modes if the currents along the microstrip lines are expanded by subdomain functions only. Therefore, using the semi-infinite traveling wave expansion functions becomes advantageous because it allows for the direct solution of the S-parameters from the method of moments. For this 2-way power divider, one can imagine an equivalent 3-port junction where three different propagating modes, single microstrip line mode, even and odd coupled microstrip line modes, propagate at port 1, port e, and port o, respectively. The corresponding scattering parameters (shown in Figs. 17–20) are investigated as a function of position of the microstrip feed line at the frequency of 20 GHz. It is noted that the scattering parameters demonstrated in the figures are normalized by the characteristic impedance of each propagating mode such that the summation of the magnitude squared of each scattering parameter is close to unity, but not exactly equal to unity because of radiation and surface-wave losses. The method for calculating characteristic impedance for single as well as coupled microstrip line modes based on a power definition is given in refs. [62,71, and 72].

4. CONCLUSIONS

A full-wave spectral-domain analysis, combined with the method of moments using various expansion functions, has been found to be a very

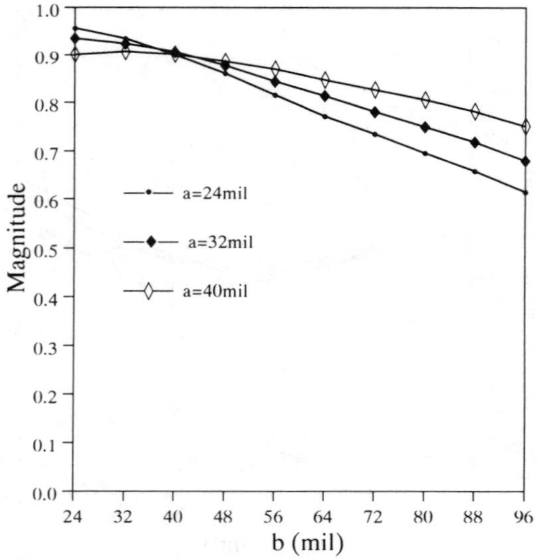

Figure 19. Magnitude of S_{e1} vs. b for a 2-way power divider (ϵ_r = 9.9, w = 24 mil, h = 25 mil).

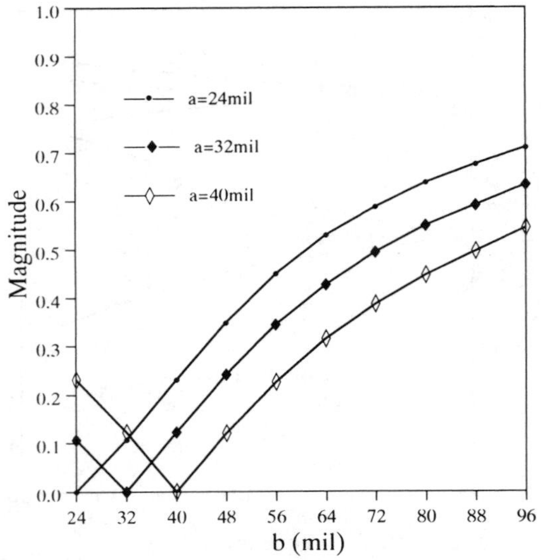

Figure 20. Magnitude of S_{o1} vs. b for a 2-way power divider (ϵ_r = 9.9, w = 24 mil, h = 25 mil).

accurate method to characterize open microstrip discontinuities of arbitrary shape. Several examples including open-end, right-angle bend, mitered-bend, basic T-, shaped T-, and 2-way power divider discontinuities were investigated to delineate different techniques in this analysis. The numerical results were further verified by comparison with available measured data for the open-end, basic T-, and shaped T-junctions. The comparison showed excellent agreement.

ACKNOWLEDGMENTS

This research was supported by U.S. Army Research Grant DAAL 03-90G-0182 and National Science Foundation Research Grant ECS 8802617.

REFERENCES

1. I. Wolff, G. Kompa, and R. Mehran, "Calculation method for microstrip discontinuities and T-junctions," *Electron. Lett.*, Vol. 8, 1972, pp. 177–179.
2. G. Kompa and R. Mehran, "Planar waveguide model for calculating microstrip components," *Electron. Lett.*, Vol. 11, 1975, pp. 459–460.
3. W. Menzel and I. Wolff, "A method for calculating the frequency dependent properties of microstrip discontinuities," *IEEE Trans. Microwave Theory Tech.*, Vol. 25, Feb. 1977, pp. 107–112.
4. R. Mehran, "Compensation of microstrip bends and Y-junctions with arbitrary angle," *IEEE Trans.*

Microwave Theory Tech., Vol. 78, June 1978, pp. 400–405.
5. R. Mehran, "Computer-aided design of microstrip filters considering dispersion loss and discontinuity effects," *IEEE Trans. Microwave Theory Tech.*, Vol. 27, March 1979, pp. 239–245.
6. R. Chadha and K. C. Gupta, "Compensation of discontinuities in planar transmission lines," *IEEE Trans. Microwave Theory Tech.*, Vol. 30, Dec. 1982, pp. 2151–2156.
7. R. Sorrentino, "Planar circuits waveguide models and segmentation method," *IEEE Trans. Microwave Theory Tech.*, Vol. 33, Oct. 1985, pp. 1057–1066.
8. K. C. Gupta, R. Garg, and I. J. Bahl, *Microstrip Lines and Slotlines,* Artech House, Dedham, MA, 1979.
9. A. Farrar and A. T. Adams, "Computation of lumped microstrip capacitances by matrix methods: Rectangular sections and end effects," *IEEE Trans. Microwave Theory Tech.*, Vol. 19, 1971, pp. 495–497.
10. A. Farrar and A. T. Adams, "Matrix methods for microstrip three-dimensional problems," *IEEE Trans. Microwave Theory Tech.*, Vol. 20, 1972, pp. 497–504.
11. M. Maeda, "An analysis of gap in microstrip transmission lines," *IEEE Trans. Microwave Theory Tech.*, Vol. 20, 1972, pp. 390–396.
12. P. Silvester and P. Benedek, "Equivalent capacitance of microstrip open circuits," *IEEE Trans. Microwave Theory Tech.*, Vol. 20, 1972, pp. 511–516.
13. P. Benedek and P. Silvester, "Equivalent capacitance for microstrip gaps and steps," *IEEE Trans. Microwave Theory Tech.*, Vol. 20, 1972, pp. 729–733.
14. P. Silvester and P. Benedek, "Microstrip discontinuity capacitances for right-angle bends, T-junctions and crossings," *IEEE Trans. Microwave Theory Tech.*, Vol. 21, 1973, pp. 341–346.
15. R. Horton, "The electrical characterization of a right-angled bend in microstrip lines," *IEEE Trans. Microwave Theory Tech.*, Vol. 21, 1973, pp. 427–429.
16. R. Horton, "Equivalent representation of an abrupt impedance step in microstrip line," *IEEE Trans. Microwave Theory Tech.*, Vol. 21, 1973, pp. 562–564.
17. Y. Rahmat-Samii, T. Itoh, and R. Mittra, "A spectral domain analysis for solving microstrip discontinuity problems," *IEEE Trans. Microwave Theory Tech.*, Vol. 22, April 1974, pp. 372–378.
18. A. F. Thomspon and A. Gopinath, "Calculation of microstrip discontinuity inductances," *IEEE Trans. Microwave Theory Tech.*, Vol. 23, 1975, pp. 648–655.
19. A. Gopinath, et al., "Equivalent circuit parameters of microstrip step change in width and cross junctions," *IEEE Trans. Microwave Theory Tech.*, Vol. 24, 1976, pp. 142–144.
20. C. Gupta and A. Gopinath, "Equivalent circuit capacitance of microstrip step change in width," *IEEE Trans. Microwave Theory Tech.*, Vol. 25, Oct. 1977, pp. 819–822.
21. N. G. Alexopoulos, J. A. Maupin, and P. T. Greiling, "Determination of the electrode capacitance matrix for GaAs FET's," *IEEE Trans. Microwave Theory Tech.*, Vol. 28, May 1980, pp. 459–466.
22. X. Zhang and K. K. Mei, "Time-domain finite difference approach to the calculation of the frequency-dependent characteristics of microstrip discontinuities," *IEEE Trans. Microwave Theory Tech.*, Vol. 36, Dec. 1988, pp. 1775–1787.
23. D. M. Sheen, S. M. Ali, M. D. Abouzahra, and J. A. Kong, "Application of the three-dimensional finite-difference time-domain method to the analysis of planar microstrip discontinuities," *IEEE Trans. Microwave Theory Tech.*, Vol. 38, July 1990, pp. 849–857.
24. A. Khebir, A. B. Kouki, and R. Mittra, "Asymptotic boundary conditions for finite element analysis of three-dimensional transmission line discontinuities," *IEEE Trans. Microwave Theory Tech.*, Vol. 38, Oct. 1990, pp. 1427–1432.
25. C. Wu, K. L. Wu, Z. Bi, and J. Litva, "Modelling of coaxial-fed microstrip patch antenna by finite difference time domain method," *Electron. Lett.*, Vol. 27(19), Sept. 1987, pp. 1991–1992.
26. N. K. Uzunoglu, N. G. Alexopoulos, and J. G. Fikioris, "Radiation properties of microstrip dipoles," *IEEE Trans. Ant. Propagat.*, Vol. 27, Nov. 1979, pp. 853–858.
27. I. E. Rana and N. G. Alexopoulos, "Current distribution and input impedance of printed dipoles," *IEEE Trans. Ant. Propagat.*, Vol. 29, Jan. 1981, pp. 99–105.
28. N. G. Alexopoulos and I. E. Rana, "Mutual impedance computation between printed dipoles," *IEEE Trans. Ant. Propagat.*, Vol. 29, Jan. 1981, pp. 106–111.
29. I. E. Rana and N. G. Alexopoulos, "Correction to 'Current distribution and input impedance of printed dipoles' and 'Mutual impedance computation between printed dipoles'," *IEEE Trans. Ant. Propagat.*, Vol. 29, Jan. 1981, pp. 99–109.
30. P. B. Katehi and N. G. Alexopoulos, "Real axis integration of Sommerfeld integrals with applications to printed circuit antennas," *J. Math. Phys.*, Vol. 24, March 1983, pp. 527–533.
31. P. B. Katehi, "Radiation losses in mm-wave open microstrip filters," *Electromagnetics,* Vol. 7, 1987, pp. 137–152; *IEEE Trans. Ant. Propagat.*, Vol. 30, July 1982, p. 822.
32. P. B. Katehi and N. G. Alexopoulos, "On the modeling of electromagnetically coupled microstrip antennas—The printed strip dipole," *IEEE Trans. Ant. Propagat.*, Vol. 32, Nov. 1984, pp. 1179–1186.

33. P. B. Katehi and N. G. Alexopoulos, "Frequency-dependent characteristics of microwave discontinuities in millimeter-wave integrated circuits," *IEEE Trans. Microwave Theory Tech.*, Vol. 33, Oct. 1985, pp. 1029–1035.
34. H. Nakano, S. R. Kerner, and N. G. Alexopoulos, "The moment method solution for printed wire antennas of arbitrary configuration," *IEEE Trans. Ant. Propagat.*, Vol. 36, Dec. 1988, pp. 1667–1674.
35. W. P. Harokopus and P. B. Katehi, "Characterization of microstrip discontinuities on multilayer dielectric substrates including radiation losses," *IEEE Trans. Microwave Theory Tech.*, Vol. 37, Dec. 1989, pp. 2058–2066.
36. C. L. Chi and N. G. Alexopoulos, "An efficient numerical approach for modeling microstrip type antennas," *IEEE Trans. Ant. Propagat.*, Vol. 38, Sept. 1990, pp. 1399–1404.
37. D. M. Pozar, "Input impedance and mutual coupling of rectangular microstrip antennas," *IEEE Trans. Ant. Propagat.*, Vol. 30, Nov. 1982, pp. 1191–1196.
38. D. R. Jackson and N. G. Alexopoulos, "Gain enhancement methods for printed circuit antennas," *IEEE Trans. Ant. Propagat.*, Vol. 33, Sept. 1985, pp. 976–987.
39. E. H. Newman and D. Forrai, "Scattering from a microstrip patch," *IEEE Trans. Ant. Propagat.*, Vol. 35, March 1987, pp. 245–251.
40. D. M. Pozar and S. M. Voda, "A rigorous analysis of a microstripline fed patch antenna," *IEEE Trans. Ant. Propagat.*, Vol. 35, Dec. 1987, pp. 1343–1350.
41. R. W. Jackson and D. M. Pozar, "Full wave analysis of microstrip open-end and gap discontinuities," *IEEE Trans. Microwave Theory Tech.*, Vol. 33, Oct. 1985, pp. 1036–1042.
42. R. H. Jansen, "The spectral-domain approach for microwave integrated circuits," *IEEE Trans. Microwave Theory Tech.*, Vol. 33, Oct. 1985, pp. 1043–1056.
43. H. Y. Yang and N. G. Alexopoulos, "A dynamic model for microstrip-slotline transition and related structures," *IEEE Trans. Microwave Theory Tech.*, Vol. 36, Feb. 1988, pp. 286–293.
44. H. Y. Yang and N. G. Alexopoulos, "Basic blocks for high frequency interconnects: Theory and experiment," *IEEE Trans. Microwave Theory Tech.*, Vol. 36, Aug. 1988, pp. 1258–1264.
45. H. Y. Yang and N. G. Alexopoulos, "Microstrip open-end and gap discontinuities in a substrate-superstrate structure," *IEEE Trans. Microwave Theory Tech.*, Vol. 37, Oct. 1989, pp. 1542–1546.
46. S. C. Wu, H. Y. Yang, N. G. Alexopoulos, and I. Wolff, "A rigorous dispersive characterization of microstrip cross and *T* junctions," *IEEE Trans. Microwave Theory Tech.*, Vol. 38, Dec. 1990, pp. 1837–1844.
47. R. W. Jackson, "Full-wave, finite element analysis of irregular microstrip discontinuities," *IEEE Trans. Microwave Theory Tech.*, Vol. 37, Jan. 1989, pp. 81–89.
48. T. Itoh, *Numerical Techniques for Microwave and Millimeter-Wave Passive Structure,* Wiley, New York, 1989.
49. A. W. Glisson and D. R. Wilton, "Simple and efficient numerical methods for problems of electromagnetic radiation and scattering from surfaces," *IEEE Trans. Ant. Propagat.*, Vol. 28, Jan. 1980, pp. 593–603.
50. J. R. Mosig and F. E. Cardiol, "Analytical and numerical techniques in the Green's function treatment of microstrip antennas and scatters," *Proc. Inst. Elec. Eng.*, Pt. H, Vol. 130, 1983, pp. 175–182.
51. J. R. Mosig, "Arbitrarily shaped microstrip structures and their analysis with a mixed potential integral equation," *IEEE Trans. Microwave Theory Tech.*, Vol. 36, Feb. 1988, pp. 314–323.
52. K. A. Michalski and D. Zheng, "Electromagnetic scattering and radiation by surfaces of arbitrary shape in layered media, part I: Theory," *IEEE Trans. Ant. Propagat.*, Vol. 38, March 1990, pp. 335–344.
53. K. A. Michalski and D. Zheng, "Electromagnetic scattering and radiation by surfaces of arbitrary shape in layered media, part II: Implementation and results for contiguous half-spaces," *IEEE Trans. Ant. Propagat.*, Vol. 38, March 1990, pp. 345–352.
54. K. A. Michalski and D. Zheng, "Analysis of microstrip resonators of arbitrary shape," *IEEE Trans. Microwave Theory Tech.*, Vol. 40, Jan. 1992, pp. 112–119.
55. T. S. Horng, S. C. Wu, H. Y. Yang, and N. G. Alexopoulos, "A generalized method for distinguishing between radiation and surface-wave losses in microstrip discontinuities," *IEEE Trans. Microwave Theory Tech.*, Vol. 38, Dec. 1990, pp. 1800–1807.
56. J. R. Mosig and F. E. Gardiol, "A dynamic radiation model for microstrip structures," *Adv. Electron. Eectron Phys.*, Vol. 59, 1982, pp. 139–237.
57. D. M. Pozar, "Improved computational efficiency for the method of moments solution of printed dipoles and patch," *Electromagnetics,* Vol. 3, 1983, pp. 299–309.
58. D. R. Jackson and N. G. Alexopoulos, "An asymptotic extraction technique for evaluating Sommerfeld-type integrals," *IEEE Trans. Ant. Propagat.*, Vol. 34, Dec. 1986, pp. 1467–1470.
59. S. M. Rao, D. R. Wilton and A. W. Glisson, "Electromagnetic scattering by surfaces of arbitrary shape," *IEEE Trans. Ant. Propagat.*, Vol. 30, May 1982, pp. 409–418.
60. T. Itoh and R. Mittra, "Spectral-domain approach for calculating dispersion characteristics of micro-

strip lines," *IEEE Trans. Microwave Theory Tech.*, Vol. 21, 1973, pp. 496–498.

61. T. Itoh, "Spectral domain immitance approach for dispersion characteristics of generalized printed transmission lines," *IEEE Trans. Microwave Theory Tech.*, Vol. 28, July 1980, pp. 733–736.
62. O. Fordham, "Two layer microstrip transmission lines," Master's thesis, University of California at Los Angeles, 1987.
63. N. G. Alexopoulos and D. R. Jackson, "Fundamental superstrate (cover) effects on printed circuit antennas," *IEEE Trans. Ant. Propagat.*, Vol. 34, Dec. 1986, pp. 1430–1438.
64. A. A. Oliner, "Leakage from higher modes on microstrip line with application to antennas," *Radio Sci.*, Vol. 22, Nov. 1987, pp. 907–912.
65. R. F. Harrington, *Field Computation by Moment Method*, Macmillan, New York, 1968.
66. B. Houshmand, W. C. Chew, and S. W. Lee, "Fourier transform of a linear distribution with triangular support and its applications in electromagnetics," *IEEE Trans. Ant. Propagat.*, Vol. 39, Feb. 1991, pp. 252–254.
67. M. Davidovitz and Y. T. Lo, "Input impedance of a probe-fed circular microstrip antenna with thick substrate," *IEEE Trans. Ant. Propagat.*, Vol. 34, July 1986, pp. 905–911.
68. G. Gronau and I. Wolff, "A simple broad-band device de-embedding method using an automatic network analyzer with time-domain option," *IEEE Trans. Microwave Theory Tech.*, Vol. 37, March 1989, pp. 479–483.
69. T. S. Horng, W. E. McKinzie, and N. G. Alexopoulos, "Full-wave spectral-domain analysis of compensation of microstrip discontinuities using triangular subdomain functions," to be presented at the *1992 IEEE-MTT's Int. Symp.*, Albuquerque, NM.
70. W. A. Johnson, D. R. Wilton, and R. M. Sharpe, "Modeling scattering from and radiation by arbitrary shaped objects with the electric field integral equation triangular surface patch code," *Electromagnetics*, Vol. 10, 1990, pp. 41–63; *IEEE Trans. Ant. Propagat.*, Vol. 39, Feb. 1991, pp. 252–254.
71. J. B. Knorr and A. Tufekcioglu, "Spectral-domain calculation of microstrip characteristic impedance," *IEEE Trans. Microwave Theory Tech.*, Vol. 23, Sept. 1975, pp. 725–728.
72. R. H. Jansen, "Unified user-oriented computation of shielded, covered and open planar microwave and millimeter-wave transmission-line characteristics," *Microwave Opt. Acoust.*, Vol. 3, Jan. 1979, pp. 14–22.

Chapter 2

Green's Function Approach for Planar Components with Regular Shapes

M. D. Abouzahra
M.I.T. Lincoln Laboratory

K. C. Gupta
University of Colorado at Boulder

2.1 Introduction

An accurate and reliable characterization of microwave and millimeter-wave circuits is one of the basic prerequisites for a successful computer-aided design (CAD) approach. The degree of accuracy to which the performance of microwave and millimeter-wave integrated circuits can be simulated depends largely upon the accuracy of the characterization and on the analysis technique used. In the first chapter, several techniques for analyzing and characterizing planar circuits were discussed. These techniques are based on finding a formal solution to the governing wave equation and associated boundary conditions. The basic approach for finding such a solution is either analytical, such as Green's function and the integral equation approaches, or numerical, such as the finite-difference and finite-element methods. The selection of one technique over another is primarily driven by the desire to obtain (1) an accurate solution, (2) compact analytical expressions if possible, and/or (3) a highly efficient computational algorithm. The Green's function approach to be discussed in this chapter is an analytical approach that offers an accurate as well as a compact analytical solution; however, the utilization of this approach is restricted to planar circuit components with regular shapes. Typical examples of the regular shapes that can be analyzed by this approach are shown in Fig. 2.1. Using the Green's functions that are already available in the literature [2.1–2.4], the impedance matrices for these planar segments can be obtained (for specified port locations).

This chapter describes the main features of the Green's function approach [2.5] for two-dimensional planar circuits. The wave equation that governs planar circuits and the associated boundary conditions are discussed first. This is followed by a brief discussion of the various analytical techniques that may be employed to obtain a solution for the wave equation; a particular technique for obtaining a solution to the wave equation in terms of Green's function is described in detail. By exploiting the fundamental voltage-current relationship, an expression for the elements of the generalized impedance matrix that characterizes planar circuits is then derived. This expression forms the basis for characterizing planar circuits. Two different techniques for obtaining Green's functions for planar circuits with regular shapes are then described. In the last section, expressions for the impedance matrix elements of several microstrip, open-boundary, planar circuits are listed. Finally, a list of Green's functions for regularly shaped planar segments with open boundaries is compiled in Appendix A. Recently developed single summation Green's function expressions are included.

2.2 The Wave Equation

Figure 2.2 depicts the basic structure of a typical, arbitrarily shaped microstrip multiport planar circuit. This structure has n-coupling ports distributed arbitrarily along its periphery. Some of the coupling ports are used to excite the electromagnetic energy inside the structure; the rest of the ports are utilized to couple the energy out of the structure. Depending on the shape of the planar structure and the locations of the coupling ports, useful characteristics can be obtained. Chapters 7 through 11 of this book describe various types of planar components.

In the planar structure illustrated in Fig. 2.3, the electromagnetic wave propagates in the dielectric medium between the two ground planes and the center conductor (planar segment). The medium of propagation is assumed to be linear, homogeneous, and isotropic. Furthermore, it is assumed that the field quantities have a time variation given by $e^{-j\omega t}$, where ω is the angular frequency in radian.

The coordinate axes in Fig. 2.3 are chosen such that the planar element lies in the xy-plane and is perpendicular to the z-axis. Thus, although the dimensions along the x- and

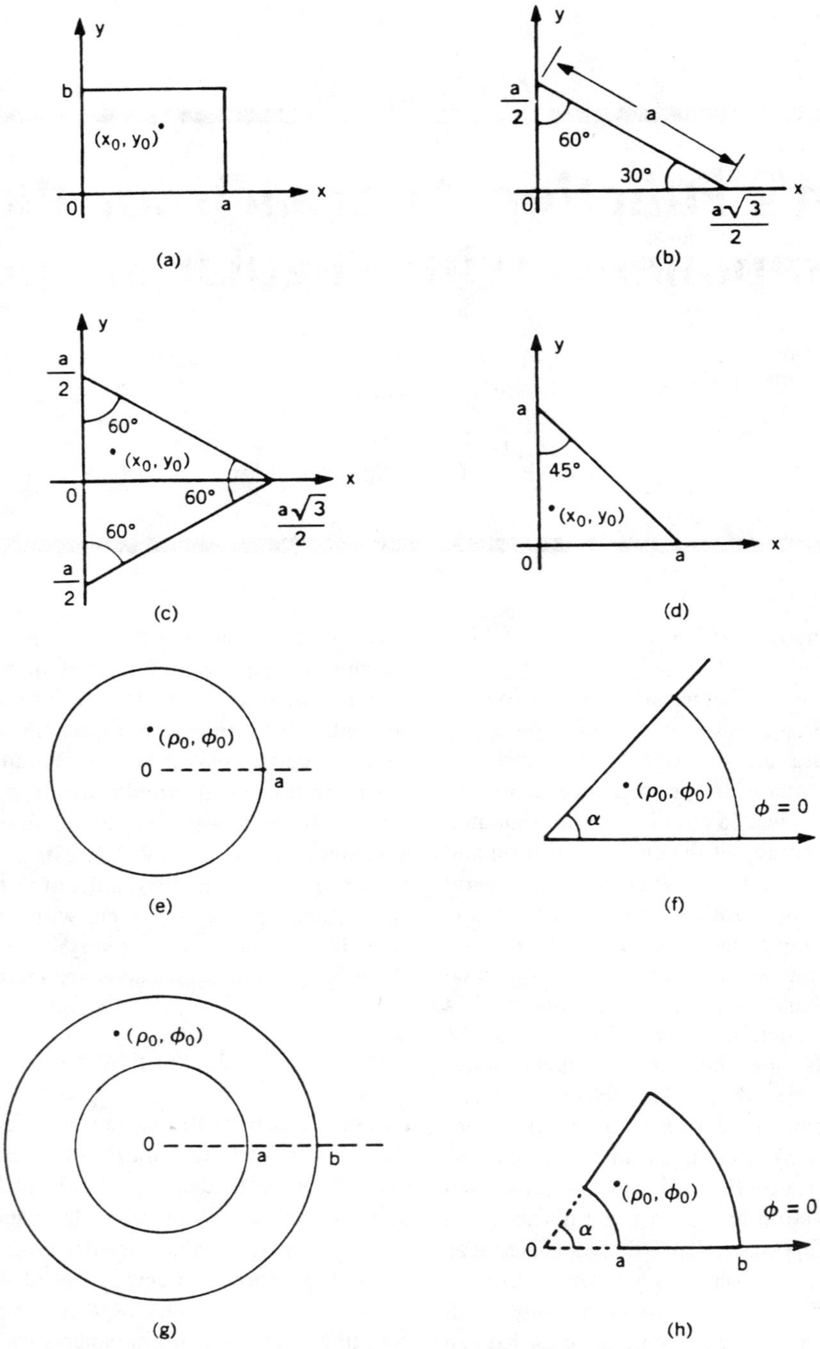

Fig. 2.1 Various planar circuit configurations for which Green's functions are available.

y-coordinates are comparable to the wavelength, the thickness along the z-direction is much smaller than the wavelength. Therefore, the fields inside the circuit can be assumed to be uniform along the z-direction, and hence we may set $\partial/\partial z = 0$ and $H_z = E_x = E_y = 0$. Given these conditions, and recalling that the dielectric material is taken to be linear, homogeneous, and isotropic, Maxwell's equations may be written as

$$\frac{\partial H_y}{\partial x} - \frac{\partial H_x}{\partial y} = j\omega\varepsilon E_z \tag{2.1}$$

$$\frac{\partial E_z}{\partial y} = -j\omega\mu H_x \tag{2.2}$$

$$\frac{\partial E_z}{\partial x} = j\omega\mu H_y \tag{2.3}$$

Using Eqs. (2.1)–(2.3), the wave equation (Helmholtz equation) that governs the electromagnetic fields in this source-free structure can be written in terms of E_z as

$$(\nabla_T^2 + k^2)E_z = 0 \tag{2.4}$$

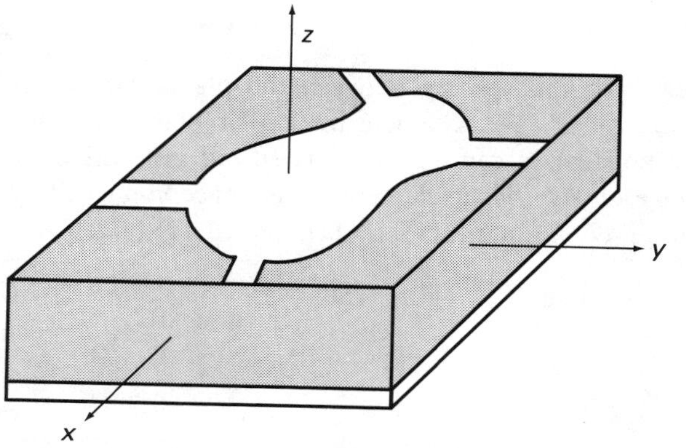

Fig. 2.2 Schematic of an arbitrarily shaped microstrip multiport planar structure.

where

$$\nabla_T^2 = \frac{\partial^2}{\partial x^2} + \frac{\partial^2}{\partial y^2} \tag{2.5}$$

$$k = \omega\sqrt{\mu\varepsilon} \tag{2.6}$$

and ε and μ denote, respectively, the permittivity and the permeability of the spacing material, ω is the angular frequency, and k denotes the wave number in the spacing material.

The characteristics of a planar circuit are determined by solving the two-dimensional Helmholtz equation outlined by (2.4) and applying the proper boundary conditions. In general, there exist two boundary conditions: one at the periphery points where no external ports are present and the other along the periphery points where external ports are connected. Prior to deriving these boundary conditions, the behavior of the surface current along the planar circuit will be investigated.

2.3 Surface Current Density on a Planar Circuit

In the absence of any external coupling ports, the surface current at the periphery C, or any other point on the planar element shown in Fig. 2.3, can be expressed as

$$\mathbf{J}_s = \hat{\mathbf{n}} \times (\mathbf{H}_1 - \mathbf{H}_2) \tag{2.7}$$

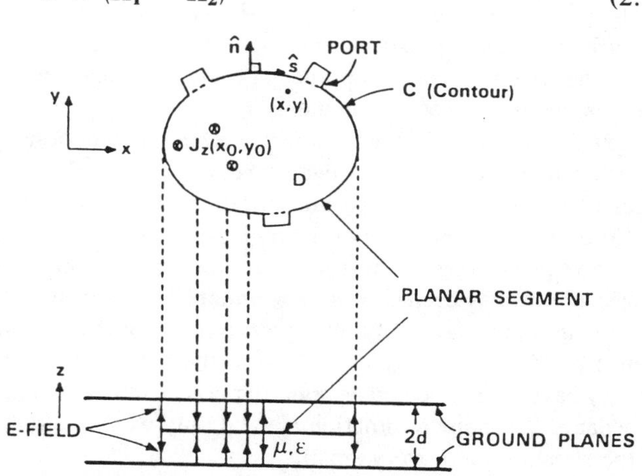

Fig. 2.3 A triplate-type planar circuit. Location of exciting current sources and the field point (x, y) on the circuit.

where $\hat{\mathbf{n}}$ is a unit vector normal to the planar element and $\mathbf{H}_1$ and $\mathbf{H}_2$ are the magnetic fields on the two opposite sides of the conducting segment. In the case of the stripline planar circuit configuration shown in Fig. 2.3, the magnetic fields on both sides of the planar element are related by $\mathbf{H}_1 = -\mathbf{H}_2$, thus reducing Eq. (2.7) to

$$\mathbf{J}_s = -2\hat{\mathbf{n}} \times \mathbf{H}_2 \tag{2.8}$$

The magnetic field $\mathbf{H}_2$, or simply $\mathbf{H}$, in Eq. (2.8) can be expressed via Maxwell's equations (2.2) and (2.3) in terms of the longitudinal electric field component E_z as follows,

$$\mathbf{H} = \frac{1}{j\omega\mu}\left(-\frac{\partial E_z}{\partial y}\hat{\mathbf{a}}_x + \frac{\partial E_z}{\partial x}\hat{\mathbf{a}}_y\right) \tag{2.9}$$

where $\hat{\mathbf{a}}_x$ and $\hat{\mathbf{a}}_y$ are unit vectors along the x- and y-coordinates, respectively. Substituting Eq. (2.9) into (2.8) yields

$$\mathbf{J}_s = \frac{2}{j\omega\mu}\left(-\frac{\partial E_z}{\partial x}\hat{\mathbf{a}}_x + \frac{\partial E_z}{\partial y}\hat{\mathbf{a}}_y\right) \tag{2.10}$$

The expression of $\mathbf{J}_s$ in (2.10) is valid at all points on the central patch of the stripline type of the planar circuit including the periphery.

For points on the periphery C, $\mathbf{J}_s$ can be written in terms of components that are tangential and normal to the periphery C as follows

$$\mathbf{J}_s = \frac{2}{j\omega\mu}\left(\frac{\partial E_z}{\partial s}\hat{\mathbf{s}} + \frac{\partial E_z}{\partial n}\hat{\mathbf{n}}\right) \tag{2.11}$$

where $\hat{\mathbf{s}}$ and $\hat{\mathbf{n}}$ are, respectively, unit vectors tangential and normal to the periphery as shown in Fig. 2.3. For a microstrip-type planar circuit, the upper ground plane of Fig. 2.3 is not present; hence there is no magnetic field above the central patch. This, of course, results in suppressing the factor of 2 appearing in (2.8), (2.10), and (2.11). To express Eq. (2.11) in a format applicable to both microstrip and stripline circuits, it is often written as

$$\mathbf{J}_s = \frac{p}{j\omega\mu}\left(\frac{\partial E_z}{\partial s}\hat{\mathbf{s}} + \frac{\partial E_z}{\partial n}\hat{\mathbf{n}}\right) \tag{2.12}$$

where $p = 1$ for microstrip-type circuits and $p = 2$ for stripline-type circuits. Having described the behavior of the surface current density on the planar element, the boundary conditions will be discussed next.

2.4 Boundary Condition Where No External Ports Are Present

Excluding the periphery points where external ports are connected, the planar circuit periphery is open (infinite impedance), shorted (zero wall impedance), or terminated by an impedance wall of finite value. For a planar circuit with open boundary, the normal component of the surface current at the periphery must be zero, which translates from (2.12) to

$$\frac{\partial E_z}{\partial n} = 0 \tag{2.13}$$

This condition is widely referred to as the *magnetic wall boundary condition.* For planar circuits with short-circuited periphery, the tangential component of the electric field at the periphery must be zero:

$$E_z = 0 \tag{2.14}$$

This condition is known as the *electric wall boundary condition.* Finally, a planar circuit whose periphery is terminated by an arbitrary impedance Z_w must satisfy the following *wall impedance boundary condition*

$$Z_z = \frac{E_z}{H_s} \tag{2.15}$$

2.5 Boundary Condition at External Ports

When external coupling ports are present, a boundary condition different from Eqs. (2.13), (2.14), and (2.15) is met. This boundary condition relates the excitation current at the coupling port to the differential of the electric field component E_z. The electromagnetic fields inside the planar structure are usually excited either by an aperture located at the periphery (edge fed) or by some internal current sources (coaxial line). The former case is normally encountered in MIC applications; the latter is often observed in antenna applications. However, both cases can be treated in the same fashion.

In an edge-fed planar circuit (microstrip or stripline type), the excitation current at the coupling port is normal to the periphery of the planar element. This situation results in simplifying (2.12), to become

$$\mathbf{J}_s = \frac{p}{j\omega\mu}\frac{\partial E_z}{\partial n}\,\hat{\mathbf{n}} \tag{2.16}$$

The RF current flowing into a coupling port can be obtained by integrating over the width of the port W to get

$$i = -\frac{p}{j\omega\mu}\int_W \frac{\partial E_z}{\partial n}\,ds \tag{2.17}$$

where ds is the incremental distance along the periphery. The negative sign in (2.17) implies that the current i flows inward whereas the unit vector $\hat{\mathbf{n}}$ points outward.

2.6 Circuit Characteristics in Terms of RF Voltage and Current

In practice it is more desirable to characterize planar circuits in terms of RF voltage v and current density $\mathbf{J}_s$ instead of electric and magnetic fields. This is done primarily to facilitate the introduction of sources (as it will be seen later), loads, and lumped impedances, as well as curved boundary conditions [2.6, 2.7]. The electric field component E_z is related to the RF voltage v between the planar circuit and the ground plane by

$$v = -E_z d \tag{2.18}$$

where d is the spacing between the planar circuit and the ground plane. The characterization of planar circuits with open boundaries can now be carried out in terms of the RF voltage on the planar element. This is accomplished by substituting (2.18) into (2.4), (2.13), and (2.17) to get

$$(\nabla_T^2 + k^2)\,v = 0 \tag{2.19}$$

with

$$\frac{\partial v}{\partial n} = 0 \tag{2.20}$$

for points on the periphery where there are no coupling ports and

$$i = \frac{p}{j\omega\mu d}\int_W \frac{\partial v}{\partial n}\,ds \tag{2.21}$$

at the coupling ports. The solution (2.19), with (2.20) and (2.21) as the boundary conditions, leads to the characterization of open-boundary-type, multiple-port planar circuits. The governing equations for other types of planar circuits can be obtained using similar procedures.

2.7 Analysis Techniques

A formal solution for the wave equation (2.19) and the associated boundary conditions (2.20) and (2.21) can be obtained by using various analytical and numerical techniques. The choice of any particular technique over another is a function of the geometrical shape of the planar circuit under consideration. Planar circuits with simple or regular geometrical shapes (e.g., rectangular, circular, sectorial, annular, and triangular) are conveniently analyzed by the Green's function approach [2.8], the resonant-mode approach [2.9], or the mode-matching approach [2.10, 2.11]. The Green's function approach and the resonant-mode approach are essentially equivalent, simply because the Green's function itself is expressed in terms of a resonant-mode expansion [2.12].

Of the three just-mentioned analytical techniques, the Green's function approach is the most attractive. This is because the Green's function approach provides compact analytical expressions that lend themselves to more efficient computer analyses. Furthermore, the Green's function approach remains to be useful even when field perturbing devices (such as shorting pins) are introduced at the periphery [2.13] or any other position of the regularly shaped planar circuit. The placement of shorting pins along the periphery of planar circuits has been shown to be an effective technique for improving and/or favorably altering the planar circuit performance [2.14]. Shorting pins located at the planar circuit periphery have been modeled as dummy ports terminated in zero impedance. Such approximation is valid only when the radius of the shorting pin is very small.

When the shorting pin has an arbitrary radius and is also arbitrarily located on the planar circuit, another analysis technique may be used [2.15]. This technique is based on the mode-matching approach, where the peripheral magnetic and electric fields are matched with the magnetic and electric fields beneath the planar circuit. The peripheral electric and magnetic fields in this approach are expressed as a Fourier series.

2.8 Green's Function Approach [2.5]

As stated earlier, Green's function technique is usually employed when the shape of the planar circuit under consideration is regular. Due to the simplicity of the planar circuit shape Green's function is obtained analytically using one of two methods: (1) the method of images or (2) the eigenfunction expansion. In deriving the Green's function, one obtains the voltage at any point on the planar circuit as a response to a given unit current source excitation located elsewhere on the circuit. Together with the precise knowledge of the locations of the ports, the Green's function can be used to obtain an impedance matrix characterizing the planar circuit under consideration, as we shall see in detail next. The two different methods for deriving the Green's function will be described later in this chapter.

The excitation of planar circuits may take place either through some internal current sources J_z (i.e., coaxial feeding) or through some apertures produced at the circuit periphery to couple the planar circuit to the external circuit (i.e., edge-fed microstrip). When the planar circuit is excited by an internal current source whose density is J_z located at any arbitrary point (x, y) inside the circuit periphery and directed in the z direction, as shown in Fig. 2.3, the wave equation can be written as

$$(\nabla_T^2 + k^2)\, v = -j\omega\mu d J_z \tag{2.22}$$

where ∇_T and k have been defined in (2.5) and (2.6), respectively. On the other hand, when the planar circuit is edge-fed by a microstrip line, the current density at the coupling port is in reality not directed along the z-axis but can be equivalently considered as such (i.e., along the z-direction) by imposing the magnetic wall condition $\partial v/\partial n = 0$ along the periphery. The equivalent fictitious surface current is given by

$$\mathbf{J}_s = \frac{p}{j\omega\mu d}\frac{\partial v}{\partial n}\,\hat{\mathbf{a}}_z \tag{2.23}$$

Thus, regardless of the method of excitation, planar circuits are considered to be fed by z-directed line currents located at the coupling ports. This is, of course, in addition to imposing the magnetic wall boundary condition along the whole periphery.

A formal solution to (2.22) can be obtained in terms of the Green's function $G(r/r_0)$, which is obtained by applying a z-directed unit line current source $\delta(r - r_0)$ located at $r = r_0$. The constructed Green's function is a solution of

$$(\nabla_T^2 + k^2)\, G\left(\frac{r}{r_0}\right) = -j\omega\mu\delta\,(r - r_0) \tag{2.24}$$

and satisfies the following boundary condition at the periphery,

$$\frac{\partial G}{\partial n} = 0 \tag{2.25}$$

Accordingly, the voltage at any point on the planar element can be written as

$$v(x, y) = \iint_D G(x, y|x_0, y_0)\, J_z(x_0, y_0)\, dx_0 dy_0 \tag{2.26}$$

where $J_z(x_0, y_0)$ denotes the excitation current injected normally into the circuit and D is the two-dimensional region of the planar element enclosed by magnetic walls as shown in Fig. 2.3

When the source current is introduced only at the periphery, the voltage v at the periphery can be expressed as

$$v(s) = \int_C G(s|s_0) J_s(s_0)\, ds_0 \tag{2.27}$$

where $J_s(s_0)$ is the line current source oriented in the z-direction and is given by (2.23), s and s_0 are distances measured along the periphery C, and the integration is over the entire periphery C. Due to the fact that the line current $J_s(s_0)$ is injected only at discrete number of points along the periphery, (2.27) may be written as

$$v(s) = \sum_j \int_{W_j} G(s|s_0) J_s(s_0)\, ds_0 \tag{2.28}$$

where the summation is carried over all the coupling ports existing along the periphery and W_j denotes the width of the jth coupling port. From (2.21) and (2.23), however, one may express the current i_j fed into the jth port in terms of a z-oriented equivalent current line to obtain

$$i_j = p \int_{W_j} J_s(s_0)\, ds_0 \tag{2.29}$$

If the widths of the coupling ports are assumed to be small so that the line current density J_s is constant (i.e., no transversal variations) over the width of the port, the integration in (2.29) can be carried out to get

$$J_s|_{\text{for } j\text{th port}} = \frac{i_j}{pW_j} \tag{2.30}$$

Upon substituting (2.30) into (2.28), the RF voltage $v(s)$ at any point on the planar element periphery becomes

$$v(s) = \sum_j \frac{i_j}{pW_j} \int_{W_j} G(s|s_0)\, ds_0 \tag{2.31}$$

The average voltage v_i over the width of the ith coupling port can be expressed in terms of $v(s)$ as follows:

$$v_i = \frac{1}{W_i} \int_{W_i} v(s)\, ds \tag{2.32}$$

By combining (2.31) and (2.32) we obtain

$$v_i = \sum_i \frac{i_j}{pW_iW_j} \int_{W_i} \int_{W_j} G(s|s_0)\, ds_0\, ds \quad (2.33)$$

Expressing the relationship between current and voltage in terms of an impedance (i.e., $z_{ij} = v_i/i_j$) yields

$$z_{ij} = \frac{1}{pW_iW_j} \int_{W_i} \int_{W_j} G(s|s_0)\, ds_0\, ds \quad (2.34)$$

where $p = 1$ for microstrip-type planar circuits and $p = 2$ for stripline-type planar circuits. Using (2.34) to calculate Z_{ij} for various i and j, the impedance matrix **Z** of the planar circuit can be constructed. The corresponding scattering matrix **S** can be determined by using the relation

$$\mathbf{S} = \sqrt{\mathbf{Y_0}}\,(\mathbf{Z} - \mathbf{Z_0})\,(\mathbf{Z} + \mathbf{Z_0})^{-1}\sqrt{\mathbf{Z_0}} \quad (2.35)$$

where $\mathbf{Z_0}$, $\sqrt{\mathbf{Z_0}}$, $\sqrt{\mathbf{Y_0}}$ are diagonal matrices with diagonal elements given by $Z_{01}, Z_{02}, \ldots, Z_{0n}$; $\sqrt{Z_{01}}, \sqrt{Z_{02}}, \ldots, \sqrt{Z_{0n}}$; and $1/\sqrt{Z_{01}}, 1/\sqrt{Z_{02}}, \ldots, 1/\sqrt{Z_{0n}}$, respectively. The matrix elements $Z_{01}, Z_{02}, \ldots, Z_{0n}$ represent the normalizing impedances at the various ports of the planar circuits. The impedance matrix elements for several planar circuits with regular shapes and open boundaries are listed in Sec. 2.10 of this chapter.

In the foregoing analysis it was assumed that the widths W of the coupling ports are small (when compared to the wavelength and the dimensions of the planar segment) such that the injected current is constant or uniform over the width of the port; however, in some cases this condition may not be true. To enforce this condition, each wide coupling port is divided into a number of parallel subports such that the current over each subport is constant. A larger impedance matrix of the planar circuit is then constructed with all the subports included. Prior to computing the scattering matrix, the multiple subports at each coupling port are combined [2.16].

The procedure to combine bordering subports is based on the fact that these subports have the same voltage but different currents. This means that the subports are effectively in parallel and hence can be combined by adding their admittances. The admittances of the various subports are obtained by inverting the planar circuit impedance matrix derived above. If ports i and j are divided into n and m subports, respectively, such that $i = \{i_1, i_2, \ldots, i_n\}$ and $j = \{j_1, j_2, \ldots, j_m\}$, then the term Y_{ij} of the overall admittance matrix is given as

$$Y_{ij} = \sum_{k=1}^{n} \sum_{l=1}^{m} Y_{kl} \quad (2.36)$$

where Y_{kl} are the elements of the admittance matrix with multiple subports. The overall scattering matrix can then be obtained from the impedance matrix using (2.35) or directly from the admittance matrix, employing

$$\mathbf{S} = \sqrt{\mathbf{Z_0}}\,(\mathbf{Y_0} - \mathbf{Y})\,(\mathbf{Y_0} + \mathbf{Y})^{-1}\sqrt{\mathbf{Y_0}} \quad (2.37)$$

where $\mathbf{Y_0}$ is a diagonal matrix of normalizing admittances at various ports with diagonal elements given by $1/Z_{01}$, $1/Z_{02}, \ldots, 1/Z_{0n}$.

To recapitulate, when the geometrical shape of the planar circuit is simple, the corresponding Green's function can be obtained analytically; consequently, the equivalent circuit parameters can be derived by using (2.34). In the following section, two methods for deriving the Green's functions of planar circuits will be described.

2.9 Green's Function Derivation

The prior discussion shows that Green's functions are very useful in characterizing planar circuits provided they can be found. In this section, two different methods for obtaining Green's functions for planar circuits with regular shapes are discussed. The two methods are (1) the method of images and (2) the eigenfunctions expansion.

2.9.1 Method of Images [2.17]

The method of images is an effective way for solving differential equations of the form given by (2.24). A solution to an inhomogeneous differential equation can be obtained by the method of images, provided that the term on the right-hand side of the equation is a periodic function [2.18] and hence can be expressed as a Fourier series. If this requirement is satisfied, the Green's function can then be expressed as an infinite series of summations of the functions (these are the eigenfunctions) appearing in the Fourier series expansion. The coefficients of the series summation of the Green's function can then be obtained by substituting the series summation into the inhomogeneous differential equation to be solved.

In the case of (2.24) a nonperiodic single source $\delta(r - r_0)$ exists at $r = r_0$ inside the region of the planar component; however, additional sources of the same type can be obtained by taking multiple images of the linear source at r_0 with respect to the various magnetic walls or electric walls of the planar component. Thus, the source term on the right-hand side of (2.24) can be modified to become periodic. The boundary condition will be satisfied by the voltage v produced by the real source and its images. It should be emphasized that the additional sources are all located outside the region of the planar component; hence the solution G still represents the Green's function for the geometrical shape of the planar circuit under consideration. For a rectangular planar component, the locations of the images are shown in Fig. 2.4.

The method of images, however, has its limitations. It can be applied only to planar circuits with shapes enclosed by boundaries that are straight lines. This is because only plane mirrors give a point image of a point source. Nonetheless, the straight-line boundary requirement is necessary but not sufficient. For instance, in the case of polygonal shapes, images are not uniquely specified in the two-dimensional space unless the internal angle at each vertex of the polygon is a submultiple of π. Thus this method is applicable only to shapes with straight regular edges and angles at each vertex equal to submultiples of π [2.19].

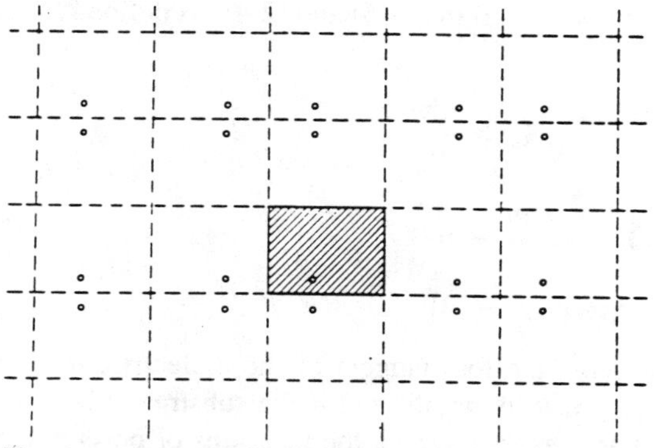

Fig. 2.4 Line current source and its images for calculation of the Green's function for a rectangular component.

2.9.2 Expansion of Green's Functions in Eigenfunctions [2.18]

This method is less restrictive than the method of images, and it is appropriate for any planar circuit with a regular shape. A regular shape is defined as being a configuration that is circular, annular, sectorial, or polygonlike, provided that any existing internal angle is a submultiple of π. In the Green's function method [2.18], the Green's function is expanded in terms of a complete and orthonormal set of eigenfunctions. These eigenfunctions ϕ_n, as well as their corresponding eigenvalues k_n^2, must satisfy an equation analogous to (2.19),

$$\nabla_T^2\phi_n + k_n^2\phi_n = 0 \tag{2.38}$$

where n denotes all the needed indices defining a certain ϕ_n. Given that the eigenfunctions are orthonormal, they must satisfy

$$\iint_D \phi_n^*\phi_n \, dx\, dy = \begin{cases} 1, & \text{if } n = m \\ 0, & \text{otherwise} \end{cases} \tag{2.39}$$

where (*) symbolizes a complex conjugate and the region D is surrounded by the periphery of the planar component. Furthermore, along the periphery the eigenfunction ϕ_n must satisfy boundary conditions similar to those satisfied by $G(r|r_0)$ and the total electric field; that is, $\partial\phi_n/\partial n = 0$ for an open boundary and $\phi_n = 0$ for a shorted boundary.

Expanding the Green's function $G(r|r_O)$ in a series of the eigenfunctions ϕ_n yields

$$G(r|r_0) = \sum_m A_m\phi_m(r) \tag{2.40}$$

where A_m are unknown coefficients yet to be determined. Substituting (2.40) into (2.24), we obtain

$$\sum_m A_m(k^2 - k_m^2)\phi_m(r) = -j\omega\mu d\delta\,(r - r_0) \tag{2.41}$$

The unknown coefficients A_m can be determined by, first, multiplying both sides of (2.41) by $\phi_n^*(r)$; second, integrating over the two-dimensional region D; and finally, applying the orthonormal property (2.39) of the eigenfunctions to get

$$A_n(k^2 - k_n^2) = -j\omega\mu d\phi_n^*(r_0) \tag{2.42}$$

By solving for A_n and substituting its value in (2.40), we obtain

$$G(r|r_0) = j\omega\mu d \sum_n \frac{\phi_n(r)\phi_n^*(r_0)}{k_n^2 - k^2} \tag{2.43}$$

where for lossless planar circuits, ϕ_n is real and hence the complex conjugate in (2.43) is not needed. Equation (2.43) reveals the basic structure of the Green's function. It is expressed as a double or a single series depending on whether or not one of the series can be summed analytically. The Green's function also involves the resonant modes of the planar structure. When we are concerned with the circuit behavior at frequencies close to a resonance, it is often sufficient to consider only the predominant resonant term. However, an accurate understanding of the planar circuit characteristics still requires the inclusion of many terms in the series.

As explained earlier, the eigenfunctions ϕ_n are determined by the shape of the planar component under consideration. Figure 2.1 illustrates the various shapes for which Green's functions are available. Appendix A to this chapter contains these Green's functions. As the reader may notice, the majority of these expressions involve a single summation and hence appear to be different from those reported earlier in [2.1]. The primary difference between the expressions outlined in Appendix A and those reported in [2.1] is that one of the series has been summed. This step has resulted in significantly improving the computational efficiency of the impedance matrix computation. Figure 2.5 illustrates this dramatic increase in computational efficiency. The two curves illustrate the percentage error in the input impedance of a rectangular microstrip patch using a single-series expression and a double-series expression [2.20]. Similar improvement in the computational efficiency has been demonstrated for triangular segments [2.21]. Analogous results should be obtained for circular and annular segments [2.2].

2.10 Z-Matrix Evaluation

The basic characteristics of a multiport planar circuit are described by (2.34). This expression evaluates the elements of the generalized impedance matrix that characterizes the planar segment under consideration. After determining the Green's function and substituting it in (2.34), the impedance matrix of the planar segment can be determined. The double integrals in (2.34) are then carried out over the widths of the appropriate ports, and an expression defining the impedance matrix elements is obtained. A scattering matrix representation for the planar circuit may be obtained by using the conversion relation given in (2.35).

Until recently, the computations of Z-matrices has been numerically involved. This is primarily due to the double infinite series that are introduced by the two-dimensional Green's function. The slow convergence of these series has dictated that a large number of terms be used (typically on the order of 100 × 100) to obtain sufficiently accurate results. However,

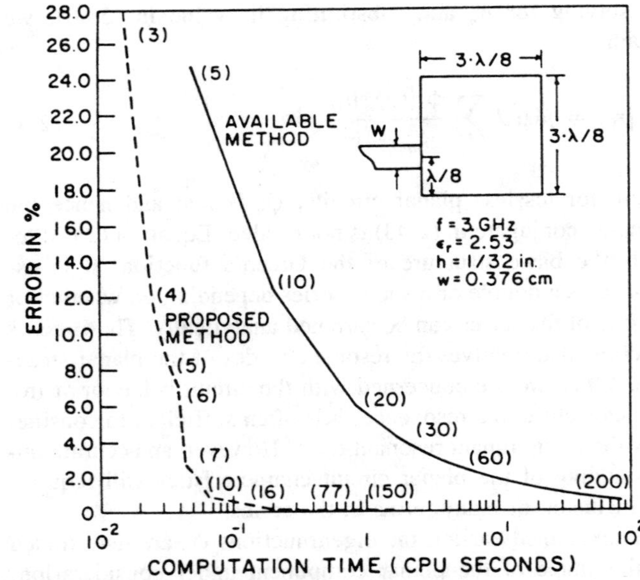

Fig. 2.5 Comparison of two methods for computation of Z-matrix of rectangular planar segments.

this criterion has led some researchers to view the Green's function approach as time consuming and hence has discouraged its widespread application.

Recently, however, remarkable success in improving the computational efficiency of most of the Z-matrices has been reported [2.2, 2.3, 2.4, 2.20–22]. The improvement in the computational efficiency has been accomplished by analytically summing up one of the series appearing in the expressions, hence reducing the double infinite series into a single infinite series. Single-series Z-matrices are now available in the literature for most of the regularly shaped planar segments. In this section a listing of the newly developed single-series impedance matrix expressions will be given. For the sake of completeness the double summation impedance matrix expressions of the circular and annular sectors will be listed as well.

2.10.1 Rectangular Segment

In [2.1] the impedance matrix elements of a rectangular microstrip planar segment has been reported to be given by

$$Z_{pq} = \frac{j\omega\mu d}{ab} \cdot \frac{\sum_{m=0}^{\infty}\sum_{n=0}^{\infty} \sigma_m\sigma_n\phi_{mn}(x_p, y_p)\phi_{mn}(x_q, y_q)}{k_x^2 + k_x^2 - k^2} \tag{2.44}$$

where ϕ_{mn}, for ports oriented along the y-direction, is

$$\phi_{mn}(x, y) = \cos k_x x \cos k_y y \; sinc\,\frac{k_y W}{2} \tag{2.45}$$

and for ports oriented along the x-direction is

$$\phi_{mn}(x, y) = \cos k_x x \cos k_y y \; sinc\left(\frac{k_x W}{2}\right) \tag{2.46}$$

The function $sinc(z)$ in (2.45) and (2.46) is defined as $\sin(z)/z$, and

$$k_x = \frac{m\pi}{a}, \qquad k_y = \frac{n\pi}{b} \tag{2.47}$$

$$\sigma_m = \begin{cases} 1, & m = 0 \\ 2, & m \neq 0 \end{cases} \tag{2.48}$$

$$k^2 = \omega^2\mu\varepsilon_0\varepsilon_r(1 - j\delta) \tag{2.49}$$

with δ being the loss tangent of the dielectric, a the rectangle's length, b its width, and d the substrate's height. Points (x_p, y_p) and (x_q, y_q) denote the locations of ports p and q, respectively.

It has been shown in [2.20] that the doubly infinite series in (2.44) can be reduced to a single infinite series by summing the inner sum. The choice of summation over n or m depends on the relative locations of ports p and q and also on the aspect ratio of the rectangular segment, as will be shown next.

Case I. Ports p and q Are Oriented Along the Same Direction

When both ports (p and q) are oriented along the same direction (i.e., both ports along x or y), we obtain

$$\begin{aligned} Z_{pq} = {} & -\frac{j\omega\mu d}{ab}\sum_{l=0}^{L} F\,\sigma_l \cos(k_u u_p)\cos(k_u u_q) \\ & \cdot \frac{\cos(\gamma_l z_>)\cos(\gamma_l z_<)\; sinc(k_u W_p/2) sinc(k_u W_q/2)}{\gamma_l \sin(\gamma_l F)} \\ & + \frac{\omega\mu d}{ab}\sum_{l=L+1}^{\infty} F \cos(k_u u_q)\cos(k_u u_p)\sin\left(\frac{k_u W_p}{2}\right) \\ & \cdot sinc\left(\frac{k_u W_q}{2}\right)\frac{\exp[-j\gamma_l(v_> - v_<)]}{\gamma_l} \end{aligned} \tag{2.50}$$

where

$$(v_>, v_<) = \begin{cases} (y_>, x_<), & l = m \\ (x_>, x_<), & l = n \end{cases} \tag{2.51}$$

When the two ports are oriented along the y-direction, $l = n$, and when they are oriented along the x-direction, $l = m$. Also

$$F = \begin{cases} b, & l = m \\ a, & l = n \end{cases} \tag{2.52}$$

$$(u_p, u_q) = \begin{cases} (x_p, x_q), & l = m \\ (y_p, y_q), & l = n \end{cases} \tag{2.53}$$

$$\gamma_l = \pm\sqrt{k^2 - k_u^2} \tag{2.54}$$

$$k_u = \begin{cases} \dfrac{m\pi}{a}, & l = m \\ \dfrac{n\pi}{b}, & l = n \end{cases} \tag{2.55}$$

$$(z_>, z_<) = \begin{cases} (y_> - b, y_<), & l = m \\ (x_> - a, x_<), & l = n \end{cases} \tag{2.56}$$

The sign of γ_l is chosen such that $\text{Im}(\gamma_l)$ is negative. W_p and W_q denote the widths of ports p and q, respectively. Additionally, the notation used in (2.51) is defined by

$$y_> = \max(y_p, y_q) \qquad y_< = \min(y_p, y_q) \tag{2.57}$$

Similar notation applies for $x_>$ and $x_<$ when $l = n$. The choice of the integer L in (2.51) becomes a trade-off between computational speed and accuracy. A compromise is to select L such that $\gamma_l F \leq 100$.

Case II. Ports p and q Are Oriented Along Different Directions

When the two ports (p and q) are oriented along different directions (x and y), various elements of the Z-matrix may be written as

$$\begin{aligned} Z_{pq} = &-\frac{j\omega\mu d}{ab}\sum_{l=0}^{L} F\,\sigma_l \cos(k_u u_p)\cos(k_u u_q)\cos(\gamma_l z_<) \\ &\cdot \cos(\gamma_l z_>)\frac{sinc(k_u W_i/2)\; sinc(\gamma_l W_j/2)}{\gamma_l \sin(\gamma_l F)} \\ &-\frac{j\omega\mu d}{ab}\sum_{l=L+1}^{\infty} F\cos(k_u u_p)\cos(k_u u_q) \\ &\cdot sinc\left(\frac{k_u W_i}{2}\right)\frac{\exp[-j\gamma_l(v_> - v_< - W_j/2)]}{\gamma_l^2 W_j} \end{aligned} \tag{2.58}$$

The choice of l is determined by the convergence of the summation in (2.58). This is realized when

$$v_> - v_< - \frac{W_j}{2} > 0 \tag{2.59}$$

The index of the inner summation is chosen so that this condition is satisfied. This condition can be written more explicitly as

$$l = m, \quad \text{if } \{\max(y_p, y_q) - \min(y_p, y_q) - W_j/2\} > 0 \tag{2.60}$$

and

$$l = n, \quad \text{if } \{\max(x_p, x_q) - \min(x_p, x_q) - W_j/2\} > 0 \tag{2.61}$$

When both these conditions are satisfied, any choice of l will ensure convergence.

If $l = n$, W_i corresponds to the port oriented along the y-direction, and W_j corresponds to the port oriented along the x-direction. On the other hand, if $l = m$, W_j is used for the port along the x-direction and W_j for the port along the y-direction.

2.10.2 30°–60°–90° Triangular Segment

Until recently, the Green's function G for a 30°–60°–90° triangular segment was obtained by summing up the double series for each of 24 line sources [2.17]. By deriving a single-series expression for each of the 24 line sources, Lee, Benalla, and Gupta have arrived at a new and computationally more efficient expression for G [2.21]. To derive the new expression, Lee and his coworkers computed a new single series Z_{pq} expression for rectangular segments with magnetic walls and with ports oriented arbitrarily *inside* the rectangle. In the previous section a Z_{pq} expression for a rectangular segment with ports oriented *parallel* to its sides (not arbitrarily oriented) was given.

The impedance matrix elements for a 30°–60°–90° triangular segment with all magnetic walls is given by

$$Z_{pq}^T = \sum_{i=l}^{24} Z_{pq}^{Rm}(x_p, y_p, \theta_p; x_i, y_i, \theta_i) \tag{2.62}$$

where $Z_{pq}^{Rm}(x_p, y_p, \theta_p; x_i, y_i, \theta_i)$ is the impedance matrix for a rectangular segment with magnetic walls, a p-port located at (x_p, y_p) with orientation θ_p, and an image port i located at (x_i, y_i) with an orientation angle of θ_i. For a given line source in a triangle, the image locations and their corresponding angles of orientation are obtainable from the geometric considerations, as described in [2.17]. The value of Z_{pq}^{Rm} is given by [2.21]

$$Z_{pq} = -\frac{j\omega\mu d}{ab}\sum_{l=0}^{\infty} F\frac{\sigma_l \Phi(t)\,\Psi(t)}{\gamma_l \sin(\gamma_l F)} \tag{2.63}$$

where

$$\begin{aligned} \Phi(t) = \frac{1}{2}\Big\{&\cos[(A - C)t + (B - D)] \\ &\cdot sinc\left[(A - C)\frac{W_\theta}{2}\right] \\ &+ \cos[(A + C)t + B + D] \\ &\cdot sinc\left[(A + C)\frac{W_\theta}{2}\right]\Big\} \end{aligned} \tag{2.64}$$

where $sinc(z) = \sin(z)/z$ and W_θ is the projected length of a port on the x- or y-axis. The expression for $\Psi(t)$ is obtained from (2.64) by replacing D with D'. The constants A, B, C, D, and D' depend on the value of the index l, the orientation angle θ, and the locations of the ports. These constants are given by

$$A = k_m \alpha_p = \frac{m\pi}{a}\cot(\theta_p) \tag{2.65}$$

$$B = (x_{po} - y_{po}\alpha_p)\frac{m\pi}{a} \tag{2.66}$$

$$C = \gamma_m = \sqrt{k^2 - \left(\frac{m\pi}{a}\right)^2} \tag{2.67}$$

$$D = -\gamma_m b \tag{2.68}$$

when $l = m$ and

$$A = k_n \alpha_p = \frac{n\pi}{b}\tan(\theta_p) \tag{2.69}$$

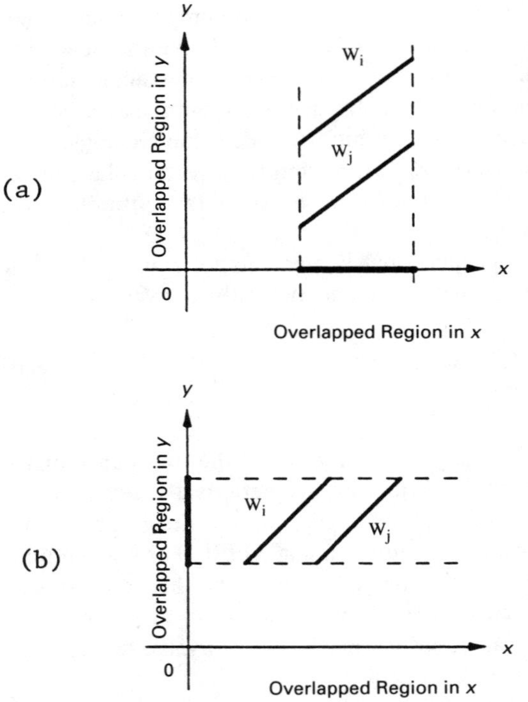

Fig. 2.6 Examples of ports with overlapping regions (a) along the x-axis; (b) along the y-axis

$$B = (y_{po} - x_{po}\alpha_p)\frac{n\pi}{b} \tag{2.70}$$

$$C = \gamma_n = \sqrt{k^2 - \left(\frac{n\pi}{b}\right)^2} \tag{2.71}$$

$$D' = -\gamma_n a \tag{2.72}$$

when $l = n$. The term σ_m is defined by (2.48), (x_{po}, y_{po}) is the coordinate of the midpoint at the port p, and $\alpha_p = \cot(\theta_p)$ is the slope of the port. The orientation angle θ_p is referenced to the x-axis when measured. The choice of $l = m$ or $l = n$ depends on *the situation;* whether the two ports have an overlapping region along the x-axis ($l = m$) or along the y-axis ($l = n$), as indicated in Fig. 2.6.

For the self-impedance $p = q$, the expression for Z_{pq} is obtained by setting $l = m$ and integrating with respect to the y-variable,

$$Z_{pq} = -\frac{j\omega\mu d}{a}\sum_{m=0}^{\infty}\frac{\sigma_m I_{pq}(y_o)}{\gamma_m \sin(\gamma_m b)} \tag{2.73}$$

where

$$\begin{aligned} I_{pq}(y_0) = {} & \frac{1}{4W_\theta}\cdot\frac{-2C}{A^2 - C^2}\cdot\sin(\gamma_m b) \\ & \cdot\{\cos[2Ay_o + 2B]\cdot sinc[AW_\theta] + 1\} \\ & + \frac{1}{4W_\theta}\Gamma\left(A, B, C, D, \frac{W_\theta}{2}\right) \\ & \cdot\{\Lambda(A - C, B - D') + \Lambda(A + C, B + D')\} \\ & - \frac{1}{4W_\theta}\Gamma\left(A, B, C, D', -\frac{W_\theta}{2}\right) \\ & \cdot\{\Lambda(A - C, B - D) + \Lambda(A + C, B + D)\} \end{aligned} \tag{2.74}$$

with

$$\begin{aligned} \Gamma\left(A, B, C, D, \frac{W_\theta}{2}\right) & \\ = {} & \frac{\sin[(A - C)(y_o + W_\theta/2) + B - D]}{(A - C)} \\ & + \frac{\sin[(A + C)(y_o + W_\theta/2) + B - D]}{(A + C)} \end{aligned} \tag{2.75}$$

and

$$\begin{aligned} \Gamma(A - C, B - D') & \\ = {} & \cos[(A - C)y_o + B - D'] \\ & \cdot sinc\left[(A - C)\frac{W_\theta}{2}\right] \end{aligned} \tag{2.76}$$

Figure 2.7 offers a comparison of the accuracy and the computational efficiency between the single-summation expression and the double-summation expression of the self-impedance Z_{pp}. For this comparison the port was located at the middle of the hypotenuse of the 30°–60°–90° triangular segment.

2.10.3 Equilateral Triangular Segment

The impedance matrix elements Z_{pq} for an equilateral triangular segment is given by

$$Z_{pq} = \frac{1}{2}(Z_{pq}^{Te} + Z_{pq}^{Tm}) \tag{2.77}$$

where Z_{pq}^{Te} and Z_{pq}^{Tm} denote, respectively, the impedance matrix elements for a 30°–60°–90° triangular segment with electric and magnetic walls. The value of Z_{pq}^{Te} is given by (2.63) through (2.76), while Z_{pq}^{Te} is given by

$$Z_{pq}^{Te} = \sum_{i=1}^{24} Z_{pi}^{Re}(x_p, y_p, \theta_p; x_i, y_i, \theta_i) \tag{2.78}$$

where $Z_{pi}^{Re}(x_p, y_p, \theta_p; x_i, y_i, \theta_i)$ is the impedance matrix element for a rectangular segment with two electric walls, a port p located at (x_p, y_p) with an orientation angle θ_p, and an image

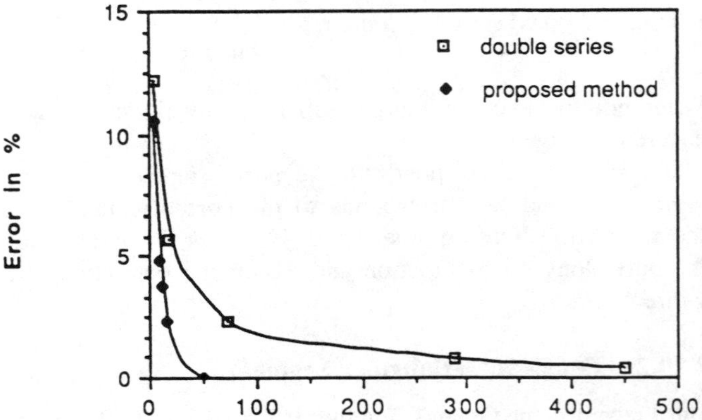

Fig. 2.7 Comparison of error as a function of CPU time for a 30°–60°–90° triangle using two approaches. (From [2.21]. Copyright © 1992. Reprinted by permission of John Wiley & Sons, Inc.)

port i located at (x_i, y_i) with an orientation angle θ_i. It should be noted that some of the images have negative values and cancel in the summation process in (2.78).

The impedance matrix elements for a rectangular segment with two electric narrow walls are given by

$$Z_{pq} = -\frac{j\omega\mu d}{ab}\sum_{l=0}^{\infty} F\frac{\sigma_l \Phi(t)\,\Psi(t)}{\gamma_l \sin(\gamma_l F)} \tag{2.79}$$

where σ_m is given by (2.48) and $\Phi(t)$ is given by

$$\Phi(t) = \frac{1}{2}\Big\{\sin[(A - C)\,t + B - D] \cdot sinc\Big[(A - C)\frac{W_\theta}{2}\Big] + \sin[(A + C)\,t + B + D] \cdot sinc\Big[(A + C)\frac{W_\theta}{2}\Big]\Big\} \tag{2.80}$$

where $sinc(z) = \sin(z)/z$ and $\Psi(t)$ is given by the same expression as (2.80) but with D replaced by D'.

When $p = q$, the self-impedance terms in the impedance matrix are given by

$$Z_{pq} = -\frac{j\omega\mu d}{b}\sum_{n=0}^{\infty}\frac{\sigma_n I_{pq}(x_0)}{\gamma_n \sin(\gamma_n a)} \tag{2.81}$$

where

$$\begin{aligned} I_{pq}(x_o) = {} & \frac{1}{W_\theta}\frac{2C}{A^2 - C^2}\cdot\sin[D - D'] \\ & \cdot\{1 - \cos[2Ax_o + 2B]\cdot sinc[AW_\theta]\} \\ & - \frac{1}{4W_\theta}\Pi\Big(A, B, C, D, \frac{W_\theta}{2}\Big) \\ & \cdot\{\Omega(A - C, B - D') + \Omega(A + C, B + D')\} \\ & + \frac{1}{4W_\theta}\Pi\Big(A, B, C, D', -\frac{W_\theta}{2}\Big) \\ & \cdot\{\Omega(A - C, B - D) + \Omega(A + C, B + D)\} \end{aligned} \tag{2.82}$$

with

$$\Pi\Big(A, B, C, D, \frac{W_\theta}{2}\Big) = \frac{\cos[(A - C)(x_o + W_\theta/2) + B - D]}{(A - C)} + \frac{\sin[(A + C)(x_o + W_\theta/2) + B + D]}{(A + C)} \tag{2.83}$$

and

$$\Omega(A - C, B - D') = \sin[(A - C)x_o + B - D']\cdot sinc\Big[(A - C)\frac{W_\theta}{2}\Big] \tag{2.84}$$

Figure 2.8 shows a comparison of the error versus computational time for the equilateral triangular segment. The two plots are computed by using the double-series and the single-series expressions of the self-impedance Z_{pp}. As is evident from Figs. 2.7 and 2.8, the single-series expressions are much more accurate and offer significant improvement in the computational efficiency.

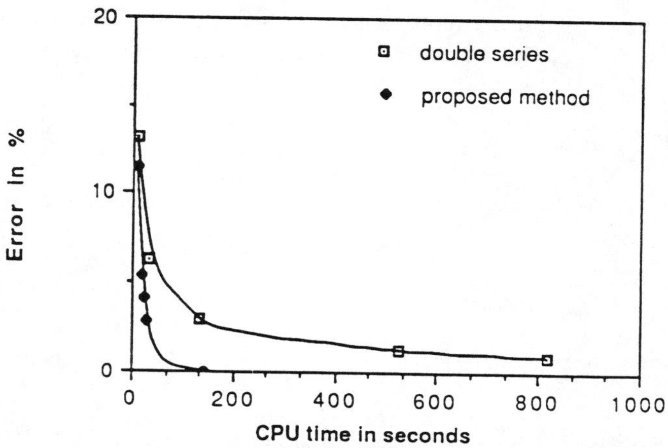

Fig. 2.8 Comparison of error as a function of CPU time for an equilateral triangle using two approaches. (From [2.21]. Copyright © 1992. Reprinted by permission of John Wiley & Sons, Inc.)

2.10.4 Isosceles Right-Angled Triangular Segment [2.4]

For a right-angled triangular segment, the elements Z_{pq} of the impedance matrix are given by single-series expression as well; however, the value of Z_{pq} is dependent on the location of the ports. Referring to Fig. 2.1(d), the following expressions apply.

a. Both Ports Are Located Along OA *or Along* OB

When both parts are located along OA or OB, the impedance matrix element Z_{pq} is given by

$$Z_{pq} = -\frac{j\omega\mu_o d}{2a}(Z_1 + Z_2) \tag{2.85}$$

where

$$Z_1 = \sum_{m=0}^{\infty} 2D_m \cos(\gamma_m a)\cos(k_x u_p)\,sinc\Big(\frac{k_x W_i}{2}\Big) + \cos(k_x u_q)\,sinc\Big(\frac{k_x W_j}{2}\Big) \tag{2.86}$$

and Z_2 is

$$Z_2 = \sum_{m=0}^{\infty} 2D_m(-1)^m \cos(k_x u_p)\,sinc\Big(\frac{k_x W_p}{2}\Big)\cdot\cos(\gamma_m u_q)\,sinc\Big(\frac{k_x W_q}{2}\Big) \tag{2.87}$$

if $u_p \geq u_q$ or

$$Z_2 = \sum_{m=0}^{\infty} 2D_m(-1)^m \cos(k_x u_q)\,sinc\Big(\frac{k_x W_q}{2}\Big)\cdot\cos(\gamma_m u_p)\,sinc\Big(\frac{k_x W_i}{2}\Big) \tag{2.88}$$

if $u_q \geq u_p$. Furthermore, $sinc(z) = \sin(z)/z$, and the parameters k_x, γ_m, D_m, k^2, and (u_p, u_q) are defined, respectively, by

$$k_x = \frac{m\pi}{a} \tag{2.89}$$

$$\gamma_m = \pm\sqrt{k^2 - k_x^2} \tag{2.90}$$

$$D_m = \frac{\sigma_m}{\gamma_m \sin(\gamma_m a)} \tag{2.91}$$

$$k^2 = \omega^2 \mu \varepsilon_o \varepsilon_r (1 - j\delta_e) \tag{2.92}$$

$$(u_p, u_q) = \begin{cases} (x_p, x_q) & \text{for } i, j \text{ along } OA \\ (y_p, y_q) & \text{for } i, j \text{ along } OB \end{cases} \tag{2.93}$$

As m becomes large, the imaginary part of the arguments of the complex trigonometric functions $\sin(\gamma_{m*})$ and $\cos(\gamma_{m*})$ appearing in (2.86), (2.87), and (2.88) will become very large and give rise to numerical complications. To avoid this numerical problem, the trigonometric functions are replaced by their large argument approximations, which are given by

$$\sin(\gamma_{m*}) = \frac{1}{2j}\exp(j\gamma_{m*}) \tag{2.94}$$

$$\cos(\gamma_{m*}) = \frac{1}{2}\exp(j\gamma_{m*}) \tag{2.95}$$

Thus, using (2.94) along with (2.95), the single-series expressions for Z_1 and Z_2 in (2.86) through (2.88) can be rewritten as

$$Z_1 = \sum_{m=0}^{L} 2D_m \cos(\gamma_m a)\cos(k_x u_p)\, sinc\left(\frac{k_x W_p}{2}\right) \cdot \cos(k_x u_q)\, sinc\left(\frac{k_x W_q}{2}\right) + \sum_{m=L+1}^{\infty} \frac{2j\sigma_m}{\gamma_m}\cos(k_x u_p)\, sinc(k_x W_p) \cdot \cos(k_x u_q)\, sinc\left(\frac{k_x W_q}{2}\right) \tag{2.96}$$

The expression for Z_2, if $u_p \geq u_q$, is

$$Z_2 = \sum_{m=0}^{L} 2D_m(-1)^m \cos(k_x u_p)\, sinc\left(\frac{k_x W_p}{2}\right) \cdot \cos(\gamma_m u_q)\, sinc\left(\frac{k_x W_q}{2}\right) + \sum_{m=L+1}^{\infty} \frac{2\sigma_m(-1)^m}{\gamma_m^2 W_q} \cdot \exp\left[-j\gamma_m\left(a - x_q - \frac{W_q}{2}\right)\right] \cdot \cos(k_x u_p)\, sinc\left(\frac{k_x W_p}{2}\right) \tag{2.97}$$

and if $u_q > u_p$, then

$$Z_2 = \sum_{m=0}^{L} 2D_m(-1)^m \cos(k_x u_q)\, sinc\left(\frac{k_x W_q}{2}\right) \cdot \cos(\gamma_m u_p)\, sinc\left(\frac{k_x W_p}{2}\right) + \sum_{m=L+1}^{\infty} \frac{2\sigma_m(-1)^m}{\gamma_m^2 W_p} \cdot \exp\left[-j\gamma_m\left(a - x_p - \frac{W_p}{2}\right)\right] \cdot \cos(k_x u_q)\, sinc\left(\frac{k_x W_q}{2}\right) \tag{2.98}$$

where the parameters appearing in these equations have been defined earlier in (2.48) and (2.89) through (2.93). The integer L in (2.96) through (2.98) is chosen such that $\gamma_m a \leq 50$; this choice is a compromise between computational speed and accuracy.

b. One Port Along OA *and the Other Along* OB

Upon replacing the trigonometric functions by their large argument approximations [i.e., (2.94) and (2.95)], the expression for the impedance matrix elements can be written as

$$\begin{aligned} Z_{pq} = -\frac{j\omega\mu d}{2a}\Bigg\{ &\sum_{m=0}^{L} 2D_m \cos(k_x u_p)\, sinc\left(\frac{k_x W_p}{2}\right) \cdot \cos[\,(\gamma_m(u_q - a)\,]\, sinc\left(\frac{k_x W_q}{2}\right) \\ &+ \sum_{m=L+1}^{\infty} \frac{2\sigma_m}{\gamma_m^2 W_q} \cdot \exp\left[-j\gamma_m\left(u_q - \frac{W_q}{2}\right)\right] \cdot \cos(k_x u_p)\, sinc\left(\frac{k_x W_p}{2}\right) \\ &+ \sum_{m=0}^{L} 2D_m(-1)^m \cos(k_x u_p)\, sinc\left(\frac{k_x W_p}{2}\right) \cdot \cos(k_x u_q)\, sinc\left(\frac{k_x W_q}{2}\right) \\ &+ \sum_{m=L+1}^{\infty} \frac{4j\sigma_m(-1)^m}{\gamma_m} \cdot \exp(-j\gamma_m a)\cos(k_x u_p)\, sinc\left(\frac{k_x W_p}{2}\right) \cdot \cos(k_x u_q)\, sinc\left(\frac{k_x W_q}{2}\right)\Bigg\} \end{aligned} \tag{2.99}$$

where W_p is the width of the port located along OA and W_q is the width of the port located along OB. The rest of the parameters are defined in Eqs. (2.89) through (2.93).

c. When Both Ports Are Located Along AB

If the locations of port p and port q are not the same (i.e., $p \neq q$), then the impedance matrix elements are given by

$$Z_{pq} = -\frac{j\omega\mu d}{2a} \tag{2.100}$$
$$\Bigg(\sum_{m=0}^{L} D_m \Bigg\{ \cos[(\gamma_m - k_x) y_> - \gamma_m a] \cdot sinc\left[(\gamma_m - k_x)\frac{W_>}{2\sqrt{2}}\right] + \cos[(\gamma_m + k_x) y_> - \gamma_m a] \cdot sinc\left[(\gamma_m + k_x)\frac{W_>}{2\sqrt{2}}\right]\Bigg\}$$
$$\cdot \Bigg\{\cos[(\gamma_m - k_x) y_<] \cdot sinc\left[(\gamma_m - k_x)\frac{W_<}{2\sqrt{2}}\right] + \cos[(\gamma_m + k_x) y_<] \cdot sinc\left[(\gamma_m + k_x)\frac{W_<}{2\sqrt{2}}\right]\Bigg\}$$
$$+ \sum_{m=L+1}^{\infty} \frac{-j\sigma_m}{\gamma_m} \cdot \exp\left[-j\gamma_m\left(y_> - \frac{W_>}{2\sqrt{2}} - y_< - \frac{W_<}{2\sqrt{2}}\right)\right]$$
$$\cdot \left\{\frac{\exp[jk_x(y_> - W_>/2\sqrt{2})]}{W_>(\gamma_m - k_x)} + \frac{\exp[-jk_x(y_> - W_>/2\sqrt{2})]}{W_>(\gamma_m + k_x)}\right\}$$
$$\cdot \left\{\frac{\exp[jk_x(y_< - W_>/2\sqrt{2})]}{W_<(\gamma_m - k_x)} + \frac{\exp[-jk_x(y_< - W_>/2\sqrt{2})]}{W_<(\gamma_m + k_x)}\right\}\Bigg)$$

where the term $W_>$ corresponds to the width of the port $y_>$, $W_<$ corresponds to the width of the port $y_<$, and

$$y_> = \max(x_p, x_q) \tag{2.101}$$

$$y_< = \min(x_p, x_q) \tag{2.102}$$

The rest of the parameters are defined in (2.89) through (2.93).

If the locations of the two ports are the same (i.e., $p = q$), then the expression for Z_{pq} becomes

$$Z_{pq} = -\frac{j\omega\mu d}{2a}(Z_1 + Z_2) \tag{2.103}$$

where

$$Z_{pq} = \sum_{m=0}^{L} 2D_m \Bigg\{\frac{1}{4} sinc^2\left[(\gamma_m - k_x)\frac{W}{2\sqrt{2}}\right] \cos[(\gamma_m - k_x)2x - \gamma_m a] \tag{2.104}$$
$$+ \frac{1}{2} sinc\left[(\gamma_m - k_x)\frac{W}{2\sqrt{2}}\right] \cdot sinc\left[(\gamma_m - k_x)\frac{W}{2\sqrt{2}}\right] \cdot \cos[\gamma_m(2x - a)]$$
$$+ \frac{1}{4} sinc^2\left[(\gamma_m + k_x)\frac{W}{2\sqrt{2}}\right] \cdot \cos[(\gamma_m + k_x)2x - \gamma_m a]$$
$$+ \frac{\sqrt{2}}{2W(\gamma_m - k_x)} \sin\left[(\gamma_m - k_x)\frac{W}{2\sqrt{2}} - \gamma_m a\right] \cdot sinc\left[(\gamma_m - k_x)\frac{W}{2\sqrt{2}}\right]$$
$$+ \frac{\sqrt{2}}{2W(\gamma_m - k_x)} \sin\left[(\gamma_m - k_x)\frac{W}{2\sqrt{2}} - \gamma_m a\right] \cdot sinc\left[(\gamma_m + k_x)\frac{W}{2\sqrt{2}}\right] \cdot \cos(2k_x x)$$
$$+ \frac{\sqrt{2}}{2W(\gamma_m - k_x)} \sin\left[(\gamma_m + k_x)\frac{W}{2\sqrt{2}} - \gamma_m a\right] \cdot sinc\left[(\gamma_m - k_x)\frac{W}{2\sqrt{2}}\right] \cdot \cos(2k_x x)$$
$$+ \frac{\sqrt{2}}{2W(\gamma_m - k_x)} \sin\left[(\gamma_m + k_x)\frac{W}{2\sqrt{2}} - \gamma_m a\right] \cdot sinc\left[(\gamma_m + k_x)\frac{W}{2\sqrt{2}}\right]$$
$$+ \frac{2\gamma_m}{W(\gamma_m^2 - k_x^2)} \sin(\gamma_m a) \cdot \left[1 + \cos(2k_x)\, sinc\left(\frac{k_x W}{\sqrt{2}}\right)\right]\Bigg\}$$

and, if $(2x - a) \geq 0$, then

$$Z_2 = \sum_{m=L+1}^{\infty} \frac{-j\sigma_m}{\gamma_m W^2} \tag{2.105}$$
$$\cdot \exp\left[-j\gamma_m\left(2a - 2x - \frac{W}{\sqrt{2}}\right)\right] \cdot \left[\frac{\exp[-jk_x(2x + W/\sqrt{2})]}{(\gamma_m - k_x)^2} + \frac{\exp[-jk_x(2x + W/\sqrt{2})]}{(\gamma_m + k_x)^2}\right]$$
$$+ \frac{-2j\sigma_m}{\gamma_m W^2(\gamma_m^2 - k_x^2)} \cdot exp\left[-j\gamma_m\left(2a - 2x - \frac{W}{\sqrt{2}}\right)\right]$$

$$+\frac{2j\sigma_m}{\gamma_m W^2}\left[\frac{1}{(\gamma_m - k_x)^2} - \frac{1}{(\gamma_m + k_x)^2}\right]$$
$$+\frac{j\sigma_m}{\gamma_m W^2}\left\{\frac{\exp[jk_x(2x + W/\sqrt{2})]}{(\gamma_m^2 - k_x^2)} + \frac{\exp[jk_x(2x - W/\sqrt{2})]}{(\gamma_m^2 - k_x^2)}\right\}$$
$$+\frac{2\sqrt{2}\sigma_m}{W(\gamma_m^2 - k_x^2)} \cdot \left[1 + \cos(2k_x x)\, sinc\left(\frac{k_x W}{\sqrt{2}}\right)\right]$$

whereas if $(2x - a) < 0$,

$$Z_2 = \sum_{m=L+1}^{\infty} \frac{-j\sigma_m}{\gamma_m W^2} \exp\left[-j\gamma_m\left(2x - \frac{W}{\sqrt{2}}\right)\right] \cdot \left\{\frac{\exp[jk_x(2x - W/\sqrt{2})]}{(\gamma_m - k_x)^2} + \frac{\exp[-jk_x(2x - W/\sqrt{2})]}{(\gamma_m + k_x)^2}\right\}$$
$$+\frac{-2j\sigma_m}{\gamma_m W^2(\gamma_m^2 - k_x^2)} \exp\left[-j\gamma_m\left(2x - \frac{W}{\sqrt{2}}\right)\right]$$
$$+\frac{2j\sigma_m}{\gamma_m W^2}\left[\frac{1}{(\gamma_m - k_x)^2} - \frac{1}{(\gamma_m + k_x)^2}\right]$$
$$+\frac{j\sigma_m}{\gamma_m W^2}\left\{\frac{\exp[jk_x(2x - W/\sqrt{2})]}{(\gamma_m^2 + k_x^2)} + \frac{\exp[jk_x(2x - W/\sqrt{2})]}{(\gamma_m^2 - k_x^2)}\right\}$$
$$+\frac{2\sqrt{2}\sigma_m}{W(\gamma_m^2 - k_x^2)}\left[1 + \cos(2k_x x)\, sinc\left(\frac{k_x W}{\sqrt{2}}\right)\right] \tag{2.106}$$

where σ_m, D_m, k_x, and γ_m have been defined earlier. The port's location is $x = x_p = x_q$; the port's width is $W = W_p = W_q$.

d. When One Port Is Located Along OA *and the Other Along* AB

The impedance matrix elements in this case are given by

$$Z_{pq} = -\frac{j\omega\mu d}{2a}\left(\sum_{m=0}^{L} 2D_m \cos(k_x u_p)\, sinc\left(\frac{k_x W_p}{2}\right) \cdot \left\{\cos[(\gamma_m - k_x)u_q]\, sinc\left[(\gamma_m - k_x)\frac{W_q}{2\sqrt{2}}\right] + \cos[(\gamma_m + k_x)\, u_q]\, sinc\left[(\gamma_m + k_x)\frac{W_q}{2\sqrt{2}}\right]\right\}\right.$$
$$+\sum_{m=L+1}^{\infty} \frac{2\sqrt{2}\sigma_m}{\gamma_m W_q} \exp\left[-j\gamma_m\left(a - u_q - \frac{W_q}{2\sqrt{2}}\right)\right] \cdot \cos(k_x u_p)$$
$$\left. \cdot\, sinc\left(\frac{k_x W_p}{2}\right)\left\{\frac{\exp[jk_x(u_q - W_q/2\sqrt{2})\,]}{(\gamma_m - k_x)} + \frac{\exp[-jk_x(u_q - W_q/2\sqrt{2})]}{(\gamma_m + k_x)}\right\}\right) \tag{2.107}$$

where

$$(u_p, u_q) = \begin{cases}(x_p, x_q) & \text{if } p \text{ along } OA,\ q \text{ along } AB \\ (x_q, x_p) & \text{if } p \text{ along } AB,\ q \text{ along } OA\end{cases} \tag{2.108}$$

In the preceding expression W_p is the width of the port located along OA, and W_q is the width of the port located along AB; the rest of the parameters were defined earlier.

e. When One Port Is Located Along OB *and the Other Along* AB

The impedance matrix elements for this configuration are given by

$$Z_{pq} = -\frac{j\omega\mu d}{2a}\left(\sum_{m=0}^{L} 2D_m(-1)^m \cos(k_x u_p) \cdot sinc\left(\frac{k_x W_p}{2}\right) \cdot \left\{\cos[(\gamma_m - k_x)\, u_q - \gamma_m a] \cdot sinc[(\gamma_m - k_x)\, W_q/2\sqrt{2}] + \cos[(\gamma_m + k_x)\, u_q - \gamma_m a] \cdot sinc\left[(\gamma_m + k_x)\frac{W_q}{\sqrt{2}}\right]\right\}\right. \tag{2.109}$$
$$+\sum_{m=L+1}^{\infty} \frac{2\sqrt{2}\,\sigma_m(-1)^m}{\gamma_m W_q} \cdot \exp\left[-j\gamma_m\left(u_q - \frac{W_q}{2\sqrt{2}}\right)\right] \cos(k_x u_p)$$
$$\left. \cdot\, sinc\left(\frac{k_x W_p}{2}\right)\left\{\frac{\exp[jk_x(u_q - W_q/2\sqrt{2})\,]}{(\gamma_m - k_x)} + \frac{\exp[-jk_x(u_q - W_q/2\sqrt{2})\,]}{(\gamma_m + k_x)}\right\}\right)$$

where

$$(u_p, u_q) = \begin{cases}(y_p, x_q) & \text{if } p \text{ along } OB,\ q \text{ along } AB \\ (y_q, y_p) & \text{if } p \text{ along } BA,\ q \text{ along } OB\end{cases} \tag{2.110}$$

W_p is the width of the port located along OB, and W_q is the width of the port located along AB. The rest of the parameters have been defined earlier.

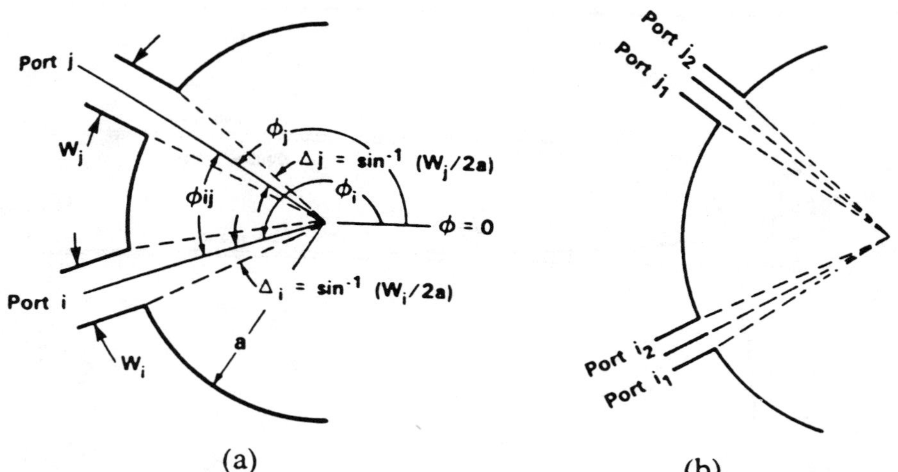

Fig. 2.9 (a) Parameters for ports at the circumference of a circular disc; (b) division of a wide port into subports.

2.10.5 Circular Disc Planar Segments

Figure 2.9 depicts the geometry of a multiport circular segment. As in the case of triangular and rectangular segments, the double-series expression of the impedance matrix elements Z_{ij} has been reduced to a single-series expression [2.2]. The new single-series expression is given by

$$Z_{ij} = \frac{j\omega\mu d}{16} \sum_{n=0}^{\infty} \cdot \frac{\sigma_n J_n(kd_i) f_n(kd_j, ka) \cos(n\phi_{ij}) \sin(n\Delta_i) \sin(n\Delta_j)}{J_n'(ka)[(nW_i/2d_i)(nW_j/2d_j)]} \tag{2.111}$$

where

$$\Delta_i = \arcsin\left(\frac{W_i}{2d_i}\right) \tag{2.112}$$

$$f_n(x, y) = Y_n'(y)J_n(x) - J_n'(y)Y_n(x) \tag{2.113}$$

$$\phi_{ij} = \phi_i - \phi_j \tag{2.114}$$

$$\sigma_n = \begin{cases} 1, & n = 0 \\ 2, & n \neq 0 \end{cases} \tag{2.115}$$

$$0 \le d_i \le d_j \le a, \qquad d_j \neq 0 \tag{2.116}$$

$$k^2 = \omega^2 \mu \varepsilon \tag{2.117}$$

The impedance matrix element expression given by (2.111) is valid for any port location along the periphery or elsewhere. When the ports are located only around the periphery, Eq. (2.111) can be simplified further to become

$$Z_{ij} = \frac{j\omega\mu d}{8\pi ka} \sum_{n=0}^{\infty} \cdot \frac{\sigma_n J_n(ka) \cos(n\phi_{ij}) \sin(n\Delta_i) \sin(n\Delta_j)}{J_n'(ka)[(nW_i/2a)(nW_j/2a)]} \tag{2.118}$$

where d stands for the substrate thickness, a is the effective radius of the disc [2.13], and Δ_i is the half-angle corresponding to port i [2.13] (see Fig. 2.8).

Figure 2.10 shows a comparison between results obtained by (2.112) and the double-series expression. These results are for Z_{11} of a four-port circular disc of radius 15 mm; the port width was 4.4 mm. The frequency of operation is 4.5 GHz, the substrate height is 1.524 mm, and the substrate permittivity is $\varepsilon_r = 2.5$. Figure 2.11 shows the time ratios (time taken by double series/time taken by single series) plotted against the number of modes used for the circular disc (and the annular ring, which follows).

2.10.6 Annular Ring Planar Segments

The elements of the impedance matrix for a multiport annular ring of an inner radius a and an outer radius b can also be obtained in a single-series form. This expression is given by [2.2],

$$Z_{ij} = \frac{j\omega\mu d}{16} \sum_{n=0}^{\infty} \frac{\sigma_n f_n(d_i k, ak) f_n(d_j k, bk) \cos(n\phi_{ij})}{f_n'(bk, ak)\,[(nW_i/2d_i)(nW_j/2d_j)]} \cdot \sin(n\Delta_i)\sin(n\Delta_j) \tag{2.119}$$

where

$$a \le d_i \le d_j \le b \tag{2.120}$$

$$f_n'(bk, ak) = Y_n'(ak)J_n'(bk) - J_n'(ak)Y_n'(bk) \tag{2.121}$$

$$\Delta_i = \arcsin\left(\frac{W_i}{2b}\right) \tag{2.122}$$

with $f_n(x, y)$, ϕ_{ij}, σ_n, and k^2 being defined earlier by (2.113), (2.114), (2.115), and (2.117), respectively.

For the case where the ports are located around the outside periphery of the ring, Eq. (2.119) reduces to

$$Z_{ij} = \frac{j\omega\mu d}{8\pi kb} \sum_{n=0}^{\infty} \cdot \frac{\sigma_n f_n(bk, ak) \cos(n\phi_{ij}) \sin(nW_i) \sin(nW_j)}{f_n'(bk, ak)[(nW_i/2b)(nW_j/2b)]} \tag{2.123}$$

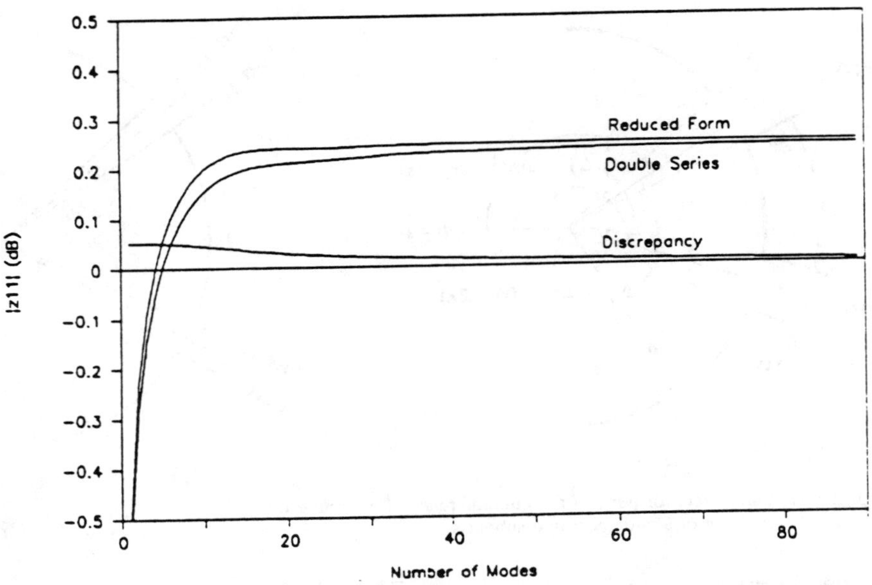

Fig. 2.10 Z_{11} for a circular disc computed by single- and double-series expressions.

Figure 2.12 shows the values of Z_{11} for an annular ring for both the single-series and the double-series expressions; it also shows the discrepancy between them. This calculation was carried out for a four-port annular ring of outer and inner radii 15 mm and 5 mm, respectively. The substrate was assumed to be 1.524 mm thick and have a relative permittivity of $\varepsilon_r = 2.5$. Furthermore, Fig. 2.11 shows the time ratios (time taken by double series/time taken by single series) plotted against the number of modes used for the annular ring and the circular disc.

2.10.7 Circular Sectorial Planar Segments

The impedance matrix elements for a circular sectorial segment (with sector angle $\alpha = \pi/l$, l being an integer) have been reported earlier [2.1, 2.14] as a double-series expression. Unlike the previous cases, this expression has not been reduced to a single form. The impedance matrix elements for this case are expressed in four different forms depending on the location of the ports (see Fig. 2.12).

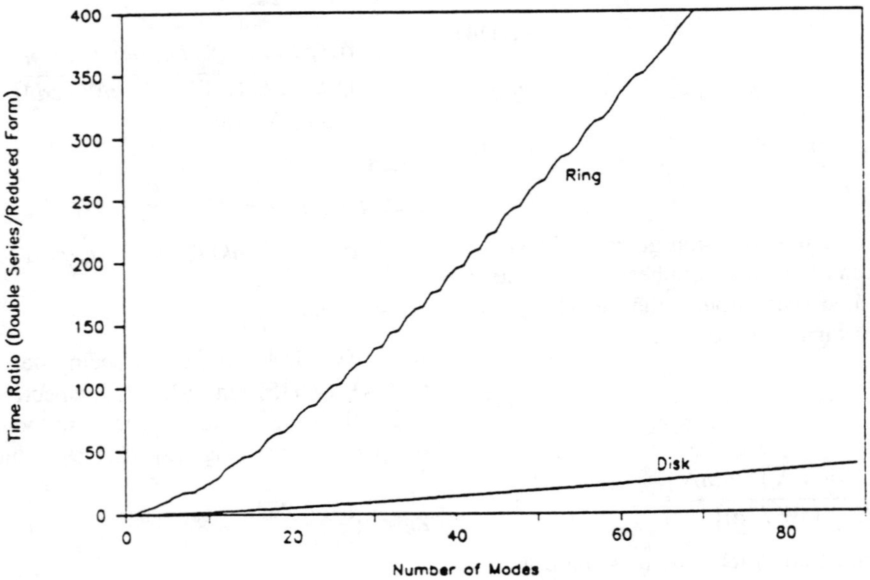

Fig. 2.11 Comparison of relative computing time for disk and ring.

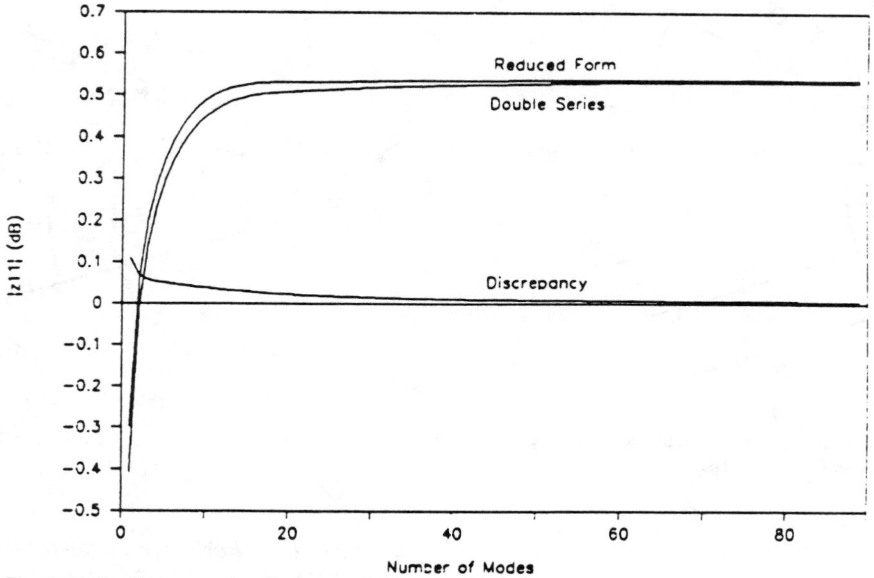

Fig. 2.12 Z_{11} for an annular ring computed by single- and double-series expressions.

a. When Both Ports Are Located Along the Curved Edge

In this case the impedance matrix elements are given by

$$Z_{ij} = \frac{2j\omega\mu d}{\alpha} \sum_{n=0}^{\infty} \sum_{m=1}^{\infty} \frac{\sigma_n \cos(n_s\phi_i)\cos(n_s\phi_j)\sin(n_s W_i/2a)\sin(n_s W_j/2a)}{[(n_s W_i/2a)(n_s W_j/2a)](a^2 - n_s^2/k_{mn_s}^2)(k_{mn_s}^2 - k^2)} \tag{2.124}$$

where the eigenvalues k_{mn_s} are solutions of

$$J_n'(k_{mn_s}\rho)\big|_{\rho=a} = 0 \tag{2.125}$$

The derivative in (2.125) is relative to ρ. In addition, $n_s = n\pi/\alpha$, $l = \pi/\alpha$, with W_i being the curvilinear width of the port measured along the circumference.

b. When Both Ports Are Located Along the Same Radial Edge

When both ports are located along the same radial edge at $\phi = \alpha$, the self- and transfer impedance elements are determined from

$$Z_{rt} = \frac{2j\omega\mu d}{\pi\alpha W_r W_t} \sum_{n=0}^{\infty} \sum_{m=1}^{\infty} \frac{\sigma_n I(r) I(t)}{k_{mn_s}^2 (a^2 - n_s^2/k_{mn_s}^2)(k_{mn_s}^2 - k^2) J_{n_s}^2(k_{mn_s}a)} \tag{2.126}$$

where

$$I(r) = \begin{cases} \displaystyle\int_{t_1}^{t_2} J_0(t)dt - 2 \sum_{k=0}^{(n_s-2)/2} [J_{2k+1}(t_2) - J_{2k+1}(t_1)], & \text{for even } n_s \\ \displaystyle J_0(t_1) - J_0(t_2) + 2 \sum_{k=1}^{(N_s-1)/2} [J_{2k}(t_1) - J_{2k}(t_2)], & \text{for odd } n_s \end{cases} \tag{2.127}$$

with

$$t_1 = k_{mn_s}\left(x_r - \frac{\Delta_r}{2}\right) \tag{2.128}$$

and

$$t_2 = k_{mn_s}\left(x_r + \frac{\Delta_r}{2}\right) \tag{2.129}$$

The term x_r in (2.128) and (2.129) represents the radial distance between port r and the origin, as illustrated in Fig. 2.13. In addition, Δ_r represents the linear width of port r measured along the radial edge of the sector. Finally, the integral appearing in (2.127) cannot be evaluated in closed form and thus will have to be computed numerically. The term $I(t)$ in (2.126) is obtained by replacing the r with t in (2.128) and (2.129).

If both ports were located along the radial edge $\phi = 0$, the resulting self- and transfer impedance elements of the Z-matrix are determined from (2.126).

c. When Ports r *and* p *Are Located Along Two Different Radial Edges*

In this case we may have port p located at the $\phi = 0$ edge and port r at the $\phi = \alpha$ edge. Such geometry yields the following expression,

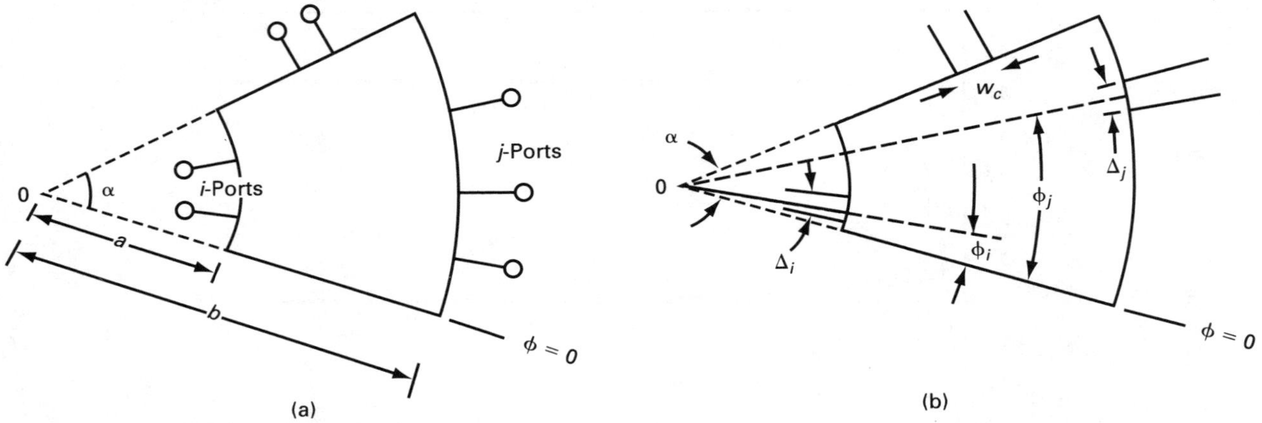

Fig. 2.13 Geometry and nomenclature for circular sectorial segments with radial and circumferential ports.

$$Z_{rp} = \frac{2j\omega\mu l d}{\pi W_r W_p} \sum_{n=0}^{\infty} \sum_{m=1}^{\infty} \frac{\sigma_n(-1)^n I(r) I(p)}{k^2_{mn_s}(a^2 - n_s^2/k^2_{mn_s})\,(k^2_{mn_s} - k^2) J^2_{n_s}(k_{mn_s}a)} \tag{2.130}$$

where all the parameters have been defined earlier.

d. When Port i *Is Located Along a Radial Edge and Port* r *Is Located Along the Curved Edge*

When port i is located along the radial edge $\phi = \alpha$ and port r is located along the curved edge at $\rho = a$, then the transfer impedance element Z_{ir} is given by

$$Z_{ir} = \frac{4j\omega\mu l d a}{\pi W_i W_r} \sum_{n=0}^{\infty} \sum_{m=1}^{\infty} \frac{\sigma_n(-1)^n \cos(n_s\phi_i)\,\sin(n_s\Delta_i/2)\,I(r)}{n_s k_{mn_s}\,(a^2 - n_s^2/k^2_{mn_s})\,(k^2_{mn_s} - k^2) J_{n_s}(k_{mn_s}a)} \tag{2.131}$$

If the i port were located along the curved edge and the second port p were located along the straight radial edge $\phi = 0$, then the impedance matrix elements of the Z-matrix are determined from (2.131) by setting $\alpha = 0$.

2.10.8 Annular Ring Sectorial Planar Segments

The geometry of an annular sector planar segment is shown in Fig. 2.14. There are three different expressions in this case, depending on the ports locations [2.1, 2.4]. These expressions correspond to the cases where (1) both ports are located along the radial edges, (2) both ports are located along the curved edges, and (3) one port is located along the radial edge and the other is located along the curved edge.

a. When Both Ports Are Located Along the Radial Edges

In this case there are three possible geometries: (1) both ports are located along the $\phi = 0$ edge, (2) both ports are located along the $\phi = \alpha$ edge, and (3) one port is located along the $\phi = 0$ edge while the other port is located along the

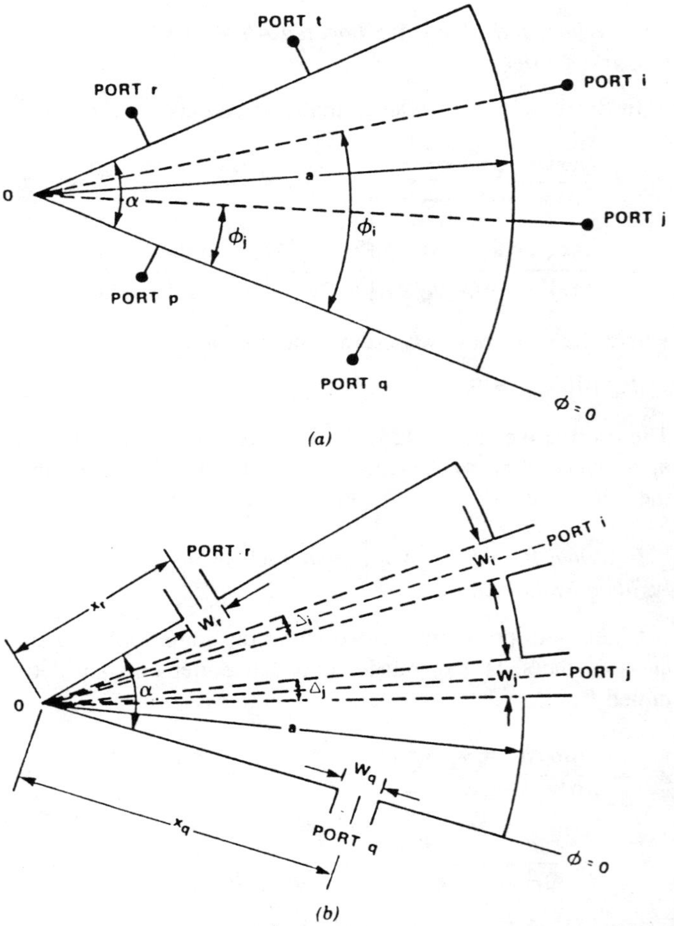

Fig. 2.14 Geometry and nomenclature for annular sectorial segments with radial and circumferential ports.

$\phi = \alpha$ edge. For all three cases, the impedance matrix elements are determined by

$$Z_{ij} = \frac{-2j\omega\mu d}{\alpha k^2(b^2 - a^2)} + \frac{2j\omega\mu d}{\alpha W_i W_j} \sum_{n=0}^{\infty} \sum_{m=1}^{\infty} \cdot \frac{\sigma_n \cos(n_s\phi_i)\cos(n_s\phi_j)}{(k_{mn_l}^2 - k^2)k_{mn_l}^2} \cdot \frac{W(i)W(j)}{[(b^2 - n_l^2/k_{mn_l}^2)F_{mn_l}^2(b)] - [(a^2 - n_l^2/k_{mn_l}^2)F_{mn_l}^2(a)]} \tag{2.132}$$

where

$$W(*) = N_{n_l}'(k_{mn_l}a)I^J(*) - J_{n_l}'(k_{mn_l}a)I^N(*) \tag{2.133}$$

$$F_{mn_l}(.) = N_{n_l}'(k_{mn_l}a)J_{n_l}(k_{mn_l}.) - J_{n_l}'(k_{mn_l}a)N_{n_l}(k_{mn_l}.) \tag{2.134}$$

d is the substrate height, a is the inner radius of the annular sector, b is its outer radius, and $n_l = nl$, where $l = \pi/\alpha$, $k^2 = \omega^2\mu\varepsilon\varepsilon_0$, and σ_n is given by (2.115). Furthermore, $I^J(*)$ is defined from (2.127) through (2.129), whereas $I^N(*)$ is defined as

$$I^N(*) = \begin{cases} \int_{t_1}^{t_2} N_0(t)dt + 2 \sum_{k=0}^{(nl-2)/2} [N_{2K+1},(t_1) - N_{2k+1}(t_2)], \text{ for even } n_l \\ N_0(t_1) - N_0(t_2) + 2 \sum_{k=1}^{(nl-2)/2} [N_{2k}(t_1) - N_{2k}(t_2)], \text{ for odd } n_l \end{cases} \tag{2.135}$$

where

$$t_1 = k_{mn_l}\left(\rho - \frac{W}{2}\right) \tag{2.136}$$

$$t_2 = k_{mn_l}\left(\rho + \frac{W}{2}\right) \tag{2.137}$$

with ρ being the radial location of the port, W its linear width, and k_{mn_l} the solutions of

$$J_{n_l}'(k_{mn_l}a)N_{n_l}'(k_{mn_l}b) - J_{n_l}'(k_{mn_l}b)N_{n_l}'(k_{mn_l}a) = 0 \tag{2.138}$$

The terms $J_{n_l}(.)$ and $N_{n_l}(.)$ in (2.138) represent, respectively, Bessel's function of the first and second kind (Neumann function) of integer order n_l. The factor $J_{n_l}'(k_{mn_l}.)$ is the derivative of the Bessel's function with respect to its argument (.). Similarly, the factor $N_{n_l}'(k_{mn_l}.)$ in (2.138) denotes the derivative of the Neumann function with respect to the argument (.). The subscript m in k_{mn_l} denotes the mth root of (2.138). For the case when $n_l = 0$, the index m can be zero, which makes the value of k_{mn_l} equal to zero. This, in turn, implies that the function F_{mn_l} is unity throughout the sectorial segment; hence the first term in (2.132) is obtained.

b. When Both Ports Are Located Along the Curved Edges

There are also three possible configurations in this case: (1) both ports are located along the $\rho = a$ inner curved edge, (2) both ports are located along the $\rho = b$ outer curved edge, and (3) one port is located along the $\rho = a$ inner edge while the other is located along the $\rho = b$ outer edge. In all these cases, the elements of the impedance matrix are given by

$$Z_{ij} = -\frac{2j\omega\mu d}{\alpha k^2(b^2 - a^2)} + \frac{2j\omega\mu d(2a)^2}{\alpha W_i W_j} \cdot \sum_{n=0}^{\infty} \sum_{m=1}^{\infty} \frac{\sigma_n F_{mn_l}(\rho_i) F_{mn_l}(\rho_j)}{n_l^2(k_{mn_l}^2 - k^2)} \cdot \frac{\cos(n_l\phi_i)\sin(n_l\Delta_i)\cos(n_l\phi_j)\sin(n_l\Delta_j)}{[(b^2 - n_l^2/k_{mn_l}^2)F_{mn_l}^2(b)] - [(a^2 - n_l^2/k_{mn_l}^2)F_{mn_l}^2(a)]} \tag{2.139}$$

where the parameters d, a, b, α, σ_n, n_l, k^2, ϕ_i, ϕ_j, Δ_i, Δ_j, $F_{mn_l}(.)$, W_i, W_j, and k_{mn_l} have been defined in the previous sections.

Recently, Alhargan and Judah [2.22] derived new single-series expressions for the impedance matrix elements of an annular sector of *arbitrary angle*. The following expressions are valid for an annular sector having one port located at the inner periphery $\rho = a$ and the other port somewhere along the outer periphery $\rho = b$:

$$Z_{11} = \frac{j\omega\mu dl}{2} \cdot \sum_{p=0}^{\infty} \frac{\sigma_p f_{\nu_p}(ak, ak) f_{\nu_p}(bk, ak) \cos^2(\nu_p\alpha/2)\sin^2(\nu_p\Delta_1)}{f_{\dot{\nu}_p}(bk, ak)} \cdot \left(\frac{2a}{W_1\nu_p}\right)^2 \tag{2.140}$$

$$Z_{1i}|_{i\neq 1} = \frac{j\omega\mu dl}{2} \sum_{p=0}^{\infty} \frac{\sigma_p f_{\nu_p}(ak, ak) f_{\nu_p}(bk, bk)}{f_{\dot{\nu}_p}(bk, ak)} \cdot \cos(\nu_p\phi_i) \cdot \sin(\nu_p\Delta_i)\cos\left(\nu_p\frac{\alpha}{2}\right)\sin(\nu_p\Delta_1)\left(\frac{2a}{W_1\nu_p}\right)\left(\frac{2b}{W_i\nu_p}\right) \tag{2.141}$$

$$Z_{ij}|_{i\neq 1, j\neq 1} = \frac{i\omega\mu dl}{2} \sum_{p=0}^{\infty} \frac{\sigma_p f_{\nu_p}(ak, bk) f_{\nu_p}(bk, bk)}{f_{\dot{\nu}_p}(bk, ak)} \cdot \cos(\nu_p\phi_i)\sin(\nu_p\Delta_i)\cos(\nu_p\phi_j) \cdot \sin(\nu_p\Delta_j)\left(\frac{2b}{W_i\nu_p}\right)\left(\frac{2b}{W_j\nu_p}\right) \tag{2.142}$$

where $\Delta_1 = \sin^{-1}(W_1/2a)$, $\Delta_i = \sin^{-1}(W_i/2b)$, and $a = W_1/2 \sin(\alpha/2)$. In addition, $\nu_p = pl$, $l = p/\alpha$, $p = 0, 1, 2, \ldots$, and $\sigma_p = 1$ for $p = 0$ and 2 otherwise.

c. When One Port Is Located Along the Radial Edge and the Other Along the Curved Edge

In this case there are four possible combinations, depending on the locations of the ports. The first two combinations are when one port is located along the $\phi = 0$ edge and the other is along the curved edge ($\rho = a$ or $\rho = b$). The last two combinations are when one port is located along the $\phi = \alpha$ edge and the other port is located along one of the curved edges

($\rho = a$ or $\rho = b$). In all four cases the impedance matrix elements are given by

$$Z_{ij} = -\frac{2j\omega\mu d}{\alpha k^2(b^2 - a^2)} + \frac{4j\omega\mu da}{\alpha W_i' W_C'} \cdot \sum_{n=0}^{\infty} \sum_{m=1}^{\infty} \frac{\sigma_n W(*) F'}{n_l(k_{mn_l}^2 - k^2)k_{mn_l}} \cdot \frac{\cos(n_l\phi_i)\cos(n_l\phi_j)\sin(n_l W_C'/2a)}{[(b^2 - n_l^2/k_{mn_l}^2)F_{mn_l}^2(b)] - [(a^2 - n_l^2/k_{mn_l}^2)F_{mn_l}^2(a)]} \tag{2.143}$$

If port i is located along the radial edge ($\phi = 0$ or $\phi = \alpha$) and port j is along the curved edge ($\rho = a$ or $\rho = b$), then $W_r' = W_i$; $W_c' = W_j$; $W(*) = W(i)$, where $W(i)$ is defined by (2.133); and $F' = F_{mn_l}(j)$, where $F_{mn_l}(j)$ is defined by (2.134). On the other hand, if port i is located along the curved edge and port j is located along the radial edge, then $W_r' = W_j$; $W_c' = W_i$; $W(*) = W(j)$, where $W(j)$ is defined by (2.133); and $F' = F_{mn_l}(i)$, where $F_{mn_l}(i)$ is defined by (2.134). In both cases W_r' denotes the linear width of the associated port, ϕ denotes the angular location of the associated port, and k_{mn_l} is the solution of (2.138).

2.10.9 Elliptical Disc Segment [2.3]

Figure 2.15 illustrates the configuration of a multiport elliptical segment. The elements of the impedance matrix for this segment are given by

$$Z_{ij} = Z_{ij}^e + Z_{ij}^o \tag{2.144}$$

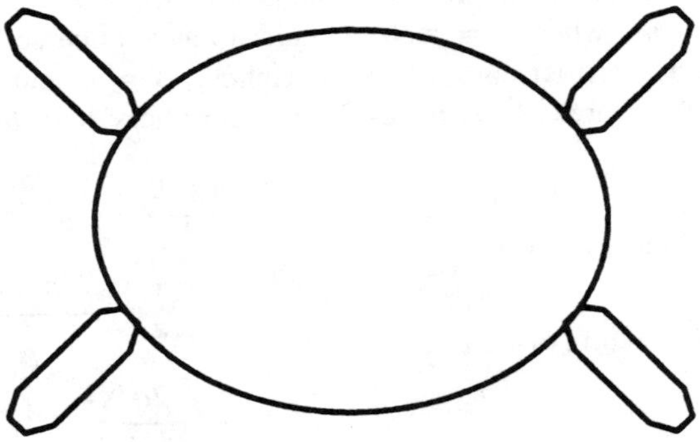

(a)

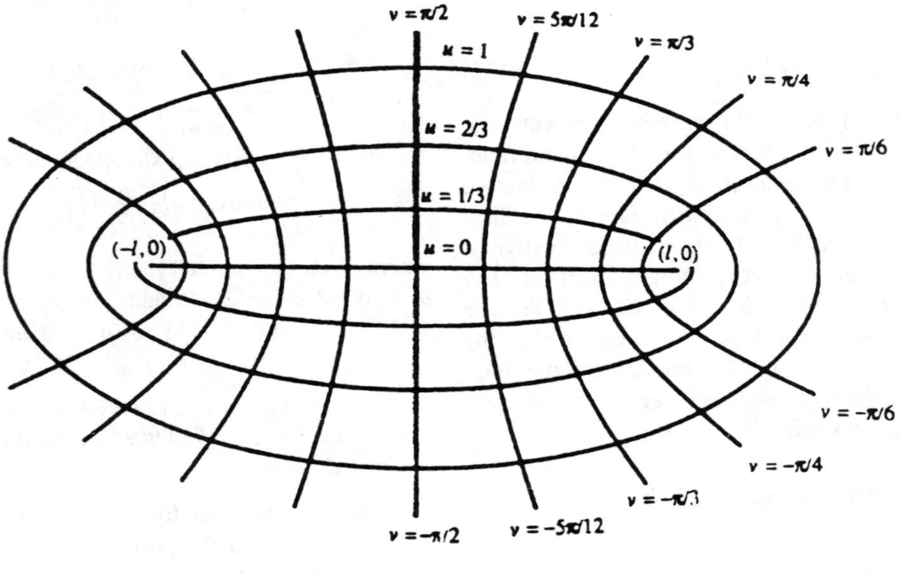

(b)

Fig. 2.15 Multiport elliptical disc and its coordinates. (a) Four port elliptic patch; (b) elliptic coordinate system, $x = l\cosh(v)\cos(v)$, $y = l\sinh(u)\sin(v)$.

with

$$Z_{ij}^{e} = \frac{j\omega\mu\, dl^2}{W_i W_j} \sum_{n=0}^{\infty} \frac{Je_n(h, \cosh u_i) fe_n(h, u_1, u_j)}{M_n^e(h) Je_n'(h, \cosh u_1)} \cdot ISe_n(h, v_i, \Delta_i, u_i) ISe_n(h, v_j, \Delta_j, u_j) \quad (2.145)$$

$$Z_{ij}^{o} = \frac{j\omega\mu\, dl^2}{W_i W_j} \sum_{n=1}^{\infty} \frac{Jo_n(h, \cosh u_i) fo_n(h, u_1, u_j)}{M_n^o(h) Jo_n'(h, \cosh u_1)} \cdot ISO_n(h, v_i, \Delta_i, u_i) ISO_n(h, v_j, \Delta_j, u_j) \quad (2.146)$$

where $h = lk = l\omega/c = 2\pi l/\lambda$, $2l$ is the focal distance of the elliptic segment, k is the wave number, and u and v are elliptic coordinates related to the cartesian coordinates via

$$x = l \cosh(u) \cos(v) \quad (2.147)$$

$$y = l \sinh(u) \sin(v) \quad (2.148)$$

Furthermore,

$$0 \le u \le u_0 \le u_1 \quad (2.149)$$

$$fe_n(h, u_1, u_0) = Ye_n'(h, \cosh u_1) Je_n(h, \cosh u_0) - Je_n'(h, \cosh u_1) Ye_n(h, \cosh u_0) \quad (2.150)$$

$$fo_n(h, u_1, u_0) = Yo_n'(h, \cosh u_1) Jo_n(h, \cosh u_0) - Jo_n'(h, \cosh u_1) Yo_n(h, \cosh u_0) \quad (2.151)$$

$$ISe_n(h, v_i, \Delta_i, u_i) = \int_{v_i-\Delta_i}^{v_i+\Delta_i} Se_n(h, \cos v) \cdot [\sinh^2 u_i + \sin^2 v]^{1/2} dv \quad (2.152)$$

$$ISO_n(h, v_i, \Delta_i, u_i) = \int_{v_i-\Delta_i}^{v_i+\Delta_i} So_n(h, \cos v) \cdot [\sinh^2 u_i + \sin^2 v]^{1/2} dv \quad (2.153)$$

$$\Delta_i \simeq \arcsin\left(\frac{W_i}{2l\sqrt{\sinh^2 u_i + \sin^2 v_i}}\right) \quad (2.154)$$

where W_i represents the port width, $Je_n(h, \cosh u)$ is the even radial Mathieu function of the first kind, $Ye_n(h, \cosh u)$ is the even radial Mathieu function of the second kind, $Se_n(h, \cos v)$ is the first kind even circumferential Mathieu function (also referred to as associated Mathieu functions), $So_n(h, \cos v)$ is the first kind odd circumferential Mathieu function, and $M_n^e(h)$ is a normalization constant.

During the course of the computation [2.3], the following expressions for Mathieu functions [2.23] were found to be highly convergent, more so than those given in [2.24]:

First kind even circumferential Mathieu function:

$$Se_{2n+p}(h, \cos v) = \sum_{m=0}^{\infty} A_{2m+p} \cos[(2m + p) v] \quad (2.155)$$

First kind odd circumferential Mathieu function:

$$So_{2n+p}(h, \cos v) = \sum_{m=0}^{\infty} B_{2m+p} \sin[(2m + p) v] \quad (2.156)$$

First kind even radial Mathieu function:

$$Je_{2n+p}(h, \cosh u) = \sqrt{\frac{\pi}{2}} \sum_{m=0}^{\infty} \cdot (-1)^{n-m} A_{2m+p} J_{2m+p}(h, \cosh u) \quad (2.157)$$

First kind odd radial Mathieu function:

$$Jo_{2n+p}(h, \cosh, u) = \sqrt{\frac{\pi}{2}} \tanh u \sum_{m=0}^{\infty} \cdot (-1)^{n-m} (2m + 1) B_{2m+p} J_{2m+p}(h, \cosh u) \quad (2.158)$$

Second kind even radial Mathieu function:

$$Ye_{2n}(h, \cosh u) = \frac{\sqrt{\pi/2}}{A_0} \sum_{m=0}^{\infty} \cdot (-1)^{n-m} A_{2m} Y_m\left(\frac{1}{2} he^{u}\right) J_m\left(\frac{1}{2} he^{-u}\right) \quad (2.159)$$

$$Ye_{2n+1}(h, \cosh u) = \frac{\sqrt{\pi/2}}{A_1} \sum_{m=0}^{\infty} \cdot (-1)^{n-m} A_{2m}\left[Y_{m+1}\left(\frac{1}{2} he^{u}\right) J_m\left(\frac{1}{2} he^{-u}\right) + Y_m\left(\frac{1}{2} he^{u}\right) J_{m+1}\left(\frac{1}{2} he^{-u}\right)\right] \quad (2.160)$$

Second kind odd radial Mathieu function:

$$Yo_{2n+p}(h, \cosh u) = \frac{\sqrt{\pi/2}}{B_{2-p}} \sum_{m=0}^{\infty} (-1)^{n-m} B_{2m-p} \cdot \left[Y_{m+1}\left(\frac{1}{2}he^{u}\right) J_{m-1+p}\left(\frac{1}{2}he^{-u}\right) - Y_{m-1+p}\left(\frac{1}{2}he^{u}\right) J_{m+1}\left(\frac{1}{2}he^{-u}\right)\right] \quad (2.161)$$

where $p \in \{0, 1\}$ and J_m, Y_m are the conventional Bessel functions.

The normalization constants in (2.145) and (2.146) are given by

$$M_{2n}^e(h) = \int_0^{2\pi} [Se_{2n}(h, \cos v)]^2 dv = 2\pi \sum_{m=0}^{\infty} \frac{A_{2m}^2}{\sigma_m} \quad (2.162)$$

where

$$\sigma_m = \begin{cases} 1, & m = 0 \\ 2, & m \neq 0 \end{cases} \quad (2.163)$$

$$M_{2n+1}^e(h) = \int_0^{2\pi} [Se_{2n+1}(h, \cos v)]^2 dv = \pi \sum_{m=0}^{\infty} A_{2m+1}^2 \quad (2.164)$$

$$M^0_{2n+p} = \int_0^{2\pi} [So_{2n+p}(h, \cos v)]^2 dv \tag{2.165}$$

$$= \pi \sum_{m=1}^{\infty} B^2_{2m+p}$$

Appendix 2.A
Green's Functions for Planar Circuits with Regular Shapes

Reference [2.1] gives an extensive list of Green's functions for regularly shaped planar segments with open, shorted, and mixed boundaries. Each function is expressed as a double infinite series. To this date most of these expressions remain valid, except for those that have been reduced to a single series. The geometries whose Green's functions have been reduced to a single series happened to be planar segments of the open-boundary type. In this appendix we shall list only those Green's functions that correspond to the open-boundaries case. Those readers who are interested in obtaining Green's functions for planar segments with shorted and mixed boundaries should consult [2.1].

By substituting the new Green's functions into (2.34) and performing the double integration, the corresponding impedance matrix elements given in Sec. 2.9 can be obtained. In the following expressions, σ_i is defined as

$$\sigma_i = \begin{cases} 1, & \text{if } i = 0 \\ 2, & \text{otherwise} \end{cases} \tag{2.A.1}$$

For planar segments with open boundaries, some of the Green's function expressions listed here involve single summation; hence they are different from those reported in [2.1].

A.1 Planar Segments with Open Boundaries

(a) Rectangle: The appropriate Green's function for the rectangle shown in Fig. 2.1(a) is given by [2.20]

$$G(x_p, y_p | x_q, y_q) \tag{2.A.2}$$

$$= -C \sum_{l=0}^{\infty} F\sigma_l \cos(k_u u_p) \cos(k_u u_q) \cdot \frac{\cos(\gamma_l z_>) \cos(\gamma_l z_<)}{\gamma_l \sin(\gamma_l F)}$$

where

$$C = \frac{j\omega\mu d}{ab} \tag{2.A.3}$$

$$F = \begin{cases} a, & l = m \\ b, & l = n \end{cases} \tag{2.A.4}$$

$$(u_p, u_q) = \begin{cases} (x_p, x_q), & l = m \\ (y_p, x_q), & l = n \end{cases} \tag{2.A.5}$$

$$\gamma_l = \pm\sqrt{k^2 - k_u^2} \tag{2.A.6}$$

$$k_u = \begin{cases} \dfrac{m\pi}{a}, & l = m \\ \dfrac{n\pi}{b}, & l = n \end{cases} \tag{2.A.7}$$

and

$$(z_>, z_<) = \begin{cases} (y_> - b, y_<), & l = m \\ (x_> - a, x_<), & l = n \end{cases} \tag{2.A.8}$$

with

$$y_< = \min(y_p, y_q) \qquad y_> = \max(y_p, y_q) \tag{2.A.9}$$

where l is a dummy variable that could be m or n and the sign of γ_l is chosen such that $\text{Im}(\gamma_l)$ is negative.

(b) 30°–60°–90° Triangle: The Green's function for the right-angled triangle shown in Fig. 2.1(b) is in the form of double summation, and it is given by [2.21]

$$G(x, y | x_0, y_0) = 8j\omega\mu d \sum_{m=-\infty}^{\infty} \sum_{n=-\infty}^{\infty} \cdot \frac{T_1(x_0, y_0) T_1(x, y)}{16\sqrt{3}\pi^2(m^2 + mn + n^2) - 9\sqrt{3}a^2k^2} \tag{2.A.10}$$

where

$$T_1(x < y) = (-1)^l \cos\left(\frac{2\pi l x}{\sqrt{3}a}\right) \cos\left[\frac{2\pi(m - n)y}{3a}\right] + (-1)^m \cos\left(\frac{2\pi m x}{\sqrt{3}a}\right) \cos\left[\frac{2\pi(n - l)y}{3a}\right] + (-1)^n \cos\left(\frac{2\pi n x}{\sqrt{3}a}\right) \cos\left[\frac{2\pi(l - m)y}{\sqrt{3}a}\right] \tag{2.A.11}$$

with the condition that

$$l = -(m + n) \tag{2.A.12}$$

It should be noted that although a single-summation Green's function is not available, an impedance matrix expression with one finite sum and one infinite series has been obtained [2.21]. Benalla and Gupta [2.21] applied the method of images and used the single-series formulation for the impedance matrix of a rectangular planar segment to obtain the result.

(c) Equilateral Triangle: As in the previous case the Green's function of the equilateral triangle depicted in Fig. 2.1(c) remains to be expressed as a double series and is given by [2.21]

$$G(x, y | x_0, y_0) = 4j\omega\mu d \sum_{m=-\infty}^{\infty} \sum_{n=-\infty}^{\infty} \cdot \frac{T_1(x_0, y_0) T_1(x, y) + T_2(x_0, y_0) T_2(x, y)}{16\sqrt{3}\pi^2(m^2 + mn + n^2) - 9\sqrt{3}a^2k^2} \tag{2.A.13}$$

where $T_1\ (x, y)$ is given by (2.A.4) and

$$T_2(x, y) = (-1)^l \cos\left(\frac{2\pi l x}{\sqrt{3}a}\right)\sin\left[\frac{2\pi(m-n)y}{3a}\right] + (-1)^m \cos\left(\frac{2\pi m x}{\sqrt{3}a}\right)\sin\left[\frac{2\pi(n-l)y}{3a}\right] + (-1)^n \cos\left(\frac{2\pi n x}{\sqrt{3}a}\right)\sin\left[\frac{2\pi(l-m)y}{3a}\right] \quad (2.A.14)$$

As for $T_1(x, y)$, the integer 1 in $T_2(x, y)$ is given by (2.A.12). As in the previous case, simplified expressions for the impedance matrix elements have been reported [2.21].

(d) Right-Angled Isosceles Triangle: The Green's function for this geometry has been reduced to a single infinite series and is given by [2.4]

$$G(x, y|x_0, y_0) = -\frac{j\omega\mu d}{2a}(G_1 + G_2) \quad (2.A.15)$$

where G_1 is given by

$$G_1 = \sum_{m=0}^{\infty} 2D_m \cos(k_x x_i)\cos(k_x x_j) \cdot \cos[\gamma_m(y_{1>} - a)]\cos(\gamma_m y_{1<}) \quad (2.A.16)$$

when $(x_i + x_j) \geq (y_i + y_j)$, and by

$$G_1 = \sum_{m=0}^{\infty} 2D_m \cos(k_x y_i)\cos(k_x y_j) \cdot \cos[\gamma_m(y_{4>} - a)]\cos(\gamma_m y_{4<}) \quad (2.A.17)$$

if $(y_i + y_j > (x_i + x_j)$. Similarly, G_2 is given by

$$G_2 = \sum_{m=0}^{\infty} 2D_m (-1)^m \cos(k_x x_i)\cos(k_x y_j)\cos(\gamma_m y_{2>})\cos(\gamma_m y_{2<}) \quad (2.A.18)$$

if $(x_i + y_j) \geq (y_i + x_j)$ and by

$$G_2 = \sum_{m=0}^{\infty} 2D_m (-1)^m \cos(k_x x_i)\cos(k_x x_j)\cos(\gamma_m y_{3>})\cos(\gamma_m y_{3<}) \quad (2.A.19)$$

when $(y_i + x_j) > (x_i + y_j)$.

Furthermore,

$$D_m = \frac{\sigma_m}{\gamma_m \sin(\gamma_m a)} \quad (2.A.20)$$

$$y_{1>} = \max(y_i, y_j) \quad (2.A.21)$$

$$y_{1<} = \min(y_i, y_j) \quad (2.A.22)$$

$$y_{2>} = \max(y_i, x_j) \quad (2.A.23)$$

$$y_{2<} = \min(y_i, x_j) \quad (2.A.24)$$

$$y_{3>} = \max(x_i, y_j) \quad (2.A.25)$$

$$y_{3<} = \min(x_i, y_j) \quad (2.A.26)$$

$$y_{4>} = \max(x_i, x_j) \quad (2.A.27)$$

$$y_{4<} = \min(x_i, x_j) \quad (2.A.28)$$

$$\gamma_m = \pm\sqrt{k^2 - k_x^2} \quad (2.A.29)$$

where σ_m is given by (2.A.1), $k^2 = \omega^2\mu\varepsilon$, $k_x = m\pi/a$, and the sign of γ_m is selected such that $\mathrm{Im}(\gamma_m)$ is negative.

(e) Circular Disc: The Green's function for the circular disc shown in Fig. 2.1(e) was originally reported in the form of doubly infinite series [2.1]. It was later demonstrated in [2.1] that the Z-matrix expression derived from this Green's function can indeed be reduced to a single series. Using a similar approach, Alhargan and Judah [2.2] showed that the Green's function expression itself can be simplified to

$$G(\rho, \phi|\rho_0, \phi_0) = \frac{j\omega\mu d}{4}\sum_{n=0}^{\infty} \frac{\sigma_n J_n(\rho k) f_n(k\rho_0, ka)\cos[n(\phi - \phi_0)]}{J_n'(ka)} \quad (2.A.30)$$

where $0 \leq r \leq r_0 \leq a$, $r_0 \neq 0$, and

$$f_n(\rho k, ak) = Y_n'(ak)J_n(\rho k) - J_n'(ak)Y_n(\rho k) \quad (2.A.31)$$

with

$$\sigma_n = \begin{cases} 1, & n = 0 \\ 0, & n \neq 0 \end{cases} \quad (2.A.32)$$

The wave number k is equal to $\omega\sqrt{\mu\varepsilon}$.

(f) Circular Sector: Until recently [2.22], Green's function for a circular sector was available as a double series and only when the sector angle α is a submultiple of π. For the circular sector shown in Fig. 2.1(f) for which $\alpha = \pi/l$, the double infinite series Green's function is given by [2.1, 2.25]

$$G(\rho, \phi|\rho_0, \phi_0) = \frac{2d}{j\omega\varepsilon\alpha a^2} + \frac{2j\pi\omega\mu d}{\alpha}\sum_{n=0}^{\infty}\sum_{m=1}^{\infty} \frac{\sigma_n J_{n_i}(k_{mn_i}\rho_0)J_{n_i}(k_{mn_i}\rho)\cos(n_i\phi_0)\cos(n_i\phi)}{\pi(a^2 - n_i^2/k_{mn_i}^2)(k_{mn_i}^2 - k^2)J_{n_i}^2(k_{mn_i}a)} \quad (2.A.33)$$

when $n_i = nl$ and the k_{mn_i} are given by

$$\frac{\partial}{\partial\rho}J_{n_i}(k_{mn_i}\rho)\,\big|_{\rho=a} = 0 \quad (2.A.34)$$

The recently derived [2.22] single-series Green's function for a circular sector of arbitrary angle is given by

$$G(\rho, \phi|\rho_0, \phi_0) = \frac{j\omega\mu d l}{2}\sum_{p=0}^{\infty} \frac{\sigma_p J_{\nu_p}(\rho k) f_{\nu_p}(ak, \rho_0 k)}{J_{\nu_p}'(ak)} \cdot \cos(\nu_p\phi)\cos(\nu_p\phi_0) \quad (2.A.35)$$

where $0 \le \rho \le \rho_0 \le a$, $\rho_0 \ne 0$, $0 \le \phi \le \alpha$, $0 \le \phi_0 \le \alpha$, and

$$f_{\nu_p}(ak, \rho_0 k) = Y_{\nu_p}'(ak)J_{\nu_p}(\rho_0 k) - J_{\nu_p}'(ak)Y_{\nu_p}(\rho k) \quad (2.A.36)$$

where $\nu = pl$, $l = \pi/\alpha$, $p = 0, 1, 2, \ldots$, and $\sigma_p = 1$ for $p = 0$ and 2 otherwise.

(g) Annular Ring: The Green's function for the annular ring shown in Fig. 2.1(g) was originally reported as a doubly infinite series [2.1, 2.25]. This expression was later simplified in [2.2] into a single-series expression. The simplified expression is given by

$$G(\rho, \phi|\rho_0, \phi_0) = \frac{j\omega\mu d}{4} \sum_{n=0}^{\infty} \frac{\sigma_n f_n(\rho k, ak) f_n(\rho_0 k, bk) \cos[n(\phi - \phi_0)]}{f_n'(bk, ak)} \quad (2.A.37)$$

where $a \le r \le r_0 \le b$, f_n is given by (2.A.31), and

$$f_n'(bk, ak) = Y_n'(ak)J_n'(bk) - J_n'(ak)Y_n'(bk) \quad (2.A.38)$$

(h) Annular Sector: The annular sector geometry is similar to the circular sector case in the sense that two Green's functions have been reported. The first Green's function is a double-series expression valid for a sector whose angle α is equal to a submultiple of π. For this configuration (see Fig. 2.1(h)), the Green's function is given by [2.1, 2.25]

$$G(\rho, \phi|\rho_0, \phi_0) = \frac{2d}{j\omega\varepsilon\alpha\,(b^2 - a^2)} + \frac{2j\omega\mu d\pi}{\alpha} \sum_{n=0}^{\infty} \sum_{m=1}^{\infty} \frac{\sigma_n F_{mn_i}(\rho) F_{mn_i}(\rho_0) \cos(n_i\phi_0) \cos(n_i\phi)}{[\pi(b^2 - n_i^2/k_{mn_i}^2)F_{mn_i}^2(b)]\,[(a^2 - n_i^2/k_{mn_i}^2)F_{mn_i}^2(a)\,(k_{mn_i}^2 - k^2)]} \quad (2.A.39)$$

where $n_i = nl$ and $l = \pi/\alpha$. Furthermore, the term $F_{mn_i}(\rho)$ is defined by

$$F_{mn_i}(\rho) = N_n'(k_{mn}a)J_n(k_{mn}\rho) - J_n'(k_{mn}a)N_n(k_{mn}\rho) \quad (2.A.40)$$

where k_{mn} are the solutions of

$$\frac{J_n'(k_{mn}a)}{N_n'(k_{mn}a)} = \frac{J_n'(k_{mn}b)}{N_n'(k_{mn}b)} \quad (2.A.41)$$

In the foregoing expressions, $N_n(.)$ denotes Neumann's function of order n, and $J_n'(.)$ and $N_n'(.)$ denote first derivatives with respect to the arguments.

The second Green's function that has been derived recently by Alhargan and Judah [2.22] is expressed in terms of a single series and is valid for an annular sector with an arbitrary angle. This expression is given by

$$G(\rho, \phi|\rho_0, \phi_0) = \frac{j\omega\mu dl}{2} \sum_{p=0}^{\infty} \frac{\sigma_p f_{\nu_p}(ak, \rho k) f_{\nu_p}(bk, \rho_0 k)}{f_{\nu_p}'(bk, ak)} \cdot \cos(\nu_p\phi)\cos(\nu_p\phi_0) \quad (2.A.42)$$

where $a \le \rho \le \rho_0 \le b$, $0 \le \phi \le \alpha$, and $0 \le \phi_0 \le \alpha$, with $\nu_p = pl$, $l = \pi/\alpha$, $p = 0, 1, 2, \ldots$, and $\sigma_p = 1$ for $p = 0$ and 2 otherwise.

References

[2.1] Gupta, K. C., and M. D. Abouzahra. "Planar circuit analysis." In T. Itoh, ed., *Numerical Techniques for Microwave and Millimeter-Wave Passive Structures*, Chapter 4, pp. 214–333. New York: John Wiley & Sons, 1989.

[2.2] Alhargan, F. A., and Sunil R. Judah. "Reduced form of the Green's functions for disks and annular rings." *IEEE Trans. Microwave Theory Tech.*, Vol. MTT-39, No. 3, pp. 601–604, March 1991.

[2.3] Alhargan, F. A., and Sunil R. Judah. "Frequency response characteristics of the multiport planar elliptic patch." *IEEE Trans. Microwave Theory Tech.*, Vol. MTT-40, No. 40, pp. 1726–1730, August 1992.

[2.4] Maramis, H. J., and K. C. Gupta. "Planar model characterization of compensated microstrip bends." *MIMICAD Technical Report No. 1*, University of Colorado (Boulder), March 1989.

[2.5] Okoshi, T. *Planar Circuits for Microwave and Lightwaves*, pp. 10–42. New York: Springer-Verlag, 1985.

[2.6] Mroczkowski, C., and W. K. Gwarek. "Microwave circuits described by two-dimensional vector wave equation and their analysis by FD-TD method." *Proc. 21st European Microwave Conf.*, pp. 866–871, Stuttgart, Germany, September 9–12, 1991.

[2.7] Gwarek, W. K. "Analysis of an arbitrarily-shaped planar circuit—A time domain approach." *IEEE Trans. Microwave Theory Tech.*, Vol. MTT-33, No. 10, pp. 1067–1072, October 1985.

[2.8] Okoshi, T., and T. Miyoshi. "The planar circuit—An approach to microwave integrate circuitry." *IEEE Trans. Microwave Theory Tech.*, Vol. MTT-20, No. 4, pp. 245–252, April 1972.

[2.9] Civalleri, P. P., and R. Sidella. "Impedance and admittance of distributed three-layer N-ports." *IEEE Trans. Circuit Theory Tech.*, Vol. CT-17, No. 8, pp. 392–398, August 1970.

[2.10] Helszajn, J. *Non-Reciprocal Microwave Junctions and Circulators*, pp. 170–176. New York: John Wiley & Sons, 1975.

[2.11] Ohta, I., I. Hagino, and T. Kaneko. "Improved circular disk 3 dB hybrids." *Electronics and Communication in Japan*, Pt. 2, Vol. 70, No. 12, pp. 66–77, 1987.

[2.12] Sorrentino, R. "Planar circuits, waveguides models, and segmentation method." *IEEE Trans. Microwave Theory Tech.*, Vol. MTT-33, No. 10, pp. 1057–1066, October 1985.

[2.13] Abouzahra, M. D., and K. C. Gupta. "Multi-port power divider/combiner circuits using circular microstrip disk configurations." *IEEE Trans. Microwave Theory Tech.*, Vol. MTT-35, No. 12, pp. 1296–1302, December 1987.

[2.14] Abouzahra, M. D., and K. C. Gupta. "Multi-way unequal power divider circuits using sector-shaped planar components." In *1989 IEEE MTT-S Int'l. Microwave Symp. Dig.*, pp. 321–324, Long Beach, California, June 13–15, 1989.

[2.15] Judah, S. R., and M. J. Page. "An analysis of an N-port microstrip planar disk device with an arbitrarily located short circuit post of arbitrary radius." *IEEE Trans. Microwave Theory Tech.*, Vol. MTT-38, No. 12, pp. 1823–1830, December 1990.

[2.16] Chadha, R., and K. C. Gupta. "Segmentation method using impedance matrices for analysis of planar microwave circuits." *IEEE Trans. Microwave Theory Tech.*, Vol. MTT-29, No. 1, pp. 71–74, January 1981.

[2.17] Chadha, R., and K. C. Gupta. "Green's functions for triangular segments in planar microwave circuits." *IEEE Trans. Microwave Theory Tech.*, Vol. MTT-28, No. 10, pp. 1139–1143, October 1980.

[2.18] Morse, P. M., and H. Feshbach. *Methods of Theoretical Physics*, Chapter 7. New York: McGraw-Hill, 1953.

[2.19] Gupta, K. C., R. Garg, and R. Chadha. *Computer-Aided Design of Microwave Circuits*, pp. 243–244. Dedham, MA: Artech House, 1981.

[2.20] Benalla, A., and K. C. Gupta. "Faster computation of Z-matrices for rectangular segments in planar microstrip circuits." *IEEE Trans. Microwave Theory Tech.*, Vol. MTT-34, No. 6, pp. 733–736, June 1986.

[2.21] Lee, S. H., A. Benalla, and K. C. Gupta. "Faster computation of Z-matrices for triangular segments in planar circuits." *Int'l. J. Millimeter-wave Computer-Aided Engineering*, Vol. 2, No. 2, pp. 98–107, April 1992.

[2.22] Alhargan, F. A., and S. R. Judah. "Circular and annular sector planar components of arbitrary angle for N-way power dividers/combiners." *IEEE Trans. Microwave Theory Tech.*, 1994 (in press).

[2.23] Morse, P. M., and H. Feshbach. *Methods of Theoretical Physics*. New York: McGraw-Hill, 1953.

[2.24] McLachlan, N. W. *Theory and Applications of Mathieu Functions*. Oxford: Oxford University Press, 1947.

[2.25] Chadha, R., and K. C. Gupta. "Green's functions for circular sectors, annular rings, and annular sectors in planar microwave circuits." *IEEE Trans. Microwave Theory Tech.*, Vol. MTT-29, No. 1, pp. 68–71, January 1981.

Chapter 3

Segmentation and Desegmentation Techniques

K. C. Gupta
University of Colorado at Boulder

M. D. Abouzahra
M.I.T. Lincoln Laboratory

3.1 Introduction

GREEN'S function approach for analysis of two-dimensional planar components with regular shapes has been presented in Chapter 2. The impedance matrix for a planar circuit with a regular geometrical shape can be obtained from the Green's function when port locations are specified. On the other hand, analysis of components with completely arbitrary shapes, for which a Green's function cannot be obtained, is discussed in the reprints included in Chapter 3.

Between these two extreme cases, there is a class of planar components in which the shape of the planar circuit is a composite of simple configurations. In these cases, the two-dimensional composite shapes can be decomposed into either (1) all regular shapes or (2) a combination of some regular shape(s) and some arbitrary shape(s). Examples of these two types of composite planar circuit configurations are shown in Fig. 3.1(a) and Fig. 3.1(b), respectively. This process of breaking down a composite shape into simpler shapes may be called *segmentation.* [3.1]

The term "segmentation" as used in this chapter (as in the analysis procedures for planar components in general) refers to the network theoretic method for characterizing a combination of multiport components when the individual characteristics of each of the individual components is known. The implementation of this "segmentation" method may be carried out in terms of S-parameters, Z-parameters, or Y-parameters. These three procedures are discussed in Sec. 3.2.

The segmentation method has been used extensively for the analysis of two-dimensional planar components. Recently, this approach has also been applied to three-dimensional electromagnetic problems [3.2]. Such problems include full-wave analysis of via-hole grounds in microstrip [3.2]. In these cases, the structure is divided into various regions sharing common interfaces, where each region is bounded by either electric or magnetic walls. A network characterization of each region is obtained in terms of generalized admittance matrices.

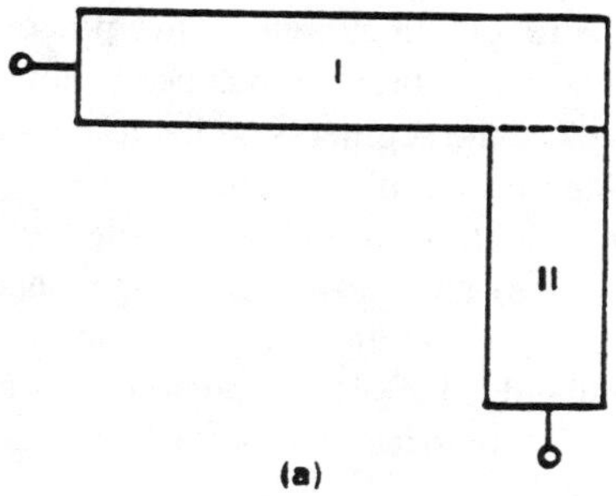

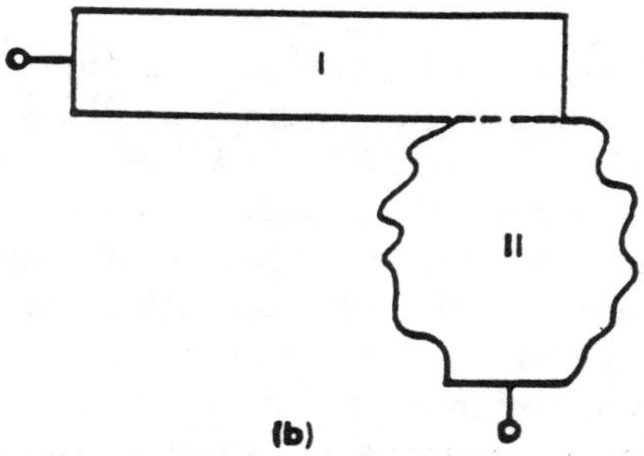

Fig. 3.1 Planar circuit of composite configurations: (a) a configuration that can be decomposed into two regularly shaped segments; (b) a configuration that can be decomposed into a regular and an arbitrary segment.

These admittance matrices are then connected together by the segmentation procedure to enforce the continuity of currents boundary condition that corresponds to the continuity of the tangential components of the magnetic fields at the interfaces. Section 3.3 presents details of the three-dimensional segmentation approach used along with the mode-matching method.

The underlying basis of the segmentation approach is the transformation of the field matching along the interface between two regions into an equivalent network connection problem. Matching of the tangential component of magnetic field is accomplished by the current continuity condition en-

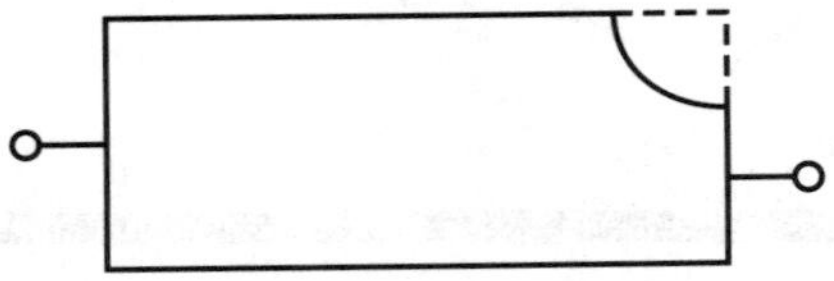

Fig. 3.2 Planar circuit configuration suitable for analysis by the desegmentation method.

forced by Kirchoff's current law. The continuity of the tangential component of electric field is assured by enforcing equivalence of voltages at the connected ports. Thus the electromagnetic field matching problem is solved in terms of an equivalent network interconnection problem.

In addition to the segmentation method, there is also a complementary process called desegmentation [3.3], which is suitable for analyzing some planar circuit configurations. An example of the latter is shown in Fig. 3.2. This configuration cannot be decomposed into regular shapes, but if a right-angled circular sector (a regular shape for which the Green's function is available) is added to the shape shown in Fig. 3.2, we end up with a regular rectangle. This process has been termed "desegmentation" [3.3]. It has been shown that if the impedance matrices of the segments added to the original configuration (right-angled circular section in the present case) and the impedance matrix of the augmented configuration (rectangle in this case) are known, the impedance matrix of the original planar segment can be obtained by using "desegmentation" formulas [3.3, 3.4]. Segmentation procedures in terms of S-, Z- and Y-matrices are described in Sec. 3.2.

3.2 Segmentation Procedures

The name segmentation has been given to this network analysis method for planar (two-dimensional) microwave circuits by Okoshi and his colleagues [3.1, 3.5, 3.6]. The basic idea is to divide a single complicated planar circuit configuration into simpler "segments" that have regular shapes and can therefore be characterized relatively easily. An example of such a segmentation is shown in Fig. 3.1(a), where a microstrip corner geometry is broken down into two rectangular segments for which Green's functions are available. Essentially, the segmentation method gives us the overall characterization or performance of the composite network, when the characterization of each of the segments is known. Originally, the segmentation method was formulated [3.1] in terms of S-matrices of individual segments; however, it was found subsequently [3.7] that a Z-matrix formulation is more efficient for microwave planar circuits (also for microstrip antennas). In this section, we will describe the segmentation procedures based on S-, Z-, and Y-matrices.

3.2.1 Segmentation Using S-Matrices [3.8]

This procedure can be illustrated by considering the interconnection of the two components shown in Fig. 3.3. The components are characterized by their S-parameters as:

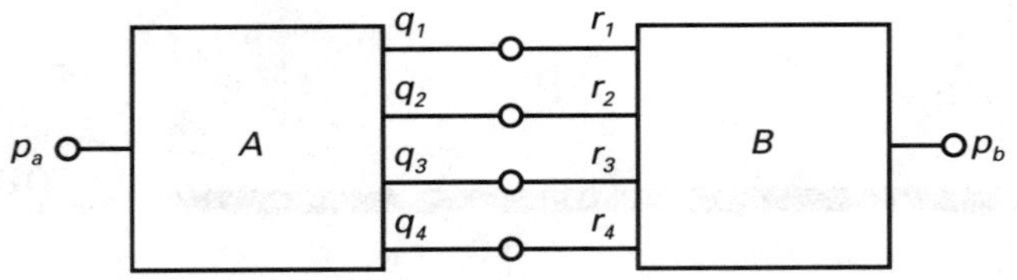

Fig. 3.3 Port nomenclature use in the segmentation formula.

$$b_A = [S_A]a_A \quad \text{and} \quad b_B = [S_B]a_B \tag{3.1}$$

where

$$b_A = \begin{bmatrix} b_{p_a} \\ b_q \end{bmatrix}, \quad a_A = \begin{bmatrix} a_{p_a} \\ a_q \end{bmatrix}, \quad b_B = \begin{bmatrix} b_{p_b} \\ b_r \end{bmatrix}, \quad \text{and} \quad a_B = \begin{bmatrix} a_{p_b} \\ a_r \end{bmatrix} \tag{3.2}$$

The port nomenclature is shown in Fig. 3.3: p_a and p_b are external ports of segments A and B, respectively, and q_1, $q_2, \ldots$ and $r_1, r_2 \ldots$ are interconnected ports numbered such that q_1 is connected to r_1, q_2 to r_2, . . . , and so on. Separating out external and connected ports, we may write

$$\begin{bmatrix} b_{p_a} \\ b_q \end{bmatrix} = \begin{bmatrix} S_{p_a} & S_{p_a q} \\ S_{qp_a} & S_{qq} \end{bmatrix} \begin{bmatrix} a_{p_a} \\ a_q \end{bmatrix} \quad \text{and} \quad \begin{bmatrix} b_{p_b} \\ b_r \end{bmatrix} = \begin{bmatrix} S_{p_b} & S_{p_b r} \\ S_{rp_b} & S_{rr} \end{bmatrix} \begin{bmatrix} a_{p_b} \\ a_r \end{bmatrix} \tag{3.3}$$

Taking

$$b_p = \begin{bmatrix} b_{p_a} \\ b_{p_b} \end{bmatrix} \quad \text{and} \quad a_p = \begin{bmatrix} a_{p_a} \\ a_{p_b} \end{bmatrix} \tag{3.4}$$

we write Eq. (3.3) as

$$\begin{bmatrix} b_p \\ b_q \\ b_r \end{bmatrix} = \begin{bmatrix} S_{pp} & S_{pq} & S_{pr} \\ S_{qp} & S_{qq} & 0 \\ S_{rp} & 0 & S_{rr} \end{bmatrix} \begin{bmatrix} a_p \\ a_q \\ a_r \end{bmatrix} \tag{3.5}$$

where

$$S_{pp} = \begin{bmatrix} S_{p_a} & 0 \\ 0 & S_{p_b} \end{bmatrix}, \quad S_{pq} = \begin{bmatrix} S_{p_a q} \\ 0 \end{bmatrix}, \quad S_{pr} = \begin{bmatrix} 0 \\ S_{p_b r} \end{bmatrix} \tag{3.6}$$

and

$$S_{qp} = S^t_{pq} = [S_{qp_a}, 0], \quad S_{rp} = S^t_{pr} = [0, S_{rp_b}] \tag{3.7}$$

The superscript t indicates transpose of a matrix and 0 denotes a null matrix of appropriate dimensions. From Eq. (3.5) we may write

$$b_p = S_{pp}a_p + S_{pq}a_q + S_{pr}a_r \tag{3.8}$$

$$b_q = S_{qp}a_p + S_{qq}a_q \tag{3.9}$$

$$b_r = S_{rp}a_p + S_{rr}a_r \tag{3.10}$$

In this case, the excitation will be the incident wave at the external ports, or a_p. Thus we desire to obtain the outgoing wave variable b_p in terms of a_p. The interconnection conditions in the case of S-parameters are expressed in terms of wave variables a's and b's. It is assumed that connected ports have been referenced to the same impedance such that

$$a_r = b_q \tag{3.11}$$

and

$$b_r = a_q \tag{3.12}$$

Introducing Eqs. (3.11) and (3.12) into Eq. (3.10) gives

$$b_r = S_{rp}a_p + S_{rr}(S_{qp}a_p + S_{qq}a_q) \tag{3.13}$$

or

$$b_r = (S_{rp} + S_{rr}S_{qp})a_p + S_{rr}S_{qq}a_q \tag{3.14}$$

which with Eq. (3.12) may be written as

$$(I - S_{rr}S_{qq})a_q = (S_{rp} + S_{rr}S_{qp})a_p \tag{3.15}$$

so

$$a_q = (I - S_{rr}S_{qq})^{-1}(S_{rp} + S_{rr}S_{qp})a_p \tag{3.16}$$

where $[I]$ is the identity matrix. Introducing Eqs. (3.12) and (3.10) into Eq. (3.9) gives

$$b_q = S_{qp}a_p + S_{qq}(S_{rp}a_p + S_{rr}a_r) \tag{3.17}$$

or

$$b_q = (S_{qp} + S_{qq}S_{rp})a_p + S_{qq}S_{rr}a_r \tag{3.18}$$

which with Eq. (3.11) may be written as

$$(I - S_{qq}S_{rr})a_r = (S_{qp} + S_{qq}S_{rp})a_p \tag{3.19}$$

so

$$a_r = (I - S_{qq}S_{rr})^{-1}(S_{qp} + S_{qq}S_{rp})a_p \tag{3.20}$$

Now, using Eqs. (3.16) and (3.20) in Eq. (3.8), we write

$$\begin{aligned} b_p = {} & S_{pp}a_p + S_{pq}(I - S_{rr}S_{qq})^{-1}(S_{rp} + S_{rr}S_{qp})a_p \\ & + S_{pr}(I - S_{qq}S_{rr})^{-1}(S_{qp} + S_{qq}S_{rp})a_p \end{aligned} \tag{3.21}$$

So the resultant matrix S_{AB} for the combination of components A and B is given by

$$\begin{aligned} S_{AB} = {} & S_{pp} + S_{pq}(I - S_{rr}S_{qq})^{-1}(S_{rp} + S_{rr}S_{qp}) \\ & + S_{pr}(I - S_{qq}S_{rr})^{-1}(S_{qp} + S_{qq}S_{rp}) \end{aligned} \tag{3.22}$$

Equation (3.22) is the segmentation formula expressed in terms of S-matrices.

3.2.2 Segmentation Using Z-Matrices

For two-dimensional components of regular shapes, Z-matrix characterization is obtained directly from the Green's function approach discussed in Chapter 2. Similarly, the Z-matrices for arbitrary shaped components may be obtained by using the contour-integral approach presented in Chapter 4. Since Z-matrix characterizations of several components are thus available, it becomes computationally efficient to develop a segmentation procedure in terms of Z-parameters [3.7].

Segmentation using Z-matrices may also be illustrated by considering the interconnection of two multiport segments shown in Fig. 3.3. As a result of the interconnections of q- and r-ports, we can write

$$V_q = V_r \quad \text{and} \quad i_q = -i_r \tag{3.23}$$

The Z-matrices of segments A and B may be written as

$$Z_A = \begin{bmatrix} Z_{p_a} & Z_{p_aq} \\ Z_{qp_a} & Z_{qq} \end{bmatrix}, \quad Z_B = \begin{bmatrix} Z_{p_b} & Z_{p_br} \\ Z_{rp_b} & Z_{rr} \end{bmatrix} \tag{3.24}$$

where Z_{p_a}, Z_{p_aq}, Z_{qp_a}, Z_{qq}, Z_{p_b}, Z_{p_br}, Z_{rp_b}, and Z_{rr} are submatrices of appropriate dimensions. When A and B are reciprocal components,

$$Z_{p_aq} = (Z_{qp_a})^t \quad \text{and} \quad Z_{p_br} = (Z_{rp_b})^t \tag{3.25}$$

The Z-matrices of the segments A and B can be written together as

$$\begin{bmatrix} V_p \\ V_q \\ V_r \end{bmatrix} = \begin{bmatrix} Z_{pp} & Z_{pq} & Z_{pr} \\ Z_{qp} & Z_{qq} & 0 \\ Z_{rp} & 0 & Z_{rr} \end{bmatrix} \begin{bmatrix} i_p \\ i_q \\ i_r \end{bmatrix} \tag{3.26}$$

where

$$V_p = \begin{bmatrix} V_{p_a} \\ V_{p_b} \end{bmatrix}, \quad i_p = \begin{bmatrix} i_{p_a} \\ i_{p_b} \end{bmatrix}$$

and

$$Z_p = \begin{bmatrix} Z_{p_a} & 0 \\ 0 & Z_{p_b} \end{bmatrix}, \quad Z_{pq} = \begin{bmatrix} Z_{p_aq} \\ 0 \end{bmatrix}, \quad Z_{pr} = \begin{bmatrix} 0 \\ Z_{p_br} \end{bmatrix}$$

It may be noted that the interconnection conditions stated in (3.23) have not been used for writing Eq. (3.26), which represents a rearrangement of individual matrices Z_A and Z_B given in (3.24). Eq. (3.23) can now be substituted into (3.26) to eliminate V_p, V_q, i_q, and i_r. The resulting expression may be written as $V_p = [Z_{AB}]i_p$, where

$$\begin{aligned} [Z_{AB}] = {} & \begin{bmatrix} Z_{p_a} & 0 \\ 0 & Z_{p_b} \end{bmatrix} \\ & + \begin{bmatrix} Z_{p_aq} \\ -Z_{p_b} \end{bmatrix} [Z_{qq} + Z_{rr}]^{-1}[-Z_{qp_a}, Z_{rp_b}] \end{aligned} \tag{3.27}$$

It may be noted that the size of Z_{AB} is $(p_a + p_b) \times (p_a + p_b)$. The second term on the right-hand side is a product of three matrices of the sizes: $(p_a + p_b) \times q$, $q \times q$, and $q \times (p_a + p_b)$, respectively. From the computational point of view, the most time-consuming step is the evaluation of the inverse of a matrix of size $(q \times q)$, where q is the number of interconnected ports.

To illustrate the foregoing procedure for combining Z-matrices of two segments together, let us consider an example of two lumped resistive networks connected together as shown in Fig. 3.4(a). The Z-matrices of the individual components A and B may be written as

$$\begin{aligned} Z_A = \begin{bmatrix} Z_{11} & Z_{13} & Z_{14} \\ Z_{31} & Z_{33} & Z_{34} \\ Z_{41} & Z_{43} & Z_{44} \end{bmatrix} &= \begin{bmatrix} 4 & 3 & 3 \\ 3 & 7 & 3 \\ 3 & 3 & 5 \end{bmatrix} \\ &= \begin{bmatrix} Z_{p_a} & Z_{p_aq} \\ Z_{qp_a} & Z_{qq} \end{bmatrix} \end{aligned} \tag{3.28}$$

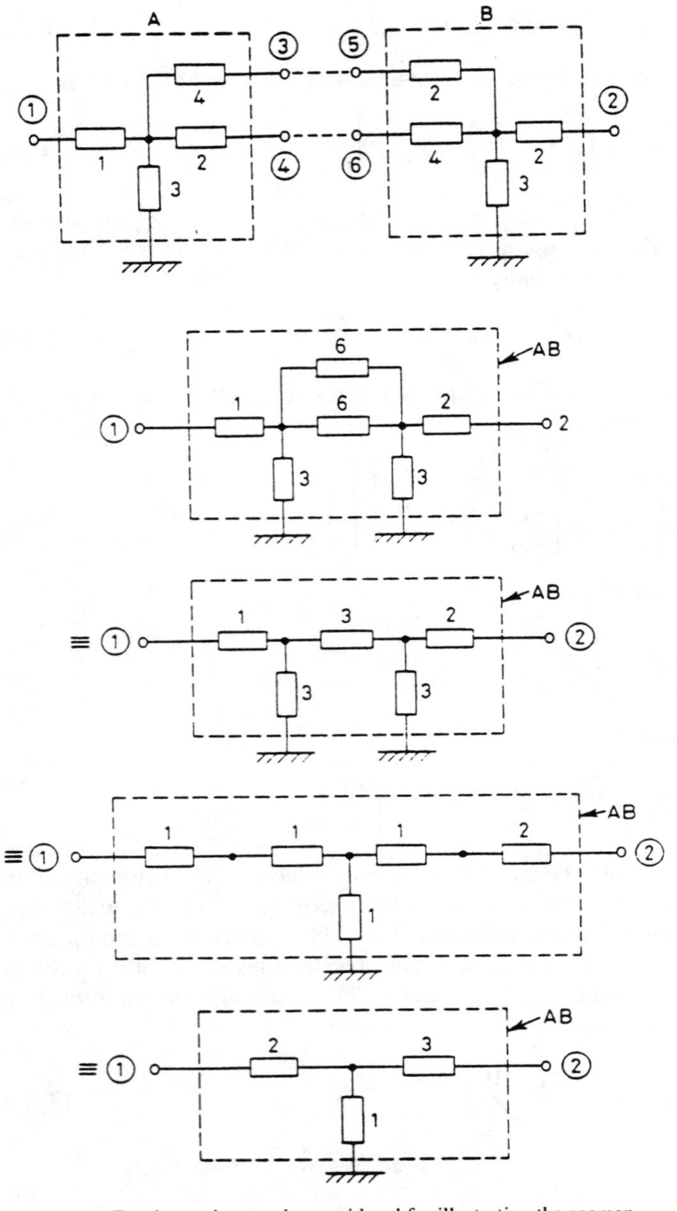

Fig. 3.4 (a) Two lumped networks considered for illustrating the segmentation procedure; (b) circuit simplification for writing Z-matrix of the combination of the networks A and B.

$$Z_B = \begin{bmatrix} Z_{22} & Z_{25} & Z_{26} \\ Z_{25} & Z_{55} & Z_{56} \\ Z_{62} & Z_{65} & Z_{66} \end{bmatrix} = \begin{bmatrix} 5 & 3 & 3 \\ 3 & 5 & 3 \\ 3 & 3 & 7 \end{bmatrix} = \begin{bmatrix} Z_{p_b} & Z_{p_b r} \\ Z_{rp_b} & Z_{rr} \end{bmatrix} \tag{3.29}$$

In terms of the notations of Eq. (3.26), we have

$$Z_{pp} = \begin{bmatrix} Z_{11} & 0 \\ 0 & Z_{22} \end{bmatrix} = \begin{bmatrix} 4 & 0 \\ 0 & 5 \end{bmatrix}$$

$$Z_{pq} = \begin{bmatrix} Z_{p_a} \\ 0 \end{bmatrix} = \begin{bmatrix} Z_{13} & Z_{14} \\ 0 & 0 \end{bmatrix} = \begin{bmatrix} 3 & 3 \\ 0 & 0 \end{bmatrix}$$

$$Z_{pr} = \begin{bmatrix} 0 \\ Z_{p_b r} \end{bmatrix} = \begin{bmatrix} 0 & 0 \\ Z_{25} & Z_{26} \end{bmatrix} = \begin{bmatrix} 0 & 0 \\ 3 & 3 \end{bmatrix}$$

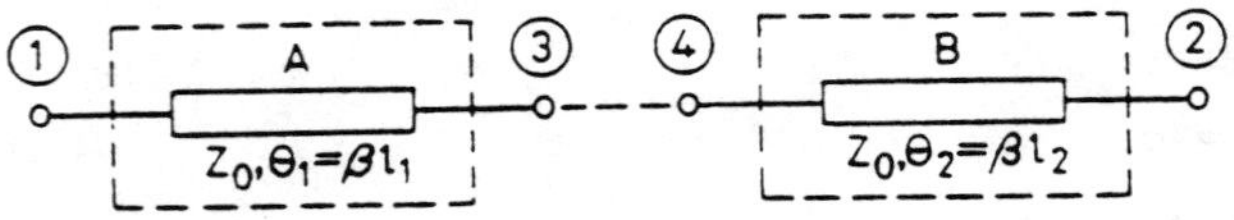

Fig. 3.5 Segmentation as applied to two sections of a transmission line.

$$Z_{qq} = \begin{bmatrix} Z_{33} & Z_{34} \\ Z_{43} & Z_{44} \end{bmatrix} = \begin{bmatrix} 7 & 3 \\ 3 & 5 \end{bmatrix}$$

$$Z_{rr} = \begin{bmatrix} Z_{55} & Z_{56} \\ Z_{65} & Z_{66} \end{bmatrix} = \begin{bmatrix} 5 & 5 \\ 3 & 7 \end{bmatrix}$$

$$Z_{qp} = [Z_{qp_a}, 0] = \begin{bmatrix} Z_{31} & 0 \\ Z_{41} & 0 \end{bmatrix} = \begin{bmatrix} 3 & 0 \\ 3 & 0 \end{bmatrix}$$

$$Z_{rp} = [0, Z_{rp_b}] = \begin{bmatrix} 0 & Z_{52} \\ 0 & Z_{62} \end{bmatrix} = \begin{bmatrix} 0 & 3 \\ 0 & 3 \end{bmatrix}$$

Substituting all these submatrices in Eq. (3.27), we get

$$[Z_{AB}] = \begin{bmatrix} 4, & 0 \\ 0, & 5 \end{bmatrix} + \begin{bmatrix} 3, & 3 \\ -3, & -3 \end{bmatrix} \cdot \begin{bmatrix} 7+5, & 3+3 \\ 3+3, & 5+7 \end{bmatrix}^{-1} \begin{bmatrix} -3, & 3 \\ -3, & 3 \end{bmatrix}$$

which may be evaluated as

$$[Z_{AB}] = \begin{bmatrix} 4, & 0 \\ 0, & 5 \end{bmatrix} + \begin{bmatrix} 3, & 3 \\ -3, & -3 \end{bmatrix} \frac{1}{18} \begin{bmatrix} 2, & -1 \\ -1, & 2 \end{bmatrix} \begin{bmatrix} -3, & 3 \\ -3, & 3 \end{bmatrix} = \begin{bmatrix} 4, & 0 \\ 0, & 5 \end{bmatrix} \begin{bmatrix} -1, & +1 \\ -1, & +1 \end{bmatrix} = \begin{bmatrix} 3, & 1 \\ 1, & 4 \end{bmatrix} \tag{3.30}$$

The resultant matrix Z_{AB} in Eq. (3.30) may be verified by rearranging the circuit shown in Fig. 3.4(a) as the one shown in Fig. 3.4(b).

Let us consider another example. Two transmission line sections of electrical lengths θ_1 and θ_2 are connected in cascade as shown in Fig. 3.5. Z-matrices for the individual sections A and B are given by

$$Z_A = (-jZ_0/\sin\theta_1) \begin{bmatrix} \cos\theta_1 & 1 \\ 1 & \cos\theta_1 \end{bmatrix} \tag{3.31}$$

and

$$Z_B = (-jZ_0/\sin\theta_2) \begin{bmatrix} \cos\theta_2 & 1 \\ 1 & \cos\theta_2 \end{bmatrix} \tag{3.32}$$

In terms of notations of Eq. (3.27), we have

$$Z_{pp} = \begin{bmatrix} Z_{p_a} & 0 \\ 0 & Z_{p_b} \end{bmatrix} = \begin{bmatrix} Z_{11} & 0 \\ 0 & Z_{22} \end{bmatrix} = \begin{bmatrix} z_1\cos\theta_1 & 0 \\ 0 & z_2\cos\theta_2 \end{bmatrix}$$

$$Z_{pq} = \begin{bmatrix} Z_{p_a q} \\ 0 \end{bmatrix} = z_1 \begin{bmatrix} 1 \\ 0 \end{bmatrix}, \quad z_1 = -jZ_0/\sin\theta_1$$

$$Z_{pr} = \begin{bmatrix} 0 \\ Z_{p_b r} \end{bmatrix} = z_2 \begin{bmatrix} 0 \\ 1 \end{bmatrix}, \quad z_2 = -jZ_0/\sin\theta_2$$

$$Z_{qq} = Z_{33} = z_1\cos\theta_1$$

$$Z_{rr} = Z_{44} = z_2\cos\theta_2$$

$$Z_{qp} = [Z_{qp_a}, 0] = [z_1, 0]$$

$$Z_{rp} = [0, Z_{rp_b}] = [0, z_2]$$

Substituting all these submatrices in Eq. (3.27), we get

$$Z_{AB} = \begin{bmatrix} z_1 \cos\theta_1 & 0 \\ 0 & z_2\cos\theta_2 \end{bmatrix} + \begin{bmatrix} z_1 \\ -z_2 \end{bmatrix} [z_1\cos\theta_1 + z_2\cos\theta_2]^{-1}[-z_1, z_2] \tag{3.33}$$

$$= \begin{bmatrix} z_1 \cos\theta_1 & 0 \\ 0 & z_2\cos\theta_2 \end{bmatrix} + \frac{1}{(z_1\cos\theta_1 + z_2\cos\theta_2)} \begin{bmatrix} z_1 \\ -z_2 \end{bmatrix} [-z_1, z_2]$$

$$= \begin{bmatrix} z_1 \cos\theta_1 & 0 \\ 0 & z_2\cos\theta_2 \end{bmatrix} + \frac{1}{(z_1\cos\theta_1 + z_2\cos\theta_2)} \begin{bmatrix} -z_1^2 & z_1 z_2 \\ z_1 z_2 & -z_2^2 \end{bmatrix}$$

Substituting for z_1 and z_2 and using trigonometric formulas for $\sin(\theta_1 + \theta_2)$ and $\cos(\theta_1 + \theta_2)$, Eq. (3.33) may be expressed

$$Z_{AB} = -jZ_0/\sin(\theta_1 + \theta_2) \begin{bmatrix} \cos(\theta_1+\theta_2) & 1 \\ 1 & \cos(\theta_1+\theta_2) \end{bmatrix} \tag{3.34}$$

which is a Z-matrix for a uniform transmission line of length $(\theta_1 + \theta_2)$. This example illustrates again the validity of Eq. (3.27).

When the segmentation method is applied to the multiport network model of microstrip antennas (discussed in Chapter 11), we are interested in the Z-matrix with respect to external ports (1 and 2 in Fig. 11.13) and also in the voltages at the ports connecting the edge admittance network(s) to the patch network. This voltage distribution at the radiating edges is expressed in terms of an equivalent line source of magnetic current. The radiation field (and associated characteristics like beamwidth, side lobe level, etc.) are obtained from the magnetic current distribution.

Referring to Fig. 3.3, voltages at the connected ports (q-ports) may be obtained by [3.6], p. 357].

$$V_q = [Z_{qp} + Z_{qq}(Z_{qq} + Z_{rr})^{-1}(Z_{rp} - Z_{qp})]i_p \tag{3.35}$$

where i_p is the current vector specifying the input current(s) at the external port(s) of the antenna.

3.2.3 Segmentation Using Y-Matrices

In several electromagnetic analysis problems it is more convenient to evaluate the Y-matrices of individual components rather than their Z- or S-matrices. Examples of such problems include multiport network modeling of external coupling between microstrip discontinuities in circuits [3.9], determining the edge admittance and mutual coupling in microstrip patch antennas (discussed in Chapter 11), and calculating the generalized admittance matrices for individual regions for use in the full-wave three-dimensional mode-matching technique [3.2] for modeling of via-hole grounds in microstrip circuits. It is convenient to use a Y-matrix formulation of the segmentation procedure in these cases.

To illustrate the Y-matrix–based segmentation procedure, consider again the configuration of Fig. 3.3 for two components A and B being connected together. In terms of Y-matrices, we can write

$$I_A = Y_A V_A \quad \text{and} \quad I_B = Y_B V_B \tag{3.36}$$

where

$$I_A = \begin{bmatrix} I_{p_a} \\ I_q \end{bmatrix}, \quad V_A = \begin{bmatrix} V_{p_a} \\ V_q \end{bmatrix}, \quad I_B = \begin{bmatrix} I_{p_b} \\ I_r \end{bmatrix}, \quad \text{and} \quad V_B = \begin{bmatrix} V_{p_b} \\ V_r \end{bmatrix} \tag{3.37}$$

So

$$\begin{bmatrix} I_{p_a} \\ I_q \end{bmatrix} = \begin{bmatrix} Y_{p_a} & Y_{p_a q} \\ Y_{qp_a} & Y_{qq} \end{bmatrix} \begin{bmatrix} V_{p_a} \\ V_q \end{bmatrix} \quad \text{and} \quad \begin{bmatrix} I_{p_b} \\ I_r \end{bmatrix} = \begin{bmatrix} Y_{p_b} & Y_{p_b r} \\ Y_{rp_b} & Y_{rr} \end{bmatrix} \begin{bmatrix} V_{p_a} \\ V_r \end{bmatrix}. \tag{3.38}$$

Taking

$$I_p = \begin{bmatrix} I_{p_a} \\ I_{p_b} \end{bmatrix} \quad \text{and} \quad V_p = \begin{bmatrix} V_{p_a} \\ V_{p_b} \end{bmatrix} \tag{3.39}$$

we write Eq. (3.38) as

$$\begin{bmatrix} I_p \\ I_q \\ I_r \end{bmatrix} = \begin{bmatrix} Y_{pp} & Y_{pq} & Y_{pr} \\ Y_{qp} & Y_{qq} & 0 \\ Y_{rp} & 0 & Y_{rr} \end{bmatrix} \begin{bmatrix} V_p \\ V_q \\ V_r \end{bmatrix} \tag{3.40}$$

where

$$Y_{pp} = \begin{bmatrix} Y_{p_a} & 0 \\ 0 & Y_{p_b} \end{bmatrix}, \quad Y_{pq} = \begin{bmatrix} Y_{p_a q} \\ 0 \end{bmatrix}, \quad Y_{pr} = \begin{bmatrix} 0 \\ Y_{p_b r} \end{bmatrix} \tag{3.41}$$

and

$$Y_{qp} = Y_{pq}^t = [Y_{qp_a}, 0] \quad \text{and} \quad Y_{rp} = Y_{rp}^t = [0, Y_{rp_b}] \tag{3.42}$$

From Eq. (3.40) we may write

$$I_p = Y_{pp}V_p + Y_{pq}V_q + Y_{pr}V_r \tag{3.43}$$

$$I_q = Y_{qp}V_p + Y_{qq}V_q \tag{3.44}$$

$$I_r = Y_{rp}V_p + Y_{rr}V_r \tag{3.45}$$

To obtain the external port currents I_p in terms of the excitation V_p, and the interconnection voltages V_r in terms of V_p, the conditions

$$I_r = -I_q \tag{3.46}$$

and

$$V_r = V_q \tag{3.47}$$

are applied to Eqs. (3.43)–(3.45) as follows. Introducing Eq. (3.46) into Eqs. (3.44) and (3.45) gives

$$(Y_{qp}V_p + V_{qq}V_q) = -(Y_{rp}V_p + Y_{rr}V_r) \tag{3.48}$$

or

$$(Y_{qp} + Y_{rp})V_p = -Y_{rr}V_r - Y_{qq}V_q \tag{3.49}$$

which with Eq. (3.47) may be written as

$$-(Y_{qq} + Y_{rr})V_q = (Y_{rp} + Y_{qp})V_p \tag{3.50}$$

so

$$V_q = -(Y_{qq} + Y_{rr})^{-1}(Y_{rp} + Y_{qp})V_p \tag{3.51}$$

which is one of the desired results. We go on to consider Eq. (3.43) making use of Eqs. (3.47 and (3.51), giving

$$I_p = Y_{pp}V_p - (Y_{pq} + Y_{pr})(Y_{qq} + Y_{rr})^{-1} (Y_{rp} + Y_{qp})V_p \tag{3.52}$$

or

$$I_p = [Y_{pp} - (Y_{pq} + Y_{pr})(Y_{qq} + Y_{rr})^{-1} (Y_{rp} + Y_{qp})]V_p \tag{3.53}$$

So the resultant matrix for component C is given by

$$Y_c = Y_{pp} - (Y_{pq} + Y_{pr})(Y_{qq} + Y_{rr})^{-1}(Y_{rp} + Y_{qp}) \tag{3.54}$$

Equation (3.54) provides the segmentation formulation in terms of Y-matrices.

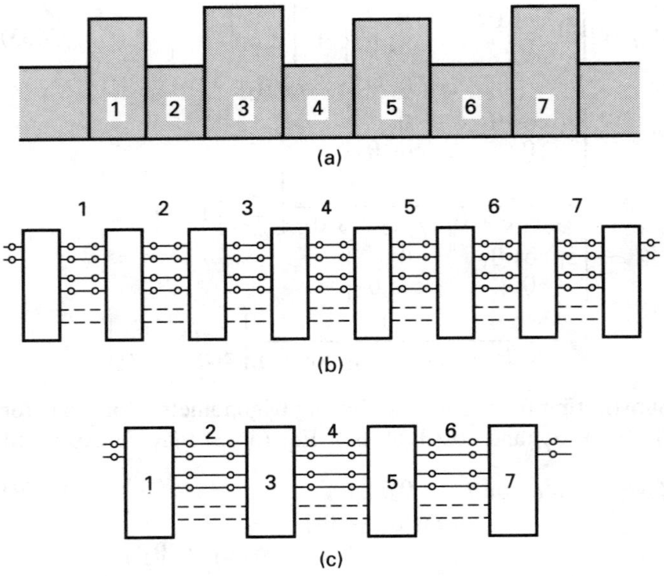

Fig. 3.6 Two different ways (b and c) of segmenting the four-stub filter (a).

3.3 Segmentation Applied with Mode-Matching Method

The segmentation approach has been used extensively for planar circuits and antennas [3.10]. This approach is based on network matching the fields across the interfaces between the various segments. When applied to three-dimensional electromagnetic problems as it has been recently reported [3.2, 3.11, 3.12], the segmentation approach is based on the mode matching between various regions. In these cases [3.2], a network theory based segmentation approach is used in place of the conventional mode-matching procedure.

3.3.1 Various Strategies for Segmenting

The generalized scattering-matrix approach discussed in [3.13] may be viewed as a network-based implementation of the mode-matching method. This approach can be illustrated by considering the case of a waveguide E-plane stub-loaded phase shifter circuit [3.12] shown in Fig. 3.6(a). This circuit consists of a rectangular waveguide section loaded with four E-plane stubs. The generalized equivalent circuit of this fixed phase shifter is shown in Fig. 3.6(b). The structure is regarded as a cascade of eight waveguide discontinuities of alternating E-plane enlargement and reduction type. Analysis of this structure by the generalized scattering method (GSM) consists of associating a generalized multiport network to each discontinuity. The resulting equivalent network (Fig. (3.6(b)) is then analyzed by applying the S-parameter segmentation formula (Eq. (3.22)) to two adjacent multiport networks at a time.

An alternative approach for segmenting the structure in Fig. 3.6(a) is to view it as being comprised of four cells or cavities (regions 1, 3, 5, and 7) connected by three waveguide sections. This is obtained by grouping each enlargement-type step with the successive reduction-type step. Each cell then constitutes a microwave circuit that can be represented by a multiport network. The overall equipment circuit is shown in Fig. 3.6(c). This segmentation procedure has been called the cellular segmentation [3.1] and is computationally more efficient than the multiport network model shown in Fig. 3.6(b).

Another more efficient approach for segmenting this structure is called transverse segmentation. This approach is shown in Fig. 3.7(a) and is based on segmenting the geometry into five cells, where the segment 1 shown has a uniform height. Segments 2, 3, 4, and 5 are connected to the segment 1 at a large number of interconnecting ports. These ports correspond to the various higher-order modes present at these interfaces. Again, we formulate an equivalent multiport network for each cell and use segmentation formulas for interconnecting the various cells together. The schematic of an equivalent multiport network for segment 1 is shown in Fig. 3.7(b). At the external ports i and 0, only the two-terminals (single port) model is used. These ports are taken to be away from discontinuities so that all the higher-order modes decay out and only the dominant waveguide mode is present.

3.3.2 Y-Matrix Representation of Various Segments

It has been shown [3.12] that multiport networks in Figs. 3.6(c) and 3.7(a) can be analyzed efficiently when a generalized admittance matrix representation is used for each of the waveguide segments. The formulation of generalized admittance matrices is discussed in detail in [3.12]. An overview of this work is included in this section.

Consider an arbitrary microwave circuit R_0 shown in Fig. 3.8. This circuit is enclosed by a perfectly conducting surface S except at the location of N_p physical external ports that extend over the surfaces S_i. The EM-field in R_0 is determined uniquely by the knowledge of the tangential electric field over the surfaces S_i. Following the treatment of [3.14], the magnetic field in R_0 can be expressed as:

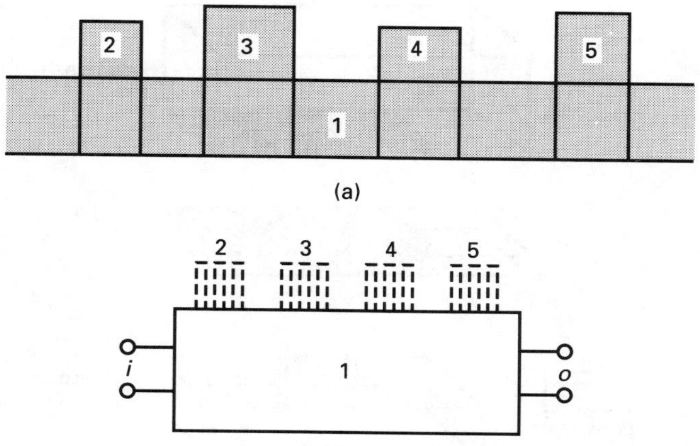

Fig. 3.7 (a) More efficient segmentation of the structure shown in Fig. 3.6(a); (b) multiport network modeling of segment 1 in Fig. 3.7(a).

$$\mathbf{H}(\mathrm{r}) = j\omega\varepsilon \oint_S \mathbf{n} \times \mathbf{E}(\mathrm{r}') \cdot \underline{\underline{\mathbf{G}}}(\mathrm{r}, \mathrm{r}')dS' \qquad (3.55)$$

$$= j\omega\varepsilon \sum_{i=1}^{N_p} \int_{S_i} \mathbf{n} \times \mathbf{E}(\mathrm{r}') \cdot \underline{\underline{\mathbf{G}}}(\mathrm{r}, \mathrm{r}')dS'$$

where the integral over S has been reduced to the portions of S_i where n $\times$ **E** is not zero. In Eq. (3.55), the quantity n $\times$ **E** can be interpreted as a magnetic surface current flowing on S_i. The quantity $\underline{\underline{\mathbf{G}}}$ in (3.55) represents the dyadic Green's function for the magnetic field in R_0, which satisfies

$$\nabla \times \nabla \times \underline{\underline{\mathbf{G}}} - k^2\underline{\underline{\mathbf{G}}} = \underline{\underline{I}}\delta(r - r') \qquad \text{in} \quad R_0 \qquad (3.56)$$

and

$$\mathrm{n} \times \nabla \times \underline{\underline{\mathbf{G}}} = 0 \qquad \text{on} \quad S \qquad (3.57)$$

where $\underline{\underline{I}}$ is the unit dyadic, δ is the Dirac delta function, and k is given by $k^2 = \omega^2\mu\varepsilon$.

Upon expanding the tangential E- and H-fields at the surface S_i into a suitable set of vector basis functions, we obtain

$$\mathrm{n} \times \mathbf{E}^{(i)} = \sum_{k=1}^{N_i} \mathbf{V}_k^{(i)}\phi_k^{(i)} \qquad (3.58)$$

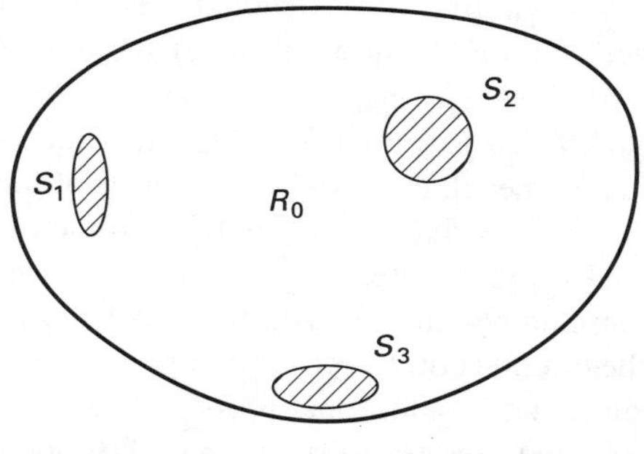

Fig. 3.8 A general configuration considered for admittance matrix formulation (from [3.12]).

$$\mathbf{H}^{(i)} = \sum_{k=1}^{N_i} \mathbf{I}_k^{(i)}\phi_k^{(i)} \qquad (3.59)$$

where N_i is the number of basis functions taken into account on S_i. The two sets of vector basis functions selected in (3.58) and (3.59) need not to be the same. This choice, however, leads to a Galerkin-type formulation that has variational properties. The basis functions $\phi_k^{(i)}$'s are selected such that they are complete and orthonormal. They also should be the modal eigenvectors of a waveguide that has S_i as its cross section. The series in (3.58) and (3.59) are truncated to N_i terms so that summations can be numerically calculated.

The expansion coefficients $\mathbf{V}_k^{(i)}$ and $\mathbf{I}_k^{(i)}$ in (3.58) and (3.59) represent, respectively, the equivalent voltage and the equivalent current on the kth electrical port of the surface S_i. Using the orthogonal properties of the selected basis functions $\phi_k^{(i)}$, the expansion coefficients $V_k^{(i)}$ and $I_k^{(i)}$ can be related to the respective field quantities on S_i by

$$\mathbf{V}_k^{(i)} = \int_{S_i} \mathbf{n} \times \mathbf{E}(\mathrm{r}) \cdot \phi_k^{(i)}(\mathrm{r})\, dr \qquad (3.60)$$

$$\mathbf{I}_k^{(i)} = \int_{S_i} \mathbf{H}(\mathrm{r}) \cdot \phi_k^{(i)}(\mathrm{r})\, dr \qquad (3.61)$$

Upon combining (3.61) with (3.55) and (3.58) we obtain

$$\mathbf{I}_m^{(j)} = \sum_{i=l}^{N_p} \sum_{k=l}^{N_i} \mathbf{Y}_{mk}^{(ji)}\mathbf{V}_k^{(i)} \qquad (3.62)$$

where $Y_{mk}^{(ji)}$ is the generalized admittance matrix of R_0 and is defined by

$$\mathbf{Y}_{mk}^{(ji)} = \int_{S_j}\int_{S_i} \phi_m^{(j)}(\mathrm{r}) \cdot \underline{\underline{\mathbf{G}}}(\mathrm{r}, \mathrm{r}') \cdot \phi_k^{(i)}(\mathrm{r}')\, dr\, dr' \qquad (3.63)$$

This quantity represents the amplitude of the mth current component driving the jth opening when a unit kth component of the voltage is applied at the ith opening, with all the other voltage components being zero; that is, all other openings are short circuited. Green's functions for rectangular cells may be obtained either in terms of the cavity resonant modes or in terms of the waveguide modes [3.12]. The latter choice yields expressions with only a single series over m or n.

3.3.3 Application to Full-Wave Modeling of Via-Hole Grounds in Microstrip

The segmentation approach for mode matching has also been recently applied to the analysis of via holes in microstrip lines [3.2]. The cylindrical via structure is enclosed in a metallic box of suitable size (Fig. 3.9(a)), such that it does not perturb the reactive fields in the proximity of the via discontinuity. To reduce the complexity of the problem, the via shape is treated as a cylinder of rectangular cross section. Taking advantage of the symmetry, only one quarter of the structure (as shown in Fig. 3.9(b)) is analyzed. The analysis is carried out by

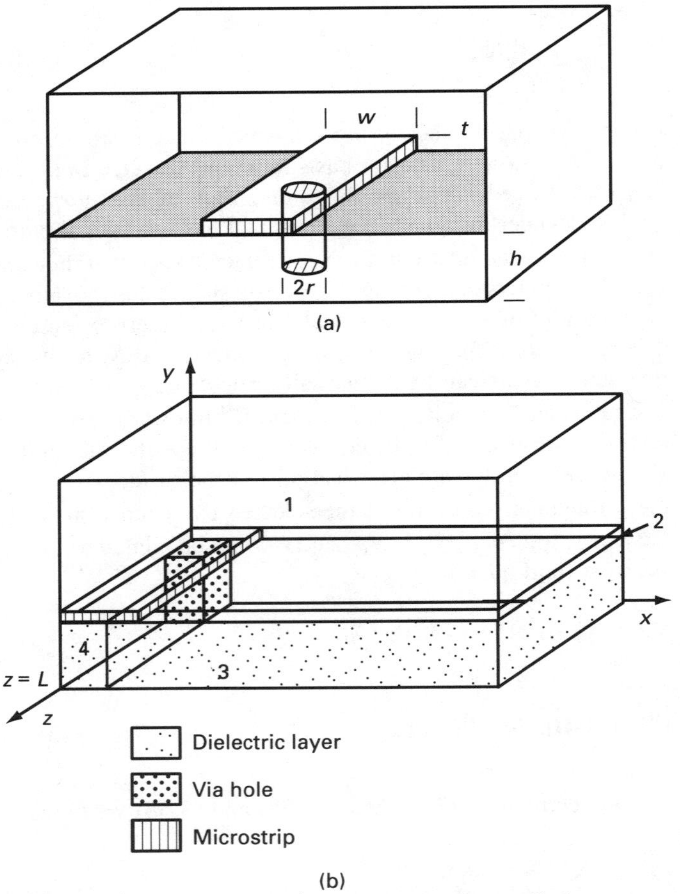

Fig. 3.9 (a) A cylindrical microstrip via enclosed in a metallic box; (b) quarter of the structure used for analysis.

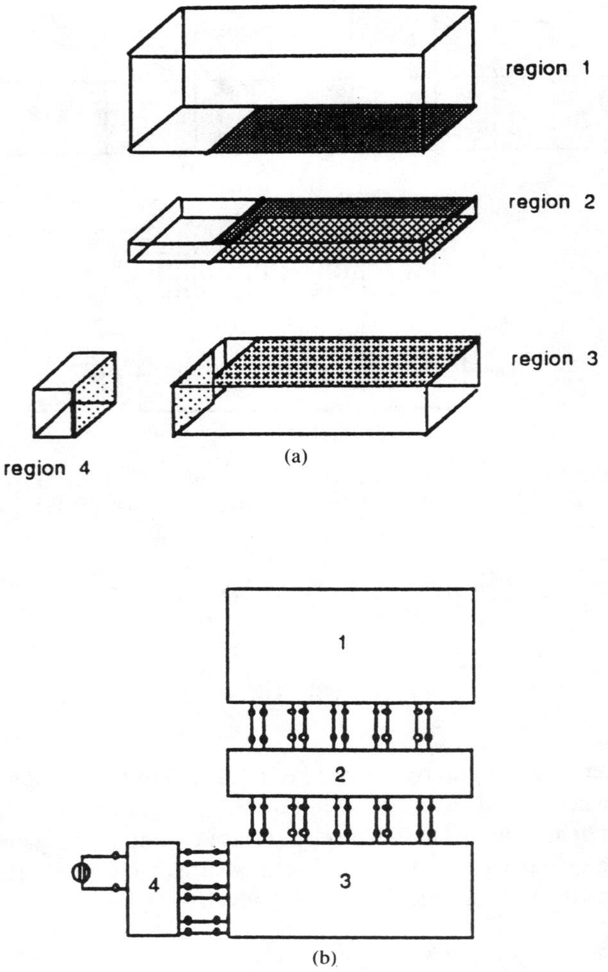

Fig. 3.10 (a) Four segments of the via-hole structure in Fig. 3.9(b); (b) equivalent multiport network model of the via-hole structure in Fig. 3.9(b).

using even-odd excitation, thus placing a magnetic-electric wall at $z = 0$. For symmetry reasons, the wall at $x = 0$ is a magnetic one.

The structure of Fig. 3.9(b) is segmented into four regions shown in Fig. 3.10(a). Region 1 corresponds to the region above the strip. Region 2 is the volume of the thickness of the strip. Note that the finite thickness of the metal is thus taken into account. Regions 3 and 4 are in the dielectric layer. Each of these regions is characterized by a generalized admittance matrix as discussed in Sec. 3.3.2. An equivalent multiport network representation is shown in Fig. 3.10(b). The excitation of the field in the structure is accomplished by an electric field distribution impressed on the front side ($z = L$) of region 4 (corresponding to the generator in Fig. 3.10(b)). The scattering parameters of the microstrip via-hole discontinuity are then readily computed by evaluating the E-field region along z in the central region. Results obtained by this segmentation approach are shown to be in good agreement with experimental measurements [3.2].

3.4 Desegmentation Procedure

The desegmentation method is a complementary technique used for two-dimensional components and microstrip antennas. This method is applicable in cases where the addition of one or more *simple* shapes to the shape to be analyzed yields another *simple* shape. A few examples of this type are shown in Fig. 3.11. As in the case of segmentation method, desegmentation method also involves replacement of continuous connections between various shapes by the multiple number of ports along the interconnection. To determine the characteristics of the given circuit using this procedure, it is necessary that the shape to be added has ports other than those along the common boundary.

Let α, as shown in Fig. 3.11, denote the shape of the circuit to be analyzed. On the addition of one or more *simple* shapes (denoted by β), a *simple* shape γ is obtained. The characteristics of β and γ can be obtained using the Green's function method. In the nomenclature used for the ports of the circuit to be analyzed, the ports that are not on the common boundaries between α and β are denoted as *p*-ports. The ports of α that are on the common boundaries with β are called *q*-ports. If the *q*-ports on the circuit α do not cover completely all the boundaries between α and β, additional ports are added to α so that the common boundaries are fully covered. The ports on β that are connected to *q*-ports are called *r*-ports. It should be noted

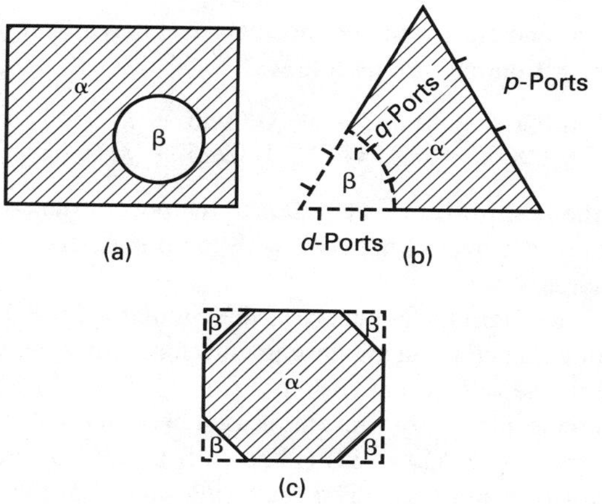

Fig. 3.11 Examples of planar circuits suitable for analysis by desegmentation method.

that higher accuracy can be obtained by increasing the number of ports on the common boundary. The ports on β that are not on the common boundary are called *d*-ports. As explained later, there must be at least as many *d*-ports on each of the segments in β as the number of *r*-ports. The *d*-ports need not be on the periphery of the segment, and can be located inside the periphery, as well. The desegmentation method enables us to obtain the characteristics of the α-shaped circuit from the characteristics of β and γ circuits. This can be done in terms of either S-matrices or Z-matrices.

3.4.1 Desegmentation Using S-Matrices [3.15]

The governing equations for the ports on β and γ can be written as

$$\begin{bmatrix} b_r \\ b_d \end{bmatrix} = \begin{bmatrix} S_{rr} & S_{rd} \\ S_{dr} & S^{\beta}_{dd} \end{bmatrix} \begin{bmatrix} a_r \\ a_d \end{bmatrix} \tag{3.64}$$

and

$$\begin{bmatrix} b_p \\ b_d \end{bmatrix} = \begin{bmatrix} S^{\gamma}_{pp} & S_{pd} \\ S_{dp} & S^{\gamma}_{dd} \end{bmatrix} \begin{bmatrix} a_p \\ a_d \end{bmatrix} \tag{3.65}$$

where the b's and a's denote the outgoing and incoming wave variables, respectively, S^{γ} is the scattering matrix of γ, and S^{β} can be directly obtained from the scattering matrices of β segment(s). To analyze the circuit, we use

$$\begin{bmatrix} b_p \\ b_q \end{bmatrix} = \begin{bmatrix} S^{\alpha}_{pp} & S_{pq} \\ S_{qp} & S_{qq} \end{bmatrix} \begin{bmatrix} a_p \\ a_q \end{bmatrix} \tag{3.66}$$

where S^{α} is the unknown S-matrix of α. The ports in q and r are ordered in the same way as for the segmentation method. Under these conditions, we have

$$a_q = b_r \tag{3.67a}$$

and

$$b_q = a_r. \tag{3.67b}$$

Using segmentation method, the S-matrix of α and those of β segments can be combined to give the S-matrix of γ. Eliminating a_q, a_r, b_q, and b_r from Eqs. (3.65)–(3.67), S^{γ} can be expressed as

$$S^{\gamma} = \begin{bmatrix} S^{\alpha}_{pp} + S_{pq} M_{11} S_{qp} & S_{pq} M_{12} S_{rd} \\ S_{dr} M_{21} S_{qp} & S^{\beta}_{dd} + S_{dr} M_{22} S_{rd} \end{bmatrix} \tag{3.68}$$

where

$$M_{12} = (I - S_{rr} S_{qq})^{-1} \tag{3.69a}$$

$$M_{21} = (I - S_{qq} S_{rr})^{-1} \tag{3.69b}$$

$$M_{11} = S_{rr} M_{21} \tag{3.69c}$$

and

$$M_{22} = S_{qq} M_{12} \tag{3.69d}$$

In the segmentation method, S^{α} is obtained from S^{β} and S^{γ} using these equations. For this purpose, the number of d-ports must be at least equal to the number of q- and r-ports. The lower-right submatrices in (3.65) and (3.68) can be equated to give

$$S^{\gamma}_{dd} = S^{\beta}_{dd} + S_{dr} S_{qq} (I - S_{rr} S_{qq})^{-1} S_{rd} \tag{3.70}$$

which results in

$$S_{qq} = (N\, S_{rr} + I)^{-1} N \tag{3.71}$$

where

$$N = (S^{t}_{dr}\, S_{dr})^{-1}\, S^{t}_{dr} (S^{\gamma}_{dd} - S^{\beta}_{dd}) S^{t}_{rd} (S_{rd}\, S^{t}_{rd})^{-1} \tag{3.72}$$

The other submatrices S^{α} can be expressed as

$$S^{\alpha} = \begin{bmatrix} S^{\gamma}_{pp} - S_{pq} M_{11} S_{qp} & S_{pd} S^{t}_{rd} (S_{rd} S^{t}_{rd})^{-1} (I - S_{rr} S_{qq}) \\ (I - S_{qq} S_{rr})(S^{t}_{dr} S_{dr})^{-1} S^{t}_{dr} S_{dp} & (N\, S_{rr} + I)^{-1} N \end{bmatrix} \tag{3.73}$$

To evaluate S^{α} with this procedure, first the submatrix S_{qq} is computed using (3.71); then S_{pq} and S_{qp} are evaluated using (3.73). Submatrix M_{11} is evaluated using (3.69c) and finally S^{α}_{pp} is obtained using (3.73).

In the case where $d = q$, the matrix N in (3.72) reduces to

$$N = S^{-1}_{dr} (S^{\gamma}_{dd} - S^{\beta}_{dd}) S^{-1}_{rd} \tag{3.74}$$

and the expression for S^{α} reduces to

$$S^{\alpha} = \begin{bmatrix} S^{\gamma}_{pp} - S_{pq} M_{11} S_{qp} & S_{pd} S^{-1}_{rd} (I - S_{rr} S_{qq}) \\ (I - S_{qq} S_{rr}) S^{-1}_{dr} S_{dp} & (N\, S_{rr} + I)^{-1} N \end{bmatrix} \tag{3.75}$$

where M_{11} is obtained from (3.69c) and the submatrices in S^{α} are evaluated in the same order as for $d > q$.

The scattering matrix S^{α} corresponds to the segment α on which additional ports at the boundaries between α and β might have been added. Let the ports of α be divided into o-ports, for which characterization is desired, and e-ports, which were added for desegmentation. The incoming and outgoing wave variables on the segment α are related as

$$\begin{bmatrix} b_o \\ b_e \end{bmatrix} = \begin{bmatrix} S^{\alpha}_{oo} & S^{\alpha}_{oe} \\ S^{\alpha}_{eo} & S^{\alpha}_{ee} \end{bmatrix} \begin{bmatrix} a_o \\ a_e \end{bmatrix} \tag{3.76}$$

where S^{α}_{oo}, S^{α}_{oe}, S^{α}_{eo}, and S^{α}_{ee} are obtained by reordering the rows and columns on S^{α} and a_o, b_o and a_e, b_e are the wave variables at the o-ports and e-ports, respectively. Characterization is needed only for the o-ports, and thus the open circuit boundary condition can be applied to e-ports, giving

$$a_e = b_e \tag{3.77}$$

From (3.76) and (3.77), we obtain

$$b_0 = [S^{\alpha}_{oo} + S^{\alpha}_{oe}(I - S^{\alpha}_{ee})^{-1} S^{\alpha}_{eo}] a_o \tag{3.78}$$

or the desired scattered matrix S_p can be expressed as

$$S_p = S^{\alpha}_{oo} + S^{\alpha}_{oe}(I - S^{\alpha}_{ee})^{-1} S^{\alpha}_{eo} \tag{3.79}$$

In the foregoing relations, I denotes an identity matrix of appropriate size.

The desegmentation method presented here can be used as a method for deembedding of multiport networks. The following example illustrates this formulation applied to deembedding a three-port network.

Example [3.15]

Consider the case of a three-port circulator with identical connectors at each port as shown in Fig. 3.12. The S-parameters for the overall network (including the connectors) are measured and are obtained as

$$S^{\gamma} = \begin{bmatrix} 0.0148 + j0.0493 & -0.134 + j0.0624 & -0.897 + j0.185 \\ -0.897 + j0.187 & 0.0148 + j0.0493 & -0.134 + j0.0624 \\ -0.134 + j0.0624 & -0.897 + j0.187 & 0.0148 + j0.0493 \end{bmatrix} \tag{3.80}$$

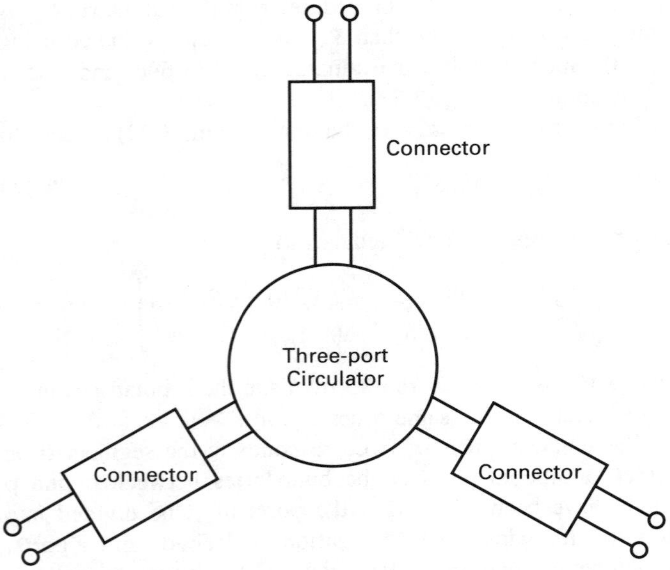

Fig. 3.12 A three-port circulator with connectors considered as an example for deembedding.

The S-parameters for the connectors are known from a separate calibration and are as follows:

$$S = \begin{bmatrix} 0.0568 - j0.0374 & -j0.0966 - j0.9544 \\ -0.0966 - j0.9544 & 0.0245 + j0.0364 \end{bmatrix} \tag{3.81}$$

where the first row in (3.81) refers to the port of the connector external to the overall network with the other port connected to the circulator.

One can obtain the S-matrix of the circulator (excluding the effect of connectors) using desegmentation with γ-network referring to the overall network and β-network referring to the three connectors only. In this case, we have $p = 0$ and $q = r = d = 3$. $S^{\gamma}_{dd} (= S^{\gamma})$ is known from (3.80) and S^{β} can be obtained from (3.81). The scattering matrix of the circulator $S^{\alpha} (= S^{\alpha}_{qq})$ is obtained using (3.71) as

$$S^{\alpha} = \begin{bmatrix} 0 & 0.1 & 0.995 \\ 0.995 & 0 & 0.1 \\ 0.1 & 0.995 & 0 \end{bmatrix} \tag{3.82}$$

3.4.2 Desegmentation Using Z-Matrices [3.3]

For this approach, the governing equations are in terms of the impedance matrices that characterize β and γ. These relations can be expressed as

$$\begin{bmatrix} v_r \\ v_d \end{bmatrix} = \begin{bmatrix} Z_{rr} & Z_{rd} \\ Z_{dr} & Z^{\beta}_{dd} \end{bmatrix} \begin{bmatrix} i_r \\ i_d \end{bmatrix} \tag{3.83}$$

and

$$\begin{bmatrix} v_p \\ v_d \end{bmatrix} = \begin{bmatrix} Z^{\gamma}_{pp} & Z_{pd} \\ Z_{dp} & Z^{\gamma}_{dd} \end{bmatrix} \begin{bmatrix} i_p \\ i_d \end{bmatrix} \tag{3.84}$$

where v's and i's are the voltage and current variables, Z^{γ} is the impedance matrix of γ, and Z^{β} can be directly obtained from the impedance matrices of the β segments. For the circuit to be analyzed, we have

$$\begin{bmatrix} v_p \\ v_q \end{bmatrix} = \begin{bmatrix} Z^{\alpha}_{pp} & Z_{pq} \\ Z_{qp} & Z_{qq} \end{bmatrix} \begin{bmatrix} i_p \\ i_q \end{bmatrix} \tag{3.85}$$

where Z^{α} is to be determined from Z^{β} and Z^{γ}. The interconnection constraints resulting from connecting the q-ports to the corresponding r-ports can be written as

$$v_q = v_r \tag{3.86a}$$

and

$$i_q + i_r = 0 \tag{3.86b}$$

Using the segmentation procedure, Z^{γ} can be expressed in terms of Z^{α} and Z^{β} as

$$Z^{\gamma} = \begin{bmatrix} Z^{\alpha}_{pp} - Z_{pq}(Z_{qq} + Z_{rr})^{-1} Z_{qp} & Z_{pq}(Z_{qq} + Z_{rr})^{-1} Z_{rd} \\ Z_{dr}(Z_{qq} + Z_{rr})^{-1} Z_{qp} & Z^{\beta}_{dd} - Z_{dr}(Z_{qq} + Z_{rr})^{-1} Z_{rd} \end{bmatrix}. \tag{3.87}$$

Carrying out steps similar to those followed in the S-matrix version of the desegmentation method, we obtain

$$Z_{qq} = -Z_{rr} - N \tag{3.88}$$

where

$$N = Z_{rd}Z^t_{rd}[Z^t_{dr}\,(Z^\gamma_{dd} - Z^\beta_{dd})\,Z^t_{rd}]^{-1}\,Z^t_{dr}Z_{dr} \tag{3.89}$$

Other submatrices in Z^α are

$$Z_{pq} = -Z_{pd}\,Z^t_{rd}(Z_{rd}\,Z^t_{rd})^{-1}\,N \tag{3.90}$$

$$Z_{qp} = -N(Z^t_{dr}\,Z_{dr})^{-1}\,Z^t_{dr}\,Z_{dp} \tag{3.91}$$

and

$$Z^\alpha_{pp} = Z^\gamma_{pp} - Z_{pq}\,N^{-1}\,Z_{qp} \tag{3.92}$$

For the case in which $d = q$, the expressions for Z^α reduce to

$$N = Z_{rd}(Z^\gamma_{dd} - Z^\beta_{dd})^{-1}\,Z_{dr} \tag{3.93}$$

$$Z_{qq} = -Z_{rr} - N \tag{3.94}$$

$$Z_{pq} = -Z_{pd}(Z^\gamma_{dd} - Z^\beta_{dd})^{-1}\,Z_{dr} \tag{3.95}$$

$$Z_{qp} = -Z_{rd}(Z^\gamma_{dd} - Z^\beta_{dd})^{-1}\,Z_{dp} \tag{3.96}$$

and

$$Z^\alpha_{pp} = Z^\gamma_{pp} - Z_{pd}(Z^\gamma_{dd} - Z^\beta_{dd})^{-1}\,Z_{dp} \tag{3.97}$$

The impedance matrix Z^α corresponds to the segment α on which additional ports at the boundaries between α and β might have been added. One can apply the open circuit condition on the added ports and delete from Z^α the rows and columns that correspond to these ports. The reduced impedance matrix corresponds to the segment α with the desired ports only.

Example

This example demonstrates how the Z-matrix version of the desegmentation method is used to obtain the Z-matrix of a resistive subnetwork α. The Z-matrices of the overall network and another subnetwork β shown in Fig. 3.13 are given.

The Z-matrices of the γ-network and the β subnetwork are given by

$$Z^\gamma = \begin{array}{c} \\ p \\ d \end{array}\!\!\begin{array}{cc} p & d \\ \multicolumn{2}{c}{\begin{pmatrix} 3 & 1 \\ 1 & 6 \end{pmatrix}} \end{array} \tag{3.98}$$

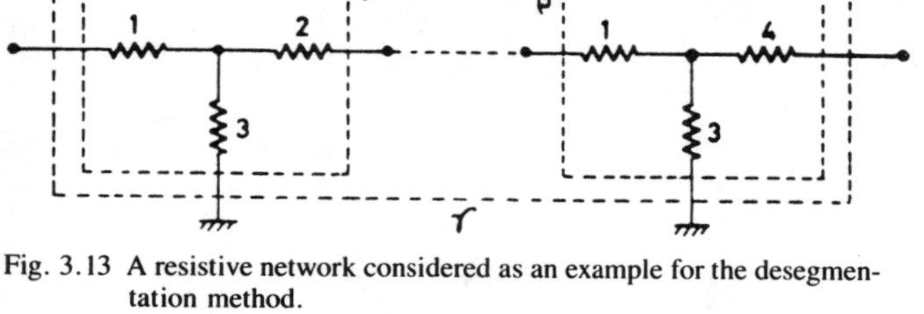

Fig. 3.13 A resistive network considered as an example for the desegmentation method.

and

$$Z^\beta = \begin{array}{c} \\ r \\ d \end{array}\!\!\begin{array}{cc} r & d \\ \multicolumn{2}{c}{\begin{pmatrix} 4 & 3 \\ 3 & 7 \end{pmatrix}} \end{array} \tag{3.99}$$

The matrix Z^α can be obtained by substituting (3.98) and (3.99) into (3.93) through (3.97) as follows.

$$N = 3(6 - 7)^{-1}\,3 = -9$$

$$Z_{qq} = -4 - (-9) = 5$$

$$Z_{pq} = -1(6 - 7)^{-1}\,3 = 3$$

$$Z_{qp} = -3(6 - 7)^{-1}\,1 = 3$$

and

$$Z_{pp} = 3 - 1(6 - 7)^{-1}\,1 = 4$$

The matrix Z^α can thus be written as

$$Z^\alpha = \begin{bmatrix} 4 & 3 \\ 3 & 5 \end{bmatrix} \tag{3.100}$$

which is seen to be the Z-matrix for the α-subnetwork.

Reference [3.4] describes an alternative formulation for the desegmentation method when Z-matrices are used. This approach becomes computationally more efficient when some of the external ports of the α-segment are located at the interface between α- and β-segments (see Fig. 3.11). In this case, the computations of Z^α_{pp} from (3.97) are not sufficient, and one needs to evaluate other submatrices of Z^α, namely, Z_{qq} from (3.94) and Z_{pq} from (3.95), also. The method given in [3.4] simplifies the computations by considering the external ports at the common interface (between α and β) to be distinct from the interconnected q- and r-ports. This calls for artificially adding q-ports even at the location where external p-port(s) of α-segment may be present. As shown in [3.4], the overall computational effort is reduced considerably.

The segmentation and desegmentation methods discussed in this chapter provide efficient tools for the analysis and design of planar circuits.

References

[3.1] Okoshi, T., and T. Takeuchi. "Analysis of planar circuits by segmentation method." *Electronics and Communication in Japan*, Vol. 58-B, pp. 71–79, August 1975.

[3.2] Alessandri, F., M. Mongiardo, and R. Sorrentino. "Full-wave modeling of via hole grounds in microstrip by three-dimensional mode matching technique." *1992 IEEE MTT-S Symp. Dig.*, pp. 1237–1240.

[3.3] Sharma, P. C., and K. C. Gupta. "Desegmentation method for analysis of two-dimensional planar microwave circuits." *IEEE Trans. Microwave Theory Tech.*, Vol. MTT-29, pp. 1094–1098, October 1981.

[3.4] Sharma, P. C. and K. C. Gupta. "An alternative procedure for implementing desegmentation method." *IEEE Trans. Microwave Theory Tech.*, Vol. MTT-32, pp. 1–4, January 1983.

[3.5] Okoshi, T. and T. Miyoshi. "The planar circuit—An approach to microwave integrated circuitry." *IEEE Trans. Microwave Theory Tech.*, Vol. MTT-20, pp. 245–252, April 1972.

[3.6] Gupta, K. C., R. Garg, and R. Chadha, *Computer-Aided Design of Microwave Circuits*. Dedham, MA: Artech House, 1981.

[3.7] Chadha, R., and K. C. Gupta. "Segmentation method using impedance matrices for analysis of planar microwave circuits." *IEEE Trans. Microwave Theory Tech.*, Vol. MTT-29, pp. 71–74, January 1981.

[3.8] Prothe, A. and K. C. Gupta. "Modeling and sensitivity analysis of parasitic coupling in microstrip circuits." MIMICAD Tech. Rept. No. 10, University of Colorado, Dept. of Electr. and Comptr. Eng., August 1991.

[3.9] Sabban, A., and K. C. Gupta. "Multiport network method for evaluating radiation loss and spurious coupling between discontinuities in microstrip circuits." *1989 IEEE MTT-S Int'l. Microwave Symp. Dig.*, Vol. II, pp. 707–710, June 1989.

[3.10] Gupta, K. C., and M. D. Abouzahra. "Planar circuit analysis." In T. Itoh, ed., *Numerical Techniques for Microwave and Millimeter-Wave Passive Structures*, Chapter 4, pp. 214–333. New York: John Wiley & Sons, 1989.

[3.11] Alessandri, F., M. Mongiardo, and R. Sorrentino. "Transverse segmentation: A novel technique for the efficient CAD of 2N-port branch-line couplers." *IEEE Microwave Guided Wave Letters*, Vol. 1, No. 8, pp. 204–207, August 1991.

[3.12] Alessandri, F., M. Mongiardo, and R. Sorrentino. "A technique for the fullwave automatic synthesis of waveguide components: Application to fixed phase shifters." *IEEE Trans. Microwave Theory Tech.*, Vol. MTT-40, No. 7, pp. 1484–1495, July 1992.

[3.13] Itoh, T. "Generalized scattering matrix technique." In T. Itoh, ed., *Numerical Techniques for Microwave and Millimeter-Wave Passive Structures*, Chapter 10, pp. 622–636. New York: John Wiley & Sons, 1989.

[3.14] Collin, R. E. *Field Theory of Guided Waves*, 2nd ed., Section 2.18. New York: IEEE Press, 1991.

[3.15] Sharma, P. C. and K. C. Gupta. "A generalized method for deembedding of multi-port networks." *IEEE Trans. Instrumentation and Measurements*, Vol. IM-30, pp. 305–307, 1981.

Chapter 4

Boundary-Element or Contour-Integral Method

THE Green's function approach discussed in Chapter 2 and the segmentation and desegmentation techniques described in Chapter 3 represent planar circuit analysis techniques whose applications are limited to components of regular shapes and composites of regular shapes. For components of arbitrary shapes, more general techniques such as that presented in this chapter are needed. This approach has been known to different groups by different names: contour-integral approach, boundary-integral method, and boundary-element method. The first two names are synonyms since they imply an integral equation formulation along a contour (or boundary) in the case of two-dimensional problems and along the enclosing boundary surface for three-dimensional problems. Often, the integral is handled numerically by dividing the boundary into small sections. The name boundary-element method emphasizes the discretization of the boundary into elements, and in that sense, the method is a special case of the finite-element method [4.1] with "elements" located only on the boundary.

Although the boundary-integral or boundary-element method is a rather well-established numerical technique in engineering [4.2, 4.3], it has not been widely used in the analysis of microwave components. Perhaps the most widespread applications have been reported by Okoshi and his colleagues [4.4] under the name of contour-integral method. This method has been reviewed at other places [4.5–4.7] also. The applications of this method to three-dimensional problems have been relatively few (for example, see [4.8 and 4.9]).

This chapter is a collection of six reprints on the contour-integral or boundary-element method. These contributions are arranged chronologically. In Paper 4.1, the contour-integral representation of the two-dimensional wave equation is formulated for arbitrarily shaped shorted-boundary circuits. The height of the component need not be small compared to the wavelength as required by the definition of the planar circuit. The analysis reported in this paper is applicable to TE_{10}-mode waveguide circuitry (provided no transverse E-field is present). Results for several waveguide discontinuities (thick inductive window, corners, and posts) are compared with experimental results available in literature. Thin waveguide circuits have also been dealt with by Grüner [4.10], who has employed a conformal mapping synthesis procedure for planar circuits having an impedance matrix with prescribed poles.

The synthesis of open-boundary arbitrarily shaped planar circuits is discussed in Paper 4.2, which describes an iterative synthesis procedure for planar circuits having an impedance matrix with prescribed poles and residues. Geometry synthesis has been implemented in two different ways. In the first implementation, the circuit pattern is represented by a number of sectors of different angles and radii. In the second way, the circuit pattern parameters are based on conformal mapping by a regular function. Examples of different synthesis processes are included. When the number of the prescribed eigenvalues and external ports is relatively small, results produced by this method are reported to be satisfactory both in computing time and accuracy.

The third paper (Paper 4.3) presents a general formulation of the boundary-element method and its application to two-dimensional electromagnetic field problems. It is emphasized that the boundary-element method is a boundary method, and therefore, if the region to be analyzed is homogeneous, then it requires nodes, necessary for analysis, on its boundary only. Thus the problem can be treated with one less dimension. Specialized formulations for waveguides with discontinuities, multimedia problems, and electromagnetic scattering problems are presented. Examples include open-ended parallel-plane waveguides, a symmetrical H-plane y-junction, a waveguide with a dielectric cylinder placed at its open end, and electromagnetic scattering from a dielectric circular cylinder.

A proposal for a boundary-integral method without using Green's function is the topic of Paper 4.4 by Kishi and Okoshi, in which the authors describe a modification of the conventional boundary-element method (BEM) by using the reciprocity theorem derived from Green's identity, thus making the use of Green's function unnecessary. This modified method has been applied to both Dirichlet- and Neumann-type boundary-value problems. Numerical results comparing this technique with the conventional BEM and variational methods are presented. One important advantage of the method is the absence of the singular point, as Green's function is not used.

Paper 4.5 describes an algorithm applicable to the analysis of planar circuits with only part of the boundary being regularly shaped. When this condition is met, the circuit can then be enclosed in rectangular or circular resonators. Consequently, the electric and magnetic fields are derived from the Green's functions of this resonator by integrating over the periphery of the circuit not coinciding with the regular shape. In a sense, the approach is a hybrid of the Green's function approach discussed in Chapter 2 and the contour-integral method of this chapter.

The last reprint (Paper 4.6) in this chapter presents an application of the contour-integral method for developing a general procedure for design and analysis of two-dimensional Rotman-type beam-forming lenses. In this paper, the geometry under consideration is divided into a central lens region and tapered regions connecting the lens region to input and output microstrip lines. Z-matrix representations of these regions are computed separately, and the segmentation method discussed in Chapter 3 is then used for evaluating the overall characterization of the lens. Thus, this paper presents an example of what could be called segmented boundary-element method (SBEM) wherein various segments of a given configuration are characterized separately (using boundary-element method) and then connected together using multiport network-type formulation.

Because of the obvious computational advantages of the boundary-element method over the finite-element method (with volume elements), it is expected that BEM will find increasing applications in the area of microwave and millimeter-wave planar circuits. Some other recent applications are reported in [4.11–4.13].

References

[4.1] Davies, J. B. "The Finite Element Method." in T. Itoh, ed., *Numerical Techniques for Microwave and Millimeter-Wave Passive Structures,* Chapter 2, p. 86. New York: John Wiley & Sons, 1989.

[4.2] Brebbia, C. A. and S. Walker. *Boundary Element Techniques in Engineering.* London: Butterworth, 1980.

[4.3] Brebbia, C. A. *The Boundary Element Method for Engineers.* London: Pentech Press, 1978.

[4.4] Okoshi, T. *Planar Circuits for Microwaves and Lightwaves.* Berlin: Springer-Verlag, 1985. See Chapter 2 on "Analysis of Planar Circuits Having Arbitrary Shapes" and Chapter 3 on "Short-Boundary Planar Circuits."

[4.5] Gupta, K. C. et al. *Computer-Aided Design of Microwave Circuits.* Dedham, MA: Artech House, 1981. See Chapter 8, Section 8.5, on "Numerical Method for Arbitrary Shapes," pp. 253–256.

[4.6] Gupta, K. C. and M. Abouzahra. "Contour Integral Approach." In T. Itoh, ed., *Numerical Techniques for Microwave and Millimeter-Wave Passive Structures.* New York: John Wiley & Sons, 1989. See Chapter 4, Section 5, pp. 247–255.

[4.7] Morita, N. "The BEM." In E. Yamashita, ed., *Analysis Methods for Electromagnetic Wave Problems,* Chapter 2. Dedham, MA: Artech House, 1990.

[4.8] Kosha, M. and Suzuki, M. "Application of the boundary-element method to waveguide discontinuities." *IEEE Trans. Microwave Theory Tech.,* Vol. MTT-34, No. 2, pp. 301–307, February 1986.

[4.9] Li, K. and Y. Fujii. "Indirect boundary element method applied to generalized microstripline analysis with application to side proximity effect in MMIC's." *IEEE Trans. Microwave Theory Tech.,* Vol. MTT-40, pp. 237–244, February 1992.

[4.10] Grüner, K. "Method of synthesizing non-uniform waveguides." *IEEE Trans. Microwave Theory Tech.,* Vol. MTT-22, pp. 317–322, March 1974.

[4.11] Tonye, E. and H. Baudrand. "Multimode S-parameters of planar multiport junctions by BEM." *Electronic Letters,* Vol. 20, No. 19, pp. 799–802, September 13, 1984.

[4.12] Zhu, L. and E. Yamashita. "New method for the analysis of dispersion characteristics of various planar transmission lines with finite metallization thickness." *IEEE Microwave and Guided Wave Letters,* Vol. 1, No. 7, pp. 164–166, July 1991.

[4.13] Wu, K. L., C. Wu, and J. Litva. "Characterizing the microwave planar circuits using the coupled finite-boundary element method." *IEEE Trans. Microwave Theory Tech.,* Vol. MTT-40, No. 10, pp. 1963–1966, October 1992.

Paper 4.1

Computer Analysis of Short-Boundary Planar Circuits

TAKANORI OKOSHI, MEMBER, IEEE, AND SEIKO KITAZAWA

***Abstract*—A method of computer analysis of planar (two-dimensional) circuits having an arbitrarily shaped short boundary is proposed. The proposed method is based upon the contour integral representation of the two-dimensional wave equation. Results of the computer analyses for simple circuits are compared with analytical solutions to show the validity and accuracy of the proposed method. Some examples of analyses of practical circuits are also presented.**

I. INTRODUCTION

THE planar circuit is a circuit concept proposed by one of the authors in 1969. It is the two-dimensional circuit that should be positioned between the distributed-constant (one-dimensional) circuit and the waveguide (three-dimensional) circuit; it is defined as an electrical circuit having dimensions comparable to the wavelength in two directions but much less thickness in one direction. The planar circuit can be classified into three types: the triplate type, open (or asymmetric) type, and short-boundary type [1].

In the past five years, the authors and their co-workers in Japan have been concentrating principally upon the analysis and design of the triplate-type planar circuits, which have open-circuit boundaries. It was because the investigation of the triplate type seemed most urgent in connection with the development of the microwave IC technology.

This paper proposes a method of the computer analysis of an arbitrarily shaped, short-boundary planar circuit. Some examples of analyses of practical circuits are also presented.

The computer-analysis technique described in this paper enables us to know the precise characteristics of circuits such as are shown in Fig. 1(a)–(c). Moreover, in the case of Fig. 1(b), the height of the waveguide need not be small compared to the wavelength as required by the definition of the planar circuit; the present analysis can also be applied to the ordinary TE_{10}-mode waveguide circuitry provided that no transverse electric field is present. Therefore the advantages of the computer-analysis technique of the short-boundary-type planar circuit extend to the conventional waveguide technology.

II. PRINCIPLE OF ANALYSIS

The computer-analysis technique developed for the triplate type may be modified to its "dual" form for its application to the short-boundary type. The most important change stems from the fact that the coupling ports

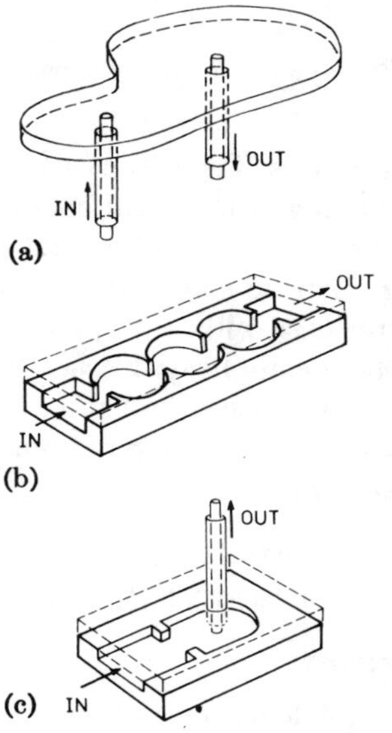

Fig. 1. Examples of the short-boundary planar circuit. (a) Coaxial-coupled type. (b) Waveguide-coupled type. (c) Mixed type.

are of entirely different form. For example, when a planar circuit is coupled to the external circuits through waveguides as shown in Fig. 1(b), a computational process is required to provide the "match" between the electromagnetic field in the planar circuit and the proper fields in the waveguides at properly selected reference planes. The coaxial ports as shown in Fig. 1(a) and (c) also require a similar computational process. In any case, the basic equation for the open-boundary planar circuit can be utilized in the earlier stage of the following analysis.

In [1], it was shown that by using Weber's solution of the two-dimensional wave equation [2], the RF voltage at a point upon the periphery of an arbitrarily shaped, homogeneous two-dimensional wave medium is given as

$$2jV(s) = \oint_c [k \cos \theta H_1^{(2)}(kr) V(s_0) - j\omega\mu d\, H_0^{(2)}(kr) i_n(s_0)]\, ds_0. \quad (1)$$

In this equation, $H_0^{(2)}$ and $H_1^{(2)}$ are the zeroth-order and first-order Hankel functions of the second kind, respectively, i_n denotes the current density flowing outwards along the periphery, s and s_0 denote the distance along the periphery C. The variable r denotes distance between points (s) and (s_0), and θ denotes the angle made by the straight line from point (s) to point (s_0) and the normal at point (s_0), as shown in Fig. 2. When i_n is given, (1) is a second-kind Fredholm integral equation with respect to the RF voltage V.

To avoid useless confusions, we restrict the following

Manuscript received May 6, 1974; revised July 25, 1974.
The authors are with the Department of Electronic Engineering, University of Tokyo, Tokyo, Japan.

Reprinted from *IEEE Trans. Microwave Theory Tech.*, vol. MTT-23, no. 3, pp. 299–306, March 1975.

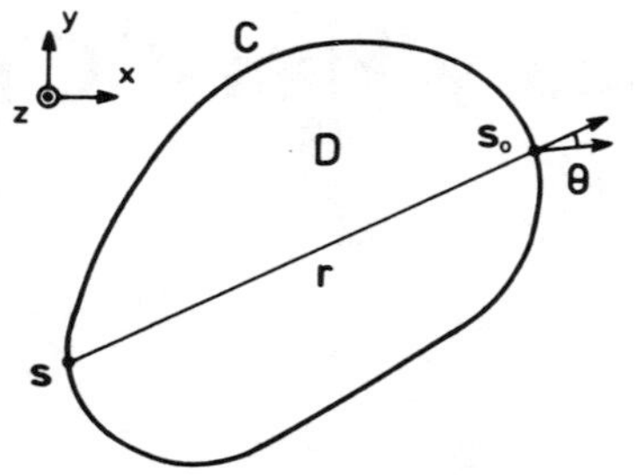

Fig. 2. Symbols used in the basic equation.

discussion to two cases. The first case is the short-boundary planar circuit having two coaxial coupling ports as shown in Fig. 1(a). The second case is that having two waveguide coupling ports as shown in Fig. 1(b). The description of the latter case will be emphasized for its practical importance. More complicated cases such as that shown in Fig. 1(c) and those circuits having three or more ports will be dealt with by modifying or combining the analyses for the previous two cases.

III. SHORT-BOUNDARY PLANAR CIRCUIT HAVING TWO COAXIAL COUPLING PORTS

A. Basic Equation

For numerical calculation we divide the periphery of a circuit as shown in Fig. 1(a) into M incremental sections, and provide M sampling points at the center of each section as shown in Fig. 3. We assume that current flows uniformly in each section. The peripheries of the coupling conductors are also divided into m and n incremental sections, and sampling points are provided. Those $m + n + M$ sampling points are numbered as follows:

1) conductor 1: $i = 1 \sim m$;
2) conductor 2: $i = (m + 1) \sim (m + n)$;
3) circuit periphery: $i = (m + n + 1) \sim (m + n + M)$.

It was shown in [1] that if we rewrite (1) into an incremental form, we obtain a matrix equation

$$2j\begin{bmatrix} V_1 \\ \cdot \\ \cdot \\ \cdot \\ V_N \end{bmatrix} = k[G_{ij}]\begin{bmatrix} V_1 \\ \cdot \\ \cdot \\ \cdot \\ V_N \end{bmatrix} + j\omega\mu\, d[F_{ij}]\begin{bmatrix} I_1 \\ \cdot \\ \cdot \\ \cdot \\ I_N \end{bmatrix} \quad (2)$$

$$G_{ij} = \begin{cases} \cos\theta_{ij} H_1^{(2)}(kr_{ij}) W_j, & (i \neq j) \\ 0, & (i = j) \end{cases}$$

$$F_{ij} = \begin{cases} H_0^{(2)}(kr_{ij}), & (i \neq j) \\ 1 - (2j/\pi)[\log(kW_i/4) - 1 + \gamma], & (i = j) \end{cases}$$

where γ denotes Euler's constant ($=0.5772$), W_i, W_j, and $I_i(= -i_n W_i)$ denote the widths of the ports (i and j) and the current flowing into port i, respectively, and $N \triangleq m + n + M$.

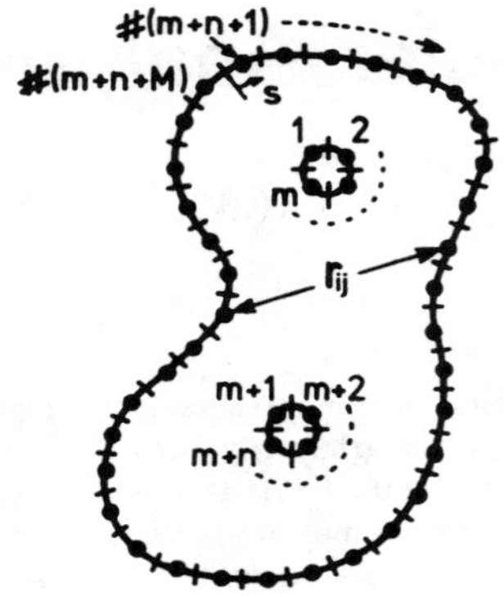

Fig. 3. Symbols used in the computer analysis—I (case of the coaxial-coupled type).

(Equation (1) is derived in [1] for the case in which there is no "hole" in the circuit, in other words when the circuit pattern is singly connected. However, (1) is applicable also to a multiply connected pattern such as is shown in Fig. 3; the proof is given in the Appendix.)

If we further define

$$[U] = 2j[E] - k[G], \qquad (E\text{: unity matrix}) \quad (3)$$

$$[H] = j\omega\mu\, d[F] \quad (4)$$

we obtain, immediately from (1), a simple equation

$$[U_{ij}]\begin{bmatrix} V_1 \\ V_2 \\ \cdot \\ \cdot \\ \cdot \\ V_N \end{bmatrix} = [H_{ij}]\begin{bmatrix} I_1 \\ I_2 \\ \cdot \\ \cdot \\ \cdot \\ I_N \end{bmatrix}. \quad (5)$$

B. Simplification of the Basic Equation

For simplicity we assume the following conditions for the position and size of the coupling conductors.

1) The radius of the conductors R is much less than the wavelength ($R \ll \lambda$, or $kR \ll 1$).

2) If we denote the distance from the center of the conductor to the nearest spot upon the circuit periphery by $r_{\min}$, $R \ll r_{\min}$ holds.

Then we may assume that the voltage V and current density i_n are both uniform along the periphery of the conductor.[1] Therefore, if the voltages and currents of two terminals p and q (see Fig. 3) are denoted by V_p, V_q, I_p, and I_q, respectively, the voltages and currents in each section around the conductors are given as V_p, V_q, I_p/m, and I_q/n. On the other hand, along the circuit periphery $V = 0$ holds. Hence, we may write

[1] We may remove this assumption if we consider $(m - 1)$ and $(n - 1)$ higher order modes in coaxial waveguides 1 and 2, respectively. Then we may take into account the (probably) reactive line impedances for those higher order modes, and we no longer need the reduction of the constraints performed to obtain (7) and (8). A longer computer time will be required, however.

$$
[U_{ij}]\begin{bmatrix} V_p \\ \vdots \\ V_p \\ V_q \\ \vdots \\ V_q \\ 0 \\ \vdots \\ 0 \end{bmatrix}
\begin{matrix} \Big\} m \\ \\ \Big\} n \\ \\ \Big\} M \end{matrix}
= [H_{ij}]\begin{bmatrix} I_p/m \\ \vdots \\ I_p/m \\ I_q/n \\ \vdots \\ I_q/n \\ I_{m+n+1} \\ \vdots \\ I_{m+n+M} \end{bmatrix}
\begin{matrix} \Big\} m \\ \\ \Big\} n \\ \\ \Big\} M \end{matrix} . \qquad (6)
$$

Equation (6) consists of $N(=m+n+M)$ scalar equations, whose number is greater than the number of variables $(M+2)$. To decrease the number of constraints, we define reduced matrices with $(M+2) \times (M+2)$ elements:

$$
[U'] = \left[\begin{array}{cc|ccc}
\sum_{i=1}^{m}\sum_{j=1}^{m} U_{ij} & \sum_{i=1}^{m}\sum_{j=m+1}^{m+n} U_{ij} & \sum_{i=1}^{m} U_{i(m+n+1)} & \cdots & \\
\sum_{i=m+1}^{m+n}\sum_{j=1}^{m} U_{ij} & \sum_{i=m+1}^{m+n}\sum_{j=m+1}^{m+n} U_{ij} & \sum_{i=m+1}^{m+n} U_{i(m+n+1)} & \cdots & \\
\hline
\sum_{j=1}^{m} U_{(m+n+1)j} & \sum_{j=m+1}^{m+n} U_{(m+n+1)j} & U_{m+n+1\ m+n+1} & \cdots & \cdot \\
\vdots & \vdots & \vdots & & \vdots \\
\cdot & \cdot & \cdot & \cdots & U_{m+n+M\ m+n+M}
\end{array}\right] \qquad (7)
$$

$$
[H'] = (\text{similar to the above}). \qquad (8)
$$

Then we may further rewrite (6) as

$$
[U']\begin{bmatrix} V_p \\ V_q \\ 0 \\ \vdots \\ 0 \end{bmatrix}
\begin{matrix} \Big\} 2 \\ \\ \Big\} M \end{matrix}
= [H']\begin{bmatrix} I_p/m \\ I_q/n \\ I_{m+n+1} \\ \vdots \\ I_{m+n+M} \end{bmatrix}
\begin{matrix} \Big\} 2 \\ \\ \Big\} M \end{matrix} . \qquad (9)
$$

The previous simplification implies that "each of the m (or n) sampling points is equally weighted."

C. Derivation of Admittance Parameters

We may derive the admittance parameters Y_{pp}, Y_{pq}, Y_{qp}, and Y_{qq} directly from H' and U'. First, we temporarily consider that all the $(M+2)$ sampling points are coupling terminals and that the planar circuit is represented by an $(M+2)$-port equivalent circuit. The admittance matrix Y of such a circuit is given from (9) as

$$
[Y] = [H']^{-1}[U']. \qquad (10)
$$

The desired parameters Y_{pp}, Y_{pq}, Y_{qp}, and Y_{qq} will be found in the top left corner of the matrix Y. (This method can readily be applied to cases in which the circuit has three or more ports.)

However, practically, the previous computation requires rather long computer time. When the circuit has only two ports, we have a simpler alternative which will be described in the following subsection.

D. Derivation of Transfer Parameters

We assume that the terminals p and q are driving and load terminals, respectively, and impedances Z_p and Z_q are connected to them. Then Z_p must have a negative real part, and must be equal to the driving point impedance multiplied by -1, provided that a stable oscillation exists in the circuit. Since

$$
Z_p = -V_p/I_p \qquad (11)
$$

$$
Z_q = -V_q/I_q \qquad (12)
$$

holds, (9) is rewritten as

$$
\{[H'] + mZ_p[X] + nZ_q[W]\}\begin{bmatrix} I_p/m \\ I_q/n \\ I_{m+n+1} \\ \vdots \\ I_{m+n+M} \end{bmatrix} = 0 \qquad (13)
$$

where X and W are again matrices determined by the shape of the circuit:

$$
[X] = \left[\begin{array}{c|ccc}
\overset{1}{U_{11}'} & 0 & \cdots & 0 \\
U_{21}' & \cdot & & \cdot \\
\vdots & \vdots & & \vdots \\
U_{(M+2)1}' & 0 & \cdots & 0
\end{array}\right] \qquad (14a)
$$

$$
[W] = \left[\begin{array}{c|c|ccc}
0 & \overset{2}{U_{12}'} & 0 & \cdots & 0 \\
\cdot & U_{22}' & \cdot & & \cdot \\
\vdots & \vdots & \vdots & & \vdots \\
0 & U_{(M+2)2}' & 0 & \cdots & 0
\end{array}\right]. \qquad (14b)
$$

In order that (13) has a nontrivial solution, i.e., a steady field exists in the circuit,

$$\det\left[H' + mZ_pX + nZ_qW\right] = 0 \tag{15}$$

must hold. This equation leads directly to a bilinear relation between $-Z_p$, the driving point impedance, and Z_q, the load impedance, as

$$-Z_p = \frac{A'Z_q + B'}{C'Z_q + D'} \tag{16}$$

where A', B', C', and D' are given as the following determinants

$$A' = n \det \begin{bmatrix} H_{11}' & \overset{2}{U_{12}'} & H_{13}' \cdots H_{1N'}' \\ & & \cdot \quad\quad \cdot \\ H_{21}' & U_{22}' & \cdot \quad\quad \cdot \\ \cdot & \cdot & \cdot \quad\quad \cdot \\ \cdot & \cdot & \cdot \quad\quad \cdot \\ \cdot & \cdot & \cdot \quad\quad \cdot \\ H_{N'1}' & U_{N'2}' & H_{N'3}' \cdots H_{N'N'}' \end{bmatrix} \tag{17a}$$

$$B' = \det\left[H_{ij}'\right] \tag{17b}$$

$$C' = mn \det \begin{bmatrix} \overset{1}{U_{11}'} & \overset{2}{U_{12}'} & H_{13}' \cdots H_{1N'}' \\ & & \cdot \quad\quad \cdot \\ U_{21}' & U_{22}' & \cdot \quad\quad \cdot \\ \cdot & \cdot & \cdot \quad\quad \cdot \\ \cdot & \cdot & \cdot \quad\quad \cdot \\ \cdot & \cdot & \cdot \quad\quad \cdot \\ U_{N'1}' & U_{N'2}' & H_{N'3}' \cdots H_{N'N'}' \end{bmatrix} \tag{17c}$$

$$D' = m \det \begin{bmatrix} \overset{1}{U_{11}'} & H_{12}' \cdots H_{1N'}' \\ & \cdot \quad\quad \cdot \\ U_{21}' & \cdot \quad\quad \cdot \\ \cdot & \cdot \quad\quad \cdot \\ \cdot & \cdot \quad\quad \cdot \\ \cdot & \cdot \quad\quad \cdot \\ U_{N'1}' & H_{N'2}' \cdots H_{N'N'}' \end{bmatrix} \tag{17d}$$

where $N' \triangleq M + 2$.

Equation (16) shows that A', B', C', and D' are quantities proportional to the so-called transfer parameters A, B, C, and D of the equivalent two-port circuit. In order that the reciprocity condition ($(AD - BC)^{1/2} = 1$) holds, we should divide A', B', C', and D' by $(A'D' - B'C')^{1/2}$ to get A, B, C, and D, respectively, as

$$\begin{bmatrix} A & B \\ C & D \end{bmatrix} = (A'D' - B'C')^{-1/2} \begin{bmatrix} A' & B' \\ C' & D' \end{bmatrix}. \tag{18}$$

IV. SHORT-BOUNDARY PLANAR CIRCUIT HAVING TWO WAVEGUIDE COUPLING PORTS

A. Basic Equation

We now consider a waveguide-coupled circuit as shown in Fig. 1(b), but having a more arbitrary peripheral shape. The periphery of the circuit is again divided and numbered, as shown in Fig. 4, as follows:

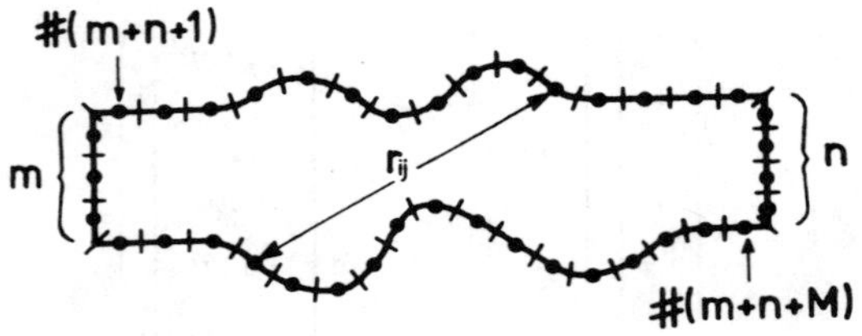

Fig. 4. Symbols used in the computer analysis—II (case of the waveguide-coupled type).

1) input port 1: $i = 1 \sim m$;
2) output port 2: $i = (m + 1) \sim (m + n)$;
3) circuit periphery: $i = (m + n + 1) \sim (m + n + M)$;

and $m + n + M (= N)$ sampling points are provided. The derivation of the $N \times N$ matrix equation (5) follows entirely the same process as was described in the preceding section.

B. Simplification of the Basic Equation

We now assume for simplicity that straight waveguide sections with appropriate length exist on both sides of the input and output planes, so that only the TE_{10}-mode exists at those planes.[2] Under this assumption we may relate $V_1 \sim V_m$ and $I_1 \sim I_m$, $V_{m+1} \sim V_{m+n}$ and $I_{m+1} \sim I_{m+n}$ with simple proportionality formulas. We use representative values of those voltages and currents defined as V_p and V_q: voltages at the center of the input and output reference planes, respectively; and I_p and I_q: currents flowing across the entire widths of the input and output reference planes, respectively. Since the variation of the voltage and current across the reference planes are both sinusoidal,

$$\begin{cases} V_1 = \alpha_1 V_p \\ V_2 = \alpha_2 V_p \\ \quad\vdots \\ V_m = \alpha_m V_p \end{cases} \qquad \begin{cases} I_1 = \alpha_1 I_p/m \\ I_2 = \alpha_2 I_p/m \\ \quad\vdots \\ I_m = \alpha_m I_p/m \end{cases} \tag{19}$$

where

$$\alpha_i = \sin\left(\frac{2i - 1}{2m}\pi\right),$$

$$\begin{cases} V_{m+1} = \beta_1 V_q \\ V_{m+2} = \beta_2 V_q \\ \quad\vdots \\ V_{m+n} = \beta_n V_q \end{cases} \qquad \begin{cases} I_{m+1} = \beta_1 I_q/n \\ I_{m+2} = \beta_2 I_q/n \\ \quad\vdots \\ I_{m+n} = \beta_n I_q/n \end{cases} \tag{20}$$

[2] We may remove half of this assumption, i.e., the presence of the straight waveguide sections "inside" the reference planes, if we consider $(m - 1)$ and $(n - 1)$ higher order modes in the input and output waveguides, respectively. Then we may take into account the (probably) reactive waveguide impedances for those higher order modes, and again no longer need the reduction of the constraints performed to obtain (22) and (23).

where

$$\beta_i = \sin\left(\frac{2i-1}{2n}\pi\right).$$

Using the previous relations, we may rewrite (5) as

$$[U_{ij}]\begin{bmatrix} \alpha_1 V_p \\ \cdot \\ \cdot \\ \cdot \\ \alpha_m V_p \\ \beta_1 V_q \\ \cdot \\ \cdot \\ \cdot \\ \beta_n V_q \\ 0 \\ \cdot \\ \cdot \\ \cdot \\ 0 \end{bmatrix} \begin{matrix} \left.\right\} m \\ \\ \left.\right\} n \\ \\ \left.\right\} M \end{matrix} = [H_{ij}]\begin{bmatrix} \alpha_1 I_p/m \\ \cdot \\ \cdot \\ \cdot \\ \alpha_m I_p/m \\ \beta_1 I_q/n \\ \cdot \\ \cdot \\ \cdot \\ \beta_n I_q/n \\ I_{m+n+1} \\ \cdot \\ \cdot \\ \cdot \\ I_{m+n+M} \end{bmatrix} \begin{matrix} \left.\right\} m \\ \\ \left.\right\} n \\ \\ \left.\right\} M \end{matrix} \cdot \quad (21)$$

Following the procedure described in the preceding section, we again define the following $(M+2) \times (M+2)$ matrices to decrease the number of constraints, as

$$[U'] = \left[\begin{array}{cc|c} \sum_{i=1}^{m}\sum_{j=1}^{m} \alpha_j U_{ij} & \sum_{i=1}^{m}\sum_{j=m+1}^{m+n} \beta_{j-m} U_{ij} & \sum_{i=1}^{m} U_{i(m+n+1)} \cdots \\ \sum_{i=m+1}^{m+n}\sum_{j=1}^{m} \alpha_j U_{ij} & \sum_{i=m+1}^{m+n}\sum_{j=m+1}^{m+n} \beta_{j-m} U_{ij} & \sum_{i=m+1}^{m+n} U_{i(m+n+1)} \cdots \\ \hline \sum_{j=1}^{m} \alpha_j U_{(m+n+1)j} & \sum_{j=m+1}^{m+n} \beta_{j-m} U_{(m+n+1)j} & U_{\substack{m+n+1 \\ m+n+1}} \cdots\cdots \\ \cdot & \cdot & \cdot \qquad \cdot \\ \cdot & \cdot & \cdot \qquad \cdot \\ \cdot & \cdot & \cdot \qquad \cdot \\ \cdot & \cdot & \cdot \qquad U_{\substack{m+n+M \\ m+n+M}} \\ \cdot & \cdot & \cdots\cdots \end{array}\right] \quad (22)$$

$$[H'] = \text{(similar to the above)}. \quad (23)$$

Using the previous matrices, we may rewrite (21) in a form identical to (9). The transfer parameters are also derived in the same way.

V. EXAMPLES OF NUMERICAL ANALYSIS

A. Short-Circuited Radial Line

To show the validity and accuracy of the proposed method of analysis, two examples of numerical analysis whose results can be compared with analytical solutions will first be described. The first example is of the coaxial-coupled type: a thin radial line terminated by a short-circuit wall as shown in Fig. 5.

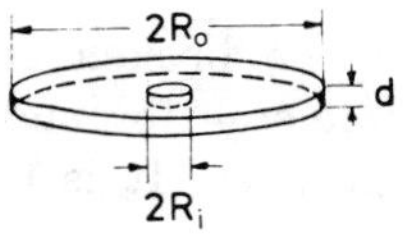

Fig. 5. A radial line.

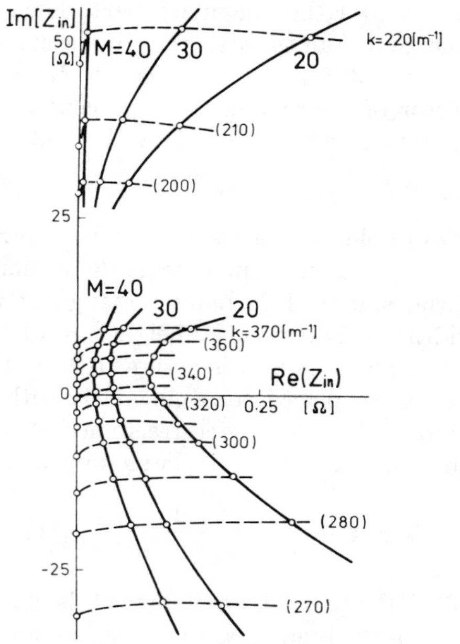

Fig. 6. The computed input impedance of a radial line. Note that the real part (abscissa) is exaggerated to show the computational error.

For such a one-port circuit, the driving-point impedance is given [1], instead of (16), by

$$-Z_p = B'/D'. \quad (24)$$

This driving-point impedance has been computed for dimensions (see Fig. 5) $R_i = 1$ mm, $R_o = 10$ mm, and $d = 0.5$ mm, at 21 frequencies ranging between $k(=2\pi f/c) = 200 \sim 400[\text{m}^{-1}]$. Various numbers of divisions have been used to obtain the estimate of the computation error.

The results of the cases for $m = 10$ are shown in Fig. 6. The abscissa and the ordinate show the real and imaginary parts of the driving-point impedance, respectively. Note, however, that the abscissa is expanded by a factor of 100 to exaggerate the computation error. Small circles upon the ordinate show the theoretical values [3]

$$-Z_p = -j\frac{Z_{0i}d}{2\pi R_i}\frac{\sin(\theta_i - \theta_o)}{\cos(\varphi_i - \theta_o)} \quad (25)$$

where subscripts i and o represent the inner and outer boundaries, respectively, θ and φ are the quantities defined by

$$H_0^{(1)}(x) = J_0(x) + jN_0(x) = G_0(x)\exp(j\theta(x))$$

$$jH_1^{(1)}(x) = -N_1(x) + jJ_1(x) = G_1(x)\exp(j\varphi(x))$$

where $x = kr$, and Z_{0i} denotes the wave impedance at the inner boundary, defined as

$$Z_{0i} = 120\pi\{G_0(kR_i)/G_1(kR_i)\}[\Omega].$$

The bracketed numerals in Fig. 6 denote the wavenumber in m^{-1}. Some of the broken curves (k = 200,210, 220[m^{-1}]) seem to converge for increasing M to values different from analytic ones. It is natural because m is finite.

The estimate of the error obtained from Fig. 6 is several-tenth ohms for the real part, and several ohms (in most cases below 2 Ω) for the imaginary part. The series resonance frequency obtained with the numerical analysis is 15.68 GHz (k = 328 m^{-1}) for m = 10, M = 40, which shows an error of approximately 1 percent as compared with the analytical value 15.87 GHz (k = 332 m^{-1}).

B. Uniform Waveguide Section

Another example, whose results can be compared with analytical ones, is a uniform waveguide section. A finite section of the standard X band rectangular waveguide with a (width) = 22.9 mm, b (height) = 10.2 mm, and l (length) = $2a$ = 45.8 mm is assumed to be terminated by a resistive sheet placed perpendicularly with respect to the waveguide axis; the surface resistance of the sheet is assumed to be equal to the waveguide impedance

$$Z_0 = 120\pi \frac{b}{a} \frac{1}{(1 - (\lambda/2a)^2)^{1/2}} [\Omega] \tag{26}$$

at f = 10.00 GHz (k = 209 m^{-1}), which is 222.2[Ω].

The results of the numerical computation of the driving-point impedance are shown in Fig. 7(a) with small dots, for $m = n$ = 10, M = 40. In the same figure, the theoretical value

$$Z_{\text{in}} = Z_0 \frac{(Z_L/Z_0) + j \tan (2\pi l/\lambda_g)}{1 + j(Z_L/Z_0) \tan (2\pi l/\lambda_g)} \tag{27}$$

is also plotted with small crosses where Z_L, Z_0, and λ_g denote the load impedance (222.2[Ω]), the waveguide impedance at the given frequency, and the wavelength in the guide, respectively. The complicated part (below, right) of Fig. 7(a) is shown in Fig. 7(b) in an enlarged scale.

C. Waveguide Section Including a Thick Inductive Window

In the following three subsections, more practical circuits are analyzed numerically; the obtained results are compared with the approximate analyses and experimental data. In those numerical analyses, for convenience of comparison, the obtained transfer parameters are once converted into equivalent T-circuit parameters; this conversion process will first be described.

We consider a waveguide section including a symmetrical obstacle, a thick inductive window as in the example shown in Fig. 8(a), and represent the entire section by three equivalent circuits: those of two straight sections and an equivalent T-circuit of the obstacle itself.

If we denote the transfer parameters of the entire section, the straight sections and the obstacle by F_c, F_L, and F_T, respectively,

$$[F_c] = [F_L][F_T][F_L]. \tag{28}$$

If the straight sections are lossless,

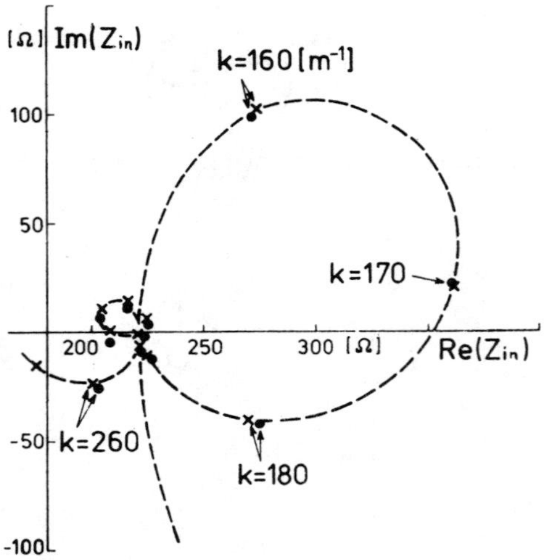

(a)

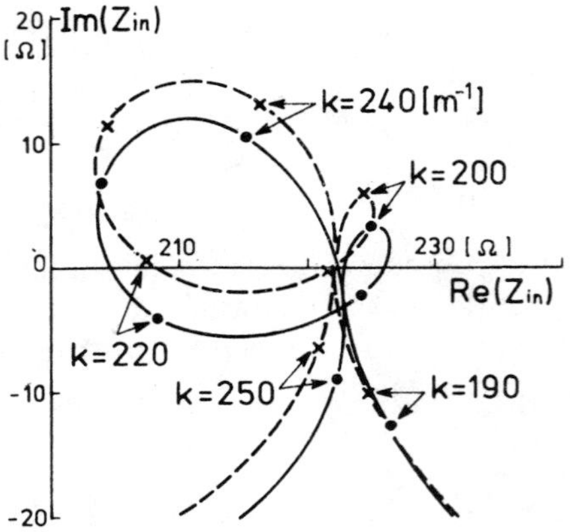

(b)

Fig. 7. The computed input impedance of a waveguide section. (a) The overall frequency characteristics. (b) An enlarged figure of a portion.

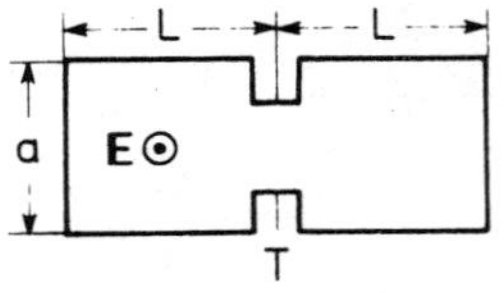

(a)

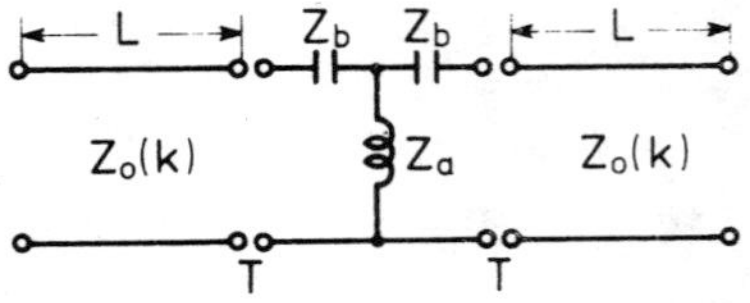

(b)

Fig. 8. Equivalent circuit representation of a symmetrical waveguide obstacle. (a) A thick inductive window (an example). (b) The equivalent T-circuit representation.

$$[F_L] = \begin{bmatrix} \cos\left(\frac{2\pi L}{\lambda_g}\right) & jZ_0 \sin\left(\frac{2\pi L}{\lambda_g}\right) \\ j\left(\frac{1}{Z_0}\right) \sin\left(\frac{2\pi L}{\lambda_g}\right) & \cos\left(\frac{2\pi L}{\lambda_g}\right) \end{bmatrix}. \tag{29}$$

From the elements of F_c computed numerically and those of F_L, the elements of F_T are obtained as

$$[F_T] = \begin{bmatrix} A_T & B_T \\ C_T & D_T \end{bmatrix} = [F_L]^{-1}[F_c][F_L]^{-1}. \quad (30)$$

On the other hand, F_T is expressed by the T-circuit parameters as

$$[F_T] = (1/Z_a) \begin{bmatrix} Z_a + Z_b & 2Z_aZ_b + Z_b^2 \\ 1 & Z_a + Z_b \end{bmatrix}. \quad (31)$$

Hence, the normalized T-circuit parameters are given in terms of the computed elements of F_T as

$$\frac{Z_a}{Z_0} = \frac{1}{C_T Z_0} \left(\triangleq j \frac{X_a}{Z_0} \right) \quad (32)$$

$$\frac{Z_b}{Z_0} = \frac{A_T - 1}{C_T Z_0} \left(\triangleq -j \frac{X_b}{Z_0} \right). \quad (33)$$

Fig. 9 shows the results of computer analyses of the frequency characteristics of thick inductive windows. The ordinate gives the normalized T-circuit parameters X_a/Z_0 and X_b/Z_0. The dimensions of the circuit are assumed to be

Case 1:

$a = 22.9$ mm $\quad d'/2 = 2.29$ mm $\quad l = 2.29$ mm $\quad L = 24.045$ mm

Case 2:

$a = 22.9$ mm $\quad d'/2 = 2.29$ mm $\quad l = 4.58$ mm $\quad L = 25.19$ mm.

The total sampling-point number is 66 in case 1, and 68 in case 2.

The dots in Fig. 9 show the results of the computer analyses. The crosses show the approximate theoretical values described in [4]. The difference between dots and crosses is found to be less than 0.02. (The errors in X_b/Z_0 might seem rather large; however, note that the scales of the ordinate for X_b/Z_0 and X_a/Z_0 are different.)

D. Waveguide Sections Including Corners

The dots in Fig. 10 show the results of the computer analyses of 30° and 60° waveguide corners. The ordinate again shows X_a/Z_0 and X_b/Z_0. The circuit dimensions are $a = 22.9$ mm and $L = 22.9$ mm, and the total sampling-point number $N = 66$ for $\alpha = 30°$ and $N = 72$ for $\alpha = 60°$. Small crosses in the figure show the experimental data found in [4].

E. Waveguide Sections Including Post

The dots in Fig. 11 show the results of the computer analyses of waveguide sections including post. The circuit dimensions are $a = 22.9$ mm and $L = 22.9$ mm, and the total sampling-point number $N = 72$, including 12 points around the post. Small crosses show the experimental data found in [4].

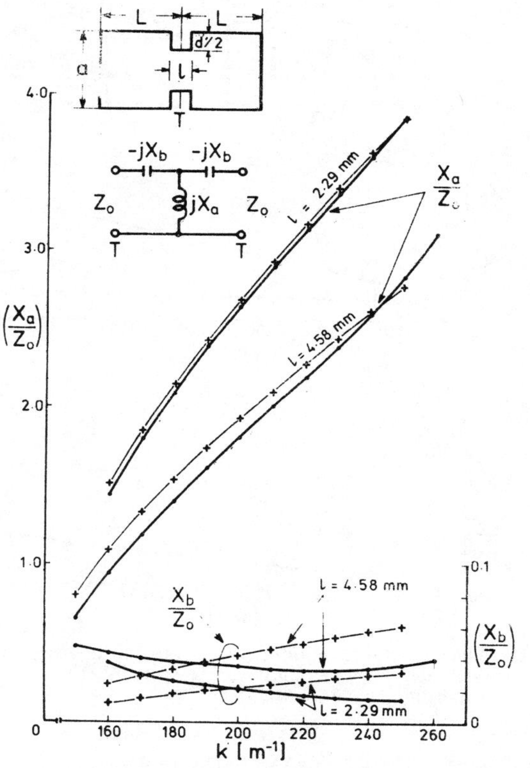

Fig. 9. The computed equivalent-circuit parameters of thick inductive windows. Crosses show the results of the approximate analysis described by Marcuvitz [4].

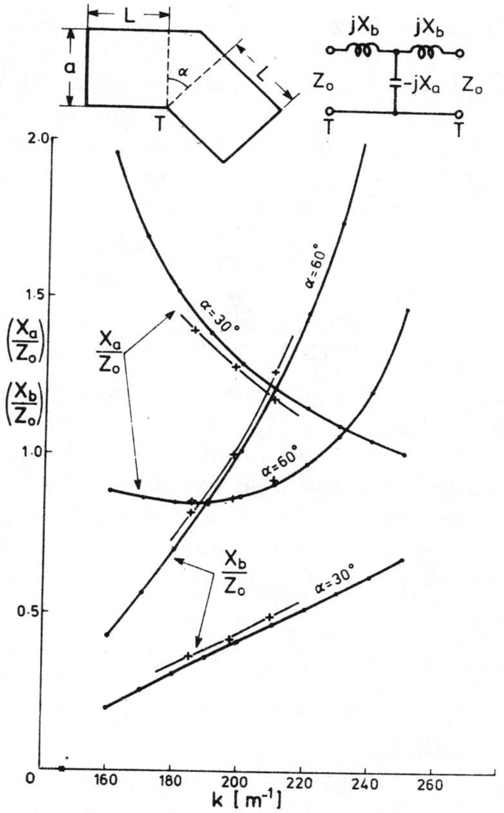

Fig. 10. The computed equivalent-circuit parameters of waveguide corners. Crosses show the experimental data described by Marcuvitz [4].

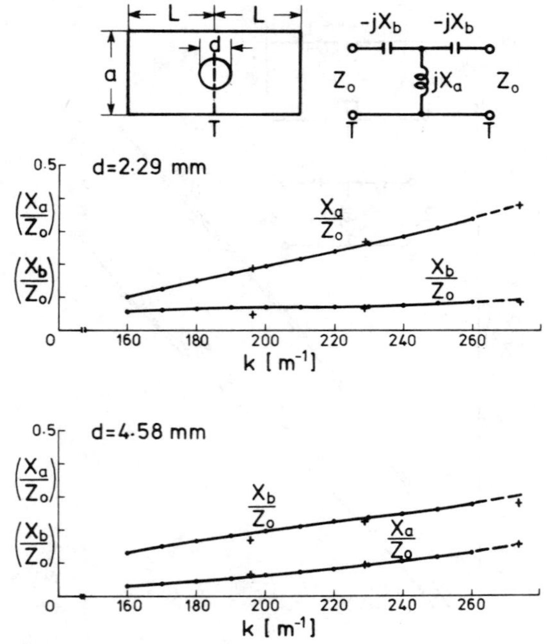

Fig. 11. The computed equivalent-circuit parameters of a post in a waveguide. Crosses show the results of the approximate analysis described by Marcuvitz [4].

VI. CONCLUSION

The basic program for the computer analysis of arbitrarily shaped, short-boundary planar circuits has been completed. The validity and error of the proposed method of analysis has been shown through comparison with theories. The required computer time is still long; for example, about 10 min are required to obtain all the dots in Fig. 9 using HITAC-8800, one of the standard Japanese high-speed computers. The improvement of the program toward shorter computer time and better accuracy is left for further efforts.

APPENDIX

PROOF OF (1) FOR A MULTIPLY CONNECTED CIRCUIT PATTERN

We now consider a planar circuit pattern having a hole as shown in Fig. 12(a). We connect the outer and inner boundaries, C_1 and C_2, with contours C_3 and C_4 which are infinitesimally separated, and give a direction of integration, as shown in Fig. 12(b), to define θ. For the entire contour $C_1 + C_2 + C_3 + C_4$, obviously (1) holds. However, for the corresponding points upon C_3 and C_4 [see Fig. 12(c)],

$$[V(s_0)]_{C3} = [V(s_0)]_{C4} \tag{34}$$

$$[i_n(s_0)]_{C3} = -[i_n(s_0)]_{C4} \tag{35}$$

must be satisfied.

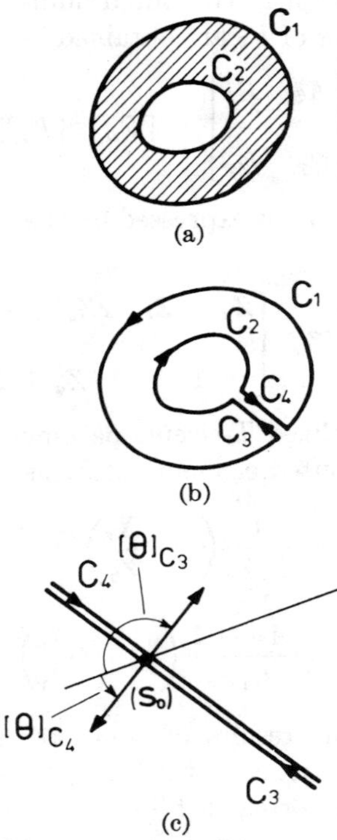

Fig. 12. Proof of (1) for a multiply connected planar circuit pattern.

The voltage $V(s)$ in the left-hand side of (1) in the text needs to be known only upon C_1 and C_2. If we consider cases in which s is located somewhere upon C_1 or C_2 and s_0 is located somewhere upon C_3 and C_4, we obtain for the corresponding points upon C_3 and C_4

$$[\theta]_{C3} = [\theta]_{C4} + \pi. \tag{36}$$

Hence,

$$[\cos\theta]_{C3} = -[\cos\theta]_{C4}. \tag{37}$$

Putting (34), (35), and (37) into the right-hand side of (1), we find that the integrals along C_3 and C_4 cancel each other. Hence, we may apply (1) to a pattern like Fig. 12(a) provided that the direction of integration along C_1 and C_2 be carefully defined.

ACKNOWLEDGMENT

The authors wish to thank S. Oda of Hitachi Ltd. for his help in the early stage of this work.

REFERENCES

[1] T. Okoshi and T. Miyoshi, "The planar circuit—An approach to microwave integrated circuitry," *IEEE Trans. Microwave Theory Tech.*, vol. MTT-20, pp. 245–252, Apr. 1972.
[2] J. A. Stratton, *Electromagnetic Theory.* New York: McGraw-Hill, 1941, p. 460.
[3] S. Ramo and J. R. Whinnery, *Fields and Waves in Modern Radio.* New York: Wiley, 1953.
[4] N. Marcuvitz, *Waveguide Handbook.* New York: McGraw-Hill, 1951.

Paper 4.2

Computer-Aided Synthesis of Planar Circuits

FUMIO KATO, MASAO SAITO, MEMBER, IEEE, AND TAKANORI OKOSHI, MEMBER, IEEE

Abstract—**This paper presents a fully computer-oriented iterative synthesis of an open-boundary planar circuit having an impedance matrix with prescribed poles and residues. A starting circuit pattern is given first, and it is represented by a finite number of parameters. Those parameters (and hence, the pattern) are then iteratively modified by using the Newton–Raphson method to realize the prescribed circuit characteristics. When the numbers of given poles and coupling ports are relatively small, the results are satisfactory both in the computing time and accuracy. Some numerical examples are given.**

I. Introduction

THE PLANAR (two-dimensional) circuit is a circuit category that should be positioned between the distributed-constant circuit and the waveguide circuit [1]. As this concept was originally proposed for rigorous analysis and design of microwave and millimeter-wave IC's, its synthesis (determination of the circuit pattern giving prescribed circuit characteristics) is an important technical task.

However, although a number of methods have been presented for the analysis [1]–[4], rather few papers have ever dealt with the synthesis. Okoshi *et al.* described a trial-and-error synthesis of a ladder-type 3-dB hybrid consisting of wide striplines [5]; their synthesis, however, is never of a general nature. Grüner dealt with the conformal-mapping synthesis of a thin waveguide section [6]. Since a thin waveguide section can be regarded as a short-boundary planar circuit, his method will also be applicable, if appropriately modified, to open-boundary planar circuit. However, Grüner described synthesis of only poles of transmission characteristics. Synthesis of residues must also be performed to make the synthesis complete.

This paper presents a fully computer-oriented iterative synthesis of an open-boundary planar circuit having an impedance matrix with prescribed poles and residues. Basically, this paper's aims are similar to Grüner's [6]; the differences are: 1) open-boundary problems are considered, 2) prescribed residues are realized in addition to poles, and 3) the mapping technique has been improved.

II. Basic Equations and Principle of Analysis

A. Circuit Equations and Impedance Matrix

Consider a planar circuit as shown in Fig. 1. It consists of a conductive (circuit) plate with an arbitrary shape, an isotropic dielectric layer, and a ground plate.[1] The boundary of the circuit pattern is denoted by Γ, and its interior is henceforth called the circuit pattern and is denoted by S. The spacing d between the circuit plate and the ground plate is assumed to be much less than the wavelength. The circuit is assumed to be lossless.

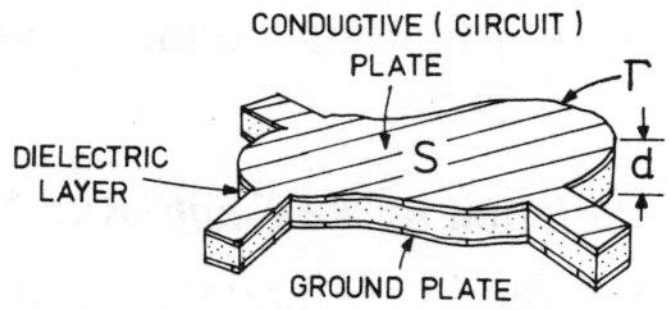

Fig. 1. An open-boundary planar circuit.

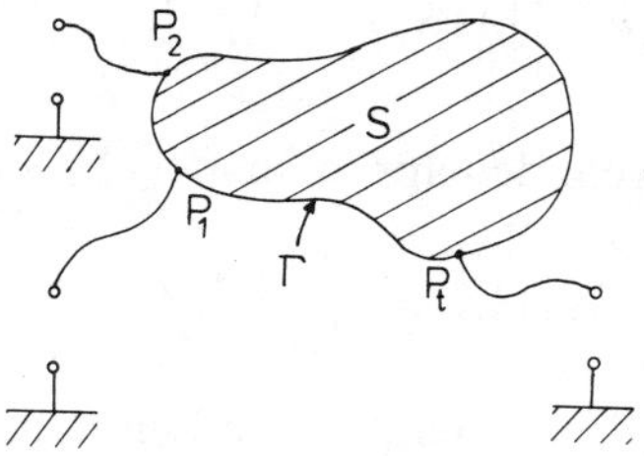

Fig. 2. Schematic representation of a multiport planar circuit.

The steady-state circuit equation is then given by the following two-dimensional Helmholtz equation:

$$\nabla^2 v(x,y) + \lambda v(x,y) = 0 \tag{1}$$

where $\lambda = \omega^2 \varepsilon\mu$, ε and μ are the permittivity and permeability of the spacing material, respectively, and ∇^2 denotes the Laplacian [1]. Upon Γ where no coupling port is present, the boundary condition is given as

$$\frac{\partial v}{\partial n} = 0 \tag{2}$$

where $\partial/\partial n$ denotes the derivative normal (outward) to Γ. Equations (1) and (2) give a countably infinite number of eigenvalues λ_i and associated normalized eigenfunctions $v_i(x,y)$.

When t ports are connected to the circuit plate, as shown in Fig. 2, at positions $P_1, P_2, \cdots, P_t$ and assumed to have negligible widths as compared with the wavelength, the (m,n)

[1] In the earlier papers [1]–[4], triplate-type planar circuits have been mainly dealt with. In this paper a two-plate circuit as shown in Fig. 1 is considered for its simplicity. Mathematically the above two types are equivalent, except that in the former the impedance level is reduced by one half because currents flow in both the upper and lower surfaces of the circuit plate.

Reprinted from *IEEE Trans. Microwave Theory Tech.*, vol. MTT-25, no. 10, pp. 814–819, Oct. 1977.

element of the impedance matrix of the t-port circuit is expressed in terms of λ_i and v_i as [1]

$$Z_{mn} = \sum_{i=0}^{\infty} \frac{-j\omega\mu d}{\lambda - \lambda_i} v_i(P_m) v_i(P_n). \tag{3}$$

We should note that λ_i's correspond to poles of Z_{mn} and that $v_i(P_m)$ and $v_i(P_n)$ determine the residue matrix associated with λ_i.

B. Approximate Solution by Rayleigh–Ritz Method

An approximate solution (λ_i and v_i) of (1) can be obtained from the stationary condition of the functional

$$I(v,\lambda) = \iint_S \left\{ \left(\frac{\partial v}{\partial x}\right)^2 + \left(\frac{\partial v}{\partial y}\right)^2 - \lambda v^2 \right\} dS. \tag{4}$$

Let $v(x,y)$ be expanded approximately by a truncated series

$$v(x,y) = \sum_{k=1}^{M} a_k f_k(x,y) \tag{5}$$

where $\{f_k(x,y)\}$ is a system of functions which is complete in the region S.

We rewrite the variational problem, (4), into a form which is more suitable to the computer analysis. By using

$$A_{kl} = \iint_S \nabla f_k \cdot \nabla f_l \, dS \tag{6a}$$

$$B_{kl} = \iint_S f_k f_l \, dS, \qquad k, l = 1, 2, \cdots, M \tag{6b}$$

$$A = \begin{bmatrix} A_{11} & \cdots & A_{1M} \\ \vdots & & \vdots \\ A_{M1} & \cdots & A_{MM} \end{bmatrix} \qquad B = \begin{bmatrix} B_{11} & \cdots & B_{1M} \\ \vdots & & \vdots \\ B_{M1} & \cdots & B_{MM} \end{bmatrix} \tag{7}$$

(4) is reduced to a matrix eigenvalue problem

$$(\boldsymbol{A} - \lambda \boldsymbol{B})\boldsymbol{a} = 0 \tag{8}$$

where $\boldsymbol{a}$ denotes a column vector consisting of the coefficients $a_1, a_2, \cdots, a_M$ in (5). Thus, from the nontrivial condition of (8), the eigenvalues can be obtained as the roots of the characteristic equation

$$\det(\boldsymbol{A} - \lambda \boldsymbol{B}) = 0. \tag{9}$$

The corresponding eigenvector $\boldsymbol{a}$ is then obtained from (8) except for a constant multiplier, which is determined by using the normalizing condition

$$\iint_S v^2(x,y) \, dS = 1. \tag{10}$$

In the actual computation, the constant multiplier of $\boldsymbol{a}$ is determined by the matrix equivalence of (10):

$$\boldsymbol{a}^T \boldsymbol{B} \boldsymbol{a} = 1 \tag{11}$$

where superscript T denotes transposition. Finally, v is given in terms of $\boldsymbol{a}$ by using (5).

III. Method of Synthesis

In the following synthesis, we must represent the circuit pattern and port locations with a finite set of parameters. However, whatever parameters are employed, the essential process of the synthesis remains unchanged. Hence, we begin this section with a general description of the synthesis without specifying any particular type of parameters. In Section III-D, two practical choices of the parameters will be presented.

A. Modified Newton–Raphson Method

We first describe the iterative process used in the synthesis. Consider a system of n equations to be solved:

$$\boldsymbol{F}(\boldsymbol{X}) = \begin{bmatrix} F_1(\boldsymbol{X}) \\ \vdots \\ F_n(\boldsymbol{X}) \end{bmatrix} = \boldsymbol{0} \tag{12}$$

where $\boldsymbol{X}$ denotes a column vector whose elements are $N (N \geq n)$ unknown quantities representing the circuit pattern, that is, $\boldsymbol{X} = (X_1, X_2, \cdots, X_N)^T$. The Jacobian matrix of $\boldsymbol{F}(\boldsymbol{X})$ with respect to $\boldsymbol{X}$ is expressed as

$$\boldsymbol{J}(\boldsymbol{X}) = \begin{bmatrix} J_{11}(\boldsymbol{X}) & \cdots & J_{1N}(\boldsymbol{X}) \\ \vdots & & \vdots \\ J_{n1}(\boldsymbol{X}) & \cdots & J_{nN}(\boldsymbol{X}) \end{bmatrix} \tag{13}$$

where

$$J_{ij}(\boldsymbol{X}) = \frac{\partial F_i(\boldsymbol{X})}{\partial X_j}, \qquad i = 1, \cdots, n; \; j = 1, \cdots, N. \tag{14}$$

We first assume a vector $\boldsymbol{X}^{(0)}$ as the initial pattern, and modify it to obtain the solution of (12). For this purpose, the $(h+1)$th solution should be obtained from the hth solution as [7]

$$\boldsymbol{X}^{(h+1)} = \boldsymbol{X}^{(h)} - \boldsymbol{J}^-(\boldsymbol{X}^{(h)})\boldsymbol{F}(\boldsymbol{X}^{(h)}) \tag{15}$$

where $\boldsymbol{J}^-$ represents a generalized inverse matrix of $\boldsymbol{J}$. Then $\boldsymbol{X}^{(h)}$ should finally converge to $\boldsymbol{X}^*$ which satisfies

$$\boldsymbol{J}^T(\boldsymbol{X}^*)\boldsymbol{F}(\boldsymbol{X}^*) = \boldsymbol{0}. \tag{16}$$

In particular, when rank $\{\boldsymbol{J}(\boldsymbol{X}^*)\} = n$ as is usually the case,

$$\boldsymbol{F}(\boldsymbol{X}^*) = \boldsymbol{0} \tag{17}$$

holds, which is the condition to be satisfied.[2] If we add, in computing $\boldsymbol{X}^{(h)}$, an additional condition (the minimum norm condition)

$$|\boldsymbol{X}^{(h+1)} - \boldsymbol{X}^{(h)}| = \text{minimum} \tag{18}$$

then the modification algorithm, (15), is rewritten as [7]

$$\boldsymbol{X}^{(h+1)} = \boldsymbol{X}^{(h)} - \boldsymbol{J}^T(\boldsymbol{J}\boldsymbol{J}^T)^{-1}\boldsymbol{F}(\boldsymbol{X}^{(h)}). \tag{19}$$

[2] In order that $\boldsymbol{X}$ converges to the solution, $\boldsymbol{X}^{(0)}$ must satisfy some conditions, which have been discussed in detail by Ben-Israel [7]. However, the convergence is guaranteed if (15) is modified so that the convergence is decelerated [8]. Such a modified equation is used in the computation described in Section IV.

B. Synthesis of Prescribed Poles

For some practical purposes such as the design of a bandpass filter or the suppression of spurious modes in a resonator, synthesis of the poles (eigenvalues) disregarding the residues is useful to some extent [6]. In this subsection, to begin with, eigenfunctions are not considered; a circuit pattern having several prescribed eigenvalues will be synthesized.

We assume that a set of the smallest n eigenvalues $\boldsymbol{\lambda} = (\lambda_1, \lambda_2, \cdots, \lambda_n)^T$ $(0 = \lambda_0 < \lambda_1 < \lambda_2 < \cdots < \lambda_n; n \leq N)$ are given. From (9), those must satisfy

$$F_i(X) = \det \{\boldsymbol{A}(X) - \lambda_i \boldsymbol{B}(X)\} = 0, \qquad i = 1, \cdots, n. \quad (20)$$

As discussed in the preceding subsection, X is successively modified by the Newton–Raphson method to arrive finally at the solution satisfying $\boldsymbol{F}(X^*) = \boldsymbol{0}$. We assume that $F_i(X) \neq 0$ and write

$$C^i_{kl}(X) = A_{kl}(X) - \lambda_i B_{kl}(X) \quad (21a)$$

$$\boldsymbol{C}_i(X) = \boldsymbol{A}(X) - \lambda_i \boldsymbol{B}(X). \quad (21b)$$

Matrix $\boldsymbol{C}_i(X)$ is then regular, and J_{ij} in (14) may be rewritten as

$$J_{ij}(X) = F_i(X) \sum_{k=1}^{M} \sum_{l=1}^{M} \frac{\partial C^i_{kl}}{\partial X_j} [C_i]^{-1}_{kl} \quad (22)$$

where $[C_i]^{-1}_{kl}$ represents the (k,l) element of the inverse matrix of $\boldsymbol{C}_i$. In actual synthesis of a circuit, each modification should be as small as possible to avoid an unreasonable circuit pattern and oscillation of the solution. Therefore we use (19) rather than (15).

The initial circuit pattern $X^{(0)}$ must be chosen with some care. We may reconcile at least one eigenvalue to the desired value merely by multiplying $X^{(0)}$ by a positive constant. Empirically, the convergence seems to be improved by choosing the initial pattern $X^{(0)}$ so that its nth (highest) eigenvalue is equal to the prescribed value.

C. Synthesis of Prescribed Impedance Matrix

We assume t coupling ports along the boundary. The residues of the impedance matrix ($\mu\, dv_i(P_m)v_i(P_n)$ in (3)) are now considered as well as the poles. Let the positions of the ports be denoted by $P_1, \cdots, P_t$; their parameter representations, by $\delta_1, \cdots, \delta_t$. Since the values of eigenfunctions at those ports must be adjusted as well as the eigenvalues, each step of the synthesis must include shifts of the port locations $P_1, \cdots, P_t$. Therefore, the parameters representing a circuit pattern are now

$$X = (X_1, \cdots, X_N, \delta_1, \cdots, \delta_t)^T. \quad (23)$$

The input data now comprise the number of ports t, the prescribed eigenvalues λ_i $(i = 1, \cdots, n)$, and the values of associated eigenfunctions at the kth port q_{ik} $(i = 1, \cdots, n; k = 1, \cdots, t)$. Let $\lambda_i(X)$ denote the ith eigenvalue of the circuit pattern X; $v_{ik}(X)$, the value of the corresponding eigenfunction at the kth port. Then we may apply the Newton–Raphson method to the following equations:

$$\boldsymbol{F}(X) = \begin{bmatrix} F_1(X) \\ \vdots \\ F_n(X) \\ F_{n+1}(X) \\ \vdots \\ F_{n(t+1)}(X) \end{bmatrix} = \begin{bmatrix} \lambda_1(X) - \lambda_1 \\ \vdots \\ \lambda_n(X) - \lambda_n \\ v_{11}(X) - q_{11} \\ \vdots \\ v_{nt}(X) - q_{nt} \end{bmatrix} = \boldsymbol{0}. \quad (24)$$

The Jacobian matrix of $\boldsymbol{F}(X)$ is then obtained from (5), (8), (11), and (24).

D. Parameters Representing Circuit Pattern

In Sections III-B and -C, the synthesis process has been formulated without referring to the implication of the parameters representing the circuit pattern. In this subsection, two choices of the parameters are described.

1) Sector Method: We assume an origin O in the region S and N sampling points along the boundary Γ. The sampling points are chosen so that the angle at the origin subtended by any adjacent two sampling points is equal to $2\pi/N$ (see Fig. 3). We denote the distance between the origin and each sampling point by $X_1, X_2, \cdots, X_N$; the circuit pattern is then represented approximately by $X = (X_1, X_2, \cdots, X_N)^T$. When we finally obtain X_j's satisfying

$$X_j > 0, \qquad j = 1, \cdots, N \quad (25)$$

we may consider X to be physically realizable.

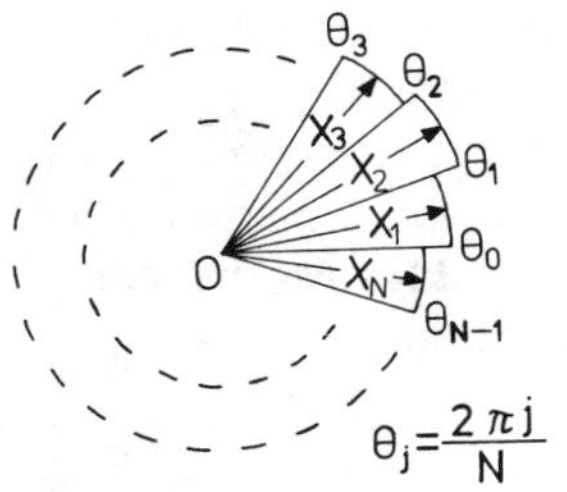

Fig. 3. Representation of the circuit pattern in the sector method.

In the polar coordinates, A_{kl} is given, from (6a), as

$$A_{kl} = \sum_{j=1}^{N} \int_0^{X_j} \int_{\theta_{j-1}}^{\theta_j} \nabla f_k \cdot \nabla f_l r \, dr \, d\theta, \qquad k, l = 1, \cdots, M \quad (26)$$

and its derivative with respect to X_j as

$$\frac{\partial A_{kl}}{\partial X_j} = \int_{\theta_{j-1}}^{\theta_j} \nabla f_k \cdot \nabla f_l r \Big|_{r = X_j} d\theta. \quad (27)$$

Expressions for B_{kl} and its derivative may be given in similar forms.

We may use the above parameters in either of the syntheses described in Sections III-B and -C. For convenience those combinations will hereafter be called synthesis processes I and II, respectively. In process II, however, we find some difficulty in considering the external ports because of an inherent discontinuous nature of the boundary consisting of many sectors (see Fig. 3).

2) *Conformal-Mapping Method:* Any simply connected region S in the complex z plane can be mapped from the interior of a unit circle Ω in another complex ζ plane (see Fig. 4) by a regular function

$$z = x + jy = \phi(\zeta) \tag{28}$$

which we express by a truncated Maclaurin series as[3]

$$\phi(\zeta) = \sum_{k=1}^{m} (\alpha_k + j\beta_k)\zeta^k, \qquad \alpha_k \text{ and } \beta_k \text{ are real.} \tag{29}$$

Thus the pattern S can be represented by α_k and β_k ($k = 1, \cdots, m$). Further, we define a vector X as

$$X = (X_1, X_2, \cdots, X_{N-1}, X_N)^T = (\alpha_1, \beta_1, \cdots, \alpha_m, \beta_m)^T \tag{30}$$

where

$$N = 2m \tag{31a}$$

$$X_{2k-1} = \alpha_k \qquad X_{2k} = \beta_k, \qquad k = 1, \cdots, m. \tag{31b,c}$$

Equation (1) expressed in the z plane is transformed, in the ζ plane, to

$$\nabla_\zeta^2 v + \lambda |\phi'(\zeta)|^2 v = 0 \left(\nabla_\zeta^2 = \frac{\partial^2}{\partial \xi^2} + \frac{\partial^2}{\partial \eta^2} \right). \tag{32}$$

The boundary condition, (2), remains unchanged because a mapping by a regular function is conformal. In accordance with the above equation, we define A_{kl} and B_{kl} as

$$A_{kl} = \iint_\Omega \nabla f_k \cdot \nabla f_l \, d\xi \, d\eta \tag{33}$$

$$B_{kl} = \iint_\Omega f_k f_l |\phi'(\zeta)|^2 \, d\xi \, d\eta \tag{34}$$

with which the matrix eigenvalue problem is expressed in the same form as (8). Quantities A_{kl} and B_{kl} and their derivatives with respect to X_j can be computed easily because the region of integrals in (33) and (34) is a unit circle. Furthermore, A_{kl}'s are independent of $\phi(\zeta)$; hence

$$\frac{\partial A_{kl}}{\partial X_j} = 0, \qquad j = 1, \cdots, N; \quad k, l = 1, \cdots, M \tag{35}$$

holds.

Let those points in the ζ plane corresponding to the positions of ports $P_1, \cdots, P_t$ be denoted by $Q_1, \cdots, Q_t$, respectively. Those points are all located on the unit circle; hence, their arguments $\delta_1, \cdots, \delta_t$ can be used as the parameters representing the port locations (see Fig. 4).

[3] Instead of (29), Grüner [6] used an expansion of the form $|\phi'(\zeta)|^2 = \exp\{\sum_{k=1}^{N} c_k f_k(\xi,\eta)\}$ where $f_k(\xi,\eta)$ is a harmonic function. Therefore, a differential equation must be solved to determine the mapping function $\phi(\zeta)$ from $|\phi'(\zeta)|^2$. Expansion (29) seems better than Grüner's because mapping function $\phi(\zeta)$ is directly obtained from calculated coefficients as a regular function.

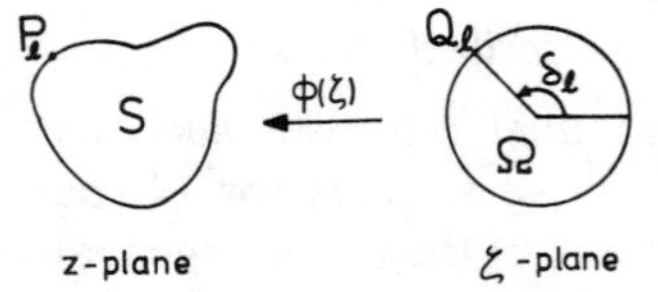

Fig. 4. Conformal mapping from the ζ plane into the z plane.

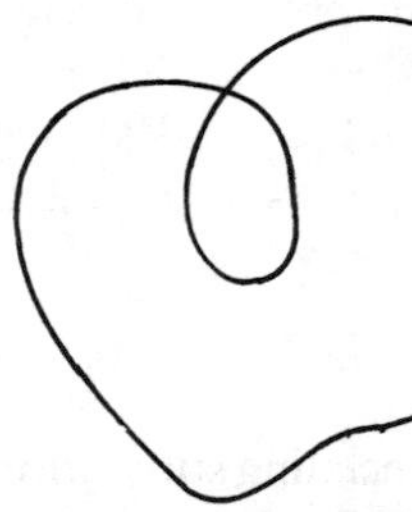

Fig. 5. An example of the circuit pattern without physical realizability.

TABLE I
FUNDAMENTAL FUNCTIONS USED IN THE SYNTHESIS

1	r	r^2	....	r^{n_r}
	r sin θ	r^2 sin θ	....	r^{n_r} sin θ
	r cos θ	r^2 cos θ	....	r^{n_r} cos θ
	⋮	⋮		⋮
	r sin $n_\theta\theta$	r^2 sin $n_\theta\theta$	....	r^{n_r} sin $n_\theta\theta$
	r cos $n_\theta\theta$	r^2 cos $n_\theta\theta$	....	r^{n_r} cos $n_\theta\theta$

We may apply the above conformal-mapping method to either of the syntheses described in Sections III-B and -C. They will be called synthesis processes III and IV, respectively. A problem in these processes is that the multivalent region, shown in Fig. 5, sometimes appears as the result of mapping. To prevent this, at least

$$\phi'(\zeta) \neq 0 \tag{36}$$

must hold in Ω. This condition is not always satisfied easily.

IV. EXAMPLES OF SYNTHESIS

Examples of synthesis based upon processes I–IV will be described. The fundamental functions used are shown in Table I, where $n_r = 4$ and $n_\theta = 2$.

A. Synthesis Processes I and II

The given parameters are: $N = 32$, $n = 3$, $\lambda_1 = 3.4$, $\lambda_2 = 7.0$, and $\lambda_3 = 7.4$. The successive modification of the circuit is shown in Fig. 6 for process I and in Fig. 7 for process II. Fig. 8(a) and (b) depicts how the lowest three eigenvalues of the circuit approach the prescribed values as the analysis and modification are iterated for process I and II, respectively. In Fig. 8 the horizontal solid lines indicate the prescribed eigenvalues.

These examples show that process I produces a smoother pattern than process II, but much more iteration is needed to

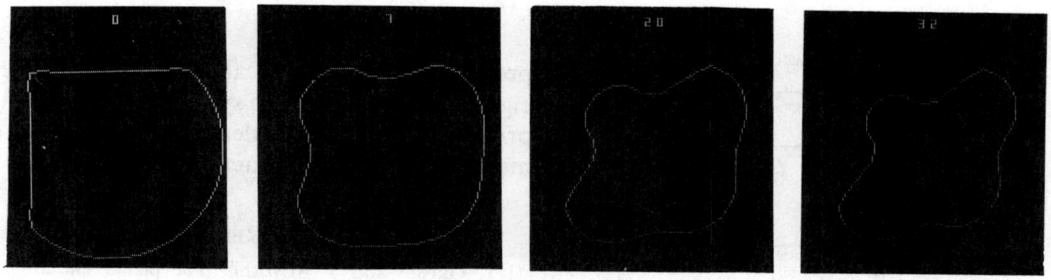

Fig. 6. Circuit patterns for iteration numbers $h = 0$, 7, 20, and 32 by process I (graphic output of the computer).

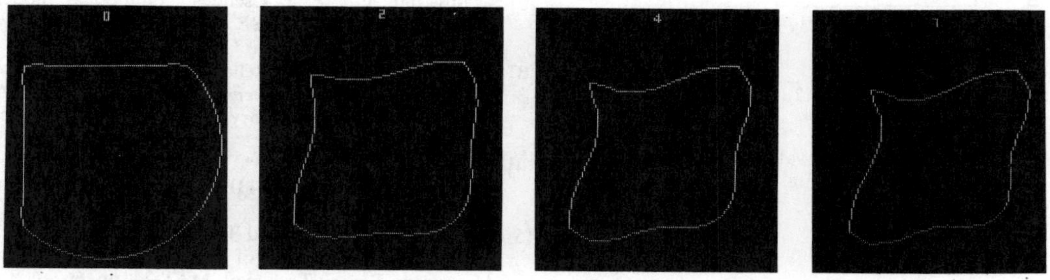

Fig. 7. Circuit patterns for iteration numbers $h = 0$, 2, 4, and 7 by process II (graphic output of the computer).

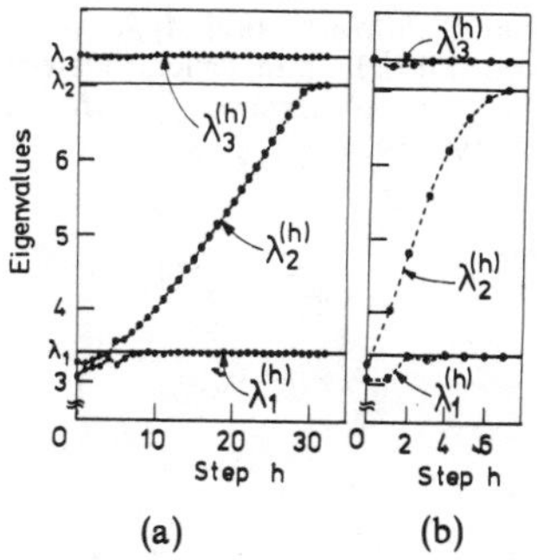

Fig. 8. The behavior of eigenvalues. (a) The case of Fig. 6. (b) The case of Fig. 7.

reach the prescribed eigenvalues. The above statement is valid for almost all cases. The computer time required for one cycle of analysis and modification was about 2 s in both processes when HITAC-8800 was used.

B. Synthesis Processes III and IV

Synthesis processes III and IV were applied to a similar problem, in which $n = 3$, $\lambda_1 = 3.5$, $\lambda_2 = 5.0$, and $\lambda_3 = 7.0$. The degree of the mapping function $m = 16$, that is, $N = 32$. The details of the modification processes are omitted for the sake of brevity.

The computer time for one cycle was about 2.5 s in both processes. The differences in the final pattern and in the speed of convergence were not remarkable for the two processes. However, processes III and IV are inferior to processes I and II in that the behavior of the convergence is somewhat oscillatory. Besides, in some cases, a univalent mapping could not be found. These methods, therefore, should better be employed only when external coupling ports must be considered.

Therefore, in this subsection, synthesis of an impedance matrix of a two-port planar circuit using process IV will be described in more detail. Two eigenvalues $\lambda_1 = 4.0$ and $\lambda_2 = 5.0$ are specified, and the prescribed values of the associated eigenfunctions at the ports are given as $q_{11} = 0.95$, $q_{12} = 0.95$, $q_{21} = -0.7$, and $q_{22} = 1.6$ (see (24)). Figs. 9 and 10 show the modification process and the convergence of parameters, respectively. In Fig. 9, P_1 and P_2 denote the port locations. The computer time required for one step was a little less than 2 s.

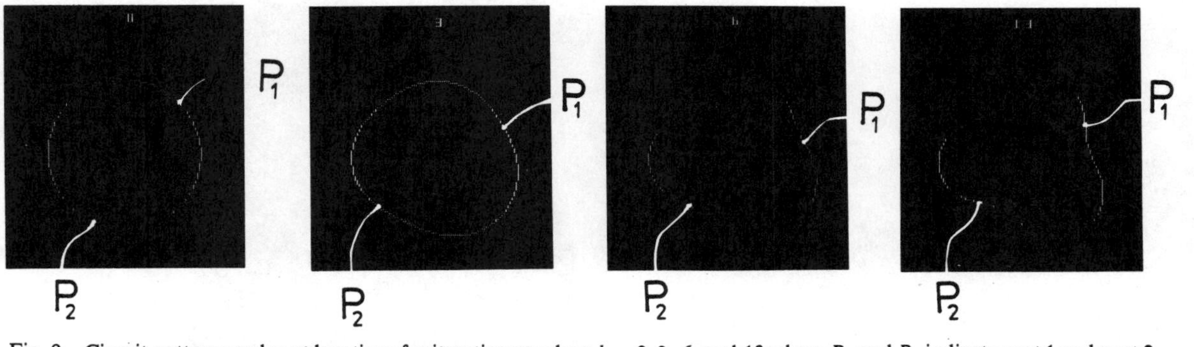

Fig. 9. Circuit patterns and port locations for iteration numbers $h = 0$, 3, 6, and 13 where P_1 and P_2 indicate port 1 and port 2, respectively (graphic output of the computer).

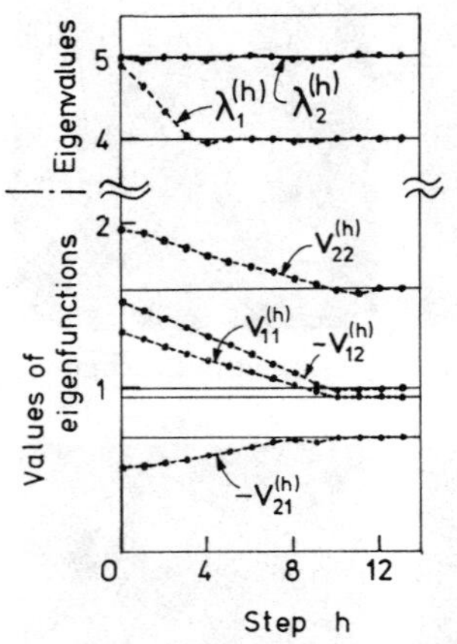

Fig. 10. The behavior of eigenvalues and the values of eigenfunctions at the ports in the case of Fig. 9.

V. Conclusion

A basic algorithm and numerical examples of the synthesis of planar circuits have been presented. When the number of the prescribed eigenvalues and external ports is relatively small, the results are satisfactory both in the computing time and accuracy. However, research is still at a primitive stage; further efforts are needed to make the proposed method practical, especially for larger numbers of eigenvalues and ports. In synthesis processes III and IV, the problem of the multivalent region must be overcome to make those methods practical.

References

[1] T. Okoshi and T. Miyoshi, "The planar circuit—An approach to microwave integrated circuitry," *IEEE Trans. Microwave Theory Tech.*, vol. MTT-20, pp. 245–252, Apr. 1972.

[2] S. Silvester and Z. J. Csendes, "Numerical modeling of passive microwave devices," *IEEE Trans. Microwave Theory Tech.*, vol. MTT-22, pp. 190–201, Mar. 1974.

[3] P. P. Civalleri and S. Ridella, "Analysis of a three-layer rectangular structure," in *Network Theory*, T. Boite, Ed. (Proc. Advanced Study Institute on Network Theory, Belgium, Sept. 1969.) London, England: Gordon and Breach, 1972.

[4] T. Okoshi and M. Saito, "Analysis and design of distributed planar circuits," presented at 1974 IEEE Int. Conf. Circuits and Systems, Apr. 23–25.

[5] T. Okoshi, Y. Uehara, and T. Takeuchi, "Segmentation method—An approach to the analysis of microwave planar circuit," *IEEE Trans. Microwave Theory Tech.*, vol. MTT-24, pp. 662–668, Oct. 1976.

[6] K. Grüner, "Method of synthesizing nonuniform waveguides," *IEEE Trans. Microwave Theory Tech.*, vol. MTT-22, pp. 317–322, Mar. 1974.

[7] A. Ben-Israel, "A Newton–Raphson method for the solution of systems of equations," *J. Math. Anal. Appl.*, vol. 15, pp. 243–252, 1966.

[8] O. Yu. Kul'chitskii and L. I. Shimelevich, "Determination of the initial approximation for Newton's method," *USSR Comp. Math. Phys.*, vol. 14, no. 4, pp. 188–190, 1974.

Paper 4.3

Application of Boundary-Element Method to Electromagnetic Field Problems

SHIN KAGAMI AND ICHIRO FUKAI

Abstract—This paper proposes an application of the boundary-element method to two-dimensional electromagnetic field problems. By this method, calculations can be performed using far fewer nodes than by the finite-element method, and unbounded field problems are easily treated without special additional consideration. In addition, the results obtained have fairly good accuracy. In this paper, analyzing procedures of electromagnetic field problems by the boundary-element method, under special conditions, are proposed and several examples are investigated.

I. Introduction

At present, the finite-element method is widely used in many fields. The main reason may be that, by the finite-element method, it is easy to handle inhomogeneities and complicated structures. However, it requires a large computer memory and long computing time to solve the final matrix equation. In addition, unbounded field problems need some additional techniques [1], [2].

Recently, the boundary-element method has been proposed, which is interpreted as a combination technique of the conventional boundary-integral equation method and a discretization technique, such as the finite-element method, and which has merits of both the above methods, i.e., the required size of the computer memory being small and the obtained results having fairly good accuracy [3], [4]. Namely, the boundary-element method is a boundary method and, therefore, if the region to be analyzed is homogeneous, then it requires nodes, necessary for calculation, on its boundary only. So the problem can be treated with one less dimension. Moreover, it can handle unbounded field problems easily, so that it is suitable for the electromagnetic field analysis which often includes unbounded regions [5], [6].

In this paper, a formulation of two-dimensional electromagnetic field problems by the boundary-element method and its application to several interesting cases, such as the problem of electromagnetic waveguide discontinuities, multi-media problems, and electromagnetic wave scattering problems [6]. In addition, several examples are analyzed and the results obtained with the boundary-element method are compared with rigorous ones, and solutions of the other numerical methods. The propriety of our analyzing procedure of the boundary-element method is verified.

Manuscript received April 15, 1983; revised October 12, 1983.
S. Kagami is with the Department of Electrical Engineering, Asahikawa Technical College, Asahikawa, Japan 070.
I. Fukai is with the Department of Electrical Engineering, Faculty of Engineering, Hokkaido University, Sapporo, Japan 060.

II. General Formulation

A two-dimensional region R enclosed by a boundary B, as illustrated in Fig. 1, is considered. In the region R, Helmholtz's equation

$$(\nabla^2 + k^2)u = 0 \tag{1}$$

holds, where u is the potential used for analysis, we write its outward normal derivative as q, and k denotes the wavenumber in free space. The boundary condition on B is

$$u = \bar{u} \tag{2}$$

or

$$q = \bar{q} \tag{3}$$

where "$^-$" means a known value. Here, Green's function

$$u^* = -\frac{j}{4} H_0^{(2)}(kr) \tag{4}$$

is introduced, where $H_0^{(2)}$ is the Hankel function of the second kind and order zero. By the method of weighted residuals [3], [4] or Green's formula, the following equation is obtained:

$$u_i + \int_B uq^* \, dc = \int_B qu^* \, dc. \tag{5}$$

In (5), the suffix "i" means an arbitrary point in the region R and q^* is the outward normal derivative of u^*

$$\begin{aligned} q^* &= \frac{j}{4} H_1^{(2)}(kr) \frac{\partial (kr)}{\partial n} \\ &= \frac{j}{4} k H_1^{(2)}(kr) \cos\angle(\boldsymbol{r}; \boldsymbol{n}). \end{aligned} \tag{6}$$

In (6), $H_1^{(2)}$ is the Hankel function of the second kind and order one. For the case where the point i is placed on the boundary B, the singular point of Green's function appears and special consideration is necessary. Now, we adopt the integration path going round the node i as shown in Fig. 2. Then, (5) is rewritten as follows:

$$u_i + \lim_{\epsilon \to 0} \int_{B'} uq^* \, dc + \lim_{\epsilon \to 0} \int_{B''} uq^* \, dc = \lim_{\epsilon \to 0} \int_{B'} qu^* \, dc + \lim_{\epsilon \to 0} \int_{B''} qu^* \, dc. \tag{7}$$

Here, we estimate the integration over the boundary B'' as follows:

$$\lim_{\epsilon \to 0} \int_{B''} uq^* \, dc = \lim_{\epsilon \to 0} \int_{B''} u \frac{j}{4} k H_1^{(2)}(k\epsilon) \, dc$$

Reprinted from *IEEE Trans. Microwave Theory Tech.*, vol. 32, no. 4, pp. 455–461, April 1984.

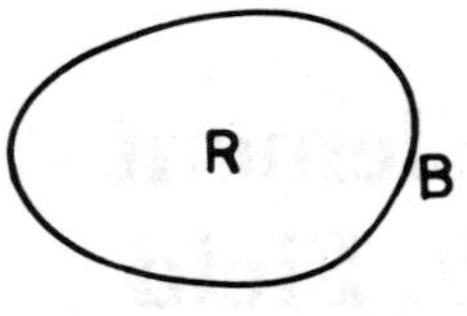

Fig. 1. Two-dimensional region R.

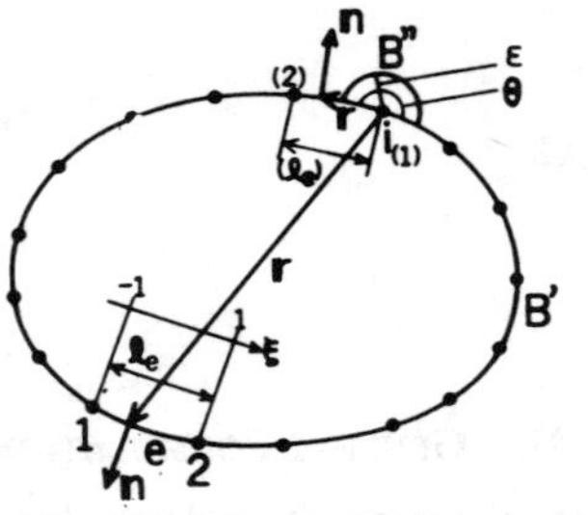

Fig. 2. Integration on each element.

$$
\begin{aligned}
&= \lim_{\epsilon \to 0} u \frac{j}{4} k H_1^{(2)}(k\epsilon)\epsilon\theta \\
&= \frac{j}{4} k\theta \lim_{\epsilon \to 0} \left[\epsilon \left\{ \frac{k\epsilon}{2} - j\left(-\frac{2}{\pi}\frac{1}{k\epsilon} \right) \right\} \right] \\
&= -\frac{\theta}{2\pi} u_i
\end{aligned}
\tag{8}
$$

$$
\begin{aligned}
\lim_{\epsilon \to 0} \int_{B''} qu^* \, dc &= \lim_{\epsilon \to 0} \int_{B''} q\left\{ -\frac{j}{4} H_0^{(2)}(k\epsilon) \right\} dc \\
&= \lim_{\epsilon \to 0} \left[q\left(-\frac{j}{4} \right) \left\{ 1 - j\frac{2}{\pi}(\ln k\epsilon + \gamma - \ln 2) \right\} \epsilon\theta \right] \\
&= 0 \\
\gamma &= 0.5772 \cdots \quad \text{(Euler's number)}.
\end{aligned}
\tag{9}
$$

From (5), (8), and (9), we derive the next equation as follows:

$$
c_i u_i + \oint_B qu^* \, dc = \oint_B qu^* \, dc,
$$

$$
c_i = 1 - \frac{\theta}{2\pi} \qquad \oint_B = \lim_{\epsilon \to 0} \int_{B'}.
\tag{10}
$$

In (10), $\oint$ denotes the Cauchy's principal value of integration. Next, the boundary B is approximated by the connection of line elements, on each of which u and q are assumed to vary linearly. Then, (10) is discretized as follows:

$$
c_i u_i + \sum_{e=1}^{n} [h_1 h_2]_e \begin{Bmatrix} u_1 \\ u_2 \end{Bmatrix}_e = \sum_{e=1}^{n} [g_1 g_2]_e \begin{Bmatrix} q_1 \\ q_2 \end{Bmatrix}_e
\tag{11}
$$

In (11), n is the number of elements and u_1, u_2, q_1, and q_2 are the potentials and the potential derivatives on the two nodes of the eth element and h_1, h_2, g_1, and g_2 are contributions of the eth element to integration. When the point i does not belong to the eth element, they are calculated with Gaussian integration as

$$
\begin{Bmatrix} h_1 \\ h_2 \end{Bmatrix} = \int_{-1}^{1} \begin{Bmatrix} \phi_1 \\ \phi_2 \end{Bmatrix} \frac{1}{4} jk H_1^{(2)}(kr) \cos\angle(\boldsymbol{r}; \boldsymbol{n}) \, d\xi \cdot \frac{1}{2} l_e
\tag{12}
$$

$$
\begin{Bmatrix} g_1 \\ g_2 \end{Bmatrix} = \int_{-1}^{1} \begin{Bmatrix} \phi_1 \\ \phi_2 \end{Bmatrix} \left\{ -\frac{1}{4} j H_0^{(2)}(kr) \right\} d\xi \cdot \frac{1}{2} l_e
\tag{13}
$$

where ξ is a normalized coordinate defined on the eth element, l_e is the length of the element, and ϕ_1 and ϕ_2 are the interpolation functions given as follows:

$$
\begin{Bmatrix} \Phi_1 \\ \Phi_2 \end{Bmatrix} = \frac{1}{2}(1 \mp \xi).
\tag{14}
$$

When the point i belongs to the eth element, calculations of g_1, g_2, h_1, and h_2 involve the limitation of $\epsilon \to 0$. In this case, the vector $\boldsymbol{r}$ is at right angles to the outward normal vector $\boldsymbol{n}$ (see Fig. 2), so that $\cos(\boldsymbol{r}; \boldsymbol{n})$ becomes equal to zero and

$$
h_1 = h_2 = 0.
\tag{15}
$$

The remainders are g_1 and g_2. Here, we consider the case where the point i coincides with the first node of the eth element

$$
\begin{aligned}
g_1 &= \lim_{\epsilon \to 0} \int_{-1+\epsilon}^{1} \frac{1}{2}(1-\xi) \left\{ -\frac{1}{4} j H_0^{(2)}(kr) \right\} d\xi \cdot \frac{1}{2} l_e \\
&= \frac{j l_e}{16} \lim_{\epsilon \to 0} \int_{-1+\epsilon}^{1} (\xi - 1) H_0^{(2)}\left\{ k(\xi+1)\frac{1}{2} l_e \right\} d\xi \\
&= \frac{j l_e}{16} \lim_{\epsilon \to 0} \int_{\epsilon}^{2} \left\{ \xi H_0^{(2)}\left(k\xi \cdot \frac{1}{2} l_e \right) - 2 H_0^{(2)}\left(k\xi \cdot \frac{1}{2} l_e \right) \right\} d\xi \\
&= \frac{j l_e}{16} \lim_{\epsilon \to 0} \left\{ \left[\frac{2\xi}{k l_e} H_1^{(2)}\left(\frac{1}{2} k l_e \xi \right) \right]^2 - 2\int^{2} H_0^{(2)}\left(\frac{1}{2} k l_e \xi \right) d\xi \right\}.
\end{aligned}
\tag{16}
$$

In the second term of (16), the Hankel function is expanded to infinite series and is integrated by term

$$
\begin{aligned}
g_1 = \frac{1}{4} j l_e \Bigg[& \frac{1}{k l_e} H_1^{(2)}(k l_e) + j\frac{2}{\pi} \left\{ \ln\left(\frac{1}{2} k l_e \right) + \gamma - 1 - \left(\frac{1}{k l_e} \right)^2 \right\} - 1 \\
& - \sum_{s=1}^{\infty} \frac{(-1)^s (k l_e)^{2s}}{(s!)^2 2^{2s} (2s+1)} \left\{ 1 - j\frac{2}{\pi} \left(\ln\frac{k l_e}{2} + \gamma - h_s - \frac{1}{2s+1} \right) \right\} \Bigg] \\
h_s = 1 + & \frac{1}{2} + \frac{1}{3} + \cdots + \frac{1}{s}.
\end{aligned}
\tag{17}
$$

From (16) and (17), we obtain the next result for g_2

$$
g_2 = -\frac{1}{4} j l_e \left[\frac{1}{k l_e} H_1^{(2)}(k l_e) - j\frac{2}{\pi} \left(\frac{1}{k l_e} \right)^2 \right]
\tag{18}
$$

If the point i is the second node of the eth element, then g_1 is estimated with (18) and g_2 with (17). The infinite series in (17) is approximated by the first few terms, since the length of the eth element l_e is chosen as $l_e < \lambda/10$, i.e., $k l_e < 2\pi/10$, so that the series rapidly converges, where λ is the wavelength in free space.

In the above calculations, the estimation of c_i is important for the Dirichlet boundary condition giving the nonzero potential. Because c_i is the coefficient of u_i, any value of c_i is permitted for the boundary condition giving the zero potential; but for the case of the nonzero potential, c_i must be calculated precisely [7].

In the matrix notation, (11) is rewritten as follows:

$$
\boldsymbol{Hu} = \boldsymbol{Gq}.
\tag{19}
$$

On each node, the potential u or its outward normal derivative q must be given as the boundary condition. Then, all the remaining u's and q's can be calculated from (19). The matrices produced

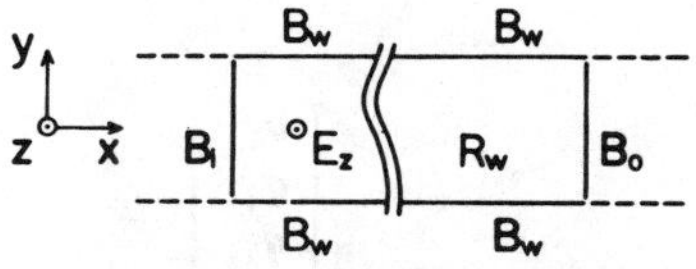

Fig. 3. A parallel-plane waveguide with discontinuities.

by the boundary-element formulation are much smaller in size than the finite-element method.

III. The Case of Waveguides with Discontinuities

For a typical example, a parallel-plane waveguide with discontinuities is considered, and the mode having the z-component of the electric field, which is chosen as the analyzed potential u, is assumed. In this case, a closed region R_w, as shown in Fig. 3, is chosen as the analyzed model, which is enclosed by the boundary of the waveguide wall B_w and pseudo-boundaries at the power supply side and the opposite load side, which we call the input-side boundary B_i and the output-side boundary B_o, respectively.

From (19), the following equation is obtained for the region R_w:

$$[H_i H_o H_w]\begin{Bmatrix} u_i \\ u_o \\ u_w \end{Bmatrix} = [G_i G_o G_w]\begin{Bmatrix} q_i \\ q_o \\ q_w \end{Bmatrix}, \quad \text{in } R_w. \tag{20}$$

In (20), suffixes i, o, and w show the quantities corresponding to the boundaries B_i, B_o, and B_w, respectively. On the waveguide wall, the electric-field component parallel to it vanishes, so that the following boundary condition is taken:

$$u_w = 0, \quad \text{on } B_w. \tag{21}$$

But on the remaining boundaries B_i and B_o, any specified value of u or q cannot be given. If they are given, the phase relation between the field components on the input- and the output-side boundaries are also given, and, therefore, the problem has already been solved. This is a contradiction.

Now, we adopt the following procedure. First, we place the two pseudo-boundaries B_i and B_o at the position where it is considered that the reflecting electromagnetic wave, generated at the discontinuites, attenuates and almost vanishes. The TE_{10}-mode field distribution is assumed on them. Then, on only B_i, the boundary condition, in complex form, is placed

$$u_i = \bar{u}_i, \quad \text{on } B_i. \tag{22}$$

Next, on B_o, we introduce the TE_{10}-mode propagation constant β, and the electric-field component is written as

$$u_o = E_{z0}\exp(-j\beta y), \quad \text{on } B_o \tag{23}$$

and its outward normal derivative q_o is also written as

$$q_o = \frac{du_o}{dy} = -j\beta E_{z0}\exp(-j\beta y), \quad \text{on } B_o. \tag{24}$$

So, on B_o, the next relationship between u_o and q_o is obtained as follows:

$$q_o = -j\beta u_0, \quad \text{on } B_o. \tag{25}$$

From (20), (21), (22), and (25), the following equation, to be solved finally, is obtained:

$$[G_i - j\beta F_o - H_o G_w]\begin{Bmatrix} q_i \\ u_o \\ q_w \end{Bmatrix} = H_i \bar{u}_i, \quad \text{in } R_w. \tag{26}$$

In the right-hand side of (26), all quantities are known and q_i, u_o, and q_w are obtained as a solution. Then, q_o is given by (25).

IV. Multi-Media Problems

For multi-media cases, any boundary method requires to make up equations for each homogeneous sub-domain constructed of one media. So it is generally said that boundary methods are weak in multi-media problems and the finite-element method is superior to the boundary-element method in such a case. However, the authors have verified that a bit of effort on the design of the computer program makes this fault of the boundary-element method negligible, and they propose a procedure of programming, for handling multi-media problems, in a slightly different style from those of the references [3], [4].

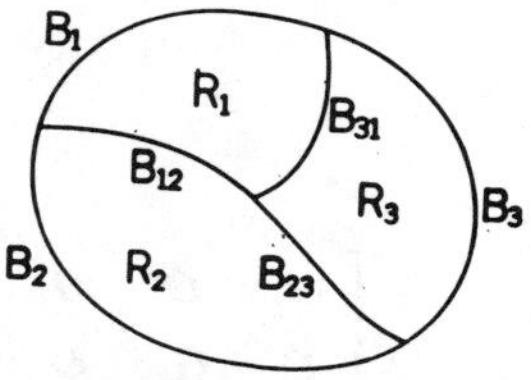

Fig. 4. Two-dimensional region constructed of three media.

Consider a two-dimensional region constructed of three different media, as shown in Fig. 4, where R_1, R_2, and R_3 are homogeneous sub-domains, B_1, B_2, and B_3 are the boundaries belonging to only R_1, R_2, and R_3, respectively, and B_{12}, B_{23}, and B_{31} are the interfaces between two adjacent sub-domains. The ordinary boundary-element technique leads to the following equations for each homogeneous sub-domain:

$$[G_1 G_{12}^{(1)} G_{31}^{(1)}]\begin{Bmatrix} q_1 \\ q_{12}^{(1)} \\ q_{31}^{(1)} \end{Bmatrix} = [H_1 H_{12}^{(1)} H_{31}^{(1)}]\begin{Bmatrix} u_1 \\ u_{12}^{(1)} \\ u_{31}^{(1)} \end{Bmatrix}, \quad \text{in } R_1 \tag{27}$$

$$[G_2 G_{23}^{(2)} G_{12}^{(2)}]\begin{Bmatrix} q_2 \\ q_{23}^{(2)} \\ q_{12}^{(2)} \end{Bmatrix} = [H_2 H_{23}^{(2)} H_{12}^{(2)}]\begin{Bmatrix} u_2 \\ u_{23}^{(2)} \\ u_{12}^{(2)} \end{Bmatrix}, \quad \text{in } R_2 \tag{28}$$

$$[G_3 G_{31}^{(3)} G_{23}^{(3)}]\begin{Bmatrix} q_3 \\ q_{31}^{(3)} \\ q_{13}^{(3)} \end{Bmatrix} = [H_3 H_{31}^{(3)} H_{13}^{(3)}]\begin{Bmatrix} u_3 \\ u_{31}^{(3)} \\ u_{13}^{(3)} \end{Bmatrix}, \quad \text{in } R_3. \tag{29}$$

In (27)–(29), superscript (i) implies the quantity defined in R_i. The boundry conditions on the interfaces are as follows:

$$u_{12}^{(1)} = u_{12}^{(2)} = u_{12}, \quad q_{12}^{(1)} = -q_{12}^{(2)} = q_{12}, \quad \text{on } B_{12} \tag{30}$$

$$u_{23}^{(2)} = u_{23}^{(3)} = u_{23}, \quad q_{23}^{(2)} = -q_{23}^{(3)} = q_{23}, \quad \text{on } B_{23} \tag{31}$$

$$u_{31}^{(3)} = u_{31}^{(1)} = u_{31}, \quad q_{31}^{(3)} = -q_{31}^{(1)} = q_{31}, \quad \text{on } B_{31}. \tag{32}$$

In the above conditions, the minus sign of q originates from the outward normal directions of the adjacent two subregions opposite each other. From (27)–(32), we obtain the next equation to be solved finally.

$$\begin{bmatrix} G_1 & O & O & G_{12}^{(1)} & -H_{12}^{(1)} & O & O & -G_{31}^{(1)} & -H_{31}^{(1)} \\ O & G_2 & O & -G_{12}^{(2)} & -H_{12}^{(2)} & G_{23}^{(2)} & -H_{23}^{(2)} & O & O \\ O & O & G_3 & O & O & -G_{23}^{(3)} & -H_{23}^{(3)} & G_{31}^{(3)} & H_{31}^{(3)} \end{bmatrix} \begin{bmatrix} q_1 \\ q_2 \\ q_3 \\ q_{12} \\ u_{12} \\ q_{23} \\ u_{23} \\ q_{31} \\ u_{31} \end{bmatrix} \quad (33)$$

$$= \begin{bmatrix} H_1 & O & O \\ O & H_2 & O \\ O & O & H_3 \end{bmatrix} \begin{bmatrix} u_1 \\ u_2 \\ u_3 \end{bmatrix}, \qquad \text{in } R_1 + R_2 + R_3.$$

V. Electromagnetic Scattering Problems

In this section, the procedure analyzing the problem of electromagnetic scattering by dielectric bodies is developed. This is considered to be the problem of multi-media and the unbounded field. Therefore, the procedure developed for multi-media problems can be utilized for the scattering problem in the form extended a little.

It is assumed that the incident wave is the E-wave and propagates in the direction parallel to the $x-y$ plane. All quantities are uniform in the z-direction. The analyzed region is constructed of two subregions, as shown in Fig. 5. Subregion R_I is inside the dielectric cylinder and subregion R_O is outside of it. The latter is an unbounded field, but at infinity, the radiation condition can be taken for the scattering wave, so that the boundary-element equations are considered on only the dielectric surface B_d

$$G_I q_I = H_I u_I, \qquad \text{in } R_I$$

$$u_I = u_{I,sc} + u_{I,in} \quad (34)$$

$$q_I = q_{I,sc} + q_{I,in}$$

$$G_O q_{O,sc} = H_O u_{O,sc}, \qquad \text{in } R_O. \quad (35)$$

Here, subscripts I and O imply the quantities defined in R_I and R_O, respectively, and sc and in indicate the scattering and incident waves, respectively. In R_I, the ruling equation is defined using the total wave, i.e., the incident wave plus the scattering wave. On the contrary, in R_O, the radiation condition can't be applied for the incident wave, so that the equation is defined by only the scattering wave.

On B_d, the following boundary conditions are taken:

$$u_{I,sc} = u_{O,sc} = u_{sc}, \qquad \text{on } B_d \quad (36)$$

$$q_{I,sc} = -q_{O,sc} = q_{sc}, \qquad \text{on } B_d. \quad (37)$$

In addition, the incident wave is described as follows:

$$u_{in} = u_{I,in} = E_{z_0,in} \exp(jk\rho). \quad (38)$$

Here, the ρ coordinate is chosen in the direction of the incidence, as in Fig. 5. For $q_{I,in}$, we derive the following equation:

$$q_{I,in} = \frac{du_{I,in}}{dn_I} = -jk\boldsymbol{n}_I \cdot \boldsymbol{n}_\rho u_{I,in}$$

$$= x\cos\theta + y\sin\theta \quad (39)$$

where θ is the incident angle. In matrix notation, the above relation is rewritten as

$$q_{I,in} = B_I u_{I,in}. \quad (40)$$

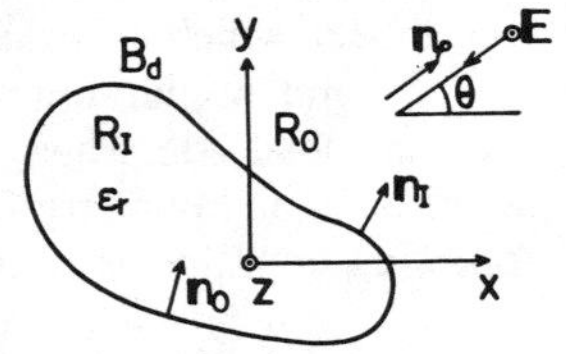

Fig. 5. Electromagnetic scattering by a dielectric cylinder.

From (34)–(40), the following equation is obtained for $R_I + R_O$:

$$\begin{bmatrix} -G_I & H_I \\ G_O & -H_O \end{bmatrix} \begin{bmatrix} u_{sc} \\ q_{sc} \end{bmatrix} = \begin{bmatrix} H_I - G_I B_I \\ O \end{bmatrix} u_{I,in}. \quad (41)$$

By (41), the scattering field is calculated on the boundary B_d provided that the incident wave is given there.

The procedure proposed here is very powerful for the case where the dielectric body has a much larger dimension than the wavelength of the incidence or has a large dielectric constant. The finite-element method is weak because it requires a large computer memory.

VI. Analyzed Results

A. Open-Ended Parallel-Plane Waveguides

We analyzed the reflection coefficients of three kinds of waveguides, i.e., flanged, unflanged, and flared, consisting of two parallel planes. These are unbounded field problems, and closed boundaries extending to infinity should be chosen. But at infinity, the radiation condition can be introduced, and at the point several wavelengths distant from the open-end of the waveguide, the field value becomes negligibly small, so that the contribution to the integration can be neglected. The finite models are chosen as in Fig. 6, which are half-models, because each of them has the symmetry axis parallel to the direction of electromagnetic wave propagation. Fig. 7 denotes the results obtained with the boundary-element method for the above three cases. They have good agreement with Lee's solution by ray theory [8], Vaynshteyn's by the Wiener-Hopf technique [9], and the results by the finite-element method extended to unbounded fields [2]. For the flared waveguide, the boundary-element method results are compared with those of the boundary integral equation method [10]. All of the above boundary-element calculations are performed with 40 nodes, while the corresponding finite-element calculations need at least 400 nodes. The problems analyzed here are unbounded ones, for which the boundary-element method seems to be the most suitable method.

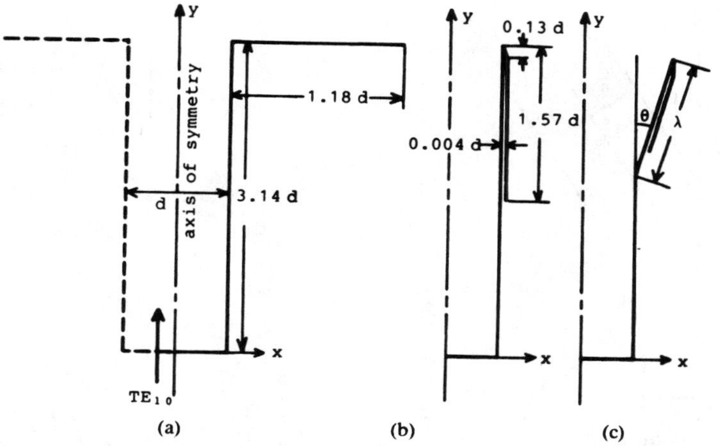

Fig. 6. Analyzed models of open-ended parallel-plane waveguides. (a) Flanged, (b) unflanged, and (c) flared.

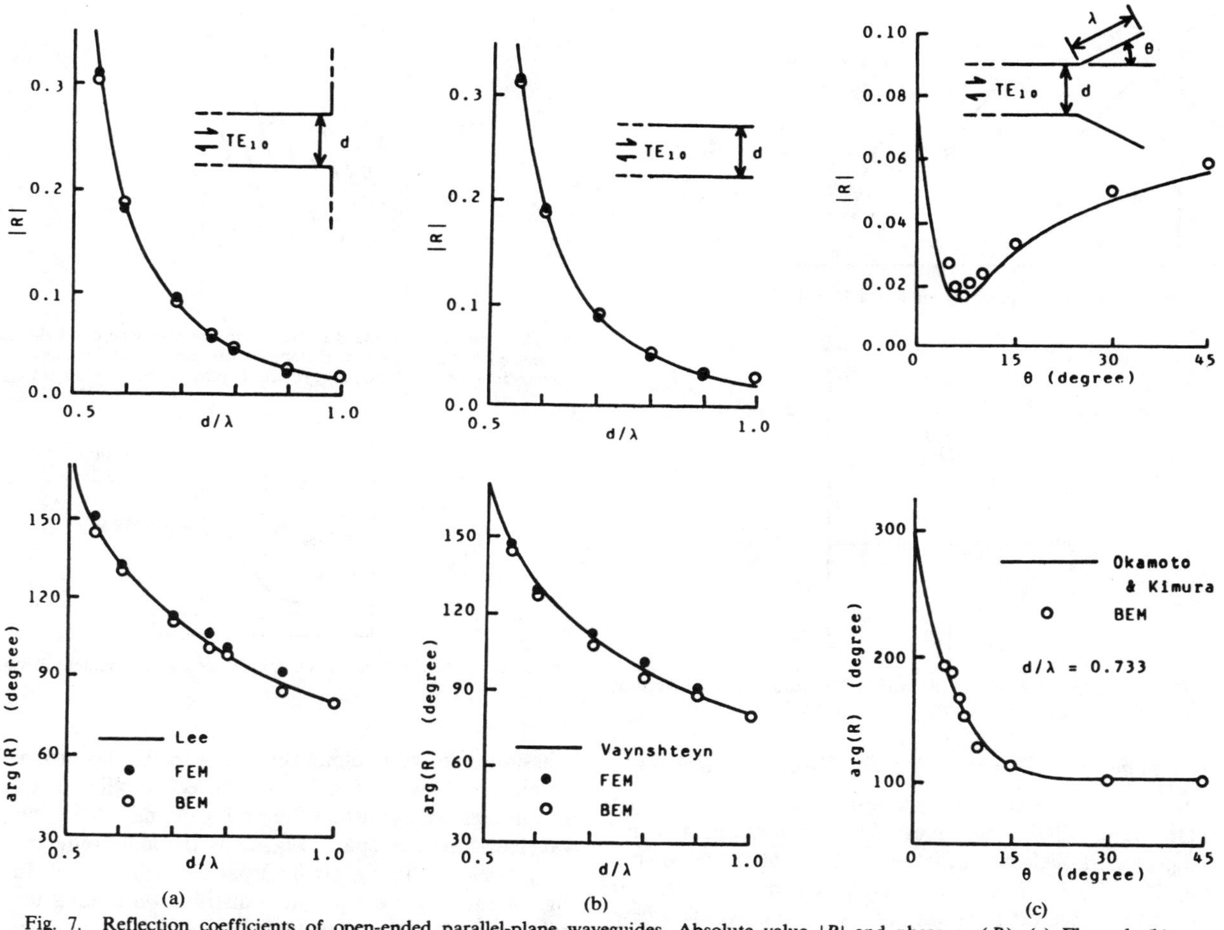

Fig. 7. Reflection coefficients of open-ended parallel-plane waveguides. Absolute value $|R|$ and phase $\arg(R)$. (a) Flanged, (b) unflanged, and (c) flared.

B. A Symmetrical H-Plane Y-Junction

The electric-field distribution in an *H*-plane Y-junction is analyzed. The electromagnetic wave is assumed to propagate along the *y*-axis, as in Fig. 8, which also indicates the analyzed model. The right half of the model is considered for the symmetry. On the input- and output-side boundaries, the TE_{10} mode is assumed. Fig. 9 shows the electric-field distribution, represented by the standing-wave on the centerlines. Details of their positions are shown in Fig. 8. From the results in Fig. 8, the reflection coefficient due to this Y-junction is calculated as 0.18. This value leads to the calculated amplitude of the electric field at the output side of the junction, $\sqrt{(1-0.18^2)/2} = 0.6956$, where that of the input side is assumed to be 1.0. The value, 0.6956, agrees well with the boundary-element result of the electric standing-wave amplitude in Fig. 9.

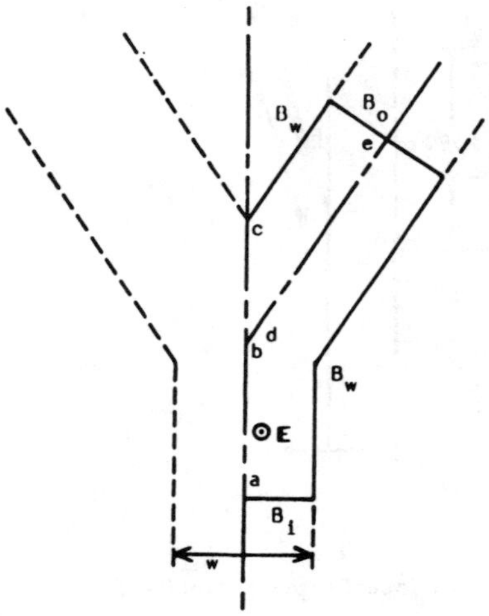

Fig. 8. Analyzed model of H-plane Y-junction.

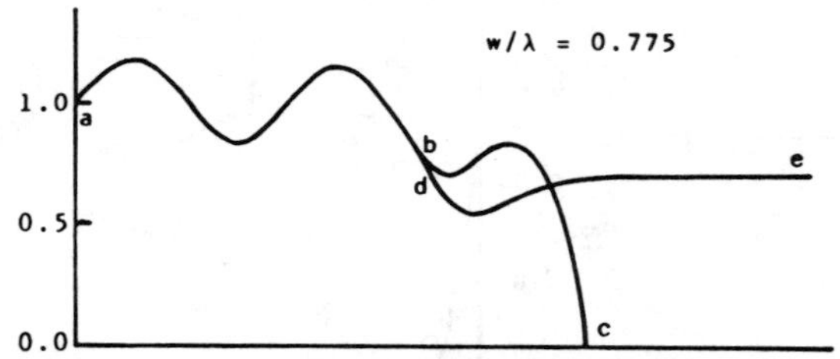

Fig. 9. The electric-field distribution in a H-plane Y-junction.

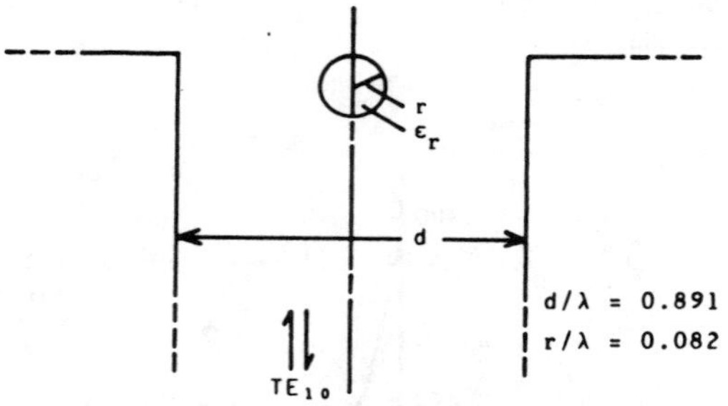

Fig. 10. Flanged open-ended waveguide with a dielectric cylinder placed at its open-end.

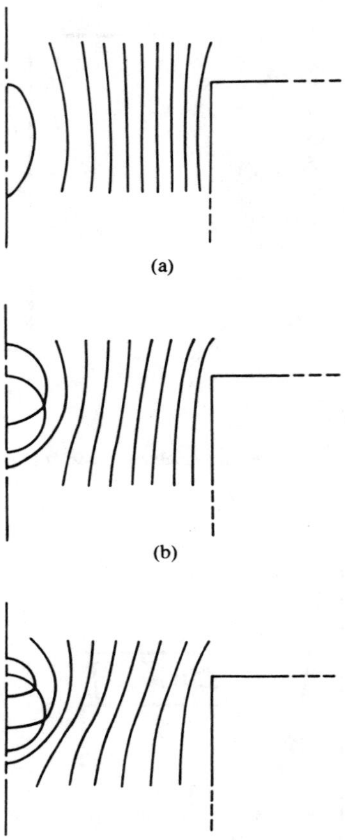

Fig. 11. The electric-field distribution in the case where a dielectric cylinder is placed at the open-end of flanged waveguides. (a) Dielectric constant of the cylinder $\epsilon_r = 1.0$, i.e., no cylinder is placed. (b) $\epsilon_r = 2.0$. (c) $\epsilon_r = 3.0$.

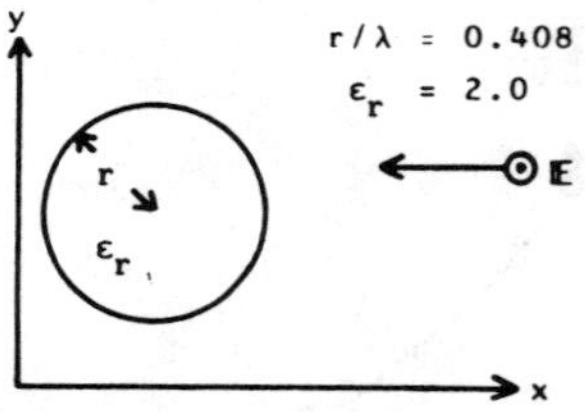

Fig. 12. Dielectric circular cylinder and the incident E-wave.

C. A Waveguide with a Dielectric Cylinder Placed at its Open-end

Here, we analyze the electromagnetic-field distribution of the case where the parallel-plane waveguide is open-ended with infinite flanges and a dielectric cylinder is placed at the open-end, as in Fig. 10. Fig. 11 shows the electric-field distributions for three values of dielectric constant, 1.0, 2.0, and 3.0. The case where $\epsilon_r = 1.0$ implies a homogeneous case, i.e., no cylinder is placed. This model brings the reflection coefficient, 0.029, which agrees with that obtained by the computer program for the homogeneous case (cf. Fig. 7(a)) and clarifies the propriety of the multi-media case analyzing procedure. From Fig. 11, it is found that the larger dielectric constant becomes, the denser the electric-field concentrates to the dielectric cylinder.

D. Electromagnetic Scattering of Dielectric Circular Cylinder

As the last example, the plane electromagnetic wave scattering by a dielectric circular cylinder is investigated. The incident wave is assumed to travel along the x-axis in the negative x-direction. Details are shown in Fig. 12. The dielectric cylinder is assumed to have a dielectric constant 2.0 and a radius 0.408λ, where λ is a wavelength in free space. Fig. 13 is the amplitude and phase of the E-wave scattering far-field pattern in this case. In addition, Fig. 14 shows the electric-field distribution around the dielectric cylinder. In Fig. 13, the results obtained by the boundary-element method are compared with analytical ones and show good accuracy. The calculations are done with only 24 nodes. For smaller cylinders, more accurate results have been obtained. For larger cylinders, no actual analysis has yet been performed, but it seems that the boundary-element method is more suitable than the finite-element method.

E. Computer Implementation

Boundary-element calculations require only a small computer memory, so that a microcomputer can handle them. The techniques discussed in this paper have been implemented by a

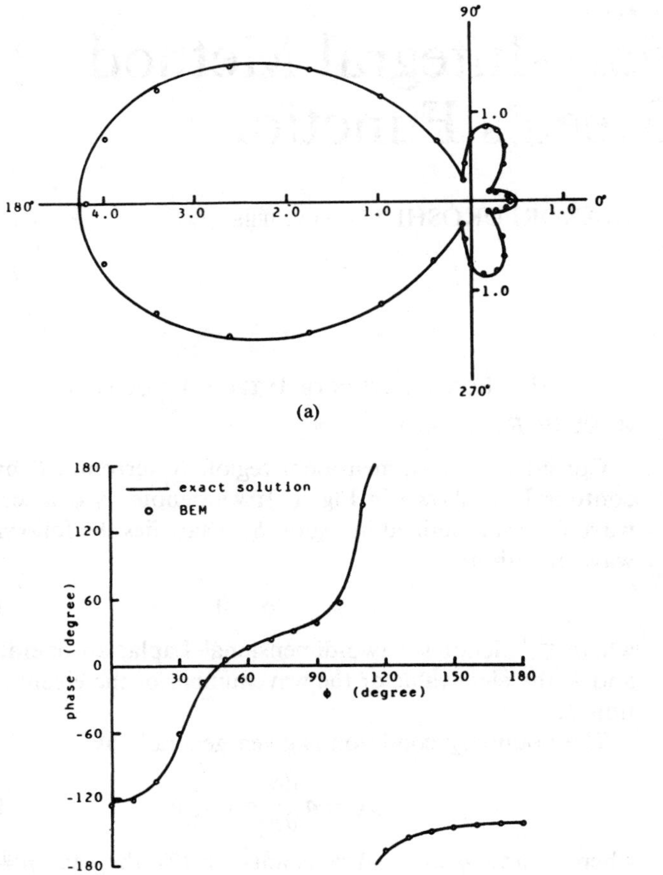

Fig. 13. *E*-wave scattering far-field pattern for dielectric circular cylinder in the case where *E*-wave incidents along *x*-axis in the negative *x*-direction. (a) Amplitude. (b) Phase.

FORTRAN program on a microcomputer, whose main CPU is MC68000 (8 MHz) and whose operating system is the UCSD *p*-system. Typically, the case of a parallel-plane waveguide having 40 nodes took about 20 m of CPU time.

VII. Conclusions

Application of the boundary-element method to electromagnetic-field problems was proposed. Several analyzing procedures for interesting cases were also given. The results obtained show that the boundary-element method is a very powerful numerical method for electromagnetic-field problems. Namely, by using the boundary-element method, far fewer nodes than by the finite-element method bring good accuracy, and unbounded field problems can be treated without any additional technique.

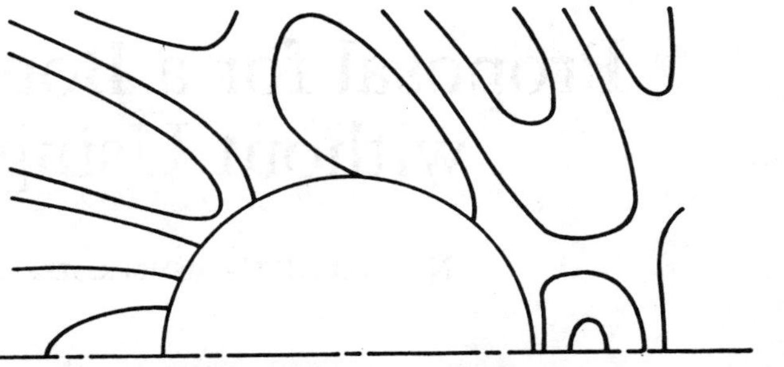

Fig. 14. The electric-field distribution around the dielectric circular cylinder.

References

[1] B. H. McDonald and A. Wexler, "Finite-element solution of unbounded field problems," *IEEE Trans. Microwave Theory Tech.*, vol. MTT-20, pp. 841–847, Dec. 1972.

[2] S. Washisu, I. Fukai, and M. Suzuki, "Extension of finite-element method to unbounded field problems," *Electron. Lett.*, vol. 15, pp. 772–774, Nov. 1979.

[3] C. A. Brebbia, *The Boundary Element Method for Engineers*. London: Pentech Press, 1978.

[4] C. A. Brebbia and S. Walker, *Boundary Element Techniques in Engineering*. London: Butterworth, 1980.

[5] S. Washisu and I. Fukai, "An analysis of electromagnetic unbounded field problems by boundary element method," *Trans. Inst. Electron. Commun. Eng. Jpn.*, vol. J64-B, pp. 1359–1365, Dec. 1981.

[6] S. Bilgen and A. Wexler, "Spline boundary element solution of dielectric scattering problems," in *12th Eur. Microwave Conf.*, (Helsinki, Finland), Sept. 1982, pp. 372–377.

[7] C. G. Williams and G. K. Cambrell, "Efficient numerical solution of unbounded field problems," *Electron. Lett.*, vol. 8, pp. 247–248, May 1972.

[8] S. W. Lee, "Ray theory of diffraction by open-ended waveguides. Part I," *J. Math. Phys.*, vol. 11, pp. 2830–2850, Sept. 1970.

[9] L. A. Vaynshteyn, *The Theory of Diffraction and the Factorization Method*. Boulder, CO: Golem, 1969.

[10] N. Okamoto and T. Kimura, "Radiation properties of *H*-plane two-dimensional horn antennas of arbitrary shape," *Trans. Inst. Electron. Commun. Eng. Jpn.*, vol. J59-B, pp. 25–32, Jan. 1976.

Proposal for a Boundary-Integral Method without Using Green's Function

NAOTO KISHI, MEMBER, IEEE, AND TAKANORI OKOSHI, FELLOW, IEEE

Abstract —**A new method for solving electromagnetic boundary-value problems is presented. The new method is a modification of the conventional boundary-element method; the conventional method is modified by using the reciprocity theorem derived from Green's identity, making the use of Green's function unnecessary. To confirm the validity of the new method, numerical analyses are presented for Dirichlet- and Neumann-type boundary-value problems of a two-dimensional scalar wave equation.**

I. Introduction

VARIOUS NUMERICAL methods are now available for the analysis of electromagnetic boundary-value problems [1]. In particular, since the early 1970's, integral-equation methods [2]–[7] have become popular because of the relatively short computer time required. In the past 15 years, of the integral-equation methods, the so-called boundary-element method (sometimes called the counter-integral method) has been used frequently to solve electromagnetic boundary-value problems [3]–[7].

In the boundary-element method, the boundary-integral equation is obtained first from Green's identity by using Green's function and is then solved by a discretization procedure similar to the finite-element method. As the Green's function for a two-dimensional scalar wave equation, various zero-order Bessel or Hankel functions, or their combinations, have been used [3], [4], [6], [7].

However, to formulate the boundary-integral equation, Green's function is not always necessary. The use of Green's function is sometimes even harmful because the integral equation will have a singular point when Green's function is used; the singular point makes the numerical analysis more complicated.

In this paper, we present a new boundary-integral method without using Green's function. In the new method, a homogeneous solution of the wave equation is used instead of Green's function, as has been done in [2] to solve scattering problems. We apply this method to solve Dirichlet- and Neumann-type boundary-value problems of a two-dimensional scalar wave equation. The moment method [10] is used in the computation.

Manuscript received November 28, 1986; revised June 11, 1987. This work was supported by a Grant-in-Aid for Scientific Research from the Ministry of Education, Science and Culture.

The authors are with the Department of Electronic Engineering, University of Tokyo, Tokyo, Japan 113.

IEEE Log Number 8716524.

II. Formulation of Integral Equation

A. Basic Equations

Consider a two-dimensional region S surrounded by a contour Γ as shown in Fig. 1. If we denote by ϕ a scalar wave function defined in region S, ϕ satisfies the following wave equation:

$$\nabla_t^2\phi + k^2\phi = 0 \qquad (1)$$

where ∇_t^2 denotes a two-dimensional Laplacian operator, and k the eigenvalue or the wavenumber of the eigenfunction ϕ.

The boundary condition is given generally as

$$p\phi + q\frac{\partial \phi}{\partial n} = 0 \qquad (2)$$

where p and q are real constants. In (2), the case $p \neq 0$, $q = 0$ and the case $p = 0$, $q \neq 0$ represent Dirichlet- and Neumann-type boundary conditions, respectively. The operator $(\partial/\partial n)$ denotes the derivative in the outward normal direction on Γ.

B. Conventional Boundary-Element Method (BEM Formulation)

In conventional BEM analyses [3]–[7], the wave equation (1) is first converted to a boundary-integral equation by using the two-dimensional Green's theorem (see [3] for the derivation) as

$$\phi(\mathbf{r}') = 2\oint_\Gamma \left(G\frac{\partial \phi}{\partial n} - \phi\frac{\partial G}{\partial n}\right) dl \qquad (3)$$

where $\phi(\mathbf{r}')$ denotes the wave function at $\mathbf{r}'$ upon the contour (see Fig. 1). Function G is the two-dimensional Green's function in free space, which satisfies

$$\nabla_t^2 G + k^2 G = -\delta(\mathbf{r}-\mathbf{r}') \qquad (4)$$

where δ denotes a Dirac's delta function.

In [3], [4], [6], and [7], Green's function has been chosen as

$$G(R) = \begin{cases} jH_0^{(2)}(kR)/4 & \text{(in [3], [6])} \\ C_0 J_0(kR) - Y_0(kR)/4 & \text{(in [4])} \\ -Y_0(kR)/4 & \text{(in [7])} \end{cases} \qquad (5)$$

Reprinted from *IEEE Trans. Microwave Theory Tech.*, vol. MTT-35, no. 10, pp. 887–892, Oct. 1987.

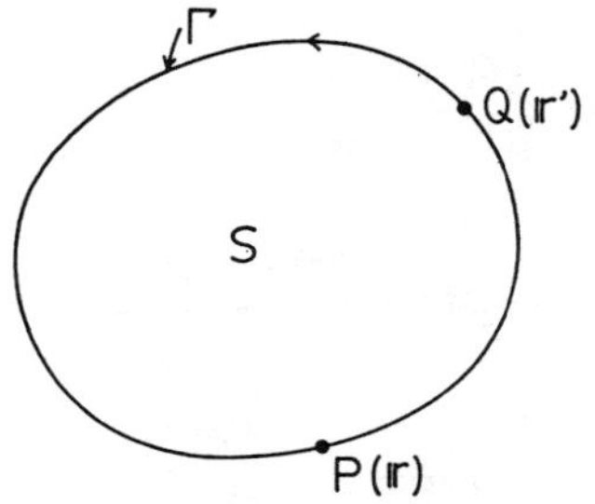

Fig. 1. A two-dimensional region S surrounded by a contour Γ.

where

$$R = |\mathbf{r} - \mathbf{r}'|$$

and

$H_0^{(2)}(kR)$	zero-order Hankel function of the second kind,
$J_0(kR)$	zero-order Bessel function of the first kind,
$Y_0(kR)$	zero-order Bessel function of the second kind,
C_0	complex constant.

In the conventional BEM, (3) is solved to obtain the solution for (1) and (2).

C. Formulation without Green's Function

We note that (3) is a boundary-integral equation; that is, it is an equation in terms of the ϕ values and their normal derivatives only on the boundary Γ. Our present concern is whether it is possible to use a homogeneous solution of (4), in other words a solution of (1), instead of G, which is the inhomogeneous solution of (4). Hereafter, we denote the homogeneous solution by ψ and call it the weight function.

If we substitute the wave function ϕ and the weight function ψ into the two-dimensional Green's theorem:

$$\int_S (\psi\nabla_t^2\phi - \phi\nabla_t^2\psi)\, dS = \oint_\Gamma \left(\psi\frac{\partial\phi}{\partial n} - \phi\frac{\partial\psi}{\partial n}\right) dl \tag{6}$$

we obtain, instead of (3), the following scalar reciprocal theorem [8], [9]:

$$\oint_\Gamma \left(\psi\frac{\partial\phi}{\partial n} - \phi\frac{\partial\psi}{\partial n}\right) dl = 0 \tag{7}$$

because both ϕ and ψ are solutions of (1).

We can solve (7) instead of (3) to obtain the solution for (1) and (2). This is the principle of the proposed method.

III. Formulation for Boundary-Element Analysis

A. Discretization of the Boundary

We apply discretization similar to that in conventional BEM analysis in solving the integral equation (7). We assume first that the region S has an elliptical or rectangular shape, with symmetries with respect to the x and y axes, and that the eigenfunction ϕ also has such symmetries. Thus, we need to consider only the first quadrant.

The boundary Γ is then approximated by N straight-line segments, which are called elements in the BEM analysis

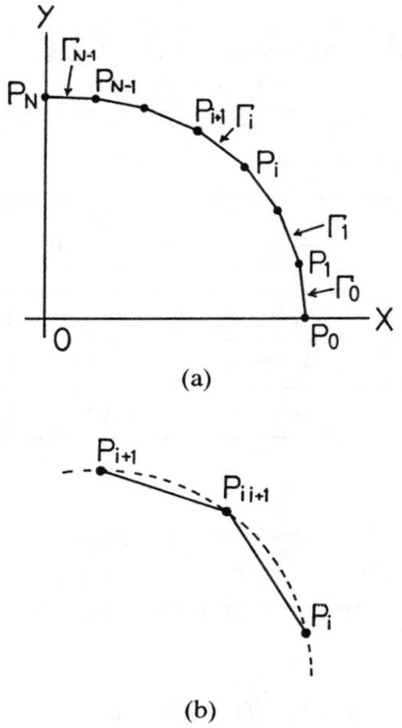

Fig. 2. Polygonal approximation of the boundary. (a) Type-1 approximation. (b) Type-2 approximation.

(see Fig. 2(a)). The function ϕ and its derivative $(\partial\phi/\partial n)$ are now defined upon the node points P_i between the two elements Γ_{i-1} and Γ_i shown in Fig. 2(a).

Another method for approximating the boundary is shown in Fig. 2(b). In this method, a segment Γ_i consists of two line elements. Therefore, the shape of the approximated boundary looks apparently the same as in Fig. 2(a) except that the number of elements N is doubled. Note, however, that in this second method the value of ϕ is defined only on old node points P_i, and not on the new inflection points P_{ii+1}. In the following, we call the two discretization methods (Fig. 2(a) and (b)) type 1 and type 2, respectively.

In both of type-1 and type-2 approximations, the value of function ϕ along element Γ_i is obtained by linear interpolation between values at node points P_i and P_{i+1} as

$$\phi = \frac{L-t}{L}\phi_i + \frac{t}{L}\phi_{i+1} \tag{8}$$

where L denotes the length of element Γ_i in type 1 and the sum of the length of two line elements $\overline{P_iP_{ii+1}}$ and $\overline{P_{ii+1}P_{i+1}}$ in type 2, and t denotes a variable expressing the distance from the node point P_i along Γ_i. The normal derivative $(\partial\phi/\partial n)$ is also linearly interpolated in a manner similar to that in (8).

B. Weight Function

We choose the weight function ψ as circular harmonics, i.e., the product of a Bessel function of the first kind and a trigonometric function

$$\psi = J_j(kr)\cos(j\theta + \rho) \tag{9}$$

where the order j and the phase ρ of the trigonometric part are chosen as shown in Table I so as to match the symmetry condition of the eigenfunction ϕ to be obtained.

TABLE I
SELECTION OF THE ORDER j AND PHASE ρ OF THE WEIGHT FUNCTION ψ ACCORDING TO THE SYMMETRY OF THE EIGENFUNCTION ϕ

symmetry about x or y-axis		$\Psi = J_j(kr)\cos(j\theta+\rho)$	
x-axis	y-axis	j	ρ
symmetric	symmetric	even	0
symmetric	antisymmetric	odd	0
antisymmetric	symmetric	odd	$\pi/2$
antisymmetric	antisymmetric	even	$\pi/2$

Thus, we can restrict the contour integral of (7) to the first quadrant, because the integrals along the x and y axes vanish owing to the symmetry of ϕ and its normal derivative.

C. Matrix Equation and Its Solutions

Using the above discretization and weight function, we now solve (7) by using the method of moments. We need $N+1$ linearly independent weight functions ψ, because the number of unknown values of ϕ is $N+1$ (see Fig. 2(a)). These $N+1$ ψ functions can be obtained by letting $j=0,1,\cdots,N$ in (9). Using these, (7) is now rewritten as

$$\sum_{i=0}^{N-1}\int_{\Gamma_i}\phi\frac{\partial\psi}{\partial n}\,dl=\sum_{i=0}^{N-1}\int_{\Gamma_i}\psi\frac{\partial\phi}{\partial n}\,dl \tag{10}$$

or, in matrix form, as

$$[A]\begin{pmatrix}\phi_0\\ \vdots\\ \phi_N\end{pmatrix}=[B]\begin{pmatrix}\dfrac{\partial\phi_0}{\partial n}\\ \vdots\\ \dfrac{\partial\phi_N}{\partial n}\end{pmatrix} \tag{11}$$

where $[A]$ and $[B]$ are square matrices of order $(N+1)$.

Either the value of ϕ or its normal derivative $(\partial\phi/\partial n)$ is zero on the boundary in Dirichlet- and Neumann-type problems, respectively. Therefore, either the left-hand side or the right-hand side of (11) vanishes, and the eigenvalue equation is reduced to

$$\det[B]=0 \quad \text{for a Dirichlet-boundary condition} \tag{12}$$

$$\det[A]=0 \quad \text{for a Neumann-boundary condition.} \tag{13}$$

These equations are the equations to be solved.

Generally, however, (11) relates the values of a function and its normal derivative $(\partial\phi/\partial n)$ on the boundary. Equation (11) can be applied, therefore, not only to Dirichlet- and Neumann-type boundary-value problems but also to general electromagnetic boundary-value problems in composite media (such as dielectric waveguides) or with composite boundary conditions.

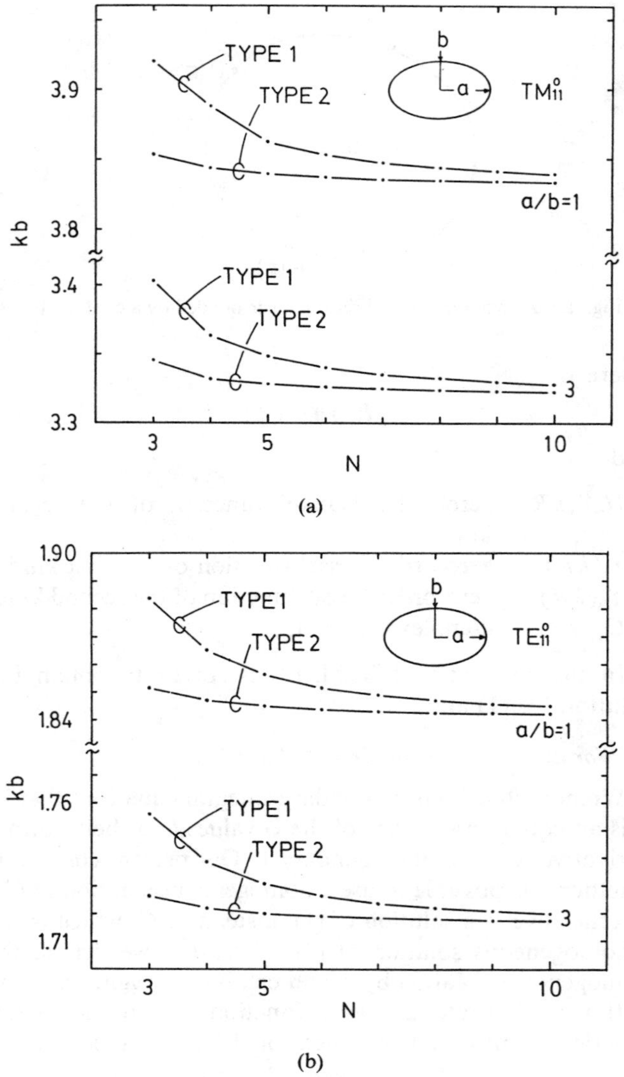

Fig. 3. Convergence characteristics for elliptical boundaries. (a) TM mode. (b) TE mode.

IV. NUMERICAL RESULTS

Some numerical solutions based upon (11)–(13) are shown in Figs. 3–7. In these figures, eigenmodes for the Dirichlet- and Neumann-type boundary conditions are called the TM and TE modes, respectively. This is because the obtained eigenfunctions ϕ give the electromagnetic field components in the direction of propagation of the TM and TE modes propagated in a metallic waveguide. The mode numbers (subscripts) are given according to those in a circular or rectangular metallic waveguide. The superscripts e and o denote, respectively, the even and odd modes with respect to the x axis, respectively.

Figs. 3 and 4 show how the calculated wavenumber kb converges as the number of elements N increases.

Fig. 3(a) and (b) shows the convergence for the elliptical boundary for the TM and TE modes, respectively. Here, TYPE 1 and TYPE 2 indicate the two methods for approximating the shape of boundary described in Section III-A.

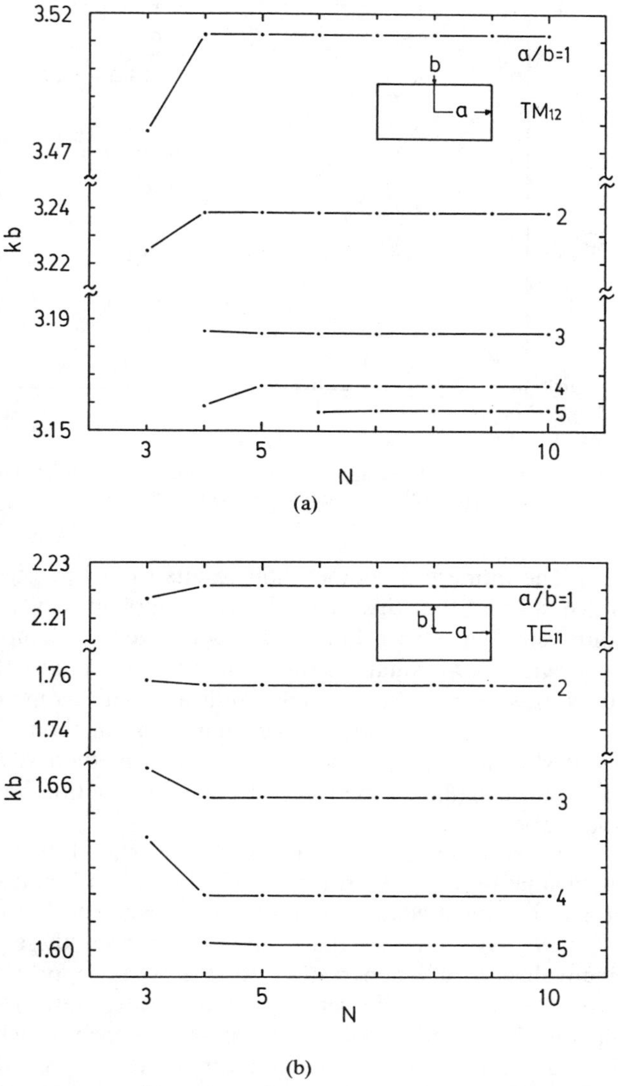

Fig. 4. Convergence characteristics for rectangular boundaries. (a) TM mode. (b) TE mode.

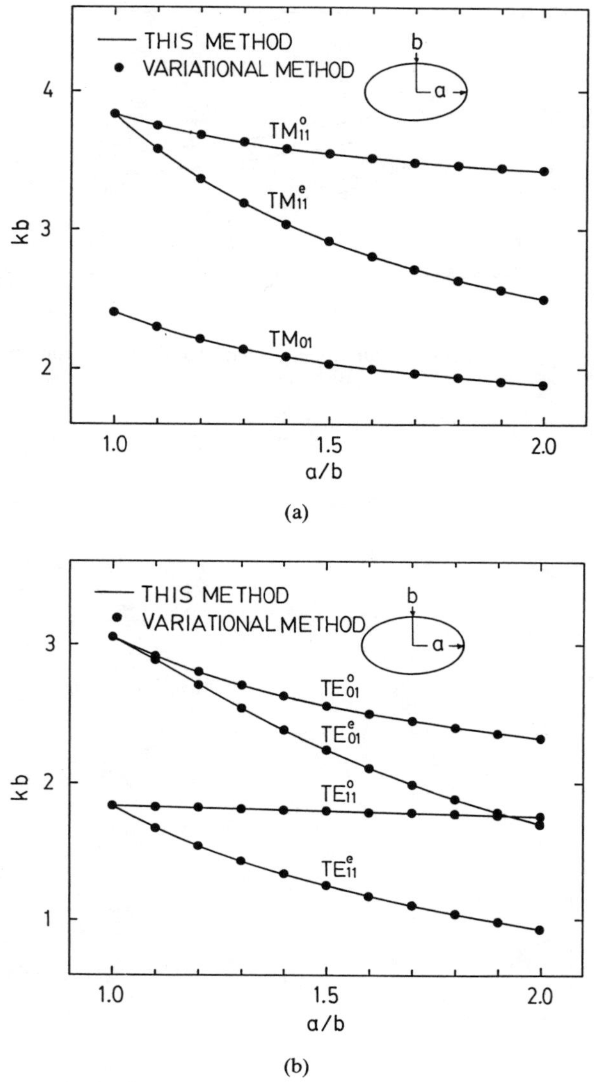

Fig. 5. Eigenvalues for elliptical boundaries as a function of the aspect ratio a/b. (a) TM mode. (b) TE mode.

When the aspect ratio a/b is unity (upper trace in Fig. 3(a) and (b)), i.e., with a circular boundary, the eigenvalues are found to converge to their analytical solutions, which are the first zeros of the Bessel functions of the first kind and their derivatives. Comparing the two types of approximations of boundary (types 1 and 2), we find that the convergence is better in type 2 than in type 1. It is also found that an eigenvalue obtained with the type-2 approximation at certain N is almost equal to that with type 1 but at twice N. This means that the shape of the boundary is more important than the node number in achieving high accuracy.

Fig. 4(a) and (b) shows the convergence for the rectangular boundary. Eigenvalues converge to analytical solutions, which are $\pi/2\sqrt{(b/a)^2+4}$ and $\pi/2\sqrt{(b/a)^2+1}$ for the TM_{12} and TE_{11} modes, respectively. An accuracy of the order of magnitude of 10^{-10} is obtained at $N=10$ independent of the aspect ratio a/b.

The solid curve in Fig. 5(a) and (b) shows the eigenvalue for the elliptical boundary as a function of the aspect ratio a/b. The results of the variational method analyses using polynomials of x and y as trial functions [11] are also shown for comparison (dots). The solid curve and the dots show good agreement.

Fig. 6 shows the difference between the eigenvalues obtained with the proposed method and the conventional BEM analyses using the integral equation (eq. (3)), both using the same number of segments N. The boundary shape is assumed as a circle and eigenmodes are TE modes. The difference is found for practical purposes to be sufficiently small and to be almost independent of the number of elements N. This means that the convergence characteristics of the proposed method are almost identical with those of the conventional BEM analysis.

Fig. 7(a) and (b) shows the values of ϕ or $(\partial\phi/\partial n)$ at nodes points upon the boundary calculated by using (11)

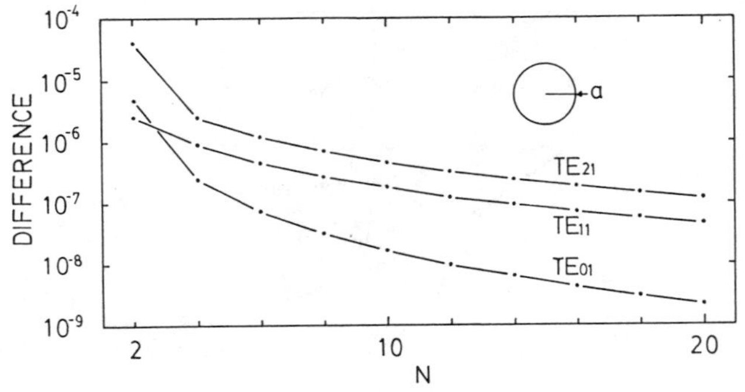

Fig. 6. Comparison of convergence characteristics with conventional BEM analysis.

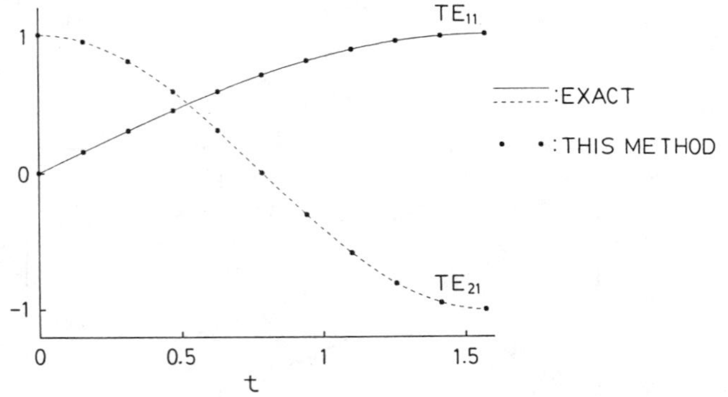

(a)

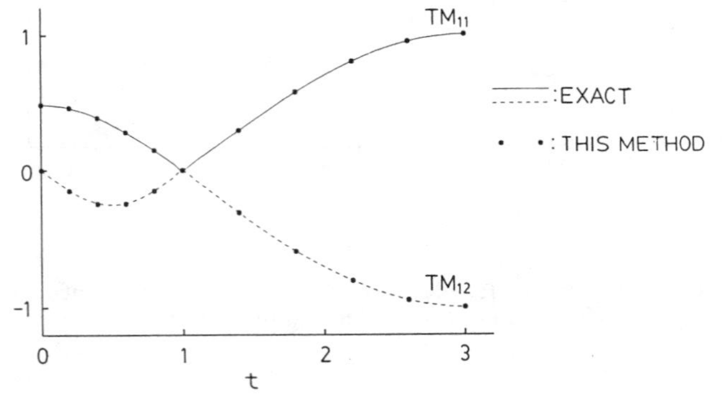

(b)

Fig. 7. Values of ϕ or its normal derivatives $(\partial\phi/\partial n)$ upon the node points on the boundary. (a) ϕ of circular boundary. (b) $(\partial\phi/\partial n)$ of rectangular boundary ($a/b = 2$).

for circular and rectangular boundary shapes, respectively. The analytical solutions are also shown as solid and dotted curves. The boundary values calculated with this method show good agreement with exact ones.

V. Discussion

1) In this paper we have presented a new boundary-integral formulation without using Green's function and have applied it to boundary-element analyses of Dirichlet- and Neumann-type boundary-value problems. From the foregoing numerical results, it is found that this method is valid and useful for solving these problems.

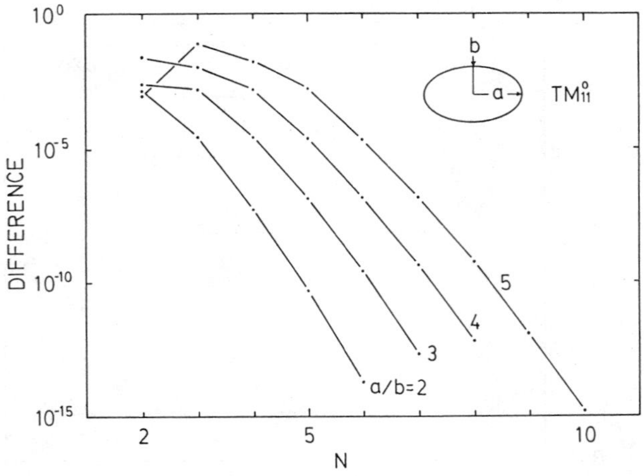

Fig. 8. Difference of the approximation of boundary between type 1 and type 2. Number of node points is that of type 2.

2) The numerical computation results shown in Figs. 3 and 4 suggest that this method can be applied to various boundary shapes with fairly rapid convergence. Comparisons with variational method and conventional BEM analyses, shown in Figs. 5 and 6, indicate that this method compares favorably with other numerical method. The result shown in Fig. 7 indicates that this method is also useful for calculating the value of the wave function on the boundary.

3) In connection with Fig. 3, it was stated that the eigenvalue obtained with the type-2 approximation at certain N is almost equal to that obtained with type-1, but at twice N. Fig. 8 shows the difference between these two eigenvalues as a function of N for the type-2 approximation. The difference is very small; therefore, we can emphasize the effectiveness of the type-2 approach, in which the approximation of the boundary shape is improved, whereas the number of node points, and hence the number of unknown variables, are kept unchanged.

4) The above statement is also supported by the fact that the convergence of the eigenvalue for the rectangular boundary (Fig. 4) is more rapid than that for the elliptical boundary (Fig. 3). We understand that this is because the approximation of the boundary shape is not necessary in the rectangular case.

5) The features of the proposed method can be summarized in comparison with the conventional BEM and other numerical methods as follows:

i) Formulation of the problem is simple and easy because Green's function is not used; hence the singular point is absent.
ii) Consideration of the modal symmetry can be done easily by proper choice of the weight function ψ.
iii) The numerical accuracy can be improved by simply improving the approximation of the boundary shape.
iv) It does not contain any spurious solutions in the numerical analysis in this paper.

VI. Conclusions

The proposed method, the boundary-integral method without Green's function, is found to be an efficient tool for solving the boundary-value problems of two-dimensional scalar wave equations.

Acknowledgment

The authors thank Dr. N. Morita of the University of Osaka and Prof. M. Hashimoto of Osaka Electro-Communications University for their helpful comments.

References

[1] S. M. Saad, "Review of numerical method for the analysis of arbitrarily shaped microwave and dielectric waveguides," *IEEE Trans. Microwave Theory Tech.*, vol. MTT-33, pp. 894–899, Oct. 1985.

[2] N. Okamoto, "Matrix formulation of scattering by a homogeneous gyrotropic cylinder," *IEEE Trans. Antennas Propagat.*, vol. AP-18, pp. 642–649, Sept. 1970.

[3] T. Okoshi and T. Miyoshi, "The planar circuit—An approach to microwave integrated circuitry," *IEEE Trans. Microwave Theory Tech.*, vol. MTT-20, pp. 245–252, Apr. 1972.

[4] Y. Ayasli, "Analysis of wide-band stripline circulators by integral equation technique," *IEEE Trans. Microwave Theory Tech.*, vol. MTT-28, pp. 200–209, Mar. 1980.

[5] C. A. Brebbia, *Boundary Element Method for Engineers.* London: Pentech Press, 1978.

[6] S. Kagami and I. Fukai, "Application of boundary-element method to electromagnetic field problems," *IEEE Trans. Microwave Theory Tech.*, vol. MTT-32, pp. 455–461, Apr. 1984.

[7] M. Matsubara, "Boundary element analysis of polarization holding fibers," (in Japanese), *Trans. IECE Japan*, vol. J67-B, pp. 968–973, Sept. 1984.

[8] E. A. N. Whitehead, "The theory of parallel-plate media for microwave lenses," *Proc. Inst. Elec. Eng.*, vol. 98, pt. III, pp. 133–140, Jan. 1951.

[9] T. Itoh and R. Mittra, "An analytical study of the echelette grating with application to open resonators," *IEEE Trans. Microwave Theory Tech.*, vol. MTT-17, pp. 319–328, June 1969.

[10] R. F. Harrington, *Field Computation by Moment Method.* New York: Macmillan, 1968.

[11] R. M. Bulley and J. B. Davies, "Computation of approximate polynomial solutions to TE modes in arbitrary shaped waveguide," *IEEE Trans. Microwave Theory Tech.*, vol. MTT-17, pp. 440–446, Aug. 1969.

A New Algorithm for the Wide-Band Analysis of Arbitrarily Shaped Planar Circuits

PAOLO ARCIONI, MARCO BRESSAN AND GIUSEPPE CONCIAURO, MEMBER, IEEE

Abstract —**A new algorithm for the wide-band analysis of the two-dimensional model of a planar circuit is described. The planar circuit is considered to be enclosed in a regularly shaped (rectangular or circular) resonator, and the electric and magnetic fields are derived from the Green's functions of this resonator by integrating over the periphery of the circuit not coinciding with the regular shape. The special form used for the Green's functions makes it possible to derive the Z parameters in a special form, similar to Foster's series, but converging much more rapidly. The calculation requires the determination of a reduced number of resonances of the planar circuit, which are obtained by an integral equation approach leading to a linear eigenvalue problem. The algorithm was implemented in an efficient CAD routine, named ANAPLAN, which is briefly described.**

I. Introduction

IN COMPARISON with usual line elements, strip and microstrip planar elements allow a greater flexibility in MIC and MMIC design. The possibility of considering a virtually infinite variety of shapes may lead to interesting solutions in the design of many circuit components, such as directional couplers, filters, and chokes. Since the publication of the early papers on planar circuits [1]–[4], interest in this subject has increased continuously. An extensive bibliography on both methodological and applicative aspects can be found in [5] and in a recent survey paper by Sorrentino [6].

The design flexibility inherent in planar circuit philosophy can be fully exploited only if CAD tools suitable for the wide-band analysis and optimization of arbitrary shapes are available. Fast algorithms for analysis are of paramount importance, as trial-and-error optimum design techniques require a significant number of sequential analyses.

A simple model used in the analysis of planar circuits assumes that the circuit is laterally bounded by a magnetic wall, except at the ports (Fig. 1(a)). Since in this model the (z-directed) electric field and the (transversal) magnetic field are z-independent, the field analysis is two-dimensional and it is much simpler than the three-dimensional one which would be required for taking into account fringing field and radiation effects rigorously. The use of this model is well established in the analysis of strip (triplate) planar circuits, where the fringing field effects are taken into account by a slight enlargement of the transversal circuit dimensions [5]. The same model was used for the analysis of microstrip planar resonators [7], [8] and circuits [9], [10]. In these last cases the fringing field effect was taken into account by considering an enlarged circuit pattern and an effective permittivity. Though the two-dimensional model is less accurate in microstrip than in triplate circuit analysis, it is, however, a useful starting point in the design of such circuits.

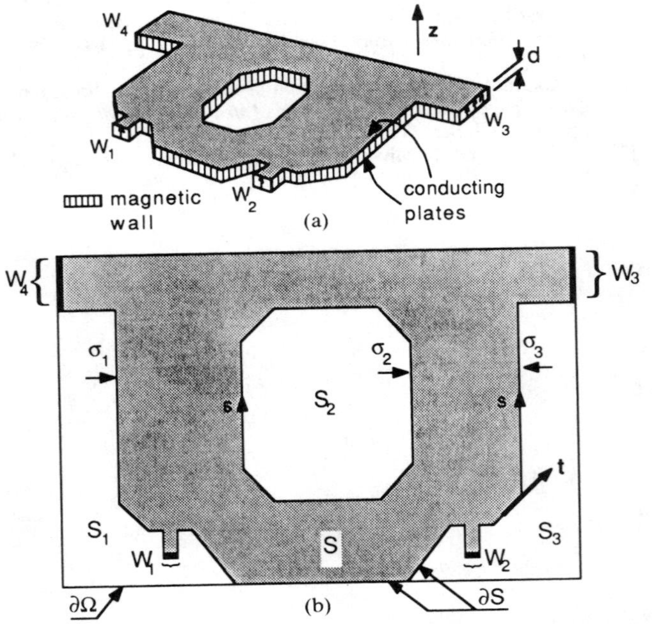

Fig. 1. (a) Two-dimensional model of a planar circuit. (b) Planar circuit enclosed inside a rectangular resonator Ω. In this example $\Omega = S + S_1 + S_2 + S_3$; $\sigma = \sigma_1 + \sigma_2 + \sigma_3$.

Thus far, three kinds of methods have been proposed for the analysis of planar elements of arbitrary shapes: i) the time-domain approach [11]; ii) the contour integral method [4]; and iii) the eigenfunction expansion method [12]. The time-domain approach has the advantage of permitting the analysis of circuits including nonlinear elements, but it leads to computer times that are prohibitive for CAD applications. The contour integral method is very efficient for single-frequency calculations, but it may require a very

Manuscript received April 15, 1987; revised December 12, 1987, and May 18, 1988. This work was supported by CNR, Progetto Finalizzato MADESS, under contract 86.02188.61.
The authors are with the Dipartimento di Elettronica, Università di Pavia, 27100 Pavia, Italy.
IEEE Log Number 8822911.

Reprinted from *IEEE Trans. Microwave Theory Tech.*, vol. MTT-36, no. 10, pp. 1426–1437, Oct. 1988.

long frequency-by-frequency analysis when used in wide-band design.

In principle the eigenfunction expansion method is well suited for wide-band analyses, as it supplies directly the poles and the residues necessary to determine the Y parameters on the basis of their Foster representation. This method requires the determination of the resonating modes of the circuit when the ports are shorted, since each pair of poles is given by the resonating frequency of a mode and the pertinent residues depend on the modal field. Due to the relatively slow convergence of the Foster series, in order to achieve a good accuracy in a wide band, the number of modes to be considered must be much larger than the number of modes occurring in the band of interest. This is a serious drawback, because the modes must be determined numerically using, for example, the finite element technique. For this reason, even though this method is the most suitable for wide-band analyses, its practical use requires a very long computer time.

In this paper we present a new algorithm, well suited for wide-band analyses of planar circuits of arbitrary shapes, represented by their two-dimensional model. The algorithm leads to the determination of all the unknown coefficients included in the following representation of the Z parameters:

$$Z_{ij}(k) = \frac{\eta d}{jkS} + jk\eta d\Lambda_{ij} + j\frac{k^3\eta}{d}\sum_{q=1}^{Q}\frac{V_{iq}V_{jq}}{k_q^2(k_q^2-k^2)} \quad (i,j=1,\cdots,N). \quad (1)$$

In this expression N is the number of the ports, S is the area of the planar circuit (see Fig. 1(b)), d is the thickness of the dielectric, $\eta=\sqrt{\mu/\epsilon}$ and $k=\omega\sqrt{\epsilon\mu}$ are the characteristic impedance and the wavenumber, respectively, Λ_{ij} are real frequency-independent coefficients, and the quantities k_q are the first Q resonating wavenumbers of the circuit when its ports are open. The coefficients V_{iq} are related to the normalized electric field $E^{(q)}$ of the qth resonant mode by

$$V_{iq} = -\frac{d}{W_i}\int_{W_i} E^{(q)}\,ds \quad (2)$$

where s is a coordinate taken along the boundary ∂S, and W_i denotes both the ith port and its width. The first term in (1) dominates at very low frequencies and represents the contribution of the parallel-plate capacitance.

Like the eigenfunction expansion method, our algorithm also involves the calculation of resonant modes. Their number, however, is strongly reduced because of the quite good convergence of the series contained in (1). In order to appreciate this point (1) should be compared with the usual Foster-type representation of the Z parameters, obtained by the Green's function approach [5]:

$$Z_{ij}(k) = \frac{\eta d}{jkS} + j\frac{\eta k}{d}\sum_{q=1}^{\infty}\frac{V_{iq}V_{jq}}{k_q^2-k^2}. \quad (3)$$

It is realized that (1) represents a modification of (3), obtainable from it by adding and subtracting from the series its low-frequency approximation (corresponding to $d^2\Lambda_{ij}$) and by retaining in the remaining series the first Q terms. The extraction of $d^2\Lambda_{ij}$ improves the convergence of the series, due to the appearance of the factor k_q^2 in the denominator of its terms. Since our algorithm leads to the direct determination of the coefficients Λ_{ij} (i.e., independently of their series representation), the number of modes to be calculated is strongly reduced. In practice it does not much exceed the number of resonances occurring in the band of interest (see Section V).

The two basic equations of the theory are derived from the integral representation of the field in terms of its boundary value. In setting up these equations the planar circuit is considered as enclosed inside a two-dimensional circular or rectangular resonator Ω, bounded by a magnetic wall too (see Fig. 1(b)). This unusual configuration is considered in order to represent the field by integrals involving the Green's functions of the resonator Ω, which can be approximated very well by rational functions of k. This causes the basic equations to assume a particular form which makes it possible to deduce (1) and to set up the algorithm for the calculation of all the coefficients involved therein. It is noted that this particular form makes it possible to determine the resonating wavenumbers very efficiently by solving a linear eigenvalue problem, following a procedure similar to that employed by Conciauro *et al.* [13] for the determination of waveguide modes.

Since it is based on an integral equation approach, our method shares some of the merits of the contour integral method (in particular with regard to the relatively small order of the involved matrices), without the drawback of requiring a frequency-by-frequency analysis. Furthermore, in cases where a part of the boundary ∂S coincides with $\partial\Omega$ (as in the case of Fig. 1(b)), the boundary condition has to be imposed only on the other part, so that the number of variables involved in the solution of the equations and the order of the matrices may be reduced significantly.

It is worth noting that a similar algorithm can be developed for the analysis of planar waveguide circuits, the main difference being the electrical wall condition on the lateral bounary. Actually, in that case a physical reasoning makes it possible to derive the first two terms of (1) straightforwardly, with a significant simplification [14].

II. Basic Equations

Let $\boldsymbol{z}$ be the unit vector along the z axis and $\boldsymbol{t}=\boldsymbol{t}(s)$ the tangent unit vector along the boundary ∂S, and let $\boldsymbol{E}=\boldsymbol{z}E(s)$ denote the electric field thereat. Moreover let $I_1, I_2,\cdots, I_N$ be the currents impressed at the N ports, which are z-directed and uniformly distributed on sheets having widths W_i. Due to the equivalence theorem, the field inside the region S, may be considered as generated by an (unknown) equivalent magnetic current sheet $-\boldsymbol{t}E(s)$ flowing along ∂S and by the electric current sheet

$$\boldsymbol{z}\sum_{i=1}^{N}\frac{I_i}{W_i}u_i(s)$$

where $u_i(s)=1$ at the ith port and zero elsewhere. Note that the electric current is zero outside the ports due to the magnetic-wall condition at the boundary of the planar circuit.

Since the equivalence theorem establishes that the field produced by these currents is zero outside the region S, it is unimportant to consider them as radiating in free space or inside a cylindrical impedance wall $\partial\Omega$, bounding a region Ω which includes S (Fig. 1(b)). Therefore we are permitted to assume that the currents act inside an outer *rectangular* or *circular* two-dimensional resonator Ω bounded by a magnetic wall. One of the advantages of this assumption can be appreciated at this point: in fact, since it permits ∂S and $\partial\Omega$ to coincide partially (for particular shapes of S), the unknown magnetic current can be reduced to that flowing in the noncommon portion of the boundaries (magnetic currents backed by magnetic wall have no effect). In the following the portion of ∂S not coinciding with $\partial\Omega$ will be denoted by σ. It is understood that σ coincides with ∂S in cases where ∂S and $\partial\Omega$ have no common parts.

The field generated by the current sheets in the resonator Ω is given by

$$E(\boldsymbol{r}) = -j\eta k \sum_{i=1}^{N} \frac{I_i}{W_i}\int_{W_i} g_{11}(\boldsymbol{r},\boldsymbol{r}',k)\,ds' + \int_{\sigma} \boldsymbol{g}_{12}(\boldsymbol{r},\boldsymbol{r}',k)\cdot\boldsymbol{t}'E(s')\,ds' \tag{4a}$$

$$\boldsymbol{H}(\boldsymbol{r}) = -\sum_{i=1}^{N} \frac{I_i}{W_i}\int_{W_i} \boldsymbol{g}_{21}(\boldsymbol{r},\boldsymbol{r}',k)\,ds' + \frac{jk}{\eta}\int_{\sigma} \overline{\boldsymbol{G}}_{22}(\boldsymbol{r},\boldsymbol{r}',k)\cdot\boldsymbol{t}'E(s')\,ds' \tag{4b}$$

where $\boldsymbol{r}$ and $\boldsymbol{r}'=\boldsymbol{r}'(s')$ are the observation and the source points, respectively, $\boldsymbol{t}'=\boldsymbol{t}'(s')$ denotes the tangent vector at the source point, and $g_{11}, \boldsymbol{g}_{12}, \boldsymbol{g}_{21}, \overline{\boldsymbol{G}}_{22}$ are Green's functions for the resonator Ω. These functions may be represented as eigenfunction expansions, using the eigenfunctions and the eigenvalues of the problems:

$$\nabla^2\phi_m + h_m'^2\phi_m = 0 \qquad (\phi_m = 0 \text{ at } \partial\Omega) \tag{5a}$$

$$\nabla^2\psi_m + h_m^2\psi_m = 0 \qquad \left(\frac{\partial\psi_m}{\partial\nu} = 0 \text{ at } \partial\Omega\right) \tag{5b}$$

$$\int_{\Omega}|\phi_m|^2\,d\Omega = 1 \qquad \int_{\Omega}|\psi_m|^2\,d\Omega = 1. \tag{5c}$$

The expressions for $g_{11}, \boldsymbol{g}_{12}, \boldsymbol{g}_{21}, \overline{\boldsymbol{G}}_{22}$ are

$$g_{11}(\boldsymbol{r},\boldsymbol{r}',k) = \sum_m \frac{\psi_m(\boldsymbol{r})\psi_m(\boldsymbol{r}')}{h_m^2-k^2} - \frac{1}{k^2\Omega} \tag{6a}$$

$$\boldsymbol{g}_{12}(\boldsymbol{r},\boldsymbol{r}',k) = -\boldsymbol{z}\times\nabla' g_{11}(\boldsymbol{r},\boldsymbol{r}',k) \tag{6b}$$

$$\boldsymbol{g}_{21}(\boldsymbol{r},\boldsymbol{r}',k) = -\boldsymbol{g}_{12}(\boldsymbol{r}',\boldsymbol{r},k) \tag{6c}$$

$$\overline{\boldsymbol{G}}_{22}(\boldsymbol{r},\boldsymbol{r}',k) = -\frac{1}{k^2}\nabla\nabla'\sum_m \frac{\phi_m(\boldsymbol{r})\phi_m(\boldsymbol{r}')}{h_m'^2} + \sum_m \frac{\boldsymbol{e}_m(\boldsymbol{r})\boldsymbol{e}_m(\boldsymbol{r}')}{h_m^2-k^2} \tag{6d}$$

where

$$\boldsymbol{e}_m = -\frac{\boldsymbol{z}\times\nabla\psi_m}{h_m}. \tag{7}$$

Formulas (4)–(7) may be deduced in many ways. The shortest one, probably, consists in determining them by duality from the general representation of fields in cylindrical regions, bounded by electric walls [15, sec. 13.2]. This derivation is straightforward, as it needs only to particularize the general formulas given in [15] to time-harmonic and z-independent sources. When Ω is circular or rectangular $\psi_m, \phi_m, h_m, h_m'$ are known, and their expressions can be found in many textbooks [e.g. 16]. It is worth noting that the last term in (6a) corresponds to the zero frequency eigenvalue occurring in the Neumann's problem (5b) ($h_0=0$, $\psi_0=\Omega^{-1/2}$).

As stated by the equivalence theorem, $\boldsymbol{E}$ and $\boldsymbol{H}$ differ from zero inside S and are zero outside. Therefore the fields at the boundary of the planar circuit must be deduced from (4a) and (4b) as limits for $\boldsymbol{r}$ tending to the boundary from the inside of S. For this reason the boundary condition on the magnetic field is

$$\boldsymbol{t}\cdot\lim_{\boldsymbol{r}\to\boldsymbol{r}_0}\boldsymbol{H}(\boldsymbol{r}) = -\sum_{i=1}^{N}\frac{I_i}{W_i}u_i(s) \qquad (\boldsymbol{r}\in S, \forall\boldsymbol{r}_0\in\sigma). \tag{8}$$

It is stressed that the discontinuities in surface Green's integrals depend on the singularities of the Green's functions when $R=|\boldsymbol{r}-\boldsymbol{r}'|\to 0$. Therefore, if series (6) are truncated, one obtains poor approximations of $\boldsymbol{E}$ and $\boldsymbol{H}$ near σ, since the truncation destroys the singularities. On the other hand, (8) plays a fundamental role in setting up our algorithm, so that a good approximation of the field at σ is very important. For this reason it is mandatory to modify (6), extracting the singular terms from the Green's functions and expressing them is closed form. The extraction requires rewriting the fundamental functions g_{11} and $\overline{\boldsymbol{G}}_{22}$ as follows:

$$g_{11}(\boldsymbol{r},\boldsymbol{r}',k) = -\frac{1}{k^2\Omega} + g_{11}^0(\boldsymbol{r},\boldsymbol{r}') + k^2\sum_m \frac{\psi_m(\boldsymbol{r})\psi_m(\boldsymbol{r}')}{h_m^2(h_m^2-k^2)} \tag{9a}$$

$$\overline{\boldsymbol{G}}_{22}(\boldsymbol{r},\boldsymbol{r}',k) = -\frac{1}{k^2}\nabla\nabla' g_{22}^0(\boldsymbol{r},\boldsymbol{r}') + \overline{\boldsymbol{G}}_{22}^0(\boldsymbol{r},\boldsymbol{r}') + k^2\sum_m \frac{\boldsymbol{e}_m(\boldsymbol{r})\boldsymbol{e}_m(\boldsymbol{r}')}{h_m^2(h_m^2-k^2)} \tag{9b}$$

where

$$g_{11}^0 = \sum_m \frac{\psi_m(\boldsymbol{r})\psi_m(\boldsymbol{r}')}{h_m^2} \qquad g_{22}^0 = \sum_m \frac{\phi_m(\boldsymbol{r})\phi_m(\boldsymbol{r}')}{h_m'^2}$$

$$\overline{\boldsymbol{G}}_{22}^0 = \sum_m \frac{\boldsymbol{e}_m(\boldsymbol{r})\boldsymbol{e}_m(\boldsymbol{r}')}{h_m^2}. \tag{10}$$

TABLE I
g_{11}^0, g_{22}^0, AND $\overline{G}_{22}^0$ FOR A TWO-DIMENSIONAL RESONATOR OF CIRCULAR CROSS SECTION

$$g_{11}^{o}(r,\phi,r',\phi') = \frac{2(r^2+r'^2)-3a^2}{8\pi a^2} - \frac{1}{2\pi}\ln\frac{R(r'R_i/a)}{a^2}$$

$$g_{22}^{o}(r,\phi,r',\phi') = \frac{1}{2\pi}\ln\frac{r'R_i}{aR}$$

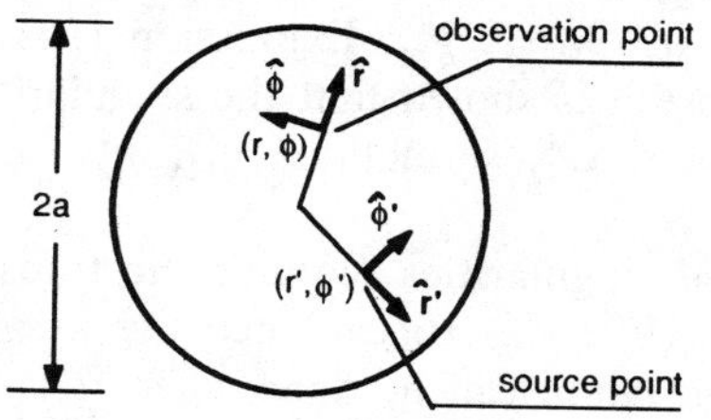

$$\overline{G}_{22}^{o}(r,\phi,r',\phi') = \frac{\hat{r}\hat{r}'}{8\pi}\left[2C\ln\frac{r'R_i}{aR} + \frac{(r^2+r'^2)C-2rr'}{R^2} + \frac{(r^2+r'^2+4a^2)C-2rr'}{(r'R_i/a)^2} - \frac{(a^2+r^2)(a^2+r'^2)}{r^2r'^2}\left(LC-AS+\frac{rr'}{(r'R_i/a)^2}\right)\right] +$$

$$+\frac{\hat{\phi}\hat{r}'}{8\pi}\left[-2S\ln\frac{r'R_i}{aR} + S\frac{r^2-r'^2}{R^2} + S\frac{r^2+r'^2-2a^2}{(r'R_i/a)^2} - \frac{(a^2-r^2)(a^2+r'^2)}{r^2r'^2}(LS+AC)\right] +$$

$$+\frac{\hat{r}\hat{\phi}'}{8\pi}\left[2S\ln\frac{r'R_i}{aR} + S\frac{r^2-r'^2}{R^2} - S\frac{r^2+r'^2-2a^2}{(r'R_i/a)^2} + \frac{(a^2+r^2)(a^2-r'^2)}{r^2r'^2}(LS+AC)\right] +$$

$$+\frac{\hat{\phi}\hat{\phi}'}{8\pi}\left[2C\ln\frac{r'R_i}{aR} - \frac{(r^2+r'^2)C-2rr'}{R^2} + \frac{(r^2+r'^2-2a^2)(rr'/a^2-C)}{(r'R_i/a)^2} - \frac{rr'}{a^2} - \frac{(a^2-r^2)(a^2-r'^2)}{r^2r'^2}\left(LC-AS+\frac{rr'}{a^2}\right)\right]$$

$$R = \sqrt{r^2+r'^2-2rr'C} \qquad R_i = \sqrt{r^2+a^4/r'^2-2ra^2C/r'}$$

$$C = \cos(\phi-\phi') \qquad S = \sin(\phi-\phi') \qquad L = \ln\frac{r'R_i}{a^2} \qquad A = \mathrm{tg}^{-1}\frac{rr'S}{a^2-rr'C}$$

TABLE II
g_{11}^0, g_{22}^0, AND $\overline{G}_{22}^0$ FOR A TWO-DIMENSIONAL RESONATOR OF RECTANGULAR CROSS SECTION

$$g_{11}^{o}(x,y,x',y') = \frac{a}{3b} + \frac{x^2+x'^2}{2ab} - \frac{|X_0^0|+|X_0^1|}{2\pi} - \frac{1}{4\pi}\sum_{m=-\infty}^{+\infty}\sum_{p,q=0}^{1}\ln T_m^{pq}$$

$$g_{22}^{o}(x,y,x',y') = -\frac{1}{4\pi}\sum_{m=-\infty}^{+\infty}\sum_{p,q=0}^{1}(-1)^{p+q}\ln T_m^{pq}$$

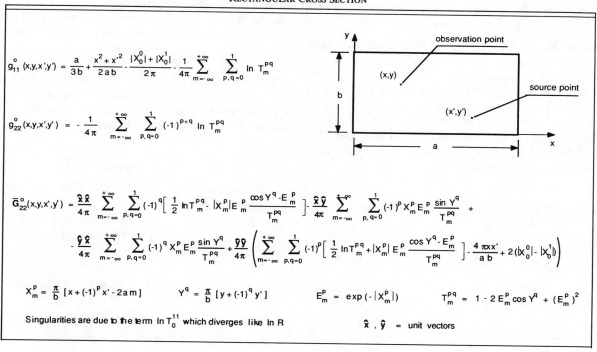

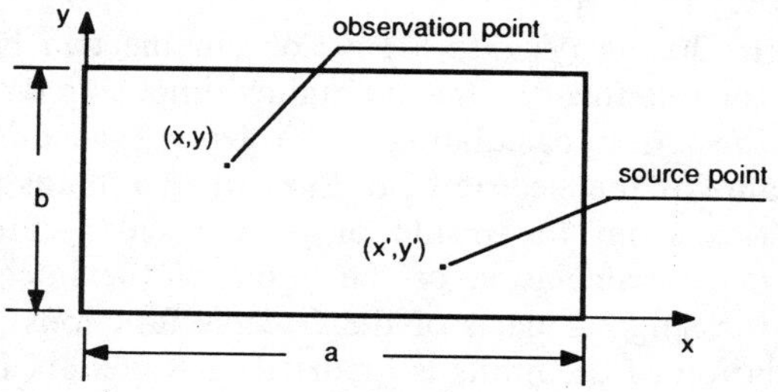

$$\overline{G}_{22}^{o}(x,y,x',y') = \frac{\hat{x}\hat{x}}{4\pi}\sum_{m=-\infty}^{+\infty}\sum_{p,q=0}^{1}(-1)^q\left[\frac{1}{2}\ln T_m^{pq} - |X_m^p|E_m^p\frac{\cos Y^q - E_m^p}{T_m^{pq}}\right] - \frac{\hat{x}\hat{y}}{4\pi}\sum_{m=-\infty}^{+\infty}\sum_{p,q=0}^{1}(-1)^p X_m^p E_m^p\frac{\sin Y^q}{T_m^{pq}} +$$

$$-\frac{\hat{y}\hat{x}}{4\pi}\sum_{m=-\infty}^{+\infty}\sum_{p,q=0}^{1}(-1)^q X_m^p E_m^p\frac{\sin Y^q}{T_m^{pq}} + \frac{\hat{y}\hat{y}}{4\pi}\left(\sum_{m=-\infty}^{+\infty}\sum_{p,q=0}^{1}(-1)^p\left[\frac{1}{2}\ln T_m^{pq} + |X_m^p|E_m^p\frac{\cos Y^q - E_m^p}{T_m^{pq}}\right] - \frac{4\pi xx'}{ab} + 2(|X_0^0| - |X_0^1|)\right)$$

$$X_m^p = \frac{\pi}{b}[x+(-1)^p x' - 2am] \qquad Y^q = \frac{\pi}{b}[y+(-1)^q y'] \qquad E_m^p = \exp(-|X_m^p|) \qquad T_m^{pq} = 1 - 2E_m^p\cos Y^q + (E_m^p)^2$$

Singularities are due to the term $\ln T_0^{11}$ which diverges like $\ln R$

$\hat{x}$, $\hat{y}$ = unit vectors

It can be easily verified that g_{11} and g_{11}^0 are the Green's functions for the scalar wave equation and for the Poisson equation in two dimensions, respectively. As they exhibit the same singularity, it is evident that the only singular term on the r.h.s. in (9a) is g_{11}^0. Expression (9b) is discussed in [17], where it is shown that the singularities are contained in g_{22}^0 and $\overline{G}_{22}^0$. Functions g_{11}^0, g_{22}^0, and $\overline{G}_{22}^0$ diverge as ln R.

The extraction of singularities requires the transformation of the series (10) so as to evidence the logarithmic terms. The transformation can be done when Ω is rectangular or circular, the cases we are interested in. As the procedure is very cumbersome, for the sake of brevity we limit ourselves to the results in Tables I and II. We note that: i) $g_{11}^0, g_{22}^0, \overline{G}_{22}^0$ are given in closed form (Ω circular) or in the form of very rapidly converging series (Ω rectangular); ii) as a consequence of the extraction of the singular terms, the series in (9a) and (9b) represent continuous functions and converge quite rapidly [17], [18], so that they can be truncated without worries; and iii) after truncation expressions (9a) and (9b) become rational functions of k. This last fact constitutes a further advantage of considering S as embedded in the outer resonator Ω. In fact, using Green's functions which are rational functions of k permits the algebraic manipulations, developed in the next section, that lead to the wide-band expression of the Z parameters.

In the following the number of terms retained in the summations in (9a) and (9b) will be denoted by M. As discussed in Section V, in our application the value of M is reasonably small since, in order to have a good accuracy, it is sufficient that h_M exceeds only slightly the value of k at the maximum frequency of interest.

After introducing (9) into (4) we obtain the two basic equations for this theory. One is obtained imposing condition (8), the other calculating $E(s)$ from (4a) as the limiting value of the electric field $E(\boldsymbol{r})$ when $\boldsymbol{r}$ tends to a point on ∂S from the inside of S. Particular care is required in the calculation of the limits of the integrals involving the singular parts of the Green's functions [19]. The calculation of the limits is reported in Appendix I.

The two basic equations are

$$\frac{\partial}{\partial s}\int_\sigma g_{22}^0(s,s')\frac{\partial}{\partial s'}E(s')\,ds' + k^2\int_\sigma \boldsymbol{t}\cdot\overline{G}_{22}^0(s,s')\cdot\boldsymbol{t}'E(s')\,ds' + k^2\sum_{m=1}^{M}\frac{a_m}{h_m^2}\boldsymbol{t}\cdot\boldsymbol{e}_m(s)$$
$$= j\eta k\sum_{i=1}^{N}\frac{I_i}{W_i}\left[\frac{u_i(s)}{2} - \fint_{W_i}\boldsymbol{t}\cdot\boldsymbol{z}\times\nabla g_{11}^0(s',s)\,ds' + k^2\sum_{m=1}^{M}\frac{\boldsymbol{t}\cdot\boldsymbol{e}_m(s)}{h_m(h_m^2-k^2)}\int_{W_i}\psi_m(s')\,ds'\right] \quad (11)$$

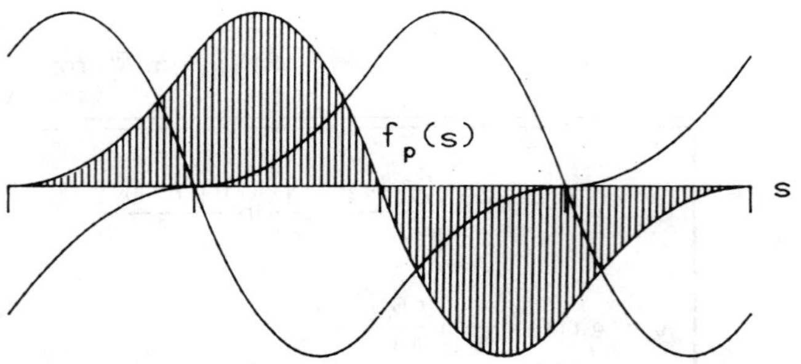

Fig. 2. A generic zero-mean base function.

for values of s corresponding to points of σ and

$$E(s) = j\eta k\sum_{i=1}^{N}\frac{I_i}{W_i}\left[\frac{W_i}{k^2\Omega} - \int_{W_i} g_{11}^0(s,s')\,ds' - k^2\sum_{m=1}^{M}\frac{\psi_m(s)}{h_m^2(h_m^2-k^2)}\int_{W_i}\psi_m(s')\,ds'\right] + \frac{E(s)}{2}u_\sigma(s) - \fint_\sigma \boldsymbol{z}\times\nabla' g_{11}^0(s,s')\cdot\boldsymbol{t}'E(s')\,ds' + \sum_{m=1}^{M}\frac{a_m}{h_m}\psi_m(s) \quad (12)$$

for any value of s, where $u_\sigma(s) = 1$ over σ, $u_\sigma(s) = 0$ elsewhere, and

$$a_m = \frac{k^2}{h_m^2 - k^2}\int_\sigma \boldsymbol{t}\cdot\boldsymbol{e}_m(s)E(s)\,ds. \quad (13)$$

In (11)–(13) the source and the observation points are indicated by their coordinates s, s' to put into evidence that both points are located on the boundary. The dash on the integral symbol denotes the "principal value."

III. Determination of the Z Matrix

For a given set of currents, (11) permits the determination of the unknown function $E(s)$ over σ. Introducing this function on the r.h.s. of (12), it is possible to find the electric field all over ∂S. Substituting this into

$$V_i = -\frac{d}{W_i}\int_{W_i} E(s)\,ds \quad (14)$$

the voltages $V_1, V_2, \cdots, V_N$ at the ports are obtained as function of the currents.

Assuming that, in general, σ consists of K separated parts $\sigma_1, \sigma_2, \cdots, \sigma_K$ (see Fig. 1(b)), the unknown function is expressed as

$$E(s) = \sum_{k=1}^{K} b_k' f_k'(s) + \sum_{p=1}^{P} b_p f_p(s) \qquad (\text{over } \sigma) \quad (15)$$

where f_k' and f_p are base functions and b_k', b_p are unknown coefficients. Functions f_k' are defined as follows:

$$f_k' = 1 \quad \text{for} \quad s\in\sigma_k; \qquad f_k' = 0 \quad \text{elsewhere}. \quad (16)$$

Any of the functions f_p has zero mean-value (see Fig. 2) and its support belongs to only one of the lines σ_k.

TABLE III

EXPRESSIONS OF THE ELEMENTS OF THE MATRICES USED IN (17) AND (22)

$C_{pq} = \int_\sigma \int_\sigma \frac{\partial f_p(s)}{\partial s} g^o_{22}(s,s') \frac{\partial f_q(s')}{\partial s'} \, ds\, ds'$	$F_{mi} = \frac{1}{h_m W_i} \int_{W_i} \psi_m(s)\, ds$
$L'_{pq} = \int_\sigma \int_\sigma f_p(s)\, \mathbf{t} \cdot \overline{\mathbf{G}}^o_{22}(s,s') \cdot \mathbf{t}'\, f_q(s')\, ds\, ds'$	$L''_{kq} = \int_{\sigma_k} \int_\sigma \mathbf{t} \cdot \overline{\mathbf{G}}^o_{22}(s,s') \cdot \mathbf{t}'\, f_q(s')\, ds\, ds'$
$Q'_{pi} = \frac{1}{W_i} \int_\sigma \left[\oint_{W_i} \mathbf{t} \cdot \mathbf{z} \times \nabla g^o_{11}(s',s)\, f_p(s)\, ds' - \frac{u_i(s)\, f_p(s)}{2} \right] ds =$	$Q''_{ki} = \frac{1}{W_i} \int_{\sigma_k} \left[\oint_{W_i} \mathbf{t} \cdot \mathbf{z} \times \nabla g^o_{11}(s',s)\, ds' - \frac{u_i(s)}{2} \right] ds =$
$= \frac{1}{W_i} \int_{W_i} \left[\oint_\sigma \mathbf{t}' \cdot \mathbf{z} \times \nabla' g^o_{11}(s,s')\, f_p(s')\, ds' - \frac{f_p(s)}{2} \right] ds$	$= \frac{1}{W_i} \int_{W_i} \left[\oint_{\sigma_k} \mathbf{t}' \cdot \mathbf{z} \times \nabla' g^o_{11}(s,s')\, ds' - \frac{f'_k(s)}{2} \right] ds = \frac{S_k}{\Omega}$
$R'_{pm} = \frac{1}{h_m^2} \int_\sigma f_p(s)\, \mathbf{t} \cdot \mathbf{e}_m(s)\, ds$	$R''_{km} = \frac{1}{h_m^2} \int_{\sigma_k} \mathbf{t} \cdot \mathbf{e}_m(s)\, ds$
$S_{hk} = \int_{\sigma_h} \int_{\sigma_k} \mathbf{t} \cdot \overline{\mathbf{G}}^o_{22}(s,s') \cdot \mathbf{t}'\, ds\, ds' = S_h \delta_{hk} - \frac{S_h S_k}{\Omega}$	$(\mathbf{S}^{-1})_{hk} = \frac{\delta_{hk}}{S_h} + \frac{1}{S}$
$T_{ij} = \frac{1}{W_i W_j} \int_{W_i} \int_{W_j} g^o_{11}(s,s')\, ds\, ds'$	

$i,j = 1,2,\ldots,N$ $h,k = 1,2,\ldots,K$ $m,n = 1,2,\ldots,M$ $p,q = 1,2,\ldots,P$ δ_{hk} = Kronecker symbol

Introducing (15) into (11) and (13), and applying the Galerkin's procedure, the following system of matrix equations is obtained:

$$-k^2\boldsymbol{R}'\boldsymbol{a} + (\boldsymbol{C} - k^2\boldsymbol{L}')\boldsymbol{b} - k^2\boldsymbol{L}''_T\boldsymbol{b}' = j\eta k(\boldsymbol{Q}' - k^2\boldsymbol{R}'\Delta\boldsymbol{F})\boldsymbol{i} \tag{17a}$$

$$-k^2\boldsymbol{R}''\boldsymbol{a} - k^2\boldsymbol{L}''\boldsymbol{b} - k^2\boldsymbol{S}\boldsymbol{b}' = j\eta k(\boldsymbol{Q}'' - k^2\boldsymbol{R}''\Delta\boldsymbol{F})\boldsymbol{i} \tag{17b}$$

$$(\boldsymbol{U} - k^2\boldsymbol{D}')\boldsymbol{a} - k^2\boldsymbol{R}'_T\boldsymbol{b} - k^2\boldsymbol{R}''_T\boldsymbol{b}' = \boldsymbol{0} \tag{17c}$$

where $\boldsymbol{U}$ is the $M \times M$ unit matrix, $\boldsymbol{a} = (a_1, a_2, \cdots, a_M)$, $\boldsymbol{b} = (b_1, b_2, \cdots, b_P)$,

$$\boldsymbol{b}' = (b'_1, b'_2, \cdots, b'_K), \qquad \boldsymbol{i} = (I_1, I_2, \cdots, I_N),$$

$$\boldsymbol{D}' = \mathrm{diag}\left[h_1^{-2}, h_2^{-2}, \cdots, h_M^{-2}\right]$$

$$\Delta = \mathrm{diag}\left[h_1^2/(h_1^2 - k^2), \cdots, h_M^2/(h_M^2 - k^2)\right] \tag{18}$$

and the elements of the other matrices are defined in Table III. The subscript T denotes the transpose. Note that all the matrices listed in Table III are k-independent and that the elements of $\boldsymbol{S}$ and $\boldsymbol{Q}''$ are related to the areas $S, S_1, S_2, \cdots, S_K$, defined in Fig. 1(b) (see Appendix II).

From (17b) we obtain

$$\boldsymbol{b}' = -\boldsymbol{S}^{-1}[\boldsymbol{R}''\boldsymbol{a} + \boldsymbol{L}''\boldsymbol{b}] - \frac{j\eta}{k}\boldsymbol{S}^{-1}\boldsymbol{Q}''\boldsymbol{i} + j\eta k\boldsymbol{S}^{-1}\boldsymbol{R}''\Delta\boldsymbol{F}\boldsymbol{i}. \tag{19}$$

Explicit expressions of the elements of $\boldsymbol{S}^{-1}$ are easily found (see Table III and Appendix II).

On substitution of (19) into (17a) and (17c) the following matrix equation is obtained:

$$\left(\begin{bmatrix} \boldsymbol{U} & \boldsymbol{0} \\ \boldsymbol{0} & \boldsymbol{C} \end{bmatrix} - k^2 \begin{bmatrix} \boldsymbol{D} & \boldsymbol{R}_T \\ \boldsymbol{R} & \boldsymbol{L} \end{bmatrix}\right) \begin{bmatrix} \boldsymbol{a} \\ \boldsymbol{b} \end{bmatrix} = j\eta k \left(\begin{bmatrix} \boldsymbol{H} \\ \boldsymbol{Q} \end{bmatrix} - k^2 \begin{bmatrix} \boldsymbol{D} - \boldsymbol{D}' \\ \boldsymbol{R} \end{bmatrix} \Delta\boldsymbol{F} \right) \boldsymbol{i} \tag{20}$$

where

$$\boldsymbol{D} = \boldsymbol{D}' - \boldsymbol{R}''_T\boldsymbol{S}^{-1}\boldsymbol{R}'' \tag{21a}$$

$$\boldsymbol{R} = \boldsymbol{R}' - \boldsymbol{L}''_T\boldsymbol{S}^{-1}\boldsymbol{R}'' \tag{21b}$$

$$\boldsymbol{L} = \boldsymbol{L}' - \boldsymbol{L}''_T\boldsymbol{S}^{-1}\boldsymbol{L}'' \tag{21c}$$

$$\boldsymbol{H} = -\boldsymbol{R}''_T\boldsymbol{S}^{-1}\boldsymbol{Q}'' \tag{21d}$$

$$\boldsymbol{Q} = \boldsymbol{Q}' - \boldsymbol{L}''_T\boldsymbol{S}^{-1}\boldsymbol{Q}''. \tag{21e}$$

Before discussing the solution of (20), we observe that substituting (12) into (14) and using (15) and (19), the voltage vector $\boldsymbol{v}=(V_1,V_2,\cdots,V_N)$ can be represented as

$$\boldsymbol{v}=\frac{\eta d}{jk}\left[\frac{1}{\Omega}\boldsymbol{I}+\boldsymbol{Q}_T''\boldsymbol{S}^{-1}\boldsymbol{Q}''\right]\boldsymbol{i}$$
$$+j\eta kd\left[\boldsymbol{T}+\left(k^2\boldsymbol{F}_T\boldsymbol{D}'-\boldsymbol{H}_T\right)\Delta\boldsymbol{F}\right]\boldsymbol{i}$$
$$+d\begin{bmatrix}\boldsymbol{H}-\boldsymbol{F}\\ \cdots\cdots\\ \boldsymbol{Q}\end{bmatrix}_T\begin{bmatrix}\boldsymbol{a}\\ \cdots\\ \boldsymbol{b}\end{bmatrix} \tag{22}$$

where $\boldsymbol{I}$ is an $N\times N$ matrix having unit elements, and $\boldsymbol{T}$ is defined in Table III.

To solve (20) we observe that (see appendix IV)

$$\left(\begin{bmatrix}\boldsymbol{U}&\vdots&\boldsymbol{0}\\ \cdots&\vdots&\cdots\\ \boldsymbol{0}&\vdots&\boldsymbol{C}\end{bmatrix}-k^2\begin{bmatrix}\boldsymbol{D}&\vdots&\boldsymbol{R}_T\\ \cdots&\vdots&\cdots\\ \boldsymbol{R}&\vdots&\boldsymbol{L}\end{bmatrix}\right)^{-1}$$
$$=\begin{bmatrix}\boldsymbol{A}\\ \cdots\\ \boldsymbol{B}\end{bmatrix}\operatorname{diag}\left(\frac{1}{\kappa_r^2-k^2}\right)\begin{bmatrix}\boldsymbol{A}\\ \cdots\\ \boldsymbol{B}\end{bmatrix}_T \tag{23}$$

where the quantities κ_r are the eigenvalues of the problem

$$\left(\begin{bmatrix}\boldsymbol{U}&\vdots&\boldsymbol{0}\\ \cdots&\vdots&\cdots\\ \boldsymbol{0}&\vdots&\boldsymbol{C}\end{bmatrix}-\kappa_r^2\begin{bmatrix}\boldsymbol{D}&\vdots&\boldsymbol{R}_T\\ \cdots&\vdots&\cdots\\ \boldsymbol{R}&\vdots&\boldsymbol{L}\end{bmatrix}\right)\begin{bmatrix}\boldsymbol{a}\\ \cdots\\ \boldsymbol{b}\end{bmatrix}_r=0$$
$$(r=1,2,\cdots,P+M) \tag{24}$$

and $(\boldsymbol{A},\boldsymbol{B})$ denotes the matrix whose columns are the eigenvectors $(\boldsymbol{a},\boldsymbol{b})_r$ normalized according to (A3). Since the matrices

$$\begin{bmatrix}\boldsymbol{U}&\vdots&\boldsymbol{0}\\ \cdots&\vdots&\cdots\\ \boldsymbol{0}&\vdots&\boldsymbol{C}\end{bmatrix}\qquad\begin{bmatrix}\boldsymbol{D}&\vdots&\boldsymbol{R}_T\\ \cdots&\vdots&\cdots\\ \boldsymbol{R}&\vdots&\boldsymbol{L}\end{bmatrix} \tag{25}$$

are positive definite (see Appendix III), the eigenvectors are real and the eigenvalues are real positive. Using (23) we obtain, from (20),

$$\begin{bmatrix}\boldsymbol{a}\\ \cdots\\ \boldsymbol{b}\end{bmatrix}=\begin{bmatrix}\boldsymbol{A}\\ \cdots\\ \boldsymbol{B}\end{bmatrix}\operatorname{diag}\left(\frac{1}{\kappa_r^2-k^2}\right)\begin{bmatrix}\boldsymbol{A}\\ \cdots\\ \boldsymbol{B}\end{bmatrix}_T$$
$$\cdot j\eta k\left(\begin{bmatrix}\boldsymbol{H}\\ \cdots\\ \boldsymbol{Q}\end{bmatrix}-k^2\begin{bmatrix}\boldsymbol{D}-\boldsymbol{D}'\\ \cdots\cdots\\ \boldsymbol{R}\end{bmatrix}\Delta\boldsymbol{F}\right)\boldsymbol{i}. \tag{26}$$

On substitution into (22) and after some manipulation involving formulas (A1) and (A5)–(A9), we obtain the voltage/current relationship in the form $\boldsymbol{v}=\boldsymbol{Z}\boldsymbol{i}$, where the Z matrix is given by the expression

$$\boldsymbol{Z}=\frac{1}{jk}\frac{\eta d}{S}\boldsymbol{I}+jk\eta d\boldsymbol{\Lambda}+jk^3\frac{\eta}{d}\boldsymbol{V}\operatorname{diag}\left(\frac{1}{\kappa_r^2\left(\kappa_r^2-k^2\right)}\right)\boldsymbol{V}_T \tag{27}$$

where

$$\boldsymbol{\Lambda}=\boldsymbol{T}+\boldsymbol{Q}_T\boldsymbol{C}^{-1}\boldsymbol{Q}+\boldsymbol{H}_T\boldsymbol{H}-\boldsymbol{F}_T\boldsymbol{H}-\boldsymbol{H}_T\boldsymbol{F} \tag{28}$$
$$\boldsymbol{V}=d(\boldsymbol{H}_T-\boldsymbol{F}_T)\boldsymbol{A}+d\boldsymbol{Q}_T\boldsymbol{B}. \tag{29}$$

IV. Discussion

The eigenvalue problem (24) is equivalent to the solution of (20) in the case $\boldsymbol{i}=0$. Equation (15), together with (19) with $\boldsymbol{i}=0$, yields a field distribution $E^{(r)}(s)$ for each eigenvector $(\boldsymbol{a},\boldsymbol{b})_r$. These fields represent the eigensolutions of the homogeneous equation obtained from (11) when the exciting currents are zero. The eigensolutions occur at some particular values of k, corresponding to the eigenvalues κ_r.

It is realized that solving the homogeneous equation means finding the resonances of the planar circuit when the ports are terminated by magnetic walls. On the other hand it is pointed out that the solution of the homogeneous equation also yields the resonances occurring in the outer regions $S_1, S_2,\cdots,S_K$. In fact it could be verified that, if the determination of the field in these regions were of interest, the above theory should be applied, with the only difference of a sign reversal in $\boldsymbol{t}$ and a different procedure of limit in the representation the boundary condition (8) ($\boldsymbol{r}$ should tend to σ from *the outside of* S rather than from the inside). When the currents are zero, an integrodifferential equation identical to (11) should be obtained. For this reason eigenvalues κ_r and eigenvectors $(\boldsymbol{a},\boldsymbol{b})_r$ correspond to resonances occurring either inside or outside the region S.

The voltages at the rth resonance are obtained from (22), putting $\boldsymbol{i}=0$ and introducing the rth eigenvector. This yields

$$\boldsymbol{v}_r=d(\boldsymbol{H}_T-\boldsymbol{F}_T)\boldsymbol{a}_r+d\boldsymbol{Q}_T\boldsymbol{b}_r. \tag{30}$$

These vectors are the columns of the matrix $\boldsymbol{V}$ (see. (29)), so that the generic element V_{ir} represents the voltage at the ith port, for the rth resonance. Apart from the approximations involved in the numerical algorithm, the voltage vectors corresponding to spurious resonances are zero as the electric field for such resonances is zero inside S. Therefore the spurious resonances have no influence in the calculation of the Z parameters, so that the voltage vectors and the elements in the diagonal matrix corresponding to them can be completely disregarded in (27). Actually, for spurious resonances the numerical algorithm yields nonzero vectors which, however, are easily detected since they are characterized by very small elements. After the identification and elimination of the spurious resonances, the dimensions of the matrix $\boldsymbol{V}$ are reduced from $N\times(P+M)$ to $N\times Q$, by retaining the Q significant columns only. At the same time the number of elements in the diagonal matrix is reduced to Q by retaining the terms containing the Q significant eigenvalues only. These eigenvalues represent the resonating wavenumbers of the open-circuited planar circuit, i.e., the quantities denoted by k_q in the Introduction. The elements of the matrix $\boldsymbol{Z}$ given by (27) have the form (1) and the coefficients involved in the summation have the meaning given in the Introduction.

V. Numerical Implementation and Examples

The algorithm described in the preceding sections was implemented in a user-oriented computer program, named ANAPLAN, whose distinguishing feature is the short computer time. In the following an outline of the program is given. Details of the program can be found in [20].

As a first step in the computing procedure, the boundary ∂S of the planar circuit is generated interactively through a graphic module, starting from a rectangular or a circular shape. The modifications of the starting contour are represented by polygonals, which constitute the discretization of the lines σ_k.

In the implementation of the algorithm we use base functions consisting of interlaced, piecewise-parabolic functions, each supported by four adjacent segments of σ_k and going to zero, together with their derivative, at the extremes of its support (see Fig. 2). Different functions, supported by three segments only, are used in proximity to the extremes of those lines σ_k that depart from $\partial\Omega$. These functions are piecewise parabolic too, but they differ from zero at the extreme coincident with $\partial\Omega$. These functions make it possible to represent $E(s)$ at the extremes of each line σ_k, where it might differ from zero. Since parabolic base functions are used, a small number of them is sufficient for an accurate representation of the field. Actually, it was observed that very good results are obtained, provided the length of no segment of the polygonals exceeds a quarter wavelength at the maximum frequency of interest. This condition is automatically verified since ANAPLAN checks the segmentation and subdivides any segments that are too long. A more dense segmentation is provided close to the possible edges of the lines σ_k, to allow for the rapid variation of the field which can occur there.

ANAPLAN automatically finds the number M of terms retained in the modal series (9) following the rule of thumb that the highest eigenvalue h_M must be about two times larger than the maximum value of k in the band of interest. We observed experimentally that this rule ensures a good accuracy in the evaluation of the Green's functions throughout the whole frequency band.

The next step is the calculation of all the coefficients of the matrices defined in Table III. The contributions to the integrals arising from the singular terms contained in $g_{11}^0, g_{22}^0, \overline{G}_{22}$ are calculated analytically. Contributions to the integrals coming from the regular parts of the Green's functions, as well as the other integrals, are evaluated numerically.

Once matrices (25) have been calculated, the eigenvectors and eigenvalues of (24) are determined, using standard library routines. Finally matrices (28) and (29) are evaluated and the Z parameters (or the S parameters derived from them) are calculated using (1).

We report the results of the wide-band analysis of the planar circuit of Fig. 3. This circuit is a 3-dB hybrid coupler and has the same dimensions as the one snalyzed by Okoshi [5, pp. 113–117] using the contour integral method. Our analysis was performed in the 0 ~ 10 GHz band. In this example, the starting contour is the external circle and the line σ is the inner contour. $P = 20$ base functions were used to represent the field at the inner contour. The value $M = 11$ was chosen by ANAPLAN. The obtained results are reported in Fig. 4, which shows

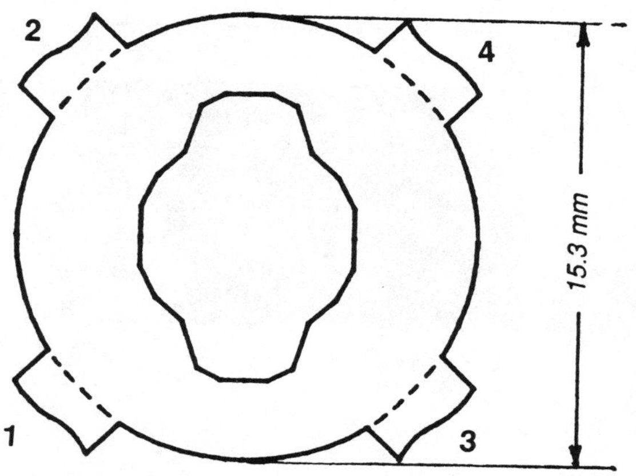

Fig. 3. The 3-dB hybrid coupler used in the test example. A more accurate definition of dimensions is given in [5, fig. 7.7]. Relative permittivity is $\epsilon_r = 2.35$.

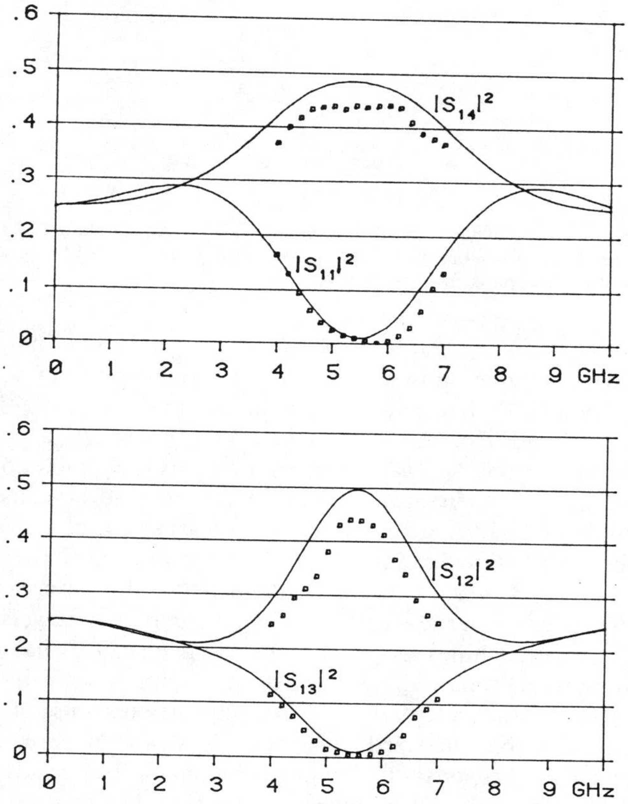

Fig. 4. Magnitudes of the S parameters of the circuit of Fig. 3 calculated by ANAPLAN. Squares represent experimental data reported in [5].

the squared magnitudes of the four scattering parameters (continous lines). The CPU time was about 5 s on a Digital VAX 8500 computer. About 3.5 s was required for the calculation of all the matrix elements listed in Table III, and a remaining 1.5 s was required for the solution of the eigenvalue problem and the calculation of the elements of the matrices Λ and V. The results of our analysis are practically coincident with the ones reported by Okoshi (differences on the plots are inappreciable). On the same figures are represented the experimental data reported by Okoshi [5, fig. 7.10(a)]. Differences are due to the fact that

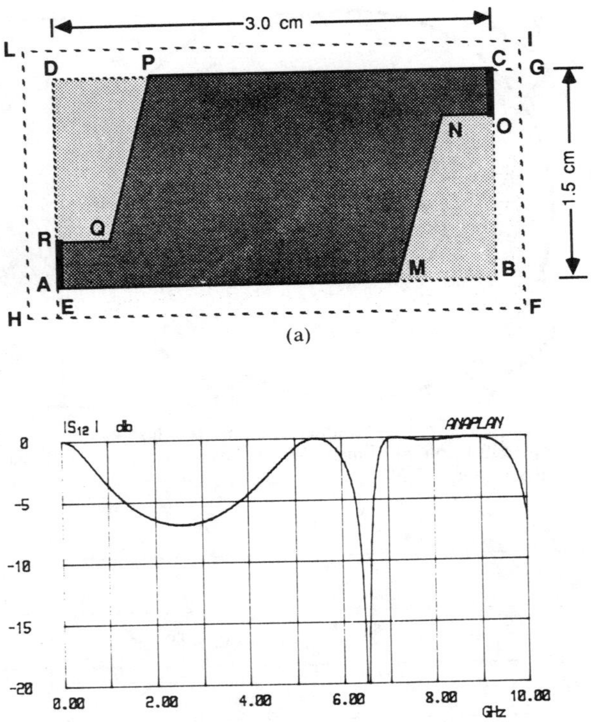

Fig. 5. (a) The geometry of the planar circuit enclosed in three different rectangular resonators. Relative permittivity is $\epsilon_r = 2.35$. (b) Magnitude of S_{12} computed by ANAPLAN.

losses are ignored in the analyses. It is pointed out that the quoted CPU time concerns the analysis in the *whole* $0 \sim 10$ GHz band. For the same circuit, Okoshi reports a computer time of 4 s *per frequency* on a HITAC 8800 computer, and 50 frequency points at least should be considered for analyzing the circuit in the same band. Though these CPU times cannot be compared exactly, as they refer to different machines, they make it possible to appreciate the speed of our algorithm, since the computing powers of the two machines are comparable. It is observed that the computing time required by our algorithm is scarcely affected by the irregularities in the frequency response of the circuitto be analyzed, whereas in cases of irregular frequency responses the number of points to be considered in a frequency-by-frequency analysis should increase dramatically.

A further example concerns the circuit in Fig. 5(a), which was analyzed in the band $0 \sim 10$ GHz, considering it as embedded in three different rectangles (ABCD, EFGD, HFIL). This example makes it possible to realize the advantage of having a part of the boundary coinciding with the external boundary $\partial\Omega$. The noncoincident part of the boundary (σ) consists of the lines MNO + PQR in the first case, of AMNOC + PQR in the second case, and of the whole boundary in the third case. In the three cases we had: $P = 12$, $M = 27$; $P = 20$, $M = 33$; and $P = 28$, $M = 37$. The increase in M derives from the increase in the size of Ω, which causes an increasing number of resonances of Ω to occur inside the band of interest. The results of the analyses, represented in Fig. 5(b), are indistinguishable in the three cases; this emphasizes that the accuracy of our method is unaffected by the choice of the outer contour $\partial\Omega$. This choice, on the contrary, has a large impact on the computing time, which in the three cases was 2.3 s, 6.3 s, and 12.2 s, respectively. Such results show the important time saving which is obtained in the analysis of shapes slightly differing from Ω. On the other hand, the maximum time of 12.2 s suggests the rapidity of the algorithm in cases where the circuit and the external resonator have no common boundary.

Appendix I
Calculation of Some Limits of Integrals Involving Singular Functions Occurring in the Derivation of (11) and (12)

All the following limits are calculated letting the observation point $\boldsymbol{r}$ tend to a point $\boldsymbol{r}_0$ of ∂S (or σ) from the inside of the region S.

The first limit is

$$L_1 = \boldsymbol{t} \cdot \lim_{\boldsymbol{r} \to \boldsymbol{r}_0} \int_\sigma \nabla \nabla' g_{22}^0(\boldsymbol{r}, \boldsymbol{r}') \cdot \boldsymbol{t}' E(s')\, ds'$$

$$= \boldsymbol{t} \cdot \lim_{\boldsymbol{r} \to \boldsymbol{r}_0} \int_\sigma \nabla \frac{\partial}{\partial s'} g_{22}^0(\boldsymbol{r}, \boldsymbol{r}') E(s')\, ds'$$

$$= -\boldsymbol{t} \cdot \lim_{\boldsymbol{r} \to \boldsymbol{r}_0} \int_\sigma \nabla g_{22}^0(\boldsymbol{r}, \boldsymbol{r}') \frac{\partial}{\partial s'} E(s')\, ds'$$

$$(\boldsymbol{r}_0 = \boldsymbol{r}_0(s) \in \sigma)$$

where the transformation was performed integrating by parts and observing that σ (or its component parts) is either a closed line or a line with extremes on $\partial\Omega$, where g_{22}^0 is zero. As we have $g_{22}^0 = -\ln R/2\pi +$ regular function (see Tables I, II), we obtain

$$\nabla g_{22}^0(\boldsymbol{r}, \boldsymbol{r}') = -\frac{\boldsymbol{r} - \boldsymbol{r}'}{2\pi R^2} + \text{regular function}.$$

Then, denoting by Δ an infinitesimal element of σ centered at $\boldsymbol{r}_0 = \boldsymbol{r}_0(s)$, we have

$$L_1 = \frac{\partial E(s)}{\partial s} \boldsymbol{t} \cdot \lim_{\boldsymbol{r} \to \boldsymbol{r}_0} \int_\Delta \frac{\boldsymbol{r} - \boldsymbol{r}'}{2\pi R^2}\, ds'$$

$$- \boldsymbol{t} \cdot \fint_\sigma \nabla g_{22}^0(\boldsymbol{r}_0, \boldsymbol{r}') \frac{\partial E(s')}{\partial s'}\, ds'$$

where

$$\fint_\sigma = \int_{\sigma - \Delta}.$$

Calculating the integral over Δ it is discovered that it is normal to $\boldsymbol{t}$. Therefore this integral does not contribute to L_1. Extracting ∇ from the last integral and observing that the logarithmic singularity of g_{22}^0 is integrable, we obtain the result

$$L_1 = -\frac{\partial}{\partial s}\int_\sigma g_{22}^0(\boldsymbol{r}_0, \boldsymbol{r}')\frac{\partial E(s')}{\partial s'}\,ds'.$$

The second limit is

$$L_2 = \boldsymbol{t}\cdot\lim_{\boldsymbol{r}\to\boldsymbol{r}_0}\int_{W_i} \boldsymbol{z}\times\nabla g_{11}^0(\boldsymbol{r}', \boldsymbol{r})\,ds' \qquad (\boldsymbol{r}_0 = \boldsymbol{r}_0(s)\in\sigma).$$

As we have $g_{11}^0 = -\ln R/2\pi + \text{regular function}$ (see Tables I, II), we obtain

$$\nabla g_{11}^0(\boldsymbol{r}', \boldsymbol{r}) = \frac{\boldsymbol{r}-\boldsymbol{r}'}{2\pi R^2} + \text{regular function}.$$

When $\boldsymbol{r}_0 \in W_i$, we have

$$L_2 = \boldsymbol{t}\cdot\lim_{\boldsymbol{r}\to\boldsymbol{r}_0}\int_\Delta \boldsymbol{z}\times\frac{\boldsymbol{r}-\boldsymbol{r}'}{2\pi R^2}\,ds' + \int\!\!\!\!\!-_{W_i} \boldsymbol{t}\cdot\boldsymbol{z}\times\nabla g_{11}^0(\boldsymbol{r}', \boldsymbol{r}_0)\,ds'$$

$$= \frac{1}{2} + \int\!\!\!\!\!-_{W_i} \boldsymbol{t}\cdot\boldsymbol{z}\times\nabla g_{11}^0(\boldsymbol{r}', \boldsymbol{r}_0)\,ds'.$$

When $\boldsymbol{r}_0 \notin W_i$ the term $1/2$ is missing and the integral is of the usual type.

The third limit is

$$L_3 = \lim_{\boldsymbol{r}\to\boldsymbol{r}_0}\int_\sigma \boldsymbol{t}'\cdot\boldsymbol{z}\times\nabla' g_{11}^0(\boldsymbol{r}, \boldsymbol{r}')E(s')\,ds'$$

$$(\boldsymbol{r}_0 = \boldsymbol{r}_0(s)\in\partial S).$$

The calculation is similar to the previous one. When $\boldsymbol{r}_0 \in \sigma$ we obtain

$$L_3 = -\frac{E(s)}{2} + \int\!\!\!\!\!-_\sigma \boldsymbol{t}'\cdot\boldsymbol{z}\times\nabla' g_{11}^0(\boldsymbol{r}_0, \boldsymbol{r}')E(s')\,ds'.$$

When $\boldsymbol{r}_0 \notin \sigma$ the term $-E/2$ is missing and the integral is of the usual type.

The last limit is

$$L_4 = \boldsymbol{t}\cdot\lim_{\boldsymbol{r}\to\boldsymbol{r}_0}\int_\sigma \overline{\boldsymbol{G}}_{22}^0(\boldsymbol{r}, \boldsymbol{r}')\cdot\boldsymbol{t}'E(s')\,ds'$$

$$= \int_\sigma \boldsymbol{t}\cdot\overline{\boldsymbol{G}}_{22}^0(\boldsymbol{r}_0, \boldsymbol{r}')\cdot\boldsymbol{t}'E(s')\,ds'$$

since the integral is continuous across σ, due to the weakness of the logarithmic singularity of $\overline{\boldsymbol{G}}_{22}^0$.

Appendix II

A. Expression for S_{hk} and $(\boldsymbol{S}^{-1})_{hk}$

Using the modal representation of $\overline{\boldsymbol{G}}_{22}^0$ we have

$$S_{hk} = \int_{\sigma_h}\int_{\sigma_k} \boldsymbol{t}\cdot\sum_m \frac{\boldsymbol{e}_m(s)\boldsymbol{e}_m(s')}{h_m^2}\cdot\boldsymbol{t}'\,ds\,ds'$$

$$= \oint_{\partial S_h}\oint_{\partial S_k} \boldsymbol{t}\cdot\sum_m \frac{\boldsymbol{e}_m(s)\boldsymbol{e}_m(s')}{h_m^2}\cdot\boldsymbol{t}'\,ds\,ds'$$

where the integrals are extended to the whole contours ∂S_h, ∂S_k, since $\boldsymbol{t}\cdot\boldsymbol{e}_m = 0$ on $\partial\Omega$. Using Stokes theorem and observing that $\nabla\times\boldsymbol{e}_m = \boldsymbol{z}h_m\psi_m$ we obtain

$$S_{hk} = \int_{S_h}\int_{S_k}\sum_m \psi_m(\boldsymbol{r})\psi_m(\boldsymbol{r}')\,dS_h\,dS_k = \delta_{hk}S_h - \frac{S_hS_k}{\Omega}$$

due to the completeness relation:

$$\delta(\boldsymbol{r}-\boldsymbol{r}') = \frac{1}{\Omega} + \sum_m \psi_m(\boldsymbol{r})\psi_m(\boldsymbol{r}').$$

The validity of the expression of $(\boldsymbol{S}^{-1})_{hk}$ given in Table III is verified directly using $\boldsymbol{S}\boldsymbol{S}^{-1} = \boldsymbol{U}$ and the geometrical relation $S = \Omega - S_1 - S_2 - \cdots - S_K$.

B. Expression for Q''_{ki}

Using the modal expansion of g_{11}^0 (see (10)) the second integral expression of Q''_{ki} given in Table III can be rewritten as

$$Q''_{ki} = \frac{1}{W_i}\int_{W_i}\left[\lim_{\boldsymbol{r}\to\boldsymbol{r}_0}\int_{\sigma_k} \boldsymbol{t}'\cdot\boldsymbol{z}\times\nabla'\sum_m \frac{\psi_m(\boldsymbol{r})\psi_m(\boldsymbol{r}')}{h_m^2}\,ds'\right]ds$$

$$(\boldsymbol{r}_0\in W_i,\ \boldsymbol{r}'\in\sigma_k).$$

Due to the boundary condition satisfied by ψ_m the integral over σ_k is transformed into a line integral over ∂S_k. Then, using Stokes theorem and observing that

$$\boldsymbol{z}\cdot\nabla'\times\nabla'\times\boldsymbol{z}\psi_m = -\nabla'^2\psi_m = h_m^2\psi_m$$

we obtain

$$Q''_{ki} = -\frac{1}{W_i}\int_{W_i}\left[\lim_{\boldsymbol{r}\to\boldsymbol{r}_0}\int_{S_k}\sum_m \psi_m(\boldsymbol{r})\psi_m(\boldsymbol{r}')\,dS'_k\right]ds$$

$$= -\frac{1}{W_i}\int_{W_i}\left[\lim_{\boldsymbol{r}\to\boldsymbol{r}_0}\int_{S_k}\left(\delta(\boldsymbol{r}-\boldsymbol{r}') - \frac{1}{\Omega}\right)dS'_k\right]ds = \frac{S_k}{\Omega}$$

as $\boldsymbol{r}$ is external to S_k.

Moreover, starting from the expression of $(\boldsymbol{S}^{-1})_{hk}$ given in Table III, it is easily verified that the following relation holds:

$$(\boldsymbol{Q}''_T\boldsymbol{S}^{-1}\boldsymbol{Q}'')_{ij} = \frac{1}{S} - \frac{1}{\Omega}. \tag{A1}$$

Appendix III

Positive Definiteness of Matrices (25)

The quadratic form associated to the first matrix is

$$f_1 = \sum_{m=1}^{M} a_m^2 + \sum_{p,q=1}^{P} b_pC_{pq}b_q.$$

Introducing the expression of C_{pq} given in Table III and using the modal expansion of g_{22}^0 (see (10)), after simple manipulations we obtain

$$f_1 = \sum_{m=1}^{M} a_m^2 + \sum_m\left[\sum_{p=1}^{P}\int_\sigma \frac{\partial f_p}{\partial s}\frac{\phi_m}{h'_m}\,ds\right]^2$$

which is always positive.

The quadratic form associated to the second matrix is

$$f_2 = \boldsymbol{a}_T\boldsymbol{D}\boldsymbol{a} + \boldsymbol{a}_T\boldsymbol{R}_T\boldsymbol{b} + \boldsymbol{b}_T\boldsymbol{R}\boldsymbol{a} + \boldsymbol{b}_T\boldsymbol{L}\boldsymbol{b}.$$

Introducing (21a)–(21c) we obtain

$$f_2 = \boldsymbol{a}_T\boldsymbol{D}'\boldsymbol{a} + \boldsymbol{a}_T\boldsymbol{R}'_T\boldsymbol{b} + \boldsymbol{a}_T\boldsymbol{R}''_T\boldsymbol{b}' + \boldsymbol{b}_T\boldsymbol{R}'\boldsymbol{a} + \boldsymbol{b}_T\boldsymbol{L}'\boldsymbol{b} + \boldsymbol{b}_T\boldsymbol{L}''_T\boldsymbol{b}' + \boldsymbol{b}'_T\boldsymbol{R}''\boldsymbol{a} + \boldsymbol{b}'_T\boldsymbol{L}''\boldsymbol{b} + \boldsymbol{b}'_T\boldsymbol{S}\boldsymbol{b}'$$

where

$$\boldsymbol{b}' = -\boldsymbol{S}^{-1}(\boldsymbol{R}''\boldsymbol{a} + \boldsymbol{L}''\boldsymbol{b}).$$

From the definition of the elements of the matrices $\boldsymbol{L}', \boldsymbol{L}'', \boldsymbol{S}, \boldsymbol{R}', \boldsymbol{R}''$, and considering the modal expansion of $\overline{G}^0_{22}$ (see (10)) it is shown that

$$L'_{pq} = \sum_m h_m^2 R'_{mp} R'_{mq} \qquad L''_{kq} = \sum_m h_m^2 R''_{mk} R'_{mq}$$

$$S_{hk} = \sum_m h_m^2 R''_{mh} R''_{mk}.$$

On substitution into the last expression of f_2 after some manipulations we obtain

$$f_2 = \sum_{m=1}^{M} \left[\frac{a_m}{h_m} + h_m \sum_{p=1}^{P} \left(R'_{pm} b_p + R''_{pm} b'_p \right) \right]^2 + \sum_{m=M+1}^{\infty} h_m^2 \left(R'_{pm} b_p + R''_{pm} b'_p \right)^2$$

which is always positive.

Appendix IV

Some Useful Relations

Due to their positive definiteness (see Appendix III), matrices (25) can be simulataneously diagonalized using the matrix $(\boldsymbol{A}, \boldsymbol{B})$ having as columns the eigenvectors of the problem (24) [21, p. 106]. We have

$$\begin{bmatrix} \boldsymbol{A} \\ \boldsymbol{B} \end{bmatrix}_T \begin{bmatrix} \boldsymbol{U} & \boldsymbol{0} \\ \boldsymbol{0} & \boldsymbol{C} \end{bmatrix} \begin{bmatrix} \boldsymbol{A} \\ \boldsymbol{B} \end{bmatrix} = \operatorname{diag}\left(\kappa_1^2, \kappa_2^2, \cdots, \kappa_{P+M}^2\right) \qquad \text{(A2)}$$

$$\begin{bmatrix} \boldsymbol{A} \\ \boldsymbol{B} \end{bmatrix}_T \begin{bmatrix} \boldsymbol{D} & \boldsymbol{R}_T \\ \boldsymbol{R} & \boldsymbol{L} \end{bmatrix} \begin{bmatrix} \boldsymbol{A} \\ \boldsymbol{B} \end{bmatrix} = \boldsymbol{U}. \qquad \text{(A3)}$$

Equation (A3) specifies the normalization of the eigenvectors. Using these expressions, (23) is verified easily.

Expression (A2) may be rewritten as

$$\boldsymbol{A}_T\boldsymbol{A} + \boldsymbol{B}_T\boldsymbol{C}\boldsymbol{B} = \operatorname{diag}\{\kappa_1^2, \kappa_2^2, \cdots, \kappa_{M+P}^2\} \qquad \text{(A4)}$$

From the same expression it is obtained:

$$\begin{bmatrix} \boldsymbol{U} & \boldsymbol{0} \\ \boldsymbol{0} & \boldsymbol{C} \end{bmatrix}^{-1} = \begin{bmatrix} \boldsymbol{A} \\ \boldsymbol{B} \end{bmatrix} \operatorname{diag}\{\kappa_1^{-2}, \kappa_2^{-2}, \cdots, \kappa_{M+P}^{-2}\} \begin{bmatrix} \boldsymbol{A} \\ \boldsymbol{B} \end{bmatrix}_T$$

or

$$\begin{bmatrix} \boldsymbol{U} & \boldsymbol{0} \\ \boldsymbol{0} & \boldsymbol{C} \end{bmatrix} \begin{bmatrix} \boldsymbol{A} \\ \boldsymbol{B} \end{bmatrix} \operatorname{diag}\{\kappa_1^{-2}, \kappa_2^{-2}, \cdots, \kappa_{M+P}^{-2}\} \begin{bmatrix} \boldsymbol{A} \\ \boldsymbol{B} \end{bmatrix}_T = \begin{bmatrix} \boldsymbol{U} & \boldsymbol{0} \\ \boldsymbol{0} & \boldsymbol{U} \end{bmatrix}.$$

This last expression yields the following useful relations:

$$\boldsymbol{A}\operatorname{diag}\{\kappa_1^{-2}, \kappa_2^{-2}, \cdots, \kappa_{M+P}^{-2}\}\boldsymbol{A}_T = \boldsymbol{U} \qquad \text{(A5)}$$

$$\boldsymbol{B}\operatorname{diag}\{\kappa_1^{-2}, \kappa_2^{-2}, \cdots, \kappa_{M+P}^{-2}\}\boldsymbol{A}_T = \boldsymbol{0}. \qquad \text{(A6)}$$

Analogously, starting from (A3) we obtain

$$(\boldsymbol{D}\boldsymbol{A} + \boldsymbol{R}_T\boldsymbol{B})\boldsymbol{A}_T = \boldsymbol{U} \qquad (\boldsymbol{D}\boldsymbol{A} + \boldsymbol{R}_T\boldsymbol{B})\boldsymbol{B}_T = \boldsymbol{0}.$$

Postmultiplying the latter of these equations by $\boldsymbol{C}\boldsymbol{B}$, using the expression of $\boldsymbol{B}_T\boldsymbol{C}\boldsymbol{B}$ deduced from (A4), and introducing the former, we obtain

$$(\boldsymbol{D}\boldsymbol{A} + \boldsymbol{R}_T\boldsymbol{B})\operatorname{diag}(\kappa_1^2, \kappa_2^2, \cdots, \kappa_{M+P}^2) = \boldsymbol{A}. \qquad \text{(A7)}$$

Furthermore, in the derivation of (27), the following identities are used:

$$\operatorname{diag}\left(\frac{h_m^2}{h_m^2 - k^2}\right) = \boldsymbol{U} + k^2 \operatorname{diag}\left(\frac{1}{h_m^2 - k^2}\right) \qquad \text{(A8)}$$

$$\operatorname{diag}\left(\frac{1}{\kappa_r^2 - k^2}\right) = \operatorname{diag}\left(\frac{1}{\kappa_r^2}\right) + k^2 \operatorname{diag}\left(\frac{1}{\kappa_r^2(\kappa_r^2 - k^2)}\right). \qquad \text{(A9)}$$

References

[1] G. Biorci and S. Ridella, "On the theory of distributed three-layer N-port networks," *Alta Frequenza*, vol. XXXVIII, pp. 615–622, Aug. 1969.

[2] B. Bianco and P. P. Civalleri, "Basic theory of three-layer N-port," *Alta Frequenza*, vol. XXXVIII, pp. 623–631, Aug. 1969.

[3] P. P. Civalleri and S. Ridella, "Impedance and admittance matrices of distributed three-layer N-port," *IEEE Trans. Circuit Theory*, vol. CT-17, pp. 392–398, Aug. 1970.

[4] T. Okoshi and T. Miyoshi, "The planar circuit—An approach to microwave integrated circuitry," *IEEE Trans. Microwave Theory Tech.*, vol. MTT-20, pp. 245–252, Apr. 1972.

[5] T. Okoshi, *Planar Circuits for Microwave and Lightwaves*. Berlin: Springer-Verlag, 1985.

[6] R. Sorrentino, "Planar circuits, waveguide models, and segmentation method," *IEEE Trans. Microwave Theory Tech.*, vol. MTT-33, pp. 1057–1066, Oct. 1985.

[7] I. Wolff and N. Knoppik, "Rectangular and circular disc capacitors and resonators," *IEEE Trans. Microwave Theory Tech.*, vol. MTT-22, pp. 857–864, Oct. 1974.

[8] I. Wolff and V. K. Tripathi, "The microstrip open-ring resonator," *IEEE Trans. Microwave Theory Tech.*, vol. MTT-32, pp. 102–107, Jan. 1984.

[9] F. Giannini, R. Sorrentino, and J. Vrba, "Planar circuit analysis of microstrip radial stub," *IEEE Trans. Microwave Theory Tech.*, vol. MTT-32, pp. 1652–1655, Dec. 1984.

[10] K. C. Gupta and M. D. Abouzahra, "Analysis and design of four-port and five-port microstrip disc circuits," *IEEE Trans. Microwave Theory Tech.*, vol. MTT-33, pp. 1422–1428, Dec. 1985.

[11] W. K. Gwarek, "Analysis of an arbitrarily-shaped planar circuit—A time-domain approach," *IEEE Trans. Microwave Theory Tech.*, vol. MTT-33, pp. 1067–1072, Oct. 1985.

[12] P. Silvester, "Finite element analysis of planar microwave networks," *IEEE Trans. Microwave Theory Tech.*, vol. MTT-21, pp. 104–108, Feb. 1973.

[13] G. Conciauro, M. Bressan, and C. Zuffada, "Waveguide modes via an integral equation leading to a linear matrix eigenvalue problem," *IEEE Trans. Microwave Theory Tech.*, vol. MTT-32, pp. 1495–1504, Nov. 1984.

[14] P. Arcioni, M. Bressan, and G. Conciauro, "Wideband analysis of planar waveguide circuits," *Alta Frequenza*, June 1988.

[15] J. Van Bladel, *Electromagnetic Fields*. Washington: Hemisphere Publishing Corp., 1985.

[16] N. Marcuvitz, *Waveguide Handbook*. Cambridge MA: Radiation Lab. MIT, 1951.

[17] M. Bressan and G. Conciauro, "Singularity extraction from the electric Green's functions in two-dimensional resonators of circular or rectangular cross-section," *Alta Frequenza*, vol. LII, pp. 188–190, Mar. 1983.

[18] M Bressan and G. Conciauro, "Singularity extraction from the electric Green's function for a spherical resonator," *IEEE Trans. Microwave Theory Tech.*, vol. MTT-33, pp. 407–414, May 1985.

[19] D. S. Jones, *Methods in Electromagnetic Wave Propagation*. Oxford: Clarendon Press, 1979.

[20] P. Arcioni and M. Bressan, "ANAPLAN: a package for the wide-band analysis of planar circuits," (in Italian) *Proc. Sesta Riunione Nazionale di Elettromagnetisimo Applicato* (Trieste), Oct. 22–24, 1986, pp. 305–308.

[21] J. N. Franklin, *Matrix Theory*. Englewood Cliffs, NJ: Prentice-Hall, 1968.

Two-Dimensional Field Analysis for CAD of Rotman-Type Beam-Forming Lenses

P.C. Sharma,[1]* K.C. Gupta,[1] C.M. Tsai,[1] J.D. Bruce,[2] and R. Presnell[2]

[1]MIMICAD Center, Department of Electrical and Computer Engineering,
University of Colorado at Boulder, Boulder, Colorado 80309-0425

[2]ESL, Inc., Sunnyvale, California 94086

Received February 8, 1991; revised October 9, 1991.

ABSTRACT

A general procedure for design and analysis of a Rotman-type beam-forming lens using 2-dimensional field analysis is proposed. Evaluation of reflection coefficients, coupling between ports, main beam direction, 3-dB beam width and side lobe level are described. The results for a sample case are in agreement with the experimental results. The procedure developed is suitable for CAD implementation leading to optimization of Rotman lens layout configuration.

1. INTRODUCTION

Rotman-type lenses, such as the one shown in Figure 1, are used extensively in multiple beam-forming radiating systems [1–2]. The classical methods [3]–[6] for analysis and design of these lens structures are based on ray optics of electromagnetic waves. In this approach, the geometry of the lens and the lengths of the lines connecting the array contour of the lens to the antenna elements are chosen such that the ray paths from a feed port to the corresponding antenna elements (through the lens and connecting lines) produce the desired phase distribution (for radiation in a specific direction) along the length of the array. This approach, however, does not incorporate the effect of (i) mutual coupling between ports; (ii) multiple scattering between contours; and (iii) discontinuity reactances at the junctions between the lens and the transmission lines at the input–output ports. Thus, the experimental performance does not always meet the theoretical design predictions. Experimental iterations, which are expensive and time consuming, thus become necessary for meeting the given specifications.

The field analysis of this type of lens has not been reported extensively in the literature. In the only published studies available [7,8], the field distribution at port apertures, around the lens periphery, is described by a contour integral solution of the wave equation using the method of moments.

The present article reports an alternative 2-dimensional field analysis approach based on the contour integral method [9] to compute the Z-matrix for the multiport model of the lens and then evaluate the S-matrix. The S-matrix yields reflection coefficients, coupling coefficients, and voltage distribution on the array elements excited by transmission lines connected to the output ports of the lens. In a microstrip-stripline configuration of the lens, the height (thickness) is much smaller than the wavelength at the operating frequency, and consequently, there is no variation of the fields along the height of the substrate. Such

*On leave from Shri G.S. Institute of Technology and Science, Indore (MP) India.

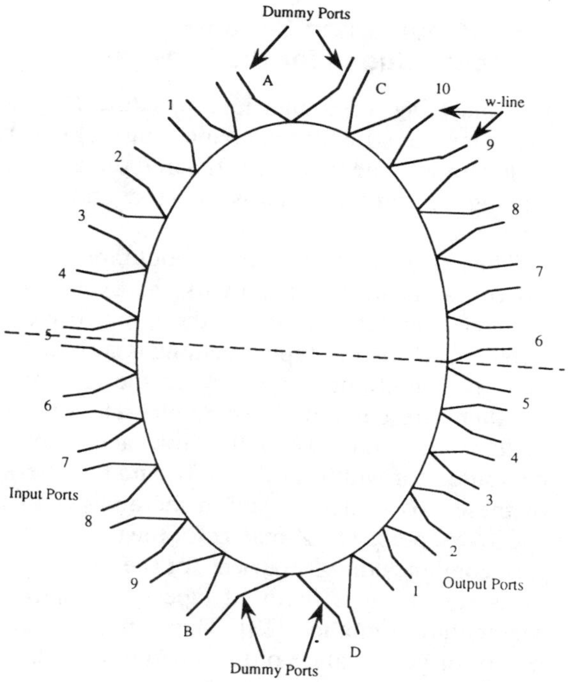

Figure 1. Rotman lens configuration.

a planar structure can be analyzed by 2-dimensional circuit analysis methods [10]. For the purpose of analyzing a Rotman-type lens, out of the several methods of 2-dimensional analysis available [10], the contour integral method [9] is suitable because of the arbitrary geometrical shape of the lens. This method yields voltage at any point on the periphery of the planar component, for a given input current at a specified feed point. The Z-matrix and, therefore, S-matrix can thus be evaluated.

The S-matrix yields reflection coefficients, coupling coefficients among various ports, and the voltage distribution at the antenna elements of the array. The radiation characteristics are then computed. The method has been used to analyze a typical Rotman lens configuration, and the analysis results are compared with the experimental data obtained from ref. 11. The analytical and experimental results are in good agreement. Details of the method of analysis are given in the following section.

2. ANALYSIS APPROACH

2.1. Two-Dimensional Analysis

As mentioned in the preceding section, the Rotman-type lens (in the present case, a microstrip version) can be analyzed by 2-dimensional field analysis approach. In view of the arbitrary geometry of lens, the contour integral method [9] is suitable for analyzing the lens. The method is discussed in the following.

2.1.1. Formulation of Contour Integral Method. Consider an arbitrary planar microstrip configuration shown in Figure 2. Using Weber's solution for cylindrical waves, the potential $V(s)$ at a point upon the periphery of the circuit geometry is found to satisfy the integral equation [9]

$$V(s) = \frac{1}{2j} \oint_c [k \cos \theta H_1^{(2)}(kr) V(s_0) - j\omega\mu d H_0^{(2)}(kr) I_n(s_0)] ds_0 \quad (1)$$

where k is the wave number, $H_0^{(2)}$ and $H_1^{(2)}$ are the zero-order and first-order Hankel functions, respectively, of the second kind. The flowing current at a point s_0 on the periphery is denoted by $I_n(s_0)$. The variable r denotes the distance between points m and n on the periphery, d is the substrate thickness, and s and s_0 are the locations of the field and source points (respectively) along the periphery. The angle between the line joining the points m and n and the normal at point n (as in Fig. 2) is denoted by θ_{mn}. The integral in eq. (1) is a contour integral over the periphery of the planar component, and is the reason for the name of this method. A detailed derivation of this integral expression is available in ref. 9. Eq. (1) gives the

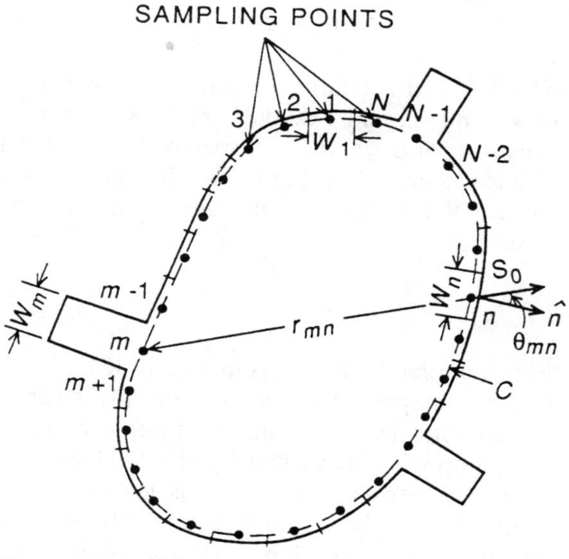

Figure 2. Arbitrary planar configuration analyzed by the contour integral approach.

relation between the rf voltage and rf current distribution along the periphery. To solve eq. (1) numerically, the circuit periphery is divided in sections numbered as 1, 2, . . . , N. The width of the mth section is indicated by W_m.

When the widths W_m of various sections are small, magnetic and electric fields may be assumed uniform over each section (port) width and eq. (1) yields a system of matrix equations:

$$\sum_{n=1}^{N} u_{mn} V_n = \sum_{n=1}^{N} h_{mn} I_n \qquad (m = 1, 2, \ldots, N) \quad (2)$$

where

$$u_{mn} = \delta_{mn} - \frac{k}{2j} \int_{W_n} \cos \theta H_1^{(2)}(kr) ds_0 \quad (3)$$

$$h_{mn} = \frac{\omega\mu d}{2} \frac{1}{W_n} \int_{W_n} H_0^{(2)}(kr) ds_0 \qquad m \neq n$$

$$= \frac{\omega\mu d}{2} \left[1 - \frac{2j}{\pi} \left[\log_e\left(\frac{kW_m}{4.0}\right) - 1 + \gamma \right] \right] \qquad (\text{for } m = n) \quad (4)$$

In eqs. (3) and (4), d is the thickness of the substrate, $\gamma = 0.45772$ and is Euler's constant, and δ_{mn} is 1 (for $m = n$), and zero otherwise. Solving eq. (2), we obtain the rf voltage on each sampling point as

$$V = U^{-1}HI \quad (5)$$

where V and I denote column vectors consisting of V_m and I_m, and U and H are $N \times N$ matrices consisting of U_{mn} and h_{mn}, respectively. The matrix U^{-1} is the inverse of the matrix U. The Z-matrix for the N-port network of Figure 2 is, therefore, given by

$$Z = U^{-1}H \quad (6)$$

In computing the Z-matrix using eqs. (3), (4), and (6), the integration in eqs. (3) and (4) is carried out numerically. The number of ports, N, in this computation, is governed by the field variation along the periphery of the lens geometry. The value of N is chosen large enough (to keep the electrical width of each port small enough) so that there is negligible field variation over each port width.

2.2. Z-Matrix and S-matrix Representations for the Lens

For analyzing a Rotman lens structure (Fig. 1), the complete structure is divided into (i) lens region without tapers; and (ii) tapered sections connecting the lens to microstrip feed and output lines.

The periphery of the lens without tapers is considered as a multiport network, as explained in the preceding subsection, and the Z-matrix is computed. Each of the taper sections is also considered as a multiport network and the Z-matrices for them are computed separately. The effective dielectric constant [12] values for each taper are computed for widths at the ends, and the average of these two values is used in the analysis of respective tapers. The Z-matrix for the overall structure combines the Z-matrices for the tapered sections and the lens (without tapers) by using the segmentation method [13]. The S-matrix is computed for given values of terminating impedances (50 Ω in the present case).

2.3. Performance of the Overall Configuration

The values of S_{mm} at various ports give the reflection coefficients. The elements S_{mn} of the S-matrix give the coupling coefficients between ports m and n. For input at the port m, the value of S_{mn} gives excitation of the antenna element connected to the port n. The radiation characteristics are computed using the voltage distribution, the given spacing between array elements and the dimensions of the aperture of the array elements (which, in the present case, are rectangular horns). The array factor of the array with isotropic radiating elements is multiplied with unit pattern of rectangular horn to compute the radiation characteristics.

3. RESULTS OF ANALYSIS AND COMPARISON WITH EXPERIMENT

3.1. Specific Configuration

The specific Rotman lens geometry, that has been analyzed using the method discussed in the preceding section, consists of 23 ports [11 ports on the feed contour and 12 ports on the array (output) contour]. Four of these ports (A, B, C, and D in Fig. 1) are terminated in 50-Ω dummy loads. This leaves 9 input ports and 10 output ports as active ports. The 10 output ports are connected to 10 antenna elements (which form the array) through 50-Ω lines. The complete lens structure is fabricated in microstrip version [11] on a substrate thickness of 1/16 in. and dielectric constant of 2.2. The loss tangent of 0.001 has been used in the analysis.

3.2. Computational Details

The common interface between the periphery of the lens and each of the taper sections is divided into 10 ports. As there are 11 taper sections on the feed contour and 12 tapers on the array contour, the feed contour has 110 ports and the array contour has 120 ports. The lens is thus represented by a 2-dimensional multiport network model with 230 ports, and the Z-matrix for the model is a 230×230 matrix.

Since the interface between a taper and the lens periphery is divided into 10 ports, there are 10 ports on each taper-end connected to the lens periphery. The port of each taper connected to a 50-Ω line or dummy load is also divided into 10 subports, each subport forming a port in computing the Z-matrix. Each of the tapered lengths of a taper section is divided into 50 ports. Thus, there are 120 ports around the periphery of the taper sections. Out of the Z-matrix of order 120×120 pertaining to these 120 ports, the Z-parameters corresponding to the 2 ends of the taper are selected to obtain a 20×20 Z-matrix for each taper. In view of the geometrical symmetry of the complete lens configuration, Z-matrices for only 12 tapers need to be computed.

As mentioned above, the Z-matrix of the lens without tapers and the Z-matrices for the tapers are combined to obtain the Z-matrix for the overall lens configuration utilizing the segmentation method [13]. The resultant Z-matrix is of the order 230×230 as there are 10 subports at each of the taper ports connected to 50-Ω lines or dummy loads. These 10 subports can then be connected

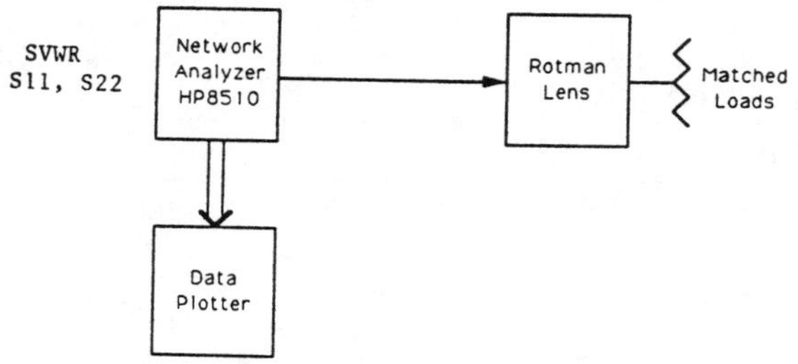

Figure 3. Experimental set-up for S_{mm} measurement.

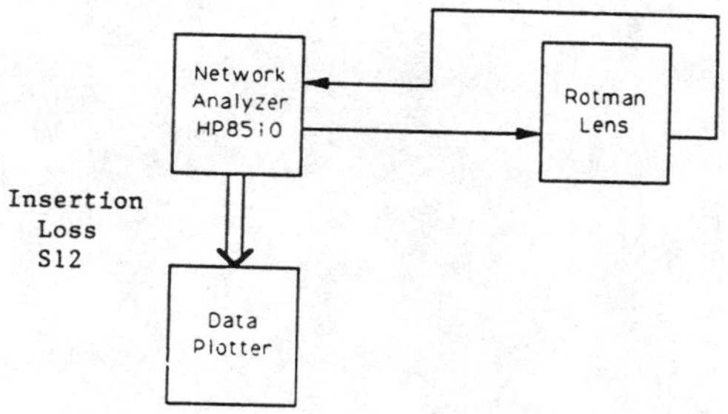

Figure 4. Experimental set-up for S_{mn} measurement.

in parallel, and the 230×230 Z-matrix is reduced to a 23×23 Z-matrix. The S-matrix is then computed with impedances of 50-Ω terminating each port. The effect of line lengths connecting the tapers to antenna elements is also taken into account.

The S-parameters so computed yield reflection coefficients, coupling coefficients between ports, and voltage distribution on the array. The array factor is computed using the field distribution, so obtained, and the given spacing between the array elements (= 1.2 in. in the present case). The array factor is multiplied with the element pattern of one of the array units. In the present case, the array elements are rectangular horns of aperture 0.75 (in.) × 4.2 (in.).

3.3. Comparison of Analysis and Experimental Results

3.3.1. Radiation Characteristics. Experimental measurement of S-parameters of the Rotman lens ports was made by direct measurement using a HP8510 network analyzer and data plotter. S_{11} and S_{22} measurements were made by terminating all ports of the lens except the one under test as shown in Figure 3. S_{12} measurements were made

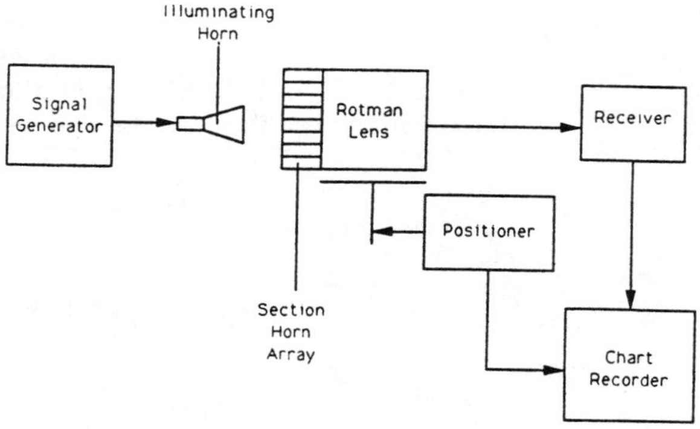

Figure 5. Experimental set-up for pattern measurements.

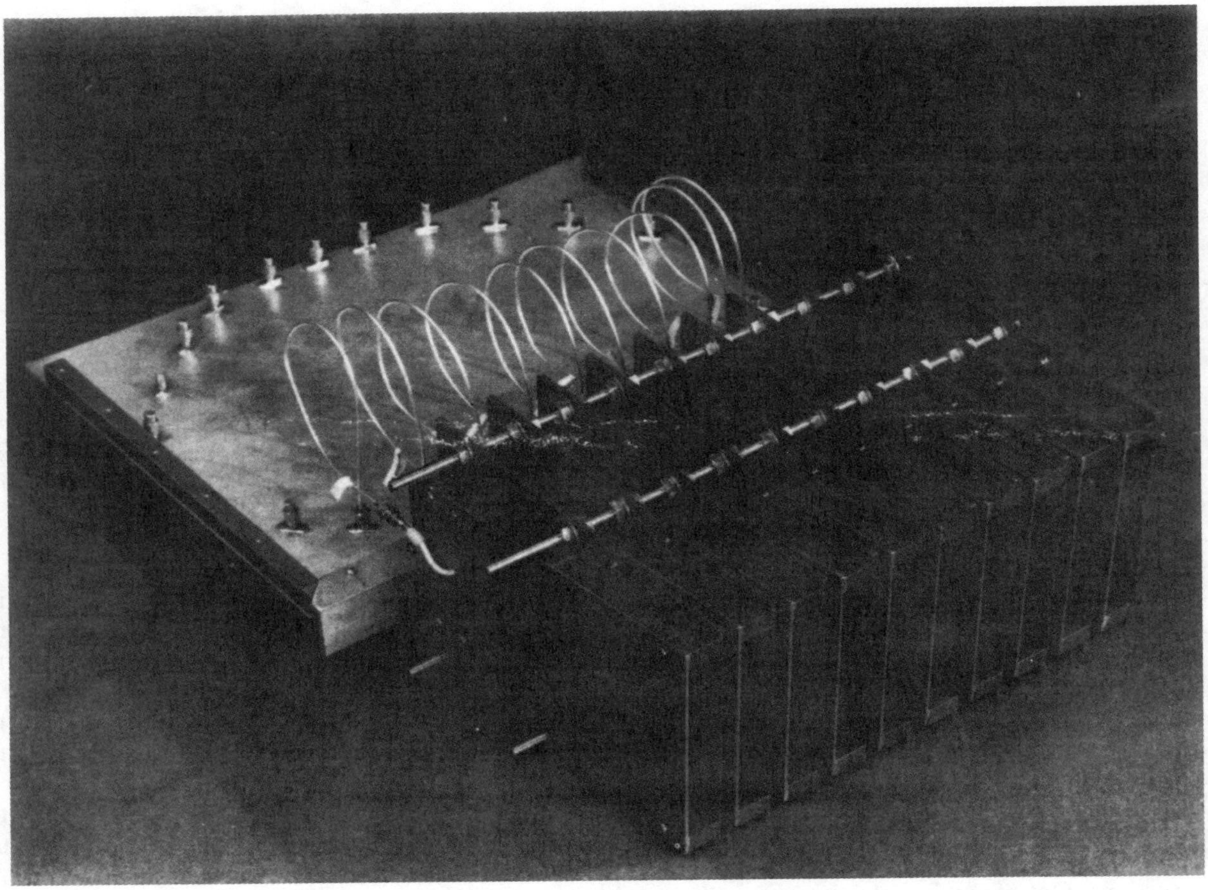

Figure 6. A photograph of Rotman lens and the array.

by terminating all ports of the lens, except the two under measurement shown in Figure 4.

Antenna pattern measurements were made using an indoor anechoic chamber equipped with rotating platform, illuminating horn antenna, and generating, receiving, and recording electronics, as shown in Figure 5. Again, all ports were terminated, except for the beam port under measurement. A photograph of the Rotman lens and the array is shown in Figure 6.

The results of analysis and the experimental measurements at 2.0 GHz are presented in this section.

The main beam directions for input at different feed ports as obtained by the analysis and as measured experimentally are given in Table I. The

TABLE I. Values of Direction of Main Beam, 3-dB Beam-Width, and Side Lobe Level for Different Feed Port Locations as Obtained by 2-Dimensional Analysis and Experiment

Feed Port (#5 is Broad Side Feed)	Main Beam Direction		3-dB Beam Width (Degree)		Side Lobe Level (dB)	
	2-D	Expt.	2-D	Expt.	2-D	Expt.
5	0.00	0.00	24.50	23.25	−12.22	−14.00
4	9.00	9.00	25.25	24.25	−11.82	−12.25
3	17.75	20.00	26.25	26.00	−11.49	−12.00
2	29.00	30.00	27.50	24.25	−10.40	−9.00
1	39.00	38.00	30.00	23.25	−9.49	−8.25

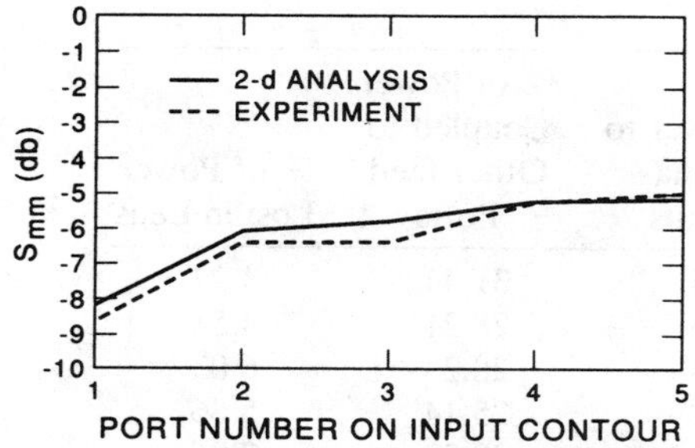

Figure 7. Comparison of analysis and experimental results for reflection coefficients at input ports.

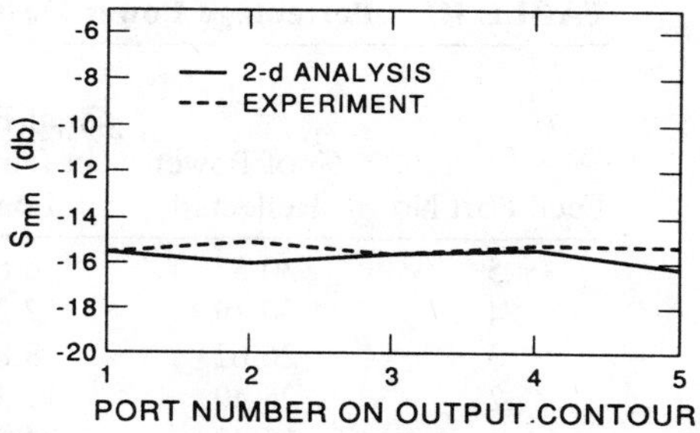

Figure 8. Comparison of analysis and experimental results for transmission coefficients from the broadside input port to various antenna ports.

ports are numbered as shown in Figure 1. It is observed that the results for the main beam direction are in excellent agreement. Theoretical values of the 3-dB beam widths increase as we move away from the broadside. Experimental values are in fair agreement, but for positions 2 and 1 (extreme position) beam width decreases, perhaps because of interaction between the lens and the array. At higher frequencies (4 and 6 GHz) beam width values increase monotonically as one moves away from the broadside. For the side lobe level, it is observed that the experimental results are within 1.8 dB of the analytical results.

3.3.2. Reflection Coefficients. The reflection coefficients at the input ports are shown in Figure 7. It is observed that the analytical and experimental results are in good agreement. In view of the geometrical symmetry of the lens structure, the values (Fig. 7) are given only for ports on one half of the lens geometry.

TABLE II. Coupling between Pairs of Ports

Port No. *m–n* on Output Contour	Coupling Coefficient (db)	Port No. on Input Contour	Coupling Coefficient (db)
1–2	−10.57	1–2	−10.63
2–3	−9.43	2–3	−12.23
3–4	−8.26	3–4	−10.38
4–5	−8.78	4–5	−11.52
5–6	−8.96		
1–3	−14.64	1–3	−15.37
2–4	−13.51	2–4	−16.17
3–5	−14.04	3–5	−13.99
4–6	−15.69	4–6	−14.97
1–4	−17.84	1–4	−16.38
2–5	−18.87	2–5	−17.11
3–6	−18.24	3–6	−16.59
4–7	−17.17		
1–5	−22.30	1–5	−17.40
2–6	−20.48	2–6	−21.77
3–7	−17.90	3–7	−21.71
1–6	−21.10	1–6	−19.60
2–7	−20.40	2–7	−18.72
3–8	−20.29	3–8	−18.70
1–7	−21.69	1–7	−15.60
2–8	−25.85	2–8	−14.87
1–8	−23.94	1–8	−16.79
2–9	−22.18		
1–9	−19.37	1–9	−16.56
1–10	−18.80		

TABLE III. Percentage Power Distribution in Ports

Feed Port No.	% of Power Reflected	% of Power in Dummy Loads	% of Power to Antenna Elements	% of Power Coupled to Other feed Ports	% of Power Lost in Lens
5	30.37	6.69	27.24	31.44	4.27
4	30.19	7.24	29.85	28.21	4.51
3	26.62	8.84	30.28	28.23	6.03
2	25.40	11.31	32.29	25.14	5.86
1	15.43	24.57	31.81	20.23	7.96
Avg. %	25.602	11.730	30.294	26.650	5.726

3.3.3. Voltage Distribution Along the Array. The values of S_{mn} at array elements are plotted in Figure 8 for a feed at broadside. The experimental and analytical results are in reasonable agreement.

3.3.4. Coupling Coefficients and Efficiency Considerations. The experimental values of coupling coefficients between ports are presently not available. However, the results of analysis are presented in Table II. The power distribution in various ports is given in Table III. It is observed that the maximum value of the input power transferred to antenna elements is 32.29%, and the minimum value is 27.24%. The average value of efficiency (average for various beam locations) equal to 30.29% is achieved with the present design.

4. CONCLUDING REMARKS

An alternative 2-dimensional field analysis method for analysis of Rotman-type beam forming lenses is described. The proposed method is used to analyze a given lens geometry. The results of analysis, namely reflection coefficients, main beam direction, 3-dB beam width, side lobe level, coupling coefficients, and power distribution in ports, are presented. The analysis results are compared with the experimental results presently available. The experimental and analysis results are in fair agreement. The results indicate the validity of the proposed method. The proposed analysis approach is well-suited for modification of the lens design for achieving improved performance, and for implementation of CAD for the design of beam-forming lenses.

ACKNOWLEDGMENTS

Initial work on this project by Mr. D. Jaisson is gratefully acknowledged. This work was partially sponsored by ESL Inc. (a subsidiary of TRW, Inc.) through a research contract at the University of Colorado.

REFERENCES

1. D. Archer, "Lens-fed multiple-beam arrays," *Microwave J.*, Sept. 1984, pp. 171–195.
2. D. T. Thomas, "Multiple beam synthesis of low sidelobe patterns in lens fed arrays," *IEEE Trans. Antennas Propagat.*, Vol. AP-26, 1978, pp. 883–886.
3. G. D. M. Peeler, "Lens antennas," in R. C. Johnson and H. Jasik (eds.), *Antenna Engineering Handbook* (2nd Ed.), McGraw-Hill, New York, 1984, pp. 16.19–16.23.
4. W. Rotman and R. F. Turner, "Wide-angle microwave lens for line source applications," *IEEE Trans. Antennas Propagat.*, Vol. AP-11, Nov. 1963, pp. 623–632.
5. M. L. Kales and R. M. Brown, "Design considerations for two-dimensional symmetric bootlace lenses," *IEEE Trans. Antennas Propagat.*, July 1965, pp. 521–528.
6. J. P. Shelton, "Focusing characteristics of symmetrically configured bootlace lenses," *IEEE Trans. Antennas Propagat.*, Vol. AP-26, No. 4, July 1978, pp. 513–518.
7. K. K. Chan, "Planar waveguide model of Rotman lens," *IEEE AP-S Symp. Dig.*, 1989, pp. 651–654.
8. K. K. Chan and W. Tam, "Field analysis of planar bootlace lens feeds," *Int. Conf. on Radar,* April 24–28, 1989, Paris, France, Vol. I, pp. 273–278.
9. T. Okoshi, "Planar circuits for microwaves and lightwaves," *Springer Series in Electrophysics,* Vol. 18, 1984, pp. 44–51.
10. K. C. Gupta, et al., *Computer-Aided Design of*

Microwave Circuits, Artech House, Dedham, MA, 1981, pp. 229–260.
11. R. Presnell, "Design of Rotman lens," Report #TR-849-1, ESL, Inc., Sunnyvale CA, October 1989.
12. K. C. Gupta, et al., *Computer-Aided Design of Microwave Circuits*, Artech House, Dedham, MA, 1981, pp. 60–62.
13. R. Chadha and K. C. Gupta, "Segmentation method using impedance matrices for analysis of planar microwve circuits," *IEEE Trans. Microwave Theory Tech.*, Vol. MTT-29, Jan. 1981, pp. 71–74.

Chapter 5

Integral Equation Approach for Three-Dimensional Analysis

THE most commonly used method for full-wave analysis of planar microwave components makes use of an integral equation formulation for current distribution on the conductors. In this method [5.1–5.4] the integral equation is invariably reduced to a matrix equation by using Galerkin's approach and/or the method of moments [5.5 and 5.6], and quite often the approach is known as "method of moments."

Various variations of this integral equation approach have been developed and implemented by different research groups around the world. A version known as electric field integral equation (EFIE) is usually formulated in spectral domain [5.7–5.10] The canonical building block in this approach is the plane-wave spectral representation of the grounded dielectric slab Green's function representing the electric field at any point due to an infinitesimal electric current element at any arbitrary point on the conductor of the planar circuit. Thus the problem is formulated in the spectral-domain approach, and then solved for the transform of the current distribution on the top conductors of the planar circuit.

An alternative approach to the EFIE is the mixed-potential integral equation (MPIE) approach introduced by Harrington [5.5] and extensively used earlier in the analysis of wire antennas. In this case, the integral equation is formulated [5.4] in terms of both electric currents and charges on the surface of conductor(s) in planar circuits. MPIE is frequently solved in the space domain using the Green's functions [5.8], which correspond physically to the potentials created by unit point sources. Determination of Green's function can be done in the ρz-plane, where z is the direction normal to the substrate. For planar circuits with top metallization in a single plane, once the Green's functions are computed and stored for the case $z = 0$, one can get rid of the z-coordinate and perform all the subsequent calculations in the x–y-plane (continuing the conductor pattern). Thus the integral equation techniques (EFIE or MPIE, spectral domain, or space domain) reduce a single three-dimensional (x-y-z) problem to two two-dimensional (ρ-z *and* x-y) problems, thus resulting in a numerically efficient approach. Consequently, going from a free-space configuration to multilayered dielectric substrates requires only a modification of the Green's function, but the two-dimensional problem in x–y-plane remains unchanged.

It may be pointed out that the use of the Green's functions for grounded dielectric substrates causes the resulting integrals to be of Sommerfeld type [5.11]. Surface waves (TE and TM) appear as poles of the functions to be integrated. Efficient numerical evaluation of these integrals calls for quite sophisticated techniques. Various implementations by different research groups differ in this regard.

This chapter is a collection of seven articles selected from numerous papers published in the past 10 years on the integral equation approach for three-dimensional analysis of planar microwave and millimeter-wave circuits. The first paper (Paper 5.1) is authored by Jansen and is an invited review of the spectral-domain approach for microwave integrated circuits published in a special issue of *MTT Transactions on Numerical Method*. This special issue was developed out of a workshop on "Critical Inspection of Field-Theoretic Methods for Microwave Problems" organized in conjunction with the 1984 MTT Symposium in San Francisco. This paper by Jansen surveys both the analytical and numerical aspects of the spectral domain method for shielded-, covered-, and open-type problems involving two- and three-dimensional strip- and slot-type fields. An exhaustive bibliography (some 110 references) is included. The approach presented in this paper has been developed into a commercially available software for simulation of microwave planar circuits [5.12]. The spectral domain approach was developed by Itoh and Mittra in a pioneering paper [5.13].

The second paper (Paper 5.2) in this chapter describes a space-domain method of moments (Galerkin's technique) for shielding microstrip circuits. The analysis was implemented in a Pascal program on an IBM-PC and was later transported to a VAX computer. Dynamic arrays were used extensively in developing a complex vector data type that was used to vectorize the software. Further developments of this implementation have also resulted in a commercially available code from Sonnet Software [5.14].

The next paper (Paper 5.3) is also a space-domain formulation, but in this case MPIE is applied to open microstrip structures. This paper appeared in IEEE-MTT Transaction special issue on *Computer-Aided Design*. This technique uses Green's functions associated with the scalar and vector potentials, which are calculated by using stratified media theory and are expressed as Sommerfeld integrals. Three different combinations of basis and test functions are described. This formulation is applicable to both microstrip circuits and microstrip antennas. Also, it is shown that the MPIE approach includes previously published static and quasi-static integral equations.

The fourth paper (Paper 5.4) in this chapter describes a different integral equation formulation for analyzing shielded thin microstrip discontinuities and circuits. The integral equation is derived by applying the reciprocity theorem and is solved by the method of moments. In this derivation, a coaxial aperture is modeled with an equivalent magnetic current and is used as the excitation mechanism for generating the microstrip currents. This formulation for excitation is more realistic than either the gap generator excitation method or the cavity resonance technique used in other integral equation formulations for full-wave analysis. Numerical results from this analysis and comparison with measured data are presented in a separate paper [5.4].

Paper 5.5 is an extension of an earlier work reported by Jansen [5.16], Rautio (Paper 5.2), and Dunleavy (Paper 5.4). The emphasis in this paper is on efficient formulation of the system matrix in the moment method procedure, which allows the derivation of the elements of any large matrix by a linear combination of elements of a precomputed index table. The table is obtained from a two-dimensional discrete fast Fourier transform. Further work by Hill has contributed to the development of a commercially available full-wave analysis code from Compact Software [5.17].

This is followed by Paper 5.6, which combines the spectral-domain approach (SDA) and the method of lines (MOL) [5.18, 5.19] for three-dimensional planar circuit discontinuities. The MOL is most efficient when only one spatial variable needs to be discretized and, similarly, the SDA is most efficient when only one-dimensional basis functions are used. The combination space-spectral–domain approach (SSDA) discussed in this paper combines the one-dimensional SDA (which is used to describe only the plane transverse to the propagation direction) with the one-dimensional MOL (which describes the circuit in the propagation direction). This paper appeared in a special issue of the *IEEE Transactions on Microwave Theory Techniques* (July 1992) on process-oriented microwave CAD and modeling.

The last paper (Paper 5.7) in this chapter describes a spatial-domain mixed-potential integral equation approach developed for the analysis of microstrip discontinuities and antennas of arbitrary shape. Conceptually, the procedure is similar to that discussed by Mosig (Paper 5.3). However, the algorithm developed in Paper 5.7 utilizes roof-top basis functions on a rectangular and triangular mixed grid and includes analytical evaluations of the quadruple moment integrals. The algorithm has been successfully implemented into an efficient program that is being developed for commercial release [5.20].

References

[5.1] Mosig, J. R. "Integral equation technique." In T. Itoh, ed., *Numerical Techniques for Microwave and Millimeter-Wave Passive Structures,* Chapter 3, pp. 133–213. New York: John Wiley & Sons, 1989.

[5.2] Sorrentino, R. "Numerical methods for passive microwave and millimeter wave structures." *IEEE Press Reprint Volume,* Part 7 on "Method of Moments" and Part 10 on "Spectral-Domain Approach," Piscataway, NJ: IEEE PRESS, 1989.

[5.3] Wong, J. J. H. *Generalized Moment Methods in Electromagnetics.* New York: John Wiley & Sons, 1991.

[5.4] Mosig, J. R., and F. E. Gardiol. "A dynamical radiation model for microstrip structures." In *Advances in Electronics and Electron Physics,* Vol. 59, pp. 139–237. New York: Academic Press, 1982.

[5.5] Harrington, R. F. "Field Computation by Moment Methods," New York: Macmillan, 1968; reprinted Melbourne, FL: R. E. Krieger, 1982.

[5.6] Ney, M. M. "Method of moments as applied to electromagnetic problems." *IEEE Trans. Microwave Theory Tech.,* Vol. MTT-33, No. 10, pp. 972–980, October 1985.

[5.7] Katehi, P. B., and N. G. Alexopoulos. "Frequency-dependent characteristics of microstrip discontinuities in millimeter-wave integrated circuits." *IEEE Trans. Microwave Theory Tech.,* Vol. MTT-33, No. 10, pp. 1029–1035, October 1985.

[5.8] Jackson, R. W., and D. M. Pozar. "Full-wave analysis of microstrip open-end discontinuities." *IEEE Trans. Microwave Theory Tech.,* Vol. MTT-33, No. 10, pp. 1036–1042, October 1985.

[5.9] Becks, T., and I. Wolff. "Improvements of spectral domain analysis techniques for arbitrary planar circuits." In H. L. Bertoni and L. B. Felsen, eds., *Directions in Electromagnetic Wave Modeling,* pp. 339–346. New York: Plenum Press, 1991.

[5.10] Itoh, T. "The spectral domain method." In E. Yamashita, ed., *Analysis Methods for EM Wave Problems,* Chapter 11. Dedham, MA: Artech House, 1990.

[5.11] Mosig, J. R., and F. E. Gardiol. "Analytical and numerical techniques in the Green's function treatment of microstrip antennas and scatterers." *Proc. Inst. Elec. Eng.,* Pt. H, Vol. 130, pp. 175–182, 1983.

[5.12] User's Manual for *LINMIC+ and SFPMIC+*. Ratingen, Germany: Jansen Microwave, 1989.

[5.13] Itoh, T., and R. Mittra. "Spectral-domain approach for calculating the dispersion characteristics of microstrip lines." *IEEE Trans. Microwave Theory Tech.,* Vol. 21, pp. 496–499, July 1973.

[5.14] *Em.* Sonnet Software, Inc., Liverpool, New York, 1991.

[5.15] Dunleavy, L. P., and P. B. Katehi. "Shielding effects in microstrip discontinuities." *IEEE Trans. Microwave Theory Tech.,* Vol. MTT-36, No. 12, pp. 1767–1774, December 1988.

[5.16] Jansen, R. H. "Hybrid mode analysis of end effects of planar microwave and millimeter wave transmission lines." *Proc Inst. Elec. Eng.,* Pt. H, Vol. 128, No. 2, pp. 77–86, April 1981.

[5.17] *Explorer.* Compact Software, Inc., Paterson, New Jersey, 1992.

[5.18] Worm, S. B., and R. Pregla. "Hybrid mode analysis of arbitrarily shaped planar microwave structures by the method of lines." *IEEE Trans. Microwave Theory Tech.,* Vol. MTT-32, pp. 191–196, February 1984.

[5.19] Pregla, R., and W. Pascher. "The method of lines." In T. Itoh, ed., *Numerical Techniques for Microwave and Millimeter Wave Passive Structures,* Chapter 6, pp. 381–446. New York: John Wiley & Sons, 1989.

[5.20] Childs, W. H. "Integration of Electromagnetic Simulation into the Microwave Circuit Environment." *MIMICAD Newsletter,* Vol. 5, No. 2, University of Colorado (Boulder), Ocober 1992.

Paper 5.1

The Spectral-Domain Approach for Microwave Integrated Circuits

ROLF H. JANSEN, SENIOR MEMBER, IEEE

(*Invited Paper*)

Abstract—A survey is given of the so-called spectral-domain approach, an analytical and numerical technique particularly suited for the solution of boundary-value problems in microwave and millimeter-wave integrated circuits. The mathematical formulation of the analytical part of this approach is described in a generalized notation for two- and three-dimensional strip- and slot-type fields. In a similar way, the numerical part of the technique is treated, keeping always in touch with the mathematical and physical background, as well as with the respective microwave applications. A discussion of different specific aspects of the approach is presented and outlines the peculiarities of shielded-, covered-, and open-type problems, followed by a brief review of the progress achieved in the last decade (1975–1984). The survey closes with considerations on numerical efficiency, demonstrating that spectral-domain computations can by speeded up remarkably by analytical preprocessing. The presented material is based on ten years of active involvement by the author in the field and reveals a variety of contributions by West German researchers previously not known to the international microwave community.

I. Introduction

GENERALLY SPEAKING, the term spectral-domain approach (SDA) refers to the application of integral transforms, such as the Fourier and Hankel transforms, to the solution of boundary-value and initial-value problems. As becomes obvious from the overview book and associated bibliography by Sneddon [1], this approach has been applied to mechanical and electromagnetic problems for at least a century. It provides an elegant tool for the reduction of the partial differential equations of mathematical physics to ordinary ones, which in many cases are amenable to further analytical processing. During the last 15 years, the spectral-domain approach has received considerably more interest together with the growing importance of printed circuits for very high frequencies, namely conventional and monolithic microwave and millimeter-wave integrated circuits (MIC's) fabricated by planar photolithographic technology. The actual and potential range of application of this technique implies hybrid thin- and thick-film circuits, monolithic MIC's on gallium arsenide, planar resonators and antennas, as well as multiconductor multilayer interconnections in high-speed computers. These circuits and components typically operate at frequencies between 0.1 and 100 GHz, and the main intention of using the approach has been the derivation of accurate, particularly frequency-dependent, design information.

Manuscript received February 5, 1985; revised May 31, 1985.

The author is with the University of Duisburg, Department of Electrical Engineering, FB9/ATE, Bismarckstrasse 81, D-4100 Duisburg 1, West Germany.

Already by 1957, Wu [2] had considered it an "obvious thing to do" to apply a Fourier transform in the analysis of microstrip lines. From the end of the 1960's on, several authors began to implement more and more of those steps which are characteristic for what today is denoted the spectral-domain approach for MIC's. Yamashita and Mittra [3], for example, solved Poisson's equation in the transform domain and computed microstrip line capacitance from a variational expression under application of parseval's theorem. Denlinger [4] in the United States and Schmitt and Sarges [5] in West Germany both derived an approximate solution to the microstrip dispersion problem in terms of the transformed strip current density. Itoh and Mittra [6], on the other hand, applied a spectral-domain approach in essentially the form it is still used today to the computation of slotline dispersion characteristics. Two years later, the same authors explicitly used the notation "spectral domain approach" for the specific technique (Galerkin's method in the transform domain) employed in one of their microstrip contributions [7]. Recent analyses still follow the basic outlines of this technique and the notation has been adopted by the microwave community.

In the initial research phase, a variety of fundamental applications and modifications of the spectral-domain approach and related methods had been reported within a few years. Coupled microstrip dispersion and characteristic impedances were computed by Kowalski and Pregla [8] and by Krage and Haddad [9]. Also, guided higher order modes in open microstrip lines were treated by Van de Capelle and Luypaert [10]. Itoh and Mittra [11] extended the spectral-domain formalism to shielded microstrip lines while Jansen [12] treated the same problem making use of a least-square criterion instead of Galerkin's method in the final step of the solution. As a first application to microstrip discontinuity problems, Rahmat-Samii *et al.* performed a quite general static spectral-domain analysis [13]. The first full-wave analyses of hybrid-mode microstrip resonator problems were reported by Itoh [14] and by Jansen [15], [16] in 1974, including rectangular, disk, ring, and concentric coupled shapes. Along the guidelines having emerged in this way, the spectral-domain approach has been used extensively for the characterization and analysis of elementary structures frequently appearing in MIC's. These structures can be classified as conducting thin patterns in one or more interfaces of a multilayer stratified dielectric medium. Therefore, the associated electromag-

Reprinted from *IEEE Trans. Microwave Theory Tech.*, vol. MTT-33, no. 10, pp. 1043–1056, Oct. 1985.

netic boundary-value problems lend themselves ideally to an SDA treatment. The partial differential equations considered are mainly the wave equation or, where small dimensions compared to wavelength prevail, the Laplace and the Poisson equation. Specific problems frequently tackled by the spectral domain approach are:

1) the static or frequency-dependent characterization of printed microwave transmission lines (a two-dimensional electromagnetic field problem).
2) the static or frequency-dependent analysis of problems concerning strip and slot transmission-line discontinuities, junctions and resonators, respectively and patch antennas (three-dimensional electromagnetic fields).

The contribution given here outlines the basic features of the analytical formulation of the spectral-domain approach as it applies to the above-mentioned problems. It is shown how for printed planar structures of arbitrary connected and disconnected shape embedded in a multilayer dielectric medium a single closed-form integral equation emerges from the application of the analytical steps of the SDA. As a result of explicit construction of that portion of the solution which depends on the vertical coordinate, this integral equation comes out reduced by one dimension compared to the original partial differential equation. From the beginning of the analysis, a considerable reduction in complexity is achieved and reduces the expense for the subsequent numerical part of the approach. This provides one of the important arguments for the superiority of the spectral-domain approach compared to other techniques.

In a discussion of the numerical procedure usually employed to solve the derived integral equation, the peculiarities of eigenvalue-type and deterministic MIC problems are treated briefly. There are arguments to prefer a Galerkin solution with certain symmetry properties for the former, while the latter do not generally result in a symmetric; respectively, Hermitian system of equations. From the obtained solutions, most of the quantities required in the characterization and analysis of MIC's can be obtained directly in the transform domain. Only one of the methods recently applied to MIC's shares several of the advantages of the SDA: the differential-difference approach (DDA), also called the method of lines [17]. A comparison with this, therefore, deserves a brief discussion.

After presenting these general features, the different aspects of the spectral-domain approach are outlined which have to be considered for shielded structures, laterally open structures, and configurations which are completely open electromagnetically. A specific implementation of the approach recently developed for the systematic frequency-dependent analysis of discontinuities and junctions in MIC's is described. It is discussed further as to how the radiation condition can be incorporated into the SDA formulation by proper choice of the integration path prevailing for the basic integral equation. To round out the

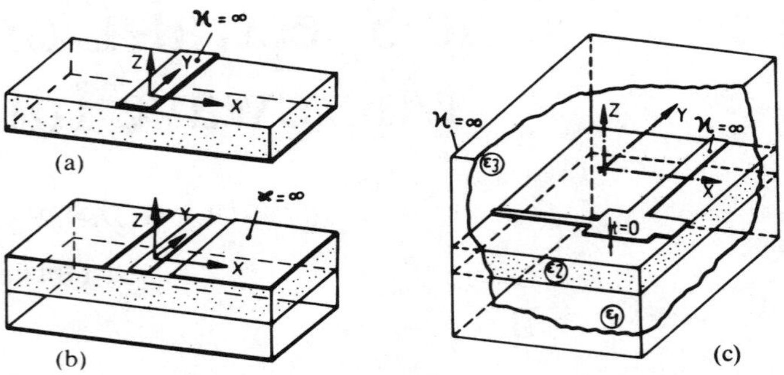

Fig. 1. (a) Microstrip line and (b) coplanar waveguide as examples for MIC transmission lines and (c) generalized MIC structure together with coordinate system used in the discussion.

picture given of the spectral-domain approach, the more important results achieved by its application are summarized in a subsequent section, and the state-of-the-art is described.

The last section of the paper is a discussion of the advantages and disadvantages of the spectral-domain approach. Emphasis is placed on the hybrid character of this technique which requires (and allows!) a certain amount of analytical preprocessing to achieve high efficiency. It is shown, further, how most of the disadvantages of the technique can be removed and to what degree, typically, a specific class of problems can be speeded up.

Remarks on the numerical problems associated with the development of user-oriented SDA packages are made and critical parts of these are illuminated. Finally, the main characteristics of the spectral-domain approach are summarized in a brief conclusion.

II. Mathematical Formulation

Some elements of the analytical steps necessary to apply the spectral-domain approach to specific problems, particularly the characterization of MIC transmission lines, have already been described in overview books [18], [19]. The treatment given here generalizes the formulation as far as possible and emphasizes those features which the different classes of MIC problems all have in common. For a visualization of the physical construction of the configurations to be considered, Fig. 1 shows (a) an open microstrip line, (b) a coplanar waveguide suspended above the ground plane of a circuit environment, (c) and a quite general shielded structure. The latter serves for the following discussion and could as well be laterally open or completely open. It provides an idealized view of the basic construction of MIC's indicating that the passive portions of these consist mainly of a thin conductor metallization in one or more interfaces of a double or multilayer dielectric medium; for an overview, see [20].

In agreement with common microwave practice, the formulation is in terms of time-harmonic electromagnetic fields, namely, a time dependence of $\exp(j\omega t)$. Vector quantities, like the electric field $\underline{E}$, are written by single underlining, matrices by double underlining. The involved conductors are assumed to have ideal conductivity κ and negligible thickness t. This is a very realistic assumption in hybrid MIC's, where strip and slot widths are usually large compared to conductor thickness. In monolithic MIC's on gallium-arsenide, this is not valid with the same generality. Here, the assumption is mainly a matter of convenience and simplification of the treatment. The consideration of finite thickness in the SDA formulation can be achieved by treating the thick metallization as a separate layer, see for example [21]–[23]. In addition, it is convenient in most cases to assume lossless dielectric media since this allows a ral number arithmetic for the SDA algorithm, except for cases where radiation or surface-wave excitation is involved. Loss parameters are usually introduced by perturbation methods subsequent to a numerical solution neglecting loss. This also applies to the evaluation of conductor loss, which can be taken into account if the asymptotic behavior of the field derived for zero metallization thickness is appropriately modified [24]–[27]. In each of the layers $i=1,2\ldots L$ of a general configuration like that of Fig. 1, the electromagnetic field is best formulated in terms of scalar LSM and LSE wave potentials [28], [29]. This is equivalent to the use of vector wave potentials having only one component in the z-direction, i.e., perpendicular to the stratified circuit medium. It allows a completely decoupled and, therefore, particularly simple analytical treatment of all classes of MIC problems [30]–[36] independent of the number of dielectric layers involved. This specific choice naturally leads to what Itoh [35] has named the spectral-domain inmittance approach. Just recently, Omar and Schünemann [37] have shown that only coupling of the LSE and the LSM contributions to the electromagnetic field occurs and is required in satisfying the edge condition with the last step of the solution. The scalar LSM and LSE wave potentials are denoted f and g here. They are subject to the homogeneous Helmholtz equation

$$(\Delta+k_i^2)f_i=(\Delta+k_i^2)g_i=0 \qquad (1)$$

in each of the layers $i=1,2\ldots L$ of the circuit medium, with k_i denoting the wave number associated with the ith layer. It should be stressed that the homogeneous form of (1) applies even in the case of excitation problems. With the spectral-domain approach, sources are introduced in a natural way as impressed current densities or electric fields only into the interfaces between the layers [38]–[41]. Instead of considering the space-domain form of the Helmholtz equation (1) directly, its spectral-domain equivalent is used. Without loss of generality, the scalar potentials may be written in the form of inverse two-dimensional Fourier transforms, for example,

$$f_i(x,y,z)=\int_{C_x}\int_{C_y}\tilde{f}_i(k_x,k_y,z)\cdot\exp\left(-j(k_x x+k_y y)\right)dx\,dy. \qquad (2)$$

For configurations of circular symmetry, the use of Hankel transforms is an adequate choice [1], [42]–[46]. In transmission-line problems, (2) is reduced by one dimension since these problems can be formulated in terms of the cross-sectional field alone postulating a longitudinal dependency of the form $\exp(-jk_y y)$. Alternatively, for these cases, the spectral wave potential $\tilde{f}_i$ may be viewed as factorized and containing a y-dependent factor in the form of a Dirac distribution, so that the integration in (2) reduces to one dimension. In the high majority of SDA contributions published, the integration paths C_x and C_y have been chosen to coincide with the respective coordinate axes. The spectral variables k_x and k_y may be interpreted each as a measure of spatial oscillation of the described field which is useful for later convergence considerations. This is immediately obvious if (2) implies a finite Fourier transform [11] and the spectral wave numbers k_x and/or k_y form an infinite numerable sequence. In that case, $f_i(k_{xm},k_{yn},z)$ describes the Fourier series coefficients of $f_i(x,y,z)$ and these values of k_{xm}, k_{yn} are chosen in such a way that the boundary conditions on a lateral shielding parallel to the z-axis are satisfied. Further generalizing, one may consider such coefficients as being associated with any two suitably chosen complete orthogonal sets of solutions of the Helmholtz equation which are TM and TE with respect to the z-direction and satisfy the lateral boundary conditions [30], [31]. At the same time, this reveals how a suitable finite integral transform can be constructed for a given cross section of the shielding. In addition, this generalization makes clear that the mathematical formulation can be discussed completely independent of the special cross section or even the existence of a lateral shielding. The Helmholtz operator of (1), if applied in the transform domain, always appears as

$$\tilde{\Delta}+k_i^2=\frac{\partial^2}{\partial z^2}+k_i^2-k_x^2-k_y^2=\frac{\partial^2}{\partial z^2}+k_{zi}^2 \qquad (3)$$

which is an ordinary differential operator. Due to the simple layered planar construction of MIC's, the transformed wave potentials $\tilde{f}_i$ and $\tilde{g}_i$ can be determined analytically and adopt the general form

$$\tilde{f}_i(k_x,k_y,z)=a_i(k_x,k_y)\cdot\cos\left(k_{zi}(z-z_i)\right)+b_i(k_x,k_y)\cdot\sin\left(k_{zi}(z-z_i)\right),\qquad i=1\cdots L. \qquad (4)$$

The functions a_i and b_i are spectral distributions weighting the elementary plane-wave constituents with respect to the z-axis. The parameter z_i is arbitrary and is, for example, introduced to allow convenient satisfaction of the boundary conditions at the conducting ground plane and cover shielding usually existing in MIC's. With the relations

$$j\omega\epsilon_i\tilde{E}_{zi}=k_\rho^2\tilde{f}_i,\; j\omega\mu_i\tilde{H}_{zi}=k_\rho^2\tilde{g}_i,\; k_\rho^2=k_x^2+k_y^2 \qquad (5)$$

it becomes clear that homogeneous boundary conditions prevail for $\tilde{f}_i$ and $\tilde{g}_i$ identical to those for the transformed field components $\tilde{E}_{zi}$ and $\tilde{H}_{zi}$. Further, since the transform defined by (2) affects only the x and y coordinates, all conditions specified for the spatial electromagnetic field in planes of constant values of z can be directly transferred into the spectral domain. Therefore, in a configuration involving the layers $i=1,2\ldots L$, the potential $\tilde{f}_i$ at the ground and top plane of the multilayer medium is

$$\tilde{f}_i(k_x, k_y, z) = a_i(k_x, k_y)\cos(k_{zi}(z - z_i)), \qquad i = 1, L \tag{6}$$

if z_1 and z_L describe the positions of the ground and cover shielding. In complete analogy, $\tilde{g}_i$ is proportional to the respective sine function for $i = 1$ and $i = L$ as a consequence of the vanishing of $H_{zi}(x, y, z)$ at $z = z_1$ and $z = z_L$. For vertically open structures like antennas and open microstrip, the potentials $\tilde{f}_L$ and $\tilde{g}_L$ both have an exponential z-dependence. The complete transformed electromagnetic field in all of the layers $i = 1, 2 \ldots L$ is derived by application of the spectral-domain equivalent $\tilde{\nabla}$ of the ∇ operator, namely, by

$$\underline{\tilde{E}}_i = \frac{1}{j\omega\epsilon_i}\tilde{\nabla} \times \tilde{\nabla} \times (\tilde{f}_i\underline{u}_z) - \tilde{\nabla} \times (\tilde{g}_i\underline{u}_z)$$

$$\underline{\tilde{H}}_i = \frac{1}{j\omega\mu_i}\tilde{\nabla} \times \tilde{\nabla} \times (\tilde{g}_i\underline{u}_z) + \tilde{\nabla} \times (\tilde{f}_i\underline{u}_z). \tag{7}$$

These relations have the same form as their spatial counterparts [28], [29] and result by substituting the algebraic multiplicators jk_x and jk_y for the respective partial differential operators. Together with the foregoing discussion they show that, in a circuit medium of L layers, the total electromagnetic field can be described by $4(L-1)$ independent spectral LSM and LSE distributions $a_i(k_x, k_y)$, $b_i(k_x, k_y)$. For the dielectric interfaces between the different layers, exactly the same number of continuity conditions can be formulated in the spectral domain, i.e.,

$$\tilde{E}_{xi} - \tilde{E}_{xi+1} = 0 \qquad \tilde{E}_{yi} - \tilde{E}_{yi+1} = 0$$

$$\tilde{H}_{xi} - \tilde{H}_{xi+1} = -\tilde{J}_{yi} \quad \tilde{H}_{yi} - \tilde{H}_{yi+1} = +\tilde{J}_{xi} \tag{8}$$

for $i = 1 \ldots L-1$ and z fixed at its interface value for each subscript i. They mirror the continuity of the electric field $\underline{E}_t$ tangential to the interfaces independent of the presence of a thin metallization. At the same time, they describe the discontinuity of the tangential magnetic field at such a metallization in terms of a surface current density $\underline{J}_t$. In interfaces which do not contain conductors, $\underline{J}_t$ is defined to be zero, enforcing the magnetic field continuity there. By analytical processing of the relations (8), all the unknown distributions a_i and b_i can be eliminated or expressed by the spectral-domain current density components $\tilde{J}_{xi}$ and $\tilde{J}_{yi}$. The latter may exist in only one of the interfaces or in several of these. In this stage of the analysis, the only boundary conditions which remain to be satisfied are those of the electric field $\underline{E}_t$ tangential to the conductor metallization and of the surface current density $\underline{J}_t$ to vanish in complementary regions. How the spectral-domain relations resulting from the analytical evaluation of (8) have to be arranged for further processing, therefore, depends on whether the metallized interfaces can be characterized as strip-type ($i = ist$) or slot-type ($i = isl$), respectively. In the general case, an algebraic spectral-domain equation results linking the mixed-type vectors

$$(\cdots \tilde{E}_{xist}, \tilde{E}_{yist} \cdots \tilde{J}_{xisl}, \tilde{J}_{yisl} \cdots)^T$$

and

$$(\cdots \tilde{J}_{xist}, \tilde{J}_{yist} \cdots \tilde{E}_{xisl}, \tilde{E}_{yisl} \cdots)^T \tag{9}$$

by a spectral immittance matrix; see for example [22], [35], [47]. T denotes transposition and is used only for convenience of writing. The lower one of the two vectors shown is put onto the right side of the described algebraic relation because its elements are better suited for expansion into known functions. These elements are typically confined to a small portion of the affected interfaces. For a similar reason, the upper one of the vectors is arranged onto the left side of the spectral-domain equation. By this, it is described by the other vector and needs to satisfy boundary conditions on only a small portion of its region of existence. The spectral algebraic relation between the vectors (9) is equivalent to a single integral equation which results by application of the transform inherent in (2). Since the whole discussion has been performed without recourse to particular shapes of the metallization pattern, this is true for arbitrary planar connected and disconnected conductors. From the procedure outlined, the kernel of the integral equation is available in analytical form. Further, this integral equation comes out reduced by one dimension compared to the original field problem associated with (1). It is one-dimensional for transmission-line problems and two-dimensional for discontinuities, resonators, and so on. As (4) shows, the z-dependency of the field is described in analytical form. In most cases of practical interest, the MIC configurations analyzed are either strip-type or slot-type exclusively, with only one layer of metallization. Under this presumption, the relation between the vectors (9) reduces to the simpler form

$$(\tilde{E}_{xist}, \tilde{E}_{yist})^T = \underline{\underline{\tilde{Z}}}(p) \cdot (\tilde{J}_{xist}, \tilde{J}_{yist})^T$$

or

$$(\tilde{J}_{xisl}, \tilde{J}_{yisl})^T = \underline{\underline{\tilde{Y}}}(p) \cdot (\tilde{E}_{xisl}, \tilde{E}_{yisl})^T. \tag{10}$$

The quantity p has been introduced to remind one of the fact that the elements of $\underline{\underline{\tilde{Z}}}$ and $\underline{\underline{\tilde{Y}}}$ depend in a known, analytical form on physical parameters defining the considered MIC problem (for example, vertical geometry, shielding dimensions, or operating frequency). Obviously, the spectral immittance matrix $\underline{\underline{\tilde{Y}}}$ is the inverse of $\underline{\underline{\tilde{Z}}}$ if both equations in (10) refer to the same problem and dielectric interface. For this reason, we may also write in the space domain

$$\underline{E}_t = L_\infty(p) \cdot \underline{J}_t \quad \text{or} \quad \underline{J}_t = L_\infty^{-1}(p) \cdot \underline{E}_t \tag{11}$$

where $\underline{E}_t = (E_x, E_y)^T$ and $\underline{J}_t = (J_x, J_y)^T$ are specialized to denote electric field and current density in the plane of the circuit metallization. The integral operators L_∞ and L_∞^{-1} are linear with respect to the vectors they operate on and both have a dyadic kernel defined by the spectral-domain Green's immittance matrices $\underline{\underline{\tilde{Z}}}$ and $\underline{\underline{\tilde{Y}}}$. The constituents of this kernel can be determined in a particularly elegant way by the so-called spectral-domain immittance approach [32], [35]. It has been shown by the author for shielded configurations that by suitable choice of an orthonormal vectorial function basis in (2), the kernel can be described by a single infinite scalar set of wave impedances $\tilde{Z}_n$ in conjunction with the elements of the function basis [30], [31]. In that case, the discrete impedance elements $\tilde{Z}_n$ are equal to the modal input wave impedances of the cylindrical shielding as seen in the plane of metallization. Independent of these details, there is always a duality between the strip-type and the slot-type formulation as visible in (10) and (11). For the former, the tangential electric field $\underline{E}_t$ has to vanish on the strip metallization F_{st}, while for the latter, the surface current density $\underline{J}_t$ does not exist in slot regions F_{sl}, i.e., outside of the metallization. Splitting up the right-hand quantities of (11) into an excited term (subscript $t$ex) and a source or impressed term (subscript $t$im) further yields

$$\underline{E}_t = \underline{0} = L_\infty(p)\cdot(\underline{J}_{t\text{ex}} + \underline{J}_{t\text{im}}) \text{ for } x, y \in F_{st}$$

$$\underline{J}_t = \underline{0} = L_\infty^{-1}(p)\cdot(\underline{E}_{t\text{ex}} + \underline{E}_{t\text{im}}) \text{ for } x, y \in F_{sl}. \quad (12)$$

These final integral equations are written here as in the formulation of a scattering problem [38]–[41]. They define, at the same time, the electric field $\underline{E}_t$ and current density $\underline{J}_t$ in the complementary regions F_{st}^{-1} and F_{sl}^{-1}, i.e., outside of strips and on the metallization around slots. If sources $\underline{J}_{t\text{im}}$, $\underline{E}_{t\text{im}}$ are note present, like in transmission-line and resonator problems, the equations in (12) each define a so-called nonstandard eigenvalue problem [48]–[50]. This notation applies since, without sources, (12) can be solved only for specific values of the parameter p (the eigenvalue), which is contained in the integral equation kernels in nonlinear, usually transcendental form. Which physical quantity is chosen in a problem as the parameter p is to some degree arbitrary. In transmission-line problems, the usual choice is $p = \beta$, i.e., the propagation constant, or $p = \beta^2$, the square of it. Resonator problems are conveniently treated in terms of $p = \omega_0$, the resonance frequency, or $p = l_0$, a dimension of the resonator.

III. Numerical Solution

The standard procedure applied in most computer solutions of the integral equations (12) today is Galerkin's method in the spectral domain, particularly for the eigenvalue problem. This is a preferable choice resulting from the self-adjointness of the involved integral operators and following the argumentation by Harrington [51]. The stationarity of such solutions has been discussed in an early contribution by the author in comparison to a least-squares alternative [31]. It has recently been shown by Lindell in a general context for the eigenvalue parameter p with respect to the trial field [48]–[50] which is $J_{t\text{ex}}$ for strip problems and $E_{t\text{ex}}$ for slot-type configurations. To simplify the discussion, restriction to strip-type problems is allowed without loss in generality. The numerical procedure is best understood if the equations prevailing in the spectral and the space domain are considered in parallel, i.e., writing briefly

$$\underline{E}_t = L_\infty(p)\cdot(\underline{J}_{t\text{ex}} + \underline{J}_{t\text{im}}) \quad \tilde{\underline{E}}_t = \tilde{\underline{\underline{Z}}}(p)\cdot(\tilde{\underline{J}}_{t\text{ex}} + \tilde{\underline{J}}_{t\text{im}}). \quad (13)$$

In the space domain, the physical vectors $\underline{E}_t$ and $\underline{J}_t = \underline{J}_{t\text{ex}} + \underline{J}_{t\text{im}}$ are different from zero in the complementary regions F_{st}^{-1} and F_{st}. The unknown surface current density $\underline{J}_{t\text{ex}}$ is expanded into a suitable, preferably complete set of expansion functions defined on F_{st} and vanishing outside. By this, continuity of the magnetic field outside the metallization is achieved at the same time. The expansion of $\underline{J}_{t\text{ex}}$ is actually performed in the original, spatial domain since this provides the best physical insight for a good choice. It depends on the specific problem under investigation whether the functions $\underline{J}_{tk}$ chosen should be easily transformable into the spectral domain or not. For the application of Galerkin's method, the set of testing functions necessary to enforce the vanishing of $\underline{E}_t$ on the conductor region F_{st} is the same as the expansion used, say $\underline{J}_{tj}$. The scalar product employed in the testing process is commonly defined by integration over F_{st} without a specific weighting factor and expressed here using parentheses (,). Making use of the linearity of the integral operator involved, the standard process of testing [51] finally yields

$$\sum_k \alpha_k(\underline{J}_{tj}, L_\infty(p)\cdot\underline{J}_{tk}) = -(\underline{J}_{tj}, L_\infty(p)\cdot\underline{J}_{t\text{im}})$$

or

$$\sum_k \alpha_k(\tilde{\underline{J}}_{tj}, \tilde{\underline{\underline{Z}}}(p)\cdot\tilde{\underline{J}}_{tk}) = -(\tilde{\underline{J}}_{tj}, \tilde{\underline{\underline{Z}}}(p)\cdot\tilde{\underline{J}}_{t\text{im}}). \quad (14)$$

The second alternative and mathematically identical equation applies as a consequence of Parseval's theorem [1], which also serves for a unique definition of the associated scalar product (,) in the spectral domain. In eigenvalue problems, the right-hand side of (14) vanishes and a nontrivial solution $\underline{J}_{t\text{ex}}$ exists only if the determinant of the respective linear system of equations is zero. This provides the nonlinear eigenvalue equation for the unknown parameter p and, subsequent to an iterative evaluation of p, the associated surface current distribution $\underline{J}_t = \underline{J}_{t\text{ex}}$. The electric field outside of the metallization is found from the application of (13). For MIC excitation problems, (14) is deterministic and can be solved in a single step for a prescribed value of the parameter p. The main difficulty in such cases is a realistic formulation of the excitation term $\underline{J}_{t\text{im}}$ such that it well describes the physical situation. Also, the introduction of such a source term may complicate the satisfaction of boundary conditions in its spatial vicinity as compared to an equivalent eigenvalue formulation [36]. As long as the source chosen has finite support in the $x-y$ plane, the field region is finite and the expansion functions are chosen properly, Galerkin's method can still be applied, for example, if the source is a slit voltage generator or a strip current sheet [38], [39]. However, if the field is excited by a transmission-line mode coming in from infinity and a reflected wave is involved, Galerkin's method cannot be applied any more since the associated scalar products become unbounded. In that case, which is a good description of practical MIC excitation problems, another version of moment methods has to be employed [40], [41] enforcing existence of the scalar products. Expansion and test functions have to be different then with the consequence that the final system of equations (14) is not symmetric or Hermitian any longer.

It should have become obvious from the discussion that interpreting the numerical procedure as "Galerkin's method in the transform domain" is too restrictive not only because of the last-mentioned details. Actually, it does not make a mathematical difference which one of the equations (14) is considered if the presumptions necessary for the application of Parseval's theorem are satisfied. As a rule of thumb, in laterally open problems, evaluation of the scalar products in (14) is alleviated if the spectral quantities are used. In these cases, expansion functions should be selected with explicitly available analytical transforms. On the other hand, for shielded configurations, it may have advantages to perform the scalar product operation in the spatial domain, particularly if a suitable orthogonal set of functions can be constructed for the description of the electromagnetic field [30], [31]. Also, it is generally easier to construct complete sets of expansion functions for conductors of complex shape in the space domain. So, the major advantage of the so-called spectral-domain approach is that it allows one to shift between the space and the transform domains in essentially all steps of the processing. The same applies to the computation of MIC design quantities from solutions obtained by the approach. Quantities like quality factors, dielectric and magnetic loss, conductor loss, and power transported in the cross section of transmission lines can with advantage be computed in either of the domains depending on the shielding situation and the specific problem. The evaluation of such design data involves volume or surface integration over the products

$$\underline{E}_i \cdot \underline{E}_i^* \quad \underline{H}_i \cdot \underline{H}_i^* \text{ and } \underline{E}_i \times \underline{H}_i^* \tag{15}$$

where the asterisk denotes complex conjugate. Integration over the vertical z-coordinate is always performed analytically due to the simple trigonometric dependencies associated with the layered MIC structure. Along the other coordinates, Parseval's theorem again allows a choice. Care has to be taken in conductor loss computations for metallizations of zero thickness. Because of the order of the edge singularity of the field for conductors of vanishing thickness [52], the square of the magnetic field tangential to the metallization is not integrable. However, this can be repaired to achieve a good approximation of conductor loss by proper modification of the asymptotic behavior of the transform of the magnetic field [24], [25]. The idea behind this is that, except for the immediate vicinity of the conductor edges, the field does not change noticeably if a small, finite thickness is introduced. The modification may also be performed in the spatial domain if a strip-type problem prevails for which the original current density distribution is available in closed form. Also, longitudinal strip current or transverse slot voltage may be evaluated in the space domain for the same reason. These quantities are often used in the calculation of characteristic impedances of strip and slot transmission lines [8], [9], [24], [34].

Similar advantageous properties as those described for the spectral-domain approach are shared to some degree by one of the methods recently applied to MIC's. This is named the differential-difference approach (DDA) here and is also called the method of lines [17], [53]–[55]. The fundamental similarity to the SDA formulation consists in the fact that it reduces the original Helmholtz equation (1) to a system of ordinary differential equations which can be solved explicitly. In contrast to the spectral-domain approach, the reduction in complexity and presumption for further analytical processing is achieved by discretization of the Helmholtz operator, writing, for example,

$$\left(\frac{\partial^2}{\partial z^2} + k_i^2\right) f_i^{m,n} + \frac{f_i^{m-1,n} - 2f_i^{m,n} + f_i^{m+1,n}}{h_x^2} + \frac{f_i^{m,n-1} - 2f_i^{m,n} + f_i^{m,n+1}}{h_y^2} = 0. \tag{16}$$

This implies a two-dimensional finite-difference representation of the field for each plane of constant coordinate z, i.e., a mesh of points m, n with $m = 1 \ldots M, n = 1 \ldots N$. It describes a band-structured system of coupled ordinary differential equations with a total number of $2MN$ unknowns for two scalar wave potentials. The system can be decoupled, i.e., brought into diagonal form, leading to the same number of discrete transformed potentials, say $\tilde{f}_i^{m,n}$ and $\tilde{g}_i^{m,n}$. For these, the z-dependency in the layered MIC structure can be described in analytical form including the boundary conditions at the ground and top planes. For the associated discretized tangential electric field and current density in the plane of metallization, the boundary conditions are formulated pointwise. This cannot be performed in terms of the transformed quantities $\tilde{f}_i^{m,n}$, $\tilde{g}_i^{m,n}$ and, therefore, requires a back-transformation into the original domain. As in the spectral-domain approach, the last step in the DDA procedure is the solution of a determinantal equation depending on one of the physical parameters p of the problem or the solution of a deterministic linear system of equations for prescribed values of p.

The method has been applied only to shielded structures so far, which is a consequence of the spatial discretization that makes it difficult to extend it to open regions. Several interesting similarities between the SDA and the DDA become plausible if one recalls that the application of finite integral transforms means a discretization in the spectral domain. Extension to open problems with the SDA is straightforward since fields of finite and infinite spatial support both have contributions over the infinite spectral domain. One of the advantages of the method of lines is that it exhibits a comparatively low numerical expense for the generation of each of the elements of the final matrix equation. In addition, it can in a very flexible way be used for the analysis of different conductor shapes and does not require a choice of expansion functions. On the other hand, application of the DDA to three-dimensional field problems results in very high matrix orders. If, for example, in a strip-type problem, a rectangular shielding with ground plane F and conductor area F_{st} is assumed, the order of the final DDA matrix is approximately

$$Q = 2M \cdot N \cdot \frac{F_{st}}{F}. \tag{17}$$

With growing complexity of the conductor shape, the spatial resolution $M \cdot N$ has to be increased and the number of floating-point operations in the differential-difference approach is proportional to Q^3. About the same resolution is achieved by an SDA treatment using $2MN$ Fourier coefficients, which, however, determines only the linear number of summations necessary to construct a matrix element. The order of the final SDA matrix is not directly related to $M \cdot N$, but mainly a question of the intelligent choice or systematic generation of expansion functions. It can be made extremely low, which makes the SDA a preferable technique for the repeated generation of MIC design information. From this point of view, it is an advantage that it allows the choice of expansion functions. Furthermore, the spectral-domain approach is specifically suited to analytical preprocessing and speedup measures as will be shown in the last section of this paper.

IV. Specific Aspects

In the analysis of MIC configurations by the spectral-domain approach three classes of structures have to be distinguished: shielded, covered, and open types. These are shown in Fig. 2 for the cross section of a single microstrip line. The shielded-type has been used extensively by contributors to the SDA in transmission-line and resonator problems. It presents a good description of real-life MIC structures only if radiation and surface-wave excitation from an adequate open structure are negligible. This applies, for example, for the technically used fundamental modes of printed strip and slot transmission lines under normal operating conditions [56]–[58] and to high-Q resonators with properly chosen, not too large, shielding dimensions [14]–[16], [30], [31]. However, practical MIC shielding cases are usually large in dimensions compared to the enclosed circuit elements, with the exception of finlines and related millimeter-wave components [59], [60]. Therefore, care has to be taken in the interpretation of data derived from a shielded-type analysis if these shall be used for MIC design purposes. With respect to this point of view, the use of the covered, i.e., laterally open-type of analysis seems to present a better choice for the characterization of MIC structures in general. A cover shielding can always be specified in the design of a practical circuit and, therefore, taken into account properly. The assumption of lateral openness is believed to provide the most realistic one if design quantities have to be computed for general applicability in the CAD of MIC's or as a basis for the generation of mathematical models. Nevertheless, nearly all of the SDA contributions to the analysis of covered-type configurations neglect energy leakage into the lateral direction. They are equivalent, therefore, to shielded-type SDA formulations with side walls removed left and right to infinity. Only recently, lateral energy leakage has been included in covered MIC analyses [40], [41], [61] and is considered a prerequisite to the description of dynamic coupling mechanisms. The open-type analysis of MIC structures, such as the one shown in Fig. 2(c), is applicable to nonradiating transmission-line modes, but mainly reserved to problems where radiation into free space is of dominant interest.

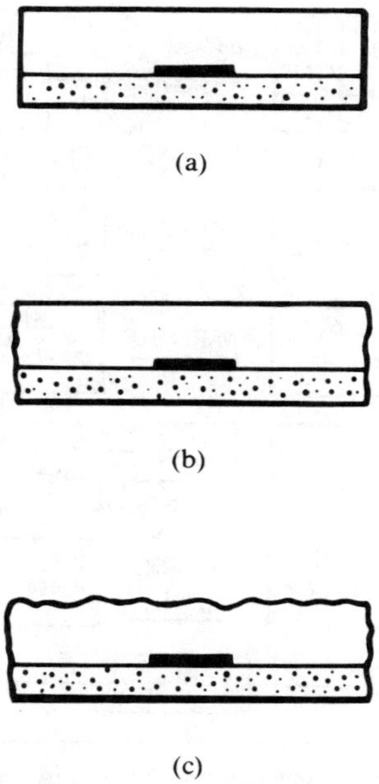

Fig. 2. Microstrip cross sections representing three different classes of MIC structures, (a) shielded, (b) covered, and (c) open type.

Using the shielded-type formulation, a systematic spectral-domain technique for the hybrid-mode characterization of MIC discontinuities and junctions has been developed by the author a few years ago [36], [62]. It is represented pictorially in the flow diagram of Fig. 3 to show how the SDA can be applied to derive design information in a very general way. The technique avoids the necessity of specifying sources and has, meanwhile, been applied successfully to a variety of strip- and slot-type problems [63]–[67]. It mainly refers to operating conditions where energy leakage into the volume field is not noticeable, but can, however, be extended to be valid without that restriction. The main idea behind the technique shown in Fig. 3 is a generalization of the Weissfloch or tangent method [28] in conjunction with a three-dimensional SDA resonator analysis. This generalization can be performed and becomes practicable here because the total electromagnetic field and current density is available from the analysis which would not be accessible or practicable in an equivalent measurement situation. On top of the left column of Fig. 3, the physical n-port investigated is shown in a shielding box with the field volume subdivided into short-circuited transmission-line sections (stubs) of length l_i and the n-port junction. The circuit representation using scattering parameters is depicted on the right-hand side with the respective reference planes RP_i, $i = 1 \ldots n$. For

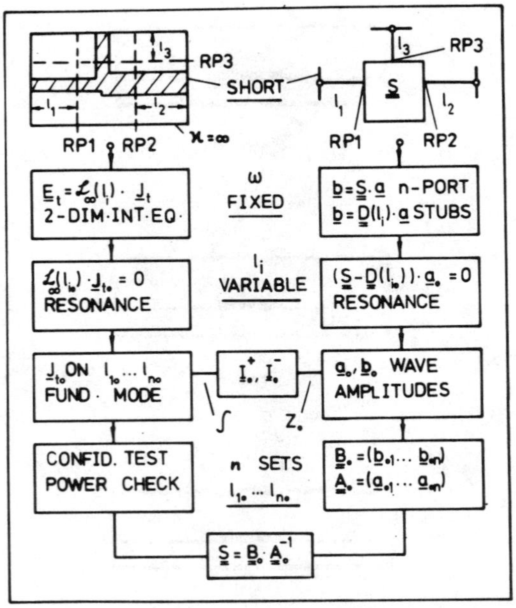

Fig. 3. A specific implementation of the SDA used for the frequency-dependent characterization of discontinuities in strip- and slot-type MIC's.

fixed operating frequency ω, the configuration is analyzed in terms of successively interchanged resonant lengths l_{i0} exactly n times. By introduction of precomputed strip current density distributions into the expansion functions used to describe $\underline{J}_t$, all the boundary conditions except those in the n-port region itself can be satisfied *a priori*. The resonant lengths and the stub current density amplitudes result from the numerical description of the n successive hypothetical resonator experiments. They are processed to obtain the complex amplitudes of the longitudinal strip currents or electric fields of the stubs. Then, using a power-related definition of characteristic stub impedance, the complex wave amplitudes associated with each of the n experiments are computed and assembled into matrices $\underline{A}_0$ and $\underline{B}_0$. The scattering matrix of the investigated n-port results from this easily. As a confidence test for the validity of the results it is checked in parallel that the power balance for the lossless n-port is satisfied to a good approximation. The technique has the advantage of providing phase information which is stationary with respect to the current density and field distribution, respectively [36]. It has its limitations if radiation mechanisms in MIC's are involved to a noticeable degree.

To understand leakage mechanisms in MIC's, the mathematical structure of the spectral immittance matrices of (10) has to be considered. Independent of the degree of openness and the number of dielectric layers prevailing in a specific problem, the spectral impedances can always be written in the form [31], [34], [36], [38], [41]

$$\underline{\underline{\tilde{Z}}}(p)=\begin{bmatrix} k_x^2 Z_{FE}+k_y^2 Z_{FH} & k_x k_y (Z_{FE}-Z_{FH}) \\ k_x k_y (Z_{FE}-Z_{FH}) & k_y^2 Z_{FE}+k_x^2 Z_{FH} \end{bmatrix} \quad (18)$$

with

$$Z_{FE}=Z_{FE}(k_\rho^2, p) \quad Z_{FH}=Z_{FH}(k_\rho^2, p).$$

The admittance matrix for slot-type problems follows from inversion of (18) and of the impedance elements Z_{FE}, Z_{FH}. It has exactly the same structure. Due to this duality, it is again sufficient to discuss the strip-type case for simplicity. The quantities Z_{FE} and Z_{FH} are the total LSM and LSE modal input wave impedances as seen into the medium below and above the plane of metallization. Thinking in terms of a transverse resonance approach [28], [29], therefore, makes clear that $1/Z_{FE}$ and $1/Z_{FH}$ have the properties of radial wave eigenfunctions in the layered circuit medium ($1/Y_{FE}$ and $1/Y_{FH}$ for slot-type problems). So, the elements Z_{FE} and Z_{FH} have poles for those values $k_\rho = k_{\rho p}$ of the radial wavenumber which are propagation constants of the LSM and LSE modes in the inhomogeneous parallel-plate medium between the ground and top planes if the conductor metallization is not present and they represent surface waves for open-type structures [36], [41]. The maximum discrete value of $k_{\rho p}^2$ corresponds to the dominant LSM_0 radial wave in the circuit medium which is the main cause for dynamic coupling in MIC's since this has zero cutoff frequency. With

$$k_{xp}=k'_{xp}+jk''_{xp}=\pm\left(k_{\rho p}^2-k_y^2\right)^{1/2} \quad (19)$$

the associated poles in the complex k_x-plane are all off the real k'_x-axis as long as k_y^2 is larger than the value of $k_{\rho p}^2$ of the LSM_0 mode. Physically, this means, for example, that MIC transmission-line modes with propagation constants k_y larger than that of the LSM_0 mode are nonradiating. This has already been discussed by Pregla in an early contribution [56]; however, his analysis has not been extended into the radiation region.

Higher order modes on covered and open MIC transmission lines do not generally exist in nonradiating form. Also, the respective MIC problems involving three-dimensional fields are always affected by energy leakage [40], [41] even if this may not be of practical concern at low frequencies. The SDA formulation can be extended in application to these cases by writing the scalar products of the final equations (14) in the form

$$\left(\tilde{\underline{J}}_{tj}, \underline{\underline{\tilde{Z}}}(p)\tilde{\underline{J}}_{tk}\right)_x=\int_{C_x}\underline{\underline{\tilde{Z}}}(k_x,p)\tilde{\underline{J}}_{tk}(k_x)\cdot\tilde{\underline{J}}_{tj}(-k_x)\,dk_x \quad (20)$$

which is a consequence of Parseval's theorem [56] and by proper choice of the integration path C_x. The encountered immittance elements are meromorphic functions with respect to k_x in the covered case. The evaluation of (20) is achieved by residue calculus techniques [68]. In three-dimensional problems, in addition, integration in the proper related sheet of the complex k_y plane is involved. The simple principle of extension into the radiation region is a further fundamental advantage of the spectral-domain approach and allows rigorous treatment of complex MIC problems.

For the general rules of evaluating SDA integrals of the type (20) and a discussion of the physical background, a

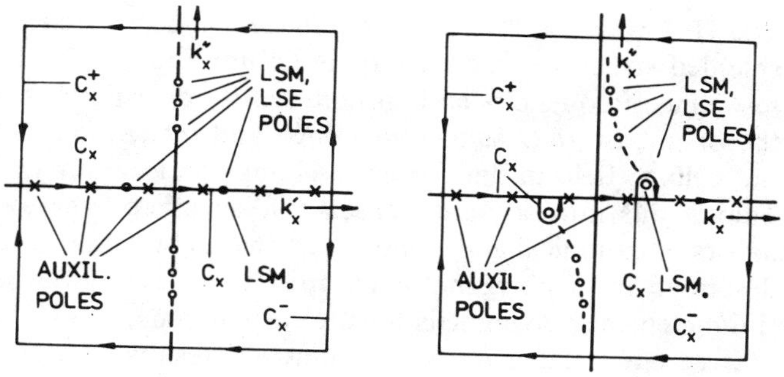

(a) (b)

Fig. 4. Integration paths C_x in the SDA treatment of covered strip-type MIC problems, energy leakage (a) neglected and (b) taken into account, respectively.

covered transmission-line configuration such as that of Fig. 2(b) is considered. By introduction of the factor

$$1=\frac{\exp(jck_xw)}{2\cos(ck_xw)}+\frac{\exp(-jck_xw)}{2\cos(ck_xw)}, \qquad c>1 \quad (21)$$

the integrals in (20) can be split up into a sum of two contributions for each of which the integration path C_x can be closed in the complex k_x-plane at infinity [39], [41]. This is shown in Fig. 4(a) for an analysis which does not include energy leakage, and in Fig. 4(b) where this mechanism is properly accounted for. In both cases, the zeros of the auxiliary cosine function in (21) introduce additional, nonphysical poles onto the real k_x'-axis. According to the relation (19) and neglecting material loss, the LSM and LSE wave poles are located on the axes or not, depending on whether the leakage mechanism is incorporated into the solution (square of propagation constant k_y^2 complex) or not (k_y^2 a real number). The quantity w is a suitable normalization width. The positive real constant c is to some degree arbitrary and can be utilized for numerical check purposes and in convergence considerations. Integrating along the real k_x'-axis across the single poles artificially introduced by (21), Cauchy principal values are taken [68].

If, in Fig. 4(a), a nonradiating mode would be considered, i.e., with a propagation constant larger than that of the LSM radial wave, the LSM_0 pole would be located on the imaginary k_x''-axis. In that case, the dominant contributions to the integrals in (20) would come from the discrete, regular set of auxiliary poles, say k_{xm}, on the real axis. With the constant c in the factor (21) chosen sufficiently large, the set of k_{xm} becomes very dense and the problem could be described in terms of this set alone. In the limit of $c\rightarrow\infty$, this is nothing else than numerical integration along the real axis of the k_x-plane. However, describing an MIC problem under radiation conditions, as actually prevailing with the pole locations of Fig. 4(a), this becomes more complicated. Numerical integration along the k_x'-axis and across the LSM_0-pole (Cauchy principle value) now means introducing a discrete standing plane contribution k_{xp} into the electric-field distribution [61]. This is equivalent to the presence of a lateral shielding far away from the MIC configuration, reflecting the radiated LSM_0-field. The same type of result is achieved if one applies a transverse resonance approach to covered structures under operating conditions in the radiation region [57], [58].

The leaky character of higher order MIC transmission-line modes, discontinuities, and junctions of the covered-type are correctly described by the integration path shown in Fig. 4(b). This can be concluded from an investigation of the migration paths of the LSM and LSE wave poles in a slightly lossy dielectric medium [41], [61]. Those physical poles which in the lossless case of Fig. 4(a) are located on the real axis (here only the LSM_0-pole) are just below the k_x'-axis for a small dielectric tangent different from zero. They migrate across the real axis of the k_x-plane if an additional radiation mechanism is involved. Therefore, the original integration path C_x (the k_x'-axis) for nonradiating situations has to be distorted in the way indicated in Fig. 4(b). Otherwise, solutions would not pass over continuously into the radiation region of operation. As Pregla has already pointed out in his early study [56], the transition from one state of a solution to another, i.e., when the LSM_0-pole appears at the origin, does not present problems since the associated residues vanish then. The same arguments and integration path discussed here are valid also for the evaluation of integrals (scalar products) with respect to the k_y-variable in the SDA solution of three-dimensional field problems. However, depending on which integration is performed first, additional branch cuts have to be regarded either in the complex k_y- or k_x-plane. A broad and thorough treatment of the spectral-domain approach for a variety of representative leaky MIC problems has been elaborated in a recent dissertation by Boukamp [41].

V. Progress: 1975–1984

To round out the view given so far for the spectral-domain approach, a review is given of the improvements of the technique and the more important results achieved by its application during the last ten years. Emphasis is placed on frequency-dependent solutions since these become more and more important with the development of practical MIC's in the millimeter-wave region. Many of the contributions mentioned do not use the SDA in its pure form but deviate from it in the one step or another of the analysis. The reader may get an impression, therefore, that a high degree of flexibility is inherent in the details of the SDA formulation. The discussion is subdivided according to three different groups of MIC structures, namely transmission lines, resonators and antennas, and, finally, discontinuities.

Considering printed microwave and millimeter-wave transmission lines first, there has been a clear tendency since 1975 to treat this class of problems in a generalized way, allowing additional dielectric layers and more complicated strip and slot configurations [24], [26], [32], [34], [35], [59], [60], [69]–[78]. The inclusion of characteristic impedance data becomes standard in computer analysis programs, and also dielectric and conductor losses are considered frequently [24], [26], [34], [59], [60], [69], [72]–[78], [79]–[80]. This makes visible the beginning orientation of the SDA towards direct application in the design of MIC's. Together with this trend, computer time and storage requirements for the analysis programs become a more important point of view [80]. However, from an appraisal of methods applied to the microstrip dispersion problem published by Kuester and Chang in 1979 [81], it can be concluded that the majority of respective computer packages at that time still involved some problems. There was significant progress, therefore, when not only the number of applications of the SDA increased further, but in addition some conceptual simplifications, modifications, and basic numerical considerations were reported. As elegant and simple concepts, for example, the transfer matrix approach [32] and the very similar spectral-domain immittance concept [35] have been presented. Also, El-Sherbiny [82], [83] provided interesting aspects to the mathematical and physical background of the SDA and applied a modified Wiener–Hopf technique in the final step of solution. Some of the rules to be regarded in order to obtain stable, accurate solutions and a unified treatment of shielded, covered, and open strip and slot structures have been reported by the author [34]. The introduction of finite conductor thickness into the SDA formalism is mainly the result of Kitazawa's work [21]–[23], [84]. Its effect on coplanar waveguide properties, for example, is an increase of guided wavelength and a decrease of characteristic impedance, respectively.

With growing experience in the use of the SDA, application of the technique shifted to more involved transmission-line problems. Coupled strip–slot structures have been studied by various authors with regard to coupler design and an extension of the range of characteristic impedances achievable in microstrip [22], [35], [47], [71], [77]. Also, an increasing portion of SDA work on transmission-line structures with anisotropic media has been reported. Borburgh [85], [86] seems to have been the first to apply the technique to microstrip on a magnetized ferrite substrate and related analyses followed [87], [88]. A variety of authors have treated printed transmission lines in single- and double-layered anisotropic dielectric media [89]–[92]. Only recently, slow-wave MIS coplanar waveguide has been studied with respect to MMIC application [93]. Beyond this, the computation of the stopband properties of several periodic structures by the spectral-domain approach has been reported [88], [94]–[96]. The last-mentioned reference also contains some numerical results on Podell-type microstrip couplers. As a further example for inhomogeneous structures, an analysis of tapered MIC transmission lines combining the SDA for uniform lines with coupled-mode theory has been presented [97]. Finally, a very efficient hybrid-mode spectral-domain approach for conductor arrays has been used by Jansen and Wiemer [98] in the design of MIC interdigital couplers and lumped elements on small computers.

Results achieved for resonators and antennas are considered together here since a large class of planar antennas makes use of resonating open patch elements. The information given on patch antennas, however, is by far incomplete, as the emphasis is placed on MIC's in this paper. The first full-wave analyses of resonators concentrated on shielded structures and gave quite accurate results for the resonance frequencies and current density distributions of the open case if Q-factors were high and interaction with the volume field in the chosen shielding box low [30], [31]. Taking this into account, Jansen studied microstrip resonators of canonical and complicated shapes, with the latter described by a polygonal contour in terms of high-order finite-element polynomials for the current density [99].

Resonator shapes for which numerical results have been generated are rectangle, circular disk and ring, concentric coupled disk-ring and double-ring structure, stretched hexagon, and regular octagon. This work has recently been supplemented by Knorr [100] who analyzed a shielded short-circuited slotline resonator and by Sharma and Bhal [101], [102] who provided shielded-type results for the triangular shape and interacting rectangular microstrip structures.

Already by 1978, Pregla [43] had investigated open resonating microstrip rings including radiation using a Hankel transform and formulating the problem in terms of complex eigenfrequency. With increasing interest in microstrip antennas, further related analyses of circular shapes were performed in the years following [44]–[46], [103]–[105]. Itoh and Menzel presented a full-wave SDA treatment of open rectangular microstrip patches in 1981 [106] with clear emphasis on antenna applications. There is also direct antenna design work, for example, contributions by Bailey and Deshpande [107], [108] and by Newman *et al.* [109], which performs only part of the computational steps in the spectral domain. Numerical integration along the real axes in the spectral domain is the dominant choice in these papers; however, singularities near the integration path may cause problems (see, for example, Newman's remarks [109]). The first results for covered MIC geometries including the excitation of LSM and LSE waves in the layered circuit medium have been provided by Boukamp and Jansen [40], [41], [61]. The main intention of this research work was to study the mathematical and physical background and prepare the way for an extension of the SDA in application to dynamic MIC coupling problems. One of the practical results achieved in this context is, for example, that lateral leakage in MIC's can be minimized if the circuit cover height is chosen slightly lower than a value which would correspond to the onset of the first LSE mode.

The same covered-type spectral-domain approach has also been used to study the simplest case of a leaky microstrip discontinuity problem, i.e., the open end, with an excitation formulation [40], [41]. The motivation was again to provide a basis for the analysis of more complicated geometries. Interestingly, the numerical results in comparison with a resonator formulation indicate that there is a noticeable coupling effect between the open ends of half-wavelength resonators, such as those used, for example, in coupled line filters. The explanation is that, on the alumina substrate investigated, the distance between the respective open microstrip ends is small compared to the wavelength of the involved LSM_0-mode. Beyond this very elementary but rigorous example, MIC discontinuities have been computed using the SDA only for the static limit [13] and by the frequency-dependent shielded-type SDA implementation outlined in the foregoing section [36], [62]–[67], [110]. Due to the three-dimensional electromagnetic fields and relatively complicated geometries involved, this sparsity of results prevails for other methods to an even larger extent. Systematic and quite extensive design data have been published for open-circuited microstrip and suspended substrate lines, as well as short-circuited slots [36] and for the symmetrical and asymmetrical gap in microstrip and suspended substrate lines [63], [65]. Also, the inductive strip discontinuity in unilateral finlines, which is the related slot-type structure, has been treated [64]. Very recent work by Koster and Jansen provided a variety of microstrip impedance step data for use in MIC design [67], [110].

VI. Efficiency Considerations

The spectral-domain approach is a hybrid technique in the sense that it requires (and allows!) a certain amount of analytical preprocessing in order to achieve high computational efficiency for a specific problem or class of problems. One of its main disadvantages is the relatively high numerical expense which has to be spent to evaluate the coefficients of the final system of equations (14). These are improper integrals or infinite series with only moderate rate of convergence. The order of the final system, on the other hand, can be held extremely small compared to other techniques. This is achieved, for example, by regarding several criteria in the choice of the expansion functions [80]. Briefly summarizing, the set of expansion functions as a whole should be twice continuously differentiable in the interior of the region on which it is defined, so that it is in the domain of the original Helmholtz operator (1). Mathematical arguments and numerical experience indicate that this avoids the existence of spurious, nonphysical solutions [80]. Further, expansion functions in MIC problems should satisfy the edge condition, i.e., have the correct order of singularity at the boundary of the conductor metallization. This is a prerequisite to obtaining accurate solutions with a low number of terms or, equivalently, with a low order of the final system of equations. The set of functions used should be complete in order to allow convergence checks and investigation. It should be chosen with all the physical insight that is available for the specific problems, from static considerations, from idealizations such as the planar magnetic-wall waveguide model [19] and so on. The main rule is not to leave work to the computer for the evaluation of what is known in advance of the physical solution or can be obtained easier. This also implies the precomputation of expansion functions by a transmission-line SDA portion (two-dimensional fields) in computer programs for the SDA solution of three-dimensional field problems [36]. Finally, the use of static together with stationary precomputed information can provide a means to generate vector expansion functions based on the continuity equation. The analytical and programming expense required on the side of the investigator may be considerable, which mirrors the hybrid character of the SDA. However, in the way outlined, very efficient CAD tools can be developed by its application.

To come to a quantification of the numerical expense associated with the SDA, the number of point operations which have to be performed in the solution of typical MIC problems is estimated. Also, the possibilities of reducing this figure shall be discussed. Let us assume a not too elementary MIC transmission-line case, in parallel, a resonator problem formulated in Cartesian coordinates. The final system of SDA equations (14) is dense and has to be generated repeatedly in the iterative localization of the zeros of its determinant as a function of the eigenvalue parameter p. Even with an intelligently chosen start value of p, this has to be done about 10 times. Under the assumption of a reasonable choice of expansion functions, the number of point operations necessary to obtain the numerical value of the final SDA determinant is usually small compared to the expense investigated for its generation. This is a consequence of the fact that the number of summations required to compute a single coefficient (integral or series) of (14) is typically much larger than the order of the system matrix. The latter may be

$$Q = 2 \cdots 10 \text{ and } Q = 4 \cdots 100 \tag{22}$$

for the transmission-line and resonator problem, respectively.

For example, $Q = 4$ could apply to a simple, rectangular half-wavelength microstrip resonator [14], [15], [36]. The number of summations or discretization points to evaluate each single scalar product may be $100 \cdots 500$ for the two-dimensional and $100^2 \cdots 500^2$ for the three-dimensional case. In particular situations, this may be even higher [34] depending on the spectral distribution of the involved fields. The complexity of the immittance functions encountered depends only on the number of dielectric layers considered and may be characterized by a figure of at least $10 \cdots 100$ point operations. On the whole, this amounts to a total count of point operations of about

$$TC = 2 \cdot 10^4 \cdots 25 \cdot 10^6 \text{ and } TC = 8 \cdot 10^6 \cdots 125 \cdot 10^{10} \tag{23}$$

for the two cases considered (symmetric SDA matrices). This looks quite high, particularly for the very right-hand side. However, one has to keep in mind that the spatial resolution assumed there is equivalent to a mesh of $25 \cdot 10^4$ points in the plane of the MIC metallization. As a rough estimate, the matrix order in a respective DDA treatment would be $Q = 5000$, the matrix itself dense, and had to be processed repeatedly about 10 times.

A reduction of numerical expense in SDA solutions is achieved first by an optimization of the expansion functions. This is performed according to the outlined criteria and with some experience from a preliminary, crude version. It can be done with a relatively small amount of reprogramming and produces a typical speedup factor of

$5 \cdots 10$ for nonelementary two- and three-dimensional problems. Also, about a factor of 10 may be gained by choosing an excitation formulation instead of solving an eigenvalue problem which applies, however, only to the three-dimensional case. The estimated speedup results since the source formulation avoids repeated generation of the final system (14). An additional reduction in computer time can be obtained by splitting off asymptotic spectral contributions from the coefficients of (14) and integrating or summing up these by analytical techniques (a factor of 10). In eigenvalue problems, it is advisable to substitute the spectral immittances by accurate one-dimensional interpolants [36] and optimize CP-time at the cost of storage requirements [31] for that part of the computation which does not depend on the eigenvalue p (a factor of 10). SDA computer programs developed for regular industrial use in MIC design justify even more expense in analytical preprocessing. For these, the normal mode of application is the repeated solution of the same problem for several different operating frequencies. Therefore, a high speedup factor compared to the first solution can be achieved if this is employed to provide for the subsequent ones a very compact low-order set of expansion functions (tested by the author for the transmission-line problem described in [98]). Thus, average CP-time is further reduced by about a factor of 5. By a combination of such analytical measures, the total count of point operations may be brought down to

$$TC = 2 \cdot 10^3 \cdots 5 \cdot 10^5 \text{ and } TC = 8 \cdot 10^5 \cdots 25 \cdot 10^7 \quad (24)$$

which is hardly achievable by other techniques. However, great care has to be taken in properly designing the employed integration algorithms, i.e., choosing a correct spectral representation. This particularly effects cases where tight coupling is involved. For loose and multiple coupled situations, a sufficiently stable matrix inversion algorithm has to be chosen.

VII. Conclusion

The spectral-domain approach allows an elegant and closed-form integral equation formulation for a broad class of MIC problems which is reduced by one dimension compared to the original field problem. It results in a particularly low-order linear system of equations and provides design-relevant parameters in both the spectral and the space domain. In so far as it may require a considerable amount of analytical preprocessing to achieve highest efficiency, it is a hybrid method. A preference of the SDA for MIC problems is to some extent confirmed by the fact that the majority of rigorous frequency-dependent MIC design information has been generated using this technique. The survey presented here further confirms this preference; however, the advice should be deduced from the discussion not to apply the SDA in a crude and schematic way.

References

[1] I. N. Sneddon, *The Use of Integral Transforms*. New York: McGraw-Hill, 1972.

[2] Tai Tsun Wu, "Theory of the microstrip," *J. Appl. Phys.*, vol. 28, pp. 299–302, 1975.

[3] E. Yamashita and R. Mittra, "Variational method for the analysis of microstrip lines," *IEEE Trans. Microwave Theory Tech.*, vol. MTT-16, pp. 251–256, 1968.

[4] E. J. Denlinger, "A frequency dependent solution for microstrip transmission lins," *IEEE Trans. Microwave Theory Tech.*, vol. MTT-19, pp. 30–39, 1971.

[5] H. J. Schmitt and K. H. Sarges, "Wave propagation in microstrip," *Nachrichtentech. Z.*, vol. 5, pp. 260–264, 1971.

[6] T. Itoh and R. Mittra, "Dispersion characteristics of slot lines," *Electron. Lett.*, vol. 7, pp. 364–365, 1971.

[7] T. Itoh and R. Mittra, "Spectral-domain approach for calculating the dispersion characteristics of microstrip lines," *IEEE Trans. Microwave Theory Tech.*, vol. MTT-21, pp. 496–499, 1973.

[8] G. Kowalski and R. Pregla, "Dispersion characteristics of single and coupled microstrips," *Arch. Elek. Übertragung*, vol. 26, pp. 276–280, 1972.

[9] M. K. Krage and G. I. Haddad, "Frequency-dependent characteristics of microstrip transmission lines," *IEEE Trans. Microwave Theory Tech.*, vol. MTT-20, pp. 678–688, 1972.

[10] A. R. Van de Capelle and P. J. Luypaert, "Fundamental- and higher-order modes in open microstrip lines," *Electron. Lett.*, vol. 9, pp. 345–346, 1973.

[11] T. Itoh and R. Mittra, "A technique for computing dispersion characteristics of shielded microstrip lines," *IEEE Trans. Microwave Theory Tech.*, vol. MTT-22, pp. 896–898, 1974.

[12] R. H. Jansen, "A modified least-squares boundary residual (LSBR) method and its application to the problem of shielded microstrip dispersion," *Arch. Elek. Übertragung*, vol. 28, pp. 275–277, 1974.

[13] Y. Rahmat-Samii *et al.*, "A spectral-domain analysis for solving microstrip discontinuity problems," *IEEE Trans. Microwave Theory Tech.*, vol. MTT-22, pp. 372–378, 1974.

[14] T. Itoh, "Analysis of microstrip resonators," *IEEE Trans. Microwave Theory Tech.*, vol. MTT-22, pp. 946–951, 1974.

[15] R. H. Jansen, "Shielded rectangular microstrip disc resonators," *Electron. Lett.*, vol. 10, pp. 299–300, 1974.

[16] R. H. Jansen, "Computer analysis of edge-coupled planar structures," *Electron. Lett.*, vol. 10, pp. 520–522, 1974.

[17] U. Schulz, "The method of lines—A new technique for the analysis of planar microwave structures" (in German), Ph.D. dissertation, Univ. of Hagen, West Germany, 1980.

[18] R. Mittra, Ed., *Computer Techniques for Electromagnetics* (Int. Series of Monographs in El. Eng.). New York: Pergamon Press, 1973.

[19] K. C. Gupta *et al.*, *Microstrip Lines and Slotlines*. Dedham, MA: Artech House, 1979.

[20] K. C. Gupta and A. Singh, *Microwave Integrated Circuits*. New Delhi: Wiley Eastern Private Ltd., 1974.

[21] T. Kitazawa *et al.*, "A coplanar waveguide with thick metal-coating," *IEEE Trans. Microwave Theory Tech.*, vol. MTT-24, pp. 604–608, 1976.

[22] R. H. Jansen, "Microstrip lines with partially removed ground metallization-theory and applications," *Arch. Elek. Übertragung*, vol. 32, pp. 485–492, 1978.

[23] T. Kitazawa *et al.*, "Analysis of the dispersion characteristics of slot line with thick metal coating," *IEEE Trans. Microwave Theory Tech.*, vol. MTT-28, pp. 387–392, 1980.

[24] R. H. Jansen, "Fast accurate hybrid mode computation of non symmetrical coupled microstrip characteristics," in *Proc. 7th Eur. Microwave Conf.*, 1977, pp. 135–139.

[25] W. Schumacher, "Current distribution in the ground plane of covered microstrip and the effect on conductor losses" (in German), *Arch. Elek. Übertragung*, vol. 33, pp. 207–212, 1979.

[26] D.Mirshekar-Syakhal and J. B. Davies, "Accurate solution of microstrip and coplanar structures for dispersion and for dielectric and conductor loss," *IEEE Trans. Microwave Theory Tech.*, vol. MTT-23, pp. 694–699, 1979.

[27] R. Pregla, "Determination of conductor losses in planar waveguide structures," *IEEE Trans. Microwave Theory Tech.*, vol. MTT-28, pp. 433–434, 1980.

[28] R. E. Collin, *Field Theory of Guided Waves*. New York: McGraw-Hill, 1960.

[29] R. R. Harrington, *Time-Harmonic Electromagnetic Fields*. New York: McGraw-Hill, 1961.

[30] R. H. Jansen, "Computer analysis of shielded microstrip structures" (in German), *Arch. Elek. Übertragung*, vol. 29, pp. 241–247, 1975.

[31] R. H. Jansen, "Numerical computation of the resonance frequencies and current density distributions of arbitrarily shaped microstrip structures" (in German), Ph.D. dissertation, *Univ.* Aachen (RWTH), West Germany, 1975.
[32] J. B. Davies and D. Mirshekar-Syakhal, "Spectral domain solution of arbitrary coplanar transmission line with multilayer substrate," *IEEE Trans. Microwave Theory Tech.*, vol. MTT-25, pp. 143–146, 1977.
[33] L. J. van der Pauw, "The radiation of electromagnetic power by microstrip configurations," *IEEE Trans. Microwave Theory Tech.*, vol. MTT-25, pp. 719–725, 1977.
[34] R. H. Jansen, "Unified user-oriented computation of shielded, covered and open planar microwave and millimeterwave transmission-line characteristics," *IEE J. Microwaves, Opt., Acoust.*, vol. MOA-3, pp. 14–22, 1979.
[35] T. Itoh, "Spectral domain immitance approach for dispersion characteristics of generalized printed transmission lines," *IEEE Trans. Microwave Theory Tech.*, vol. MTT-28, pp. 733–736, 1980.
[36] R. H. Jansen, "Hybrid mode analysis of end effects of planar microwave and millimeter-wave transmission lines, *Proc. Inst. Elec. Eng.*, pt. H, vol. 128, pp. 77–86, 1981.
[37] A. S. Omar and K. Schünemann, "Space domain decoupling of LSE and LSM fields in generalized planar guiding structures," in *IEEE MTT Symp. Dig.*, 1984, pp. 59–61.
[38] L. J. Van der Pauw, "The radiation and propagation of electromagnetic power by a microstrip transmission line," *Philips Res. Rep.*, vol. 31, pp. 35–70, 1976.
[39] R. H. Jansen, unpublished results and computer program—SFPMIC—, "Source formulation approach to planar microwave integrated circuits," Univ. of Duisburg, West Germany, 1979/80.
[40] J. Boukamp and R. H. Jansen, "The high-frequency behaviour of microstrip open ends in microwave integrated circuits including energy leakage," in *Proc. 14th Eur. Microwave Conf.*, 1984, pp. 142–147.
[41] J. Boukamp, "Spectral domain techniques for the analysis of radiation effects in microwave integrated circuits" (in German), Ph.D. dissertation, Univ. of Aachen (RWTH), West Germany, 1984.
[42] T. Itoh and R. Mittra, "Analysis of a microstrip disk resonator," *Arch. Elek. Übertragung*, vol. 27, pp. 456–458, 1973.
[43] S. G. Pintzos and R. Pregla, "A simple method for computing the resonant frequencies of microstrip ring resonators," *IEEE Trans. Microwave Theory Tech.*, vol. MTT-26, pp. 809–813, 1978.
[44] W. C. Chew and J. A. Kong, "Resonance of the axial-symmetric modes in microstrip disk resonators," *J. Math. Phys.*, vol. 21, pp. 582–591, 1980.
[45] W. C. Chew and J. A. Kong, "Resonance of nonaxial symmetric modes in circular microstrip disk antenna," *J. Math. Phys.*, vol. 21, pp. 2590–2598, 1980.
[46] K. Araki and T. Itoh, "Hankel transform domain analysis of open circular microstrip radiating structures," *IEEE Trans. Antennas Propagat.*, vol. AP-29, pp. 84–89, 1981.
[47] D. Mirshekar-Syakhal and J. B. Davies, "Accurate analysis of coupled strip-finline structure for phase constant, characteristic impedance, dielectric and conductor losses," *IEEE Trans. Microwave Theory Tech.*, vol. MTT-30, pp. 906–910, 1982.
[48] I. V. Lindell, "Variational methods for nonstandard eigenvalue problems in waveguide and resonator analysis," *IEEE Trans. Microwave Theory Tech.*, vol. MTT-30, pp. 1194–1204, 1982.
[49] G. J. Gabriel and I. V. Lindell, "Comments on Variational methods for nonstandard eigenvalue problems in waveguide and resonator analysis," *IEEE Trans. Microwave Theory Tech.*, vol. MTT-31, pp. 786–789, 1983.
[50] G. J. Gabriel and I. V. Lindell, "Further comments on Variational methods for nonstandard eigenvalue problems in waveguide and resonator analysis," *IEEE Trans. Microwave Theory Tech.*, vol. MTT-32, pp. 474–476, 1984.
[51] R. F. Harrington, *Field Computation by Moment Methods*. New York: Macmillan, 1968.
[52] R. Mittra and S. W. Lee, *Analytical Techniques in the Theory of Guided Waves*. New York: Macmillan, 1971.
[53] S. B. Worm, "Analysis of planar microwave structures of arbitrary shape" (in German), Ph.D. dissertation, Univ. of Hagen, West Germany, 1983.
[54] H. Diestel and S. B. Worm, "Analysis of hybrid field problems by the method of lines with nonequidistant discretization," *IEEE Trans. Microwave Theory Tech.*, vol. MTT-32, pp. 633–638, 1984.
[55] S. B. Worm and R. Pregla, "Hybrid-mode analysis of arbitrarily shaped planar microwave structures by the method of lines," *IEEE Trans. Microwave Theory Tech.*, vol. MTT-32, pp. 191–196, 1984.
[56] G. Kowalski and R. Pregla, "Dispersion characteristics of single and coupled microstrips with double-layer substrates," *Arch. Elek. Übertragung*, vol. 27, pp. 125–130, 1973.
[57] H. Ermert, "Guided modes and radiation characteristics of covered microstrip lines," *Arch. Elek. Übertragung*, vol. 30, pp. 65–70, 1976.
[58] H. Ermert, "Guiding characteristics and radiation characteristics of planar waveguides," in *Proc. 8th Eur. Microwave Conf.*, 1978, pp. 94–98.
[59] H. Hofmann, "Dispersion of planar waveguides for millimeter-wave application," *Arch. Elek. Übertragung*, vol. 31, pp. 40–44, 1977.
[60] L. P. Schmidt and T. Itoh, "Spectral domain analysis of dominant and higher order modes in fin lines," *IEEE Trans. Microwave Theory Tech.*, vol. MTT-28, pp. 981–985, 1980.
[61] J. Boukamp and R. H. Jansen, "Spectral domain investigation of surface wave excitation and radiation by microstrip lines and microstrip disk resonators," *Proc. 13th Eur. Microwave Conf.*, 1983, pp. 721–726.
[62] R. H. Jansen, "A new unified numerical approach to the frequency dependent characterization of strip, slot and coplanar MIC components for CAD purposes," Univ. of Duisburg, West Germany, Res. Rep. FB9/ATE, pp. 29–33, 1980/1981.
[63] R. H. Jansen and N. H. L. Koster, "A unified CAD basis for the frequency dependent characterization of strip, slot and coplanar MIC components," in *Proc. 11th Eur. Microwave Conf.*, 1981, pp. 682–687.
[64] N. H. L. Koster and R. H. Jansen, "Some new results on the equivalent circuit parameters of the inductive strip discontinuity in unilateral fin lines," *Arch. Elek. Übertragung*, vol. 35, pp. 497–499, 1981.
[65] N. H. L. Koster and R. H. Jansen, "The equivalent circuit of the asymmetrical series gap in microstrip and suspended substrate lines," *IEEE Trans. Microwave Theory Tech.*, vol. MTT-30, pp. 1273–1279, 1982.
[66] R. H. Jansen and N. H. L. Koster, "New aspects concerning the definition of microstrip characteristic impedance as a function of frequency," in *IEEE MTT-Symp. Dig.*, 1982, pp. 305–307.
[67] N. H. L. Koster and R. H. Jansen, "The microstrip discontinuity: A revised description," *IEEE Trans. Microwave Theory Tech.*, vol. MTT-33, 1985.
[68] G. B. Arfken, *Mathematical Methods for Physicists*. New York: Academic Press, 1970.
[69] J. B. Knorr and K. D. Kuchler, "Analysis of coupled slots and coplanar strips on dielectric substrate," *IEEE Trans. Microwave Theory Tech.*, vol. MTT-23, pp. 541–548, 1975.
[70] N. Samardzija and T. Itoh, "Double-layered slot line for millimeter wave integrated circuits," *IEEE Trans. Microwave Theory Tech.*, vol. MTT-24, pp. 827–831, 1976.
[71 T. Itoh, "Generalized spectral domain method for multiconductor printed lines and its application to tunable suspended microstrips," *IEEE Trans. Microwave Theory Tech.*, vol. MTT-26, pp. 983–987, 1978.
[72] J. B. Knorr and P. M. Shayda, "Millimeter-wave fin line characteristics," *IEEE Trans. Microwave Theory Tech.*, vol. MTT-28, pp. 737–743, 1980.
[73] L. P. Schmidt *et al.*, "Characteristics of unilateral fin-line structures with arbitrarily located slots," *IEEE Trans. Microwave Theory Tech.*, vol. MTT-29, pp. 352–355, 1981.
[74] D. Mirshekar-Syakhal and J. B. Davies, "An accurate unified solution to various fin-line structures, of phase constant, characteristic impedance, and attenuation," *IEEE Trans. Microwave Theory Tech.*, vol. MTT-30, pp. 1854–1861, 1982.
[75] D. Mirshekar-Syakhal, "An accurate determination of dielectric loss effect in MMIC's including microstrip and coupled microstrip lines," *IEEE Trans. Microwave Theory Tech.*, vol. MTT-31, pp. 950–954, 1983.
[76] A. K. Sharma and W. J. R. Hoefer, "Propagation in coupled unilateral and bilateral finlines," *IEEE Trans. Microwave Theory Tech.*, vol. MTT-31, pp. 489–502, 1983.
[77] W. Schumacher, "Computation of guided wavelengths of planar transmission lines in two- and three-layered media using a moment method" (in German), Ph.D. dissertation, Univ. of Karlsruhe, West Germany, 1979.

[78] W. Schuhmacher, "Hybrid modes and field distributions in unsymmetrical planar transmission lines" (in German), *Arch. Elek. Übertragung*, vol. 34, pp. 445–453, 1980.

[79] J. B. Knorr and A. Tufekcioglu, "Spectral domain calculation of microstrip characteristic impedance," *IEEE Trans. Microwave Theory Tech.*, vol. MTT-23, pp. 725–728, 1975.

[80] R. H. Jansen, "High speed computation of single and coupled microstrip parameters including dispersion, high-order modes, loss and finite strip thickness," *IEEE Trans. Microwave Theory Tech.*, vol. MTT-26, pp. 75–82, 1978.

[81] E. F. Kuester and D. C. Chang, "An appraisal of methods, for computation of the dispersion characteristics of open microstrip," *IEEE Trans. Microwave Theory Tech.*, vol. MTT-27, pp. 691–694, 1979.

[82] A. M. A. El-Sherbiny, "Exact analysis of shielded microstrip lines and bilateral fin lines," *IEEE Trans. Microwave Theory Tech.*, vol. MTT-29, pp. 669–675, 1981.

[83] A. M. A. El-Sherbiny, "Millimeter-wave performance of shielded slot-linss," *IEEE Trans. Microwave Theory Tech.*, vol. MTT-30, pp. 750–756, 1982.

[84] T. Kitazawa and R. Mittra, "Analysis of finline with finite metallization thickness," *IEEE Trans. Microwave Theory Tech.*, vol. MTT-32, pp. 1484–1487, 1984.

[85] J. Borburgh, "Theoretical investigation of the dispersion and field distribution of guided modes on a microstrip line with gyrotropic substrate" (in German), Ph.D. dissertation, Univ. of Erlangen, West Germany, 1976.

[86] J. Borburgh, "The behaviour of guided modes on the ferrite-filled microstrip line with the magnetization perpendicular to the ground plane," *Arch. Elek. Übertragung*, vol. 31, pp. 73–77, 1977.

[87] Y. Hayashi and R. Mittra, "An analytical investigation of finlines with magnetized ferrite substrate," *IEEE Trans. Microwave Theory Tech.*, vol. MTT-31, pp. 495–498, 1983.

[88] C. Surawatpunya *et al.*, "Bragg interaction of electromagnetic waves in a ferrite slab periodically loaded with metal strips," *IEEE Trans. Microwave Theory Tech.*, vol. MTT-32, pp. 689–695, 1984.

[89] A. M. A. El-Sherbiny, "Hybrid mode analysis of microstrip lines on anisotropic substrates," *IEEE Trans. Microwave Theory Tech.*, vol. MTT-29, pp. 1261–1266, 1981.

[90] H. Lee and V. K. Tripathi, "Spectral domain analysis of frequency dependent propagation characteristics of planar structures on uniaxial medium," *IEEE Trans. Microwave Theory Tech.*, vol. MTT-30, pp. 1188–1193, 1982.

[91] T. Kitazawa and Y. Hayashi, "Propagation characteristics of striplines with multilayered anisotropic media," *IEEE Trans. Microwave Theory Tech.*, vol. MTT-31, pp. 429–433, 1983.

[92] M. Horno and R. Marques, "Coupled microstrips on double anisotropic layers," *IEEE Trans. Microwave Theory Tech.*, vol. MTT-32, pp. 467–470, 1984.

[93] Y. Fukuoka *et al.*, "Analysis of slow-wave coplanar waveguide for monolithic integrated circuits," *IEEE Trans. Microwave Theory Tech.*, vol. MTT-31, pp. 567–573, 1983.

[94] K. Ogusu, "Propagation properties of a planar dielectric waveguide with periodic metallic strips," *IEEE Trans. Microwave Theory Tech.*, vol. MTT-29, pp. 16–21, 1981.

[95] T. Kitazawa and R. Mittra, "An investigation of striplines and finlines with periodic stubs," *IEEE Trans. Microwave Theory Tech.*, vol. MTT-32, pp. 684–688, 1984.

[96] F. J. Glandorf, "Numerical solution of the electromagnetic eigenvalue problem of periodically inhomogenous microstrip lines" (in German), Ph. dissertation, Univ. of Duisburg, West Germany, 1982.

[97] D. Mirshekar-Syakhal and J. B. Davies, "Accurate analysis of tapered planar transmission lines for microwave integrated circuits," *IEEE Trans. Microwave Theory Tech.*, vol. MTT-29, pp. 123–128, 1981.

[98] R. H. Jansen and L. Wiemer, "Multiconductor hybrid-mode approach for the design of MIC couplers and lumped elements including loss, dispersion and parasitics," in *Proc. 14th Eur. Microwave Conf.*, 1984, pp. 430–435.

[99] R. H. Jansen, "High-order finite element polynomials in the computer analysis of arbitrarily shaped microstrip resonators," *Arch. Elek. Übertragung*, vol. 30, pp. 71–79, 1976.

[100] J. B. Knorr, "Equivalent reactance of a shorting septum in a fin line: Theory and experiment," *IEEE Trans. Microwave Theory Tech.*, vol. MTT-29, pp. 1196–1202, 1981.

[101] A. K. Sharma and B. Bhat, "Analysis of triangular microstrip resonators," *IEEE Trans. Microwave Theory Tech.*, vol. MTT-30, pp. 2029–2031, 1982.

[102] A. K. Sharma and B. Bhat, "Spectral domain analysis of interacting microstrip resonant structures," *IEEE Trans. Microwave Theory Tech.*, vol. MTT – 31, pp. 681–685, 1983.

[103] K. Kawano and H. Tomimuro, "Spectral domain analysis of an open slot ring resonator," *IEEE Trans. Microwave Theory Tech.*, vol. MTT-30, pp. 1184–1187, 1982.

[104] K. Araki *et al.*, "A study on circular disk resonators on a ferrite substrate," *IEEE Trans. Microwave Theory Tech.*, vol. MTT-30, pp. 147–154, 1982.

[105] S. M. Ali *et al.*, "Vector Hankel transform analysis of annular-ring microstrip antenna," *IEEE Trans. Antennas Propagat.*, vol. AP-30, pp. 637–644, 1982.

[106] T. Itoh and W. Menzel, "A full-wave analysis method for open microstrip structures," *IEEE Trans. Antennas Propagat.*, vol. AP-29, pp. 63–68, 1981.

[107] M. C. Bailey and M. D. Deshpande, "Integral equation formulation of microstrip antennas," *IEEE Trans. Antennas Propagat.*, vol. AP-30, pp. 651–656, 1982.

[108] M. D. Deshpande and M. C. Bailey, "Input impedance of microstrip antennas," *IEEE Trans. Antennas Propagat.*, vol. AP-30, pp. 645–650, 1982.

[109] E. H. Newman *et al.*, "Mutual impedance computation between microstrip antennas," *IEEE Trans. Microwave Theory Tech.*, vol. MTT-31, pp. 941–945, 1983.

[110] N. H. L. Koster, "Frequency-dependent characterization of discontinuities in planar microwave transmission lines" (in German), Ph.D. dissertation, Univ. of Duisburg, West Germany, 1984.

Paper 5.2

An Electromagnetic Time-Harmonic Analysis of Shielded Microstrip Circuits

JAMES C. RAUTIO, MEMBER, IEEE, AND ROGER F. HARRINGTON, FELLOW, IEEE

Abstract —A Galerkin analysis of microstrip circuits of arbitrary planar geometry enclosed in a rectangular conducting box is described. The technique entails a time-harmonic electromagnetic analysis evaluating all fields and surface currents. This analysis is suitable for the accurate verification of microstrip designs prior to fabrication.

A computer program implementing the analysis has been written in Pascal on a personal computer. Agreement with measurements of several microstrip structures suggests a high degree of accuracy.

I. Introduction

THIS PAPER DESCRIBES an electromagnetic analysis of arbitrary microstrip (i.e., planar) circuits contained in a rectangular conducting box. The analysis proceeds by subdividing the microstrip circuit metallization into small rectangular subsections. An explicit surface current distribution is assumed to exist in each subsection. We evaluate the tangential electric fields due to the current in each subsection and then adjust the magnitude of the current in all subsections such that the weighted residual of the total tangential electric field goes to zero on all metallization. All surface currents are determined and the problem is solved. The *N*-port circuit parameters follow immediately.

The assumed surface current distribution in a subsection is called an expansion function. With the integral of the resulting electric field weighted by the same function, we have a special case of the method of moments known as a Galerkin technique [2], [3]. The magnitude of the current in each subsection is "adjusted" by matrix inversion.

The fields due to current in an individual subsection are represented by a sum of homogeneous rectangular waveguide modes. Thus, this technique is closely related to the spectral domain approach [7]. The technique described here [4]–[6] was originally developed as an extension of an analysis of planar waveguide probes described in section 8-11 of [1].

Manuscript received November 11, 1986; revised March 26, 1987. This work was supported in part by a fellowship from the General Electric Company, Electronics Laboratory, Syracuse, NY, and in part by the Office of Naval Research, Arlington, VA, under Contract N00014-85-K-0082.

The authors are with the Department of Electrical and Computer Engineering, Syracuse University, Syracuse, NY 13210

IEEE Log Number 8715142.

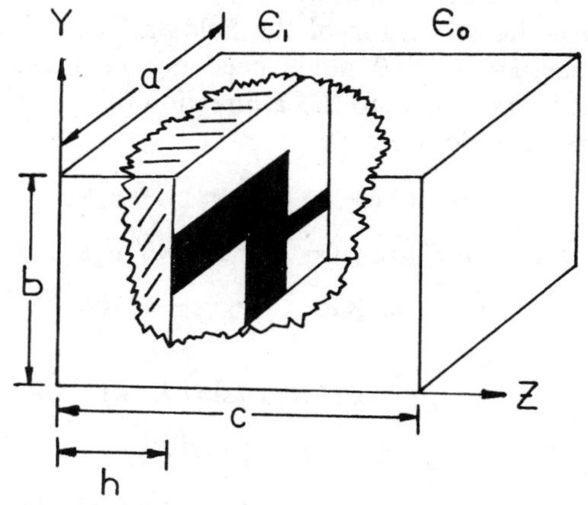

Fig. 1. The microstrip circuit is completely contained in a shielding, conducting rectangular box. The coordinate system is oriented so as to emphasize the fact that the fields are represented as a sum of homogeneous rectangular waveguide modes with the waveguide tube along the z axis.

II. Method of Analysis

The rectangular conducting box is treated as two separate waveguides joined at $z = h$ (Fig. 1) with the indicated regions and dielectric constants. Note that region 0 is usually but not necessarily restricted to free space. The tangential (or transverse to z) fields in a given region due to current on a single subsection is expressed as a sum of homogeneous waveguide modes. Expressions for the tangential fields are written as a weighted sum of these modes:

$$\boldsymbol{E}_t^1 = \sum V_i \frac{\sin\left(K_{iz}^1 z\right)}{\sin\left(K_{iz}^1 h\right)} \boldsymbol{e}_i$$

$$\boldsymbol{H}_t^1 = \sum V_i Y_i^1 \frac{\cos\left(K_{iz}^1 z\right)}{\sin\left(K_{iz}^1 h\right)} \boldsymbol{h}_i$$

$$\boldsymbol{H}_t^1 = \sum V_i Y_i^1 \frac{\cos\left(K_{iz}^1 z\right)}{\sin\left(?K_{iz}^1 h\right)} \boldsymbol{h}_i$$

$$\boldsymbol{E}_t^0 = \sum V_i \frac{\sin\left[K_{iz}^0 (c - z)\right]}{\sin\left[K_{iz}^0 (c - h)\right]} \boldsymbol{e}_i$$

$$\boldsymbol{H}_t^0 = \sum V_i Y_i^0 \frac{\cos\left[K_{iz}^0 (c - z)\right]}{\sin\left[K_{iz}^0 (c - h)\right]} \boldsymbol{h}_i \qquad (1)$$

Reprinted from *IEEE Trans. Microwave Theory Tech.*, vol. MTT-35, no. 8, pp. 726–730, Aug. 1987.

where V_i is the modal coefficient (amplitude) of the ith mode, and Y_i is the admittance of the ith (m, n) mode as follows:

$$Y_i^{1\,\mathrm{TE}} = jK_{iz}^1/(\omega\mu_1) \qquad Y_i^{1\,\mathrm{TM}} = j\omega\epsilon_1/K_{iz}^1$$

$$Y_i^{0\,\mathrm{TE}} = -jK_{iz}^0/(\omega\mu_0) \qquad Y_i^{0\,\mathrm{TM}} = -j\omega\epsilon_0/K_{iz}^0$$

$$K_{iz}^1 = -\sqrt{K_1^2 - K_x^2 - K_y^2} \qquad K_{iz}^0 = +\sqrt{K_0^2 - K_x^2 - K_y^2}.$$

Note that the modal admittances are the admittances of the standing wave modes rather than those of the usual traveling wave modes (they differ by the constant j). The $\boldsymbol{e}_i$ and $\boldsymbol{h}_i$ are the orthonormal mode vectors which form a basis for the expansion of the fields in each region. Note that the $m = 0, n = 0$ mode need not be included as all current is transverse to the z direction [8], [9]. For rectangular waveguide, we have

$$\boldsymbol{e}_i^{\mathrm{TE}}(x, y) = N_1 g_1 \boldsymbol{u}_x - N_2 g_2 \boldsymbol{u}_y$$

$$\boldsymbol{e}_i^{\mathrm{TM}}(x, y) = N_2 g_1 \boldsymbol{u}_x + N_1 g_2 \boldsymbol{u}_y$$

$$\boldsymbol{h}_i = \boldsymbol{u}_z \times \boldsymbol{e}_i \qquad \boldsymbol{e}_i = -\boldsymbol{u}_z \times \boldsymbol{h}_i$$

where

$$g_1 = \cos(K_x x)\sin(K_y y)$$

$$g_2 = \sin(K_x x)\cos(K_y y).$$

The N_1 and N_2 are normalizing constants dependent on the mode numbers and waveguide dimensions. If a different geometry is selected for the waveguide shield, only the above mode vectors need be changed.

Given a specific current distribution on the surface of the substrate, we must determine the modal coefficients, the V_i, of the field generated by that surface current. This is accomplished by setting the discontinuity in magnetic field equal to the assumed surface current. Then, using the orthogonality of the modal vectors, we may determine the V_i of the field generated by the current

$$V_i = -\hat{Z}_i \int\int \boldsymbol{J}_s \cdot \boldsymbol{e}_i \, ds$$

$$\hat{Y}_i = Y_i^0 \operatorname{ctn}\left[K_{iz}^0(c-h)\right] - Y_i^1 \operatorname{ctn}\left(K_{iz}^1 h\right)$$

$$\hat{Z}_i = 1/\hat{Y}_i.$$

The admittance $\hat{Y}_i$ is the parallel connection of the admittances of the two shorting planes at $z = 0$ and $z = c$ transformed back to the substrate surface $z = h$. Multilayered geometries need only modify $\hat{Y}_i$.

Substitution of V_i into (1) yields the tangential fields everywhere in the waveguide. Specialization of V_i to a delta function for $\boldsymbol{J}_s$ provides the Green's function in the "spatial" domain for current on the surface of the substrate. The Green's function is a cosine and sine series in two dimensions with the coefficients of the series representing the Green's function in the "spectral" domain.

Evaluation of the V_i requires the evaluation of surface integrals of the current distribution dotted with a mode vector. We use the "rooftop" distribution [10], which is separable with respect to x and y. One component of current, either x or y, is evaluated at a time. The distribution has a triangle function dependence in the direction of current flow and a rectangle function dependence in the lateral direction. This is shown in Fig. 2, where the rectangular base of the three-dimensional figure represents the rectangular subsection, and the height above the base is proportional to current density. Fig. 3 shows how several rooftop functions can be placed on overlapping subsections to provide a piecewise linear approximation to the current in the direction of current flow and a step approximation in the lateral direction.

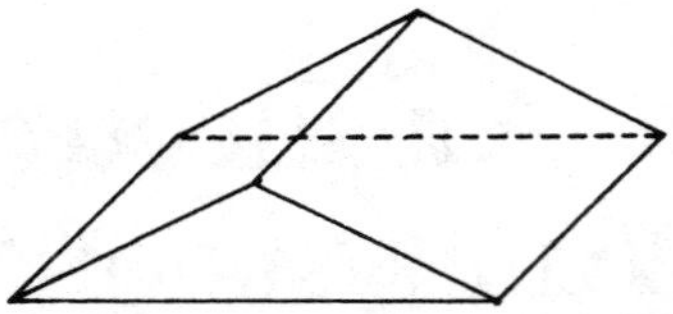

Fig. 2. The product of a triangle function in one direction by a rectangle function in the lateral direction gives a rooftop function which will be used as an expansion function.

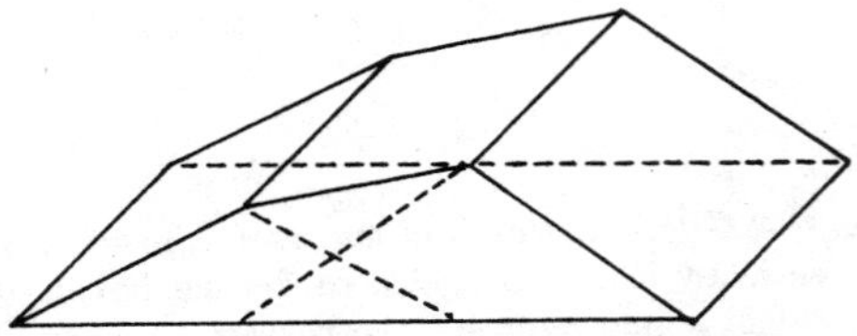

Fig. 3. Two rooftop functions placed on overlapping rectangles give a piecewise linear approximation to the current in the direction of current flow. Additional rooftop functions placed side by side will provide a step approximation to the surface current in the direction lateral to current flow.

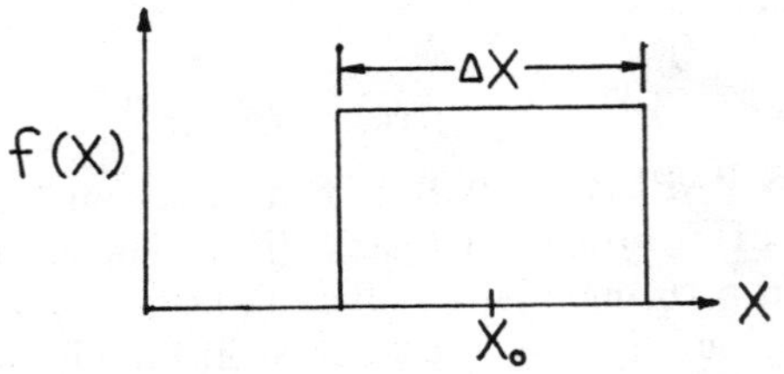

Fig. 4. The rectangular pulse function is used to represent the current density on a subsection in the direction lateral to current flow.

Since the rooftop function is separable, the integral for V_i reduces to the product of two one-dimensional integrals. The simplest integral to evaluate is the integral involving the rectangle function, Fig. 4. We require the evaluation of

$$F_c = \int f(x)\cos(Kx)\,dx \quad \text{and} \quad F_s = \int f(x)\sin(Kx)\,dx.$$

The constant K is the wavenumber corresponding to the variable of integration, for example, $M\pi/a$. Evaluation of the integrals yields

$$F_c = \frac{2}{K}\sin(K\Delta x/2)\cos(Kx_0), \qquad K \neq 0$$

$$F_s = \frac{2}{K}\sin(K\Delta x/2)\sin(Kx_0), \qquad K \neq 0$$

$$F_c = \Delta x \quad \text{and} \quad F_s = 0, \qquad K = 0.$$

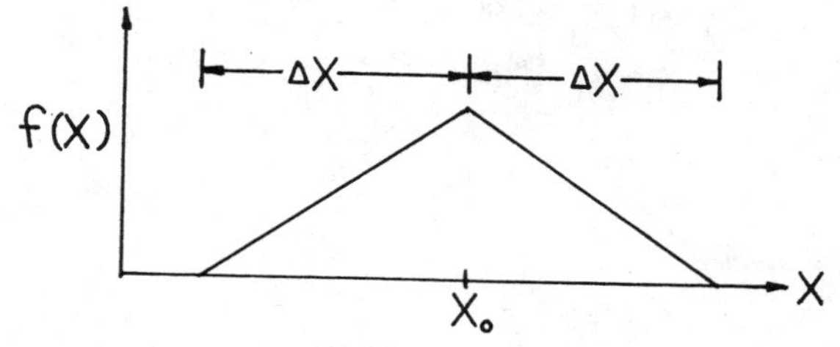

Fig. 5. The triangular pulse function represents the current density on a subsection in the direction of current flow. Thus, no line charges are generated.

Note that the term which depends on the subsection dimensions is the same for both cases. We refer to that term as $G(\Delta x)$. The integrand may be a function of y, in which case we have $G(\Delta y)$. For the rectangular pulse, we have

$$G(\Delta x) = \frac{2}{K} \sin(K\Delta x/2), \qquad K \neq 0$$
$$= \Delta x, \qquad K = 0.$$

We use F_c, F_s, $G(\Delta x)$, and $f(x)$ as a generic notation. The functions which they represent depend on the pulse being considered. For the triangle function, Fig. 5, we have

$$F_c = G(\Delta x)\cos(Kx_0) \quad \text{and} \quad F_s = G(\Delta x)\sin(Kx_0)$$

with

$$G(\Delta x) = \frac{2}{\Delta x K^2}(1 - \cos(K\Delta x)), \qquad K \neq 0$$
$$= \Delta x, \qquad K = 0.$$

Thus, to evaluate the integral for V_i we need only evaluate the ith modal vector at the midpoint of the subsection and multiply by the appropriate $G(\Delta x)$ and $G(\Delta y)$.

In the following equations, $f(x)$ denotes a pulse function which is a function of x, and $f(y)$ denotes a pulse function which is a function of y. $G(\Delta x)$ is the constant, derived above, which is obtained when $f(x)$ is multiplied by a sine or cosine and integrated over the domain of $f(x)$. The same holds for $G(\Delta y)$. We indicate distinct $f(x)$ and $G(\Delta x)$ functions by subscripts. Quantities relating to a source subsection are indicated by a prime. Quantities relating to a field subsection remain unprimed. A primed modal function indicates that the modal vector is to be evaluated at the center of the source subsection, e.g., $g_1' = g_1(x_0, y_0)$. An unprimed modal function is to be evaluated at the field point (or center of the field subsection).

In general, we consider a current distribution of the form $\boldsymbol{J}_s = J_x f_1(x) f_2(y) \boldsymbol{u}_x + J_y f_3(x) f_4(y) \boldsymbol{u}_y$. The pulse functions f_1 and f_4 are triangle functions while f_2 and f_3 are rectangle functions. Other functions could be used [4]. The pulse functions are centered on the subsection under consideration. Calculation of the V_i and substitution into (1) to calculate the tangential electric field at the substrate surface yield

$$\begin{aligned} E_x = \sum_{m,n} &\left[-G_1'(\Delta x) G_2'(\Delta y) g_1' g_1 \left(N_1^2 \hat{Z}_{mn}^{\mathrm{TE}} + N_2^2 \hat{Z}_{mn}^{\mathrm{TM}}\right)\right] J_x \\ &+ \left[G_3'(\Delta x) G_4'(\Delta y) N_1 N_2 g_2' g_1 \left(\hat{Z}_{mn}^{\mathrm{TE}} - \hat{Z}_{mn}^{\mathrm{TM}}\right)\right] J_y \\ E_y = \sum_{m,n} &\left[G_1'(\Delta x) G_2'(\Delta y) N_1 N_2 g_1' g_2 \left(\hat{Z}_{mn}^{\mathrm{TE}} - \hat{Z}_{mn}^{\mathrm{TM}}\right)\right] J_x \\ &+ \left[-G_3'(\Delta x) G_4'(\Delta y) g_2' g_2 \left(N_2^2 \hat{Z}_{mn}^{\mathrm{TE}} + N_1^2 \hat{Z}_{mn}^{\mathrm{TM}}\right)\right] J_y. \end{aligned} \tag{2}$$

This equation is similar to (18) in [7], illustrating the similarity between this technique and the spectral-domain approach.

For a Galerkin implementation, we need the integral of the electric field weighted by a rooftop function, say $f_1(x) f_2(y)$, at a field subsection. This integration is effected by multiplying each term in the summation by $G_1(\Delta x) G_2(\Delta y)$. These weighted integrals (reactions) of the electric field are also used in the evaluation of the N-port circuit parameters.

A. Implementation of the Galerkin Solution

The technique is implemented by subdividing the metallization into small, overlapping rectangles. We need two sets of rectangles, one for x-directed and a second for y-directed current. The centers of the two sets of subsections need to be offset with respect to each other; otherwise the microstrip edges will not properly align. More importantly, a subsection of x-directed current cannot induce current in a collocated y-directed subsection. This situation gives incorrect results.

Both problems are solved by offsetting one set of subsections as in [10]. The current densities on the subsections form a set of dependent variables in a system of equations. The weighted integrals of the electric field on the subsections form a set of independent variables related to the dependent variables by an impedance matrix whose elements are calculated above. Select one (or more) subsections as a source; set the integral of the electric field on that subsection equal to one and all the others to zero (zero tangential electric field on a conductor). Matrix inversion provides the solution. Techniques for the efficient calculation of the matrix elements have been developed [4], [5].

B. The Source Model

Microstrip circuit inputs and outputs are usually taken at the edge of the substrate by means of a coaxial cable penetrating the shielding sidewall at $z = h$. The coax shield is connected to the microstrip shield, and the coax center conductor is attached to a microstrip conductor.

The coax aperture can be modeled by a conductor-backed circulating magnetic current. We assume that the aperture is small and that the aperture current has negligible effect. When we compare measured data with calculated data [4], [6], we find that the contribution from the aperture field is important and that it can be modeled as a small fringing capacitance in shunt with the connector.

We model the current injected by the coax center conductor as a subsection of current directed perpendicularly to the sidewall and centered on the sidewall. The port subsection uses the same roof-top current distribution. This facilitates the transition from the port subsection to microstrip subsections. In the analysis, we set the tangential electric field to a constant value on all port subsections and to zero on all other subsections.

C. Evaluation of Input Admittance

We initially discuss the input admittance of a one port circuit. Quantities associated with that port are designated by subscript 1. Elements in the admittance matrix of the entire microstrip system have double numerical subscripts.

We use the usual variational expression [1, pp. 348–349]

$$Y_i = -\frac{I_1^2}{\iint \boldsymbol{E}\cdot\boldsymbol{J}\,ds}.$$

Since $\boldsymbol{E}$ or $\boldsymbol{J}$ is zero everywhere except at the port subsection, we need only consider the port subsection. The weighted integral of the electric field on that subsection is equal to one by definition. The current on the subsection is proportional to that same weighting function, the constant of proportionality being $Y_{11} = J_1$. Thus, the denominator of the above expression is just Y_{11}. The input current is the input current density multiplied by the width of the input, Δw, usually either Δx or Δy. Thus,

$$Y_1 = -(Y_{11}\Delta w)^2/Y_{11} = -Y_{11}(\Delta w)^2.$$

In a like manner, the transfer admittance between any two ports, say port a and port b, of an N-port circuit may be determined by

$$Y_{\text{tran}} = -Y_{ab}\Delta w_a \Delta w_b.$$

The sign of a transfer admittance depends on circuit geometry. This is because we define positive current in the direction of the positive axis, while circuit theory defines positive current as directed into the body of the multiport.

III. Software Implementation

The analysis was implemented in a Pascal program on an IBM-PC and later transported to a VAX computer. Dynamic arrays (a data type available in Pascal) were used extensively in developing a complex vector data type which was used to vectorize the software.

On the IBM-PC a small circuit (a dozen subsections) can be analyzed in a few minutes. Larger circuits (100 subsections) require several hours per frequency. The VAX version of the analysis provides a factor of ten improvement. The software has not been optimized.

A mouse-based microstrip geometry capture program has also been written. A five-section low-pass filter was subdivided into 611 subsections in less than an hour using this program. The output of the program (a text file containing the coordinates of the center of each subsection) is used directly as input to the analysis program.

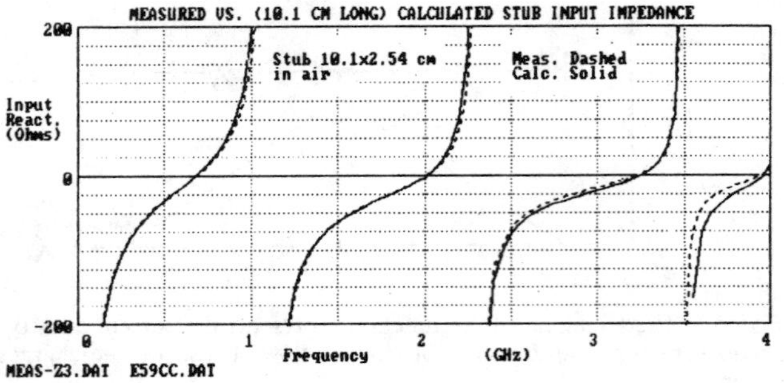

Fig. 6. Measurements (dashed line) of a large open-circuited microstrip stub agree well with calculations (solid line), suggesting that a high level of accuracy has been realized.

Fig. 6 shows a sample analysis, a comparison between measured and calculated data for a microstrip open circuited stub. The stub is 10.1 cm long and 2.54 cm wide. The agreement between the measured and calculated data is typical. Space does not permit more detailed results here. The interested reader is referred to [4] and [6].

IV. Conclusions

A technique for the analysis of shielded microstrip circuits has been presented. The technique is a Galerkin implementation of the method of moments and is closely related to the spectral-domain approach. The analysis is a complete time-harmonic electromagnetic analysis of microstrip.

The analysis may be used in the evaluation of individual microstrip discontinuities or, with faster computers, in the evaluation of entire microstrip circuits.

While the analysis is numerically intensive, it is sufficiently efficient that results for simple circuits can be obtained in reasonable time even with a small personal computer.

References

[1] R. F. Harrington, *Time-Harmonic Electromagnetic Fields*. New York: McGraw-Hill, 1961.

[2] R. F. Harrington, "Matrix methods for field problems," *Proc. IEEE*, vol. 55, pp. 136–149, 1967.

[3] R. F. Harrington, *Field Computation by Moment Methods*. New York: Macmillan, 1968. Reprinted Melbourne, FA.: Krieger, 1982.

[4] J. C. Rautio, "A time-harmonic electromagnetic analysis of shielded microstrip circuits," Ph.D. dissertation, Syracuse University, Syracuse, NY, 1986.

[5] J. C. Rautio and R. F. Harrington, "Efficient evaluation of the system matrix in an electromagnetic analysis of shielded microstrip circuits," to be published.

[6] J. C. Rautio and R. F. Harrington, "Preliminary results of a time-harmonic electromagnetic analysis of shielded microstrip circuits," in *27th ARFTG Conf. Dig.*, spring 1986, pp. 121–134.

[7] R. H. Jansen, "The spectral-domain approach for microwave integrated circuits," *IEEE Trans. Microwave Theory Tech.*, vol. MTT-33, pp. 1043–1056, 1985.

[8] T. Vu Khac and C. T. Carson, "$m = 0$ and $n = 0$ mode and rectangular waveguide slot discontinuity," *Electron. Lett.*, vol. 9, pp. 431–432, 1973.

[9] J. Van Bladel, "Contribution of the ψ = constant mode to the modal expansion in a waveguide," *Proc. Inst. Elec. Eng.*, vol. 128, pp. 247–251, 1981.

[10] A. W. Glisson and D. R. Wilton, "Simple and efficient numerical methods for problems of electromagnetic radiation and scattering from surfaces," *IEEE Trans. Antennas Propagat.*, vol. AP-28, pp. 593–603, 1980.

Paper 5.3

Arbitrarily Shaped Microstrip Structures and Their Analysis with a Mixed Potential Integral Equation

JUAN R. MOSIG, MEMBER, IEEE

Abstract —This paper gives a comprehensive description of the mixed potential integral equation (MPIE) as applied to microstrip structures. This technique uses Green's functions associated with the scalar and vector potential which are calculated by using stratified media theory and are expressed as Sommerfeld integrals. Several methods of moments allowing the study of irregular shapes are described. It is shown that the MPIE includes previously published static and quasi-static integral equations. Hence, it can be used at any frequency ranging from dc to higher order resonances. Several practical examples including an L-shaped patch have been numerically analyzed and the results are found to be in good agreement with measurements.

I. Introduction

THE PRACTICAL advantages of microstrip structures have been discussed in many papers and are now too well known to be repeated here. On the other hand, it is perhaps worthwhile to point out that such structures are very well suited for mathematical modeling. This seldom mentioned "theoretical" advantage is mostly due to the relatively simple geometry of microstrip structures and has certainly contributed to their popularity. Indeed, every analytical technique commonly used in electromagnetics has been applied to microstrip, giving rise to a surprisingly great number of different and apparently unrelated models.

In this paper, we will put the emphasis on the analysis of microstrip structures having upper conductors of arbitrary shape. The general term *microstrip structures* includes here patches of finite size and discontinuities obtained by interconnecting several microstrip lines through a patch.

The purpose of many microstrip models is to provide an equivalent circuit for a given structure. The simplest models yield lumped *LC* circuits, valid at low frequencies. Improvements of these models introduce ohmic losses (a series resistance) and dielectric losses (a parallel conductance). At higher frequencies, the microstrip structure can no longer be represented by a classical *RLC* circuit. More accurate models are then employed that take into account the dynamic behavior of the fields and yield scattering or impedance matrices with, in general, complex elements varying with frequency. If the field analysis is made for an open microstrip structure, the port matrices (S or Z) should include real terms accounting for the radiation losses (surface and space waves).

Equivalent capacitances and inductances of a microstrip structure can be obtained by solving, respectively, a static integral equation [1], [2] and a quasi-static integral equation [3], [4]. But these models are restricted to low frequencies, where the real part of any impedance parameter is negligible and the imaginary part behaves like the reactance of a classical *LC* circuit.

A more accurate model including dispersion is the waveguide/cavity model which leads directly to the scattering matrix of the structure [5]. However, this model neglects fringing fields as well as radiation and surface waves. Therefore it is restricted to electrically thin substrates. The same restriction applies to models based upon the planar circuit concept [6]–[8].

The most general and rigorous treatment of microstrip structures is given by the well-known electric field integral equation (EFIE) technique, usually formulated in the spectral domain [10], [11]. In this paper, we use a modification of the EFIE, called the mixed potential integral equation (MPIE), and we solve it in the space domain. The MPIE is numerically stable and can be solved with efficient algorithms [12]. Working in the space domain helps to keep a good physical insight of the problem.

The MPIE was introduced by Harrington [13] and has been extensively used for the analysis of wire antennas. Here, the MPIE will be applied to microstrip introducing specific kernels which account for dielectric layers and for the ground plane. It is shown that the MPIE models can be used at any frequency, from the static case up to the determination of higher order modes in a resonant patch. Several specializations to particular frequency ranges are discussed in detail. Also, this formulation includes coupling, dispersion, radiation losses, and surface waves and therefore provides a powerful and flexible technique for the study of microstrip structures.

II. The MPIE for Microstrip Structures

A. Initial Assumptions

We consider here microstrip structures where the substrate and the ground plane have infinite transverse dimensions (Fig. 1a). Theoretical developments are given here for

Manuscript received May 4, 1987; revised October 19, 1987.
The author is with the Laboratoire d'Electromagnétisme et d'Acoustique, École Polytechnique Fédérale de Lausanne, CH-1007 Lausanne, Switzerland.
IEEE Log Number 8718674.

Reprinted from *IEEE Trans. Microwave Theory Tech.*, vol. 36, no. 2, pp. 314–323, Feb. 1988.

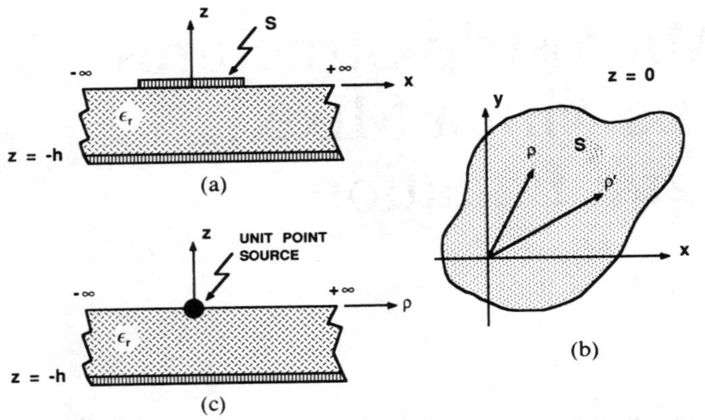

Fig. 1. (a) $y-z$ plane cut and (b) $x-y$ plane cut of a microstrip structure and (c) associated problem for the evaluation of the Green functions.

a single-layer substrate. Modifications needed to account for multiple layers will be mentioned later on. Ohmic losses in the upper conductor are included by introducing a surface impedance equal to the ratio of the tangential electric field to the surface current density. If standard thin-film technology is used, the upper conductor thickness (25–50 μm) can be neglected against substrate thickness, but is still many times greater than the skin depth at microwave frequencies (6 μm for copper at 1 GHz). Hence the upper conductor is modeled as an electric current sheet and the surface impedance can be estimated as

$$Z_s = (1+j)\sqrt{\pi f \mu_0/\sigma}. \tag{1}$$

Dielectric losses will be accounted for by introducing, as customary, a complex dielectric constant:

$$\epsilon_r = \epsilon_r'(1-j\tan\delta). \tag{2}$$

Ohmic losses in the ground plane can be taken into account by modifying the Green's functions of the problem [14]. A simplified procedure keeps the Green's functions associated with an ideal ground plane and doubles the resistivity of the upper conductor.

As mentioned in the introduction, radiation and surface waves are automatically included in this formulation. Finally, the convention $\exp(j\omega t)$ will be used throughout the paper.

B. The Integral Equation

To set up an integral equation for the currents and the charges, we start with the boundary condition associated with the tangential electric field in the surface S of the upper conductor (Fig. 1(b)):

$$\boldsymbol{e}_z \times [\boldsymbol{E}^{(e)} + \boldsymbol{E}^{(s)}] = Z_s \boldsymbol{J}_s. \tag{3}$$

Here, $\boldsymbol{E}^{(e)}$ is the excitation field and $\boldsymbol{E}^{(s)}$ is the scattered field, which can be derived from a scalar potential V and a vector potential $\boldsymbol{A}$ as

$$\boldsymbol{E}^{(s)} = -j\omega \boldsymbol{A} - \nabla V. \tag{4}$$

These potentials are in turn expressed, by means of the corresponding Green's functions $G_V, \overline{\overline{G}}_A$, as superposition integrals of the charge and current densities q_s and $\boldsymbol{J}_s$:

$$V(\boldsymbol{\rho}) = \int_S ds' G_V(\boldsymbol{\rho}|\boldsymbol{\rho}') q_s(\boldsymbol{\rho}') \tag{5}$$

$$\boldsymbol{A}(\boldsymbol{\rho}) = \int_S ds' \overline{\overline{G}}_A(\boldsymbol{\rho}|\boldsymbol{\rho}') \cdot \boldsymbol{J}_s(\boldsymbol{\rho}'). \tag{6}$$

Finally, the MPIE can be written as

$$\boldsymbol{e}_z \times \boldsymbol{E}^{(e)}(\boldsymbol{\rho}) = \boldsymbol{e}_z \times \left[j\omega \int_S ds' \overline{\overline{G}}_A(\boldsymbol{\rho}|\boldsymbol{\rho}') \cdot \boldsymbol{J}_s(\boldsymbol{\rho}') + \nabla \int_S ds' G_V(\boldsymbol{\rho}|\boldsymbol{\rho}') q_s(\boldsymbol{\rho}') + Z_s \boldsymbol{J}_s(\boldsymbol{\rho}) \right] \tag{7}$$

where charge and current densities are related through the continuity equation

$$\nabla \cdot \boldsymbol{J}_s + j\omega q_s = 0.$$

Rigorously, the MPIE is a Fredholm integral of the second kind, but the term $Z_s \boldsymbol{J}_s$ is usually small and the MPIE behaves numerically as a Fredholm integral to the first kind. Prior to solving (7) numerically with a method of moments, the Green's functions must be evaluated. These functions correspond physically to the potentials created by unit point sources (Fig. 1(c)). Hence, their determination can be done in the $\rho-z$ plane. Once the Green's functions are computed and stored for the case $z=0$, we can get rid of the z coordinate and perform all the subsequent calculations in the $x-y$ plane of Fig. 1(b). In general, integral equation techniques reduce one 3-D problem ($x-y-z$) to two 2-D problems ($\rho-z$ and $x-y$) resulting in a numerically efficient approach. For instance, going from free-space to multilayered substrates requires only a modification of the Green's functions, but the 2-D problem in the $x-y$ plane remains unchanged.

C. The Green's Functions

A thorough study of the Green's functions for microstrip structures can be found in previous works by the author [15]. For sake of completeness, we will mention here the final results for a single layer and for $z=0$.

For the dyadic $\overline{\overline{G}}_A$, we consider only its x, y components, since neither the z component of the current nor the z component of the electric field need to be considered in (7). By symmetry considerations, we have

$$G_V(\boldsymbol{\rho}|\boldsymbol{\rho}') = G_V(\boldsymbol{\rho}-\boldsymbol{\rho}'|0) = g_V(|\boldsymbol{\rho}-\boldsymbol{\rho}'|) \tag{8}$$

$$G_A^{st}(\boldsymbol{\rho}|\boldsymbol{\rho}') = G_A^{st}(\boldsymbol{\rho}-\boldsymbol{\rho}'|0) = \begin{cases} g_A(|\boldsymbol{\rho}-\boldsymbol{\rho}'|) & st = xx, yy \\ 0 & st = xy, yx. \end{cases} \tag{9}$$

The Green's functions g_V and g_A show translational invariance in the $x-y$ plane and therefore are only functions of the source–observer distance $R = |\boldsymbol{\rho}-\boldsymbol{\rho}'|$. Their expressions in terms of Sommerfeld integrals are [15]:

$$2\pi\epsilon_0 g_V(R) = \int_0^\infty d\lambda\, J_0(\lambda R) \frac{\lambda(u_0 + u\tanh uh)}{D_{\mathrm{TE}} D_{\mathrm{TM}}} \tag{10}$$

$$(2\pi/\mu_0) g_A(R) = \int_0^\infty d\lambda\, J_0(\lambda R) \frac{\lambda}{D_{\mathrm{TE}}} \tag{11}$$

where

$$D_{TE} = u_0 + u \coth uh \qquad D_{TM} = \epsilon_r u_0 + u \tanh uh$$

and

$$u_0 = \sqrt{\lambda^2 - k_0^2} \qquad u = \sqrt{\lambda^2 - \epsilon_r k_0^2}$$

with λ being the dummy spectral variable.

TE (TM) surface waves appear as zeros of D_{TE} (D_{TM}) and hence as poles of the functions to be integrated in (10) and (11). It is worth pointing out that the result (9) stems from the fact that the traditional approach of Sommerfeld is used. However, it has been recently shown [16] that this approach should be modified if the conductors are not restricted to horizontal planes, in order to keep the unicity of the scalar Green's function G_V.

Stratified-media theory [17] allows the generalization of (9) and (10) to multilayered substrates. The two-layer case has been recently investigated in connection with microstrip radome problems [18].

Efficient numerical evaluation of the integrals (10) and (11) calls for quite sophisticated techniques [15]. Nevertheless, since the integrals are functions of only one space variable, namely the source–observer distance, they can be precomputed for a given range of values and stored as

D. Space and Spectral Domains

The MPIE deserves the qualification of the space-domain technique because once the Green's functions have been computed, we get rid of the spectral variable λ and the integral equation is effectively solved in the $x-y$ plane. On the contrary, the spectral domain EFIE [10], [11] keeps the calculations in the spectral plane until the final steps. But it must be emphasized that both models are physically equivalent and their differences are purely numerical. Indeed, they are both rigorously derived from Maxwell's equations and if no approximations are made in their numerical treatment, they should provide identical results.

results.

III. Specializations of the MPIE

Fig. 2(a) shows a microstrip structure with two ports. The excitation fields are produced by an ac generator connected to the input port, while the output port is loaded by an arbitrary impedance. Surface currents and charges exist in the upper conductor and from them the port impedance matrix can be determined.

From a circuit point of view, two particular cases deserve consideration. In the first one (Fig. 2(c)) the generator is a dc battery and the load is an open circuit. No current flow exists, and the sole unknown is the charge density, whose determination allows the computation of the capacitance of the microstrip structure. In the second case, we have a low-frequency current generator at the input port and a short circuit at the output port (Fig. 2(b)). A divergenceless surface current flows through the closed circuit. There is no surface charge and $I_1 = -I_2$. From the surface current, an inductance associated with the microstrip structure can be determined.

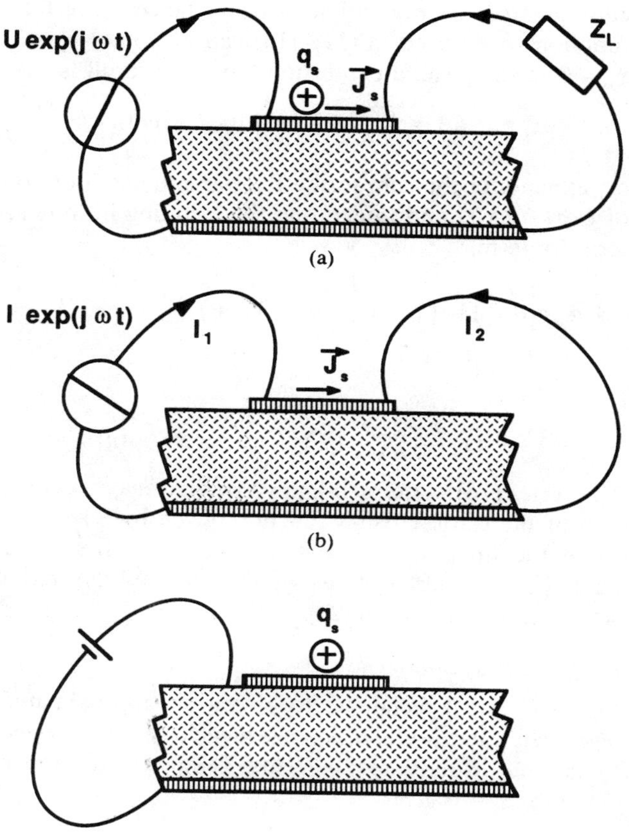

Fig. 2. Three possible excitations of a microstrip circuit: (a) dynamic (time-harmonic), (b) quasi-static and (c) static.

Both cases are included in the MPIE model and give rise, respectively, to static and quasi-static specializations of the MPIE.

A. The Static Case

In the absence of currents, (7) becomes

$$\boldsymbol{e}_z \times \nabla \int_S ds' G_V(\boldsymbol{\rho}|\boldsymbol{\rho}') q_s(\boldsymbol{\rho}') = \boldsymbol{e}_z \times \boldsymbol{E}^{(e)}(\boldsymbol{\rho}). \quad (12)$$

In many practical situations, it is customary to assume that the excitation field is created by some charge distribution $q_s^{(e)}$ via the same Green's function. Then, (12) can be rewritten as

$$\boldsymbol{e}_z \times \nabla \int_S ds' G_V(\boldsymbol{\rho}|\boldsymbol{\rho}') \left[q_s^{(e)}(\boldsymbol{\rho}') + q_s(\boldsymbol{\rho}') \right] = 0 \quad (13)$$

which implies, by integration over the tangential coordinates, that

$$\int_S ds' G_V(\boldsymbol{\rho}|\boldsymbol{\rho}') \left[q_s^{(e)}(\boldsymbol{\rho}') + q_s(\boldsymbol{\rho}') \right] = \text{constant} = U. \quad (14)$$

Instead of starting with an excitation charge, solving (13) for the "scattered charge" q_s, and finally computing the voltage U with (14), it will frequently be easier to start by assuming the voltage U known and considering (14) as an integral equation for the total charge $q_s^{(e)} + q_s$. This last approach follows closely the circuit representation of Fig. 2(c), and corresponds to the well-known static integral equation for the evaluation of capacitances. The Green's function to be used in (12)–(14) can be found by setting $k_0 = 0$ in the general expression (10). The result is

$$2\pi\epsilon_0 g_V(R) = \int_0^{\infty} d\lambda\, J_0(\lambda R)(1+\epsilon_r \coth \lambda h)^{-1} \quad (15)$$

or, expanding the sum inside the parentheses into powers of $\exp(-2\lambda h)$ and integrating the resulting infinite series term by term,

$$4\pi\epsilon_0 g_V(R) = (1-\eta)\left[\frac{1}{R_0} - (1+\eta)\sum_{n=1}^{\infty}(-\eta)^{n-1}\frac{1}{R_n}\right] \quad (16)$$

with

$$\eta = (\epsilon_r - 1)/(\epsilon_r + 1) \qquad R_n^2 = R^2 + 4n^2h^2.$$

The series (16) is the well-known partial image representation of the static Green's function, given by Silvester [19], while the integral representation (15) was first used by Patel [20]. Generalizations of (15) to multilayered substrates can be found in [21].

B. The Quasi-Static Case

The classical technique to obtain an approximated integral equation useful at low frequencies implies neglecting losses and displacement currents. Taking the divergence of (7) with $Z_s = 0$ gives

$$j\omega \boldsymbol{e}_z\cdot\nabla\times\int_S ds'\,\overline{\overline{G}}_A(\rho|\rho')\cdot\boldsymbol{J}_s(\rho') = \boldsymbol{e}_z\cdot\nabla\times\boldsymbol{E}^{(e)}(\rho)$$

$$= -j\omega\mu_0 \boldsymbol{e}_z\cdot\boldsymbol{H}^{(e)}(\rho) \quad (17)$$

where the equivalence $\nabla\cdot(\boldsymbol{e}_z\times\boldsymbol{X}) = \boldsymbol{e}_z\cdot(\nabla\times\boldsymbol{X})$ has been used. Introducing now the del operator under the integration sign leads to

$$\boldsymbol{e}_z\cdot\int_S ds'\,\overline{\overline{G}}_H(\rho|\rho')\cdot\boldsymbol{J}_s(\rho') + \boldsymbol{e}_z\cdot\boldsymbol{H}^{(e)}(\rho) = 0 \quad (18)$$

where the dyadic Green function associated with the magnetic field is $\mu_0\overline{\overline{G}}_H = \nabla\times\overline{\overline{G}}_A$. According to (9), its relevant components are given by

$$\mu_0 G_H^{zx} = -\partial g_A/\partial y \qquad \mu_0 G_H^{zy} = \partial g_A/\partial x. \quad (19)$$

Equation (18) simply expresses the fact that the total normal magnetic field must vanish on the surface of lossless conductors at any frequency. However, since (18) is a scalar equation, it does not suffice in general to determine the two components of the surface current. The second scalar equation is obtained by neglecting the displacement current in Maxwell's equations. Then $\nabla\times\boldsymbol{H} = \boldsymbol{J}$ and, consequently, $\boldsymbol{J}_s$ is solenoidal, i.e.,

$$\nabla\cdot\boldsymbol{J}_s = 0. \quad (20)$$

The set of equations (18)–(20) defines the quasi-static model.

As in the static case, it will sometimes be convenient to introduce an excitation current $\boldsymbol{J}_s^{(e)}$. Then, (18) is transformed into

$$\boldsymbol{e}\cdot\int_S ds'\,\overline{\overline{G}}_H(\rho|\rho')\cdot\left[\boldsymbol{J}_s(\rho') + \boldsymbol{J}_s^{(e)}(\rho')\right] = 0 \quad (21)$$

and the system of equations (20) and (21) is solved taking into account the additional condition

$$\int_C dl\,\boldsymbol{e}_n\cdot\boldsymbol{J}_s = I \quad (22)$$

which relates the excitation surface current to the total current entering the structure in Fig. 2(b).

Since displacement currents are neglected, the current distribution satisfies a static Poisson equation. Consequently, to ensure the internal coherence of the model, the Green's functions arising in (17), (18), and (21) must be static too. At zero frequency, (11) becomes

$$(4\pi/\mu_0) g_A(R) = 2\int_0^{\infty} d\lambda\, J_0(\lambda R)(1+\coth\lambda h)^{-1}$$

$$= \left(\frac{1}{R} - \frac{1}{\sqrt{R^2+4h^2}}\right) \quad (23)$$

which is the solution to the problem of a point source above a ground plane. Therefore, the quasi-static model is independent on the substrate permitivity.

IV. The Method of Moments

In order to apply the MPIE to irregular microstrip shapes, we need a very flexible numerical technique. The most frequent choice is a method of moments with subsectional basis functions [13]. In this approach, the upper conductor is divided into elementary domains (cells) and the basis functions defined over each cell. We have chosen the rectangular cell as the simplest shape still able to provide good approximations for many practical structures. More sophisticated shapes for the cells, such as triangles [22] and quadrangles, have been used in scattering problems and could also be applied to microstrip problems.

We also need to select the basis functions. In general, each component of the surface current will depend on the two coordinates x, y, but it is possible to use basis functions which are, inside each cell, constant along the transverse coordinate. This yields expansions for J_{sx} and J_{sy} which are discontinuous along, respectively, y and x, but the associated charge is still nonsingular. Basis functions ensuring continuity of the current in any direction, such as bilinear expansions, may be used, but the improved accuracy of the results is balanced by the increased difficulty of the computations.

The choice of test functions is also a crucial matter. To illustrate this, three possible combinations of basis and test functions will be described (Fig. 3). Other possibilities are given in [25].

Case A) Rooftop and Galerkin

An interesting possibility is using overlapping rooftop functions for the two components of the surface current (Fig. 3(a)). Then, according to the continuity equation, the basis functions for q_s are 2-D pulse doublets. The MPIE is tested by using the same rooftop functions and this yields a Galerkin procedure.

Define $\boldsymbol{T}_i$ as the vector rooftop function associated with two adjacent cells S_i^+ and S_i^- (Fig. 4(a)). The union of

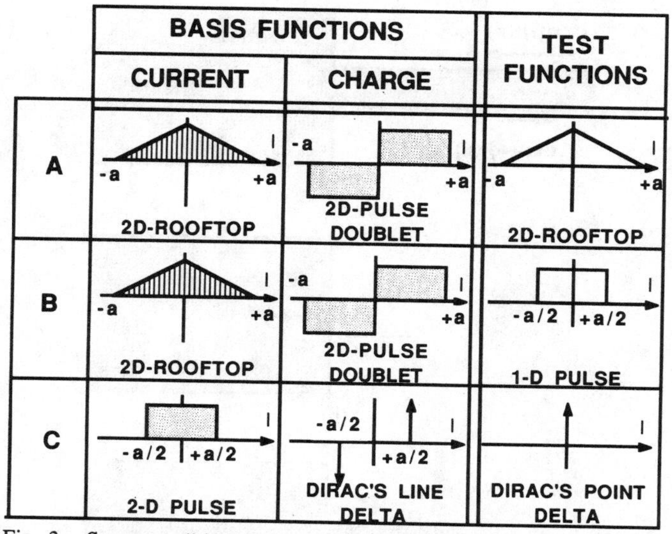

Fig. 3. Some possible choices for the basis and test functions defined over rectangular domains. All the two-dimensional functions considered are independent of the transverse coordinate.

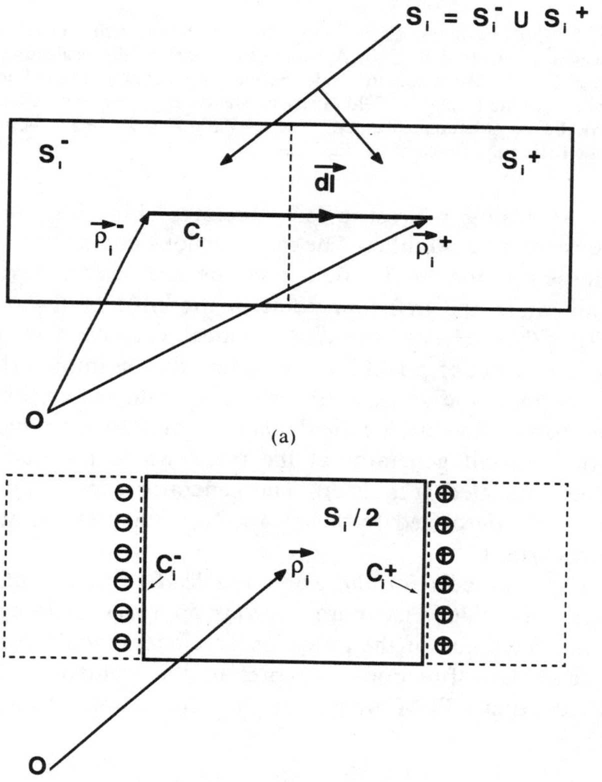

Fig. 4. (a) Longitudinal testing segments C_i linking the centers of adjacents cells (S_i^- and S_i^+) and (b) transverse segments (C_i^- and C_i^+) containing the line charge densities in the point-matching approach.

these two cells will be simply denoted by S_i. In general, we need to consider N_x x-directed functions and N_y y-directed functions, the total number being $N = N_x + N_y$. Therefore,

$$T_i = \begin{cases} e_x T_{ix} & i = 1, 2 \cdots N_x \\ e_y T_{iy} & i = N_x + 1 \cdots N. \end{cases} \tag{24}$$

The current and the charge are expanded as

$$J_s = \sum_{i=1}^{N} \alpha_i T_i \qquad j\omega q_s = \sum_{i=1}^{N} \alpha_i \Pi_i \tag{25}$$

where the α_i are unknown coefficients and the functions $\Pi_i = -\nabla \cdot T_i$ correspond to the pulse doublets.

Standard application of the method of moments yields a matrix equation with the elements of the matrix given by

$$z_{ij} = a_{ij} + v_{ij} + l_{ij} \tag{26}$$

where the contribution of $\overline{A}, V$, and the ohmic losses are, respectively,

$$a_{ij} = j\omega \int_{S_i} ds\, T_i(\rho) \cdot \int_{S_j} ds' \overline{\overline{G}}_A(\rho|\rho') \cdot T_j(\rho') \tag{27}$$

$$v_{ij} = \frac{1}{j\omega} \int_{S_i} ds\, \Pi_i(\rho) \int_{S_j} ds' G_V(\rho|\rho') \Pi_j(\rho') \tag{28}$$

$$l_{ij} = Z_s \int_{S_i} ds\, T_i(\rho) \cdot T_j(\rho). \tag{29}$$

Notice that a_{ij} vanishes if T_i is perpendicular to T_j, while $l_{ij} = 0$ if there is no intersection between S_i and S_j. In general the computation of each matrix element requires a fourfold integral. Even if two integrals can be evaluated analytically through an adequate change of variables, this approach remains cumbersome and simpler possibilities must be investigated.

Case B) Rooftop and Testing Along Segments

This modification has been suggested by Glisson and Wilton [12] and successfully applied to microstrip resonators and antennas [23].

The basis functions are the same as in A) but testing is done along the segment C_i linking the centers of cells S_i^+ and S_i^- (Fig. 4(a)). Thus we get, instead of (27)–(29),

$$a_{ij} = j\omega \int_{C_i} dl \cdot \int_{S_j} ds' \overline{\overline{G}}_A(\rho|\rho') \cdot T_j(\rho') \tag{30}$$

$$v_{ij} = \frac{1}{j\omega} \int_{S_j} ds' \left[G_V(\rho_i^+|\rho') - G_V(\rho_i^-|\rho') \right] \Pi_j(\rho') \tag{31}$$

$$l_{ij} = Z_s \int_{C_i} dl \cdot T_j(\rho) \tag{32}$$

where ρ_i^+, ρ_i^- denote the centers of the cells S_i^+, S_i^-. These expressions, simpler than (27)–(29), can be brought to effective numerical evaluation [23].

In Section III, we have mentioned the fact that the MPIE remains valid at low frequency and tends to the static integral equation. However, the condition of the matrix of moments worsens when the frequency decreases, thus preventing accurate results. This drawback can be removed by testing along the segments belonging to an open tree and replacing the remaining segments by closed loops [24]. According to Faraday's law, a null circulation of the electric field along closed loops is equivalent to enforcing a zero average value of the normal magnetic field inside the loop. Hence, the quasi-static integral equation (21) is included in the MPIE.

Case C) 2-D Pulses and Point Matching

The simplest, but still meaningful, combination of basis and test functions expands the components of the current over a set of 2-D pulses. In order to approximately satisfy the appropriate edge conditions on the surface current, these pulses are defined over domains which do not coincide with the original cells. Rather, each domain, symbolically denoted by $S_i/2$, is a combination of two cell's halves and can be considered as a two-dimensional extension of the segment C_i (Fig. 4(b)).

The associated charges are now line charges (Dirac's delta functions) distributed along two segments C_i^+ and C_i^- (Fig. 4(b)). Testing the MPIE is performed by point matching at the centers of segments C_i. Only the component of the electric field parallel to the segment is tested. A general matrix element is still given by (26), but now we have

$$a_{ij} = j\omega \boldsymbol{e}_i \cdot \int_{S_j/2} ds' \overline{\overline{G}}_A(\boldsymbol{\rho}_i|\boldsymbol{\rho}') \cdot \boldsymbol{e}_j \tag{33}$$

$$v_{ij} = \frac{1}{j\omega} \int_{C_j^+} dl' G_V(\boldsymbol{\rho}_i|\boldsymbol{\rho}') - \frac{1}{j\omega} \int_{C_j^-} dl' G_V(\boldsymbol{\rho}_i|\boldsymbol{\rho}') \tag{34}$$

$$l_{ij} = Z_s \delta_{ij} \tag{35}$$

where δ_{ij} is the Kronecker symbol and $\boldsymbol{e}_i(\boldsymbol{e}_j)$ is a unit vector parallel to $C_i(C_j)$.

The Numerical Integration Problem

The differences between the several combinations of basis and test functions disappear if an inaccurate numerical integration is used. For instance, it is meaningless to apply a Galerkin approach of type A) and then perform the integrations in (27)–(29) by using the mean-value theorem, because the resulting algorithm will be more like a point-matching technique. In this sense, the technique B) can be considered a particular version of A) using a rather loose integration technique.

A simplification of technique B) uses for the current 2-D pulses instead of rooftop functions, while keeping the 2-D pulse doublets for the charge [12], [13]. The continuity equation is no longer satisfied, but the approach can be justified on numerical grounds as being technique B) with an approximate surface integration.

V. Mathematical Treatment of the Excitation

The excitation fields are seldom known in a direct way, except in a few cases, such as exciting with a plane wave or with a series voltage gap generator (very unpractical in microstrip). Therefore, the excitation fields must usually be computed from a given distribution of currents and charges. The simplest model for the excitation is a vertical filament of unit current (Dirac's delta) acting on some point of the upper conductor. This model is a first-order approximation of real-world coaxial pins but can only be used with the method of moments of the type A), where the testing integrations suffice to smooth out the delta's singularity. For techniques B) and C), a more accurate model of the coaxial probe has been developed in [23].

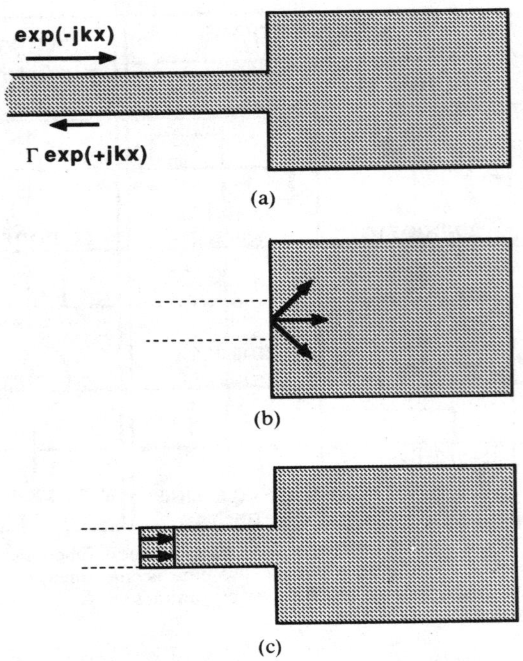

Fig. 5. (a) Microstrip discontinuity showing the incident and reflected waves in the feed line. (b) Approximate model totally neglecting the feed line. (c) Approximate model partially neglecting the feed line. In both (b) and (c) the field analysis yields Z_{in}, and the reflection coefficient is estimated as $(Z_{in} - Z_c)/(Z_{in} + Z_c)$, Z_c being the characteristic impedance of the feed line.

Concerning microstrip-fed structures (Fig. 5(a)), there are several possibilities. The most obvious one neglects the microstrip line in the field analysis and uses a vertical filament at the insertion point in the edge of the patch (Fig. 5(b)). Hence, the mathematical excitation is $\boldsymbol{J}_s = \boldsymbol{e}_z \delta(z)$. A better possibility, including discontinuity effects in the insertion zone, is to include a finite section of the microstrip line in the field analysis and to introduce a series current generator at the point where the line has been truncated (Fig. 5(c)). The generator can be mathematically described by a half-rooftop function bearing a unit current.

These models for the excitation lead to values of the input impedance. A more rigorous approach yielding directly the value of the reflection coefficient would require special basis functions to represent the incident and reflected quasi-TEM waves on the semi-infinite feed line [11].

VI. Numerical Results

A. The Linear Resonator

In order to study the convergence of the results with the number of longitudinal cells, we consider first an open-circuited microstrip line resonator with aspect ratio $L/w = 37.4$ and $\epsilon_r = 1$ (Fig. 6). Since this is a very narrow patch, only longitudinal currents are considered. The numerical algorithms of Section IV, labeled as before A), B), and C), are applied with one cell along the transverse direction and N cells along the longitudinal coordinate. For a fixed N, the modulus of the determinant of the moments' matrix shows a sharp minimum at the resonance. Fig. 6 gives the

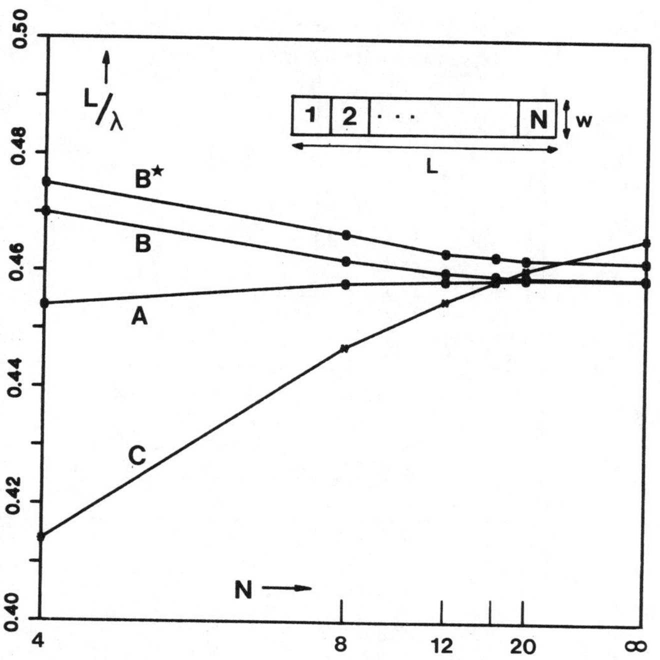

Fig. 6. The narrow linear resonator: convergence of the resonant frequency as a function of the number N of cells for the three techniques of Section IV.

normalized resonant frequency as a function of $1/N$. This allows graphical extrapolation for the case $N = \infty$.

Techniques B) and A) both converge quickly to an extrapolated frequency $L/\lambda_0 = 0.459$, which can be considered "numerically exact." The relative error for $N = 8$ is 0.9 percent for the testing-along segments algorithm B) and only 0.1 percent for the Galerkin algorithm A). On the other hand, point matching C) converges rather slowly and the extrapolated value for infinity is slightly different ($L/\lambda_0 = 0.466$). From the point of view of computation time, B) is three times slower and A) seven times slower than C). Hence the algorithm B) represents a good compromise and cells of length 0.05λ ensure accuracy of 1 percent.

Fig 6 also gives results on a modification of algorithm B) which allows for a transverse variation of the longitudinal current. This modification, denoted B*), accounts for edge effects, with a dependence of the type $[1-(2y/w)^2]^{1/2}$. The predicted resonant frequency changes only by 0.7 percent, and this difference becomes even smaller if more than one cell is allowed in the transverse direction.

B. The Rectangular Patch

The second test case is a wide rectangular patch of length $L = 150$ mm and aspect ratio $L/w = 2$ (Fig. 7). The substrate parameters are $h = 3.175$ mm, $\epsilon_r = 2.56$, and $\tan\delta = 0.0015$.

The patch is excited by a coaxial probe at $x = 58.33$ mm, $y = 37.5$ mm. To study the relevance of the number of transverse cells, numerical tests were made with a number of cells fixed along x ($N = 9$) and variable along y ($M = 3, 5, 7$). Results for the input impedance near the resonance are presented in the Smith chart of Fig. 7. The

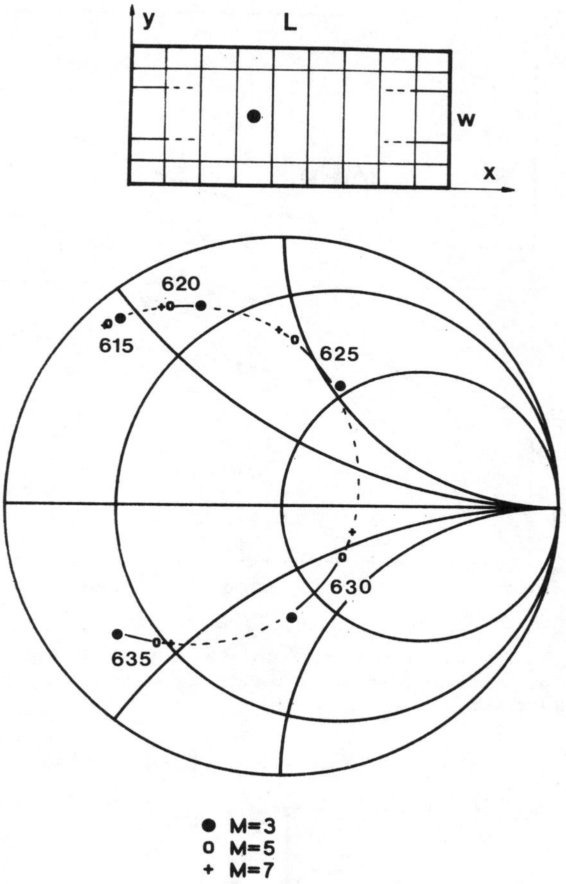

Fig. 7. The rectangular patch: convergence of the input impedance as a function of the number of cells in the transverse direction (615–635 MHz).

extrapolated ($M = \infty$) value of the resonant frequency is 628.9 MHz. The error is 0.2 percent for $M = 7$ and still only 0.7 percent for $M = 3$. This shows that the boundary condition imposing infinite values for the current density at the lateral edges can be neglected in the numerical treatment of wide patches, without appreciable loss of accuracy. It is also worth mentioning that all the points in Fig. 7 are almost on the same curve. This means that the impedance level is almost independent of the number of transverse cells, $M > 3$ being enough for engineering accuracy.

The rectangular patch was also analyzed well below the resonance for frequencies ranging from 50 to 500 MHz. The input impedance $R + jX$ normalized to 50 Ω is plotted in Fig. 8. In addition to a small real part, which accounts for ohmic losses, dielectric losses, and radiation, there is a reactance whose limiting value at low frequency is $-1/\omega C_0$ (dashed line), C_0 being the static capacitance of the patch [12].

C. The L-Shaped Patch

To illustrate the performance of the MPIE when dealing with irregular shapes, we have selected an L-shaped patch (Fig. 9(a)). Its dimensions are $a = b = 56$ mm, and $c = d = 28$ mm. The substrate parameters are $\epsilon_r = 4.34$, $\tan\delta = 0.02$,

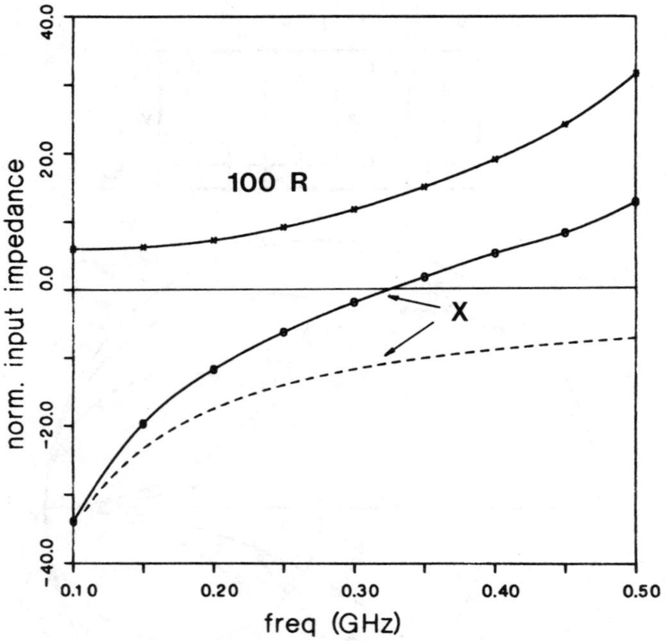

Fig. 8. Normalized input impedance $Z = (R + jX)/50\ \Omega$ of a rectangular patch. The static approximation $Z \cong 1/j\omega C_0$ is given by the dashed line.

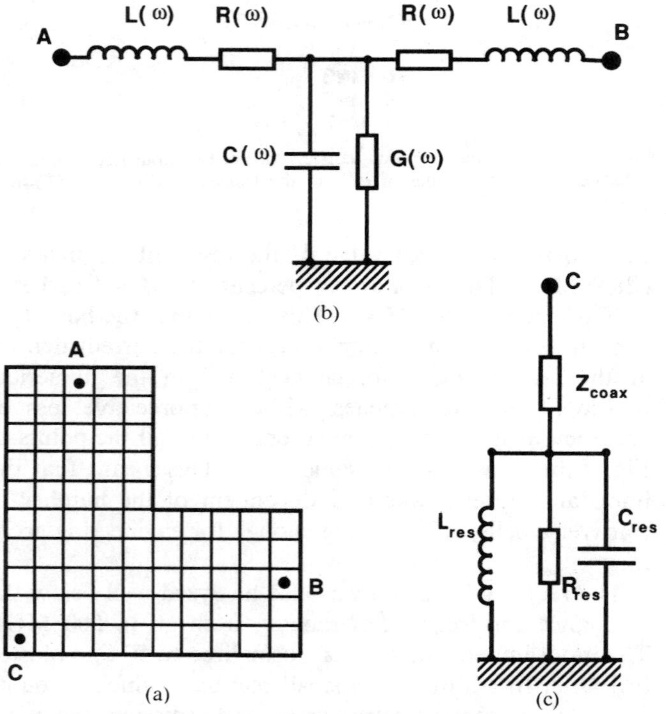

Fig. 9. The L-shaped microstrip patch. (a) Decomposition into elementary cells showing the coaxial-fed ports A, B, C. (b) Equivalent circuit at low frequency for a two-port excitation (A and B). (c) Equivalent circuit near resonance for an one-port excitation (C).

and $h = 0.8$ mm, and the patch is divided into 75 square cells.

We looked first for higher order resonances, exciting the patch with a coaxial probe C located at $x = 2.8$ mm, $y = 2.8$ mm. Two resonances were found, at 1.555 GHz and 2.536 GHz. A first resonance, at 0.998 GHz, is missed

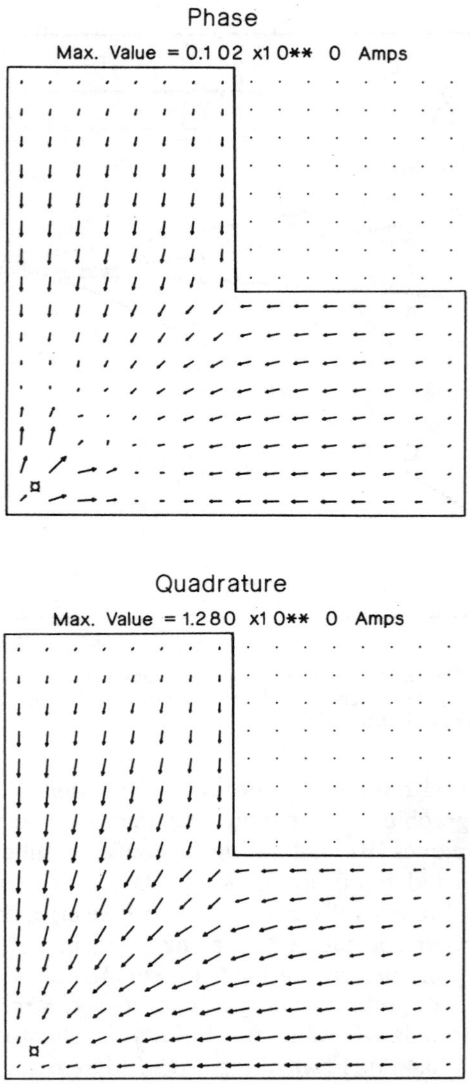

Fig. 10. Real (in-phase) and imaginary (in quadrature) parts of the surface current existing in the L-patch at the second resonance. The maximum value of the current corresponds to the longest arrow. Excitation current in the coaxial is $1 + j0$ A.

due to the symmetrical location of the coaxial probe. Fig. 10 gives a vector representation for the real and imaginary parts of the surface current density when the total excitation current entering the patch is normalized to $1 + j0$ A. As in any resonating situation, the imaginary part is stronger and its pattern is independent of the coaxial position. On the other hand, the real part, neglected in many microstrip models, corresponds to near-field effects created by the coaxial probe. This real part can modify greatly the input reactance values, mainly in weak resonances.

The input impedance at the second resonance is given in Fig. 11 and compared with measurements. The theoretical predictions are very good, with an error of only 1 percent in the resonant frequency, 4 percent in the maximum resistance, and a slight difference in the reactance values. The patch behaves as a parallel resonant circuit with a small series reactance due to the probe (Fig. 9(c)). The

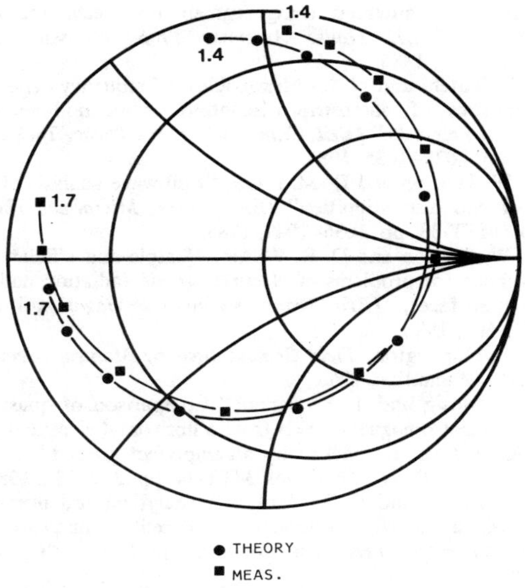

Fig. 11. Input impedance of the L-shaped patch near the second resonance (1.4–1.7 GHz): ● = theory, ■ = measurements.

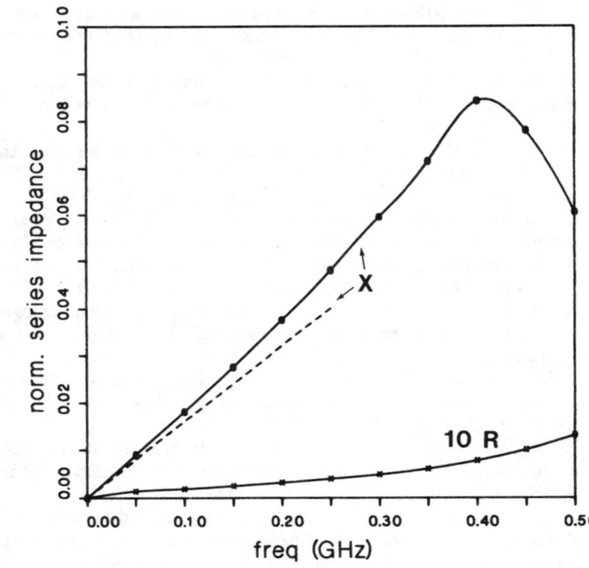

Fig. 12. Normalized series impedance $Z_s = (R + jX)/50\ \Omega$ of an L-shaped patch below the first resonance (see the equivalent circuit of Fig. 9(b)). The dashed line represents the quasi-static approximation $Z_s \cong j\omega L_0/50\ \Omega$.

elements of the equivalent circuit are easily obtained from the input impedance values of Fig. 12.

We have also considered the L-shaped patch as a two-port network with coaxial excitations at points A ($x_A =$ 8.4 mm, $y_A = 47.6$ mm) and B ($x_B = 47.6$ mm, $y_B =$ 8.4 mm). At low frequency, the patch behaves as a microstrip bend discontinuity and we can assume the equivalent circuit of Fig. 9(b). The normalized values of the series impedance $R(\omega) + jX(\omega)$ and of the parallel admittance $G(\omega) + jB(\omega)$ are given in Figs. 12 and 13. Again, the MPIE predicts correctly the frequency behavior of the structure. In particular, as the frequency goes to zero the reactance and susceptance values tend toward, respectively, the quasi-static values ωL_0 and ωC_0, obtained with (14) and (21).

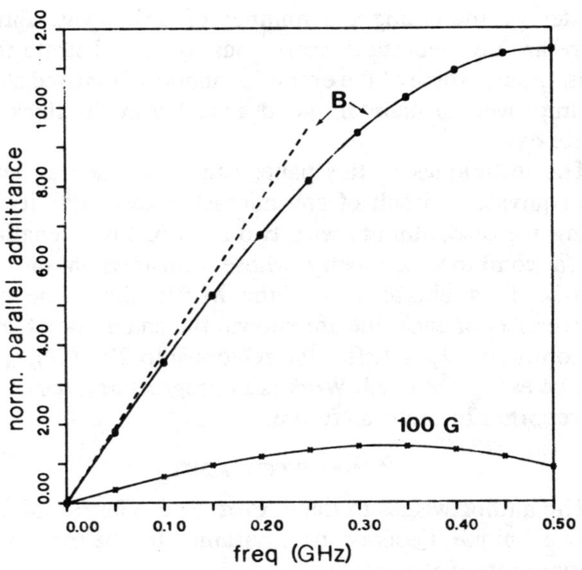

Fig. 13. Normalized parallel admittance $Y_p = (G + jB) \times 50\ \Omega$ of an L-shaped patch below the first resonance (see the equivalent circuit of Fig. 9(b)). The dashed line represents the static approximation $Y_p \cong j\omega C_0 \times 50\ \Omega$.

VII. Concluding Remarks

The mixed potential integral equation has been found to be a very convenient tool for studying microstrip structures. Combined with a method of moments using subsectional basis, this technique can easily analyze conductors of irregular shape. Also, the MPIE remains valid at any frequency and can be used for studying higher order resonances as well as for characterizing microstrip discontinuities well below the first resonance. Thus, the techniques described in this paper are particularly useful for problems where the frequency is too high for assuming a quasi-static situation, but still too low for computing the fields as expansions over the resonant modes.

In this paper, we have also pointed out the connections existing between the MPIE and other models used for microstrip. In particular the well-known static and quasi-static integral equations are embedded in the MPIE, and this explains why the proposed algorithms are successful in providing first-order corrections to static capacitances and steady-state inductances.

Convergence studies have shown that cells of linear dimensions 0.05 λ already give good results. Under this condition, accurate theoretical values are obtained for resonant frequencies, quality factors, and input impedances of patch resonators. For discontinuities, the optimum cell size to obtain an accurate equivalent circuit is mainly related to the geometry of the upper conductor.

The MPIE includes surface waves and radiation. Multilayered substrates can be accommodated by suitable modifications of the Green's functions. Handling multiple conductors at different levels (stacked patches) is only a matter of increasing the number of unknowns. Finally, there are no theoretical restrictions to the substrate thickness, though some of the excitation models discussed should be improved to maintain good accuracy in the thick substrate case.

The techniques of this paper can be applied to obtain the equivalent circuit of any microstrip discontinuity. Exciting the discontinuity with two lines of finite length, we get a combined geometry whose transmission or chain matrix T is obtained with the MPIE. Since the chain matrices T_0 of each line are known, the chain matrix of the discontinuity T_D satisfies the relationship $T = T_0 T_D T_0$ and can be easily obtained. Work is in progress and results will be reported in the near future.

Acknowledgment

The author wishes to thank Prof. F. E. Gardiol of Ecole Polytechnique Fédérale de Lausanne for helping in the preparation of the manuscript.

References

[1] A. Farrar and A. T. Adams, "Matrix methods for microstrip three-dimensional problems," *IEEE Trans. Microwave Theory Tech.*, vol. MTT-20, pp. 497–504, 1972.

[2] P. Benedek and P. Silvester, "Capacitance of parallel rectangular plates separated by a dielectric sheet," *IEEE Trans. Microwave Theory Tech.*, vol. MTT-20, pp. 504–510, 1972.

[3] A. F. Thomson and A. Gopinath, "Calculation of microstrip discontinuities inductances," *IEEE Trans. Microwave Theory Tech.*, vol. MTT-23, pp. 648–655, 1975.

[4] P. Anders and F. Arndt, "Microstrip discontinuities capacitances and inductances for double steps, mitered bends with arbitrary angle and asymmetric right-angle bends," *IEEE Trans. Microwave Theory Tech.*, vol. MTT-28, pp. 1213–1217, 1980.

[5] T. S. Chu and T. Itoh, "Generalized scattering matrix method for analysis of cascaded and offset microstrip step discontinuities," *IEEE Trans. Microwave Theory Tech.*, vol. MTT-34, pp. 280–284, 1986.

[6] R. Chadha and K. C. Gupta, "Segmentation methods using impedance matrices for the analysis of planar microwave circuits," *IEEE Trans. Microwave Theory Tech.*, vol. MTT-29, pp. 71–74, 1981.

[7] R. Sorrentino, "Planar circuits waveguide models and segmentation method," *IEEE Trans. Microwave Theory Tech.*, vol. MTT-33, pp. 1057–1066, 1985.

[8] Y. Suzuki and T. Chiba, "Computer analysis method for arbitrarily shaped microstrip antenna with multiterminals," *IEEE Trans. Antennas Propagat.*, vol. AP-32, pp. 585–590, 1984.

[9] V. Palanisamy and R. Garg, "Analysis of arbitrarily shaped microstrip patch antennas using segmentation technique and cavity model," *IEEE Trans. Antennas Propagat.*, vol. AP-34, pp. 1208–1213, 1986.

[10] P. B. Katehi and N. G. Alexopoulos, "Frequency-dependent characteristics of microstrip discontinuities in millimeter-wave integrated circuits," *IEEE Trans. Microwave Theory Tech.*, vol. MTT-33, pp. 1029–1035, 1985.

[11] R. W. Jackson and D. M. Pozar, "Full-wave analysis of microstrip open-end discontinuities," *IEEE Trans. Microwave Theory Tech.*, vol. MTT-33, pp. 1036–1042, 1985.

[12] A. W. Glisson and D. R. Wilton, "Simple and efficient numerical methods for problems of electromagnetic radiation and scattering from surfaces," *IEEE Trans. Antennas Propagat.*, vol. AP-29, pp. 593–603, 1980.

[13] R. F. Harrington, *Field Computations by Moment Methods.* New York: Macmillan, 1968.

[14] J. R. Mosig and T. K. Sarkar, "Comparison of quasi-static and exact electromagnetic fields from a horizontal electric dipole above a lossy dielectric backed by an imperfect ground," *IEEE Trans. Microwave Theory Tech.*, vol. MTT-34, pp. 379–387, 1986.

[15] J. R. Mosig and F. E. Gardiol, "Analytical and numerical techniques in the Green's function treatment of microstrip antennas and scatterers," *Proc. Inst. Elec. Eng.*, pt. H, vol. 130, pp. 175–182, 1983.

[16] K. A. Michalski, "On the scalar potential of a point charge associated with a time-harmonic dipole in a layered medium," to be published in *IEEE Trans. Antennas Propagat.*

[17] J. R. Wait, *Electromagnetic Waves in Stratified Media.* Oxford: Pergamon Press, 1962.

[18] N. G. Alexopoulos and D. R. Jackson, "Fundamental superstrate (cover) effects on printed circuit antennas," *IEEE Trans. Antennas Propagat.*, vol. AP-32, pp. 807–816, 1984.

[19] P. Silvester and P. Benedek, "Electrostatics of the microstrip: Revisited," *IEEE Trans. Microwave Theory Tech.*, vol. MTT-20, pp. 756–758, 1972.

[20] P. D. Patel, "Calculation of capacitance coefficients for a system of irregular finite conductors on a dielectric sheet," *IEEE Trans. Microwave Theory Tech.*, vol. MTT-19, pp. 862–869, 1971.

[21] R. Crampagne, M. Ahmadpanah, and J. L. Guiraud, "A simple method for determining the Green's function for a large class of MIC lines having multilayered dielectric structures," *IEEE Trans. Microwave Theory Tech.*, vol. MTT-26, pp. 82–87, 1978.

[22] S. M. Rao, D. R. Wilton, and A. W. Glisson, "Electromagnetic scattering by surfaces of arbitrary shape," *IEEE Trans. Antennas Propagat.*, vol. AP-30, pp. 409–418, 1982.

[23] J. R. Mosig and F. E. Gardiol, "General integral equation formulation for microstrip antennas and scatterers," *Proc. Inst. Elec. Eng.*, pt. H, vol. 132, pp. 424–432, 1985.

[24] D. R. Wilton and A. W. Glisson, "On improving the stability of electric field integral equation at low frequency," presented at IEEE AP-S Int. Symp., Los Angeles, CA, June 1981.

[25] J. R. Mosig and F. E. Gardiol, "A dynamical radiation model for microstrip structures," in *Advances in Electronics and Electron Physics*, vol. 59. New York: Academic Press, 1982, pp. 139–237.

Paper 5.4

A Generalized Method for Analyzing Shielded Thin Microstrip Discontinuities

LAWRENCE P. DUNLEAVY AND PISTI B. KATEHI, MEMBER, IEEE

Abstract —A new integral equation method is described for the accurate full-wave analysis of shielded thin microstrip discontinuities. The integral equation is derived by applying the reciprocity theorem, then solved by the method of moments. In this derivation, a coaxial aperture is modeled with an equivalent magnetic current and is used as the excitation mechanism for generating the microstrip currents. Computational aspects of the method have been explored extensively. A summary of some of the more interesting conclusions is included.

I. Introduction

THE NEED FOR more accurate microstrip circuit simulations has become increasingly apparent with the advent of monolithic microwave integrated circuits (MMIC's), as well as the increased interest in millimeter-wave and near-millimeter-wave frequencies. The development of more accurate microstrip discontinuity models, based on full-wave analyses, is of the utmost importance in improving high-frequency circuit simulations and reducing lengthy design cycle costs. Further, in most applications the microstrip circuit is enclosed in a shielding cavity (or housing) as shown in Fig. 1. There are two main conditions where shielding effects are significant: 1) when the frequency approaches or is above the cutoff frequency f_c for higher order modes and 2) when the metal enclosure is physically close to the circuitry. A full-wave analysis is required to accurately model these effects.

Although shielding effects have been studied to some extent in the past (e.g. [1]), the treatment has been incomplete, particularly for more complicated structures such as coupled line filters. Further, shielding effects are not accurately accounted for in the discontinuity models of most available microwave CAD software. To address these inadequacies, this paper develops an accurate method for analyzing thin strip discontinuities in shielded microstrip. The method presented is based on an integral equation approach. The integral equation is derived by an application of the reciprocity theorem and is then solved by the method of moments.

Manuscript received April 19, 1988; revised August 29, 1988. This work was supported primarily by the National Science Foundation under Contract ECS-8602530. Support was also provided by the Army Research Office under Contract DAAL03-87-K-0088 and by the Microwave Products Division of Hughes Aircraft Company.

L. P. Dunleavy was with the Department of Electrical Engineering and Computer Science, University of Michigan, Ann Arbor, MI. He is now with the Industrial Electronics Group, Hughes Aircraft Company, Torrance, CA.

P. B. Katehi is with the Department of Electrical Engineering and Computer Science, University of Michigan, Ann Arbor, MI 48109-2122.

IEEE Log Number 8824528.

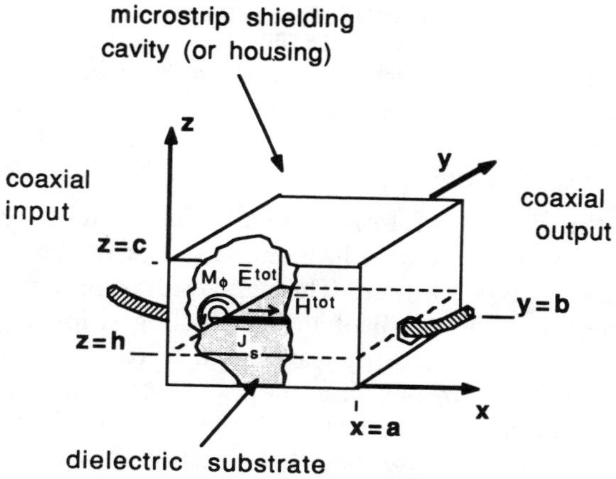

Fig. 1. Basic shielded microstrip geometry.

To derive a realistically based formulation, a coaxial excitation mechanism is used. To date, all full-wave analyses of microstrip discontinuities use either a gap generator excitation method [2]–[4] or a cavity resonance technique [5], [6]. Both of these techniques are purely mathematical tools. The former has no physical basis relative to an actual circuit. The latter is also abstract, since in any practical circuit some form of excitation is present. In fact, one of the most common excitations in practice comes from a coaxial feed (Fig. 1). A magnetic current model for such a feed is used in the present treatment as the excitation.

In addition to developing the theory, computational aspects of the solution are explored extensively. This is an important area that has been largely neglected in the presentation of numerical solutions of this nature. Most significantly, it is shown that an optimum sampling range may be specified that dictates how to divide the conducting strip for best computational accuracy. The method developed in this paper has been applied to study the effect of shielding on the characteristics of discontinuities of the type shown in Fig. 2. Numerical results from this study are presented in a companion paper [7] and are seen to be in excellent agreement with measured data.

II. Theoretical Formulation

The details of the theoretical derivation for the present method are given in [8]. Hence, only a summary of the key steps is described below.

Reprinted from *IEEE Trans. Microwave Theory Tech.*, vol. 36, no. 12, pp. 1758–1766, Dec. 1988.

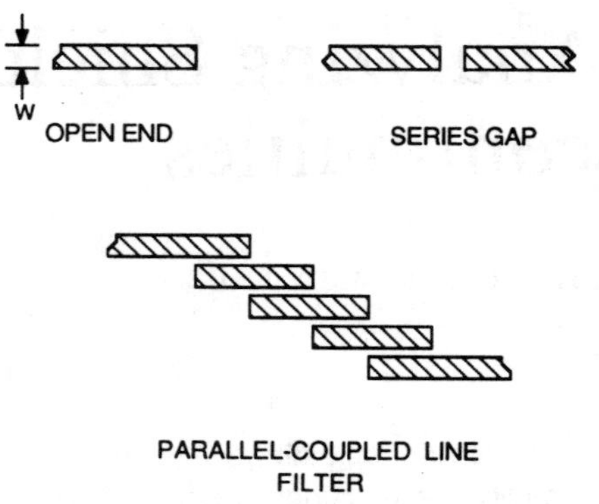

Fig. 2. Discontinuity structures addressed in the present research.

A. Integral Equation

In the theoretical formulation, a few simplifying assumptions are made to reduce unnecessary complexity and excessive computer time. Throughout the analysis, it is assumed that the width of the conducting strips is small compared to the microstrip wavelength λ_g (the "thin-strip" approximation). In this case, the transverse component of the current may be neglected. While substrate losses are accounted for, it is assumed that the strip conductors and the walls of the shielding box are lossless and that the strip has infinitesimal thickness. These assumptions are valid for the high-frequency analysis of the microstrip structures of Fig. 2, provided good conductors are used in the metalized areas.

Consider the geometry of Fig. 1. In most cases the coaxial feed, or "launcher," is designed to allow only transverse electromagnetic (TEM) propagation, and the feed's center conductor is small compared to a wavelength ($kr_a \ll 1$). In these cases, the radial electric field will be dominant in the aperture and we can replace the feed by an equivalent magnetic surface current $\overline{M}_s$ [9]. This current is sometimes called a frill current. The source $\overline{M}_s$ induces the current distribution $\overline{J}_s$ on the conducting strip and produces the total electric field $\overline{E}^{\text{tot}}$ and the total magnetic field $\overline{H}^{\text{tot}}$ inside the cavity as indicated in Fig. 1.

Now consider a cavity geometry similar to Fig. 1, with the strip conductors as well as the coaxial input and output removed. Assume a test current $\overline{J}_q$ existing on a small subsection of the area which was occupied by the strip. The fields inside this new geometry are denoted by $\overline{E}_q$ and $\overline{H}_q$. Using the reciprocity theorem, the two sets of sources ($\overline{M}_s$, $\overline{J}_s$; and $\overline{J}_q$) are related according to

$$\iiint_V \left(\overline{J}_s \cdot \overline{E}_q - \overline{H}_q \cdot \overline{M}_s\right) dv = \iiint_V \overline{J}_q \cdot \overline{E}^{\text{tot}} dv \tag{1}$$

where V represents the volume of the interior of the cavity.

Note that the reciprocity theorem has been widely used for developing integral equations similar to (1) for application to antenna and scattering problems [10]–[12]. Since $\overline{J}_q \cdot \overline{E}^{\text{tot}}$ is zero everywhere inside the cavity, the right-hand side of (1) vanishes. Reducing the remaining volume integrals in (1) to surface integrals results in

$$\iint_{S_{\text{strip}}} \overline{E}_q(z=h) \cdot \overline{J}_s \, ds = \iint_{S_f} \overline{H}_q(x=0) \cdot \overline{M}_s \, ds \tag{2}$$

where S_{strip} is the surface of the conducting strip and S_f is the surface of the coaxial aperture(s). For one-port discontinuities, S_f represents the surface of the feed on the left-hand side of Fig. 1, while for two-port discontinuities S_f represents both feed surfaces. An integral equation similar to (2) can be derived for the case of gap generator excitation by setting $\overline{M}_s = 0$ and assuming that E_x is nonzero at one point on the strip [8].

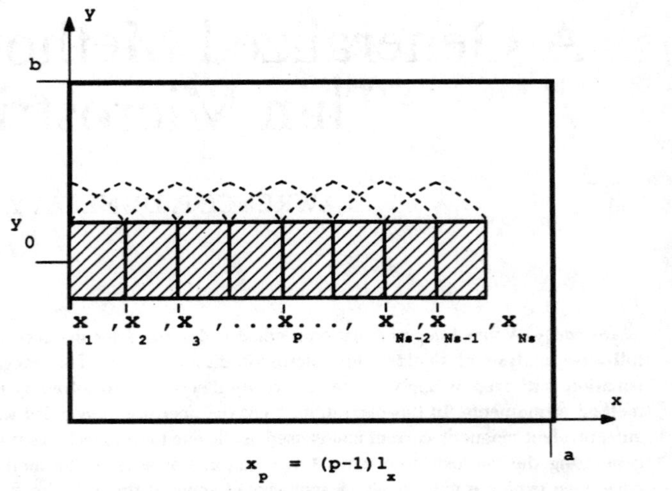

Fig. 3. Strip geometry for expansion of longitudinal current into overlapping sinusoidal basis functions.

In order to solve the integral equation (2), the current distribution $\overline{J}_s$ is expanded into a series of orthonormal functions as follows[1]:

$$\overline{J}_s = \psi(y) \sum_{p=1}^{N_s} I_p \alpha_p(x) \hat{x} \tag{3}$$

where I_p are unknown current coefficients and N_s is the number of sections considered on the strip (Fig. 3). The function $\psi(y)$ describes the transverse variation of the current and is given by [2], [13]

$$\psi(y) = \begin{cases} \dfrac{\dfrac{2}{\pi W}}{\sqrt{1 - \left[\dfrac{2(y - Y_0)}{W}\right]^2}}, & Y_0 - W/2 \leqslant y \leqslant Y_0 + W/2 \\ 0, & \text{otherwise} \end{cases} \tag{4}$$

where W is the width of the microstrip line and Y_0 is the y coordinate of the center of the strip with respect to the origin in Fig. 1.

[1]The assumed time dependence is $e^{j\omega t}$.

The basis functions $\alpha_p(x)$ are described by

$$\alpha_p(x) = \begin{cases} \dfrac{\sin[K(x_{p+1}-x)]}{\sin(Kl_x)}, & x_p \leqslant x \leqslant x_{p+1} \\ \dfrac{\sin[K(x-x_{p-1})]}{\sin(Kl_x)}, & x_{p-1} \leqslant x \leqslant x_p \\ 0, & \text{otherwise} \end{cases} \tag{5}$$

for $p \neq 1$, and

$$\alpha_1(x) = \begin{cases} \dfrac{\sin[K(l_x-x)]}{\sin(Kl_x)}, & 0 \leqslant x \leqslant l_x \\ 0, & \text{otherwise} \end{cases} \tag{6}$$

for $p=1$, where K is a scaling factor, taken to be equal to the wave number in the dielectric, x_p is the x coordinate of the pth subsection ($=(p-1)l_x$), and l_x is the subsection length ($l_x = x_{p+1} - x_p$). For computation, all of the geometrical parameters are normalized with respect to the dielectric wavelength (λ_d); hence the normalized scaling factor is equal to 2π.

The integral equation (2) can now be transformed into a matrix equation by substituting the expansion of (3) for the current $\bar{J}_s$. The result may be put in the form

$$[\boldsymbol{Z}][\boldsymbol{I}] = [\boldsymbol{V}]. \tag{7}$$

In the above, $[\boldsymbol{Z}]$ is an $N_s \times N_s$ impedance matrix, $[\boldsymbol{I}]$ is a vector composed of the unknown current coefficients I_p, and $[\boldsymbol{V}]$ is the excitation vector. The individual elements of the impedance matrix are given by

$$Z_{qp} = \iint_{S_p} \bar{E}_q(z=h) \cdot \hat{x}\psi(y)\alpha_p(x)\, ds \tag{8}$$

where S_p is the area of the two subsections on either side of the point x_p. The elements of the excitation vector are found according to

$$V_q = \iint_{S_f} \bar{H}_q \cdot \bar{M}_s\, ds. \tag{9}$$

Once the elements of the impedance matrix and excitation vector have been computed, the current distribution is found by solving (7) as follows:

$$[\boldsymbol{I}] = [\boldsymbol{Z}]^{-1}[\boldsymbol{V}]. \tag{10}$$

B. Evaluation of Impedance Matrix Elements

Before evaluating the elements of the impedance matrix, the Green's function associated with the electric current $\bar{J}_q$ is derived. To do this the cavity is divided into two regions: region 1 consists of the volume contained within the substrate ($z<h$), while region 2 is the volume above the substrate surface ($z>h$).

The integral form of the electric field is given in terms of the Green's function by

$$\bar{E}_q^i = -j\omega\mu_0 \iiint_V \left[\left(\bar{\bar{I}} + \frac{1}{k_i^2}\overline{\nabla}\overline{\nabla}\right)\cdot\left(\bar{\bar{G}}^i\right)^T\right]\cdot \bar{J}\, dv' \tag{11}$$

where $k_i^2 = \omega^2\mu_0\epsilon_i$. The index i indicates that the above holds in each region (i.e., for $i=1,2$).

In (11), $\bar{\bar{G}}^i$ is a dyadic Green's function [14] satisfying the following equation:

$$\nabla^2\bar{\bar{G}}^i + k_i^2\bar{\bar{G}}^i = -\bar{\bar{I}}\delta(\bar{r}-\bar{r}') \tag{12}$$

where $\bar{\bar{I}}$ is the unit dyadic ($=\hat{x}\hat{x}+\hat{y}\hat{y}+\hat{z}\hat{z}$), $\bar{r}$ is the position vector of a field point anywhere inside the cavity, and $\bar{r}'$ is the position vector of an infinitesimal current source.

Because of the existence of an air–dielectric interface and the assumption of a unidirectional current, the dyadic Green's function will have the form

$$\bar{\bar{G}}^i = G_{xx}^i\hat{x}\hat{x} + G_{xz}^i\hat{x}\hat{z}. \tag{13}$$

The dyadic components of (13) are found by applying appropriate boundary conditions at the walls: $x=0, a$; $y=0, b$; and $z=0, c$; and at the air–dielectric interface [8]. These components may be expressed as

$$G_{xx}^{(1)} = \sum_{m=1}^{\infty}\sum_{n=0}^{\infty} A_{mn}^{(1)} \cos k_x x \sin k_y y \sin k_z^{(1)} z \tag{14}$$

$$G_{xz}^{(1)} = \sum_{m=1}^{\infty}\sum_{n=0}^{\infty} B_{mn}^{(1)} \sin k_x x \sin k_y y \cos k_z^{(1)} z \tag{15}$$

$$G_{xx}^{(2)} = \sum_{m=1}^{\infty}\sum_{n=0}^{\infty} A_{mn}^{(2)} \cos k_x x \sin k_y y \sin k_z^{(2)} (z-c) \tag{16}$$

$$G_{xz}^{(2)} = \sum_{m=1}^{\infty}\sum_{n=0}^{\infty} B_{mn}^{(2)} \sin k_x x \sin k_y y \cos k_z^{(2)} (z-c) \tag{17}$$

where

$$k_x = n\pi/a \tag{18}$$

$$k_y = m\pi/b \tag{19}$$

$$k_z^{(1)} = \sqrt{k_1^2 - k_x^2 - k_y^2} \tag{20}$$

$$k_z^{(2)} = \sqrt{k_0^2 - k_x^2 - k_y^2} \tag{21}$$

$$k_1 = \omega\sqrt{\mu_0\epsilon_1} \tag{22}$$

$$k_0 = \omega\sqrt{\mu_0\epsilon_0} \tag{23}$$

and

$$A_{mn}^{(1)} = \frac{-\varphi_n \cos k_x x' \sin k_y y' \tan k_z^{(2)}(h-c)}{abd_{1mn} \cos k_z^{(1)} h} \tag{24}$$

$$A_{mn}^{(2)} = \frac{-\varphi_n \cos k_x x' \sin k_y y' \tan k_z^{(1)} h}{abd_{1mn} \cos k_z^{(2)}(h-c)} \tag{25}$$

$$B_{mn}^{(1)} = \frac{-\varphi_n(1-\epsilon_r^*) k_x \cos k_x x' \sin k_y y' \tan k_z^{(1)} h \tan k_z^{(2)}(h-c)}{abd_{1mn} d_{2mn} \cos k_z^{(1)} h} \tag{26}$$

$$B_{mn}^{(2)} = \frac{-\varphi_n(1-\epsilon_r^*) k_x \cos k_x x' \sin k_y y' \tan k_z^{(1)} h \tan k_z^{(2)}(h-c)}{abd_{1mn} d_{2mn} \cos k_z^{(2)}(h-c)}. \tag{27}$$

In (24)–(27), ϵ_r^* is the complex dielectric constant of the substrate and

$$\varphi_n = \begin{cases} 2 & \text{for } n = 0 \\ 4 & \text{for } n \neq 0 \end{cases} \tag{28}$$

$$d_{1mn} = k_z^{(2)} \tan k_z^{(1)} h - k_z^{(1)} \tan k_z^{(2)} (h - c) \tag{29}$$

$$d_{2mn} = k_z^{(2)} \epsilon_r^* \tan k_z^{(2)} (h - c) - k_z^{(1)} \tan k_z^{(1)} h. \tag{30}$$

In view of (11)–(30), the elements of the impedance matrix may be put in the following form[2]:

$$Z_{qp} = \frac{j\omega\mu_0 K^2 l_x^4}{16ab \sin^2 Kl_x} \zeta_q \zeta_p \sum_{n=0}^{NSTOP} \varphi_n \cos k_x x_q \cos k_x x_p \cdot [\operatorname{sinc} R_{1n} \quad \operatorname{sinc} R_{2n}]^2 LN(n) \tag{31}$$

with $LN(n)$ given by the series

$$LN(n) = \sum_{m=1}^{MSTOP} L_{mn}. \tag{32}$$

The series elements L_{mn} are given by

$$L_{mn} = \frac{\varphi_n \left[\sin(k_y Y_0) J_0\left(\frac{k_y W}{2}\right)\right]^2 \tan k_z^{(1)} h \tan k_z^{(2)} (h-c)}{\left[k_z^{(2)} \tan k_z^{(1)} h - k_z^{(1)} \tan k_z^{(2)} (h-c)\right]} \cdot \frac{\left[k_z^{(2)} \epsilon_r^* \left(1 - \frac{k_x^2}{k_1^2}\right) \tan k_z^{(2)} (h-c) - k_z^{(1)} \left(1 - \frac{k_x^2}{k_0^2}\right) \tan k_z^{(1)} h\right]}{\left[k_z^{(2)} \epsilon_r^* \tan k_z^{(2)} (h-c) - k_z^{(1)} \tan k_z^{(1)} h\right]} \tag{33}$$

where Y_0 is the y coordinate of the center of the strip, and

$$\operatorname{sinc}(t) = \begin{cases} \frac{\sin t}{t} & \text{for } t \neq 0 \\ 1 & \text{for } t = 0 \end{cases} \tag{34}$$

$$\zeta_q = \begin{cases} 2 & \text{for } q = 1 \\ 4 & \text{otherwise} \end{cases} \tag{35}$$

$$R_{1n} = \frac{1}{2}(K + k_x) l_x \tag{36}$$

$$R_{2n} = \frac{1}{2}(K - k_x) l_x. \tag{37}$$

C. Evaluation of the Excitation Vector Elements

The formulation for the excitation vector elements for the one-port case will now be carried out. The case for two-port excitation is a straightforward extension [8].

To evaluate the excitation vector elements according to (9), we need to find the magnetic field $\overline{H}_q$ and the frill current $\overline{M}_s = M_\phi \hat{\phi}$. An approximate expression for the frill current is given by [9]

$$\overline{M}_s = -\frac{V_0}{\rho \ln\left(\frac{r_b}{r_a}\right)} \hat{\phi} \tag{38}$$

where V_0 is the complex voltage applied by the coaxial line at the feed point, r_b is the radius of the coaxial feed's outer conductor, r_a is the radius of the coaxial feed's inner conductor, and ρ, ϕ are cylindrical coordinates referenced to the feed's center.

Substituting from (38) into (9) yields (with $ds = \rho\, d\rho\, d\phi$)

$$V_q = -\frac{V_0}{\ln\left(\frac{r_b}{r_a}\right)} \iint_{S_f} H_{q\phi}^i (x=0)\, d\rho\, d\phi = -\frac{V_0}{\ln\left(\frac{r_b}{r_a}\right)} \left[\iint_{S_f^{(1)}} H_{q\phi}^{(1)} (x=0)\, d\rho\, d\phi + \iint_{S_f^{(2)}} H_{q\phi}^{(2)} (x=0)\, d\rho\, d\phi\right] \tag{39}$$

where $S_f^{(1)}$ is the portion of the feed surface below the substrate–air interface ($z'' = \rho \sin\phi \leqslant -t$); $S_f^{(2)}$ is the portion of the feed surface above the substrate ($z'' = \rho \sin\phi \geqslant -t$); and $H_{q\phi}^{(1)}(x=0)$ and $H_{q\phi}^{(2)}(x=0)$ are the $\hat{\phi}$ components of the magnetic field in regions 1 and 2, respectively, evaluated on the plane of the aperture.

After solving for the magnetic fields $H_{q\phi}^i(x=0)$ and substituting the resulting expressions into (39), the following formulation is produced for excitation vector elements:

$$V_q = \frac{-V_0 \zeta_q K l_x^2}{\ln\left(\frac{r_b}{r_a}\right) 4ab \sin K l_x} \sum_{n=0}^{NSTOP} \cos k_x x_q \cdot \operatorname{sinc} R_{1n} \quad \operatorname{sinc} R_{2n} \, [MN(n)] \tag{40}$$

where $MN(n)$ is expressed in terms of the series given by

$$MN(n) = \sum_{m=1}^{MSTOP} M_{mn}. \tag{41}$$

[2] The expression given here for the impedance matrix elements, and that given shortly for the excitation vector elements, apply to the case of an open-end or series gap. Slight modifications are necessary for analysis of parallel coupled line filters.

The series elements M_{mn} are given by the following integral:

$$M_{mn} = \iint_{S_f} \mathscr{M}^{i}_{mn}\, d\rho\, d\phi$$

$$= \iint_{S_f^{(1)}} \mathscr{M}^{(1)}_{mn}\, d\rho\, d\phi + \iint_{S_f^{(2)}} \mathscr{M}^{(2)}_{mn}\, d\rho\, d\phi. \quad (42)$$

The above integrations are performed numerically, with the integrands $\mathscr{M}^{i}_{mn}$ given by

$$\mathscr{M}^{(1)}_{mn} = \cos\phi\, c^{(1)}_{zmn} \cos k_y(\rho\cos\phi + Y_c)\sin k_z^{(1)}(\rho\sin\phi + h_c)$$
$$- \sin\phi\, c^{(1)}_{ymn} \sin k_y(\rho\cos\phi + Y_c)\cos k_z^{(1)}(\rho\sin\phi + h_c) \quad (43)$$

for ρ and ϕ in region 1, and

$$\mathscr{M}^{(2)}_{mn} = \cos\phi\, c^{(2)}_{zmn} \cos k_y(\rho\cos\phi + Y_c)$$
$$\cdot \sin k_z^{(2)}(\rho\sin\phi - c + h_c) - \sin\phi\, c^{(2)}_{ymn} \sin k_y(\rho\cos\phi + Y_c)$$
$$\cdot \cos k_z^{(2)}(\rho\sin\phi - c + h_c) \quad (44)$$

for ρ and ϕ in region 2. In (43) and (44) Y_c and h_c are the y and z coordinates of the coaxial feed, and

$$c^{(1)}_{ymn} = \frac{c^{(1)}_{zmn}}{k_y d_{2mn}} \Big\{ k_z^{(1)} k_z^{(2)} \epsilon_r^* \tan k_z^{(2)}(h-c)$$
$$- \left[\left(k_z^{(1)}\right)^2 + k_x^2(1-\epsilon_r^*) \right] \tan k_z^{(1)} h \Big\} \quad (45)$$

$$c^{(1)}_{zmn} = \frac{\varphi_n k_y \tan k_z^{(2)}(h-c)}{d_{1mn}\cos k_z^{(1)} h} \sin k_y Y_0\, J_0\!\left(k_y \frac{W}{2}\right) \quad (46)$$

$$c^{(2)}_{ymn} = \frac{c^{(2)}_{zmn}}{k_y d_{2mn}} \Big\{ k_z^{(1)} k_z^{(2)} \tan k_z^{(1)} h$$
$$- \left[\left(k_z^{(2)}\right)^2 \epsilon_r^* - k_x^2(1-\epsilon_r^*) \right] \tan k_z^{(2)}(h-c) \Big\} \quad (47)$$

$$c^{(2)}_{zmn} = \frac{\varphi_n k_y \tan k_z^{(1)} h}{d_{1mn}\cos k_z^{(2)}(h-c)} \sin k_y Y_0\, J_0\!\left(k_y \frac{W}{2}\right). \quad (48)$$

The above outlines the theory for computing the current distribution on the conducting strips of shielded microstrip discontinuities. The next step is to use the current distribution to derive the network parameters of the discontinuity under consideration. However, since the methods used to derive network parameters are described elsewhere [2], [8], [15], only a brief summary is given in the Appendix.

The theoretical method developed above has been implemented in a Fortran program. The remainder of the paper addresses computational aspects of the solution for the current distribution and discontinuity network parameters.

III. Computation of Current Distribution

To gain insight into the nature of the computations, we will now examine plots of a typical impedance matrix, excitation vector, and current distribution for an open-ended microstrip line.

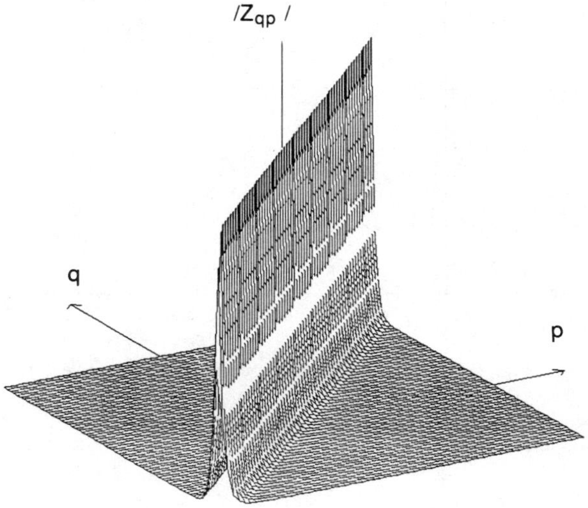

Fig. 4. Impedance matrix for an open end.

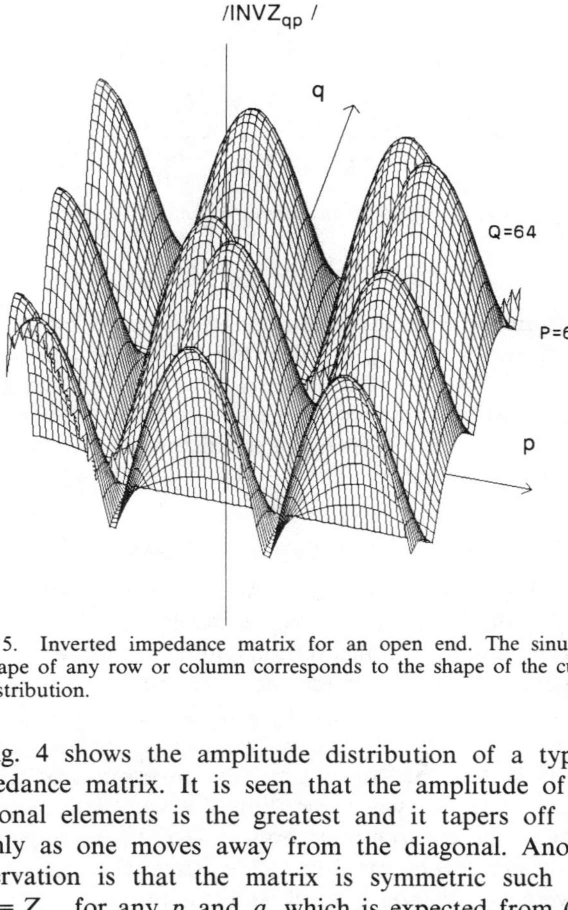

Fig. 5. Inverted impedance matrix for an open end. The sinusoidal shape of any row or column corresponds to the shape of the current distribution.

Fig. 4 shows the amplitude distribution of a typical impedance matrix. It is seen that the amplitude of the diagonal elements is the greatest and it tapers off uniformly as one moves away from the diagonal. Another observation is that the matrix is symmetric such that $Z_{qp} = Z_{pq}$ for any p and q, which is expected from (31). When the impedance matrix of Fig. 4 is inverted, the amplitude distribution is as shown in Fig. 5. The inverted impedance matrix shows a sinusoidal shape for any given row or column.

Fig. 6 shows the amplitude distribution for the excitation vector. The amplitude is highest over the subsection closest to the feed and then tapers off smoothly. In contrast, the excitation vector for the gap generator method has only one nonzero value, at the position of the source.

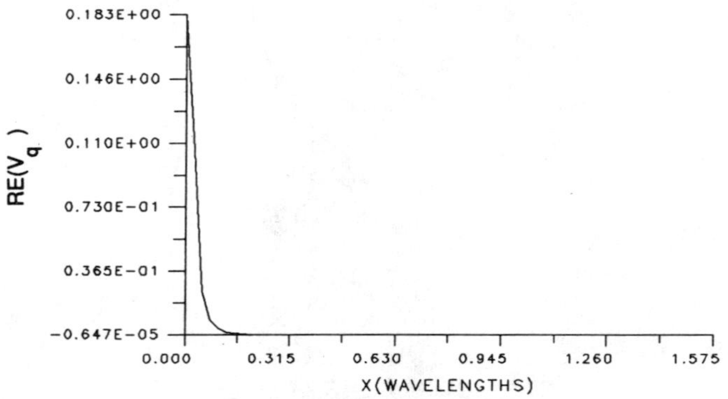

Fig. 6. Amplitude distribution of the excitation vector.

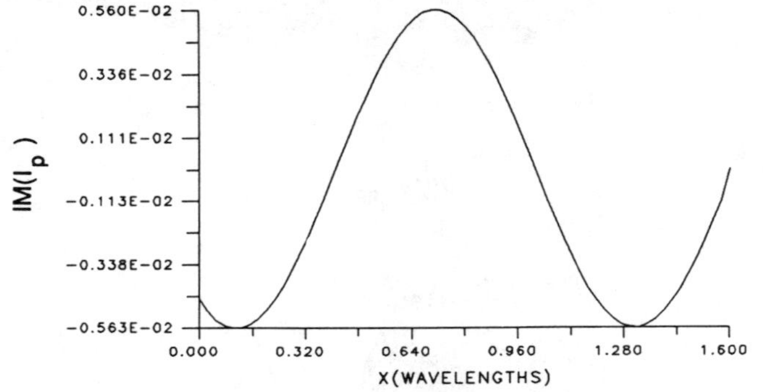

Fig. 7. Imaginary part of the current distribution for an open-ended line.

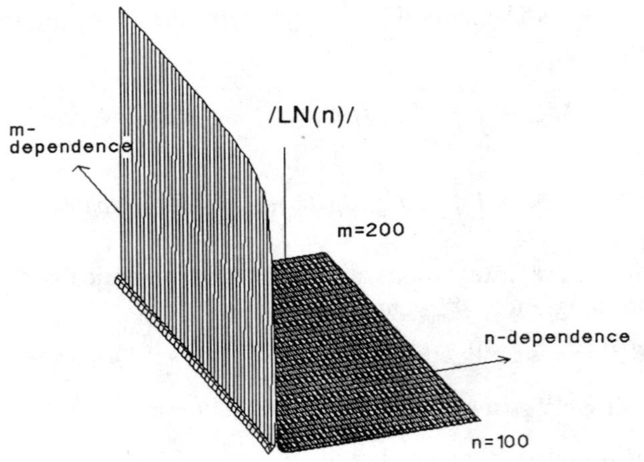

Fig. 8. Three dimensional plot of $LN(n)$ versus summation indices.

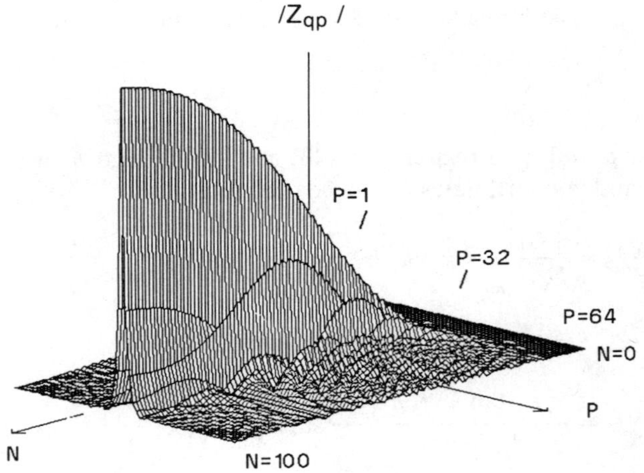

Fig. 9. Convergence of impedance matrix elements. A row ($q = 32$) of the matrix is seen to be well formed after adding 100 terms on n.

Multiplying the inverted impedance matrix by the excitation vector of Fig. 6 yields the current distribution of Fig. 7. It can be seen that the shape of the current is similar to that exhibited by the first column of the inverted impedance matrix. This is not surprising given the shape of the excitation vector.

IV. Convergence of Z_{qp} and V_q

In the expressions of (31) and (40) for the impedance matrix and excitation vector elements, the summations over m and n are theoretically infinite. The number of elements included in these series depends on the convergence behavior of Z_{qp} and V_q with the summation indices.

As seen from (31), the convergence of the impedance matrix is described mainly by the convergence of $LN(n)$ Fig. 8 shows the typical variation of $LN(n)$ with m and n. Most of the contributions from $LN(n)$ to the impedance matrix are concentrated in the first several n values. The convergence over m is good, and it appears that performing the computations out to $m = 200$ may be sufficient. Note, however, that the allowable truncation points for the summations over m and n vary with the geometry. The values quoted here are for illustration purposes only.

The computation of Z_{qp} over n is illustrated for a typical impedance matrix in Fig. 9. Shown is the convergence behavior for one row ($q = 32$) of the 64×64 element impedance matrix of Fig. 4. This behavior is representative of that for any row. After only a few terms the diagonal element ($p = q = 32$) rises above the others, and after adding 100 terms the amplitude distribution is well formed.

Similar conclusions can be drawn for the convergence of the excitation vector elements with respect to the summation indices m and n.

V. Convergence of Network Parameters

The convergence behavior of the elements of the impedance matrix and excitation vector is important to examine; yet the more relevant question remains: how are the final results affected by various convergence-related parameters?

To answer this question, a series of numerical experiments were carried out, and the main results are presented here. As illustrated in Fig. 10, an open-end discontinuity can be represented by either an effective length extension L_{eff} or an equivalent capacitance c_{op}. The microstrip effective dielectric constant ϵ_{eff} is calculated from the distance between two adjacent maxima of the open-end current distribution (Fig. 7).

The experiments investigated the convergence behavior of L_{eff} and ϵ_{eff} with respect to the sampling rate $N_x(= 1/l_x)$ and the truncation points $NSTOP$, $MSTOP$ for the

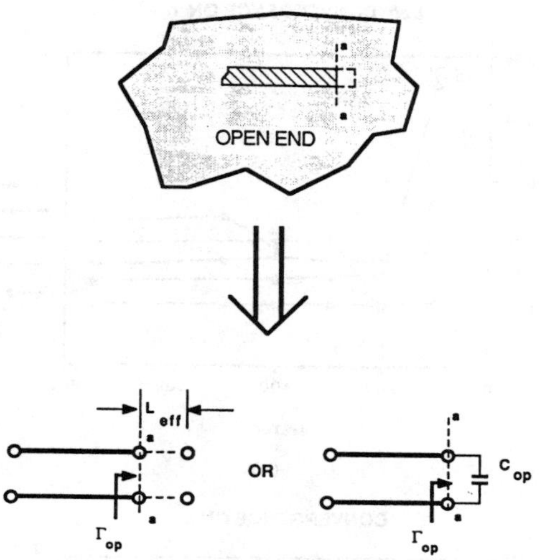

Fig. 10. Representation of a shielded microstrip open end.

summations over n and m, respectively. These numerical experiments have been grouped into three separate categories, each exploring a different aspect of the convergence behavior.[3]

A. Effect of K Value

Using the program mentioned above, data were generated to plot L_{eff} with ϵ_{eff} versus N_x for several different values of the normalized scaling factor K of (5) and (6). Fig. 11(a) shows the convergence behavior of L_{eff} for a typical case. It is seen that a relatively flat convergence region exists for all the K values between about 40 and 100 samples per wavelength. Outside this region the convergence behavior depends on K.

At first glance, it appears that the best convergence is achieved for higher K values (e.g. $K = 8\pi$); however, quite the opposite conclusion results from examining the ϵ_{eff} computation. As can be seen from Fig. 11(b), the best convergence for ϵ_{eff} is obtained for low K values.

Based on these and other observations [8], it was determined that a value of $K = 2\pi$ gives the best overall convergence behavior for the L_{eff} and ϵ_{eff} computations.

B. L_{eff}, ϵ_{eff} Convergence on n and m

To investigate the convergence of the network parameter computations with the summation index n, several program runs were executed for different values of $NSTOP$, with $MSTOP$ fixed at 1000. Data were generated to plot L_{eff} and ϵ_{eff} versus n for several l_x values. Fig. 12(a) shows that for all the l_x values, good convergence on n is achieved after 500 terms. The same can be said for the convergence of ϵ_{eff}.

In examining the convergence behavior with n it was found that, for a given subsection length l_x, cavity length a, and truncation point $NSTOP$, a maximum sampling limit exists beyond which the computed current becomes

[3]The parameters used for the plots shown in this section are as follows: $\epsilon_r = 9.7$, $W = h = 0.025$ in, $a = 3.5$ in, $b = c = 0.25$ in, $f = 18$ GHz.

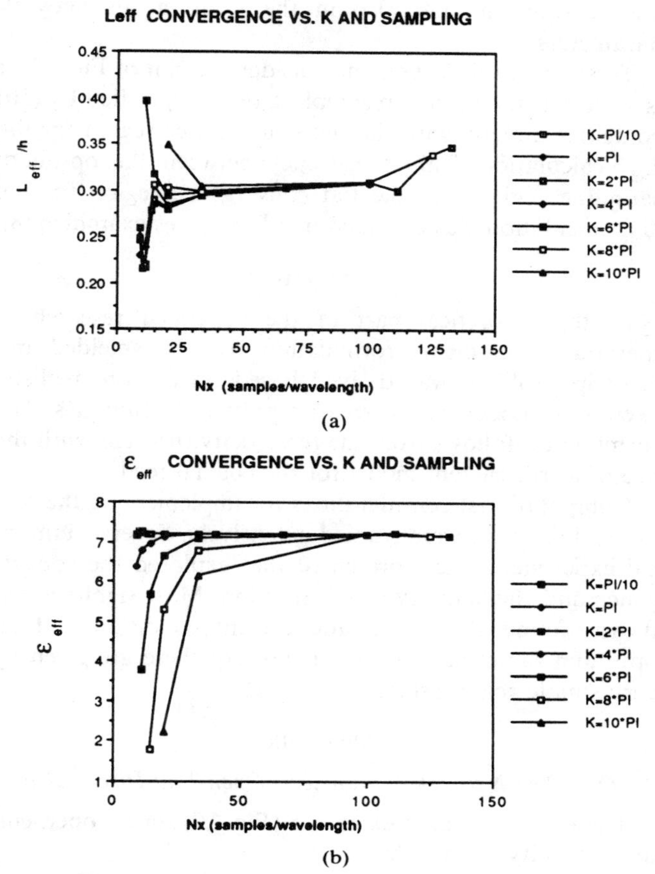

Fig. 11. Convergence of L_{eff} and ϵ_{eff} versus sampling.

completely erratic. This is called the erratic current condition and is given by the following simple relationship:

$$NSTOP * l_x < a \text{ or } N_x > \frac{NSTOP}{a}. \tag{49}$$

Outside of the region defined by (49), the numerical solution appears to be completely stable. To investigate the convergence behavior with respect to the summation index m, $NSTOP$ was fixed at 500, and the program was run for different values of $MSTOP$. Fig. 12(b) shows that L_{eff} converges well on m after about 500 terms. The convergence behavior of ϵ_{eff} on m was found to be similar to that for L_{eff}.

C. Optimum Sampling Range

In this last numerical experiment, the effect of varying l_x on the numerical accuracy of the matrix solution was examined. This was done by studying the variation of the matrix condition number [16] with respect to l_x for a fixed matrix size. After studying several cases it was found that an *optimum sampling range* may be defined by the following choice of subsection length l_x:

$$\frac{1.5a}{NSTOP} \leqslant l_x \leqslant \frac{4a}{NSTOP}. \tag{50}$$

Sampling within this range automatically avoids the erratic current condition and provides the best accuracy in the matrix solution, and also in the solution for network parameters.

To support this last claim, consider the plot of Fig. 13. It is seen that the optimum sampling region specified by (50) coincides directly with the flat convergence region for the L_{eff} calculation. This consistency between the optimum sampling region and the flat convergence region for the L_{eff} calculation was observed in all the cases examined [8].

VI. Summary

In the theoretical part of the presented research, a method of moments formulation for the shielded microstrip problem was derived based on a more realistic excitation model than used with previous techniques. The formulation follows from the reciprocity theorem, with the use of a frill current model for the coaxial feed.

Computational considerations for implementing the theoretical solution were studied extensively. Several numerical experiments were presented that explored the convergence and the stability of the solution. Most significantly, it was found that an erratic current condition and an optimum sampling range exist; both of these are given by very simple relationships.

Appendix

A. One-Port Network Parameters (Open-End Discontinuity)

The effective length extension (Fig. 10) for an open-end discontinuity is given by

$$L_{\text{eff}} = \frac{\lambda_g}{4} - d_{\max} \tag{A1}$$

where $d_{\max}$ is the distance from the end of the line to a current maximum.

The normalized equivalent capacitance (Fig. 10) can be expressed as

$$c_{op} = \frac{\sin 2\beta_g d_{\max}}{\omega(1-\cos 2\beta_g d_{\max})} = \frac{\sin 2\beta_g L_{\text{eff}}}{\omega(1+\cos 2\beta_g L_{\text{eff}})}. \tag{A2}$$

In the above, β_g is the phase constant of microstrip transmission line.

B. Two-Port Network Parameters (Gap Discontinuity, Coupled Line Filters)

For the computation of two-port network parameters, the strip geometry is assumed to be physically symmetric with respect to the center of the cavity (in both the x and y directions of Fig. 1). The network parameters are determined by analyzing the current from the even- and odd-mode excitations as discussed in [2], [8], [15].

The normalized impedance parameters are given by

$$z_{11} = \frac{z_{\text{IN}}^e + z_{\text{IN}}^o}{2} \tag{A3}$$

$$z_{12} = \frac{z_{\text{IN}}^o - z_{\text{IN}}^e}{2} \tag{A4}$$

where z_{IN}^e and z_{IN}^o are the input impedances of the even- and odd-mode networks. The scattering parameters for the network may be derived using the following relations:

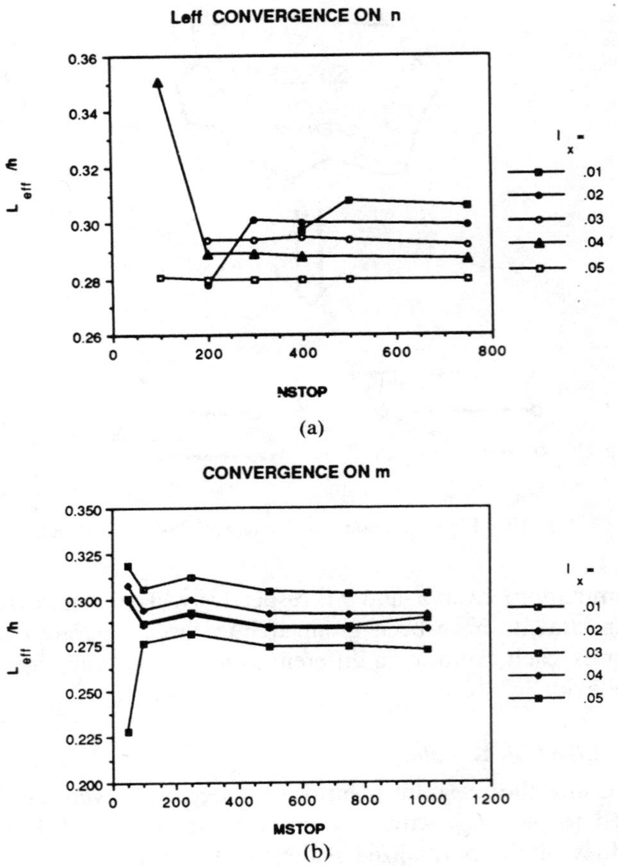

Fig. 12. Convergence of L_{eff} on n and m.

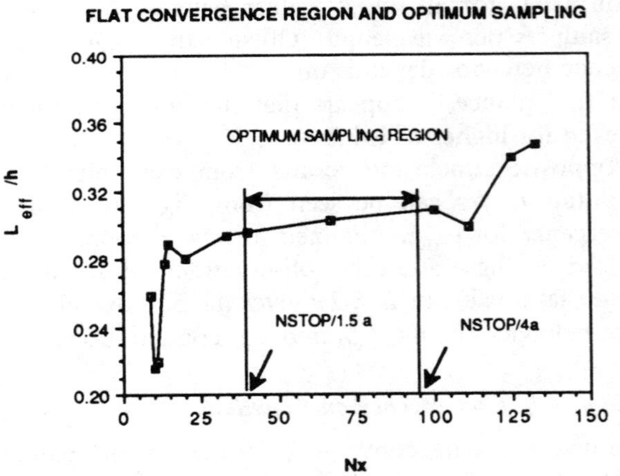

Fig. 13. Illustration of optimum sampling range which is seen to correspond directly with the flat convergence region for the L_{eff} computation.

$$S_{11} = S_{22} = \frac{z_{11}^2 - 1 - z_{12}^2}{D} \tag{A5}$$

$$S_{12} = S_{21} = \frac{2z_{12}}{D} \tag{A6}$$

where

$$D = z_{11}^2 + 2z_{11} - z_{12}^2. \tag{A7}$$

Acknowledgment

The authors thank E. Watkins, J. Schellenberg, and M. Tutt for their contributions to this work.

References

[1] R. H. Jansen and N. H. L. Koster, "Accurate results on the end effect of single and coupled lines for use in microwave circuit design," *Arch. Elek. Ubertragung.*, vol. 34, pp. 453–459, 1980.

[2] P. B. Katehi and N. G. Alexopoulos, "Frequency-dependent characteristics of microstrip discontinuities in millimeter-wave integrated circuits," *IEEE Trans. Microwave Theory Tech.*, vol. MTT-33, pp. 1029–1035, Oct. 1985.

[3] R. H. Jansen, and W. Wertgen, "Modular source-type 3D analysis of scattering parameters for general discontinuities, components and coupling effects in (M)MICs," in *Proc. 17th European Microwave Conf.* (Rome), 1987, pp. 427–432.

[4] J. C. Rautio, "An electromagnetic time-harmonic analysis of shielded microstrip circuits," *IEEE Trans. Microwave Theory Tech.*, vol. MTT-35, pp. 726–729, 1987.

[5] R. H. Jansen, "Hybrid mode analysis of end effects of planar microwave and millimeter-wave transmission lines," *Proc. Inst. Elec. Eng.*, vol. 128, pp. 77–86, Apr. 1981.

[6] T. Itoh, "Analysis of microstrip resonators," *IEEE Trans. Microwave Theory Tech.*, vol. MTT-22, pp. 946–951, 1974.

[7] L. P. Dunleavy and P. B. Katehi, "Shielding effects in microstrip discontinuities," pp. 1767–1774, this issue.

[8] L. P. Dunleavy, "Discontinuity characterization in shielded microstrip: A theoretical and experimental study," Ph.D. dissertation, University of Michigan, Apr. 1988.

[9] R. F. Harrington, *Time-Harmonic Electromagnetic Fields.* New York: McGraw-Hill, 1961, pp. 111–112.

[10] C. Chi and N. G. Alexopoulos, "Radiation by a probe through a substrate," *IEEE Trans. Antennas Propagat.*, vol. AP-34, pp. 1080–1091, Sept. 1986.

[11] N. N. Wang, J. H. Richmond, and M. C. Gilreath "Sinusoidal reaction formulation for radiation and scattering from conducting surfaces," *IEEE Trans. Antennas Propagat.*, vol. AP-23, pp. 376–382, May 1975.

[12] E. H. Newman and D. H. Pozar, "Electromagnetic modeling of composite wire and surface geometries," *IEEE Trans. Antennas Propagat.*, vol. AP-26, pp. 784–789, Nov. 1978.

[13] J. C. Maxwell, *A Treatise on Electricity and Magnetism*, 3rd. ed., vol. 1. New York: Dover, 1954, pp. 296–297.

[14] C. T. Tai, *Dyadic Green's Functions in Electromagnetic Theory.* Scranton, PA: Intext Educational Publishers, 1971.

[15] P. B. Katehi, "Radiation losses in mm-wave open microstrip filters," *Electromagnetics*, vol. 7, pp. 137–152, 1987.

[16] G. H. Golub and C. F. Van Loan, *Matrix Computations.* Baltimore, MD: John Hopkins University Press, 1983, pp. 26–27.

An Efficient Algorithm for the Three-Dimensional Analysis of Passive Microstrip Components and Discontinuities for Microwave and Millimeter-Wave Integrated Circuits

Achim Hill and Vijai K. Tripathi, *Senior Member, IEEE*

Abstract **—A numerical technique for the full-wave analysis of shielded, passive microstrip components on a two-layer substrate is presented. The distinct feature of the technique is a novel, efficient formulation for establishing the system matrix in the moment method procedure which allows the derivation of the elements of any large matrix by a linear combination of elements in a precomputed index table. The table is obtained from a two-dimensional discrete fast Fourier transform. In the moment method procedure, the two-dimensional surface current is represented by locally defined rooftop functions. The effect of the resonant modes associated with the metallic enclosure on the numerical procedure is examined. In order to demonstrate the features and the accuracy of the technique, numerical results for microstrip open end and for a right-angle bend with and without the compensated corner are computed by using the resonant technique and are compared with other published computational and experimental data.**

I. Introduction

A considerable amount of work has been done in recent years on the frequency-dependent characterization and modeling of microstrip components and discontinuities, resulting in several useful numerical techniques [1]. These include solutions based on waveguide and two-dimensional cavity models, use of the method of lines, finite difference and finite element techniques, and the solution based on an integral equation formulation in real space and the Fourier transform domain [1]–[31]. All the accurate methods are, in general, computationally intensive; devising techniques to improve the efficiency of various methods continues to be a challenging task.

The integral equation formulation in real space and the spectral domain has become a promising technique for the analysis and simulation of components and discontinuities in microwave and millimeter-wave circuits. The simulation of passive (M)MIC structures can be classified into three categories, namely open, shielded, and partly shielded configurations. Each shielding type requires a somewhat different numerical treatment. Efficient computational methods have been derived for the open [21] and partly shielded structures [23]. However, the shielded configuration still requires intensive computational treatment. The insufficient development of full-wave simulators which accurately incorporate housing effects has also been pointed out recently by Jansen and Wiemer [25].

In this paper a technique which allows for an efficient numerical treatment of the shielded circuit structure is presented. It leads to reasonable computation times for the analysis of 3-D microstrip structures for the case when 2-D locally defined surface currents are employed in the moment method. The analysis represents an extension of the previous work reported by Jansen [7], Koster [8], Rautio and Harrington [12], [13] and Dunleavy and Katehi [27], [28].

Manuscript received October 23, 1989; revised July 24, 1990.

A. Hill was with the Department of Electrical and Computer Engineering, Oregon State University, Corvallis, OR 97331. He is now with Compact Software Inc., Paterson, NJ 07504.

V. K. Tripathi is with the Department of Electrical and Computer Engineering, Oregon State University, Corvallis, OR 97331.

IEEE Log Number 9040562.

II. Matrix Equation Formulation

Consider the boxed structure shown in Fig. 1. The source-free medium consists of three homogeneous, isotropic dielectric layers and is bounded by a box of perfectly conducting metal. Each layer r ($r = 1,2,3$) of thickness H_r is characterized by its relative dielectric constant ϵ_r. The box extends from $x = 0$ to $x = a$, $y = 0$ to $y = b$ with bottom and cover plates at $z = 0$ and $z = c$. The microstrip metallization of zero thickness and infinite conductivity is located at the interface.

The tangential electric field components on the interface (E_x, E_y) are expressed in terms of surface currents (J_x, J_y):

$$
\begin{aligned}
E_x &= \sum_m \sum_n F_{mn} XY_{mn} \int J_x(x', y') \cos k_{xm}x' \sin k_{yn}y' \, dx' \, dy' \\
&\quad \cdot \cos k_{xm}x \sin k_{yn}y \\
&\quad + \sum_m \sum_n F_{mn} RZ_{mn} \int J_y(x', y') \sin k_{xm}x' \cos k_{yn}y' \, dx' \, dy' \\
&\quad \cdot \cos k_{xm}x \sin k_{yn}y \\
E_y &= \sum_m \sum_n F_{mn} RZ_{mn} \int J_x(x', y') \cos k_{xm}x' \sin k_{yn}y' \, dx' \, dy' \\
&\quad \cdot \sin k_{xm}x \cos k_{yn}y \\
&\quad + \sum_m \sum_n F_{mn} YX_{mn} \int J_y(x', y') \sin k_{xm}x' \cos k_{yn}y' \, dx' \, dy' \\
&\quad \cdot \sin k_{xm}x \cos k_{yn}y. \qquad (1)
\end{aligned}
$$

Reprinted from *IEEE Trans. Microwave Theory Tech.*, vol. 39, no. 1, pp. 83–91, Jan. 1991.

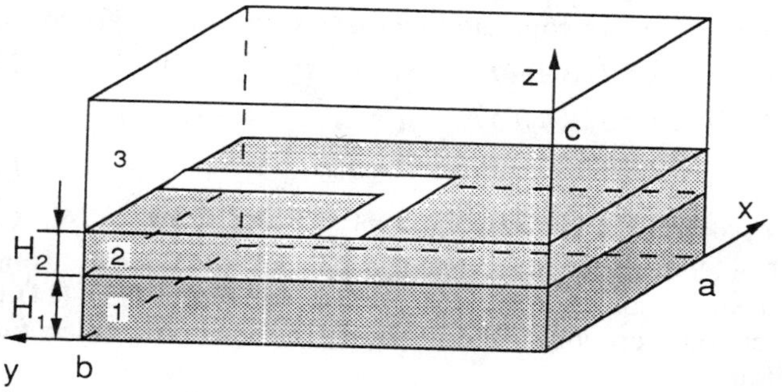

Fig. 1. Microstrip discontinuity in shielded box.

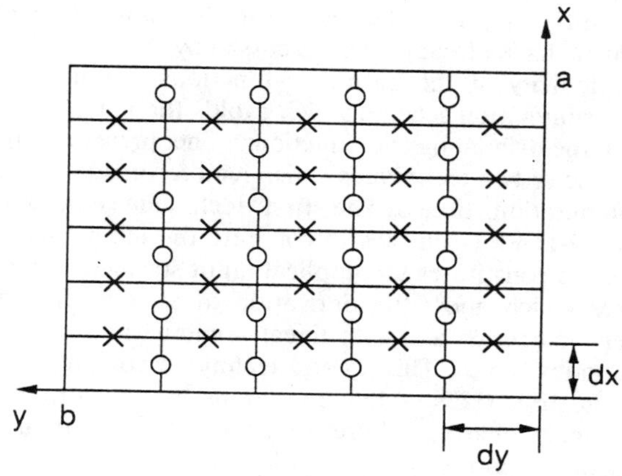

Fig. 2. Discretization of conductor surface. Crosses represent center of x-directed current; circles represent centers of y-directed currents.

The expressions $F_{mn}XY_{mn}$, $F_{mn}R_{mn}$, and $F_{mn}YX_{mn}$ represent the components of the spectral-domain Green's dyadic. These have been derived by Jansen and Koster [7], [8] and they are summarized in the Appendix for convenience. The discrete spectral variables in the above equation are given by $k_{xm} = m\pi/a$ and $k_{yn} = n\pi/b$.

Equation (1), a Fredholm integral equation of the first kind, maps the surface current into an electric field on the interface. Following the moment method procedure, the current is expanded into basis functions and substituted in (1). In a subsequent step (1) is tested with suitable functions which in the present case were chosen to be identical to the basis functions leading to a Galerkin implementation of the method. The present method employs rooftop functions as the basis elements [13], [15]–[17]. Fig. 2 shows the discretization of the metallized surface with discretization Δx and Δy in the x and y directions, respectively. The center of x-directed currents is marked with a cross, and the center of y-directed currents is marked with a circle. Note that the x- and y-directed currents are offset by $(\Delta x/2, \Delta y/2)$ in order to ensure edge conditions and generate correct results. A detailed discussion for the current offset can be found in [12]. The total current is approximated by

$$J = \sum_k a_{xk} J_{xx}(x, x_k) J_{xy}(y, y_k) + \sum_i a_{yi} J_{yx}(x, x_i), J_{yy}(y, y_i) \tag{2}$$

where

$$J_{uu}(u, u_k) = \begin{cases} \dfrac{u - u_k}{\Delta u} + 1, & u_k - \Delta u < u < u_k \\ \dfrac{u_k - u}{\Delta u} + 1, & u_k < u < u_k + \Delta u \\ 0, & \text{otherwise} \end{cases} \tag{3}$$

with $u = x$ or y and

$$J_{uv}(v, v_k) = \begin{cases} 1, & v_k - \Delta v/2 < v < v_k + \Delta v/2 \\ 0, & \text{otherwise} \end{cases} \tag{4}$$

with $uv = xy$ or yx.

After applying the testing procedure, the linear system of equations can be summarized as follows:

$$\begin{bmatrix} V_x \\ V_y \end{bmatrix} = \begin{bmatrix} P_{xx} & P_{xy} \\ P_{yx} & P_{yy} \end{bmatrix} \begin{bmatrix} A_x \\ A_y \end{bmatrix}. \tag{5}$$

The left-hand side represents the scalar product of the electric field and a testing function at the ith subsection. These vanish on the metal except for subsections where sources are defined. The P matrix contains the testing products and the Green's dyadic of the associated boundary value problem. Vector A is formed by the expansion coefficients for current as given in (2), i.e., $A_{xi} = a_{xi}$ and $A_{yi} = a_{yi}$. In the above equation,

$$V_{xi} = \iint E_x J_{xi}\, dx\, dy$$

$$V_{yi} = \iint E_y J_{yi}\, dx\, dy \tag{6}$$

$$P_{xx}^{ij} = \sum_m \sum_n G_{mn}^{xx} \cos k_{xm} x_{jx} \sin k_{yn} y_{jx} \cos k_{xm} x_{ix} \sin k_{yn} y_{ix}$$

$$P_{xy}^{ij} = \sum_m \sum_n G_{mn}^{xy} \sin k_{xm} x_{jy} \cos k_{yn} y_{jy} \cos k_{xm} x_{ix} \sin k_{yn} y_{ix}$$

$$P_{yx}^{ij} = \sum_m \sum_n G_{mn}^{yx} \cos k_{xm} x_{jx} \sin k_{yn} y_{jx} \sin k_{xm} x_{iy} \cos k_{yn} y_{iy}$$

$$P_{yy}^{ij} = \sum_m \sum_n G_{mn}^{yy} \sin k_{xm} x_{jy} \cos k_{yn} y_{jy} \sin k_{xm} x_{iy} \cos k_{yn} y_{iy} \tag{7}$$

and

$$\begin{aligned} G_{mn}^{xx} &= F_{mn} XY_{mn} G_t^2(\Delta x) G_r^2(\Delta y) \\ G_{mn}^{xy} &= F_{mn} R_{mn} G_t(\Delta y) G_r(\Delta x) G_t(\Delta x) G_r(\Delta y) \\ G_{mn}^{yx} &= F_{mn} R_{mn} G_t(\Delta x) G_r(\Delta y) G_t(\Delta y) G_r(\Delta x) \\ G_{mn}^{yy} &= F_{mn} YX_{mn} G_t^2(\Delta y) G_r^2(\Delta x) \end{aligned} \tag{8}$$

where

$$G_t(\Delta u) = \begin{cases} \dfrac{2}{\Delta u k_u^2}(1 - \cos(k_u \Delta u)), & k_u \neq 0 \\ \Delta u, & k_u = 0 \end{cases} \tag{9}$$

$$G_r(\Delta u) = \begin{cases} \dfrac{2}{k_u}(\sin(k_u \Delta u/2), & k_u \neq 0 \\ 0, & k_u = 0. \end{cases} \tag{10}$$

III. Enhanced Algorithm

The evaluation of the elements P^{ij} of the moment matrix is time consuming owing to the two-dimensional summation and the repeated computation of the Green's dyadic with associated harmonic functions, and the time requirement for the solution of the linear system (9) is negligible compared with the formation of the matrix.

In addition to the possible use of fast Fourier transform (FFT), efficiency considerations for the summation procedure in (7) have already been discussed by Rautio [12], where the periodicity of the harmonic functions was used to develop a summation scheme which avoids the periodic evaluation of the trigonometric functions. The present approach makes use of two techniques which lead to drastic reductions of computation times. The first technique employs customized 2-D FFT routines to compute the index tables, and the second constitutes the application of specialized indexing routines which allow the derivation of all elements of the moment matrix by a simple linear combination of elements of the index tables. This second technique completely eliminates the IEX$\cdot$(IEX$-1)/2$ operations for the formation of the moment matrix, where IEX is the total number of expansion functions.

In order to employ FFT subroutines the representation of the moment matrix has to be transformed into a suitable form. In addition, the interface is uniformly discretized in x and y directions such that

$$\begin{aligned}
x_{ix} &= p_{ix}\,\Delta x, & p_{ix} &= 0,1,\cdots,M\\
y_{iy} &= s_{iy}\,\Delta y, & s_{iy} &= 0,1,\cdots,N\\
x_{iy} &= \left(p_{iy}+\tfrac{1}{2}\right)\Delta x, & p_{iy} &= 0,1,\cdots,M-1\\
y_{ix} &= \left(s_{ix}+\tfrac{1}{2}\right)\Delta y, & s_{ix} &= 0,1,\cdots,N-1.
\end{aligned} \tag{11}$$

Since the respective operator of the Fredholm integral equation (1) is self-adjoint, the moment matrix is symmetric and P_{yx}, which is equal to P_{xy}, is no longer considered in the numerical treatment. After using trigonometric identities and substituting (11) into (7), the moment matrix P can be rewritten as

$$\begin{aligned}
P_{xx}^{ij} = {} & f_{xx}(p_{jx}-p_{ix}, s_{jx}-s_{ix}) - f_{xx}(p_{jx}-p_{ix}, s_{jx}+s_{ix})\\
& + f_{xx}(p_{jx}+p_{ix}, s_{jx}-s_{ix}) - f_{xx}(p_{jx}+p_{ix}, s_{jx}+s_{ix})
\end{aligned} \tag{12}$$

$$\begin{aligned}
P_{xy}^{ij} = {} & f_{xy}(p_{jy}+p_{ix}, s_{jy}+s_{ix}) - f_{xy}(p_{jy}+p_{ix}, s_{jy}+s_{ix})\\
& + f_{xy}(p_{jy}-p_{ix}, s_{jy}+s_{ix}) - f_{xy}(p_{jy}-p_{ix}, s_{jy}-s_{ix})
\end{aligned} \tag{13}$$

$$\begin{aligned}
P_{yy}^{ij} = {} & f_{yy}(p_{jy}-p_{iy}, s_{jy}-s_{iy}) + f_{yy}(p_{jy}-p_{iy}, s_{jy}+s_{iy})\\
& - f_{yy}(p_{jy}+p_{iy}, s_{jy}-s_{iy}) - f_{yy}(p_{jy}+p_{iy}, s_{jy}+s_{iy})
\end{aligned} \tag{14}$$

with

$$\begin{aligned}
f_{xx}(u,v) &= \sum_m \sum_n G_{mn}^{xx} \cos\frac{m\pi u}{M} \cos\frac{n\pi v}{N}\\
f_{xy}(u,v) &= \sum_m \sum_n G_{mn}^{xy} \sin\frac{m\pi(u+1/2)}{M} \sin\frac{n\pi(v+1/2)}{N}\\
f_{yy}(u,v) &= \sum_m \sum_n G_{mn}^{yy} \cos\frac{m\pi u}{M} \cos\frac{n\pi v}{N}.
\end{aligned} \tag{15}$$

The expressions for P^{ij}, (12)–(14), are now in a form which is suited for the application of 2-D discrete FFT's. If we were to compute each element of the moment matrix by using the FFT algorithm, we would require IT = IEX$*$(IEX$-1)/2$ evaluations for a matrix with dimensions IEX, e.g. IT = 4485 if IEX = 300. The technique presented here allows us to reduce the IT evaluations to IT = three evaluations for any matrix size. The elements P^{ij} are obtained from a linear combination of components of the FFT's with respect to the transform variable, as will be outlined in the following. Instead of computing (12)–(14) for each $P_{xx}^{ij}, P_{xy}^{ij}, P_{yy}^{ij}$, only three FFTs are computed and stored in suitable arrays:

$$P_{xx}(u,v) = f_{xx}(u,v), \qquad u=0,\cdots,M;\ v=0,\cdots,N \tag{16}$$

$$P_{xy}(u,v) = f_{xy}(u,v), \qquad u=0,\cdots,M-1;\ v=0,\cdots,N-1 \tag{17}$$

$$P_{yy}(u,v) = f_{yy}(u,v), \qquad u=0,\cdots,M;\ v=0,\cdots,N. \tag{18}$$

The elements of the moment matrix are then derived from a linear combination of (16)–(18), as implied in (12)–(14) under consideration of the periodicy of the trigonometric functions.

The double infinite summations have to be truncated at a suitable bound such that a certain convergence criterion is fulfilled. This upper bound can exceed the period of the respective FFT's. In other words, the sampling ratio, defined as the number of spatial frequency samples per discretization length, is larger than unity. To circumvent this problem, the functions G_{mn} in (20) are presampled at periodic intervals to form a summation which is then submitted to the FFT routines. The idea of such a first-stage summation has been used before, by Rautio [12]; however, the implementation was different. As a consequence of the first-stage summation scheme, each of the three FFT's in (21)–(23) is split into four FFT's, which yields a total number of 12 FFT's for the evaluation of a moment matrix. To incorporate an arbitrary number of spatial frequencies in the FFT algorithm, the three basic FFT formulations given in (16)–(18) are decomposed as follows.

$$\begin{aligned}
\sum_m \sum_n G_{mn}^{xx} \cos\frac{m\pi p}{M} \cos\frac{n\pi s}{N}
= {} & \sum_m \sum_n G_{mn}^{xxee} \cos\frac{m\pi p}{M} \cos\frac{n\pi s}{N}\\
& + (-1)^s \sum_m \sum_n G_{mn}^{xxeo} \cos\frac{m\pi p}{M} \cos\frac{n\pi s}{N}\\
& + (-1)^p \sum_m \sum_n G_{mn}^{xxoe} \cos\frac{m\pi p}{M} \cos\frac{n\pi s}{N}\\
& + (-1)^{(s+p)} \sum_m \sum_n G_{mn}^{xxoo} \cos\frac{m\pi p}{M} \cos\frac{n\pi s}{N}
\end{aligned} \tag{19}$$

$$\begin{aligned}
\sum_m \sum_n G_{mn}^{xy} \sin\frac{m\pi(p+1/2)}{M} \sin\frac{n\pi(s+1/2)}{N}
= {} & \sum_m \sum_n G_{mn}^{xyee} \sin\frac{m\pi(p+1/2)}{M} \sin\frac{n\pi(s+1/2)}{N}\\
& + (-1)^s \sum_m \sum_n G_{mn}^{xyeo} \sin\frac{m\pi(p+1/2)}{M} \cdot \cos\frac{n\pi(s+1/2)}{N}\\
& + (-1)^p \sum_m \sum_n G_{mn}^{xyoe} \cos\frac{m\pi(p+1/2)}{M} \cdot \sin\frac{n\pi(s+1/2)}{N}\\
& + (-1)^{(s+p)} \sum_m \sum_n G_{mn}^{xyoo} \cos\frac{m\pi(p+1/2)}{M} \cdot \cos\frac{n\pi(s+1/2)}{N}.
\end{aligned} \tag{20}$$

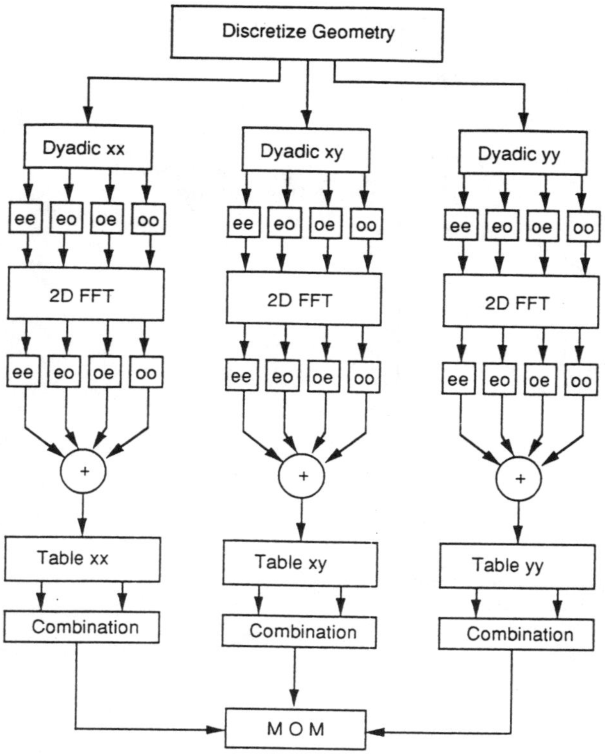

Fig. 3. Enhanced algorithm for establishing the moment matrix (MOM).

The quantity

$$\sum_{m}\sum_{n} G_{mn}^{yy}\cos\frac{m\pi p}{M}\cos\frac{n\pi s}{N}$$

is identical to (24) with G_{mn}^{xx} replaced by G_{mn}^{yy}. The abbreviations in (24) and (25) are defined as follows:

$$G_{mn}^{xxpq}=\sum_{i_p}\sum_{j_q}(m+i_pM,n+j_qN)$$

$$G_{mn}^{yypq}=\sum_{i_p}\sum_{j_q}(m+i_pM,n+j_qN)$$

$$G_{mn}^{xypq}=\sum_{i_p}\sum_{j_q}G^{xy}(m+i_pM,n+j_qN)(-1)^{(i_p/2+j_q/2)} \quad (21)$$

with $p=e$ or O and $q=e$ or O and $i_e,j_e=0,2,4,\cdots$; $i_o,j_o=1,3,5,\cdots$. The formation of the moment matrix is summarized in Fig. 3. After discretizing the geometry, each Green's dyadic together with the associated spectral representation of the basis functions is presampled as suggested in (26). Next, 2-D FFT's are applied as outlined in (16) through (18). Results of the FFT's are then stored to form index tables. Now the moment matrix of any geometry on the discretized interface can be built by a suitable linear combination of the elements of the three index tables as implied in (12) through (14).

IV. Numerical Results

The algorithm described above is applicable to both deterministic and eigenvalue problems [8], [12], [27]. The deterministic procedure as discussed in [12] and [27] allows for the derivation of port impedances for given excitations at the terminal ports. These impedance values are then used for the determination of the network matrix. The eigenvalue formulation requires the evaluation of the resonance frequency of the microstrip structure in the shielding box, from which discontinuity parameters are extracted as shown in [7], [8], and [32]. We have used the deterministic approach for the first example only and the eigenvalue computation approach for all the subsequent examples.

As an initial confidence test for the presented technique, the input impedance of a microstrip stub was computed and compared with results obtained by Rautio and Harrington [13]. The box was 4 cm long, 2 cm wide, and 5 cm high. The strip line was 1 cm wide and 2.81 cm long and was deposited on a 1-cm-thick substrate with a relative dielectric constant of 10. The results are shown in Fig. 4 and the agreement is obvious.

The open end discontinuity of a microstrip line can be characterized by an effective length extension, l_{eff}, or an equivalent terminating capacitance which accounts for the fringing fields at the open end. Dynamic simulation of the open end effect in a shielded environment have recently been developed by Jansen and Koster [30] and Dunleavy and Katehi [26], and these are included for comparison in Fig. 5. The shielding geometry and strip dimensions were chosen to be identical to those used in [26]. The box was 7.747 mm wide and 5.08 mm high. The frequency behavior of the effective length computed in this work behaves in the same manner as described by Koster [8]. For low frequencies the length decreases then passes through a minimum and increases for higher frequencies. The box used in the presented analysis exhibits a resonance around 24 GHz for the specified dimensions, which explains the deviation of the effective length at this point from the results obtained in [26] and [29].

The S parameters for typical right-angle-bend geometries having various W/H ratios are shown in Fig. 6 for an alumina substrate of height 0.635 mm. All bends are analyzed in a square box of 12.7 mm side length and 3.81 mm height. For the nominal 50 Ω line ($W/H\simeq 1$) S-parameter simulations are compared with design formulas from Kirschning *et al.* [30], which were derived from measurements in the frequency range between 2 and 14 GHz for an open structure. The deviation in magnitude of S_{11} is within 3% for the frequency range up to 14 GHz. However, larger differences are observed for the phase beyond 4 GHz. These differences are attributed to the effect of the shielding box used in these computations.

Next, a bend is simulated in the frequency range between 50 and 120 GHz to demonstrate the effect of parasitic box resonances. The bend of 100 μm width is deposited on a 100 μm GaAs substrate with a 5 μm oxide layer. The box dimensions were $a=1.9$ mm and $c=1.05$ mm. Fig. 7 shows the associated S parameters. To show the effect of box resonances, the height of the box is reduced to $c=200$ μm. The oxide layer is removed in this example. Investigations of the first LSM_{10} parametric waveguide mode predict a resonance in the vicinity of 80 GHz, which is verified by the S-parameter computation shown in Fig. 8.

As a last example, the S parameters of a right-angle bend are compared for a compensated corner. The compensation is obtained by cutting out a square at the corner of the bend. The form of the compensation is shown in the inset of

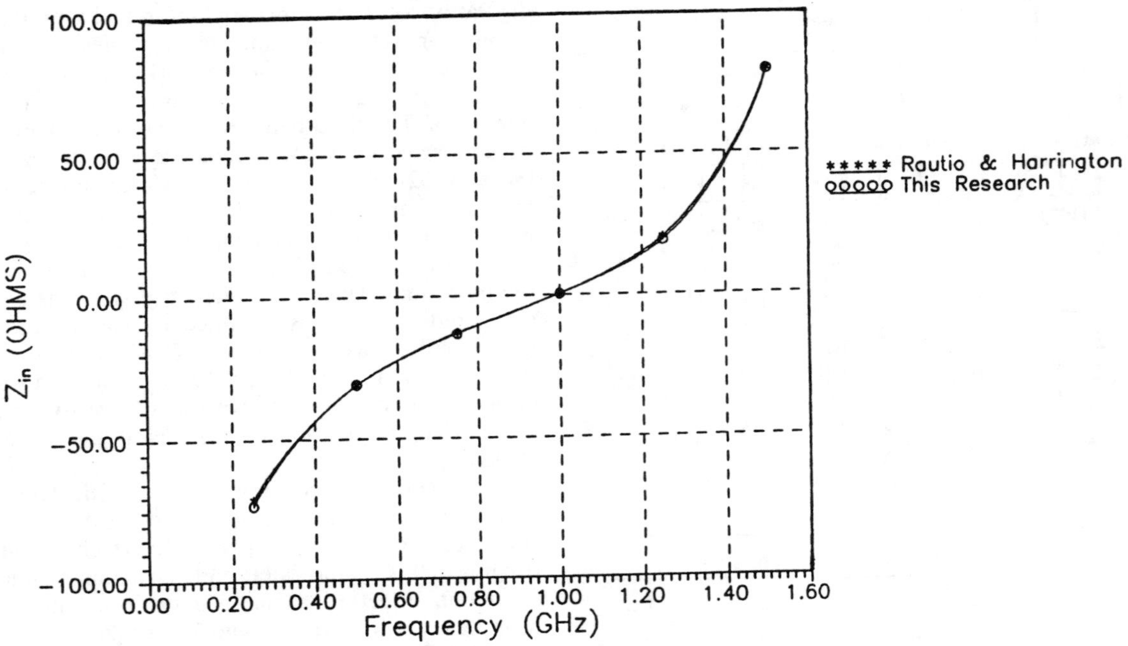

Fig. 4. Stub input impedance ($a = 2$ cm, $b = 4$ cm, $c = 5$ cm, $H_1 = 1$ cm, $\epsilon_1 = 10$).

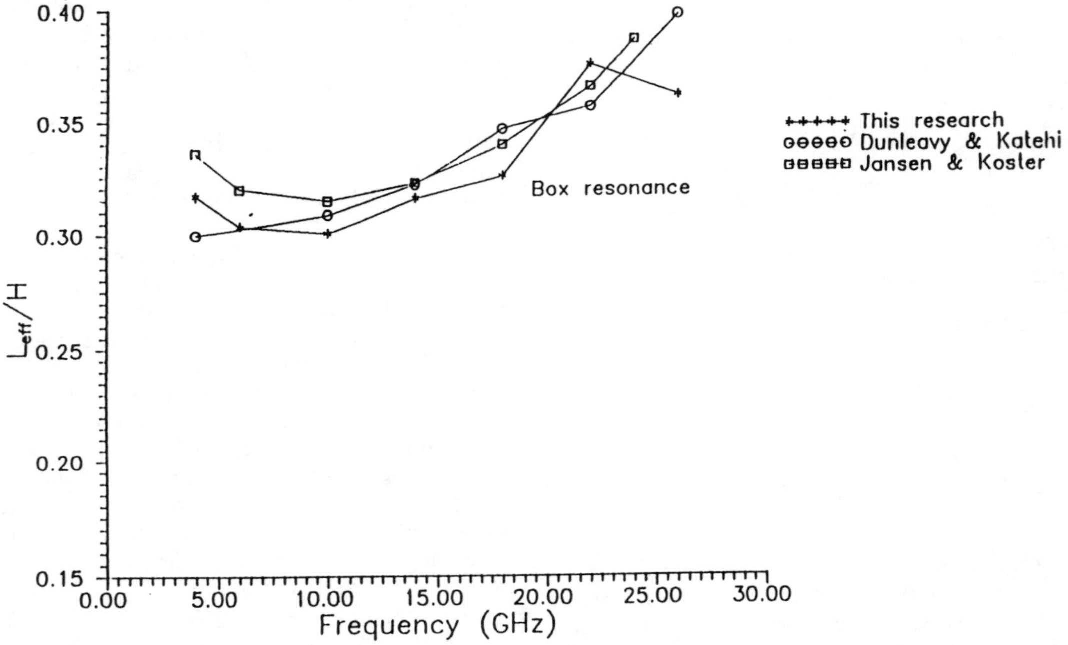

Fig. 5. Effective length of open microstrip end ($a = 7.747$ mm, $c = 5.08$ mm, $H_1 = 0.635$ mm, $\epsilon_1 = 9.6$, $W/H_1 = 1.57$).

Fig. 9(a). The configuration of the first example for right-angle bends is used with a W/H ratio of 2. The S parameters are compared for a compensation ratio of $s/W = 0.83$. Fig. 9(a) shows the magnitude of S_{11}. For the compensated case a considerable reduction of S_{11} is achievable. The ratio of S_{11} for the compensated case and S_{11} for the uncompensated case is about 0.125 at the low frequency end at 6 GHz and 0.27 at the high frequency end at 20 GHz. Comparing the deviation in the phase of S_{11} (Fig. 9(b)) yields a change of capacitive to inductive loading, which is due to the increased current crowding effect in the corner region of the compensated edge.

V. Conclusion

An enhanced algorithm for the full-wave analysis of microstrip discontinuities on a double-layered substrate has been presented. The procedure is based on the use of index tables that are computed from 2-D discrete FFT routines. Elements of the associated moment matrix are then derived

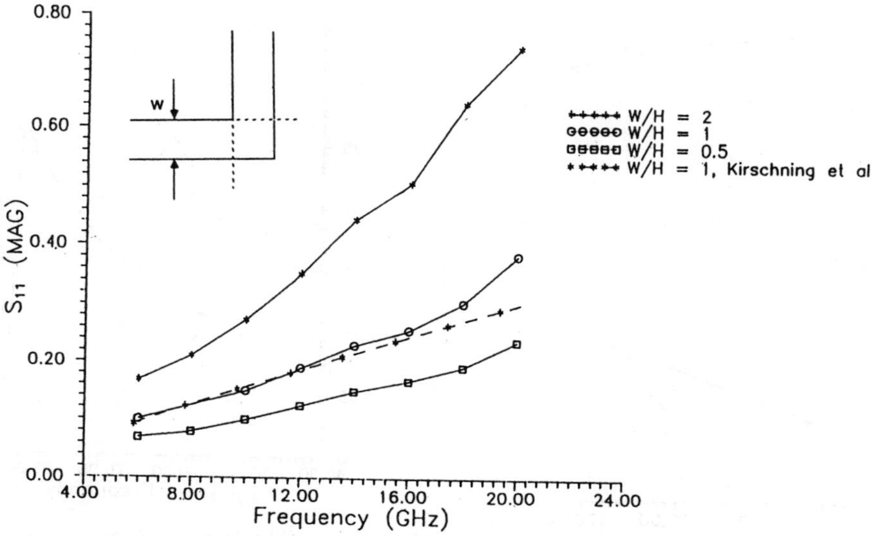

(a)

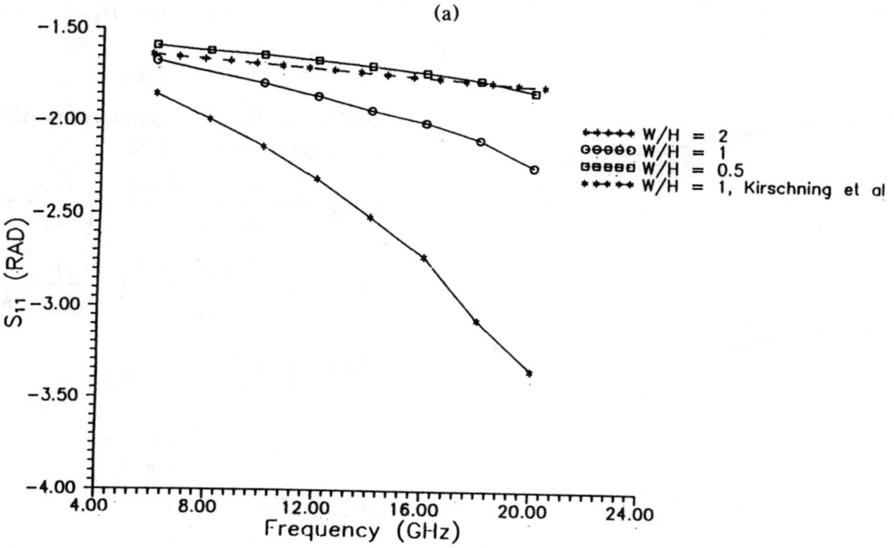

(b)

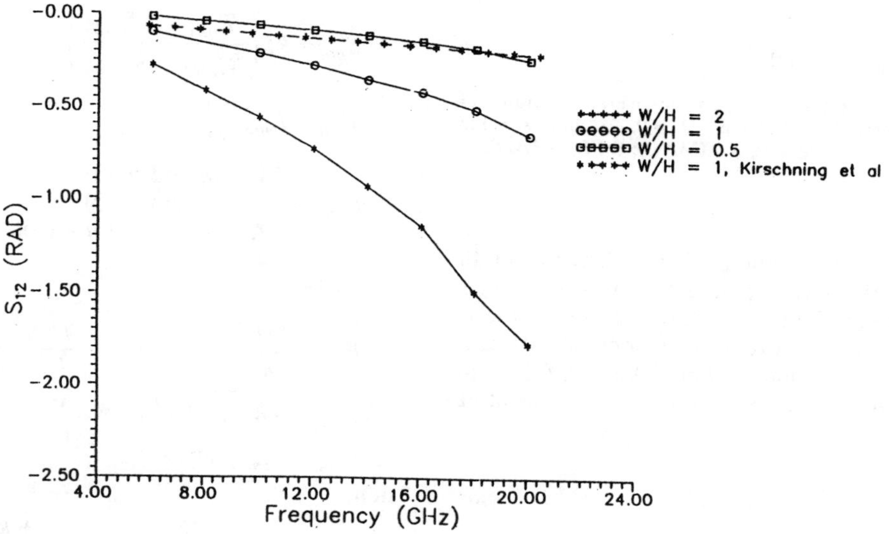

(c)

Fig. 6. S parameter of right-angle bend for various W/H_1 ratios ($a = b = 12.7$ mm, $c = 3.18$ mm, $H_1 = 0.635$ mm, $\epsilon_1 = 9.8$): (a) S_{11} (MAG); (b) S_{11} (RAD); (c) S_{12} (RAD). Dotted lines in the inset represent reference planes.

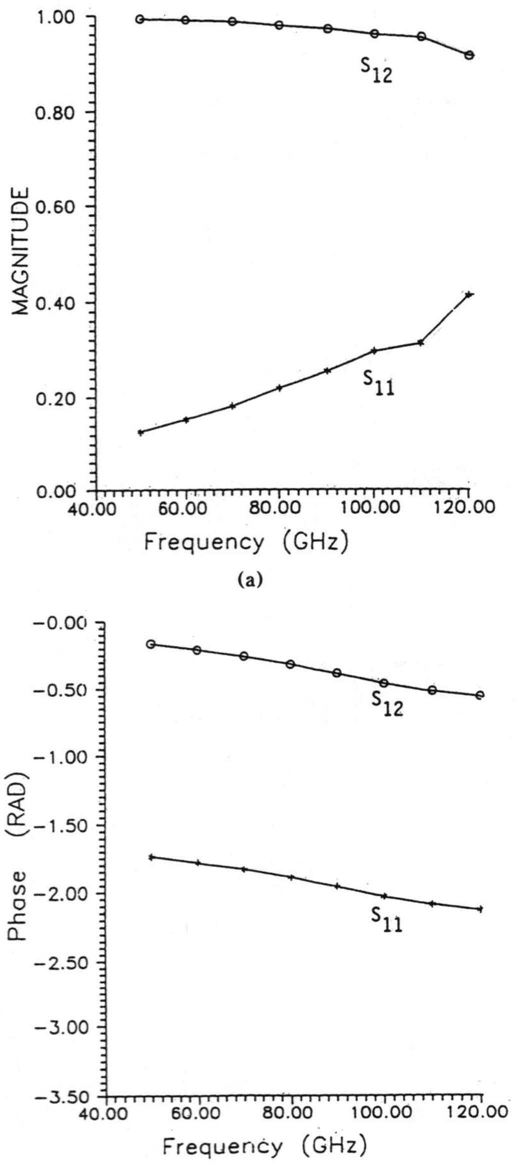

(b)

Fig. 7. S parameter of right-angle bend, double-layered substrate ($a = b = 1.9$ mm, $c = 1.05$ mm, $H_1 = 0.1$ mm, $H_2 = 0.005$ mm, $\epsilon_1 = 12.9$, $\epsilon_2 = 3.9$, $W = 0.1$ mm): (a) S_{11} (MAG), S_{12} (MAG); (b) S_{11} (RAD), S_{12} (RAD).

from a simple linear combination of the elements in the index table. The method was applied to simulate resonant modes for various discontinuity structures, including the effect of the shielding walls, in order to demonstrate its capabilities. The presented technique should be helpful in the characterization of single and coupled microstrip discontinuities for (M)MICs.

Acknowledgment

The authors gratefully acknowledge the helpful correspondence and discussions with Dr. N. H. L. Koster of Duisburg University, Federal Republic of Germany, and A. Skirevik of Ecole Polytechnique, Luasanne.

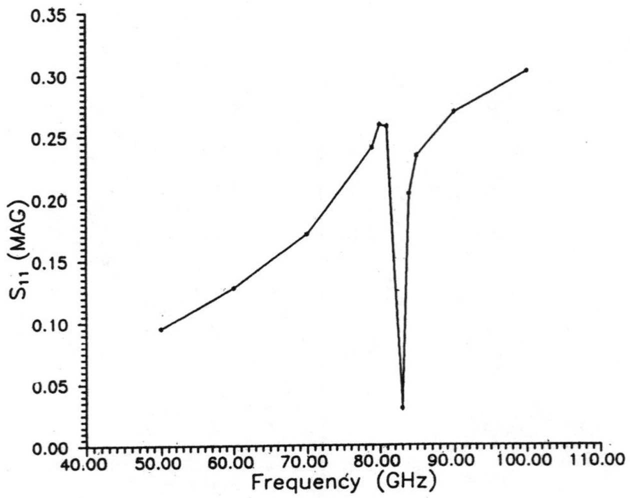

Fig. 8. S_{11} (MAG) of right-angle bend at box resonances $a = b = 1.9$ mm, $c = 0.2$ mm, $H_1 = 0.1$ mm, $W/H_1 = 1$).

Appendix

Expressions for the components of the Green's dyadic in (1) are summarized. Define

$$Umn = 1 - \frac{\epsilon_2}{\epsilon_1}\frac{T_{mn1}T_{mn2}}{k_{zmn2}^2} + \left(\frac{T_{mn1}}{\epsilon_1} + \frac{T_{mn2}}{\epsilon_2}\right)\frac{\epsilon_3}{T_{mn3}} \tag{A1}$$

$$V_{mn} = 1 - \frac{T_{mn1}T_{mn2}}{k_{zmn1}^2} + \left(\frac{T_{mn1}}{k_{zmn1}^2} + \frac{T_{mn2}}{k_{zmn2}^2}\right)\frac{k_{zmn3}^2}{T_{mn3}} \tag{A2}$$

$$X_{mn} = \frac{R_{mn}}{\left(k_{xmn}^2 + k_{ymn}^2\right)U_{mn}} \tag{A3}$$

$$Y_{mn} = \frac{S_{mn}}{\left(k_{xmn}^2 + k_{ymn}^2\right)V_{mn}} \tag{A4}$$

$$Z_{mn} = X_{mn} - Y_{mn} \tag{A5}$$

$$R_{mn} = \frac{T_{mn1}}{\epsilon_1} + \frac{T_{mn2}}{\epsilon_2} \tag{A6}$$

$$S_{mn} = k_0^2\left(\frac{T_{mn1}}{k_{zmn1}^2} + \frac{T_{mn2}}{k_{zmn2}^2}\right) \tag{A7}$$

$$F_{mn} = L_{mn}\frac{j\omega\epsilon_0}{ab} \tag{A8}$$

$$L_{mn} = \begin{cases} 4, & m \text{ and } n > 0 \\ 0, & m \text{ and } n = 0 \\ 2, & m \neq 0 \text{ and } n = 0 \text{ or } n \neq 0 \text{ and } m = 0 \end{cases} \tag{A9}$$

$$k_0^2 = \frac{\omega^2}{v^2} \tag{A10}$$

$$k_{xm} = \frac{m\pi}{a} \qquad k_{yn} = \frac{n\pi}{b} \tag{A11}$$

$$k_{zmni} = \sqrt{k_0^2\epsilon_i - \left(k_{xm}^2 + k_{yn}^2\right)} \tag{A12}$$

$$T_{mni} = k_{zmni}\tan\left(k_{zmni}H_i\right). \tag{A13}$$

Then,

$$XY_{mn} = k_{xm}^2 X_{mn} + k_{yn}^2 Y_{mn} \tag{A14}$$

$$YX_{mn} = k_{yn}^2 X_{mn} + k_{xm}^2 Y_{mn} \tag{A15}$$

$$RZ_{mn} = k_{xm}k_{yn}Z_{mn}. \tag{A16}$$

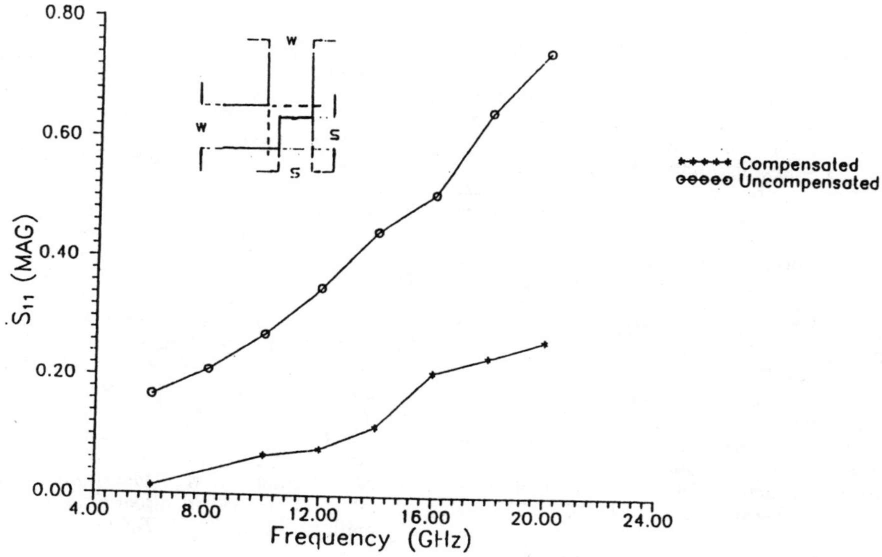

(a)

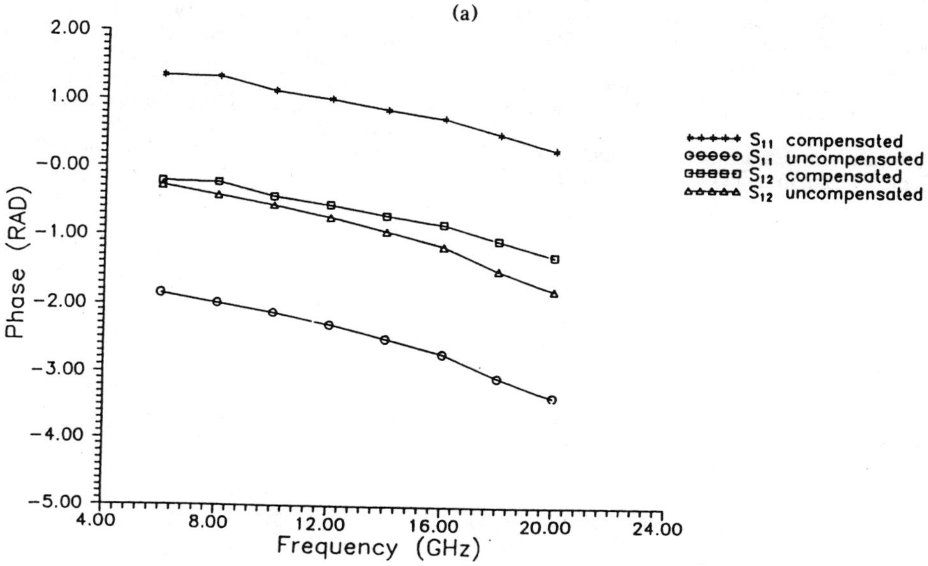

(b)

Fig. 9. Comparison of S parameter for compensated bend and right-angle bend ($a = b = 12.7$ mm, $c = 3.18$ mm, $H_1 = 0.635$ mm, $\epsilon_1 = 9.8$, $W/H_1 = 2$): (a) S_{11} (MAG); (b) S_{11} (RAD), S_{12} (RAD). Dotted lines in the inset represent reference planes.

References

[1] T. Itoh, Ed., *Numerical Techniques for Microwave and Millimeter Wave Passive Structures*. New York: Wiley, 1989.

[2] I. Wolff *et al.*, "Calculation method for microstrip discontinuities and T-junctions," *Electron. Lett.*, vol. 8, pp. 177–179, 1972.

[3] T. S. Chu and T. Itoh, "Analysis of microstrip discontinuity by the modified residue calculus technique," *IEEE Trans. Microwave Theory Tech.*, vol. MTT-33, pp. 1024–1028, Oct. 1985.

[4] N. H. L. Koster and R. H. Jansen, "The microstrip step discontinuity: A revised description," *IEEE Trans. Microwave Theory Tech.*, vol. MTT-34, pp. 213–222, Feb. 1986.

[5] P. B. Katehi and N. G. Alexopoulos, "Frequency-dependent characteristics of microstrip discontinuities in millimeter-wave integrated circuits," *IEEE Trans. Microwave Theory Tech.*, vol. MTT-33, pp. 1029–1035, Oct. 1985.

[6] R. W. Jackson and D. M. Pozar, "Full-wave analysis of microstrip open-end and gap discontinuity," *IEEE Trans. Microwave Theory Tech.*, vol. MTT-33, pp. 1036–1042, Oct. 1985.

[7] R. H. Jansen, "Hybrid mode analysis of end effects of planar microwave and millimetrewave transmission lines," *Proc. Inst. Elec. Eng.*, vol. 128, pt. H, no. 2, pp. 77–86, Apr. 1981.

[8] N. H. L. Koster, "Zur charakterisierung der frequenzabhaengigen Eigenschaften von Diskontinuitaeten in planaren Wellenleitern," Ph.D. thesis, Universitaet Duisburg, 1984.

[9] R. Sorrentino and T. Itoh, "Transverse resonance analysis of finline discontinuities," *IEEE Trans. Microwave Theory Tech.*, vol. MTT-32, pp. 1633–1638, December 1984.

[10] L. P. Schmidt, "Zur feldtheoretischen Berechnung fon transversalen Diskontinuitaeten in Mikrostrip-Leitungen," Ph.D. thesis, RWTH Aachen, West Germany, 1979.

[11] W. L. Chang, "Filterelemente and Resonatoren aus geschirmten Streifenleitungen mit sprunghafter Breitenaenderung," Ph.D. thesis, TH Darmstadt, West Germany, 1977.

[12] J. C. Rautio, "A time-harmonic electromagnetic analysis of shielded microstrip circuits," Ph.D. thesis, Syracuse University, Syracuse, NY, 1986.

[13] J. C. Rautio and R. F. Harrington, "An electromagnetic time-harmonic analysis of shielded microstrip circuits," *IEEE Trans. Microwave Theory Tech.*, vol. MTT-35, pp. 726–730, Aug. 1987.

[14] R. H. Jansen, "The spectral-domain approach for microwave integrated circuits," *IEEE Trans. Microwave Theory Tech.*, vol. MTT-33, pp. 1043–1056, Oct. 1985.

[15] A. W. Glisson and D. R. Wilton, "Simple and efficient numerical methods for problems of electromagnetic radiation and scattering from surfaces," *IEEE Trans. on Antennas and Propagation*, vol. AP-28, pp. 593–603, September 1980.

[16] R. W. Jackson, "Full-wave, finite element analysis of irregular microstrip discontinuities," *IEEE Trans. Microwave Theory Tech.*, vol. 37, pp. 81–89, Jan. 1989.

[17] A. Skrivervik and J. R. Mosig, "Equivalent circuits of microstrip discontinuities including radiation effects," in *IEEE MTT-S Int. Microwave Sump. Dig.*, 1989, pp. 1147–1150.

[18] Z. J. Cendes and J. Lee, "The transfinite element method for modeling MMIC devices," *IEEE Trans. Microwave Theory Tech.*, vol. 36, pp. 1639–1649, Dec. 1988.

[19] W. P. Harokopus and P. B. Katehi, "An accurate characterization of open microstrip discontinuity including radiation losses, in *IEEE MTT-S Int. Microwave Symp. Dig.*, 1989, pp. 231–234.

[20] B. S. Worm and R. Pregla, "Hybrid-mode analysis of arbitrarily shaped planar microwave structures by the method of lines, *IEEE Trans. Microwave Theory Tech.*, vol. MTT-32, pp. 191–196, Feb. 1984.

[21] J. R. Mosig, "Arbitrarily shaped microstrip structures and their analysis with a mixed potential integral equation," *IEEE Trans. Microwave Theory Tech.*, vol. 36, pp. 314–323, Feb. 1988.

[22] Z. Chen and B. Gao, "Deterministic approach to full-wave analysis of discontinuities in MICs using the method of lines," *IEEE Trans. Microwave Theory Tech.*, vol. 37, pp. 606–611, Mar. 1989.

[23] A. Nakatani, S. A. Maas, and J. Castaneda, "Modeling of high frequency MMIC passive components," in *IEEE MTT-S Int. Microwave Symp. Dig.*, 1989, pp. 1139–1142.

[24] W. Wertgen and R. H. Jansen, "Spectral iterative techniques for the full-wave 3D analysis of (M)MIC structures," in *IEEE MTT-S Int. Microwave Symp. Dig.*, 1988, pp. 709–712.

[25] R. J. Jansen and L. Wiemer, "Full-wave theory based development of mm-wave circuit models for microstrip open end, gap, step, bend and tee," in *IEEE MTT-S Int. Microwave Symp. Dig.*, 1989, pp. 779–782.

[26] L. P. Dunleavy and P. B. Katehi, "A generalized method for analyzing shielded thin microstrip discontinuities," *IEEE Trans. Microwave Theory Tech.*, vol. 36, pp. 1758–1766, Dec. 1988.

[27] L. P. Dunleavy and P. B. Katehi, "Shielding effects in microstrip discontinuities," *IEEE Trans. Microwave Theory Tech.*, vol. 36, pp. 1767–1774, Dec. 1988.

[28] L. P. Dunleavy, "Discontinuity characterization in shielded microstrip: A theoretical and experimental study," Ph.D. thesis, University of Michigan, 1988.

[29] R. H. Jansen and N. H. L. Koster, "Accurate results on the end effect of single and coupled lines for use in microwave circuit design," *Arch. Elek. Übertragung.*, Band 34, pp. 453–459, 1980.

[30] M. Kirschning *et al.*, "Measurement and computer aided modelling of microstrip discontinuities by an improved resonator method," in *IEEE MTT-S Int. Microwave Symp. Dig.*, 1983, pp. 495–497.

[31] R. R. Jansen and W. Wertgen, "Modular source-type 3D analysis of scattering parameters for general discontinuities, components and coupling effects in (M)MIC's," in *Proc. 17th European Microwave Conf.* (Rome), Sept. 1987, pp. 427–437.

[32] A. Hill, "Quasi-TEM and full wave numerical methods for the characterization of microstrip discontinuities," Ph.D. thesis, Oregon State University, Corvallis, OR, 1989.

Paper 5.6

Rigorous Analysis of 3-D Planar Circuit Discontinuities Using the Space-Spectral Domain Approach (SSDA)

Ke Wu, *Member, IEEE,* Ming Yu, *Student Member, IEEE,* and Ruediger Vahldieck, *Senior Member, IEEE*

***Abstract*—A new method, the Space-Spectral Domain Approach (SSDA), has been developed to determine scattering parameters for arbitrarily shaped multilayered planar MIC/MMIC discontinuities. Although the basic framework of the SSDA has been introduced previously, only resonant frequencies of planar circuit discontinuities could be calculated. The SSDA as presented in this paper is not only significantly extended, but it also introduces the new concept of self-consistent hybrid boundary conditions to replace the modal source concept in the feed line. Furthermore, a general error function is derived to provide a direct assessment of the discretization accuracy. The convergence behavior of this new method is investigated, and current standing-wave profiles along microstrip throughlines with matched, open and short-circuited conditions are given. Finally, *S*-parameters for several microstrip discontinuities with abrupt and smooth transition are illustrated to demonstrate the flexibility of this new approach.**

Introduction

ACCURATE characterization of planar discontinuities is the basis for industrial applications of computer-aided design of monolithic microwave integrated circuits (MMIC) and miniature hybrid microwave integrated circuits (MHMIC). In general, these circuits are composed of cascaded planar transmission lines which are interconnected by circuit discontinuities. The difficulties in describing the scattering parameters of these discontinuities are accentuated by the possibility of an irregularly shaped contour and the presence of a multilayered substrate topology.

Hitherto known full-wave techniques applicable to arbitrarily shaped 3-D discontinuities are mostly based on spatial discretization of the structure (i.e., FDTD [1]–[2], TLM [3], [4], FEM [5], [6]). Although these techniques are very flexible, they provide accurate results only at the expense of memory space and CPU time. This is in particular true when very thin substrate layers are involved (i.e., insulating layers in semiconductor based transmission lines). In this case, techniques which also discretize the space transverse to the propagation direction need a very fine resolution to accommodate these layers. This may require the use of supercomputer power to obtain results in a reasonable time. Other techniques which are known to be computationally very efficient, like the spectral domain approach (SDA) [7]–[9] and other alternative methods [10]–[12], lose some of their advantages when applied to spatial 3-D discontinuities. In particular when these discontinuities are arbitrarily shaped, convergence of the basis functions (SDA) becomes generally a problem.

To avoid difficulties associated with 2-D basis functions or 3-D spatial discretization, the authors recently [18], [19] have introduced a novel combination of two different modeling techniques, the method of lines (MOL) [13]–[16] and the SDA, to form the space spectral domain approach (SSDA). In this technique, the disadvantages associated with each of the methods when applied individually to 3-D discontinuities can be largely eliminated. This is so, because the MOL is most efficient when only one spatial variable needs to be discretized and similarly, the SDA is most efficient when only 1-D basis functions are used. Therefore, the SSDA combines the 1-D SDA (which is used to describe only the plane transverse to the propagation direction) with the 1-D MOL (which describes the circuit in propagation direction). This combination takes advantage of the flexibility of the MOL to model arbitrary discontinuities and at the same time adds the computational efficiency of the SDA. In addition, the SSDA accounts automatically for the singularity of fields (or currents) along the edges of slots (or strips).

So far the SSDA has only been capable of analyzing resonant structures. In this paper the SSDA is extended to calculate the s-parameters of discontinuities. This extended version of the SSDA employs the concept of self-consistent inhomogeneous (or hybrid) boundary conditions at the end of feedlines which are connected to either side of the discontinuity.

This approach makes it possible to simulate the whole structure via an eigenvalue equation in which the solution

Manuscript received August 19, 1991; revised February 12, 1992. This work was supported by Microtel Pacific Research Ltd. (MPR), Vancouver, and the Natural Sciences and Engineering Research Council (NSERC) of Canada.

K. Wu was with the Department of Electrical and Computer Engineering, University of Victoria, P.O. Box 3055, Victoria, BC, Canada V8W 3P6. He is presently with the Département de génie électrique École Polytechnique, Case postale 6079, succursale A. Montréal, PQ, Canada H3C 3A7.

M. Yu and R. Vahldieck are with the Department of Electrical and Computer Engineering, University of Victoria, P.O. Box 3055, Victoria, BC, Canada V8W 3P6.

IEEE Log Number 9108319.

Reprinted from *IEEE Trans. Microwave Theory Tech.*, vol. 40, no. 7, pp. 1475–1483, July 1992.

is the reflection coefficient of the discontinuity. The hybrid boundary conditions have been used before in [21] and [22] but in the first case to model the forward and reflected waves individually and in the second case to find the total field at the launching point by using a modal source approach. In the method presented here, the reflection coefficient (or s_{11}) is obtained directly.

Another contribution resulting from this work is that error functions are derived based on a comparison between the differential and difference operators in the inhomogeneous boundary conditions. These functions are useful in determining the discretization accuracy and the error introduced. At the same time, a limiting criterion is derived which indicates when and how the discretization size should be changed.

THEORY

In the following the equidistant discretization scheme is used for simplicity. The scattering parameter analysis of a 3-D planar discontinuity problem with arbitrary contour and multilayered substrates is shown in Fig. 1. The electromagnetic field in each dielectric region is described by two scalar potential functions, ψ^e and ψ^h, which satisfy the Helmholtz equation and the boundary conditions. Both potential functions are z-oriented and hence correspond to the TM and TE modes in the guided structure. Since the principal analytical steps involved in the space-spectral domain approach have been well explained in [19], the emphasis in the following analysis is on how to simulate the 3-D scattering problems by the self-consistent inhomogeneous boundary conditions implemented in the SSDA algorithm.

Instead of discretizing the 3-D planar structure in the x and z directions as required by the conventional 2-D MOL, the structure is discretized in the z-direction only. This step corresponds to slicing the structure in the x-y planes for each of the two scalar potential functions separately. The distance between two slices is determined by the discretization size h. Using the Fourier transform, the two scalar potential functions are written in the spectral domain along the x-direction. This step means that a set of continuous expansion functions are assigned to each discrete line. Considering a structure with open bilateral boundaries leads to infinite integrals, these can be approximated by the integration over a finite space (0, a) [23]:

$$\tilde{\psi}^{e,h}(\alpha, y, z) = \int_{-\infty}^{+\infty} \psi^{e,h}(x,y,z)e^{j\alpha x}\,dx \approx \int_0^a \psi^{e,h}(x,y,z)e^{j\alpha x}\,dx. \quad (1)$$

A. Inhomogeneous (Hybrid) Boundary Conditions

It is assumed that at some distance from Port 1 of the discontinuity there will be a standing wave of the fundamental mode only consisting of incident and reflected

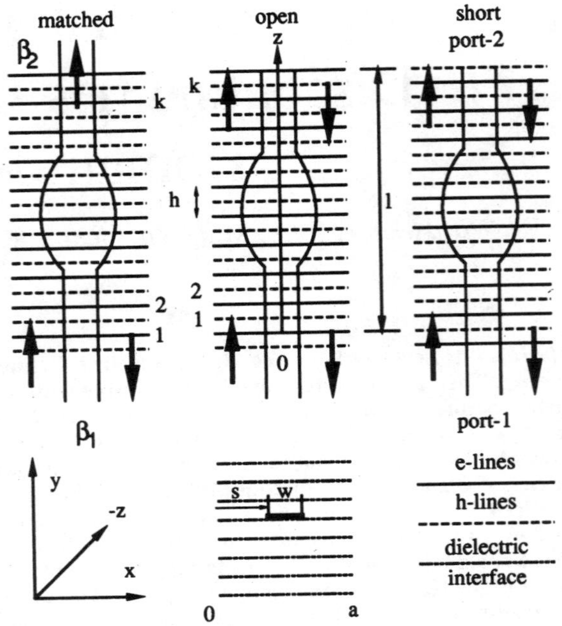

Fig. 1. Illustration of an arbitrary 3-D planar multilayered circuits with matched, open and short circuited Port 2.

waves:

$$\psi^e = \psi_0^e(e^{-j\beta_1 z} - re^{j\beta_1 z})$$
$$\psi^h = \psi_0^h(e^{-j\beta_1 z} + re^{j\beta_1 z}) \quad (2)$$

where β_1 is the propagation constant at the boundary of Port 1 calculated separately by using the SDA, r is the voltage reflection coefficient and ψ_0^e, ψ_0^h are the incident TE/TM potentials at $z = 0$. The inhomogeneous boundary conditions can be derived independently without considering the spectral domain factors. With reference to the matched, open- and short-circuited conditions, as illustrated in Fig. 1, three different cases for the boundaries exist, these are the Dirichlet, Neumann and hybrid boundary conditions. For the matched condition at Port 2 there are two choices for the discretization scheme depending on whether to assign an e or h line as the first line. In the following, the discretization scheme begins with an h-line (open-circuit).

In case of the matched and open-circuit conditions, the hybrid boundary condition at $z = 0$ for ψ^e can be expressed as

$$\psi^e|_{z=0} = \psi_1^e \qquad \text{(Dirichlet kind)} \quad (3)$$

and at $z = 0.5\,h$ for ψ^h:

$$\left.\frac{\partial \psi^h}{\partial z}\right|_{z=0.5h} = \frac{\partial \psi_1^h}{\partial z} = -j\beta_1 \frac{1 - j\tau tg(0.5\beta_1 h)}{\tau - jtg(0.5\beta_1 h)}\psi_1^h \qquad \text{(Neumann kind)} \quad (4)$$

in which $\tau = (1 + r)/(1 - r)$. The voltage reflection coefficient is thus explicitly involved in the hybrid boundary conditions. At Port 2 the matched and open-circuit conditions correspond to:

$$\left.\frac{\partial \psi^e}{\partial z}\right|_{z=L-h} = \frac{\partial \psi_k^e}{\partial z} = -j\beta_2 \psi_k^e$$

(matched condition) (5)

and

$$\left.\frac{\partial \psi^e}{\partial z}\right|_{z=L-h} = \frac{\partial \psi_k^e}{\partial z} \approx \frac{\Delta \psi_k^e}{\Delta z} = -\frac{1}{h}\psi_k^e$$

(open-circuited condition) (6)

respectively, where β_2 is the propagation constant at Port 2 if a two-port circuit is considered. The propagation constants β_1 and β_2 can be derived from the 1-D SDA or MOL. Note that the matched condition corresponds to the discretization scheme of the open-circuit condition. In a similar way, the hybrid boundary conditions obtained for the short-circuit situation is as follows:

$$\left.\frac{\partial \psi^e}{\partial z}\right|_{z=0.5h} = \frac{\partial \psi_1^e}{\partial z} = -j\beta_1 \frac{\tau - jtg(0.5\beta_1 h)}{1 - j\tau tg(0.5\beta_1 h)} \psi_1^e$$

(at Port 1)

$$\left.\frac{\partial \psi^h}{\partial z}\right|_{z=L-h} = \frac{\partial \psi_k^h}{\partial z} \approx \frac{\Delta \psi_k^h}{\Delta z} = -\frac{1}{h}\psi_k^h$$

(at Port-2) (7)

Obviously, the potential functions and their first derivatives constitute the characteristic solutions of the whole circuit. It is interesting to see that the complex functions of the inhomogeneous boundary conditions at the input described in (4) and (7) are not only expressed in terms of the propagation constant β_1, but also the discretization interval h and the unknown voltage reflection coefficient r (or s_{11}). In other words, the inhomogeneous boundary conditions are no longer "static" and strongly depend on the unknown scattering parameter, which in turn depends on the geometry of the structure of interest as well as the operating frequency. This is why the inhomogeneous boundary conditions are said to be self-consistent.

B. Error Functions and Limiting Conditions of Discretization

Judging from the inhomogeneous boundary conditions, the discretization size h is involved and plays an important role in the analysis. Intuitively speaking, the smaller the interval h is, the more accurate the numerical results become. However, it is not advisable to chose a very fine discretization scheme since this leads not only to a time-consuming algorithm but also deteriorates its efficiency and stability. So far, there is no detailed analysis treating this problem. In the following, analytical error functions are introduced to provide some criteria on the limiting conditions of the discretization. These criteria are useful for gaining insight into the error magnitude introduced in the analysis due to the discretization.

To begin with, the finite difference operator is applied to approximate the differential operator in dealing with the inhomogeneous boundary conditions at both Port 1 and Port 2. In view of the matched or open-circuit condition, as shown in Fig. 1, a simple analytical expression is obtained from (2):

$$\left.\frac{\Delta \psi^h}{\Delta z}\right|_{z=0.5h} = \frac{\psi_1^h - \psi_0^h}{h} = -j\beta_1 \frac{\sin(0.5\beta_1 h)}{0.5\beta_1 h} \cdot \frac{(1-r)}{e^{-j0.5\beta_1 h} + re^{j0.5\beta_1 h}} \psi_1^h. \quad (8)$$

In comparison with (4), the error function at Port 1 can be defined by the difference operator over its differential counterpart:

$$\xi_1 = 1 - \left| \frac{\left.\frac{\Delta \psi^h}{\Delta z}\right|_1}{\left.\frac{\partial \psi^h}{\partial z}\right|_1} \right| \quad (9)$$

with

$$\frac{\left.\frac{\Delta \psi^h}{\Delta z}\right|_1}{\left.\frac{\partial \psi^h}{\partial z}\right|_1} = \frac{tg(0.5\beta_1 h)}{0.5\beta_1 h} \frac{1}{1 - j\tau tg(0.5\beta_1 h)}. \quad (10)$$

Similarly, the error function is obtained for the short-circuit condition, which essentially is the same expression as (10) after replacing τ by $1/\tau$:

$$\xi_1 = 1 - \left| \frac{\left.\frac{\Delta \psi^e}{\Delta z}\right|_1}{\left.\frac{\partial \psi^e}{\partial z}\right|_1} \right| \quad (11)$$

with

$$\frac{\left.\frac{\Delta \psi^e}{\Delta z}\right|_1}{\left.\frac{\partial \psi^e}{\partial z}\right|_1} = \frac{tg(0.5\beta_1 h)}{0.5\beta_1 h} \frac{\tau}{\tau - jtg(0.5\beta_1 h)}. \quad (12)$$

The error function at Port 2 can also be derived based on the same definition as in (9) and (11) if the matched condition is considered:

$$\xi_2 = 1 - \left| \frac{\left.\frac{\Delta \psi^e}{\Delta z}\right|_k}{\left.\frac{\partial \psi^e}{\partial z}\right|_k} \right| \quad (13)$$

with

$$\frac{\left.\frac{\Delta \psi^e}{\Delta z}\right|_k}{\left.\frac{\partial \psi^e}{\partial z}\right|_k} = \frac{\sin(0.5\beta_2 h)}{0.5\beta_2 h} e^{-j0.5\beta_2 h} \tag{14}$$

Apparently, the error functions described in (9)–(14) have the same characteristic behavior as the function $\sin(x)/x$. The minimum point of the error function requires that $x(=\beta h)$ be equal to zero which is impossible in practical applications. Therefore, an error term is inevitably introduced into the analysis. As indicated in (10), (12), and (14), the error function may consist of magnitude and phase, but only the magnitude part is considered here for brevity. Note that although the error function is seemingly defined only at the input, it is virtually valid throughout the line as long as the discretization and fundamental mode are concerned. This is because the differential operations of Maxwell's and Helmholtz' equations are approximated by the corresponding finite difference operation at any location of the line. In general, minimizing the error function is to restrict the product $x(=\beta h)$ within a certain margin close to zero such that the function $\sin(x)/x$ approaches unity. To do so, the following special criteria (3 dB criterion) can be defined:

$$\frac{\sin(0.5\beta h)}{0.5\beta h} \geq 0.707. \tag{15}$$

This is the limiting condition of the discretization, in which β should be max (β_1, β_2). Solving (15) leads to the following expression:

$$\frac{h}{\lambda_g} \leq 0.22 \tag{16}$$

where λ_g is the smallest guiding wavelength along the line, Equations (15) and (16) mean that the interval size h should be smaller than one fifth of the guiding wavelength. Although there is no lower limit of the discretization steps, an adequate choice should be made to guarantee both accuracy and efficiency of the algorithm. In view of the required accuracy in practical applications, it is necessary to choose at least one tenth of the guiding wavelength. On the other hand, the error functions defined at Port 1 are dependent on the unknown voltage reflection coefficient, and subsequently on the structure itself. It is believed that such a criterion is not limited to the present method and is also applicable to other approaches employing discretization like TLM, finite-difference technique and even FEM.

C. The Space-Spectral Domain Approach and the Determinant Equation

This section describes the determinant equation derived from the SSDA procedure. The solution of this determinant equation is the unknown reflection coefficient r. The matched condition is taken as an example in the following analysis. The inhomogeneous boundary conditions are

$$\left.\frac{\partial \psi^h}{\partial z}\right|_{z=0.5h} = u\psi_1^h$$
$$\left.\frac{\partial \psi^e}{\partial z}\right|_{z=L-h} = -v\psi_k^e \tag{17}$$

in which u and v are the coefficients defined in (4) and (5). In order to maintain the essential transformation properties (known from the MOL procedure), symmetric second-order finite-difference operators are required to deal with the Helmholtz equation and, in particular, the field equations tangential to the interfaces. Using the concept and algorithm described in [21], the electric and magnetic potential vectors in the original discrete domain are normalized by quasi-complex diagonal matrices [15]–[17]:

$$\bar{\psi}^e = \bar{\bar{r}}^e \bar{\phi}^e$$
$$\bar{\psi}^h = \bar{\bar{r}}^h \bar{\phi}^h \tag{18}$$

with

$$\bar{\bar{r}}^e = \begin{bmatrix} \sqrt{uh} & & & \\ & 1 & & \\ & & \ddots & \\ & & & 1 \end{bmatrix} \tag{19a}$$

$$\bar{\bar{r}}^h = \begin{bmatrix} 1 & & & \\ & \ddots & & \\ & & 1 & \\ & & & \sqrt{vh} \end{bmatrix} \tag{19b}$$

Therefore, the first derivatives of the potential functions are approximated by

$$\bar{\bar{r}}^{h^{-1}}\left(h\frac{\partial \bar{\psi}^e}{\partial z}\right) \Rightarrow \bar{\bar{r}}^h \bar{\bar{D}} \bar{\bar{r}}^e \bar{\phi}^e = \bar{\bar{D}}_z \bar{\phi}^e$$
$$\bar{\bar{r}}^{e^{-1}}\left(h\frac{\partial \bar{\psi}^h}{\partial z}\right) \Rightarrow -\bar{\bar{r}}^e \bar{\bar{D}}^t \bar{\bar{r}}^h \bar{\phi}^h = -\bar{\bar{D}}_z^t \bar{\phi}^h \tag{20}$$

where superscript t denotes the transposed matrix and D is the bidiagonal matrix which has been formulated in [13]–[16]. The second derivatives of the potential vectors are transformed to:

$$\bar{\bar{r}}^{e^{-1}}\left(h^2\frac{\partial^2 \psi^e}{\partial z^2}\right) \Rightarrow \bar{\bar{D}}_z^t \bar{\bar{D}}_z \bar{\phi}^e = -\bar{\bar{D}}_{zz}^{ee} \bar{\phi}^e$$
$$\bar{\bar{r}}^{h^{-1}}\left(h^2\frac{\partial^2 \psi^h}{\partial z^2}\right) \Rightarrow -\bar{\bar{D}}_z \bar{\bar{D}}_z^t \bar{\phi}^h = -\bar{\bar{D}}_{zz}^{hh} \bar{\phi}^h. \tag{21}$$

Note that the unknown voltage reflection coefficient is directly involved with the first element of $\bar{\bar{r}}^e$ and its related matrices.

Helmholtz' equations for ψ^e and ψ^h can now be transformed to uncouple the differential equations in the space-spectral domain via the complex transformation matrices $\overline{\overline{T}}^{e,h}$, which can be obtained numerically from an eigenvalue analysis [21]:

$$\frac{d^2\overline{V}^e}{dy^2} - \left(\alpha^2 + \frac{\overline{\overline{T}}^{e^{-1}}\overline{\overline{D}}_{zz}^{ee}\overline{\overline{T}}^e}{h^2} - \epsilon_r k_0^2\right)\overline{V}^e = 0$$

$$\frac{d^2\overline{V}^h}{dy^2} - \left(\alpha^2 + \frac{\overline{\overline{T}}^{h^{-1}}\overline{\overline{D}}_{zz}^{hh}\overline{\overline{T}}^h}{h^2} - \epsilon_r k_0^2\right)\overline{V}^h = 0 \quad (22)$$

with

$$\overline{\phi}^{e,h} = \overline{\overline{T}}^{e,h}\overline{V}^{e,h}$$

where α is the Fourier transform factor along the x-direction. As mentioned in [21], sometimes the columns of the complex transformation matrices $\overline{\overline{T}}^{e,h}$ have to be suitably rearranged such that the elementary matrix $\overline{\overline{\delta}} = \overline{\overline{T}}^{h^t}\overline{\overline{D}}_z\overline{\overline{T}}^e$ retains quasi-diagonal properties [13]–[16]. This is usually done by sorting the absolute eigenvalues. It is worthwhile noting that the matrices $\overline{\overline{T}}^{e,h}$ are unique once the longitudinal boundary conditions are given and they are totally independent of the metallic contour of the discontinuity. The conductor circuit is only involved in form of the basis functions which will be explained later. The solution to (22) simply describes the wave propagation in the y-direction and can be written as a set of inhomogeneous transmission line equations which gives a relationship for $\overline{V}^{e,h}$ and its derivatives in the bottom and top boundaries of one dielectric layer [19].

Applying the continuity condition at each dielectric interface leads to a matrix relationship between the tangential field components of two adjacent subregions in the interface plane. Next, by successively utilizing the continuity condition and multiplying the resulting matrices by the transmission line matrices associated with the multilayer subregions, the boundary conditions from the top and bottom walls can be transformed into the interface plane of the discontinuity. This leads to a kind of space-spectral Green's function in the transformed domain which must be transformed back into the original domain [19]. This step can be performed by the conventional MOL and SDA procedures independently. From the mathematical viewpoint there is no difference which procedure is applied first. However, applying the MOL first leads to a better physical understanding and easier mathematical treatment. Since the planar conductors continuously extend over the entire surface of the circuit, the discretization lines intersecting the conductor section are equal to the total number ($2 \times k$) of the potential lines. As a result, the matrix elements of the resulting Green's function in the space-spectral domain are once again coupled to each other through the reverse transformation back into the original domain:

$$\begin{pmatrix} \overline{\overline{Z}}_{xx}(\alpha) & \overline{\overline{Z}}_{xz}(\alpha) \\ \overline{\overline{Z}}_{zx}(\alpha) & \overline{\overline{Z}}_{zz}(\alpha) \end{pmatrix} \begin{pmatrix} \overline{i}_x(\alpha) \\ j\overline{i}_z(\alpha) \end{pmatrix} = \begin{pmatrix} \overline{e}_x(\alpha) \\ j\overline{e}_z(\alpha) \end{pmatrix} \quad (23)$$

This equation is now subject to the SDA technique from which the eigensolution can be obtained directly in the spectral domain. To do so, the Galerkin's technique is used together with an appropriate choice of basis functions defined on the conductor surface for each slicing line in the z direction. This leads to a characteristic matrix equation system which must be solved for the zeros of its determinant. Whereby the determinant is a function of the reflection coefficient r:

$$\overline{\overline{F}}(r)\begin{bmatrix}\overline{a}\\ \overline{b}\end{bmatrix} = 0. \quad (24)$$

In contrast to the 3-D SDA, only one-dimensional basis functions are needed here. In order to achieve a fast algorithm, the following trigonometric functions combined with the edge condition are used:

$$i_{xp}^{m^h}(x) = \frac{a_p^{m^h}\sin\left\{\dfrac{p\pi(x-s)}{w}\right\}}{\sqrt{1-\left\{\dfrac{2}{w}(x-s)-1\right\}^2}},$$

$$p = 1, 2, 3, \cdots \quad \text{with } m^h = 1, 2, \cdots, k$$

$$i_{zq}^{m^e}(x) = \frac{b_q^{m^e}\cos\left\{\dfrac{q\pi(x-s)}{w}\right\}}{\sqrt{1-\left\{\dfrac{2}{w}(x-s)-1\right\}^2}},$$

$$q = 0, 1, 2, \cdots \quad \text{with } m^e = 1, 2, \cdots, k \quad (25)$$

where $a_p^{m^h}$ and $b_q^{m^e}$ denote the unknown modal current coefficients, which must be determined for each line. Note that e and h refer to electric and magnetic potential lines, respectively. Subscripts p and q denote the number of the basis functions. For irregularly shaped discontinuities, the geometric parameters w and s become a function of the z-coordinate and therefore are different for each line. In general, this does not complicate the analysis of planar structures at all, as long as the circuit contour can be described mathematically or by a set of coordinates. It should be emphasized that, due to the flexibility in handling arbitrary circuit topology, the SSDA is very well suited for contour-driven CAD software. In addition, singularities of the circuit in the x direction are automatically considered in the formulation of the basis functions.

Once the voltage reflection coefficient r (s_{11}) is known, an arbitrary constant for the first element of the x-oriented current coefficients can be assumed. Applying a singular value decomposition technique to (24) yields all the current coefficients for the chosen basis functions assigned to each discrete line. Therefore, the total surface current across the line can be obtained by a simple integration. It is worthwhile pointing out that infinite summation of the spectral terms, when constructing the characteristic equation $\overline{\overline{F}}$ in (24) must be truncated at a suitable value N for practical calculations. N can be different for each line. However, for simplicity, only equal numbers for each line will be considered in the following calculations.

Results and Discussion

First of all, the influence of the voltage reflection coefficient $r(s_{11})$ on the error function is examined by assuming that the 3 dB criterion defined in (15) and (16) is satisfied for two cases: $h/\lambda_g = 0.2$ and $h/\lambda_g = 0.05$. It is important to note that such a function is related only to the discretization error and cannot be regarded as the overall accuracy criterion although both errors are related to each other to some extent. Fig. 2 and Fig. 3 display the magnitude of the error function versus the phase of s_{11} in degrees at Port 1 which varies from 0° to 360° with different voltage reflection coefficients. It is obvious that choosing a fine discretization significantly reduces the error term. On the other hand, maximum and minimum error may occur at different locations of the phase. In case of $h/\lambda_g = 0.05$, for example, maximum error points are quasi-symmetrically located at two sides of one minimum location around $\angle s_{11} = 175°$ while in case of $h/\lambda_g = 0.2$ two minimum locations exist which are close to $\angle s_{11} = 160°$ and $\angle s_{11} = 320°$, respectively. Another observation is that the error for small reflections appears to be smaller than that of larger reflections, which can be explained by the fact that a higher reflection yields a distinct variation of the standing-wave pattern and consequently causes higher discretization error of the differential operators. This is in particular true if the reference plane of discretization coincides with the position on the line where a strong variation of the waveform occurs. In other words, choosing different lengths of the feed line and/or terminal line results in different phase terms of the voltage reflection coefficient at the reference plane with no change in the magnitude of r.

To demonstrate the SSDA, three simple examples of through-lines with matched, open- and short-circuit conditions at the boundary of Port 2 are given. Since only propagation of the fundamental mode is considered along the uniform transmission lines, one basis function of each J_x and J_z component is needed to provide enough accuracy. Fig. 4 shows a convergence test of $\angle s_{11}$ for a short-circuited through microstrip line as a function of the truncated spectral term along the x-direction with different discretization size. It can be seen that the convergence is quite beyond $N = 75$ for three discretization sizes.

Fig. 5 displays the surface current distribution of J_x and J_z components along the longitudinal direction of the line under matched condition. There is a negligible standing-wave ($|r| = 0.022$) on the line which should not be the case if the matching were perfect. This phenomenon can be explained by the fact that the matched condition implemented in this theory is a necessary condition which does not provide a complete match due to the error of discretization and the truncated terms of the infinite spectrum. This has also been reported in [24]. Increasing the

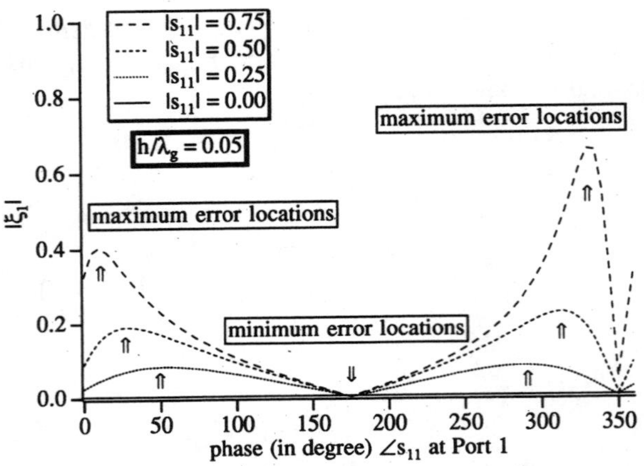

Fig. 2. Magnitude of the error function at Port 1 as a function of the angle of r with different magnitudes of r in case of fine discretization.

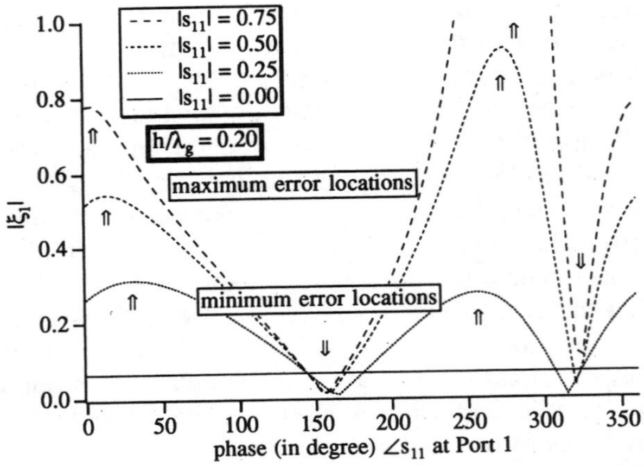

Fig. 3. Magnitude of the error function at Port 1 as a function of the angle of r with different magnitudes of r in case of rough discretization.

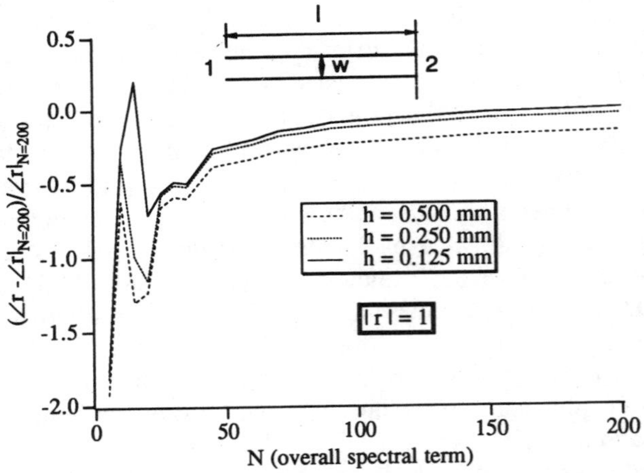

Fig. 4. Relative convergence behavior of the voltage reflection coefficient $\angle r$ versus the truncated spectral term N with different discretization size for a shorted through microstrip line. $|r| = 1$ is always obtained. The parameters used in the calculation: $l = 9$ mm, $w = 1$ mm, $h = 0.25$ mm, $a = 10$ mm, $s = 4.5$ mm, $f = 12$ GHz, $\epsilon_r = 10$, $\epsilon_{\text{eff}} = 8.0474$ and substrate thickness is 0.25 mm.

spectral term N can improve the mismatch (i.e., $|s_{11}| = 0.059$ for $N = 100$ to 0.022 for $N = 150$). In addition, the magnitude of J_x tends to vanish because of its antisymmetry of the current distribution on the conductor.

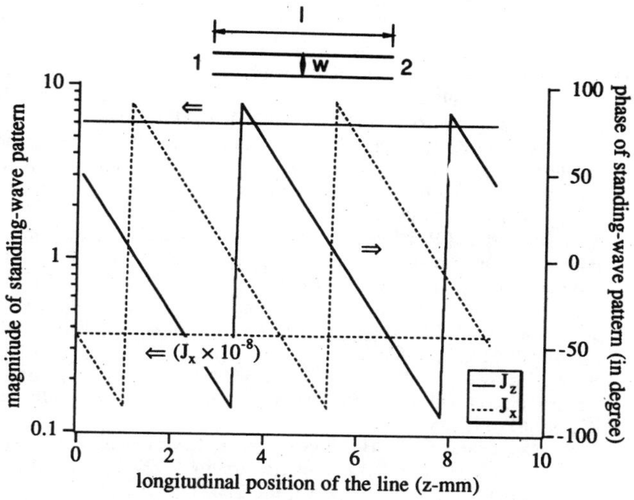

Fig. 5. Magnitude and phase distributions along the z-direction of the through microstrip line as described in Fig. 4 with the matched condition.

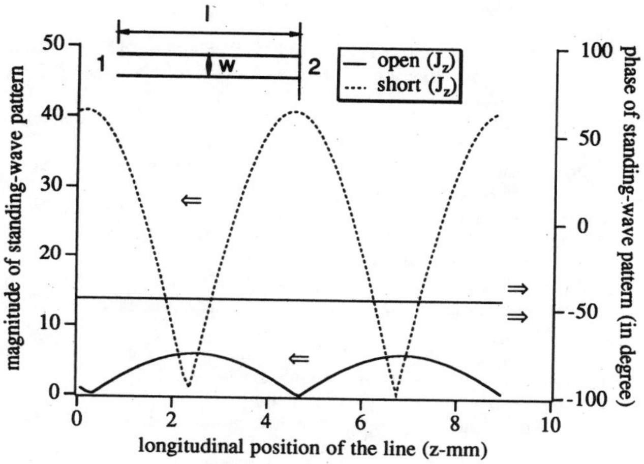

Fig. 6. Standing-wave profiles of the J_z component along the through microstrip line as described in Fig. 4 with the short and open-circuit termination.

Fig. 6 illustrates the complete standing-wave of the J_z component along the line with the open and short at Port 2 ($|r|$ at $z = 0$ is 1.0 for both open and short circuits), which agrees well with the physical perception. As expected, the maximum and minimum points for the open and short are alternatively located, and the open and short points of the line are clearly indicated by the magnitude of the standing-waves.

Fig. 7 shows a comparison of the parameter s_{11} obtained by this method and by others (i.e. [21], [25]) for a microstrip step discontinuity. A good agreeent can be observed over the frequency range up to 25 GHz, while a small discrepancy of numerical results appears beyond that frequency range. This may be due to different dimensions of the shielding box.

Transmission characteristics of two closely spaced microstrip step discontinuities are shown in Fig. 8. It is evident that there is a strong interaction between both steps since the separation of both steps is less than half the guided wavelength. Interestingly, a tighter coupling of both steps leads to a lower reflection coefficient over the frequency. The phase of the step discontinuity is shown in Fig. 8(b). These results suggest that a strong inter-

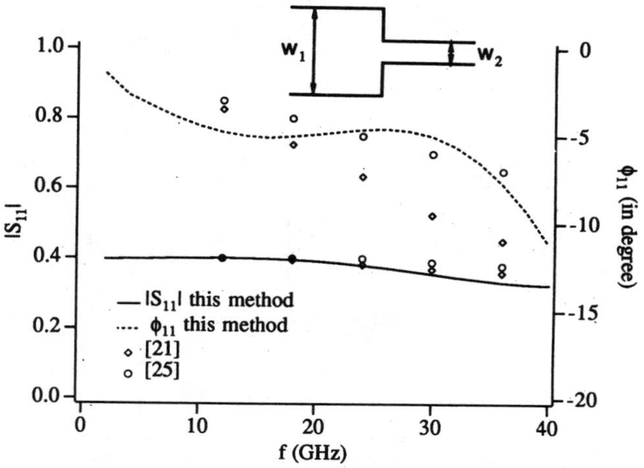

Fig. 7. Frequency-dependent reflection characteristics (s_{11}) of a microstrip step discontinuity. $w_1 = 1.00$ mm, $w_2 = 0.25$ mm, $\epsilon_r = 10.0$, $a = 10$ mm and thickness of the dielectric substrate $t = 0.25$ mm.

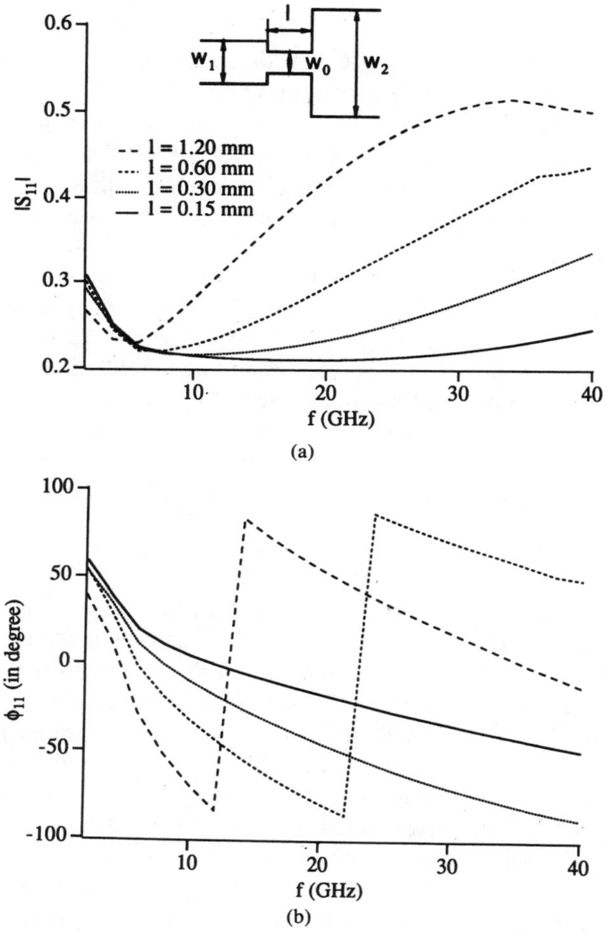

Fig. 8. S-parameters for a cascaded step discontinuity separated by a transmission line of length 1. $w_1 = 0.4$ mm, $w_0 = 0.2$ mm, $w_2 = 0.8$ mm, $\epsilon_r = 3.8$, $a = 5$ mm and $t = 0.25$ mm. (a) Magnitude of s_{11}. (b) Phase of s_{11}.

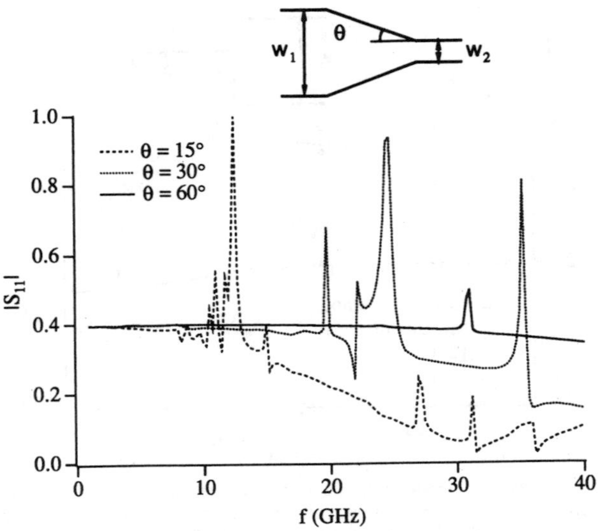

Fig. 9. Frequency response of a linear taper with a variable angle θ. Dimension of the structure is the same as that used in Fig. 7.

action between two single step discontinuities is not limited to only cases of short interconnecting stubs.

The frequency-dependent reflection characteristic of a linear microstrip taper is presented in Fig. 9 to demonstrate the flexibility and efficiency of this full-wave approach when arbitrary discontinuity contours are considered. The limiting case of the taper is $\theta = 90$ degree in which the taper is reduced to an abrupt step discontinuity analyzed in Fig. 7. An oscillating behavior of s_{11} appears in cases of long tapers ($\theta = 30$ degree and $\theta = 15$ degree) which indicates some kind of resonance effect. For the $\theta = 15$ degree taper this resonance effect occurs at higher frequencies than for the one with $\theta = 30$ degree. Similar characteristics of linear microstrip tapers have been obtained by the planar circuit approach in [26].

Conclusion

A new approach using the Space-Spectral Domain Approach (SSDA) has been presented to calculate scattering parameters and field/current distributions for three-dimensional discontinuity problems in MIC/MMIC circuits. The theory presented in this paper demonstrates how to implement self-consistent hybrid boundary conditions. Analytical error functions are introduced for the first time to estimate the error magnitude due to the discretizatior scheme used in this method. The convergence behavior of the method is illustrated as a function of the truncated spectral terms and with different discretization sizes. Surface current standing-wave profiles along microstrip through-lines with matched, open- and short-circuit conditions are calculated. A comparison with results from other methods validates this new approach. Some practical discontinuities including the linear microstrip taper have been analyzed, to demonstrate the efficiency and flexibility of this technique in treating arbitrary planar circuit contours frequently found in M(H)MIC's.

References

[1] K. S. Yee, "Numerical solution of initial boundary value problems involving Maxwell's equations in isotropic media," *IEEE Trans. Antennas Propagat.*, vol. AP-14, pp. 302–307, May 1966.

[2] X. Zhang and K. K. Mei, "Time-domain finite difference approach to the calculation of the frequency-dependent characteristics of microstrip discontinuities," *IEEE Trans. Microwave Theory Tech.*, vol. 36, pp. 1775–1785, Dec. 1988.

[3] S. Akhtarzad and P. B. Johns, "Three-dimensional transmission-line matrix computer analysis of microstrip resonators," *IEEE Trans. Microwave Theory Tech.*, vol. MTT-23, pp. 990–997, Dec. 1975.

[4] W. J. R. Hoefer, "The transmission-line matrix method—theory and applications," *IEEE Trans. Microwave Theory Tech.*, vol. MTT-33, pp. 882–893, Oct. 1985.

[5] R. W. Jackson, "Full wave finite element analysis of irregular microstrip discontinuities," *IEEE Trans. Microwave Theory Tech.*, vol. 37, pp. 81–89, Jan. 1989.

[6] J. C. Rautio and R. F. Harrington, "An electromagnetic time-harmonic analysis of shielded microstrip circuits," *IEEE Trans. Microwave Theory Tech.*, vol. MTT-35, pp. 726–730, Aug. 1987.

[7] T. Itoh and R. Mittra, "A technique for computing dispersion characteristics of shielded microstrip lines," *IEEE Trans. Microwave Theory Tech.*, vol. MTT-21, pp. 896–989, Oct. 1976.

[8] R. H. Jansen, "The spectral domain approach for microwave integrated circuits," *IEEE Trans. Microwave Theory Tech.*, vol. MTT-33, pp. 1043–1056, 1985.

[9] M. Helard, J. Citerne, O. Picon, and V. F. Hanna, "Theoretical and experimental investigation of finline discontinuities," *IEEE Trans. Microwave Theory Tech.*, vol. MTT-33, pp. 994–1003, Oct. 1985.

[10] W. Wertgen and R. H. Jansen, "Iterative, monotonically convergent hybrid mode simulation of complex, multiply branched (M)MIC conductor geometries," in *1990 IEEE MTT-S Int. Microwave Symposium Dig.*, pp. 559–562.

[11] R. Sorrentino and T. Itoh, "Transverse resonance analysis of finline discontinuities," *IEEE Trans. Microwave Theory Tech.*, vol. MTT-32, pp. 1632–1638, Dec. 1984.

[12] I. Wolff, G. Kompa and R. Mehran, "Calculation method for microstrip discontinuities and T-junctions," *Electron. Lett.*, vol. 8, pp. 177–179, Apr. 1972.

[13] U. Schulz and R. Pregla, "A new technique for the analysis of the dispersion characteristics of planar waveguides," *Arch. Elek. Ubertragung.*, *Band 34*, pp. 169–173, 1980.

[14] S. B. Worm and R. Pregla, "Hybrid mode analysis of arbitrarily shaped planar microwave structures by the method of lines," *IEEE Trans. Microwave Theory Tech.*, vol. MTT-32, pp. 191–196, Feb. 1984.

[15] R. Pregla and W. Pascher, "The method of lines," in *Numerical Techniques for Microwave and Millimeter Wave Passive Structures*, ch. 6, T. Itoh, Ed., New York: Wiley, 1989, pp. 381–446.

[16] K. Wu and R. Vahldieck, "Comprehensive MoL analysis of a class of semiconductor-based transmission lines suitable for microwave and optoelectronic application," *Int. J. of Numerical Modelling*, Special Issue, T. Itoh Ed., vol. 4, New York: Wiley, 1991, pp. 45–62.

[17] H. Diestel and S. B. Worm, "Analysis of hybrid field problems by the method of lines with nonequidistant discretization," *IEEE Trans. Microwave Theory Tech.*, vol. MTT-32, pp. 633–638, June 1984.

[18] K. Wu and R. Vahldieck, "A novel space-spectral domain approach and its application to three-dimensional MIC/MMIC circuits," in *Proc. 19th EuMC*, London, Sept. 1989, pp. 751–756.

[19] ——, "A new method of modeling three-dimensional MIC/MMIC circuits: the Space-Spectral Domain Approach," *IEEE Trans. Microwave Theory Tech.*, vol. 38, pp. 1309–1318, Sept. 1990.

[20] R. W. Jackson, "Full wave, finite element analysis of irregular microstrip discontinuities," *IEEE Trans. Microwave Theory Tech.*, vol. 37, pp. 81–89, Jan. 1989.

[21] Z. Q. Chen and B. X. Gao, "Deterministic approach to full-wave analysis of discontinuities in MIC's using the method of lines," *IEEE Trans. Microwave Theory Tech.*, vol. 37, pp. 606–611, Mar. 1989.

[22] S. B. Worm, "Full-wave analysis of discontinuities in planar waveguides by the method of lines using a source approach," *IEEE Trans. Microwave Theory Tech.*, vol. 38, pp. 1510–1514, Oct. 1990.

[23] T. Uwano and T. Itoh, "Spectral domain approach," in *Numerical Techniques for Microwave and Millimeter Wave Passive Structures*, ch. 5, T. Itoh, Ed., New York: Wiley, 1989, pp. 334–380.

[24] M. Drissi, V. Fouad Hanna, and J. Citerne, "Theoretical and exper-

imental investigation of open microstrip gap discontinuity,'' in *Proc. 18th EuMC* Sept. 1988, pp. 203–209.

[25] N. H. L. Koster and R. H. Jansen, ''The microstrip step discontinuity: a revised description,'' *IEEE Trans. Microwave Theory Tech.*, vol. MTT-34, pp. 213–223, Feb. 1986.

[26] R. Chadha and K. C. Gupta, ''Compensation of discontinuities in planar transmission lines,'' in *1982 IEEE MTT-S Int. Microwave Symposium Dig.*, pp. 308–310.

[27] W. J. R. Hoefer, ''A contour formula for compensated microstrip steps and open ends,'' in *IEEE MTT-S Int. Microwave Symp. Dig.*, Boston, 1983, pp. 524–526.

[28] D. Raicu, ''Universal taper for compensation of step discontinuities in microstrip lines,'' *IEEE Microwave Guided Wave Lett.*, vol. 1, pp. 249–251, 1991.

Paper 5.7

Electromagnetic Modeling of Passive Circuit Elements in MMIC

David C. Chang, *Fellow, IEEE,* and Jian X. Zheng, *Student Member, IEEE*

***Abstract*—A spatial-domain mixed-potential integral equation method is developed for the analysis of microstrip discontinuities and antennas of arbitrary shape. The algorithm is based on roof-top basis functions on a rectangular and triangular mixed grid and analytical evaluation of the quadruple moment integrals involved. The algorithm is successfully implemented into an accurate, efficient and versatile computer program. The numerical results agree with the measured ones very well.**

I. Introduction

It is commonly accepted that electromagnetic modeling and CAD are much needed to achieve a first-pass design for monolithic microwave/millimeter-wave integrated circuits (MMIC). As the operating frequency and functionality of these chips continue to increase, our inability to accurately model the effect of junction discontinuities and parasitic coupling among circuit elements is quickly becoming a critical bottleneck in ensuing a successful design. For operating frequencies beyond 20 GHz for a typical circuit, traditional quasi-static [1], [2] and other waveguide methods [3], [4] can no longer be expected to yield accurate results. In this paper, we shall present the algorithmic development of a full-wave method and its application to microstrip structures of general shape. We believe that this method is not only computationally efficient, but also yields a physical interpretation compatible to the physical picture as to how the current waves should behave on a microstrip, particularly near a junction region.

The algorithm we developed, called P(seudo)-mesh, is derived from the application of moment method to a mixed-potential integral equation (MPIE) in spatial domain for finding both current and charge distributions on the microstrip surface. The formulation of the integral equation itself can be traced back to the well-known work of Harrington [5]. For planar structures, it has at least two distinctive advantages when compared with a typical electric field integral equation (EFIE): one is that the Green's functions involved in the kernel of a MPIE are scalar functions of electric and magnetic types and they can be represented typically by one-dimensional Sommerfeld integrals. The other is that the singularity in the Green's functions of both types is of the order of $1/R$, where $R = |\bar{r} - \bar{r}'|$ is the distance between the source and observation points; the moment integrals associated with this singular term are, in fact, known analytically in closed-form. For a planar structure s as shown in Fig. 1, the MPIE can be written down as

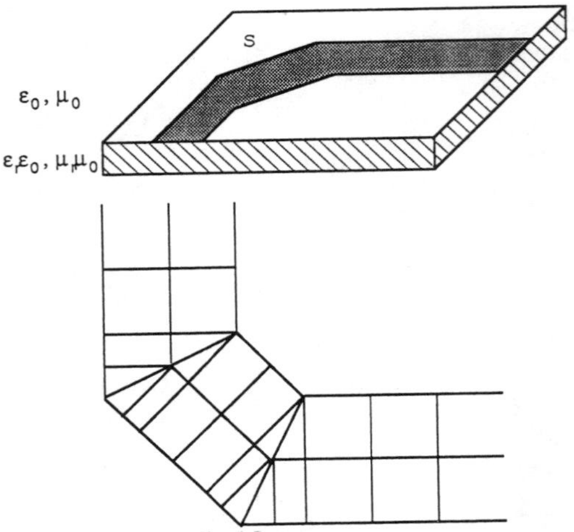

Fig. 1. A microstrip circuit and its gridded structure.

$$\frac{j\omega\mu_0}{4\pi}\int_s ds \int_s ds' \left[G_m(\bar{r}, \bar{r}')\bar{T}(\bar{r}) \cdot \bar{J}(\bar{r}') - \frac{1}{k_0^2} G_e(\bar{r}, \bar{r}') \nabla \cdot \bar{T}(\bar{r}) \nabla' \cdot \bar{J}(\bar{r}') \right] = \int_s ds\bar{E}_i(\bar{r}) \cdot \bar{T}(\bar{r}) \quad (1)$$

where G_e and G_m are the Green's functions of electric and magnetic types; $\bar{J}(\bar{r}')$ and $\bar{T}(\bar{r})$ are, respectively, the current distribution and the test function, which satisfy the boundary condition on the structure; k_0 and μ_0 are, respectively, the wavenumber and permeability in free space, ω is the angular frequency; and $\bar{E}_i$ is the impressed electric field on the structure. When the source point (x', y', z') and the field point (x, y, z) are on the same plane,

Manuscript received August 7, 1990; revised January 14, 1992. This work was supported by the Center for Microwave/Millimeter-Wave Computer-Aided Design, University of Colorado at Boulder, Boulder, CO 80309.

The authors are with the MIMICAD Center, Department of Electrical and Computer Engineering, University of Colorado at Boulder, Boulder, CO 80309-0425.

IEEE Log Number 9201710.

Reprinted from *IEEE Trans. Microwave Theory Tech.*, vol. 40, no. 9, pp. 1741–1747, Sept. 1992.

the Green's functions for microstrip structures are [6]

$$G_m(\rho) = \int_0^\infty 2J_0(\lambda\rho) \frac{\mu_r \lambda}{\mu_r u_0 + u_n \coth(u_n h)} d\lambda \tag{2}$$

$$G_e(\rho) = \int_0^\infty 2J_0(\lambda\rho) \frac{\lambda[u_0 + \mu_r u_r \tanh(u_n h)]}{[\epsilon_r u_0 + u_n \tanh(u_0 h)][\mu_r u_0 + u_n \coth(u_n h)]} d\lambda \tag{3}$$

where h is the thickness of the substrate; $\epsilon_0 \epsilon_r$ and $\mu_0 \mu_r$ are the dielectric permittivity and permeability, respectively; $J_0(\lambda\rho)$ is the zero-th order Bessel function.

$$\rho = \sqrt{(x - x')^2 + (y - y')^2} \tag{4}$$

$$u_0 = \sqrt{\lambda^2 - 1}; \quad \text{Re}(u_0) \geq 0, \text{Im}(u_0) \geq 0 \tag{5}$$

$$u_n = \sqrt{\lambda^2 - \epsilon_r \mu_r}; \quad \text{Re}(u_n) \geq 0, \text{Im}(u_n) \geq 0. \tag{6}$$

The MPIE formulation has been previously adopted by several authors to model microstrip structures, both in the form of microstrip patch antennas and microstrip circuit discontinuities [7]–[8]. For instance, in the works by Wu, et al. [7], a microstrip structure is divided into two sets of rectangular cells and pulse basis functions are used to approximate the charge and current distribution separately. Special form of linear ''roof-top'' basis functions is used to approximate the current distribution by Mosig in [8], again for a set of rectangular cells. These methods obviously are most appropriate when the structure under investigation can be, in fact, naturally divided into rectangular cells, but would not be as efficient when cells of reasonable size (about 20 cells per waveguide wavelength) cannot be fitted into the boundaries of a structure. On the other hand, a full implementation of roof-top basis functions for triangular cells has been reported recently [15]. The current distribution in a given cell is expressed in terms of the nodal currents at its vertices. Boundary conditions at the edge of a microstrip structure are difficult to enforce in this case, particularly when non-rectangular corners are encountered. Furthermore, it has been demonstrated in [15] that, for a given structure, the rate of convergence for such a scheme may depend upon the particular orientation of the cells selected.

The new algorithm presented here uses a combination of rectangular and triangular cells in a self-consistent manner in order to take into account the regularity in shape over the major portion of a microstrip structure, while still preserving the flexibility to model junctions of arbitrary shape locally. It embodies the advantages of the methods described above but without their respective disadvantages. As it will become clear later, it also has a very attractive physical interpretation, which in turn lends itself naturally to the choice of cells for a given geometry. Mixed use of rectangular cells and triangular cells and the derivation can be found in finite element analysis in the solution of scalar integral or differential equations [16]. Special consideration has to be taken for our solution of vector current distribution, whose normal component, instead of itself, is continuous on cell boundaries, on a microstrip structure.

II. Roof-Top Basis Functions

As stated in the introduction, what distinguishes the P-mesh from other similar use of roof-top basis functions to approximate the current distribution on a microstrip structure is that we are able to mix rectangular cells and triangular cells in a self-consistent manner as shown in Fig. 1. This self-consistency is derived from the observation that in order to avoid the unphysical occurence of a δ-function charge density in the numerical process, only the normal component of the current density, but *not* the current density itself, is required to be continuous across a cell boundary. Thus, instead of expressing the current distribution in terms of *vector* nodal currents at the three vertices of a triangular cell [15], we can implement a modified version in which the current distribution is expressed in terms of the normal components on the three sides [17]. In order to solve for the current uniquely, we further impose an additional requirement that these normal components have to remain constant across their respective boundaries. In the case of a rectangular cell, two of the four additional conditions can be shown as redundant and thus the number of equations is again reduced to six for the six unknown coefficients.

2.1 *Roof-Top Functions on Rectangular Cells*

Denote the side formed by nodes i and j as side (i, j). We can express the current density distribution $\bar{J}_\alpha(x, y)$ in rectangle α in terms of the normal component $I_\alpha^{i,j}$ on the sides, where the subscript α and the superscripts i, j mean the side (i, j) of cell α:

$$\bar{J}_\alpha(x, y) = \sum_{i=1}^{4} I_\alpha^{i,i+1} \bar{D}_\alpha^{i,i+1}(x, y) \tag{7}$$

where $\bar{D}_\alpha^{i,i+1}$ is the expression for the corresponding roof-top function to the side $(i, i + 1)$ of cell α (see Fig. 2). Since a rectangular cell has four vertices or nodes, we can consider i as a cyclic number so that $i = i - 4$ for $i > 4$ and $i = i + 4$ for $i < 1$. It is not difficult to show that $\bar{D}_\alpha^{i,i+1}$, which has a magnitude of 1 on side $(i, i + 1)$ and vanishes on the opposite side, i.e. side $(i + 2, i + 3)$ as shown in Fig. 2, is given by

$$\bar{D}_\alpha^{i,i+1}(x, y) = \frac{[(y_{i+1} - y_i)(x - x_{i-1}) - (x_{i+1} - x_i)(y - y_{i-1})]}{\Delta_{i-1,i,i+1}} \cdot \frac{(x_{i-1} - x_i)\hat{x} + (y_{i-1} - y_i)\hat{y}}{d_{i-1,i}}; \quad (x, y) \in \text{rectangle } \alpha \tag{8}$$

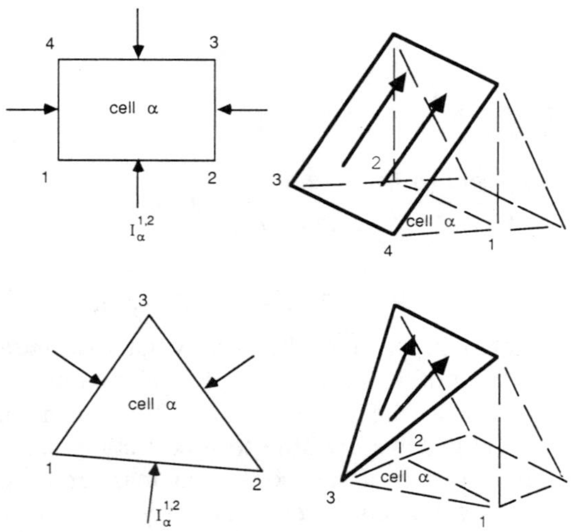

Fig. 2. The roof-top functions on a rectangular cell and a triangular cell.

$$d_{i,j} = \sqrt{(x_i - x_j)^2 + (y_i - y_j)^2)} \tag{9}$$

$$\Delta_{i-1,i,i+1} = \begin{vmatrix} 1 & x_{i-1} & y^{i-1} \\ 1 & x_i & y_i \\ 1 & x_{i+1} & y_{i+1} \end{vmatrix} \tag{10}$$

and $\hat{x}$, $\hat{y}$ are the unit vectors, respectively, in the x and y-directions. The divergence of the cell current density distribution can be written as

$$\nabla \cdot \bar{J}_\alpha(x, y) = \sum_{i=1}^{4} I_\alpha^{i,i+1} Q_\alpha^i; \qquad (x, y) \in \text{rectangle } \alpha \tag{11}$$

where

$$Q_\alpha^i = -\frac{1}{d_{i-1,i}} \tag{12}$$

2.2 Roof-Top Functions on Triangular Cells

Like in the case of a rectangular cell, the current density on a triangular cell α is given by

$$\bar{J}_\alpha(x, y) = \sum_{i=1}^{3} I_\alpha^{i,i+1} \bar{D}_\alpha^{i,i+1}(x, y);$$

$$(x, y) \in \text{triangle } \alpha \tag{13}$$

$$\nabla \cdot \bar{J}_\alpha(x, y) = \sum_{i=1}^{3} I_\alpha^{i,i+1} Q_\alpha^i; \qquad (x, y) \in \text{triangle } \alpha \tag{14}$$

where the roof-top functions $\bar{D}_\alpha^{i,i+1}$ are now

$$\bar{D}_\alpha^{i,i+1}(x, y) = -\frac{d_{i,i+1}}{|\Delta_{i-1,i,i+1}|} \cdot [(x - x_{i-1})\hat{x} + (y - y_{i-1})\hat{y}] \tag{15}$$

$$Q_\alpha^i = -\frac{2d_{i,i+1}}{|\Delta_{i-1,i,i+1}|} \tag{16}$$

and $i = i - 3$ for $i > 3$, or $i = i + 3$ for $i < 1$.

Unlike that for a rectangle, the roof-top basis function for a triangle changes direction at different locations. For triangle α as shown in Fig. 2, the roof-top function for side $(i, i + 1)$ is a vector parallel to side $(i - 1, i)$ at node i, parallel to side $(i - 1, i + 1)$ at node $(i + 1)$ and vanished at node $(i - 1)$. The incoming normal component is defined as 1 on side $(i, i + 1)$.

2.3 Pseudo-Mesh Current Distribution Representation

Since for both rectangular and triangular cells we can now express the current by the normal current density across the cell boundary and since each of these normal current densities is assumed to be constant along the boundary, we can now characterize the current in the cell by the *total current flow* into and out of each cell. Topologically, this is the same as replacing a microstrip structure by equivalent meshes, and the current distribution on the surface area of a cell by the current flow along corresponding meshes as shown in Fig. 3. Unlike a real mesh structure, however, the net amount of total current flows into and out of a "junction" does not follow the conventional Kirchhoff's law. In fact, the difference between the incoming current and the outgoing current contributes to the charge distribution on the cell. The requirement that the normal component of current must vanish at the edges of a microstrip circuit can be easily implemented by "opening" the corresponding meshes connecting the edges. We should note that the use of triangular "meshes" fully captures the physical phenomenon of a current flow round the corner of a bend. Thus, one of the advantages for the P-mesh representation is that it can be constructed according to the physical intuition a designer has, and such intuition usually results in fast convergence of the computational process.

2.4 The Global Expression for Current Distribution

In Section II, we have discussed the roof-top basis functions on individual cells. To complete the P-mesh development, we still need to integrate the individual current unknown, i.e. $I_\alpha^{i,i+1}$ for the cell α into a global set of "mesh" current I_m, $m = 1, 2, \cdots, M$, where M is the total number of the interconnecting meshes. As we mentioned earlier, meshes at the boundary of a microstrip structure are "disconnected" since the normal current at the edge of a boundary cell is zero. For adjacent cells α and α' as shown in Fig. 4, at the common boundary describable either by $(\alpha; i, i + 1)$ or $(\alpha'; i', i' + 1)$, the unknown current across this boundary is now expressed in terms of the mesh current I_m so that

$$I_m = I_\alpha^{i,i+1} = -I_{\alpha'}^{i',i'+1}. \tag{17}$$

The roof-top basis function corresponding to this unknown current is

$$\bar{H}_m = \bar{D}_\alpha^{i,i+1} - \bar{D}_{\alpha'}^{i',i'+1}. \tag{18}$$

It should be noted that $\bar{D}_\alpha^{i,i+1}$ is defined only in cell α, while $\bar{D}_{\alpha'}^{i',i'+1}$ is defined only in cell α'. The divergence

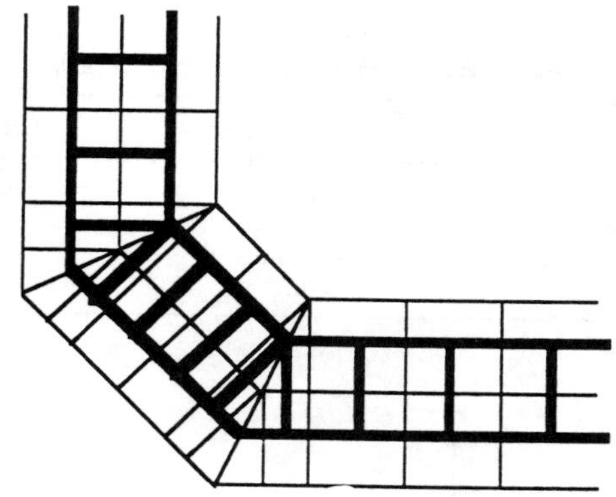

Fig. 3. The current flow in a wire mesh.

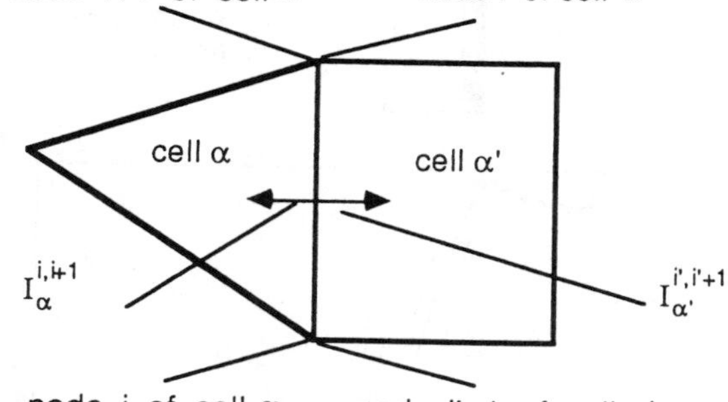

Fig. 4. The side current for two adjacent cells.

of the roof-top basis function is

$$P_m = Q_\alpha^i - Q_{\alpha'}^{i'}. \tag{19}$$

Therefore, the P-mesh current distribution on a structure can finally be expressed as

$$\bar{J}(x, y) = \sum_{m=1}^{M} I_m \bar{H}_m(x, y) \tag{20}$$

where M is the number of the unknown interconnecting "meshes."

III. Matrix Solution to the MPIE

Using each of the $\bar{H}_m(x, y)$ as the test functions in (1) and substituting (20) into (1) yield a matrix equation:

$$\sum_{m'=1}^{M} I_{m'} Z_{m,m'} = V_m; \qquad m = 1, 2, \cdots, M \tag{21}$$

where

$$Z_{m,m'} = \frac{j\omega\mu_0}{4\pi} \int_{s_m} ds \int_{s_{m'}} ds' K_{m,m'}(x, y; x', y') \tag{22}$$

$$K_{m,m'}(x, y; x', y') = G_m(\rho)\bar{H}_m(x, y) \cdot \bar{H}_{m'}(x', y') - \frac{1}{k_0^2} G_e(\rho) P_m P_{m'} \tag{23}$$

$$V_m = \int_{s_m} ds\, \bar{H}_m(x, y) \cdot \bar{E}_i(x, y) \tag{24}$$

$$\rho = \sqrt{(x - x') + (y - y')^2}. \tag{25}$$

The surface integration on s_m has to be carried out over the two adjacent cells, s_α and $s_{\alpha'}$ which share a common boundary or "mesh" m, i.e. $m = (\alpha; i, i+1) = (\alpha'; i', i'+1)$. The matrix element $Z_{m,m'}$ consists of quadruple integrals of the form

$$\int_{\text{cell}\,\alpha} ds \int_{\text{cell}\,\alpha'} ds' G_{m,e}(\rho) x^{\mu} y^{\nu} x'^{\mu'} y'^{\nu'};$$

$$\mu, \nu, \mu', \nu' \geq 0 \;\&\; 0 \leq \mu + \nu, \mu' + \nu' \leq 1 \tag{26}$$

For planar structures, the $G_{m,e}$ in (26) are Sommerfeld integrals (see (2) and (3)), and cannot be solved analytically. It is noticed that the Green's functions $G_{e,m}$ are only one-dimensional functions of ρ. We can evaluate them numerically first and then curve-fit them into polynomials over the range of ρ determined by the maximum and minimum distance between two cells α and α' [19]:

$$G_{m,e}(\rho) = \sum_{p=-1}^{N_p} C_p^{m,e} \rho^p \tag{27}$$

where $C_p^{m,e}$ are the coefficients from curve-fitting, N_p is the order of the polynomials. With these semi-analytical expressions of the Green's functions, the integrals in (26) are simplified as

$$Q(\alpha, \alpha', \mu, \nu, \mu', \nu', p) = \int_{\text{cell}\,\alpha} ds \int_{\text{cell}\,\alpha'} ds' \rho^p x^{\mu} y^{\nu} x'^{\mu'} y'^{\nu'};$$

$$p = -1, 0, \cdots, N_p. \tag{28}$$

Analytical solutions to the integrals with even p in (28) can be found in many books on finite elements [20]. But no formulation has been found to deal with the integrals with odd p. The integrals with odd p are solved analytically in [19]. The derivation is very complicated and it will not be included in this paper.

IV. De-Embedding of Network Parameters

In practical applications, we assume a gap voltage source at the far end of a feed line as the excitation. When the feed line is long enough, the current distribution is very close to a sinusoidal function just $0.1 \sim 0.2\lambda_g$ away from junctions, a measure of current standing wave at each port leads directly to the scattering matrix of the microstrip structure [19]. Thus, unlike those methods which use the input admittance, or the current at the voltage source [13], the scattering matrix does not require the knowledge, nor the confusion arising from the definition of the characteristic impedance of a microstrip line. Several de-embedding techniques have been developed during the course of this work, and the most accurate scheme is a three-point curve-fitting scheme. The current distribution at three uniformly-spaced points is detected to provide three equations

$$z = -z_0: J_1 = a \exp(\gamma z_0) - b \exp(-\gamma z_0) \tag{29}$$

$$z = 0: J_2 = a - b \tag{30}$$

$$z = z_0: J_3 = a \exp(-\gamma z_0) - b \exp(\gamma z_0) \tag{31}$$

where a and b are the amplitudes of the incident wave and reflected wave at $z = 0$, respectively. Summation of (29) and (31) yields

$$2(a - b) \cosh (\gamma z_0) = (J_1 + J_3). \quad (32)$$

Substituting (30) into (32) gives

$$\cosh (\gamma z_0) = \frac{J_1 + J_3}{2J_2}. \quad (33)$$

Unique γ can be solved from (33) as long as $Im(\gamma) z_0 < (\pi/2)$. Then, the incident and reflected waves can be obtained from either two of (29), (30) and (31).

The advantage of the de-embedding technique over the typical VSWR method used in experimental procedure is that we don't need to care about high standing waves since we always deal with the real part and imaginary part instead of the magnitude of current. This feature makes it very convenient to solve a multi-port network problem by just changing the excitation states instead of using matched loads.

V. Numerical Results

The P-mesh algorithm has been well implemented into a versatile computer code. It has been used to analyze various kinds of microstrip circuits and antennas. Before we proceed to discuss some examples, it is necessary to define some commonly used notations:

ϵ_r	substrate dielectric constant;
h	substrate thickness;
σ	microstrip conductivity;
w	microstrip line width;
λ_g	waveguide wavelength;
$\gamma = \alpha + j\beta$	complex propagation constant of a microstrip line;
N_w	number of cells per waveguide wavelength;
N_t	number of cells in transverse direction;
N_c	number of cells in a structure.

No obvious convergence but an error bound is observed. To give an idea what the accuracy is of the P-mesh code with $N_t = 1$ and $N_w = 20$ when the microstrip width is less than 5% λ_g, we list the error bounds for the cases of large values and small values of the main parameters in Table I (Note: small values are approximately defined as $|S_{i,j}| < -30$ dB and $\angle S_{i,j} < 5°$, and large values are approximately defined as $|S_{i,j}| >> 0.1$ and $\angle S_{i,j} >> 5°$). In the following analysis, we always use $N_t = 1$ and $N_w = 20$.

The first example is the double-stub structure as shown in Fig. 5. It was previously fabricated and measured for the purpose of determining the parasitic coupling of two parallel stubs [21]. Two local minima are observed in the response of $|S_{2,1}|$ (see Fig. 5). It happens even the stub separation is as large as $\lambda_g/4$. Since the two stubs are identical in length, only one minimum is expected if we break the structure into two single stubs and connect them without considering the coupling effect. But, when we

TABLE I
The Error Bounds for the Main Parameters

Parameter	Small Value	Large Value
λ_g	<0.5%	<0.5%
Resonant frequency	<0.5%	<0.5%
$\|S_{i,j}\|$	<2.0 dB	<1.5%
$\angle S_{i,j}$	<0.2°	<1.5%

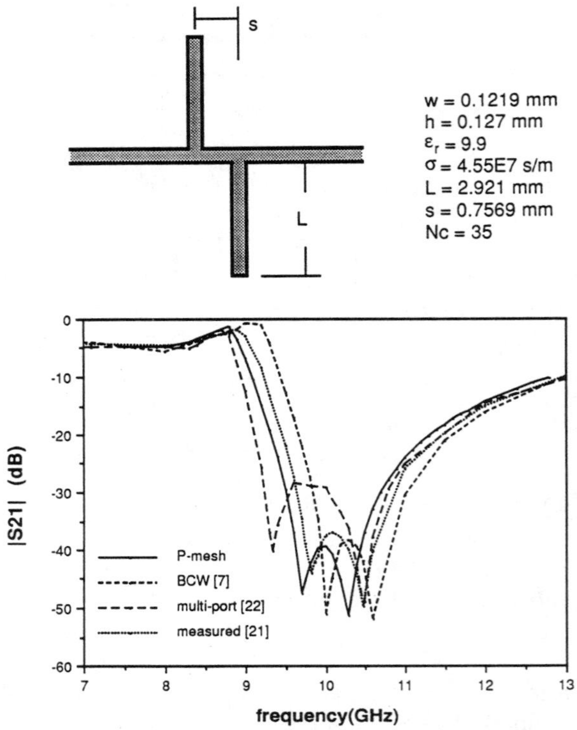

Fig. 5. A double-stub structure and its frequency response.

consider it as a four-port network first, and then terminate the two ports with open-ends, we can always get the double minima [23]. Obviously, the double minimum response is not caused by the radiation from the open-ends, but created by the mutual coupling between the two junctions and stubs. We can also see from Fig. 5 that the theoretical results agree with the measured ones very well. There is a 1.5% difference in the resonant frequency between the numerical and experimental results, but it is within the fabrication error bound (2%).

The other example is a serpentile line formed by three cascaded U-bends as shown in Fig. 6. Fig. 7 shows the frequency responses. It is predicted that $|S_{1,1}| < 0.2$ and $|S_{2,1}| > 0.966$ for a single U-bend, while $|S_{1,1}| < 0.05$ for a double-bend portion (see Fig. 6). Obviously, the accumulation of the reflections from the junctions has degraded the performance of the circuit. In fact, the circuit performance is also affected by the mutual coupling in the circuit. The effect of mutual coupling can easily be demonstrated by comparing the frequency responses of the three cascaded U-bends calculated in two different ways: (1) three U-bends analyzed as a whole entity (three U-bends 1); (2) three U-bends analyzed as the cascading

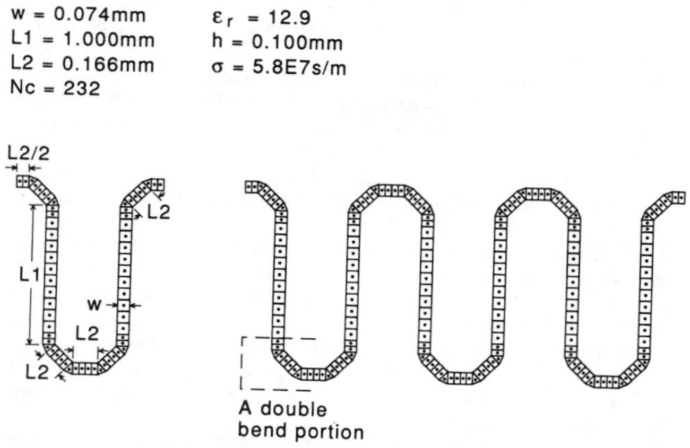

Fig. 6. A serpentile line formed by three U-bends.

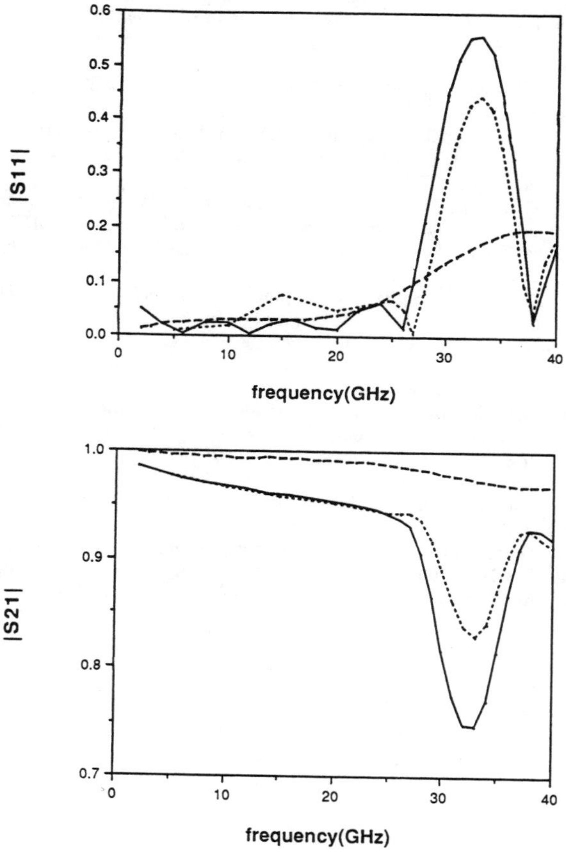

Fig. 7. The frequency responses of the U-bends.

of three individual U-bends (three U-bends 2). A strong resonance is predicted at 32 GHz resulting from multiple reflections. It is noticed that there are substantial differences in the frequency responses of the three U-bends, especially at the resonance, between the two cases. When the coupling is included, we have $|S_{1,1}|^2 = 0.31$ and $|S_{2,1}|^2 = 0.57$. But when the coupling is omitted, $|S_{1,1}|^2 = 0.19$ and $|S_{2,1}|^2 = 0.69$ are predicted. Obviously, the differences account for the mutual coupling among the U-bends. It has been found that the radiation loss is not significant in a typical MMIC circuit [18]. The power loss in this case is $1 - |S_{1,1}|^2 - |S_{2,1}|^2 = 0.12$ and attributes to the conductor loss resulting from imperfect conductor,

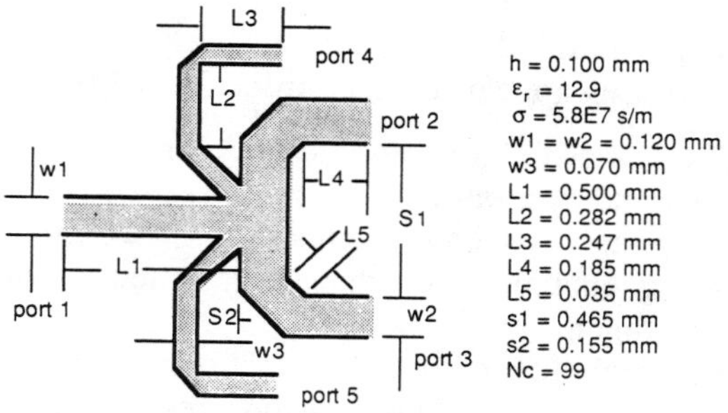

Fig. 8. A complex five-port junction.

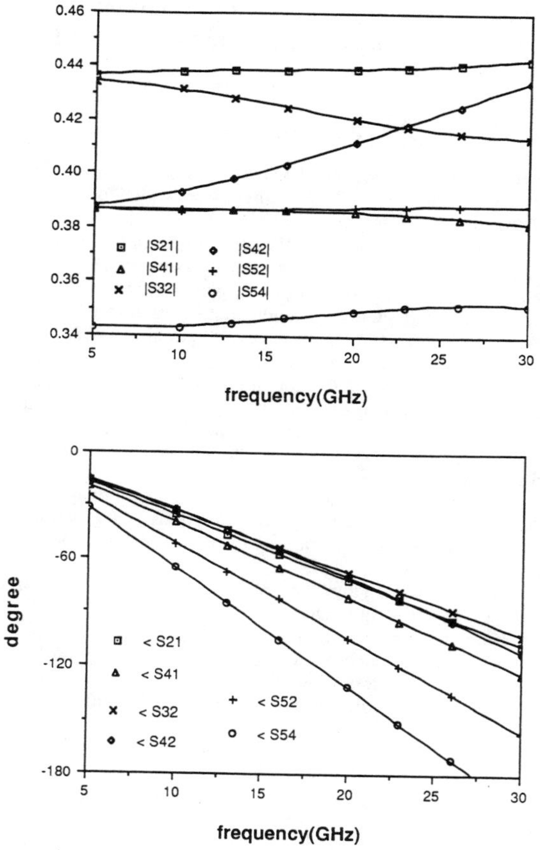

Fig. 9. The frequency responses of the five-port junction.

which is modeled as an impedance boundary condition (the value of the impedance is half of that used in [8] when the strip thickness is much larger than the skin depth).

To demonstrate the ability of the P-mesh algorithm and the code, we provide a brief analysis of a five-port junction (see Fig. 8). Static theory tells that $|S_{2,1}| = |S_{3,2}|$ and $|S_{4,1}| = |S_{4,2}| = |S_{5,2}|$. This is true at low frequency. But the high frequency responses change very much (see Fig. 9). More complicated circuits can be solved in the same way.

VI. Conclusions

We have demonstrated the P-mesh algorithm and its applications in the analysis of MMIC circuits. It can also be applied to analyze microstrip antennas [24]. In fact, almost any planar structures can be analyzed using the P-mesh algorithm efficiently and accurately. The following conclusions appear to be in order for MMIC circuits:

1. The P-mesh algorithm and code are versatile, efficient and accurate.
2. Coupling between elements in a MMIC circuit may strong enough to change the frequency responses very much.
3. Accumulation of small reflections from junctions and bends may result in strong resonances and the resonances may cause serious power loss in a MMIC circuit.
4. Metallic loss might be a very important loss factor, especially when resonances are established.

References

[1] A. Farrar and A. T. Adams, "Matrix methods for microstrip three-dimensional problems," *IEEE Trans. Microwave Theory Tech.*, vol. MTT-20. pp. 497-504, Aug. 1972.

[2] P. Silvester and P. Benedek, "Microstrip discontinuity capacitance for right-angle bends, T-junctions and crossings," *IEEE Trans. Microwave Theory Tech.*, vol. MTT-21, pp. 341-346, May 1973.

[3] R. Chadha and K. C. Gupta, "Segmentation method using impedance matrices for analysis of planar microwave circuits," *IEEE Trans. Microwave Theory Tech.*, vol. MTT-29, pp. 71-74, Jan. 1981.

[4] T. S. Chu and T. Itoh, "Comparative study of mode matching formulations of microstrip discontinuity problems," *IEEE Trans. Microwave Theory Tech.*, vol. MTT-33, pp. 1018-1023, Oct. 1985.

[5] R. F. Harrington, *Field Computations by Moment Methods*. New York: Macmillan, 1968, Ch. 4.

[6] B. L. Brim and D. C. Chang, "Accelerated numerical computation of the spatial domain dyadic Green's functions of a grounded dielectric slab," *National Science Meeting Dig.*, p. 164, Boulder, CO, Jan. 1987.

[7] D. I. Wu, D. C. Chang, and B. I. Brim, "Accurate numerical modelling of microstrip junctions and discontinuities," *Int. J. Microwave Millimeter-wave Computer Aided Engineering*, to be published.

[8] J. R. Mosig, "Arbitrarily shaped microstrip structures and their analysis with a mixed potential integral equation," *IEEE Trans. Microwave Theory Tech.*, vol. 36, pp. 314-323, Feb. 1988.

[9] B. J. Rubin, "Modeling of arbitrarily shaped signal lines and discontinuities," *IEEE Trans. Microwave Theory Tech.*, vol. 37, pp. 1057-1060, June 1989.

[10] W. Wertgen and R. H. Jansen, "Efficient direct and iterative electrodynamic analysis of geometrically complex MIC and MMIC structures," *Int. J. Numerical Modelling: Electronic Networks, Devices and Fields*, vol. 2, pp. 153-186, Feb. 1989.

[11] R. W. Jackson, "Full-wave, finite element analysis of irregular microstrip discontinuities," *IEEE Trans. Microwave Theory Tech.*, vol. 37, pp. 81-89, Jan. 1989.

[12] J. C. Rautio and R. F. Harrington, "Preliminary results of a time-harmonic electromagnetic analysis of shielded microstrip circuits," in *27th Automatic RF Techniques Group Conf. Dig.*, Baltimore, June 1986, pp. 121-134.

[13] P. B. Katehi and N. G. Alexopoulos, "Frequency-dependent characteristics of microstrip discontinuities in millimeter-wave integrated circuits," *IEEE Trans. Microwave Theory Tech.*, vol. MTT-33, pp. 1029-1035, Oct. 1985.

[14] R. W. Jackson and D. M. Pozar, "Full-wave analysis of microstrip open-end and gap discontinuities," *IEEE Trans. Microwave Theory Tech.*, vol. MTT-33, pp. 1036-1042, Oct. 1985.

[15] J. X. Zheng and D. C. Chang, "Convergence of the numerical solution for a microstrip junction based upon a triangular cell expansion," in *National Science Meeting Dig.*, p. 212, Boulder, CO, Jan. 1990.

[16] L. J. Segerlind, *Applied Finite Element Analysis*. New York: Wiley, 1984, pp. 27-99.

[17] D. C. Chang and J. X. Zheng, "Numerical modeling of planar circuits with pseudo meshes," *National Science Meeting Dig.*, Boulder, CO, Jan. 1990, p. 265.

[18] J. X. Zheng and D. C. Chang, "Numerical modeling of chamfered bends and other microstrip junctions of general shape in MMICs," *IEEE MTT Int. Microwave Symp. Dig.*, Dallas, May 1990, pp. 709-712.

[19] J. X. Zheng, "Electromagnetic Modeling of Microstrip Circuit Discontinuities and Antennas of Arbitrary Shape," Ph.D. dissertation, University of Colorado at Boulder, 1990, ch. 4.

[20] T. Y. Yang, *Finite Element Structural Analysis*. Englewood Cliffs, NJ: Prentice-Hall, pp. 288-296.

[21] C. Goldsmith, Texas Instrument, private communication.

[22] A. Sabban and K. C. Gupta, "A planar-lumped model for coupled microstrip line discontinuities," *IEEE Trans. Microwave Theory Tech.* Special Issue, vol. 38, Dec. 1990.

[23] K. Larson and J. Dunn, University of Colorado at Boulder, private communication.

[24] J. X. Zheng and D. C. Chang, "Computer-aided design of electromagnetically-coupled and tuned, wide band microstrip patch antennas," in *IEEE AP-Symp. Dig.*, Dallas, May 1990, pp. 1120-1123.

Chapter 6

Finite-Difference, Finite-Element, and TLM Methods

IN the past several years there has been increasing interest in applying three-dimensional full-wave methods to the analysis of microwave and millimeter-wave planar circuits. The integral equation approach discussed in Chapter 5 is very convenient when electric currents in the planar circuit are confined to one or more two-dimensional planes. To account for vertically oriented currents (such as in via holes and air bridges), a variety of three-dimensional full-wave methods have been reported in the literature. Of these methods, three particular methods have become the most popular, namely, the finite-difference time-domain (FD-TD) method, the finite-element method (FEM), and the transmission line matrix (TLM) method. Each of these methods has its own unique characteristics. In the FD-TD method, Maxwell's equations are discretized in space and time over a finite volume, and the derivatives are approximated by finite differences. By appropriately selecting the points at which the various field components are to be evaluated, the resulting set of finite-difference equations can then be solved, and a solution that satisfies the boundary conditions can be obtained [6.1–6.3].

The TLM method exploits the analogy between field propagation in space and voltage-current propagation in a spatial transmission line network. In the TLM method Maxwell's equations are discretized in space and represented by a transmission line matrix. The numerical calculations are performed by exciting the matrix at specific points by voltage or current impulses. The propagation of these impulses over the matrix are then followed as they are scattered by the nodes and bounce at boundaries. The output that is observed at a chosen point consists of a series of impulses separated by constant time intervals. The Fourier transform of this output function leads to the matrix response to a sinusoidal input. The transient characteristics of this method ensures that the whole frequency spectrum, up to the matrix cutoff, is included in the final answer [6.4–6.7]. A good description of the TLM method is given in [6.8].

It is quite clear from the foregoing discussion that the TLM and the FD-TD methods are similar. In a recent paper Hoefer [6.9] compared the two methods with each other and with other time-domain electromagnetic simulation algorithms as well. In this interesting paper Hoefer stated that as a general rule, the TLM and the FD-TD methods produce practically identical results. Hoefer also stated that for each TLM scheme there exists an equivalent FD-TD formulation, and vice versa, which suggests that one method can be derived from the other [6.10]. This relationship between the TLM method and the FD-TD method was demonstrated mathematically earlier by P. B. Johns [6.11] but only under certain conditions. Additional discussions on the relationship between the FD-TD technique and the TLM method are available in [6.12–6.16].

The finite-element method and its variants are somewhat different from the TLM and the FD-TD methods. This method was developed initially in mechanical and civil engineering applications and has been of increasing importance in the area of electromagnetics. A general treatment of the method itself as a numerical technique can be found in many books [6.17–6.19]. In addition, a very enlightening description of the finite-element method and its application to electromagnetic problems can be found in [6.20]. Basically, the FEM is based on dividing the region of interest into a number of nonoverlapping surface or volume subregions (depending on whether a two- or three-dimensional structure is being considered) called finite elements. Regardless of the shape of the element, the unknown function within each element (which may be scalar potential or a vectorial field component) is then approximated by a polynomial function. Each of these functions must satisfy some particular boundary conditions such as continuity. Application of the Rayleigh-Ritz procedure then transforms the functional minimization into a linear system of equations. The equations to be solved are usually expressed not in terms of the field variables but in terms of an integral-type functional. The functional is chosen such that the field solution makes the functional stationary. The total functional is the sum of the integral over each element.

All three methods, the FD-TD, the TLM, and the FEM, have become well-established techniques for characterizing passive microwave components. Eight representative reprint papers describing the applications of these techniques to planar components are included in this chapter. Papers 6.1 through 6.4 are FD-TD application papers. Paper 6.1 describes the suitability of the FD-TD method for characterizing microstrip discontinuities. The validity of this approach was demonstrated by applying it to five different symmetric microstrip discontinuities. Paper 6.2 shows, for the first time, the efficiency and the accuracy of the FD-TD approach when applied to the analysis of two-dimensional passive microstrip components. The input impedance of a line-fed rectangular patch antenna and the frequency-dependent scattering parameters of two commonly used components (a low-pass filter and a branch line coupler) are calculated. The predicted results are found to be in excellent agreement with the measurements.

Paper 6.3 presents an overview on the application of the FD-TD technique to the simulation and design of planar microwave integrated circuits. It is also displayed how three-dimensional microwave circuit structures and coplanar microwave circuits are modeled by the FD-TD approach. FD-TD analyses of widely used planar microstrip and coplanar structures, like radial lines, curved stubs, microstrip meander lines, band-reject filters, and coplanar T-junction are presented. Furthermore, a demonstration of how the FD-TD approach can be combined with semiconductor modeling to analyze nonlinear microwave integrated circuits is given.

In Paper 6.4 a two-dimensional finite-difference time-domain algorithm is used advantageously to analyze a class of shielded, fully or partially open millimeter-wave and optical waveguide structures. The paper shows how this two-dimensional FD-TD algorithm could lead to a significant reduction in computational time and storage requirements as compared with the conventional three-dimensional approach. The effectiveness of this approach is demonstrated by applying it to a variety of integrated circuit structures such as dielectric waveguide, insulated and trapped image guides, coplanar lines, and microstrip lines.

Additional applications of the FD-TD method to the analysis of planar circuits can be found in [6.21–6.34]. In [6.34] the conventional FD-TD algorithm is adapted to nonorthogonal computational grids and is used to characterize three-dimensional discontinuity problems. With this algorithm, an accuracy better than is possible with the staircase approach (conventionally used in the FD-TD algorithm) has been accomplished.

Paper 6.5 presents an overview of the TLM method of analysis, discusses its historical development, and also describes in detail its versatility and flexibility. An impressive list of potential applications as well as a list of references describing specific two-dimensional and three-dimensional electromagnetic applications are also given. Paper 6.6 discusses the application of a numerical method called Bergeron's method to the analysis of microstrip components. This method was first developed in Japan [6.35]. The formulation of the Bergeron method is described as being fundamentally equivalent to that of the TLM method. Both methods are based on the property of the traveling wave that is formulated as the general solution of one-dimensional wave equation by d'Alembert. The Bergeron's method, however, is characterized [6.35] as being more convenient because it provides direct handling of the electromagnetic field variables and the characteristics of the medium, in contrast with the TLM method in which each variable is divided into an incident and reflected components. Paper 6.6 illustrates the capability and validity of the Bergeron's method by calculating the frequency characteristics of a side-coupled microstrip filter. The calcuated scattering parameters for this planar component were compared with measured results as well as results produced by the FEM method.

The application of the FEM method to the analysis of planar components is discussed and illustrated in Papers 6.7 and 6.8. A new numerical procedure called the transfinite-element method [6.36] is applied in Paper 6.7 to the analysis of planar microstrip components. This new technique is based on using the FEM method in conjunction with the planar waveguide model. The effectiveness of this technique was demonstrated by employing it to analyze several widely encountered microstrip- and stripline-type planar components. The use of a full-wave finite-element technique to analyze microstrip discontinuities located on open microstrip is illustrated by Webb in Paper 6.8. This technique is described as being capable of taking into account all electromagnetic phenomena, including radiation and surface wave effects, as well as coupling due to closely spaced junctions. To verify its accuracy, this technique was used to analyze a step discontinuity, a straight rectangular stub, and a bent rectangular stub. The calculated characteristics for these components were found to be in reasonable agreement with the measurements. Additional discussion on the application of the FEM method to the analysis of planar components can be found in [6.35–6.41]. A treatment of the absorbing boundary conditions for the FEM analysis of planar circuits can be found in [6.42].

References

[6.1] Yee, K. S. "Numerical solution of initial boundary value problems involving Maxwell's equations in isotropic media." *IEEE Trans. Antennas Propagat.*, Vol. AP-14, pp. 302–307, May 1966.

[6.2] Mur, G. "Absorbing boundary conditions for the finite-difference approximation of the time-domain electromagnetic-field equations." *IEEE Trans. Electromagnetic Compatibility,* Vol. EMC-23, pp. 377–381, November 1981.

[6.3] Choi, D. H. and W. J. R. Hoefer. "The finite-difference time-domain method and its application to eigenvalue problems." *IEEE Trans. Microwave Theory Tech.*, Vol. MTT-34, pp. 1464–1470, December 1986.

[6.4] Johns, P. B. and R. L. Beurle. "Numerical solution of 2-dimensional scattering problems using a transmission-line matrix." *Proc. IEE*, Vol. 118, pp. 1203–1208, September 1971.

[6.5] Johns, P. B. "The solution of inhomogeneous waveguide problems using a transmission-line matrix." *IEEE Trans. Microwave Theory Tech.*, Vol. MTT-22, pp. 209–215, March 1974.

[6.6] Akhtarzad, S. and P. B. Johns. "Solution of Maxwell's equations in three space and time by the t.l.m. method of numerical analysis." *Proc. IEE*, Vol. 122, pp. 1349–1352, December 1975.

[6.7] Hoefer, W. J. R. "Huygens and the computer—A powerful alliance in numerical electromagnetics." *Proc. IEEE,* Vol. 79, pp. 1459–1471, October 1991.

[6.8] Hoefer, W. J. R. "The transmission line matrix (TLM) method." In T. Itoh, ed., *Numerical Techniques for Microwave and Millimeter Wave Passive Structures,* Chapter 8, pp. 496–591. New York: John Wiley & Sons, 1989.

[6.9] Hoefer, W. J. R. "Time domain electromagnetic simulation for microwave CAD applications." *IEEE Trans. Microwave Theory Tech.,* Vol. MTT-40, pp. 1517–1527, July 1992.

[6.10] Chen, Z., M. Ney, and W. J. R. Hoefer. "A new finite-difference time-domain formulation and its equivalence with the TLM symmetrical condensed node." *IEEE Trans. Microwave Theory Tech.,* Vol. MTT-39, pp. 2160–2169, December 1991.

[6.11] Johns, P. B. "On the relationship between TLM and finite-difference methods for Maxwell's equations." *IEEE Trans. Microwave Theory Tech.,* Vol. MTT-35, pp. 60–61, January 1987.

[6.12] Gwarek, W. K. "Comments on the relationship between TLM and finite-difference methods for Maxwell's equations." *IEEE Trans. Microwave Theory Tech.,* Vol. MTT-35, pp. 872–873, September 1987.

[6.13] Celuch-Marcysiak, M. and W. K. Gwarek. "Formal equivalence and efficiency comparison of the FD-TD, TLM, SN methods in application to microwave CAD programs." *Proc. XXI European Microwave Conference,* pp. 199–204, Stuttgart, Germany, September 9–11, 1991.

[6.14] Simons, N. R. S. and E. Bridges. "Equivalence of propagation characteristics for the transmission line matrix and finite-difference time-domain methods in two dimensions." *IEEE Trans. Microwave Theory Tech.,* Vol. MTT-39, pp. 354–357, February 1991.

[6.15] Celuch-Marcysiak, M. and W. K. Gwarek. "Comments on: A new finite-difference time-domain formulation and its equivalence with the TLM symmetrical condensed node." *IEEE Trans. Microwave Theory Tech.,* Vol. MTT-41, pp. 168–170, January 1993.

[6.16] Chen, Z., M. M., Ney, and W. J. R. Hoefer. "Reply to comments on: A new finite-difference time-domain formulation and its equivalence with the TLM symmetrical condensed node." *IEEE Trans. Microwave Theory Tech.,* Vol. MTT-41, pp. 170–172, January 1993.

[6.17] Silvester, P. P. and R. L. Ferrari. *Finite Elements for Electrical Engineers.* New York: Cambridge University Press, 1983.

[6.18] Davies, A. J. *The Finite Element Method: A First Approach.* Oxford: Oxford University Press, 1987.

[6.19] Booton, R. C., Jr. *Computational Methods for Electromagnetics and Microwaves.* New York: John Wiley & Sons, 1992.

[6.20] Davies, J. B. "The Finite Element Method." In T. Itoh, ed., *Numerical Techniques for Microwave and Millimeter Wave Passive Structures,* Chapter 2, pp. 33–132. New York: John Wiley & Sons, 1989.

[6.21] Gwarek, W. W. "Analysis of an arbitrarily-shaped planar circuit—A time-domain approach." *IEEE Trans. Microwave Theory Tech.,* Vol. MTT-33, pp. 1067–1072, October 1985.

[6.22] Gwarek, W. K. "Analysis of arbitrarily-shaped two-dimensional microwave circuits by finite-difference time-domain method." *IEEE Trans. Microwave Theory Tech.,* Vol. MTT-36, pp. 738–744, April 1988.

[6.23] Zhang, X., J. Fang, K. K. Mei, and Y. Liu. "Calculations of the dispersive characteristics of microstrips by the time-domain finite difference method." *IEEE Trans. Microwave Theory Tech.,* Vol. MTT-36, pp. 263–267, February 1988.

[6.24] Reineix, A. and B. Jecko. "Analysis of microstrip patch antennas using finite difference time domain method." *IEEE Trans. Antennas Propagat.,* Vol. AP-37, pp. 1361–1369, November 1989.

[6.25] Rittweger, M. and I. Wolff. "Analysis of complex passive M(MIC)-components using the finite difference time-domain approach." *1990 IEEE MTT-S International Microwave Symp.,* Dallas, Texas, pp. 1147–1150, May 8–10, 1990.

[6.26] Gwarek, W. K. "Inhomogeneous two-dimensional model for analysis of microstrip discontinuities." *20th European Microwave Conference,* pp. 1222–1227, September 10–13, Budapest, Hungary, 1990.

[6.27] Maeda, S., T. Kashiwa, and I. Fukai. "Analysis of crosstalk between parallel microstrips using finite-difference time-domain method." *Electronics and Communications in Japan,* Pt. 2, Vol. 74, No. 11, pp. 23–30, 1991.

[6.28] Paul, D. L., E. M. Daniel, and C. J. Railton. "Fast finite difference time domain method for the analysis of planar microstrip circuits." *Proc. 21st European Microwave Conf.,* pp. 303–308, Stuttgart, Germany, September 9–11, 1991.

[6.29[Gwarek, W. K. and C. Mroczkowski. "An inhomogeneous two-dimensional model for the analysis of microstrip discontinuities." *IEEE Trans. Microwave Theory Tech.,* Vol. MTT-39, pp. 1655–1658, September 1991.

[6.30] Buchanan, W. J. and N. K. Gupta. "Simulation of three-dimensional finite-difference time domain method on limited memory systems." *IEEE Int'l. Conf. on Computation in Electromagnetics,* Conf. Publication #350, London, November 25–27, 1991.

[6.31] Ko, W. L. and R. Mittra. "A combination of FD-TD and Prony's method for analyzing microwave integrated circuits." *IEEE Trans. Microwave Theory Tech.,* Vol. MTT-39, pp. 2176–2181, December 1991.

[6.32] Shorthouse, D. B. and C. J. Railton. "The incorporation of static field solution into the finite difference time domain algorithm." *IEEE Trans. Microwave Theory Tech.,* Vol. MTT-40, pp. 986–994, May 1992.

[6.33] Haffa, S., D. Hollmann, and W. Wiesbeck. "The finite difference method for S-parameter calculation of arbitrary three-dimensional structures." *IEEE Trans. Microwave Theory Tech.,* Vol. MTT-40, pp. 1602–1610, August 1992.

[6.34] Lee, J. F., R. Palendech, and R. Mittra. "Modelling three-dimensional discontinuities in waveguides using non-orthogonal FD-TD algorithm." *IEEE Trans. Microwave Theory Tech.,* Vol. MTT-40, pp. 346–352, February 1992.

[6.34] Yoshida, N. and I. Fukai. "Transient analysis of a stripline having a corner in three-dimensional space." *IEEE Trans. Microwave Theory Tech.,* Vol. MTT-32, pp. 491–498, May 1984.

[6.35] Lee, J. F. and Z. J. Cendes. "Transfinite elements: A highly efficient procedure for modeling open field problems." *J. Appl. Phys.,* Vol. 61, pp. 3913–3915, April 1987.

[6.36] Silvester, P. "Finite element analysis of planar microwave networks." *IEEE Trans. Microwave Theory Tech.,* Vol. MTT-21, pp. 104–108, February 1973.

[6.37] Garcia, P. and J. P. Webb. "Optimization of planar devices by the finite element method." *IEEE Trans. Microwave Theory Tech.,* Vol. MTT-38, pp. 48–53, January 1990.

[6.38] Lee, J. F. "Analysis of passive microwave devices by using three-dimensional tangential vector finite elements." *Int'l. J. Numerical Modelling: Electronic Networks, Devices and Fields,* Vol. 3, No. 4, pp. 235–246, 1990.

[6.39] Azizur Rahman, B. M., F. A. Fernandez, and J. B. Davies. "Review of finite element methods for microwave and optical waveguides." *Proc. IEEE,* Vol. 79, pp. 1442–1448, October 1981.

[6.40] Lo, J. O. Y., A. Konrad, J. L. Coulomb, and J. C. Sabonnadiere. "Microstrip discontinuity analysis by the time domain finite elements." *9th Annual Review of Progress in Applied Computational Electromagnetics,* pp. 846–855, Monterey, California, March 1993.

[6.41] Konrad, A. and J. O. Y. Lo. "Time domain solution of planar circuits." *J. Electromagnetic Waves and Applications,* Vol. 7, No. 1, pp. 77–92, 1993.

[6.42] Webb, J. P. "Absorbing boundary conditions for the finite-element analysis of planar devices." *IEEE Trans. Microwave Theory Tech.,* Vol. MTT-38, pp. 1328–1332, September 1990.

Paper 6.1

Time-Domain Finite Difference Approach to the Calculation of the Frequency-Dependent Characteristics of Microstrip Discontinuities

XIAOLEI ZHANG AND KENNETH K. MEI, FELLOW, IEEE

Abstract —The frequency-dependent characteristics of the microstrip discontinuities have previously been analyzed using several full-wave approaches. The time-domain finite different (TD–FD) method presented in this paper is another independent approach and is relatively new in its application for obtaining the frequency-domain results for microwave components [26]. The purpose of this paper is to establish the validity of the TD–FD method in modeling circuit components for MMIC CAD applications.

I. Introduction

MICROSTRIP discontinuities (Fig. 1) are the basic constituent elements of microstrip integrated circuits. The accurate modeling of these discontinuities using different numerical approaches is one of the most important topics in microwave CAD. The current cut-and-try cycles in the design of microstrip integrated circuits will be greatly reduced if the frequency-dependent characteristics of the discontinuities can be obtained with certainty. Using network concepts, various microstrip resonators, couplers, and filters can be directly analyzed from the interconnection of microstrip discontinuities and microstrip line segments.

The study of microstrip discontinuities started in the early 1960's. For nearly a decade, the analyses were mostly quasi-static in nature [1]–[12]. The first accurate full-wave frequency-dependent analysis appeared around 1975 [13]–[15]. This approach began with the use of a waveguide model with electric-wall top and bottom planes and magnetic-wall sides planes to characterize the microstrip. The effective dielectric constant of the filling and the width of the guide are assumed to be frequency dependent and are determined in such a way that the model and the actual microstrip line have the same frequency-dependent propagation constant and characteristic impedance. Using the waveguide model to represent the original microstrip, the fields at the region of the discontinuities are expanded into waveguide modes, and the modes of different regions are matched at intersection planes. From the matching coefficients the S matrix for different propagation modes can thus be calculated. The waveguide model approach is efficient and has reasonable accuracy for calculating the magnitude of the S parameters in the lower frequency range, but it is not able to take into account the radiation effect (since the model is a closed one) and the surface wave generation. Besides, the mode-matching step will also introduce error due to the fact that the actual modes excited in the microstrip discontinuities are not the same as those used in the model and accordingly will not match in exactly the same way. There is also an obvious limitation on the kinds of structures this method can be applied to. It cannot, for example, be used to analyze the microstrip open-end structure where one side of the discontinuity is not connected to a microstrip and where the radiation and surface waves are present.

A full-wave approach to the microstrip open end problem was first proposed by James and Henderson in 1979 [16]. The analysis on the far end of the microstrip open end, where the surface wave and the radiation wave are the constitutents of the fields, is carried out using an analytic mode-expansion technique. On the microstrip side, a TEM wave is taken as the dominant mode incident field, and the semiempirical results for the propagation constant and the characteristic impedance are used for this incident wave. The fields at both sides are matched at the interface and a variational step is taken to reduce the error introduced by the assumption of the TEM field pattern where the electric field has a vertical constant value under the strip and is zero elsewhere in the transverse plane. Mainly due to the roughness of the field pattern assumed, the results of this method are not very accurate, but the analysis did provide valuable physical insight.

Another important method which has been used by several investigators to model the microstrip discontinuities is the spectral–domain approach [17]–[19]. In using this method to analyze the shielded or covered structures, the fields and currents involved are Fourier transformed (with respect to the space variables) into the so-called spectral domain. The shape of the current on the microstrip is assumed to be close to the actual current distribution and is easily Fourier-transformable. The spectral-domain components of the fields and currents are

Manuscript received May 3, 1988; revised July 29, 1988. This work was supported by the Office of Naval Research Under Contract N00014-86-K-0420.

The authors are with the Department of Electrical Engineering and Computer Sciences, University of California at Berkeley, Berkeley, CA 94720.

IEEE Log Number 8823766.

Reprinted from *IEEE Trans. Microwave Theory Tech.*, vol. 36, no. 12, pp. 1775–1787, Dec. 1988.

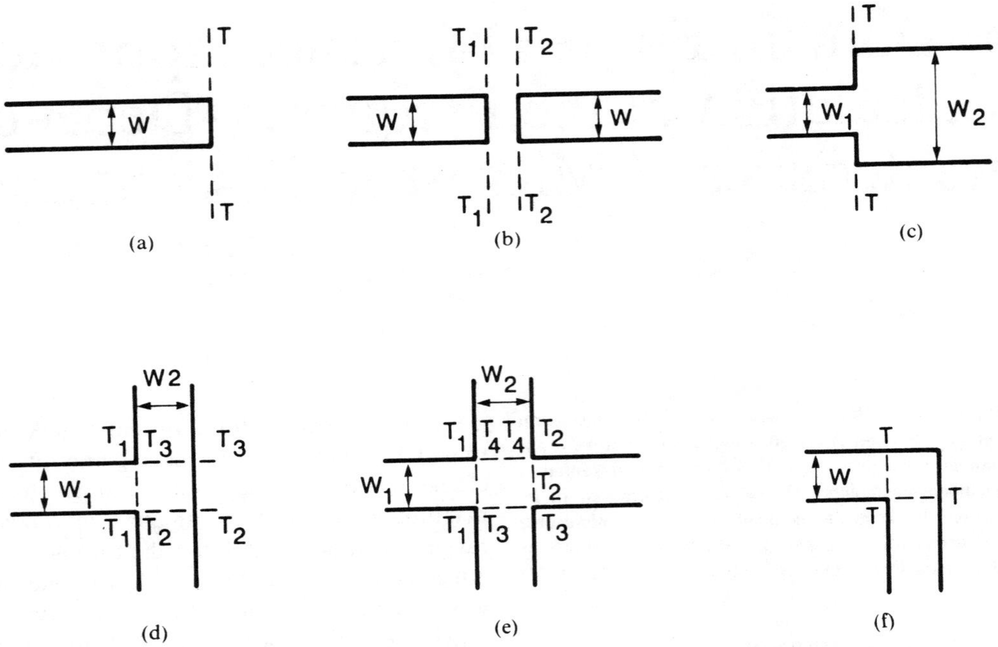

Fig. 1. Microstrip discontinuities. (a) Open-end. (b) Gap. (c) Step-in-width. (d) T junction. (e) Cross-junction. (f) Bend.

related according to the field continuity and boundary conditions and thus set up a system of equations for the variables. The inverse-transformed field solutions are used to calculate the S parameters.

Although a relatively accurate method for the type of components it is capable of calculating, the spectral-domain approach depends strongly on the current distributions assumed, which in many cases are hard to specify with high accuracy; thus it is limited in its applications. Besides, the frequency range which can be dealt with by this method is limited due to the difficulties which arise near the cutoff frequency of the higher order mode of the microstrip.

In recent years, the moment method has also been used by several investigators [20], [21] on the discontinuity problems. This method could in principle be an accurate one with wide applications, but due to the complexity of the Green's functions for the microstrip configurations it is not economical to make very fine numerical divisions to the microstrip for accurate results. In fact in many cases only a rational function form is used to represent the transverse current distribution on the microstrip, which may not correspond to the actual current distribution up to a certain frequency.

All the above-mentioned investigations are done in frequency domain; that is, the data for the whole frequency range are calculated one frequency at a time. It is an expensive task when the results of a wide frequency range are sought. This led us to seek an alternative way of calculating the frequency-domain data. Since a pulse response contains all the information of a system for the whole frequency range, it is a natural approach to use a pulse in the time domain to excite the microstrip structures, and from the time-domain pulse response to extract the frequency-domain characteristics of the system via the Fourier transform [26].

One numerical scheme which can be used to calculate the time-domain fields is the time-domain finite difference (TD–FD) method. It was first proposed by K. S. Yee in 1966 [22] and has been used by many investigators to solve electromagnetic scattering problems. Other numerical methods which can be used to solve this type of initial boundary value problem include the TLM method and Bergeron's method. Among these methods the TD–FD method is the most direct from a mathematical point of view, and is especially suitable for the accurate calculation of the microstrip fields, the reasons for which will be explained below.

Early investigators used the time-domain methods as a tool to obtain qualitative results that graphically illustrate the field propagation rather than to obtain design data via the Fourier transform of time domain results [24], [25]. In the process of our investigation, it has been found [26] that the Fourier transform of the time-domain results is very sensitive to numerical errors, notably those resulting from the imperfect treatment of the absorbing boundary conditions used to truncate the numerical computations of an open structure. Thus, even though the time-domain results may be reasonably accurate, the frequency-domain results obtained from their Fourier transform may not be acceptable as useful data.

The present available absorbing boundary conditions for the discretized wave equations are either not good enough in quality or require impractically large computer memories. Recently, a new type of absorbing boundary algorithm has been developed [29], [30] which can greatly improve the quality of the local absorbing boundary conditions. Using this new boundary treatment, together with a

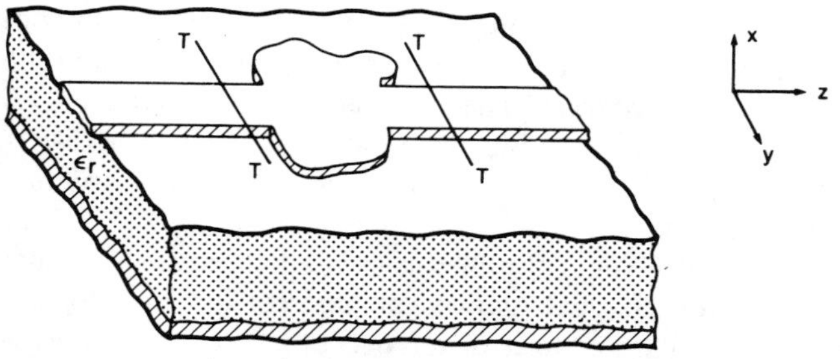

Fig. 2. A generalized microstrip discontinuity.

large enough computation domain, an accurate time-domain field can be obtained which can be used in the following Fourier transform step.

The calculated frequency-domain design data were compared with the available published results. These comparisons further demonstrate that the TD–FD approach is a viable method for modeling microstrip components.

II. Formulation of the Problem and the Numerical Method

A. General Formulation of the Problem

The generalized microstrip discontinuity under investigation is shown in Fig. 2 (a more general one will be an N-port structure instead of a two-port), where the strip and the bottom plane are made of a perfect conductor ($\sigma = \infty$) and the substrate has a relative dielectric constant of ϵ_r. The structure is assumed to be in an open environment, that is, above the dielectric and the metal strip surface, free space is assumed to extend to infinity; in the horizontal direction, apart from the discontinuity region, the substrate-ground structure also extends uniformly into infinity.

The Maxwell equations governing the solution of this problem are

$$\frac{\partial \vec{E}}{\partial t} = \frac{1}{\epsilon_i} \nabla \times \vec{H}$$

$$\frac{\partial \vec{H}}{\partial t} = -\frac{1}{\mu_0} \nabla \times \vec{E} \qquad (1)$$

where $i = 1,2$ represents the substrate and the free-space region, respectively. At the interface of the two regions, the field continuity conditions are enforced.

For the uniqueness of the solution of these Maxwell equations, the following conditions must be satisfied:

a) The initial condition for the fields must be specified on the whole domain of interest; that is, $\vec{E}(\vec{r}, t = 0)$ and $\vec{H}(\vec{r}, t = 0)$ must be given everywhere inside the computation domain.
b) The tangential components of $\vec{E}$ and $\vec{H}$ on the boundary of the domain of interest must be given for all $t > 0$. For the boundary at infinity, Sommerfeld's radiation condition must be satisfied; that is, the wave at infinity must be of an outgoing type.

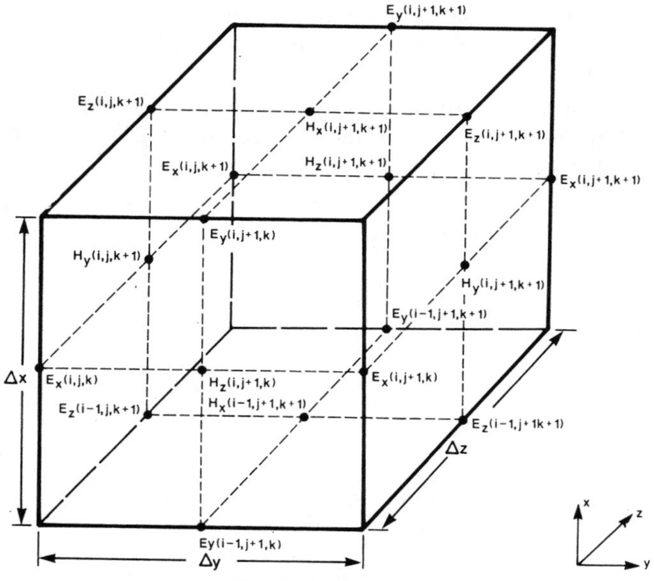

Fig. 3. Yee's mesh.

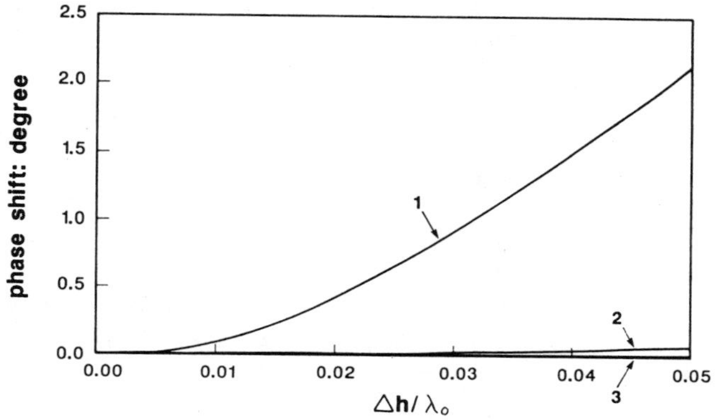

Fig. 4. Comparison of the numerical and microstrip dispersions (phase shift per time step): 1: microstrip dispersion ($w/h = 1.5$, $\epsilon_r = 13$); 2: leapfrog scheme, $c\Delta t/\Delta h = 1/7$; 3: a fourth-order scheme, $c\Delta t/\Delta h = 1/7$ [27].

B. The Time-Domain Finite Difference Algorithm

There are many ways to solve the system of Maxwell equations in (1) numerically. The TD–FD algorithm is one of the most suitable schemes for the purpose of this investigation.

To simulate the wave propagation in three dimensions, Yee [22] arranged the spatial nodal points, where different components of $\vec{E}$ and $\vec{H}$ are to be calculated, as in Fig. 3. The repetitive arrangement of the cells of Fig. 3 fills the computation domain with a finite difference mesh. Every component of $\vec{H}$ can be obtained by the loop integral of $\vec{E}$ using the four surrounding $\vec{E}$ nodal values according to Maxwell's curl equation for $\vec{E}$. A similar approach holds for the calculation of $\vec{H}$.

In this algorithm, not only the placement of the $\vec{E}$ and $\vec{H}$ nodes are off in space by half a space step, but the time instants when the $\vec{E}$ or $\vec{H}$ fields are calculated are also off by half a time step. To be more specific, if the components of $\vec{E}$ are calculated at $n\Delta t$, where Δt is the discretization unit in time, or the time step, and n is any nonnegative integer, the components of $\vec{H}$ are calculated at $(n+1/2)\Delta t$. For this reason, this algorithm is also called the leapfrog method.

To summarize, for a homogeneous region of space, the discretization of the Maxwell curl equations (1) leads to the following:

$$E_x^{n+1}(i,j,k) = E_x^n(i,j,k) + \frac{\Delta t}{\epsilon} \cdot \left[\frac{H_z^{n+1/2}(i,j+1,k) - H_z^{n+1/2}(i,j,k)}{\Delta y} - \frac{H_y^{n+1/2}(i,j,k+1) - H_y^{n+1/2}(i,j,k)}{\Delta z} \right]$$

$$E_y^{n+1}(i,j,k) = E_y^n(i,j,k) + \frac{\Delta t}{\epsilon} \cdot \left[\frac{H_x^{n+1/2}(i,j,k+1) - H_x^{n+1/2}(i,j,k)}{\Delta z} - \frac{H_z^{n+1/2}(i+1,j,k) - H_z^{n+1/2}(i,j,k)}{\Delta x} \right]$$

$$\cdot E_z^{n+1}(i,j,k) = E_z^n(i,j,k) + \frac{\Delta t}{\epsilon} \cdot \left[\frac{H_y^{n+1/2}(i+1,j,k) - H_y^{n+1/2}(i,j,k)}{\Delta x} - \frac{H_x^{n+1/2}(i,j+1,k) - H_x^{n+1/2}(i,j,k)}{\Delta y} \right]$$

$$H_x^{n+1/2}(i,j,k) = H_x^{n-1/2}(i,j,k) - \frac{\Delta t}{\mu} \cdot \left[\frac{E_z^n(i,j,k) - E_z^n(i,j-1,k)}{\Delta y} - \frac{E_y^n(i,j,k) - E_y^n(i,j,k-1)}{\Delta z} \right]$$

$$H_y^{n+1/2}(i,j,k) = H_y^{n-1/2}(i,j,k) - \frac{\Delta t}{\mu} \cdot \left[\frac{E_x^n(i,j,k) - E_x^n(i,j,k-1)}{\Delta z} - \frac{E_z^n(i,j,k) - E_z^n(i-1,j,k)}{\Delta x} \right]$$

$$H_z^{n+1/2}(i,j,k) = H_z^{n-1/2}(i,j,k) - \frac{\Delta t}{\mu} \cdot \left[\frac{E_y^n(i,j,k) - E_y^n(i-1,j,k)}{\Delta x} - \frac{E_x^n(i,j,k) - E_x^n(i,j-1,k)}{\Delta y} \right] \quad (2)$$

where Δx, Δy, and Δz are the space discretization units in the x, y, and z directions, respectively, and Δt is the time discretization interval. The numbering of the different $\vec{E}$ and $\vec{H}$ components is illustrated in Fig. 3, which is different from that in Yee's original paper due to programming considerations.

The TD–FD algorithm has several advantages over other schemes for the calculation of microstrip time-domain fields. First, the central difference nature of the leapfrog method makes it a relatively accurate method (second-order accuracy in both time and space), compared to other first-order schemes. Second, there is no need for special treatment of the edge of the microstrip if the tangential $\vec{E}$ and vertical $\vec{H}$ components are arranged on the metal strip and only the parallel components of the electric field are arranged on the edge of the strip. Finally, the leapfrog algorithm has the unique characteristics that the numerical scheme has no dissipation (amplitude increase or decrease for any frequency component) and only a small amount of dispersion [27]. It has been shown [28], [29] that the numerical dispersion is negligible compared to the physical dispersion of the microstrip structure, as seen in Fig. 4 for comparing the microstrip dispersion and numerical dispersions for the leapfrog scheme and for a fourth-order finite difference scheme used to solve the one-dimensional wave equation. (The actual frequency range of interest for the microstrip discontinuity problems corresponds to $\Delta h/\lambda_0 \approx 0$–$0.03$.) Thus, no higher order finite difference is needed for the accurate modeling of the microstrip structures.

For any finite difference scheme, a stability condition must be found which guarantees that the numerical error generated in one step of the calculation does not accumulate and grow. The stability criterion of Yee's algorithm is the Courant condition [23]:

$$v_{\max} \cdot \Delta t \leqslant \frac{1}{\sqrt{\dfrac{1}{\Delta x^2} + \dfrac{1}{\Delta y^2} + \dfrac{1}{\Delta z^2}}}. \quad (3)$$

For the special case of $\Delta x = \Delta y = \Delta z = \Delta h$, (3) becomes

$$v_{\max} \cdot \Delta t \leqslant \frac{1}{\sqrt{3}} \cdot \Delta h. \quad (3a)$$

where $v_{\max}$ is the maximum signal phase velocity in the configuration being considered.

The stability of the absorbing boundary condition cannot be achieved exactly due to the imperfection of nearly all the presently available absorbing boundary conditions for the numerical solution of wave equations (there will always be some unrealistic reflection wave going back to the computation domain due to the boundary treatment). But since the computation lasts for only a limited time, one can always minimize the influence of the unrealistic reflections by making the computation domain sufficiently large and stopping the computation after the useful information has been obtained.

C. Choice of the Excitation Pulse

The excitation pulse used in this investigation has been chosen to be Gaussian in shape. A Gaussian pulse has a smooth waveform in time, and its Fourier transform (spectrum) is also a Gaussian pulse centered at zero frequency. These unique properties make it a perfect choice for investigating the frequency-dependent characteristics of the microstrip discontinuities via the Fourier transform of the pulse response.

An ideal Gaussian pulse which propagate in the $+z$ direction will have the following expression:

$$g(t,z) = \exp\left[-\frac{\left(t - t_0 - \dfrac{z - z_0}{v}\right)^2}{T^2}\right] \tag{4}$$

where v is the velocity of the pulse in the specific medium, and the pulse has its maximum at $z = z_0$ when $t = t_0$.

The Fourier transform of the above Gaussian pulse has the form

$$G(f) \propto \exp\left[-\pi^2 T^2 f^2\right]. \tag{5}$$

The choices of the parameters T, t_0, and z_0 are subject to two requirements. The first is that after the space discretization interval Δz has been chosen fine enough to represent the smallest dimension of the structure and the time discretization interval Δt has been chosen small enough to meet the stability criterion (3), the Gaussian pulse must be wide enough to contain enough space divisions for a good resolution. And at the same time, the spectrum of the pulse must be wide enough (or the pulse must still be narrow enough) to maintain a substantial value within the frequency range of interest. If these last two conditions cannot be satisfied simultaneously, Δz has to be rechosen to be even smaller.

The pulse width W chosen in this work is approximately 20 space steps. We define the pulse width to be the width between the two symmetric points which have 5 percent of the maximum value of the pulse. Therefore, T is determined from

$$\exp\left[-\frac{\left(\dfrac{W}{2}\right)^2}{(vT)^2}\right] = \exp(-3)(\approx 5\%) \tag{6}$$

or

$$T = \frac{1}{\sqrt{3}} \cdot \frac{10\Delta z}{v}. \tag{7}$$

By making this choice of T, the maximum frequency which can be calculated is

$$f_{\max} = \frac{1}{2T} \qquad \left(G\left(\frac{1}{2T}\right) \approx 0.1\right) \tag{8}$$

$$= \frac{1}{2} \cdot \frac{\sqrt{3}\,v}{10\Delta z} \tag{9}$$

which, with the specific Δz chosen, is high enough to cover the entire frequency range of interest, as will be shown below in the discussion of the numerical results.

The second requirement is that the choice of z_0 and t_0 be made such that initial "turn on" of the excitation will be small and smooth.

Another consideration in excitation is the specification of the spatial distribution of the field on the excitation plane. Ideally, the use of the dominant mode distribution is preferred. But this distribution is generally not known with high enough accuracy. By the use of our knowledge of the modes in the microstrip structure, a very simple field distribution can be specified at the excitation plane which serves our purposes almost as well.

The tangential electric field to be specified on the excitation plane is assumed to have only the E_x component, which is distributed uniformly under the strip and is zero elsewhere. This is not the exact dominant mode field distribution, although in the latter case the energy is also concentrated under the strip. But for frequencies under the cutoff frequency of the first waveguide type higher order mode and below the strong coupling frequency of the substrate modes (the lowest one of them will be referred to below as the inflection frequency), the only mode which can propagate down the microstrip is the dominant mode. The substrate modes which do not cut off at those frequencies have imaginary propagation (phase) constants with respect to the major wave propagation direction due to the presence of the metal strip and thus can only propagate sidewards. After the inflection frequency, there will be modes other than the dominant mode which can propagate down the line. But those modes will not contaminate our lower frequency results after the Fourier transform. This is due to the fact that the physical model and the numerical method we used are both linear; thus the modes at different frequencies will not couple energy from each other.

Therefore, as long as we allow a certain distance for the Gaussian pulse to propagate out of the excitation plane, and thus allow the unwanted substrate modes at lower frequencies to leave the central region of the microstrip, the low-frequency component of the pulse will consist of the dominant mode only. Graphically, the pulse pattern in the transverse direction gradually becomes stabilized (to its actual physical form) and a prominent edge effect manifests itself as the calculation goes on and the pulse is seen to propagate down the line.

Also, for the reasons mentioned above, caution must be taken when interpreting the Fourier transformed results after the inflection frequency, as the time-domain method does not have the capability of distinguishing between modes. These results do not correspond to the exact dominant mode results in general, although how great an influence the higher order modes have on the dominant mode results is still left to be determined.

D. Dielectric–Air Interface Treatment and the Artificial Absorbing Boundary Conditions

Fig. 5 shows the finite difference computation domain used for the discontinuity problems (in this case, a mi-

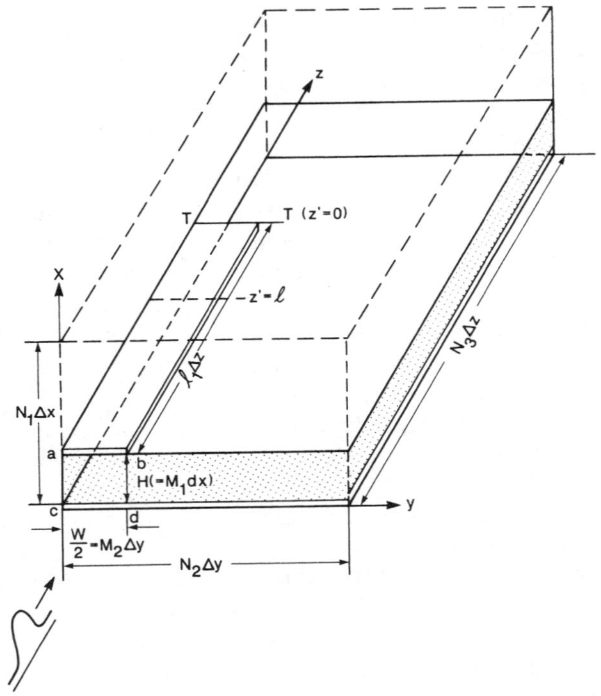

Fig. 5. Finite difference computation domain for microstrip open-end.

crostrip open end is shown as an example). Due to symmetry, only half of the structure is placed in the mesh domain with a magnetic wall at the plane of symmetry.

The finite difference form (2) of Maxwell's equations is derived in the uniform region of the medium and therefore cannot be applied to the nodal points on the dielectric–air interface or on the boundary planes of the finite difference mesh. All these points require special treatment.

The field components which lie on the dielectric–air interface are the tangential components of $\vec{E}$ (E_y and E_z) and the vertical component of $\vec{H}$ (H_x). In calculating H_x, (2) can still be used because the value of μ does not change across the boundary, and the E_y and E_z components used to calculate H_x are the tangential components with respect to the interface and are thus continuous across the boundary. To calculate E_y and E_z, however, a finite difference formulation other than (2) must be derived from the field continuity conditions across the boundary. The derivation is given in the Appendix, which is similar to that of Lin [31] for a two-dimensional finite element scheme, and the result is that E_y can be obtained by the discretization form of

$$\frac{\epsilon_1+\epsilon_2}{2}\cdot\frac{\partial E_y}{\partial t}=\frac{\partial H_x}{\partial z}-\frac{\Delta H_z}{\Delta x} \tag{10}$$

and E_z can be obtained through

$$\frac{\epsilon_1+\epsilon_2}{2}\cdot\frac{\partial E_z}{\partial t}=\frac{\Delta H_y}{\Delta x}-\frac{\partial H_x}{\partial y}. \tag{11}$$

In other words, the average value of ϵ is used in (2) for the calculation of the interface E_y and E_z nodes.

The values of E_y, E_z, and H_x vanish on the metal strip because of the assumption of a perfectly conducting surface. This also holds for the ground plane.

Since the computation domain cannot include the whole space, the finite difference mesh must be truncated to accommodate the finite computer memories. In solving our problems, the truncation planes are the side, top, and end surfaces (Fig. 5). The numerical algorithm on the truncation planes must simulate the propagation of the outgoing waves; this is known as the artificial absorbing (or radiation) boundary condition.

The perfect absorbing boundary conditions are usually global in nature, which makes them quite expensive to implement and requires excessively large computer memories. The local absorbing boundary conditions, which make use of only the fields at the neighboring space and time nodes, are relatively inexpensive to implement.

There are quite a few local absorbing boundary conditions available, but most of them are not "absorbing" enough for the purpose of this investigation. As mentioned in the Introduction, the Fourier transform of the time-domain results is very sensitive to the reflection errors. A small amount of reflection may not visibly influence the time-domain fields, but the transformed results could be far off.

To improve the local absorbing boundary conditions, a new approach based on the "local cancellation of the leading order errors" has been developed and shown to provide substantial improvement of absorbing qualities. The boundary treatment used in this investigation will be discussed below.

Consider the end boundary ($z=N_3\Delta z$) first. In most of the cases under consideration, the end surface will contain one end of the microstrip with its other end connected to the microstrip discontinuity. For most high-dielectric-constant substrates, owing to the guiding nature of the metal strip, the major direction of the power flow is in the $+z$ direction, or nearly normal incident. The sideways leakage and radiation are small. This is quite similar to a one-dimensional propagation case, and a natural choice of the boundary condition is to use the field values a few Δz before and a few Δt earlier for the present boundary fields.

In a microstrip structure, the existence of the dielectric substrate makes the wave velocity in the main propagation direction $+z$ be less than the velocity of light c in free space. Denote this velocity as v (usually v is some fraction of c and is also weakly dependent on frequency due to the dispersive nature of the structure). Assume $a_1\cdot v=c$, where a_1 is a constant. (Rigorously speaking, it will be a weak function of frequency. Here for the purpose of the boundary treatment only its low-frequency value is adapted). For the stability criterion (3a) to be satisfied, we must have (choose $\Delta x=\Delta y=\Delta z=\Delta h$)

$$v_{\max}\Delta t=c\,\Delta t=a_1 v\,\Delta t\equiv k\,\Delta h \tag{12}$$

where k is a certain constant satisfing

$$k\leqslant\frac{1}{\sqrt{3}}$$

and therefore

$$v \cdot \Delta t = \frac{k}{a_1} \cdot \Delta h. \tag{13}$$

From (13) it is clear that we can always make the wave travel a certain integral number of space steps in some integral number of time steps by choosing a specific k (better to be less than but close to $1/\sqrt{3}$ for high accuracy), thus avoiding the need for interpolation, and the stability condition (3a) or (13) is still satisfied.

Take the 50 Ω line on the alumina substrate ($\epsilon_r = 9.6$, $W/H = 1.0$) as an example. Here $\epsilon_{reff}(f=0) \approx 6.6$; then $v = 1/\sqrt{6.6} \cdot c$, or $a_1 = \sqrt{6.6}$. If we choose $k = 0.514$, then k/a_1 will be approximately $1/5$; that is, the wave will travel one space step in approximately five time steps.

After choosing the parameters as above, the boundary value of the fields can now be specified as the value of the inner nodes at several time steps earlier. Again, for the alumina case, the boundary condition of the fields will be

$$E_i(N_3 \Delta z, n\Delta t) = E_i[(N_3 - 1)\Delta z, (n-5)\Delta t] \tag{14}$$

where $E_i(N_3 \Delta z, n\Delta t)$ is any electric field component which lies on the boundary of the computation domain (actually only the tangential components of the fields are needed for later calculations). By using this treatment, a storage of the next-to-boundary nodal fields for several time steps is needed. Since this is a storage of two-dimensional data, it will not increase the memory requirement significantly, compared to the major three-dimensional storage.

Usually, merely applying the boundary operation as above will still leave a visible (3–5 percent) amount of reflection. This is partly due to the fact that the true wave propagation is not one-dimensional, and partly because the velocity of the wave is not a constant but rather a function of frequency.

To improve the earlier boundary treatment, it is found that if we apply the same kind of boundary condition on the tangential $\vec{H}$ field next to the boundary, i.e. (for the alumina case),

$$H_i[(N_3 - 1/2)\Delta z, (n+1/2)\Delta t] = H_i[(N_3 - 3/2)\Delta z, (n + 1/2 - 5)\Delta t] \tag{15}$$

and compare it with those $\vec{H}$ values calculated from the loop integration of $\vec{E}$ fields (here the boundary $\vec{E}$'s used are obtained in the previous computation step by using boundary condition (14)), these two $\vec{H}$ fields will always have the property that the errors contained in them due to the imperfect treatment of the boundary condition will have opposite signs and the magnitudes of these errors will maintain a known ratio. Therefore by a weighted average (1 : 5 in this case) of these two $\vec{H}$ fields, we can get an error cancellation effect and the resulting boundary operation will have a much improved quality. This kind of error cancellation approach can actually be applied to any kind of linear local boundary conditions. A general discussion of it for the one-dimensional case can be found in [30].

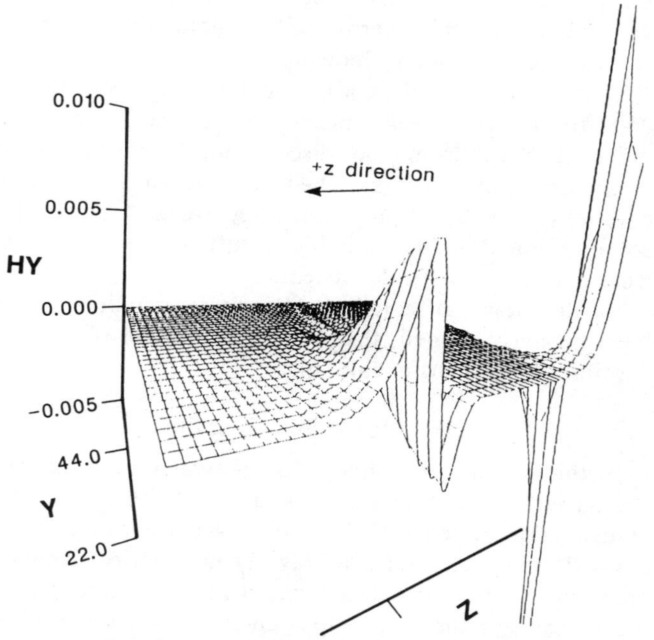

Fig. 6. dc magnetic field on the front surface due to the electric wall boundary treatment.

The treatment of the side boundary is similar to that of the end. But in this case the normal incident approximation is no longer valid. The direction of incidence of the wave changes with both time and position. An accurate and economic treatment of the side boundary condition has not been found so far. As a compromise, the boundary tangential $\vec{E}$ fields will always be given the value of their inner neighbors one time step earlier. A similar operation is done for the $\vec{H}$ nodes $1/2$ space step from the boundary, and the error cancellation algorithm still works in this case.

On the top plane, only the boundary condition for $\vec{E}$ fields is applied in the same way as for the side plane. The cancellation scheme does not work for the top plane, since the field is of the evanescent type in that direction.

The front surface needs some special treatment. During the time when the Gaussian pulse is excited, under the strip on plane *abcd* of Fig. 5, the vertical field is given the value of the Gaussian pulse. Elsewhere on the front surface the electric fields are fixed to be zero. This is equivalent to an electric wall boundary condition (the magnetic wall or symmetric boundary condition turned out not to be able to give a clear tail of the pulse). Following the passing of the pulse with part of it reflected back from the discontinuities, the front surface should now behave in a "transparent" way, as in the real case. This means that from the moment the reflected wave reaches the front surface a radiation type of boundary condition must be "switched on." It is found that the early enforced electric wall boundary condition induced a dc current or tangential magnetic field on the front surface and nearby (Fig. 6). This local dc field, although it has no influence on the traveling pulse, does cause trouble in the boundary treatment. That is, if we switch on the radiation condition on this wall, the numerical errors will accumulate very rapidly and the solutions soon "blow up."

To solve this problem, what actually has been done is that after the pulse leaves the source plane and before it is reflected back from the discontinuities, the radiation boundary condition is switched on at a surface which is parallel to the source plane but a few space steps into the computation domain, and this is sufficient to avoid the trouble caused by the dc current.

At this stage, after all the boundary conditions have been properly treated, the numerical solution of the discontinuity problems is quite direct.

III. Numerical Results

In this investigation, five kinds of symmetric microstrip discontinuities on alumina substrate ($\epsilon_r = 9.6$) for a 50 Ω transmission system ($W/H = 1.0$, $W = 0.6$ mm) have been studied. They are (refer to Fig. 1) microstrip open-end, cross and T junctions, step-in-width and gap. Among these discontinuities, the microstrip open-end case has been given the most detailed discussion and compared with all the available published results.

A. Microstrip Open-End Terminations

A microstrip open-end on alumina substrate ($\epsilon_r = 9.6$) as shown in Fig. 5 is studied first. The parameters of the structure are as follows:

thickness of substrate: $H = 0.6$ mm
width of metal strip: $W = 0.6$ mm
thickness of metal strip: $t = 0.0$.

To accommodate the structural details of the microstrip, the mesh parameters have been chosen to be

space interval: $\Delta h = H/10 = 0.06$ mm;
$\Delta x = \Delta y = \Delta z = \Delta h$;
$N_1 = 40,\ N_2 = 120,\ N_3 = 190$;
$l_1 = 120,\ M_1 = 10,\ M_2 = 5$;
$l = 20, 30, 40, 50\ (\Delta h)$ have all been used to calculate the results for comparison;
time step $\Delta t = k \cdot \Delta h / c$ s, where c is the velocity of light in air and k is a constant restricted by the stability criterion (3);
$k = 0.514$ in this calculation.

A Gaussian pulse excitation is used at the front surface. It is uniform under the strip (in plane *abcd* of Fig. 4) and has only the E_x component with the following specified value:

$$E_x(t) = \exp\left[-\frac{(t-t_0)^2}{T^2}\right] \tag{16}$$

where $t_0 = 350\,dt$ and $T = 40\,dt$; elsewhere on the front surface, set $E_x = E_y = 0$. The pulse width in space is about $20dh$, which is wide enough to obtain good resolution. The frequency spectrum of this pulse is from dc to about 100 GHz.

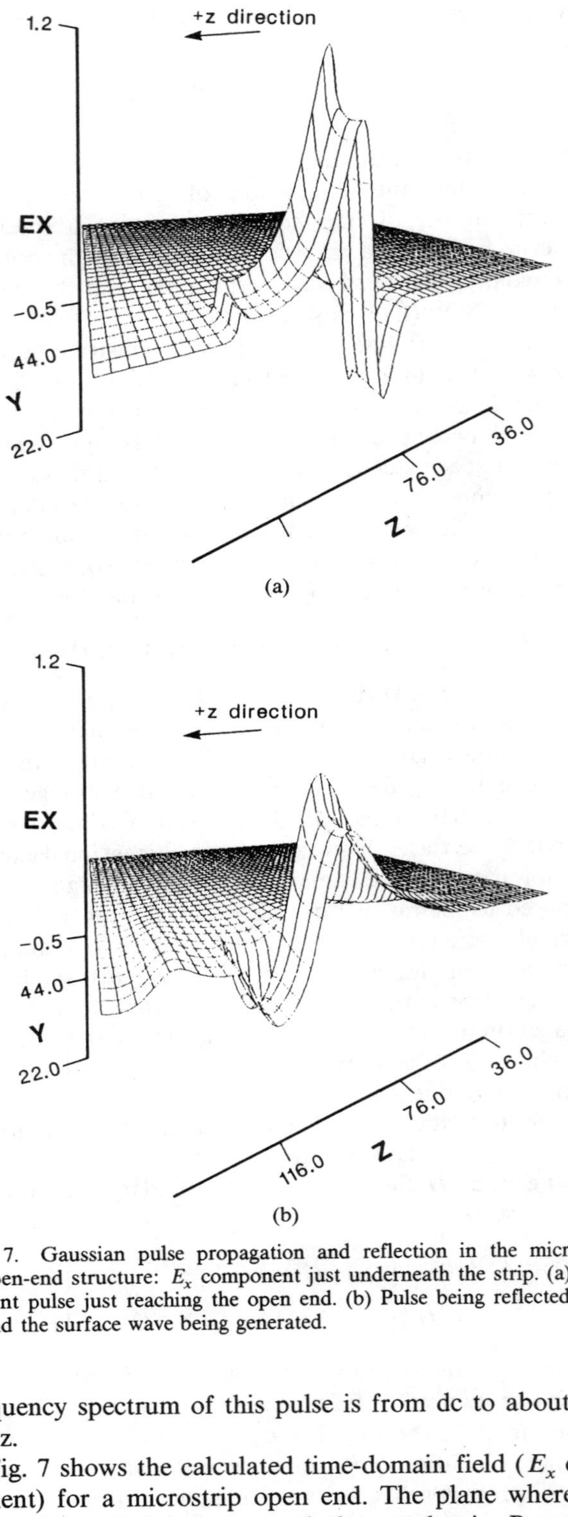

Fig. 7. Gaussian pulse propagation and reflection in the microstrip open-end structure: E_x component just underneath the strip. (a) Incident pulse just reaching the open end. (b) Pulse being reflected back and the surface wave being generated.

Fig. 7 shows the calculated time-domain field (E_x component) for a microstrip open end. The plane where the plot is drawn is just underneath the metal strip. Part (a) is the field distribution at the moment when the Gaussian pulse just reaches the open end and is reflected. The reflected wave is seen to have the same sign as the incident wave and is added to the incident wave. Part (b) shows the reflected wave and a small amount of traveling surface wave.

The microstrip open end structure is a one-port network (Fig. 5). Its scattering matrix has only one element, that is,

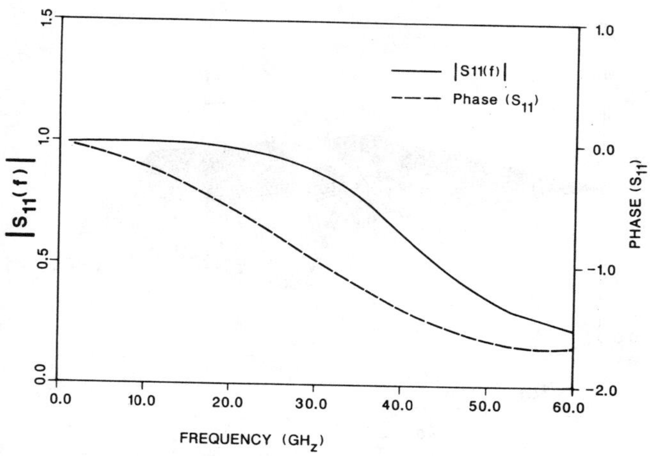

Fig. 8. Frequency-dependent S parameter of open end: magnitude and phase.

S_{11} or the reflection coefficient. S_{11} is defined as

$$S_{11}(f)=\frac{V_{\text{ref}}(f)}{V_{\text{inc}}(f)} \tag{17}$$

where $V_{\text{ref}}(f)$ is the transformed reflection voltage at the input plane (i.e., the reference plane $T-T$ in Fig. 5) of the one-port, and V_{inc} is the transformed incident voltage at the same position. In this calculation, the incident field is obtained from that of an infinitely long microstrip, and the reflected field from the open end is obtained from the difference between the total open-end field and the incident field.

It is common practice in microwave network calculation for S_{11} to be calculated away from the discontinuity through the transmission line formula (refer to Fig. 5.)

$$S_{11}(f)=\frac{V_{\text{ref}}(f,z'=l)\cdot e^{\gamma(f)l}}{V_{\text{inc}}(f,z'=l)\cdot e^{-\gamma(f)l}} = \frac{V_{\text{ref}}(f,z'=l)}{V_{\text{inc}}(f,z'=l)}\cdot e^{2\gamma(f)l}. \tag{18}$$

This is done here to allow the higher order, evanescent modes which were generated near the discontinuity to die out. They are not included in the calculation.

Fig. 8 shows the calculated results of the magnitude and phase of $S_{11}(f)$ for the open end under consideration. The uniqueness of the solution to the use of either the field at one point under the metal strip or the voltage between the strip and the ground, when substituted into (18) for the V's there, has been checked and found to be well satisfied. This is expected to be true once the mode distribution is well established on the line.

The equivalent circuit as shown in Fig. 9 with frequency-dependent circuit parameters is used to model the microstrip open end. Here

$$Y(f)=G(f)+j2\pi fC(f) \tag{19}$$

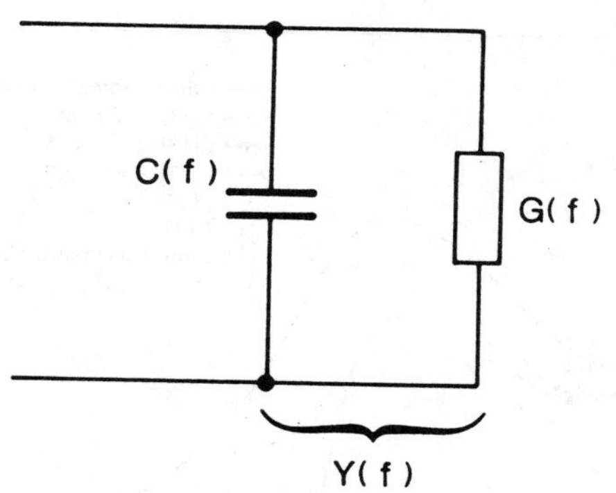

Fig. 9. Equivalent circuit of the microstrip open end.

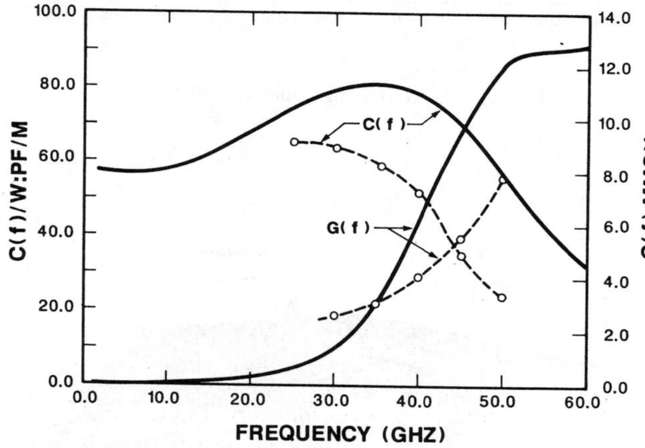

Fig. 10. Frequency-dependent equivalent circuit parameter $C(f)/W$ and $G(f)$ for open end. Solid lines denote time-domain results; dashed lines denote the results of Katehi and Alexopoulos [21].

and $Y(f)$ is related to $S_{11}(f)$ through

$$Y(f)=\frac{1-S_{11}(f)}{1+S_{11}(f)}\cdot\frac{1}{Z_0(f)}. \tag{20}$$

Here $Z_0(f)$ is the characteristic impedance of the microstrip, which is calculated using the ratio of voltage and current [26].

The calculated $C(f)$ and $G(f)$ are plotted in Fig. 10 together with the results presented by Katehi and Alexopoulos [21] for the same structure. The comparison shows quite an amount of discrepancy, especially for higher frequencies. Although both models should be questioned, there is an obvious question in the result of [21] for $G(f)$ in that it does not exhibit a trend to go smoothly down to zero as the frequency goes to dc. For $C(f)$, an equivalent parameter $\Delta l(f)$ can be derived from it and has been compared with extensive published results below.

The parameter which can also be used to account for the capacitive characteristic of the open end is the effective increase in length Δl. It is related to C through [32]

$$\frac{C(f)}{W}=\frac{\Delta l(f)}{h}\sqrt{\epsilon_{\text{eff}}}\,\frac{1}{Z_0(f)}\,\frac{h}{W} \tag{21}$$

where ϵ_{eff} is the effective dielectric constant of the microstrip [32].

The calculated $\Delta l(f)$, together with the comparison with several other published results, is given in Fig. 11.

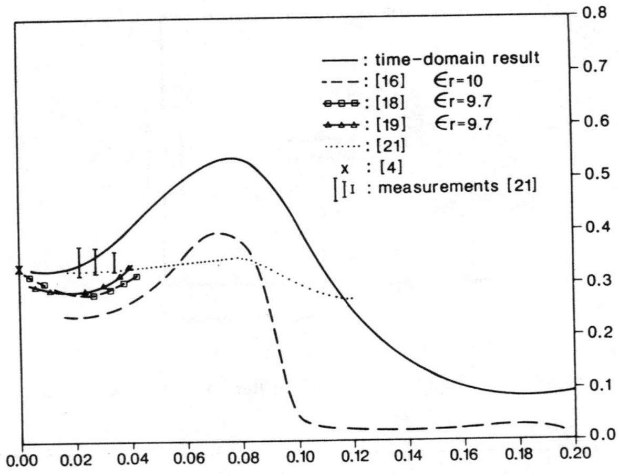

Fig. 11. Effective length increase $\Delta l(f)/H$.

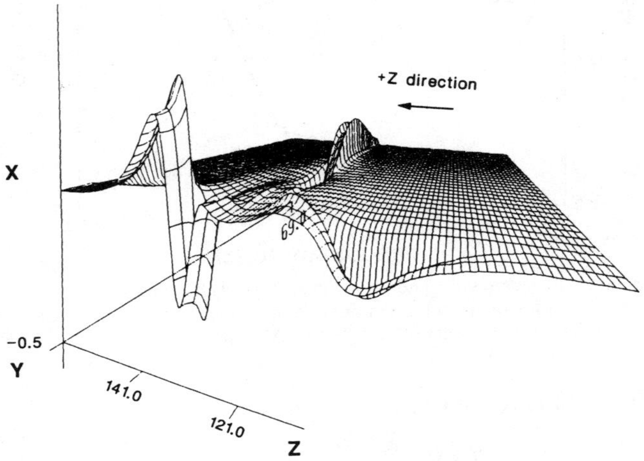

Fig. 12. Gaussian pulse propagation and reflection in the microstrip cross-junction: E_x component just underneath the strip.

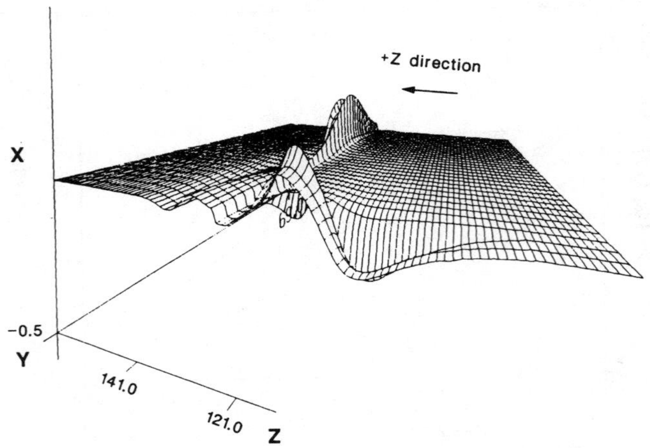

Fig. 13. Gaussian pulse propagation and reflection in the microstrip T junction: E_x component just underneath the strip.

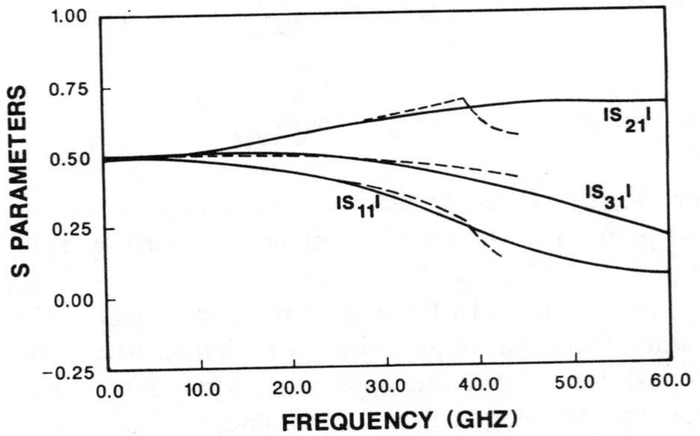

Fig. 14. Frequency-dependent S parameters of the microstrip cross-junction. Solid: time-domain result, dash: Mehran, $\epsilon_r = 9.7$, $h = 0.635$ mm, $W_1 = W_2 = 0.56$ mm [14].

The normalized frequency range used here corresponds to the frequencies from zero to 100 GHz for the structure concerned. The dc result of the time-domain calculation is very close to the quasi-static result of [4], the low frequency part is close to the experimental result of Edward ($W/H = 0.9$) [21], and the shape of the curve is similar to that of James and Henderson [16], which has too low a dc value though. The peak in the curve which appeared near the cutoff frequency of the first TE-type higher order mode of the microstrip has been predicted in [3].

B. Microstrip Cross-Junctions and T Junctions

Using a computation mesh domain similar to that for the open end (this time with $N_1 = 30$, $N_2 = 65$, $N_3 = 180$, and $\Delta h = H/8$), the symmetric microstrip cross-junctions and T junctions ($W_1 = W_2 = W = 0.6$ mm) under symmetric excitations are studied. Here the side boundary is given the same type of boundary treatment as for the end, since in this case sideways flow becomes another major direction of power flow.

Figs. 12 and 13 show the time-domain field distributions for the cross-junction and T junction. The Gaussian pulse which travels into the cross-junction is seen to split four ways after it hits the cross-junction. For the case of the T junction, a small amount of surface wave is observed to travel past the junction, as most of the energy is either reflected backward or transmitted sideways.

Figs. 14 and 15 plot the calculated S parameters for the cross- and T junctions (magnitudes only), together with the results of Mehran [14] as comparison. These results are calculated directly from the definition of each S parameter using the transformed time-domain fields in a way similar to the open-end case. The reference planes for the networks which represent the discontinuities are indicated in Fig. 1.

The independent S parameters for the cross-junction are S_{11}, S_{21}, and S_{31}. From Fig. 14, it is seen that these three S parameters all acquire the value 0.5 at dc and very low frequencies, indicating that the four branches of the cross-junction each get an equal 1/4 share of the total field energy; thus the U condition for the S matrix is well satisfied, and no (detectable) radiation occurs at those frequency ranges. The same is not true for higher frequency ranges.

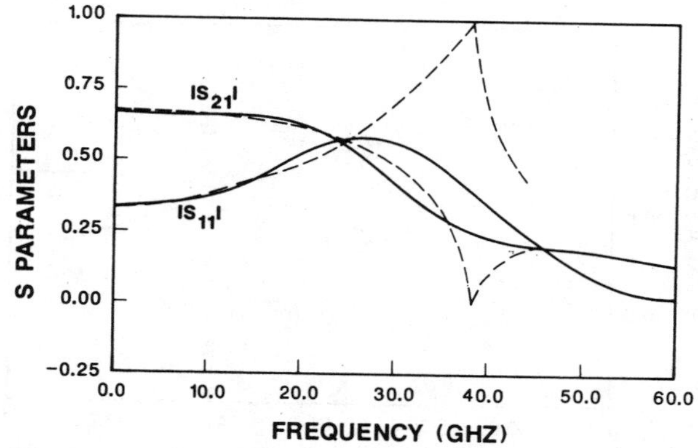

Fig. 15. Frequency-dependent S parameters of the microstrip T junction. Solid: time-domain result, dash: Mehran, $\epsilon_r = 9.7$, $h = 0.635$ mm, $W_1 = W_2 = 0.56$ mm [14].

The independent S parameters S_{11} and S_{21} for the T junction under symmetric excitation are plotted in Fig. 15. Again the U condition is checked to be satisfied in the very low frequency range.

From Figs. 14 and 15, it is seen that the time-domain results and the results of Mehran have very good agreement at dc and the lower frequency range. The discrepancies which occurred at very high frequencies are believed to be partly due to the fact that the waveguide model approach used in [14] is not able to take into account the radiation and surface wave generation effects which happened at higher frequency, and partly due to the fact that the time-domain results will not represent the exact dominant mode parameters after the inflection frequency, which is around 35 GHz for this configuration. The fact that the T junction results have earlier and larger discrepancy further confirmed this point, because here the surface wave becomes another reason for the waveguide model to fail, adding to the radiation loss.

C. Microstrip Step-in-Width and Gaps

Figs. 16 and 17 show the calculated time-domain fields and S parameters for the microstrip step-in-width ($W_1/H = 1.0$, $W_2/W_1 = 2.0$) using a mesh with size similar to that of T and cross-junction calculations. The S parameters for the step-in-width are defined as

$$S_{11} = \frac{V_{1\,\mathrm{ref}}(f)}{V_{1\,\mathrm{inc}}(f)} \tag{22}$$

$$S_{21} = \frac{\dfrac{V_{2\,\mathrm{trans}}(f)}{\sqrt{Z_{02}(f)}}}{\dfrac{V_{1\,\mathrm{inc}}(f)}{\sqrt{Z_{01}(f)}}} \tag{23}$$

$$S_{22} = \frac{V_{2\,\mathrm{ref}}(f)}{V_{2\,\mathrm{inc}}(f)} \tag{24}$$

$$S_{12} = S_{21} \tag{25}$$

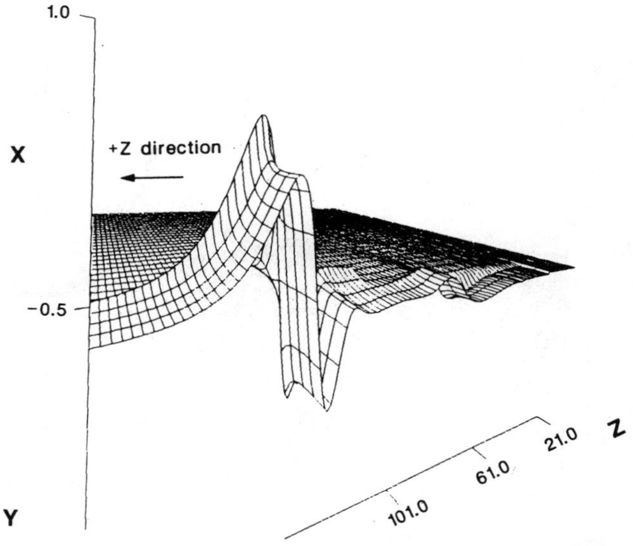

Fig. 16. Gaussian pulse propagation and reflection in the microstrip step-in-width structure: E_x component just underneath the strip.

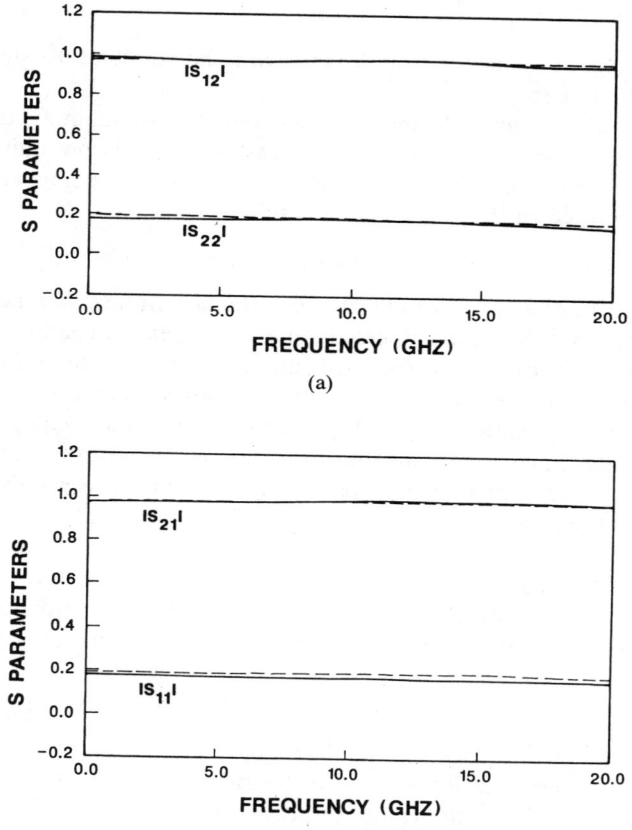

Fig. 17. Frequency-dependent S parameters of the microstrip step-in-width. Solid: time-domain result, dash: Koster and Jansen, $\epsilon_r = 10$ [33].

where $Z_{01}(f)$ and $Z_{02}(f)$ are the characteristic impedances of the microstrip lines connected to port 1 and 2 of the step, respectively, and are calculated in the same way as shown in [26].

The calculated frequency dependence of the S parameters of the step-in-width is quite flat over a large frequency

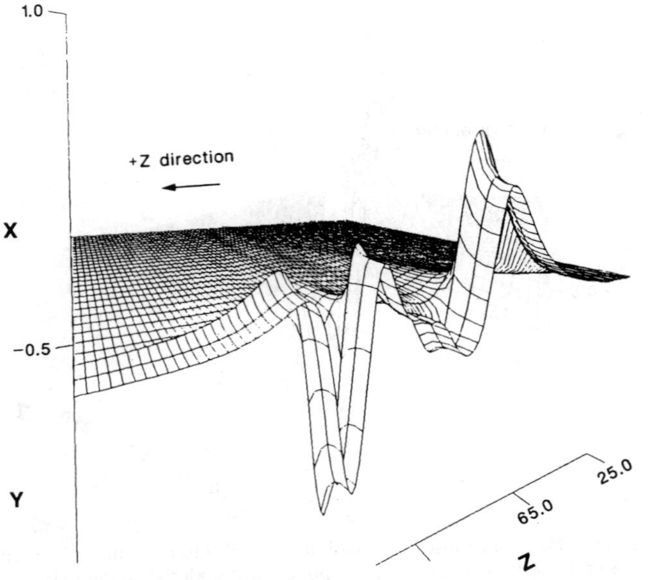

Fig. 18. Gaussian pulse propagation and reflection in the microstrip gap: E_x component just underneath the strip.

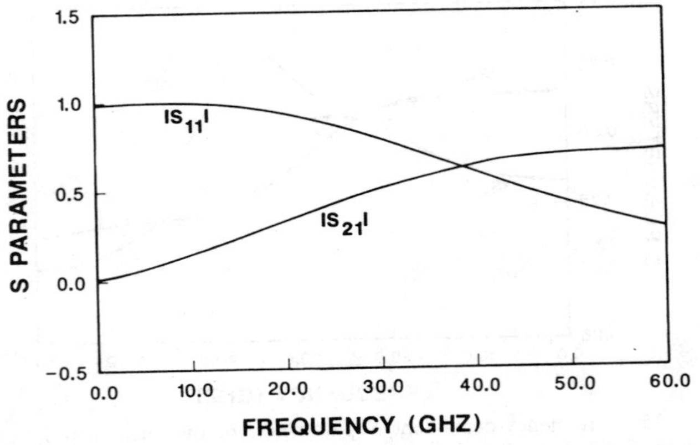

Fig. 19. Frequency-dependent S parameters of the microstrip gap.

range. It is in good agreement with the result of Koster and Jansen [33].

Figs. 18 and 19 are the calculated time-domain fields and S parameters for the microstrip gap discontinuity ($S/H = 0.5$). The dc block characteristics of the gap are exhibited in the S parameter results.

IV. Conclusion

It has been shown that the time-domain finite difference approach is capable of calculating the dispersion characteristics of the microstrip discontinuities over a large frequency range. It is a very general method and can find wide applications in modeling various microwave components. Further investigations on different computer-memory-saving schemes will make this method more suitable for CAD purposes.

Appendix

To calculate the E_y and E_z components on the dielectric–air interface, we start with the Maxwell equation

$$\frac{\partial \vec{E}}{\partial t} = \frac{1}{\epsilon_i} \nabla \times \vec{H} \tag{A1}$$

where $i = 1$, 2 denotes the dielectric constant in the substrate and the air region, respectively. Taking the calculation of E_y as an example, we have, from (A1),

$$\frac{\partial E_y}{\partial t} = \frac{1}{\epsilon_i}\left(\frac{\partial H_x}{\partial z} - \frac{\partial H_z}{\partial x}\right). \tag{A2}$$

Since E_y, H_x and $\partial H_x/\partial z$ are continuous across the interface, it is very obvious that $\partial H_z/\partial x$ is discontinuous across the boundary; i.e., we can get from (A2)

$$\frac{1}{\epsilon_2}\left(\frac{\partial H_z}{\partial x}\right)_2 - \frac{1}{\epsilon_1}\left(\frac{\partial H_z}{\partial x}\right)_1 = \left(\frac{1}{\epsilon_2} - \frac{1}{\epsilon_1}\right)\frac{\partial H_x}{\partial z} \neq 0. \tag{A3}$$

Therefore on each side of the interface and at positions very close to the boundary, we can write

$$\epsilon_1 \frac{\partial E_y}{\partial t} = \frac{\partial H_x}{\partial z} - \left(\frac{\partial H_z}{\partial x}\right)_1$$
$$\epsilon_2 \frac{\partial E_y}{\partial t} = \frac{\partial H_x}{\partial z} - \left(\frac{\partial H_z}{\partial x}\right)_2. \tag{A4}$$

Approximate $(\partial H_z/\partial_x)_1$ and $(\partial H_z/\partial_x)_2$ by

$$\left(\frac{\partial H_z}{\partial x}\right)_1 \approx \frac{H_z(m) - H_z(m-1/2)}{\frac{\Delta x}{2}}$$
$$\left(\frac{\partial H_z}{\partial x}\right)_2 \approx \frac{H_z(m+1/2) - H_z(m)}{\frac{\Delta x}{2}} \tag{A5}$$

where m is assumed to be the position of the interface, and $m+1/2$ and $m-1/2$ denote the positions a half step above and below the interface, respectively.

Substituting (A5) into (A4), we get

$$H_z(m) \approx \frac{\epsilon_1}{\epsilon_1+\epsilon_2} H_z(m+1/2) + \frac{\epsilon_2}{\epsilon_1+\epsilon_2} H_z(m-1/2) + \frac{\epsilon_2-\epsilon_1}{\epsilon_1+\epsilon_2}\frac{\partial H_x}{\partial z}\Delta x/2. \tag{A6}$$

Substituting the $H_z(m)$ value of (A6) back into (A5), substituting the resulting (A5) into the two expressions in (A4), and adding the two together, we get

$$\frac{\epsilon_1+\epsilon_2}{2}\cdot\frac{\partial E_y}{\partial t} = \frac{\partial H_x}{\partial z} - \frac{H_z(m+1/2) - H_z(m-1/2)}{\Delta x} = \frac{\partial H_x}{\partial z} - \frac{\Delta H_z}{\Delta x}. \tag{A7}$$

Similarly, we get

$$\frac{\epsilon_1+\epsilon_2}{2}\cdot\frac{\partial E_z}{\partial t} = \frac{\Delta H_y}{\Delta x} - \frac{\partial H_x}{\partial y}. \tag{A8}$$

Thus we can further discretize (A7) and (A8) to calculate the tangential $\vec{E}$ components on the dielectric–air interface.

It is straightforward to show that this boundary approximation is of first order in space and second order in time.

References

[1] J. J. Campbell, "Application of the solutions of certain boundary value problems to the symmetrical four-port junction and specially truncated bends in parallel-plate waveguides and balanced strip-transmission lines," *IEEE Trans. Microwave Theory Tech.*, vol. MTT-16, pp. 165–176, Mar. 1968.

[2] L. S. Napoli and J. J. Hughes, Foreshortening of microstrip open circuits on alumina substrate," *IEEE Trans. Microwave Theory Tech.*, vol. MTT-19, pp. 559–561, June 1971.

[3] D. S. James and S. H. Tse, "Microstrip end effects," *Electron. Lett.*, vol. 8, pp. 46–47, Jan. 27, 1972.

[4] P. Silvester and P. Benedek, "Equivalent capacitances of microstrip open circuits," *IEEE Trans. Microwave Theory Tech.*, vol. MTT-20, pp. 511–516, Aug. 1972.

[5] T. Itoh, R. Mittra, and R. D. Ward, "A method for computing edge capacitance of finite and semi-infinite microstrip lines," *IEEE Trans. Microwave Theory Tech.*, pp. 847–849, Dec. 1972.

[6] C. Gupta and A. Gopinath, "Equivalent circuit capacitance of microstrip step change in width," *IEEE Trans. Microwave Theory Tech.*, vol. MTT-25, pp. 819–822, Oct. 1977.

[7] A. A. F. Thomson and A. A. Gopinath, "Calculation of microstrip discontinuity inductances," *IEEE Trans. Microwave Theory Tech.*, vol. MTT-23, pp. 648–655, Aug. 1975.

[8] P. Stouten, "Equivalent capacitances of T junctions," *Electron. Lett.*, vol. 9, pp. 552–553, Nov. 1973.

[9] I. Wolff, A. G. Kompa, and R. Mehran, "Calculation method for microstrip discontinuities and T-junctions," *Electron. Lett.*, vol. 8, pp. 177, Apr. 1972.

[10] P. Silvester and P. Benedek, "Microstrip discontinuity capacitances for right-angle bends, T junctions, and crossings," *IEEE Trans. Microwave Theory Tech.*, vol. MTT-21, pp. 341–346, May 1973.

[11] M. Maeda, "An analysis of gap in microstrip transmission lines," *IEEE Trans. Microwave Theory Tech.*, vol. MTT-20, pp. 390–396, June 1972.

[12] A. Farrar and A. T. Adams, "Matrix method for microstrip three-dimensional problems," *IEEE Trans. Microwave Theory Tech.*, vol. MTT-20, pp. 497–504, Aug. 1972.

[13] G. Kompa and R. Mehran, "Planar waveguide model for calculating microstrip components," *Electron. Lett.*, vol. 11, pp. 459–460, Dept. 1975.

[14] R. Mehran, "The frequency-dependent scattering matrix of microstrip right-angle bends, T-junctions and crossings," *Arch. Elek. Übertragung.*, vol. 29, pp. 454–460, Nov. 1975.

[15] W. Menzel and I. Wolff, "A method for calculating the frequency dependent properties of microstrip discontinuities," *IEEE Trans. Microwave Theory Tech.*, vol. MTT-25, pp. 107–112, Feb. 1977.

[16] J. R. James and A. Henderson, "High-frequency behavior of microstrip open-circuit terminations," *IEE J. Microwave Opt. Acoust.*, vol. 3, pp. 205–211, Sept. 1979.

[17] R. H. Jansen, "The spectral-domain approach for microwave integrated circuits," *IEEE Trans. Microwave Theory Tech.*, vol. MTT-33, pp. 1043–1056, Oct. 1985.

[18] R. H. Jansen, "Hybrid mode analysis of end effects of planar microwave and millimeterwave transmission lines," *Proc. Inst. Elec. Eng.*, vol. 128, pt. H, no. 2, pp. 77–86, Apr. 1981.

[19] J. S. Hornsby, "Full-wave analysis of microstrip resonator and open-circuit end effect," *Proc. Inst. Elec. Eng.*, vol. 129, pt. H, no. 6, pp. 338–341, Dec. 1982.

[20] R. W. Jackson and D. M. Pozar, "Full-wave analysis of microstrip open-end and gap discontinuities," *IEEE Trans. Microwave Theory Tech.*, vol. MTT-33, pp. 1036–1042, Oct. 1985.

[21] P. B. Katehi and N. G. Alexopoulos, "Frequency-dependent characteristics of microstrip discontinuities in millimeter-wave integrated circuit," *IEEE Trans. Microwave Theory Tech.*, vol. MTT-33, pp. 1029–1035, Oct. 1985.

[22] K. S. Yee, "Numerical solution of initial boundary value problems involving Maxwell's equations in isotropic media," *IEEE Trans. Antennas Propagat.*, vol. AP-14, pp. 302–307, May 1966.

[23] A. Taflove and M. E. Brodwin, "Numerical solution of steady-state electromagnetic scattering problems using the time-dependent Maxwell's equations," *IEEE Trans. Microwave Theory Tech.*, vol. MTT-23, pp. 623–630, 1975.

[24] S. Koike, N. Yoshida, and I. Fukai, "Transient analysis of microstrip gap in three-dimensional space," *IEEE Trans. Microwave Theory Tech.*, vol. MTT-33, pp. 726–730, Aug. 1985.

[25] S. Koike, N. Yoshida, and I. Fukai, "Transient analysis of coupling between crossing lines in three-dimensional space," *IEEE Trans. Microwave Theory Tech.*, vol. MTT-35, pp. 67–71, Jan. 1987.

[26] X. Zhang, J. Fang, K. K. Mei, and Y. Liu, "Calculation of the dispersive characteristics of microstrips by the time-domain finite difference method," *IEEE Trans. Microwave Theory Tech.*, vol. MTT-36, pp. 263–267, Feb. 1988.

[27] L. Lapidus and G. H. Pinder, *Numerical Solution of Partial Differential Equations in Science and Engineering.* New York: Wiley, 1982.

[28] J. Fang, X. Zhang, K. K. Mei, and Y. Liu, "Time domain computation and numerical dispersion," presented at XXIInd General Assembly of the International Union of Radio Science, Tel Aviv, Israel, Aug. 24–Sept. 2, 1987.

[29] X. Zhang, "Time-domain finite difference calculation of the frequency-dependent characteristics of the microstrip discontinuities," master's report, Dept. EECS, University of California, Berkeley, Dec. 1987.

[30] K. K. Mei and J. Fang, "A super-absorption boundary algorithm for one dimensional wave equations," submitted to *J. Comput. Phys.*

[31] C. C. Lin, "Numerical modeling of two-dimensional time domain electromagnetic scattering by underground inhomogeneities," Ph.D dissertation, Dept. EECS, University of California, Berkeley, 1985.

[32] K. C. Gupta, R. Garg, and I. J. Bahl, *Microstrip Lines and Slotlines.* Dedham, MA: Artech House, 1979.

[33] N. H. L. Koster and R. H. Jansen, "The microstrip step discontinuity: A revised description," *IEEE Trans. Microwave Theory Tech.*, vol. MTT-34, pp. 213–223, Feb. 1986.

Paper 6.2

Application of the Three-Dimensional Finite-Difference Time-Domain Method to the Analysis of Planar Microstrip Circuits

DAVID M. SHEEN, SAMI M. ALI, SENIOR MEMBER, IEEE, MOHAMED D. ABOUZAHRA, SENIOR MEMBER, IEEE, AND JIN AU KONG, FELLOW, IEEE

***Abstract*—A direct three-dimensional finite-difference time-domain (FDTD) method is applied to the full-wave analysis of various microstrip structures. The method is shown to be an efficient tool for modeling complicated microstrip circuit components as well as microstrip antennas. From the time-domain results, the input impedance of a line-fed rectangular patch antenna and the frequency-dependent scattering parameters of a low-pass filter and a branch line coupler are calculated. These circuits are fabricated and the measurements are compared with the FDTD results and shown to be in good agreement.**

I. Introduction

FREQUENCY-domain analytical work with complicated microstrip circuits has generally been done using planar circuit concepts in which the substrate is assumed to be thin enough that propagation can be considered in two dimensions by surrounding the microstrip with magnetic walls [1]–[6]. Fringing fields are accounted for by using either static or dynamic effective dimensions and permittivities. Limitations of these methods are that fringing, coupling, and radiation must all be handled empirically since they are not allowed for in the model. Also, the accuracy is questionable when the substrate becomes thick relative to the width of the microstrip. To fully account for these effects, it is necessary to use a full-wave solution.

Full-wave frequency-domain methods have been used to solve some of the simpler discontinuity problems [7], [8]. However, these methods are difficult to apply to a typical printed microstrip circuit.

Modeling of microstrip circuits has also been performed using Bergeron's method [9], [10]. This method is a modification of the transmission line matrix (TLM) method, and has limitations similar to the finite-difference time-domain (FDTD) method due to the discrete modeling of space and time [11], [12]. A unique problem with this method is that the dielectric interface and the perfectly conducting strip are misaligned by half a space step [12].

The FDTD method has been used extensively for the solution of two- and three-dimensional scattering problems [13]–[17]. Recently, FDTD methods have been used to effectively calculate the frequency-dependent characteristics of microstrip discontinuities [18]–[21]. Analysis of the fundamental discontinuities is of great importance since more complicated circuits can be realized by interconnecting microstrip lines with these discontinuities and using transmission line and network theory. Some circuits, however, such as patch antennas, may not be realized in this way. Additionally, if the discontinuities are too close to each other the use of network concepts will not be accurate due to the interaction of evanescent waves. To accurately analyze these types of structures it is necessary to simulate the entire structure in one computation. The FDTD method shows great promise in its flexibility in handling a variety of circuit configurations. An additional benefit of the time-domain analysis is that a broad-band pulse may be used as the excitation and the frequency-domain parameters may be calculated over the entire frequency range of interest by Fourier transform of the transient results.

In this paper, the frequency-dependent scattering parameters have been calculated for several printed microstrip circuits, specifically a line-fed rectangular patch antenna, a low-pass filter, and a rectangular branch line coupler. These circuits represent resonant microstrip structures on an open substrate; hence, radiation effects can be significant, especially for the microstrip antenna. Calculated results are presented and compared with experimental measurements.

The FDTD method has been chosen over the other discrete methods (TLM or Bergeron's) because it is extremely efficient, its implementation is quite straightforward, and it may be derived directly from Maxwell's equations. Many of the techniques used to implement this

Manuscript received September 29, 1989; revised March 21, 1990. This work was supported by NSF Grant 8620029-ECS, the Joint Services Electronics Program (Contract DAAL03-89-C-0001), RADC Contract F19628-88-K-0013, ARO Contract DAAL03-88-J-0057, ONR Contract N00014-89-J-1019, and the Department of the Air Force.

D. M. Sheen, S. M. Ali, and J. A. Kong are with the Department of Electrical Engineering and Computer Science and the Research Laboratory of Electronics, Massachusetts Institute of Technology, Cambridge, MA 02139.

M. D. Abouzahra is with MIT Lincoln Laboratory, Lexington, MA 02173.

IEEE Log Number 9036153.

Reprinted from *IEEE Trans. Microwave Theory Tech.*, vol. 38, no. 7, pp. 849–857, July 1990.

method have been demonstrated previously [18]–[20]; however, simplification of the method has been achieved by using a simpler absorbing boundary condition [22]. This simpler absorbing boundary condition yields good results for the broad class of microstrip circuits considered by this paper. Additionally, the source treatment has been enhanced to reduce the source effects documented in [18]–[20].

II. Problem Formulation

The FDTD method is formulated by discretizing Maxwell's curl equations over a finite volume and approximating the derivatives with centered difference approximations. Conducting surfaces are treated by setting tangential electric field components to 0. The walls of the mesh, however, require special treatment to prevent reflections from the mesh termination.

A. Governing Equations

Formulation of the FDTD method begins by considering the differential form of Maxwell's two curl equations which govern the propagation of fields in the structures. For simplicity, the media are assumed to be piecewise uniform, isotropic, and homogeneous. The structure is assumed to be lossless (i.e., no volume currents or finite conductivity). With these assumptions, Maxwell's curl equations may be written as

$$\mu \frac{\partial \boldsymbol{H}}{\partial t} = -\nabla \times \boldsymbol{E} \tag{1}$$

$$\epsilon \frac{\partial \boldsymbol{E}}{\partial t} = \nabla \times \boldsymbol{H}. \tag{2}$$

In order to find an approximate solution to this set of equations, the problem is discretized over a finite three-dimensional computational domain with appropriate boundary conditions enforced on the source, conductors, and mesh walls.

B. Finite-Difference Equations

To obtain discrete approximations to these continuous partial differential equations the centered difference approximation is used on both the time and space first-order partial differentiations. For convenience, the six field locations are considered to be interleaved in space as shown in Fig. 1, which is a drawing of the FDTD unit cell [13]. The entire computational domain is obtained by stacking these rectangular cubes into a larger rectangular volume. The $\hat{x}$, $\hat{y}$, and $\hat{z}$ dimensions of the unit cell are Δx, Δy, and Δz, respectively. The advantages of this field arrangement are that centered differences are realized in the calculation of each field component and that continuity of tangential field components is automatically satisfied. Because there are only six unique field components within the unit cell, the six field components touching the shaded upper eighth of the unit cell in Fig. 1 are considered to be a unit node with subscript indices i, j, and k corresponding to the node numbers in the $\hat{x}$, $\hat{y}$, and $\hat{z}$ directions. This notation implicitly assumes the $\pm 1/2$ space indices and thus simplifies the notation, rendering the formulas directly implementable on the computer. The time steps are indicated with the superscript n. Using this field component arrangement, the above notation, and the centered difference approximation, the explicit finite difference approximations to (1) and (2) are

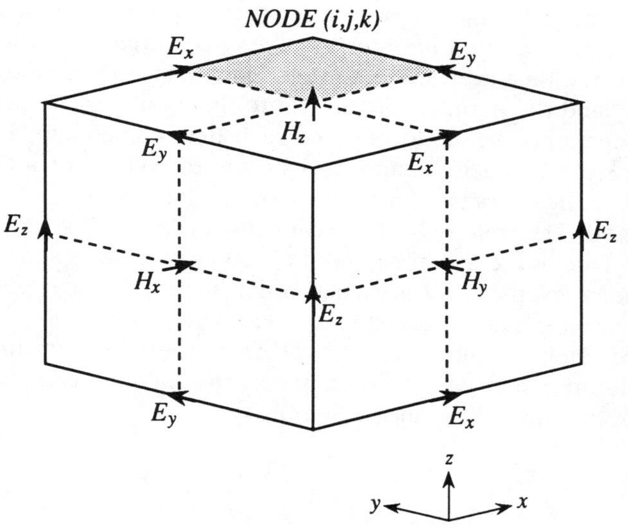

Fig. 1. Field component placement in the FDTD unit cell.

$$H^{n+1/2}_{x\,i,j,k} = H^{n-1/2}_{x\,i,j,k} + \frac{\Delta t}{\mu \Delta z}\left(E^n_{y\,i,j,k} - E^n_{y\,i,j,k-1}\right) - \frac{\Delta t}{\mu \Delta y}\left(E^n_{z\,i,j,k} - E^n_{z\,i,j-1,k}\right) \tag{3}$$

$$H^{n+1/2}_{y\,i,j,k} = H^{n-1/2}_{y\,i,j,k} + \frac{\Delta t}{\mu \Delta x}\left(E^n_{z\,i,j,k} - E^n_{z\,i-1,j,k}\right) - \frac{\Delta t}{\mu \Delta z}\left(E^n_{x\,i,j,k} - E^n_{x\,i,j,k-1}\right) \tag{4}$$

$$H^{n+1/2}_{z\,i,j,k} = H^{n-1/2}_{z\,i,j,k} + \frac{\Delta t}{\mu \Delta y}\left(E^n_{x\,i,j,k} - E^n_{x\,i,j-1,k}\right) - \frac{\Delta t}{\mu \Delta x}\left(E^n_{y\,i,j,k} - E^n_{y\,i-1,j,k}\right) \tag{5}$$

$$E^{n+1}_{x\,i,j,k} = E^n_{x\,i,j,k} + \frac{\Delta t}{\epsilon \Delta y}\left(H^{n+1/2}_{z\,i,j+1,k} - H^{n+1/2}_{z\,i,j,k}\right) - \frac{\Delta t}{\epsilon \Delta z}\left(H^{n+1/2}_{y\,i,j,k+1} - H^{n+1/2}_{y\,i,j,k}\right) \tag{6}$$

$$E^{n+1}_{y\,i,j,k} = E^n_{y\,i,j,k} + \frac{\Delta t}{\epsilon \Delta z}\left(H^{n+1/2}_{x\,i,j,k+1} - H^{n+1/2}_{x\,i,j,k}\right) - \frac{\Delta t}{\epsilon \Delta x}\left(H^{n+1/2}_{z\,i+1,j,k} - H^{n+1/2}_{z\,i,j,k}\right) \tag{7}$$

$$E^{n+1}_{z\,i,j,k} = E^n_{z\,i,j,k} + \frac{\Delta t}{\epsilon \Delta x}\left(H^{n+1/2}_{y\,i+1,j,k} - H^{n+1/2}_{y\,i,j,k}\right) - \frac{\Delta t}{\epsilon \Delta y}\left(H^{n+1/2}_{x\,i,j+1,k} - H^{n+1/2}_{x\,i,j,k}\right). \tag{8}$$

The half time steps indicate that $\boldsymbol{E}$ and $\boldsymbol{H}$ are alternately calculated in order to achieve centered differences for the time derivatives. In these equations, the permittivity and the permeability are set to the appropriate values depending on the location of each field component. For the electric field components on the dielectric–air interface the average of the two permittivities, $(\epsilon_0 + \epsilon_1)/2$, is used. The validity of this treatment is explained in [20].

Due to the use of centered differences in these approximations, the error is second order in both the space and time steps; i.e., if Δx, Δy, Δz, and Δt are proportional to Δl, then the global error is $O(\Delta l^2)$. The maximum time step that may be used is limited by the stability restriction of the finite difference equations,

$$\Delta t \leqslant \frac{1}{v_{\max}} \left(\frac{1}{\Delta x^2} + \frac{1}{\Delta y^2} + \frac{1}{\Delta z^2} \right)^{-1/2} \tag{9}$$

where $v_{\max}$ is the maximum velocity of light in the computational volume. Typically, $v_{\max}$ will be the velocity of light in free space unless the entire volume is filled with dielectric. These equations will allow the approximate solution of $\boldsymbol{E}(\boldsymbol{r},t)$ and $\boldsymbol{H}(\boldsymbol{r},t)$ in the volume of the computational domain or mesh; however, special consideration is required for the source, the conductors, and the mesh walls.

C. Source Considerations

The volume in which the microstrip circuit simulation is to be performed is shown schematically in Fig. 2. At $t = 0$ the fields are assumed to be identically 0 throughout the computational domain. A Gaussian pulse is desirable as the excitation because its frequency spectrum is also Gaussian and will therefore provide frequency-domain information from dc to the desired cutoff frequency by adjusting the width of the pulse.

In order to simulate a voltage source excitation it is necessary to impose the vertical electric field, E_z, in a rectangular region underneath port 1 as shown in Fig. 2. The remaining electric field components on the source plane must be specified or calculated. In [18]–[20] an electric wall source is used; i.e., the remaining electric field components on the source wall of the mesh are set to 0. An unwanted side effect of this type of excitation is that a sharp magnetic field is induced tangential to the source wall. This results in some distortion of the launched pulse. Specifically, the pulse is reduced in magnitude due to the energy stored in the induced magnetic field and a negative tail to the pulse is immediately evident. An alternative excitation scheme is to simulate a magnetic wall at the source plane. The source plane consists only of E_x and E_z components, with the tangential magnetic field components offset $\pm \Delta y/2$. If the magnetic wall is enforced by setting the tangential magnetic field components to zero just behind the source plane, then significant distortion of the pulse still occurs. If the magnetic wall is enforced directly on the source plane by using image theory (i.e., $\boldsymbol{H}_{\tan}$ outside the magnetic wall is equal

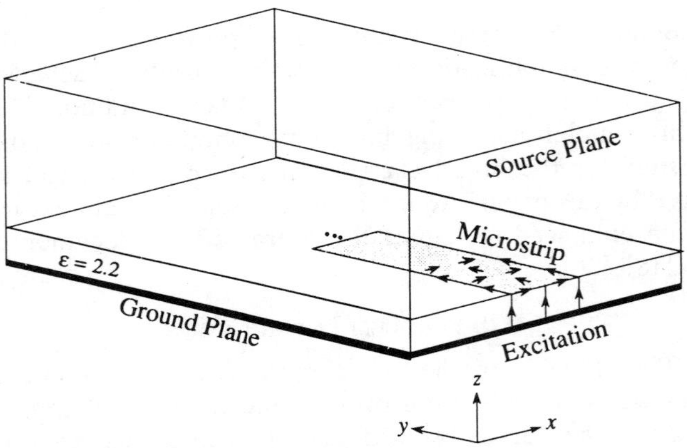

Fig. 2. Computational domain.

to $-\boldsymbol{H}_{\tan}$ inside the magnetic wall), then the remaining electric field components on the source plane may be readily calculated using the finite-difference equations. Using this excitation, only a minimal amount of source distortion is apparent. The launched wave has nearly unit amplitude and is Gaussian in time and in the $\hat{y}$ direction:

$$E_z = f_s(t) = e^{-(t-t_0)^2/T^2}. \tag{10}$$

It is assumed that excitation specified in this way will result in the fundamental mode only propagating down the microstrip in the frequency range of interest.

The finite-difference formulas are not perfect in their representation of the propagation of electromagnetic waves. One effect of this is numerical dispersion; i.e., the velocity of propagation is slightly frequency dependent even for uniform plane waves. In order to minimize the effects of numerical dispersion and truncation errors, the width of the Gaussian pulse is chosen for at least 20 points per wavelength at the highest frequency represented significantly in the pulse.

D. Conductor Treatment

The circuits considered in this paper have a conducting ground plane and a single dielectric substrate with metallization on top of this substrate in the ordinary microstrip configuration. These electric conductors are assumed to be perfectly conducting and have zero thickness and are treated by setting the electric field components that lie on the conductors to zero. The edge of the conductor should be modeled with electric field components tangential to the edge lying exactly on the edge of the microstrip as shown in Fig. 2.

E. Absorbing Boundary Treatment

Due to the finite capabilities of the computers used to implement the finite-difference equations, the mesh must be limited in the $\hat{x}$, $\hat{y}$, and $\hat{z}$ directions. The difference equations cannot be used to evaluate the field components tangential to the outer boundaries since they would require the values of field components outside of the mesh. One of the six mesh boundaries is a ground plane and its tangential electric field values are forced to be 0.

The tangential electric field components on the other five mesh walls must be specified in such a way that outgoing waves are not reflected using the absorbing boundary condition [22], [23]. For the structures considered in this paper, the pulses on the microstrip lines will be normally incident to the mesh walls. This leads to a simple approximate continuous absorbing boundary condition, which is that the tangential fields on the outer boundaries will obey the one-dimensional wave equation in the direction normal to the mesh wall. For the $\hat{y}$ normal wall the one-dimensional wave equation may be written

$$\left(\frac{\partial}{\partial y} - \frac{1}{v}\frac{\partial}{\partial t}\right)\boldsymbol{E}_{\tan} = 0. \tag{11}$$

This equation is Mur's first approximate absorbing boundary condition and it may be easily discretized using only field components on or just inside the mesh wall, yielding an explicit finite difference equation [22],

$$E_0^{n+1} = E_1^n + \frac{v\Delta t - \Delta y}{v\Delta t + \Delta y}\left(E_1^{n+1} - E_0^n\right) \tag{12}$$

where E_0 represents the tangential electric field components on the mesh wall and E_1 represents the tangential electric field components one node inside of the mesh wall. Similar expressions are immediately obtained for the other absorbing boundaries by using the corresponding normal directions for each wall. It should be noted that the normal incidence assumption is not valid for the fringing fields which are propagating tangential to the walls; therefore the sidewalls should be far enough away that the fringing fields are negligible at the walls. Additionally, radiation will not be exactly normal to the mesh walls. Second-order absorbing boundary conditions [22], [23] which account for oblique incidence will not work on the mesh walls where the microstrip is incident because these absorbing boundary conditions are derived in uniform space.

The results presented show that the first-order absorbing boundary treatment is sufficiently accurate and particularly well suited to the microstrip geometry. It is possible to obtain more accurate normal incidence absorbing boundary conditions [24], which have been used in the FDTD calculation of microstrip discontinuities [20]. Due to the more dynamic, resonant behavior of the circuits considered, Mur's first approximate absorbing boundary condition is used and this allows for accurate simulation of the various microstrip circuits.

F. Time Marching Solution

The finite difference equations, (3)–(8), are used with the above boundary and source conditions to simulate the propagation of a broad-band Gaussian pulse on the microstrip structure. The essential aspects of the time-domain algorithm are as follows:

- Initially (at $t = n = 0$) all fields are 0.
- The following are repeated until the response is ≈ 0:
  - Gaussian excitation is imposed on port 1.
  - $\boldsymbol{H}^{n+1/2}$ is calculated from FD equations.
  - $\boldsymbol{E}^{n+1}$ is calculated from FD equations.
  - Tangential $\boldsymbol{E}$ is set to 0 on conductors.
  - Save desired field quantities.
  - $n \rightarrow n+1$.
- Compute scattering matrix coefficients from time-domain results.

One additional consideration is that the reflections from the circuit will be reflected again by the source wall. To eliminate this, the circuit is placed a sufficient distance from the source, and after the Gaussian pulse has been fully launched, the absorbing boundary condition is switched on at the source wall.

G. Frequency-Dependent Parameters

In addition to the transient results obtained naturally by the FDTD method, the frequency-dependent scattering matrix coefficients are easily calculated.

$$[V]^r = [S][V]^i \tag{13}$$

where $[V]^r$ and $[V]^i$ are the reflected and incident voltage vectors, respectively, and $[S]$ is the scattering matrix. To accomplish this, the vertical electric field underneath the center of each microstrip port is recorded at every time step. As in [20], it is assumed that this field value is proportional to the voltage (which could be easily obtained by numerically integrating the vertical electric field) when considering propagation of the fundamental mode. To obtain the scattering parameter $S_{11}(\omega)$, the incident and reflected waveforms must be known. The FDTD simulation calculates the sum of incident and reflected waveforms. To obtain the incident waveform, the calculation is performed using only the port 1 microstrip line, which will now be of infinite extent (i.e., from source to far absorbing wall), and the incident waveform is recorded. This incident waveform may now be subtracted from the incident plus reflected waveform to yield the reflected waveform for port 1. The other ports will register only transmitted waveforms and will not need this computation. The scattering parameters, S_{jk}, may then be obtained by simple Fourier transform of these transient waveforms as

$$S_{jk}(\omega) = \frac{\mathscr{F}\mathscr{T}\{V_j(t)\}}{\mathscr{F}\mathscr{T}\{V_k(t)\}}. \tag{14}$$

Note that the reference planes are chosen with enough distance from the circuit discontinuities to eliminate evanescent waves. These distances are included in the definition of the circuit so that no phase correction is performed for the scattering coefficients. For all of the circuits considered, the only unique coefficients are in

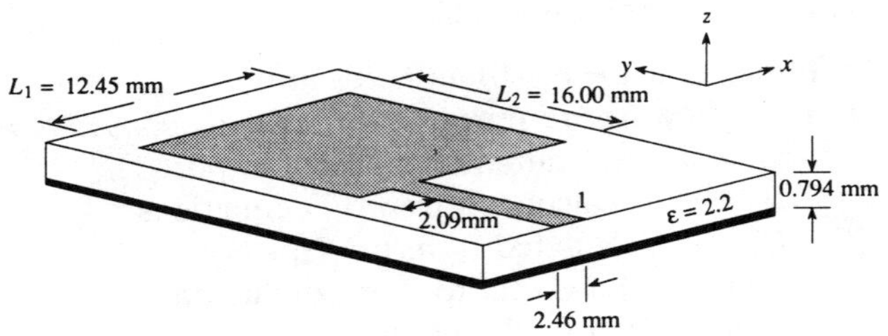

Fig. 3. Line-fed rectangular microstrip antenna detail.

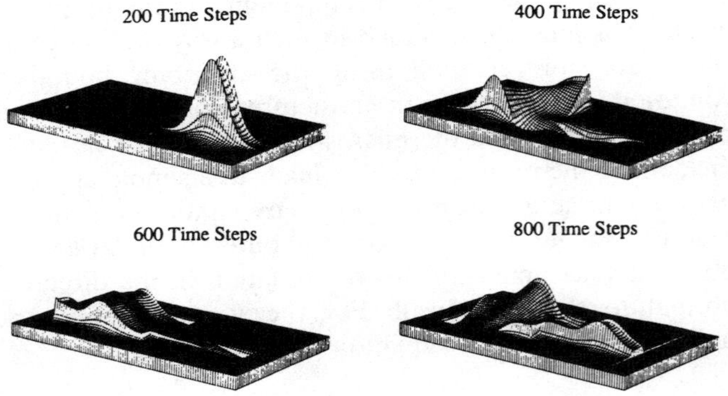

Fig. 4. Rectangular microstrip antenna distribution of $E_z(x, y, t)$ just underneath the dielectric interface at 200, 400, 600, and 800 time steps.

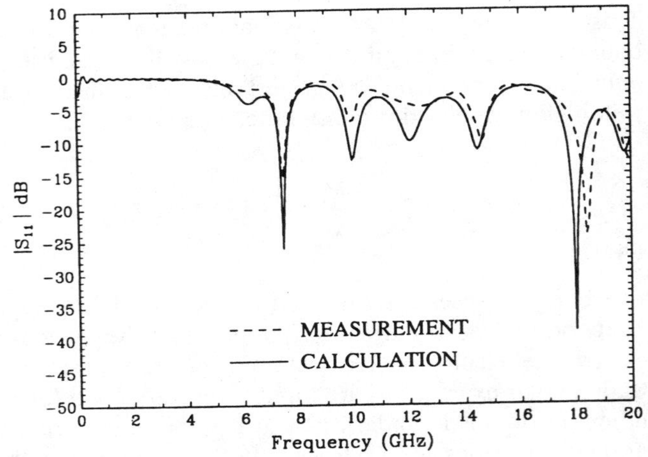

Fig. 5. Return loss of the rectangular antenna.

the first column of the scattering matrix (i.e., $S_{11}(\omega), S_{21}(\omega), S_{31}(\omega), \cdots$).

III. Numerical Results

Numerical results have been computed for three configurations, a line-fed rectangular patch antenna, a low-pass filter, and a branch line coupler. These circuits have dimensions on the order of 1 cm, and the frequency range of interest is from dc to 20 GHz. The operating regions of all of these circuits are less than 10 GHz; however, the accuracy of the computed results at higher frequencies is examined. These circuits were constructed on Duroid substrates with $\epsilon = 2.2$ and thickness of 1/32 inch (0.794 mm). Scattering matrix coefficients were measured using an HP 8510 network analyzer, which is calibrated to 18 GHz, but provides measurement to 20 GHz.

A. Line-Fed Rectangular Microstrip Antenna

The actual dimensions of the microstrip antenna analyzed are shown in Fig. 3. The operating resonance approximately corresponds to the frequency where $L_1 =$ 1.245 mm $= \lambda/2$. Simulation of this circuit involves the straightforward application of the finite-difference equations, source, and boundary conditions. To model the thickness of the substrate correctly, Δz is chosen so that three nodes exactly match the thickness. An additional 13 nodes in the $\hat{z}$ direction are used to model the free space above the substrate. In order to correctly model the dimensions of the antenna, Δx and Δy have been chosen so that an integral number of nodes will exactly fit the rectangular patch. Unfortunately, this means the port width and placement will be off by a fraction of the space step. The sizes of the space steps are carefully chosen to minimize the effect of this error.

The space steps used are $\Delta x = 0.389$ mm, $\Delta y = 0.400$ mm, and $\Delta z = 0.265$ mm, and the total mesh dimensions are $60 \times 100 \times 16$ in the $\hat{x}$, $\hat{y}$, and $\hat{z}$ directions respectively. The rectangular antenna patch is thus $32\,\Delta x \times 40\,\Delta y$. The length of the microstrip line from the source plane to the edge of the antenna is $50\,\Delta y$, and the reference plane for port 1 is $10\,\Delta y$ from the edge of the patch. The microstrip line width is modeled as $6\,\Delta x$.

The time step used is $\Delta t = 0.441$ ps. The Gaussian half-width is $T = 15$ ps, and the time delay t_0 is set to be $3T$ so the Gaussian will start at approximately 0. The simulation is performed for 8000 time steps, somewhat longer than for other circuits, due to the highly resonant behavior of the antenna. The computation time for this circuit is approximately 12 h on a VAXstation 3500 workstation.

The spatial distribution of $E_z(x, y, t)$ just beneath the microstrip at 200, 400, 600, and 800 time steps is shown in Fig. 4, where the source Gaussian pulse and subsequent propagation on the antenna are observed. Notice that the absorbing boundary condition for the source has been implemented several nodes from the source plane to eliminate any undesirable effects of switching from source to absorbing boundary conditions. Three-dimensional properties of the propagation are observed including enhancement of the field near the edges of the microstrip.

The scattering coefficient results, shown in Fig. 5, show good agreement with the measured data. The operating resonance at 7.5 GHz is almost exactly shown by both theory and measurement. This result is a significant advancement over planar circuit techniques, which without empirical treatment will allow only $|S_{11}| = 1.0$ because they do not allow for radiation. Also, the resonance frequency calculated using planar circuit concepts will be sensitive to errors in the effective dimensions of the patch. Additional resonances are also in good ageement with experiment, except for the highest resonance near 18 GHz, which is somewhat shifted.

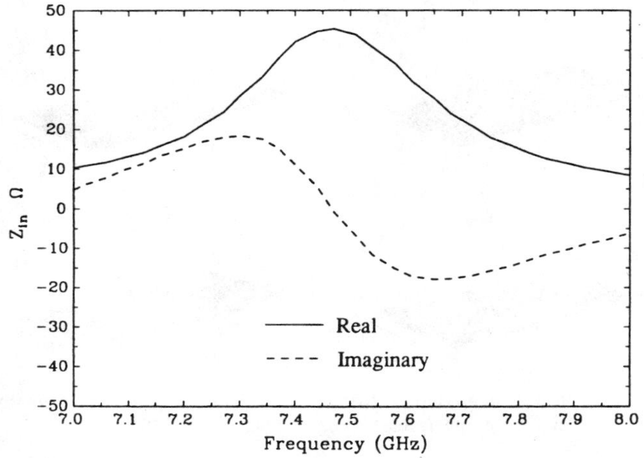

Fig. 6. Input impedance of the rectangular antenna near the operating resonance of 7.5 GHz as calculated from the FDTD results.

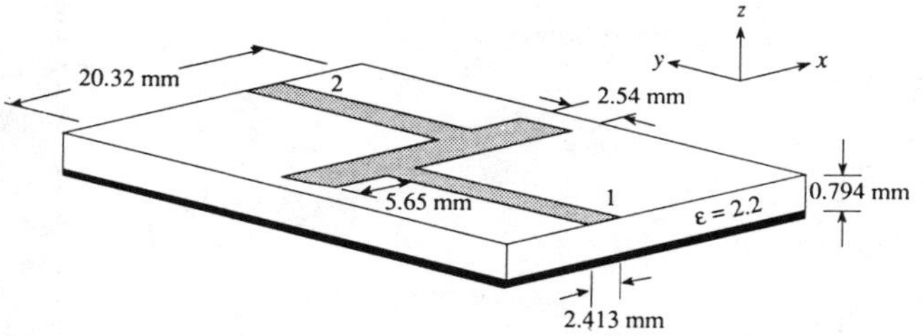

Fig. 7. Low-pass filter detail.

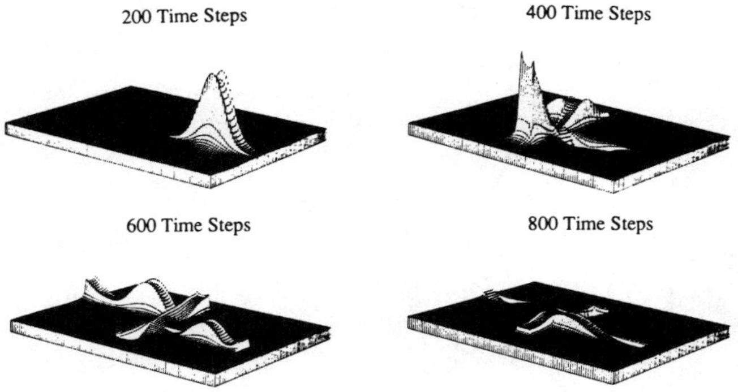

Fig. 8. Low-pass filter distribution of $E_z(x, y, t)$ just underneath the dielectric interface at 200, 400, 600, and 800 time steps.

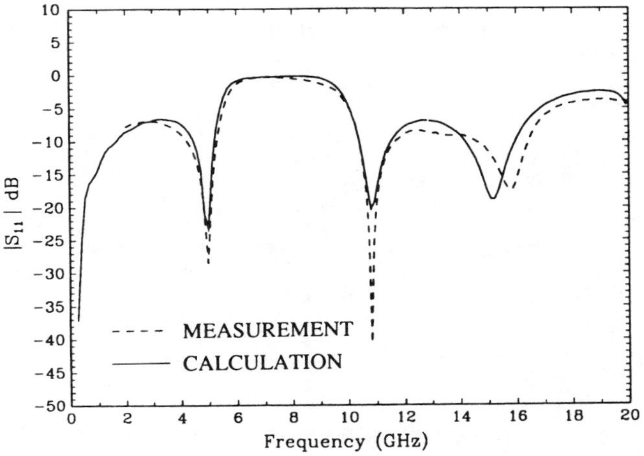

Fig. 9. Return loss of the low-pass filter.

Input impedance for the antenna may be calculated from the $S_{11}(\omega)$ calculation by transforming the reference plane to the edge of the microstrip antenna,

$$Z_{in} = Z_0 \frac{1 + S_{11} e^{j2kL}}{1 - S_{11} e^{j2kL}} \tag{15}$$

where k is the wavenumber on the microstrip, L is the length from the edge of the antenna to the reference plane ($10 \Delta y$), and Z_0 is the characteristic impedance of the microstrip line. For simplicity of the Z_{in} calculation, the microstrip is assumed to have a constant characteristic impedance of 50 Ω, and an effective permittivity of 1.9 is used to calculate the wavenumber. Results for the input impedance calculation near the operating resonance of 7.5 GHz are shown in Fig. 6.

B. Microstrip Low-Pass Filter

The low-pass filter analyzed is one designed according to the criteria established by [2] and is shown in Fig. 7. As in the rectangular antenna, Δx, Δy, and Δz are carefully chosen to fit the dimensions of the circuit. The space steps Δx and Δy are chosen to exactly match the dimensions of the rectangular patch; however, the locations and widths of the ports will be modeled with some error.

The space steps used are $\Delta x = 0.4064$ mm, $\Delta y = 0.4233$ mm, and $\Delta z = 0.265$ mm, and the total mesh dimensions are $80 \times 100 \times 16$ in the $\hat{x}$, $\hat{y}$, and $\hat{z}$ directions respectively. The long rectangular patch is thus $50 \Delta x \times 6 \Delta y$. The distance from the source plane to the edge of the long patch is $50 \Delta y$, and the reference planes for ports 1 and 2 are $10 \Delta y$ from the edges of the patch. The strip widths of ports 1 and 2 are modeled as $6 \Delta x$.

The time step used is $\Delta t = 0.441$ ps. The Gaussian half-width is $T = 15$ ps and the time delay t_0 is set to be $3T$. The simulation is performed for 4000 time steps to allow the response on both ports to become nearly 0. The computation time for this circuit is approximately 8 h on a VAXstation 3500 workstation.

The spatial distribution of $E_z(x, y, t)$ just beneath the microstrip at 200, 400, 600, and 800 time steps is shown in Fig. 8. The scattering coefficient results, shown in Figs. 9 and 10, again show good agreement in the location of the response nulls. The desired low-pass filter performance is seen in the sharp S_{21} roll-off beginning at approximately 5 GHz. There is again some shift near the high end of the frequency range. In the S_{21} results, the stopband for the calculated curve is somewhat narrower than the measured results. Some experimentation with planar circuit techniques has led to the conclusion that this narrowing is caused predominantly by the slight misplacement of the ports inherent in the choice of Δx and Δy.

C. Microstrip Branch Line Coupler

The branch line coupler, shown in Fig. 11, is used to divide power equally between ports 3 and 4 from ports 1 or 2. This occurs at the frequency where the center-to-center distance between the four lines is a quarter wave-

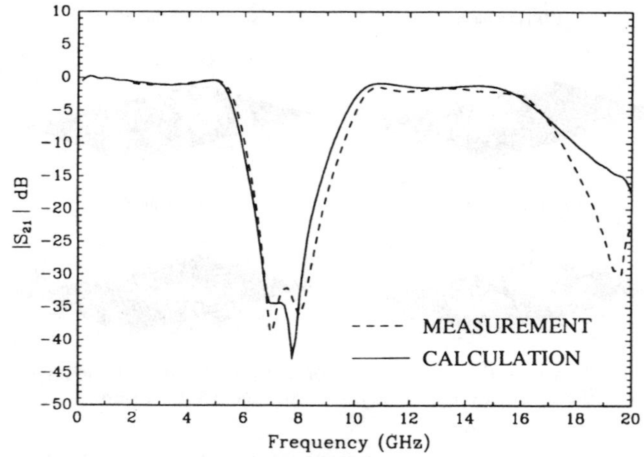

Fig. 10. Insertion loss of the low-pass filter.

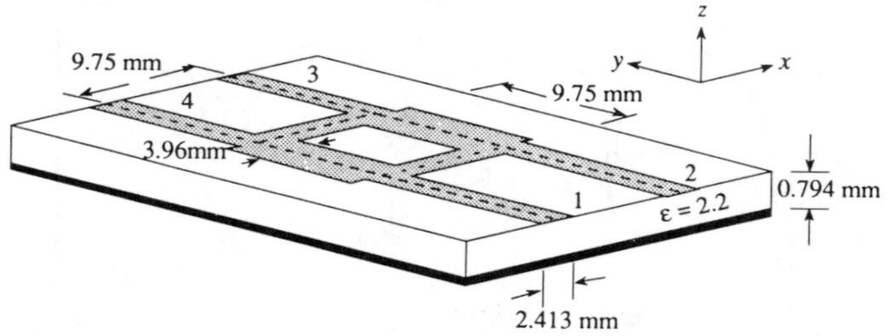

Fig. 11. Branch line coupler detail.

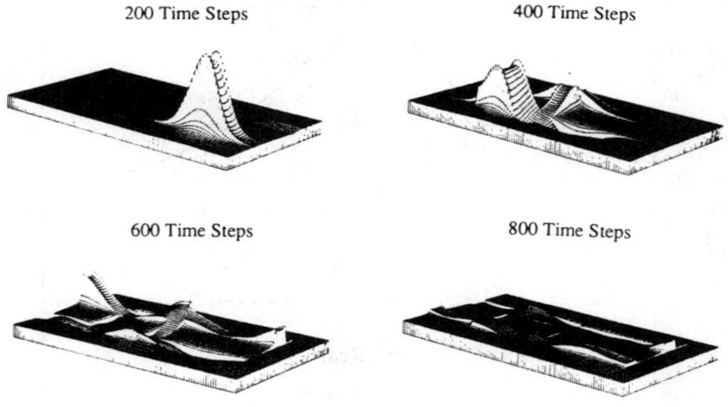

Fig. 12. Branch line coupler distribution of $E_z(x, y, t)$ just underneath the dielectric interface at 200, 400, 600, and 800 time steps.

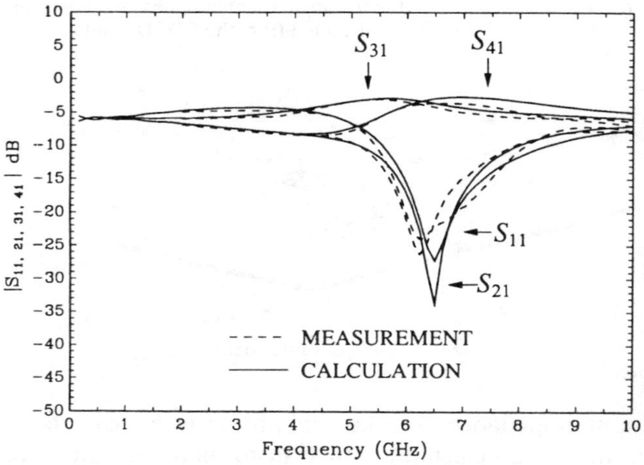

Fig. 13. Scattering parameters of the branch line coupler.

length. Also, the phase difference between ports 3 and 4 is 90° at this frequency. To model this circuit, Δx, Δy, and Δz are chosen to match the dimensions of the circuit as effectively as possible. The space step, Δz, is chosen to match the substrate thickness exactly. The space steps, Δx and Δy, are chosen to match the center-to-center distance (9.75 mm) exactly. Again, small errors in the other $\hat{x}$ and $\hat{y}$ dimensions occur.

The space steps used are $\Delta x = 0.406$ mm, $\Delta y = 0.406$ mm, and $\Delta z = 0.265$ mm and the total mesh dimensions are $60 \times 100 \times 16$ in the $\hat{x}$, $\hat{y}$, and $\hat{z}$ directions, respectively. The center-to-center distances are $24\Delta x$ and $24\Delta y$. The distance from the source plane to the edge of the coupler is $50\Delta y$, and the reference planes for ports 1 through 4 are $10\Delta y$ from the edges of the coupler. The strip widths of ports 1 through 4 are modeled as $6\Delta x$. The wide strips in the coupler are modeled as $10\Delta x$ wide.

The time step used is $\Delta t = 0.441$ ps. The Gaussian half-width is $T = 15$ ps and the time delay t_0 is set to be $3T$. The simulation is performed for 4000 time steps to allow the response on all four ports to become nearly zero. The computation time for this circuit is approximately 6 h on a VAXstation 3500 workstation.

The spatial distribution of $E_z(x, y, t)$ just beneath the microstrip at 200, 400, 600, and 800 time steps is shown in Fig. 12. The scattering coefficient results, shown in Fig. 13, again show good agreement in the location of the response nulls and crossover point. The desired branch line coupler performance is seen in the sharp S_{11} and S_{21} nulls which occur at approximately the same point (6.5 GHz) as the crossover in S_{31} and S_{41}. At this crossover point S_{31} and S_{41} are both approximately -3 dB, indicating that the power is being evenly divided between ports 3 and 4. The nulls in S_{11} and S_{21} at the operating point indicate that little power is being transmitted by ports 1 and 2. The phase difference between S_{31} and S_{41} is verified to be approximately 90° at the operating point (≈ 6.5 GHz) in both the calculated and measured coefficients, as shown in Fig. 14. Some shift in the location of the nulls is again observed and is likely due to the slight modeling errors in the widths of the lines.

D. Error Discussion

In general, agreement between measured and calculated results has been good; however, there are several reasons to explain the small discrepancies in the results. The modeling error occurs primarily in the inability to match all of the circuit dimensions. The space steps, Δx, Δy, and Δz, may be freely chosen; this allows exact matching of only one circuit dimension (or two edges) in each of the $\hat{x}$, $\hat{y}$, and $\hat{z}$ directions. Another small source of error is the exclusion of dielectric and conductor loss in the FDTD calculation. This causes the calculated S parameters to be shifted up in amplitude from the measured data at the higher frequencies. Measurement errors occur because of the microstrip-to-coaxial transitions, which are

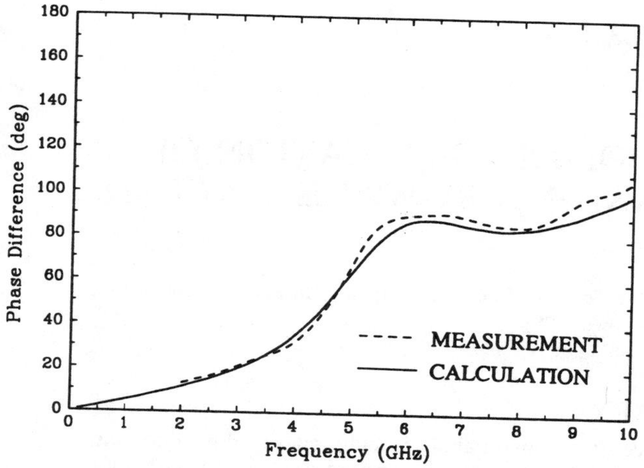

Fig. 14. Phase difference of port 3 relative to port 4 of the branch line coupler.

not de-embedded in the measurement, and because the network analyzer and its connectors are rated only to 18 GHz.

IV. Conclusions

The finite-difference time-domain method has been used to perform time-domain simulations of pulse propagation in several printed microstrip circuits. In addition to the transient results, frequency-dependent scattering parameters and the input impedance of a line-fed rectangular patch antenna have been calculated by Fourier transform of the time-domain results. These results have been verified by comparison with measured data. The versatility of the FDTD method allows easy calculation of many complicated microstrip structures. With the computational power of computers increasing rapidly, this method is very promising for the computer-aided design of many types of microstrip circuit components.

References

[1] G. D'Inzeo, F. Giannini, C. Sodi, and R. Sorrentino, "Method of analysis and filtering properties of microwave planar networks," *IEEE Trans. Microwave Theory Tech.*, vol. MTT-26, pp. 462–471, July 1978.

[2] G. D'Inzeo, F. Giannini, and R. Sorrentino, "Novel microwave integrated low-pass filters," *Electron. Lett.*, vol. 15, no. 9, pp. 258–260, Apr. 26, 1979.

[3] K. C. Gupta, R. Garg, and R. Chadha, *Computer-Aided Design of Microwave Circuits*. Dedham, MA: Artech House, 1981.

[4] T. Okoshi, *Planar Circuits for Microwaves and Lightwaves*. Berlin: Springer-Verlag, 1985.

[5] W. K. Gwarek, "Analysis of arbitrarily shaped two-dimensional microwave circuits by finite difference time domain method," *IEEE Trans. Microwave Theory Tech.*, vol. 36, pp. 738–744, Apr. 1988.

[6] K. C. Gupta and M. D. Abouzahra, "Planar circuit analysis," in *Numerical Techniques for Microwave and Millimeter-Wave Passive Structures*, T. Itoh, Ed. New York: Wiley, 1989, pp. 214–333.

[7] P. B. Katehi and N. C. Alexopoulos, "Frequency dependent characteristics of microstrip discontinuities in millimeter wave integrated circuits," *IEEE Trans. Microwave Theory Tech.*, vol. MTT-33, pp. 1029–1035, Oct. 1985.

[8] R. W. Jackson, "Full-wave, finite element analysis of irregular microstrip discontinuities," *IEEE Trans. Microwave Theory Tech.*, vol. 37, pp. 81–89, Jan. 1989.

[9] S. Koike, N. Yoshida, and I. Fukai, "Transient analysis of microstrip line on anisotropic substrate in three-dimensional space," *IEEE Trans. Microwave Theory Tech.*, vol. 36, pp. 34–43, Jan. 1988.

[10] T. Shibata, T. Hayashi, and T. Kimura, "Analysis of microstrip circuits using three-dimensional full-wave electromagnetic field analysis in the time domain," *IEEE Trans. Microwave Theory Tech.*, vol. 36, pp. 1064–1070, June 1988.

[11] S. Akhtarzad and P. B. Johns, "Solution of Maxwell's equations in three space dimensions and time by the T. L. M. method of numerical analysis," *Proc. Inst. Elec. Eng.*, vol. 122, no. 12, pp. 1344–1348, Dec. 1975.

[12] W. J. R. Hoefer, "The transmission-line matrix method—Theory and applications," *IEEE Trans. Microwave Theory Tech.*, vol. MTT-33, pp. 882–893, Oct. 1985.

[13] K. S. Yee, "Numerical solution of initial boundary value problems involving Maxwell's equations in isotropic media," *IEEE Trans. Antennas Propagat.*, vol. AP-14, pp. 302–307, May 1966.

[14] A. Taflove and M. E. Brodwin, "Numerical solution of steady state electromagnetic scattering problems using the time dependent Maxwell's equations," *IEEE Trans. Microwave Theory Tech.*, vol. MTT-23, pp. 623–630, Aug. 1975.

[15] A. Taflove, "Application of the finite difference time domain method to sinusoidal steady state electromagnetic penetration problems," *IEEE Trans. Electromagn. Compat.* vol. EMC-22, pp. 191–202, Aug. 1980.

[16] A. Taflove and K. R. Umashankar, "The finite difference time domain (FD–TD) method for electromagnetic scattering and interaction problems," *J. Electromagn. Waves and Appl.*, vol. 1, no. 3, pp. 243–267, 1987.

[17] A. C. Cangellaris, C. C. Lin, and K. K. Mei, "Point matched time domain finite element methods for electromagnetic radiation and scattering," *IEEE Trans. Antennas Propagat.*, vol. AP-35, pp. 1160–1173, Oct. 1987.

[18] X. Zhang, J. Fang, K. K. Mei, and Y. Liu, "Calculations of the dispersive characteristics of microstrips by the time-domain finite difference method," *IEEE Trans. Microwave Theory Tech.*, vol. 36, pp. 263–267, Feb. 1988.

[19] X. Zhang and K. K. Mei, "Time domain finite difference approach for the calculation of microstrip open-circuit end effect," *IEEE Trans. Microwave Theory Tech.*, in *IEEE MTT-S Int. Microwave Symp. Dig.*, 1988, pp. 363–366.

[20] X. Zhang and K. K. Mei, "Time domain finite difference approach to the calculation of the frequency dependent characteristics of microstrip discontinuities," *IEEE Trans. Microwave Theory Tech.*, vol. 36, pp. 1775–1787, Dec. 1988.

[21] D. H. Choi and W. J. R. Hoefer, "The finite-difference time-domain method and its application to eigenvalue problems," *IEEE Trans. Microwave Theory Tech.*, vol. MTT-34, pp. 1464–1470, Dec. 1986.

[22] G. Mur, "Absorbing boundary conditions for the finite difference approximation of the time domain electromagnetic field equations," *IEEE Trans. Electromagn. Compat.*, vol. EMC-23, pp. 377–382, Nov. 1981.

[23] B. Enquist and A. Majda, "Absorbing boundary conditions for the numerical simulation of waves," *Math. Comput.*, vol. 31, no. 139, pp. 629–651, July 1977.

[24] K. K. Mei and J. Fang, "A super-absorption boundary algorithm for one dimensional wave equations," submitted to *J. Comput. Phys.*

Paper 6.3

FINITE DIFFERENCE TIME-DOMAIN SIMULATION OF ELECTROMAGNETIC FIELDS AND MICROWAVE CIRCUITS

INGO WOLFF

Department of Electrical Engineering and Sonderforschungsbereich 254, Duisburg University, Bismarckstrasse 81, D-4100 Duisburg, Germany

SUMMARY

An overview is given on the application of finite difference time-domain analysis techniques to the simulation of electromagnetic wave phenomena and the design of planar microwave integrated circuits. In particular the application to the analysis of three-dimensional discontinuities in coplanar microwave circuits is described and numerical and experimental results are presented. Finally the design of planar microwave circuits containing non-linear active components using the finite difference time-domain method in combination with a harmonic balance technique is demonstrated using an example of a LUFET circuit.

1. INTRODUCTION

For the design of planar microwave integrated circuits up to 1985 mainly analysis techniques in the frequency domain have been used. With the requirements for new and flexible tools in the design of planar circuits, for example, with closely coupled elements, alternative techniques had to be studied. One of these techniques is the finite difference time-domain (FDTD) analysis which in principle has been known for a long time. Yee[1] already in 1966 proposed this technique for the analysis of electromagnetic boundary value problems.

For a long time the FDTD technique was only used to qualitatively demonstrate electromagnetic field solutions in the time domain. Only the introduction of absorbing walls[2–4] made this technique into a powerful tool for application in real circuit design. This was soon realized by several authors who presented first results for the time-domain analyses of eigenvalue problems,[5] planar circuits,[6] microstrip lines and discontinuities[7–11] and to arbitrary two-dimensional circuits.[8] A combination of time-domain analysis techniques with the method of lines was presented.[10]

In this paper an overview shall be given on finite difference time-domain simulations of electromagnetic fields and the application of this method to a realistic microwave circuit design. It will be shown that this technique enables us to model three-dimensional circuit structures in microstrip and coplanar microwave circuits and that it is a powerful analysis technique for non-linear microwave integrated circuit design by combining semiconductor modelling with the finite difference time-domain analysis of passive circuit elements if, for reasons to be discussed later, a harmonic balance technique is used for this combination.

2. THE FINITE DIFFERENCE TIME-DOMAIN METHOD

The finite difference time-domain technique (FDTD) is a numerical method for the solution of electromagnetic field problems which has a large numerical but a low analytical expense. Despite the large numerical expense it is believed to be one of the most efficient techniques, because basically it stores only the field distribution at one moment in memory instead of working with a large equation system matrix. The field solution for each other time then is determined from Maxwell's equations and is calculated using a timestepping procedure based on the finite difference formulation of Maxwell's equations in space and time. The 'leapfrog algorithm',[1] which will be described below, fits very well on modern computer architectures, so that the data required to describe a three-dimensional field distribution can be handled in a reasonable time. Therefore the method can efficiently be implemented on vector or on parallel computers. Sufficiently accurate

Reprinted with permission from "Finite Difference Time-Domain Simulation of Electromagnetic Fields and Microwave Circuits," I. Wolff, *Int. Jour. of MIMICAE*, vol. 5, no. 3, pp. 163–182, © 1992. Reprinted with permission of John Wiley & Sons, Inc.

results can be obtained by using a single precision floating-point expression requiring only four bytes, as has been shown recently.[11]

The electromagnetic field in the space which has to be analysed can be excited in different ways, as will be discussed later. A transient analysis, where a pulse in space and time, e.g. in the form of a Gaussian pulse, is excited inside the circuit or component, is very advantageous. The reflexions and transmissions of the pulse inside the component or circuit are computed in the time domain and finally the time-domain signals at defined ports are transformed back to the frequency domain using a fast Fourier transform. This transient analysis delivers a broadband frequency response in one single computational run.

If the space is discretized into $N \times M \times L$ cells, a cube of $6NML$ memory places is needed for storage in the finite difference time-domain analysis. If typically the discretization in the plane of a planar microwave circuit is 30×40 elements and about 15 discretization cells are needed perpendicular to the substrate plane (these are the lowest numbers for the analysis, for example, of a simple microstrip component), the storage requirement is about 100,000 to 120,000 memory places.

It will be shown that the FDTD-method offers the possibility of a flexible problem formulation with an analytical expense which is much lower than that of frequency-domain methods like, for example, the spectral-domain analysis technique. Arbitrarily shaped planar structures like, for example, radial stubs or curved lines can easily be approximated in a rectangular mesh. Even very rough mesh sizes lead to an acceptable accuracy in the results.

2.1. *The leapfrog algorithm*

The fundamental idea of the FDTD method is to discretize space and time into finite differences and thereby to reduce Maxwell's differential equations

$$\frac{\partial}{\partial t}\mathbf{E} = \frac{\partial}{\varepsilon}\operatorname{curl}\mathbf{H}, \qquad \frac{\partial}{\partial t}\mathbf{H} = \frac{1}{\mu}\operatorname{curl}\mathbf{E} \tag{1}$$

to finite difference equations. If the space cell shown in Figure 1, which first was used by Yee,[1]

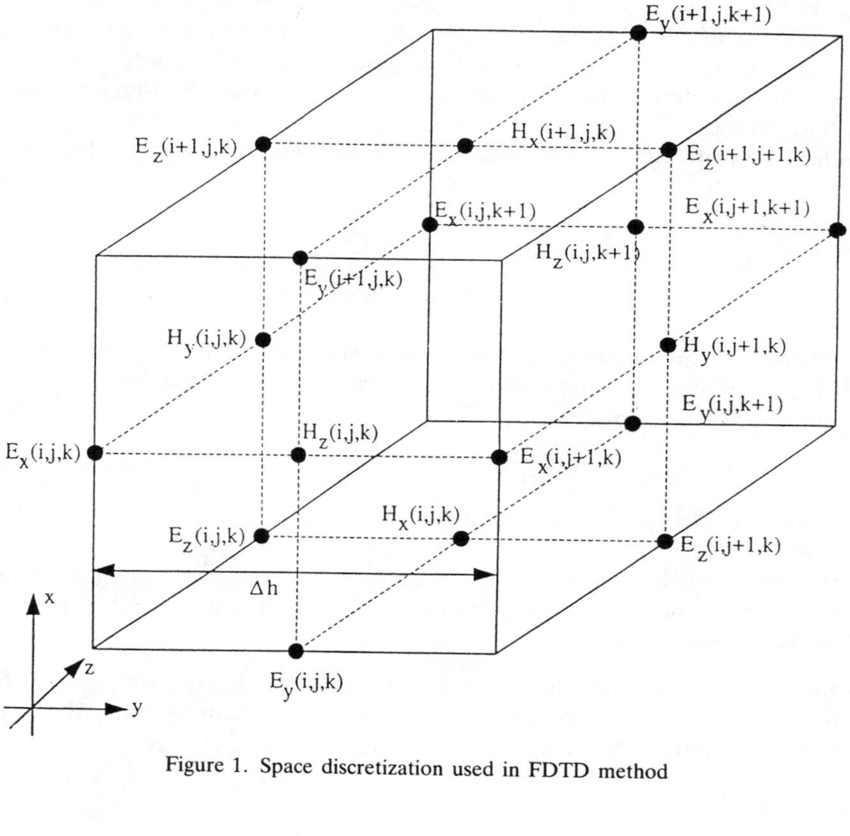

Figure 1. Space discretization used in FDTD method

is applied to calculate the electromagnetic field components in the defined nodes, the resulting finite difference equations are of central difference form and of second-order accuracy. They are given by:

$$E_x^{n+1}(i,j,k) = E_x^n(i,j,k) + \frac{\Delta t}{\varepsilon \Delta h}(H_z^{n+1/2}(i,j,k) - H_z^{n+1/2}(i,j-1,k) - H_y^{n+1/2}(i,j,k) + H_y^{n+1/2}(i,j,k-1))$$

$$E_y^{n+1}(i,j,k) = E_y^n(i,j,k) + \frac{\Delta t}{\varepsilon \Delta h}(H_x^{n+1/2}(i,j,k) - H_x^{n+1/2}(i,j,k-1) - H_z^{n+1/2}(i,j,k) + H_z^{n+1/2}(i-1,j,k))$$

$$E_z^{n+1}(i,j,k) = E_z^n(i,j,k) + \frac{\Delta t}{\varepsilon \Delta h}(H_y^{n+1/2}(i,j,k) - H_y^{n+1/2}(i-1,j,k) - H_x^{n+1/2}(i,j,k) + H_x^{n+1/2}(i,j-1,k))$$

$$H_x^{n+1/2}(i,j,k) = H_x^{n-1/2}(i,j,k) - \frac{\Delta t}{\mu \Delta h}(E_z^n(i,j+1,k) - E_z^n(i,j,k) - E_y^n(i,j,k+1) + E_y^n(i,j,k)) \quad (2)$$

$$H_y^{n+1/2}(i,j,k) = H_y^{n-1/2}(i,j,k) - \frac{\Delta t}{\mu \Delta h}(E_x^n(i,j,k+1) - E_x^n(i,j,k) - E_z^n(i+1,j,k) + E_z^n(i,j,k))$$

$$H_z^{n+1/2}(i,j,k) = H_z^{n-1/2}(i,j,k) - \frac{\Delta t}{\mu \Delta h}(E_y^n(i+1,j,k) - E_y^n(i,j,k) - E_x^n(i,j+1,k) + E_x^n(i,j,k))$$

The indices ,i, j, k define the position of the field node, the distances between the nodes are $\Delta x = \Delta y = \Delta z = \Delta h$. Using the node arrangement of Figure 1, the time differentiation can also be expressed by a central difference formulation, if the electric fields are calculated at the time $n\Delta t$ and the magnetic fields at the time $(n+0{\cdot}5)\Delta t$, respectively. In (2) the time-steps are characterized by the upper index of the field components.

The schematic way in which the electric field is calculated from the magnetic field and then again the magnetic field from the electric field, is called the 'leapfrog' algorithm and was first used in a similar way by Yee.[1] It shall be explained using, for example, the last equation of (2) and the corresponding nodes in Figure 1 which are found in the front plane of the rectangular node-mesh. The field component H_z at time step $n+0{\cdot}5$, which is represented by the central node of the rectangular front node-mesh, is calculated (using the induction law) from its value at time-step $n-0{\cdot}5$ and the values of the E_x- and E_y-field components at time-step n in the nodes surrounding the central node in the front plane of Figure 1. A similar argumentation holds for all other field components. In this way time-step by time-step the field components in all space elements are calculated by an iteration process. Because the differences which approximate the differentiations are of second-order accuracy, this method is accurate enough to deliver an accurate time-dependent solution.

Furthermore, the convergence of the leapfrog method can be assured, if the 'stability condition'

$$v\Delta t \leqslant \frac{1}{\sqrt{\dfrac{1}{\Delta x^2} + \dfrac{1}{\Delta y^2} + \dfrac{1}{\Delta z^2}}} \quad (3)$$

is fulfilled, where v is the maximum phase velocity of the electromagnetic waves in the analysed structure. If (3) is fulfilled, the numerical error in the stepwise calculation does not increase from step to step during the calculations.

2.2. *The excitation of the electromagnetic field*

The electromagnetic field in the structure which is to be analysed must be 'excited', i.e. at the beginning of the calculation a certain field structure must be defined in the grid. Three possible methods shall be shortly discussed:

(a) At the time t_0 (starting time) the electromagnetic field is defined within the total space of the structure considered. This method has been used, for example, in [5] for the analysis of microwave resonators. The field distribution used at t_0 depends on the experience of the program user who, for example, may define the field distribution similar to that of a special resonant mode to determine the eigenvalues of this mode.

(b) The second possibility is to define a harmonic field oscillation on a boundary of the structure. Again the transversal distribution of the electromagnetic field in the boundary may depend on the experience of the programmer. This method is only applicable in cases where no reflections occur in the circuit which is to be analysed, because using a harmonic excitation, incident and reflected waves cannot be divided on the structure and, for example, the scattering parameters cannot be estimated.

(c) The third form of excitation is to use an electromagnetic field pulse with a finite length in space and time. Using this kind of excitation, in principle arbitrary n-ports and their transmission properties can be analysed, if the incident, transmitted and reflected pulses can be separated. Because such a pulse contains a broad spectrum of frequencies, the transmission properties in a desired frequency range can be determined in one analysis cycle.

The spatial transversal distribution of the pulse again must be defined by the user, but very crude assumptions are good enough if only enough space is left between the excitation plane and the reference plane of the structure which is to be analysed, so that electromagnetic fields which are real solutions of Maxwell's equations can be established in the space between the source and the structure. So, for example, in the case of a microstrip line the assumption of a constant electric field across the strip width and zero field outside the strip leads to good results, if the microstrip line is long enough to build up the typical electromagnetic wave on a microstrip line with, for example, large electric field values at the edges of the strip while propagating along the line section.

The pulse length in time must be so short that the spectrum contains all desired frequencies. The length in space must be so short that incident and reflected pulses can be exactly separated. On the other hand, the node distances of the grid used must be smaller than the space length of the pulse to guarantee a proper calculation of the spatial field distribution.

Version (c) of the discussed excitations is very suitable for the analysis of planar microwave circuits. As the exciting pulse a Gaussian pulse described by $f(t) = \exp\{-(t-t_0)^2/T\}$ is used.[9] Its length in time is determined by T. Because, for example, the microstrip line is dispersive, the pulse form is changed when propagating along the line.

From the calculated electromagnetic field, time-dependent voltages and currents can be calculated at the ports of the circuits. Of course the same problems exist as in the case of the frequency-domain calculations that, for example, the voltage is no longer defined exactly in a line structure with two different dielectric matcrials (in the frequency domain: modes are no longer TEM but hybrid modes), but these problems need not to be discussed here. From the time-dependent signals the equivalent values in the frequency domain can be determined using a Fourier transform, so that the scattering parameters at defined ports or reference planes can be calculated from these results.

2.3. *The absorbing walls*

If the transient analysis is to be applied, magnetic or electric walls cannot be used in the boundaries of a problem which shall be solved using the FDTD-technique, because reflections from these boundaries would be superimposed upon the reflected and transmitted pulses in the circuit and therefore would disturb the information on the circuit properties. To allow the separation of the pulses which are scattered at the component or circuit without disturbance by other signals, absorbing boundary conditions have to be used, which enclose the structure to be analysed.

An ideal absorbing boundary can be defined by an infinite space behind this boundary in which a wave penetrates without exciting a reflected wave at the boundary. Such an ideal absorbing wall which exactly simulates the radiation condition leads to a very high computational expense, but an approximate solution for an absorbing wall can be found very easily,[3,4,9] if the field in the boundary always has the same value as at a distance Δh from the boundary at a time $n_p \Delta t$ before the actual time. $n_p \Delta t$ is the time the wave needs to propagate along the distance Δh. Using the stability condition (3) this means that the condition

$$\frac{\Delta h}{v\Delta t} = 1 + (\sqrt{3\varepsilon_{\text{eff}}}) \tag{4}$$

where the parenthetical term defines the integer value, must hold. If $E^n_{\tan}(I,j,k)$ is the tangential component of the electric field in the absorbing wall at a time step $n\Delta t$ and $E^{n-n_p}_{\tan}(I-1,j,k)$ the tangential electric field in the distance Δh perpendicular before the wall at a time $(n-n_p)\Delta t$, the condition for the absorbing wall can be formulated as follows:

$$E^n_{\tan}(I,j,k) = E^{n-n_p}_{\tan}(I-1,j,k) \tag{5}$$

Equation (5) is only a good approximation for the real absorbing boundary behaviour, if the phase velocity of the wave is dispersionless and if the wave is purely transversal and propagates into a direction perpendicular to the boundary. The microstrip line, as is well known, is dispersive. Therefore (5) leads to a wall with a small reflection coefficient (3 per cent to 5 per cent). But using a similar model for the tangential magnetic fields, a description of the wall can be found with an error of similar magnitude but with a changed sign. The weighted superposition of both processes leads to a good approximation of the absorbing wall,[4,9] as long as the wave propagates perpendicularly to the wall. The same model can also be used for walls parallel to the line structure, if the wall is far enough from the line, so that the incident fields are to a first approximation transversal fields. In the case of a wall above the line structure (5) must be used without any error compensation.

It can be shown that the described boundary algorithm, which fulfils the wave equation exactly in the case of a non-dispersive and single direction wave propagation, also allows wave excitation in the opposite direction in the air-filled area for a frequency f_{osc} corresponding to the period

$$T_{\text{osc}} = \frac{\Delta h}{v_{\text{ph}}} + \frac{\Delta h}{c} \tag{6}$$

which is the time the pulse needs to travel one space step on the line structure, plus the time, the pulse needs to travel one space step in air. So, the stability of the absorbing boundary is not given for f_{osc}, because an oscillation can be built up in the boundary because of possible reflections and the two possible propagation directions.

Figure 2 shows a typical result of such a boundary oscillation for an example of a radial microstrip stub. The figure shows the time-domain signals at the input and the output port of the radial stub, which are defined on the input and output microstrip line, far away from the stub-structure. After more than 3000 time iterations, suddenly an instability arises in the time-domain solution, owing to the wall oscillation as described above.

Alternative absorbing boundary conditions without this disadvantage normally have higher reflections or they are (as mentioned above) numercially too expensive. But owing to the fact, that the absorbing boundary algorithm described above offers a number of field-samples over the actual moment in time, together with buffered samples of passed moments, digital filter techniques even with non-causal systems can be used to suppress the oscillation of the absorbing boundary.

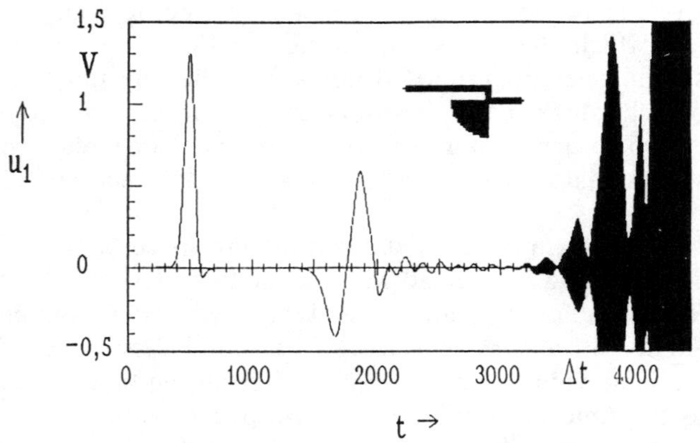

Figure 2. Unstable time-domain solution resulting from an absorbing boundary oscillation

The simplest method is to build an average of some neighbouring samples in the time domain which means nothing else but the convolution of a rectangular function with a width T_{osc} or multiplying the boundary field with a real lowpass filter function in the frequency domain. This will not influence the signals at microwave frequencies in the time-domain pulse very much, but it improves the stability of the absorbing boundary drastically. The introduction of such a boundary finally reduces the computation domain in which the electromagnetic field must be calculated. Thereby the storage requirements and the computation time are reduced.

2.4. *Matched sources*

As has already been mentioned above, the excitation of the pulse inside the circuit is not a critical process, but a certain space must be available so that, for example, a constant electric field distribution across a microstrip line can change to a real electromagnetic field solution of the microstrip line. Such an exact field solution is needed at the input port of the component or circuit which is to be analysed. To avoid long lines at the input of the circuit (because they need additional computational space in the computer), other techniques must be developed which provide an electromagnetic field distribution which is a valid solution on the input line. Such a solution shall be called a matched source. Its principle is shown in Figure 3.

In Reference 9 the problem of matched sources was solved by separating the stimulated pulse and the reflected field by using a transmission line connected to the component under test, which is long compared to the pulse width. After the pulse has left the excitation plane, the reflected pulse propagates through an absorbing boundary which is switched on at this time. There has to be a distance between the excitation plane and the absorbing wall mentioned, because the homogeneous pulse excitation (see above) induces a dc-offset in the tangential H-field owing to the violation of Maxwell's equations.

Accuracy recommends the use of an excitation pulse whose duration in space is sufficiently long. Arbitrarily long pulse duration can be used if matched sources are available. They can be simulated by separating the calculation domain into two parts. The first part contains only a transmission line which can be excited in the conventional way. The transmission line is terminated with an absorbing boundary at the other end and it should be long enough to stabilize the pulse form after excitation and furthermore to let the mentioned dc-offset of the H-field vanish.

The field close to the termination of this line, which is a real physical solution, is used to excite the second part of the computation domain containing the component connected to short transmission lines. In this area superposition of pulse propagation in both directions is allowed. The absorbing boundary algorithm can be used for the reflected part as before, because separation of the transmitted and reflected part of the pulse is possible, using the information of the first computation domain.

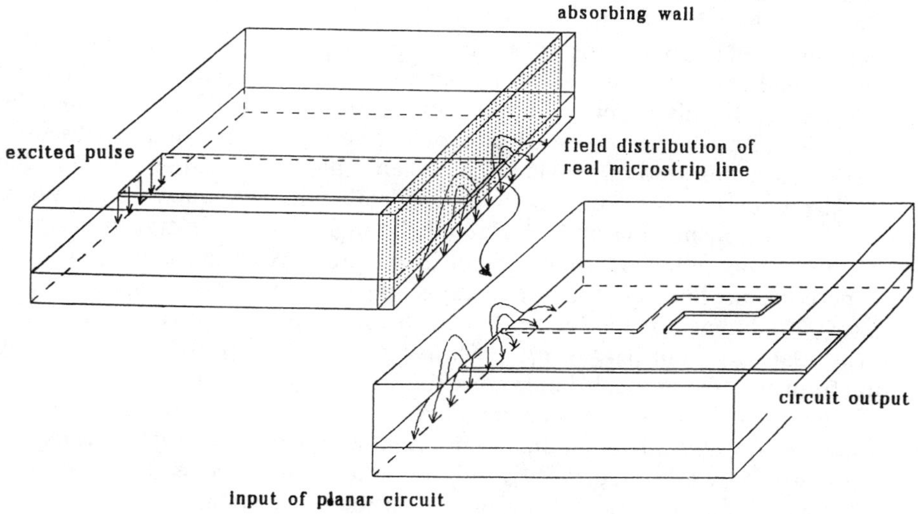

Figure 3. Principle of a 'matched source' for exciting field pulses in microstrip circuits

3. APPLICATIONS

In a FDTD program, working on the above-described principles, the metallization structure can be defined using a graphical editor. Input and output ports of the structures are uniform microstrip lines (e.g. 50 Ω lines for comparison with measurements) or other planar lines like, for example, coplanar lines.

As described before, the structure is enclosed in a 'shielding' of absorbing walls. The following data are typical for an analysis of a simple microstrip line. The total space is discretized into 40 grid elements of width Δh parallel to the line length, 100 elements parallel to the width of the line and 30 elements perpendicular to the substrate material. The time length Δt typically is chosen so that the wave propagates along the grid length in five time-steps. This time is calculated using a static approximation of the phase velocity. The spatial width of the exciting field pulse is about $40\Delta h$. From these data all other values are determined. If the pulse width is defined as the time or space where it has decreased to 0·05 of its maximum value, the maximum frequency which can be considered, using the mentioned pulse data, is about 50 GHz.

If t_0 is chosen to be $300\Delta t$, the pulse is 'switched on' at a moment when its value is still very small (2×10^{-12} of the maximum value). The fields at the ports are evaluated at the absorbing boundaries for the electric fields and $0{\cdot}5\Delta h$ in front of the boundaries for the magnetic fields. About 1000 to 5000 time-steps (depending on the complexity of the structure) are used for the analysis; however, 2^{13} time points are used for the Fourier transform to get accurate results in the frequency domain. The values for time-steps higher than the calculated ones are filled with zero.

The FDTD-method offers the possibility of a flexible problem formulation with an analytical expense which is much lower than, for example, that of the spectral-domain analysis technique. Arbitrarily shaped planar line discontinuities and components and even three-dimensional structures, such as components with metallizations in two or more dielectric layers, can be analysed using this technique.

In a first example a meander line, which consists of multiple coupled line sections and coupled discontinuities is shown (Figure 4). This structure has been discretized into 212 elements in the x-direction, 120 elements in the y-direction and 16 elements in the z-direction. Only four grid elements are needed over the line width and the height of the substrate material. The theoretical frequency range of the analysis is 35 GHz. Figure 4(a) shows the meander line structure, Figure 4(b) the propagation of the wave through the structure at different times, Figure 4(c) the time-domain signal at the input and the output ports and Figure 4(d) the scattering parameters in comparison to measurements. The agreement between theory and experiment is good. In particular the good agreement of the scattering parameter phases over a large frequency range must be mentioned here. Only a small frequency shift in the resonant frequency near 23 GHz can be recognized. This is a discretization error which can be avoided by reducing the mesh size.

It is interesting to recognize, that the discretization error does not result in a magnitude error, but only in a frequency offset, as already has been mentioned in Reference 9. It has been found that the accuracy of the solution at higher frequencies can be improved by assuming only one more grid point over the width of the microstrip line.

Several other interesting applications to purely planar microwave structures like radial stubs or spiral inductors, together with a discussion of possible errors and how to avoid them, can be found in Reference 12. In this paper the application of the FDTD method to three-dimensional discontinuity problems, as they often occur in coplanar circuit design, shall be discussed.

As a first example for this kind of problem the coplanar band-reject filter shown in Figure 5(a), is considered. The filter has been designed for application in a frequency multiplier.[13] It was designed to have zero transmission at 18 GHz and the highest possible transmission at 36 GHz. The size of the filter was reduced, using a bend in the stub. With the assumption of negligible dispersion of the coplanar lines and open end capacitances the effective length of the bent stub, considering also the influence of the bends, has been found to be 1780 μm on a 635 μm Al_2O_3-substrate with a relative permittivity of $\varepsilon_r = 9{\cdot}8$ using the FDTD program. This is smaller than it would result from the geometrical dimensions shown in Figure 5(a).

Furthermore the band-reject filter has been realized using a symmetrical stub to avoid additional bond-wires which are needed across the main line, if an asymmetrical structure is used. Using the symmetrical structure shown in Figure 5(a), only two bond wires across the stubs are needed to avoid the excitation of the fundamental odd mode on the coplanar line.

A filter with and without bond-wires has been analysed using the FDTD method. The bond-

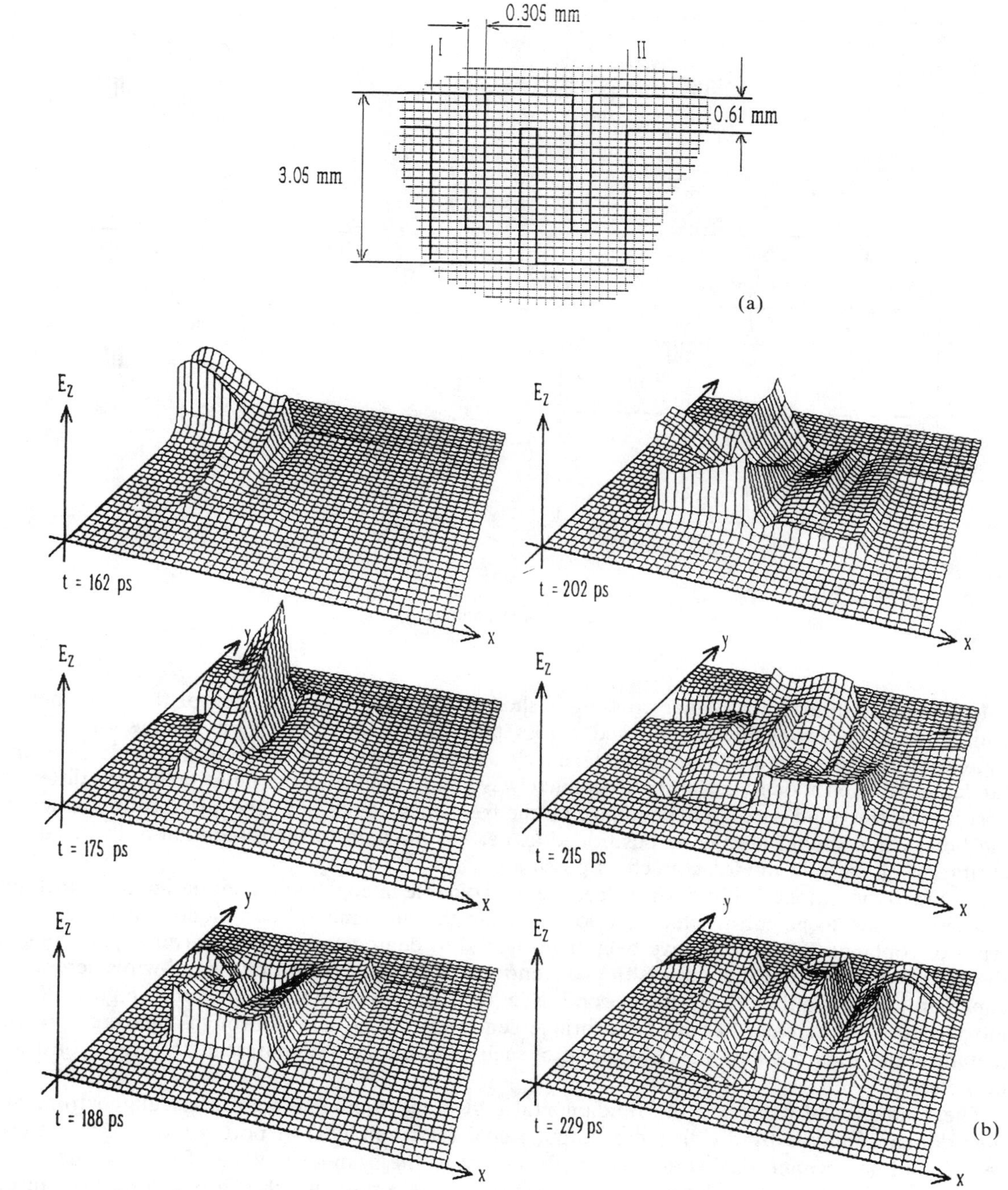

Figure 4. (a) Microstrip meander line structure, substrate Al_2O_3, ϵ_r = 9·978 (measured), height = 635 μm, (b) wave propagation through the meander line at different times, (c) time-domain signals at input and output port of the meander line, (d) scattering parameters of the meander line (solid line, calculated; dashed line, measured)

wires are simulated as a real three-dimensional conducting structure. Figures 5(b) and (c) show the influence of the bond-wires on the filter characteristics. As the two figures show, there is almost no similarity between the transfer functions of the two structures. (1) The filter with bond-wires behaves as wanted. The agreement between theory and measurement is good. Only the resonances near 26 GHz are much sharper in the simulation than in the measurement. This is because the losses have been neglected in the FDTD analysis. (2) The results shown in Figure 5(c) for the filter without bond-wires again show quite good agreement between theory and measurement, but the structure does not have the desired transfer characteristic. Apparently, the structure now has the function of two symmetrical, coupled slotlines. From this result it may be recognized how essential the correct simulation of three-dimensional bond-wires or airbridge structure in coplanar waveguide technique is.

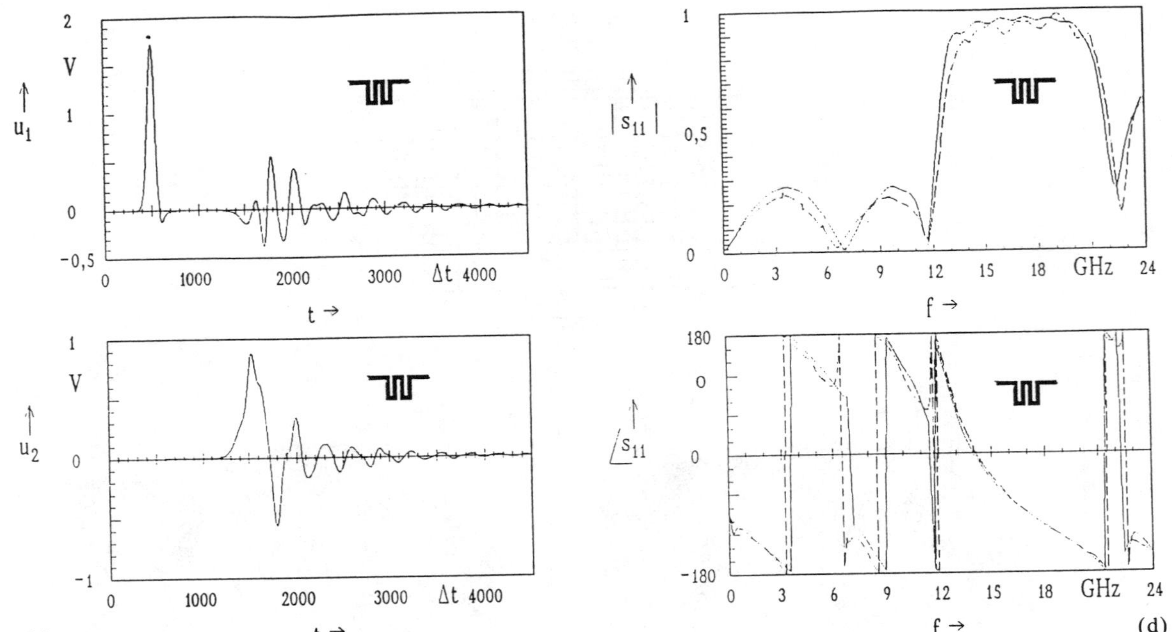

(c) (d)

Figure 4. Continued

In Figure 6 (insert) a coplanar crossing is shown. To avoid the excitation of the fundamental odd mode, bond-wires are again used across the second line. The figure shows the FDTD simulation of the excited odd mode voltage u_3^{odd} at port 3 compared to the excited voltage of the fundamental even mode (the normal coplanar waveguide mode) at the input port 1 for different positions of the bond-wires in dependence on the frequency. As can be recognized from the figure, the bond-wires should always be positioned as near as possible to the discontinuity to avoid the excitation of the odd mode as much as possible.

As already mentioned above, airbridges are unavoidable in coplanar circuit design if T-junctions or crossings are to be constructed that do not excite the fundamental odd mode intensively. Two types of coplanar T-junctions have been investigated to demonstrate the powerful applications of the FDTD method: the first type is the conventional T-junction with three bond-wires across the coplanar lines (Figure 7(a)); the second is a modified 'airbridge T-junction' (Figure 7(b)), introduced by Reference 14 with an airbridge centre conductor and a through-connected ground metallization. Figure 8 shows a SEM photo of such an airbridge to demonstrate this new construction clearly.

The fundamental even and the fundamental odd mode S-parameters are calculated from the time-domain analysis by dividing the output signal (slot voltages) at port 3 into a symmetrical (even) and an asymmetrical (odd) part with respect to the symmetry plane of the coplanar line. From these signals a generalized scattering matrix of the 3-port, which is a 6×6 matrix can be calculated. The 6×6 matrix includes the scattering parameters of the even and the odd mode as well as the mode conversion occurring from the even to the odd mode and vice versa. If port 1 is excited by an even mode voltage pulse, the odd mode excitation at the other ports can be calculated easily. From the designer's point of view especially, the conversion from the even mode, which is the conventional coplanar mode, into the unwanted odd mode is important.

Two T-junctions with different geometrical sizes have been analysed to demonstrate the different effects which may occur within these components. Moreover, measurements have been provided for comparison, using a modern on-wafer measurement equipment.[15] Measurement of the odd mode signal is not possible because the network analyser has no facilities to deal with different modes which have different characteristic impedances simultaneously. So measurements are available only for the even mode signals.

The first T-junction is based on coplanar lines with $w = 75$ μm centre conductor linewidth and $s = 50$ μm gap width between the centre conductor and the ground plane. These dimensions are typical for hybrid microwave circuits, for example, on alumina substrate. The second T-junction

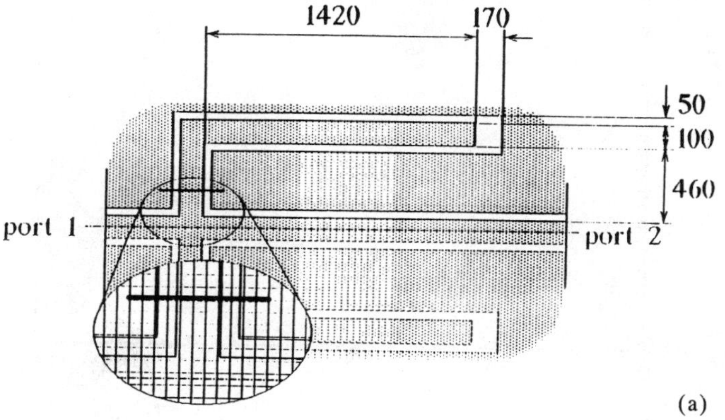

(a)

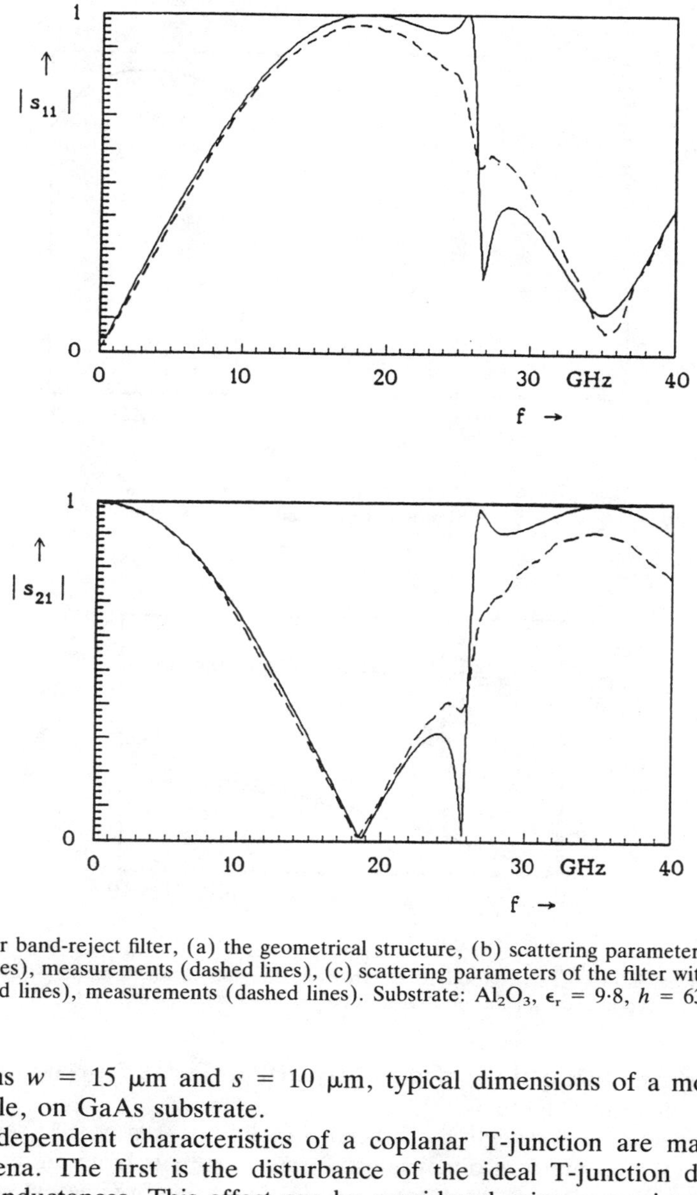

(b)

Figure 5. The coplanar band-reject filter, (a) the geometrical structure, (b) scattering parameters of the filter with bond wires, theory (solid lines), measurements (dashed lines), (c) scattering parameters of the filter without bond wires, theory (solid lines), measurements (dashed lines). Substrate: Al_2O_3, $\epsilon_r = 9{\cdot}8$, $h = 635$ μm

has the dimensions $w = 15$ μm and $s = 10$ μm, typical dimensions of a monolithic integrated circuit, for example, on GaAs substrate.

The frequency-dependent characteristics of a coplanar T-junction are mainly based on two different phenomena. The first is the disturbance of the ideal T-junction due to the parasitic capacitances and inductances. This effect can be considered using a quasistatic approach.[16] The second reason is the mode conversion. Owing to the small size of the coplanar T-junction compared to the wavelength, the frequency-dependent characteristics generally are weak. Many more problems are caused by the excited odd mode. Accordingly, suppression of the odd mode excitation is of more importance than minimization of, for example, the parasitic capacitance of the T-

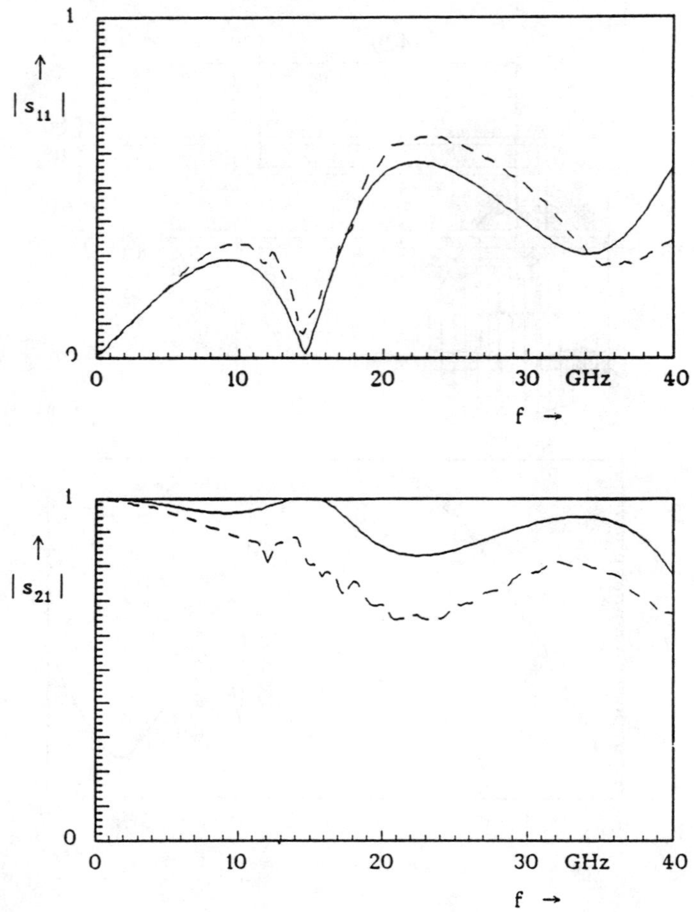

(c)

Figure 5. Continued

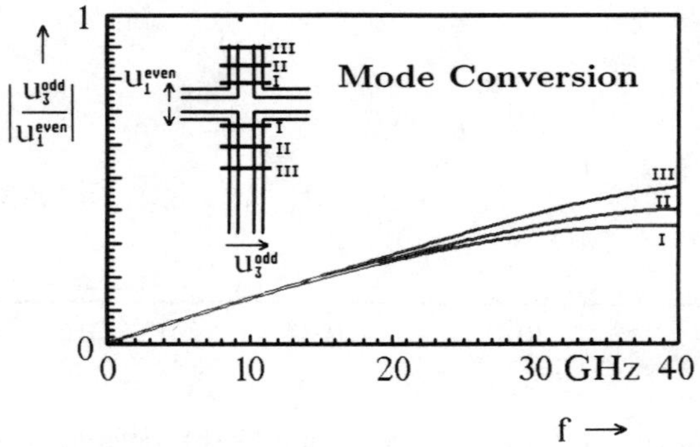

Figure 6. Calculated voltage ratio of excited odd mode and fundamental incident mode on a coplanar crossing in dependence on the position of applied bond-wires (airbridges)

junction. Indeed, the capacitance of the modified T-junction (Figure 7(a)) normally is higher than that of the conventional one (Figure 7(b)).

Figure 9 shows the calculated reflection coefficients (solid lines) at port 1 (see insert) of the T-junctions with the larger geometrical dimensions. Because of their large dimensions they have a recognizable frequency-dependence of the (even-mode) scattering parameters which easily can be compared to measurements (dashed lines). As mentioned above, the reflection coefficients of the modified T-junction increase faster than that of the conventional one. Owing to the small ratio of airbridge height to centre conductor width (3 μm to 75 μm), the parasitic capacitance is absolutely dominant. However, even in this case, the behaviour of the new T-junction with respect

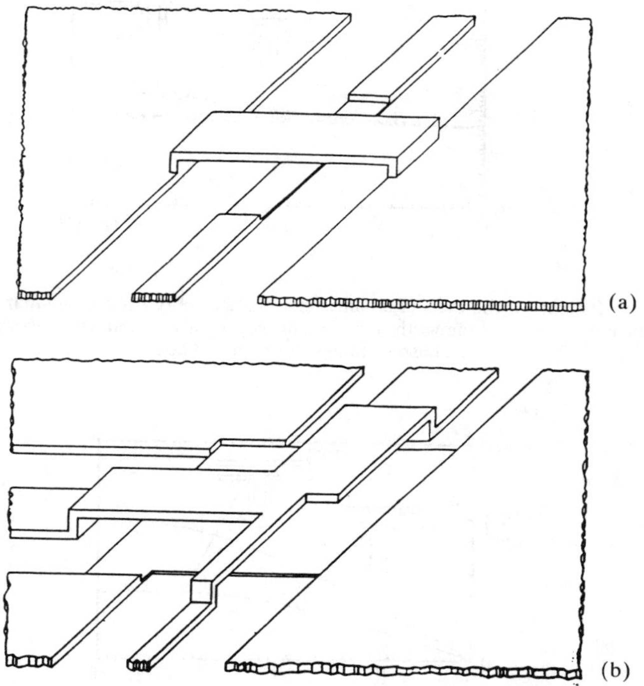

Figure 7. (a) The conventional airbridge used for T-junctions; (b) the new developed airbridge T-junction

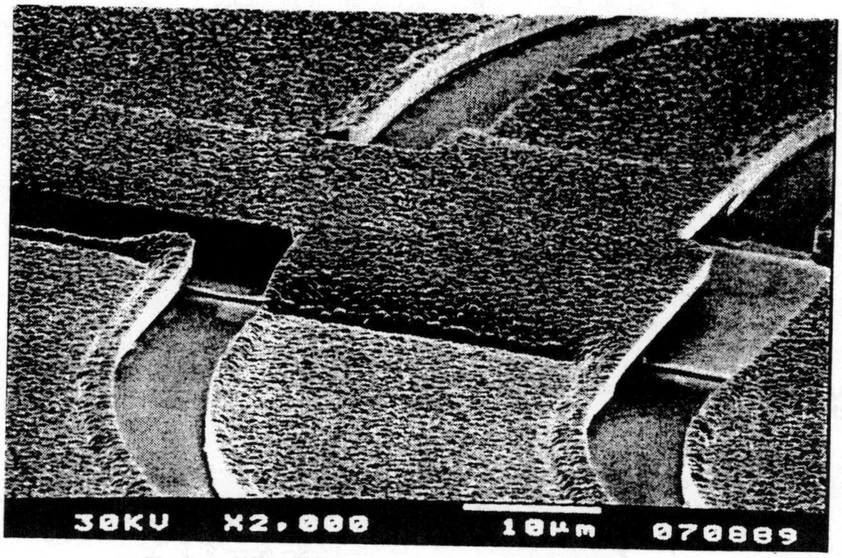

Figure 8. SEM photo of the coplanar airbridge T-junction

to the odd mode excitation is much better, compared to that of the conventional one (Figure 10). Again the mode conversion from port 1 to port 3 is shown in Figure 10 as an example, because it offers the most significant frequency-dependent characteristic.

For the more realistic size of the small T-junction this advantage of the modified type is even more significant, as is shown in Figure 11. The excitation of the odd mode in the modified T-junction is negligibly small. With the dimensions of this T-junction (airbridge height 3 μm, centre conductor width 15 μm), the frequency-dependence of the even mode S-parameters is mainly caused by the odd mode excitation, as pointed out in Figure 12. The conventional T-junction has a reverse frequency characteristic, which cannot be explained by the parasitic capacitance. Measurements performed for these small T-junctions show no frequency characteristic up to 40 GHz, as is also predicted by the theoretical results in Figure 12.

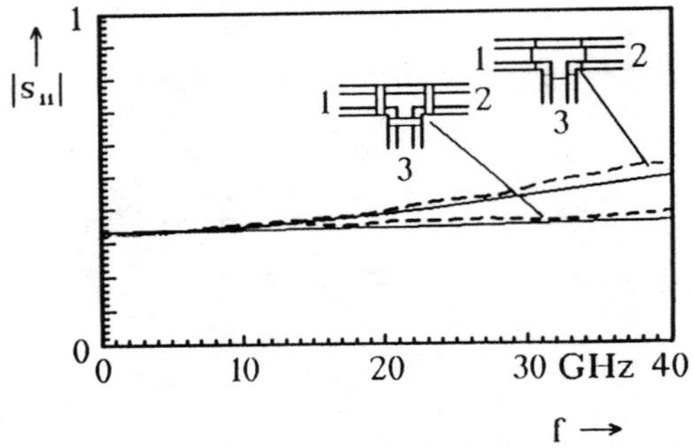

Figure 9. The reflection coefficients of the even (fundamental) mode at port 1 for a coplanar T-junction with conventional airbridges and the airbridge T-junction. Linewidth w = 75 μm; gap width s = 50 μm, theory (solid lines), measurements (dashed lines). Substrate: GaAs

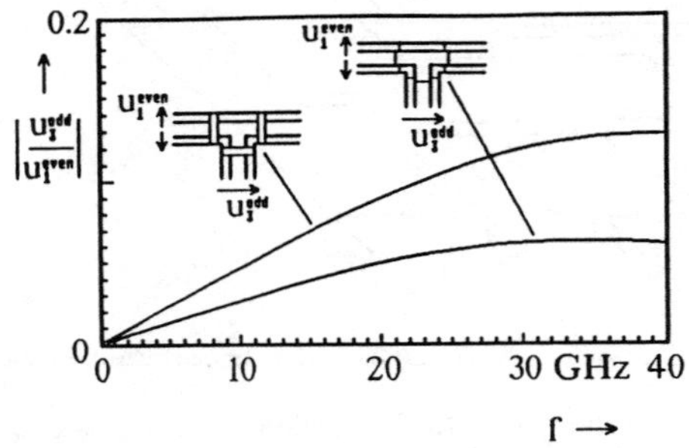

Figure 10. Voltage ratio describing the mode conversion from the incident fundamental mode at port 1 into the odd mode at port 3. Geometrical parameters as in Figure 9

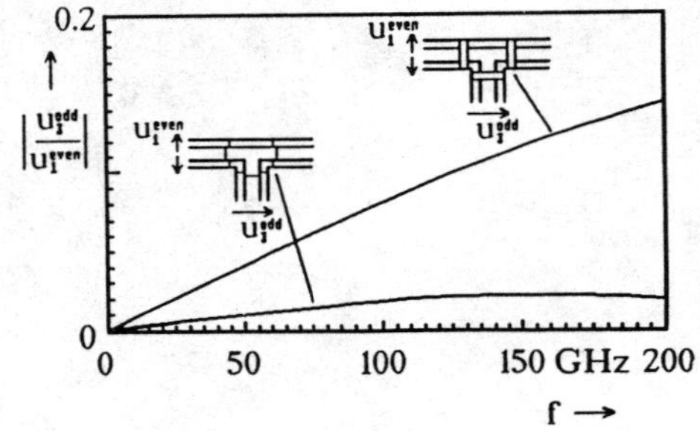

Figure 11. Voltage ratio describing the mode conversion from the incident fundamental mode at port 1 into the odd mode at port three for T-junctions with geometrical dimensions w = 15 μm and s = 10 μm. Substrate: GaAs

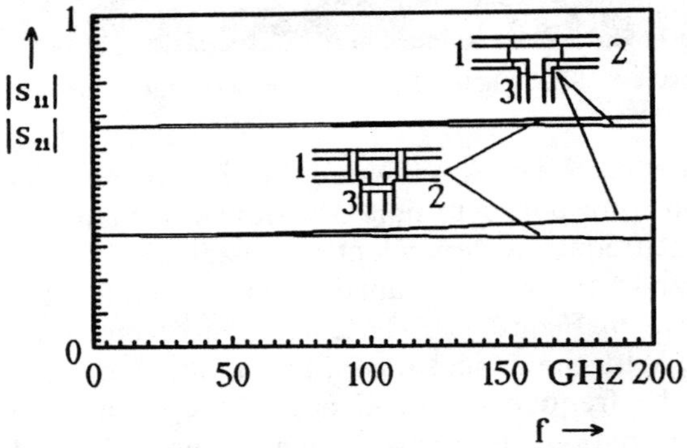

Figure 12. Magnitude of selected scattering parameters of the conventional coplanar T-junction and the airbridge T-junction in dependence on the frequency, w = 15 μm; s = 10 μm. Substrate: GaAs

4. ERRORS

The results achieved confirm that the assumptions made for the application of the FDTD method, i.e. perfect conductors and no material losses, make it possible to simulate wave propagation in the presented (M)MIC-components with acceptable accuracy. Radiation effects are dominant compared to metallization losses, as can be concluded from the fact, that the attenuation of the components is described accurately. Discretization errors do not result in a magnitude error, but in a frequency offset, as mentioned above. This is essentially caused by the following phenomena, which are very similar to those of the transmission line matrix (TLM) method,[17] and which are due to the discretized nature of these methods:

In the case of equidistant discretization the leapfrog algorithm used is able to simulate wave propagation without increasing or decreasing the amplitudes up to a cutoff frequency

$$f_c = \frac{1}{\pi \Delta t} \arcsin\left(\frac{\Delta t}{(\varepsilon\mu)^{0\cdot5}\Delta h}\right) \tag{7}$$

which corresponds to a wavelength of a few space elements depending on the difference factors used. Only a small dispersion, which means a weakly frequency-dependent phase-velocity occurs. This dispersion is a little bit higher for wave propagation in the main axis direction than in arbitrary directions. The frequency shift for the calculated examples is less than 0·3 per cent. An error correction after simulation is possible by correcting the frequencies after Fourier transforms.

High non-uniform fields, for example, at the edges of the metallized structures cannot be resolved by the used grid. This so called coarseness error is reported in Reference 11. Its result is, that the metallized contour is assumed to be too small, if it lies exactly on the grid nodes. This means, that the FDTD method as applied straightforwardly (as has been done in the demonstrated examples) calculates structures with metallization dimensions larger than the real dimensions which have been used in measurements. This fact can explain the small shift of the resonant frequencies which, for example, can be detected in Figure 6(b). These phenomena have to be considered when the structure is discretized.

Reflections at the absorbing boundaries also cause errors as well as the truncation of the pulses before transformation into the frequency domain. The truncation error may cause ripples in the frequency response.

5. ANALYSIS OF DISTRIBUTED MICROWAVE CIRCUITS CONTAINING NON-LINEAR ACTIVE ELEMENTS

The standard approach to the analysis of microwave circuits containing non-linear components consists of dividing the circuit into several segments which may be modelled independently of each other, e.g. transmission lines, discontinuities, FETs. These models are then used for the analysis of the total circuit with some circuit analysis software. The method which has found widest acceptance for the analysis of non-linear microwave circuits is the harmonic balance method (HBM), as described, for example, in References 18 and 19. A major drawback of the segmentation approach is its inability to account for couplings in complex, narrow structures, except for the rare cases where a model for these structures is known. Moreover, the choice of the correct segmentation, if it exists, is a matter of discretion and therefore a potential source of error.

To overcome this problem, time-domain methods have gained in interest, but these methods, as described, for example, in the above sections, up to now have been primarily employed for the simulation of linear, passive components. As already has been shown[20,21] they also may be enhanced to account for active, non-linear devices as well. However, time-domain methods with non-linear enhancement are subject to specific restrictions as well. In the presence of non-linear elements, a characterization of the circuit as in the linear case is impossible. Therefore, if excitations change even slightly, the analysis has to be performed entirely again, including the computationally expensive linear space domain. Hence the effort for n analyses is as high as n times the effort for a single analysis, because with time-domain methods there is no way of utilizing the results of the former analyses as a solution estimate, as is possible in the HBM case. Multi-tone excitations (e.g. of mixer circuits) cannot easily be handled with time-domain methods.

5.1. *The compression approach*

To circumvent these restrictions, this work proposes a technique which combines time-domain methods for the simulation of linear structures with the HBM for non-linear analyses,[24] resulting in several advantages:

- Non-linear components embedded in complex structures may be considered.
- All field-theoretical aspects, which, for example, time-domain methods are able to account for, are fully integrated into the non-linear analysis.
- All advantages of the HBM with respect to multiple analyses are exploited.
- All analysis options of the HBM, especially the possibility to analyse multitone excitations, are preserved.
- The characterization of the linear part has to be done only once and is independent of the actual circuit simulation.
- In contrast to other approaches for the characterization of linear parts as, for example, the Green's function-based John's Matrix concept,[20] the compression approach yields as a result the minimum of required information in a highly condensed form.

Basically, the circuit is considered to be a superposition of a single linear space domain governed by Maxwell's equations, and a number of non-linear regions made up by non-linear elements. An interface between both domains is introduced by defining ports in the space domain to which the non-linear elements are connected. The non-linear elements are assumed to be described by standard CAD models. As characterization of the linear part, the entire passive, linear space domain, including the regions where the physical non-linear devices are located, is described by a single matrix $\mathbf{S}_c$ of scattering parameters (which is termed the compression matrix) plus appropriate reference resistances.

It is important to note that this is an essential difference to the segmentation method, where the linear part is divided into several components, which are considered to be independent—with respect to field couplings—from each other. This prevents segmentation methods from utilizing the accuracy benefits of, for example, time-domain methods.

The resulting equivalent circuit, consisting of the non-linear elements connected to the ports of the linear part (Figure 13), is then analysed with an HBM simulator.

A crucial point of this approach is the necessity to consider 'inner' ports, i.e. ports, which are located within complex structures, being connected to non-linear elements but not to any reference line. Special attention has to be paid to these ports with respect to the definition of port variables and reference resistances.

5.2. *Definition of inner ports*

Port voltage is defined to be the integral of the electric field strength along a path inside the non-linear element, under the assumption that there is approximately no dependence of this value on the actual integration path. Current is taken to be the closed-curve integral of magnetic field strength along a curve which closely surrounds the current path.

Each port belongs to a 'local group', where all ports of a local group are located in such a close distance that Kirchhoff's laws can be assumed to be applicable within that group (many groups

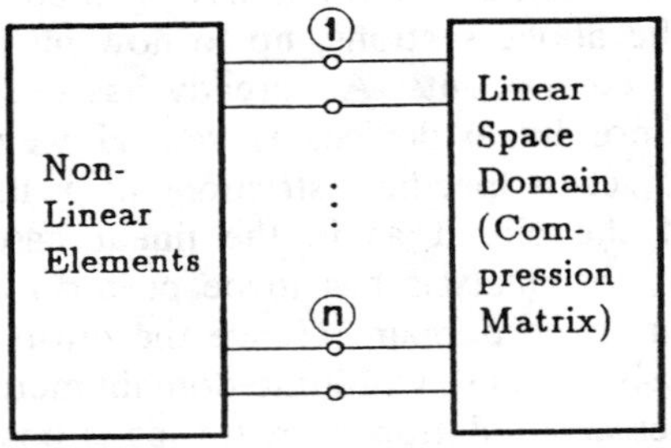

Figure 13. Interface between non-linear elements and the surrounding linear space domain. Non-linear elements interact by means of the compression matrix

will contain only one port). All ports of a given local group share a virtual ground. Owing to the general non-existence of electrical potential, voltage between ports of different groups is undefined. This implies firstly that connecting ports of different groups (e.g. virtually by means of a circuit simulation software) will yield meaningless results and secondly that components which are both non-linear and distributed have to be broken into lumped non-linear elements, each in its own local group (e.g. a long gate width FET may be broken into a number of FET 'slices'). Ports of different groups are connected exclusively by means of the compression matrix, thus reflecting the physical structure.

Since most standard non-linear CAD models relate voltages and currents, the appropriate description of the linear circuitry, in term of scattering parameters, takes the form:

$$(\mathbf{I} - \mathbf{S}')\mathbf{u} - (\mathbf{I} + \mathbf{S}')\mathbf{Z}\mathbf{i} = \mathbf{0} \tag{8}$$

where $\mathbf{S}' = (s'_{ij})$, $s'_{ij} = s_{ij}\,(Z_i/Z_j)^{1/2}$, $\mathbf{Z} = \mathrm{diag}(Z_i)$. s_{ij}, Z_i, $\mathbf{u}$ and $\mathbf{i}$ are scattering parameters, reference resistances and vectors of port voltages and currents, respectively. If scattering parameters are determined from voltages and currents, equation (8) exactly reproduces the relation of these quantities for whatever reference resistances are chosen. For 'non-physical' reference resistances the corresponding scattering parameters may not be interpreted as to relate the actual wave quantities any longer. However, at inner ports, where there are *a priori* no physically meaningful reference resistances available, there is no need to consider wave quantities at all, because equation (8) contains all the information required to interface non-linear models. Therefore, at inner ports, the reference resistances may be thought of as being temporary scaling factors which can be chosen arbitrarily. However, using proper reference resistances at outer (i.e. associated with a transmission line) ports enables the interpretation of the corresponding scattering parameters in terms of wave quantities. Thus the primary reason for using scattering parameters as characterization of the linear part is that equation (8) provides a convenient interface to non-linear models as well as professional circuit simulation software.

5.3. *Determination of the compression matrix*

The straightforward way to determine the compression matrix is to perform p analyses with p linear independent excitations $\mathbf{a}_i$ yielding p responses $\mathbf{b}_i$, where the wave quantities are determined from port voltages and currents. The compression matrix is then given by $\mathbf{S}_c = \mathbf{B}\mathbf{A}^{-1}$ where $\mathbf{B} = (\mathbf{b}_1, \mathbf{b}_2 \ldots \mathbf{b}_p)$, $\mathbf{A} = (\mathbf{a}_1, \mathbf{b}_2 \ldots \mathbf{a}_p)$ and p is the number of ports considered. However, some methods for field computations are limited to excitation via reference lines. In this case the linear independency can be achieved by using a few (outer) ports as major excitation ports while introducing varying terminations at all inner ports which are not available for direct excitation. The use of opens and shorts as terminations has proved to be sufficient, supplying these ports with an input which is +1 or −1 (any reference resistance) times the corresponding outgoing waves, respectively. For accuracy reasons it is important that short connections resemble actual current paths of non-linear components.

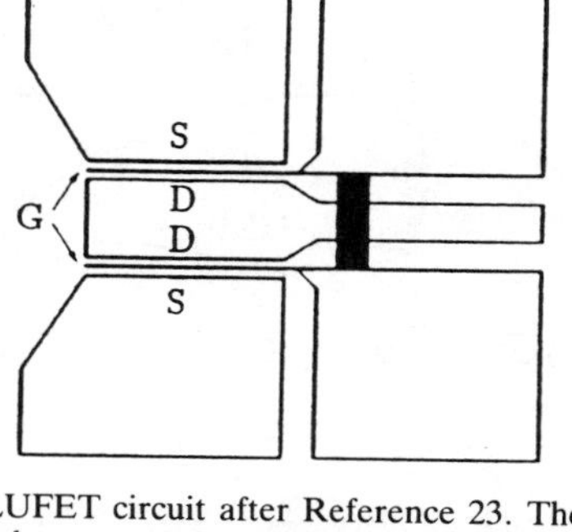

Figure 14. Schematic in-phase combiner LUFET circuit after Reference 23. The black area indicates an airbridge. G, S, and D mark gate, source and drain regions, respectively

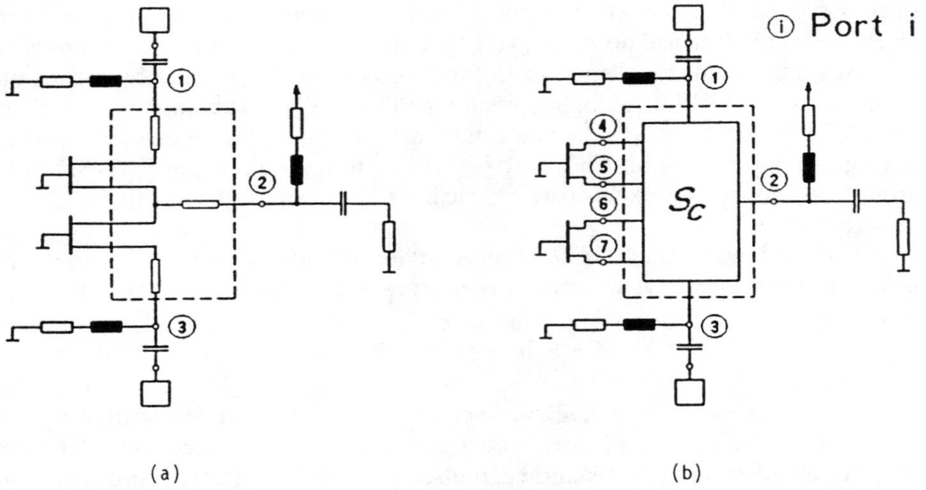

Figure 15. Schematic equivalent circuits including combiner LUFET and biasing circuitry. The area enclosed by the dashed line corresponds to the structure shown in Figure 14. (a) Segmentation method approach; (b) compression method approach

5.4. *Results*

The compression approach was applied in an in-phase combiner LUFET structure after References 22 and 23, shown in Figure 14. The resulting equivalent circuit containing the compression

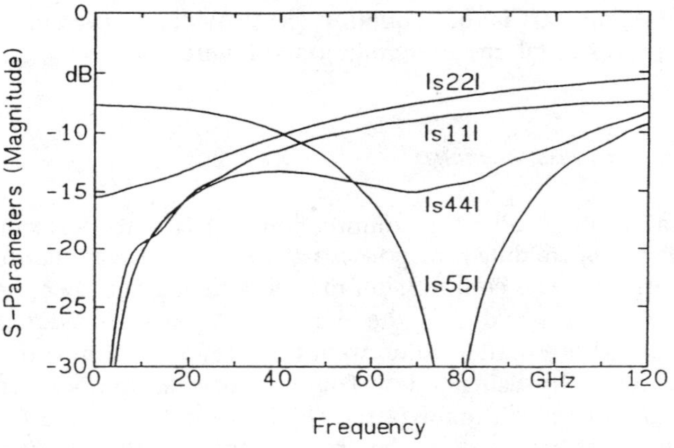

(a)

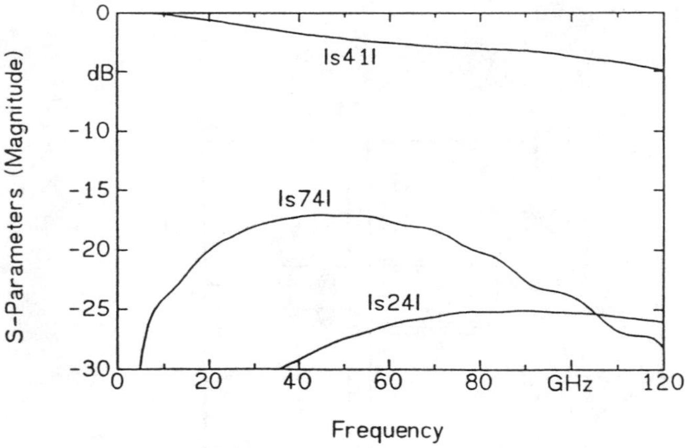

(b)

Figure 16. Calculated frequency characteristic of the linear network (compression matrix). (a) Return losses; (b) selected transmission parameters with respect to port 4

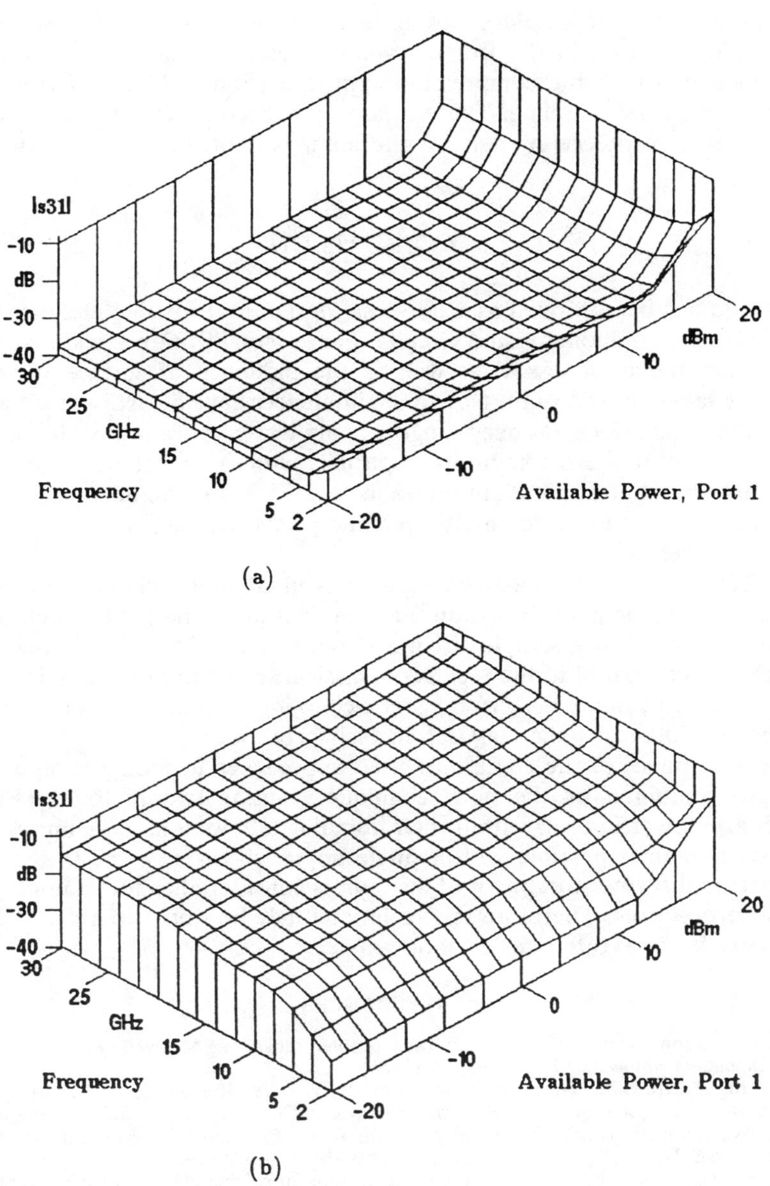

Figure 17. Calculated port isolation between ports 3 and 1 of the combiner. (a) Segmentation approach; (b) compression approach

matrix, the non-linear elements plus some biasing and power supply circuitry is shown in Figure 15(b). The area enclosed by the dashed line corresponds to the structure shown in Figure 14, which includes the passive 'capacitances' of the FET's metallization structure (the capacitances due to space charges have been accounted for in the non-linear model of the FET). The FET gate width was 200 μm. The linear space domain is represented by a seven-port compression matrix, to which each FET is connected by two ports (source-gate, drain-gate; Figure 15(b)). There are five local port groups (ports 1, 2, 3, 4–5, 6–7). The corresponding segmented equivalent circuit is shown in Figure 15(a). The slotlines as well as the coplanar lines are replaced by ideal transmission lines. Their electrical parameters were estimated from geometry and material parameters in a conventional way.

The compression matrix was determined using the three-dimensional full-wave FDTD simulation method described in section 2. The non-linear calculations were carried out with an in-house HBM software for multi-tone analysis. Figure 16 shows the magnitude of some characteristic S-parameters calculated in the described way.

Figure 17(a) and (b) show the calculated port isolation (between port 1 and port 3) for the segmentation and compression approach case, respectively. The significant difference of up to 20 dB is due to the presence of a field coupling between both FET source ports (ports 4 and 7, Figure 15(b)). A comparison of the frequency-dependence of mag(s_{74}) (Figure 16(b)) and the port isolation calculated with the compression approach (Figure 17(b)) shows that the isolation is mainly determined by this coupling. There is no way to account for this effect with segmentation methods unless there are precise models for the inner part of the combiner structure available.

6. CONCLUSIONS

The FDTD method has been proved as a universal design tool for the analysis of complex three-dimensional microwave components and circuits. Finite metallization thickness, conductor losses, dielectric losses and radiation losses, as well as the influence of surface waves on the circuit parameters, can be taken into account. Not all the influences have been discussed and demonstrated in this paper. The method does not need lengthy analytical preparation of the problems. With the increasing capabilities of modern computers, especially with the development of parallel computer systems with a large number of parallel processors, the FDTD techniques may become one of the most powerful design techniques for multilayer microwave circuits with integrated three-dimensional component structure.

Using the FDTD analysis of the passive circuit part, an analysis technique for distributed circuits which contain non-linear elements based on the combination of the FDTD method with the HBM technique was presented. For a sample circuit, a comparison of the circuit analysis results based on this approach was compared to the results of classical segmentation analysis. It was shown that the new method is able to predict electromagnetic couplings within a condensed microwave circuit which cannot be considered by the classical analysis procedure.

A number of advantages of the commonly used approaches including computational efficiency and accuracy are combined by the new technique. Joining two methods (FDTD and HBM methods) which are powerful tools within their domains, allows a more comprehensive treatment of circuit behaviour than is possible with a single approach on its own. In the author's opinion this is especially true if further treatment of a circuit is considered as for example, by optimization or noise analysis procedures. Therefore this technique may be suited to serve the increasing need of circuit designers for integrated design tools as well.

REFERENCES

1. K. S. Yee, 'Numerical solution of initial boundary value problems involving Maxwell's equations in anisotropic media', *IEEE Trans. Antennas Propagat.*, **AP-14**, 302–307 (1966).
2. A. Taflove and M. E. Brodwin, 'Numerical solution of steady-state electromagnetic scattering problems using the time-dependent Maxwell's equations', *IEEE Trans. Microwave Theory Tech.*, **MTT-23**, 623–630 (1975).
3. G. Mur, 'Absorbing boundary conditions for the finite-difference approximation of the time-domain electromagnetic-field equations', *IEEE Trans. Electromagn. Compat.*, **EMC-23**, 377–382 (1981).
4. J. Fang and K. K. Mei, 'A super-absorbing boundary algorithm for solving electromagnetic problems by time-domain finite-difference method', *1988 IEEE AP-S International Symposium Digest*, Syracuse, NY, pp. 472–475, June 1988.
5. D. H. Choi and W. J. R. Hoefer, 'The finite-difference-time-domain method and its application to eigenvalue problems', *IEEE Trans. Microwave Theory Tech.*, **MTT-34**, 1464–1470 (1986).
6. W. K. Gwarek, 'Analysis of an arbitrarily-shaped planar circuit—a time domain approach', *IEEE Trans. Microwave Theory Tech.*, **MTT-33**, 1067–1072 (1985).
7. X. Zhang, J. Fang, K. K. Mei and Y. Liu, 'Calculations of the dispersive characteristics of microstrips by the time-domain finite difference method', *IEEE Trans. Microwave Theory Tech.*, **MTT-36**, 263–267 (1988).
8. W. K. Gwarek, 'Analysis of arbitrarily shaped two-dimensional microwave circuits by finite-difference time-domain method', *IEEE Trans. Microwave Theory Tech.*, **MTT-36**, 738–744 (1988).
9. X. Zhang and K. K. Mei, 'Time-domain finite difference approach to the calculation of the frequency-dependent characteristics of microstrip discontinuities', *IEEE Trans. Microwave Theory Tech.*, **MTT-36**, 1775–1787 (1988).
10. S. Nam, H. Ling and T. Itoh, 'Time-domain method of lines applied to the uniform microstrip line and its step discontinuity', *1989 IEEE MTT-S Internat. Microwave Symposium Digest*, Long Beach, CA, pp. 997–1000, June 1989.
11. C. J. Railton and J. P. McGeehan, 'Analysis of microstrip discontinuities using the finite difference time domain technique', *1989 IEEE MTT-S Internat. Microwave Symposium Digest*, Long Beach, CA, pp. 1009–1012, June 1989.
12. I. Wolff and M. Rittweger, 'Finite difference time-domain analysis of planar microwave circuits', *Arch. Elektrotechn.*, **74**, 189–201 (1991).
13. M. Rittweger, M. Abdo and I. Wolff, 'Full-wave analysis of coplanar discontinuities considering three-dimensional bond wires', *1991 IEEE MTT-S Internat. Microwave Sympos. Digest*, Boston, MA, pp. 467–468, June 1991.
14. N. H. L. Koster, S. Kosslowski, R. Bertenburg, S. Heinen and I. Wolff, 'Investigations in air bridges used for MMICs in CPW technique', *Proc. 19th European Microwave Conf.*, London, pp. 666–671, September 1989.
15. G. Gronau and I. Wolff, 'A simple broad-band device de-embedding method using an automatic network analyzer with time-domain option', *IEEE Trans. Microwave Theory Tech.*, **MTT-37**, 479–483 (1989).
16. M. Naghed and I. Wolff, 'A three-dimensional finite-difference calculation of equivalent capacitances of coplanar waveguide discontinuities', *IEEE Trans. Microwave Theory Tech.*, **MTT-38**, 1808–1815 (1990).

17. W. J. R. Hoefer, 'The transmission-line matrix method—theory and application', *IEEE Trans. Micwoave Theory Tech.*, **MTT-33**, 882–893 (1985).
18. V. Rizzoli and A. Neri, 'State of the art and present trends in nonlinear microwave CAD techniques', *IEEE Trans. Microwave Theory Tech.*, **MTT-36**, 343–365 (1988).
19. K. S. Kundert and A. S. Vincentelli, 'Finding the steady-state response of analog and microwave circuits', *Alta Frequenza*, **LVII**, 379–387 (1988).
20. R. H. Voelker and R. J. Lomax, 'A finite-difference transmission line matrix method incorporating a nonlinear device model', *IEEE Trans. Microwave Theory Tech.*, **MTT-38**, 302–312 (1990).
21. P. Russer, M. Schwab and F. X. Kaertner, 'Numerical analysis of microwave oscillators', *1990 Internat. Workshop of West German IEEE MTT/AP Joint Chapter on Integrated Nonlinear Microwave and Millimeterwave Circuits (INMMC '90) Digest*, Duisburg, Germany, pp. 155–175, October 1990.
22. T. Tokumitsu, S. Hara, T. Takenaka and M. Aikawa, 'Divider and combiner Line-Unified FET's as basic circuit function modules—Part I', *IEEE Trans. Microwave Theory Tech.*, **MTT-38**, 1210–1217 (1990).
23. T. Tokumitsu, S. Hara, T. Takenaka and M. Aikawa, 'Divider and combiner Line-Unified FET's as basic circuit function modules—Part II', *IEEE Trans. Microwave Theory Tech.*, **MTT-38**, 1218–1226 (1990).
24. J. Kunisch, M. Rittweger, S. Heinen and I. Wolff, 'The compression approach: a new technique for the analysis of distributed circuits containing nonlinear elements', *Proc. 21st European Microwave Conf.*, Stuttgart, pp. 1296–1301, September 1991.

Paper 6.4

An Efficient Two-Dimensional Graded Mesh Finite-Difference Time-Domain Algorithm for Shielded or Open Waveguide Structures

Veselin J. Brankovic, Dragan V. Krupezevic, and Fritz Arndt, *Senior Member, IEEE*

Abstract—**A finite-difference time-domain (FD-TD) algorithm is described for the efficient full-wave analysis of a comprehensive class of millimeter-wave and optical waveguide structures. The FD-TD algorithm is based on a two-dimensional graded mesh combined with adequately formulated absorbing boundary conditions. This allows the inclusion of nearly arbitrarily shaped, fully or partially lateral open or shielded guiding structures with or without layers of finite metallization thickness. Moreover, lossy dielectrics and/or lossy conductors are included in the theory. The algorithm leads to a significant reduction in cpu time and storage requirements as compared with the conventional three-dimensional eigenvalue FD-TD mesh formulation. Dispersion characteristic examples are calculated for structures suitable for usual integrated circuits, such as insulated image guides, ridge guides, dielectric waveguides, trapped image guides, coplanar-lines and microstrip lines. The theory is verified by comparison with results obtained by other methods.**

I. Introduction

SHIELDED, open or partially open waveguide structures of the class shown in Fig. 1 have found increasing interest for integrated circuit applications in the millimeter-wave and optical frequency range [1]–[17]. As this class includes a wide variety of specially shaped waveguiding structures used, in the design of integrated circuits, it is highly desirable to dispose of a reliable computer analysis which is sufficiently general and flexible to allow dominant and higher-order mode solutions of all desired cases including open or partially open structures.

Various methods of analyzing one or several of the structures of Fig. 1 have been the subject of many papers, including different kinds of mode-matching techniques [1]–[10], [13], the finite-element method [11], the finite-difference frequency-domain method [12], [14], [15], and, more recently, the finite-difference time-domain (FD-TD) method [18]–[28]. As for the criteria of flexibility, accuracy and computational efficiency, the FD-TD method is considered to be a very appropriate candidate for solving the waveguide eigenvalue problems with particular cross-sectional shapes, such as those shown in Fig. 1. The conventional FD-TD approach, however, utilizes a three-dimensional Yee's mesh for appropriate resonating sections which may be obtained by placing shorting planes along the axis of propagation of the structure under investigation. By repeating the calculation of the resonant frequency of these resonators for different distances of the shorting planes, the dispersion characteristic is then determined step by step. Consequently, the conventional approach requires considerable cpu time, needs a relatively large memory size, and tends to inaccuracies in the near of the cutoff frequencies.

A new two-dimensional FD-TD formulation has been introduced very recently [30] which helps to alleviate the above mentioned shortcomings of the conventional FD-TD approach, similar to [29], [33] where a 2-D TLM analysis has been described. Although many similarities exist between the TLM and the FD-TD method, cf. e.g.

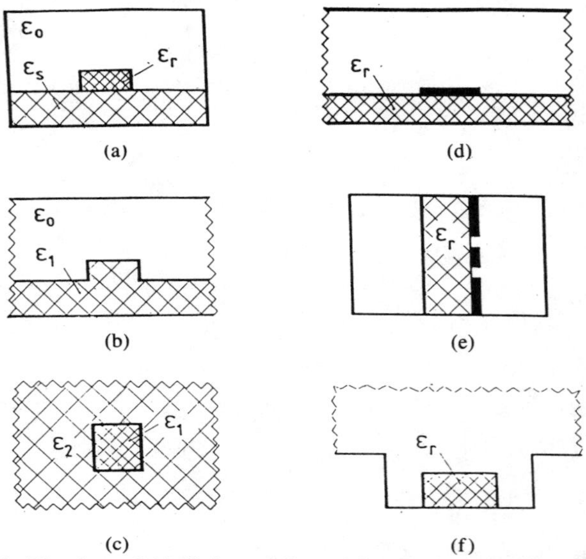

Fig. 1. The class of shielded, partially or fully open millimeter-wave and optical waveguide structures investigated with the new 2-D FD-TD method. (a) Shielded insulated image guide. (b) Lateral open ridge guide. (c) Open dielectric waveguide. (d) Lateral open microstrip line. (e) Coplanar line. (f) Open trapped image guide.

Manuscript received March 31, 1992; revised July 27, 1992.
The authors are with the Microwave Department, University of Bremen Kufsteiner Str., NW1, D-2800 Bremen, Germany.
IEEE Log Number 9203677.

Reprinted from *IEEE Trans. Microwave Theory Tech.*, vol. 40, no. 12, pp. 2272–2277, Dec. 1992.

[17], [31], the new 2-D FD-TD approach has the advantage of being nearly an order of magnitude faster than the 2-D TLM method. Moreover, the numerical difficulties in the field simulation in the vicinity of conductive edges reported in [32] for the classical TLM condensed node formulation (which is used also in [29], [33]) are inherently avoided by the 2-D FD-TD formulation presented in this paper.

The purpose of this paper is to extend the two-dimensional FD-TD method introduced in [30] by involving adequate absorbing boundary conditions, a graded mesh algorithm and an efficient calculation procedure for lossy structures. The inclusion of absorbing boundary conditions allows a simple, flexible, and versatile FD-TD solution for analyzing the complete class of shielded, partially or fully open waveguide structures of Fig. 1. The graded mesh permits the investigation of structures with realistic dimensions by making the mesh finer in regions of particular interest. Losses are an important factor in modern circuit design. Numerical results will be presented to elucidate the usefulness of the method. The theory is verified by comparison with results available from other methods.

II. Theory

A. Two-Dimensional FD-TD Mesh Formulation

The usual FD-TD method is formulated by discretizing Maxwell's curl equations over a finite volume and approximating the derivatives with centered difference approximations. This leads to the known three-dimensional Yee's mesh [18] in various modifications [19]–[28].

In order to derive the new two-dimensional FD-TD mesh formulation, we utilize the relationships between the transverse fields $\vec{E}_t$ and $\vec{H}_t$ of a guided wave travelling in $+z$ direction (Fig. 2)

$$\vec{E}_t(z \pm \Delta l) = \vec{E}_t(z) \cdot e^{\mp j\beta\Delta l}$$

$$\vec{H}_t(z \pm \Delta l) = \vec{H}_t(z) \cdot e^{\mp j\beta\Delta l}. \quad (1)$$

By applying Yee's formulations, e.g. in the form of [21] for anisotropic structures, a set of FD-TD equations for the six components $\vec{H}$ and $\vec{E}$ is then derived. The result is shown for H_x as an example,

$$\begin{aligned} H_x^{n+1/2}(i, j + 1/2) &= H_x^{n-1/2}(i, j + 1/2) \\ &\quad + s/[\mu_{xx}(i, j + 1/2) \cdot Z_{F0}] \\ &\quad \cdot \{E_y^n(i, j + 1/2) \cdot [e^{-j\beta\Delta l} - 1] \\ &\quad + E_z^n(i, j) - E_z^n(i, j + 1)\}, \quad (2) \end{aligned}$$

where the stability factor is $s = c\Delta t/\Delta w$ with $\Delta w = \Delta x = \Delta y = \Delta l$, c is the velocity of light, Z_{F0} is the characteristic impedance of free space, and μ_{xx} is the diagonal element of the relative permeability tensor. The condition for stability in free space for a uniform mesh is $s \leq 1/\sqrt{3}$ [21]. The remaining finite difference equations related to the other five field components can be similarly calculated.

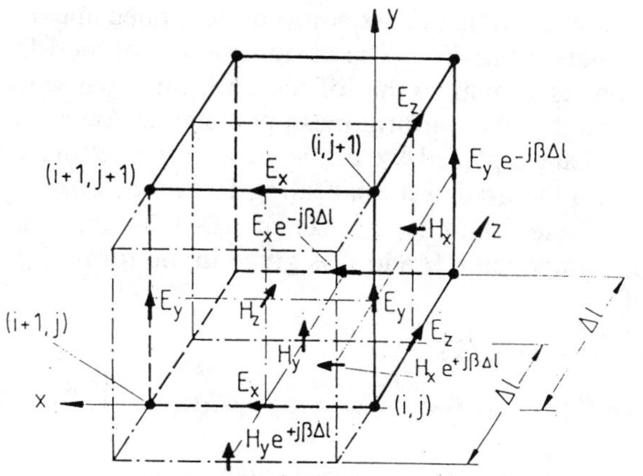

Fig. 2. New two-dimensional FD-TD mesh (i, j) for solving 2-D eigenvalue problems of waveguiding structures.

B. FD-TD Formulation of the Absorbing Boundary Conditions

For simplification, the equations of the boundary conditions are given only for the x-direction. In y-direction, the equations may be found analogously. Following Mur [20], we assume that the mesh is located in the region $x \geq 0$, that the boundary conditions are given for the plane $x = 0$, and that the outgoing wave has normal incidence. We then obtain the approximation for the boundary condition which would determine each field component W on the outer surface. This is consistent with an outgoing wave (i.e. this is absorbed), in the following form [20]:

$$\left(\partial_x - \frac{\partial_t}{v_{px}}\right) W \bigg|_{x=0} = 0, \quad (3)$$

where v_{px} is the phase velocity in x-direction.

The finite-difference formulation of (3) may be written in terms of the tangential electrical field components in the subregions 1) (outer region), and 2) (mesh region) as

$$E_{1t}^{n+1} = E_{2t}^n + \frac{v_{px}\Delta t - \Delta x}{v_{px}\Delta t + \Delta x}(E_{2t}^{n+1} - E_{1t}^n). \quad (4)$$

For the graded mesh, the following approximation of equation (4) is used

$$E_{1t}^{n+1} = E_{2t}^n + \frac{s_{xloc} - \sqrt{\epsilon_{xloc}}}{s_{xloc} + \sqrt{\epsilon_{xloc}}}(E_{2t}^{n+1} - E_{1t}^n), \quad (5)$$

where s_{xloc} is the local stability factor $s_{xloc} = c\Delta t/(\Delta w \cdot p_{loc})$, with the local graded mesh scaling factor p_{loc}, and where ϵ_{xloc} is the local relative premittivity in x-direction.

C. Graded Mesh Algorithm

In order to improve the computational efficiency, a graded mesh has turned out to be very advantageous [26]. This is particularly true for open structures where, if necessary, the enclosing absorbing boundary box is made sufficiently large so that the conditions mentioned above may be adequately satisfied. The basic structure of the FD-TD equations is similar to that of the uniform mesh with the exception that the uniform mesh parameters $\Delta w = \Delta x = \Delta y = \Delta l$ are replaced by $\Delta x = p\Delta w$, $\Delta y = q\Delta w$, $\Delta l = r\Delta w$, with the graded mesh scaling factors p, q, r.

With these relations, the set of FD-TD equations for the six components $\vec{H}$ and $\vec{E}$ is given in the form (e.g. for H_x) of

$$H_x^{n+1/2}(i, j+1/2) = H_x^{n-1/2}(i, j+1/2) + s/[\mu_{xx}(i, j+1/2) \cdot Z_{F0}] \cdot \left\{ \frac{1}{r} E_y^n(i, j+1/2) \cdot [e^{-j\beta\Delta l} - 1] + \frac{1}{q} [E_z^n(i, j) - E_z^n(i, j+1)] \right\}. \quad (6)$$

In order to maintain numerical convergence, the stability factor has to be chosen on the basis of the smallest mesh parameter. To ensure continuity of fields across the interfaces between coarse and fine meshes, space averaging [28] is applied.

D. FD-TD Formulation for Lossy Structures

For deriving the FD-TD equations for lossy structures, two approaches are possible: The first approach differs between a) lossy dielectric regions (with the conductivity $\sigma_D << \omega\epsilon$) and b) lossy metal regions (with the finite conductivity $\sigma_M >> \omega\epsilon$). 2) The second, more general approach includes both cases simultaneously. In our calculations we used the general approach which has the advantage that only one formulation is necessary for the whole lossy structure.

By applying the Maxwell equations and Yee's formulations, we obtain for the component E_y, as an example:
a) In the case of a lossy dielectric region ($\sigma_D << \omega\epsilon$):

$$E_y^{n+1}(i, j+1/2) = \left[1 - \frac{\Delta t}{\epsilon_0} \frac{\sigma_D(i, j+1/2)}{\epsilon_{ry}(i, j+1/2)}\right] \cdot E_y^n(i, j+1/2) + \frac{s}{\epsilon_{ry}(i, j+1/2)} \cdot Z_{F0} \cdot DH, \quad (7)$$

where

$$DH = \frac{H_x^{n+1/2}(i, j+1/2)[1 - e^{j\beta\Delta l}]}{r} - \frac{H_z^{n+1/2}(i+1/2, j+1/2) - H_z^{n+1/2}(i-1/2, j+1/2)}{p},$$

and $\Delta x = p\Delta w$, $\Delta y = q\Delta w$, $\Delta l = r\Delta w$, s: stability factor, $\epsilon = \epsilon_r\epsilon_0$.
b) In the case of a lossy metallic region ($\sigma_M >> \omega\epsilon$):

$$E_y^{n+1}(i, j+1/2) = -E_y^n(i, j+1/2) + \frac{2}{\Delta w \sigma_M(i, j+1/2)} \cdot DH. \quad (8)$$

The more general approach is obtained, assuming

$$E_y^{n+1/2}(i, j+1/2) = \tfrac{1}{2}[E_y^{n+1}(i, j+1/2) + E_y^n(i, j+1/2)],$$

by the following equation where σ denotes the finite conductivity for lossy dielectric or metall regions

$$E_y^{n+1}(i, j+1/2) = \frac{1 - \dfrac{\Delta t}{2\epsilon_0} \cdot \dfrac{\sigma(i, j+1/2)}{\epsilon_{ry}(i, j+1/2)}}{1 + \dfrac{\Delta t}{2\epsilon_0} \cdot \dfrac{\sigma(i, j+1/2)}{\epsilon_{ry}(i, j+1/2)}} \cdot E_y^n(i, j+1/2) + \frac{\dfrac{sZ_{F0}}{\epsilon_{ry}(i, j+1/2)}}{1 + \dfrac{\Delta t}{2\epsilon_0} \cdot \dfrac{\sigma(i, j+1/2)}{\epsilon_{ry}(i, j+1/2)}} \cdot DH. \quad (9)$$

The attenuation factor α of lossy structures is then calculated via the quality factor by using the equations already outlined in [33]. For the determination of the quality factor, however, an exponential regression procedure is used, i.e., for the corresponding resonance frequency, the envelope of the signal curve is evaluated by integration over several time periods. This yields high accuracy.

The principal numerical calculation steps in the two-dimensional FD-TD algorithm are similar to those in the conventional FD-TD approach with the exception that a propagation factor β has to be selected first (the choices of ß are indicated in the figure legends). After launching an excitation pulse, waiting until the distribution of the pulse is stable and performing the Fourier transformation, the modal frequencies related to the selected propagation factor are obtained. The mesh for the pulse propagation iteration needs only to be built up in the cross section x, y dimension, the number of nodes in z direction is reduced to ± 1. Also it should be emphasized that the amplitudes of all field components in the new formulation are complex quantities.

III. Results

Good agreement between our 2-D FD-TD dispersion results and those of the FD-frequency domain method [14] may be observed in Fig. 3 for the shielded insulated image guide. Fig. 4 shows the dispersion curves for different permittivities for the lateral open dielectric ridge guide. The results of the FD-TD method for a relatively low distance of the absorbing boundaries a = 5 h (i.e. only 2 h distance from the ridge) compare well with those obtained by the FD frequency domain method [14] for the shielded

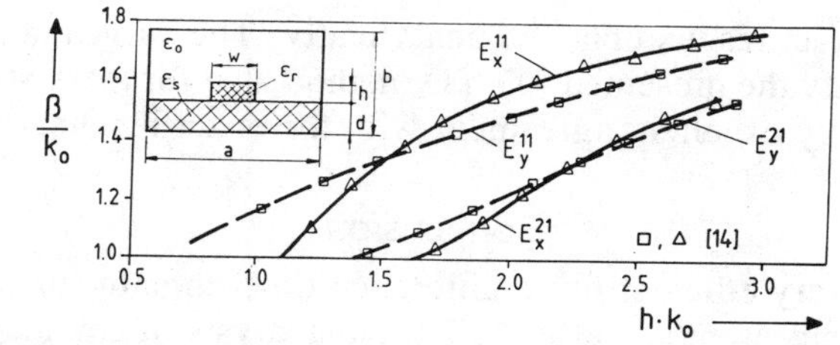

Fig. 3. Shielding insulated image guide. $w/h = 2.25$, $d/h = 0.5$, $a/h = 13.5$, $b/h = 8$, $\epsilon_r = 3.8$, $\epsilon_s = 1.5$. discretization: 54 × 64. Number of iterations $N_i = 1000$. Comparison with the FD frequency domain method [14]. (Choice of the propagation constant: $\beta = 0.5/h \cdots 5/h$, in e.g. 20 intermediate steps.)

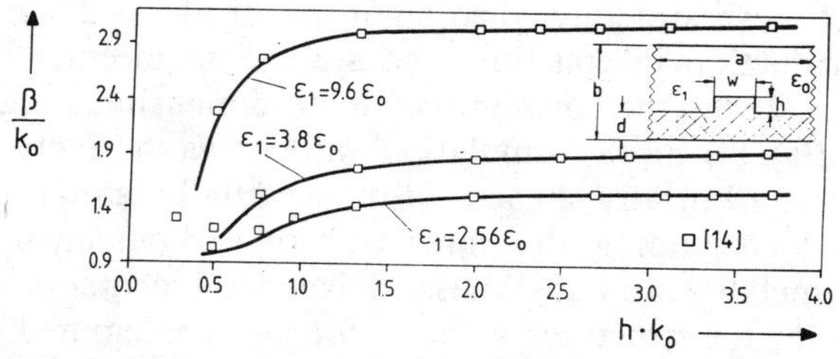

Fig. 4. Lateral open dielectric ridge guide. $w/h = 2$, $d/h = 1$, $b/h = 6$. Discretization: 60 × 72. Number of iterations $N_i = 2000$. Distance of the absorbing boundaries $a = 5$ h. Comparison with the FD frequency domain method [14] for the shielded dielectric ridge guide, lateral shield distance $a = 100$ h. (Choice of the propagation constant: $\beta = (0.3\sqrt{\epsilon_r})/h \cdots (3.8\sqrt{\epsilon_r})/h$, in e.g. 20 intermediate steps.)

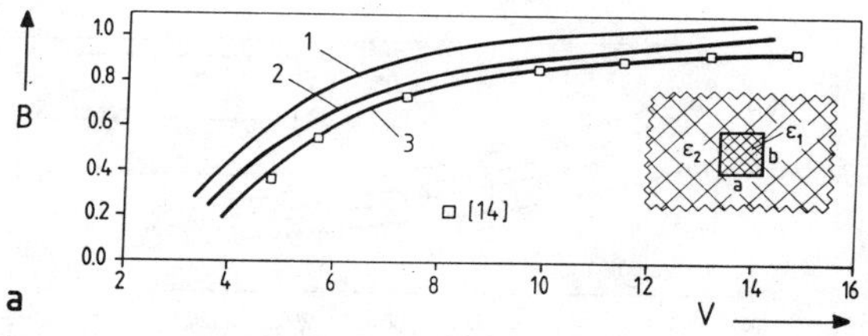

Fig. 5. Open dielectric guide $a = b$, $\epsilon_1 = 13.1\ \epsilon_0$, $\epsilon_2 = \epsilon_0$. Normalized dispersion curves of the E_{11}^y-mode, $B = [(\beta/k_0)^2 - 1]/(\epsilon_{r1} - 1)$ versus $V = k_0 a\sqrt{\epsilon_{r1} - 1}$, of the single dielectric waveguide for different discretizations: 1 (8 × 16), 2 (16 × 32), 3 (40 × 80). Comparison with the FD frequency domain method [14] for the shielded dielectric waveguide, shield distance 100a. Absorbing boundary in distance $5a/2$. Number of iterations $N_i = 2000$. Graded mesh ratio for the dielectric kernel: 1:5. (Choice of the propagation constant: $\beta = 2/a \cdots 15/a$, in e.g. 20 intermediate steps.)

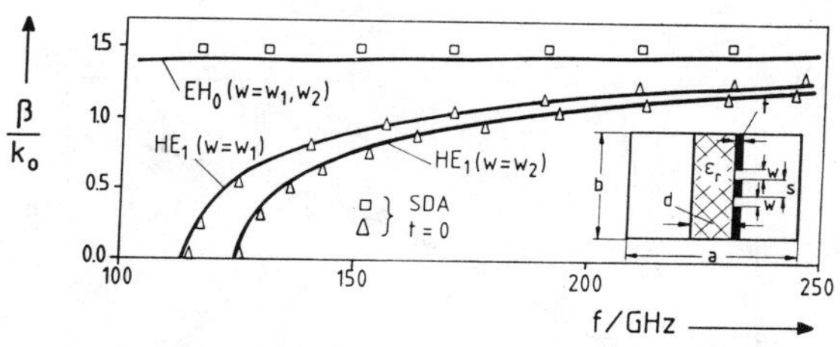

Fig. 6. EH_0- and HE_1-mode normalized propagation factor for the shielded coplanar line with finite metallization thickness t for different slot widths $w = w_1, w_2$. (For EH_0, the two curves are identical within the drawing accuracy of this figure.) Dimensions: $a = 0.8636$ mm, $b = 0.4318$ mm, $s = 0.072$ mm, $d = 0.126$ mm, $t = 0.018$ mm, $\epsilon_r = 3.7$, $w_1 = 0.09$ mm, $w_2 = 0.144$ mm. Discretization: 96 × 24. Number of iterations $N_i = 2000$. Comparison with own calculations with the Spectral Domain Approach (SDA) for $t = 0$. (Choice of the propagation constant: $\beta = 0 \cdots 7800$ 1/m, in e.g. 20 intermediate steps.)

dielectric ridge guide where the lateral shield distance is $a = 100$ h. These curves are considered to verify the absorbing boundary formulations made for the 2-D FD-TD method.

As an example for open structures, Fig. 5 presents the E_{11}^y-mode dispersion curves for a dielectric guide. A graded mesh was used with a mesh ratio for the dielectric kernel (ϵ_{r1}) to the outer region of 1 to 5. The normalized dispersion curves, $B = [(\beta/k_0)^2 - 1]/(\epsilon_{r1} - 1)$ versus $V = k_0 a\sqrt{\epsilon_{r1} - 1}$, are plotted for different discretizations: 1 (8 × 16), 2 (16 × 32), 3 (40 × 80). The absorbing boundary is located in a distance of $5a/2$. The comparison with the FD frequency domain method [14] for the shielded dielectric waveguide (shield distance $100a$) demonstrates good agreement with the discretization case number 3.

Fig. 6 shows the EH_0- and HE_1-mode normalized propagation factor as a function of frequency for a realistic mm-wave shielded coplanar line with finite metallization thickness. Due to the finite thickness included in the 2-D FD-TD method, there are slight deviations in comparison with own calculations by the spectral domain approach (SDA) for $t = 0$. Two slot widths w_1 and w_2 are included in the calculations in order to demonstrate the flexibility of the method also for varying typical structure dimensions.

In principle, the limitations of the applicability of the method—like for all time domain methods—are given by the limited storage and cpu time capabilities of the computer and by the fact of decreasing accuracy concerning the propagation factors of the higher-order modes with increasing order. The computer requirements, however, are significantly reduced by utilizing the 2-D FD-TD method described in this paper and by introducing a graded mesh.

A shielded and a lateral open microstrip line is calculated in Fig. 7. The influence of the lateral shield on the expanded normalized propagation factor ratio $B\ [\%] = 100 \cdot (\sqrt{\epsilon_r} - (\beta/k_0))/\sqrt{\epsilon_r}$ as a function of frequency is demonstrated for typical dimensions for mm-wave applications. For a distance c with more than about ten times the half strip width, the influence of the lateral shield becomes negligible and tends to values obtained with the lateral absorbing boundary (curves 4, 5).

The 2-D FD-TD method presented in this paper may be utilized to calculate the characteristic impedance. After performing the Fourier transformation at the corresponding modal frequency, in each point of the mesh, the complete set of electric and magnetic field components is obtained. Dependent from the chosen definition of the characteristic impedance, the related integration can be carried out numerically. This procedure has already successfully been tested for the shielded microstrip line by using the voltage-current definition. Because this topic would be beyond the scope of this paper, results will be presented in a further paper.

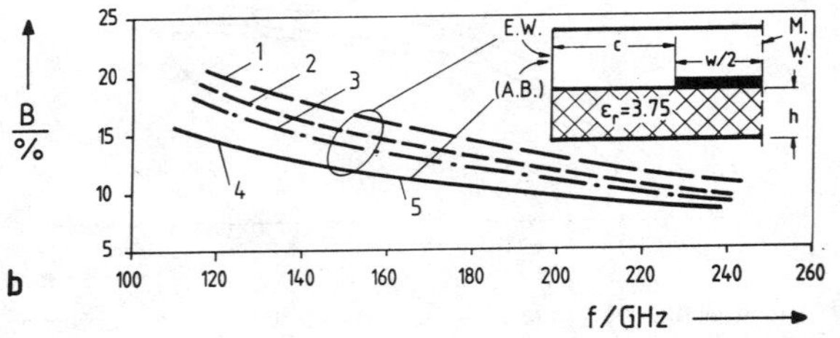

Fig. 7. Shielded and lateral open microstrip line. Influence of the lateral shield on the expanded normalized propagation factor ratio B [%] $= 100 \cdot (\sqrt{\epsilon_r} - (\beta/k_0))/\sqrt{\epsilon_r}$ as a function of frequency for a single microstrip line. Dimensions: $w = 0.16$ mm, height $b = 0.192$ mm, $\epsilon_r = 3.7$, $h = 0.128$ mm. Shield distances c: 1 ($c = 0.248$ mm $\triangleq 31\Delta l$), 2 ($c = 0.328$ mm $\triangleq 41\Delta l$), 3 ($c = 0.408$ mm $\triangleq 51\Delta l$), 4 ($c = 1.28$ mm $\triangleq 160\Delta l$). 5: Lateral open structure (curves 4 and 5 are identical within the drawing accuracy of Fig. 7.) (Choice of the propagation constant: $\beta = 3800 \cdots 9800$ 1/m, in e.g. 20 intermediate steps.)

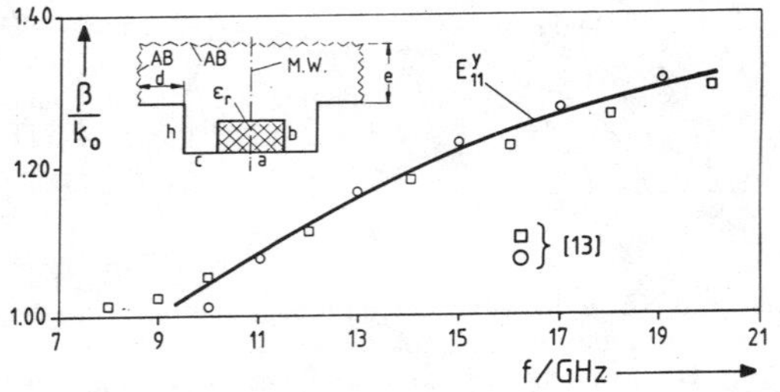

Fig. 8. Open trapped image guide. E^y_{11}-mode normalized propagation factor as a function of frequency. Dimensions: $a = b = c = 6$ mm, $h = 7$ mm, $\epsilon_r = 2.23$. Discretization: 54 × 42. Distance of the absorbing boundaries $d = 6$ mm, $e = 7$ mm. Number of iterations $N_i = 1200$. Comparison with [13]: □ □ □ combined effective dielectric constant transverse resonance method, ○ ○ ○ effective dielectric constant method. (Choice of the propagation constant: $\beta = 200 \cdots 550$ 1/m, in e.g. 20 intermediate steps.)

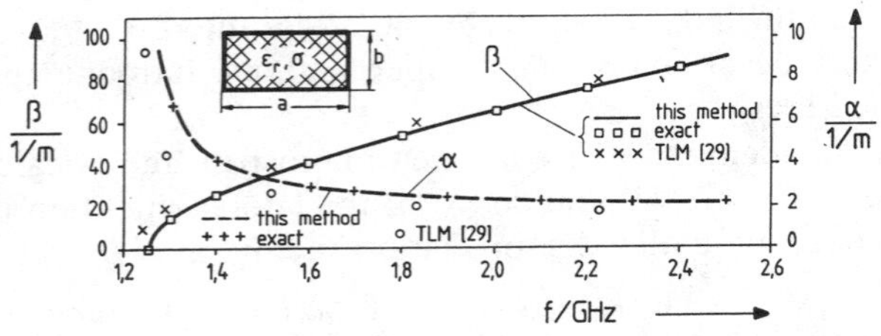

Fig. 9. Waveguide filled with a lossy dielectric. Attenuation factor α and propagation factor β as a function of frequency. Comparison with the TLM method results (○ ○ ○, × × ×) of [33] and with the exact results (+ + +, □ □ □) obtained analytically. Parameters: $a = 60$ mm, $b = 40$ mm, $\epsilon_r = 4$, $\sigma = 0.02$ S/m. Discretization: 30 × 40. Number of iterations $N_i = 8000$. (Choice of the propagation constant: $\beta = 0 \cdots 90$ 1/m, in e.g. 20 intermediate steps.)

Fig. 8 shows the E^y_{11}-mode normalized propagation factor as a function of frequency for the open trapped image guide. The results compare well with those calculated in [13] with the transverse resonance effective dielectric constant techniques.

As an example where exact results are available, a waveguide filled with a lossy dielectric is investigated in Fig. 9. The attenuation factor α and propagation factor β calculated with our method as a function of frequency are compared with the TLM method results of [33] and with the exact results obtained analytically. The curves in Fig. 9 verify the presented FD-TD method also for lossy structures by excellent agreement with the exact results.

Conclusion

A very efficient finite-difference time-domain formulation is presented for the full-wave analysis of shielded or open waveguiding structures. Lossy dielectrics and/or lossy conductors are included in the theory. This method utilizes advantageously a two-dimensional instead of the original three-dimensional Yee's mesh combined with adequately formulated absorbing boundary conditions and a graded mesh structure. The algorithm leads to a significant reduction in cpu time and storage requirements as compared with the conventional three-dimensional eigenvalue FD-TD mesh formulation and allows the inclusion of nearly arbitrarily shaped, fully or partially lateral open or shielded guiding structures with or without layers of finite metallization thickness. Dispersion characteristic examples for structures suitable for usual integrated circuits, such as insulated image guides, ridge guides, dielectric waveguides, trapped image guides, coplanar-lines, and microstrip lines elucidate the usefulness of the method. The theory is verified by comparison with results available from other methods.

References

[1] W. V. McLevidge, T. Itoh, and R. Mittra, "New waveguide structures for millimeter-wave and optical integrated circuits," *IEEE Trans. Microwave Theory Tech.*, vol. MTT-23, pp. 788–794, Oct. 1975.

[2] R. M. Knox "Dielectric waveguide microwave integrated circuits—An overview," *IEEE Trans. Microwave Tech.*, vol. MTT-24, pp. 806–814, Nov. 1976.

[3] J. A. Paul and Y.-W. Chang, "Millimeter wave image-guide integrated passive devices," *IEEE Trans. Microwave Theory Tech.*, vol. MTT-26, pp. 751–754, Oct. 1978.

[4] T. Itoh, "Open guiding structures for mmW integrated circuits," *Microwave J.*, pp. 113–126, Sept. 1982.

[5] S. E. Miller, "Integrated optics: An introduction," *Bell Syst. Tech. J.*, vol. 48, pp. 2059–2069, Sept. 1969.

[6] R. M. Knox and P. D. Toulios, "Integrated circuits for the millimeter through optical frequency range," in *Proc. Symp. Submillimeter Waves*, New York, May 1970, pp. 497–516.

[7] R. M. Knox, "Dielectric waveguides: A low-cost option for IC's," *Microwaves*, pp. 56–64, Mar. 1976.

[8] M. Ikeuchi, H. Swami, and H. Niki, "Analysis of open-type dielectric waveguides by the finite element iterative method," *IEEE Trans. Microwave Theory Tech.*, vol. MTT-29, pp. 234–239, Mar. 1981.

[9] N. Deo and R. Mittra, "A technique for analyzing planar dielectric waveguides for millimeter wave integrated circuits," *Arch. Elek. Übertragung*, vol. 37, pp. 236–244, July/Aug. 1983.

[10] U. Crombach, "Analysis of single and coupled rectangular dielectric waveguides," *IEEE Trans. Microwave Theory Tech.*, vol. MTT-29, pp. 870–874, Sept. 1981.

[11] B. M. A. Rahman and J. B. Davies, "Finite-element solution of integrated optical waveguides," *J. Lightwave Technol.*, vol. LT-2, pp. 682–687, Oct. 1984.

[12] S. M. Saad, "Review of numerical methods for the analysis of arbitrarily-shaped microwave and optical dielectric waveguides," *IEEE Trans. Microwave Tech.*, vol. MTT-33, pp. 894–899, Oct. 1985.

[13] W.-B. Zhou, and T. Itoh, "Analysis of trapped image guides using effective dielectric constant and surface impedances," *IEEE Trans. Microwave Theory Tech.*, vol. MTT-30, pp. 2163–2166, Dec. 1982.

[14] K. Bierwirth, N. Schulz, and F. Arndt, "Finite-difference analysis of rectangular dielectric waveguide structures," *IEEE Trans. Microwave Theory Tech.*, MTT-34, pp. 1104–1114, Nov. 1986.

[15] T. K. Sarkar, M. Manela, V. Narayanan, and A. R. Djordjevic, "Finite difference frequency-domain treatment of open transmission structures," *IEEE Trans. Microwave Theory Tech.*, vol. 38, pp. 1609–1616, Nov. 1990.

[16] T. Itoh, "Overview of quasi-planar transmission lines," *IEEE Trans. Microwave Theory Tech.*, vol. 37, pp. 275–280, Feb. 1989.

[17] M. Celuch-Marcysiak, W. K. Gwarek, "Formal equivalence and efficiency comparison of the FD-TD, TLM and SN methods in application to microwave CAD programs," in *Proc. European Microwave Conf.*, Stuttgart, Sept. 1991, pp. 199–204.

[18] K. S. Yee, "Numerical solution of initial boundary value problems involving Maxwell's equations on isotropic media," *IEEE Trans. Antennas Propagat.*, vol. AP-14, pp. 302–307, May 1966.

[19] A. Taflove, and M. E. Brodwin, "Numerical solution of steady state electromagnetic scattering problems using the time dependent Maxwell's equations," *IEEE Trans. Microwave Theory Tech.*, vol. MTT-23, pp. 623–630, Aug. 1975.

[20] G. Mur, "Absorbing boundary conditions for the finite-difference approximation of the time-domain electromagnetic-field equations," *IEEE Trans. on Elect. Compatibility*, vol. EMC-23, pp. 377–382, Nov. 1981.

[21] D. H. Choi, and W. J. R. Hoefer, "The finite-difference time-domain method and its applications to eigenvalue problems," *IEEE Trans. Microwave Theory Tech.* vol. MTT-34, pp. 1464–1470, Dec. 1986.

[22] W. Gwarek, "Analysis of arbitrarily shaped two-dimensional microwave circuits by finite-difference time-domain method," *IEEE Trans. Microwave Theory Tech.*, vol. 36, pp. 738–744, Apr. 1988.

[23] G.-C. Liang, Y.-W. Liu, and K. K. Mei, "Full-wave analysis of coplanar waveguide and slotline using the time-domain finite-difference method," *IEEE Trans. Microwave Theory Tech.*, vol. 37, pp. 1949–1957, Dec. 1989.

[24] D. M. Sheen, S. A. Ali, M. D. Abouzahra, and J. A. Kong, "Application of the three-dimensional time-domain method to the analysis of planar microstrip circuits," *IEEE Trans. Microwave Theory Tech.*, vol. 38, pp. 849–857, July 1990.

[25] M. Rittweger, M. Abdo, and I. Wolff, "Full-wave analysis of coplanar discontinuities considering three-dimensional bond wires," in *IEEE MTT-S Int. Symp. Dig.*, June 1991, pp. 465–468.

[26] D. H. Choi and W. J. R. Hoefer, "A graded mesh FD-TD algorithm for eigenvalue problems," in *Proc. European Microwave Conf.*, Rome, Sept. 1987, 413–417.

[27] Z. Chen, M. M. Ney, and W. J. R. Hoefer, "A new finite-difference time-domain formulation and its equivalence with the TLM symmetrical condensed node," *IEEE Trans. Microwave Theory Tech.*, vol. 39, Dec. 1991, pp. 2160–2169.

[28] I. S. Kim, and W. J. R. Hoefer, "A local mesh refinement algorithm for the time domain-finite difference method using Maxwell's curl equations," *IEEE Trans. Microwave Theory Tech.*, vol. 38, pp. 812–815, June 1990.

[29] H. Jin, R. Vahldieck, and S. Xiao, "An improved TLM full-wave analysis using a two dimensional mesh," in *IEEE MTT-S Int. Microwave Symp. Dig.*, July 1991, pp. 675–677.

[30] F. Arndt, V. J. Brankovic, and D. V. Krupezevic, "An improved FD-TD full wave analysis for arbitrary guiding structures using a two-dimensional mesh," in *IEEE MTT-S Int. Microwave Symp. Dig.*, 1992.

[31] N. R. S. Simons and E. Bridges, "Equivalence of propagation characteristics for the transmission-line matrix and finite-difference time-domain methods in two dimensions," *IEEE Trans. Microwave Theory Tech.*, vol. 39, pp. 354–357, Feb. 1991.

[32] J. S. Nielsen, and W. J. R. Hoefer, "Modification of the condensed 3-D TLM node to improve modeling of conductor edges," *IEEE Microwave Guided Wave Lett.*, vol. 2, pp. 105–110, Mar. 1992.

[33] H. Jin, R. Vahldieck, and S. Xiao, "A full-wave analysis of arbitrary guiding structures using a two-dimensional TLM mesh," in *Proc. European Microwave Conf.*, Stuttgart, pp. 205–210, Sept. 1991.

Paper 6.5

The Transmission-Line Matrix Method—Theory and Applications

WOLFGANG J. R. HOEFER, SENIOR MEMBER, IEEE

(Invited Paper)

Abstract—**This paper presents an overview of the transmission-line matrix (TLM) method of analysis, describing its historical background from Huygens's principle to modern computer formulations. The basic algorithm for simulating wave propagation in two- and three-dimensional transmission-line networks is derived. The introduction of boundaries, dielectric and magnetic materials, losses, and anisotropy are discussed in detail. Furthermore, the various sources of error and the limitations of the method are given, and methods for error correction or reduction, as well as improvements of numerical efficiency, are discussed. Finally, some typical applications to microwave problems are presented.**

I. Introduction

BEFORE THE ADVENT of digital computers, complicated electromagnetic problems which defied analytical treatment could only be solved by simulation techniques. In particular, the similarity between the behavior of electromagnetic fields, and of voltages and currents in electrical networks, was used extensively during the first half of the twentieth century to solve high-frequency field problems [2]–[4].

When modern computers became available, powerful numerical techniques emerged to predict directly the behavior of the field quantities. The great majority of these methods yield harmonic solutions of Maxwell's equations in the space or spectral domain. A notable exception is the transmission-line matrix (TLM) method of analysis which represents a true computer simulation of wave propagation in the time domain.

In this paper, the theoretical foundations of the TLM method are reviewed, its basic algorithm for simulating the propagation of waves in unbounded and bounded space is derived, and it is shown how the eigenfrequencies and field configurations of resonant structures can be determined with the Fourier transform. Sources and types of errors are discussed, and possible pitfalls are pointed out. Then, various methods of error correction are presented, and the most significant improvements to the conventional TLM approach are described. A referenced list of typical applications of the method is included as well. In the conclusion, the advantages and disadvantages of the method are summarized, and it is indicated under what circumstances it is appropriate to select the TLM method rather than other numerical techniques for solving a particular problem.

Manuscript received February 22, 1985; revised May 31, 1985.

The author is with the Department of Electrical Engineering, University of Ottawa, Ottawa, Ontario, Canada K1N 6N5.

II. Historical Background

Two distinct models describing the phenomenon of ligh were developed in the seventeenth century: the corpuscula model by Isaak Newton and the wave model by Christia Huygens. At the time of their conception, these model were considered incompatible. However, modern quantun physics has demonstrated that light in particular, an electromagnetic radiation in general, possess both granula (photons) and wave properties. These aspects are comple mentary, and one or the other usually dominates, depend ing on the phenomenon under study.

At microwave frequencies, the granular nature of elec tromagnetic radiation is not very evident, manifesting itsel only in certain interactions with matter, while the wav aspect predominates in all situations involving propagatio and scattering. This suggests that the model proposed b Huygens, and later refined by Fresnel, could form the basi for a general method of treating microwave propagatio and scattering problems.

Indeed, Johns and Beurle [5] described in 1971 a nove numerical technique for solving two-dimensional scatterin problems, which was based on Huygens's model of wav propagation. Inspired by earlier network simulation tech niques [2]–[4], this method employed a Cartesian mesh o open two-wire transmission lines to simulate two-dimen sional propagation of delta function impulses. Subsequen papers by Johns and Akhtarzad [6]–[16] extended th method to three dimensions and included the effect o dielectric loading and losses. Building upon the ground work laid by these original authors, other researcher [17]–[34] added various features and improvements such a variable mesh size, simplified nodes, error correction tech niques, and extension to anisotropic media.

The following section describes briefly the discretize version of Huygens's wave model which is suitable fc implementation on a digital computer and forms the al gorithm of the TLM method. A detailed description of thi model can be found in a very interesting paper by P. B Johns [9].

III. Huygen's Principle and Its Discretization

According to Huygens [1], a wavefront consists of number of secondary radiators which give rise to spherica wavelets. The envelope of these wavelets forms a ne

Reprinted from *IEEE Trans. Microwave Theory Tech.*, vol. MTT-33, no. 10, pp. 882–893, Oct. 1985.

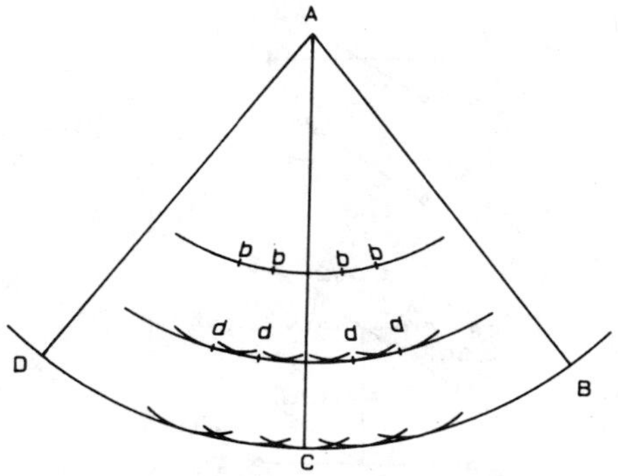

Fig. 1. Huygens's principle and formation of a wavefront by secondary wavelets.

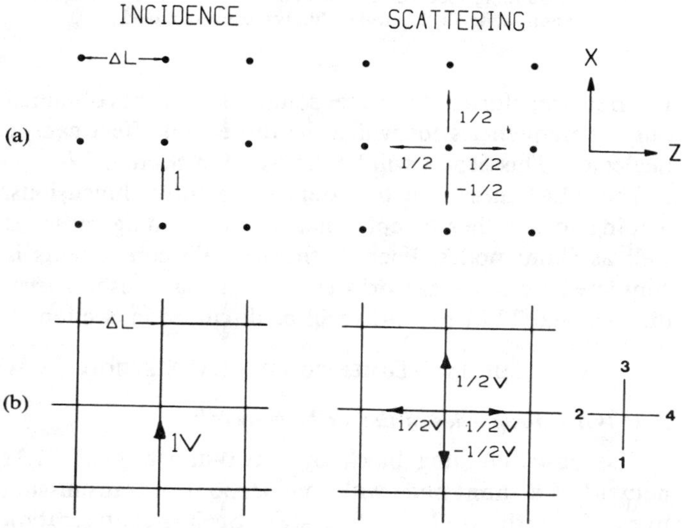

Fig. 2. The discretized Huygens's wave model (a) in two-dimensional space and (b) in an equivalent Cartesian mesh of transmission lines (after Johns [9]).

wavefront which, in turn, gives rise to a new generation of spherical wavelets, and so on (Fig. 1). In spite of certain difficulties in the mathematical formulation of this mechanism, its application nevertheless leads to an accurate description of wave propagation and scattering, as will be shown below.

In order to implement Huygens's model on a digital computer, one must formulate it in discretized form. To this end, both space and time are represented in terms of finite, elementary units Δl and Δt, which are related by the velocity of light such that

$$\Delta t = \Delta l / c. \tag{1}$$

Accordingly, two-dimensional space is modeled by a Cartesian matrix of points or nodes, separated by the mesh parameter Δl (see Fig. 2(a)). The unit time Δt is then the time required for an electromagnetic pulse to travel from one node to the next.

Assume that a delta function impulse is incident upon one of the nodes from the negative x-direction. The energy in the pulse is unity. In accordance with Huygen's principle, this energy is scattered isotropically in all four directions, each radiated pulse carrying one fourth of the incident energy. The corresponding field quantities must then be $1/2$ in magnitude. Furthermore, the reflection coefficient "seen" by the incident pulse must be negative in order to satisfy the requirement of field continuity at the node.

This model has a network analog in the form of a mesh of orthogonal transmission lines, or transmission-line matrix (Fig. 2(b)), forming a Cartesian array of shunt nodes which have the same scattering properties as the nodes in Fig. 2(a). It can be shown that there is a direct equivalence between the voltages and currents on the line mesh and the electric and magnetic fields of Maxwell's equations [5].

Consider the incidence of a unit Dirac voltage-impulse on a node in the TLM mesh of Fig. 2(b). Since all four branches have the same characteristic impedance, the reflection coefficient "seen" by the incident impulse is indeed $-1/2$, resulting in a reflected impulse of -0.5 V and three transmitted impulses of $+0.5$ V.

The more general case of four impulses being incident on the four branches of a node can be obtained by superposition from the previous case. Hence, if at time $t = k\,\Delta t$, voltage impulses ${}_kV_1^i$, ${}_kV_2^i$, ${}_kV_3^i$, and ${}_kV_4^i$ are incident on lines 1–4, respectively, on any junction node, then the total voltage impulse reflected along line n at time $(k+1)\,\Delta t$ will be

$${}_{k+1}V_n^r = \frac{1}{2}\left[\sum_{m=1}^{4} {}_kV_m^i\right] - {}_kV_n^i. \tag{2}$$

This situation is conveniently described by a scattering matrix equation [7] relating the reflected voltages at time $(k+1)\,\Delta t$ to the incident voltages at the previous time step $k\,\Delta t$

$${}_{k+1}\begin{pmatrix} V_1 \\ V_2 \\ V_3 \\ V_4 \end{pmatrix}^r = 1/2 \begin{pmatrix} -1 & 1 & 1 & 1 \\ 1 & -1 & 1 & 1 \\ 1 & 1 & -1 & 1 \\ 1 & 1 & 1 & -1 \end{pmatrix} \times {}_k\begin{pmatrix} V_1 \\ V_2 \\ V_3 \\ V_4 \end{pmatrix}^i . \tag{3}$$

Furthermore, any impulse emerging from a node at position (z, x) in the mesh (reflected impulse) becomes automatically an incident impulse on the neighboring node. Hence

$$\begin{aligned} {}_{k+1}V_1^i(z,x) &= {}_{k+1}V_3^r(z,x-1) \\ {}_{k+1}V_2^i(z,x) &= {}_{k+1}V_4^r(z-1,x) \\ {}_{k+1}V_3^i(z,x) &= {}_{k+1}V_1^r(z,x+1) \\ {}_{k+1}V_4^i(z,x) &= {}_{k+1}V_2^r(z=1,x). \end{aligned} \tag{4}$$

Consequently, if the magnitudes, positions, and directions of all impulses are known at time $k\,\Delta t$, the corresponding values at time $(k+1)\,\Delta t$ can be obtained by operating (3) and (4) on each node in the network. The impulse response of the network is then found by initially fixing the magni-

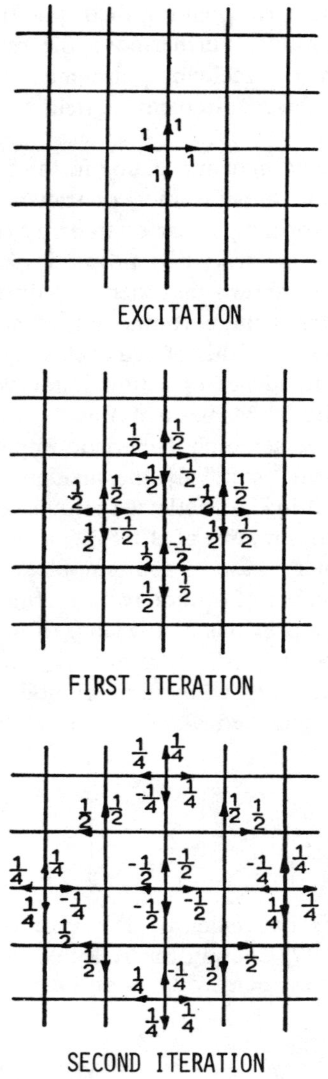

Fig. 3. Three consecutive scatterings in a two-dimensional TLM network excited by a Dirac impulse.

tudes, directions, and positions of all impulses at $t = 0$ and then calculating the state of the network at successive time intervals.

The scattering process described above forms the basic algorithm of the TLM method. Three consecutive scatterings are shown in Fig. 3, visualizing the spreading of the injected energy across the two-dimensional network.

This sequence of events closely resembles the disturbance of a pond due to a falling drop of water. However, there is one obvious difference, namely the discrete nature of the TLM mesh which causes dispersion of the velocity of the wavefront. In other words, the velocity of a signal component in the mesh depends on its direction of propagation as well as on its frequency.

In order to appreciate the importance of this dispersion, note that the process in Fig. 3 depicts a short episode of the response of the TLM network to a single impulse which contains all frequencies. Thus, harmonic solutions to a problem are obtained from the impulse response via the

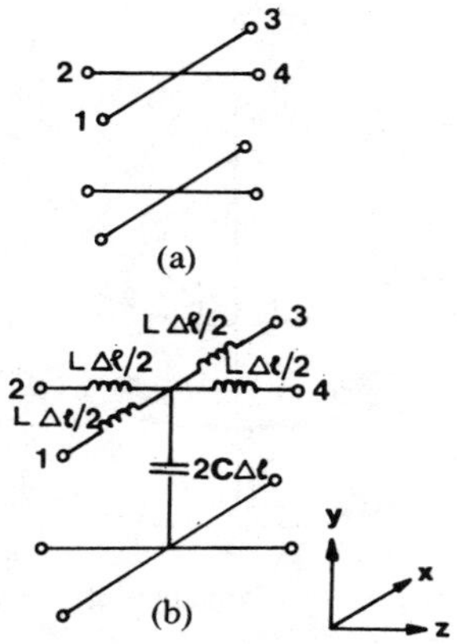

Fig. 4. The building block of the two-dimensional TLM network. (a) Shunt node. (b) Equivalent lumped-element model.

Fourier transform. Accurate solutions will be obtained only at frequencies for which the dispersion effect can be neglected. This aspect will be discussed in Section IV.

The TLM mesh can be extended to three dimensions, leading to a rather complex network containing series as well as shunt nodes. Each of the six field components is simulated by a voltage or a current in that mesh. Three-dimensional TLM networks will be discussed in Section V.

IV. The Two-Dimensional TLM Method

A. *Wave Properties of the TLM Network*

The basic building block of a two-dimensional TLM network is a shunt node with four sections of transmission lines of length $\Delta l/2$ (see Fig. 4(a)). Such a configuration can be approximated by the lumped-element model shown in Fig. 4(b). Comparing the relations between voltages and currents in the equivalent circuit with the relations between the H_z-, H_x-, and E_y-components of a TE_{m0} wave in a rectangular waveguide, the following equivalences can be established [5]:

$$E_y \equiv V_y \qquad -H_z \equiv (I_{x3} - I_{x1})$$
$$-H_x \equiv (I_{z2} - I_{z4}) \quad \mu \equiv L \quad \epsilon \equiv 2C. \tag{5}$$

For elementary transmission lines in the TLM network, and for $\mu_r = \epsilon_r = 1$, the inductance and capacitance per unit length are related by

$$1/\sqrt{LC} = 1/\sqrt{\epsilon_0 \mu_0} = c \tag{6}$$

where $c = 3 \times 10^8$ m/s.

Hence, if voltage and current waves on each transmission-line component travel at the speed of light, the complete network of intersecting transmission lines represents a medium of relative permittivity twice that of free space. The means that as long as the equivalent circuit in Fig. 4 is valid, the propagation velocity in the TLM mesh is $1/\sqrt{2}$ the velocity of light.

Note that the dual nature of electric and magnetic fields also allows us to simulate, for example, the longitudinal

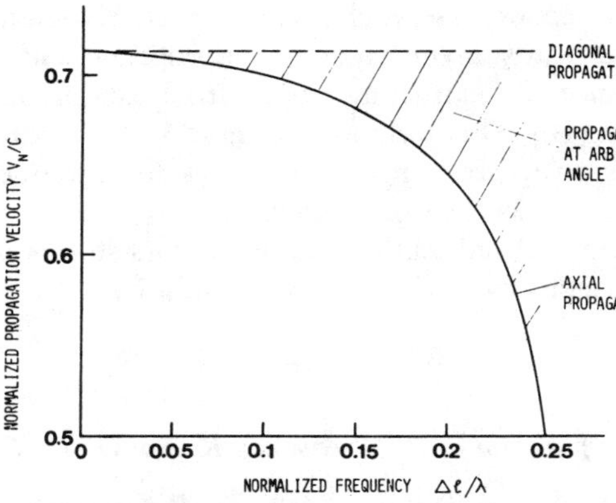

Fig. 5. Dispersion of the velocity of waves in a two-dimensional TLM network (after Johns and Beurle [5]).

magnetic field of TE modes by the network voltage, while the network currents simulate the transverse electric-field components. Whatever the relationship between field and network variables, the wave properties of the mesh, which will be discussed next, remain the same. Considering the mesh as a periodic structure, Johns and Beurle [5] calculate the following dispersion relation for propagation along the main mesh axes:

$$\sin(\beta_n \Delta l/2) = \sqrt{2}\sin[\omega \Delta l/(2c)] \tag{7}$$

where β_n is the propagation constant in the network. The resulting ratio of velocities on the matrix and in free space, $v_n/c = \omega/(\beta_n c)$, is shown in Fig. 5. It appears that a first cutoff occurs for $\Delta l/\lambda = 1/4$ (λ is the free-space wavelength). However, no cutoff occurs in the diagonal direction, where the velocity is frequency-independent, while in intermediate directions, the velocity ratio lies somewhere between the two curves shown in Fig. 5.

In conclusion, the TLM network simulates an isotropic propagating medium only as long as all frequencies are well below the network cutoff frequency, in which case the network propagation velocity may be considered constant and equal to $c/\sqrt{2}$.

B. *Representation of Lossless and Lossy Boundaries*

Electric and magnetic walls are represented by short and open circuits, respectively, at the appropriate positions in the TLM mesh. To ensure synchronism, they must be placed halfway between two nodes. In practice, this is achieved by making the mesh parameter Δl an integer fraction of the structure dimensions. Curved walls are represented by piecewise straight boundaries as shown in Fig. 6.

In the computation, the reflection of an impulse at a magnetic or electric wall is achieved by returning it, after one unit time step Δt, with equal or opposite sign to its boundary node of origin.

Lossy boundaries can be represented in the same way as lossless boundaries, with the difference that the reflection coefficient in each boundary branch is now

$$\rho = (R-1)/(R+1) \tag{8}$$

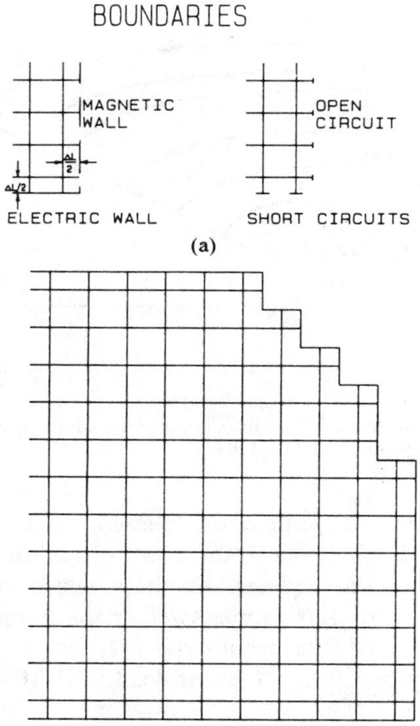

Fig. 6. Representation of boundaries in the TLM mesh. (a) Electric and magnetic walls. (b) Curved wall represented by a piecewise straight boundary.

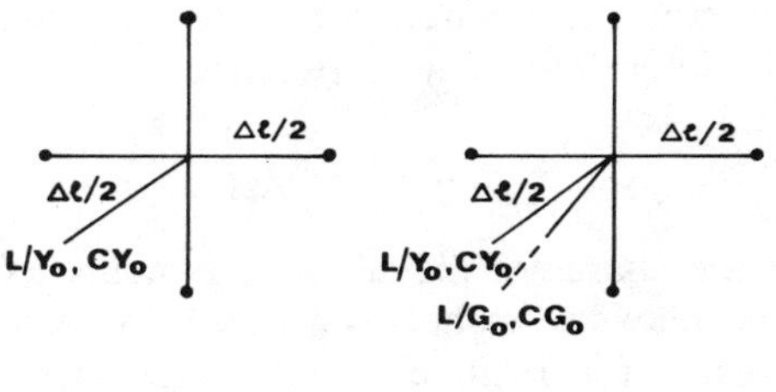

Fig. 7. Simulation of permittivity and losses. (a) Permittivity stub. (b) Permittivity stub and loss stub.

instead of unity. R is the normalized surface resistance of the boundary.

For a good but imperfect conductor of conductivity σ, the reflection coefficient ρ is approximately

$$\rho \simeq -1 + 2[\epsilon_0 \omega/(2\sigma)]^{1/2} \tag{9}$$

Note that since ρ depends on the frequency ω, the loss calculations are accurate only for that frequency which has been selected in determining ρ.

C. *Representation of Dielectric and Magnetic Materials*

The presence of dielectric or magnetic material (for example, in partial dielectric or magnetic loading of a waveguide) can be taken into account by loading inside nodes with reactive stubs of appropriate characteristic impedance and a length equal to half the mesh spacing [7], as shown in Fig. 7(a). For example, if the network voltage

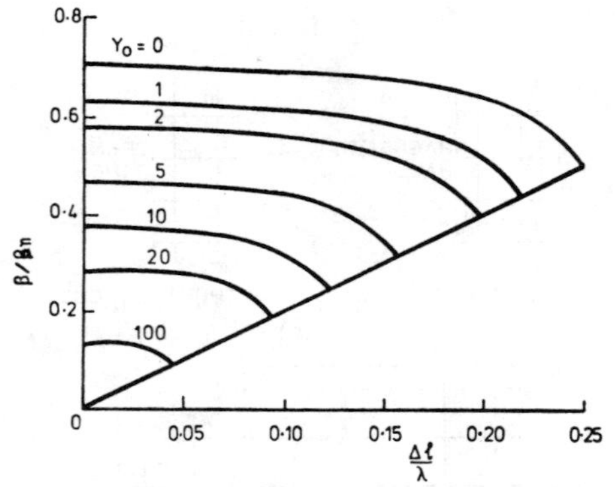

Fig. 8. Dispersion of the velocity of waves in a two-dimensional stub-loaded TLM network with the characteristic admittance of the reactive stubs as a parameter (after Johns [7]).

simulates an electric field, an open-circuited shunt stub of length $\Delta l/2$ will produce the effect of additional capacity at the nodes. This reduces the phase velocity in the structure and, at the same time, satisfies the boundary conditions at the air–dielectric interface [7]. At low frequencies, the velocity of waves in the stub-loaded TLM mesh is given by

$$v_n^2 = 0.5c^2/(1+Y_0/4) \tag{10}$$

where c is the free-space velocity, and Y_0 is the characteristic admittance of the stubs, normalized to the admittance of the main network lines. Note that the velocity in the network is now made variable by altering the single constant Y_0. The relationship between ϵ_r of the simulated space and Y_0 is

$$\epsilon_r = 2(1+Y_0/4). \tag{11}$$

The velocity characteristic along the main axes of the stub-loaded network is shown in Fig. 8 for various values of Y_0. Again, for relatively low frequencies, the mesh velocity is practically the same in all directions.

In cases where the voltage on the TLM mesh represents a magnetic field, the open shunt stubs describe a permeability. The velocity of waves in a magnetically loaded medium will be simulated correctly by such a mesh. However, the interface conditions are not satisfied, and a correction must be introduced in the form of local reflection and transmission coefficients at the interface between the media, as described in [7].

D. *Description of Dielectric Losses*

Losses in a dielectric can be accounted for in two different ways. One can either consider the TLM mesh to consist of lossy transmission lines, or one can load the nodes of a lossless mesh with so-called loss-stubs (Fig. 7(b)).

In the first case, the magnitude of each pulse is reduced by an appropriate amount while traveling from one node to the next, and the ensuing change in velocity is accounted for by increasing the time required to reach the next node [8]. This method is particularly suited for homogeneous structures.

In the second case, each node is resistively loaded with a matched transmission line of appropriate characteristic admittance G_0, extracting energy from each node at every iteration [10]. This technique is more suitable for inhomogeneous structures since it describes the interface conditions as well as the loss mechanism.

The normalized admittance of the loss-stub is related to the conductivity σ of the lossy medium by

$$G_0 = \sigma \Delta l(\mu_0/\epsilon_0)^{1/2}. \tag{12}$$

E. *Computation of the Frequency Response of a Structure*

The previous sections have described how the wave properties of two-dimensional unbounded and bounded space can be simulated by a two-dimensional mesh of transmission lines, and how the impulse response of such a mesh can be computed by iteration of (3) and (4). Any node (or several nodes) may be selected as input and/or output points. The output function is an infinite series of discrete impulses of varying magnitude, representing the response of the system to an impulsive excitation (see Fig. 12). The output corresponding to any other input may be obtained by convolving it with this impulse response.

Of particular interest is the response to a sinusoidal excitation which is obtained by taking the Fourier transform of the impulse response. Since the latter is a series of delta functions, the Fourier integral becomes a summation, and the real and imaginary parts of the output spectrum are

$$\mathrm{Re}\left[F(\Delta l/\lambda)\right] = \sum_{k=1}^{N} {}_kI \cos(2\pi k \Delta l/\lambda) \tag{13}$$

$$\mathrm{Im}\left[F(\Delta l/\lambda)\right] = \sum_{k=1}^{N} {}_kI \sin(2\pi k \Delta l/\lambda) \tag{14}$$

where $F(\Delta l/\lambda)$ is the frequency response, ${}_kI$ is the value of the output response at time $t = k\Delta l/c$, and N is the total number of time intervals for which the calculation has been made, henceforth called the "number of iterations."

In the case of a closed structure, this frequency response represents its mode spectrum. A typical example is Fig. 9(a), which shows the cutoff frequencies of the modes in a WR-90 waveguide.

Note that, as in a real measurement, the position of input and output points as well as the nature of the field component under study will affect the magnitudes of the spectral lines. For example, if input and output nodes are situated close to a minimum of a particular mode field, the corresponding eigenfrequency will not appear in the frequency response. This feature can be used either to suppress or enhance certain modes.

F. *Computation of Fields and Impedances*

Since the network voltages and currents are directly proportional to field quantities in the simulated structure, the TLM method also yields the field distribution. In order to obtain the configuration of a particular mode, its eigen-

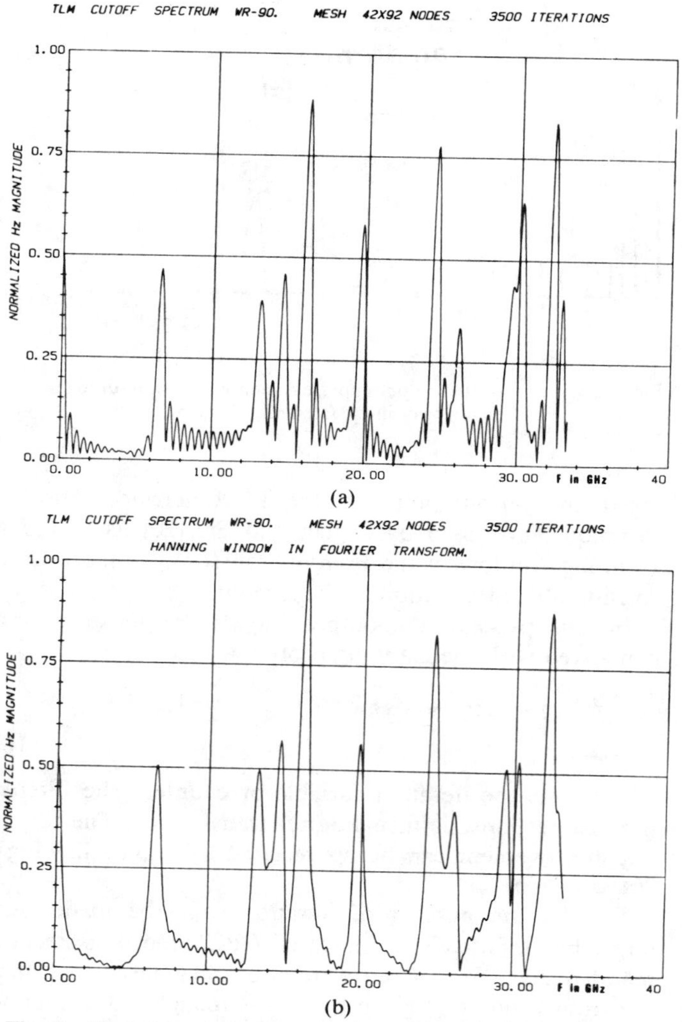

Fig. 9. Typical output from a two-dimensional TLM program. (a) Cutoff spectrum of a WR-90 waveguide. (b) The same spectrum after convolution of the output impulse function with a Hanning window (TLM) mesh: 42×92 nodes, 3500 iterations).

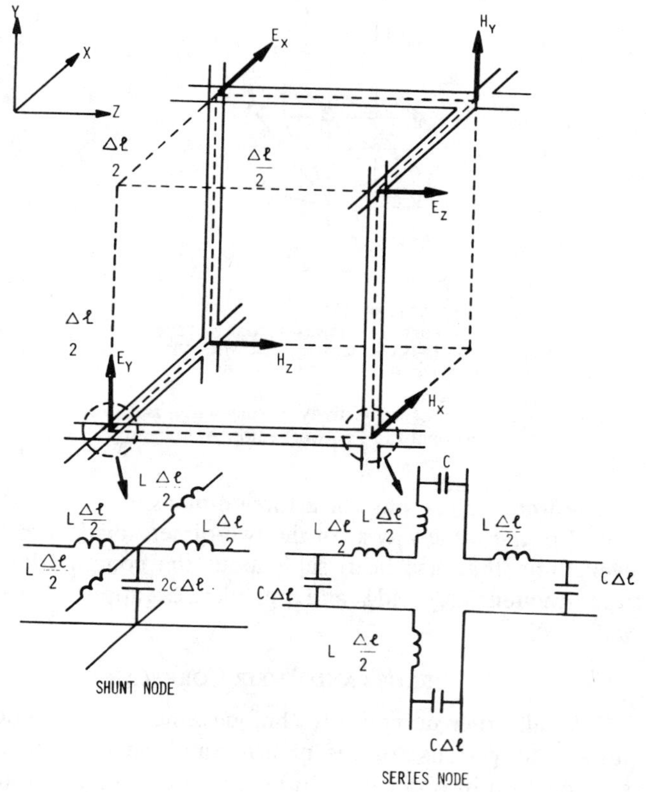

Fig. 10. The three-dimensional TLM cell featuring three series and three shunt nodes.

frequency must be computed first. Then the Fourier transform of the network variable representing the desired field component is computed at each node during a second run. In this process, (13) and (14) are computed for each node, with $\Delta l/\lambda$ corresponding to the eigenfrequency of the mode. The field between nodes can be obtained by interpolation techniques.

Impedances can, in turn, be obtained from the field quantities. Local field impedances can be found directly as the ratio of voltages and currents at a node, while impedances defined on the basis of particular field integrals (such as the voltage–power impedance in a waveguide) are computed by stepwise integration of the discrete field values. This procedure is identical to that used in finite-element and finite-difference methods of analysis.

V. The Three-Dimensional TLM Method

The two-dimensional method described above can be extended to three dimensions at the expense of increased complexity [10] to [15]. In order to simultaneously describe all six field components in three-dimensional space, the basic shunt node must be replaced by a hybrid TLM cell consisting of three shunt and three series nodes as shown in Fig. 10. The side of the cell is $\Delta l/2$. The voltages at the three series nodes represent the three electric-field components, while the currents at the series nodes represent the magnetic-field components.

The wave properties of the three-dimensional mesh are similar to that of its two-dimensional counterpart with the difference that the low-frequency velocity is now $c/2$ instead of $c/\sqrt{2}$ [15].

Boundaries are simulated by short-circuiting shunt nodes (electric wall) or open-circuiting shunt nodes (magnetic wall) situated on a boundary. Wall losses are included by introducing imperfect reflection coefficients.

Magnetic and dielectric materials may be introduced by adding short-circuited $\Delta l/2$ series stubs at the series nodes and open-circuited $\Delta l/2$ shunt stubs at the shunt nodes, respectively. Furthermore, losses are taken into account by resistively loading the shunt nodes in the network (see Fig. 11). Even anisotropic materials may be simulated by introducing at each of the three series or shunt nodes of a cell a stub with a different characteristic admittance [17]. Finally, losses as well as permittivities and permeabilities can be varied in space and in time by controlling the admittances of the dissipative and reactive stubs. The relationships between material parameters and stub admittances are the same as in the two-dimensional case.

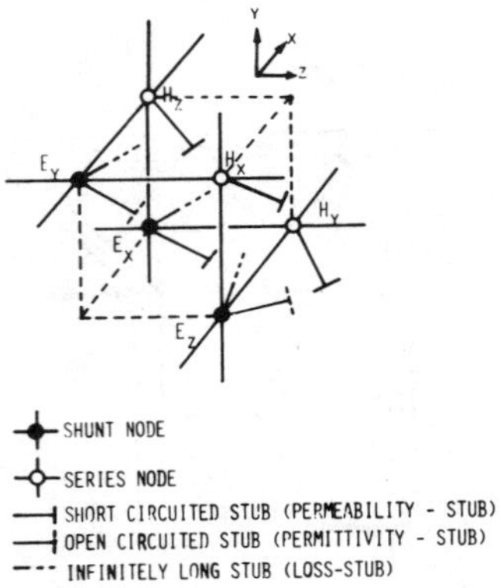

Fig. 11. Simulation of permittivity, permeability, and losses in a three-dimensional TLM network (after Akhtarzad [13]).

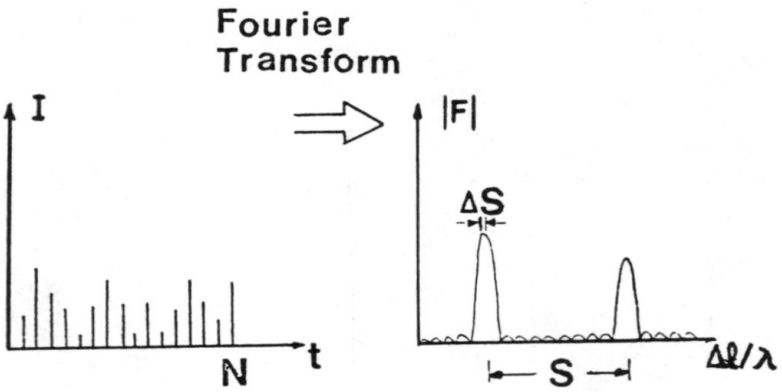

(a) (b)

Fig. 12. (a) Truncated output impulse response and (b) resulting truncation error in the frequency domain.

The impulse response of a three-dimensional network is found in the same way as in the two-dimensional case, and everything that has been said about the computation of eigenfrequencies, fields, and impedances, applies here as well.

VI. Errors and Their Correction

Like all other numerical techniques, the TLM method is subject to various sources of error and must be applied with caution in order to yield reliable and accurate results. The main sources of error are due to the following circumstances:

a) The impulse response must be truncated in time.
b) The propagation velocity in the TLM mesh depends on the direction of propagation and on the frequency.
c) The spatial resolution is limited by the finite mesh size.
d) Boundaries and dielectric interfaces cannot be aligned in the 3-D TLM model.

The resulting errors will be discussed below, and ways of eliminating or, at least, significantly reducing these errors will be described.

A. *Truncation Error*

The need to truncate the output impulse function leads to the so-called truncation error: Due to the finite duration of the impulse response, its Fourier transform is not a line spectrum but rather a superposition of $\sin x/x$ functions (Gibbs's phenomenon) which may interfere with each another such that their maxima are slightly shifted. The resulting error in the eigenfrequency, or truncation error, is given by

$$E_T \leqq \Delta S/(\Delta l/\lambda_c) = 3\lambda_c/(SN^2\pi^2\Delta l) \qquad (15)$$

where N is the number of iterations and S is the distance in the frequency domain between two neighboring spectral peaks (see Fig. 12).

This expression shows that the truncation error decreases with increasing separation S and increasing number of iterations N. It is thus desirable to suppress all unwanted modes close to the desired mode by choosing appropriate input and output points in the TLM network. Another technique, proposed by Saguet and Pic [20], is to use a Hanning window in the Fourier transform, resulting in a considerable attenuation of the sidelobes.

In this process, the output impulse response is first convolved with the Hanning profile

$$f_k(k) = 0.5(1+\cos \pi k/N), \qquad k=1,2,3,\cdots,N \qquad (16)$$

where k is the iteration variable or counter. The filtered impulse response is then Fourier transformed. The resulting improvement can be appreciated by comparing Figs. 9(a) and 9(b).

Finally, the number of iterations may be made very large, but this leads to increased CPU time. It is recommended that the number of iterations be chosen such that the truncation error given by (16) is reduced to a fraction of a percent and can be neglected.

B. *Velocity Error*

If the wavelength in the TLM network is large compared with the network parameter Δl, it can be assumed that the fields propagate with the same velocity in all directions. However, when the wavelength decreases, the velocity depends on the direction of propagation (see Fig. 5). At first glance, the resulting velocity error can be reduced only by choosing a very dense mesh, unless propagation occurs essentially in an axial direction (e.g., rectangular waveguide), in which case the error can be corrected directly using the dispersion relation (7). Fortunately, the velocity error responds to the same remedial measures as the coarseness error (which will be described next), and it therefore does not need to be corrected separately.

C. *Coarseness Error*

The coarseness error occurs when the TLM mesh is too coarse to resolve highly nonuniform fields as can be found at corners and wedges. This error is particularly cumbersome when analyzing planar structures which contain such regions. A possible but impractical measure would be to choose a very fine mesh. However, this would lead to large memory requirements, particularly for three-dimensional

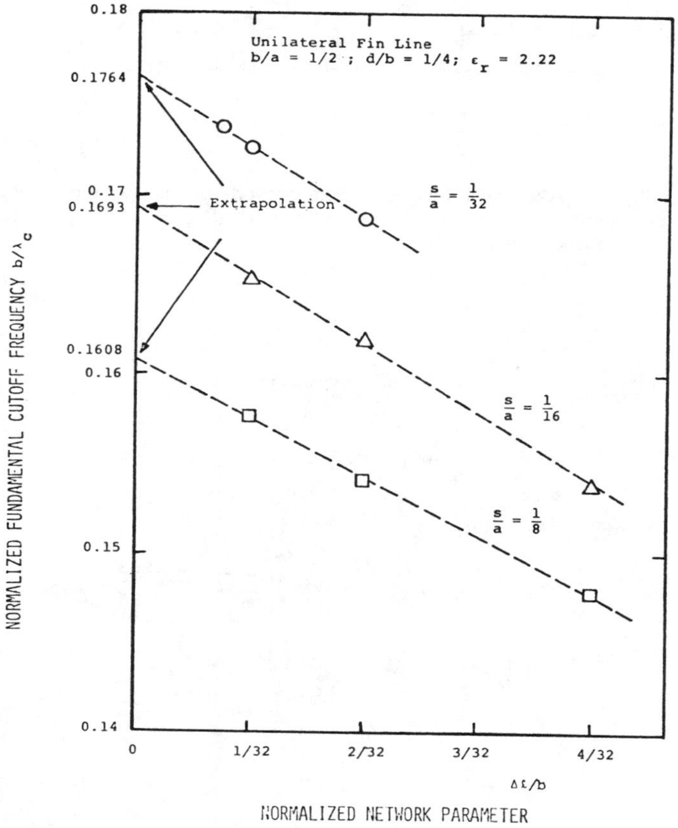

Fig. 13. Elimination of coarseness error by linear extrapolation of results obtained with TLM meshes of different parameter $\Delta l/b$ (after Shih and Hoefer [25]).

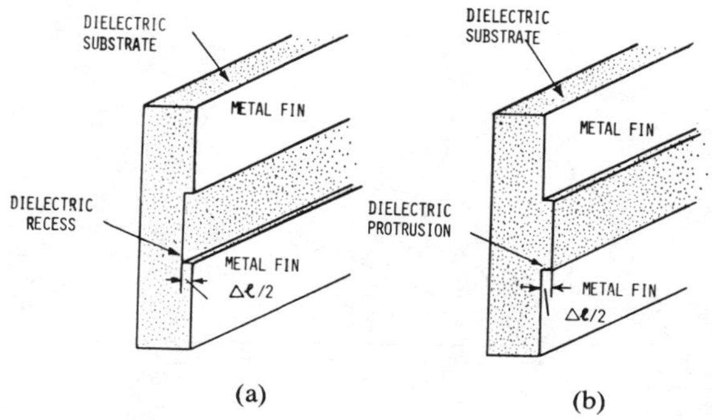

Fig. 14. Misalignment of conducting boundaries and dielectric interfaces in the three-dimensional TLM simulation of planar structures (after Shih [26]).

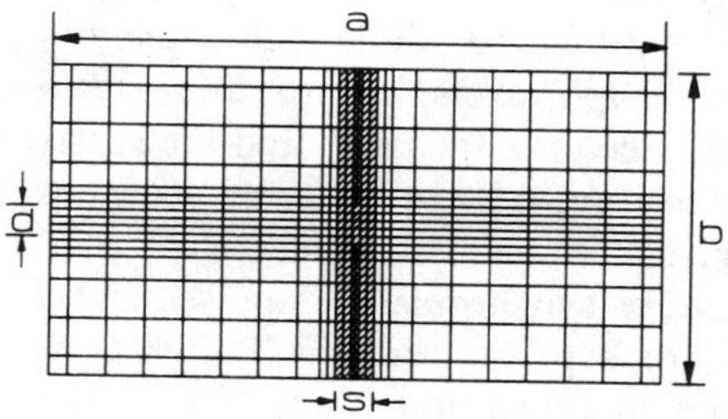

Fig. 15. Two-dimensional TLM network with variable mesh size for the computation of cutoff frequencies of finlines (after Saguet and Pic [28]).

problems. A better response is to introduce a network of variable mesh size to provide higher resolution in the nonuniform field region [27]–[29]. This approach is described in the next section; however, it requires more complicated programming. Yet another approach, proposed by Shih and Hoefer [25], is to compute the structure response several times using coarse meshes of different mesh parameter Δl, and then to extrapolate the obtained results to $\Delta l = 0$ as shown in Fig. 13.

Both measures effectively reduce the error by one order of magnitude and simultaneously correct the velocity error.

D. Misalignment of Dielectric Interfaces and Boundaries in Three-Dimensional Inhomogeneous Structures

Due to the particular way in which boundaries are simulated in a three-dimensional TLM network, dielectric interfaces appear halfway between nodes, while electric and magnetic boundaries appear across such nodes. This can be a problem when simulating planar structures such as microstrip or finline. In the TLM model, the dielectric either protrudes or is undercut by $\Delta l/2$, as shown in Fig. 14. Unless the resulting error is acceptable, one must make two computations, one with recessed and one with protruding dielectric, and take the average of the results. The problem does not occur in a variation of the three-dimensional TLM method involving an alternative node configuration proposed by Saguet and Pic [31], and described in the next section.

VII. Variations of the TLM Method

A number of modifications of the conventional TLM method have been proposed over the last few years with the aim of reducing errors, memory requirements, and CPU time. Some of them have already been mentioned, such as the introduction of a Hanning window [20], and extrapolation from coarse mesh calculations [21], [25]. Some effort has also been directed towards improving the efficiency of programing techniques [24].

In the following, three other interesting and significant innovations will be discussed briefly.

A. TLM Networks with Nonuniform Mesh

In order to ensure synchronism, the conventional TLM network uses a uniform mesh parameter throughout. This can lead to considerable numerical expenditure if the structure contains sharp corners or fins producing highly nonuniform fields and thus demands a high density mesh. Saguet and Pic [28] and Al-Mukhtar and Sitch [29] have independently proposed ways to implement irregularly graded TLM meshes which, as in the finite-element method, allow the network to adapt its density to the local nonuniformity of the fields. Fig. 15 shows such a network as proposed by Saguet and Pic [28] for the computation of cutoff frequencies in a finline. Note, however, that the size of the mesh cells is not arbitrary as in the case of finite elements; the length of each side is an odd integer multiple P of the smallest cell length in the network. To keep the

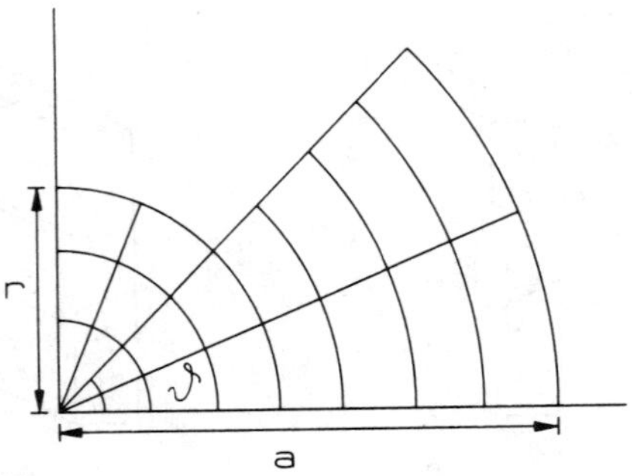

Fig. 16. A radial TLM mesh for the treatment of circular ridged waveguides (after Al-Mukhtar and Sitch [29]).

velocity of traveling impulses the same in all branches, the inductivity per unit length of the longer mesh lines is increased by a factor P while their capacity per unit length is reduced by $1/P$. This, in turn, increases their characteristic impedance by a factor P, and the scattering matrix of nodes connecting cells of different size must be modified accordingly.

To preserve synchronism, impulses traveling on longer branches are kept in store for P iterations before being reinjected at the next node.

For the configuration shown in Fig. 15, Saguet and Pic found that computing time was reduced between 3.5 and 5 times over a uniform mesh, depending on the relative size of the larger cells.

A different approach has been proposed by Al-Mukhtar and Sitch [27], [29]. They describe two possible ways to modify the characteristics of mesh elements in order to ensure synchronism, one involving the insertion of series stubs between nodes and loading of nodes by shunt stubs, the other involving modification of inductivity and capacity per unit length in such a way that propagation velocity in a branch becomes proportional to its length. The work by Al-Mukhtar and Sitch also covers the representation of radial meshes (see Fig. 16) as well as three-dimensional inhomogeneous structures. They report an economy of 45 percent in computer expenditure for a two-dimensional ridged waveguide problem, and a 40-percent reduction in storage and an 80-percent reduction in run time for a three-dimensional finline problem thanks to mesh grading.

B. A Punctual Node for Three-Dimensional TLM Networks

Conventional three-dimensional TLM networks require three shunt and three series nodes for the representation of one single cell (see Fig. 10). Saguet and Pic [31] have proposed an alternative method of interconnection. Representation of the short transmission-line sections by two rather than three lumped elements (see Fig. 17) makes it possible to realize both shunt and series connections in one point, resulting in a punctual node with 12 branches. This node is equivalent to a cell, such as that in Fig. 10, in which the inner connections have been eliminated. Losses and dielectric or magnetic loading can be simulated with stubs in the same way as discussed earlier.

This new node representation reduces, according to Saguet and Pic [31], the computation time by about 30

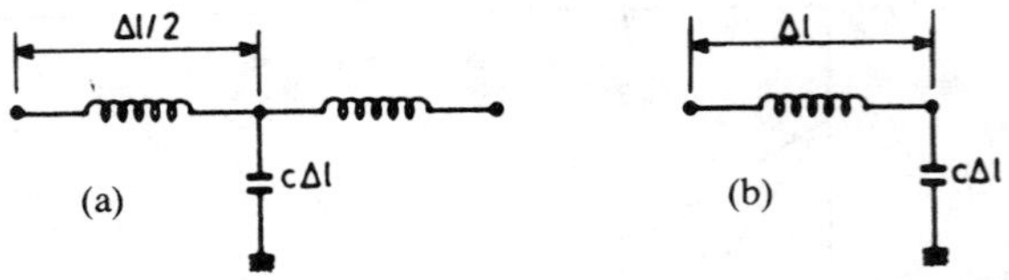

Fig. 17. Alternative lumped-element network for the two-dimensional shunt node. (a) Classical representation. (b) Representation proposed by Saguet and Pic [31].

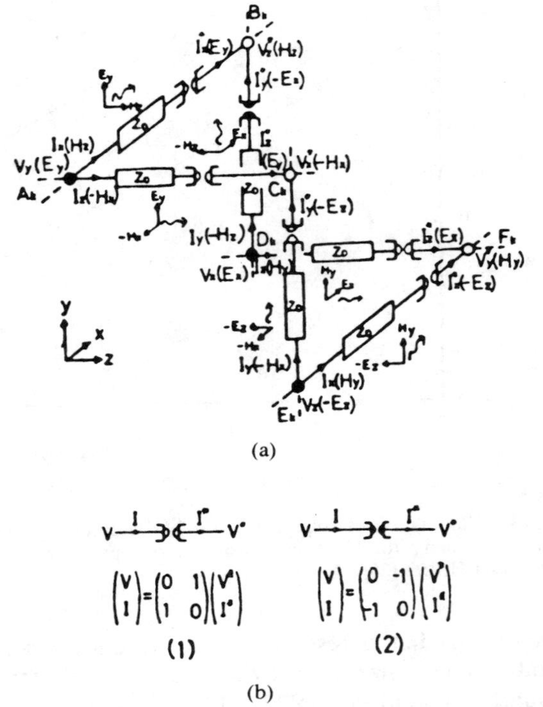

Fig. 18. Alternative network for three-dimensional TLM analysis proposed by Yoshida *et al.* [33]. (a) Equivalent circuit of an alternative three-dimensional TLM cell. (b) Definition of gyrators in (a): 1) positive gyrator, 2) negative gyrator.

percent. By employing both the punctual node and variable mesh size in a three-dimensional program, Saguet [32] has computed the resonant frequencies of a finline cavity 35 times faster than with a program based on the traditional TLM method.

C. Alternative Network Simulating Maxwell's Equations

Yoshida, Fukai, and Fukuoka [19], [23], [30], [33] have described a network similar to the TLM mesh, differing only in the way the basic cell element has been modeled. Instead of series and shunt nodes, this network contains so-called electric and magnetic nodes which are both "shunt-type nodes": while at the electric node, the voltage variable represents an electric field, it symbolizes a magnetic field at the magnetic node. The resulting ambivalence in the nature of the network voltage and current must be removed by inserting gyrators between the two types of nodes, as shown in Fig. 18. The wave properties of this network are identical with that of the conventional TLM

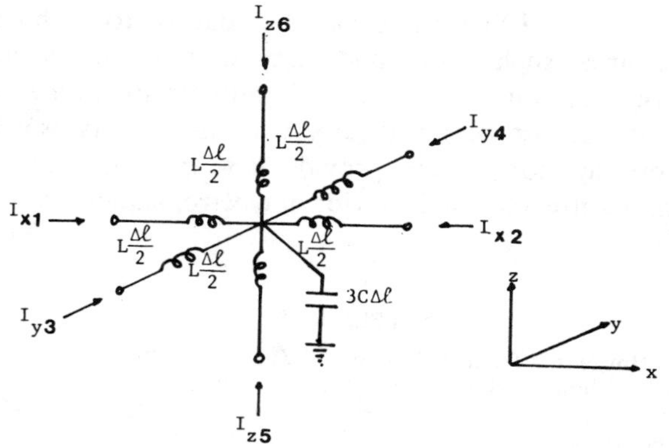

Fig. 19. Basic node of a three-dimensional scalar TLM network (after Choi and Hoefer [34]).

mesh. Errors and limitations are the same, and so are the possibilities of introducing losses and isotropic as well as anisotropic dielectric and magnetic materials.

D. The Scalar TLM Method

In those cases where electromagnetic fields can be decomposed into TE and TM modes (or LSE and LSM modes), it is only necessary to solve the scalar wave equation. Choi and Hoefer [34] have described a scalar TLM network to simulate a single field component or a Hertzian potential in three-dimensional space. The scalar TLM mesh can be thought of as a two-dimensional network to which additional transmission lines are connected orthogonally at each node as shown in Fig. 19. Such a structure could be realized in the form of a three-dimensional grid of coaxial lines.

The voltage impulses traveling across such a network represent the scalar variable to be simulated. Boundary reflection coefficients depend on both the nature of the boundary and that of the quantity to be simulated. For example, impulses will be subject to a reflection coefficient of -1 at a lossless electric wall if they represent either a tangential electric- or a normal magnetic-field component. A normal electric or a tangential magnetic field will be reflected with a coefficient of $+1$ in the same circumstances.

The slow-wave velocity in the three-dimensional scalar mesh is $c/\sqrt{3}$ as opposed to $c/2$ in the conventional TLM network. Dielectric or magnetic material as well as losses may be simulated using reactive and dissipative stubs.

The scalar method requires only 1/4 of the memory space and is seven times faster than the conventional method for a commensurate problem. However, its application is severely restricted, as it can be applied to scalar wave problems only.

VIII. Applications of the TLM Method

In the previous sections, the flexibility, versatility, and generality of the transmission-line matrix method has been demonstrated. In the following, an overview of potential applications of the method is given, and references describing specific applications are indicated. This list is not exhaustive, and many more applications can be found, not only in electromagnetism, but also in other fields dealing with wave phenomena, such as optics and acoustics.

For completeness, it should be mentioned that the TLM procedure can also be used to model and solve linear and nonlinear lumped networks [35]–[37] and diffusion problems [38]. Readers with a special interest in these applications should consult these references for more details.

Wave problems can be simulated in unbounded and bounded space, either in the time domain or—via Fourier analysis—in the frequency domain. Arbitrary homogeneous or inhomogeneous structures with anisotropic, space- and time-dependent electrical properties, including losses, can be simulated in two and three dimensions.

Below are some typical application examples.

A. Two-Dimensional Scattering Problems in Rectangular Waveguides (Field Distribution of Propagating and Evanescent Modes, Wave Impedance, Scattering Parameters)

- Open-circuited rectangular waveguide (TE_{n0}) [5].
- Bifurcation in rectangular waveguide (TE_{n0}) [5].
- Scattering by arbitrarily-shaped two-dimensional discontinuities in rectangular waveguide (TE_{n0}) including losses.

B. Two-Dimensional Eigenvalue Problems (Eigenfrequencies, Mode Fields)

- Cutoff frequencies and mode fields in homogeneous waveguides of arbitrary cross section, such as ridged waveguides [6], [8], [13], [26], [27].
- Cutoff frequencies and mode fields in inhomogeneous wavegudies of arbitrary cross section, such as dielectric loaded waveguides, finlines, image lines [7], [13], [16], [18], [21], [22], [25], [26], [28].

C. Three-Dimensional Eigenvalue and Hybrid Field Problems (Dispersion Characteristics, Wave Impedances, Losses, Eigenfrequencies, Mode Fields, Q-Factors)

- Characteristics of dielectric loaded cavities [10], [13], [14], [15], [18], [26], [27], [29], [32], [34].
- Dispersion characteristics and scattering in inhomogeneous planar transmission-line structures, including anisotropic substrate [11], [12], [13], [14], [17], [26], [27], [29], [30], [32].
- Transient analysis of transmission-line structures [19], [23], [33].

General-purpose two-dimensional and three-dimensional TLM programs can be found in Akhtarzad's Ph.D. thesis [13]. They can be adapted to most of the applications described above. If the various improvements and modifications described in Section VII are implemented in these programs, versatile and powerful numerical tools for the solution of complicated field problems are indeed obtained.

IX. Discussion and Conclusion

This paper has described the physical principle, the formulation, and the implementation of the transmission-line matrix method of analysis. Numerous features and applications of the method have been discussed, in particular the principal sources of error and their correction, the inclusion of losses, inhomogeneous and anisotropic properties of materials, and the capability to analyze transient as well as steady-state wave phenomena.

A general-purpose two-dimensional TLM program can be written in about 80 lines of FORTRAN, while a three-dimensional program is about 110 lines long. [13].

The method is limited only by the amount of memory storage required, which depends on the complexity of the structure and the nonuniformity of fields set up in it. In general, the smallest feature in the structure should at least contain three nodes for good resolution. The total storage requirement for a given computation can be determined by considering that each two-dimensional node requires five real number storage places, and an additional number equal to the number of iterations is needed to store the output impulse function. A basic three-dimensional node requires twelve number locations; if it is completely equipped with permittivity, permeability, and loss stubs, the required number of stores goes up to 26. Again, one real number must be stored per output function and per iteration. The number of iterations required varies between several hundred and several thousand, depending on the size and complexity of the TLM mesh.

As far as computational expenditure is concerned, the TLM method compares favorably with finite-element and finite-difference methods. Its accuracy is even slightly better by virtue of the Fourier transform, which ensures that the field function between nodes is automatically circular rather than linear as in the two other methods.

The main advantage of the TLM method, however, is the ease with which even the most complicated structures can be analyzed. The great flexibility and versatility of the method reside in the fact that the TLM network incorporates the properties of the electromagnetic fields and their interaction with the boundaries and materials. Hence, the electromagnetic problem need not to be reformulated for every new structure; its parameters are simply entered into a general-purpose program in the form of codes for boundaries, losses, permeability and permittivity, and excitation of the fields. Furthermore, by solving the problem through simulation of wave propagation in the time domain, the solution of large numbers of simultaneous equations is avoided. There are no problems with convergence, stability, or spurious solutions.

Another advantage of the TLM method resides in the large amount of information generated in one single computation. Not only is the impulse response of a structure obtained, yielding, in turn, its response to any excitation, but also the characteristics of the dominant and higher order modes are accessible in the frequency domain through the Fourier transform.

In order to increase the numerical efficiency and reduce the various errors associated with the method, more programing effort must be invested. Such an effort may be worthwhile when faced with the problem of scattering by a three-dimensional discontinuity in an inhomogeneous transmission medium, or when studying the overall electromagnetic properties of a monolithic circuit.

Finally, the TLM method may be adapted to problems in other areas such as thermodynamics, optics, and acoustics. Not only is it a very powerful and versatile numerical tool, but because of its affinity with the mechanism of wave propagation, it can provide new insights into the physical nature and the behavior of electromagnetic waves.

References

[1] C. Huygens, "Traité de la Lumière" (Leiden, 1690).

[2] J. R. Whinnery and S. Ramo, "A new approach to the solution of high-frequency field problems," *Proc. IRE*, vol. 32, pp. 284–288, May 1944.

[3] G. Kron, "Equivalent circuit of the field equations of Maxwell—I," *Proc. IRE*, vol. 32, pp. 289–299, May 1944.

[4] J. R. Whinnery, C. Concordia, W. Ridgway, and G. Kron, "Network analyzer studies of electromagnetic cavity resonators," *Proc. IRE*, vol. 32, pp. 360–367, June 1944.

[5] P. B. Johns and R. L. Beurle, "Numerical solution of 2-dimensional scattering problems using a transmission-line matrix," *Proc. Inst. Elec. Eng.*, vol. 118, no. 9, pp. 1203–1208, Sept. 1971.

[6] P. B. Johns, "Application of the transmission-line matrix method to homogeneous waveguides of arbitrary cross-section," *Proc. Inst. Elec. Eng.*, vol. 119, no. 8, pp. 1086–1091, Aug. 1972.

[7] P. B. Johns, "The solution of inhomogeneous waveguide problems using a transmission-line matrix," *IEEE Trans. Microwave Theory Tech.*, vol. MTT-22, pp. 209–215, Mar. 1974.

[8] S. Akhtarzad and P. B. Johns, "Numerical solution of lossy waveguides: T.L.M. computer program," *Electron. Lett.*, vol. 10, no. 15, pp. 309–311, July 25, 1974.

[9] P. B. Johns, "A new mathematical model to describe the physics of propagation," *Radio Electron. Eng.*, vol. 44, no. 12, pp. 657–666, Dec. 1974.

[10] S. Akhtarzad and P. B. Johns, "Solution of 6-component electromagnetic fields in three space dimensions and time by the T.L.M. method," *Electron. Lett.*, vol. 10, no. 25/26, pp. 535–537, Dec. 12, 1974.

[11] S. Akhtarzad and P. B. Johns, "T.L.M. analysis of the dispersion characteristics of microstrip lines on magnetic substrates using 3-dimensional resonators," *Electron. Lett.*, vol. 11, no. 6, pp. 130–131, Mar. 20, 1975.

[12] S. Akhtarzad and P. B. Johns, "Dispersion characteristic of a microstrip line with a step discontinuity," *Electron. Lett.*, vol. 11, no. 14, pp. 310–311, July 10, 1975.

[13] S. Akhtarzad, "Analysis of lossy microwave structures and microstrip resonators by the TLM method," Ph.D dissertation, Univ. of Nottingham, England, July 1975.

[14] S. Akhtarzad and P.B. Johns, "Three-dimensional transmission-line matrix computer analysis of microstrip resonators," *IEEE Trans. Microwave Theory Tech.*, vol. MTT-23, pp. 990–997, Dec. 1975.

[15] S. Akhtarzad and P. B. Johns, "Solution of Maxwell's equations in three space dimensions and time by the T.L.M. method of analysis," *Proc. Inst. Elec. Eng.*, vol. 122, no. 12, pp. 1344–1348, Dec. 1975.

[16] S. Akhtarzad and P. B. Johns, "Generalized elements for T.L.M. method of numerical analysis," *Proc. Inst. Elec. Eng.*, vol. 122, no. 12, pp. 1349–1352, Dec. 1975.

[17] G. E. Mariki, "Analysis of microstrip lines on inhomogeneous anisotropic substrates by the TLM numerical technique," Ph.D. thesis, Univ. of California, Los Angeles, June 1978.

[18] W. J. R. Hoefer and A. Ros, "Fin line parameters calculated with the TLM method," in *IEEE MTT Int. Microwave Symp. Dig.* (Orlando, FL), Apr. 28–May 2, 1979.

[19] N. Yoshida, I. Fukai, and J. Fukuoka, "Transient analysis of two-dimensional Maxwell's equations by Bergeron's method," *Trans. IECE Japan*, vol. J62B, pp. 511–518, June 1979.

[20] P. Saguet and E. Pic, "An improvement for the TLM method," *Electron. Lett.*, vol. 16, no. 7, pp. 247–248, Mar. 27, 1980.

[21] Y.-C. Shih, W. J. R. Hoefer, and A. Ros, "Cutoff frequencies in fin lines calculated with a two-dimensional TLM-program," in *IEEE MTT Int. Microwave Symp. Dig.* (Washington, DC), June 1980, pp. 261–263.

[22] W. J. R. Hoefer and Y.-C. Shih, "Field configuration of fundamental and higher order modes in fin lines obtained with the TLM method," presented at URSI and Int. IEEE-AP Symp., Quebec, Canada, June 2–6, 1980.

[23] N. Yoshida, I. Fukai, and J. Fukuoka, "Transient analysis of three-dimensional electromagnetic fields by nodal equations," *Trans. IECE Japan*, vol. J63B, pp. 876–883, Sept. 1980.

[24] A. Ros, Y.-C. Shih, and W. J. R. Hoefer, "Application of an accelerated TLM method to microwave systems," in *10th Eur. Microwave Conf. Dig.* (Warszawa, Poland), Sept. 8–11, 1980, pp. 382–388.

[25] Y.-C. Shih and W. J. R. Hoefer, "Dominant and second-order mode cutoff frequencies in fin lines calculated with a two-dimensional TLM program," *IEEE Trans. Microwave Theory Tech.*, vol. MTT-28, pp. 1443–1448, Dec. 1980.

[26] Y.-C. Shih, "The analysis of fin lines using transmission line matrix and transverse resonance methods," M.A.Sc. thesis, Univ. of Ottawa, Canada, 1980.

[27] D. Al-Mukhtar, "A transmission line matrix with irregularly graded space," Ph.D. thesis, Univ. of Sheffield, England, Aug. 1980.

[28] P. Saguet and E. Pic, "Le maillage rectangulaire et le changement de maille dans la méthode TLM en deux dimensions," *Electron. Lett.*, vol. 17, no. 7, pp. 277–278, Apr. 2, 1981.

[29] D.A. Al-Mukhtar and J. E. Sitch, "Transmission-line matrix method with irregularly graded space," *Proc. Inst. Elec. Eng.*, vol. 128, pt. H, no. 6, pp. 299–305, Dec. 1981.

[30] N. Yoshida, I. Fukai, and J. Fukuoka, "Application of Bergeron's method to anisotropic media," *Trans. IECE Japan*, vol J64B, pp. 1242–1249, Nov. 1981.

[31] P. Saguet and E. Pic, "Utilisation d'un nouveau type de noeud dans la méthode TLM en 3 dimensions," *Electron. Lett.*, vol. 18, no. 11, pp. 478–480, May 1982.

[32] P. Saguet, "Le maillage parallelepipédique et le changement de maille dans la méthode TLM en trois dimensions," *Electron. Lett.*, vol. 20, no. 5, pp. 222–224, Mar. 15, 1984.

[33] N. Yoshida and I. Fukai, "Transient analysis of a stripline having a corner in three-dimensional space," *IEEE Trans. Microwave Theory Tech.*, vol. MTT-32, pp. 491–498, May 1984.

[34] D. H. Choi and W. J. R. Hoefer, "The simulation of three-dimensional wave propagation by a scalar TLM model," in *IEEE MTT Int. Microwave Symp. Dig.* (San Francisco), May 1984.

[35] P. B. Johns, "Numerical modelling by the TLM method," in *Large Engineering Systems*, A. Wexler, Ed. Oxford: Pergamon Press, 1977.

[36] J. W. Bandler, P. B. Johns, and M. R. M. Rizk, "Transmission-line modeling and sensitivity evaluation for lumped network simulation and design in the time domain," *J. Franklin Inst.*, vol. 304, no. 1, pp. 15–23, 1977.

[37] P. B. Johns and M. O'Brien, "Use of the transmission-line modelling (T.L.M.) method to solve non-linear lumped networks," *Radio Electron. Eng.*, vol. 50, no. 1/2, pp. 59–70, Jan./Feb. 1980.

[38] P. B. Johns, "A simple explicit and unconditionally stable numerical routine for the solution of the diffusion equation," *Int. J. Num. Meth. Eng.*, vol. 11, pp. 1307–1328, 1977.

Paper 6.6

Analysis of Microstrip Circuits Using Three-Dimensional Full-Wave Electromagnetic Field Analysis in the Time Domain

TSUGUMICHI SHIBATA, MEMBER, IEEE, TOSHIO HAYASHI, AND TADAKATSU KIMURA, MEMBER, IEEE

Abstract —**Calculation of the frequency characteristics for microstrip circuits based on a three-dimensional full-wave electromagnetic field analysis in the time domain is proposed. In this method, the circuit is excited by a pulse which includes broadened frequency components. The frequency characteristics are then computed at once from the Fourier transform of the output transient responses. To evaluate the validity and capability of the method, a side-coupled microstrip filter is analyzed and the frequency characteristics are calculated. A quasi-static analysis of this filter is also presented and the results compared with measurements. The frequency characteristics calculated with the full-wave analysis in the time domain show excellent agreement with the measured values, thus demonstrating the validity and the power of the analytical method.**

I. Introduction

WITH THE DEVELOPMENT of high-speed device technology, it is becoming increasingly important to develop high-performance MMIC's which will serve as key components in compact and low-cost communication systems of the future. In MMIC's, microstrip circuits such as transmission lines (i.e., interconnect), filters, and couplers are used as circuit elements. Circuit designers are turning their attention to techniques for precisely analyzing their electrical properties. Some simple circuits have already been analyzed using quasi-static analysis. In most practical situations, however, because of the complicated electromagnetic fields that result from their three-dimensional shapes and locations, precise analysis cannot be expected from the more common analytical methods, or the so-called quasi-static analysis method. For greater accuracy, computer-aided numerical analysis of full-wave three-dimensional electromagnetic fields is required. This numerical analysis technique is thus an important key for the development of MMIC designs.

Two approaches have been taken with respect to full-wave electromagnetic field analysis. One is the time-domain analysis approach which solves the initial boundary value problem of the wave or Maxwell's equations. The other approach is a frequency-domain analysis method based on the eigenvalue problem of the vector Helmholtz equation. One of the advantages of the time-domain analysis, which is the main reason we have employed it here, is that the analysis is stable and yields a unique result, while the frequency-domain analysis suffers from spurious solutions and considerable care must be taken in the calculation process.

The numerical formulation of the time-domain analysis is given by Yee (the FD-TD method) [1], which uses finite difference equations of Maxwell's equations. An alternative method is based on the equivalent circuits of electromagnetic fields [2]. Johns *et al.* constructed an equivalent circuit with finite length transmission lines and developed a method of analysis based on the scattering matrix of each line (the T.L.M. method) [3]–[7]. These methods have been substantially modified in a number of reports [8]–[11]. Especially, Yoshida, Fukai and their coworkers have refined the T.L.M. method by expressing the medium constants by additional lumped elements to the equivalent circuit. They call their approach the Bergeron method. They also applied this method to the analysis of microstrip circuits and successfully showed the time variation of electromagnetic fields and power flow around the circuits [12]–[16]. But, for microstrip circuit design, it is necessary to calculate the frequency characteristics of the circuit. Since their analysis uses sinusoidal sources to excite the circuits, it is insufficient for this purpose.

The aim of this paper is to demonstrate calculation of the frequency characteristics for microstrip circuits and to establish the frequency-domain MMIC design method based on time-domain electromagnetic field analysis. The frequency characteristics are calculated using a raised cosine pulse source and applying the Fourier transform to the transient response of the circuit. In this way the frequency characteristics can be obtained at once over the full range of frequencies, whereas only one point of the frequency characteristic can be given using a sinusoidal source. By comparing the computed results and measured

Manuscript received October 20, 1987; revised January 19, 1988.
The authors are with NTT LSI Laboratories, 3-1 Morinosato Wakamiya, Atsugi-shi, Kanagawa, 243-01, Japan.
IEEE Log Number 8821075.

Reprinted from *IEEE Trans. Microwave Theory Tech.*, vol. 36, no. 6, pp. 1064–1070, June 1988.

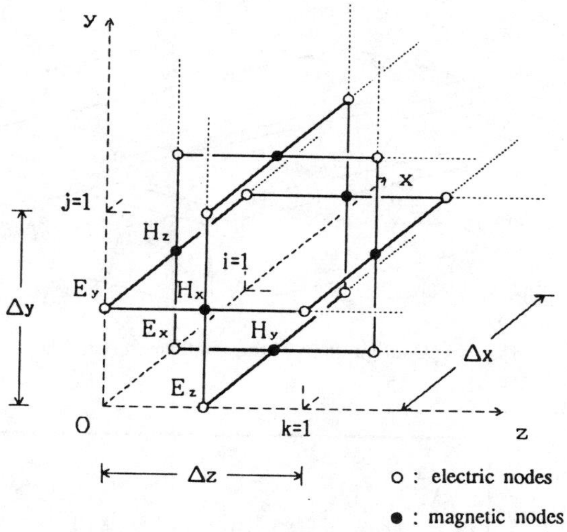

Fig. 1. The net of an equivalent circuit. Electric nodes,

$$E_x\left(i+\frac{1}{2}, j, k\right), \quad i=0,1,\cdots,L-1;\ j=0,1,\cdots,M;\ k=0,1,\cdots,N$$

$$E_y\left(i, j+\frac{1}{2}, k\right), \quad i=0,1,\cdots,L;\ j=0,1,\cdots,M-1;\ k=0,1,\cdots,N$$

$$E_z\left(i, j, k+\frac{1}{2}\right), \quad i=0,1,\cdots,L;\ j=0,1,\cdots,M;\ k=0,1,\cdots,N-1$$

and magnetic nodes,

$$H_x\left(i, j+\frac{1}{2}, k+\frac{1}{2}\right), \quad i=0,1,\cdots,L;\ j=0,1,\cdots,M-1;\ k=0,1,\cdots,N-1$$

$$H_y\left(i+\frac{1}{2}, j, k+\frac{1}{2}\right), \quad i=0,1,\cdots,L-1;\ j=0,1,\cdots,M;\ k=0,1,\cdots,N-1$$

$$H_z\left(i+\frac{1}{2}, j+\frac{1}{2}, k\right), \quad i=0,1,\cdots,L-1;\ j=0,1,\cdots,M-1;\ k=0,1,\cdots,N$$

are indicated by open circles and closed circles, respectively.

data, the time-domain analysis scheme can be verified. A side-coupled microstrip filter will be analyzed as an example to demonstrate the practicality of the scheme.

In the next section (Section II), we will briefly review the Bergeron method. The analysis of a single microstrip line is described in Section III as a preliminary to the analysis of a side-coupled microstrip filter in Section IV. In Section V, the results of the quasi-static analysis are presented and the necessity of the full-wave analysis is emphasized. In Section VI, we will discuss the accuracy of the full-wave, time-domain analysis. Finally, a few concluding remarks will be offered in Section VII.

II. The Equivalent Circuit of the Electromagnetic Field and Numerical Scheme of Analysis

In this section, we review the Bergeron method [12], which is used to solve for three-dimensional electromagnetic fields in the time domain using the equivalent circuits of the fields. Fig. 1 shows an equivalent circuit disposed in three-dimensional space. There are six kinds of nodes in the equivalent circuit: three electric nodes, E_x, E_y, and E_z, which are indicated by open circles in the figure, and three magnetic nodes H_x, H_y, and H_z, shown as closed circles. Branches consist of $\Delta l/2$ length transmission lines and gyrators, as shown in detail in Fig. 2. Values of the dielectric constant ϵ and permeability μ are expressed in terms of lumped capacitances connected to the corresponding electric and magnetic nodes, respectively. Electromagnetic field components E_x, E_y, E_z and H_x, H_y, H_z correspond to the node voltages of the equivalent circuit V_x, V_y, V_z and V_x^*, V_y^*, V_z^*, respectively, at each point (the * denoting the variables at the magnetic nodes). Electromagnetic field analysis results in the calculation of this equivalent circuit response. Fig. 3 shows the equivalent circuit at the E_x node. The following equations describe the node voltages and branch currents at the E_x node.

$$V_x(i,j,k,t)+Z_0I_{yi}(i,j,k,t) = I_{yo}^*\left(i, j-\frac{1}{2}, k, t-\Delta t\right) + Z_0V_z^*\left(i, j-\frac{1}{2}, k, t-\Delta t\right) \tag{1}$$

$$V_x(i,j,k,t)-Z_0I_{yo}(i,j,k,t) = I_{yi}^*\left(i, j+\frac{1}{2}, k, t-\Delta t\right) - Z_0V_z^*\left(i, j+\frac{1}{2}, k, t-\Delta t\right) \tag{2}$$

$$V_x(i,j,k,t)+Z_0I_{zi}(i,j,k,t) = I_{zo}^*\left(i, j, k-\frac{1}{2}, t-\Delta t\right) + Z_0V_y^*\left(i, j, k-\frac{1}{2}, t-\Delta t\right) \tag{3}$$

$$V_x(i,j,k,t)-Z_0I_{zo}(i,j,k,t) = I_{zi}^*\left(i, j, k-\frac{1}{2}, t-\Delta t\right) - Z_0V_y^*\left(i, j, k-\frac{1}{2}, t-\Delta t\right) \tag{4}$$

$$V_x(i,j,k,t)-\frac{\Delta t}{4C(i,j,k)}I_c(i,j,k,t) = V_x(i,j,k,t-\Delta t) + \frac{\Delta t}{4C(i,j,k)}I_c(i,j,k,t-\Delta t) \tag{5}$$

$$I_{yi}(i,j,k,t)-I_{yo}(i,j,k,t) + I_{zi}(i,j,k,t)-I_{zo}(i,j,k,t)-I_c(i,j,k,t)=0 \tag{6}$$

where Z_0 is the characteristic impedance of the transmission line and C is given by

$$C=(\epsilon_s-1)\epsilon_0\Delta \tag{7}$$

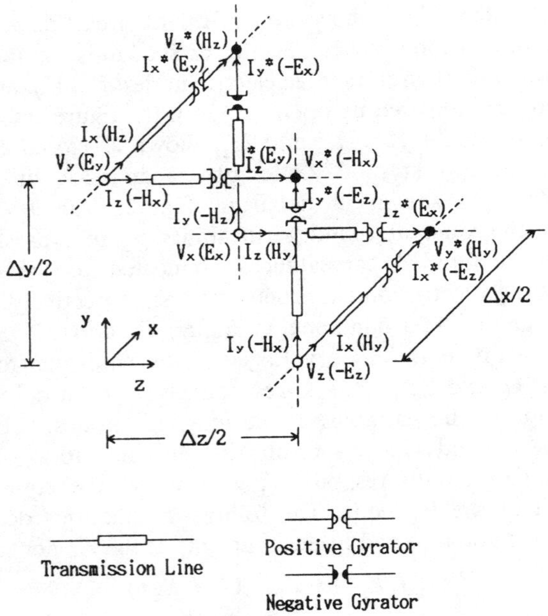

Fig. 2. Equivalent circuit elements. The equivalent circuit consists of finite length transmission lines and gyrators. $V-I$ characteristics of gyrators are

$$\begin{bmatrix} V \\ I \end{bmatrix} = \begin{bmatrix} 0 & 1 \\ 1 & 0 \end{bmatrix} \begin{bmatrix} V^* \\ I^* \end{bmatrix}$$

for the positive gyrator and

$$\begin{bmatrix} V \\ I \end{bmatrix} = \begin{bmatrix} 0 & -1 \\ -1 & 0 \end{bmatrix} \begin{bmatrix} V^* \\ I^* \end{bmatrix}$$

for the negative gyrator. Correspondence of circuit variables to electromagnetic field components is also shown in this figure.

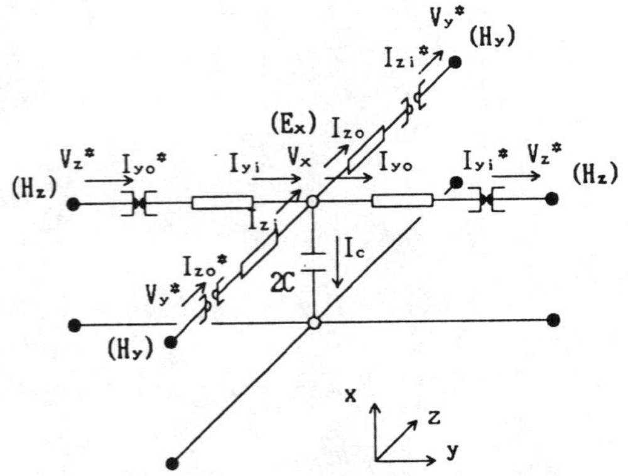

Fig. 3. Equivalent circuit at the E_x node.

at the electric nodes and

$$C = (\mu_s - 1)\mu_0 \Delta \tag{8}$$

at the magnetic nodes, where $\epsilon = \epsilon_s \epsilon_0$ and $\mu = \mu_s \mu_0$. The simulation time step Δt is related to the space distance Δ as

$$\Delta t = \frac{\Delta}{4c}. \tag{9}$$

We denoted the variables at the space point $x = i$, $y = j$, $z = k$ and at the time t as $V(i, j, k, t)$. Similar equations can be written relating to the E_y, E_z, H_x, H_y and H_z

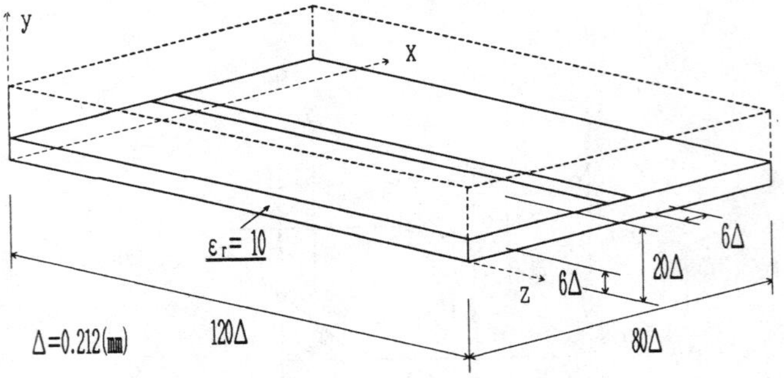

Fig. 4. Configuration of the single microstrip line.

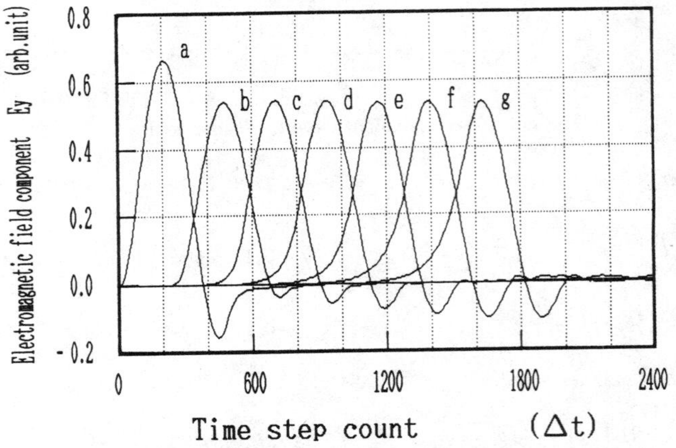

Fig. 5. The time variations of the electromagnetic field component E_y. The waves (a)–(g) are at the point $z = 0$, 20Δ, 40Δ, 60Δ, 80Δ, 100Δ, and 120Δ along the center line of the microstrip conductor in the substrate, respectively.

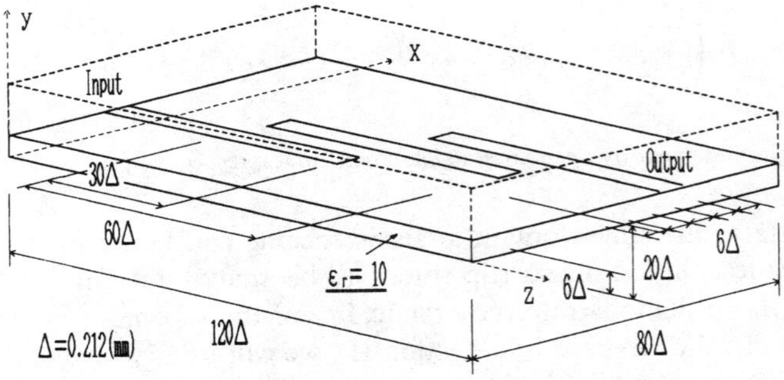

Fig. 6. Configuration of the side-coupled microstrip filter.

nodes. From these equations, we can calculate the node voltages at t from $t - \Delta t$.

III. Analysis of a Single Microstrip Line

Prior to the description of a side-coupled microstrip filter, we discuss the results of a single microstrip line analysis. The configuration of the analyzed microstrip line is shown in Fig. 4. The net size of the equivalent circuit used here is $80\Delta x \times 20\Delta y \times 120\Delta z$ $(\Delta x = \Delta y = \Delta z = \Delta)$. At the top and side faces of the equivalent circuit, irregular nodes are ended with matching impedances Z_0; thus, the infinite boundary condition is almost realized. The micro-

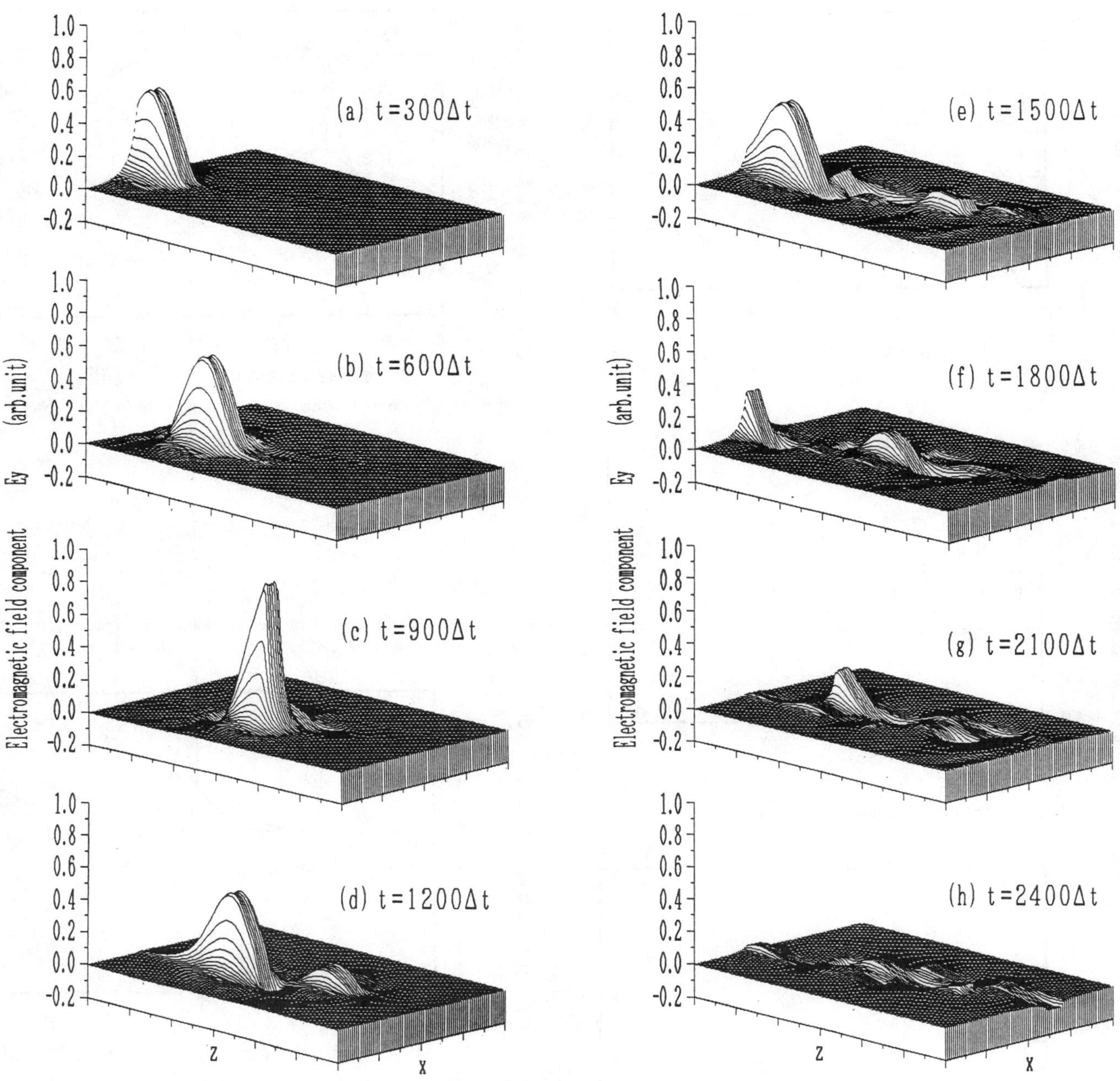

Fig. 7. The time variation of electromagnetic field component E_y at the $z-x$ plane in the substrate.

strip line parameters are as follows:

width of the microstrip conductor: $W = 6\Delta$;
thickness of the substrate: $H = 6\Delta$;
dielectric constant of the substrate: $\epsilon_r = 10.0$;
length of each transmission line in the equivalent circuit: $\Delta/2 = 0.106$ (mm).

Then, the simulation time step Δt becomes 0.176 (ps) from (9). The boundary conditions of the microstrip conductor and the bottom plane of the substrate are realized by short-circuiting the E_x and E_z nodes and open-circuiting the H_y node on the plane. As an input signal, a raised cosine pulse which equals a band-limited wave was adopted of the form

$$\begin{aligned} E_y(t) &= 1 - \cos(2\pi f_{\text{band}} t) \qquad \text{for } 0 \leqslant t < 1/f_{\text{band}} \\ &= 0 \qquad \text{for } 1/f_{\text{band}} \leqslant t \\ f_{\text{band}} &= 11.8 \text{ (GHz)}. \end{aligned} \tag{10}$$

Here, an internal impedance of Z_0 is applied to the E_y nodes under the microstrip conductor at the $z = 0$ plane, and initial values of the other field components are made zero. We applied only the E_y component under the microstrip line conductor to simplify the procedure instead of the total external electromagnetic field. The time variations of the electromagnetic field component E_y at the points under the center line of the microstrip conductor are plotted in Fig. 5.

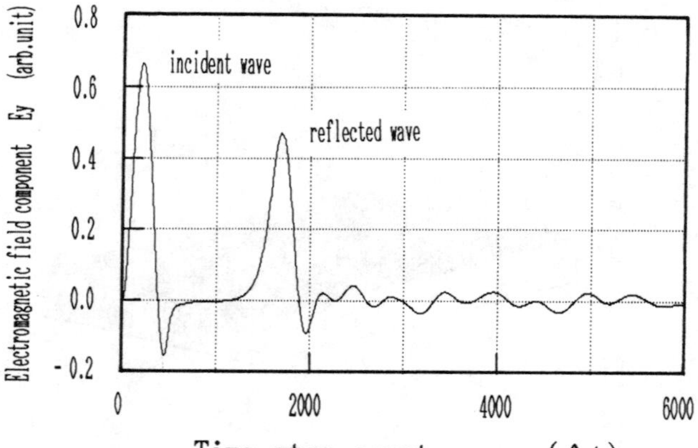

Fig. 8. The time variation of E_y at the input end.

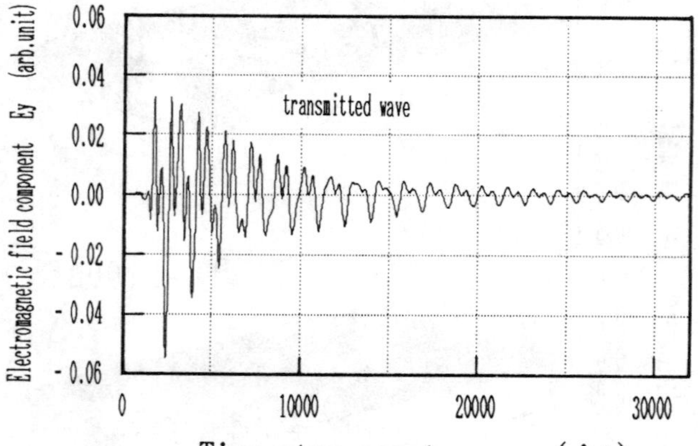

Fig. 9. The time variation of E_y at the output end.

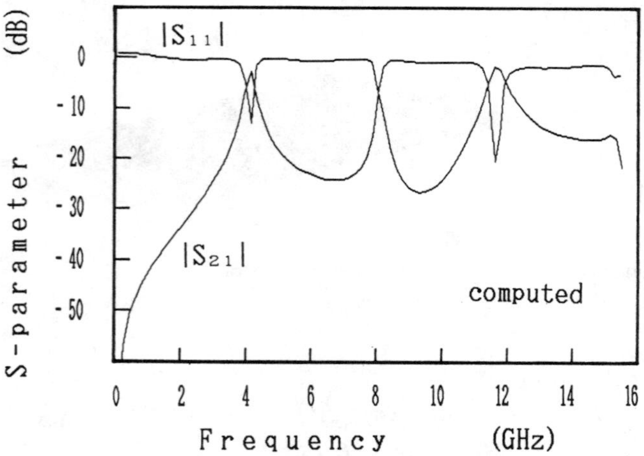

Fig. 10. Computed frequency characteristics of $|S_{21}|$ and $|S_{11}|$.

As a result of simplification of the input signal, the abrupt pulse height change from wave (a) to (b) has been reduced. The initial field component E_y was spread in space, thus creating a fringing field. The rest of the component was scattered and lost as radiation. The pulse height decreases rapidly near the input end and settles after propagating some distance (a few Δ lengths) along the microstrip in the z direction. At that point, the electromagnetic field distribution of the propagating mode is realized. We will take account of this pulse height reduction to calculate the frequency characteristics of the circuit in the next section.

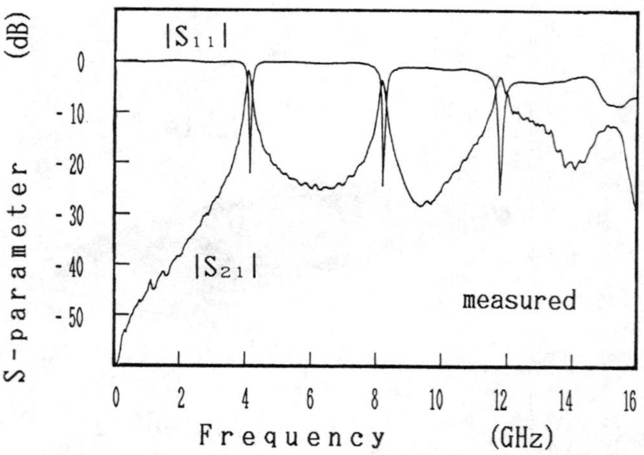

Fig. 11. Measured frequency characteristics of $|S_{21}|$ and $|S_{11}|$.

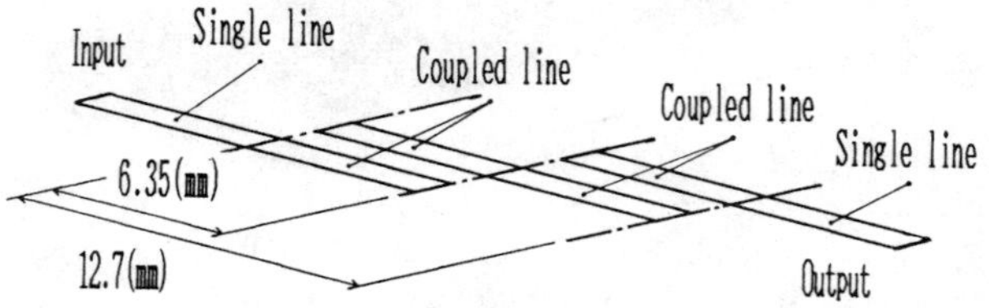

Fig. 12. Quasi-static analysis of the side-coupled microstrip filter. The filter is divided into four parts consisting of single and coupled lines.

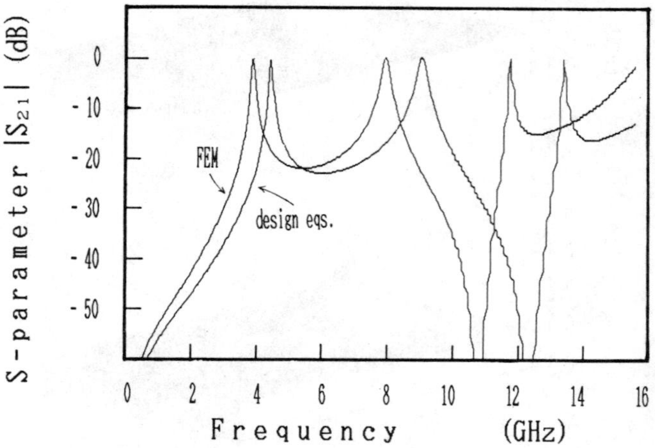

Fig. 13. The frequency characteristics of $|S_{21}|$ obtained by quasi-static analysis.

Notice also that the dispersion of the microstrip line can be simulated, which smoothes the slope of the pulse front and ripples the pulse tail because low-frequency components propagate faster than high-frequency components. Compare wave (g) with wave (b) in Fig. 5, for example.

IV. Analysis of a Side-Coupled Microstrip Filter

Here we describe the results of a side-coupled microstrip filter analysis. The configuration of the side-coupled microstrip filter is illustrated in Fig. 6. The width of the

TABLE I
PARAMETERS USED IN THE QUASI-STATIC ANALYSIS

	Single line		Coupled line			
	capacitance [pF/m]	inductance [nH/m]	self capacitance [pF/m]	mutual capacitance [pF/m]	self inductance [nH/m]	mutual inductance [nH/m]
Design equations	178.	422.	173.	11.0	420.	80.9
FEM	176.	469.	163.	11.8	572.	114.

microstrip line and the thickness and dielectric constant of the substrate are the same as the single microstrip line in the previous section. Other important size parameters are

space between microstrip conductors: $S = 6\Delta$;
length of the center microstrip conductor: $L = 60\Delta$.

The input signal given by (10) is applied at the input end and the electromagnetic field is computed step by step in time. The time variation of the electromagnetic field component E_y at the $z-x$ plane in the substrate is drawn in Fig. 7, in which (a)–(c) show that the incident wave propagates along the first microstrip conductor and is reflected at the end of the line. A part of the wave transfers to the center microstrip, while the reflected wave returns to the input end, in (d)–(e). The reflected wave fades away from view and the remainder resonates around the center microstrip in (f)–(g). The resonant component transfers to the third microstrip line and is transmitted to the output end in (h).

Figs. 8 and 9 are plots of E_y at input and output ends, respectively. From these waves, the frequency characteristics of the filter can be calculated. S parameters can be calculated from the Fourier transform of the reflected wave in Fig. 8 and the transmitted wave in Fig. 9, normalizing by the frequency spectrum of the incident wave (b) in Fig. 5 to correct the effect of the pulse height reduction discussed previously. Computed S parameters $|S_{11}|$ and $|S_{21}|$ are shown in Fig. 10.

In order to verify this computational method, we fabricated the side-coupled microstrip filter and measured $|S_{11}|$ and $|S_{21}|$ using the network analyzer. The results are shown in Fig. 11. The computed results show excellent agreement with the measured data.

V. THE RESULTS OF QUASI-STATIC ANALYSIS

In this section we describe the quasi-static analysis of the side-coupled microstrip filter which was previously analyzed using the time-domain, full-wave analysis. Then the findings based on the two different methods will be compared. Here the side-coupled microstrip filter is divided into four parts, consisting of single and coupled microstrip lines, as shown in Fig. 12. Single and coupled line parameters for a unit length can be estimated from static line capacitances on the assumption of TEM mode propagation. We calculated two sets of line capacitances from the design equations in [17] and the static field analysis using the finite element method (FEM) in [18]. The line inductance of the single line is calculated by

$$L = \frac{1}{c^2 C_{\text{air}}}. \tag{11}$$

The self- and mutual inductances of the coupled line are given by

$$L_{\text{self}} = \frac{1}{2c^2}\left(\frac{1}{C_{\text{self,air}} - C_{\text{mutual,air}}} + \frac{1}{C_{\text{self,air}} + C_{\text{mutual,air}}}\right) \tag{12}$$

$$L_{\text{mutual}} = \frac{1}{2c^2}\left(\frac{1}{C_{\text{self,air}} - C_{\text{mutual,air}}} - \frac{1}{C_{\text{self,air}} + C_{\text{mutual,air}}}\right) \tag{13}$$

where c is the velocity of an electromagnetic wave in free space, and "air" means the capacitance of the line configuration with the dielectric substrate replaced by air. Calculated parameters are summarized in Table I. We obtained the $|S_{21}|$ of the side-coupled microstrip filter by circuit simulation using SPICE. The results are shown in Fig. 13. The differences between the inductances calculated by design equations and those by our numerical calculation using FEM range from 10 to 30 percent. However, both of the $|S_{21}|$ in Fig. 13 show the same tendency except for the scale of frequencies. Also important to note is that the results of quasi-static analysis are in disagreement with measured data, especially in the frequency domain over 8 GHz. These disagreements stem from the assumption of TEM mode propagation and/or from the separation of the filter into single and coupled lines. Thus, three-dimensional, full-wave analysis is needed for more precise computation.

VI. DISCUSSION

We turn now to a consideration of the accuracy of frequency characteristic calculations using time-domain, full-wave analysis. Comparing the computed $|S_{11}|$ in Fig. 10 to the measured $|S_{11}|$ in Fig. 11, some difference in the peak level at the resonance point can be observed, resulting from the lack of frequency resolution. According to Fourier transform theory, frequency resolution Δf is given by the inverse of the observed time period T. On the other hand the time step Δt depends on the branch length $\Delta/2$, as shown in (9). Moreover, the length of Δ should be made small enough to express not only the shapes and locations of the microstrips but also the electromagnetic field distribution using the network variables. Our simulations using other discretizations indicated that the accuracy of $|S_{21}|$, especially the resonance peak level around 8 GHz, depends on how many Δ's there are between microstrip lines. We need at least 5Δ lengths between isolated conductors. For this reason, it is important to make Δ sufficiently small, although it seems wasteful or oversampled to make the computational time step Δt too small and to require many iterations in time merely to obtain the frequency characteristics up to 16 GHz. To obtain the frequency characteristics shown in Fig. 10, we have computed the electromagnetic field distribution up to the $32768\Delta t$ time step. The resulting frequency resolution of Δf becomes 173 MHz.

VII. Conclusions

We propose an analysis method using pulse excitation in the time-domain, equivalent to a full-wave electromagnetic field analysis. The characteristics of a side-coupled microstrip filter were computed and excellent agreement with the measured values was obtained, thus confirming the validity of this analytical scheme. This example essentially requires the three-dimensional electromagnetic field treatment since no other method can provide the same level of precision. Moreover, the Bergeron method is effective for various media and is expected to produce various kinds of analysis, such as full-wave analysis in lossy conductors and dielectrics. Thus, the techniques described here are also promising for microstrip circuit analysis in the case of low conductivity of lines and/or semiconducting substrate.

Acknowledgment

The authors would like to thank Professor I. Fukai and Associate Professor N. Yoshida of Hokkaido University for their helpful discussions and A. Iwata, NTT LSI Laboratories, for his encouragement throughout this work.

References

[1] K. S. Yee, "Numerical solution of initial boundary value problems involving Maxwell's equations in isotropic media," *IEEE Trans. Antennas Propagat.*, vol. AP-14, pp. 302–307, May 1966.

[2] G. Kron, "Equivalent circuit of the field equations of Maxwell-I," *Proc. I.R.E.*, pp. 289–299, May 1944.

[3] P. B. Johns and R. L. Beurle, "Numerical solution of 2-dimensional scattering problems using a transmission-line matrix," *Proc. Inst. Elec. Eng.*, vol. 118, pp. 1203–1208, Sept. 1971.

[4] P. B. Johns, "The solution of inhomogeneous waveguide problems using a transmission-line matrix," *IEEE Trans. Microwave Theory Tech.*, vol. MTT-22, pp. 209–215, Mar. 1974.

[5] S. Akhtarzad and P. B. Johns, "Solution of Maxwell's equations in three space dimensions and time by t.l.m. method of numerical analysis," *Proc. Inst. Elec. Eng.*, vol. 122, pp. 1344–1348, Dec. 1975.

[6] S. Akhtarzad and P. B. Johns, "Generalised elements for t.l.m. method of numerical analysis," *Proc. Inst. Elec. Eng.*, vol. 122, pp. 1349–1352, Dec. 1975.

[7] P. B. Johns and G. Butler, "The consistency and accuracy of the TLM method for diffusion and its relationship to existing methods," *Int. J. Numer. Methods Eng.*, vol. 19, pp. 1549–1554, 1893.

[8] Y. Shih and W. J. R. Hoefer, "The accuracy of TLM analysis of finned rectangular waveguides," *IEEE Trans. Microwave Theory Tech.*, vol. MTT-28, pp. 743–746, July 1980.

[9] D. A. Al-Mukhtar and J. E. Sitch, "Transmission-line matrix method with irregularly graded space," *Proc. Inst. Elec. Eng.*, vol. 128, pp. 299–305, Dec. 1981.

[10] G. E. Mariki and C. Yeh, "Dynamic three-dimensional TLM analysis of microstrip- lines on anisotropic substrate," *IEEE Trans. Microwave Theory Tech.*, vol. MTT-33, pp. 789–799, Sept. 1985.

[11] W. J. R. Hoefer, "The transmission-line matrix method—Theory and applications," *IEEE Trans. Microwave Theory Tech.*, vol. MTT-33, pp. 882–893, Oct. 1985.

[12] N. Yoshida and I. Fukai, "Transient analysis of a stripline having a corner in three-dimensional space," *IEEE Trans. Microwave Theory Tech.*, vol. MTT-32, pp. 491–498, May 1984.

[13] S. Koike, N. Yoshida, and I. Fukai, "Transient analysis of microstrip gap in three-dimensional space," *IEEE Trans. Microwave Theory Tech.*, vol. MTT-33, pp. 726–730, Aug. 1985.

[14] S. Koike, N. Yoshida, and I. Fukai, "Transient analysis of a directional coupler using a coupled microstrip slot line in three-dimensional space," *IEEE Trans. Microwave Theory Tech.*, vol. MTT-34, pp. 353–357, Mar. 1986.

[15] S. Koike, N. Yoshida, and I. Fukai, "Transient analysis of microstrip side-coupled filter in three-dimensional space," *Trans. IECE Japan*, vol. E69, no. 11, pp. 1199–1205, Nov. 1986.

[16] S. Koike, N. Yoshida, and I. Fukai, "Transient analysis of coupling between crossing lines in three-dimensional space," *IEEE Trans. Microwave Theory Tech.*, vol. MTT-35, pp. 67–71, Jan. 1987.

[17] K. C. Gupta, R. Garg, and I. J. Bahl, *Microstrip Lines and Slotlines*. Dedham, MA: Artech House, 1979, pp. 87–89 and 337–340.

[18] P. Silvester, "High-order polynomial triangular finite elements for potential problems," *Int. J. Eng. Sci.*, vol. 7, pp. 849–861, 1969.

Paper 6.7

The Transfinite Element Method for Modeling MMIC Devices

ZOLTAN J. CENDES, MEMBER, IEEE, AND JIN-FA LEE

Abstract —A new numerical procedure called the transfinite element method is employed in conjunction with the planar waveguide model to analyze MMIC devices. By using analytic basis functions together with finite element approximation functions in a variational technique, the transfinite element method is able to determine the fields and scattering parameters for a wide variety of stripline and microstrip devices. With minor modification, the transfinite element method can also be applied to waveguide junctions. We show that the transfinite element method can be used to treat singular points in waveguide junctions very efficiently. Examples that have been calculated by this method are a rectangular waveguide two-slot-20 dB coupler, stripline band-elimination filter, and several microstrip discontinuity problems. Good agreement of the numerical results with published values demonstrates the validity of the proposed procedure.

I. Introduction

To BE USEFUL at high frequencies, models of MMIC devices must solve the wave equations first derived by Maxwell. There are two ways to do this. One way is to solve the full vector wave equations in three-dimensions; the other is to employ the planar circuit model that approximates the fringe fields and hybrid modes of the device but maintains its essential wave and dispersion characteristics. This paper presents a new procedure to solve the second of these alternatives and shows that accurate results are obtained for several typical MMIC devices.

In the planar waveguide model, the scalar Helmholtz equation is solved in two-dimensions for the electromagnetic field distribution. The procedure is as follows [1]–[3]:

1) Approximate the actual three-dimensional MMIC device with an equivalent N-port planar waveguide model.
2) Solve for the electromagnetic fields and scattering matrix coefficients in the equivalent planar waveguide model.

We propose here a new method for the second of these steps that is considerably more efficient and more general than the existing alternatives. The method is based on the transfinite element procedure first proposed by the authors for the solution of unbounded electrostatics problems [4] and later extended to the solution of electromagnetic scattering problems [5]. Unlike the eigensolution procedure reported in [2] that requires that a set of orthonormalized eigenmodes be determined for the planar waveguide, the transfinite element method is deterministic and hence is much more efficient. And, unlike the finite difference time-domain method of [3], the transfinite element method reported here is time harmonic and thus eliminates the need for expensive numerical time integration.

Manuscript received April 14, 1988; revised August 1, 1988. This work was supported in part by the Pennsylvania Ben Franklin Partnership Fund and by the Ansoft Corporation.

The authors are with the Department of Electrical and Computer Engineering, Carnegie Mellon University, Pittsburgh, PA 15213.

IEEE Log Number 8824168.

II. The Planar Waveguide Model

Throughout this paper we will refer to "the equivalent planar waveguide model." In this model the actual three-dimensional MMIC device is transformed into a planar circuit that can be solved by two-dimensional analysis. This is accomplished by replacing the actual dimensions and material properties of the MMIC device with effective dimensions and material properties for an equivalent planar waveguide. This operation is different with striplines than it is with microstrip:

1) In stripline circuits, the dominant propagating mode is TEM, for which effective dimensions are easily calculated by using quasi-static analysis or by using the empirical formula of [1].
2) In microstrip, the dominant mode is non-TEM and the field pattern thus varies with frequency. In the low-frequency limit, the TEM approximation can be used to construct an equivalent planar waveguide model. Formulas that model the frequency dependence of the effective parameters may then be used for higher frequencies.

In this paper, the frequency-dependent effective dielectric constant in the microstrip circuits is given by [10]

$$\epsilon_{re}(f) = \epsilon_r - \frac{\epsilon_r - \epsilon_{re}(0)}{1+P}$$

$$P = \left(\frac{h}{Z_{0m}}\right)^{1.33} \left[0.43f^2 - 0.009f^3\right] \quad (1)$$

where ϵ_r is the true relative dielectric constant, h is the height of the substrate in millimeters, f is the frequency in GHz, and the characteristic impedance Z_{0m} is in ohms. The frequency-dependent effective width is modeled

Reprinted from *IEEE Trans. Microwave Theory Tech.*, vol. 36, no. 12, pp. 1639–1649, Dec. 1988.

as [10]

$$W_e(f) = W + \frac{W_e(0) - W}{1 + f/f_g}$$

$$f_g = \frac{c}{2W\sqrt{\epsilon_r}} \qquad (2)$$

where W is the true physical width of the microstrip. The values of the $\epsilon_{re}(0)$ and $W_e(0)$ are calculated by using quasi-static analysis [10]. The accuracy and frequency range of (1), (2) are described fully in [10]. These formulas work well provided that the width of the strip in the discontinuity region is easily determined. However, in some problems the width of the strip is ambiguous and reduces the correctness of this approximation.

III. The Transfinite Element Method

A. The Functional

Fig. 1 shows an equivalent planar waveguide model for a typical MMIC device. Since the fringe field is taken into account by using effective dimensions and material properties, we can write the equation for the component E_z of the electric field perpendicular to the plane of the conductor as

$$\nabla^2 E_z + k^2 \epsilon_{re} E_z = 0 \quad \text{in } \Omega$$
$$\frac{\partial E_z}{\partial n} = 0 \quad \text{on } \partial\Omega \qquad (3)$$

where Ω is the effective problem domain, $\partial\Omega$ is the boundary on the sides, and ∂n is the normal derivative.

A functional corresponding to (3) is obtained by applying Galerkin's method. The result is

$$F(E_z) = -\int_{\Omega} E_z^*(\nabla^2 E_z + k^2 \epsilon_{re} E_z)\, d\Omega \qquad (4)$$

where * represents the complex conjugate. We need to separate the solution region into two parts: let Ω_d represent the discontinuity region of the planar circuit, and Ω_i the semi-infinite ports. Equation (4) then becomes

$$F(E_z) = -\int_{\Omega_d} E_z^*(\nabla^2 E_z + k^2 \epsilon_{re} E_z)\, d\Omega - \sum_{i=1}^{P} \int_{\Omega_i} E_z^*(\nabla^2 E_z + k^2 \epsilon_{re} E_z)\, d\Omega \qquad (5)$$

where P is the number of ports. Now apply Green's theorem to the first of these integrals. This gives

$$F(E_z) = \int_{\Omega_d} (\nabla E_z^* \cdot \nabla E_z - k^2 \epsilon_{re} E_z^* E_z)\, d\Omega - \sum_{i=1}^{P} \oint_{\Gamma_i} E_z^* \frac{\partial E_z}{\partial n}\, d\Gamma - \sum_{i=1}^{P} \int_{\Omega_i} E_z^*(\nabla^2 E_z + k^2 \epsilon_{re} E_z)\, d\Omega. \qquad (6)$$

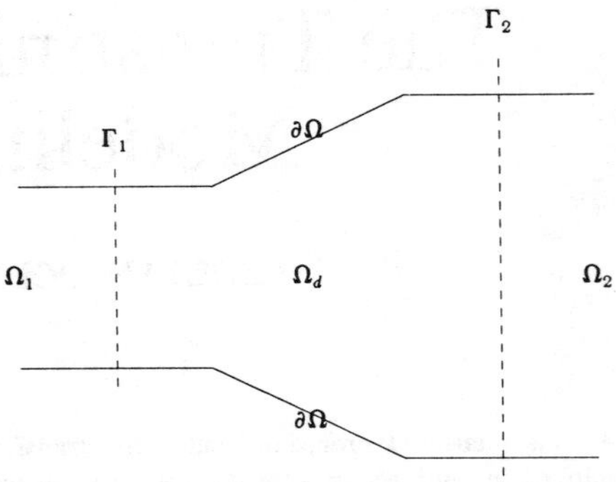

Fig. 1. A two-port planar waveguide junction.

The reason for writing the functional in this form is that the boundary integral in (6) provides the natural boundary conditions for the solution space.

B. The Solution Space

Assume that port 1 is excited by the dominant TEM mode. By modal analysis, the z component of the electric field within port i can be written as

$$E^{(i)} = \delta_{i1} \Phi_{\text{inc}} + \sum_{j=0}^{\infty} a_{ij} \Phi_{ij} \qquad (7)$$

where δ_{i1} is the Kronecker delta

$$\delta_{i1} = \begin{cases} 1 & \text{if } i = 1 \text{ (input port)} \\ 0 & \text{otherwise} \end{cases} \qquad (8)$$

the a_{ij} are unknown coefficients, and fields Φ_{inc} and Φ_{ij} are given as

$$\Phi_{\text{inc}} = \exp(\gamma_{10} y)$$
$$\Phi_{ij} = \cos\left(\frac{j\pi x}{W_i}\right) \exp(-\gamma_{ij} y). \qquad (9)$$

In these equations, W_i is the effective width of port i, and ϵ_{ri} is the effective dielectric constant of port i, and the propagation constant is

$$\gamma_{ij} = \sqrt{\left(\frac{j\pi}{W_i}\right)^2 - k^2 \epsilon_{ri}}. \qquad (10)$$

The local coordinate in port i is defined such that the $+\hat{y}$ direction is the direction of propagation of the scattered wave. The origin of the coordinate is located on the port boundary Γ_i.

The solution space is now taken to be the analytic basis functions (9) for the port regions Ω_i and finite element basis functions for the discontinuity region Ω_d. By separating the finite element nodes in Ω_d into two parts—interior nodes ϕ^I and P sets of boundary nodes ϕ^{Γ_i}, we obtain the following hybrid solution space:

$$\Lambda \equiv \left\{ \phi | \phi = \left\langle \begin{array}{ll} \tilde{\alpha}^I \phi^I + \sum_{i=1}^{P} \tilde{\alpha}^{\Gamma_i} \phi^{\Gamma_i} & \text{in } \Omega_d \\ \delta_{i1}\Phi_{\text{inc}} + \sum_{j=0}^{M} a_{ij}\Phi_{ij} & \text{in } \Omega_i \end{array} \right. \quad \phi \in C^0 \right\} \tag{11}$$

where the α are the Lagrangian interpolation polynomials, $\tilde{}$ denotes a row vector, $\underset{\sim}{}$ denotes a column vector, M is the number of modes in each port, and C^0 is the set of continuous functions. Notice that we have used two different kinds of basis functions in different regions, and also that since the function ϕ must be continuous, these two representations must be matched along the port reference planes.

Since the analytic basis functions satisfy the Helmholtz equation in the port regions, the integral over Ω_i in (6) is zero, independent of the coefficients a_{ij}. Thus the functional becomes

$$F(E_z) = \int_{\Omega_d} \left(\nabla E_z^* \cdot \nabla E_z - k^2 \epsilon_{re} E_z^* E_z \right) d\Omega - \sum_{i=1}^{P} \oint_{\Gamma_i} E_z^* \frac{\partial E_z}{\partial n} d\Gamma. \tag{12}$$

Continuity of the electric field is imposed by requiring that the field approximations in Ω_d and in Ω_i be identical at the finite element nodes along the port boundary Γ_i. This condition may be expressed as

$$\underset{\sim}{\phi}^{\Gamma_i} = \delta_{i1} \underset{\sim}{P}_{\text{inc}} + [P_i] \underset{\sim}{a}_i \tag{13}$$

where

$$\underset{\sim}{P}_{\text{inc}} = [1 \quad 1 \quad \cdots \quad 1]^T$$

$$[P_i] = [\underset{\sim}{P}_{i0} \quad \underset{\sim}{P}_{i1} \quad \cdots \quad \underset{\sim}{P}_{iM}]$$

$$\underset{\sim}{P}_{ij} = \left[\cos\left(\frac{j\pi x_1}{W_i}\right) \quad \cos\left(\frac{j\pi x_2}{W_i}\right) \quad \cdots \quad \cos\left(\frac{j\pi x_m}{W_i}\right) \right]^T$$

$$\underset{\sim}{a}_i = [a_{i0} \quad a_{i1} \quad \cdots \quad a_{iM}]^T. \tag{14}$$

Here m is the number of nodes on Γ_i, ^{T} represents the transpose, and x_i is the coordinate of node i.

Note that continuity of the derivative of the electric field is automatically provided by the natural boundary conditions of the functional (6).

C. Extremizing the Functional

By requiring that the trial functions E_z be in the solution space Λ, the boundary integral in (12) can be integrated analytically. The result is

$$\oint_{\Gamma_i} E_z^* \frac{\partial E_z}{\partial n} d\Gamma = \delta_{i1}\gamma_{10} a_{10}^* W_1 - \tilde{a}_i^* [\gamma_i] \underset{\sim}{a}_i \tag{15}$$

where

$$[\gamma_i]_{jk} = \begin{cases} 0 & j \neq k \\ W_i \gamma_{i0} & j = k = 0 \\ W_i \gamma_{ij}/2 & j = k \neq 0. \end{cases} \tag{16}$$

Finally, substituting (13)–(16) into (12), the functional can then be expressed in matrix form as

$$F(E_z) = \left[\tilde{\phi}^I \tilde{P}_{\text{inc}} + \tilde{a}_1 [P_1]^T \tilde{a}_2 [P_2]^T \right]^* \cdot \begin{bmatrix} [S_{II}] & [S_{I\Gamma_1}] & [S_{I\Gamma_2}] \\ [S_{\Gamma_1 I}] & [S_{\Gamma_1\Gamma_1}] & [S_{\Gamma_1\Gamma_2}] \\ [S_{\Gamma_2 I}] & [S_{\Gamma_2\Gamma_1}] & [S_{\Gamma_2\Gamma_2}] \end{bmatrix} \begin{bmatrix} \underset{\sim}{\phi}^I \\ \underset{\sim}{P}_{\text{inc}} + [P_1]\underset{\sim}{a}_1 \\ [P_2]\underset{\sim}{a}_2 \end{bmatrix} - a_{10}^* W_1 \gamma_{10} + \sum_{i=1}^{P} \tilde{a}_i^* [\gamma_i] \underset{\sim}{a}_i$$

where

$$[S_{ij}] = \int_{\Omega_d} \left(\nabla \tilde{\alpha}^i \cdot \nabla \underset{\sim}{\alpha}^j - k^2 \epsilon_{re} \tilde{\alpha}^i \underset{\sim}{\alpha}^j \right) d\Omega. \tag{17}$$

This equation is shown for clarity with only two ports; if there are more ports, then additional terms similar to that of port 2 need to be added.

Extremizing this with respect to ϕ^{*I} and a_i^* gives the final matrix equation

$$\begin{bmatrix} [S_{II}] & [S_{I\Gamma_1}][P_1] & [S_{I\Gamma_2}][P_2] \\ [P_1]^T[S_{\Gamma_1 I}] & [P_1]^T[S_{\Gamma_1\Gamma_1}][P_1] + [\gamma_1] & [P_1]^T[S_{\Gamma_1\Gamma_2}][P_2] \\ [P_2]^T[S_{\Gamma_2 I}] & [P_2]^T[S_{\Gamma_2\Gamma_1}][P_1] & [P_2]^T[S_{\Gamma_2\Gamma_2}][P_2] + [\gamma_2] \end{bmatrix} \begin{bmatrix} \underset{\sim}{\phi}^I \\ \underset{\sim}{a}_1 \\ \underset{\sim}{a}_2 \end{bmatrix} = \begin{bmatrix} [S_{I\Gamma_1}]\underset{\sim}{P}_{\text{inc}} \\ [P_1]^T[S_{\Gamma_1\Gamma_1}]\underset{\sim}{P}_{\text{inc}} + \gamma_{10} W_1 \underset{\sim}{\delta} \\ [P_2]^T[S_{\Gamma_2\Gamma_1}]\underset{\sim}{P}_{\text{inc}} \end{bmatrix} \tag{18}$$

where

$$\underset{\sim}{\delta} = [1 \quad 0 \quad 0 \quad \cdots \quad 0]^T.$$

Notice that since the $[S_{ij}]$ submatrices are sparse and the number of modes M required in the formulation is small, the matrix multiplications in (18) can be done very efficiently. The finally matrix equation is sparse and symmetric and can be solved by using the preconditioned conjugate gradient method (PCCG) [12].

D. The Scattering Matrix

The scattering matrix for two-port circuits is obtained in two steps

1) Take port 1 to be the input port and solve (18). Since the incident wave is assumed to have unit amplitude, the scattering coefficients are defined as

$$S_{11} = a_{10}$$
$$S_{12} = a_{20}\sqrt{\frac{\gamma_{20}W_2}{\gamma_{10}W_1}}. \qquad (19)$$

2) Change the input port to be port 2 and repeat the procedure in step 1. The scattering coefficients S_{21}, S_{22} can then be found again by using (19).

For an N-port circuit, the analysis needs to be performed N times to determine the $N \times N$ scattering matrix. From the solution of (18), one obtains not only the scattering matrix of the device but also the excitations of the higher modes on the port reference plane for each port. Chu and Itoh [11] have defined a generalized scattering matrix to characterize microstrip step discontinuities. With the present method, a generalized scattering matrix can be computed for general MMIC devices.

IV. Numerical Results for MMIC Components

A general purpose computer program has been developed to model MMIC devices using the transfinite element method [8]. To show the validity and generality of the method, we present numerical results for several MMIC devices together with the detailed descriptions of the planar waveguide model that has been used in the analysis.

A. Stripline Band-Elimination Filter

Shown in Fig. 2(a) is a two-port stripline filter with a circular disk. The characteristic impedance of both ports is 50 Ω, the substrate height is $2h = 0.64$ cm and the relative dielectric constant is 2.4. To solve this problem with the transfinite element method, we first convert to the equivalent planar waveguide model with the effective dimensions shown in Fig. 2(b). This geometry is then discretized by using triangular finite elements, and solved by means of the transfinite element method. The transmission coefficient computed by means of this procedure is plotted in Fig. 2(c). As shown, good agreement exists between these results and published experimental data [7]. It should be noted that the curve in Fig. 2(c) was produced by using the adaptive spectral response modeling procedure in [9]. The squares on the abscissa of this graph correspond to the frequencies actually employed in the computation.

We also like to point out that the zero in the transmission coefficient in Fig. 2(c) corresponds to the first resonance of the circular disk. The field intensity at the resonance frequency 2.976 GHz is plotted in Fig. 2(d) to provide more physical insight. It is apparent from the intensity plot that the transmission zero is caused by the orthogonality of the modal distribution and of the TEM field distribution at the output port at resonance.

B. Microstrip Step Discontinuity

Fig. 3(a) shows a microstrip step discontinuity with a substrate height $h = 0.635$ mm and substrate dielectric constant $\epsilon_r = 9.7$. The corresponding planar waveguide model at low frequency is shown in Fig. 3(b) together with the frequency-dependent effective parameters. Since the problem is symmetric, only half of the geometry is used to solve for the scattering coefficients. A comparison of the transmission coefficient computed by transfinite element method and by [14] is given in Fig. 3(c). This figure also gives results computed with the *generalized scattering matrix method* [11]. The discontinuity of S_{11} in the figure is due to the excitation of the second mode in the wider port.

C. Microstrip T-Junction

The microstrip T-junction shown in Fig. 4(a) has a substrate height $h = 0.65$ mm and a substrate dielectric constant $\epsilon_r = 10.1$ The quasi-static analysis to find the planar waveguide model at low frequency requires that we solve the Laplace equation twice; this performed in the following way:

- First, input the cross section of the microstrip line as shown in Fig. 4(b). Then create a finite element mesh as shown in Fig. 4(c) by using the process of the *Delaunay triangulation* [13].
- Assume that the top conductor carries constant current and solve for the magnetic vector potential distribution. The magnetic vector potential contours are shown in Fig. 4(d). From the stored energy of the system we can obtain the inductance $L = 0.3373\mu_0$, where μ_0 is the permeability of free space.
- Compute the capacitance by solving the potential distribution. The capacitance computed is $C = 20.322\epsilon_0$, where ϵ_0 is the permittivity of free space. The equal potential contours are shown in Fig. 4(e).
- The effective parameters at low frequency are given by

$$Z = \sqrt{\frac{L}{C}}$$
$$\epsilon_{re} = \frac{LC}{\epsilon_0\mu_0}$$
$$W_e = \frac{Ch}{\epsilon_{re}\epsilon_0}.$$

The characteristics thus computed are summarized as follows:

$$Z = 48.54\ \Omega$$
$$W_e'(0) = 1.927 \text{ mm}$$
$$\epsilon_{re}(0) = 6.855.$$

Also shown in Fig. 4(a) are the formulas which provide the frequency-dependent effective parameters.

The reflection and transmission coefficients computed from the transfinite element method are compared with the results by Mehran [15]. Good agreement is obtained, as

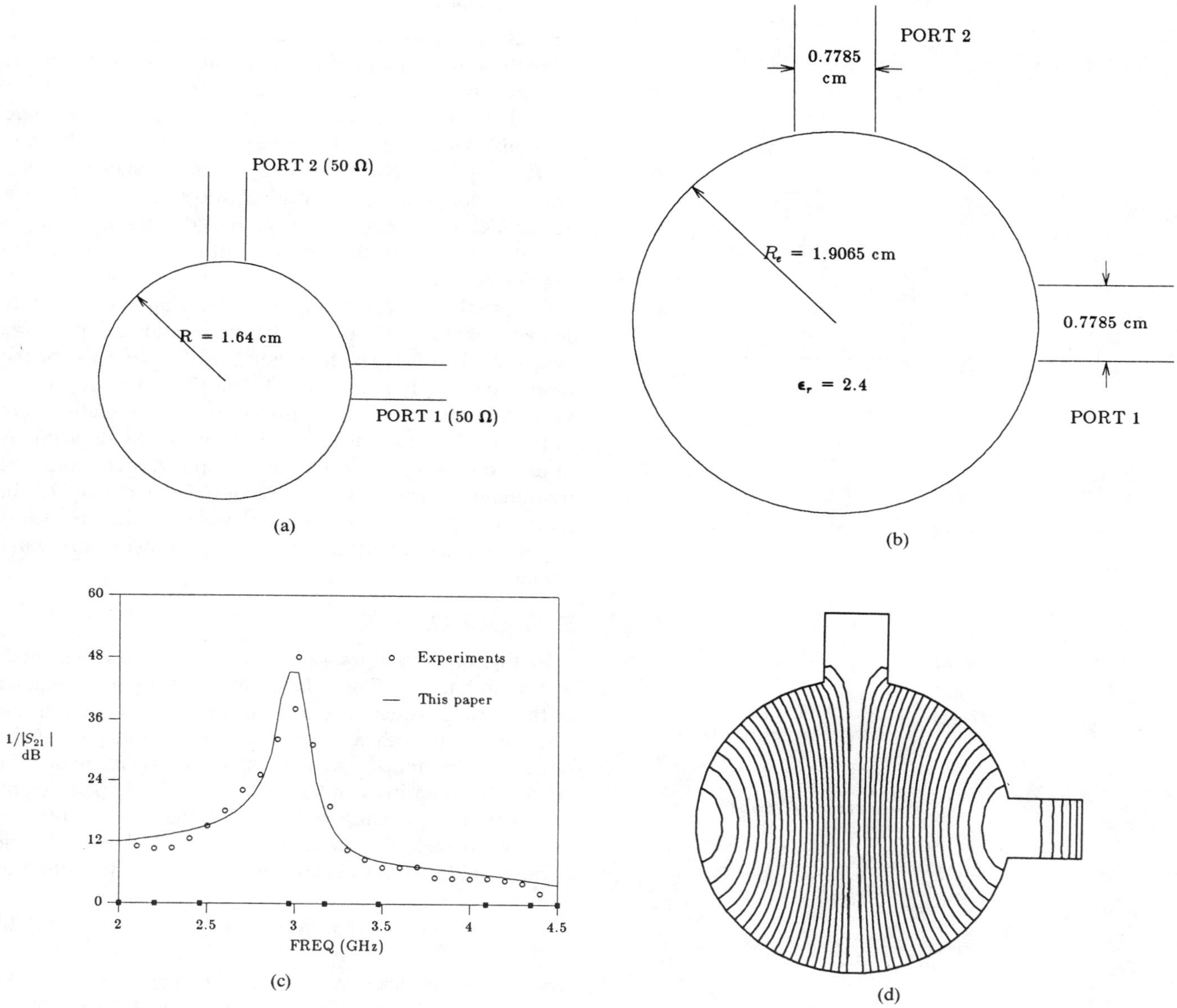

Fig. 2. (a) Geometry of a stripline disk band-elimination filter coupled to two 50 Ω striplines. (b) Planar waveguide model for the stripline disk in (a). (c) Calculation of the transmission coefficient for the band-elimination filter by the transfinite element method compared with experimental values. (d) Contours of equal field intensity for the band-elimination filter at the first transmission zero (2.976 GHz).

can be seen from Fig. 4(f). To understand the transmission zero at 33.7 GHz, we present a plot of the field intensity in Fig. 4(g). A TEM wave comes in from port 1 and excites the first resonance mode in the square region. Because of the TEM excitation in port 1, the polarization is a *cosine* distribution along ports 2 and 3, which are orthogonal to the TEM mode for output.

D. *Microstrip Radial Stubs*

Fig. 5(a) presents the geometry of a shunt-connected microstrip radial stub. This device has been analyzed by Giannini [16]; we used the formulas from [16] to construct the low-frequency planar waveguide model given in Fig. 5(b). Again, due to the symmetry only one-half of the geometry is modeled. The effective dielectric constant of the radial stub is obtained by using the filling factor for a circular disk capacitor as developed in [17]. Fig. 5(c) shows the finite element mesh with 342 unknowns that is used to obtain the results in Fig. 5(d).

The numerical results in Fig. 5(d) are computed using the model in Fig. 5(b) through the entire frequency range. This neglects the dispersion caused by the effective parameter changes in the microstrip structure. When compared to the measurements in [16], larger discrepancies at high frequencies are observed, although the results are still acceptable. Fig. 5(e) shows the plots of the real part of the field at 3, 7, 11, and 19 GHz. These plots again provide physical insight of the transmission zeros and total transmission in Fig. 5(d). The typical computation time to obtain the response at one frequency for Fig. 5(d) is less than 30 s on a Sun 3/110 workstation.

(a)

(b)

Parameter values :

$$Z_1 = 46.94\,\Omega \qquad\qquad Z_2 = 4.655\,\Omega$$

$$\epsilon_{r1}(f) = 9.7 - \frac{3.3657}{1.0 + 0.00327\left(0.43 f^2 - 0.009 f^3\right)}$$

$$\epsilon_{r2}(f) = 9.7 - \frac{0.768}{1.0 + 0.0707\left(0.43 f^2 - 0.009 f^3\right)}$$

$$W_1'(f) = 0.5 + \frac{1.524}{1.0 + \dfrac{f}{96.324}}$$

$$W_2'(f) = 15 + \frac{2.185}{1.0 + \dfrac{f}{3.211}}$$

(b)

(c)

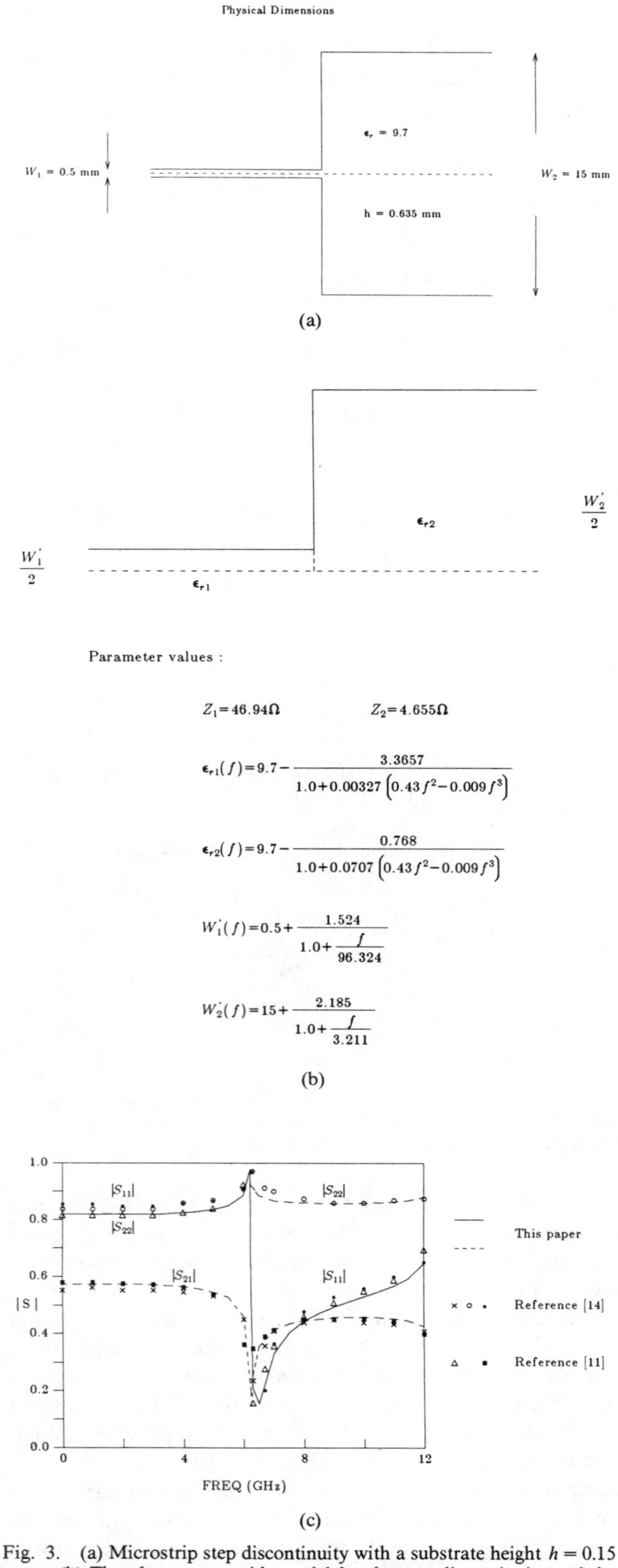

Fig. 3. (a) Microstrip step discontinuity with a substrate height $h = 0.15$ cm. (b) The planar waveguide model for the step discontinuity and the frequency-dependent effective parameters. (c) Comparison of the scattering coefficients computed by using the transfinite element method for the device in (a) with data from [14] and [11].

E. Waveguide Junction Problems

By changing the analytic basis functions in (9) from a *cosine* to a *sine* distribution within the waveguide regions, the transfinite element method is transformed into a method to model two-dimensional rectangular waveguide junctions. An example of a waveguide junction problem is the rectangular waveguide two-slot coupler shown in Fig. 6(a). The coupling of the incident wave to ports 2 and 4 are through the apertures. The current density at the edges of the slots are mathematically infinite and are called singular points.

The usual approach to model singular points with finite elements is to use more elements around the point as shown in Fig. 6(b). The total number of unknowns for the mesh shown in Fig. 6(b) is 492, and the scattering coefficients computed by the transfinite element method are plotted in Fig. 6(c), together with measured data [6]. A large discrepancy between the computed data and the measurements exists, as can be seen from the figure. In order to improve the accuracy of the numerical model at singular points, we introduce the use of transfinite singular elements.

F. Singular Elements

In Fig. 6(d) we show two conducting planes that intersect at an angle β. When $\beta > \pi$, the field becomes singular at this point. Notice in the figure that we can enclose the singular point with a circular arc with radius r_0. If the radius r_0 is much smaller than the wavelength, the Helmholtz equation can be replaced by the Laplace equation within the circular region. Hereafter, we will refer to the circular region as the *singular region* Ω_s. From [18], the electric field in the singular region can then be written as

$$E(\rho, \theta) = e_1 \rho^{\pi/\beta} \sin\left(\frac{\pi\theta}{\beta}\right) \qquad (20)$$

where ρ is the distance from the singular point, θ is the angle, and e_1 is an unknown coefficient. We have found that using even the single basis function in (20) in the transfinite element method is enough to model the solution in the singular region Ω_s.

Along the lines of the previous sections, we divide the problem domain for the waveguide junction with singularities into three parts: a discontinuity region Ω_d, waveguide regions Ω_i, and singular regions Ω_s. Accordingly, the solution space in (11) becomes

$$\Lambda \equiv \left\{ \phi \,|\, \phi = \left\{ \begin{array}{ll} \tilde{\alpha}^I \phi^I + \displaystyle\sum_{i=1}^{P} \tilde{\alpha}^{\Gamma_i} \phi^{\Gamma_i} + \tilde{\alpha}^{\Gamma_s} \phi^{\Gamma_s} & \text{in } \Omega_d \\[3mm] \delta_{i1}\Phi_{\text{inc}} + \displaystyle\sum_{j=0}^{M} a_{ij}\Phi_{ij} & \text{in } \Omega_i \quad \phi \in C^0 \\[3mm] e_1 \rho^{\pi/\beta} \sin\left(\dfrac{\pi\theta}{\beta}\right) & \text{in } \Omega_s \end{array} \right. \right\}.$$

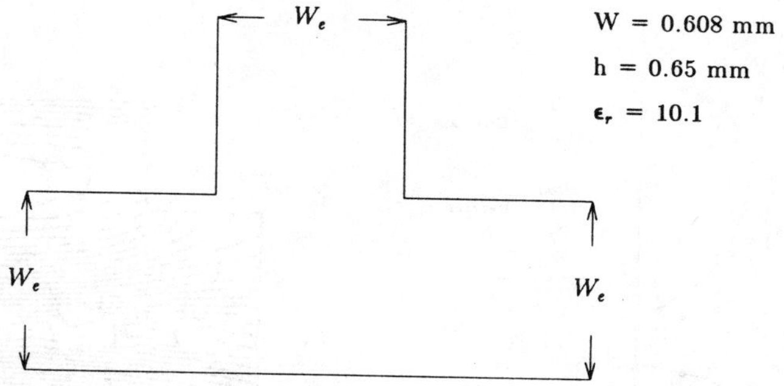

Planar Waveguide Model :

$$Z = 48.54\Omega$$

$$W'(f) = 0.608 + \frac{1.319}{1.0 + \frac{f}{77.6295}}$$

$$\epsilon_{re}(f) = 10.1 - \frac{3.245}{1.0 + 0.00307\left(0.43 f^2 - 0.009 f^3\right)}$$

(a)

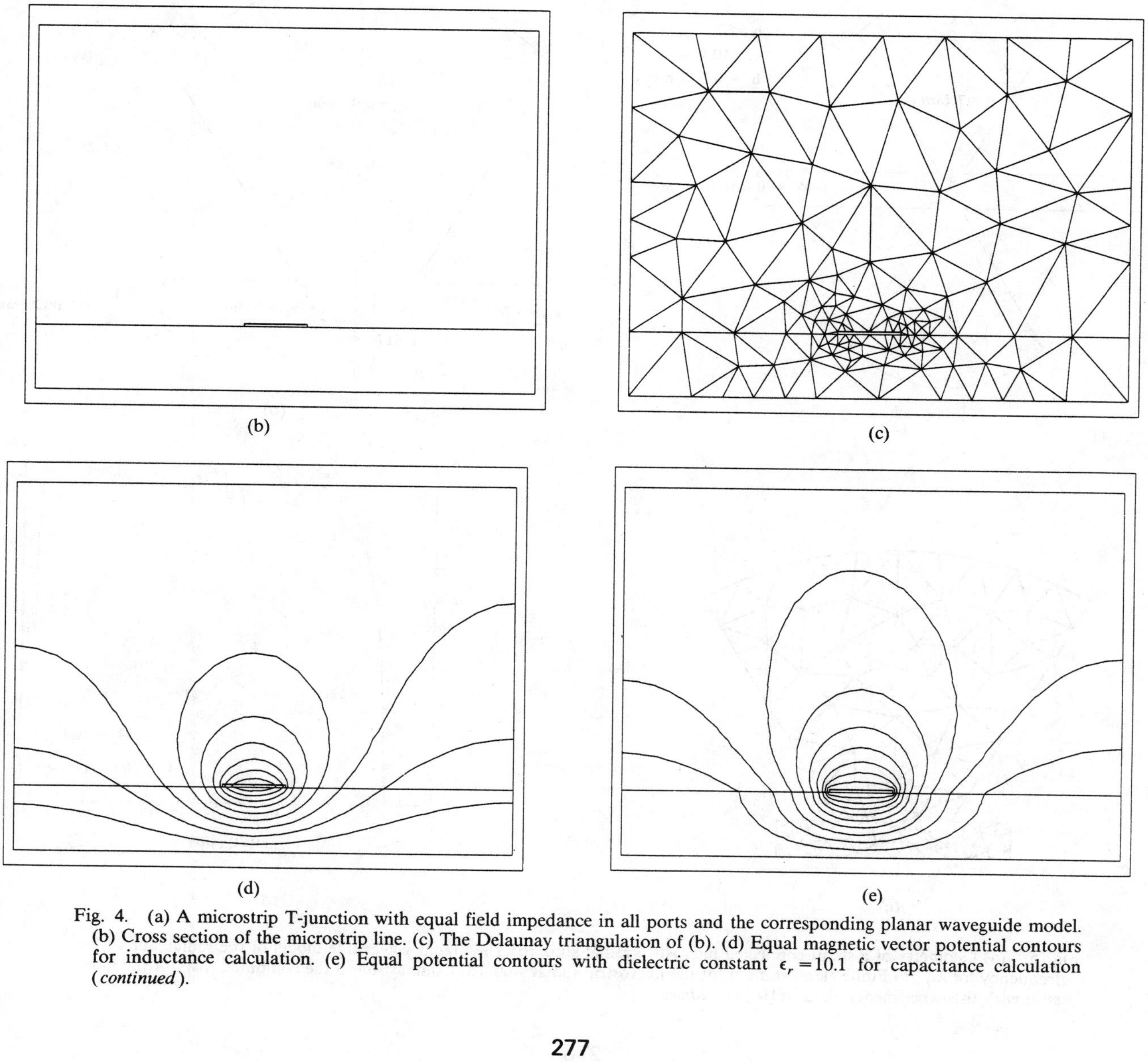

Fig. 4. (a) A microstrip T-junction with equal field impedance in all ports and the corresponding planar waveguide model. (b) Cross section of the microstrip line. (c) The Delaunay triangulation of (b). (d) Equal magnetic vector potential contours for inductance calculation. (e) Equal potential contours with dielectric constant $\epsilon_r = 10.1$ for capacitance calculation (*continued*).

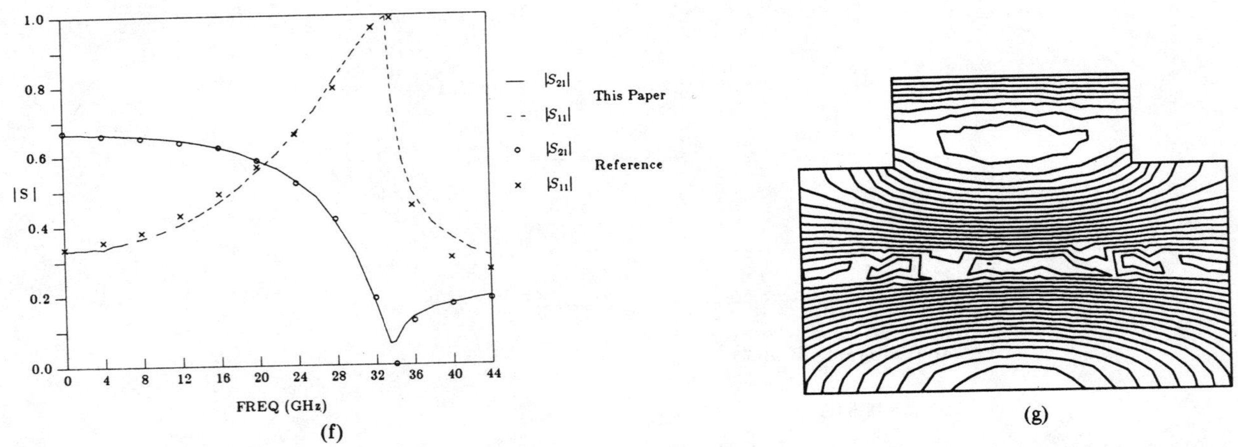

Fig. 4. (*Continued*) (f) Numerical results for the reflection and the transmission coefficients for the T-junction compared with Mehran [15]. (g) Contours of equal field intensity for the T-junction at 33.7 GHz.

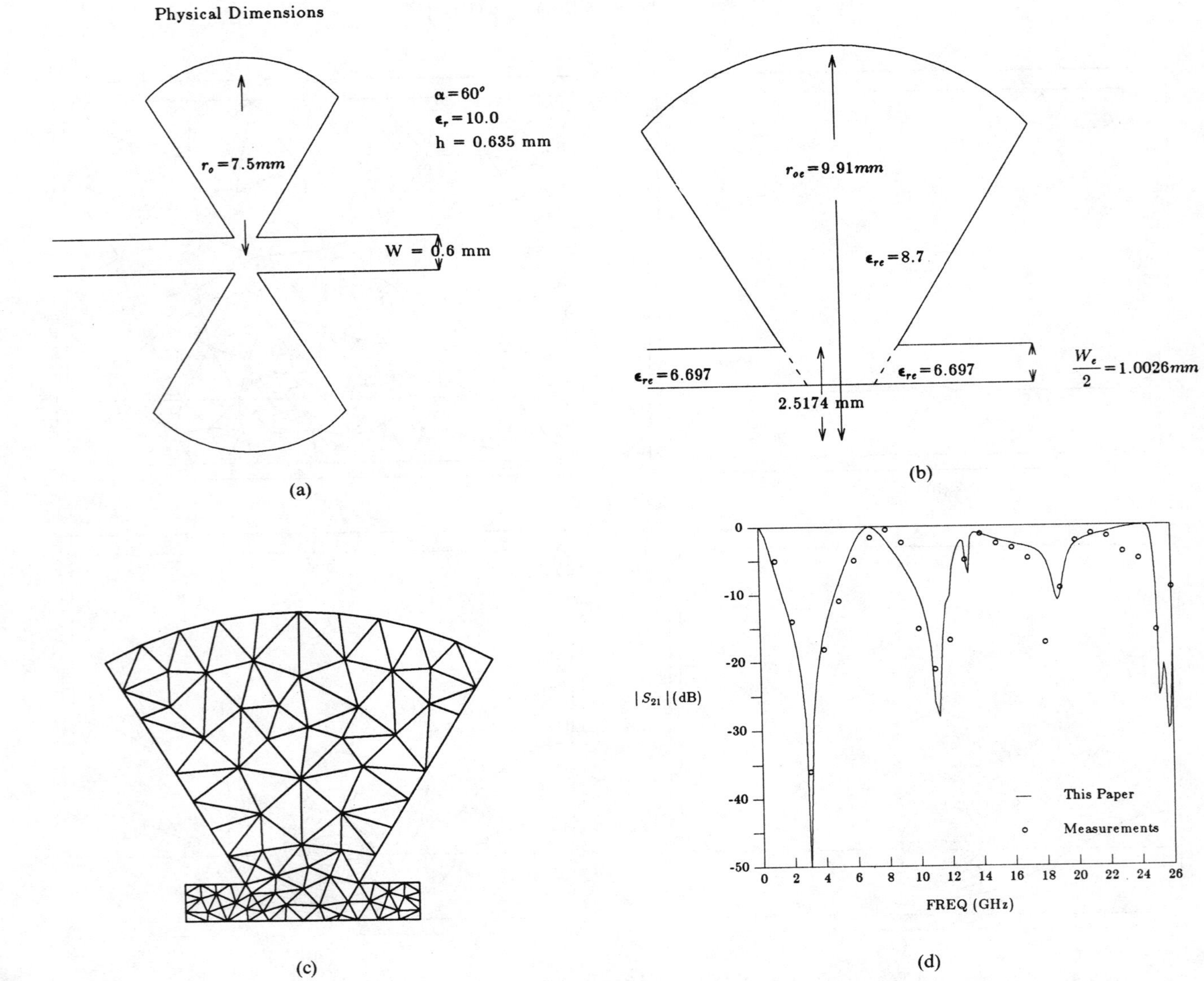

Fig. 5. (a) Geometry of a shunt-connected microstrip radial stub. (b) The planar waveguide model of the radial stub at low frequency for (a). (c) Finite element mesh for the microstrip radial stub. (d) Comparison of the computed $|S_{21}|$ for the radial stub with the experimental data in [16] (*continued*).

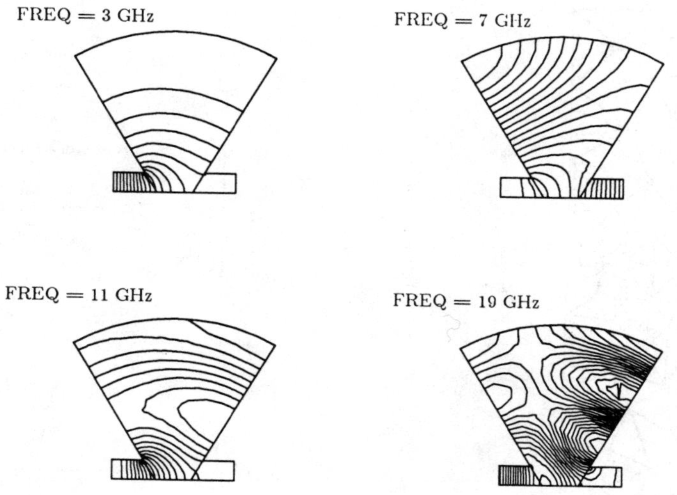

(e)

Fig. 5. (*Continued*) (e) Plots of the real part of the field at 3, 7, 11, and 19 GHz.

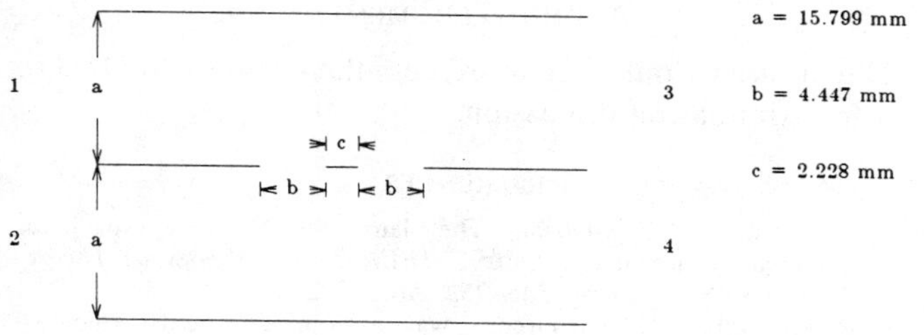

(a)

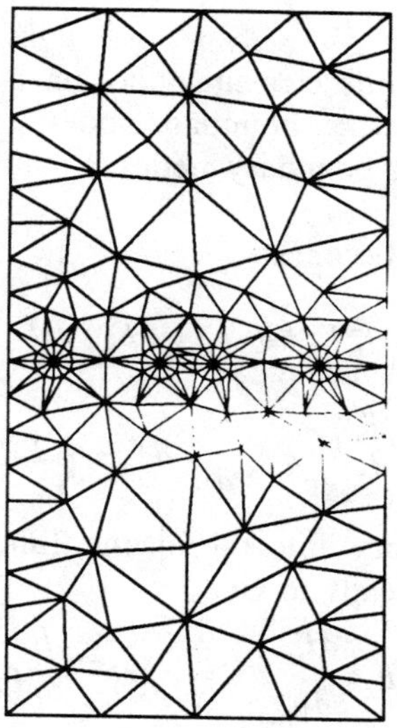

(b)

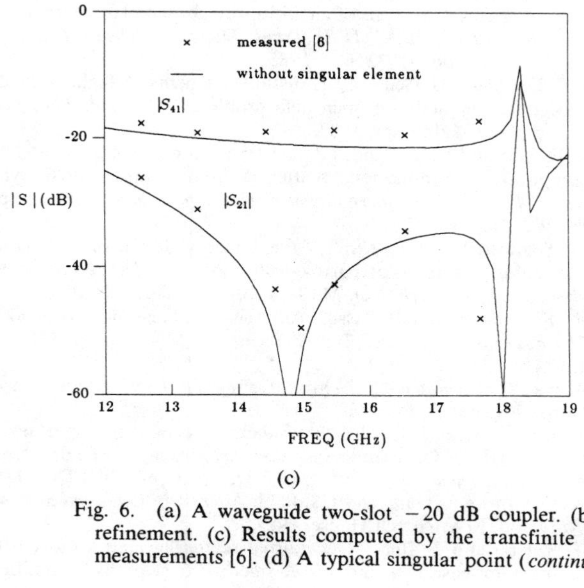

(c)

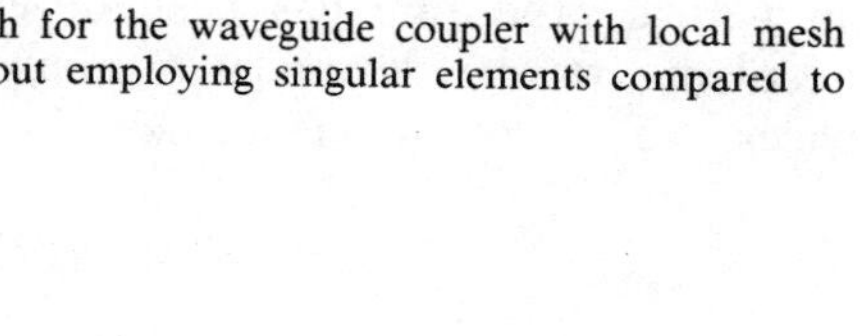

(d)

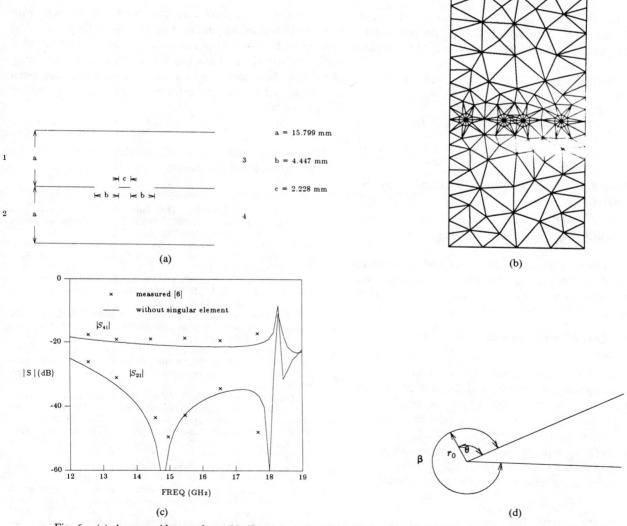

Fig. 6. (a) A waveguide two-slot −20 dB coupler. (b) Finite element mesh for the waveguide coupler with local mesh refinement. (c) Results computed by the transfinite element method without employing singular elements compared to measurements [6]. (d) A typical singular point (*continued*).

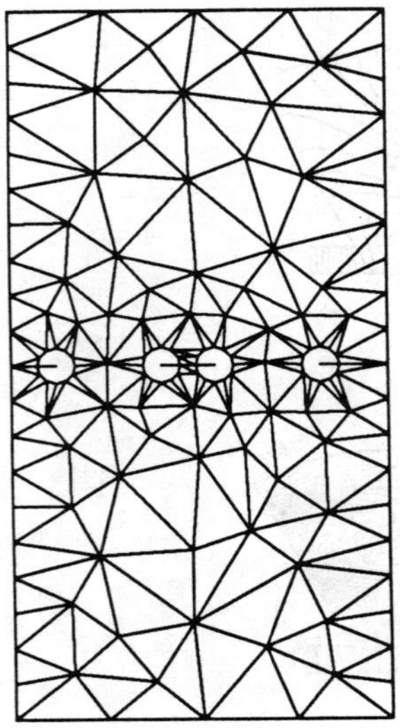

(e)

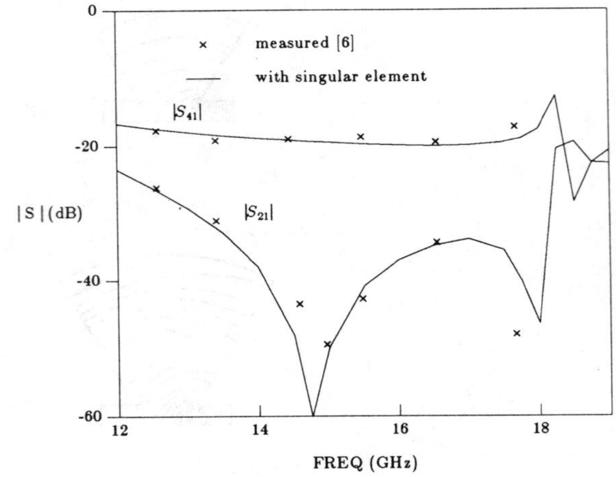

(f)

Fig. 6. (*Continued*) (e) Finite element mesh with singular elements. (f) Comparison of the results from the transfinite element method that includes singular elements with measurements [6].

Notice that the finite element nodes have been separated into: interior nodes ϕ^I, boundary nodes ϕ^{Γ_i} on port reference plane i, and boundary nodes ϕ^{Γ_s} along the circular arc Γ_s.

The functional becomes

$$F(E_z) = \int_{\Omega_d} \left(\nabla E_z^* \cdot \nabla E_z - k^2 \epsilon_{re} E_z^* E_z \right) d\Omega$$

$$- \sum_{i=1}^{P} \oint_{\Gamma_i} E_z^* \frac{\partial E_z}{\partial n} d\Gamma - \oint_{\Gamma_s} E_z^* \frac{\partial E_z}{\partial n} d\Gamma. \quad (22)$$

The new boundary integral in the functional may be integrated analytically to give

$$\oint_{\Gamma_s} E_z^* \frac{\partial E_z}{\partial n} d\Gamma = e_1^* r_0^{2\pi/\beta} \frac{\beta}{2} e_1. \quad (23)$$

Finally, imposing continuity conditions and performing the variational analysis as in the previous sections, we obtain a singular element to model waveguide junction discontinuities.

G. Scattering Coefficients Computed by Using the Singular Element

Fig. 6(e) shows the finite element mesh for the problem in Fig. 6(a) using singular elements at the slot edges. The total number of unknowns is 362 since the singular element requires fewer basis functions to model the singularity. The computed scattering coefficients S_{21}, S_{41} are plotted in Fig. 6(f). Compared to experimental values [6], much more accurate answers are obtained by using singular elements than is the case with conventional elements.

V. Conclusion

A new numerical procedure has been developed that employs the planar waveguide model to analyze passive MMIC devices. Various components have been studied using the new method and the results compared with previously published methods or experimental data; good agreement is observed in many cases. The method has also been applied to model arbitrary waveguide discontinuity problems. In these cases, a singular element has been developed to model the infinite fields that occur at reentrant corners.

Acknowledgment

The authors would like to express their thanks to Dr. D. Sun for stimulating discussion.

References

[1] T. Okoshi and T. Miyoshi, "The planar circuit—An approach to microwave integrated circuitry," *IEEE Trans. Microwave Theory Tech.*, vol. MTT-20, pp. 245–252, Apr. 1972.
[2] R. Sorrentino, "Planar circuits, waveguide models, and segmentation method," *IEEE Trans. Microwave Theory Tech.*, vol. MTT-33, pp. 1057–1066, Oct. 1985.
[3] W. K. Gwarek, "Analysis of an arbitrarily-shaped planar circuit—A time-domain approach," *IEEE Trans. Microwave Theory Tech.*, vol. MTT-33, pp. 1067–1072, Oct. 1985.
[4] J. F. Lee and Z. J. Cendes, "Transfinite elements: A highly efficient procedure for modeling open field problems," *J. Appl. Phys.*, vol. 61, pp. 3913–3915, Apr. 1987.
[5] J. F. Lee and Z. J. Cendes, "The transfinite element method for computing electromagnetic scattering from arbitrary lossy cylinders," *IEEE AP-S International Symposium Digest*, AP03–5, pp. 99–102, June 1987.
[6] H. Schmiedel and F. Arndt, "Field theory design of rectangular waveguide multiple-slot narrow-wall couplers," *IEEE Trans. Microwave Theory Tech.*, vol. MTT-34, pp. 791–797, July 1986.
[7] R. R. Bonetti and P. Tissi, "Analysis of planar disk networks," *IEEE Trans. Microwave Theory Tech.*, vol. MTT-26, pp. 471–477, July 1978.
[8] *Maxwell Microwave IC Computer Program Users Guide*, Ansoft Corp., Pittsburgh, PA, 1987.
[9] J. F. Lee and Z. J. Cendes, "An adaptive spectral response modeling procedure for multi-port microwave circuits," *IEEE Trans. Microwave Theory Tech.*, vol. MTT-35, pp. 1240–1247, Dec. 1987.
[10] K. C. Gupta, R. Garg, and I. J. Bahl, *Microstrip Lines and Slotlines*. Norwood, MA: Artech House, 1979.
[11] T. S. Chu and T. Itoh, "Generalized scattering matrix method for analysis of cascaded and offset microstrip step discontinuities,"

IEEE Trans. Microwave Theory Tech., vol. MTT-34, pp. 280–284, Feb. 1986.

[12] M. Hestenes, *Conjugate Direction Methods in Optimization*. New York: Springer–Verlag, 1980.

[13] Z. J. Cendes, D. N. Shenton, and H. Shahnasser, "Magnetic field computation using Delaunay triangulation and complementary finite element methods," *IEEE Trans. Magn.*, vol. MAG-19, pp. 2551–2554, 1983.

[14] G. Kompa, "S-matrix computation of microstrip discontinuities with a planar waveguide model," *Arch. Elek. Übertragung.*, vol. 30, pp. 58–64, Feb. 1976.

[15] R. Mehran, "Calculation of microstrip bends and Y-junctions with arbitrary angle," *IEEE Trans. Microwave Theory Tech.*, vol. MTT-26, pp. 400–405, June 1978.

[16] F. Giannini, M. Ruggieri, and J. Vrba, "Shunt-connected microstrip radial stubs," *IEEE Trans. Microwave Theory Tech.*, vol. MTT-34, pp. 363–366, Mar. 1986.

[17] I. Wolff and N. Knoppik, "Rectangular and circular microstrip disk capacitors and resonators," *IEEE Trans. Microwave Theory Tech.*, vol. MTT-22, pp. 857–864, Oct. 1974.

[18] J. D. Jackson, *Classical Electrodynamics*. New York: Wiley, 1975, pp. 75–78.

Paper 6.8

Full-Wave, Finite Element Analysis of Irregular Microstrip Discontinuities

ROBERT W. JACKSON, SENIOR MEMBER, IEEE

Abstract —**Finite element expansion currents are used to formulate a full-wave analysis of microstrip discontinuities. A rigorous analysis of fairly irregular structures is possible, including radiation and surface wave effects as well as coupling between closely spaced junctions. The step, stub, and bent-stub discontinuities are analyzed using this technique. Measurements are presented which verify stub calculations.**

I. Introduction

IN THE PAST several years, designers of microwave and millimeter-wave integrated circuits have come to depend heavily on computer-aided techniques to reduce design time and improve performance. Most CAD modeling of passive circuits centers on the microstrip circuit medium and includes models of microstrip bends, steps, tees, and other discontinuities [1], [2]. In early integrated circuit designs, these models were used with some success. In recent years, however, increases in operating frequency and higher performance requirements have made some of the earlier models (based on a quasi-static assumption) insufficiently accurate. Fully electromagnetic models are now often required in order to include effects such as dispersion, radiation, and coupling. In addition to higher frequency requirements, smaller, more densely packed circuits are being designed in order to reduce cost. This further increases coupling and makes it awkward to use many of the standard junctions. As a result, irregularly shaped junctions may be necessary to reduce crowding, and modeling of these junctions becomes important. This paper presents a technique for calculating rigorously, at high frequencies, the characteristics of somewhat irregular microstrip junctions.

A number of full-wave techniques have been published for the analysis of such simpler microstrip discontinuities as the open end [3]–[6] and step [7]. Both [5] and [6] used finite element currents (piecewise sinusoids) to model the open-end and gap discontinuities on an open substrate, but in the former case, finite elements were used only in an area local to the discontinuity and precomputed sinusoids were used elsewhere. In [3], [4], and [7], the authors also used precomputed sinusoids but modeled the discontinuity locally with entire domain currents. Some of these junctions may also be analyzed using the full-wave techniques which have been applied to finline junctions [8], [9].

For more complicated or irregular junctions, most analyses use some type of simplifying assumption, such as a quasi-static approximation [10], [11] or a magnetic wall approximation [12]. Exceptions include the use of finite element currents to model irregularly shaped microstrip resonators in a closed cavity [13] and irregularly shaped antennas [14], [15]. The method of lines has also been used to discretize the currents on irregularly shaped microstrip structures in an enclosure [16]. In that reference and in recent work by Rautio and Harrington [17], the entire modeled structures are discretized. The former de-embeds discontinuity parameters from a set of resonator calculations or periodic line calculations. The latter work uses a magnetic source current on the enclosure wall to excite an input microstrip line. The resulting current at that point is used to calculate an admittance which includes both the excitation effects and the discontinuity effects. A de-embedding must then take place.

A formulation is presented here which models microstrip junctions on an open substrate using finite element expansion currents (rooftop functions) and sinusoidal precomputed expansion currents for input and output microstrip lines. In contrast to [16] and [17], only the junction itself is modeled with a fine mesh of finite element currents. This setup allows the finite element resolution to be adjusted to the junction without using a very large number of elements to model the slowly varying input and output currents. A substantial improvement in accuracy and numerical efficiency can be achieved. Also, the junction scattering parameters are determined directly from the amplitudes of the reflected and transmitted sinusoids, and no adjustment for source effects is necessary. When resonator methods [7], [13], [16] are used to determine two-port discontinuity parameters, two or more resonators must be analyzed. This is not necessary when the technique presented in this paper is used. Whereas [5] and [6] use only x-directed currents and subdivide along the x direction, this formulation uses both x- and y-directed currents and subdivides along both directions. This subdivision makes it easier to analyze irregular structures than

Manuscript received November 6, 1987; revised June 20, 1988. This work was supported in part by the Air Force Office of Scientific Research under Contract F49620-82-C-0035.

The author is with the Department of Electrical and Computer Engineering, University of Massachusetts, Amherst, MA 01003.

IEEE Log Number 8824250.

Reprinted from *IEEE Trans. Microwave Theory Tech.*, vol. 37, no. 1, pp. 81–89, January 1989.

the methods described in [3]–[7]. Jansen and Wertgen [18] have developed a similar formulation except that the entire structure (sources, feedlines, and discontinuities) is enclosed in a conducting box.

In this analysis, neither cover plate nor sidewalls are assumed to be present [20] and therefore radiation and surface wave losses can occur. The measurements and calculations presented show the effects that these losses can have on the behavior of a common microstrip structure. Changing the analysis to include a cover plate is very easily accomplished. In that case, radiative losses would be due to the excitation of parallel-plate modes [19] instead of surface wave and space wave radiation. Similar effects should occur for discontinuities in an electrically large lossy box.

It should be noted that not including sidewalls is, in most cases, a reasonable assumption, since substrate thicknesses and junction sizes are typically very small compared to the junction's distance from a sidewall. Sidewalls may become important if the operating frequency is near a resonant frequency of the enclosure. A practical circuit is not operated near these frequencies in a high-Q enclosure, since all parts of the circuit will couple to each other through the resonant mode. If operation is required near such a frequency, absorbing material is added to the box (along the cover, for example) in order to damp the resonance. This again causes the wall effect to be small, since a parallel-plate wave traveling out from a discontinuity is substantially damped by the time it reaches a sidewall, is reflected, and returns to the discontinuity. Not including sidewalls in this analysis also generates some very useful redundancies in the numerical calculations.

In what follows, the moment method formulation is discussed very briefly, followed by a more detailed description of the expansion mode setup, along with some of the useful symmetries. Next, results are presented for the step, stub, and bent stub. Three sets of measured data are compared to calculated results for the stubs. Calculations of radiation loss of a quarter-wave stub are compared to measurements. And finally, the effect of coupling between closely spaced discontinuities is demonstrated via the bent-stub configuration.

II. Theory

The generic configuration which will be analyzed is presented in Fig. 1. A grounded dielectric slab is shown which extends to $\pm$ infinity in the x and y directions. No cover plate is assumed, although adding one would be a minor modification. Finite element currents, x- and y-directed, are located in the cross-hatched region and are excited by precomputed incident, reflected, and transmitted sinusoidal currents which overlap the finite element region. Although the configuration as shown assumes $\hat{x}$ propagation on the input/output lines, $\hat{y}$ propagating sinusoids could be added (i.e., for a 90° corner). Also, the sinusoidal input/output lines are shown to be centered around $y = 0$, but they can be offset in the transverse direction if desired. The junction which is to be analyzed is

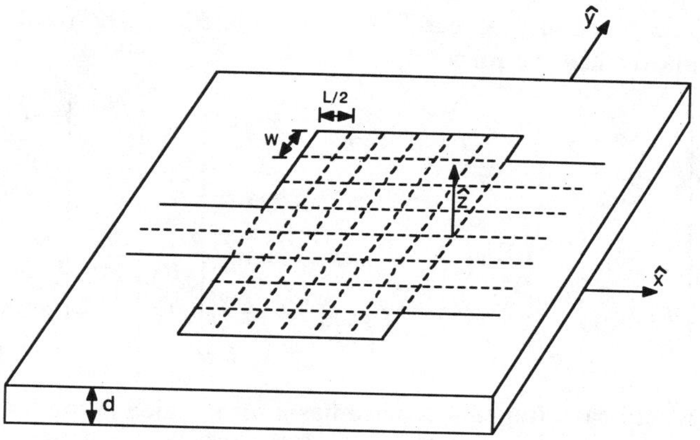

Fig. 1. Arrangement of finite element currents and input/output currents on a grounded dielectric slab. Specific discontinuities form a subset of these currents.

carved out of the finite element region, and this makes it possible to analyze many types of junctions. Examples include open ends, stubs, steps, asymmetric steps, corners, and others. The most significant limitation is that the junction must conform to certain discrete sizes.

A. Moment Method Formulation

Since the method of moments is well known, it is only discussed briefly in order to set the notation and to note some salient points. The formulation, similar to [5], begins by determining the x and y components of the electric field on the surface ($z = d$) of the grounded dielectric slab due to a surface current $\vec{J}$ on the same surface,

$$\vec{E}(x, y) = \frac{1}{(2\pi)^2} \int_{-\infty}^{\infty} \vec{\vec{Q}}(k_x, k_y) \cdot \vec{J}(k_x, k_y) e^{jk_x x} e^{jk_y y} \, dk_x \, dk_y \quad (1)$$

where $\vec{E}(x, y)$ is the two-component surface field, $\vec{J}(k_x, k_y)$ is the Fourier transform of the surface current, and $\vec{\vec{Q}}(k_x, k_y)$ is the Fourier transform of the Green's function for a current element located at the origin (see Appendix I). The current is expanded as follows.

$$\vec{J} = \sum_{j=1}^{NM} A_j \vec{J}_j^{fe} + R\vec{J}^R + T\vec{J}^T + \vec{J}^I \quad (2)$$

where $\vec{J}^R$, $\vec{J}^T$, $\vec{J}^I$ are known reflected, transmitted, and incident currents, and the $\vec{J}_j^{fe}$ are rooftop expansion currents, some of which are x-directed and some y-directed. NM is the total number of finite elements, and A_j, R, T are complex coefficients which are to be determined. Weighted averages of the tangential electric field on the microstrip surface are set to zero according to the expression

$$-\iint_{-\infty}^{\infty} \vec{W}_i(x, y) \cdot \vec{E}(x, y) \, dx \, dy = 0, \qquad i = 1, 2, 3, \cdots, NM + 2 \quad (3)$$

where $\vec{W}_i$ are the weighting functions which, except for two, are the same as the finite element expansion currents. The procedure is therefore almost Galerkin. The resulting matrix has the form

$$\begin{bmatrix} \vec{\vec{Z}} & \vec{\vec{Z}}_{xy} & \vec{Z}_{xR} & \vec{Z}_{xT} \\ \vec{\vec{Z}}_{yx} & \vec{\vec{Z}}_{yy} & \vec{Z}_{yR} & \vec{Z}_{yT} \\ (\vec{Z}_{W1x})^t & (\vec{Z}_{W1y})^t & Z_{W1R} & Z_{W1T} \\ (\vec{Z}_{W2x})^t & (\vec{Z}_{W2y})^t & Z_{W2R} & Z_{W2T} \end{bmatrix} \begin{bmatrix} \vdots \\ A_j \\ \vdots \\ R \\ T \end{bmatrix} = - \begin{bmatrix} \vec{Z}_{xI} \\ \vec{Z}_{yI} \\ Z_{W1I} \\ Z_{W2I} \end{bmatrix} \tag{4}$$

where each impedance term has a form which is similar to, for example,

$$\left(\vec{\vec{Z}}_{xy}\right)_{i,j} = \frac{-1}{(2\pi)^2} \iint_{-\infty}^{\infty} W_i^x(k_x, k_y)^* Q_{xy}(k_x, k_y) \cdot J_j^y(k_x, k_y)\, dk_x\, dk_y. \tag{5}$$

The impedance elements shown in (4) will be described more completely in the next section, but all are a result of a double spectral integration which is performed numerically. This integration includes a careful evaluation of the poles which occur in $Q_{ij}(k_x, k_y)$ [21]. These poles (in practical cases, only one) correspond to surface waves (with no cover) or parallel-plate waves (with cover). In the former case, radiation effects are also included.

After the impedances have been computed, (4) is solved to determine the transmission and reflection coefficients.

Two types of expansion functions and two types of weighting functions are used in this analysis. Rooftop functions are used for expansion and weighting in the junction area, while precomputed sinusoidal currents model the incident, reflected, and transmitted waves which extend away from the junction. Two additional weighting functions are necessary; these will be discussed later.

B. Finite Element Currents

The finite element currents are described by

$$J_j^x(x, y) = t([x - x_j]/L) \cdot s([y - y_j]/W) \tag{6a}$$

$$J_j^y(x, y) = s(2[x - \bar{x}_j]/L) \cdot t([y - \bar{y}_j]/2W) \tag{6b}$$

where

$$t(u) = \begin{cases} 1 - 2|u|, & |u| < 0.5 \\ 0, & \text{otherwise} \end{cases}$$

$$s(u) = \begin{cases} 1, & |u| < 0.5 \\ 0, & \text{otherwise.} \end{cases}$$

Referring to Fig. 1, the x-directed currents are centered on the $x =$ constant dashed lines at points midway between the $y =$ constant dashed lines, while the y-directed currents are centered on the $y =$ constant dashed lines at points midway between the $x =$ constant dashed lines. The x-directed currents overlap each other in the x direction but not in the y direction, and the reverse is true for the y-directed currents. The same approach was used in [17] except that, in the present formulation, the impedances formed between a weighting current and an expansion current depend only on the position vector separating the two. This would not be true if sidewalls were present (as in [17] and [18]), since the location of the currents with respect to the wall also has an effect when the wall is nearby. The symmetries generated by not having walls plus the symmetries generated by reciprocity make it necessary to calculate only the first row or column of $\vec{\vec{Z}}_{xx}$, $\vec{\vec{Z}}_{xy}$ and $\vec{\vec{Z}}_{yy}$. All the other elements of those matrices and the matrix $\vec{\vec{Z}}_{yx}$ can be determined from them.

C. Sinusoidal Expansion Functions

Ideally the incident, reflected, and transmitted currents would have the form

$$\vec{J}^I = [g_x^-(y)\hat{x} + jg_y^-(y)\hat{y}][\cos(\beta^- x) - j\sin(\beta^- x)] \tag{7a}$$

$$\vec{J}^R = [g_x^-(y)\hat{x} - jg_y^-(y)\hat{y}][\cos(\beta^- x) + j\sin(\beta^- x)] \tag{7b}$$

$$\vec{J}^T = [g_x^+(y)\hat{x} + jg_y^+(y)\hat{y}][\cos(\beta^+ x) - j\sin(\beta^+ x)] \tag{7c}$$

where $\vec{J}^I$ and $\vec{J}^R$ are zero for $x > 0$, and $\vec{J}^T$ is zero for $x < 0$. The functions g_x^- and g_y^- are the transverse variations of the x- and y-directed currents for $x < 0$, and g_x^+ and g_y^+ are the transverse variations for $x > 0$. As discussed in [5], truncating the cosine portion of the x-directed currents at $x = 0$ causes a longitudinal current discontinuity, and numerical difficulties result. Instead, the cosines in (7) are truncated one-quarter guide wavelength from a zero of the sine. The functions then extend away from the junction an integral number of half wavelengths before again terminating. The resulting expansion functions are

$$\vec{J}^I(x, y) = [g_{xs}^-(y)\hat{x} + jg_{ys}^-(y)\hat{y}] \cdot [f^-(\beta^- x + \pi/2) - jf^-(\beta^- x)] \tag{8a}$$

$$\vec{J}^R(x, y) = [g_{xs}^-(y)\hat{x} - jg_{ys}^-(y)\hat{y}] \cdot [f^-(\beta^- x + \pi/2) + jf^-(\beta^- x)] \tag{8b}$$

$$\vec{J}^T(x, y) = [g_{xs}^+(y)\hat{x} + jg_{ys}^+(y)\hat{y}] \cdot [f^+(\beta^+ x + \pi/2) - jf^+(\beta^+ x)] \tag{8c}$$

where

$$f^-(x) = \begin{cases} \sin x, & -n\pi < x < 0 \\ 0, & \text{otherwise.} \end{cases}$$

$$f^+(x) = -f^-(-x).$$

The functions $g_{xs}^-, g_{xs}^+, g_{ys}^-, g_{ys}^+$ model $g_x^-, g_x^+, g_y^-, g_y^+$ and are sums of the functions $s([y - y_j]/W)$ and $t([y - y_j]/2W)$, which were described previously. The exact form of these summed functions and the propagation constants β^- and β^+ are computed before beginning the impedance calculation in (4). This precomputation is fast compared to the discontinuity calculation and is described in Appendix II. Note also that the functions in (8) are normally shifted so as to terminate in the middle of the finite element grid instead of at $x = 0$ as described above.

The impedance vectors $\vec{Z}_{xI}$, $\vec{Z}_{xR}$, $\vec{Z}_{xT}$, $\vec{Z}_{yI}$, $\vec{Z}_{yR}$, and $\vec{Z}_{yT}$ are formed by taking the inner product of either W_i^x or W_i^y with one of the currents $\vec{J}^I$, $\vec{J}^R$, or $\vec{J}^T$. For example,

$$\left(\vec{Z}_{yR}\right)_i = \frac{-1}{(2\pi)^2} \iint_{-\infty}^{\infty} (W_i^y)^* \left(Q_{yx} J_x^R + Q_{yy} J_y^R\right) dk_x \, dk_y$$

forms the ith component of the vector $\vec{Z}_{yR}$. (It is understood that W_i^y, J_x^R, J_y^R are the Fourier transforms of the spatial functions described by (6) and (8).)

To determine the dimensions of the impedance matrix in (4), we note that there are $NM_x \cdot NM_y$ x-directed finite element currents, $NM_x \cdot (NM_y - 1)$ y-directed finite element currents, plus the $\vec{J}^R$ and $\vec{J}^T$ currents. So far, we have only tested with finite element functions which total $NM_x \cdot NM_y + NM_x(NM_y - 1)$. As in [4] and [5], two more testing functions are necessary, and these are chosen to be

$$\vec{W}_1(x, y) = t(x) g_{xs}^{+}(y) \hat{x} \tag{9a}$$

$$\vec{W}_2(x, y) = t(x - x_0) g_{xs}^{-}(y) \hat{x} \tag{9b}$$

and $x_0 = -(NM_x + 1)L/2$. So these testing functions straddle the lines separating the finite element region and the purely sinusoidal regions (see Fig. 1) and have a y dependence which corresponds to the input and output waveforms. The bottom two rows of (4) obviously result from forming the inner product of these two weighting functions and the various expansion functions.

D. Numerical Considerations

As mentioned previously, because each finite element has the same size and form, and because each mutual impedance depends only on the vector difference between the locations of the two currents involved, many useful redundancies occur. So only the top row of elements in, for example, $\vec{Z}_{xx}$ needs to be calculated, and the remaining terms can be determined from them. In addition, due to the shifting property of Fourier transforms, the integrands in each of the impedance elements of the top row differ only by product factors such as $\exp(-jk_x L/2)$ and/or $\exp(-jk_y W)$. Likewise for the other submatrices. Therefore, at each integration point in (k_x, k_y) the integrands of all the necessary impedances within a submatrix differ from their neighbors by one multiplication, and evaluation of the integrand proceeds quickly.

The overall setup of the software is such that *all* the currents in the grid shown in Fig. 1 are assumed to be present initially. Using the various symmetries, the impedance matrix in (4) is computed. Then, at the end of the routine, various rows and columns are deleted so as to form a specific junction out of the general gridwork in the figure. This makes it relatively easy to change the program from analysis of one type of discontinuity to analysis of another. Of course, initially one must choose the size of the input port and the size of the output port and determine the location of these ports in terms of offset and overlap within the finite element region. Several finite current elements are included on the input and output lines in the vicinity of a junction (see Fig. 3 inset) in order to model current disturbances in that area.

The principal cost of this flexibility is that only junctions and discontinuities having certain discrete sizes can be analyzed. For example, the width of the input or output port must be an integer multiple of W. In many cases this is not a problem, since interpolation can be used to determine the characteristics of a noninteger junction.

III. Results—Numerical and Experimental

The formulation described above was used to analyze a step discontinuity, a stub, and a bent stub. In this section, we compare the step discontinuity results predicted by this theory with the results predicted by Koster and Jansen. We then compare the stub results to measurements.

A. Step Discontinuity

In [7], Koster and Jansen presented the results of an analysis of the microstrip step discontinuity. The details of their formulation are not completely clear; however, they do refer to their method as having been described in a previous paper [4] on end effects. Some of the features of their analysis are that the expansion currents are entire domain in a local region near the discontinuity, the expansion currents approximate the edge condition, and the structure is enclosed in a conducting box. This is in contrast to the work presented here, which features finite element currents in an open or covered structure. These currents also approximate the edge condition, but do so with pulse and triangular functions. By not using currents which closely model the proper conditions at the edge, some accuracy is sacrificed; however, in many practical cases, this loss of accuracy is not significant. In return, a finite element approach allows the analysis of fairly complicated structures.

Using the finite element formulation, we have analyzed the step discontinuity for the substrate and frequency parameters used by Koster and Jansen. A comparison of the results shows that the S parameter magnitudes are almost identical and the phases show reasonable agreement. Fig. 2 shows a comparison of the phases calculated in this work to those calculated in [7]. The transmission phase is numerically a very stable result and differs from [7] by at most a couple of degrees. As reported in [7], the most sensitive quantity was found to be the S_{22} phase. For the results shown in Fig. 2, the size of the grid was six or ten sections in the transverse direction ($NM_y = 6$ or 10) and 28 in the longitudinal direction ($NM_x = 28$). The length of the total grid is slightly over π/β^-. More precisely, $L/2$ (defined in Fig. 1) is equal to $\pi/(\beta^-[NM_x - 2])$. Doubling NM_y has a negligible effect on the results. Increasing NM_x from 20 to 28 results in a change of slightly less than 1° for $W_2/W_1 = 3$ or 5 at $d/\lambda_0 = 0.04$. Our conclusion is that the S parameter magnitudes are very accurate and that the

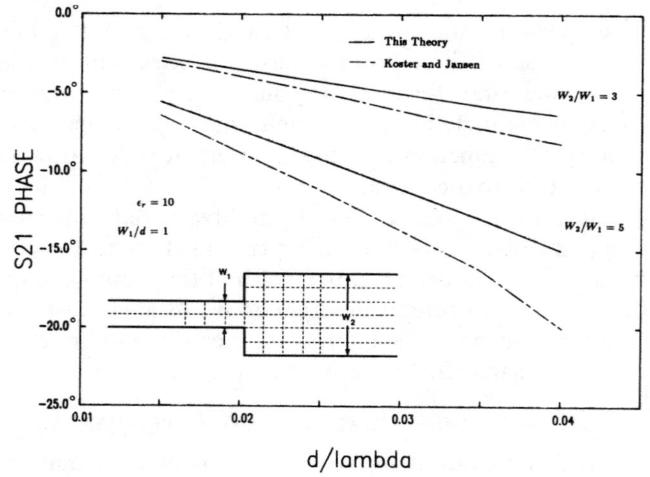

(a)

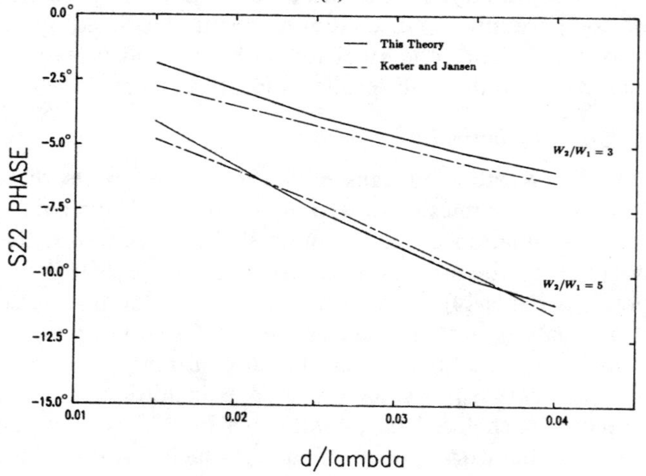

(b)

Fig. 2. Comparison of step phase calculations with those of Koster and Jansen [7]. (a) Transmission phase. (b) Reflection phase.

phases are accurate to within one or two degrees. This is sufficient for most applications. For further accuracy, the currents at the edges must be modeled more carefully.

B. Stub

Fig. 3 shows the measured and calculated phase of a single open-circuit stub attached to a transmission line. Two discontinuities are evident: a tee junction and an open end. Note that the measured and calculated resonant frequencies (where the stub is one-quarter wavelength) differ by about 1.5 percent and the phase error is less than 7° over most of the band.

The stub structure was etched on the surface of a soft substrate (Duroid 6010.2) with a dielectric constant near 11 and a thickness of 1.27 mm. Stub dimensions were measured to within ± 0.025 mm. Transmission phase and magnitude were determined using a Hewlett Packard 8510 network analyzer. In order to reliably measure the phase, the following procedure was used. First, the connector–microstrip line–stub–microstrip line–connector phase was measured and the data stored. The stub was then carefully cut away and the assembly remeasured. By subtracting the

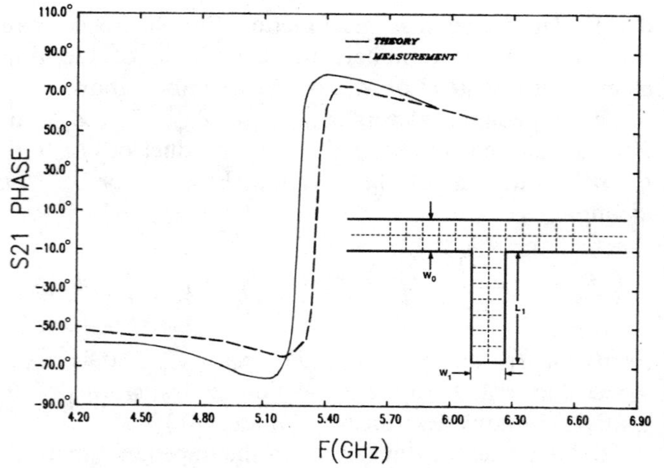

Fig. 3. Transmission phase of a single stub with measured (model) dimensions; $W_0 = 1.45$ mm (1.42), $W_1 = 1.40$ mm (1.40), $L_1 = 4.98$ mm (4.99).

phase of the second assembly from the phase of the first, we obtain the measured phase of the stub. This type of de-embedding technique is somewhat crude; however, it does not rely on the phase reproducibility of the connector transition or inhomogeneities in substrate permittivity. A reasonably good connector-to-microstrip transition is needed. In this case, each transition had a return loss of better than 25 dB. In order to accurately determine the substrate dielectric constant, the microstrip through line was further cut into several linear resonators [22]. The resonant frequency of each resonator was measured and used to determine that the dielectric constant was $\epsilon_r = 10.86 \pm 0.1$. The error is primarily due to errors in determining resonator dimensions (rough ends due to the cut).

The numerical model of the stub used nine divisions in the x direction and nine in the y. With reference to Fig. 1, both the dimensions $L/2$ and W are one-half the stub width. The input and output lines have the same width and are offset to the top of the grid. The model structure has slightly different dimensions from the actual measured structure, but these differences have an insignificant effect on resonant frequency. Due to the measurement uncertainties in the stub dimensions and in the dielectric constant (which are input to the analysis), the computed curve shown in Fig. 3 could be shifted by ± 1.5 percent in frequency from the nominal value shown. This would be a worst-case variation.

Note that a lossless ideal or quasi-static model for this junction predicts that the phase near resonance should jump from plus to minus 90°. The model presented here reproduces the measured *reduced* peak phase. This reduction may be due to radiation and conductor loss in the experiment and radiation loss in the model.

Figs. 4 and 5 illustrate some of the high-frequency characteristics which a full-wave analysis will reproduce. A shortened version of the stub shown in Fig. 3 was fabricated such that a resonance occurred near 10 GHz. This is a fairly high frequency for this substrate thickness and permittivity ($d/\lambda_0 = 0.04$, $d\sqrt{\epsilon_r}/\lambda_0 = 0.13$), and therefore

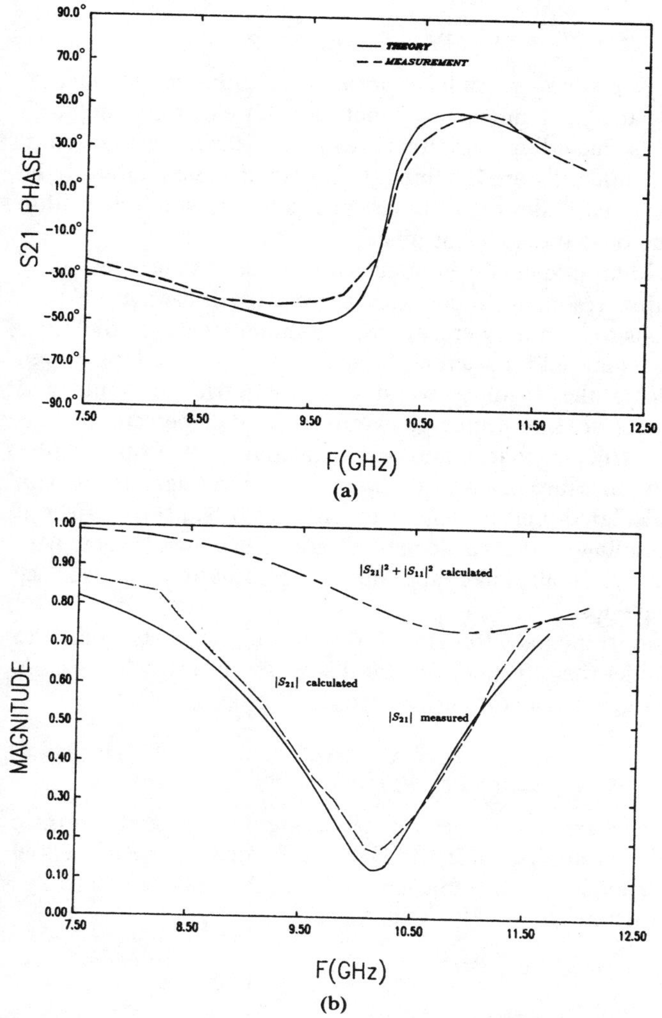

Fig. 4. Characteristics of a single stub with measured (model) dimensions; $W_0 = 1.40$ mm (1.45), $W_1 = 1.40$ mm (1.40), $L_1 = 2.16$ mm (2.17). (a) Transmission phase. (b) Transmission magnitude. $|S_{21}|^2 + |S_{11}|^2 < 1$ indicates radiation and surface wave loss.

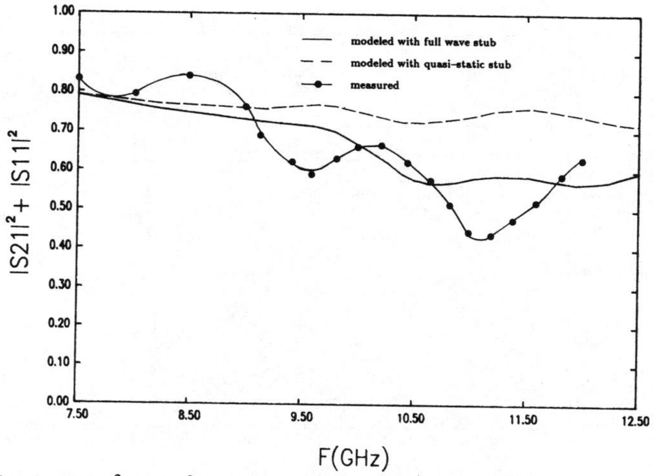

Fig. 5. ($|S_{21}|^2 + |S_{11}|^2$) measured and modeled for a single stub embedded in the middle of a 5.1 cm transmission line (see text).

significant coupling and radiation occur. To properly model the current in this case, the density of finite element modes was increased such that $L/2$ and W in Fig. 1 are one-quarter of a microstrip width (W_0).

Fig. 4 shows the measured and calculated transmission magnitude and phase. The measurements presented in Fig. 4 were performed in the manner described previously. Dielectric constant was determined to be 10.65 ± 0.15 and dimensional inaccuracies were as described previously. The uncertainty in these input quantities could cause a worst-case shift of the calculated curve by ± 1.5 percent. Agreement between measured and nominal calculation is excellent. Calculated peak phase is further reduced from the lower frequency case and measurements confirm this reduction. In addition, the calculated transmission isolation at resonance is reduced from roughly 40 dB (~ 35 dB measured) for the 5 GHz stub to 18 dB (15 dB measured) for the 10 GHz stub. A quasi-static calculation with or without loss does not show this reduced isolation.

Radiation loss can be an important effect for a substrate of this thickness. In Fig. 4b, we plot the quantity $G = |S_{21}|^2 + |S_{11}|^2$ based on the full-wave stub analysis. For a lossy junction, $(1 - G)$ is the fraction of incident power lost in the junction. The figure shows that the theoretical loss (to radiation and surface waves) peaks at about 25 percent near 11 GHz.

In order to verify the calculation, transmission and reflection measurements were made of a stub embedded between two microstrip lines, each 2.5 cm long. Fig. 5 plots the quantity G calculated from these measurements. No de-embedding was performed, and therefore the curve includes the conductor loss on the input and output lines. Measurements of the assembly with the stub cut off (a 5.0 cm through line) show a loss of 0.5 dB per inch. Incorporating this loss per unit length in a standard quasi-static model of the transmission line-stub-transmission line structure, results in a calculated G which is plotted in Fig. 5. This model includes conductor loss but not radiation loss. The stub length in the quasi-static model had to be adjusted so that the model resonant frequency coincided with the measured resonant frequency. Note that there is not a particularly good agreement with measurements above 9.0 GHz. Replacing only the quasi-static stub model with a two-port having the full-wave stub S parameters and leaving in the quasi-static lossy input/output lines results in the solid curve in the figure. This curve includes the conductor loss on the input and output lines and the radiation loss of the stub. Averaging the ripple in the measured curve (due to connector discontinuities) results in a curve which is in reasonable agreement with the full-wave stub model.

To summarize, the straight-stub full-wave analysis shows good agreement with measured data. Commercial CAD software using entirely quasi-static models shows good agreement in resonant frequency for the longer stub, but is high by about 7 percent for the shorter stub. Quasi-static models do not predict the measured reduction in peak phase variation unless conductor loss is included, and that will not account for the entire observed reduction.

The full-wave calculations in Fig. 4a include radiation and surface wave effects but not conductor loss. Since

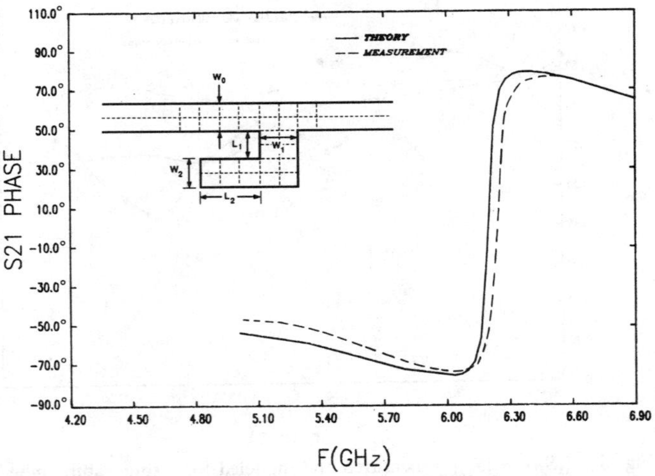

Fig. 6. Transmission phase of bent stub with measured (model) dimensions; $W_0 = 1.33$ mm (1.36), $W_1 = 1.35$ mm (1.39), $W_2 = 1.32$ mm (1.36), $L_1 = 1.38$ mm (1.36), $L_2 = 2.11$ mm (2.09).

there is good agreement with measurements, radiation and surface waves must play an important part, more so than conductor loss. Quasi-static models do not predict the reduced isolation measured at high resonant frequencies.

C. Bent Stub

Fig. 6 shows the measured and theoretical transmission phase of a bent stub. This is a very complex structure. It has three discontinuities: a tee, a 90° corner, and an open end. In addition, the stub is separated from the main line by one substrate thickness, and therefore coupling is a factor. As Fig. 4 shows, the agreement between measured and calculated results is excellent.

The measurement procedure and modeling parameters were similar to what was used on the low-frequency straight stub. A dielectric constant of 11.0 ± 0.1 was measured. The model dimensions were chosen such that the total length of the bent-stub model is the same as the total length of the measured stub. As a result of the discreteness which is inherent in the modeling procedure, the modeled widths (W_0, W_1, W_2) are slightly different from the actual widths in the measured structure. Measured and nominal calculated resonant frequency differ by 0.7 percent. Due to the uncertainty in the dielectric constant and the measured dimensions which are input to the analysis, the computed result could shift by ± 1.8 percent (worst case) in frequency.

A model of the same structure using quasi-static junction models and full-wave dispersive transmission lines gives a resonant frequency of 7.07 GHz—a 13 percent error. These results, plus the long straight-stub results (where the quasi-static model was good), indicate that coupling between different parts of the circuit is important.

It should be noted that coupling in this modeling scheme is not strong coupling (as in a Lange coupler). Strong coupling would require very accurate field calculations near interior edges.

IV. Conclusion

In this paper, we have presented a technique for analyzing irregular microstrip junctions. All electromagnetic effects, including radiation effects, are taken into account. No sidewalls are assumed, but a cover plate can easily be included if desired. Numerical results are stable to within one or two degrees of phase.

Measurements were made of three stub structures. Measured resonant frequencies agree with full-wave calculations to within 2 percent when nominal measured dielectric constant and structure dimensions were used as input. Using the absolute worst case dielectric constants and dimensions as input gives an error of 3 percent for the low-frequency stub and less than that for the other stubs. The measurements also show: (1) general agreement with calculated phase behavior to within 5–10°, (2) the effect of coupling within a structure, and (3) general agreement between calculated and measured radiation and surface wave loss.

A principal strength of this technique is its ability to model the effects of several discontinuities which are close enough to each other for coupling to occur.

Appendix I
Grounded Slab Green's Function

The tangential electric field components on the surface of a grounded dielectric slab are related to the tangential currents on the same surface via the following Green's function:

$$\vec{\vec{G}}(x, y, x_0, y_0) = \frac{1}{(2\pi)^2} \iint_{-\infty}^{\infty} dk_x \, dk_y \, \vec{\vec{Q}}(k_x, k_y) e^{jk_x(x-x_0)} e^{jk_y(y-y_0)} \quad \text{(A1)}$$

where

$$Q_{xx}(k_x, k_y) = -jZ_0 \sin(k_1 d) \cdot \frac{(\epsilon_r k_0^2 - k_x^2) k_2 \cos(k_1 d) + jk_1 (k_0^2 - k_x^2) \sin(k_1 d)}{k_0 T_e T_m} \quad \text{(A2)}$$

$$Q_{yx}(k_x, k_y) = jZ_0 k_x k_y \sin(k_1 d) \frac{k_2 \cos(k_1 d) + jk_1 \sin(k_1 d)}{k_0 T_e T_m} \quad \text{(A3)}$$

$$Q_{xy}(k_x, k_y) = Q_{yx}(k_x, k_y), \; Q_{yy}(k_x, k_y) = Q_{xx}(k_x \to k_y, k_y \to k_x)$$

where

$$k_2^2 = k_0^2 - k_x^2 - k_y^2$$
$$k_1^2 = \epsilon_r k_0^2 - k_x^2 - k_y^2$$
$$T_e = k_1 \cos(k_1 d) + jk_2 \sin(k_1 d)$$
$$T_m = \epsilon_r k_2 \cos(k_1 d) + jk_1 \sin(k_1 d). \quad \text{(A4)}$$

Appendix II
Infinite Microstrip Line

In this appendix, we present the formulation used to precompute the propagation constant β and the transverse dependency of the x- and y-directed currents on the input and output lines. The procedure is standard, but is outlined here in order to show how it fits in with the discontinuity calculation.

Assuming a current distribution of the form

$$\vec{J}(x, y) = e^{-jhx}\left[g_{xs}(y)\hat{x} + jg_{ys}(y)\hat{y}\right] \quad \text{(A5)}$$

(1) gives the following expression for $\vec{E}$:

$$\vec{E}_i = \frac{e^{-jhx}}{2\pi}\int_{-\infty}^{\infty} dk_y\left[Q_{ix}(-h, k_y)G_{xs}(k_y) + jQ_{iy}(-h, k_y)G_{ys}(k_y)\right]e^{jk_y y} \quad \text{(A6)}$$

where $i = x$ or y and G_{xs}, G_{ys} are the Fourier transforms of g_{xs}, g_{ys}.

The functions g_{xs} and g_{ys} are now expanded so that

$$g_{xs}(y) = \sum_{m=1}^{N_y} c_m g_x^m(y) \quad \text{(A7a)}$$

$$g_{ys}(y) = \sum_{m=1}^{N_y - 1} d_m g_y^m(y) \quad \text{(A7b)}$$

where

$$g_x^m(y) = s(y/W - [m - 1/2]) + s(y/W + [m - 1/2]) \quad \text{(A8a)}$$

$$g_y^m(y) = t(y/2W - m/2) - t(y/2W + m/2) \quad \text{(A8b)}$$

and the functions s and t were defined in (6). N_y and W are chosen such that the total width of the microstrip line is $2N_yW$. Equations (A7) are then Fourier transformed and combined with (A6). Following the well-known moment method procedure, functions (A8) are used to test that the electric field generated by (A6) is zero on the conductor surface. The resulting impedance matrix is

$$\begin{bmatrix} \overleftrightarrow{Z}_{xx}^{\infty}(h) & \overleftrightarrow{Z}_{xy}^{\infty}(h) \\ \overleftrightarrow{Z}_{yx}^{\infty}(h) & \overleftrightarrow{Z}_{yy}^{\infty}(h) \end{bmatrix} \begin{bmatrix} c_1 \\ \vdots \\ c_{N_y} \\ d_1 \\ \vdots \\ d_{N_y - 1} \end{bmatrix} = \begin{bmatrix} \vdots \\ 0 \\ \vdots \end{bmatrix} \quad \text{(A9)}$$

where

$$(Z_{xx}^{\infty})_{m,n} = \frac{1}{2\pi}\int_{-\infty}^{\infty} dk_y G_x^{m*}(k_y)Q_{xx}(-h, k_y)G_x^n(k_y) \quad \text{(A10a)}$$

$$(Z_{xy}^{\infty})_{m,n} = \frac{1}{2\pi}\int_{-\infty}^{\infty} dk_y G_x^{m*}(k_y)Q_{xy}(-h, k_y)jG_y^n(k_y) \quad \text{(A10b)}$$

$$(Z_{yy}^{\infty})_{m,n} = \frac{1}{2\pi}\int_{-\infty}^{\infty} dk_y G_y^{m*}(k_y)Q_{yy}(-h, k_y)G_y^n(k_y). \quad \text{(A10c)}$$

The value of h which makes the determinant of Z^{∞} zero is the propagation constant β. The values of c_i, d_j can then be determined from (A9) and combined with (A7) to generate $g_{xs}(y)$ and $g_{ys}(y)$. The coefficients $\{c_i, d_i\}$ and β are stored for use in the discontinuity calculation.

Acknowledgment

The author wishes to express his thanks to J. J. Burke for his very careful measurements.

References

[1] P. Silvester and P. Benedek, "Microstrip discontinuity capacitances for right-angle bends, *T*-junctions and crossings," *IEEE Trans. Microwave Theory Tech.*, vol. MTT-21, pp. 341–346, 1973.

[2] A. Gopinath *et al.*, "Equivalent circuit parameters on microstrip step change in width and cross junctions," *IEEE Trans. Microwave Theory Tech.*, vol. MTT-24, pp. 142–144, 1976.

[3] R. H. Jansen, "Hybrid mode analysis of end effects of planar microwave and millimeter wave transmission lines," *Proc. Inst. Elec. Eng.*, vol. 128, pt. H, pp. 77–86, Apr. 1978.

[4] J. Boukamp and R. H. Jansen, "The high frequency behavior of microstrip open ends in microwave integrated circuits including energy leakage," in *Proc. 14th European Microwave Conf.*, 1984, pp. 142–147.

[5] R. W. Jackson and D. M. Pozar, "Microstrip open-end and gap discontinuities," *IEEE Trans. Microwave Theory Tech.*, vol. MTT-33, pp. 1036–1042, Oct. 1985.

[6] P. B. Katehi and N. C. Alexopoulos, "Frequency-dependent characteristics of microstrip discontinuities in millimeter-wave integrated circuits," *IEEE Trans. Microwave Theory Tech.*, vol. MTT-33, pp. 1029–1035, Oct. 1985.

[7] N. H. L. Koster and R. H. Jansen, "The microstrip step discontinuity: A revised description," *IEEE Trans. Microwave Theory Tech.*, vol. MTT-34, pp. 213–223, Feb. 1986.

[8] R. Sorrentino and T. Itoh, "Transverse resonance analysis of finline discontinuities," *IEEE Trans. Microwave Theory Tech.*, vol. MTT-32, pp. 1633–1638, Dec. 1984.

[9] H. El-Hennawy and K. Schunemann, "Impedance transformation in fin lines," *Proc. Inst. Elec. Eng.*, vol. 129, pp. 342–350, Dec. 1982.

[10] K. C. Gupta *et al.*, *Computer-Aided Design of Microwave Circuits.* Dedham, MA: Artech House, 1981.

[11] T. Okoshi, *Planar Circuits for Microwaves and Lightwaves.* New York: Springer Verlag, 1985.

[12] G. Kompa, "*S*-matrix computation of microstrip discontinuities and a planar waveguide model," *Arch. Elek. Übertragung.*, vol. 30, pp. 58–64, 1976.

[13] R. H. Jansen, "High-order finite element polynomials in the computer analysis of arbitrarily shaped microstrip resonators," *Arch. Elek. Übertragung.*, vol. 29, pp. 241–247, 1975.

[14] J. R. Mosig and F. E. Gardiol, "General integral equation formulation for microstrip antennas and scatterers," *Proc. Inst. Elec. Eng.*, vol. 132, pt. H, no. 7, pp. 425–432, Dec. 1985.

[15] J. R. Mosig, "Arbitrarily shaped microstrip structures and their analysis with a mixed potential integral equation," *IEEE Trans. Microwave Theory Tech.*, vol. 36, pp. 314–323, Feb. 1988.

[16] S. B. Worm and R. Pregla, "Hybrid-mode analysis of arbitrarily shaped planar microwave structures by the method of lines," *IEEE Trans. Microwave Theory Tech.*, vol. MTT-32, pp. 186–191, Feb. 1984.

[17] J. C. Rautio and R. F. Harrington, "An electromagnetic time-harmonic analysis of shielded microstrip circuits," *IEEE Trans. Microwave Theory Tech.*, vol. MTT-35, pp. 726–730, Aug. 1987.

[18] R. H. Jansen, "Modular source-type 3D analysis of scattering parameters for general discontinuities, components and coupling effects in (M)MICs," in *17th European Microwave Conf. Proc.*, (Rome, Italy), 1987, pp. 427–432.

[19] R. W. Jackson, "Considerations in the use of coplanar waveguide for millimeter-wave integrated circuits," *IEEE Trans. Microwave Theory Tech.*, vol. MTT-34, pp. 1450–1456, Dec. 1986.

[20] R. H. Jansen, "The spectral domain approach for microwave integrated circuits," *IEEE Trans. Microwave Theory Tech.*, vol. MTT-33, pp. 1043–1056, 1985.
[21] D. M. Pozar, "Input impedance and mutual coupling of rectangular microstrip antennas," *IEEE Trans. Antennas Propagat.*, vol. AP-30, pp. 1191–1196, Nov. 1982.
[22] T. C. Edwards, *Foundations for Microstrip Circuit Design*. New York: Wiley, 1981, p. 189.

Chapter 7

Nonreciprocal Planar Components

THIS chapter deals with the analysis and design of nonreciprocal planar circuits. As in the case of reciprocal planar circuits, nonreciprocal planar circuits can have a regular shape (i.e., circular disc, annular ring, triangular, rectangular, sectorial, and elliptical) or an arbitrary shape. For over 30 years planar circuits with ferrite substrate have been used to design junction circulators [7.1]. The general properties of these devices have been investigated by many authors [7.2–7.5], and an impressive bibliography on this subject can be found in [7.2] and [7.3]. In addition, planar circuits with ferrite substrate have been recently found to exhibit very interesting properties as power combiners and dividers [7.6, 7.7]. The general properties of nonreciprocal planar components have been investigated by many methods, the most popular of which are the Green's function approach, the contour integral approach, and the finite-element method. The choice of one technique over another is dependent on the type and shape of the planar circuit under consideration, and also significantly influenced by the experience and preference of the investigator.

Eight reprint papers are included in this chapter. The first three deal with the Green's function–based analysis of planar circulators with regularly shaped resonators. The following two papers describe the application of the contour-integral approach to the analysis of planar circulators having resonators with simple as well as arbitrary shapes. The sixth paper presents an application of the finite-element method to the analysis of arbitrarily shaped planar circulators. The last two papers are particularly interesting; the first is concerned with using ferrite resonators to design nonreciprocal planar power dividers/combiners, whereas the second presents a very revealing and insightful discussion on using ferrite ring resonators to design three- and four-port nonreciprocal planar components with novel characteristics.

The ferrite junction circulator was in use for a number of years before its theory of operation received any attention. The versatility of this device is indicated by the fact that in addition to its use as a circulator, it can also be used as an isolator or as a switch. Paper 7.1 represents the first theoretical design and analysis of stripline circulators. This classical paper is based on using the normal modes of the center disc structure (i.e., resonator) to analyze the stripline *y*-junction circulator. A Green's function for the resonator was used to derive the circuit characteristics of the nonreciprocal planar component. Although Bosma addressed only circulators with circular disc geometry, his method of analysis can be implemented on other structures having threefold symmetry such as triangles, annular rings, and so on [7.6–7.13]. In addition to presenting a theoretical procedure for calculating the frequency characteristics of the stripline *y*-junction circulator, Bosma also discussed a general method for obtaining wider bandwidth. Paper 7.2 extends Bosma's work and gives an explanation to the circulator operation in terms of the rotation of the magnetic field pattern of the $n = 1$ mode of the resonant junction when it is magnetized by an external field. In Paper 7.3, Wu and Rosenbaum used the Green's function method, outlined by Bosma in Paper 7.1, to illustrate how the operational bandwidth of a microstrip circulator can be broadened by selecting appropriate values for the disc radius and the width of the coupling transmission lines. A thorough review of the methods used in Papers 7.1, 7.2, and 7.3 can also be found in [7.16].

The main subject of Paper 7.4 is the analysis of arbitrarily shaped ferrite planar circuits by using the contour-integral approach discussed in Chapter 4 of this book. A second analytical approach based on the Green's function is also discussed in this paper. The validity of both techniques was demonstrated by investigating the magnetic tuning characteristics of square- and triangle-shaped ferrite resonators. The authors also used both the contour-integral approach and the Green's function approach to calculate the frequency characteristics of a triangular circulator. In Paper 7.5, Ayasli used the contour-integral approach outlined in Paper 7.4 along with a new set of boundary conditions, consistent with the Green's function analysis, to examine the wide-band junction circulator designs reported earlier in Paper 7.3 and in [7.17]. The validity of the new set of boundary conditions was verified by comparing the results produced by this method with those reported in [7.17] and in Paper 7.3. Ayasli also noted in this paper that the free-space Green's function consistently used in the contour-integral approach [7.17, 7.18, Paper 7.4] is not unique, but in fact can be selected from a certain class of functions. This arbitrariness in the Green's function and its impact on the accuracy and convergence of the numerical results was also investigated. The use of the contour-integral approach to design planar circulators with broad-band characteristics can also be found in [7.19–7.21].

The application of the finite-element method to the analysis of planar circulators of arbitrary shape is discussed in Paper 7.6. In this paper, Miyoshi's eigenfunction expansion approach is used to set up the impedance matrix formulation for the problem. The variational method is then implemented to select an appropriate basis function that can be used in the expansion of the eigenfunction. Miyoshi's approach proceeds by using a polynomial expansion which describes the fields in the complete resonator. In this paper, however, the resonator region is subdivided into triangular segments and a polynomial expansion for the eigenfunction in each triangular segment is obtained. These are then assembled together to form the complex matrix eigenvalue problem. This approach is verified by applying it to planar circulators having resonators with circular, triangular, regular hexagonal, and irregular hexagonal shapes. The finite-element method has also been employed to calculate the cutoff space of elliptical and cloverleaf-shaped planar gyromagnetic resonators with electric and magnetic walls [7.22–7.23].

Paper 7.7 describes a novel and generalized analysis of nonreciprocal power dividers and combiners. This analysis leads to the requirements that are to be met by the elements of the scattering and impedance matrices of the component if nonreciprocal power division and combining properties are to be attained. The derived requirements are general and hence are independent of the geometry or the structure of the device. An example on how to use these requirements to design a microstrip nonreciprocal power divider/combiner is then presented. A ferrite circular disc resonator is used in the example. Design formulas expressing the power division–combining ratio as a function of the physical parameters and the material properties of the resonator are also given.

Finally, Paper 7.8 examines the properties of novel nonreciprocal components constructed using hollow ferrite rings. Power circulation properties as well as the power division–combining properties are discussed. This paper shows how the ring geometry provides additional design parameters and flexibility that can be used to realize multiport planar components with novel circulation and power division and combining characteristics. A very insightful description on how a four-port power divider junction (with ferrite discs or rings) works is also given.

References

[7.1] Milano, U., J. Saunders, and L. Davis. "A Y-junction stripline circulator." *IRE Trans. on Microwave Theory and Tech.*, Vol. 8, pp. 346–351, May 1960.

[7.2] Bosma, H. "Junction circulators." In *Advances in Microwaves*, Vol. 6, pp. 215–239. New York: Academic Press, 1971.

[7.3] Kumar, R. C. "Ferrite-junction circulator bibliography." *IEEE Trans. Microwave Theory Tech.*, Vol. MTT-18, pp. 524–530, September 1970.

[7.4] Hines, M. E. "Reciprocal and nonreciprocal modes of propagation in ferrite stripline and microstrip devices." *IEEE Trans. Microwave Theory Tech.*, Vol. MTT-19, pp. 442–451, May 1971.

[7.5] Schloemann, E. and R. Blight. "Broad-based stripline circulators based on YIG and Li-Ferrite single crystals." *IEEE Trans. Microwave Theory Tech.*, Vol. MTT-34, pp. 1394–1400, December 1986.

[7.6] Usachov, V. P., B. A. Gapeev, and D. I. Ogloblin. "Nonreciprocal ferrite dividers-adders." *Trudy MVTU*, No. 397, pp. 55–62, 1983.

[7.7] Vamberskii, M. V., V. P. Usachov, and S. A. Shelukhin. "Technical design of two-channel non-reciprocal microstrip line splitter-combiners." *Radioelectronics and Communication Systems*, Vol. 27, No. 12, pp. 17–20, 1984.

[7.8] Helszajn, J. and D. S. James. "Planar triangular resonators with magnetic wall." *IEEE Trans. Microwave Theory Tech.*, Vol. MTT-26, pp. 95–100, February 1978.

[7.9] Helszajn, J., D. S. James, and W. T. Nisbet. "Circulators using planar triangular and circulator resonators in microstrip." *IEEE Trans. Microwave Theory Tech.*, Vol. MTT-27, pp. 188–1193, February 1979.

[7.10] Wu, Y. S. and F. J. Rosenblum. "Modes chart for microstrip ring resonators," *IEEE Trans. Microwave Theory Tech.*, Vol. MTT-21, pp. 487–489, July 1973.

[7.11] Helszajn, J., W. T. Nisbet, and J. Sharp. "Mode charts for gyromagnetic planar ring resonators in microstrip." *Electron Lett.*, Vol. 23, No. 24, pp. 1290–1291, November 1987.

[7.12] Orlando, A. T. F. and R. M. O. Stumpf. "Annular ferrite circulator in stripline technology." *Proc. 21st European Microwave Conf.*, Vol. 2, pp. 1159–1164, Stuttgart, Germany, September 9–11, 1991.

[7.13] Borjak, A. M. and L. E. Davis. "More compact ferrite circulator junctions with predicted performance." *IEEE Trans. Microwave Theory Tech.*, Vol. MTT-40, pp. 2352–2358, December 1992.

[7.14] Davis, L. E. and V. Dimitriyev. "Nonreciprocal devices using ferrite ring resonators." *Proc. Inst. Elect. Eng.*, Pt. H, Vol. 139, No. 3, pp. 257–263, June 1992.

[7.15] Helszajn, J. "Characteristics of circulators using planar triangular and disk resonators." *IEEE Trans. Microwave Theory Tech.*, Vol. MTT-28, pp. 616–621, June 1980.

[7.16] Rosenblum, F. J. "Integrated ferrimagnetic devices." In *Advances in Microwaves*, Vol. 8, pp. 203–294. New York: Academic Press, 1974.

[7.17] Ayter, S. and Y. Ayasli. "The frequency behavior of the stripline circulator junction." *IEEE Trans. Microwave Theory Tech.*, Vol. MTT-26, pp. 197–202, March 1978.

[7.18] Okoshi, T. and T. Miyoshi. "The planar circuit—An approach to microwave integrated circuitry." *IEEE Trans. Microwave Theory Tech.*, Vol. MTT-20, pp. 245–252, April 1972.

[7.19] Miyoshi, T., and S. Miyaushi. "The design of planar circulators for wideband operation." *IEEE Trans. Microwave Theory Tech.*, Vol. MTT-28, pp. 210–215, March 1980.

[7.20] Riblet, G. P. and E. R. B. Hansson. "The use of symmetry to simplify the integral equation method with application to 6-sided circulator resonators." *IEEE Trans. Microwave Theory Tech.*, Vol. MTT-30, pp. 1219–1223, August 1982.

[7.21] Miyoshi, T. and T. Shinhama. "Fully computer-aided synthesis of a planar circulator." *IEEE Trans. Microwave Theory Tech.*, Vol. MTT-34, pp. 294–297, February 1986.

[7.22] Helszajn, J. and A. A. P. Gibson. "Cutoff spaces of elliptical gyromagnetic planar circuits and waveguides using finite elements." *IEEE Trans. Microwave Theory Tech.*, Vol. MTT-37, pp. 71–80, January 1989.

[7.23] Helszajn, J. and D. L. Lynch. "Cutoff space of cloverleaf resonators with electric and magnetic walls." *IEEE Trans. Microwave Theory Tech.*, Vol. MTT-40, pp. 1620–1629, August 1992.

Paper 7.1

On Stripline *Y*-Circulation at UHF*

H. BOSMA†

***Summary*—The simplified boundary-value problem of the circular strip-line *Y*-circulator is stated and, following experimental results, is solved approximately. The circulation parameters are calculated and discussed. The frequency characteristics are evaluated and a general method for broadbanding the device is presented. From the calculated and measured field distribution, an explanatory description of the circulation mechanism is given. Finally, other possible solutions and the features for use at UHF are discussed.**

INTRODUCTION

THE STRIPLINE *Y*-circulator (Fig. 1) polarized above the resonance field has been found to have useful properties in the UHF region and, consequently, has received much attention.[1–5] In its basic form, it is a member of a larger family of nonreciprocal devices, the junction circulators. The general properties including the existence of and the tuning procedure for the symmetrical junction circulators have been studied by Auld[6] through the consideration of their scattering matrices and associated quantities. For the stripline *Y*-circulator in particular, the same has been done by Milano,[1] *et al.*

However, as far as the present author is aware, no consistent theory is yet available about the intrinsic circulator mechanism in terms of the electromagnetic field. The same can be said about the frequency characteristics of these types of circulators. It is true that several attempts have been made to explain the junction circulator in terms of Faraday rotation,[7,8] asymmetrical diffraction[9] or field displacement[5] but, although these efforts have confirmed, more or less, the intuitive thinking about the phenomenon, they have not led to a consistent and manageable theory. In a recent paper,

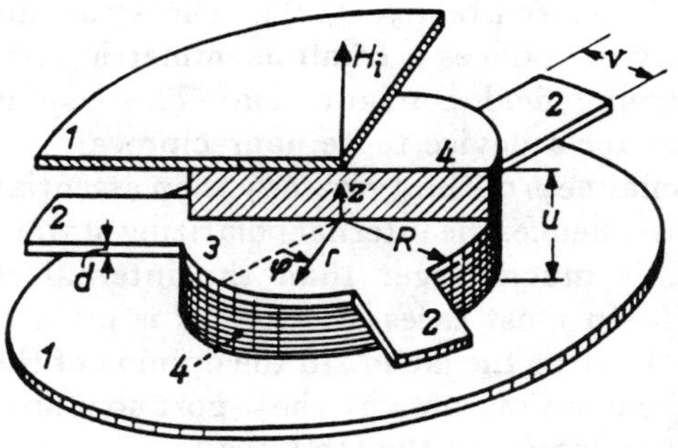

Fig. 1—The circulator configuration. (1) Outer conductors. (2) Inner conductors. (3) Center conductor. (4) Ferrite disks.

Skomal[10] explained the junction circulator in terms of two contrarotating surface waves. He evaluated some of the design parameters. However, some of his results, *e.g.*, the dependence of applied field on magnetization, do not agree with the findings of the present paper. This may be due to the fact that some of the assumptions made by Skomal are not always tenable. In particular, the circumferential distance between adjacent ports is not equal to the effective wavelength in all cases[11] (see below).

The investigation by Bosma,[5] which dealt with the 3-port stripline *Y*-circulator as this paper does, looked quite promising, but it has turned out that some of its results were not in agreement with experiment. Although the problem was stated correctly, the attempt, as is shown below, failed because a practically wrong (although mathematically nearly possible) solution was traced.

In this paper, the same problem is restated and, following the results of experiment,[12] an approximation of the problem is proposed. Then, again led by experimental results, a supposition for a practical solution is made. This solution is tested in the circulator equations and is shown to be consistent with other theoretical considerations. It also agrees well with all experimental results available.

From the solution, first order approximation, of the frequency dependence of the circulator characteristics are obtained. A general method of broadbanding the circulator is proposed. These results are also confirmed by experiment.

*Received April 5, 1963; revised manuscript received August 16, 1963.

†Philips Research Laboratories, N. V. Philips' Gloeilampenfabrieken, Eindhoven, The Netherlands.

[1] U. Milano, J. Saunders and L. Davis, "A *Y*-junction stripline circulator," IRE TRANS. ON MICROWAVE THEORY AND TECHNIQUES, vol. 8, pp. 346–351; May, 1960.

[2] S. Yoshida, "*J-B*and stripline *Y* circulator," PROC. IRE, vol. 48, p. 1664; September, 1960

[3] L. Freiberg. "Lightweight *Y*-junction stripline circulator," IRE TRANS. ON MICROWAVE THEORY AND TECHNIQUES, vol. 8, p. 672; November, 1960.

[4] G. V. Buehler and A. F. Eikenberg, "Stripline *Y*-circulator for the 100 to 400 Mc region," PROC. IRE, vol. 49, p. 518; February, 1961.

[5] H. Bosma, "On the principle of stripline circulation," *Proc. IEE*, vol. 109, pt. B, suppl. no. 21, pp. 137–146; January, 1962.

[6] B. A. Auld, "The synthesis of symmetrical waveguide circulators," IRE TRANS. ON MICROWAVE THEORY AND TECHNIQUES, vol. 7, pp. 238–247; April, 1959.

[7] P. J. Allen, "The turnstile circulator," IRE TRANS. ON MICROWAVE THEORY AND TECHNIQUES, vol. 4, pp. 223–228; October, 1956.

[8] C. Bownes, "Discussion on microwave ferrites II," PROC. IEEE, vol. 109, pt. B, suppl. no. 21, p. 163; January, 1962.

[9] V. G. Feoktistov, "Diffraction model of *Y*-circulator," *Radiotekknika i Elektronika*, vol. 7, pp. 1763–1768; October, 1962.

[10] E. N. Skomal, "Theory of operation of a 3-port *Y*-junction ferrite circulator," IEEE TRANS. ON MICROWAVE THEORY AND TECHNIQUES, vol. 11, pp. 117–123; March, 1963.

[11] J. B. Davies and P. Cohen, "Theoretical Design of Symmetrical Junction Stripline Circulators," IEEE TRANS. ON MICROWAVE THEORY AND TECHNIQUES, vol. MTT-11, pp. 506–512; November, 1963.

[12] Since this paper is dealing with the theory of the stripline circulator, the experiments and their results are not described in detail. They will probably be published elsewhere.

Reprinted from *IEEE Trans. Microwave Theory Tech.*, vol. MTT-12, pp. 61–72, Jan. 1964.

Furthermore, it will be found that circulator action is based on a slight distortion of the degeneracy of two first order resonances of the "isotropic" disk configuration caused by a small amount of gyrotropy of the disk material. Under the amount of gyrotropy, we understand the ratio κ/μ, where κ and μ are the elements of the Polder tensor [*cf.*, (12)]. The small amount of gyrotropy introduces a small asymmetric distortion of the isotropic field configuration. This asymmetry is necessary for a device to be nonreciprocal.

The smallness of the gyrotropy is an essential feature. As a consequence, the internal polarizing static magnetic field H_i is much larger than the internal resonance field[13] H_0. In most cases, the former is more than four times as large as the latter. In the opinion of the author, this is the main reason why the 3-port stripline Y-circulator is so effective in the UHF region.

Properties of the Scattering Matrix

Before the program outlined above is started, some remarks on the scattering matrix of a lossless, cyclic-symmetric 3-port may be made. The scattering matrix S may be written in the form

$$S = \begin{pmatrix} \alpha & \gamma & \beta \\ \beta & \alpha & \gamma \\ \gamma & \beta & \alpha \end{pmatrix}. \tag{1}$$

Carlin[14] has shown that a matched, lossless 3-port is a circulator and Thaxter and Heller[15] have pointed out that, if the reflection-coefficient α is small $|\alpha| \ll 1$, the equations

$$|\gamma| = |\alpha|, \qquad |\beta| = 1 - 2|\alpha|^2 \tag{2}$$

hold approximately. For a circulator which rotates in the opposite sense, γ is interchanged with β. An elegant description of these properties has been given by Butterweck.[16]

The elements of S satisfy the equations

$$|\alpha|^2 + |\beta|^2 + |\gamma|^2 = 1 \tag{3}$$

$$\alpha\beta^* + \beta\gamma^* + \gamma\alpha^* = 0. \tag{4}$$

By virtue of (3), $|\beta|$ can be expressed in terms of $|\alpha|$ and $|\gamma|$. The three terms of (4) can be interpreted as three vectors which span a triangle. Then, inequality relations like

$$|\beta|\,|\gamma| \leq |\alpha|(|\beta| + |\gamma|) \tag{5}$$

are valid. From these considerations Butterweck deduced that in the $|\alpha|$, $|\gamma|$ diagram the possible 3-ports are restricted to a region which is bounded by three ellipses. In Fig. 2 this region is indicated by the shaded area. The origin of the diagram $(|\alpha|, |\gamma|) = (0, 0)$ represents an ideal clockwise rotating circulator. Starting from this point, it is easy to show that, for $|\alpha| \ll 1$, $|\gamma|$ is given by the series

$$|\gamma| = |\alpha| + \theta|\alpha|^2 + \cdots, \quad -1 \leq \theta \leq 1. \tag{6}$$

Then, by virtue of (3), $|\beta|^2$ is given by the series

$$|\beta|^2 = 1 - 2|\alpha|^2 - 2\theta|\alpha|^3 \cdots. \tag{7}$$

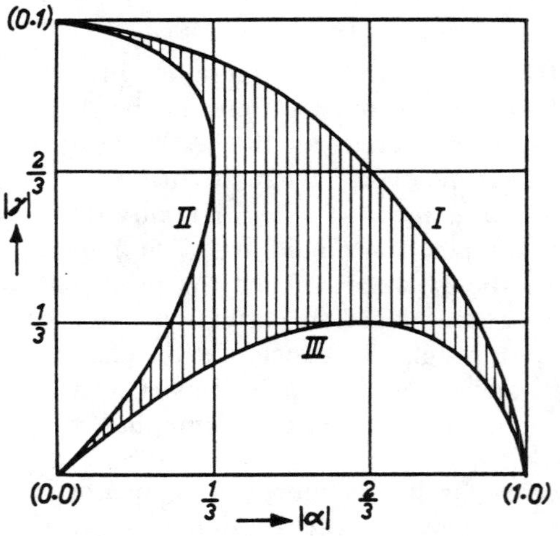

Fig. 2—The closed region of possible, lossless and cyclic-symmetric 3-ports, (Butterweck[16]).
(I) $|\alpha|^2 + |\alpha||\gamma| + |\gamma|^2 - |\alpha| - |\gamma| = 0.$
(II) $|\alpha|^2 - |\alpha||\gamma| + |\gamma|^2 + |\alpha| - |\gamma| = 0.$
(III) $|\alpha|^2 - |\alpha||\gamma| + |\gamma|^2 - |\alpha| + |\gamma| = 0.$

The 3-Port Stripline Y-Circulator

In Fig. 1, a picture of the circulator is given. The cylindrical coordinate system, (r, ϕ, z), used further on, has been drawn into it and several of the important dimensions are presented.

The two ferrite disks are placed between the two earth plates (the outer conductors) and the center conductor. They are magnetized perpendicularly to the plane of the conductors by a static magnetic field. The value of the internal polarizing field (in the ferrite) is denoted by H_i.

In the elementary case, the ferrite disks and the center conductor both have circular form and all three have the same radius R. In practice this need not be so. The disks and the center conductor may have different sizes and forms. Other forms, such as triangles or clover[4] leaves, can be used if only a 3-fold rotation symmetry is present; this, in order to maintain the 3-fold cyclic symmetry of the circulator. Moreover, dielectric inserts may be present. For mathematical simplicity only the elementary configuration is investigated in this paper.

At equal distances around the edge of the center conductor, three inner conductors of stripline waveguides

[13] For reasons of simplicity, we presume that H_0 is given by $H_0 = \omega/\gamma$, where $\gamma = 2\pi 2.8$McOe, although in the configuration in question the ferromagnetic resonance occurs at an internal bias field which can be solved from $\omega = \gamma\sqrt{H(H+4\pi M)}$, where $4\pi M$ is the magnetization of the ferrite. H_0 is used as a reference quantity only.

[14] H. J. Carlin, "Principles of gryator networks," *Proc. Symp. on Modern Advances in Microwave Techniques*, Polytechnical Institute of Brooklyn, N. Y., p. 175–204; November, 1954.

[15] J. B. Thaxter and G. S. Heller, "Circulators at 70 and 140 kMc," Proc. IRE, vol. 48, pp. 110–111; January, 1960.

[16] H. J. Butterweck, "Der Y-zirkulator," *Archiv der Elektrischen Uebertrabung*, vol. 17, p. 163–176; April, 1963.

are connected. The centers of these connections are taken at ϕ-values $-\pi/3$, $\pi/3$ and π for the input, output and decoupled line, respectively. If v is the width of the stripline, a stripline width angle Ψ is defined by (see Fig. 3)

$$\Psi = \sin^{-1}(v/2R). \tag{8}$$

The thickness d of the inner and center conductors is assumed to be zero. The distance between the two outer conductors is u, twice the thickness of one ferrite disk.

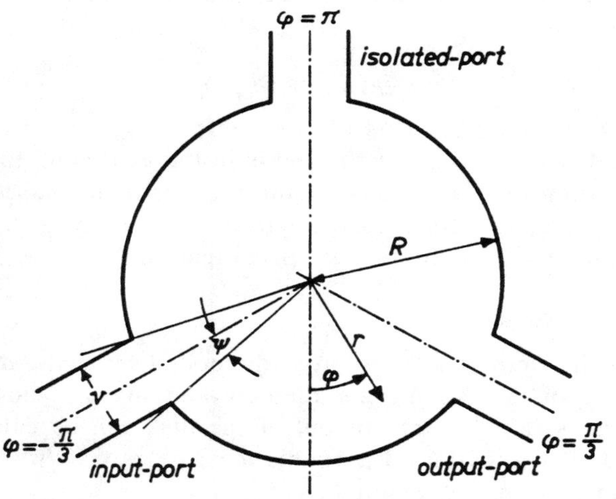

Fig. 3—The configuration of the center conductor.

Field Configuration in and Characteristic Impedance of the Striplines

In the cross section of the striplines, the field intensities are supposed not to vary over the width of the inner conductors and to be zero outside these conductors. The stray fields at the edges are neglected. However, especially in the neighborhood of the connections to the center conductor, rather large differences from the assumed ideal field configuration can be expected to exist. This supposition made about the fields in the stripline is clearly an approximation only.

The striplines carry TEM modes only. Hence, the electric field intensities are perpendicular to the conductors (in the z-direction) and the magnetic field intensities are parallel to them and perpendicular to the directions of propagation (in the ϕ-direction). Furthermore, the fields on the two sides of the inner conductors are 180° out of phase.

The characteristic impedance of the striplines is not important, because of the method of analysis presented. For convenience of matching to the other elements of the experimental setups, the characteristic impedances are made equal to 50 Ω in all practical cases. This is done by the proper choice[17] of u with respect to v (and d).

The ratio of the electric and magnetic field intensities of a traveling wave, the intrinsic wave, impedance ζ, is of much more importance for the analysis. In the simple model of the circulator, no dielectric materials other than air are placed between the conductors of the striplines. Hence, ζ is given by

$$\zeta = \sqrt{\mu_0/\epsilon_0} = 120\pi\,\Omega. \tag{9}$$

Magnetic Field Intensity at the Edge of a Disk

As is the case with the fields in the striplines, the electric as well as the magnetic field intensities on either side of the center conductor are equal and oppositely directed at any instant. Hence, the field problem need be solved only for one disk in association with the fields on the appropriate sides of the inner conductors of the striplines.

Furthermore, it is supposed that the field in a disk does not depend on the z-coordinate, so that the problem can be reduced to two of three dimensions.

Except at the connections of the striplines, no radial current can flow from the edge of the center conductor. From boundary equations, it is known that, at the connections, the tangential component of the magnetic field intensity in the ferrite at its edge H_ϕ (R, ϕ) is equal to the magnetic field intensity in the corresponding stripline at the same place. Consequently, H_ϕ (R, ϕ) is constant over the stripline widths and it is zero elsewhere.[18] This can be expressed by

$$H_\phi(R,\phi) = \begin{cases} a) & -\pi/3 - \Psi < \phi < -\pi/3 + \Psi, \\ b) & \pi/3 - \Psi < \phi < \pi/3 + \Psi, \\ c) & \pi - \Psi < \phi < \pi + \Psi, \\ d) & \text{elsewhere}. \end{cases} \tag{10}$$

The fields at the different connections need not be in phase, so that in principle a, b and c are complex numbers.

The investigation in this paper will be restricted to values of v which are not too large in comparison with R; more precisely

$$v < R. \tag{11}$$

The smaller the ratio v/R is [*cf.*, (8)], the more accurate is the assumed distribution of $H_\phi(R, \phi)$, but the larger is the disturbing influence of the stray fields of the striplines. Hence, v must not be very small.

Electric Field Intensity in the Disk

The electric field intensity in the disk is assumed to have a z-component only. It is denoted $E_z(r, \phi)$. The specific permittivity of the ferrite is denoted by ϵ and the

[17] R. W. Peters *et al.*, "Handbook of Triplate Microwave Components," Sanders Associates, Inc., Nashua, N. H. 1956.

[18] Hereafter, it is shown that the electric field intensity at the edge of the disk is not equal to zero. Thus, an electric stray field which has a radial component is present. Consequently, the supposition $H_\phi(R, \phi) = 0$ outside the connections is not exact. However, from a theoretical estimation and also from experiments, it appeared that the resultant inaccuracy is not larger than those of the other approximate suppositions.

specific gyrotropic permeability by $\|\mu\|$. The latter has a tensor character and is given by

$$\|\mu\| = \begin{Bmatrix} \mu & -i\kappa & 0 \\ i\kappa & \mu & 0 \\ 0 & 0 & 1 \end{Bmatrix}. \tag{12}$$

An effective specific permeability μ_{eff} can be introduced by

$$\mu_{eff} = (\mu^2 - \kappa^2)/\mu \tag{13}$$

and an intrinsic wave number k by

$$k^2 = \omega^2 \mu_0 \epsilon_0 \mu_{eff} \epsilon. \tag{14}$$

With these definitions made, it is shown by Bosma[5] that $E_z(r, \phi)$ satisfies the homogeneous Helmholtz equation

$$\left[\frac{\partial^2}{\partial r^2} + \frac{1}{r}\frac{\partial}{\partial r} + \frac{1}{r^2}\frac{\partial^2}{\partial \phi^2} + k^2\right] E_r(r, \phi) = 0 \tag{15}$$

and that the tangential and radial components of the magnetic field intensity in the disk are related to $E_z(r, \phi)$ by, respectively,

$$H_\phi(r, \phi) = i\left[\frac{\partial E_r}{\partial r} + i\frac{\kappa}{\mu}\frac{1}{r}\frac{\partial E_r}{\partial \phi}\right] \Big/ \omega\mu_0\mu_{eff} \tag{16}$$

$$H_r(r, \phi) = -i\left[\frac{1}{r}\frac{\partial E_r}{\partial \phi} - i\frac{\kappa}{\mu}\frac{\partial E_r}{\partial r}\right] \Big/ \omega\mu_0\mu_{eff}. \tag{17}$$

If $H_\phi(r, \phi)$ is known at the edge [*cf.*, (10)], then, by virtue of (16), $E_z(r, \phi)$ is determined by oblique boundary conditions. Then, a Green's function $G(r, \phi; r', \phi')$ can be introduced[19] such that

$$E_z(r, \phi) = \int_{-\pi}^{\pi} G(r, \phi; R, \phi') H_\phi(R, \phi') d\phi'. \tag{18}$$

In most cases, one is interested in $E_z(r, \phi)$ at the edge only and, therefore, in order to simplify the equations, one may write

$$G(R, \phi; R, \phi') = G(\phi; \phi'). \tag{19}$$

From the distribution $H_\phi(R, \phi)$ given by (10), $E_z(R, \phi)$ can be evaluated at the centers of the connections.[20] For small values of Ψ, the following approximate equations result:

$$E_z(R, -\pi/3) = A = 2\Psi[G(-\pi/3; -\pi/3)a + G(-\pi/3; \pi/3)b + G(-\pi/3, \pi)c] \tag{20}$$

$$E_z(R, \pi/3) = B = 2\Psi[G(\pi/3; -\pi/3)a + G(\pi/3; \pi/3)b + G(\pi/3, \pi)c] \tag{21}$$

$$E_z(R, \pi) = C = 2\Psi[G(\pi; -\pi/3)a + G(\pi; \pi/3)b + G(\pi; \pi)c]. \tag{22}$$

Since E_z is parallel to the cylindrical air-ferrite interface, it is the same on both sides of the edge. Therefore, A, B, and C are also the electrical field intensities in the corresponding striplines.

Eqs. (20)–(22) express the relations between the RF-em fields in the striplines (the incident and reflected waves) in terms of the properties of the field pattern in the disk. Hence, the scattering matrix elements of the 3-port circulator can be expressed in terms of $G(\phi, \phi')$.

PROPERTIES OF $G(\phi; \phi')$

The Green's function can be interpreted as a transfer impedance function for the fields in the disk configuration. Before it is evaluated for the particular case of this paper, the conditions imposed upon it by general circulator characteristics are investigated.

A. Losslessness

If the field configuration at the boundary of the disk is known, by Poynting's theorem the average power which is radiated into or out of the disk can be calculated. For a lossless 3-port this must be zero. With the complex Poynting vector $\boldsymbol{P}$,

$$\boldsymbol{P} = \tfrac{1}{2}\boldsymbol{E} \times \boldsymbol{H}^*, \tag{23}$$

in which the asterisk denotes the complex conjugate, this is expressed by

$$Re\left[\int_S \boldsymbol{P} \cdot \boldsymbol{n} dS\right] = 0, \tag{24}$$

where $\boldsymbol{n}$ is normal to the closed ferrite surface S. As the normal component of $\boldsymbol{P}$ is nonzero only on the cylindrical part of the ferrite surface, by virtue of (18), (19), (23) and (24) it is easy to show that losslessness of the disk implies that the equation

$$\int_{-\pi}^{\pi}\int_{-\pi}^{\pi} H_\phi^*(R, \phi)[G(\phi; \phi') + G^*(\phi'; \phi)] \cdot H_\phi(R, \phi') d\phi d\phi' = 0 \tag{25}$$

must be satisfied. Since this must be valid for all distributions $H_\phi(R, \phi)$, the condition for losslessness is

$$G(\phi'; \phi) = -G^*(\phi; \phi') \tag{26}$$

which is analogous to the relation $Z_{ji} = -Z_{ij}^*$ which is valid for lossless $2n$-ports in the impedance matrix theory.

B. Cyclic Symmetry

The cyclic symmetry of a 3-port requires the condition

$$G(\phi + 2\pi/3; \phi' + 2\pi/3) = G(\phi; \phi') \tag{27}$$

to be fulfilled.

[19] D. van Dantzig and H. A. Lauwerier, "The North Sea problem," *Proc. Koninklijke Nederlandse Akademie van Wetenschappen*, series A (mathematical sciences), vol. LXIII, p. 170–180; 1960.

[20] Instead of evaluating $E_z(R, \phi)$ at the centers of the striplines, the average values of $E_z(R, \phi)$ over the stripline widths could have been determined. This more complicated method has been carried out by Bosma,[5] but its results are not much more accurate.

C. Resonances of the Disk

The disk configuration with disconnected striplines has two types of resonances.

1) *Electric wall:* If the edge of the center conductor is short-circuited to the outer conductors, the resonance condition is that $E_z(R, \phi)$ is identically zero. Hence, the resonance condition is

$$G(\phi; \phi') = 0. \tag{28}$$

For the subject matter of this paper, this type of resonances is of no importance.

2) *Magnetic wall:* If the edge of the disk constitutes a magnetic wall, the resonance condition is

$$G(\phi; \phi') = \text{infinity}. \tag{29}$$

Simplified Boundary Equations

At this stage, it is advantageous to introduce two quantities ξ and Φ defined by

$$i\xi = 2\Psi G(-\pi/3; -\pi/3) \tag{30}$$

$$\Phi = 2\Psi G(-\pi/3; \pi/3). \tag{31}$$

The condition for losslessness (26) implies that ξ is real. In general, ϕ is complex. The application of the cyclic symmetry condition (27) to the coefficients of the right hand members of (20)–(22) and the substitution of (30) and (31) into these equations result in

$$A = i\xi a + \Phi b - \Phi^* c \tag{32}$$

$$B = -\Phi^* a + i\xi b + \Phi c \tag{33}$$

$$C = \Phi a - \Phi^* b + i\xi c. \tag{34}$$

Matching Condition

A good method of investigating the more special properties of the circulator is to terminate the output and the isolated port with reflectionless loads. Then, the parts of a wave incident upon the input port, which are dissipated in these terminations, are measures of the transmission coefficients. The reflected part of that wave is related to the reflection coefficient. In an earlier paper,[5] this has been called the matching condition.

In this situation, only outward traveling waves are present in the output and isolated striplines so that, by virtue of (9), the relations

$$B/b = C/c = -\zeta \tag{35}$$

hold. Substitution into (32)–(34) and introduction of the complex quantity Θ given by

$$\Theta = \zeta + i\xi \tag{36}$$

result in the relations

$$A = \left(i\xi + \frac{\Phi^3 - \Phi^{*3} + 2\Phi\Phi^*\Theta}{\Phi\Phi^* + \Theta^2}\right) a \tag{37}$$

$$b = \frac{\Phi^2 + \Phi^*\Theta}{\Phi\Phi^* + \Theta^2} a \tag{38}$$

$$c = \frac{\Phi^{*2} - \Phi\Theta}{\Phi\Phi^* + \Theta^2} a. \tag{39}$$

Scattering Matrix

The scattering matrix S of a cyclic symmetric 3-port is given by (1). If the electric and magnetic field intensities of the wave incident upon the input port are A_i and a_i, respectively, the following relations are valid:

$$A_i = \zeta a_i \tag{40}$$

$$A = (1 + \alpha) A_i \tag{41}$$

$$a = (1 - \alpha) a_i \tag{42}$$

$$\beta = B/A_i \tag{43}$$

$$\gamma = C/A_i. \tag{44}$$

After some algebraic manipulation, the following expressions are found for α, β and γ:

$$\alpha = \frac{\Phi^3 - \Phi^{*3} + \Phi\Phi^*(2\Theta - \Theta^*) - \Theta^2\Theta^*}{\Phi^3 - \Phi^{*3} + 3\Phi\Phi^*\Theta + \Theta^3} \tag{45}$$

$$\beta = -\frac{(\Phi^2 + \Phi^*\Theta)(\Theta + \Theta^*)}{\Phi^3 - \Phi^{*3} + 3\Phi\Phi^*\Theta + \Theta^3} \tag{46}$$

$$\gamma = -\frac{(\Phi^{*2} - \Phi\Theta)(\Theta + \Theta^*)}{\Phi^3 - \Phi^{*3} + 3\Phi\Phi^*\Theta + \Theta^3}. \tag{47}$$

All the elements of the scattering matrix are now expressed in terms of the Green's function $G(\phi; \phi')$ and the properties of the medium[21] of the striplines ζ.

Circulation Condition

From the considerations made above about cyclic symmetric 3-ports, such a 3-port is an ideal clockwise rotating circulator, if the conditions

$$\alpha = \gamma = 0 \tag{48}$$

$$|\beta| = 1 \tag{49}$$

are fulfilled. It will be clear that this only can be so if all media are lossless and all conductors are perfect. These conditions can be satisfied for given values of R and $4\pi M$ only with certain values of ω and H_i. Such a combination of the parameters R, ω, H_i and $4\pi M$ could be called a circulation adjustment.

The condition $\gamma=0$ is conceptually the same as the requirement $c=0$ for all a [*cf.*, (39)]. In the latter form, it was earlier called[5] the circulation condition. Since

$$\Theta + \Theta^* = 2\zeta \neq 0. \tag{50}$$

[21] The formulas given apply only if the stripline medium is lossless.

application of $\gamma=0$ to (47) yields the circulation equation

$$\Theta = \Phi^{*2}/\Phi. \tag{51}$$

Substitution into (45) and (46) shows that $\alpha=0$ and

$$\beta = -\Phi/\Phi^{*} \tag{52}$$

which satisfy the other circulation requirements posed by (48) and (49). At first sight, one might wonder why $\gamma=0$ is a sufficient condition for circulator action, but it has been shown above that this is implied by losslessness and cyclic symmetry.

Rewriting the circulation equation (51) in its real and imaginary parts yields the two circulation equations (54) and (55) derived by Bosma.[5] Therefore, the elements of the latter can be explained now in terms of the real and imaginary parts of the transfer wave impedances of the disk configuration and of the intrinsic wave impedance of the stripline medium.

Green's Function

To solve the circulation equation in a particular case, Φ and Θ must be expressed in the characterizing quantities R, Ψ, ω, H_i and $4\pi M$. To that end $G(\phi;\phi')$ will now be deduced.

In the present case, the general solution of the Helmholtz equation (15) is a series in which the general term is given by

$$E_{z,n}(r,\phi) = a_n J_n(kr)e^{in\phi}, \tag{53}$$

where J_n is the nth order Bessel function. The corresponding term of the series for $H_\phi(r,\phi)$ is, by virtue of (16),

$$H_{\phi,n}(r,\phi) = i\frac{a_n}{\zeta_{\mathrm{eff}}}\left[J_n'(kr) - \frac{\kappa}{\mu}\,\frac{nJ_n(kr)}{kr}\right]e^{in\phi}, \tag{54}$$

where ζ_{eff} is the effective intrinsic wave impedance of the ferrite as given by

$$\zeta_{\mathrm{eff}} = \sqrt{\mu_0\mu_{\mathrm{eff}}/\epsilon_0\epsilon}. \tag{55}$$

Inversely, if

$$H_{\phi,n}(R,\phi) = A_n e^{in\phi} \tag{56}$$

is the nth Fourier component of $H_\phi(R,\phi)$ at the edge, $r=R$ the corresponding term of the Bessel-Fourier series of $E_z(r,\phi)$ is given by

$$E_{z,n}(r,\phi) = iA_n\frac{\zeta_{\mathrm{eff}}J_n(kr)e^{in\phi}}{\dfrac{\kappa}{\mu}\,\dfrac{nJ_n(x)}{x} - J_n'(x)}, \qquad \text{(5}$$

where

$$x = kR = \omega\sqrt{\mu_0\epsilon_0\mu_{\mathrm{eff}}\epsilon}\,R. \qquad \text{(5}$$

Let $H_\phi(R,\phi)$ be everywhere zero at the edge except a an azimuth ϕ' over a small angle $\Delta\phi'$ where it i $H_\phi(R,\phi')$. Then,

$$A_n = (1/2\pi)H_\phi(R,\phi')e^{-in\phi'}\Delta\phi' \qquad \text{(5}$$

and the corresponding contribution to $E_z(r,\phi)$ is

$$\Delta E_z(r,\phi) = \frac{i\zeta_{\mathrm{eff}}H_\phi(R,\phi')\Delta\phi'}{2\pi}\sum_{n=-\infty}^{\infty}\frac{J_n(kr)e^{in(\phi-\phi')}}{\dfrac{\kappa}{\mu}\,\dfrac{nJ_n(x)}{x} - J_n'(x)} \qquad \text{(6}$$

Comparing this with the defining relation (18) afte some algebra, one obtains

$$G(r,\phi;R,\phi') = -\frac{i\zeta_{\mathrm{eff}}J_0(kr)}{2\pi J_0'(x)} + \frac{\zeta_{\mathrm{eff}}}{\pi}\sum_{n=1}^{\infty}\frac{\dfrac{\kappa}{\mu}\,\dfrac{nJ_n(x)}{x}\sin n(\phi-\phi') - iJ_n'(x)\cos n(\phi-\phi')}{\{J_n'(x)\}^2 - \left\{\dfrac{\kappa}{\mu}\,\dfrac{nJ_n(x)}{x}\right\}^2}J_n(kr). \qquad \text{(6}$$

Resonances

Only the resonances of the disk configuration wit magnetically short-circuited edge have importance fo the present investigation. From (29), it is seen that suc a resonance occurs if the denominator of one of th terms of the series (61) is zero,

$$J_n'(x) - \frac{\kappa}{\mu}\,\frac{nJ_n(x)}{x} = 0 \qquad \text{(6}$$

where n may be any positive or negative integer. Fro (54), it is easy to show that for these resonance $H_{\phi,n}(R,\phi)$ is identically zero. Moreover, for $n>0$ th field pattern is rotating to the right and for $n<0$ to th left. If the disk medium is isotropic, $\kappa/\mu=0$, the reso nance conditions of the clockwise and anticlockwis rotating modes are identical

$$J_n'(x) = 0 \qquad \text{(6}$$

and, hence, these resonances are degenerate. When th gyrotropy increases, $\kappa/\mu\neq 0$, such a degenerate pa splits giving different x-values. For a disk R at a pa ticular magnetic adjustment κ/μ and μ_{eff}, these tw resonances have different frequencies which can b calculated from (58). With (62) and the recurrence rel tions of Bessel functions, it is easy to deduce that for small amount of gyrotropy $\kappa/\mu \ll 1$ the difference of x-values of a pair of resonances is determined by

$$(\Delta x)_{n,j} = 2\frac{n x_{n,j}}{x_{n,j}^2 - n^2}(\kappa/\mu), \quad (64)$$

where $x_{n,j}$ is the jth solution of the nth order equation (63). n denotes the order of the resonant modes and j the number of the pair. In Fig. 4, the κ/μ-dependence has been sketched for several of these modes.

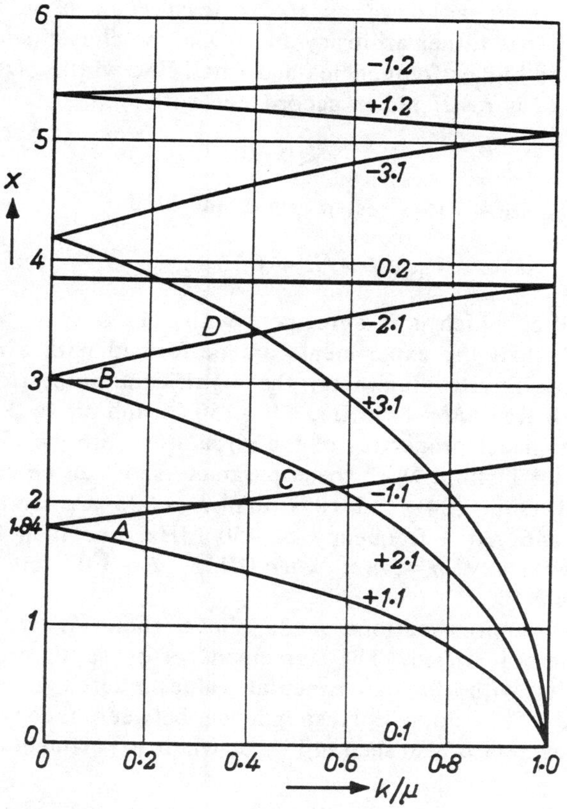

Fig. 4—The dependence on κ/μ of the x-values of several resonant disk modes. + sign denotes a clockwise rotating mode. − sign denotes an anticlockwise rotating mode. The first index refers to the order and the second one to the pair-number of a mode.

From experiments, it appeared that the practical circulation adjustment is situated between the first pair 1st-order resonances. In Fig. 4, this is in the neighborhood of A. Hence, the value of x is about equal to $x_{1,1}$. Therefore, it is supposed that the relation

$$J_1'(x_{1,1}) = 0, \qquad x_{1,1} = 1.84 \quad (65)$$

is exactly valid for the circulator.

From the measured frequency difference between the first pair 1st-order resonances at the (magnetic) circulation adjustment and the value of $x_{1,1}$ calculated from it with (58), it appeared, by virtue of (64), that $\kappa/\mu = 0.125$. The other parameters, determined experimentally, of this circulation adjustment are $f = 450$ MHz, $R = 3.14$ cm, $H_i \approx 850$ Oe, $4\pi M = 1750$ Oe and $\Delta H \approx 150$ Oe.

The experimental value of κ/μ is very small, much smaller than would have been presumed intuitively. It is the opinion of the author that this is the main reason why an explanation of the junction circulation mechanism has been awaited for so long.

Approximate Circulation Adjustment

Remembering the experimental data above, it is easy to see that the term for $n=1$ of the Green's function (61) is large compared with the other terms. Therefore, $G(r, \phi; R, \phi')$ will be approximated by retaining the term for $n=1$ only. A numerical estimation made afterwards showed that the error introduced in this way amounts to a few per cent only.

Substituting (65) into (61), by virtue of (19), for $G(\phi; \phi')$, the formula

$$G(\phi; \phi') = -\frac{x_{1,1}\zeta_{eff}}{\pi(\kappa/\mu)}\sin(\phi - \phi') \quad (66)$$

is found. Then, by virtue of (30), (31) and (36), Θ and Φ are found to be

$$\Theta = \zeta \quad (67)$$

$$\Phi = \frac{\sqrt{3}x_{1,1}\zeta_{eff}}{\pi(\kappa/\mu)}\Psi. \quad (68)$$

In this particular approximated case, both these quantities are real. Due to (9) and (55), substitution into the circulation equation (51) yields

$$\kappa/\mu = \frac{\sqrt{3}x_{1,1}}{\pi}\Psi\sqrt{\mu_{eff}/\epsilon}. \quad (69)$$

As $\omega\sqrt{\mu_0\epsilon_0} = 2\pi/\lambda$, λ being the free space wavelength, (58) yields

$$R/\lambda = \frac{x_{1,1}}{2\pi\sqrt{\mu_{eff}\epsilon}}\cdot \quad (70)$$

For small values of Ψ, this factor[22] in the right hand member of (69) can be replaced by $(v/2R)$. Then, elimination of R from (69) and (70) yields the relation

$$\frac{\kappa/\mu}{\mu_{eff}} = \sqrt{3}\,\frac{v}{\lambda}\cdot \quad (71)$$

Ferrite Properties

In order that the parameters R and H_i can be calculated at a given frequency, the electromagnetic properties of the ferrite, ϵ, κ, μ and μ_{eff} have to be known. Since the ferrite is biased far above the magnetic resonance

[22] If, for the electrical field intensities at the ports, the averaged values and not the values of $E_z(R, \phi)$ at the centers of the striplines had been taken, the factor in Ψ in (69) would have been[6] sin (Ψ). Then, the replacement by $(v/2R)$ is exact for all values of Ψ.

field $\kappa/\mu \ll 1$, the real parts of κ and μ need be considered only. These are given by

$$\mu = 1 + \frac{hm(h^2 - 1)}{(h^2 - 1)^2 - s^2} \tag{72}$$

$$\kappa = \frac{m(h^2 - 1)}{(h^2 - 1)^2 + s^2}, \tag{73}$$

where

$$m = 4\pi M/H_0 \tag{74}$$

$$h = H_i/H_0 \tag{75}$$

$$s = \Delta H/H_0. \tag{76}$$

For the circulators considered, the approximations $h^2 \gg 1$ and $h^2 \gg s$ can be made, so that (73) and (74) become

$$\mu = \mu_{eff} = (h + m)/h \tag{77}$$

$$\kappa = m/h^2. \tag{78}$$

Circulation Parameters

Insertion of (77) and (78) into (70) and (71) yields

$$R/\lambda = \frac{x_{1,1}}{2\pi\sqrt{\epsilon}}\sqrt{\frac{h}{h+m}} \tag{79}$$

$$h = \sqrt{\frac{\lambda m}{\sqrt{3}v}} - m, \tag{80}$$

while elimination of h from these equations results in

$$R/\lambda = \frac{x_{1,1}}{2\pi\sqrt{\epsilon}}\sqrt{1 - \sqrt{\sqrt{3}\,\frac{mv}{\lambda}}}\,. \tag{81}$$

From the last relation, one might conclude that small circulators can be obtained with large magnetizations and stripline widths. However, calculation of h shows that this quantity then becomes small (or even negative), which from considerations of resonance losses and magnetic saturation is not practical.

In general, it may be stated that in the UHF region h must be as large as possible. From (81), it is easily seen that h attains its maximum value for

$$m_{h_{max}} = h_{max} = (1/4\sqrt{3})\,\frac{\lambda}{v}\,. \tag{82}$$

By virtue of (81), the corresponding relative disk radius is given by

$$(R/\lambda)_{h_{max}} = \frac{x_{1,1}}{2\pi\sqrt{2\epsilon}} = 0.207e^{-1/2}. \tag{83}$$

The inverse proportionality with $\sqrt{\epsilon}$ is self-evident. Increase of h_{max} by diminution of v (and corresponding choice of m) is not unrestricted, as the resulting stray fields become undesirably large.

Eqs. (80) and (81) can be rewritten in the form

$$H_i = \sqrt{\frac{\lambda}{\sqrt{3}v}H_0 \cdot 4\pi M} - 4\pi M \tag{84}$$

$$R/\lambda = \frac{x_{1,1}}{2\pi\sqrt{\epsilon}}\sqrt{\frac{H_i}{H_i + 4\pi M}}\,. \tag{85}$$

As H_0 is inversely proportional with λ, for the same stripline width and magnetization (v and $4\pi M$ constant), H_i and R/λ do not depend on frequency. Hence, with the latter increasing, H_i and R become too small with respect to H_0 and v, respectively. Consequently, v must be chosen smaller for higher frequencies. A good estimation is obtained by the assumption that H_i must be four times as large as H_0 at least, *i.e.*, $h>4$. This number is rather arbitrary and it may be chosen smaller for the higher frequencies and small line widths. Due to (83), v is restricted in accordance with

$$v < \lambda/30, \tag{86}$$

so that for $\epsilon \approx 16$ a restriction of the kind

$$v/R < 0.75 \tag{87}$$

is valid, which is in agreement with (11).

Most of the experiments are performed with a manganese ferrite aluminate, the significant properties of which are $4\pi M = 1750$ Oe, $\Delta H = 150$ Oe and $\epsilon = 14.2$. The geometrical properties of the circulator were $v = 15$ mm and $u = 11$ mm. With these parameters, it can be calculated from (84) and (85) that $H_i = 935$ Oe and $R/\lambda = 0.046$. At a frequency of 450 MHz, one then finds theoretically $h = 5.82$, $\kappa/\mu = 0.112$, $R = 3.07$ cm and $v/R = 0.48$.

All approximations made above about h, κ/μ and v/R are justified. The agreement of these theoretical results with the experimental values quoted above is good. This good correspondence between theory and experiment is also seen in Fig. 5, where experimental and

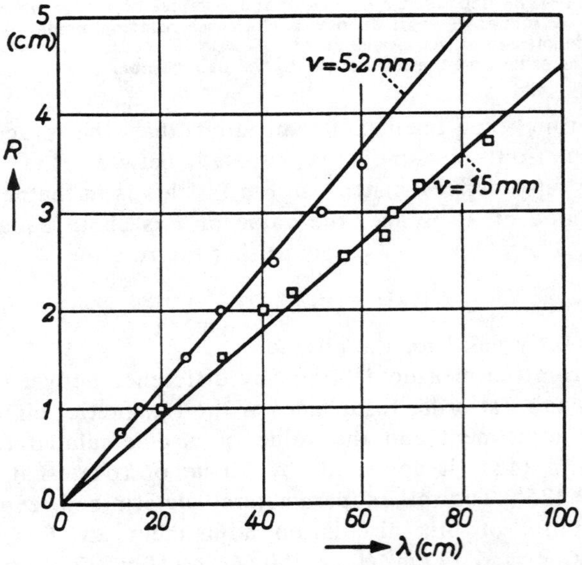

Fig. 5—Theoretical and experimental disk radii against free space wavelength with the stripline width as parameter. The drawn lines denote the theoretical results, (○) measured radii for v = 5.2 mm and (□) measured radii for v = 15.0 mm.

theoretical values of R are given as a function of λ, with values of v 15 mm and 5.2 mm, respectively. Values of H_i are not inserted into Fig. 5 as they were not measured accurately because of concentration of the static magnetic flux in the disks in most cases.

Frequency Dependence

For any ferrite disk R, there is a circulation adjustment for one frequency only. This frequency will be called the circulation frequency ω_0. If all other parameters are kept constant, the circulator characteristics are functions of frequency. Analogously, H_i can be varied in order to have a tunable circulator. The latter case will not be investigated here.

If $\omega \neq \omega_0$, $G(\phi;\phi')$ is a function of ω. Neglecting all terms for $n \neq 1$ again, from (61) it is

$$G(\phi;\phi') = \frac{Z_{eff}\left[\frac{\kappa}{\mu}\,\frac{J_1(x)}{x}\sin(\phi-\phi') - iJ_1'(x)\cos(\phi-\phi')\right]J_1(x)}{\pi\left[\{J_1'(x)\}^2 - \left\{\frac{\kappa}{\mu}\,\frac{J_1(x)}{x}\right\}^2\right]}. \tag{88}$$

For $x = x_{1,1}$ and $\omega = \omega_0$ it is easy to deduce that μ does not depend on ω and that the relations

$$\frac{d(\kappa/\mu)}{d\omega} = \frac{\kappa/\mu}{\omega_0}, \qquad \frac{dx}{d\omega} = \frac{x_{1,1}}{\omega_0} \tag{89}$$

hold. Using them together with recurrence relations of Bessel functions and (65) and (66), $G(\phi;\phi')$ can be developed into a Taylor-series. Neglecting 2nd and higher order terms of $\delta\omega$, where

$$\delta\omega = \omega - \omega_0, \tag{90}$$

this series is given by

$$G(\phi;\phi') = -\frac{\zeta_{eff}}{\pi}\left[\frac{x_{1,1}}{\kappa/\mu}\sin(\phi-\phi') + i\frac{x_{1,1}(x_{1,1}^2-1)}{(\kappa/\mu)^2\omega_0}\cos(\phi-\phi')\delta\omega\right]. \tag{91}$$

Then, by virtue of (30), (31) and (36), the quantities Θ and Φ are found to be

$$\Theta = \zeta(1 - i2\eta) \tag{92}$$

$$\Phi = \zeta(1 + i\eta), \tag{93}$$

where

$$\eta = \frac{x_{1,1}^2 - 1}{\sqrt{3}(\kappa/\mu)}\cdot\frac{\delta\omega}{\omega_0}. \tag{94}$$

Substitution into (45)–(47) yields, for the elements of the scattering matrix,

$$\alpha = \frac{-3\eta^2 + i2\eta}{(4-9\eta^2) - i6\eta} \tag{95}$$

$$\beta = \frac{(-4+6\eta^2) + i2\eta}{(4-9\eta^2) - i6\eta} \tag{96}$$

$$\gamma = \frac{b\eta^2 + i2\eta}{(4-9\eta^2) - i6\eta}. \tag{97}$$

Eqs. (3) and (4) are easily verified. For small η or, what is the same, for small frequency shifts, $\delta\omega/\omega_0 \ll 1$, α, β and γ can be approximated by

$$\alpha = \gamma = i\eta/2 = i\frac{x_{1,1}^2 - 1}{2\sqrt{3}(\kappa/\mu)}\cdot\frac{\delta\omega}{\omega_0} \tag{98}$$

$$1 - |\beta|^2 = \frac{1}{2}\eta^2 = \frac{1}{2}\left\{\frac{x_{1,1}^2 - 1}{\sqrt{3}(\kappa/\mu)}\right\}^2\left(\frac{\delta\omega}{\omega_0}\right)^2. \tag{99}$$

α and γ measure the return losses and the isolation, while $(1-|\beta|^2)$ is an expression for the forward attenuation. Due to dissipative losses, the latter is much larger in practice. The relation $\alpha = \gamma$ is easy to verify experimentally.

Bandwidth

The bandwidth of a circulator is not an unambiguous quantity. It can be defined in several ways. A practically useful method is the specification of a maximal value for the modulus of the reflection coefficient. Thus,

$$|\alpha| \le \rho_{max}. \tag{100}$$

The isolation is limited at the same time. Substitution of (98) yields, for the relative bandwidth,

$$\frac{\Delta\omega}{\omega_0} = \frac{4\sqrt{3}}{x_{1,1}^2 - 1}(\kappa/\mu)\rho_{max}. \tag{101}$$

For $\rho_{max} = 0.1$ (which is the same as a VSWR of 1.22, an isolation of 20 db and a minimal insertion loss of 0.1 db) at 450 MHz, a bandwidth of 3.2 per cent is found. Experimentally, it appeared to be 4.1 per cent. Why the experimental bandwidth is larger than the theoretical one can be understood from the fact that the dissipative losses, which tend to increase the bandwidth, are not considered in the theory.

From (101), it can be seen that the bandwidth is proportional to the amount of gyrotropy κ/μ. This can be understood from an inspection of (62) or (64). The disturbance of the original isotropic resonance is proportional to κ/μ. The larger κ/μ is, the more distant are the resonant clock- and anticlock-wise rotating modes. Hence, in the frequency region between the latter, the dependence of several quantities on frequency will be smaller, resulting in a larger bandwidth. Although, in most cases, a large bandwidth is desirable, κ/μ cannot be chosen large, as a small H_i would be the result which is not permitted because of the losses due to it. However, it is shown below that another straightforward method exists to obtain large bandwidths.

Input Impedance

Let the characteristic impedance of the striplines be Z and the output and isolated ports be reflectionlessly terminated. Then, upon neglecting higher order terms of $\delta\omega/\omega_z$ again, the input impedance Z_i is easily derived. The result is

$$Z_i = R_i + iX_i \tag{102}$$

with

$$R_i = Z \tag{103}$$

$$X_i = Z \frac{x_{1,1}^2 - 1}{\sqrt{3}(\kappa/\mu)} \cdot \frac{\delta\omega}{\omega_0}. \tag{104}$$

The real part of Z_i, R_i does not depend on frequency, but the imaginary part X_i does. Remembering that the time dependent factor of the field is exp $(-i\omega t)$, one can see that, for $\omega < \omega_0$, X_i is inductive and, for $\omega > \omega_0$, capacitive. This parallel resonant circuit behavior is confirmed by experiment.

Bandwidth Enlargement

From the dependence of the input impedance on frequency, a method of enlarging the bandwidth enormously can be proposed. For a lossless, cyclic-symmetric 3-port circulator, the isolation is related directly to the reflection coefficient [*cf.*, (7)]. When the latter is zero, the former is infinite. Now, α can be made zero over a rather large bandwidth and, hence, so can γ. This can be accomplished with a lossless series circuit of an inductance L and a capacitance C. The series impedance of such a circuit is

$$Z_{series} = i\left(\frac{1}{\omega C} - \omega L\right). \tag{105}$$

Choosing L and C so that the relations

$$\omega_0 = 1/\sqrt{LC}, \tag{106}$$

$$2\omega L_0 = Z \frac{x_{1,1}^2 - 1}{\sqrt{3}(\kappa/\mu)} \tag{107}$$

hold and connecting such resonant LC-circuits in series with the three ports, the reflections are eliminated over a rather large bandwidth centered at the circulation frequency.

In this way, with small coils and capacitors, it was found possible to make a circulator having a bandwidth of about 30 per cent. From 310 MHz up till 420 MHz, it had a maximal VSWR of 1.2, a minimal isolation of 19.5 db and a maximal insertion loss of 1.1 db. This last figure is a little high, but it is believed that it can be improved by using a better ferrite.

Field Distribution

It is easy to show that at the circulation adjustment the relations

$$b = a, \qquad c = 0 \tag{108}$$

hold [*cf.*, (10)] and that, upon substitution into (18), they yield, for the electric field intensity in the disk, the 1st order approximation

$$E_z(r, \phi) = \frac{2x_{1,1}Z_{eff} \sin \Psi J_1(kr) \sin \phi}{\pi(\kappa/\mu)J_1(x_{1,1})} a. \tag{109}$$

Insertion of this result into (16) and (17) gives for the magnetic field intensity in the disk the approximate expressions

$$H_\phi(r, \phi) = -\frac{2x_{1,1} \sin \Psi}{\pi(\kappa/\mu)J_1(x_{1,1})} \cdot \left[\frac{\kappa}{\mu} \frac{J_1(kr)}{kr} \cos \phi - iJ_1'(kr) \sin \phi\right] a \tag{110}$$

$$H_r(r, \phi) = -\frac{2x_{1,1} \sin \Psi}{\pi(\kappa/\mu)J_1(x_{1,1})} \cdot \left[\frac{\kappa}{\mu} J_1'(kr) \sin \phi + i \frac{J_1(kr)}{kr} \cos \phi\right] a. \tag{111}$$

Inspection of these formulas shows that the magnetic field intensity is anticlockwise rotating and elliptically polarized. The axis ratio is κ/μ. Calculation of the high frequency part of the magnetization shows that it is clockwise rotating and also elliptically polarized with an axis ratio $\kappa/\mu/(\mu_{eff}-1)$.

The electric field has been measured, too. The results of this experiment are sketched in Fig. 6, in which the lines of equal amplitude and the constant phase lines have been drawn. Comparison of the experimental results with the expression given above shows that the agreement is good, except at those places where the latter is small; that is, along the diameter $\phi = 0, \pi$, which is symmetrically situated between the input and the output port. This is quite understandable as in those places the neglected terms for $n \neq 1$ give a non-negligible contribution to the total electric field intensity and spread the phase lines.

Retaining 18 terms of the series involved, a solution of the same problem, as it is stated by Bosma,[5] has been determined with an electronic computer. Indeed, the field distribution so obtained agrees much better with the experimental result of Fig. 6. The circulation parameters of this "accurate" solution are $x = 1.852$, $h = 6.034$, $H_i = 970$ Oe, $4\pi M = 1750$ Oe, $\kappa/\mu = 0.109$, $f = 450$ MHz and $R = 2.90$ cm. These data, too, correspond well to those quoted above.

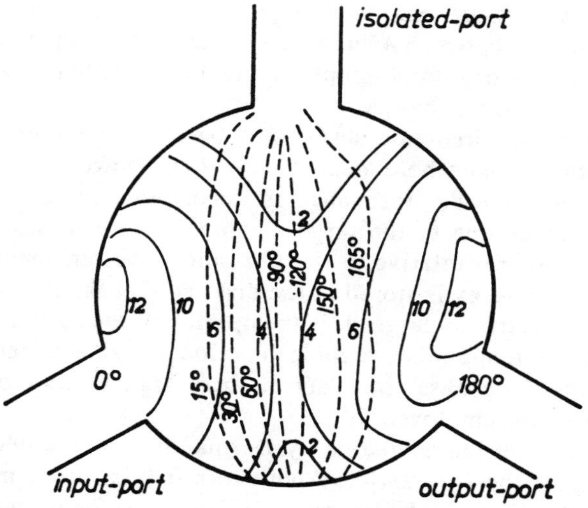

Fig. 6—Measured amplitude and phase of the electric field in the circulator disks. Drawn lines indicate the lines of equal amplitude. Broken lines are equal phase-lines. (The amplitude is given in arbitrary units.)

The Circulation Mechanism

The primary condition for circulation seems to be that the adjustment is in the neighborhood of the degeneracy of two resonances of the disk configuration (with disconnected striplines). If these resonances form a pair of the same order, as is the case with the solution found here, the implication of (64) is that κ/μ is small. This is also confirmed by Butterweck[16] who studied a model of the 3-port waveguide junction circulator, consisting of a circular-cylindrical cavity with an axially biased, concentric, thin and full height post of ferrite and three rectangular waveguides coupled weakly to the cavity at equal distances around the cylindrical wall. Butterweck also found that circulator action was possible near the degeneracy of the two first-order oppositely rotating modes. Although, in his case, κ/μ in the ferrite is not small, the effective gyrotropy, as averaged over the whole cavity volume, is.

The first order resonance of the isotropic disks has a distribution of the electric field as sketched in Fig. 7(a) at $t=0$. It is antisymmetric with respect to the symmetry diameter. A quarter of a period later $t=t/4$, the electric field is identically zero but, then, the magnetic field attains its maximum. The lines of the latter have been drawn in Fig. 7(b). If, instead of being isotropic, the disks are weakly gyrotropic, the magnetic field present at $t=T/4$ will induce a magnetic dipole distribution. This has been sketched in Fig. 8(b). As the disks are polarized perpendicularly, these dipoles rotate to the right. A quarter of a period earlier $t=0$, they were directed as it has been indicated in Fig. 8(a). The magnetic induction then being zero, they cause a demagnetizing field which, too, has been sketched in Fig. 7(a). The latter is in phase (or counterphase) with the electric field and it is symmetric with respect to the symmetry diameter. Hence, at the edge $r=R$, a real, radial power flow exists. In four regions, this power flow is alternately

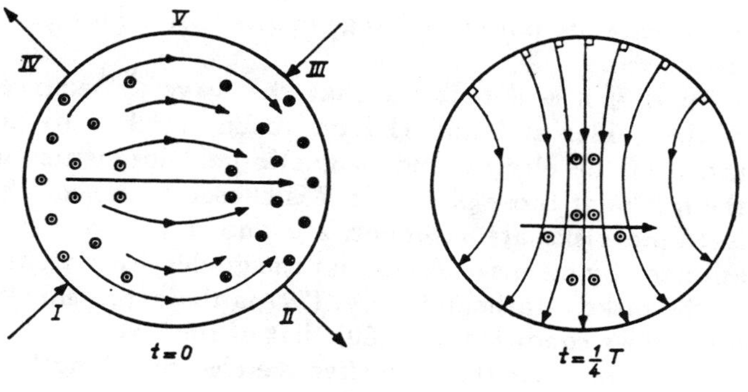

(a) (b)

Fig. 7—The electromagnetic field in the circulator disks at times (a) $t=0$ and (b) $t=T/4$, respectively. The circles denote the electric field intensity, the drawn lines the magnetic field intensity and the straight arrows the power flow.

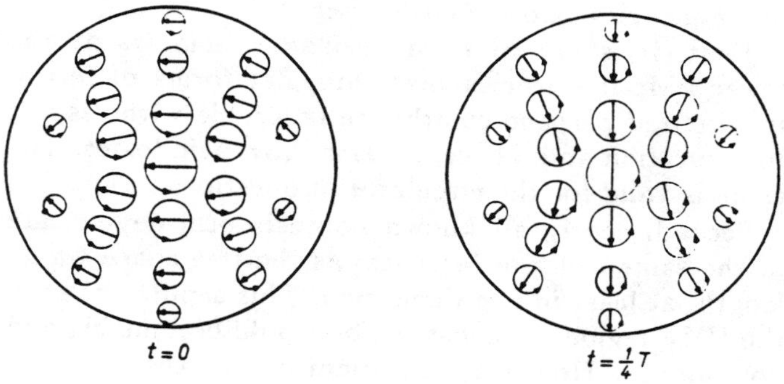

(a) (b)

Fig. 8—The magnetic dipole distribution in the circulator disks at times (a) $t=0$ and (b) $t=T/4$, respectively.

inwardly and outwardly directed. In Fig. 7(a), this has been indicated with four radial arrows. Because of the resonant character of the isotropic (symmetric) part of the magnetic field, the reactive power flow at the edge is zero.

If, in two successive regions (I and II), waveguides (striplines) are connected, the power radiated inwardly and outwardly, respectively, can be transported via these guides. If, in the other two regions (III and IV), no guides are present, no radial current can flow from the center conductor, no power can be radiated and, consequently, the azimuthal component of the magnetic field must there be zero. This can be so if only other modes are excited. As these modes are not resonant, they do not contribute substantially to the electric field at the edge, but they do near the symmetry diameter. The (small) electric field excited in this way will be in phase with the magnetic field there [Fig. 7(b)], so that across the symmetry diameter a power flow occurs which transports the energy from region I towards region II.

Due to the excitation of the other modes, $n\neq 1$ everywhere at the edge except at the regions I and II, the azimuthal magnetic field is zero. Because near V [Fig. 7(a)] the electric field is also zero, connection of a waveguide at that place does not disturb the fields and, hence, no power exchange occurs there. This guide is isolated.

Finally, it will be clear that the wave impedances of the guides at I and II have to be matched to the ratios of the electric and magnetic field intensities in the ferrite at those places. If R is chosen such that the isotropic disks are resonant, κ/μ and μ_{eff} have to be adjusted by means of H_i so that the guides are matched to the disks. Mathematically, (79) is the expression of the former condition and (80) that of the latter.

It will be clear that the effective electrical length of the circulator is equal to half a wavelength.

Application in the UHF Region

Three reasons exist why the 3-port stripline Y-circulator polarized above resonance is advantageously used at frequencies below 1000 Mc/sec.

First, it is known from resonance isolators at the lower kMc-frequencies that thin, flat forms of ferrite magnetized perpendicularly are favorable with respect to saturation and, hence, for small low-field losses. The same is valid for the circulator in question.

Second, nearly all known nonreciprocal devices are of the same order of largeness as the free space wavelength, at least in one dimension. This signifies that in the UHF region the construction would become big and voluminous. However, the diameter of the stripline Y-circulator is determined mainly by the wavelength in the ferrite. By virtue of (84) $\epsilon = 14$–16, the latter is an order of magnitude smaller than the free space wavelength as is, therefore, the circulator as well.

The last and most important reason is the fact that the circulator in question requires essentially a very high bias field $h>4$. It is much higher than it is in the case of any one of the well-known principles of nonreciprocal action. And, as has been argued above, this high bias field is very favorable for small insertion losses, which are usually difficult to obtain in the UHF region.

Other Solutions of the Problem

Any solution of the circulation equation (51) may result in conditions for circulator adjustment. Such solutions can be found in the neighborhood of degeneracies of disk resonances. This has been confirmed by Cohen and Davies[11] who computed electronically, from (54) of Bosma,[5] curves in the κ/μ, x-plane along which circulator action is possible. The straight line $x=1.84$ for $\kappa/\mu<0.25$ coincides nearly with one of these curves. Other curves join at the degeneracies B, C and D (Fig. 4) and are between the mode pairs $(-2,1$—$2,1)$, $(-1,1$—$2,1)$ and $(-2,1$—$3,1)$, respectively. To give a complete picture, for small values of κ/μ, they have also found a curve within the region ABC, which seems not to be related to a degeneracy. Moreover, the separate curves are linked crossing over a resonant mode curve somewhere between two degeneracies, *e.g.*, over the curve $-1,1$ between A and C. However, it may appear that these solutions do not apply to practical circulators and, in many cases, they do not.

In fact, circulator action has been detected experimentally somewhere near B (Fig. 4). However, using the second order resonance is not a practical proposition, since due to the large x-value the radius is large. Furthermore, relative to the transmitted power, much of its field energy is stored in the disk, so that the insertion loss is relatively large. It is the opinion of the author that many rectangular waveguide junction circulators can be improved because, in many of them higher order resonances are employed.

The solution evaluated by Bosma[5] is near the degeneracy of the first anticlockwise rotating 1st order mode and the first clockwise rotating 2nd order mode which is in the neighborhood of C in Fig. 4. The mathematical difficulties of the evaluation of this solution are much greater than of the solution presented in this paper, because the gyrotropy is not small ($\kappa/\mu \approx 0.5$) and it is not permitted to neglect all terms of the series involved except the two resonant terms. Besides, it has turned out that the approximation $J_0(x)=0$ made by Bosma[5] is not quite justified. But, although the solution near C (Fig. 4) is mathematically nearly possible and would require a small disk radius, it is not a practical proposition since κ/μ is not very small; hence, as h is not large, $h \approx 1.5$–2.0 in the UHF region, resonance (and perhaps low field) losses are not small. In fact, the resonances of the disk configuration could not be detected even at values of κ/μ larger than about 0.3, because they were masked by losses. The agreement between the theoretical and the experimental radii as a function of frequency found by Bosma[5] must be looked upon as merely an awkward coincidence. If at the time the field distribution in the disk had been measured, the practically false character of the solution would have been observed. Now, the conclusions drawn from it may be considered obsolete.

The theory developed in this paper applies also to circulators which are biased below resonance, if the section about the magnetic properties of the ferrite and its consequences are restated. Moreover, an analogous analysis can probably be carried out for the rectangular waveguide junction circulator.

Acknowledgment

The author wishes to express his gratitude to G. de Vries for his lasting and stimulating interest, for the fruitful discussions held with him; K. Kegel for the many accurately performed experiments; and H. J. C. A. Nunnink and J. Vlietstra for the trouble taken by them to find, despite the many singularities of the problem, a practical solution with an electronic computer. This solution has contributed in great measure to the attainment of the simplified analysis presented in this paper.

Operation of the Ferrite Junction Circulator

C. E. FAY FELLOW, IEEE, AND R. L. COMSTOCK MEMBER, IEEE

***Abstract*—The operation of symmetrical circulators is described in terms of the counter-rotating normal modes (fields varying as exp $n\phi$) of the ferrite-loaded circuits. The rotating modes, which are split by the applied magnetic field, form a stationary pattern which can be rotated in space to isolate one of the ports of the circulator. A detailed field theory of the strip-line *Y*-junction circulator operating with $n=1$ is presented. Experiments designed to confirm the validity of the rotating normal mode description of circulator action in the *Y*-junction circulator also are presented; these include measurements of mode frequencies and electric field patterns. The results of the field theory are used in a design procedure for quarter-wave coupled strip-line circulators. The results of the design procedure are shown to compare adequately with experimental circulators. Higher mode operation of strip-line circulators is described. The operation of waveguide cavity circulators is shown to depend on the rotating ferrite-loaded cavity modes.**

Introduction

A VERSATILE MICROWAVE DEVICE which is becoming one of the most widely used is the ferrite junction circulator. Its versatility is indicated by the fact that in addition to its use as a circulator, it also can be used as an isolator or as a switch. The three-port version of the ferrite junction circulator, usually called the *Y*-junction circulator, is most commonly used. It can be constructed in either rectangular waveguide or strip line. The waveguide version is usually an *H*-plane junction, although *E*-plane junction circulators also can be made. The strip-line ferrite junction circulator is usually made with coaxial connectors and is principally applicable to the UHF and low-microwave frequencies.

Manuscript received July 13, 1964; revised September 8, 1964.
C. E. Fay is with Bell Telephone Labs., Inc., Murray Hill, N. J.
R. L. Comstock is with Lockheed Missiles and Space Co., Research Labs., Palo Alto, Calif., He was formerly with Bell Telephone Labs., Inc., Murray Hill, N. J.

The ferrite junction circulator was in use for a number of years before its theory of operation received much attention in the literature. Early experimenters found that waveguide *T* junctions having a transversely magnetized ferrite slab suitably placed in the junction could, with proper matching and adjustment of the magnetic field, be changed into circulators. The bandwidth of such devices was very narrow. Refinements producing greater symmetry were found to increase bandwidth so that useful devices were obtained [1]–[4].

More recently a number of papers have appeared in the literature which bear on the theory of operation of the ferrite junction circulator. Auld [5] has considered the theory of symmetrical junction circulators in terms of the scattering matrix of the device. He has shown the necessary relations of the eigenvectors of the matrix, and has indicated how these relations may be obtained. Milano, Saunders, and Davis [6] have applied these concepts in the design of a *Y*-junction strip-line circulator. Bosma [7], [8], has made an analysis of the strip-line *Y*-junction circulator in terms of the normal modes of the center disk structure. In his second paper, he shows that the circulation condition is near a degeneracy of a pair of resonances of the disk structure. Butterweck [9] has considered the case of the waveguide junction circulator and has given the waveguide equivalent of Bosma's explanation. Others have attempted explanations based on field displacement, scattering from a post, or surface waves. All of these, if used with proper boundary conditions, conceivably could lead to the same conclusions.

This treatment will consist of an extension of Bosma's approach to the problem. First, we shall present a phenomenological description of the operation of the

Reprinted from *IEEE Trans. Microwave Theory Tech.*, vol. MTT-13, pp. 15–27, Jan. 1964.

strip-line Y-junction circulator followed by a development of the field theory applicable to this form of circulator, and then some experimental evidence of the validity of the approach will be submitted. Then, the theory is applied to provide a detailed design procedure for quarter-wave coupled strip-line circulators. Finally, we shall discuss operation in higher modes, operation of a four-port junction, and operation in rectangular waveguide.

I. Phenomenological Description

The strip-line Y-junction ferrite circulator consisting of two ferrite cylinders filling the space between a metallic conducting center disk and two conducting ground plates presents the simplest geometrical arrangement and, therefore, is the easiest to treat analytically. We shall present a development of the strip-line Y-circulator starting from more familiar concepts. This circulator in its basic form is illustrated in Fig. 1. The connections to the center disk are in the form of three strip-line center conductors attached to the disk at points 120° apart around its circumference. A magnetizing field is applied parallel to the axis of the ferrite cylinders.

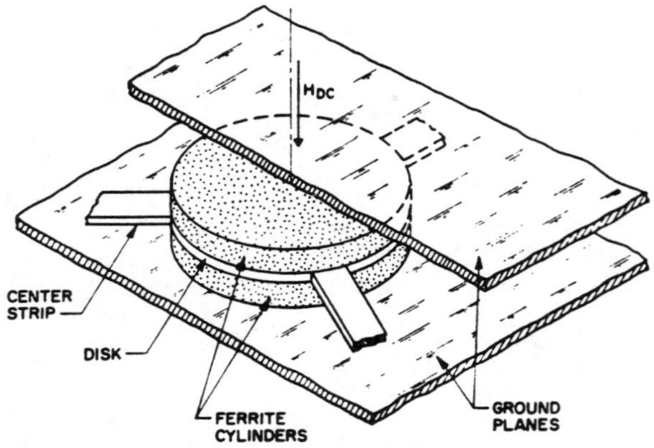

Fig. 1. Strip-line Y-junction circulator.

It was found, experimentally, that the Y-circulator had some but not all of the properties of a low-loss transmission cavity. At its resonant frequency, it was well matched but slightly undercoupled, and a standing wave existed in the structure. The maximum isolation occurred almost at the frequency at which the insertion loss was minimum. The isolation at the third port and the return loss from the input port correspond quite well as the frequency is changed. The above experimental evidence suggests a resonance of the center disk structure as being an essential feature of the operation of the circulator. The lowest frequency resonance of the circular disk structure of Fig. 1 is the dipolar mode in which the electric field vectors are perpendicular to the plane of the disk and the RF magnetic field vectors lie parallel to the plane of the disk. This mode, as excited at port 1, is illustrated in the unmagnetized case by the standing-wave pattern of Fig. 2(a). In the case of an isolated disk, the RF H lines would curve over the edge of the disk and continue back on the underside. With ferrite cylinders on each side of the disk, the bottom cylinder behaves as a mirror image of the top one, and the analysis of the device need only be concerned with one cylinder. Ports 2 and 3, if open-circuited, will see voltages which are 180° out of phase with the input voltage, and about half of the value of the input voltage. If the standing-wave pattern is rotated, as in Fig. 2(b), then port 3 is situated at the voltage null of the disk and the voltages at ports 1 and 2 are equal. The device is equivalent to a transmission cavity between ports 1 and 2, and port 3 is isolated. The standing-wave pattern of Fig. 2(b) in which all fields vary as $e^{j\omega t}$ can be generated by two counter-rotating field patterns of the same configuration. Each of these patterns would involve an

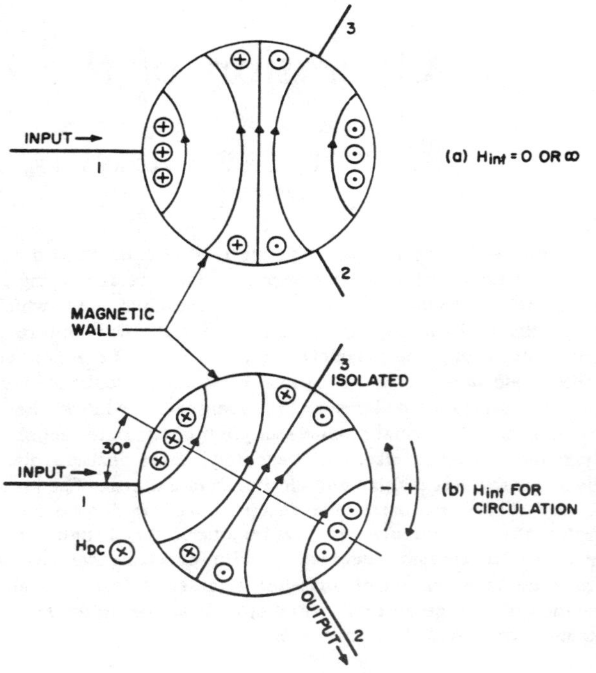

Fig. 2. (a) Dipolar mode of a dielectric disk. $H_{\text{int}}=0$ or ∞. (b) Analogous pattern of a magnetized disk. H_{int} for circulation. The pattern for the magnetized disk has been rotated to isolate port 3. The magnetic fields of higher modes than the dipolar mode are needed to achieve the pattern shown in (b).

RF magnetic field pattern which is circularly polarized at the center of the disk, becomes more elliptical as the radius increases, and is linearly polarized at the edge of the disk. If a magnetic biasing field (H_{dc}) is applied in the direction of the axis of the disk, the two counter-rotating patterns are no longer resonant at the same frequency. It is necessary when considering circulator action in the magnetized disk to include all of the normal modes, of which the dipolar field is but one, especially the nonrotating ϕ-independent mode ($n=0$ in the notation of Section II). This is the case since, as shown by Auld [5], circulator action must involve at least three normal modes. For the magnetized disk, the magnetic field pattern is a net pattern composed of many modes (the mode components of this pattern will be derived under specialized assumptions in Section II). The pattern rotating, in the same sense as the electron spins of the ferrite cylinder tend to precess under the influence of the biasing field, will have an effective scalar permeability which we shall call μ^+ at the center of the cylinder, and an effective permeability μ_e for linear polarization at the edge of the disk. The pattern rotating in the opposite sense from the electron spin precession will have an effective scalar permeability μ^- at the center of the cylinder, and an effective permeability μ_e at the edge of the cylinder. Accordingly, we shall designate these split modes as "+" (plus) or "−" (minus) depending on their sense of rotation compared to that of the electron spin precession. If we excite the system of Fig. 2(b) at a frequency intermediate between the resonant frequencies of the split modes, the impedance of the + mode, which has the higher resonant frequency, will have an inductive reactance component, and that of the − mode, which has the lower resonant frequency, will have a capacitive reactance component. If the frequency chosen is such that the capacitive reactance component of one equals the inductive reactance component of the other, the total impedance will be real at the operating frequency f_0, Fig. 3. If the degree of splitting is adjusted such that the phase angles of the impedances of the two modes are each 30° ($|X/R| = \tan 30°$) at the operating frequency, then the standing-wave pattern will be rotated 30° from that which obtains with no splitting of the modes. This is illustrated in Fig. 2(b). The + mode, which has the inductive reactance component, will have its voltage maximum leading the current (or H-field) maximum at the input port by 30° in time phase, which is also 30° in space phase since for this mode there is one revolution per cycle. The − mode will have its voltage maximum lagging the current maximum at the input port by 30°. Therefore, the two voltage maxima will coincide at a point 30° away from the input port as shown in Fig. 2(b). The rotation of the *pattern* is in the direction of rotation of the + mode. This 30° rotation brings the E-field null of the pattern to port 3 so that no voltage exists at this port. However, for the power entering at port 1, the device acts as a transmission cavity with power leaving at port 2. Thus, the direction of circulation is the direction of rotation of the − mode for $H_{dc} < H_{res}$ (ferromagnetic resonance field); for $H_{dc} > H_{res}$, the circulation is in the opposite direction. Since the device has complete symmetry, any power entering at port 2 will set up a new set of similar modes which will result in this power leaving at port 3, and port 1 will be isolated. Similarly, power entering at port 3 is transmitted to port 1, and port 2 is isolated.

The sum of the impedances of the two counter-rotating modes gives the impedance of the stationary mode. This stationary resonance will have the same Q as each of its components, and will be the one to be considered when the loaded Q and the matching of the circulator are discussed. It is shown in the next section that the approximate equivalent circuit of the resulting stationary mode is a shunt resonator as indicated in Fig. 3.

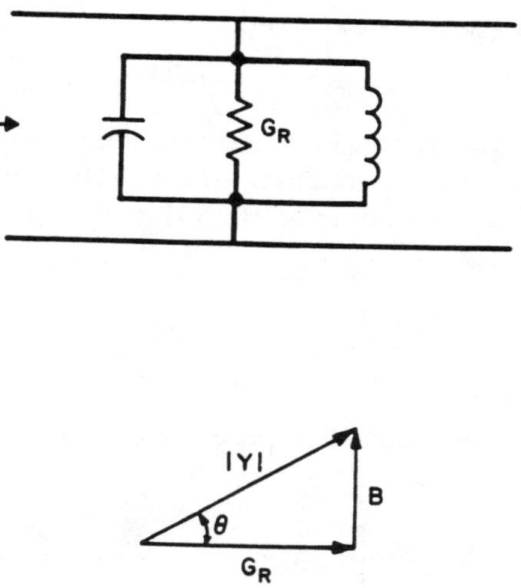

Fig. 3. Lumped element equivalent circuit of the stationary mode resonator.

We assume the resonator is lossless in this analysis so that the loaded Q, Q_L is described completely by the stored energy in the resonator and the power radiated into the connecting striplines. The field patterns of Fig. 2 are only those of the standing wave. The power transfer must be through a traveling wave whose fields are such as to result in an outgoing Poynting flux at port 2. In transmission cavities of large Q, the fields of the standing wave predominate. However, as Q_L becomes small this picture becomes inaccurate but still useful as will be shown by experimental results.

II. Field Theory of the Ferrite Junction Circulator

In the preceding section, a phenomenological description of circulator operation was presented. Here we will use the Maxwell equations, and the equations of motion of the magnetization to show the validity of the phenomenological picture and to arrive at some useful results relative to the design of this class of circulators. The boundary value problem which we shall discuss is the same as that considered by Bosma [7]. The circulator geometry consists of two ferrite cylinders separated by a disk center conductor fed by symmetrical transmission lines (see Fig. 1). The analysis is expected to predict the behavior of many similar circulators, e.g., the waveguide junction with a ferrite post.

Transverse electric (TE) waves with no variation along the dc bias field are considered. For these waves the electric field in the ferrite ($\boldsymbol{E} = \mathbf{a}_z E_z$) satisfies the homogeneous Helmholtz equation in cylindrical coordinates

$$\left[\frac{\partial^2}{\partial r^2} + \frac{1}{r}\frac{\partial}{\partial r} + \frac{1}{r^2}\frac{\partial^2}{\partial \phi^2} + k^2\right] E_z = 0 \qquad (1)$$

with time dependence $e^{j\omega t}$ and

$$k^2 = \omega^2 \epsilon\epsilon_0 \mu_{eff} \mu_0 = \omega^2 \epsilon_0 \epsilon \mu_0 (\mu^2 - \kappa^2)/\mu \quad (2)$$

where μ and κ are the Polder tensor components appropriate to a normally magnetized disk [10]. Equation (1) has solutions, depending on the value of n, given by

$$E_{zn} = J_n(x)(a_{+n}e^{jn\phi} + a_{-n}e^{-jn\phi}) \quad (3)$$

where

$$x = kr$$

Maxwell's equations then give for the ϕ-component of $\vec{H}_n$

$$H_{\phi n} = jY_{eff} \left\{ a_{+n}e^{jn\phi} \left[J_{n-1}(x) - \frac{nJ_n(x)}{x}\left(1 + \frac{\kappa}{\mu}\right)\right] + a_{-n}e^{-jn\phi}\left[J_{n-1}(x) - \frac{nJ_n(x)}{x}\left(1 - \frac{\kappa}{\mu}\right)\right]\right\} \quad (4)$$

with

$$Y_{eff} = \sqrt{\frac{\epsilon\epsilon_0}{\mu_0 \mu_{eff}}}$$

As stated in the preceding section, it has been found useful to treat the circulator as a tightly coupled resonator. With this in mind, it is then of interest to find the character of the normal modes in the uncoupled case ($\psi = 0°$, where 2ψ is the angle subtended by the strip lines, Fig. 4). Even in the uncoupled case, it is difficult to record the exact boundary conditions. Bosma [8] has argued that the normal modes resulting from the boundary condition given by

$$H_\phi(r = R) = 0 \quad (5)$$

are sufficient for describing the field in the disk in the uncoupled case (Fig. 4 for $\psi = 0°$). Justification for this boundary condition is the exponential falloff of fields outside the disk boundary. The normal modes, which result when the boundary condition given by (5) is used with (4), have resonant frequencies given by the roots of

$$+\text{ mode,} \quad J_{n-1}(kR) - \frac{nJ_n(kR)}{kR}\left(1 + \frac{\kappa}{\mu}\right) = 0$$

$$-\text{ mode,} \quad J_{n-1}(kR) - \frac{nJ_n(kR)}{kR}\left(1 - \frac{\kappa}{\mu}\right) = 0 \quad (6)$$

Several of the roots of these equations have been given by Bosma [8] in the form $(kR)_{m,n}^{\pm}$ vs. κ/μ, where m is the order of the root for a given value of n. For fixed value of m and n, there are two modes whose frequency separation depends on κ/μ and for small κ/μ is proportional to κ/μ. Fig. 5 shows a curve of κ/μ vs. σ where $\sigma = |\gamma| H_{int}/\omega$ with $p = |\gamma| 4\omega M_s/\omega$ as a parameter (H_{int} is the internal field and $4\pi M_s$ is the saturation induction). From an inspection of these curves, it is seen that $|\kappa/\mu|$ and hence the splitting is largest for a fixed value of p when $\sigma < 1$. This is shown, experimentally, to

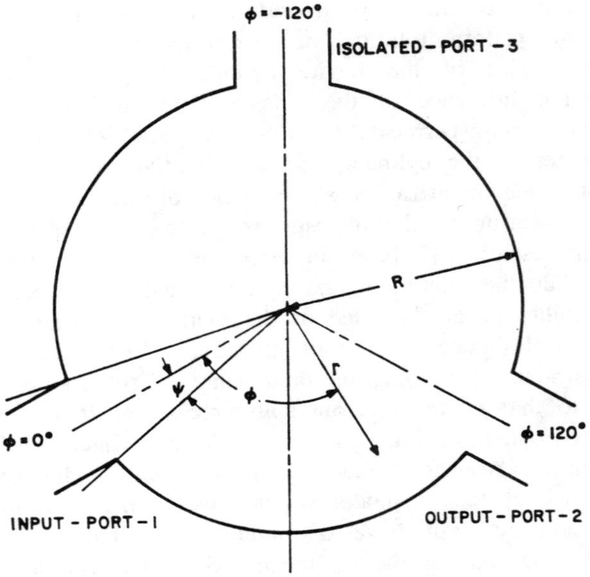

Fig. 4. Schematic diagram of strip-line circulator.

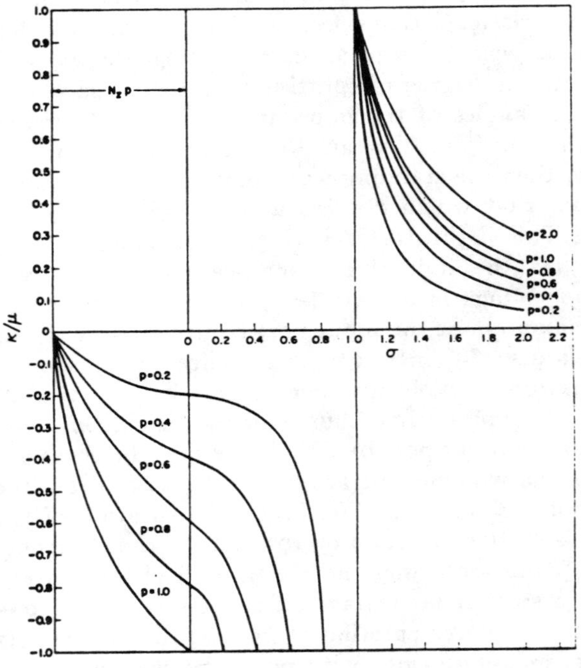

Fig. 5. κ/μ vs. σ, where $\sigma = |\gamma| H_{int}/\omega$.

be the case in Section III. Also, it is evident that the splitting will vary rapidly with p (and hence $4\pi M_s$) in this region. The lowest root of (6) will be arrived at graphically after equations are derived relating to the coupled disk.

The procedure we will use for treating the coupled case ($\psi \neq 0$), where the normal modes are driven by the TEM waves on the strip lines, is to assume initially that the system is functioning as a circulator and then to find the conditions on the rotating normal modes to satisfy the assumption. The conditions for circulator action are perhaps most easily described in terms of the scattering matrix [5] but, in the field theory discussed here, it is necessary to state the conditions in terms of the electric and magnetic fields. We follow Bosma [7] in assuming H_ϕ to be constant over the width of the strip lines, and take for the boundary conditions at $r = R$

$$\left.\begin{array}{ll} -\psi < \phi < \psi, & H_\phi = H_1 \\ 120° - \psi < \phi < \psi + 120°, & H_\phi = H_1 \\ -120° - \psi < \phi < \psi - 120°, & H_\phi = 0 \\ \text{elsewhere}, & H_\phi = 0 \end{array}\right\} \tag{7}$$

and

$$\left.\begin{array}{ll} \phi = 0, & E_z = E_1 \\ \phi = 120°, & E_z = -E_1 \\ \phi = -120°, & E_z = 0 \end{array}\right\} \tag{8}$$

It is necessary for both E_z and H_ϕ to be zero at the isolated port ($\phi = -120°$) since the boundary condition there is an admittance boundary condition, i.e., the ratio of H_ϕ to E_z at this port must be the wave admittance.

To compute the complete function $E_z(\phi, R)$, it is assumed as an approximation that E_z has a sinusoidal distribution with one period around the periphery of the disk, i.e., only the $n = 1$ mode is included for the electric field. The justification for this approximation is twofold: First, the amplitudes of the electric fields of the higher modes will be small since only the $n = 1$ mode is near resonance; and second, measurements of the electric field described in Section III conform closely to such a sinusoidal distribution, at least for narrow-band circulators. Equation (3) with the boundary conditions (8) then gives for the mode amplitudes

$$a_+ = \frac{E_1}{2J_1(kR)}\left(1 + \frac{j}{\sqrt{3}}\right)$$

and

$$a_- = \frac{E_1}{2J_1(kR)}\left(1 - \frac{j}{\sqrt{3}}\right) \tag{9}$$

so that the total electric field is in the form of a sinusoidal standing wave

$$E_z = E_1 \frac{J_1(kr)}{J_1(kR)}\left(\cos\phi - \frac{\sin\phi}{\sqrt{3}}\right) \tag{10}$$

For the magnetic field, we expand the function specified in (7) into a Fourier series, as was done previously by Bosma [7], so that

$$H = H_1\left[\frac{2\psi}{\pi} + \sum_{n=1}^{\infty} \frac{\sin n\psi}{n\pi} \cos n\phi + \sqrt{3}\frac{\sin n\psi}{n\pi} \sin n\phi\right] \tag{11}$$

which for the $n = 1$ mode can be written

$$H_{\phi 1} = H_1 \frac{\sin\psi}{2\pi}\left[(1 - j\sqrt{3})e^{j\phi} + (1 + j\sqrt{3})e^{-j\phi}\right] \tag{12}$$

However, the $n = 1$ component of H does not satisfy the circulation conditions. These conditions are satisfied only by the entire set of modes with the amplitudes given above.

Another solution for $H_{\phi 1}$ can be obtained from (4) by setting $n = 1$ and using the values of the coefficients a_+ and a_- given in (9). The result

$$H_{\phi 1} = jY_{\text{eff}} \frac{E_1}{2J_1(kR)} \cdot \left\{\left(1 + \frac{j}{\sqrt{3}}\right)\left[J_0(kR) - \frac{J_1(kR)}{kR}\left(1 + \frac{\kappa}{\mu}\right)\right]e^{j\phi} + \left(1 - \frac{j}{\sqrt{3}}\right)\left[J_0(kR) - \frac{J_1(kR)}{kR}\left(1 - \frac{\kappa}{\mu}\right)\right]e^{-j\phi}\right\} \tag{13}$$

is only compatible with (12) if

$$J_0(kR) - \frac{J_1(kR)}{kR}\left(1 - \frac{\kappa}{\mu}\right) = -\left[J_0(kR) - \frac{J_1(kR)}{kR}\left(1 + \frac{\kappa}{\mu}\right)\right]$$

or

$$J_0(kR) = \frac{J_1(kR)}{kR} = 0 \tag{14}$$

of which the first root is

$$(kR)_{1,1} = 1.84 \tag{15}$$

This result is exact under the present boundary conditions and not approximate as assumed by Bosma [8]. By comparing (15) with the normal mode resonances in the uncoupled case (6), it is seen that $(kR)_{1,1}$ lies somewhere between $(kR)_{1,1}^+$ and $(kR)_{1,1}^-$. The exact value of operating frequency must be determined from a solution of (15) with k given by (2). In Fig. 6 is shown a graph of μ_{eff} vs. σ with p as a parameter for a thin disk. This curve is useful in conjunction with (15) in evaluating the disk radius R, given M_s, H_0, and ω. To illustrate the relationship of the operating frequency to the normal mode resonant frequencies, we show in Fig. 7 a plot of (6) ($n = 1$) and (14) vs. kR.

One condition for circulation is that the operating frequency lie between the resonant frequencies for the $n = \pm 1$ modes. This is not the only condition since it is also required that the input admittance at port 1 and the output admittance at port 2 be adjusted to match the admittance at the terminals. Equating the values of $H_{\phi 1}$ given by (12) and (13), and taking account of the condition given by (14), leads to an expression for the input wave admittance of the circulator at resonance

$$Y_w = G_R = \frac{H_1}{E_1} \approx \frac{Y_{\text{eff}}\,|\kappa/\mu|}{\sin\psi} \tag{16}$$

which can be reduced to (69) of Bosma [7] for small ψ. The loaded Q of the circulator can be derived from the

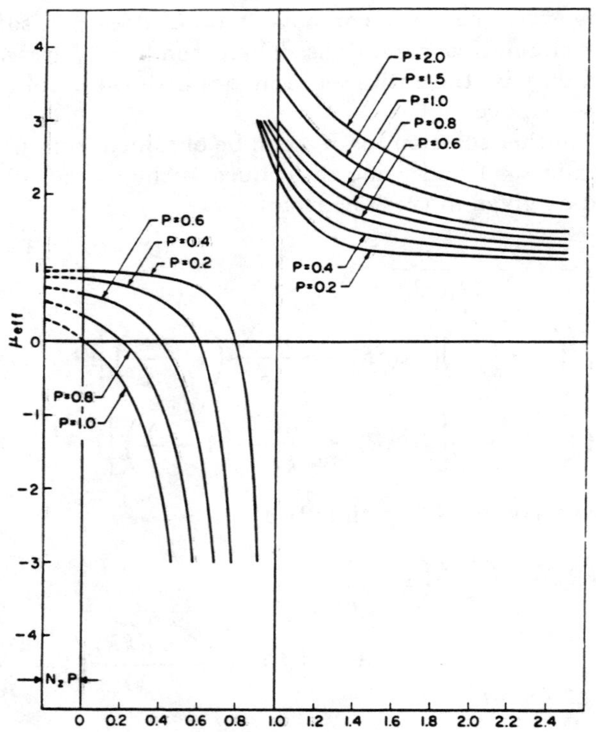

Fig. 6. μ_{eff} vs. σ

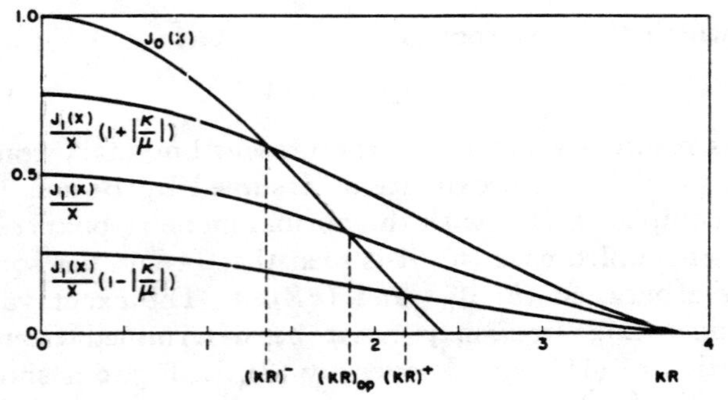

Fig. 7. Graphical solution of (9) and (17) for $n=1$. The operating frequency ($kR=1.84$) lies between the two normal mode resonances $(kR)^{\pm}$.

fields given in this section by starting from the definition

$$Q_L = \frac{\omega U}{P_{out}} \tag{17}$$

where U is the stored energy in the two cylinders and P_{out} is the power radiated out the strip lines. For U, we have

$$U = \frac{\epsilon\epsilon_0 d}{J_1^2(kR)} \int_0^{2\pi} \int_0^R E_m^2 J_1^2(kr) \cos^2 \phi' r dr d\phi'$$
$$= 1.11\, E_m R^2 d\epsilon\epsilon_0 \tag{18}$$

where E_m is the maximum electric field at the periphery of the disk, d is the cylinder height, and ϕ' has its zero reference at $\phi = -30°$. The total radiated output power is given by

$$P_{out} = d^2 E_m^2 G_R \cos^2 30° \tag{19}$$

where G_R is the conductance at the output strip lines and also the conductance looking into the resonator disk. Combining (17), (18), and (19), and evaluating the Bessel functions results in

$$Q_L = 1.48 \frac{\omega R^2 \epsilon\epsilon_0}{G_R d} \tag{20}$$

This result is physically satisfying because a large G_R is realized by making ψ large, which increases the coupling to the resonator.

It is possible to calculate the insertion loss of the ferrite cylinder and metallic disk circulator by calculating the unloaded Q, Q_0 of the structure as the result of including magnetic loss and dielectric loss. Since the expression for the resonant frequency and standing-wave pattern at the operating point involved only μ_{eff}, [(14) and (2)], if we assign a magnetic loss component to μ_{eff} so that

$$\mu_{eff} = \mu'_{eff} - j\mu''_{eff}$$

then the magnetic Q, Q_μ is just

$$\frac{\mu'_{eff}}{\mu''_{eff}}$$

From [11], we find

$$\mu_{eff} = \frac{1 - (p + \sigma + j\alpha)^2}{1 - (\sigma + j\alpha)(p + \sigma + j\alpha)} \tag{21}$$

where the loss has been introduced by substituting for σ, $\sigma + j\alpha$, where $\alpha = \gamma\Delta H/2\omega$ is the resonance linewidth in the usual reduced units. Separating the real and imaginary parts of (21) we find, taking their ratio and neglecting terms in α^4, that

$$Q_\mu = \frac{\mu'_{eff}}{\mu''_{eff}}$$
$$= \frac{(p\sigma + \sigma^2 - 1)[(p+\sigma)^2 - 1] + \alpha^2(2\sigma^2 + 3\sigma p + p^2 + 2)}{\alpha p[(p+\sigma)^2 + 1]} \tag{22}$$

In the case of $\sigma = 0$, this expression reduces to

$$Q_\mu = \frac{1 - (\kappa/\mu)^2}{[1 + (\kappa/\mu)^2]\alpha\kappa/\mu} \tag{23}$$

For most circulators operated below resonance, p and κ/μ are $\ll 1$ and the unloaded Q can be approximated by

$$Q_\mu = \frac{2\omega^2}{\gamma^2 4\pi M_5 \Delta H} \tag{24}$$

Likewise, the dielectric loss can be taken into account if we know the dielectric loss tangent of the ferrite. For this we have $Q_\epsilon = \epsilon'/\epsilon'' = 1/\tan\delta$. Then the total unloaded Q, Q_0 neglecting conductor losses is given by

$$\frac{1}{Q_0} = \frac{1}{Q_\mu} + \frac{1}{Q_\epsilon} \quad (25)$$

The insertion loss in dB is then, at band-center

$$\text{IL (dB)} = 10 \log_{10} \left(1 - \frac{Q_L}{Q_0}\right) \quad (26)$$

This represents a lower limit since the loss is assumed to be due to the fields of the standing wave only, and the conductor losses are not included.

The analysis up to this point has resulted in equations for the loaded Q of a single resonator. To relate the properties of this resonator to those of the two rotating normal modes, we express the electric field from (3) in terms of the normal mode fields, at $r = R$.

$$E_z = E^+e^{-j\phi} + E^-e^{j\phi} \quad (27)$$

where, from (9)

$$E^\pm = \frac{E_1}{2}\left(1 \pm \frac{j}{\sqrt{3}}\right)$$

When both sides of (27) are divided by H_1, the driving magnetic field, and evaluated at $\phi = 0$, we find expressions for the impedances for the two normal modes at port 1 at the operating frequency ω_{op}

$$Z^\pm = \frac{1}{2G_R}\left(1 \pm \frac{j}{\sqrt{3}}\right) \quad (28)$$

The input impedance is given by

$$Z = Z^+ + Z^- = \frac{1}{G_R} \quad (29)$$

These impedances can be represented by the lumped constant resonators shown in Fig. 8(a). The resonant frequencies of the resonators ($\omega^\pm$) are given by the roots of (6). The impedances of the + and − resonators at ω_{op} have phase angles of $\pm 30°$ as described in Section I. The operating frequency is approximately midway between the + and − mode resonant frequencies, i.e.,

$$\omega_{op} \approx \frac{\omega^+ + \omega^-}{2} \quad (30)$$

A detailed analysis of this equivalent circuit shows that in the vicinity of band-center, the two resonant circuits can be replaced by the single shunt resonator shown in Fig. 3 with the conductance on resonance and Q determined by (16) and (20).

Since the individual rotating mode resonators have 30° phase angles at band-center, a definite relation exists between the splitting $2\delta' = (\omega^+ - \omega^-)/\omega$, and the loaded Q, i.e.,

$$2\delta' = \frac{\tan 30°}{Q_L} = \frac{1}{\sqrt{3}Q_L} \quad (31)$$

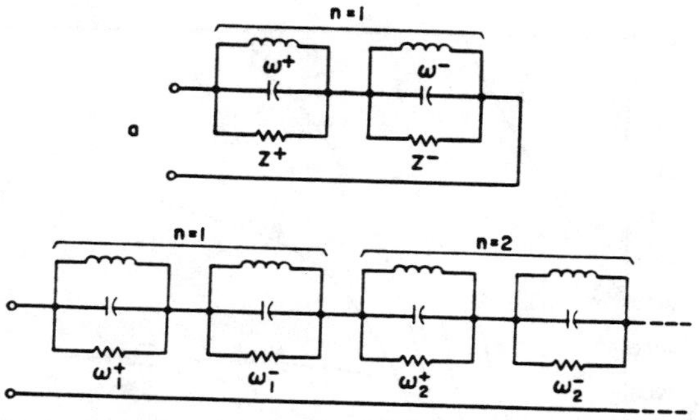

Fig. 8. Equivalent circuit of rotating normal modes ($\omega^\pm$). The $n=1$ mode resonances are shown in (a), and the more general circuit is shown in (b).

Equation (31) has been derived for the waveguide circulator by Butterweck [9] and is expected to be a general result for all three-port circulators which operate on the principle of the splitting of normal modes. A generalization of the above arguments for the higher order modes of the disk leads to the equivalent circuit shown in Fig. 8(b). In Sections III and IV it is shown that circulator action is possible between the resonances of the higher modes as represented in the equivalent circuit by the additional resonators. For example, if $\omega_{op} \approx (\omega_2^+ + \omega_2^-)/2$, the $n=1$ and all other modes are effectively short circuits, and we have the equivalent circuit similar to that shown in Fig. 8(a).

In Section IV, these results will be applied to the design of a stripline circulator. In the next section, some pertinent experimental results will be quoted.

III. Experimental

In this section, experimental evidence in support of the concepts developed in Sections I and II is presented. The experimental data have been taken on an L-band strip-line circulator which has dimensions large enough to allow probing of the electric fields at the edge of the center disk with sufficient mechanical accuracy so that relatively smooth plots are obtained.

A mode plot of the circulator is given in Fig. 9, which relates the mode resonances to the external field, H_{dc}. For the purposes of this plot, the center disk structure is lightly coupled to the 50-ohm strip lines by means of quarter-wave sections of high impedance line. This is done so that the various modes of the disk structure will be lightly loaded, and the resonances sharp and easily distinguished. The plot of resonant frequency as a function of applied magnetic field in Fig. 9 shows the $n = \pm 1$ modes, the $n = \pm 2$ modes, and parts of the $n = 0$ and $n = +3$ modes. For circulator purposes, we are interested principally in the $n=1$ modes since these involve the smallest size of the disk structure for a given frequency and, therefore, presumably the least loss in the structure. The points of circulation are shown on the plot. These occur near the degeneracies of modes of like n. The sense of circulation, also indicated, is for the H_{dc} vector pointing toward the viewer. It will be noted that the sense of circulation for a given set of modes

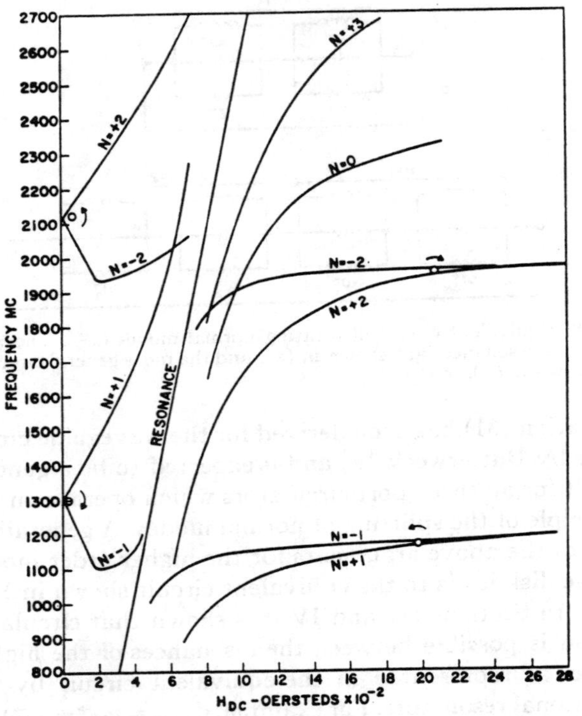

Fig. 9. Mode chart of experimental L-band circulator. Circulation is observed at the positions of the circles in the directions shown.

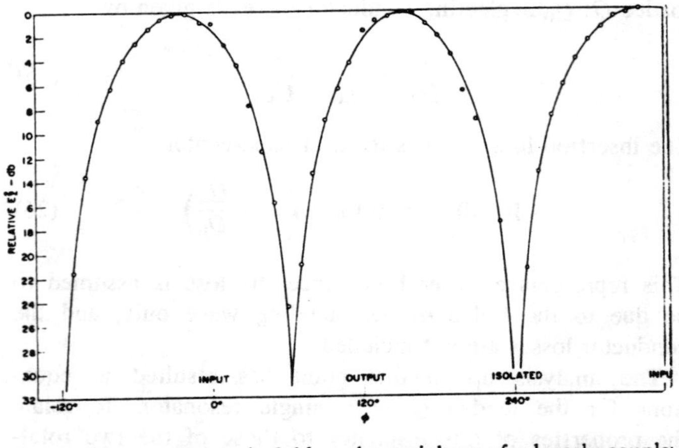

Fig. 10. Measured electric field at the periphery of a loosely couple circulator. The measured curve is predicted accurately by th function 20 log (cos ϕ+sin ϕ/tan 120°).

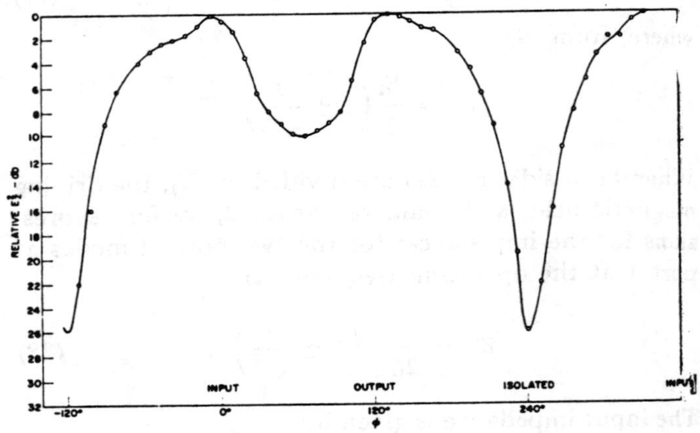

Fig. 11. Measured electric field at the periphery of a tightly coupled circulator.

above ferrimagnetic resonance is opposite to that for the same set of modes below ferrimagnetic resonance. The reason for this is that the − mode is the lower frequency one below resonance, and the higher frequency one above resonance. Therefore, the sign of the reactance of this mode at the operating frequency changes from the below resonance to the above resonance case, and hence, the sense of circulation will change if the reasoning given in Section I is followed. For the $n=2$ mode, the sense of circulation in each case is opposite to that of the corresponding $n=1$ mode case. The reason for this will be considered in the discussion of higher mode circulators.

A plot of the relative electric field, E_z, at the edge of the disk for the lightly coupled circulator is shown in Fig. 10. The circulator was adjusted for the below resonance point of circulation near the convergence of the $n=1$ modes, and the electric field probed at points around the circumference of the disk. The distribution is very nearly sinusoidal as predicted, with a slight distortion at the points where the high impedance feed lines are connected. The field maxima are seen to be approximately 30° away from the input and output ports. The circulator is very narrow-band in this case, and the required biasing field quite small, ~20 oersteds.

A more heavily coupled case is shown in Fig. 11. Here the same center structure is coupled to the 50 ohm strip lines by means of low impedance quarter-wave sections. This provides a more representative circulator which instead of being extremely narrow band, as in the previous case, has a>25 dB isolation bandwidth of approximately 15 per cent. The frequency of operation is nearly the same as before, but the required biasing field has increased to about 200 oersteds. This implies at greater separation or splitting of the modes as indicated in Fig. 9. The field plot of Fig. 11 shows a greater distortion in the vicinity of the connections to the center disk, as might be expected, since these connections are now considerably wider than in the previous case (ψ = 22°). The other result to be noted is that the minimum in the electric field between the input and output ports is much shallower than previously. Aside from the distortions discussed above, it is obvious that the standing-wave pattern in this case is maintaining the 30° rotation predicted but is requiring a greater mode separation to do so. This greater mode separation is required in order to maintain the 30° admittance phase angles with the lower loaded Q of the modes which results from the heavier coupling to the strip lines.

IV. Y-Circulator Design Considerations

The design of a strip-line circulator employing the circular disk and ferrite cylinder form can be carried out using the principles discussed in Sections I and II.

The result will yield a reasonable approximation to the desired operating characteristics.

The use of the circulation condition between the resonances of the $n = 1$ modes seems most desirable since this results in a minimum diameter of the metallic disk and ferrite cylinders. There seems to be no advantage in using the higher modes for a simple Y-circulator. The design will be based on the use of a quarter-wave transformer for coupling the circulator to the strip lines.

We must first choose the isolation bandwidth of the circulator. This is usually one of the things specified for the particular application of the circulator. Since the isolation of a Y-circulator usually corresponds quite closely to the return loss at the input port, and the return loss at the input port can be converted to VSWR (voltage standing-wave ratio), we define the following:

δ = fractional frequency deviation from f_0 (band-center) = $\frac{1}{2}$ bandwidth

VSWR = voltage standing-wave ratio at input port at frequencies $f = f_0(1 \pm \delta)$

Y_R = admittance of disk resonator at $f = f_0(1 \pm \delta)$

G_R = conductance of disk resonator

θ = phase angle of Y_R

Q_L = loaded Q of the disk resonator

δ' = fractional frequency splitting required for an admittance phase angle of 30° at the resonator

The resonator is assumed to have the equivalent circuit of Fig. 3. If a quarter-wave transformer or equivalent means is used for matching the resonator to the connecting strip lines, the susceptance of the transformer at $f = f_0(1 \pm \delta)$ can be used to cancel that of the resonator at these frequencies. For these frequencies, it can then be shown that for small δ

$$\text{VSWR} \approx \frac{|Y_R|^2}{G_R^2} = \sec^2 \theta \tag{32}$$

which allows us to determine θ. Then

$$Q_L = \frac{\tan \theta}{2\delta} \tag{33}$$

and we can find δ' from (31) using Q_L found previously. Since δ' relates to the amount of "splitting" required of the two $n = 1$ modes, using (6), the value of κ/μ can be determined. Since the relation of δ' to κ/μ is approximately linear for small splitting, we find an approximate relation

$$\frac{\kappa}{\mu} = 2.46\delta' = \frac{0.71}{Q_L} \tag{34}$$

that is useful for design purposes. The value of the constant in the above equation is appropriate for values of κ/μ in the range 0.25 to 0.5 which should cover the most practical cases. Having found the value of κ/μ for the desired bandwidth, we find from Fig. 5 that we can obtain it with the biasing field in the ferrite either above or below that required for resonance, i.e., $\sigma > 1$ for above resonance, or $\sigma < 1$ for below resonance, where $\sigma = \gamma H_{\text{int}}/\omega$. Above resonance operation requires a somewhat smaller disk diameter, a higher saturation, p, where $p = \gamma 4\pi M_s/\omega$; and a higher biasing field, hence, a larger magnet. Below resonance operation requires a minimum field and a relatively low-saturation material, ($p < 1$). Most Y-circulators in use at the present time employ the below resonance type of operation. For minimum field requirements it is convenient to operate at $\sigma = 0$ which implies that the ferrite is just saturated or $H_{dc} \approx N_z 4\pi M_s$. From Fig. 5 and 6, it can be seen that at $\sigma = 0$, $p = \kappa/\mu$ and $\mu_{\text{eff}} = 1 - (\kappa/\mu)^2$. It is now possible to calculate R, the disk radius, from (2) and (15) assuming the value of ϵ is known.

It is of course acceptable to use a lower value of p than that determined previously at the expense of increased biasing field. On the other hand, if too low a value of p is chosen, operation too close to $\sigma = 1$ may be required and resonance losses will be appreciable. If higher values of p are used, the ferrite is operated below saturation, and hence, low-field loss may be experienced.

The remaining dimension to be found is the height of the ferrite cylinder, d. Equation (20) may be rewritten as

$$G_R d = \frac{1.48\omega R^2 \epsilon\epsilon_0}{Q_L} \tag{35}$$

where everything on the right-hand side is already known. Thus, if we choose a value of G_R, the height d is determined. However, G_R cannot be specified until the matching is provided.

So far it has been assumed that a quarter-wave transformer will be used for matching and that its susceptance at the band edges will cancel the resonator susceptance at these points. In the admittance Smith chart of Fig. 12, there are shown two radii which represent the loci of admittances at $f_0(1+\delta)$ and $f_0(1-\delta)$ looking into a terminated transformer which is a quarter wavelength long at f_0, the band-center. Also shown are dotted curves representing loci of admittances having phase angles θ from (32). The example of Fig. 12 is taken for a bandwidth of 15 per cent ($\delta = 0.075$) with 30 dB isolation (VSWR = 1.065, $\theta = 14.3°$). The intersections of the radii and dotted loci define a resonator having the desired Q_L. The value of G_R and the characteristic admittance of the transformer, Y_T, now are determined if we specify the characteristic admittance, Y_0, of the strip line. Assuming for the example that $Y_0 = 0.02$ mho (50 ohm), then $Y_T = 0.048$ mho and $G_R = 0.115$ mho. The resonator admittance characteristic is seen to be wrapped into a tight loop at the end of the transformer and then transformed to a match at f_0, and to approximately the desired VSWR at the crossover point, $f_0(1 \pm \delta)$.

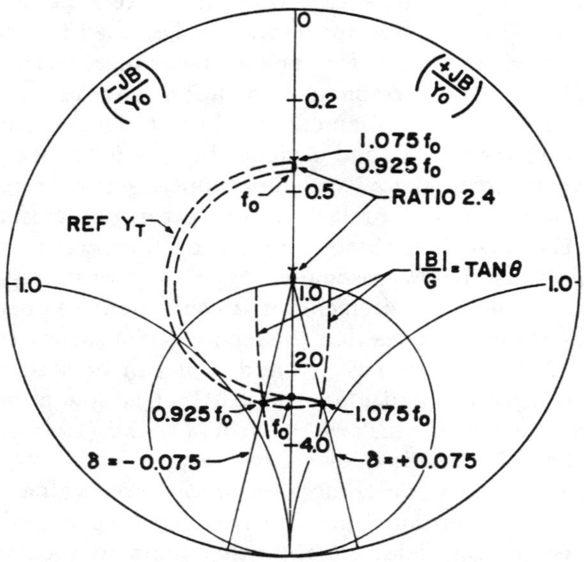

Fig. 12. Smith chart showing quarter-wavelength transformation to achieve a wide-band match. In this case the circulator is matched at midband with a 30 dB isolation bandwidth of 15 per cent.

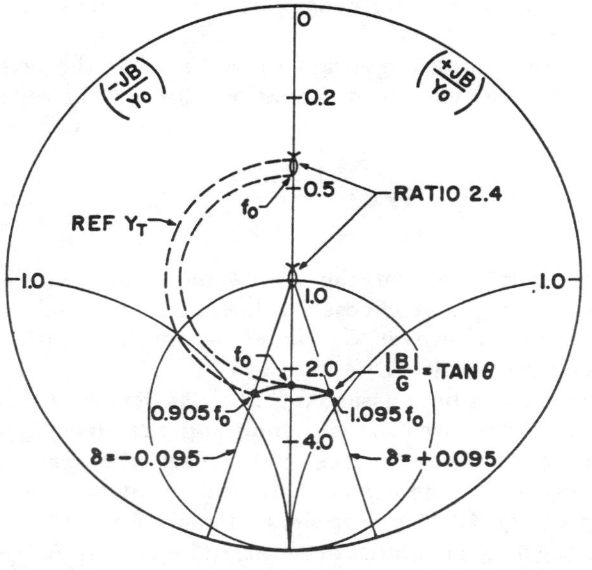

Fig. 13. Quarter-wavelength transformation for wide-band match with ripple in the isolation response. The 30 dB isolation bandwidth is 19 per cent in this case.

On inspection of Fig. 12, it will seem obvious that if either the transformer ratio or G_R of the resonator were changed slightly, the loop could be moved so that the origin was at its center, and the VSWR would be more uniform over the band. On the other hand, a larger loop could be used, still keeping all points within the specified VSWR. In Fig. 13, we show the latter case with the loop laid out so the midband point and the band edges all lie on a 1.065 VSWR circle. Using the same transformer ratio and working backward as compared to the previous procedure, it is found that G_R is now 0.108 mho and $\delta \approx 0.095$ with $\tan \theta = 0.310$, $Q_L = 1.63$. The value of G_R found in the foregoing can be used in (35) to find the thickness of the ferrite cylinder, d.

TABLE I

COMPARISON OF DESIGN AND ACTUAL CIRCULATORS

Design quantity	From fig. or eq.	*L*-Band Circulator		*C*-Band Circulator	
		19% B-W design	actual	19% B-W design	actual
Q_L	(33)	1.63	1.72	1.63	1.2
κ/μ	(34)	0.435	~0.5	0.435	~0.59
p	Fig. 5	0.435	0.52	0.435	0.28
μ_{eff}	Fig. 6	0.81	*	0.81	~0.6
ϵ		14	*	14	14
R cm	(15)	2.01	2.10*	0.655	0.76
$4\pi M_s$ gauss	$\frac{p\omega}{\gamma}$	200	240	625	400
G_R mho	Fig. 13	0.108	0.112	0.108	0.082
Y_T mho	Fig. 13	0.048	0.049	0.048	0.041
d cm	(35)	0.342	0.310	0.112	0.133
H_{dc} oers	$N_z 4\pi M_s$	145	~200	450	~870
Insertion loss dB	(26)	0.15	~0.1*	0.15	~0.17
30 dB Bandwidth	Fig. 13	0.19	0.185	0.19	0.12

* This structure has a 1.35-inch diam. ferrite cylinder inside a 1.75-inch-diam. alumina ring, hence μ_{eff} and ϵ will not be uniform and there is less material having magnetic loss than assumed in the design.

The result of the foregoing 19 per cent bandwidth design for 30 dB isolation compared to actual data on two circulators is shown in Table I.

Both circulators listed in the table were designed before the procedure outlined here was developed. The *L*-band circulator is observed to approximate fairly closely the design outlined.

V. Higher Mode Circulators and Other Forms

In this section we briefly sketch the operation of other strip-line circulators and also some waveguide circulators.

Y Circulator Using the $n = 2$ Modes

As indicated previously, a *Y* circulator can be made using the $n = 2$ modes (also $n = 4$ and higher, excluding those having threefold symmetry). The resonator pattern rotation required for the $n = 2$ mode is shown in Fig. 14. The required rotation in this case is 15° geometrically which corresponds to 30° electrically since this mode has two cycles around the circumference of the disk. The direction of rotation of the pattern is the same as the sense of circulation for this circulator.

The Four-Port Single Junction Circulator in Strip Line

The four-port single junction or *X*-junction circulator [12] seems attractive from the standpoint of compactness. An additional parameter is required [5] for this circulator as compared to the three-port junction. In this case, $n = 1$ modes are split as in the three-port circulator and, in addition, the $n = 0$ mode, which cannot split, is tuned to the operating frequency of the split $n = 1$ modes. The resonator standing-wave patterns are illustrated in Fig. 15. The $n = 1$ mode standing-wave pattern is rotated 45° toward the output port. With the

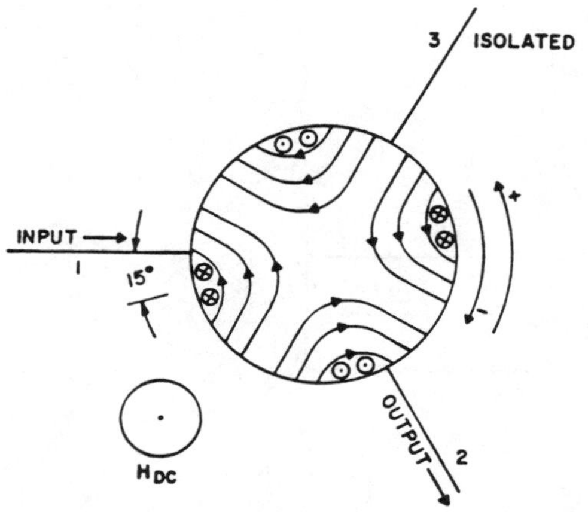

Fig. 14. Mode pattern for $n=2$ mode operation of disk circulator.

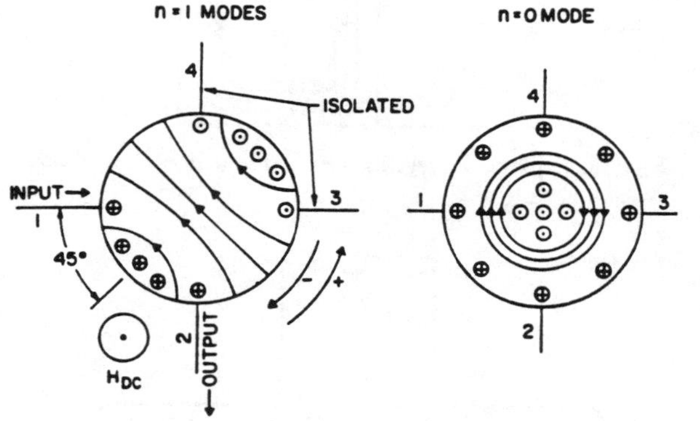

Fig. 15. Four-port circulator synthesized from $n=1$ and $n=0$ modes.

$n=0$ mode excited by the input port and with equal E-fields for the two patterns at the input port, the fields are also equal and in phase at the output, port 2. However, at ports 3 and 4, the fields are in opposite phase and will cancel. Thus we have transmission from port 1 to port 2, and ports 3 and 4 are isolated. The $n=0$ mode, which is normally about twice the frequency of the $n=1$ modes can be tuned lower in frequency by means of a capacitive post extending from the ground plane through a hole in the center of the ferrite cylinder to close proximity to the center of the metallic center disk. This post also raises the frequency of the $n=1$ modes somewhat so that the two resonant frequencies converge more rapidly. An $n=0$ mode of the cavity enclosing the ferrite strip-line assembly can be used in place of the $n=0$ disk mode. Therefore, this mode may be found to be more easily brought to the desired operating frequency.

The Y-Junction Circulator in Rectangular Waveguide—The H-Plane Junction

The three-port waveguide circulator usually consists of a 120° H-plane junction having a post of ferrite at the center of the junction as shown in Fig. 16. The biasing magnetic field is applied in the direction of the

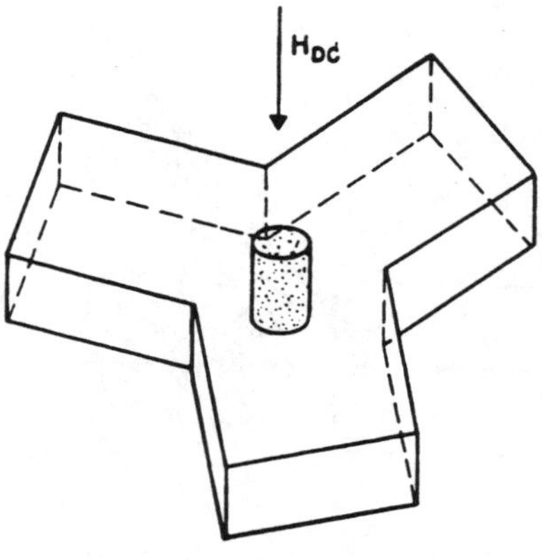

Fig. 16. Three-port H-plane waveguide circulator.

cylindrical axis of the post. With rectangular waveguide operating in the dominant TE_{10} mode, and with the ferrite post being considerably smaller in diameter than the width of the waveguide, the ferrite is excited mainly by the transverse RF H-field of the wave.

The operation of the waveguide circulator can be developed in a manner similar to that of the strip-line circulator [9]. First let us consider a lightly coupled case in which the ferrite post is contained in a cylindrical cavity which is coupled to the branching waveguides by small irises as shown in Fig. 17. If we assume this cavity is resonant at the operating frequency in the TM_{110} mode, the field pattern will be similar to that shown in the figure. Such a standing-wave pattern can be generated by two similar counter rotating patterns. The RF magnetic field will be almost circularly polarized in the ferrite for each sense of rotation. As soon as magnetic biasing field is applied to the ferrite, the resonance splits in a manner similar to that of the strip-line circulator. If the bias is adjusted so that each of the counter-rotating modes has a 30° phase angle, the resultant standing-wave pattern will have the orientation shown in Fig. 17. Here it is obvious that two ports are coupled to the standing-wave pattern and the third port lies at a null. This is equivalent to the $n=1$ mode circulation of the strip-line junction. Calculation of the frequency of resonance of this cavity is possible if small coupling is assumed. Writing down expressions for the fields in the ferrite and air portions, and equating E_z, B_r, and H_θ at the ferrite air boundary will yield an equation which could be solved for the resonant frequency. However, in the case of the actual circulator, the coupling irises are enlarged to the full size of the waveguide so that the cavity boundaries are no longer well defined. This is the equivalent of the broadbanding of the stripline circulator by tighter coupling to the feed lines. The calculation of the operating frequency becomes difficult in this case. Additional broadbanding can be obtained by impedance transforming means. One often used method is to make the circulator in reduced height, hence low-impedance waveguide, and to match to normal waveguide by stepped transformer

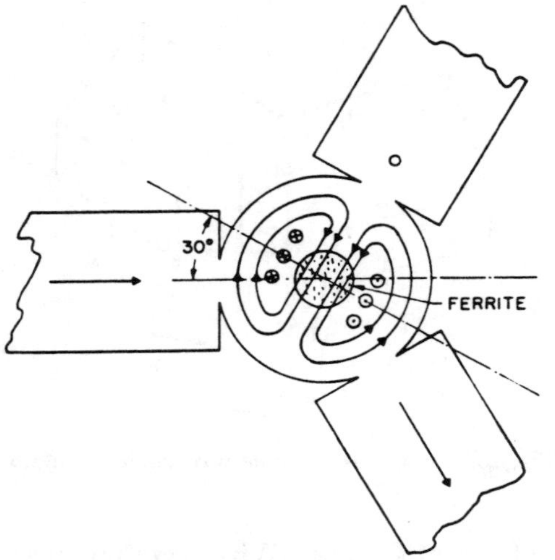

Fig. 17. Mode pattern for H-plane waveguide circulator using TM_{110} mode.

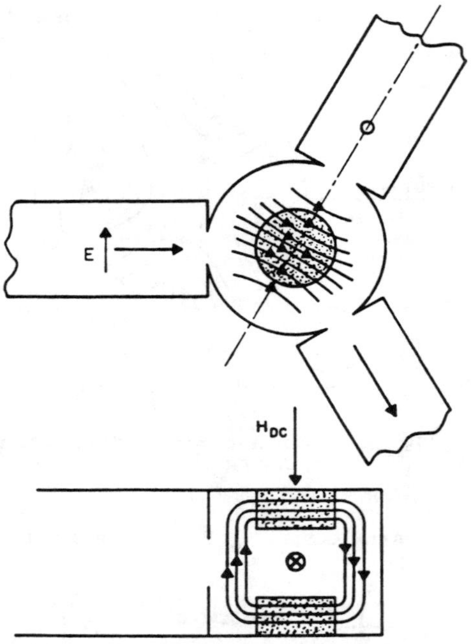

Fig. 18. E-plane waveguide circulator using TE_{111} mode.

sections. Another method is to use triangular shaped ferrites whose points are in the centers of the waveguides. The points of the triangles are probably in regions of nearly linear polarization so that they are acting more like dielectric tapers than ferrimagnetic material. A partial height ferrite post compensated for by increased diameter also may be used, as well as a reduced diameter post with dielectric sleeve to maintain the proper resonant frequency.

The waveguide circulator can operate in the $n=2$ mode also in a manner similar to the strip-line case. However, the larger ferrite required, with its probable higher loss, makes this mode unattractive.

The E-Plane Y-Junction Waveguide Circulator

An E-plane waveguide Y-junction circulator can also be made [13]. The ferrite post in this case is mounted perpendicular to the narrow wall junction and is excited by the longitudinal RF H-fields. The cavity mode involved in this case is a TE_{111} mode as illustrated in Fig. 18. It is advantageous here to remove the center portion of the ferrite post leaving only two short cylinders at the ends of the cavity, since the center of the cavity is not active magnetically and ferrite here is of little value as well as a possible source of some loss. There is no strip line equivalent to this device since there are no longitudinal H-fields in TEM mode line.

The E-plane junction circulator has not been used to any extent. It might be advantageous in high power applications since the ferrite can be confined to regions of low E-field. It has some disadvantage in that a large air gap in the magnetic biasing circuit is inherent.

The Four-Port X-Junction Circulator in Waveguide

A few four-port waveguide junction circulators or X-junction circulators, have been described [14], [15]. These operate in a manner similar to the strip-line four-port junction. The TM_{110} cavity mode is operated with a 45° rotation of the standing-wave pattern and the TM_{010} cavity mode tuned to resonate at the same frequency by means of a central metallic post as illustrated in Fig. 19. The ferrite post and the metallic post are usually each partial height.

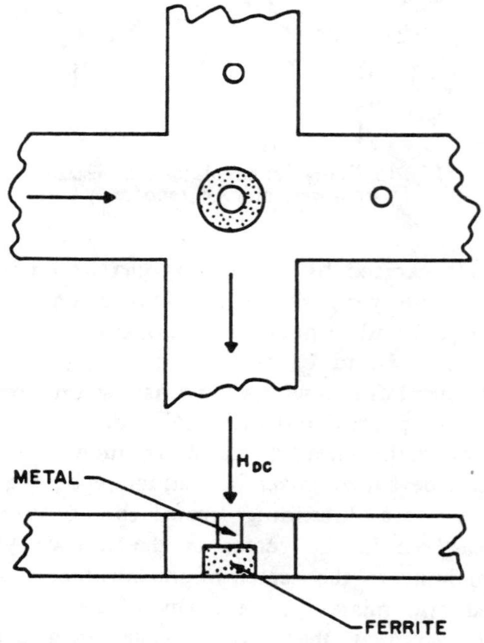

Fig. 19. Waveguide four-port circulator.

Conclusion

The operation of the ferrite junction circulator has been explained on the basis of a split resonance of the junction in which the standing-wave pattern of the resonator is rotated to produce a null or no coupling at the isolated port. The derivation of the field relations necessary to provide the above condition for the strip-line case has been carried out and experimental verification of the normal modes of the resonator and the postulated field distribution for circulation have been shown. A design procedure using the quarter-wave transformer for matching a strip-line circulator has been outlined.

While the explanation and analysis of the strip-line circulator given here has been based on a circular disk and cylindrical ferrite structure for simplicity, it seems reasonable that the same basic principles should apply to other structures having a threefold symmetry. These principles are:

1) The symmetrical microwave circuit must support two counter-rotating modes which in the absence of a magnetic biasing field are degenerate.
2) The degeneracy must be removed by a magnetic biasing field directed along the axis of symmetry.
3) Circulation then will occur at a frequency between the resonant frequencies of the nondegenerate modes.
4) The splitting of the nondegenerate modes must be adjusted to approximate the desired bandwidth, which is determined by the degree of external coupling to the microwave circuit, (31).

The basic resonator can, in principle, extend well beyond the ferrite loaded portion, since the ferrite is useful only where the RF H fields approach circular polarization. Other symmetrical junctions which have been used are simple 120° strip-line junctions without a circular disk, and triangular junctions of both ferrite and center conductor. Application of these principles to waveguide circulators has been outlined.

Acknowledgment

The authors are indebted to L. K. Anderson and B. A. Auld for many constructive suggestions and criticisms.

References

[1] Fowler, H., presented at Symposium on Mircowave Properties and Applications of Ferrites, Harvard University, Cambridge, Mass., Apr, 1956.

[2] Schauge-Pettersen, T., Novel design of a 3-port circulator, Norwegian Defense Establishment Rept. 1958.

[3] Swanson, W. E., and G. J. Wheeler, Tee circulator, *IRE WESCON Conv. Rec.*, pt. 1, 1958, pp 151–156.

[4] Chait, H. N., and T. R. Curry, A new type *Y*-circulator, *J. Appl. Phys.*, suppl. vol 30, Apr 1959, pp 152S–153S.

[5] Auld, B. A., The synthesis of symmetrical waveguide circulators, *IRE Trans. on Microwave Theory and Techniques*, vol MTT-7, Apr 1959, pp 238–246.

[6] Milano, U., J. Saunders, and L. Davis, A *Y*-junction stripline circulator, *IRE Trans. on Microwave Theory and Techniques*, vol MTT-8, May 1960, pp 346–351.

[7] Bosma, H. On the principle of strip line circulation, *Proc. IEE*, vol 109, pt B, suppl 21, Jan 1962, pp 137–146.

[8] Bosma, H., On strip line *Y*-circulation at UHF, *IEEE Trans. on Microwave Theory and Technique*, vol MTT-12, Jan 1964, pp 61–72.

[9] Butterweck, H. J., Der *Y* Zirkulator, *AEU*, vol 17, Apr 1963, pp 163–176.

[10] Lax, B., and K. J. Button, *Microwave Ferrites and Ferrimagnetics*, New York: McGraw-Hill, 1962.

[11] Suhl, H., and L. R. Walker, Topics in guided wave propagation through gyromagnetic media, *Bell Sys. Tech. J.* vol 33, May 1954, pp 579–659; Jul 1954, pp 939–986; Sep 1964, pp 1133–1194.

[12] Editorial, *Microwaves*, vol 2, p 92, Aug 1963.

[13] Yoshida, S., *E*-Type T circulator, *Proc. IRE* (*Correspondence*), vol 47, Nov 1959, p. 2018.

[14] Yoshida, S., *X* circulator, *Proc. IRE* (*Correspondence*), vol 47, Jun 1959, p. 1150.

[15] Davis, L. E., M. D. Coleman, and J. J. Cotter, Four-port crossed waveguide junction circulators, *IEEE Trans. on Microwave Theory and Techniques*, vol MTT-12, Jan 1964, pp 43–47.

Paper 7.3

Wide-Band Operation of Microstrip Circulators

Y. S. WU, MEMBER, IEEE, AND FRED J. ROSENBAUM, SENIOR MEMBER, IEEE

***Abstract*—Octave bandwidth operation of *Y*-junction stripline and microstrip circulators is predicted using Bosma's Green's function analysis. The width of the coupling transmission lines is found to be a significant design parameter. Theoretical and experimental results are presented which show that wide lines and a smaller than usual disk radius can be used to obtain wide-band operation. A microstrip circulator is reported which operates from 7–15 GHz.**

Also presented are an analysis of the input impedance and an approximate equivalent circuit for the *Y*-junction circulator which shows the relationship between Bosma's equivalent circuit and that of Fay and Comstock.

$$G(r,\theta;R,\theta') = -\frac{jZ_{\text{eff}}J_0(Sr)}{2\pi J_0'(SR)} + \frac{Z_{\text{eff}}}{\pi}\sum_{n=1}^{\infty}\frac{(\kappa/\mu)(nJ_n(SR)/(SR))\sin n(\theta-\theta') - jJ_n'(SR)\cos n(\theta-\theta')}{(J_n'(SR))^2 - [(\kappa/\mu)(nJ_n(SR)/(SR))]^2}J_n(Sr) \tag{1}$$

I. INTRODUCTION

The design of stripline and microstrip *Y*-junction circulators is generally based on the works of Bosma [1] and Fay and Comstock [2]. The former uses a Green's function approach to solve the boundary value problem of a ferrite loaded resonator coupled to three transmission lines. The latter gives a phenomenological explanation of circulator operation in terms of the rotation of the magnetic field pattern of the $n = 1$ mode of the resonant junction when it is magnetized in an external field. Experimental measurements [3]–[5] and an assumed equivalent circuit have been used to achieve broad-band operation of circulators by means of impedance matching the junction [6], [7].

Typical designs result in a specification of the center frequency, the geometry, and the ferrite parameters, such that the ratio κ/μ, which describes the anisotropic splitting of the ferrite, is usually small. This leads to narrow band operation of the junction which must then be broad banded using external tuning elements. In this paper octave bandwidth operation of *Y*-junction circulators is predicated using Bosma's method and demonstrated experimentally. The required junctions are smaller in diameter than usual and the coupling (impedance) transformers are much wider than usual. An explicit equivalent circuit is also developed which shows the equivalence of the results of Bosma and of Fay and Comstock.

Manuscript received January 15, 1974; revised May 8, 1974. This work was supported in part by the Air Force Avionics Laboratory under Contract F33615-72-C-1034, Captain M. Davis, Project Monitor.

Y. S. Wu was with the Department of Electrical Engineering, Washington University, St. Louis, Mo. 63130. He is now with the Central Research Lab., Texas Instruments Incorporated, Dallas, Tex. 75222.

F. J. Rosenbaum is with the Department of Electrical Engineering, Washington University, St. Louis, Mo. 63130.

II. INPUT IMPEDANCE OF *Y*-JUNCTION CIRCULATOR

Bosma's Green's function approach [1] gives the electric field inside two perfectly conducting parallel disks, completely filled with a ferrite magnetized normal to the disks, that is generated by a unit source $H_\theta = \delta(\theta - \theta')$ located at the edge of the disk. The Green's function for the wave equation satisfied by the electric field is

where

- μ, κ — Polder tensor elements [8] of the ferrite;
- $J_n(Sr)$ — Bessel function of the first kind with order n;
- S — $= (\omega/c)(\mu_{\text{eff}}\epsilon_f)^{1/2}$ = radial wave propagation constant;
- μ_{eff} — $= (\mu^2 - \kappa^2)/\mu$ = effective permeability of the ferrite;
- Z_{eff} — $(\mu_0\mu_{\text{eff}}/\epsilon_0\epsilon_f)^{1/2}$ = intrinsic wave impedance of ferrite;
- ϵ_f — relative dielectric constant of ferrite;
- $J_n'(Sr)$ — derivative of $J_n(Sr)$ with respect to its argument;
- r — radial coordinate;
- R — disk radius.

The magnetic field is derived from the electric field using Maxwell's equations.

Boundary conditions must now be applied. In general, it is very difficult to effect precise boundary conditions for planar microwave integrated circuits. However, because there are no radial currents at the edge of the conducting disks except at the ports (neglecting fringing), Bosma, Davies and Cohen [9], and others suggest the following boundary condition for the three-port *Y*-junction circulator shown in Fig. 1:

$$H_\theta = \begin{cases} a, & -\frac{\pi}{3} - \psi < \theta < -\frac{\pi}{3} + \psi \\ b, & \frac{\pi}{3} - \psi < \theta < \frac{\pi}{3} + \psi \\ c, & \pi - \psi < \theta < \pi + \psi \\ 0, & \text{elsewhere} \end{cases} \tag{2}$$

where a, b, and c are constants.

Reprinted from *IEEE Trans. Microwave Theory Tech.*, vol. MTT-22, no. 10, pp. 849–856, Oct. 1974.

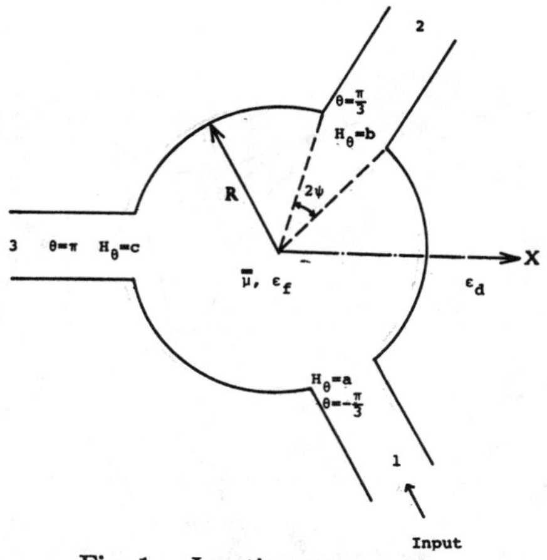

Fig. 1. Junction geometry.

From the above boundary conditions and the Green's function, the total fields can be solved by applying the superposition integral:

$$E_z(r,\theta) = \int_{-\pi}^{\pi} G(r,\theta;R,\theta')H_\theta(\theta')\, d\theta'. \tag{3}$$

Thus from E_z/H_θ, one can derive the input wave impedance Z_{in} and thereby the elements of the scattering matrix for a three-port symmetrical junction:

$$\bar{\bar{S}} = \begin{bmatrix} \alpha & \gamma & \beta \\ \beta & \alpha & \gamma \\ \gamma & \beta & \alpha \end{bmatrix}. \tag{4}$$

The input impedance is

$$Z_{\text{in}} = -Z_d - \left(\frac{j2Z_{\text{eff}}}{\pi}\right)\left(\frac{C_1^3 + C_2^3 + C_3^3 - 3C_1C_2C_3}{C_1^2 - C_2C_3}\right) \tag{5}$$

where $Z_d = 120\pi/(\epsilon_d)^{1/2}\ \Omega$ and ϵ_d is the relative dielectric constant of medium surrounding the ferrite disk. The scattering matrix elements are

$$\alpha = 1 + \frac{\pi Z_d(C_1^2 - C_2C_3)}{jZ_{\text{eff}}(C_1^3 + C_2^3 + C_3^3 - 3C_1C_2C_3)} \tag{6}$$

$$\beta = \frac{\pi Z_d(C_2^2 - C_1C_3)}{jZ_{\text{eff}}(C_1^3 + C_2^3 + C_3^3 - 3C_1C_2C_3)} \tag{7}$$

$$\gamma = \frac{\pi Z_d(C_3^2 - C_1C_2)}{jZ_{\text{eff}}(C_1^3 + C_2^3 + C_3^3 - 3C_1C_2C_3)} \tag{8}$$

where

$$A_n = J_n'(SR)$$

$$B_n = J_n(SR)$$

$$C_1 = \frac{\psi B_0}{2A_0} + \sum_{n=1}^{\infty}\left(\frac{\sin^2 n\psi}{n^2\psi}\right)\frac{A_nB_n}{A_n^2 - (n\kappa/\mu SR)^2B_n^2} - \frac{\pi Z_d}{j2Z_{\text{eff}}} \tag{9a}$$

$$C_2 = \frac{\psi B_0}{2A_0} + \sum_{n=1}^{\infty}\left(\frac{\sin^2 n\psi}{n^2\psi}\right) \cdot \frac{A_nB_n\cos(2n\pi/3) - (jn\kappa/\mu SR)B_n^2\sin(2n\pi/3)}{A_n^2 - (n\kappa/\mu SR)^2B_n^2} \tag{9b}$$

$$C_3 = \frac{\psi B_0}{2A_0} + \sum_{n=1}^{\infty}\left(\frac{\sin^2 n\psi}{n^2\psi}\right) \cdot \frac{A_nB_n\cos(2n\pi/3) + (jn\kappa/\mu SR)B_n^2\sin(2n\pi/3)}{A_n^2 - (n\kappa/\mu SR)^2B_n^2}. \tag{9c}$$

When the circulator is not perfectly matched one can define the return loss, isolation, and insertion loss by

$$\text{return loss} = 20\log_{10}|\alpha| \tag{10a}$$

$$\text{isolation} = 20\log_{10}|\beta| \tag{10b}$$

$$\text{insertion loss} = 20\log_{10}|\gamma|. \tag{10c}$$

The quantity ψ is the half-angle subtended by the coupling transmission lines where they meet the edge of the disk. The resonator is uncoupled when $\psi = 0$. If no impedance transformers are used the circulator is said to be directly coupled, with the input lines having a width $w = 2R\sin\psi$. In this theory ψ is a fundamental design parameter.

Equation (5) contains the variables C_1, C_2, and C_3, which, as indicated in (9), are infinite series. Davies and Cohen [9] consider terms up to $n = 6$ and Whiting [10] even more. In our opinion, since the actual field distribution is smoother than the assumed step function, retaining too many terms does not fit the real situation. On the other hand, retaining only the $n = 1$ mode is also erroneous because the $n = 0$ and $n = 2$ terms play important roles, especially in broad-band circulators [11].

The input wave impedance is calculated from (5), retaining terms up through $n = 3$, and the results shown in Fig. 2. In what follows, the ferrite is just saturated, the applied bias field being approximately $4\pi M_s$, the value of the saturation magnetization. In this case the elements of the Polder tensor are found to be [12], [13]

$$\mu = 1 \tag{11a}$$

$$\kappa = -\frac{\omega_m}{\omega} \tag{11b}$$

$$\omega_m = 2\pi\gamma(4\pi M_s) \tag{11c}$$

where $\gamma = 2.8$ MHz/0e and ω is the microwave radian frequency.

The input wave resistance, normalized to Z_d was calculated for several values of the coupling angle ψ (in radians). As shown in Fig. 2(a), it has two peaks, one at 8.6 GHz and another at 11 GHz for a light coupling angle $\psi = 0.1$. However, when the coupling angle increases

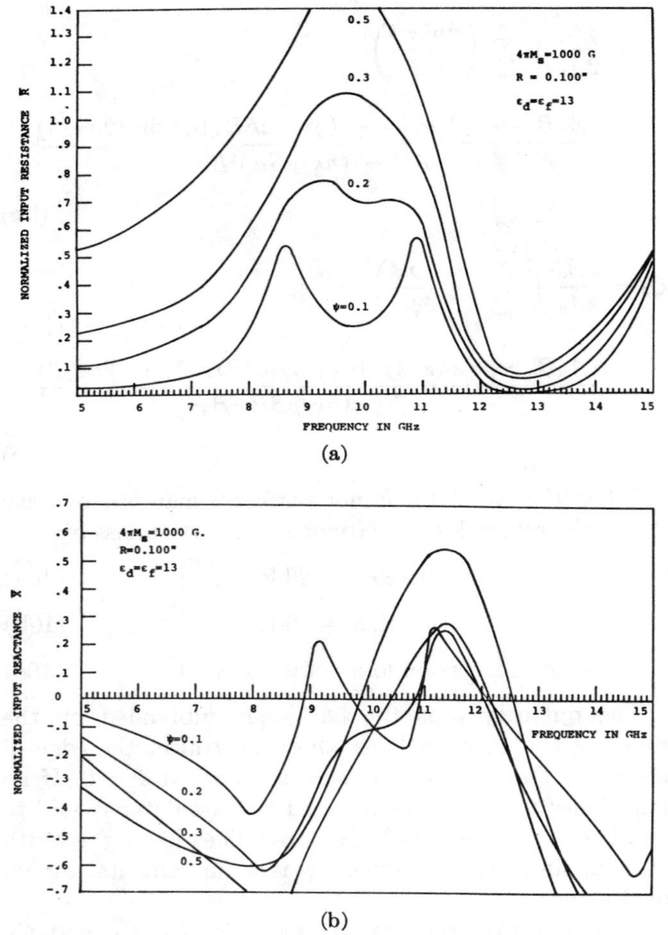

Fig. 2. (a) Normalized input wave resistance of a junction circulator with $4\pi M_s = 1000$ G for various coupling angles. (b) Normalized input wave reactance of a junction circulator with $4\pi M_s = 1000$ G for various coupling angles.

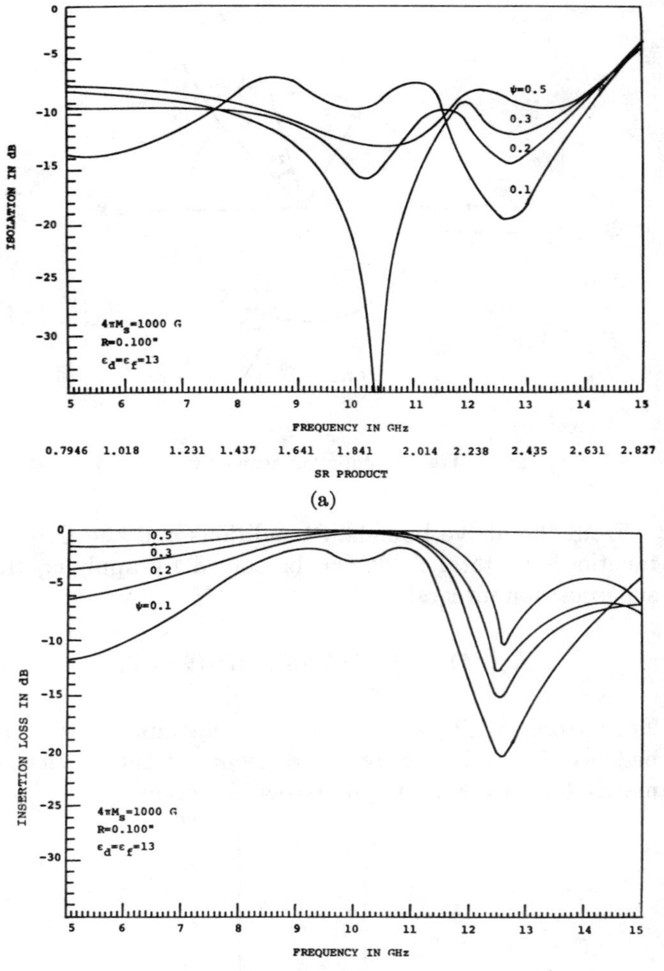

Fig. 3. (a) Calculated isolation for various coupling angles. (b) Calculated insertion loss for various coupling angles.

(tight coupling), the valley between these two peaks becomes flatter and rises. Finally, at $\psi = 0.3$, the two peaks overlap and coalesce to a single peak at $f = 10$ GHz.

Fig. 2(b) shows the corresponding normalized input wave reactances for the above circulators. For a light coupling angle ($\psi = 0.1$), the input reactance is inductive at low frequency ($f < 8.9$ GHz), then capacitive between 8.9 and 9.6 GHz, again inductive for $9.6 < f < 10.9$ GHz, and finally becomes capacitive again at high frequency $f < 10.9$ GHz. These results matched very well with Salay and Peppiatt's [4] and Simon's [3] measurements. The difference in sign of the input reactance of Fig. 2(b) and that of Salay's comes from the negative harmonic $\exp(-j\omega t)$ used here. There are three resonances for $\psi = 0.1$ in Fig. 2(b). The resonances at 8.9 and 11 GHz appear as parallel resonances and the central resonance at 10 GHz is a series resonance. As the coupling angle ψ becomes larger, the two outer resonances move closer to the central resonance. Finally, the reactance becomes that of a parallel R–L–C resonator for tightly coupled circulators.

The input impedance at the band center is of particular importance since this impedance must be matched to the coupling lines if perfect circulation ($\alpha = \beta = 0, \gamma = 1$) is required at band center. The center frequency impedance normalized to Z_d can be calculated from (5), retaining the $n = 1$ term only, by noticing that $A_1 = J_1'(x) = 0$ at the band center. It is found to be

$$\bar{Z}_{\text{in}} = \frac{2}{1 + (\psi_c/\psi)^2} \tag{12}$$

where

$$\psi_c = \frac{\pi}{\sqrt{3}(1.84)} \frac{Z_d}{Z_{\text{eff}}} \left| \frac{\kappa}{\mu} \right|$$

is the circulation angle derived by Bosma. Equation (12) shows that the normalized input wave impedance is purely resistive and increases from zero to two when the coupling angle varies from zero to $\psi \gg \psi_c$. When $\psi = \psi_c$, the circulator is matched as predicted by Bosma. When $\psi \neq \psi_c$, the circulator can be matched by applying transformers (6). However, the impedance ratio of the transformer must take (12) into account.

The theoretical performance of the circulator can be calculated from (10). Fig. 3(a) is the calculated isolation for magnetization $4\pi M_s = 1000$ G, radius $R = 0.100$ in, and dielectric constant $\epsilon_d = \epsilon_f = 13$ for various coupling angles. To get good circulation, the coupling angle should neither be too narrow nor too wide. Here the best result is obtained at $\psi = 0.3$ which is the circulation angle ψ_c. Fig. 3(b) is the insertion loss for various coupling angles. The large insertion loss at low frequencies ($f < 8$ GHz) and high frequencies ($f < 11$ GHz) is due to mismatch (reflection) at the input junction.

III. EQUIVALENT CIRCUIT FOR JUNCTION CIRCULATOR

In order to understand its input impedance and to have an analytical basis for wide-band impedance matching, it is useful to develop an equivalent circuit representation for the junction circulator. For circulators working close to the TM_{110} disk resonance the first order ($n = 1$) term dominates. Thus we may consider this mode alone for the calculation of input impedance as a first order approximation. In (5),

$$C_1 = -2M + jQ \tag{13a}$$

$$C_2 = M + jN \tag{13b}$$

$$C_3 = M - jN \tag{13c}$$

$$M = -\frac{\psi}{2}\frac{A_1B_1}{A_1^2 - (\kappa/\mu SR)^2B_1^2} \tag{13d}$$

$$N = -\frac{\psi}{2}\frac{\sqrt{3}(\kappa/\mu SR)B_1^2}{A_1^2 - (\kappa/\mu SR)^2B_1^2} \tag{13e}$$

$$Q = \frac{\pi Z_d}{2Z_{\text{eff}}}. \tag{13f}$$

Now, consider the input impedance of a series combination of two parallel R–L–C circuits as shown in Fig. 4(a). The input impedance, using $\exp(-j\omega t)$, is [11]

$$Z_{\text{in}} = \frac{1}{(1/R_c) - j(\omega C_+ - 1/\omega L_+)} + \frac{1}{(1/R_c) - j(\omega C_- - 1/\omega L_-)}. \tag{14}$$

Equation (5) is the input impedance of a circulator derived from field theory, while (14) is the impedance of the series combination of two parallel R–L–C circuits. By comparing the results given by the field and circuit approximations, one can derive the following analytic expression for the equivalent circuit elements:

$$R_c = \frac{Z_d}{2}F \tag{15}$$

$$L_\pm = \frac{\psi R\eta}{\pi c}\frac{\mu_{\text{eff}}}{(1 \pm \delta)}F = \frac{L_0}{(1 \pm \delta)}F\ \frac{h}{m} \tag{16}$$

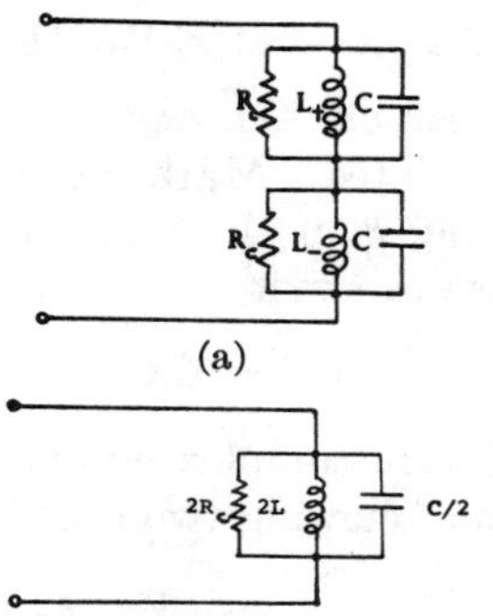

Fig. 4. Equivalent circuit of a junction circulator showing the correspondence between Fay and Comstock's two-resonator model (a) and Bosma's single resonator model (b).

$$C_\pm = \frac{1}{(1.84)^2}\frac{\pi\epsilon_f R}{\psi\eta c}\frac{1}{F} = C_0\frac{1}{F}\ \frac{f}{m} \tag{17}$$

where

- η $= 120\pi\ \Omega$ = intrinsic impedance of free space;
- R radius of ferrite disk;
- c $= 3 \times 10^8$ m/s = speed of light in vacuum;
- δ $= \left[\left(\frac{\kappa}{\mu}\right)^2 - \left(\frac{\sqrt{3}\psi SR}{2Q}\right)^2\right]^{1/2}$ = deviation factor; (18)
- F geometric factor for the microstrip or stripline characteristic impedance.

Equation (17) shows that the two capacitances are equal and essentially independent of frequency. On the other hand, the two inductances are not the same. For a fixed deviation factor δ, $L_+ < L_0$, and this resonator will have a higher resonance frequency than the L_-C_- one. Both inductances are proportional to the ferrite effective permeability. It is the parallel capacitance which gives bandwidth limitations for broad-band circulators.

For a lightly coupled circulator, ψ approaches zero, so the deviation factor tends to κ/μ, as seen in (18). Referring to Figs. 2(a) and 4(a), the input resistance has two peaks corresponding to the two parallel resonances at $f_\pm = 1/2\pi(L_\pm C_0)^{1/2}$. The higher resonance corresponds to the counterclockwise rotating wave resonance inside the ferrite disk and the lower one corresponds to the clockwise resonance. As the coupling becomes tighter, the deviation factor decreases. Finally, as ψ becomes large enough such that $\kappa/\mu = \sqrt{3}\psi_c SR/2Q$, which is the perfect circulation condition $\psi = \psi_c$, the deviation factor vanishes. In this case, we have $L_+ = L_- = L_0$, and the two resonators are identical. Thus the final equivalent circuit for a circulator at perfect circulation is simply a parallel R–L–C circuit as shown in Fig. 4(b). This is the reason Bosma suggests a single parallel R–L–C resonator for the equivalent circuit of a circulator, while Fay and Comstock suggest Fig. 4(a).

When the coupling angle becomes too large, the deviation factor becomes imaginary. Both the inductances L_+ and L_- are now complex numbers and the equivalent circuit loses its physical meaning.

IV. DESIGN OF A WIDE-BAND CIRCULATOR

To get perfect circulation, one of the three ports must be completely isolated. Mathematically, this can be achieved by setting β in the scattering matrix equal to zero. From (7) we can write

$$C_2^2 = C_1 C_3. \tag{19}$$

Since C_1, C_2, C_3 are complex numbers, (19) can be decomposed into the following two conditional equations:

$$\text{(A)} \quad P = \frac{M(M^2 - 3N^2)}{(M^2 + N^2)} \tag{20}$$

$$\text{(B)} \quad Q = \frac{N(3M^2 - N^2)}{(M^2 + N^2)} \tag{21}$$

where

$$P = \operatorname{Re}(C_1) = \frac{\psi B_0}{2A_0} + \sum_{n=1}^{\infty} \frac{\sin^2 n\psi}{n^2\psi} \frac{A_n B_n}{A_n^2 - (n\kappa/\mu SR)^2 B_n^2}$$

$$Q = \operatorname{Im}(C_1) = \frac{\pi Z_d}{2Z_{\text{eff}}}$$

$$M = \operatorname{Re}(C_2) = \frac{\psi B_0}{2A_0} + \sum_{n=1}^{\infty} \left(\frac{\sin^2 \psi}{n^2\psi}\right) \frac{A_n B_n \cos(2n\pi/3)}{A_n^2 - (n\kappa/\mu SR)^2 B_n^2}$$

$$N = \operatorname{Im}(C_2) = \sum_{n=1}^{\infty} \left(\frac{\sin^2 n\psi}{n^2\psi}\right) \frac{(n\kappa/\mu SR) B_n^2 \sin(2n\pi/3)}{A_n^2 - (n\kappa/\mu SR)^2 B_n^2}.$$

Equations (20) and (21) are the two conditions required for perfect circulation. They are essentially the same conditions as derived by Davies and Cohen [9]. As shown earlier, the quantities P, M, and N are infinite series with each term corresponding to a particular resonator mode. These infinite terms are necessary for the circulator to satisfy the particular boundary conditions. However, not all terms are important in practical application. Since most circulators operate close to the $n = 1$ resonance, many people consider only the $n = 1$ term and the above two conditions reduce to

$$SR = 1.84 \tag{22a}$$

$$\psi = \frac{\pi}{\sqrt{3}(1.84)} \frac{Z_d}{Z_{\text{eff}}} \left|\frac{\kappa}{\mu}\right|. \tag{22b}$$

Fig. 5 shows the computed perfect circulation roots from (20), retaining terms up to the third order. The roots were sought from $SR = 0$ through $SR = 3.0$. There are two sets of results. One, specified by mode 1+, is the mode 1A mentioned by Davies and Cohen. The other set is the usual circulation root which is concentrated at $SR = 1.84$ for small values of $|\kappa/\mu|$. Since the mode 1+ solutions require a larger disk and also need a large impedance ratio between the substrate and the disk [9], this mode is not practically suitable. As shown by Davies and Cohen there is a further set of roots between the 1+ and 1− modes, corresponding to the middle zero of the $\psi = 0.1$ curve in Fig. 2(b), at 9.6 GHz.

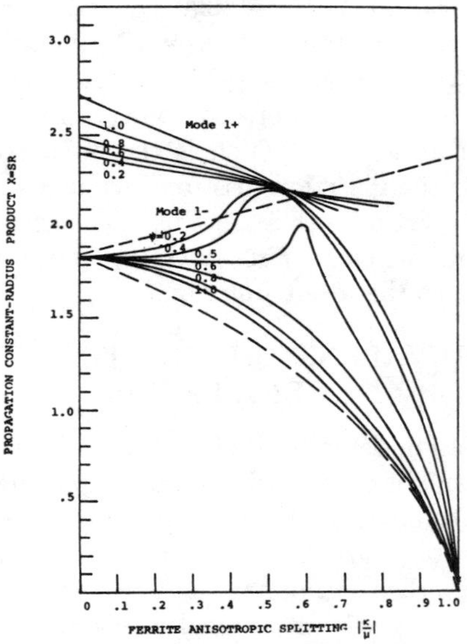

Fig. 5. Perfect circulation roots of the first circulation condition for various coupling angles (third order).

Therefore, we will pay special attention to the lower solution. The dashed curves are the roots of the uncoupled resonator. For lightly coupled circulators ($\psi < 0.4$), the roots first increase with increasing $|\kappa/\mu|$ and then decrease. On the other hand, for tightly coupled ($\psi > 0.5$) circulators, the roots start going down even for small $|\kappa/\mu|$. This phenomena is noticed by circulator designers [13], [14], but is not predicted by Davies and Cohen and Whiting. Massé [14] noticed that his broad-band circulator requires a radius about 7.5 percent less than the predicted design. He suggests that this is caused by fringing fields. In addition to this, the calculations here show that higher order modes (especially $n = 2$) have a significant effect in shrinking the required circulator radius.

Fig. 6 is the second circulation condition which is found

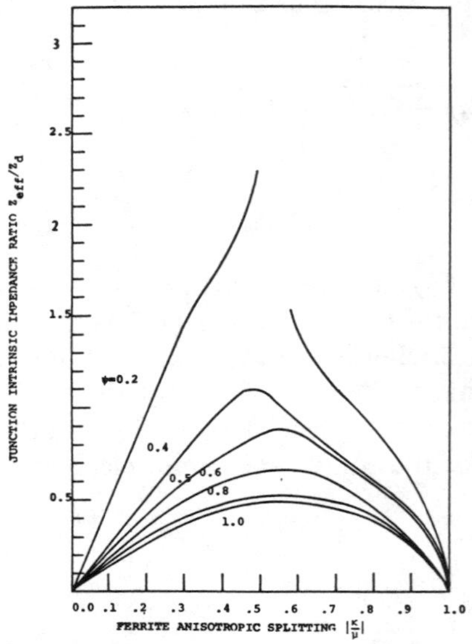

Fig. 6. Normalized junction impedance ratio as a function of anisotropic splitting factor calculated from the second circulation condition for various coupling angles.

by substituting the roots obtained from (20) into (21). The second condition gives the dielectric and disk wave impedance ratio at the junction. Both circulation conditions must be satisfied to get perfect circulation.

Since the two conditions for circulation are known, the design of a direct coupled circulator becomes a simple task. The first step is to choose the ferrite material and the dielectric medium. This is easily achieved in stripline. By using arc plasma spray (APS) [15] methods, or by inserting a ferrite puck into a hole in the substrate these choices can be made independently in microstrip. The second step is to plot the impedance ratio curve as a function of $|\kappa/\mu|$. The third step is to choose the coupling angle and get the intersection point of the impedance ratio curve and the corresponding curves in Fig. 6. Then the operating point $|\kappa/\mu|$ is specified (condition 2) from the abscissa of the intersection point. The final step is to determine the circulator radius from the corresponding point found in Fig. 5 (condition 1).

In the design of a circulator, one tries to find the intersection of the following two curves.

a) The impedance ratio Z_{eff}/Z_d of the chosen puck and dielectric material.

b) The calculated impedance ratio of the second circulation condition given in Fig. 6.

However, the condition 2 curves shown in Fig. 6 have a general property; all curves have positive slope with increasing $|\kappa/\mu|$ when $|\kappa/\mu| < 0.5$ and negative slope when $|\kappa/\mu| > 0.5$. For a weakly magnetized ferrite, operated below resonance, the impedance ratio is approximately given by

$$\frac{Z_{eff}}{Z_d} = \left(\frac{\epsilon_d}{\epsilon_f}\right)^{1/2}\left(1 - \left(\frac{\kappa}{\mu}\right)^2\right)^{1/2}. \tag{23}$$

This impedance ratio is shown for $\epsilon_d/\epsilon_f = 1$ in Fig. 7. As indicated, this curve always has a negative slope. Thus when $|\kappa/\mu|$ at the center frequency is less than 0.5, there is only one intersection near $|\kappa/\mu| \approx 0.2$ for $\psi = 0.2$. In this case a circulator with fixed radius works only in a limited frequency range. Even with quarter-wave transformers the bandwidth is still limited to about 25 percent [2], [6]. However, when $|\kappa/\mu|$ is greater than 0.5, both the impedance ratio curves and condition 2 curves have the same negative slope. Thus there can be a continuous solution if appropriate coupling angle and materials are selected.

For example, a special case is also shown in Fig. 7. The impedance ratio curve for $\epsilon_d/\epsilon_f = 1$ is plotted as a function of $|\kappa/\mu|$, along with the condition 2 curve for $\psi = 0.51$. As shown in the figure, there is no intersection for $|\kappa/\mu| < 0.5$. However, these two curves nearly overlap for $0.5 < |\kappa/\mu| < 1.0$. This means that for this particular design, the circulator not only works at $|\kappa/\mu| = 0.5$, but all the way from $|\kappa/\mu| = 0.5$ to $|\kappa/\mu| = 1.0$. Thus, as seen from Fig. 8, it can be used, in principle, from $SR = 0$ up to $SR \approx 1.84$. The importance of this phenomena is that the regular single intersection circulator can be replaced

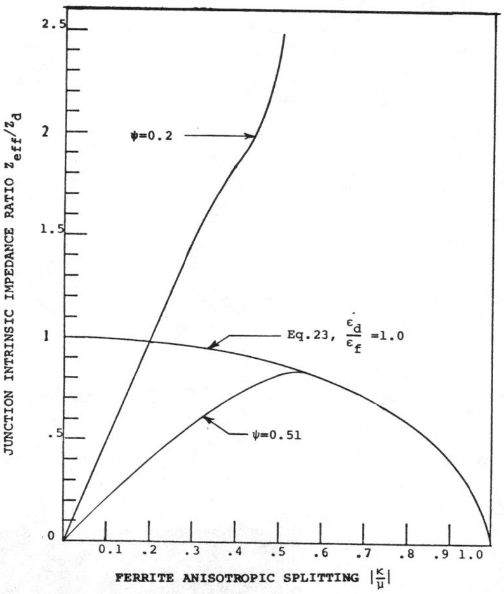

Fig. 7. Comparison of junction wave impedance ratio from the second circulation condition and from (23), as a function of $|\kappa/\mu|$.

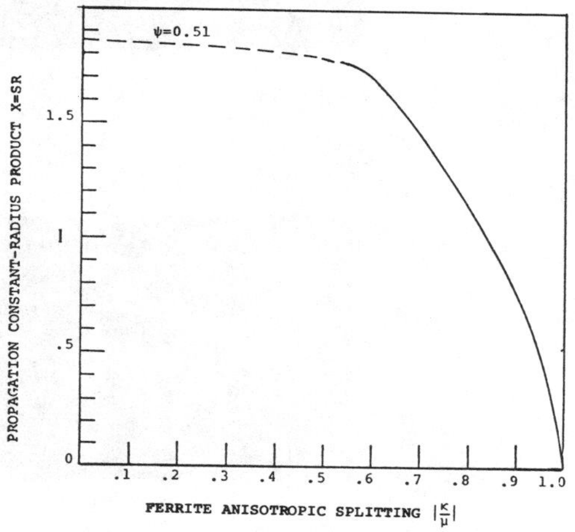

Fig. 8. Operating range of a continuous tracking circulator.

by an inherently broad-band circulator whose operating range is from $|\kappa/\mu| = |\omega_m/\omega| = 1$ or $\omega_L = \omega_m$ to $|\kappa/\mu| = |\omega_m/\omega| = 0.5$ or $\omega_H = 2\omega_m$. The bandwidth $(\omega_H - \omega_L)$ is approximately $\omega_m/2\pi = f_m = \gamma(4\pi M_s)$. An octave bandwidth circulator can be easily achieved according to this theory.

Notice that $SR = 0$ does not mean zero frequency. Since the radial wave propagation constant S is given by $S = (\omega/c)(\mu_{eff}\epsilon_f)^{1/2}$, the lower frequency which corresponds to $SR = 0$ is the one which yields $\mu_{eff} = 0$. For a ferrite just saturated, this frequency is given by $f = \gamma(4\pi M_s)$. However, because of low field losses, the actual operation band cannot reach $\mu_{eff} = 0$ or $|\kappa/\mu| = 1$.

To prove the existence of such a broad-band circulator, a microstrip device was fabricated using the structure shown in Fig. 9. The circulator radius is 0.100 in and the junction coupling angle is $\psi \simeq 0.525$ rad. The ferrite

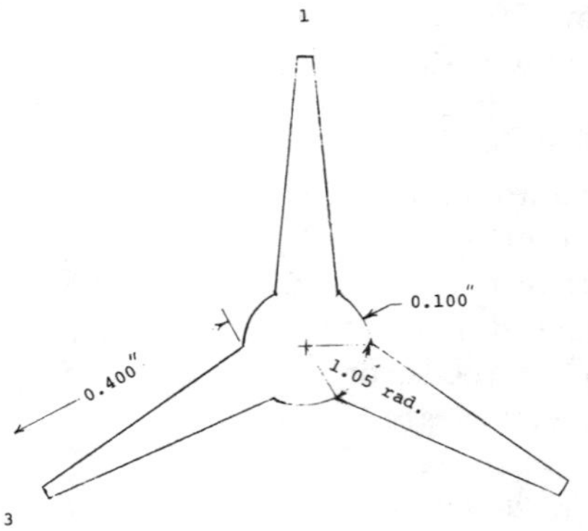

Fig. 9. Sample conductor pattern for a continuous tracking circulator.

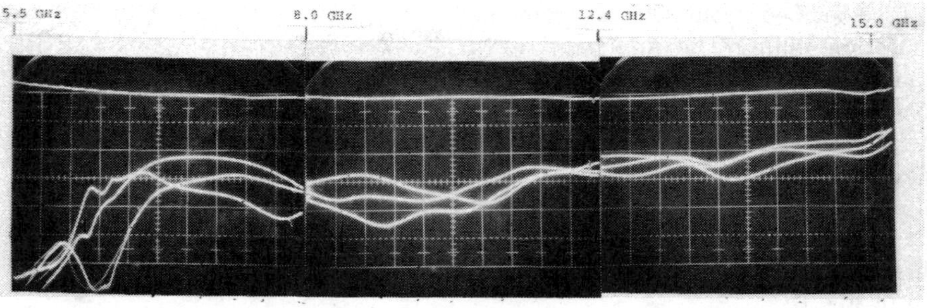

(a)

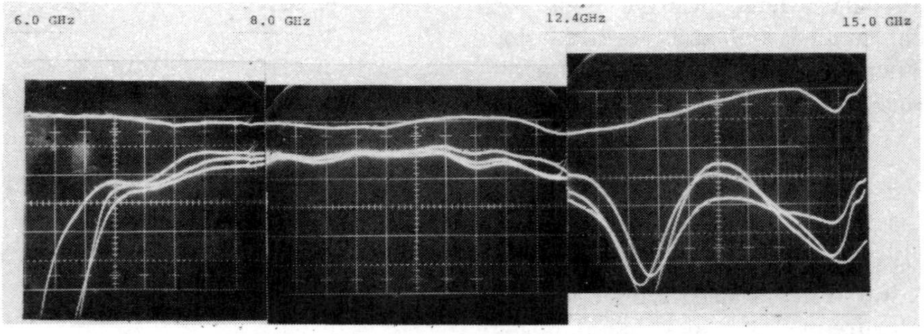

(b)

Fig. 10. Performance of an experimental continuous tracking circulator. (a) Isolation scale: 5 dB/div. (b) Insertion loss: scale: 1 dB/div.

material was TT1-390 which has a saturation magnetization of 2150 G(f_m = 6.02 GHz). The circulator is matched by a linear transformer to each of three 50-Ω microstrip lines on the same ferrite substrate. This linear transformer should be at least one wavelength long to get a good match.

Fig. 10(a) shows the experimental isolation for each of the three input ports, measured sequentially. The frequency range covered is from 5.5 to 15.0 GHz. The scale is 5 dB/div with the top line as the reference. It is seen that an average isolation of about 15 dB is obtained over the range 7–12.4 GHz. The resonance loss and cutoff region is obvious in the photograph below 6 GHz. The operating Bessel function argument SR runs from 0 up to 1.84. This follows, approximately, our prediction from Fig. 8. The center frequency is located at 9 GHz (SR = 1.2).

Although the device operates with large $|\kappa/\mu|$, the insertion loss shown in Fig. 10(b) is less than 1.0 dB over the range. The device was biased top and bottom with 0.2-in-diameter magnets, about 3/8 in tall. The large absorption shown near 13.2 GHz is caused by radiation from the top magnet.

V. CONCLUSION

These experimental results suggest that microstrip Y-junction circulators with bandwidth of the order of (2.8 MHz/Oe) × ($4\pi M_s$) can be readily fabricated simply by choosing the appropriate (wide) coupling angle and disk radius ($SR \approx 1.2$) and then suitably connecting to the 50-Ω input lines. Improved performance should be obtained through the use of a thick substrate such that the direct coupled 50-Ω lines are already as wide as the required coupling junction width. When using thin substrates, the isolation obtainable over the band will depend on the impedance characteristics of the transformers chosen. The quality of the terminations, connectors, and homogeneity of the biasing magnets are also very important as in any circulator design. This experiment gives good evidence of the existence of such a broad-band circulator.

The model presented here gives an improved understanding of the design and operation of circulators in the following ways. The input impedance of a circulator is analyzed rigorously using a Green's function approach. The equivalent circuit of the junction circulator is found to be a series combination of two parallel *RLC* circuits which reduce to a single resonator when the conditions for perfect circulation are achieved. Thus the equivalence between Bosma's model and that of Fay and Comstock is shown.

A continuous tracking technique is discovered which makes the fabrication of octave bandwidth circulators straightforward. This circulator is recognized [16] by: a) wide coupling angle $\psi \simeq 0.5$, and b) smaller than usual circulator radius; center frequency of operating band is found from $SR \approx 1.2$.

ACKNOWLEDGMENT

The authors wish to thank Dr. M. Davis, now of General Electric Company, and D. H. Harris of the Monsanto Research Corporation, Dayton, Ohio, for their interest, support, and encouragement; R. West of Trans-Tech for the substrates; and Dr. R. H. Knerr of the Bell Laboratories for metallizing and etching the substrates.

REFERENCES

[1] H. Bosma, "On stripline *Y*-circulation at UHF," *IEEE Trans. Microwave Theory Tech.* (*1963 Symposium Issue*), vol. MTT-12, pp. 61–72, Jan. 1964.
[2] C. E. Fay and R. L. Comstock, "Operation of the ferrite junction circulator," *IEEE Trans. Microwave Theory Tech.* (*1964 Symposium Issue*), vol. MTT-13, pp. 15–27, Jan. 1965.
[3] J. W. Simon, "Broadband strip-transmission line *Y*-junction circulators," *IEEE Trans. Microwave Theory Tech.*, vol. MTT-13, pp. 335–345, May 1965.
[4] S. J. Salay and H. J. Peppiatt, "Input impedance behavior of stripline circulator," *IEEE Trans. Microwave Theory Tech.* (Corresp.), vol. MTT-19, pp. 109–110, Jan. 1971.
[5] ——, "An accurate junction circulator design procedure," *IEEE Trans. Microwave Theory Tech.* (Corresp.), vol. MTT-20, pp. 192–193, Feb. 1972.
[6] L. K. Anderson, "An analysis of broadband circulators with external tuning elements," *IEEE Trans. Microwave Theory Tech.*, vol. MTT-15, pp. 42–47, Jan. 1967.
[7] E. Schwartz, "Broadband matching of resonant circuits and circulators," *IEEE Trans. Microwave Theory Tech.*, vol. MTT-16, pp. 158–165, Mar. 1968.
[8] D. Polder, "On the theory of ferromagnetic resonance," *Phil. Mag.*, vol. 40, pp. 99–115, Jan. 1949.
[9] J. B. Davies and P. Cohen, "Theoretical design of symmetrical junction stripline circulator," *IEEE Trans. Microwave Theory Tech.*, vol. MTT-11, pp. 506–512, Nov. 1963.
[10] K. Whiting, "Design data for UHF circulators," *IEEE Trans. Microwave Theory Tech.* (Corresp.), vol. MTT-15, pp. 195–198, Mar. 1967.
[11] H. Bosma, "Junction circulators," in *Advances in Microwaves*, vol. 6, L. Young, Ed. New York: Academic, 1971.
[12] E. Schlömann, "Microwave behavior of partially magnetized ferrites," *J. Appl. Phys.*, vol. 41, pp. 204–214, Jan. 1970.
[13] D. J. Massé and R. A. Pucel, "Microstrip propagation on magnetic substrates—Part II: Experiment," *IEEE Trans. Microwave Theory Tech.*, vol. MTT-20, pp. 309–313, May 1972.
[14] D. J. Massé, "Broadband microstrip junction circulators," *Proc. IEEE* (Lett.), vol. 56, pp. 352–353, Mar. 1968.
[15] D. H. Harris *et al.*, "Polycrystalline ferrite films for microwave applications deposited by arc-plasma," *J. Appl. Phys.*, vol. 41, pp. 1348–1349, Mar. 1970.
[16] Y. S. Wu and F. J. Rosenbaum, "Ferrite/dielectric high power phase shifter development—Part II on broadband *Y*-junction circulators," Air Force Avionics Lab., Air Force Systems Command, Wright-Patterson Air Force Base, Ohio, Tech. Rep. AFAL-TR-73-250, pt II.

Paper 7.4

Ferrite Planar Circuits in Microwave Integrated Circuits

TANROKU MIYOSHI, MEMBER, IEEE, S. YAMAGUCHI, AND SHINJI GOTO

Abstract—The ferrite planar circuit to be discussed in this paper is a general planar circuit using ferrite substrates magnetized perpendicular to the ground conductors. The main subject of this paper is the analysis of an arbitrarily shaped triplate ferrite planar circuit. In particular, the circuit parameters of the equivalent multiport are determined. To analyze ferrite planar circuits in general, two approaches are possible. One approach is based upon a contour-integral solution of the wave equation. In the other approach the fields in the circuit are expanded in terms of orthonormal eigenfunctions. Examples of the application of such analyses are described.

I. Introduction

THE planar circuit is defined as an electrical circuit whose thickness in one direction is much less than one wavelength and whose dimensions in the orthogonal directions are comparable to the wavelength. The concept of the planar circuit was proposed by Okoshi in 1969 [1]. Since then, its analysis [2]–[5] and synthesis [6], [7] have been investigated for many circuits using isotropic material for the spacer.

This paper will present the general treatment of a planar circuit using ferrite material for the spacer. In particular, an arbitrarily shaped ferrite planar circuit is discussed. The ferrite is magnetized in the direction perpendicular to the ground plane.

Stripline circulators [8] often used in the microwave integrated circuits and edge-guided mode devices [9], [10] are considered ferrite planar circuits. They are, strictly speaking, two-dimensional circuits because they require essentially a disk resonator, wide striplines, and tapered sections.

The main subject of this paper is the analysis of an arbitrarily shaped, triplate ferrite planar circuit. In particular, the circuit parameters of the equivalent multiport are determined. To analyze ferrite planar circuits in general, two approaches are possible. One approach is based upon a contour-integral solution of the wave equation. In the other approach the fields in the circuit are expanded in terms of orthonormal eigenfunctions. Examples of such analyses are also described.

II. Basic Equation

A ferrite planar circuit consists of an arbitrarily shaped thin center conductor sandwiched by two ferrite substrates

Manuscript received September 7, 1976; revised December 17, 1976.
The authors are with the Department of Electronic Engineering, Kobe University, Kobe, Japan.

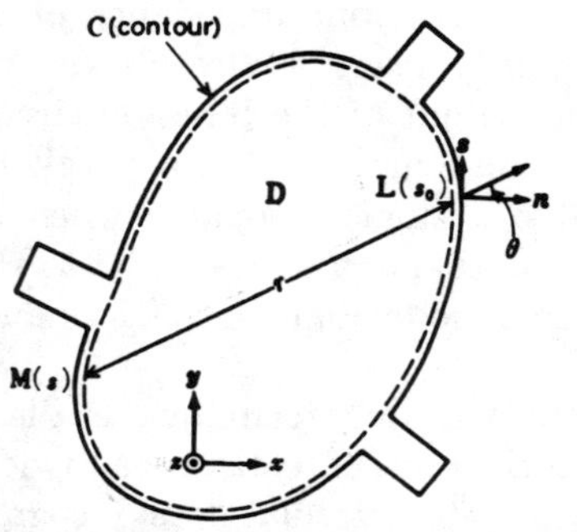

Fig. 1. Center conductor of a ferrite planar circuit and symbols used in the integral equation.

with magnetic field perpendicular to the conducting plate. It is assumed to be excited symmetrically with respect to the upper and lower ground conductors. There are several coupling ports as shown in Fig. 1 and the remainder of the periphery is assumed to be open circuited. The xy coordinates and z axis, respectively, are set parallel and perpendicular to the conductors. The bias magnetization is in the z direction. The thickness of the planar circuit is $2d$.

When the spacing d is much smaller than the wavelength and ferrite spacers are homogeneous and linear, only the field components E_z, H_x, and H_y with no variation along the z axis are considered. It is deduced directly from Maxwell's equation that the following equation governs the electromagnetic fields in the ferrite planar circuit:

$$(\nabla_T{}^2 + \omega^2 \varepsilon \mu_{\text{eff}})V = 0 \tag{1}$$

where

$$\nabla_T{}^2 = \frac{\partial^2}{\partial x^2} + \frac{\partial^2}{\partial y^2} \qquad \mu_{\text{eff}} = \frac{\mu^2 - \kappa^2}{\mu}.$$

Here V given by $E_z \times d$ denotes the RF voltage of the center conductor with respect to the ground conductors. The effective permeability μ_{eff} is given by μ and κ which are the diagonal and off-diagonal coefficients of permeability tensor for magnetization in the z direction. The sign of μ_{eff} will depend upon the frequency and the internal magnetic field.

At a coupling port, the following boundary condition given by the differential equation must apply:

$$j\frac{\kappa}{\mu}\frac{\partial V}{\partial t} + \frac{\partial V}{\partial n} = -j\omega\mu_{\text{eff}}\, di_n \tag{2}$$

where i_n is the surface current density normal to the periphery and ∂n and ∂t, respectively, are the derivative normal to

Reprinted from *IEEE Trans. Microwave Theory Tech.*, vol. MTT-25, no. 7, pp. 593–600, July 1977.

the periphery and the tangential derivative around the periphery.

Almost at the periphery where the coupling ports are absent, the current flow normal to the periphery is assumed to be zero, that is, $i_n = 0$. Actually, however, the fringing magnetic fields are always present. A simple correction for this effect is to enlarge the periphery outwards by an amount of $0.447d \times K (K = 0.4)$ in advance of the analysis. The coefficient K was determined by comparing the measured resonant frequencies for the various ferrite planar resonators with the theoretical ones, which were calculated by the Rayleigh–Ritz variational method assuming that the circuits were lossless. This will be explained later in Section IV-C.

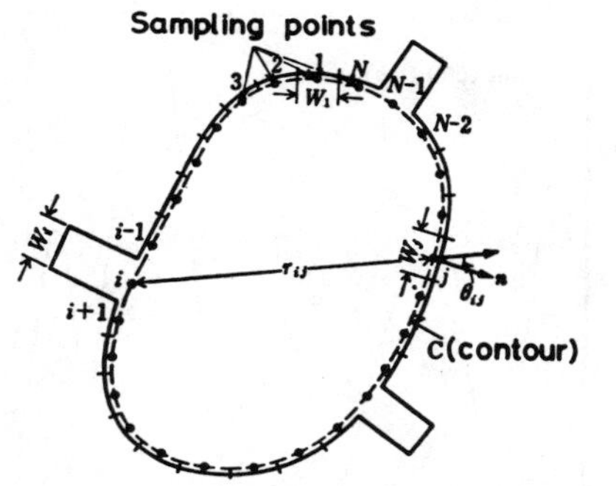

Fig. 2. Symbols used in the computer analysis.

III. Analysis Based Upon a Contour-Integral Equation

A. Integral Equation

If we introduce Green's function G for (1), the RF voltage V_p at a point P in the circuit is given by a line integral

$$V_P = \oint_c \left\{ -j\omega\mu_{\text{eff}}\, di_n G + V\left(j\frac{\kappa}{\mu}\frac{\partial G}{\partial t} - \frac{\partial G}{\partial n}\right)\right\} dt. \quad (3)^1$$

If we now use the free-space Green's function for G in (3), then we must select different types of Green's functions according to the sign of μ_{eff}.

When $\mu_{\text{eff}} > 0$, $G = H_0^{(2)}(kr)/4j$ should be used as Green's function, where $H_0^{(2)}$ is the zeroth-order Hankel function of second kind and $k = \omega\sqrt{\varepsilon\mu_{\text{eff}}}$. Then from (3), the RF voltage at a point upon the periphery is found to satisfy the following equation:

$$V_M = \frac{1}{2j}\oint_c \left\{ j\omega\mu_{\text{eff}}\, dH_0^{(2)}(kr)(-i_n) + k\left(\cos\theta - j\frac{\kappa}{\mu}\sin\theta\right) H_1^{(2)}(kr)V_L \right\} dt. \quad (4)$$

In this equation $H_1^{(2)}$ is the first-order Hankel function of second kind. The variable r denotes distance between points M and L represented by s and s_0, respectively, and θ denotes the angle made by the straight line from point M to point L and the normal at point L as shown in Fig. 1. If the current density i_n injected upon the periphery is known, (4) becomes a Fredholm integral equation of the second kind in terms of the RF voltage.

B. Computational Formulation

For a numerical calculation, we divide the periphery into N incremental sections and set N sampling points defined at the center of each section as shown in Fig. 2. When we assume that the magnetic and electric field intensities are uniform across each section, the above integral equation results in a matrix equation:

$$\sum_{j=1}^{N} u_{ij}V_j = \sum_{j=1}^{N} h_{ij}I_j, \qquad i = 1,2,\cdots,N \quad (5)$$

where

$$u_{ij} = \delta_{ij} - \frac{k}{2j}\int_{W_j}\left\{\cos\theta - j\frac{\kappa}{\mu}\sin\theta\right\} H_1^{(2)}(kr)\, dt_j$$

$$h_{ij} = \begin{cases} \dfrac{\omega\mu_{\text{eff}}\, d}{4W_j}\displaystyle\int_{W_j} H_0^{(2)}(kr)\, dt_j, & i = j \\ \dfrac{\omega\mu_{\text{eff}}\, d}{4}\left\{1 - \dfrac{2}{\pi}\left(\log\dfrac{kW_i}{4} - 1 + \gamma\right)\right\}, & i = j \end{cases} \quad (6)$$

where $\gamma = 0.5772\cdots$: Euler's constant and $I_j = -2i_nW_j$ represents the total current flowing into the jth port. The formulas u_{ij} and h_{ij} in (6) have been derived assuming that the jth section is straight. From the above relations, the impedance matrix of the equivalent N-port is given by

$$Z = U^{-1}H \quad (7)$$

whe.e U^{-1} denotes the inverse matrix to U. Then one element of the impedance matrix is given by

$$Z_{ij} = \frac{1}{\det U}\begin{vmatrix} u_{11} & \cdots & h_{1j} & \cdots & u_{1N} \\ \vdots & & \vdots & & \vdots \\ u_{N1} & \cdots & h_{Nj} & \cdots & u_{NN} \end{vmatrix}. \quad (8)$$

When the circuit has no coupling port, that is, $I_j = 0$, from the nontrivial condition of (5), we have

$$\det U = 0. \quad (9)$$

This equation gives the resonant frequency of the circuit.

When $\mu_{\text{eff}} < 0$, $G = \{K_0(hr) + j\pi I_0(hr)\}/2\pi$ is applicable, where $h = \omega\sqrt{\varepsilon|\mu_{\text{eff}}|}$ and I_0 and K_0 are the zeroth-order modified Bessel functions of the first and second kind, respectively. In this case the elements of matrices U and H in (7) are given by

$$u_{ij} = \delta_{ij} - \frac{h}{\pi}\int_{W_j}\left(\cos\theta - j\frac{\kappa}{\mu}\sin\theta\right)(K_1 - j\pi I_1)\, dt$$

---

[1] $$V_p = \oint_c \left\{G\frac{\partial V}{\partial n} - V\frac{\partial G}{\partial n}\right\} dt = \oint_c \left\{G\left(\frac{\partial V}{\partial n} + j\frac{\kappa}{\mu}\frac{\partial V}{\partial t}\right) - j\frac{\kappa}{\mu}\frac{\partial V}{\partial t}G - V\frac{\partial G}{\partial n}\right\} dt.$$

This relation and the relations (2) and $\oint_c G(\partial V/\partial t)\, dt = -\oint_c V(\partial G/\partial t)\, dt$ give (3).

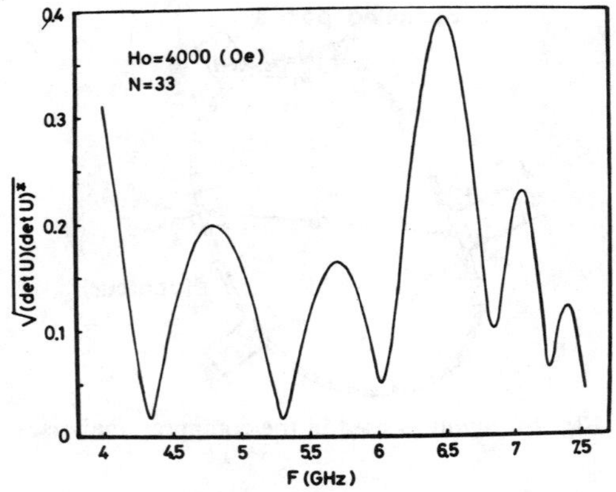

Fig. 3. The variation of $|\det U|$ as a function of frequency of a disk-shaped circuit at $H_0 = 4000$ Oe for $N = 33$.

$$h_{ij} = \begin{cases} \dfrac{j\omega\mu_{\text{eff}}\, d}{2\pi} \dfrac{1}{W_j} \displaystyle\int_{W_j} (K_0 + j\pi I_0)\, dt, & i \neq j \\ -\dfrac{j\omega\mu_{\text{eff}}\, d}{2\pi} \left\{ \left(\log \dfrac{hW_i}{4} + \gamma - 1 \right) - j\pi \right\}, & i = j. \end{cases} \tag{10}$$

C. Examples of Analysis

In all of the following examples, the ferrimagnetic material is assumed to be lossless with the saturation magnetization $4\pi M_s = 1300$ G, the dielectric constant $\varepsilon = 15.6$, and the thickness $d = 2$ mm.

As an example of the computer analysis described so far, the resonant frequencies of a disk-shaped circuit were computed first to check the computation accuracy. Since $\det U = 0$ in (9) is never realized for real frequency due to the computation error, we define the frequency which gives the minimum of $|\det U|$ as the eigenvalue. The variation of $|\det U|$ is shown as a function of frequency F (gigahertz) in Fig. 3 for $N = 33$ at $H_0 = 3300$ Oe, which shows the first ($F = 4.35$), the second ($F = 5.31$), the third ($F = 6.05$), the fourth ($F = 6.85$), and the fifth ($F = 7.26$) minima. By comparing these calculated eigenvalues with the theoretical ones, which should be given by the roots of

$$J_n'(ka) - \frac{\kappa}{\mu} \frac{nJ_n(ka)}{ka} = 0, \qquad n = 0, \pm 1, \pm 2 \cdots$$

$$I_n'(ha) - \frac{\kappa}{\mu} \frac{nI_n(ha)}{ha} = 0, \qquad n = 1, 2 \cdots$$

we found that the computation error was within 2.0 percent for the sampling number 33.

Next, the characteristics of the Y-junction stripline circulator were computed as shown in Fig. 4. Here the internal magnetic field is 3700 Oe, $N = 33$, and 50-Ω striplines are coupled to the circulator. The circulator performance in Fig. 4 is obtained above the ferrimagnetic resonance point of circulation, which is about 5.7 GHz in this case. On the other hand, the resonant frequencies of $+1$ and -1 modes, respectively, are 5.5 and 4.9 GHz, which means that the center frequency is not midway between $+1$ and -1 mode resonant frequencies but exterior to the region. This is believed to be due to operating at a frequency far from the

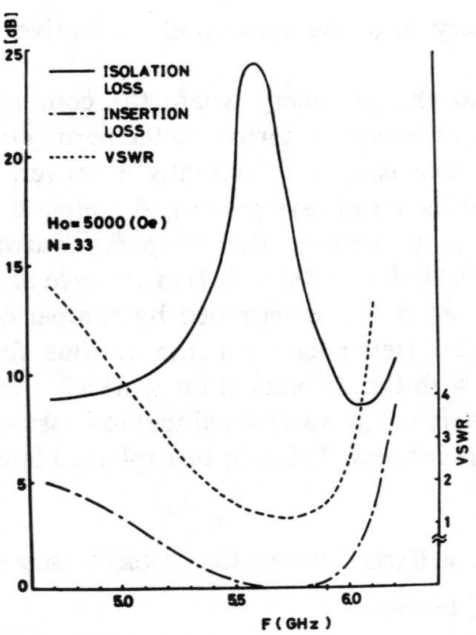

Fig. 4. Computed performance of a stripline Y-junction circulator coupled by the striplines of 50 Ω.

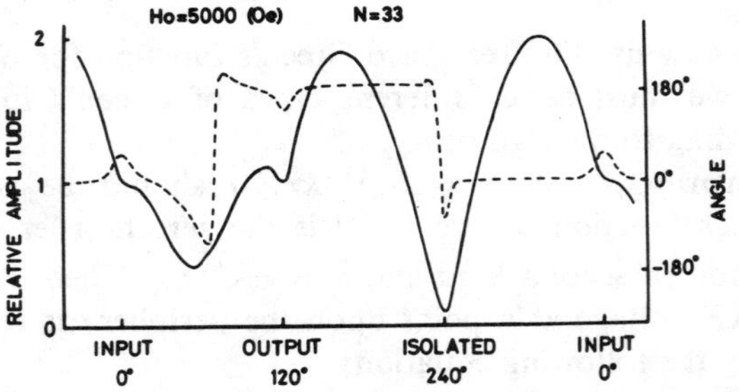

Fig. 5. Computed RF voltage distribution, amplitude (solid curve) and phase (broken curve), along the periphery of a Y-junction stripline circulator at the center frequency for $N = 33$.

degeneracy of the ± 1 modes, i.e., at a greater separation of the modes, and the strong influence of higher order modes. Fig. 5 shows the RF voltage distribution along the periphery of a Y-junction stripline circulator at the center frequency for $N = 33$. The solid and broken curves show the relative amplitude and phase of the RF voltage along the periphery, respectively. The distribution of the amplitude is not sinusoidal, as might be expected, but exhibits a shallower minimum between the input and output ports and a distortion in the vicinity of ports. This is due to the influence of higher order modes as mentioned previously, which results from the strong coupling to the stripline ports.

IV. Analysis Based Upon an Eigenfunction Expansion

A. Formulation of Circuit Parameters

Now we introduce the Green's function which satisfies the following boundary condition along the contour C in (3):

$$j \frac{\kappa}{\mu} \frac{\partial G}{\partial t} - \frac{\partial G}{\partial n} = 0. \tag{11}$$

The RF voltage at a point in the circuit is given by

$$V = j\omega\mu_{\text{eff}}\, d \oint_c G(-i_n)\, dt. \tag{12}$$

Furthermore, we expand the Green's function in terms of the complex eigenfunctions ϕ_a which derive from the eigenvalue problem defined by

$$(\nabla_T^2 + \omega_a^2 \varepsilon\mu_{\text{eff}})\phi_a = 0, \qquad (\text{in } D)$$
$$j\frac{\kappa}{\mu}\frac{\partial\phi_a}{\partial t} - \frac{\partial\phi_a}{\partial n} = 0, \qquad (\text{on } C) \tag{13}$$
$$\iint_D \varepsilon\phi_a\phi_b^*\, dS = \delta_{ab}$$

where the asterisk means a complex conjugate of ϕ_b. Then we can represent the RF voltage in the circuit, using eigenfunctions, by

$$V = j\omega\, d \oint_c \sum_{a=0}^{\infty} \frac{\phi_a\phi_a^*}{\omega_a^2 - \omega^2}(-i_n)\, dt. \tag{14}$$

Next, to calculate the circuit parameters of the equivalent multiport, we define approximately the RF voltage on a port and the total current flowing into a port, respectively, as

$$V_i = \frac{1}{W_i}\int_{W_i} V(t_i)\, dt_i \qquad I_j = \int_{W_j} \{-2i_n(t_j)\}\, dt_j. \tag{15}$$

Substituting (15) into (14), we have

$$V_i = \sum_{j=1}^{l}\left(\sum_{a=0}^{\infty} \frac{j\omega\, d}{2W_iW_j}\int_{W_i}\int_{W_j} \frac{\phi_a^*(t_j)\phi_a(t_i)}{\omega_a^2 - \omega^2}\, dt_i\, dt_j\right) I_j \tag{16}$$

where l is the number of ports coupling to the planar circuit. Thus one element of the impedance matrix of the equivalent multiport becomes

$$Z_{ij} = \sum_{a=0}^{\infty} \frac{j\omega\, d}{2W_iW_j}\int_{W_i}\int_{W_j} \frac{\phi_a^*(t_i)\phi_a(t_j)}{\omega_a^2 - \omega^2}\, dt_i\, dt_j. \tag{17}$$

It is clear from the above equation that the impedance matrix is not symmetric, i.e., $Z_{ij} \neq Z_{ji}$, because ϕ_a are generally complex eigenfunctions, but $Z_{ij} = -Z_{ji}^*$, which corresponds to the lossless condition of the circuit. To obtain the performance of a ferrite planar circuit by means of (17), we must solve the eigenvalue problem defined by (13) repeatedly at different frequencies for a given circuit and a given bias magnetic field. This is due to the fact that μ_{eff} contained in the problem is a function of the operation frequency even if the bias magnetic field is given.

B. Computational Formulation

To solve the eigenvalue problem in (13), in general, the Rayleigh–Ritz variational method, using a polynomial approximation, will be employed.

Since (13) is found to be the Euler equation of the functional I,

$$I = \iint_D (|\nabla_T\phi|^2 - \omega^2\varepsilon\mu_{\text{eff}}|\phi|^2)\, dS + j\frac{\kappa}{\mu}\oint_c \phi^* \frac{\partial\phi}{\partial t}\, dt. \tag{18}$$

Instead of solving (13), we attempt to find the approximate complex functions which minimize the functional I.

Let the function ϕ be replaced by

$$\phi = \sum_{i=1}^{M} c_i f_i \tag{19}$$

there the c_i denote the complex expansion coefficients to be determined and the f_i are the real basis functions. The stationary points of I can be selected by evaluating the M equations $\partial I/\partial c_i^* = 0$. This immediately gives the matrix eigenvalue problem

$$(A - \omega^2\varepsilon\mu_{\text{eff}}B)c = 0 \tag{20}$$

where

$$A_{ij} = \iint_D \nabla f_i \cdot \nabla f_j\, dS + j\frac{\kappa}{\mu}\oint_c f_i \frac{\partial f_j}{\partial t}\, dt,$$
$$B_{ij} = \iint_D f_i f_j\, dS.$$

Here, the values of μ, κ, and μ_{eff} are constant if both the bias magnetic field and the frequency are given. Thus the eigenvalue problem given in (13) has been approximated by the algebraic eigenvalue problem contained in (20). In (20), noting that the A matrix is Hermitian and that the B matrix is symmetric and positive definite, the eigenvalue ω_a^2 is found to be real. The problem given in (20) will be solved easily by a library program when (20) is rewritten in the usual form of the eigenvalue problem of a Hermitian matrix. To normalize the approximating eigenfunctions having arbitrary amplitude, the coefficients calculated should be multiplied by

$$\left(\varepsilon \sum_{i=1}^{M}\sum_{j=1}^{M} c_i^* c_j B_{ij}\right)^{-1/2}$$

When the ferrite planar circuit has no coupling ports, i.e., in the case of a resonator, from the nontrivial condition of (20), we also have

$$\det(A - \omega^2\varepsilon\mu_{\text{eff}}B) = 0. \tag{21}$$

However, in this case the angular frequency ω contained implicitly in μ, κ, and μ_{eff} is unknown. The resonant frequencies of the circuit are given by the roots of (21).

C. Results

In all of the following examples, a polynomial of order 5 will be used to approximate the eigenfunctions, which gives a matrix size of order 21.

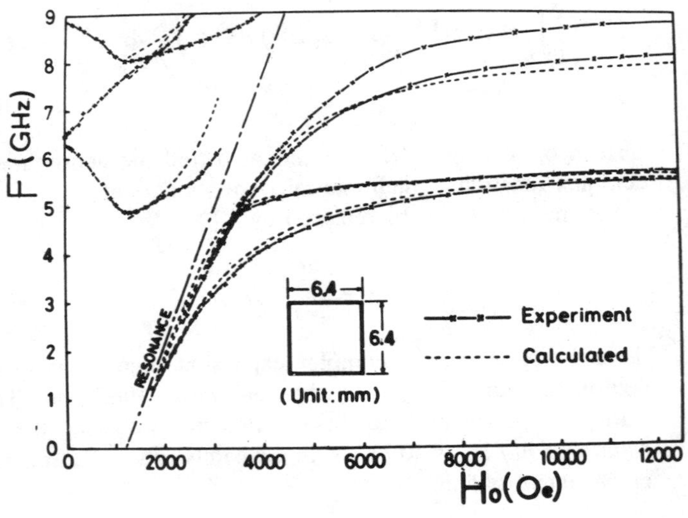

Fig. 6. Magnetic tuning characteristics of a square resonator. The broken curves were calculated taking fringing fields into account.

First, the characteristics of ferrite planar resonators, of which nothing has been reported so far except a disk-shaped circuit [11], were studied. Fig. 6 shows the magnetic tuning characteristics of a square resonator with a side 6.4 mm long. The broken curves were calculated taking the effects of fringing fields into account as mentioned in Section II, assuming that the circuits are lossless. The measured resonant frequencies shown in solid curves are found to be in good agreement with the calculated values, especially above the ferrimagnetic resonance. This is probably because the influence of the magnetic loss is smaller in this region. In the experiment, the square ferrimagnetic substrates (25×25 mm^2) with a saturation magnetization of $4\pi Ms = 1300$ G, linewidth $\Delta H = 68$ Oe, a dielectric constant of $\varepsilon = 15.6$, and thickness $d = 2$ mm were used for the spacing material.

Fig. 7 shows the computed instantaneous distribution of the RF voltage in the square resonator for the fundamental mode. Equiamplitude (upper) and phase (lower) lines are shown for (a) $\mu_{eff} > 0$ at $H_0 = 1300$ Oe, and (b) $\mu_{eff} < 0$ at $H_0 = 2300$ Oe. The fields are found to rotate clockwise as in a disk resonator. It is also found that the fields are somewhat concentrated along the periphery when $\mu_{eff} < 0$.

In the case of a triangular resonator with a side 10 mm long, the magnetic tuning characteristics and the instantaneous RF voltage distribution for the fundamental mode are shown in Figs. 8 and 9, respectively. It is found from the figures that almost the same resonant characteristics as obtained for a square resonator result. It generally follows

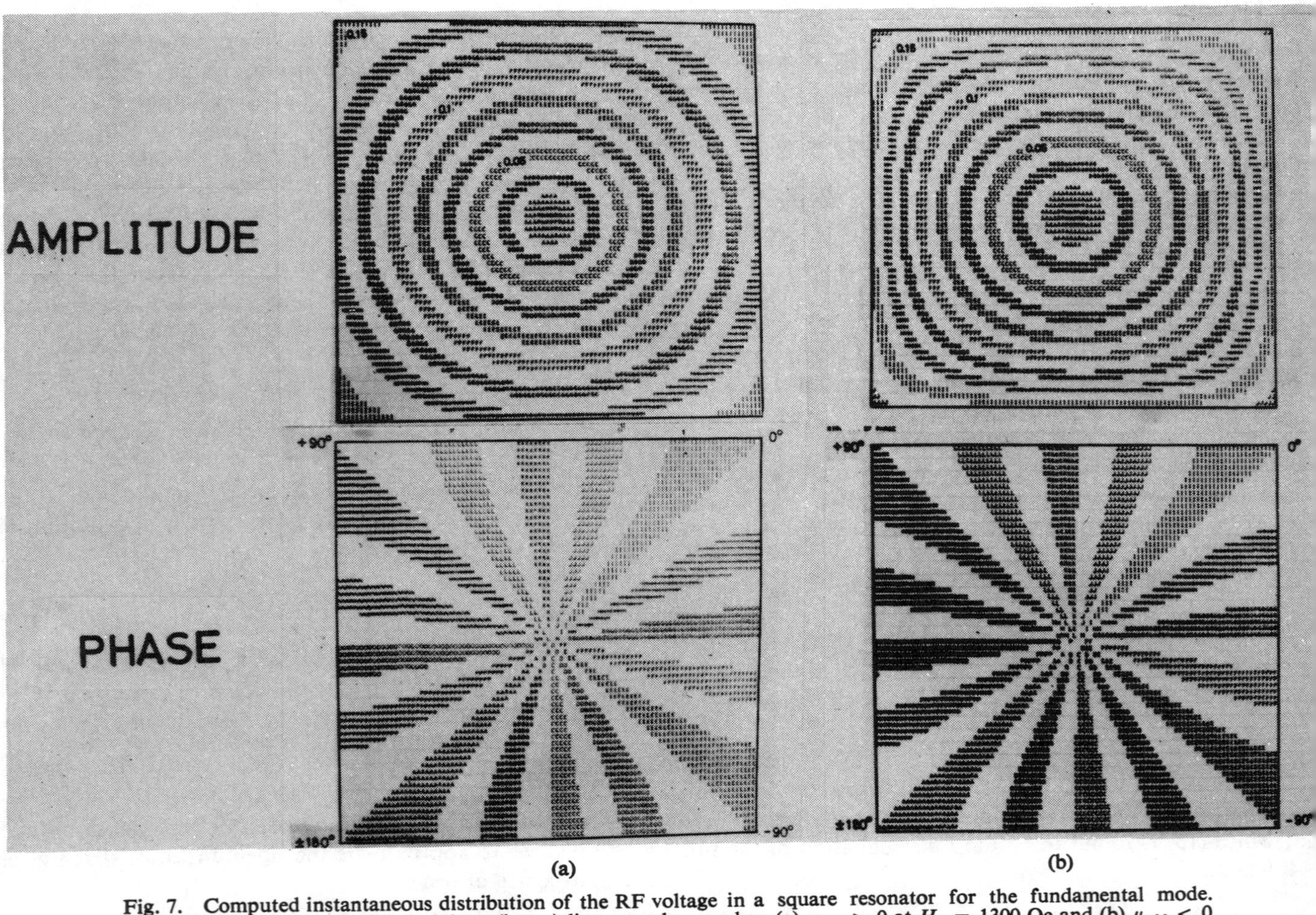

Fig. 7. Computed instantaneous distribution of the RF voltage in a square resonator for the fundamental mode. Equiamplitude (upper) and equiphase (lower) lines are shown when (a) $\mu_{eff} > 0$ at $H_0 = 1300$ Oe and (b) $\mu_{eff} < 0$ at $H_0 = 2300$ Oe.

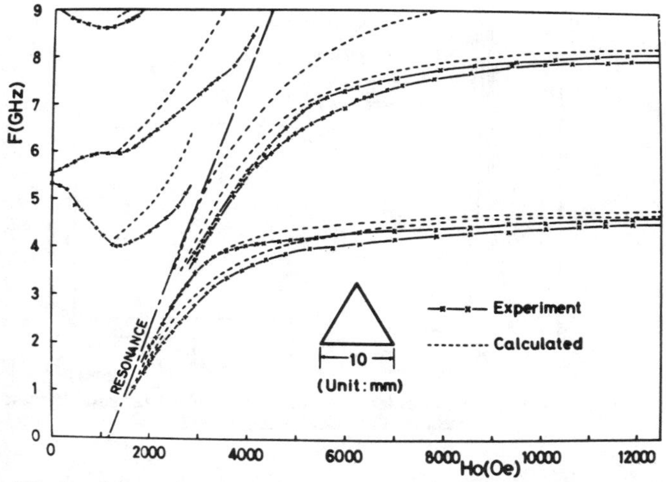

Fig. 8. Magnetic tuning characteristics of a triangular resonator.

that when $\mu_{eff} > 0$, the modes rotating both clockwise and counterclockwise are the ones resonating in ferrite planar resonators. On the other hand, when $\mu_{eff} < 0$, only the mode rotating clockwise can exist, and, furthermore, the fields in resonators are concentrated along the periphery.

Finally, several interesting applications of the analysis to a triangular ferrite planar circuit will be shown. Fig. 10 shows a center conductor plate of a three-port triangular circuit used in the calculation. The shaded portion (hexagon) in this figure is regarded as a planar circuit coupled by three striplines. We calculated the characteristics of the circuit for various applied magnetic fields and characteristic impedances Z_0 of the striplines. When the applied magnetic field is 5300 Oe, a circulator performance as shown in Fig. 11 was obtained above the ferrimagnetic resonance when $Z_0 = 30\ \Omega$. The light line curves are calculated by the method based upon the eigenfunction expansion for $i = a = 21$. The heavy line curves by the contour-integral method for $N = 33$ are also shown for the comparison. This performance can be explained by considering two fundamental rotating modes, i.e., a mode rotating clockwise and the other rotating counterclockwise. Consequently, the principle of operation is the same as for a disk-shaped stripline circulator.

When the applied magnetic field at the triangular ferrite planar circuit is 3300 Oe, μ_{eff} is negative in the frequency range between 7.19 and 9.24 GHz. In this range the performance of the so-called edge-guided mode circulator was calculated as shown in Fig. 12 for $Z_0 = 50\ \Omega$. We note that such a performance has not yet been obtained experimentally.

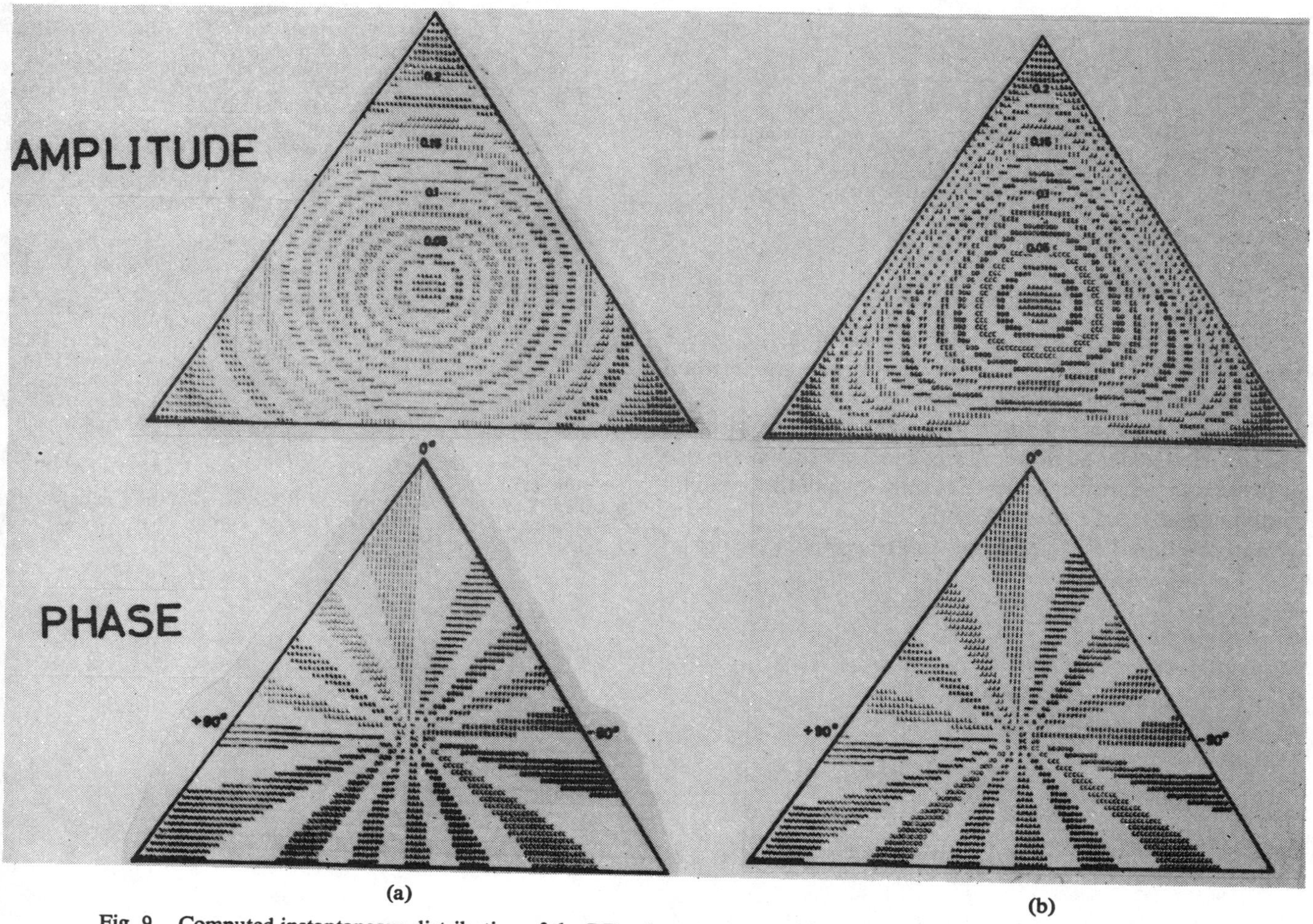

(a) (b)

Fig. 9. Computed instantaneous distribution of the RF voltage in a triangular circuit for the fundamental mode when (a) $\mu_{eff} > 0$ at $H_0 = 1300$ Oe and (b) $\mu_{eff} < 0$ at $H_0 = 2300$ Oe.

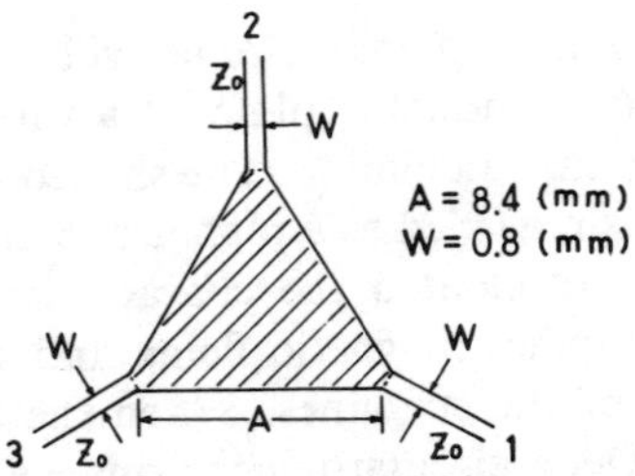

Fig. 10. Center conductor of a triangular ferrite planar circuit used in the calculation.

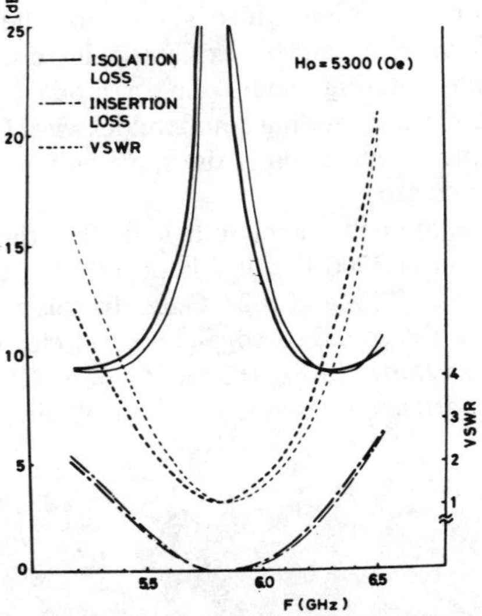

Fig. 11. Computed circulator performance of a triangular circuit at $H_0 = 5300$ Oe for $Z_0 = 30\ \Omega$. The light line curves are by the method based upon eigenfunction expansion for $i = a = 21$. The heavy line ones are by the integral equation method for $N = 33$.

V. Conclusion

We have shown that an arbitrarily shaped ferrite planar circuit can be analyzed using the contour-integral method or the eigenfunction expansion method.

Although there is no difference in the labor required to analyze repeatedly at different frequencies using these two approaches, the former is more readily adapted to a circuit with a complicated pattern than the latter.

We hope that these approaches will be useful in the design

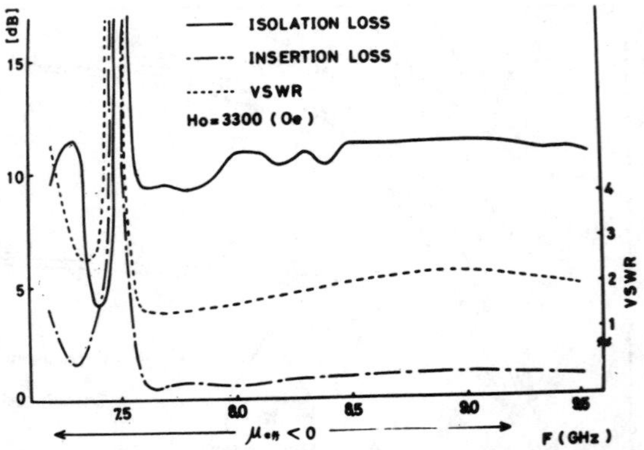

Fig. 12. Computed circulator performance of a triangular circuit at $H_0 = 3300$ Oe for $Z_0 = 50\ \Omega$.

and analysis of microwave integrated circuits on ferrite substrates.

Acknowledgment

The authors wish to thank Prof. T. Okoshi of University of Tokyo for his encouragement.

References

[1] T. Okoshi, "The planar circuit," in *Rec. of Professional Groups, IECE of Japan*, Paper SSD68-37/CT68-47, Feb. 17, 1969.

[2] P. P. Civalleri and S. Ridella, "Impedance and admittance matrices of distributed three-layer *N*-ports," *IEEE Trans. Circuit Theory*, vol. CT-17, pp. 392–398, Aug. 1970.

[3] T. Okoshi and T. Miyoshi, "The planar circuit—An approach to microwave integrated circuitry," *IEEE Trans. Microwave Theory Tech.*, vol. MTT-20, pp. 245–252, Apr. 1972.

[4] P. Silvester, "Finite element analysis of planar microwave networks," *IEEE Trans. Microwave Theory Tech.*, vol. MTT-21, pp. 104–108, Feb. 1973.

[5] H. Jui-Pang, T. Anada, and O. Kondo, "Analysis of microwave planar circuits by normal mode method," *Trans. IECE of Japan*, vol. 58-B, pp. 671–678, Dec. 1974.

[6] K. Grüner, "Methods of synthesizing nonuniform waveguides," *IEEE Trans. Microwave Theory Tech.*, vol. MTT-22, pp. 317–322, Mar. 1974.

[7] F. Kato and M. Saito, "Computer-aided design of planar circuits," *Trans. IECE of Japan*, vol. J59-A, pp. 47–54, Jan. 1976.

[8] H. Bosma, "On stripline *Y*-circulation at UHF," *IEEE Trans. Microwave Theory Tech.*, vol. MTT-12, pp. 61–72, Jan. 1964.

[9] M. E. Hines, "Reciprocal and nonreciprocal mode of propagation in ferrite stripline and microstrip devices," *IEEE Trans. Microwave Theory Tech.*, vol. MTT-19, pp. 442–451, May 1971.

[10] P. de Santis and F. Pucci, "The edge-guided wave circulator," *IEEE Trans. Microwave Theory Tech.*, vol. MTT-23, pp. 516–519, June 1975.

[11] C. E. Fay and R. L. Comstock, "Operation of the ferrite junction circulator," *IEEE Trans. Microwave Theory Tech.*, vol. MTT-13, pp. 15–27, Jan. 1965.

Paper 7.5

Analysis of Wide-Band Stripline Circulators by Integral Equation Technique

YALCIN AYASLI

***Abstract*—The analysis of wide-band *Y*-junction stripline circulators using Green's function method was reported in the literature. In this paper, similar analyses are performed using an integral equation method and the results are compared.**

The boundary conditions used in the analyses are also discussed. A new boundary condition representing the actual fields more precisely than previously is formulated and applied to the junction. The results obtained with the new boundary conditions are examined and compared with the previous theoretical and experimental results.

The current and voltage distributions that are created at the ports under the assumed boundary conditions are calculated and compared with the known stripline and junction modes.

In the formulation, it is observed that the Green's function is not unique and it can be selected from a certain class of functions. This arbitrariness is introduced into the formulation by means of a complex parameter C_0. The effect of this parameter on the numerical results is investigated and it is shown that in certain regions of the complex C_0 plane, the numerical results converge on the analytical results.

I. Introduction

WU AND Rosenbaum [1] proposed the concept of intrinsically wide-band stripline circulators and experimentally verified this with a 7–14 GHz octave band design in 1974. Later, Ayter and Ayasli [2] reported the design of another wide-band circulator at 2–4 GHz using frequency-independent design curves. They also calculated and compared the theoretical responses of both 2–4 and 7–14 GHz designs using Green's function technique.

In this paper, the theoretical performance of the two wide-band designs mentioned above are examined by the integral equation technique [3]. In this examination, the boundary conditions are chosen consistent with the Green's function analysis to allow comparison with the previous Ayter–Ayasli results [2]. The validity of these boundary conditions are then investigated and a new set which approximates the real distributions better is introduced. The new set of boundary conditions are applied to the junction and the results are compared with the previous results.

The integral equation method is especially suitable for the analysis of arbitrary shaped planar microwave circuits. In the application of the method to such circuits, the free space Green's function has been consistently used [3], [4].

Manuscript received June 6, 1979; revised October 17, 1979.

Y. Ayasli was with the Department of Electrical Engineering, Middle East Technical University, Ankara, Turkey. He is now with the Raytheon Research Division, Waltham, MA, on leave of absence from the Middle East Technical University.

In this paper it is pointed out that the Green function suitable to the integral equation method is not unique and, in fact, can be chosen from a certain class of functions. This arbitrariness in the Green's function is also investigated and its effect on the accuracy and convergence of the numerical results is examined.

II. Integral Equation Formulation

For the junction geometry shown in Fig. 1 and when the thickness of the ferrite is much less than the wavelength, the fields in the junction can be taken as independent of z. Then only the z component of the electric field can exist and it should satisfy the wave equation.

$$\frac{\partial^2 E_z}{\partial r^2}+\frac{1}{r}\frac{\partial E_z}{\partial r}+\frac{1}{r^2}\frac{\partial^2 E_z}{\partial \theta^2}+k^2E_z=0 \tag{1}$$

where

$k^2=\omega^2\epsilon_0\epsilon_f\mu_{\text{eff}}$	radial wave propagation constant
$\mu_{\text{eff}}=(\mu^2-\kappa^2)/\mu$	effective permeability of the ferrite
μ,κ	Polder tensor elements.

Following the analysis of Miyoshi *et al.* [3], the electric field E_z at some point $\bar{r}_M$ on the circumference of the junction is given by

$$E_z(\bar{r}_M)=2\oint_c j\omega\mu_{\text{eff}}G(kR)H_l(\bar{r}_0)\,dl$$
$$-2\oint_c\left(\cos\theta_0-j\frac{\kappa}{\mu}\sin\theta_0\right)\frac{\partial G(kR)}{\partial R}E_z(\bar{r}_0)\,dl. \tag{2}$$

Contour C and the quantities such as $\bar{r}_M$, $\bar{r}_0$, and θ_0 are shown in Fig. 2. If the tangential magnetic field component $H_l(\bar{r}_0)$ is assumed to be known along C, then (2) is a Fredholm integral equation in terms of E_z. $G(kR)$, in this equation is the two-dimensional Green's function and can be expressed in the most general form as

$$G(kR)=C_0J_0(kR)-\frac{1}{4}Y_0(kR) \tag{3}$$

where

C_0	a complex constant
R	$=\lvert\bar{r}_M-\bar{r}_0\rvert$
$J_0(kR)$	first kind zeroth-order Bessel function
$Y_0(kR)$	second kind zeroth-order Bessel function.

In (3), the coefficient of $Y_0(kR)$ is determined to be $-1/4$ from the singularity condition at $R=0$. Because there is no other condition imposed on $G(kR)$, the constant C_0

Reprinted from *IEEE Trans. Microwave Theory Tech.*, vol. MTT-28, no. 3, pp. 200–209, March 1980.

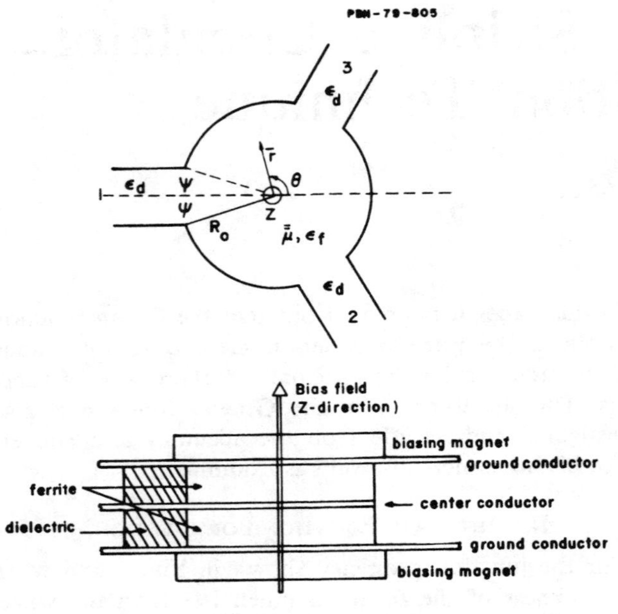

Fig. 1. Junction geometry. (a) Top view. (b) Cross section.

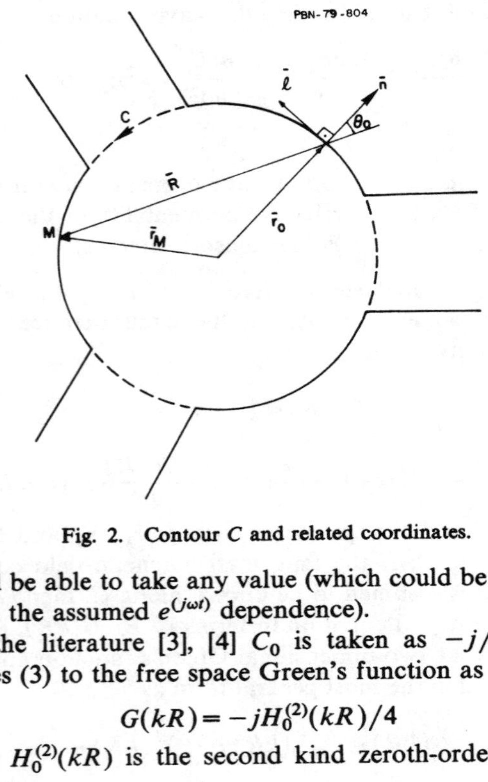

Fig. 2. Contour C and related coordinates.

should be able to take any value (which could be complex due to the assumed $e^{(j\omega t)}$ dependence).

In the literature [3], [4] C_0 is taken as $-j/4$ which reduces (3) to the free space Green's function as

$$G(kR) = -jH_0^{(2)}(kR)/4 \tag{4}$$

where $H_0^{(2)}(kR)$ is the second kind zeroth-order Hankel function.

Analytically one can readily show that the solution for $E_z(\bar{r}_M)$ is independent of C_0. However, these equations are solved numerically and not analytically. To obtain a numerical solution, the integral equation is reduced to a finite size matrix equation. In this transformation truncation errors occur due to the finite matrix size used. In addition, the usual sources of error peculiar to the numerical solutions are present. Thus the numerical solutions are only an approximation to the analytical solutions and the effect of the parameter C_0 on these numerical solutions is not obvious. The effect of C_0 on the numerical results is discussed in Appendix A.

III. Numerical Formulation

For the numerical evaluation of the integral equation (2), the contour C can be divided into N equal segments. If N is large enough, the length of the segments will be small and the fields existing in the junction can be taken as constant along each segment.

Total length of contour C is about a wavelength in the operation band and, as an example, choosing $N=36$ would correspond to sampling a sine wave at 10° intervals.

If the contour integral is thought of as the sum of N line integrals over an interval w, (2) can be written as

$$e_i = j2\omega\mu_{\text{eff}} \sum_{j=1}^{N} h_j \int_{w_j} G(kR_{ij})\,dl_j - \sum_{j=1}^{N} 2ke_j \int_{w_j} \left(\cos\theta_{ij} - j\frac{\kappa}{\mu}\sin\theta_{ij}\right) G'(kR_{ij})\,dl_j. \tag{5}$$

In this equation, e_i is the value of E_z at the ith segment and h_j is the value of H_l at the jth segment. Because e_i and h_j are taken to be constant, they are taken outside the integrations.

Equation (5) can be expressed as

$$\sum_{j=1}^{N} U_{ij} e_j = \sum_{j=1}^{N} T_{ij} h_j, \qquad i = 1, 2, \cdots, N \tag{6}$$

or as

$$[U][e] = [T][h]. \tag{7}$$

Comparing (5) and (6),

$$U_{ij} = \delta_{ij} + 2k \int_{w_j} \left(\cos\theta_{ij} - j\frac{\kappa}{\mu}\sin\theta_{ij}\right) G'(kR_{ij})\,dl_j \tag{8}$$

and

$$T_{ij} = j2\omega\mu_{\text{eff}} \int_{w_j} G(kR_{ij})\,dl_j \tag{9}$$

are found. If the h_j's are assumed to be known around the junction circumference, then the e_j's can be calculated from (6).

IV. Application of the Method to Wide-Band Designs and Boundary Conditions

Wide-band designs operate in the approximate octave band $f_z + f_m < f < 2f_m$. The designs require that the coupling angle 2Ψ is chosen as equal or wider than 60°. The angle 2Ψ is chosen as 60°, 65°, and 75° in references [1], [2], and [5], respectively. As a result of wide coupling angle, when the circulator circumference is divided into N segments, more than one sampling point falls into the input ports. As an example, for Wu–Rosenbaum [1] design, the junction circumference and the sampling points are shown in Fig. 3, if $N=36$. The boundary conditions on the junction are discussed in terms of this example.

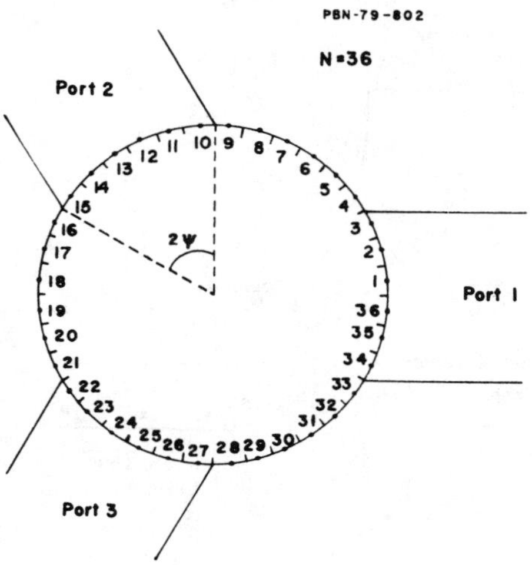

Fig. 3. Junction geometry and sampling points for Wu–Rosenbaum design and $N=36$.

A. Constant Magnetic Field at the Ports

The boundary condition used in the Green's function analysis can be expressed in terms of complex constants a, b, and c as in [6]

$$H_l=\begin{cases} a, & \text{in port 1} \\ b, & \text{in port 2} \\ c, & \text{in port 3} \\ 0, & \text{elsewhere.} \end{cases} \tag{10}$$

The electric field E_z at the ports then comes out as a function of the θ variable and by averaging it along a portwidth, an equivalent electric field is defined.

To be able to compare the results of the integral equation analysis with the results of the Green's function analysis, boundary conditions similar to the ones expressed in (10) are applied first. In terms of the tangential magnetic fields at the sampling points and referring to the numbers in Fig. 3, boundary conditions equivalent to (10) become

$$\begin{aligned} a&=h_1=h_2=h_3=h_{34}=h_{35}=h_{36} \\ b&=h_{10}=h_{11}=h_{12}=h_{13}=h_{14}=h_{15} \\ c&=h_{22}=h_{23}=h_{24}=h_{25}=h_{26}=h_{27}. \end{aligned} \tag{11}$$

The average electric fields E_1, E_2, and E_3 at the ports can then be calculated as

$$\begin{aligned} E_1&=(e_1+e_2+e_3+e_{34}+e_{35}+e_{36})/6 \\ E_2&=(e_{10}+e_{11}+e_{12}+e_{13}+e_{14}+e_{15})/6 \\ E_3&=(e_{22}+e_{23}+e_{24}+e_{25}+e_{26}+e_{27})/6. \end{aligned} \tag{12}$$

These average field quantities defined at the ports are also related through the scattering matrix

$$S=\begin{bmatrix} \alpha & \beta & \gamma \\ \gamma & \alpha & \beta \\ \beta & \gamma & \alpha \end{bmatrix} \tag{13}$$

of the junction. If it is assumed that the junction is excited from port 1 and the other ports are terminated by matched loads, the average field quantities are [6]

$$\begin{aligned} a&=(1-\alpha)a^+ \\ b&=-\beta a^+ \\ c&=-\gamma a^+ \end{aligned} \tag{14}$$

and

$$\begin{aligned} E_1&=Z_d(1+\alpha)a^+ \\ E_2&=Z_d\beta a^+ \\ E_3&=Z_d\gamma a^+. \end{aligned} \tag{15}$$

In these equations a^+ represents the incident average magnetic field H_l at port 1 and Z_d is the wave impedance of the input stripline. Without losing any generality a^+ can be taken as unity and from (14) and (15):

$$\begin{aligned} a&=2-E_1/Z_d \\ b&=-E_2/Z_d \\ c&=-E_3/Z_d. \end{aligned} \tag{16}$$

When (16) is combined with (7), N equations with the electric fields e_i as unknowns result. If this system of equations is expressed as a matrix equation in the form of

$$[U'][e]=[R] \tag{17}$$

then, for the example of Fig. 3,

$$\begin{aligned} U'(I,J)=U(I,J)+[&T(I,1)+T(I,2)+T(I,3) \\ &+T(I,34)+T(I,35)+T(I,36)]/6Z_d, \\ &J=1,2,3,34,35,36,\quad I=1,2,\cdots,36 \end{aligned} \tag{18a}$$

$$\begin{aligned} U'(I,J)=U(I,J)+[&T(I,10)+T(I,11)+T(I,12) \\ &+T(I,13)+T(I,14)+T(I,15)]/6Z_d, \\ &J=10,11,12,13,14,15,\quad I=1,2,\cdots,36 \end{aligned} \tag{18b}$$

$$\begin{aligned} U'(I,J)=U(I,J)+[&T(I,22)+T(I,23)+T(I,24) \\ &+T(I,25)+T(I,26)+T(I,27)]/6Z_d, \\ &J=22,23,24,25,26,27,\quad I=1,2,\cdots,36 \end{aligned} \tag{18c}$$

and

$$\begin{aligned} R(I)=2[&T(I,1)+T(I,2)+T(I,3)+T(I,34) \\ &+T(I,35)+T(I,36)],\qquad I=1,2,\cdots,36 \end{aligned} \tag{19}$$

can be written.

From the solution of (17), the electric field E_z sampled at 36 points along the junction circumference can be found. Using these fields, the scattering matrix elements α, β, and γ can be found as

$$\begin{aligned} \alpha&=(e_1+e_2+e_3+e_{34}+e_{35}+e_{36})/6Z_d-1 \\ \beta&=(e_{10}+e_{11}+e_{12}+e_{13}+e_{14}+e_{15})/6Z_d \\ \gamma&=(e_{22}+e_{23}+e_{24}+e_{25}+e_{26}+e_{27})Z_d. \end{aligned} \tag{20}$$

The performance of the junction as a circulator can then be calculated as

$$\begin{aligned} \text{Return loss}&=20\log_{10}|\alpha| \\ \text{Insertion loss}&=20\log_{10}|\beta| \\ \text{Isolation}&=20\log_{10}|\gamma|. \end{aligned} \tag{21}$$

B. Constant Electric Field at the Ports

For the Green's function analysis, the boundary conditions given in (10) are the most convenient to apply. But they do not truly represent the actual field distributions at the ports.

The TEM voltage and current distributions that exist on a stripline [7] are shown in Fig. 4. Since no z-variation for the fields is assumed, then H_l and E_z should have distributions similar to current and voltage distributions, respectively. In the region where the ports joint the junction, however, higher order modes exist and modify these distributions. In spite of this, the distributions in Fig. 4 suggest that taking the electric field distribution as constant and letting the magnetic field distribution free for any value it takes seems more meaningful. Thus a new and more realistic boundary condition is

$$E_z = \begin{cases} E_1, & \text{in port 1} \\ E_2, & \text{in port 2} \\ E_3, & \text{in port 3} \end{cases} \tag{22}$$

and

$$H_l = 0, \qquad \text{outside the ports.} \tag{23}$$

Using (22) and (23) instead of (10) and following a similar routine, a matrix equation of the form (17) can be obtained. Due to the mixed boundary conditions used, however, the unknowns of the problem become the tangential magnetic field at the ports and the electric field outside the ports.

V. Numerical Results

The integral equation method is used for the analysis of two wide-band circulator designs reported in the literature [1], [2]. For each design, two different sets of boundary conditions described above are separately examined for $C_0 = +300$ (see Appendix A for the choice of C_0).

The resonant frequencies of the disk-shaped circuit is calculated first to check the computational accuracy. The equation which gives the resonance frequencies is

$$\det[U] = 0. \tag{24}$$

Due to computational errors, (24) is not completely satisfied and the frequencies which give the minimum of $|\det[U]|$ can be taken as the eigenvalues [3]. The exact eigenvalues, on the other hand are the roots of

$$|J_n'(kR_0)| - \left| \frac{\kappa}{\mu} \frac{nJ_n(kR_0)}{kR_0} \right| = 0, \qquad n = 0, 1, 2, \cdots. \tag{25}$$

Thus the exact and calculated resonance frequencies can be compared giving information on the computational errors involved. From this comparison, it is found that for $N = 33$, $H_i = 3300$ Oe, $4\pi M_s = 1300$ G, $R_0 = 0.0039$ m, and $C_0 = -j/4$, the difference between the exact and calculated frequencies is less than 0.15 percent. For similar parameters, Miyoshi *et al.* [3] reports a 2-percent computational error. This improvement in the accuracy of the numerical results is achieved by using (8) and (9) instead of the approximate expressions in the calculation of diagonal terms for the $[U]$ and $[T]$ matrices.

PBN-79-801

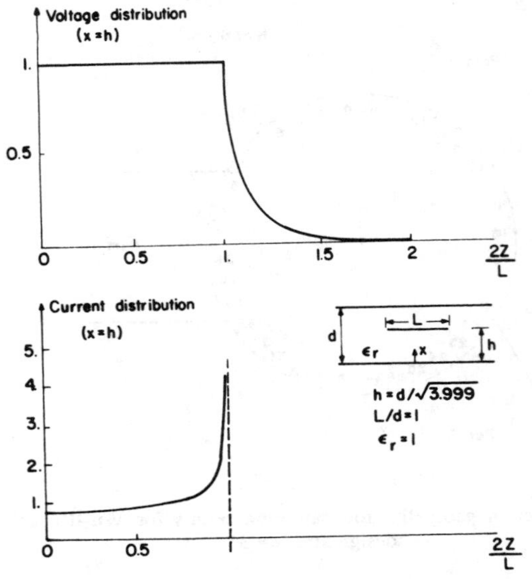

Fig. 4. Calculated voltage and current distributions on a typical stripline configuration (in relative units) [7].

A. Constant Magnetic Field at the Ports

The numerical results obtained for the Wu–Rosenbaum design [1] with constant magnetic field boundary condition at the ports are shown in Fig. 5. On the same figure the results obtained with the Green's function analysis [2] are also shown for comparison.

The results obtained by the two different methods are in good agreement. In the analysis of reference [2], the Green's function is calculated by using the first three terms of the infinite series expansion; therefore, a wider resonance is not surprising. The degradation in the performance of the junction around 13.4 GHz is present in both results.

The integral equation technique is also applied to Ayter–Ayasli design [2] and the results are shown in Fig. 6 together with the results of the Green's function analysis. In this application, the contour C is divided into 30 equal segments with 6 sampling points left for each port. This sets $\Psi = 0.628$ rads. The actual Ψ used in the design is 0.646 rad. The difference between the coupling angles Ψ used in the two methods of analysis makes point by point comparison difficult and causes the offset in the center frequencies seen in Fig. 6. The general shape of the curves are, however, in agreement.

B. Constant Electric Field at the Ports

The two wide-band designs of the previous section are also investigated using the constant electric field boundary condition. The results are shown in Figs. 7 and 8. The shift to higher frequencies seen in Fig. 8 should be due to the slightly different Ψ value used in the model.

Comparing Figs. 7, 8 and 5, 6, respectively, a general decrease in isolation and a general increase in insertion loss are observed. Although the reported experimental

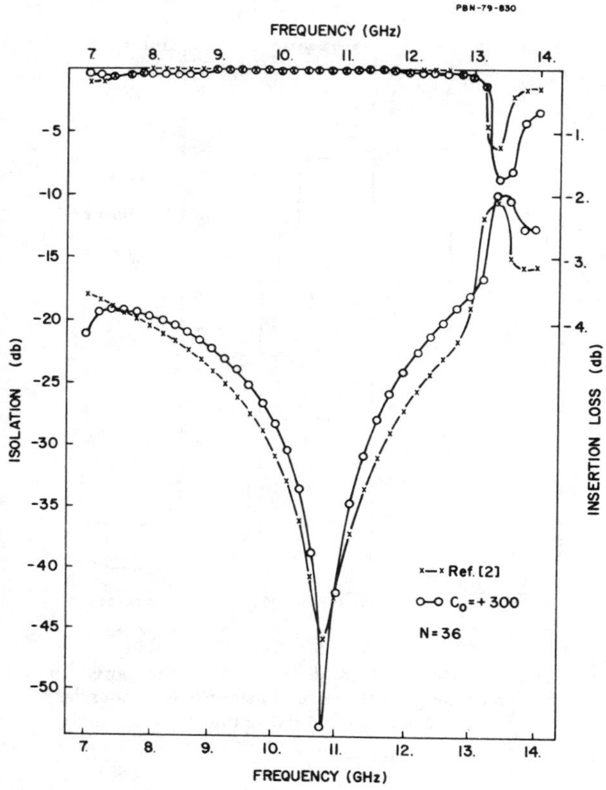

Fig. 5. Comparison of integral equation and Green's function results for Wu–Rosenbaum design with constant magnetic field boundary condition.

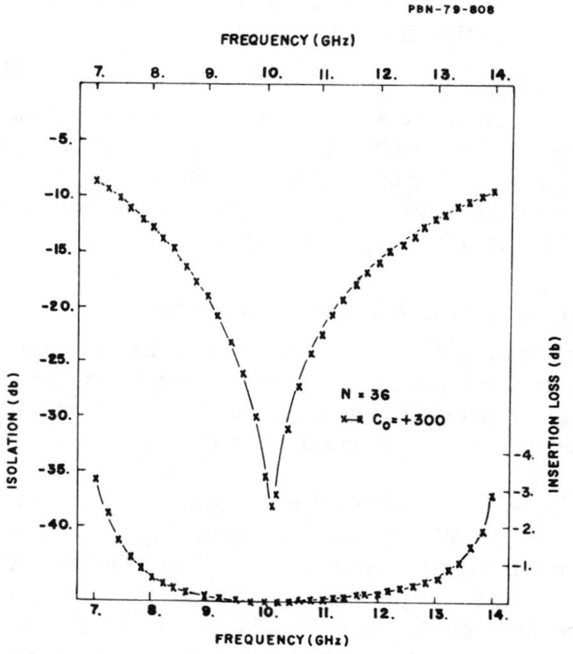

Fig. 7. Integral equation results for Wu–Rosenbaum design with constant electric field boundary condition.

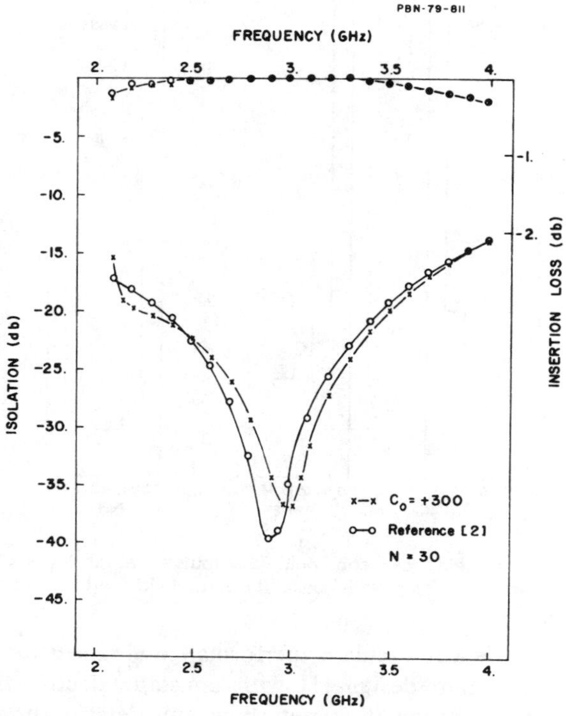

Fig. 6. Comparison of integral equation and Green's function results for Ayter–Ayasli design with constant magnetic field boundary condition.

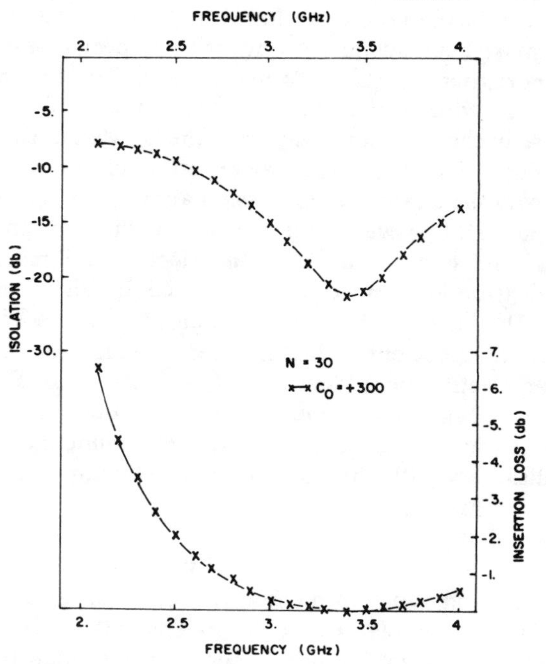

Fig. 8. Integral equation results for Ayter–Ayasli design with constant electric field boundary condition.

results [1], [2] include the effect of the matching sections and thus cannot be directly compared with the curves of Figs. 7 and 8, these observed tendencies improve the agreement between the experimental and calculated constant electric field boundary condition results.

The boundary conditions require $H_t = 0$ outside the ports, when actually fringing fields exist there. The effect of these fringing fields is included in the model by defining an effective disk radius as [3]

$$R_{eff} = R_0 + 0.1788 \times D \tag{26}$$

where D is the ferrite disk thickness. To allow comparison with the previous results [2], this correction was not made for the results of Figs. 5 and 6. If this correction were made for these figures, the center frequencies would move to 9.6 GHz and 2.65 GHz, respectively.

C. Calculated Field Distributions at the Ports

In comparing the results obtained with the two different boundary conditions, it is also instructive to examine the field distributions which form at the ports, checking whether or not they violate any of the previous assumptions.

In Fig. 9, the electric field distributions at the ports are shown for the Wu–Rosenbaum design [1] with the constant magnetic field boundary condition. On the other hand, the magnetic field distributions for the same design with constant electric field boundary condition are shown in Fig. 10. In both figures, the distributions are calculated at the center frequencies.

To be able to interpret these distributions better, the modes that exist in the junction are shown in Fig. 11 [6]. From the examination of these modes, it is clear that fields must have symmetry only with respect to the isolated port (port 3). At the other ports, the fields are not symmetric with respect to the port center, and instead increase in the direction away from the isolated port.

In Figs. 9 and 10, the calculated field distributions agree with the qualitative discussion above. In the isolated port (port 3), however, since the sum of the two junction modes results in a null field, the electric and magnetic field distributions must be determined by the stripline mode. The magnetic field distribution at port 3 in Fig. 10 is in good agreement with the current distribution in Fig. 4. The electric field distribution at port 3 in Fig. 9, however, disagrees with the voltage distribution in Fig. 4, the same way the constant magnetic field boundary condition disagrees with the current distribution in the same figure.

VI. Conclusion

The integral equation method is applied to the analysis of wide-band circulators. In this analysis, the junction is first modeled with boundary conditions similar to the conditions used in the previous Green's function analysis [2]. The results obtained by these two different methods are then compared. This comparison shows that the results are in agreement.

In the integral equation analysis, it is noted that a more realistic boundary condition can be formulated and applied to the junction. The results obtained with the different boundary conditions are then compared, showing that the new boundary condition increases the agreement between the theoretical and experimental results.

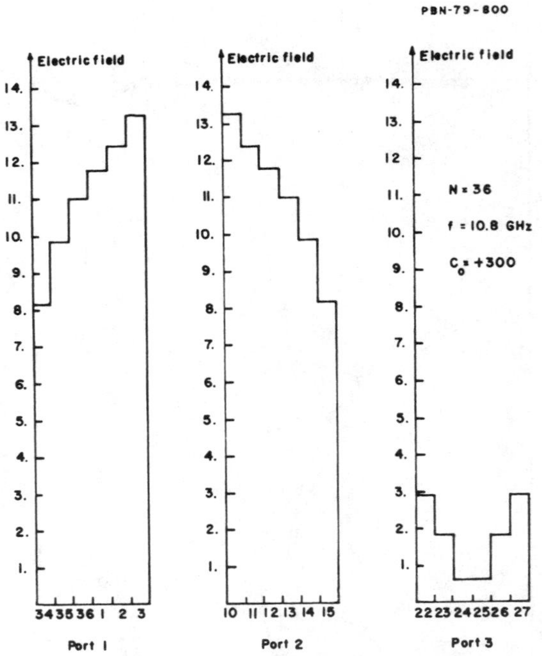

Fig. 9. Calculated electric field distribution at the ports for Wu–Rosenbaum design with constant magnetic field boundary condition (in relative units).

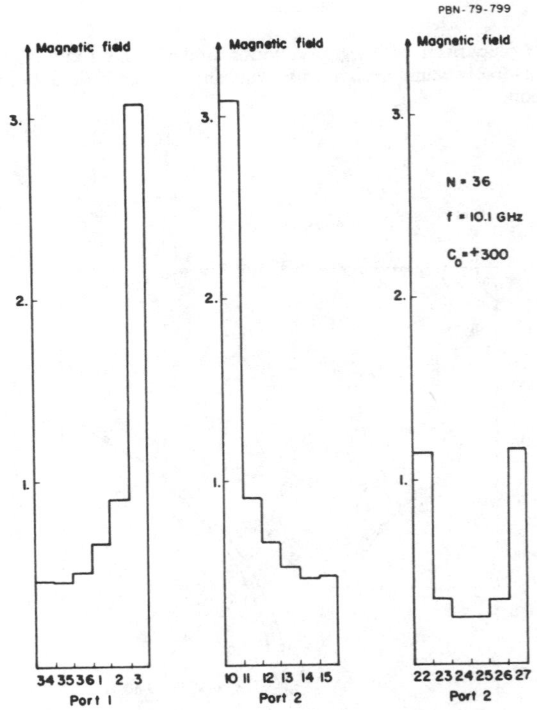

Fig. 10. Calculated magnetic field distribution at the ports for Wu–Rosenbaum design with constant electric field condition (in relative units).

The theoretical circulator performance obtained for the Wu–Rosenbaum design [1] with constant electric field boundary condition does not show any deterioration at 13.4 GHz as observed in both Green's function and integral equation results with constant magnetic field boundary condition. This means that the deterioration is an effect strictly due to the constant magnetic field boundary condition used.

The magnetic or electric field distributions created at the ports under two different boundary conditions are also calculated and compared with the modes of the junction and of the stripline. From this comparison, it is observed that only the constant electric field boundary condition at the ports does not lead to any inconsistencies.

For analyses based on the solution of integral equations in the form of (2), it is shown that an arbitrary parameter C_0 can be introduced into the Green's function used. The physical results should ideally be independent of this parameter and there are no *a priori* criteria which dictates a certain choice for it. The integral equation is then transformed to a finite size matrix equation and the effect of this parameter on the numerical solutions is investigated.

The computer results of this investigation show that by finding a region where the results are independent of C_0, the difference between the numerical and theoretical results can be decreased and a convergence between them can be obtained. In this region of convergence, the numerical results become independent of the number of equations used and thus a minimum number of equations showing a C_0 independent region in their solution can be used.

The numerical results also indicate that in the process of increasing C_0 to find a region of convergence, the stability of the matrix equations decreases linearly with C_0. However, while the condition number of the system is increasing steadily to rather large values, very little change is observed in the numerical results. Clearly, a compromise seems possible for the proper choice of C_0. The convergence obtained through the arbitrariness of C_0 is a property of the integral equation formulation and thus its application is not limited to two-dimensional problems.

Appendix A

In this appendix, the effect of the parameter C_0 on the numerical results is investigated. In Fig. 12, the calculated values of the first resonance frequency for $N=33$, $R_0=0.0039$ m, $H_i=3300$ Oe, and $4\pi M_s=1300$ G are plotted for several values of C_0 as it is varied along the $\pm$real axes. From this figure, it is seen that as C_0 is increased, the frequencies obtained from the minimum of $|\det[U]|$ converge on the exact value of 4.37080 GHz calculated from (25) for $n=1$. Six significant number accuracy is the limit of the Bessel's function expansions and the single precision calculations used.

In Section II, it is pointed out that the physical results, when obtained analytically, should be independent of C_0. But the integral equation is solved numerically by transforming it to a finite set of linear equations. As a result of this transformation, the more numerical solutions converge to the exact integral equation solutions, the less they should depend on the parameter C_0.

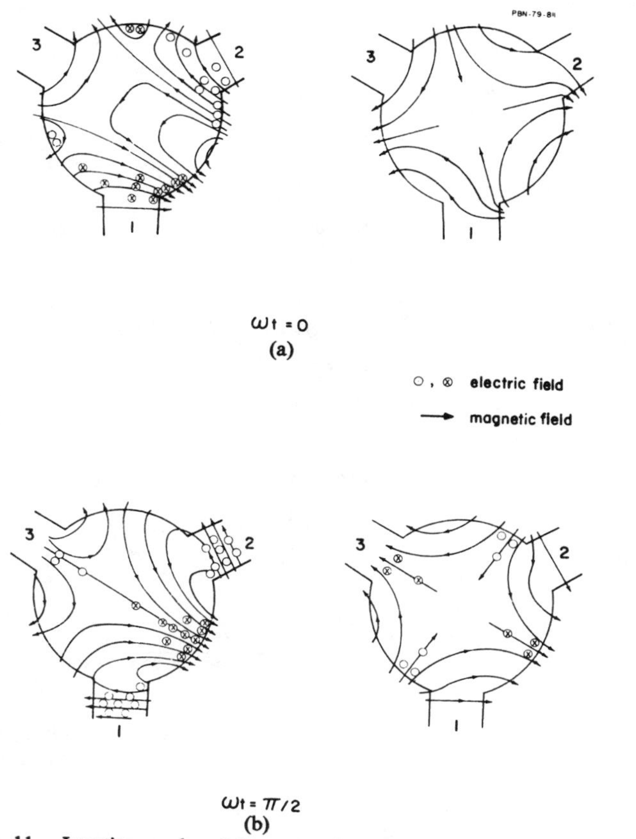

Fig. 11. Junction modes at the center frequency (to obtain isolation at port 3, the field configurations shown at (a) and (b) should be superimposed) [6].

In Fig. 13, the effect of C_0 on the diagonal element of the scattering matrix is shown for the Ayter–Ayasli design at a single frequency. This figure shows that as C_0 is increased along the $\pm$real and $\pm$imaginary axes on the complex C_0 plane, the magnitude of α converges to a certain value. Convergence occurs fastest on the real C_0 axis. After convergence is reached, as C_0 is varied in a rather large region from $+50$ to $+1000$, $|\alpha|$ changes only 2 percent.

In Fig. 14, the variation of $|\alpha|$ with C_0 as the latter is varied along the +real is shown for the four cases mentioned in Section V. For these four cases, the convergence is clearly seen for C_0 larger than $+200$. Above this value, the results are nearly independent of C_0. This independence means that in this C_0 region, the numerical results behave as expected from the analytical solutions and thus the matrix equation is modeling the junction properly.

In Figs. 13 and 14, the variation of $|\alpha|$ with the parameter C_0 are shown only at one frequency near midband. To see if the convergence obtained for certain C_0 values is sensitive to the parameters of the model, the behavior of the circulator for one of the four cases above is shown in Fig. 15 over the entire frequency band for two different values of C_0. From this figure, it is seen that the results for $C_0=+300$ and $C_0=+500$ are in close agreement over the band of operation.

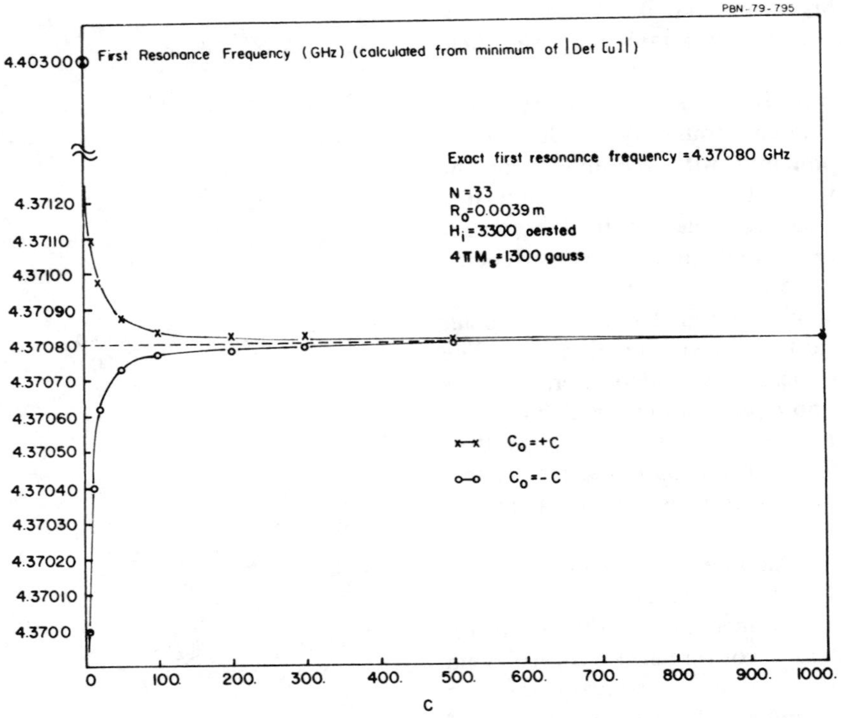

Fig. 12. Variation of the calculated first resonance frequency as the parameter C_0 is varied along the real axis.

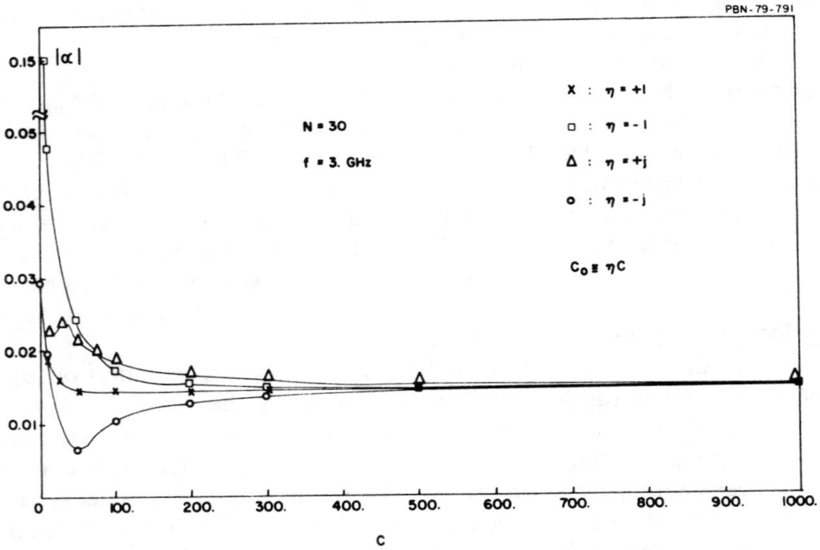

Fig. 13. Variation of $|\alpha|$ as the parameter C_0 is varied along the four axes in the complex C_0 plane for Ayter–Ayasli design and constant magnetic field boundary condition.

In the C_0 region where the results converge to a certain value, the numerical results should also be independent of N, the number of sampling points. This must be the case because if the matrix equations are modeling the integral equation properly, the physical results should not depend on N.

In Fig. 16, for Wu–Rosenbaum design and $C_0 = +300$, the circulator behavior is compared for two different values of N. The two sets of curves are in good agreement.

In Fig. 17, on the other hand, the same comparison is made for $C_0 = 0$. It is seen that the results for this value of C_0 do not converge for these two different N values.

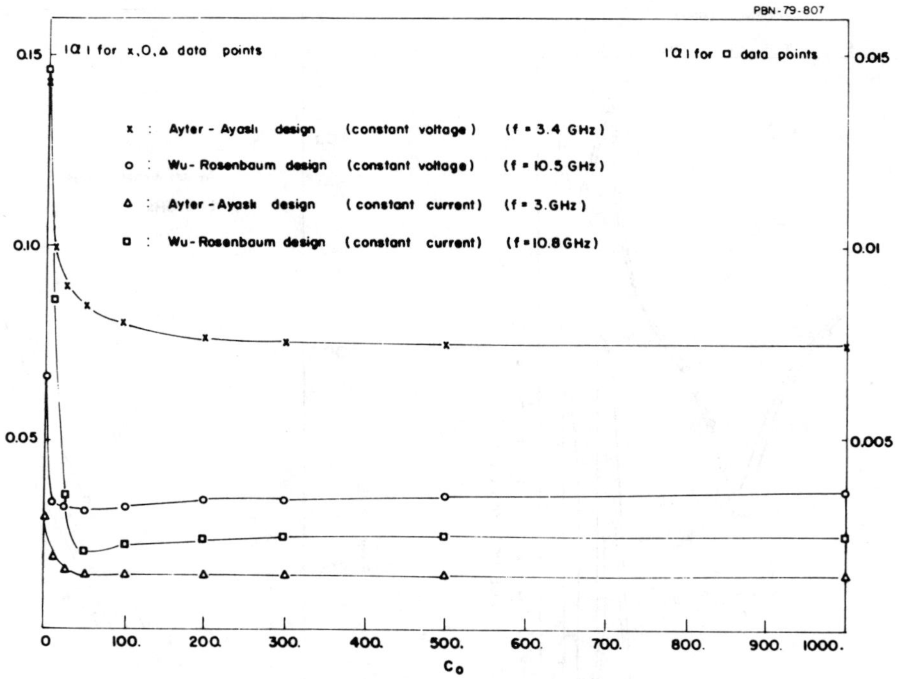

Fig. 14. Variation of $|\alpha|$ as the parameter C_0 is varied on the positive real axis for the four examples considered.

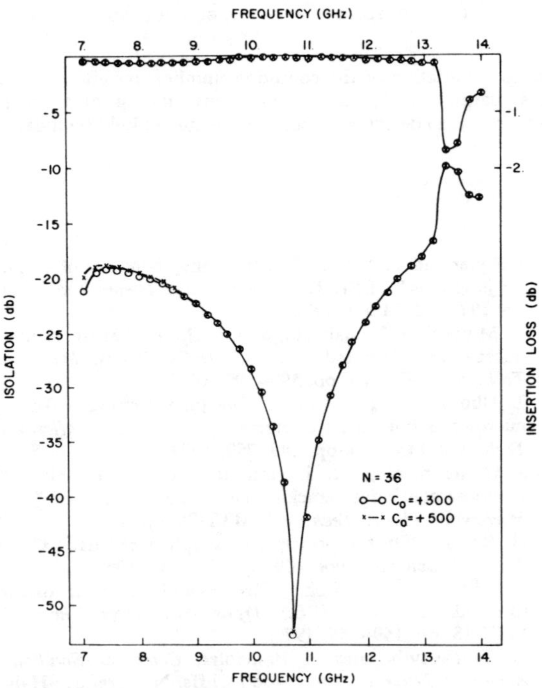

Fig. 15. Comparison of results for two different values of C_0 on Wu–Rosenbaum design with constant magnetic field boundary condition.

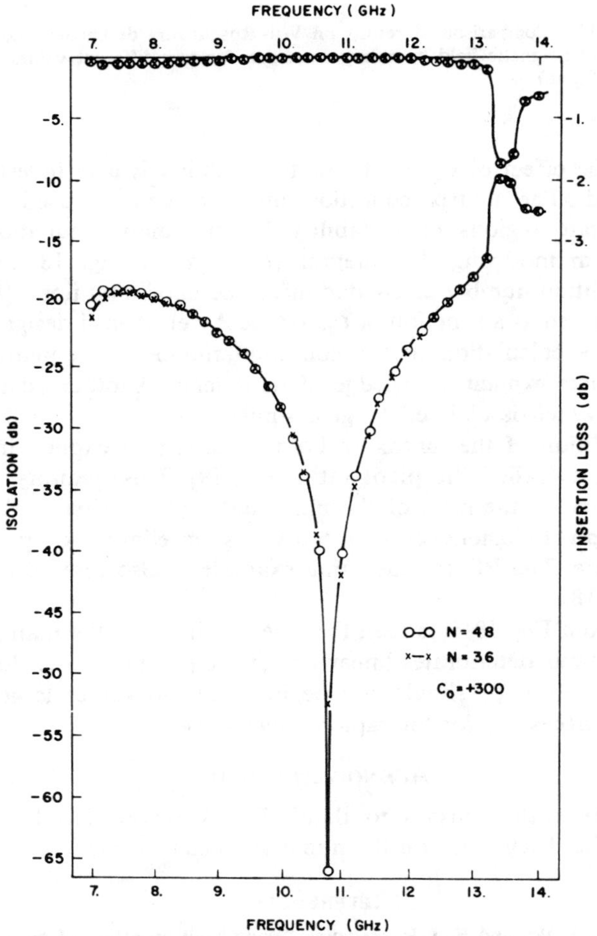

Fig. 16. Comparison of results on Wu–Rosenbaum design with constant magnetic field boundary condition for two different values of $N(C_0 = +300)$.

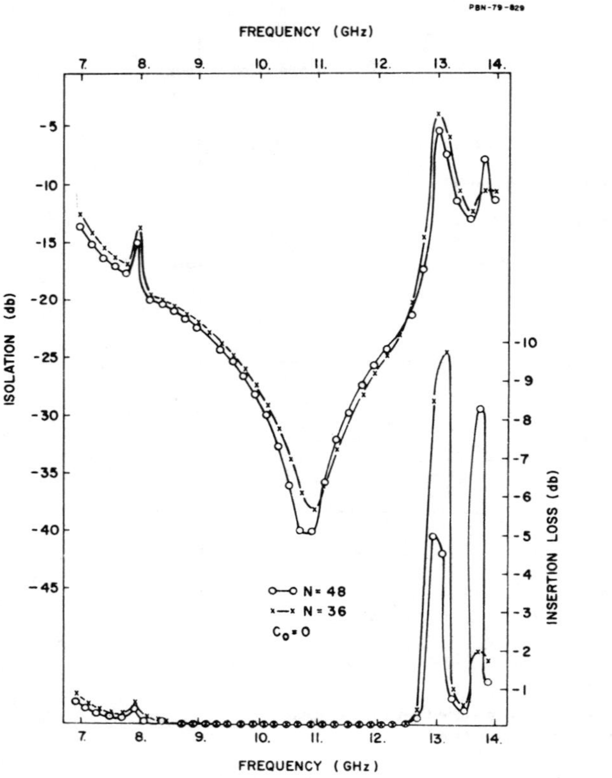

Fig. 17. Comparison of results on Wu–Rosenbaum design with constant magnetic field boundary condition for two different values of $N(C_0=0)$.

The effect of C_0 on the matrix stability is also investigated. The matrix condition numbers can be used to indicate regions of instability for the matrix equation system modeling the integral equation. In Fig. 18, the condition number calculated using the maximum norm [8] is plotted as a function of C_0 for the Ayter–Ayasli design.

The calculation of the condition number of a matrix requires explicit knowledge of its inverse. Another quantity which is claimed to give similar information on the condition of the matrix without requiring an explicit inverse is called the pivot ratio (PR) [9]. This quantity is defined as the ratio of the magnitudes of the first to the last pivot elements chosen in a Gaussian elimination procedure. The PR for the same example is also plotted on Fig. 18.

From Fig. 18, it is seen that the condition of the matrix equations deteriorates linearly with the parameter C_0. For this reason, C_0 should not be increased to values larger than necessary for the rapid convergence.

Acknowledgment

The author wishes to thank Dr. Altunkan Hizal for helpful discussions on the numerical calculations.

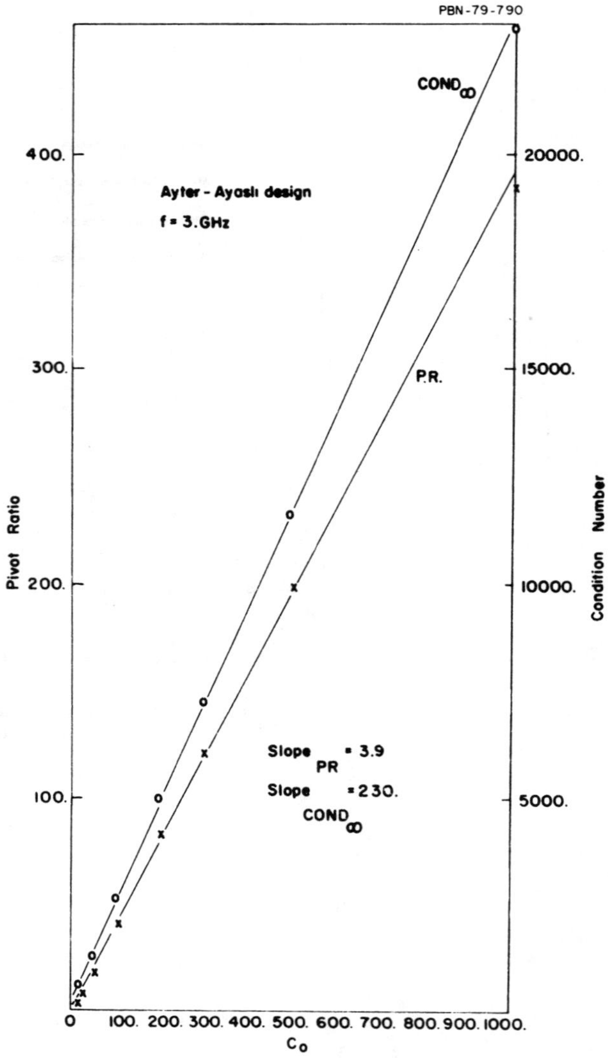

Fig. 18. Variation of the condition number (infinite norm) and PR as the parameter C_0 is varied along the positive real axis for Ayter–Ayasli design with constant magnetic field boundary condition.

References

[1] Y. S. Wu and F. J. Rosenbaum, "Wide-band operation of microstrip circulators," *IEEE Trans. Microwave Theory Tech.*, vol. MTT-22, pp. 849–856, 1974.

[2] S. Ayter and Y. Ayasli, "The frequency behavior of stripline circulator junctions," *IEEE Trans. Microwave Theory Tech.*, vol. MTT-26, pp. 197–202, March 1978.

[3] T. Miyoshi, S. Yamaguchi, and S. Goto, "Ferrite planar circuits in microwave integrated circuits," *IEEE Trans. Microwave Theory Tech.*, vol. MTT-25, pp. 593–599, 1977.

[4] T. Okoshi and T. Miyoshi, "The planar circuit—An approach to microwave integrated circuitry," *IEEE Trans. Microwave Theory Tech.*, vol. MTT-20, pp. 245–252, 1972.

[5] J. G. de Koning, R. J. Hamilton, and T. L. Hierl, "Full-band Low-loss continuous tracking circulation in K-band," *IEEE Trans. Microwave Theory Tech.*, vol. MTT-25, pp. 152–155, 1977.

[6] H. Bosma, "On the principle of stripline circulation," *Proc. IEEE Part B Supplement*, vol. 109, pp. 137–146, 1962.

[7] R. Mittra and T. Itoh, "Charge and potential distributions in shielded striplines," *IEEE Trans. Microwave Theory Tech.*, vol. MTT-18, pp. 149–156, 1970.

[8] G. E. Forsythe and C. B. Moler, *Computer Solution of Linear Algebraic Systems*. Englewood Cliffs, NJ: Prentice-Hall, 1967.

[9] R. Mittra and C. A. Klein, "The use of pivot ratios as a guide to stability of matrix equations arising in the method of moments," *IEEE Trans. Antennas Propagat.*, vol. AP-23, pp. 448–450, 1975.

Paper 7.6

A Finite Element Analysis of Planar Circulators Using Arbitrarily Shaped Resonators

RONALD W. LYON, MEMBER, IEEE AND JOSEPH HELSZAJN, MEMBER, IEEE

Abstract—A planar circulator consists, in general, of three transmission lines, connected through suitable matching networks to a magnetized ferrite resonator having three-fold symmetry. This paper describes a finite element analysis which enables the Z-matrix of a planar circulator using arbitrary shaped resonators to be calculated. This technique allows quite general computer programs to be written which permit tables of circulation solutions to be calculated. Results for junctions using disk, triangular, and irregular hexagonal resonators are included in the text. The frequency response of junction circulators using various configurations whose magnetic variables have been chosen so that they operate over the widely used tracking interval has also been evaluated. The optimum response is in each case associated with a unique coupling angle.

I. INTRODUCTION

A PLANAR junction circulator consists, in general, of a three-fold symmetric resonator of arbitrary shape to which three transmission lines are connected. A complete description of planar circulators using disk resonators has been presented in the literature [1]–[5], and some approximate analyses are available for junctions using triangular [6] and Wye [7] resonators. The general boundary value problem has been treated by Miyoshi [8], [9] who used a contour integration formulation to form the entries of the impedance matrix of the junction. Miyoshi also presented an alternative analysis in which the open circuit parameters of the junction are expanded in terms of the modes of the magnetized planar resonator. The decoupled resonator is analyzed using a variational method where the fields are expressed as a single polynomial expansion for the complete resonator. This second method requires that a mathematical derivation of the polynomial coefficients be derived analytically for each individual resonator shape required.

This paper describes a finite element approach using the variational formulation introduced by Miyoshi. The method, which reduces to that of Silvester [10]–[14] in the demagnetized case, differs from that of Miyoshi in that it permits a complicated resonator shape to be subdivided into a number of smaller elements, thus allowing a quite general computer program to be written. Both the finite element method, described in this paper, and the contour integration method used by Miyoshi, currently require the manipulation of relatively large matrices. The finite element method, however, has the advantage that it is possible to build on the work which has been found valuable in waveguide analysis. The contour integration approach requires that the matrix problem be recomputed for each coupling width chosen, whereas in the finite-element approach the coupling angle is not chosen until after the matrix manipulations are completed. This may result in a computational saving if a large number of coupling angles are to be considered.

To determine the performance of the three-port device as a circulator, the circulation boundary conditions [3] are imposed on the elements of the Z matrix. This gives the operating frequency and gyrator level of the device and, subsequently, allows the frequency response to be calculated. Circulation conditions over the complete κ/μ range for circulators using disk, irregular hexagonal, and triangular resonators are presented in this paper. As an application of this work, the frequency response of the input admittance has been evaluated with $\kappa/\mu = 0.67$ at the center frequency.

It is shown that circulators using each resonator shape can be arranged to exhibit well-behaved equivalent circuits. These would be consistent with the design of an octave-band circulator subject to the design of a suitable matching network. None of the resonators analyzed, however, exhibit characteristics which would allow any one to be designated 'ideal' for the design of octave-band circulators.

II. ELECTROMAGNETIC AND NETWORK FORMULATION FOR PLANAR JUNCTION CIRCULATORS

One description of a planar junction circulator is in terms of its impedance matrix. In order to obtain this matrix, the relationship between the electric and magnetic fields at the coupling ports must be determined. Bosma [1], [2] has obtained such a relationship for a disk circulator using Green's function techniques. A similar procedure has been utilized by Miyoshi [8] for arbitrary resonator shapes in which he derives an expansion for the open circuit parameters in terms of a series of eigenfunctions ϕ_a which

Manuscript received August 18, 1981; revised May 19, 1982.

R. W. Lyon was with Heriot–Watt University, and is now with ERA Technology Ltd., Leatherhead, Surrey, England.

J. Helszajn is with Heriot–Watt University, Department of Electrical Engineering, 31–35 Grassmarket, Edinburgh, EH1 2HT, Scotland.

Reprinted from *IEEE Trans. Microwave Theory Tech.*, vol. 30, no. 11, pp. 1964–1974, Nov. 1982.

he calculates using a variational method. In this section, this expansion due to Miyoshi is developed.

A generalized schematic diagram of a circulator is shown in Fig. 1. It consists of a three-fold symmetric, magnetized, ferrite resonator to which coupling lines are connected. These are printed onto either a dielectric or demagnetized ferrite substrate. The boundary of the resonator is designated by a contour ξ along which two unit vectors are defined, a normal vector $\hat{\boldsymbol{n}}$ and a tangential vector $\boldsymbol{t}$. The separation H between the ground plane and the center conductor is arranged to be small with respect to the wavelength in order to ensure that higher order modes which vary in the z direction are suppressed. This restriction, when applied together with the boundary conditions on the center conductor, implies that only the (E_z, H_x, H_y) field components exist.

The E_z field in a planar junction circulator satisfies the wave equation [1], [2]

$$\left(\nabla_t^2 + k_{\text{eff}}^2\right) E_z = 0 \tag{1}$$

where k_{eff}, the wave number, is given by

$$k_{\text{eff}}^2 = \omega^2 \mu_0 \mu_{\text{eff}} \epsilon_0 \epsilon_f. \tag{2}$$

ϵ_f is the relative permittivity of the ferrite medium, and the effective permeability μ_{eff} is given by

$$\mu_{\text{eff}} = \frac{\mu^2 - \kappa^2}{\mu} \tag{3}$$

where μ and κ are the diagonal and off-diagonal components of the tensor permeability of the ferrite.

On the boundary, the tangential magnetic field H_t is equal to zero and this may be expressed as a boundary condition on E_z to give

$$\frac{\partial E_z}{\partial n} + j\frac{\kappa}{\mu}\frac{\partial E_z}{\partial t} = 0 \qquad \text{on } \xi. \tag{4a}$$

At the coupling ports, H_t is not zero and E_z satisfies

$$\frac{\partial E_z}{\partial n} + j\frac{\kappa}{\mu}\frac{\partial E_z}{\partial t} = j\omega\mu_0\mu_{\text{eff}} H_t. \tag{4b}$$

In order to solve (1) in conjunction with (4a–b), it is convenient to introduce a Green's function $G(\boldsymbol{r}|\boldsymbol{r}_0)$. Two variables are defined when using a Green's function; the point $\boldsymbol{r}$ at which the E_z field is observed and the coupling port coordinate $\boldsymbol{r}_0$. These conventions are summarized in Fig. 2.

The Green's function $G(\boldsymbol{r}|\boldsymbol{r}_0)$ is defined as the solution to the equation

$$\left(\nabla_t^2 + k_{\text{eff}}^2\right) G(\boldsymbol{r}|\boldsymbol{r}_0) = -\,j\omega\mu_0\mu_{\text{eff}}\,\delta(\boldsymbol{r} - \boldsymbol{r}_0) \tag{5}$$

where $\delta(\boldsymbol{r} - \boldsymbol{r}_0)$ is the dirac delta function. The Green's function must satisfy the boundary condition

$$\frac{\partial G(\boldsymbol{r}|\boldsymbol{r}_0)}{\partial n} - j\frac{\kappa}{\mu}\frac{\partial G(\boldsymbol{r}|\boldsymbol{r}_0)}{\partial t} = 0. \tag{6}$$

The Green's function, whose units are (Ω/m), is a generalization of that used for a disk by Bosma [1] who sets

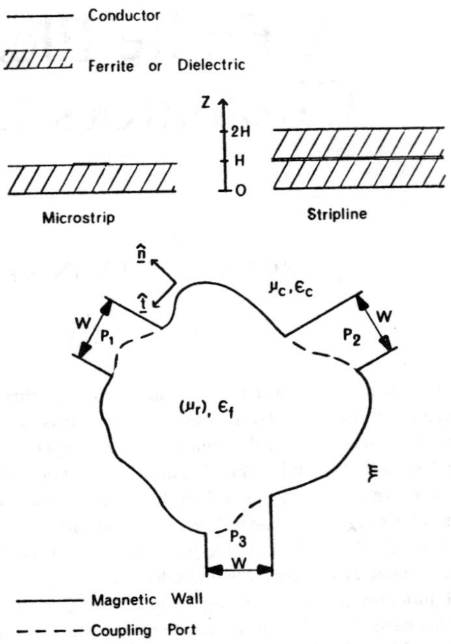

Fig. 1. Generalized schematic diagram of planar junction circulator.

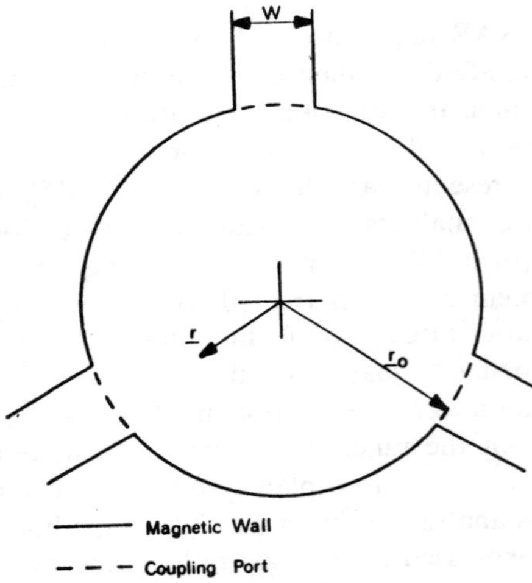

Fig. 2. Coordinate convention for Green's function.

out its properties in detail and show that

$$E_z(\boldsymbol{r}) = \sum_{i=1}^{3} \int_{P_i} G(\boldsymbol{r}|\boldsymbol{r}_0) H_t(\boldsymbol{r}_0)\, dt_0 \tag{7}$$

where the integration is carried out over the coupling port P_i. $E_z(\boldsymbol{r})$ applies at any point in the resonator including the intervals defined by the coupling points.

At this point, the derivation of Miyoshi's [8] contour integration method and variational method diverge. The contour integration method consists of discretizing the boundary and, by employing a different Green's function which satisfies the wave equation but which does not satisfy (6), reducing a contour integration whose form is similar to (7) to a set of matrix equations. In his variational approach, Miyoshi expanded the Green's function as a series of eigenfunctions ϕ_a to give

$$G(\boldsymbol{r}|\boldsymbol{r}_0) = j\omega\mu_0\mu_{\text{eff}} \sum_{a=1}^{\infty} \frac{\phi_a(\boldsymbol{r})\phi_a^*(\boldsymbol{r}_0)}{k_a^2 - k_{\text{eff}}^2}. \tag{8}$$

In practice, sufficiently accurate results are obtained when this series is truncated to around 10 terms. The eigenfunctions ϕ_a and the eigenvalues k_a are the solutions to the differential equation

$$(\nabla_t^2 + k_a^2)\phi_a(\boldsymbol{r}) = 0 \tag{9}$$

subject to the boundary condition imposed on the Green's function given in (6). The eigenfunctions are orthogonal and are normalized so that

$$\int\int_s \phi_a(\boldsymbol{r})\phi_a^*(\boldsymbol{r})\, ds = 1. \tag{10}$$

In the demagnetized planar circuit, ϕ_a is directly equivalent to the electric field of the resonant modes in the planar resonator, whereas, in the magnetized case, ϕ_a represents the complex conjugate of E_z.

Assuming that H_t is a constant over each coupling port, Miyoshi derived the relation between the average electric field at port i and the magnetic field at port j from (7) and (8) giving

$$\eta_{ij} = \frac{j\omega\mu_0\mu_{\text{eff}}}{W} \sum_{a=1}^{\infty} \frac{1}{k_a^2 - k_{\text{eff}}^2} \int_{P_i} \phi_a^*(\boldsymbol{r})\, dt \int_{P_j} \phi_a(\boldsymbol{r}_0)\, dt_0 \tag{11}$$

assuming all ports other than j are open circuited.

In order to derive a relationship for Z_{ij} from (11) it is necessary to introduce the characteristic impedance R_e of a planar transmission line of width W, substrate thickness H, and constitutive parameters ε_f and μ_{eff}. In stripline, R_e may be calculated using Richardson's technique [15] while the equivalent waveguide technique is commonly used in microstrip [17], [18]. For an n port where each of the coupling ports are of equal width W, Z_{ij} is given by

$$Z_{ij} = \frac{jR_e k_{\text{eff}}}{W} \sum_{a=1}^{\infty} \frac{1}{k_a^2 - k_{\text{eff}}^2} \int_{P_i} \phi_a^*(\boldsymbol{r})\, dt \int_{P_j} \phi_a(\boldsymbol{r})_0\, dt_0. \tag{12}$$

This analysis is a more general statement of the treatment presented by Bosma [2] for the particular case of a disk resonator.

III. Variational Solution for Eigenfunctions Using Matrix Eigenvalue Method

The Z matrix of a junction circulator can be derived provided that the eigenfunctions ϕ_a, which satisfy (9) together with the boundary conditions given by (6), are known. It is only possible to solve these equations analytically in a very small number of cases, and the most convenient method, in the general case, is to use a variational approach.

Miyoshi [8] has recognized that the trial function ϕ_a', which causes the functional

$$F(\phi_a'(\boldsymbol{r})) = \int\int_s |\nabla_t \phi_a'(\boldsymbol{r})|^2 - k_a^2 |\phi_a'(\boldsymbol{r})|^2\, ds - j\frac{\kappa}{\mu}\oint_\xi (\phi_a'(\boldsymbol{r}))^* \frac{\partial \phi'(\boldsymbol{r})}{\partial t}\, dt \tag{13}$$

to be minimized, satisfies both the differential equation and the boundary conditions for the eigenfunctions ϕ_a. When the value of κ/μ is set to zero, $F(\phi_a')$ reduces to the functional used by Silvester [10] in his analysis of arbitrarily shaped waveguides.

The trial function ϕ_a' is an approximation to the exact function ϕ_a and it is expanded as

$$\phi_a'(\boldsymbol{r}) = \sum_{i=1}^{n} u_i \alpha_i. \tag{14}$$

The terms α_i are a suitable set of real basis functions and u_i are the complex coefficients. There are n basis functions included in the expansion. In this paper, the basis functions are chosen using the finite element method.

Substituting (14) into the functional $F(\phi_a')$ in (13) and ensuring that the functional is minimized by imposing the Rayleigh–Ritz condition

$$\frac{\partial F(\phi_a'(\boldsymbol{r}))}{\partial u_i^*} \tag{15}$$

reduces the problem to a set of simultaneous equations of the form

$$[[A] - k_a^2[B]][u] = 0 \tag{16}$$

which may be recognized to be the general matrix eigenvalue problem

$$[A][u] = k_a^2[B][u]. \tag{17}$$

If the symmetric square matrix B is reduced to the product LL^T, where L is a lower triangular matrix, (17) may be reduced to the familiar eigenvalue problem

$$[L^{-1}][A][L^{-1}]^T[u] = k_a^2[u]. \tag{18}$$

A and B are square matrices

$$A_{ij} = \int\int_s \nabla_t \alpha_i \cdot \nabla_t \alpha_j\, ds - j\frac{\kappa}{\mu}\oint_\xi \alpha_i \frac{\partial \alpha_j}{\partial t}\, dt \tag{19}$$

and may be reduced to

$$[A] = [D] + j\frac{\kappa}{\mu}[C] \tag{20}$$

where

$$D_{ij} = \int\int_s \nabla_t \alpha_i \cdot \nabla_t \alpha_j\, ds \tag{21a}$$

$$C_{ij} = -\oint_\xi \alpha_i \frac{\partial \alpha_j}{\partial t}\, dt. \tag{21b}$$

The elements of the B matrix are given by

$$B_{ij} = \int\int_s \alpha_i \alpha_j\, ds. \tag{22}$$

The B matrix is reduced to LL^T computationally and, thus, an analytic expression for L is not given. Once n is selected, the matrix eigenvalue equation will yield n eigenvalues k_a^2 and column eigenvectors $[u]$, n is the number of basis functions included in (14).

Silvester, using the finite element method, has calculated both the B matrix and the D matrix (the B matrix is the T matrix and the D matrix is the S matrix in Silvester's notation) which are symmetric and are fully tabulated in [11]. The C matrix is skew-symmetric and is derived in this paper.

IV. Finite Element Analysis

In the variational method, it is possible to choose any suitable set of basis functions to expand the trial function ((14)). Miyoshi [8] uses a polynomial expansion which describes the fields in the complete resonator. This has the disadvantage that the A and B matrices given by (20) and (22) must be recalculated for every different resonator shape. In the finite element method used by Silvester [10]–[14] in his analysis of arbitrary shaped waveguides, this problem is overcome by subdividing the resonator region into triangular elements. A polynomial expansion for the eigenfunction ϕ_a is formed in each triangle in terms of $(u_i \alpha_i)$ in (14) enabling the A and B matrices to be calculated. These are then assembled together to form the complete matrix eigenvalue problem.

Using the finite element method, Silvester [11] has presented expressions for polynomial basis functions in triangular elements. These are given in terms of triangular area coordinates for each point inside a triangle. Each coordinates is defined as the ratio of the perpendicular distance to the wall opposite vertex i to the length of the altitude drawn to vertex i. From Fig. 3 it can be seen that the α_1 coordinate of point Q is given by

$$\zeta_1 = \frac{e}{d} \tag{23}$$

and the other coordinates ζ_2 and ζ_3 are defined in a similar manner. It is important to note that the three coordinates are related by [11]

$$\zeta_1 + \zeta_2 + \zeta_3 = 1. \tag{24}$$

In the finite element method described in this paper, the basis functions α_i in (14) are polynomials of degree p and these are arranged to provide a p'th-order interpolation to the eigenfunction ϕ_a over each triangle. In general, a polynomial of degree p in two coordinates will have m coefficients where

$$m = (p+1)(p+2)/2. \tag{25}$$

Over each triangular element, m points (nodes) are distributed and each basis function is arranged to take the value 1 at one node and 0 at all the others. Thus, the coefficients of the basis functions α_i in (14) represent the amplitude of the eigenfunction ϕ_a' at point i. The distribution of the nodes over a triangle for first-, second-, third-, and fourth-order polynomials are illustrated in Fig. 4. Each point is labeled with three integers, i, j, and k from which its triangular area coordinates $(\zeta_1, \zeta_2, \zeta_3)$ can be

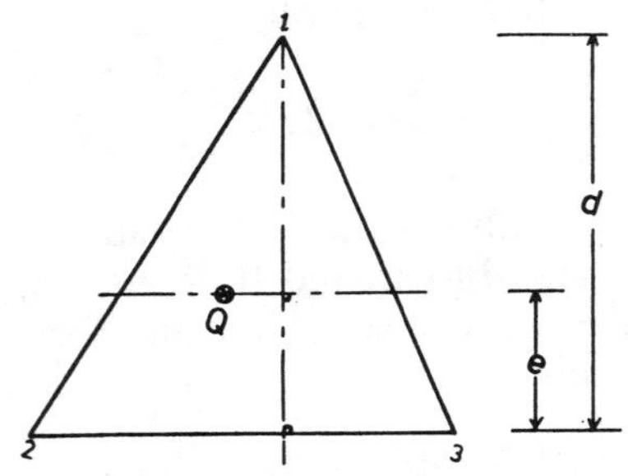

Fig. 3. Dimensions used in the definition of triangular area coordinates.

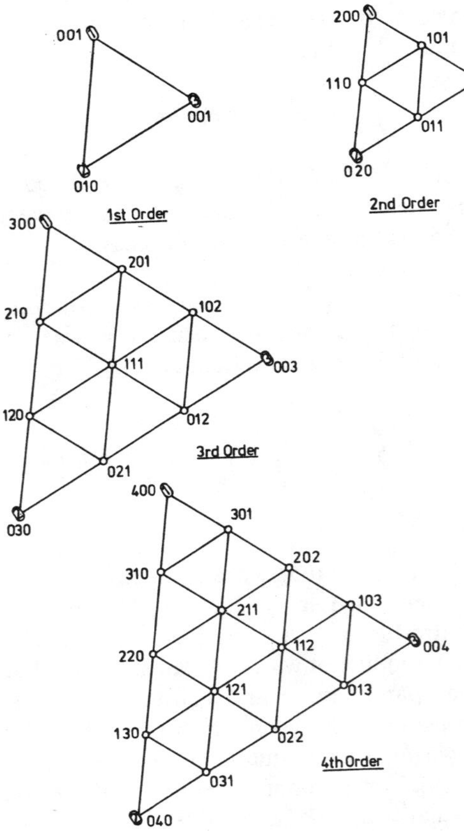

Fig. 4. Distribution of nodes over triangle for first-, second-, third-, and fourth-order polynomials.

derived using the relation

$$(\zeta_1, \zeta_2, \zeta_3) = \left(\frac{i}{p}, \frac{j}{p}, \frac{k}{p}\right). \tag{26}$$

Equation (24) is also satisfied since

$$i + j + k = p. \tag{27}$$

Associated with each point is a basis function $\alpha_{ijk}(\zeta_1, \zeta_2, \zeta_3)$ which is given by [11]

$$\alpha_{ijk}(\zeta_1, \zeta_2, \zeta_3) = P_i(\zeta_1) P_j(\zeta_2) P_k(\zeta_3) \tag{28}$$

where

$$P_r(\zeta_q) = \prod_{L=1}^{r} \left(\frac{P(\zeta_q) - L + 1}{L} \right), \qquad r \geqslant 1 \tag{29a}$$

$$= 1, \qquad r = 0 \tag{29b}$$

where q and r are dummy variables which satisfy $q \in \{1, 2, 3\}$

and $r \in \{i, j, k\}$. The basis function α_{ijk} takes the value 1 at node (i, j, k) and the value 0 at all other nodes in the triangle.

These formulas may now be substituted into (20) and (22) to calculate the A and B matrices for a triangle.

The number of terms taken in (14) is determined by the number of triangles included in a finite element division of any particular resonator shape. There is not a simple relation between the number of elements and the number of basis functions since, in essence, there will be one basis function for each node in the resonator. The number of nodes in a particular element is a function of the order of the polynomial approximation within that element. Equation (25) gives the number of nodes in each element as a function of the nodes of approximation p.

In addition to the order of polynomial approximation within each element the total number of nodes in any resonator is determined by the number and orientation of the elements in the resonator. This is demonstrated in Fig. 5, in which a triangle is split into three elements each of which, individually, have three nodes. Once they are assembled, however, certain nodes coincide leaving only four in total. Thus, it is not possible to make a simple statement of the value of n in (14). In practice, the value $90 < n < 100$ have been found to give accurate results.

The C matrix involves integration around the contour ξ and so the matrix has the value [0] if the element does not lie along the boundary. If the element has one or more of its sides lying along the boundary the C matrix for this element can be written down provided it is known for the case where an element has only one side on the boundary.

Consider the case where the first-order polynomial interpolation is to be employed (i.e., $p = 1$ in (25)). The following expressions for the basis functions $\alpha_{100}, \alpha_{010}$, and α_{001} can be derived from (28):

$$\alpha_{100} = \zeta_1$$
$$\alpha_{010} = \zeta_2$$
$$\alpha_{001} = \zeta_3. \tag{30}$$

Substituting these expressions into (21b) leads to the following expression for the C matrix when the magnetic wall lies opposite the point (1,0,0):

$$[C^{(1,0,0)}] = \begin{bmatrix} 0 & 0 & 0 \\ 0 & 1 & -1 \\ 0 & 1 & -1 \end{bmatrix}. \tag{31}$$

This matrix can be seen to be skew symmetric.

If the boundary lies opposite points (0,1,0) or (0,0,1), the C matrix may be calculated simply by re-arranging the matrix derived for a boundary opposite (1,0,0). For example, in the case where a boundary lies opposite point (0,1,0), subscript (1) → (2) subscript (2) → (3), and subscript (3) → (1). Thus, the C matrix is given by

$$[C^{(0,1,0)}] = \begin{bmatrix} -1 & 0 & 1 \\ 0 & 0 & 0 \\ -1 & 0 & 1 \end{bmatrix}. \tag{32}$$

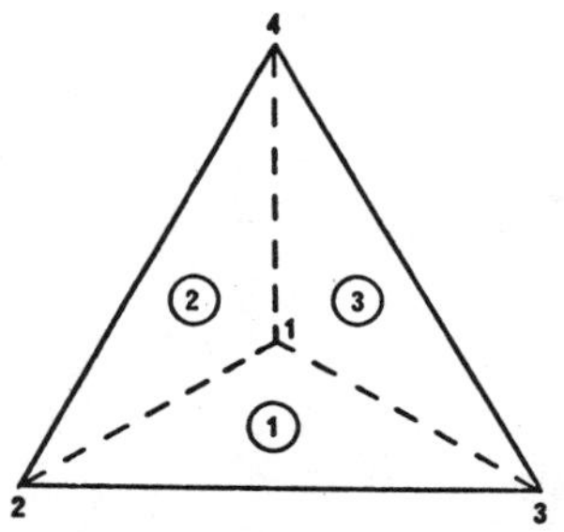

Fig. 5. Triangle split into three elements.

If the element has two of three sides which lie along the boundary, the C matrix is calculated by adding the matrices derived for elements which have magnetic walls on a single side.

The matrices given in (31) and (32) refer to a single, isolated triangular element. Several matrices must be assembled together, however, to form the complete eigenvalue problem. Consider the case illustrated in Fig. 5 where a triangle is split up into three elements in a fashion which preserves the 120° symmetry. The C matrix for the complete shape is given by

$$[C_4] = \begin{bmatrix} 0 & 0 & 0 & 0 \\ 0 & 1 & -1 & 0 \\ 0 & 1 & -1 & 0 \\ 0 & 0 & 0 & 0 \end{bmatrix} + \begin{bmatrix} 0 & 0 & 0 & 0 \\ 0 & 0 & 0 & 0 \\ 0 & 0 & 1 & -1 \\ 0 & 0 & 1 & -1 \end{bmatrix} + \begin{bmatrix} 0 & 0 & 0 & 0 \\ 0 & -1 & 0 & 1 \\ 0 & 0 & 0 & 0 \\ 0 & -1 & 0 & 1 \end{bmatrix} = \begin{bmatrix} 0 & 0 & 0 & 0 \\ 0 & 0 & -1 & 1 \\ 0 & 1 & 0 & -1 \\ 0 & -1 & 1 & 0 \end{bmatrix} \tag{33}$$

which is also skew-symmetric. In the examples considered in this paper, the A, B, and C matrices, when finally assembled are of the order 100×100.

The calculation of the C matrix for first-order polynomials is reasonably easy, but for higher order polynomials the volume of algebra becomes too large to be performed by hand. Silvester [11] encountered similar difficulties and in order to overcome the problem a computer program was written to evaluate the matrix elements analytically. By adopting a similar approach the authors have evaluated the C matrices for up to fourth-order interpolation and these are tabulated in Table I. In most cases, fourth-order interpolation is the maximum which can be used in practice as there are 15 nodes in each element and large matrices can be generated when only a few elements are incorporated. Silvester [11] presents a table contrasting the relative merits of using higher order interpolation or alternatively more elements using a lower order polynomial approximation.

V. Computation of Circulation Conditions

A suite of computer programs have been written which implement the theoretical results discussed in the previous sections. These consist, firstly, of a program which uses the finite element method to evaluate the resonant modes of a

TABLE I
TABLE OF C MATRICES FOR POLYNOMIALS UP TO FOURTH ORDER

N= 1
COMMON DENOMINATOR IS 2

0	0	0
0	1	-1
0	1	-1

N= 2
COMMON DENOMINATOR IS 6

0	0	0	0	0	0
0	0	0	0	0	0
0	0	0	0	0	0
0	0	0	3	-4	1
0	0	0	4	0	-4
0	0	0	-1	4	-3

N= 3
COMMON DENOMINATOR IS 80

0	0	0	0	0	0	0	0	0	0
0	0	0	0	0	0	0	0	0	0
0	0	0	0	0	0	0	0	0	0
0	0	0	0	0	0	0	0	0	0
0	0	0	0	0	0	0	0	0	0
0	0	0	0	0	0	0	0	0	0
0	0	0	0	0	0	40	-57	24	-7
0	0	0	0	0	0	57	0	-81	24
0	0	0	0	0	0	-24	81	0	-57
0	0	0	0	0	0	7	-24	57	-40

N= 4
COMMON DENOMINATOR IS 1890

0	0	0	0	0	0	0	0	0	0	0	0	0	0	0
0	0	0	0	0	0	0	0	0	0	0	0	0	0	0
0	0	0	0	0	0	0	0	0	0	0	0	0	0	0
0	0	0	0	0	0	0	0	0	0	0	0	0	0	0
0	0	0	0	0	0	0	0	0	0	0	0	0	0	0
0	0	0	0	0	0	0	0	0	0	0	0	0	0	0
0	0	0	0	0	0	0	0	0	0	0	0	0	0	0
0	0	0	0	0	0	0	0	0	0	0	0	0	0	0
0	0	0	0	0	0	0	0	0	0	0	0	0	0	0
0	0	0	0	0	0	0	0	0	0	0	0	0	0	0
0	0	0	0	0	0	0	0	0	0	945	-1472	804	-384	107
0	0	0	0	0	0	0	0	0	0	1472	0	-2112	1024	-384
0	0	0	0	0	0	0	0	0	0	-804	2112	0	-2112	804
0	0	0	0	0	0	0	0	0	0	384	-1024	2112	0	-1472
0	0	0	0	0	0	0	0	0	0	-107	384	-804	1472	-945

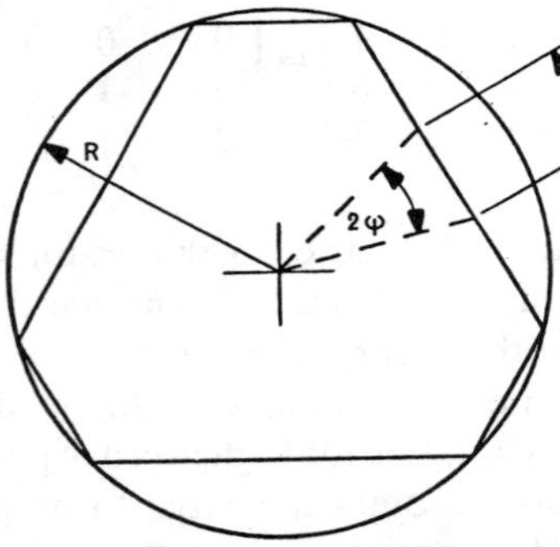

Fig. 6. Definition of coupling angle ψ.

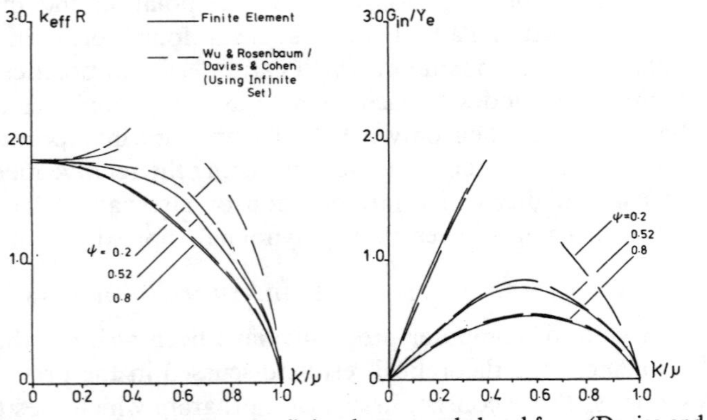

Fig. 7. Comparison between finite element and closed form (Davies and Cohen) solutions for disk resonator.

magnetized ferrite-loaded planar resonator. These resonant modes are then used by a second program to calculate the elements of the impedance matrix. This program then determines the lowest value of $k_{\text{eff}}R$ which satisfies the first circulation condition [3].

$$\operatorname{Im}(Z_{\text{in}}) = 0 \tag{34}$$

for a range of values of κ/μ and coupling angles (the angle which sustends the coupling interval W in Fig. 6). The input resistance of the circulator at this value of $k_{\text{eff}}R$ is then calculated using the second circulation condition

$$R_{\text{in}} = \operatorname{Re}(Z_{\text{in}}) \tag{35}$$

where

$$Z_{\text{in}} = Z_{11} + \frac{Z_{12}^2}{Z_{12}^*}. \tag{36}$$

In keeping with convention, the circulation data are tabulated in terms of G_{in} and B_{in} where

$$Y_{\text{in}} = G_{\text{in}} + jB_{\text{in}} = \frac{1}{Z_{\text{in}}}. \tag{37}$$

In Fig. 7(a) and (b), the results obtained using the method described in this paper are compared with those produced by previous authors for a disk [3], [5]. In calculating these results, the infinite series in (12) has been truncated to 10 terms. It can be seen that the agreement is best for larger coupling angles and less good for small values of ψ. The reason for this is that smaller coupling angles excite higher order modes more strongly, and these higher order modes tend to be computed less accurately by the finite element program.

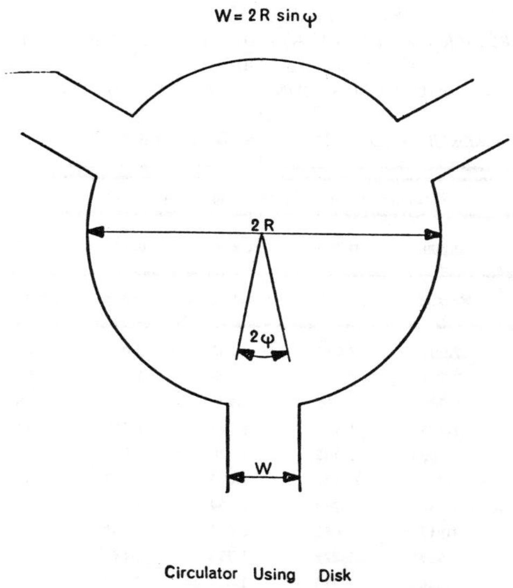

Fig. 8. Coordinate system of disk resonator.

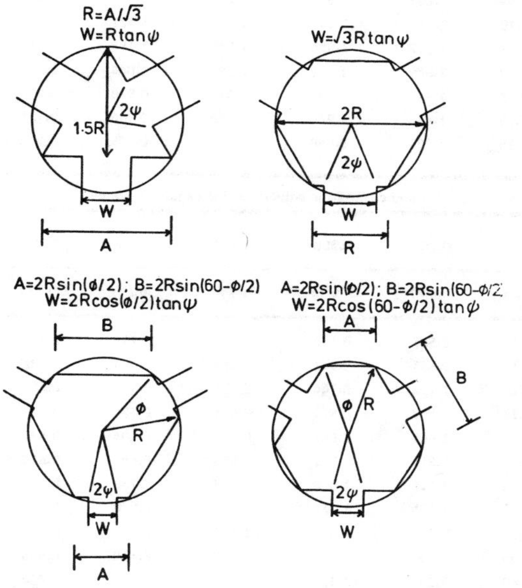

Fig. 9. Coordinate system of triangle, regular hexagonal, narrow, and broad-wall coupled irregular hexagonal resonators.

In Tables II–VI, the first and second circulation solutions are tabulated for disk, triangular, regular hexagon, narrow and broad wall coupled irregular hexagons illustrated in Figs. 8 and 9. These tables are computed retaining the first 10 eigenfunctions ϕ_a in (12). It is observed that at certain points in the tables (e.g., a triangle $0.7 < \kappa/\mu < 0.85$) the values $k_{\text{eff}}R$ and G_{in}/Y_f take the value 0. This indicates that no circulation condition was located with $k_{\text{eff}}R < 3.0$. In certain cases, G_{in}/Y_f takes a negative value. At these points, the lowest circulation condition represents rotation in the opposite direction.

The values of G_{in} and B_{in} are normalized to Y_f which is given by

$$Y_f = \sqrt{\epsilon_f}\, Y_r. \tag{38}$$

TABLE II
TABLES OF $k_{\text{eff}}R$ (TABLE A) AND G/Y_f (TABLE B) FOR CIRCULATORS USING DISK, TRIANGLE REGULAR HEXAGON, NARROW AND BROAD-WALL COUPLED IRREGULAR HEXAGON ($\phi = 50$), RESPECTIVELY

First circulation solution for disk circulator

PSI	0.200	0.300	0.400	0.500	0.600	0.700	0.800
K/U	Keff.R	Keff.R	Keff.R	Keff.R	Keff.R	Keff.R	Keff.R
0.10	1.862	1.859	1.857	1.854	1.851	1.848	1.846
0.15	1.864	1.859	1.852	1.845	1.839	1.833	1.828
0.20	1.869	1.859	1.847	1.834	1.821	1.810	1.801
0.25	1.880	1.861	1.840	1.818	1.798	1.780	1.766
0.30	1.903	1.869	1.833	1.798	1.767	1.741	1.720
0.35	1.968	1.889	1.826	1.772	1.729	1.693	1.665
0.40	2.206	1.968	1.823	1.741	1.681	1.636	1.600
0.45	2.216	2.180	1.842	1.711	1.630	1.572	1.529
0.50	2.228	2.224	2.192	1.664	1.564	1.497	1.448
0.55	0.000	0.000	0.000	1.605	1.487	1.414	1.361
0.60	2.193	2.187	2.171	1.532	1.402	1.325	1.270
0.65	2.162	2.136	2.071	1.443	1.308	1.230	1.175
0.70	2.518	2.064	1.884	1.305	1.204	1.133	1.078
0.75	2.342	1.941	1.692	1.190	1.088	1.023	0.972
0.80	2.328	1.762	1.469	1.068	0.969	0.907	0.859
0.85	2.375	1.519	1.235	0.927	0.836	0.779	0.736
0.90	1.973	1.214	0.976	0.756	0.679	0.631	0.594
0.95	0.904	0.812	0.654	0.530	0.477	0.442	0.415

Second circulation solution for disk circulator

PSI	0.200	0.300	0.400	0.500	0.600	0.700	0.800
K/U	Gin/Yf	Gin/Yf	Gin/Yf	Gin/Yf	Gin/Yf	Gin/Yf	Gin/Yf
0.10	0.495	0.335	0.255	0.210	0.181	0.162	0.149
0.15	0.750	0.505	0.383	0.314	0.271	0.242	0.223
0.20	1.010	0.676	0.511	0.417	0.359	0.320	0.295
0.25	1.282	0.850	0.637	0.517	0.443	0.394	0.363
0.30	1.580	1.029	0.760	0.612	0.522	0.463	0.427
0.35	1.980	1.223	0.882	0.701	0.593	0.526	0.484
0.40	3.011	1.500	1.001	0.780	0.656	0.580	0.534
0.45	2.685	1.728	1.079	0.837	0.706	0.623	0.572
0.50	2.828	1.803	1.350	0.893	0.749	0.660	0.606
0.55	-0.000	-0.000	-0.000	0.935	0.782	0.689	0.633
0.60	3.073	1.775	1.282	0.963	0.806	0.712	0.654
0.65	3.511	1.721	1.225	0.979	0.823	0.728	0.670
0.70	0.830	1.688	1.234	0.969	0.833	0.745	0.687
0.75	0.234	1.474	1.177	0.971	0.834	0.749	0.694
0.80	-0.429	1.356	1.154	0.972	0.839	0.755	0.700
0.85	-0.278	1.279	1.135	0.968	0.840	0.758	0.704
0.90	1.320	1.240	1.119	0.961	0.838	0.758	0.706
0.95	1.651	1.244	1.109	0.950	0.833	0.755	0.707

Y_r is the admittance of an air spaced planar transmission line whose width is defined by the coupling interval.

VI. Computation of Frequency Response

In addition to the calculation of circulation conditions, it is also necessary to investigate the frequency response of the input admittance of the junction. This is determined by a third program which evaluates the input admittance as a function of the normalized frequency variable

$$\delta = \frac{f - f_0}{f_0} \tag{39}$$

TABLE III
TABLES OF $k_{\text{eff}}R$ (TABLE A) AND G/Y_f (TABLE B) FOR CIRCULATORS USING DISK, TRIANGLE, REGULAR HEXAGON, NARROW AND BROAD-WALL COUPLED IRREGULAR HEXAGON ($\phi = 50$), RESPECTIVELY

First circulation solution for triangular circulator

PSI	0.200	0.300	0.400	0.500	0.600	0.700	0.800
K/U	Keff.R	Keff.R	Keff.R	Keff.R	Keff.R	Keff.R	Keff.R
0.10	2.455	2.450	2.444	2.438	2.432	2.428	2.426
0.15	2.451	2.438	2.423	2.409	2.396	2.397	2.382
0.20	2.448	2.421	2.392	2.366	2.346	2.331	2.323
0.25	2.458	2.402	2.351	2.312	2.283	2.262	2.250
0.30	2.504	2.386	2.301	2.247	2.210	2.183	2.166
0.35	2.536	2.383	2.241	2.174	2.127	2.089	2.061
0.40	2.524	2.518	2.175	2.095	2.040	1.996	1.961
0.45	2.493	2.492	2.101	2.007	1.946	1.897	1.857
0.50	2.448	2.445	2.021	1.916	1.849	1.795	1.751
0.55	1.993	1.910	1.845	1.790	1.743	1.703	1.671
0.60	1.990	1.837	1.756	1.691	1.638	1.593	1.557
0.65	2.263	1.767	1.662	1.586	1.525	1.475	1.431
0.70	0.000	1.703	1.559	1.473	1.407	1.353	1.305
0.75	0.000	2.042	1.444	1.352	1.284	1.230	1.182
0.80	0.000	0.000	1.309	1.209	1.142	1.090	1.046
0.85	0.000	0.000	1.154	1.054	0.990	0.941	0.900
0.90	2.690	2.508	0.948	0.861	0.808	0.766	0.731
0.95	0.631	0.634	0.865	0.931	0.583	0.547	0.518

Second circulation solution for triangular circulator

PSI	0.200	0.300	0.400	0.500	0.600	0.700	0.800
K/U	Gin/Yf	Gin/Yf	Gin/Yf	Gin/Yf	Gin/Yf	Gin/Yf	Gin/Yf
0.10	1.706	1.095	0.780	0.583	0.448	0.347	0.268
0.15	2.562	1.629	1.146	0.849	0.649	0.503	0.391
0.20	3.444	2.144	1.476	1.079	0.820	0.637	0.500
0.25	4.431	2.640	1.755	1.262	0.956	0.745	0.590
0.30	5.703	3.142	1.974	1.398	1.057	0.829	0.663
0.35	7.066	3.797	2.134	1.497	1.143	0.916	0.754
0.40	7.323	5.322	2.243	1.566	1.200	0.969	0.804
0.45	-21.798	5.174	2.310	1.604	1.236	1.004	0.841
0.50	7.199	4.965	2.348	1.629	1.262	1.033	0.870
0.55	4.847	2.835	1.960	1.489	1.197	0.993	0.836
0.60	5.556	2.915	1.998	1.518	1.222	1.016	0.861
0.65	10.533	2.979	2.014	1.535	1.242	1.042	0.893
0.70	-0.000	3.130	2.025	1.542	1.253	1.057	0.913
0.75	-0.000	5.718	2.029	1.537	1.251	1.059	0.917
0.80	-0.000	-0.000	2.067	1.556	1.265	1.069	0.925
0.85	-0.000	-0.000	2.052	1.540	1.255	1.064	0.923
0.90	1.454	1.031	1.996	1.501	1.230	1.048	0.913
0.95	3.585	2.438	2.765	2.541	1.249	1.054	0.915

TABLE IV
TABLES OF $k_{\text{eff}}R$ (TABLE A) AND G/Y_f (TABLE B) FOR CIRCULATORS USING DISK, TRIANGLE, REGULAR HEXAGON, NARROW AND BROAD-WALL COUPLED IRREGULAR HEXAGON ($\phi = 50$), RESPECTIVELY

First circulation solution using regular hexagon

PSI	0.200	0.300	0.400	0.500	0.524
K/U	Keff.R	Keff.R	Keff.R	Keff.R	Keff.R
0.10	2.004	2.003	2.002	2.000	1.999
0.15	1.997	1.994	1.990	1.985	1.984
0.20	1.985	1.981	1.974	1.965	1.963
0.25	1.973	1.965	1.953	1.938	1.934
0.30	1.960	1.946	1.928	1.905	1.899
0.35	1.945	1.924	1.896	1.863	1.856
0.40	1.930	1.897	1.857	1.813	1.803
0.45	1.917	1.866	1.811	1.754	1.742
0.50	1.913	1.829	1.755	1.686	1.671
0.55	0.000	1.783	1.688	1.608	1.592
0.60	2.348	1.726	1.610	1.522	1.504
0.65	2.267	1.653	1.519	1.426	1.407
0.70	2.151	1.551	1.413	1.321	1.302
0.75	2.012	1.854	1.276	1.184	1.174
0.80	1.849	1.691	1.166	1.064	1.051
0.85	1.660	1.484	1.028	0.928	0.914
0.90	1.464	1.218	0.838	0.764	0.753
0.95	1.093	0.880	0.602	0.539	0.530

Second circulation solution using regular hexagon

PSI	0.200	0.300	0.400	0.500	0.524
K/U	Gin/Yf	Gin/Yf	Gin/Yf	Gin/Yf	Gin/Yf
0.10	0.569	0.375	0.276	0.216	0.205
0.15	0.853	0.561	0.413	0.323	0.306
0.20	1.133	0.744	0.546	0.426	0.404
0.25	1.407	0.921	0.674	0.524	0.496
0.30	1.661	1.082	0.786	0.604	0.571
0.35	1.915	1.241	0.897	0.685	0.646
0.40	2.155	1.388	0.997	0.758	0.713
0.45	2.375	1.516	1.083	0.820	0.771
0.50	2.583	1.619	1.151	0.871	0.818
0.55	-0.000	1.693	1.199	0.910	0.856
0.60	3.079	1.739	1.229	0.938	0.884
0.65	2.881	1.759	1.245	0.957	0.903
0.70	2.739	1.753	1.248	0.968	0.915
0.75	2.462	1.720	1.231	0.955	0.910
0.80	2.374	1.648	1.239	0.961	0.915
0.85	2.350	1.617	1.237	0.964	0.918
0.90	2.314	1.623	1.223	0.960	0.916
0.95	2.018	1.577	1.237	0.962	0.914

where f is the operating frequency and f_0 is the center frequency. The ferrite material is assumed to be just saturated and the value of κ/μ is given by

$$\frac{\kappa}{\mu} = \frac{\gamma M_0}{\omega \mu_0}. \tag{40}$$

γ is the gyromagnetic ratio ($2.21 \times 10^5(\text{rad/s}/(A/m))$), μ_0 the permeability of free space ($4\pi \times 10^{-7}$ H/m), and M_0 the saturation magnetization (Telsa).

One interesting case is the class of devices which are arranged so that κ/μ is 0.67 at the center of frequency. This implies that the value of κ/μ varies from 0.5 to 1.0 over an octave frequency band. The input admittance of a junction using a disk resonator is shown in Fig. 10 for a range of coupling angles ψ between 0.45 and 0.8. For smaller coupling angles, the equivalent circuit is not well behaved over the frequency interval. The input admittance is, in general, complex except for ψ close to 0.5 where it is a nearly frequency independent conductance. This is the so-called tracking solution [5], [16].

In Fig. 11, the frequency response of a circulator using the triangular resonator is given. While G/Y_f is nearly

TABLE V

TABLES OF $k_{\mathrm{eff}}R$ (TABLE A) AND G/Y_f) TABLE B) FOR CIRCULATORS USING DISK, TRIANGLE, REGULAR HEXAGON, NARROW AND BROAD-WALL COUPLED IRREGULAR HEXAGON ($\phi = 50$), RESPECTIVELY

First circulation solution using irregular hexagon narrow wall coupled (Ø = 50)

PSI	0.200	0.300	0.400	0.436
K/U	Keff.R	Keff.R	Keff.R	Keff.R
0.10	2.004	2.003	2.002	2.002
0.15	1.997	1.995	1.993	1.992
0.20	1.988	1.985	1.980	1.978
0.25	1.976	1.971	1.964	1.961
0.30	1.953	1.955	1.944	1.939
0.35	1.949	1.937	1.920	1.912
0.40	1.934	1.916	1.891	1.881
0.45	1.919	1.892	1.857	1.843
0.50	1.903	1.862	1.815	1.797
0.55	1.893	1.823	1.761	1.738
0.60	2.542	1.778	1.697	1.670
0.65	2.505	1.731	1.622	1.592
0.70	2.441	1.617	1.523	1.499
0.75	2.556	2.084	1.365	1.347
0.80	2.564	1.854	1.256	1.217
0.85	2.570	1.633	1.122	1.073
0.90	1.856	1.372	0.933	0.892
0.95	1.173	0.999	0.671	0.635

Second circulation solution using irregular hexagon narrow wall coupled (Ø = 50)

PSI	0.200	0.300	0.400	0.436
K/U	Gin/Yf	Gin/Yf	Gin/Yf	Gin/Yf
0.10	0.424	0.280	0.207	0.188
0.15	0.635	0.418	0.308	0.281
0.20	0.844	0.555	0.408	0.371
0.25	1.050	0.689	0.505	0.459
0.30	1.252	0.819	0.598	0.542
0.35	1.449	0.944	0.686	0.621
0.40	1.639	1.063	0.769	0.695
0.45	1.815	1.172	0.844	0.761
0.50	1.972	1.263	0.906	0.816
0.55	2.117	1.331	0.951	0.857
0.60	2.683	1.385	0.984	0.888
0.65	2.463	1.428	1.010	0.912
0.70	2.357	1.439	1.025	0.929
0.75	0.036	1.363	1.024	0.931
0.80	-0.547	1.308	1.035	0.936
0.85	-0.141	1.276	1.039	0.939
0.90	1.524	1.245	1.035	0.937
0.95	1.460	1.181	1.032	0.935

TABLE VI

TABLES OF $k_{\mathrm{eff}}R$ (TABLE A) AND G/Y_f (TABLE B) FOR CIRCULATORS USING DISK, TRIANGLE, REGULAR HEXAGON, NARROW AND BROAD-WALL COUPLED IRREGULAR HEXAGON ($\phi = 50$), RESPECTIVELY

First circulation solution using irregular hexagon broad wall coupled (Ø = 50)

PSI	0.200	0.300	0.400	0.500	0.600	0.611
K/U	Keff.R	Keff.R	Keff.R	Keff.R	Keff.R	Keff.R
0.10	2.004	2.002	2.000	1.997	1.994	1.912
0.15	1.997	1.993	1.988	1.981	1.974	1.857
0.20	1.986	1.979	1.970	1.958	1.945	1.802
0.25	1.973	1.961	1.945	1.926	1.906	1.747
0.30	1.958	1.939	1.914	1.885	1.857	1.691
0.35	1.942	1.912	1.875	1.835	1.799	1.633
0.40	1.928	1.880	1.827	1.776	1.731	1.573
0.45	1.926	1.844	1.771	1.707	1.656	1.510
0.50	2.273	1.805	1.707	1.630	1.572	1.444
0.55	2.251	1.764	1.632	1.544	1.481	1.374
0.60	2.185	1.715	1.549	1.453	1.386	1.299
0.65	2.089	2.011	1.459	1.358	1.289	1.220
0.70	1.977	1.884	1.371	1.258	1.185	1.133
0.75	1.857	1.752	1.243	1.133	1.072	1.038
0.80	1.687	1.524	1.107	1.014	0.957	0.932
0.85	1.493	1.289	0.969	0.884	0.828	0.811
0.90	1.246	0.902	0.789	0.726	0.679	0.665
0.95	0.914	0.670	0.561	0.511	0.478	0.472

Second circulation solution using irregular hexagon broad wall coupled (Ø = 50)

PSI	0.200	0.300	0.400	0.500	0.600	0.611
K/U	Gin/Yf	Gin/Yf	Gin/Yf	Gin/Yf	Gin/Yf	Gin/Yf
0.10	0.719	0.473	0.348	0.271	0.219	0.414
0.15	1.075	0.706	0.518	0.403	0.324	0.467
0.20	1.425	0.934	0.683	0.529	0.424	0.493
0.25	1.768	1.154	0.841	0.648	0.516	0.505
0.30	2.099	1.364	0.987	0.755	0.597	0.509
0.35	2.416	1.558	1.119	0.850	0.667	0.509
0.40	2.720	1.733	1.234	0.931	0.726	0.507
0.45	3.022	1.886	1.329	0.999	0.777	0.503
0.50	3.875	2.014	1.406	1.057	0.822	0.498
0.55	3.760	2.116	1.461	1.103	0.861	0.492
0.60	3.563	2.163	1.482	1.126	0.886	0.485
0.65	3.327	2.278	1.479	1.131	0.898	0.477
0.70	3.079	2.128	1.479	1.130	0.901	0.468
0.75	2.870	2.026	1.445	1.101	0.893	0.461
0.80	2.832	2.023	1.413	1.094	0.899	0.452
0.85	2.781	1.994	1.396	1.090	0.901	0.444
0.90	2.776	1.921	1.369	1.077	0.897	0.437
0.95	2.680	1.950	1.400	1.092	0.893	0.426

frequency independent within specific limits, B/Y_f retains a finite slope over the whole range of coupling angles.

The result for a regular hexagon is given in Fig. 12. The smaller values of coupling, while exhibiting a small value of B/Y_f over the frequency range, cannot be described by a constant conductance. Conversely, devices with larger coupling angles, which exhibit a frequency independent G/Y_f, have a finite susceptance slope parameter. It can be seen that the solution remains well behaved for narrower coupling angles than the disk and triangle. An upper bound is placed on the maximum coupling angle which may be used by the width of the side of the hexagon to which the coupling port is connected.

It is also possible to design circulators using irregular hexagonal resonators which may be coupled through both the broad and narrow walls. The width of the narrow wall restricts the range of possible coupling angles to lower upper limit than that for the broad wall. The results for a narrow wall coupled circulator are given in Fig. 13 and for a broad-wall coupled device in Fig. 14. It is observed that

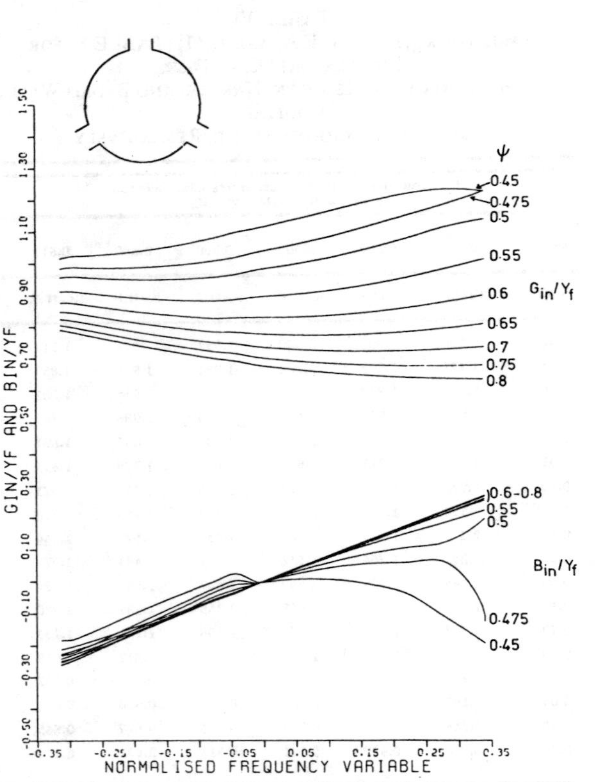

Fig. 10. Frequency response of planar circulator using disk resonator ($\kappa/\mu = 0.67$).

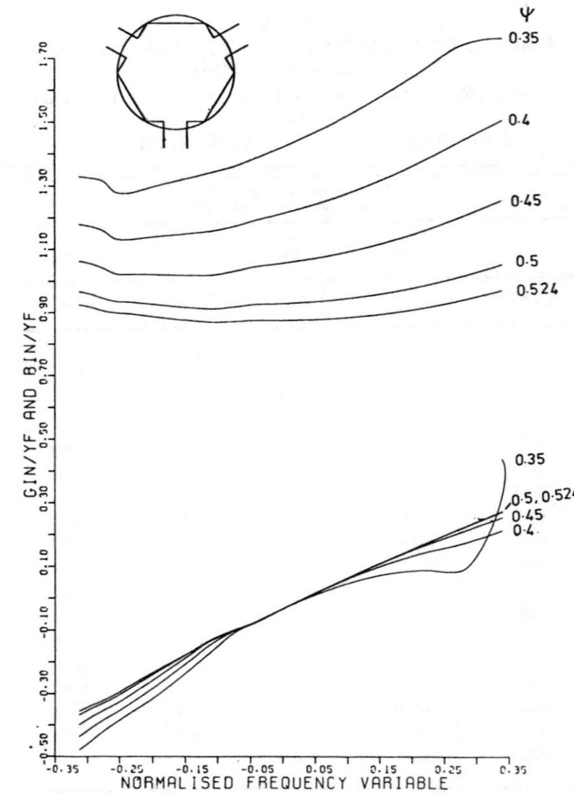

Fig. 12. Frequency response of planar circulator using regular hexagonal resonator ($\kappa/\mu = 0.67$).

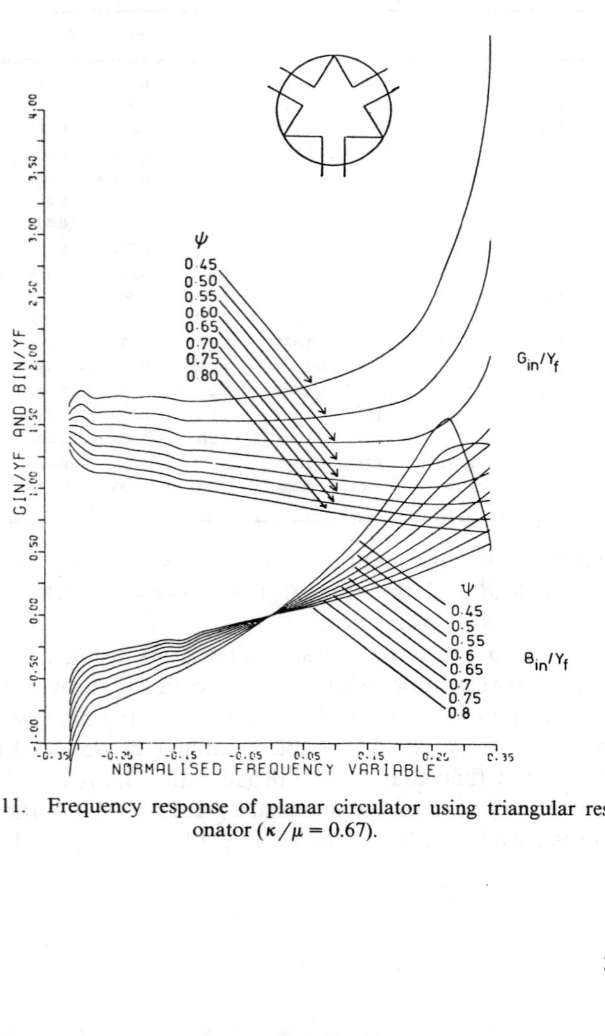

Fig. 11. Frequency response of planar circulator using triangular resonator ($\kappa/\mu = 0.67$).

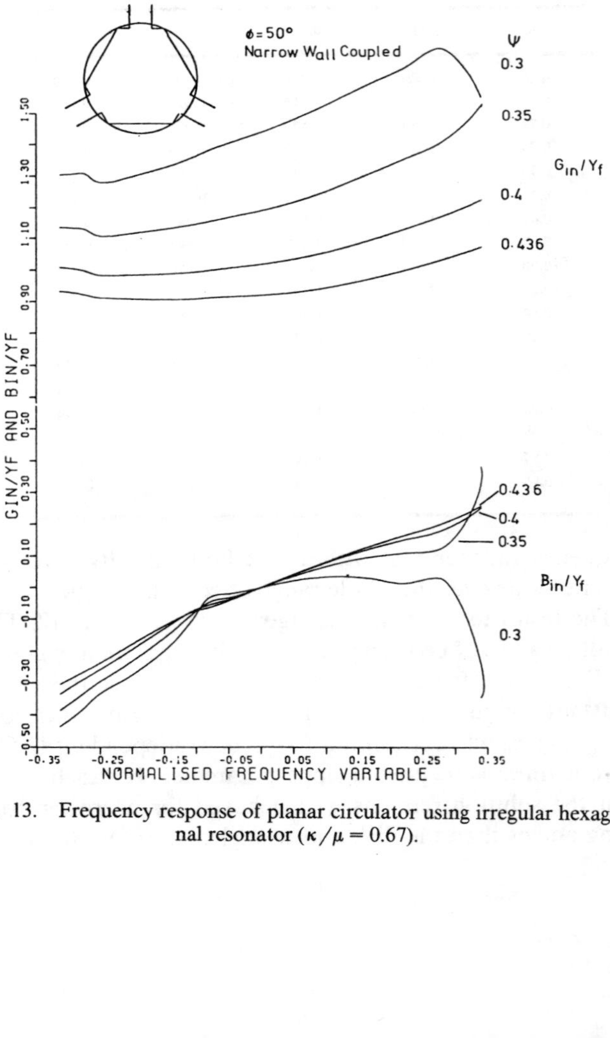

Fig. 13. Frequency response of planar circulator using irregular hexagonal resonator ($\kappa/\mu = 0.67$).

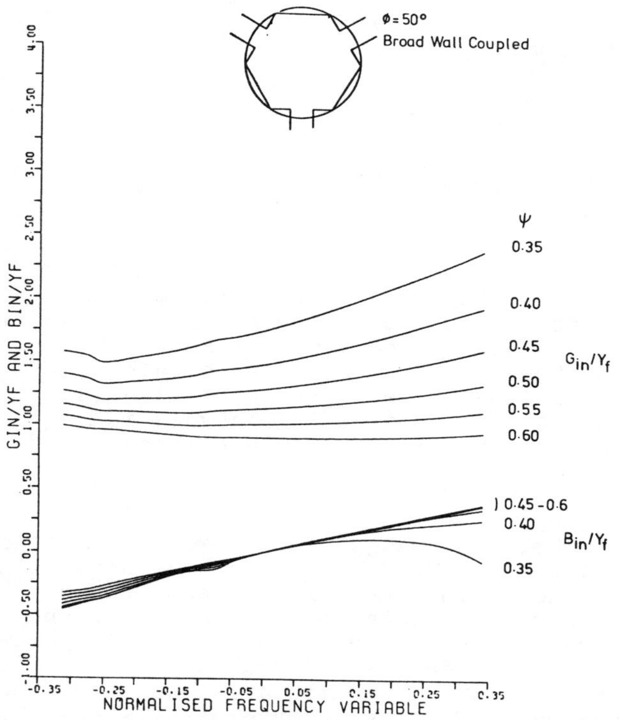

Fig. 14. Frequency response of planar circulator using irregular hexagonal resonator ($\kappa/\mu = 0.67$).

the gyrator conductance of disk, regular and irregular hexagons by and large exhibit the same values for the optimum coupling angle in the tracking region.

VII. Conclusions

This paper has described a finite element analysis of planar junction circulators. It has been used to calculate the circulation conditions of devices using disk, triangular, regular, and irregular hexagons over the range $0 < \kappa/\mu < 1$.

The method can also be used to plot the frequency response of junctions. In order to demonstrate this, the octave band defined by the so-called tracking interval, $0.5 < \kappa/\mu < 1.0$, has been studied for each of the resonators discussed previously.

Acknowledgment

The authors would like to thank the Procurement Executive, Ministry of Defense, (DCVD, UK) who sponsored this work.

References

[1] H. Bosma, "On stripline circulation at UHF," *IEEE Trans. Microwave Theory Tech.*, (1963 Symp. Issue), vol. MTT-12, pp. 61–72, Jan. 1964.

[2] H. Bosma, "Junction Circulators," *Advances in Microwaves*, vol. 6, pp. 215–239, 1971.

[3] J. B. Davies and P. Cohen, "Theoretical design of symmetrical junction circulator," *IEEE Trans. Microwave Theory Tech.*, vol. MTT-11, pp. 506–512, Nov. 1963.

[4] C. E. Fay and R. L. Comstock, "Operation of the ferrite junction circulator," *IEEE Trans. Microwave Theory Tech.*, vol. MTT-13, pp. 15–27, Jan. 1965.

[5] Y. S. Wu and F. J. Rosenbaum, "Wideband operation of microstrip circulators," *IEEE Trans. Microwave Theory Tech.*, vol. MTT-22, pp. 849–856, Oct. 1974.

[6] J. Helszajn, D. S. James, and W. T. Nisbet, "Circulators using planar triangular resonators," *IEEE Trans. Microwave Theory Tech.*, vol. MTT-27, pp. 188–193, Feb. 1979.

[7] J. Helszajn and W. T. Nisbet, "Circulators using planar wye resonators," *IEEE Trans. Microwave Theory Tech.*, vol. MTT-29, pp. 689–699, July 1981.

[8] T. Miyoshi, S. Yamaguchi, and S. Goto, "Ferrite planar circuits in microwave integrated circuits," *IEEE Trans. Microwave Theory Tech.*, vol. MTT-25, pp. 593–600, July 1977.

[9] T. Miyoshi and S. Miyaushi, "The design of planar circulators for wideband operation," *IEEE Trans. Microwave Theory Tech.*, vol. MTT-28, pp. 210–215, Mar. 1980.

[10] P. Silvester, "Finite element solution of the homogeneous waveguide problem," *Alta Frequenza*, vol. 38, pp. 593–600, May 1969.

[11] P. Silvester, "High order polynomial triangular finite elements for potential problem," *Int. J. Eng. Sci.*, vol. 7, pp. 849–861, 1969.

[12] P. Silvester, "A general high-order finite element waveguide analysis program," *IEEE Trans. Microwave Theory Tech.*, vol. MTT-17, pp. 204–210, Apr. 1969.

[13] Z. J. Csendes and P. Silvester, "Numerical solution of dielectric-loaded waveguides—Part 1: "Finite element analysis," *IEEE Trans. Microwave Theory Tech.*, vol. MTT-18, pp. 1124–1131, Dec. 1970.

[14] Z. J. Csendes and P. Silvester, "Numerical solution of dielectric-loaded waveguides—Part 2: "Modal approximation technique," *IEEE Trans. Microwave Theory Tech.*, vol. MTT-19, pp. 504–509, June 1971.

[15] J. K. Richardson, "An approximate method of calculating Z_0 of a symmetric stripline," *IEEE Trans. Microwave Theory Tech.*, vol. MTT-15, pp. 130–131, 1967.

[16] J. Helszajn, "Operation of the tracking circulator," *IEEE Trans. Microwave Theory Tech.*, vol. MTT-29, pp. 700–707, July 1981.

[17] H. A. Wheeler, "Transmission line properties of parallel strips separated by a dielectric sheet," *IEEE Trans. Microwave Theory Tech.*, vol. MTT-13, pp. 172–185 Mar. 1965.

[18] W. T. Nisbet and J. Helszajn, "Mode charts for microstrip resonators on dielectric and magnetic substrates using a transverse resonance method," *Microwave, Opt., Acoust.*, vol. 3, no. 2, pp. 69–77, Mar. 1979.

NONRECIPROCAL TWO-CHANNEL MICROWAVE DIVIDERS AND ADDERS

M. V. Vamberskii,* V. P. Usachov, and S. A. Shelukhin

Izvestiya VUZ. Radioelektronika,
Vol. 27, No. 10, pp. 16-22, 1984

UDC 621.372.832.8

Two-channel nonreciprocal divider-adders are investigated. These are devices which ensure that the outputs and the input are decoupled in the power-division mode, and that the output and inputs are decoupled in the addition mode. Normalized scattering and impedance matrices of the divider-adders are obtained, and the effect of the amplitude and phase relations between the input signals and the addition efficiency is investigated. An electrodynamic analysis is given of broadband nonreciprocal divider-adders, and expressions are obtained which enable these devices to be designed.

A serious drawback of ring connections, stub bridges and different kinds of directional couplers and other devices, which divide and add power, is the lack of decoupling between the output and input arms for arbitrary loads. In a number of cases, for example, in channels with active elements, this may lead to unstable oscillation of the apparatus. This often makes it necessary to use additional decoupling devices in power-division systems. There is therefore a need to design a device which acts both as a power distributor and as a decoupling device. Such devices are particularly necessary when designing solid-state amplifiers in the shortwave part of the microwave band, where the power of the active elements is limited, and their tendency to self-excitation is high [1].

In this paper we consider nonreciprocal microwave dividers and adders, and we generalize the individual problems involved in investigating them, particularly as described in [1,2]. As a result of a matrix analysis, carried out assuming that there are no dissipative losses, we formulate the requirements imposed on the elements of the scattering and impedance matrices of the nonreciprocal divider-adder devices. These requirements are independent of the specific construcional form of the devices, and in this sense are general. To obtain analytical relations, which enable specific types of devices to be designed (in microstrip lines), we then solve the electrodynamic boundary value problem, as a result of which the elements of the matrices are expressed in terms of the constructional parameters of the device, and the general requirements formulated above are imposed on the expressions obtained.

Matrix analysis of the nonreciprocal divider-adder. It is obvious that the minimum number of arms of a two-channel nonreciprocal divider-adder should be four. In the case of nonreciprocal division, one arm is the input arm, and two are output arms, and to ensure decoupling between the output and input arms it is necessary for the microwaves reflected from the useful loads in the output arms to be absorbed in a ballast load, connected in the fourth arm. Similar considerations also apply for nonreciprocal power addition. We will therefore consider the nonreciprocal eight-terminal network shown in Fig. 1. Without loss of generality, we will assume, for the case of division, that arm 1 is the input arm, and arms 2 and 4 are the output arms: for addition, arms 2 and 4 are the input arms and arm 3 is the output arm.

In the division mode, the power applied to arm 1 (P_{1in}), must be divided between arms 2 and 4 in the proportion d:1, i.e.,

$$P_{2out} = dP_{1in}/(d+1),\ P_{4out} = P_{1in}/(d+1). \qquad (1)$$

In the addition mode, the output in arm 3 (P_{3out}) must be equal to the sum of the powers of the electromagnetic waves applied to the arms 2 and 4 in the same proportion d:1, i.e.,

*Deceased.

Reprinted with permission from *Radioelectronics and Communication Systems*, vol. 27, no. 10, pp. 14–19, 1984.

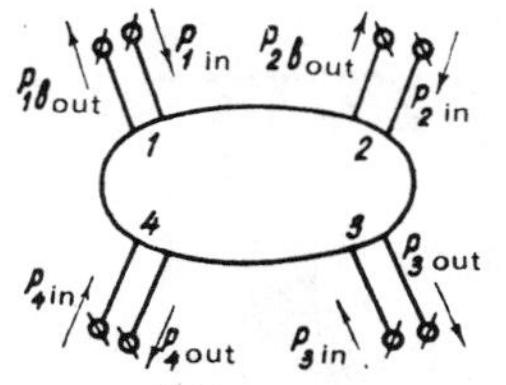

Fig. 1

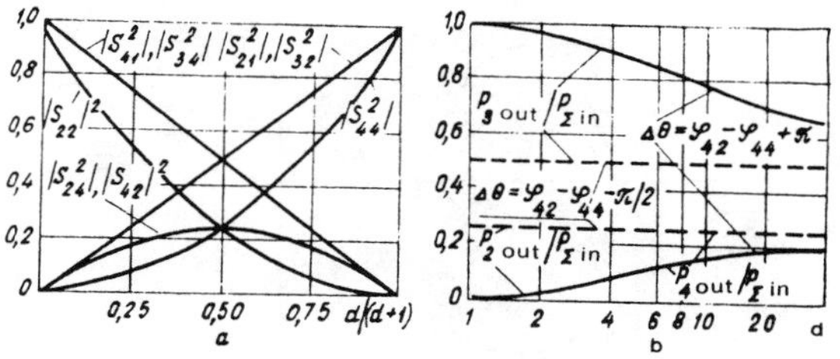

Fig. 2.

$$P_{3\text{out}} = P_{2\text{in}} + P_{4\text{in}} \quad \text{when } P_{2\text{in}}/P_{4\text{in}} = d. \tag{2}$$

Finally, the conditions for the input and output arms to be decoupled is that in the division mode, the power transfer factors from arms 2 and 4 into arm 1 must be zero, while in the addition mode, the transfer factors from arm 3 into arms 2 and 4 must be zero, i.e.,

$$S_{12} = S_{14} = S_{23} = S_{43} = 0. \tag{3}$$

The normalized scattering matrix of a loss-free passive device must be unitary

$$[\tilde{S}^*][S] = [\tilde{S}][S^*] = [1]. \tag{4}$$

Solving (4), taking into account conditions (1), (2), and (3), we obtain the normalized scattering matrix for the nonrecriprocal divider-adder in the form

$$[S] = \begin{bmatrix} 0 & 0 & e^{j\varphi_{13}} & 0 \\ d^{\frac{1}{2}}(d+1)^{-\frac{1}{2}} e^{j\varphi_{21}} & -(d+1)^{-1} e^{j(\varphi_{42}+\varphi_{21}-\varphi_{41})} & 0 & -d^{\frac{1}{2}}(d+1)^{-1} e^{j(\varphi_{44}+\varphi_{21}-\varphi_{41})} \\ 0 & -d^{\frac{1}{2}}(d+1)^{-\frac{1}{2}} e^{j(\varphi_{34}+\varphi_{42}-\varphi_{44})} & 0 & (d+1)^{-\frac{1}{2}} e^{j\varphi_{34}} \\ (d+1)^{-\frac{1}{2}} e^{j\varphi_{41}} & d^{\frac{1}{2}}(d+1)^{-1} e^{j\varphi_{42}} & 0 & d(d+1)^{-1} e^{j\varphi_{44}} \end{bmatrix}. \tag{5}$$

where $\varphi_{21}, \varphi_{13}, \varphi_{34}, \varphi_{41}, \varphi_{42}, \varphi_{44}$ are arbitrary phase angles, the value of which has no effect on the division, addition, and decoupling, and is determined by the specific construction of the device.

In Fig. 2a we show curves of the moduli of the elements of the scattering matrix (5) as a function of the division factor d. As follows from the graph, in a 3-dB nonreciprocal divider-adder (with d = 1), the decoupling that can be achieved in principle between the output arms 2 and 4 in the division mode (or the input arms in the addition mode) ignoring dissipative losses, is 6.02 dB. Then the modulus of the reflection coefficient from the nonreciprocal divider-adder when arm 2 (or 4) is fed, and when a matched load is connected to arm 4 (or 2) is not zero (the graphs of $|S_{22}|^2$ and $|S_{44}|^2$ in Fig. 2), which is a bacic property of the device. It can also be seen from Fig. 2 that when d = 0 or d = ∞, the nonreciprocal divider-adder becomes a three-arm circulator with a circulation direction 1→4→3→1 or 1→2→3→1 respectively.

When a matched load is connected to arm 3, the nonreciprocal divider-adder operates as a nonreciprocal divider (ND), whose scattering matrix can be written as follows:

$$[S]_{ND} = \begin{vmatrix} 0 & 0 & 0 \\ d^{\frac{1}{2}}(d+1)^{-\frac{1}{2}} e^{j\varphi_{21}} & -(d+1)^{-1} e^{j(\varphi_{42}+\varphi_{21}-\varphi_{41})} & -d^{\frac{1}{2}}(d+1)^{-1} e^{j(\varphi_{44}+\varphi_{21}-\varphi_{41})} \\ (d+1)^{-\frac{1}{2}} e^{j\varphi_{41}} & d^{\frac{1}{2}}(d+1)^{-1} e^{j\varphi_{42}} & d(d+1)^{-1} e^{j\varphi_{44}} \end{vmatrix} \tag{6}$$

If the load is connected to arm 1, we obtain a nonreciprocal adder (NA), whose scattering matrix has the form

$$[S]_{NA} = \begin{bmatrix} -(d+1)^{-1} e^{j(\varphi_{42}+\varphi_{21}-\varphi_{41})} & 0 & -d^{\frac{1}{2}}(d+1)^{-1} e^{j(\varphi_{44}+\varphi_{21}-\varphi_{41})} \\ -d^{\frac{1}{2}}(d+1)^{-\frac{1}{2}} e^{j(\varphi_{34}+\varphi_{42}-\varphi_{44})} & 0 & (d+1)^{-\frac{1}{2}} e^{j\varphi_{34}} \\ d^{\frac{1}{2}}(d+1)^{-1} e^{j\varphi_{42}} & 0 & d(d+1)^{-1} e^{j\varphi_{44}} \end{bmatrix} \tag{7}$$

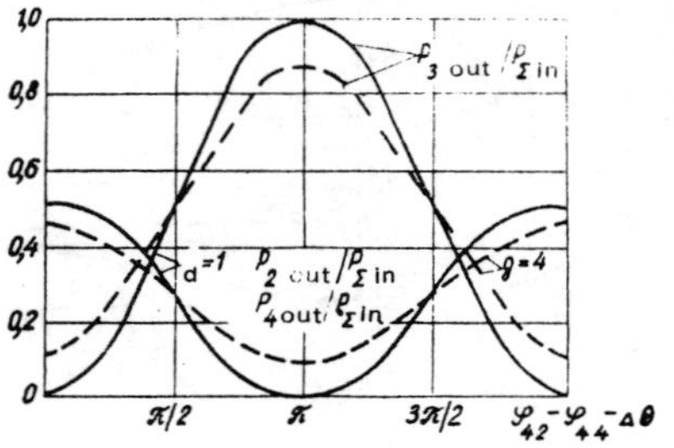

Fig. 3

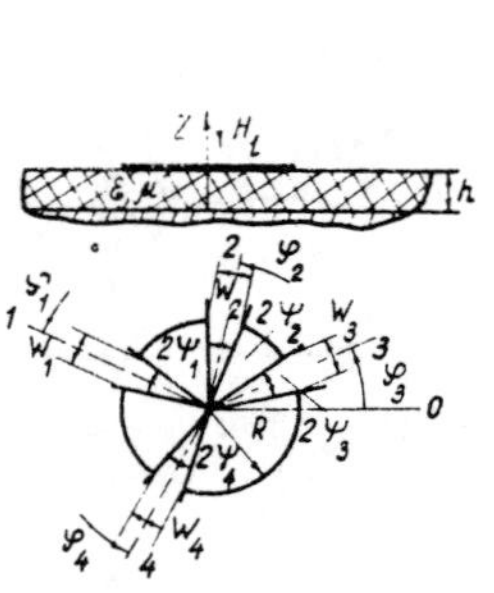

Fig. 4

When operating in the addition mode, the amplitude and phase relations between the input signals have a considerable effect on the device characteristics. Consider the operation of an adder, intended for adding equal powers (d = 1) when input arms 2 and 4 are fed with normalized waves of power a_4 and $a_2=\sqrt{g}a_4e^{-j\Delta\theta}$. The efficiency E of the summation will be represented by the quantity.

$$E = P_{3out}/(P_{2in} + P_{4in}) = |\sqrt{g}e^{j(\varphi_{42}-\varphi_{44}-\Delta\theta)} - 1|^2/(2g+2). \quad (8)$$

In Figs. 2b and 3 we show graphs illustrating the effect of the power ratio g of the input signals and their phase difference $\Delta\theta$ on the characteristics of the nonreciprocal adder. As can be seen from the graph, for the correct phase ratio between the input signals, namely $\Delta\theta=\varphi_{42}-\varphi_{44}\pm\pi$, the difference between the input powers has a comparatively small effect on the operation of the device. Even if the power of one of the signals is four times greater than the power of the other, the efficiency of the adder is 0.9. A change in the phase difference $\Delta\theta$ (Fig.3) has a much greater effect. In this connection we can conclude that when designing devices with nonreciprocal divider-adders it is necessary to pay particular attention to ensure that the optimum phase relations are obtained. To ensure that the efficiency of the adder is not less than 0.9, it is necessary, for g = 1, to satisfy the condition $1{,}2\,\pi \geqslant \Delta\theta - \varphi_{42} + \varphi_{44} \geqslant 0{,}8\,\pi$.

To simplify the further analysis, it is convenient to obtain the normalized impedance matrix of the two-channel nonreciprocal divider-adder. Bearing in mind the fact that for a loss-free device $[Z]+[\tilde{Z}^*]=0$ we obtain

$$[Z] = ([1]+[S])([1]+[S])^{-1} = \begin{bmatrix} Z_{11} & Z_{12} & Z_{13} & Z_{14} \\ -Z_{12}^* & Z_{22} & -Z_{12}e^{j\varphi_{13}} & Z_{24} \\ -Z_{13}^* & Z_{12}e^{-j\varphi_{13}} & Z_{11} & Z_{14}e^{-j\varphi_{13}} \\ -Z_{14}^* & -Z_{24}^* & -Z_{14}^*e^{j\varphi_{13}} & Z_{44} \end{bmatrix}, \quad (9)$$

where

$$Z_{11} = j\,\mathrm{ctg}\left[\frac{\theta}{2} + \mathrm{artg}\frac{d\sin\varphi_{44} - \sin(\varphi_{21}-\varphi_{41}+\varphi_{42})}{d+1+\cos(\varphi_{21}-\varphi_{41}+\varphi_{42}) - d\cos\varphi_{44}}\right],$$

$$Z_{22} = j\,\mathrm{ctg}\left[\frac{\theta}{2} + \mathrm{arctg}\frac{d\sin\varphi_{44} + \sin(\varphi_{41}+\varphi_{13}+\varphi_{34})}{d+1-d\cos\varphi_{44} - \cos(\varphi_{41}+\varphi_{13}+\varphi_{34})}\right],$$

$$Z_{44} = j\,\mathrm{ctg}\left[\frac{\theta}{2} + \mathrm{arctg}\frac{\sin(\varphi_{41}-\varphi_{21}-\varphi_{42}) - d\sin(\theta-\varphi_{44})}{d+1+\cos(\varphi_{41}-\varphi_{21}-\varphi_{42}) + d\cos(\theta-\varphi_{44})}\right],$$

$$Z_{12} = 2d^{\frac{1}{2}}(d+1)^{-\frac{1}{2}}(1-e^{-j\varphi_{44}})e^{j(\theta-\varphi_{21})}/(M-e^{j\theta}M^*),$$

$$Z_{14} = 2(d+1)^{-\frac{1}{2}}[1+e^{j(\varphi_{44}+\varphi_{21}-\varphi_{41})}]e^{j(\varphi_{13}+\varphi_{34})}/(M-e^{j\theta}M^*),$$

$$Z_{13} = 2e^{j\varphi_{13}}/(1-e^{j\theta}M^*/M),$$

$$Z_{24} = -2d^{\frac{1}{2}}(d+1)^{-1}[1-e^{j(\varphi_{41}+\varphi_{34}+\varphi_{13}-\varphi_{44})}]e^{j(\varphi_{44}+\varphi_{21}-\varphi_{41})}/(M-e^{j\theta}M^*),$$

$$\theta = \varphi_{21}+\varphi_{13}+\varphi_{42}+\varphi_{34},$$

$$M = 1+[e^{j(\varphi_{21}-\varphi_{41}+\varphi_{42})} - de^{j\varphi_{44}}]/(d+1).$$

Electrodynamic analysis of a microstrip nonreciprocal divider-adder. A microstrip two-channel nonreciprocal divider-adder is a ferrite disc resonator of radius R with four microstrip lines connected to it as shown in Fig. 4. We will assume that only T-modes propagate in the conducting lines, the substrate thickness h is small, and the fields in the disc do not vary along the x axis. Suppose the boundary conditions for the azimuthal component $H_q(R, \varphi)$ of the microwave magnetic field has the following form:

$$H_\varphi(R,\varphi) = \begin{cases} H_{\varphi_1} & \text{for } \varphi_1 - \psi_1 \leqslant \varphi \leqslant \varphi_1 + \psi_1, \\ H_{\varphi_2} & \text{»} \quad \varphi_2 - \psi_2 \leqslant \varphi \leqslant \varphi_2 + \psi_2, \\ H_{\varphi_3} & \text{»} \quad \varphi_3 - \psi_3 \leqslant \varphi \leqslant \varphi_3 + \psi_3, \\ H_{\varphi_4} & \text{»} \quad \varphi_4 - \psi_4 \leqslant \varphi \leqslant \varphi_4 + \psi_4. \\ 0 & \text{for all remaining values of } \varphi. \end{cases} \tag{10}$$

The soluton of Maxwell's equations, taking (10) into account by the method described in [3,4], enables an expression to be obtained for the electric field $E_z(R, \varphi)$ at the edge of the disc

$$E_z(R,\varphi) = j\frac{\omega\mu_0\mu_\perp}{k_\perp\pi}\sum_{m=1}^{4} H_{\varphi m}\sum_{n=-\infty}^{\infty} F_n \frac{\sin(n\psi_m)}{n} e^{jn(\varphi-\varphi_m)}, \tag{11}$$

where

$$F_n = \left[\frac{J'_n(k_\perp R)}{J_n(k_\perp R)} - \frac{n}{k_\perp R}\frac{k}{\mu}\right]^{-1}, \quad k_\perp = \omega\sqrt{\mu_0\varepsilon_0\mu_\perp\varepsilon},$$

$\mu_\perp = (\mu^2 - k^2)/\mu$, $\mu_0 = 4\pi\cdot 10^{-7}$ H/m, $\varepsilon_0 = (1/36\pi)\,10^{-9}$ F/m, ω is the angular frequency of the electromagnetic field, ε is the permittivity of the ferrite, k and μ are the components of the magnetic permeability tensor of the ferrite, and $J_n(k_\perp R)$, $J'_n(k_\perp R)$ are Bessel functions of the first kind and its derivative with respect to the argument.

We will write an expression for the flow of complex power in the terminal plane of the i-th arm: $P_{\text{compl}\,i} = 0{,}5 \int [\bar{E}\bar{H}^*]\,d\bar{S}$, where S_i is the surface of the window of the disc resonator, connected to the i-th arm. In the case considered

$$P_{\text{compl}\,i} = j\frac{\omega\mu_0\mu_\perp hR}{k_\perp\pi} H^*_{\varphi_i}\sum_{m=1}^{4} H_{\varphi_m}\sum_{n=-\infty}^{\infty} F_n\frac{\sin(n\psi_i)\sin(n\psi_m)}{n^2} e^{jn(\varphi_i-\varphi_m)}. \tag{12}$$

On the other hand, the complex power in the i-th arm of the multiterminal network is related to the voltage and the currents in this arm by the equation $P_{\text{compl}\,i} = 0{,}5\,U_i I_i^*$, or, taking the z-parameters into account,

$$P_{\text{compl}\,i} = 0{,}5 I_i^* \sum_{m=1}^{4} z_{im} I_m. \tag{13}$$

We will equate expressions (12) and (13) first substituting the relation $I_i = H_{\varphi_i} 2R\sin\psi_i$ into (13): we obtain

$$j\frac{\omega\mu_0\mu_\perp hRH^*_{\varphi_i}}{k_\perp\pi}\sum_{m=1}^{4} H_{\varphi m}\sum_{n=-\infty}^{\infty} F_n\frac{\sin(n\psi_i)\sin(n\psi_m)}{n^2 e^{-j(\varphi_i-\varphi_m)}} = H^*_{\varphi_i} 2R^2\sin\psi_i \sum_{m=1}^{4} z_{im} H_{\varphi_m}\sin\psi_m. \tag{14}$$

Since Eq. (14) must be satisfied identically for any values of $H_{\phi m}$, we obtain from this an expression for the z-parameters of the connection, which, after normalization, has the form

$$Z_{il} = j\sqrt{\mu_\perp}\frac{\sqrt{w_i w_l}}{2\pi R}\sum_{n=-\infty}^{\infty} F_n\frac{\sin(n\psi_i)\sin(n\psi_l)}{n^2\sin\psi_i\sin\psi_l} e^{jn(\varphi_i-\varphi_l)}. \tag{15}$$

Theoretical relations. In order that the four-arm connection shown in Fig. 4 should be a nonreciprocal divider-adder it is necessary to choose its constructional parameters (the geometrical dimensions, and the electrical characteristics of the ferrite) in such a way that the values of Z_{il}, calculated from (15), satisfy the previously formulated conditions (9). The solution of the system of 16 linearly independent equations with real variables obtained, enables us to obtain 16 unknowns, six phase angles, four values of the angular width of each arm, three angles characterizing the arrangement of the arms, the radius of the disc resonator, and two components of the magnetic permeability tensor An accurate solution of the system is difficult to obtain, since (15) represents z_{11} in the form of a sign-varying series with an infinite number of terms, which does not reduce to a rational function. We must therefore consider an approximate solution.

As preliminary experimental investigations of a nonreciprocal divider-adder have shown, the values of the disc radius and the magnetizing field are found in the resonance region of the ±1-st spacial harmonic, which occurs for values of $k_{\perp}R=1{,}84$. In this case, we can neglect harmonics with number n not equal to ±1, and (15) simplifies to

$$Z_{il}=[(k_{\perp}\sqrt{\mu_{\perp}}/\pi)\sin(\varphi_i-\varphi_l)]/(k/\mu). \tag{16}$$

Taking this approximation into account, the set of equations connecting the constructional parameters of the connection and the z-parameters of the nonreciprocal divider-adder can be converted to the following form:

$$\left.\begin{aligned}
&\frac{2}{3}10^{-8}f\sqrt{\varepsilon}\sqrt{w_1w_3}\,\frac{\mu_{\perp}}{k/\mu}\sin(\varphi_1-\varphi_3)=\pm 1\\
&\frac{2}{3}10^{-8}f\sqrt{\varepsilon}\sqrt{w_1w_2}\,\frac{\mu_{\perp}}{k/\mu}\sin(\varphi_1-\varphi_2)=\pm\sqrt{\frac{d}{d+1}}\\
&\frac{2}{3}10^{-8}f\sqrt{\varepsilon}\sqrt{w_1w_4}\,\frac{\mu_{\perp}}{k/\mu}\sin(\varphi_1-\varphi_2)=\pm\frac{1}{\sqrt{d+1}}\\
&\varphi_4=\varphi_2+\pi
\end{aligned}\right\}. \tag{17}$$

The two solutions of this system correspond to the two possible forms of the nonreciprocal divider-adder: scheme *a* in which the arms are arranged clockwise (1, 2, 3, 4), and scheme b in which the order of the arms is 1, 4, 2, 3. The values of the phase angles of the normalized scattering matrix for scheme *a*: $\varphi_{13}=\varphi_{21}=\varphi_{44}=\varphi_{42}=\pi$ $\varphi_{41}=\varphi_{34}=0$; and for scheme b: $\varphi_{13}=\varphi_{21}=\varphi_{34}=0$; $\varphi_{44}=\varphi_{42}=\varphi_{41}=\pi$.

As a result of the approximate solution of system (17) we obtain the following equations which enable the construction of the nonreciprocal divider-adder to be calculated:

$$\left.\begin{aligned}
&w_2=w_4d\\
&\varphi_4=\varphi_2+\pi\\
&\sin\varphi_1=\frac{\pm 15\cdot 10^7k/\mu}{f\sqrt{\varepsilon}\sqrt{w_1w_3}\mu_{\perp}}\\
&\sin\varphi_2=\mp\sqrt{\frac{d}{d+1}}\,\frac{15\cdot 10^7k/\mu}{f\sqrt{\varepsilon}\sqrt{w_2w_3}\mu_{\perp}}\\
&\sin(\varphi_1-\varphi_2)=\mp\sqrt{\frac{d}{d+1}}\,\frac{15\cdot 10^7k/\mu}{f\sqrt{\varepsilon}\sqrt{w_1w_2}\mu_{\perp}}\\
&k_{\perp}R=1{,}84
\end{aligned}\right\}. \tag{18}$$

Here the angle ϕ_3 is measured from the origin $\phi_3 = 0$.

A nonreciprocal ferrite two-channel powered divider-adder, designed using the approximate formulas (18), should have the following normalized scattering matrix:

$$[S]=\begin{bmatrix}
0 & 0 & \pm 1 & 0\\
\pm d^{\frac{1}{2}}(d+1)^{-\frac{1}{2}} & -(d+1)^{-1} & 0 & -d^{\frac{1}{2}}(d+1)^{-1}\\
0 & -d^{\frac{1}{2}}(d+1)^{-\frac{1}{2}} & 0 & (d+1)^{-\frac{1}{2}}\\
\mp(d+1)^{-\frac{1}{2}} & -d^{\frac{1}{2}}(d+1)^{-1} & 0 & -d(d+1)^{-1}
\end{bmatrix}.$$

Relations (18) form the basis of an engineering procedure for designing nonreciprocal divider-adders, the practicability of which was confirmed in practice.

REFERENCES

1. M. V. Vamberskii, S. A. Shelukhin, and V. P. Usachov, "Integrated centimeter band transistor amplifiers," Trudy MVTU, no. 305, pp. 149-154, 1979.

2. V. P. Usachov, B. A. Gapeev, and D. I. Ogloblin, "Nonreciprocal ferrite dividers-adders," Trudy MVTU, no. 397, pp. 55-62, 1983.
3. M. V. Vamberskii and V. I. Kazantsev, "Practical design of strip microcirculators," Radiotekhnika, no. 11, pp. 24-30, 1965.
4. T. V. Shelukhin, M. V. Vamberskii, and V. P. Abramov, "Multiarm integrated circulators," in: Physics of Magnetic Phenomena [in Ukrainian], Turkmensk. Gos. Univ., Ashkhabad, 1973.

Paper 7.8

Nonreciprocal devices using ferrite ring resonators

V.A. Dmitriyev, DrSc
L.E. Davis, PhD, CEng, FIEE

Indexing terms: Ferrite discs, Nonreciprocal devices, Three-port/Four-port components

Abstract: Novel nonreciprocal components based on hollow ferrite discs, i.e. ferrite ring resonators, are discussed. Some three-port and four-port components are analysed, and some design guidelines are presented. It is shown that a ring structure requires a smaller diameter and ferrite thickness than a disc structure for the same frequency. A ring also provides another variable in design and permits ports to be connected to the inside as well as the outside. Novel components and increased component density may therefore be possible at the expense of bandwidth.

1 Introduction

Three-port circulators have been extensively discussed in the literature, and components with good performance in stripline, waveguide and microstrip are widely available commercially. Papers by Bosma [1], Fay and Comstock [2] established the principles using an approach based on disc resonators. Triangular, hexagonal and other shapes have also been used successfully [3]. Wu and Rosenbaum [4] described a 'tracking' design procedure for broad bandwidth circulators which has been extended by others [5–8]. Nonsymmetrical circulators and nonreciprocal dividers and combiners using ferrite discs have been investigated by Sheluchin *et al.* [15, 16].

This paper discusses the properties of novel nonreciprocal components based on hollow ferrite discs, i.e. ferrite ring resonators. The behaviour of a ferrite ring has been investigated before [9–12], but not in the context described below, and components based on ferrite ring resonators have been investigated by Dmitriyev *et al.* [13–16] and also by Nagao who has discussed double-frequency circulation with, and without, a central conducting pin [17, 18]. It will be shown here that ring structures require smaller dimensions than disc structures, and it is proposed that output ports can be placed inside the ring as well as outside. A 'hole' in the centre of the disc provides additional variables, flexibility, and opportunities for novel multiport components. Active components or loads could perhaps be put inside, as well as outside, the ring thereby allowing increased component density. The purpose of this paper is to outline the analysis of some three-port and four-port components and to give some design guidelines.

Paper 8769H (E12), first received 16th July 1990 and in final revised form 16th January 1992

V.A. Dmitriyev is with the Moscow State Technical University named after Bauman (MSTU), Moscow 107005, Russia

L.E. Davis is with the Department of Electrical Engineering and Electronics, UMIST, PO Box 88, Manchester M60 1QD, United Kingdom

2 Junction configurations

Fig. 1 shows the ring structure to be discussed with inner and outer radii of r and R, respectively, and thickness h.

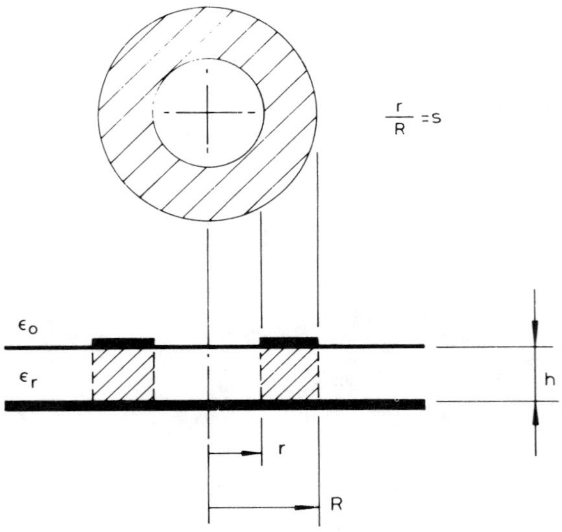

Fig. 1 *Microstrip ring resonator*
Idealised ferrite volume with magnetic walls is shown shaded

We define the ratio of the radii $r/R = s$, and, as is customary for simplicity, it will be assumed that the inner and outer curved surfaces of the ferrite ring are magnetic walls. If R is fixed, as the radius of the hole is increased, s increases, and, as the value of s approaches unity, the structure approaches that of a thin ring resonator. Near the centre of a disc the elliptically polarised RF magnetic field is coupled to the ferrite medium by the applied static field and the resulting angular field displacement gives rise to circulation [1, 2]. Therefore, the removal of the central region of the disc to create a ring with an inner magnetic wall, in addition to the outer magnetic wall, can be expected to reduce the bandwidth over which circulation occurs. However, the presence of a hole in the centre of a circulator disc provides not only an additional variable (parameters) but also, perhaps, an opportunity for ports to be located externally or internally on the ring, or to have mixed locations. Other angular locations also

Reprinted with permission from *IEE Proc.*, Part H, vol. 139, no. 3, V. A. Dmitriyev and L. E. Davis, "Nonreciprocal Devices Using Ferrite Ring Resonators," pp. 257–263, June 1992, IEE.

provide interesting properties. Some multiport combinations are indicated in Fig. 2. The external port combinations shown in Figs. 2*a–c* can be used on discs as well as rings, whereas Figs. 2*d–g* shows internal as well as

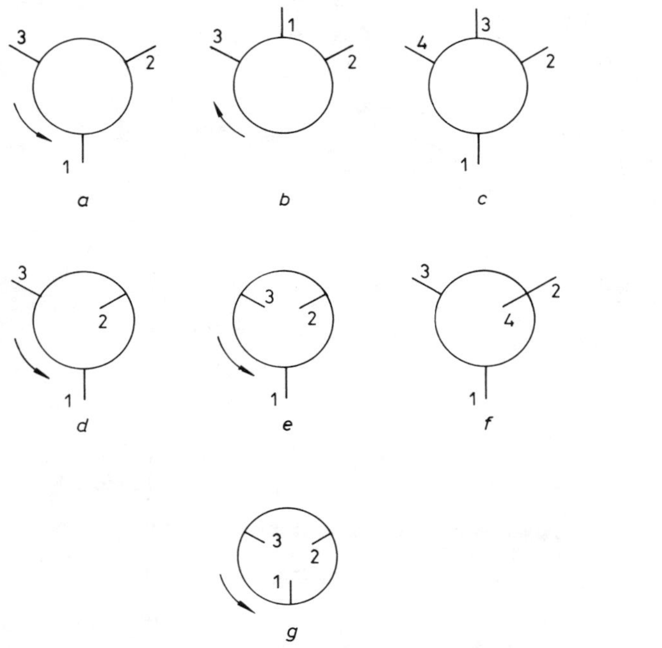

Fig. 2 *Three-port circulator and four-port power-divider junctions with ferrite discs or rings*

external ports and can only be realised with rings. Fig 2*b* suggests a compact circulator with all ports on the same side of the function, and Fig. 2*d* may be a compact iso-circulator if a matched load can be attached at the internal port 2. Also, Fig. 2*d* could be used with an amplifier or diode at port 2 to provide a phase shifter.

Figs. 2*c* and 2*f* are nonreciprocal four-port power-divider/combiners and a phenomenological description of the behaviour of these two structures can be given in terms of the dominant mode TM_{110} configurations [1, 2] as follows. Fig. 3*a* shows the well known idealised field pattern in the disc for perfect circulation between ports 1 and 2, and the zero of electric field at port 3 causes that port to be isolated. The direction of the angular displacement of the pattern with respect to the input port depends upon the direction of the static field and whether the junction is biased above or below gyromagnetic resonance. If the field pattern in Fig. 3*a* is superimposed directly onto the 4-port structure shown in Fig. 2*c*, corresponding to an input at port 1, it is clear that the signal will emerge from ports 2 and 3 with zero signal at port 4 (Fig. 4*a*). If the mode pattern is rotated anticlockwise through 120° this corresponds to an input at port 2, and the signal will emerge from port 4, with ports 1 and 3 isolated, as shown in Fig. 4*b*. If this procedure is repeated by rotating the field pattern again through 60°, and then a further 60°, corresponding to inputs at ports 3 and 4 respectively, the behaviour shown in Figs. 4*c* and 4*d* is obtained. We see that in Figs. 4*a,c,d* the junction has power-divider action. The behaviour in Fig. 4 can be obtained with a disc or with a ring resonator. The behaviour shown in Fig. 5 is associated with an internal port and therefore can only be obtained with a ring. In Fig. 5*a* the input at port 1 emerges at ports 2 and 4; in Fig. 5*b* the input at port 2 emerges from ports 3 and 4; in Fig. 5*c* the input at port 3 emerges from port 1; in Fig. 5*d* the input at port 4 emerges from ports 2 and 3. Again, Figs. 5*a,b,d* exhibit power-divider action, and power combiner action can be envisaged.

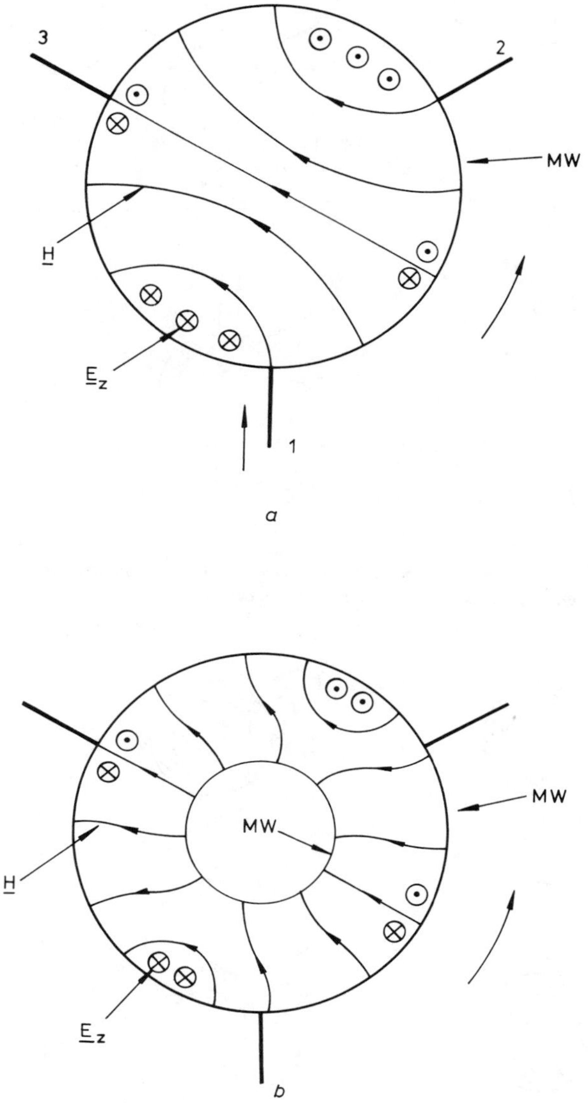

Fig. 3 *Idealised TM_{110}-mode configuration in magnetic wall resonator showing perfect circulation*
a Disc
b Ring

Using an inner port of a ring resonator, double-sided microstrip circuits can be considered. For example, two microstrip circulators of the type shown in Fig. 2*d* could be stacked as shown in Fig. 6, or connected in other ways such as by microstrip–slotline coupling, to make a four-port circulator using only one magnet system. Assuming the inner ferrite surface is a magnetic wall, the central pin (or via) does not affect the junction behaviour and the junction on the underside can be positioned at any angle with respect to that on the topside. For example, with the arrangement shown in Fig. 6, the inner port on the underside is diametrically opposite that on the topside, and consequently ports 3 and 4 must be positioned as shown. However, the set of ports on the underside can be rotated through any angle that is convenient for the layout of circuit 1 (topside) and circuit 2 (underside), e.g. through 180°, so that ports 3 and 4 are immediately beneath ports 1 and 2, respectively.

The analysis in the next Section permits a detailed investigation of the phenomenological descriptions presented above.

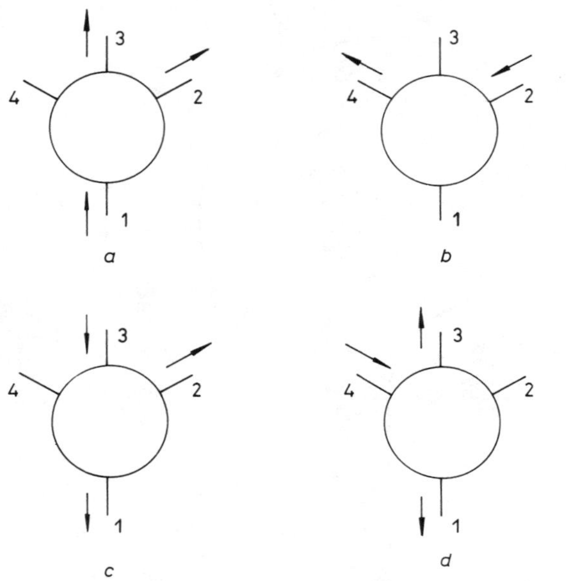

Fig. 4 *Behaviour of nonreciprocal power-divider junction (disc or ring) with external ports and input at each port in turn*

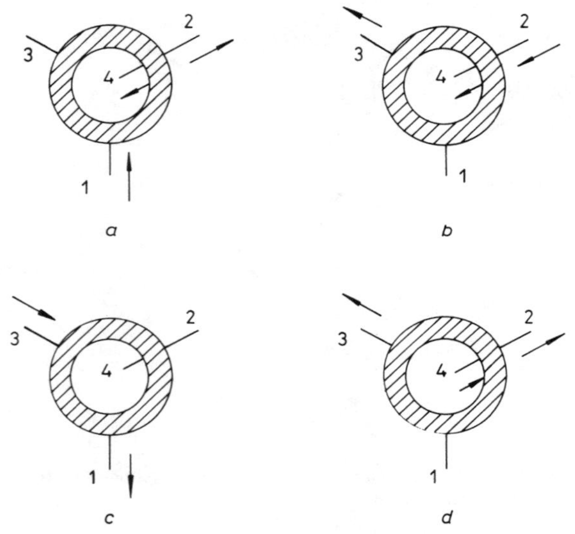

Fig. 5 *Behaviour of nonreciprocal ring power-divider junction with one internal port and input at each port in turn*

3 Analysis

3.1 Impedance matrix

The co-ordinate system for the ferrite ring resonator is shown in Fig. 7. The inner and outer radii are at $\rho = r, R$, and the centre lines of the four strips which form the ports are at ϕ_1, ϕ_2, ϕ_3 and ϕ_4. The strips subtend half angles of ψ_1, ψ_2, ψ_3 and ψ_4 and the fourth strip is on the inside of the ring. The ferrite ring is magnetised perpendicular to the plane of the ring. Consistent with the usual magnetic wall assumption, H_ϕ is specified to be zero on the inner and outer curved surfaces of the ring except over the angles subtended by the strips. In those regions, H_ϕ is assumed to be a complex constant. Using Fourier series expansions, expressions can be developed for H_ϕ and E_z and, in turn, for a voltage and current associated with each port.

The impedance matrix for a general nonreciprocal four-port junction is expressed by the relationship

$$\begin{Bmatrix} V_1 \\ V_2 \\ V_3 \\ V_4 \end{Bmatrix} = \begin{bmatrix} Z_{11} & Z_{12} & Z_{13} & Z_{14} \\ Z_{21} & Z_{22} & Z_{23} & Z_{24} \\ Z_{31} & Z_{32} & Z_{33} & Z_{34} \\ Z_{41} & Z_{42} & Z_{43} & Z_{44} \end{bmatrix} \begin{Bmatrix} I_1 \\ I_2 \\ I_3 \\ I_4 \end{Bmatrix} \quad (1)$$

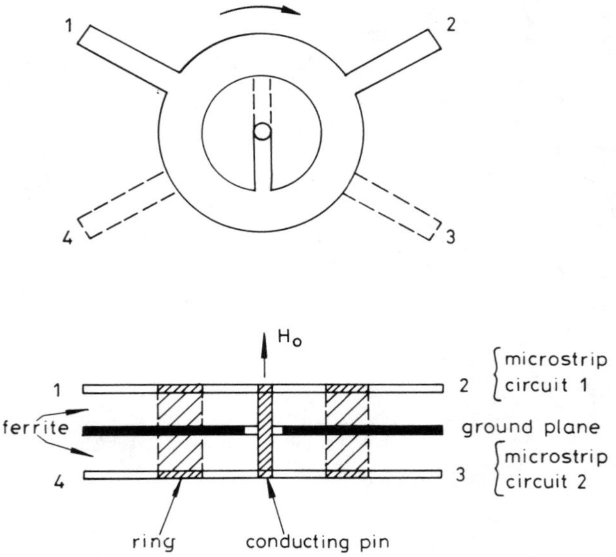

Fig. 6 *Four-port ring circulator using two stacked three-port microstrip circulators with axial interconnect, and one magnet system (not shown)*

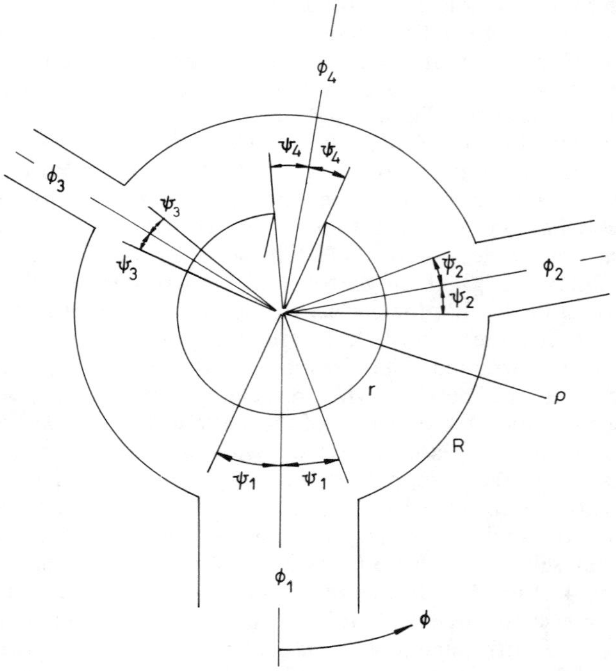

Fig. 7 *Co-ordinate system for nonreciprocal four-port power-divider junction*

It is usual to normalise the impedance matrix, and this is also necessary to permit the scattering matrix [S] to be obtained from [Z]. The elements of the unnormalised impedance matrix [Z] can be normalised with respect to

$$Z_{0i} = \frac{120\pi h}{2R\psi_i} (\mu_d/\varepsilon_d)^{1/2} \quad i = 1, 2, 3 \quad (2)$$

and

$$Z_{04} = \frac{120\pi h}{2r\psi_4} (\mu_d/\varepsilon_d)^{1/2} \tag{3}$$

so that the elements of the normalised matrix are

$$Z_{li} = \hat{Z}_{li}/(Z_{0l} Z_{0i})^{1/2} \tag{4}$$

These elements can be expressed as follows

$$Z_{li} = \frac{j\eta(\psi_l \psi_i)^{1/2}}{\pi} \sum_{n=-\infty}^{\infty} \frac{\sin(n\psi_l) \sin(n\psi_i)}{n^2 \psi_l \psi_i \Delta} [J_n(x_2) B_n(x_1) - Y_n(x_2) A_n(x_1)] \exp[jn(\phi_i - \phi_l)] \quad (i, l = 1, 2, 3) \tag{5}$$

$$Z_{l4} = \frac{j\eta(\psi_l \psi_4)^{1/2}}{\pi} \left(\frac{R}{r}\right)^{1/2} \sum_{n=-\infty}^{\infty} \frac{\sin(n\psi_l) \sin(n\psi_4)}{n^2 \psi_l \psi_4 \Delta} \times [J_n(x_2) B_n(x_2) - Y_n(x_2) A_n(x_2)] \times \exp[(jn(\phi_4 - \phi_l)] \quad (l = 1, 2, 3) \tag{6}$$

$$Z_{4,i} = \frac{j\eta(\psi_4 \psi_i)^{1/2}}{\pi} \left(\frac{r}{R}\right)^{1/2} \sum_{n=-\infty}^{\infty} \frac{\sin(n\psi_4) \sin(n\psi_i)}{n^2 \psi_4 \psi_i \Delta} \times [J_n(x_1) B_n(x_1) - Y_n(x_1) A_n(x_1)] \exp[jn(\phi_i - \phi_4)] \quad (i = 1, 2, 3) \tag{7}$$

$$Z_{44} = \frac{j\eta\psi_4}{\pi} \sum_{-\infty}^{\infty} \frac{\sin^2(n\psi_4)}{n^2 \psi_4^2 \Delta} [J_n(x_1) B_n(x_2) - Y_n(x_1) A_n(x_2)] \tag{8}$$

where $\eta = Z_{\text{eff}}/Z_d = (\mu_{\text{eff}}\, \varepsilon_d/\varepsilon_f\, \mu_d)^{1/2}$, and ε_d, μ_d are the constants of the dielectric medium surrounding the strips. The terms Δ, x_1, and x_2 are defined in eqns. 20–24.

The factor $(\psi_l \psi_i)^{1/2}$ in eqns. 5–7 has been taken outside the summation because the summation often depends only weakly on ψ, and use can be made of the $(\sin x)/x$ form within the summation. Also, it is sometimes convenient to calculate $Z_{\text{eff}}\, \psi/Z_d$.

If required, the scattering matrix $[S]$ can be obtained from the following transformation:

$$[S] = \frac{[Z] - [I]}{[Z] + [I]} \tag{9}$$

where $[I]$ is the unit matrix.

3.2 Special cases

3.2.1 Symmetrical 3-port ring circulator with three outer ports

This component is modelled if $\phi_1 = 0$, $\phi_2 = 2\pi/3$, $\phi_3 = 4\pi/3$ and $\psi_1 = \psi_2 = \psi_3$. As there is 3-fold symmetry, we need to consider only one port. The fourth row and column are deleted from eqn. 1, and with $V_3 = I_3 = 0$ the input impedance of the circulator is given by

$$Z_{in} = \frac{V_1}{I_1} = Z_{11} - \frac{Z_{12} Z_{31}}{Z_{32}} \tag{10}$$

For ideal circulation, we require the imaginary part to be zero and the real part to be unity, i.e.

$$Im(Z_{in}) = 0 \tag{11a}$$

$$Re(Z_{in}) = 1 \tag{11b}$$

Later in the paper, eqns. 11a, b will be referred to as the first and second circulation conditions.

3.2.2 Asymmetrical 3-port ring circulator with three outer ports (*Fig. 8*)

In this case we have $\phi_1 = 0$, $\phi_2 \neq \phi_3 - \phi_2 \neq 2\pi/3$, and $\psi_1 \neq \psi_2 \neq \psi_3$. As in Section 3.2.1, the fourth row and

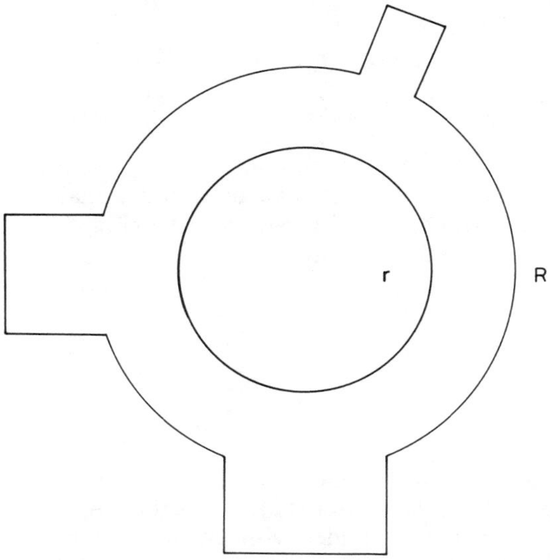

Fig. 8 *Asymmetrical three-port ring circulator with three external ports*

column are deleted from eqn. 1 and the input impedances must satisfy conditions similar to those of eqn. 11, namely

$$Im(Z_{in})_i = 0 \tag{12a}$$

$$Re(Z_{in})_i = 1 \tag{12b}$$

but we must consider both the input and the output ports. The input and output lines may have different characteristic impedances and the circulator must be matched at the input and output, therefore we have six pairs of equations to be satisfied to obtain circulation. However, any lossless junction has the property that

$$[Z] = -[Z]_T^* \tag{13}$$

where the asterisk * means complex conjugate and the subscript T means transpose. Using eqn. 13 it can be shown that at the 'ith' port

$$(Z_{in})_i = -(Z_{\text{out}}^*)_i \tag{14}$$

i.e. the imaginary parts are equal but the real parts differ in sign depending on whether the ith port is an input or an output. The effect of eqn. 14 is to reduce the conditions for circulation from six pairs of equations to three pairs. For example, $(Z_{in})_1$ is given by eqn. 10 and

$$(Z_{in})_2 = Z_{22} - \frac{Z_{23} Z_{12}}{Z_{13}} \tag{15}$$

$$(Z_{in})_3 = Z_{33} - \frac{Z_{31} Z_{23}}{Z_{21}} \tag{16}$$

3.2.3 Nonreciprocal 4-port power divider–combiner

If the input is at port 1, and the outputs at port 2 and 4 (Fig. 5*a*), with port 3 isolated, we have $V_3 = I_3 = 0$, and eqn. 1 yields

$$\begin{aligned} V_1 &= Z_{11} I_1 + Z_{12} I_2 + Z_{14} I_4 \\ V_2 &= Z_{21} I_1 + Z_{22} I_2 + Z_{24} I_4 \\ 0 &= Z_{31} I_1 + Z_{32} I_2 + Z_{34} I_4 \\ V_4 &= Z_{41} I_1 + Z_{42} I_2 + Z_{44} I_4 \end{aligned} \tag{17}$$

From this system of eqns. $(Z_{in})_1 = V_1/I_1$ can be obtained, and, as before, for the ideal component eqn. 11 must be satisfied. If there are inputs at ports 2 and 4, port 1 is isolated and eqn. 1 becomes

$$\begin{aligned} 0 &= Z_{12}I_2 + Z_{13}I_3 + Z_{14}I_4 \\ V_2 &= Z_{22}I_2 + Z_{23}I_3 + Z_{24}I_4 \\ V_3 &= Z_{32}I_2 + Z_{33}I_3 + Z_{34}I_4 \\ V_4 &= Z_{42}I_2 + Z_{43}I_3 + Z_{44}I_4 \end{aligned} \tag{18}$$

From eqn. 18 $(Z_{in})_2$ and $(Z_{in})_4$ can be found. Finally, if the input is at port 3 and the output at port 1 (Fig. 5*c*) we have $V_2 = I_2 = V_4 = I_4 = 0$ and then

$$\begin{aligned} V_1 &= Z_{11}I_1 + Z_{13}I_3 \\ 0 &= Z_{21}I_1 + Z_{23}I_3 \\ V_3 &= Z_{31}I_1 + Z_{33}I_3 \\ 0 &= Z_{41}I_1 + Z_{43}I_3 \end{aligned} \tag{19}$$

From eqn. 19 $(Z_{in})_3$ can be found.

Inspection of Fig. 5*a* shows that, when the ring junction is used as a power divider with an input at port 1 and outputs at ports 2 and 4, port 1 remains isolated from the effects of any mismatched terminations in ports 2 and 4 because any signal entering port 2 or port 4 emerges from port 3.

The case shown in Fig. 5*c* is the most straightforward to investigate because ports 2 and 4 are isolated. Therefore it is similar to the symmetrical ring with three outer ports, and to the symmetrical ring with one inner port, with the appropriate ports isolated.

4 Computed results

It is well known that the dominant mode (TM_{110}) resonance of a dielectric disc of radius R when its curved surface is a magnetic wall is given by $kR = 1.84$ [1, 2]. As described in the Introduction, it can be expected that, if a hole with a magnetic wall is made in the disc, the eigenmode resonance frequencies will decrease as the hole, radius r, is made larger while R is constant. Resonance in the magnetised ring will occur when

$$\Delta = A_n(x_2)B_n(x_1) - A_n(x_1)B_n(x_2) = 0 \tag{20}$$

where

$$A_n(x) = J_n'(x) - \frac{n}{x}\cdot\frac{\kappa}{\mu}J_n(x) \tag{21}$$

$$B_n(x) = Y_n'(x) - \frac{n}{x}\cdot\frac{\kappa}{\mu}Y_n(x) \tag{22}$$

and

$$x_1 = k_{eff}\,r \quad \text{and} \quad x_2 = k_{eff}\,R \tag{23}$$

$$k_{eff} = \omega(\varepsilon_0\varepsilon_f\mu_0\mu_{eff})^{1/2}, \quad \mu_{eff} = (\mu^2 - \kappa^2)/\mu \tag{24}$$

When the ring is unmagnetised, $\kappa/\mu = 0$, and the solutions of eqn. 20 are shown in Fig. 9 in the form of a graph of x_2 versus s, where $s = r/R = x_1/x_2$. It can be seen that the value of x_2 decreases from 1.84 to unity as s approaches unity, as expected.

With a ferrite disc, the degeneracy of the $n = \pm 1$ solutions for the TM_{110} mode is split when a static magnetic field is applied perpendicular to the plane of the disc, i.e. when $\kappa/\mu > 0$. The computed result is shown in Fig. 10 and compared with solutions for a magnetised ring with a small hole ($s = 0.2$) and a larger hole ($s = 0.5$). The two solutions when $\kappa/\mu = 0$ are obviously the same as those shown in Fig. 9, and we see that the split in the solutions for the $n = \pm 1$ modes of the disc ($s = 0$) increases as κ/μ increases. The solution for the $n = +1$ mode increases almost linearly while that for the $n = -1$ mode decreases

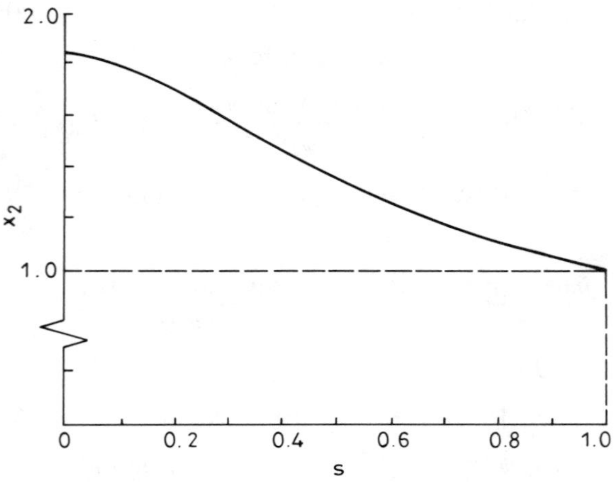

Fig. 9 *Computed decrease in x_2 of dielectric ring resonator ($\kappa/\mu = 0$) as s increases*

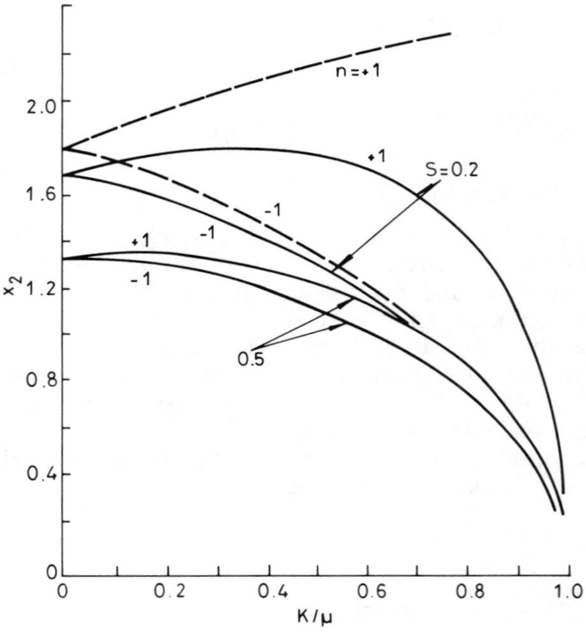

Fig. 10 *Gyromagnetic splitting of disc ($s = 0$), thick ring ($s = 0.2$), and thinner ring ($s = 0.5$) as κ/μ increases*

---- Disc

and approaches zero more steeply as κ/μ approaches unity [1, 2]. It is immediately noticeable that the $n = +1$ curve for a *ring* does not behave like that for a disc; it decreases to zero at $\kappa/\mu = 1$, and there is a maximum splitting at $\kappa/\mu \sim 0.7$. The convergence of the $n = \pm 1$ curves for both rings is due to the fact that the resonance frequencies become nearly equal, and also both modes are cut off at $\kappa/\mu = 1$.

The way in which the splitting decreases as the hole is made larger for various values of κ/μ up to $\kappa/\mu = 0.5$ is shown in Fig. 11. The splitting for each structure at a specific value of κ/μ is normalised to the average value at that value of κ/μ. Thus the ordinate in Fig. 11 is given by

$$(\Delta x_2)_{av} = \frac{2[(x_2)_{+1} - (x_2)_{-1}]}{[(x_2)_{+1} + (x_2)_{-1}]} \tag{25}$$

and this is plotted as a function of $s = x_1/x_2$. It can be seen that the splitting decreases sharply as s is increased, i.e. as the ring is made thinner, and, if $s > 0.5$, the normalised splitting is less than 0.1 and a circulator with reasonable bandwidth would be difficult to achieve.

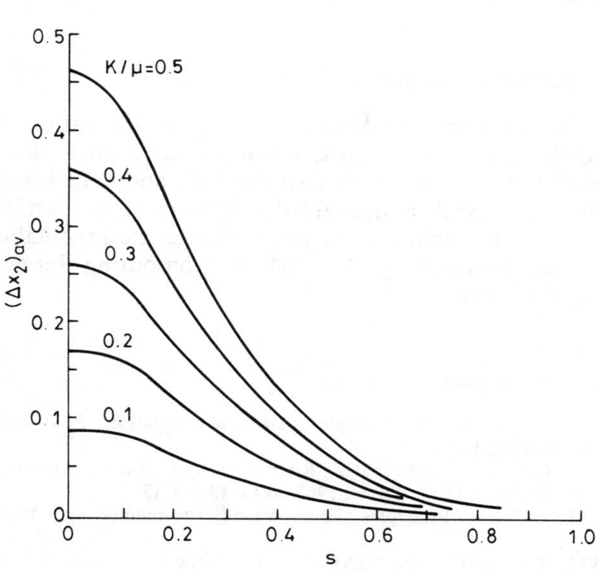

Fig. 11 *Magnitude of average splitting $(\Delta x_2)_{av}$ versus s for various values of κ/μ*

The conditions for circulation for a symmetrical ferrite ring with three outer ports as described in Section 3.2.1 have been computed. An investigation of convergence in the equations used showed that, even with large values of κ/μ, only 11 terms ($n = \pm 5$) are required for satisfactory results.

The results for a ring with $s = 0.5$, are shown in Figs. 12 and 13. It is noteworthy that Fig. 12 shows that the first circulation condition is practically independent of ψ, i.e. the strip width. Fig. 13 indicates that the effect of the hole is to reduce the ratio Z_{eff}/Z_d compared to that required with a disc. For example, from Fig. 9, with $s = 0.5$, it can be seen that $x_2 = 1.35$, whereas with a disc, $x_2 = 1.84$. To consider the possibility of a 'tracking' circulator, a graph of the impedance ratio

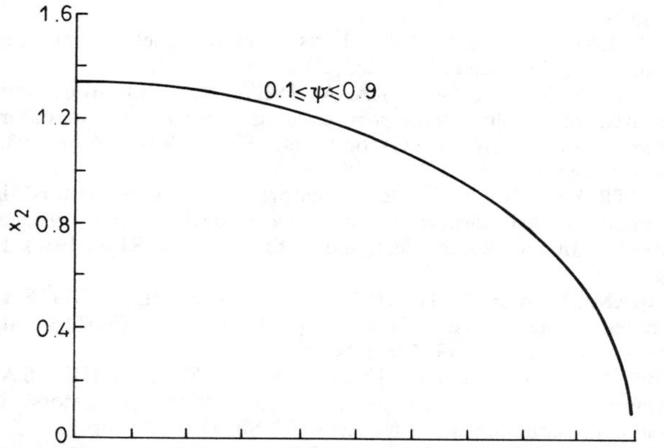

Fig. 12 *Roots of first circulation condition versus κ/μ for ring ($s = 0.5$) with various coupling angles*

$$\frac{Z_{eff}}{Z_d} = \left(\frac{\varepsilon_d}{\varepsilon_f}\right)^{1/2}\left[1 - \left(\frac{\kappa}{\mu}\right)^2\right]^{1/2} \qquad (26)$$

with $\varepsilon_d/\varepsilon_f = 1$, as proposed [4] eqn. 23), is superimposed on the second circulation condition (Fig. 13). It can be seen that 'tracking' is only possible in the range $0.7 < \kappa/\mu < 1.0$. However, if $\kappa/\mu = 0.2$ and $\psi = 0.2$, it can be seen that $Z_{eff}/Z_d \sim 0.13$ whereas, with a disc, $Z_{eff}/Z_d \sim 0.9$.

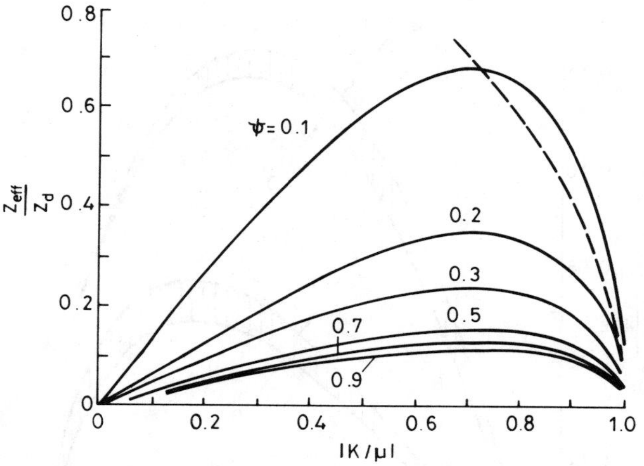

Fig. 13 *Normalised junction impedance of ring ($s = 0.5$) versus κ/μ, obtained from second circulation condition, with various coupling angles*
---- eqn. 26

With equal values of Z_{eff}, it follows that, under these circumstances, the height of the ferrite required for the ring would be approximately one-seventh of that required for a disc.

A tracking ring circulator, but with a narrower bandwidth than that of a disc, can be obtained with $0.1 < \psi < 0.2$ and $0.7 < \kappa/\mu < 1.0$, i.e. with narrow strip widths. This results suggests an interesting improvement over conventional disc circulators. Disc circulators require $\psi \geqslant 0.51$ for tracking circulation, and this, in turn, requires impedance transformers. However, with a ring circulator, we see that it is possible to select a small value of ψ. Therefore, with a suitable choice of s, it may be possible to design the value of ψ to correspond to that required to match directly to the input lines. This would obviate the need for transformers and reduce the size of the circulator.

The circulation conditions for a symmetrical 'inverted' ring circulator, i.e. with all three ports on the inside of the ring ($s = 0.5$) have been comuted, but are not shown for brevity. The general features are similar to those shown in Figs. 12 and 13 except that the required Z_{eff}/Z_d is larger with the 'inverted' circulator. A tracking circulator is possible with $0.1 < \psi < 0.2$ and $0.6 < \kappa/\mu < 1.0$.

Further computations show that a hole with half the diameter of the disc ($s = 0.5$) is the largest that can reasonably be allowed. It has also been noted that x_2 is only very weakly dependent on ψ; in fact x_2 is virtually independent of ψ when $s = 0.5$. This leads to a potentially useful simplfication in design because the results also show that the product $\psi(Z_{eff}/Z_d)$ is also only weakly dependent on ψ. An example is given in Fig. 14 for $s = 0.5$ and we see that $\psi Z_{eff}/Z_d$ varies by less than 15% when ψ varies in the range 0.1–0.5 with κ/μ constant. This means that ring circulator designs can be obtained using Fig. 12 and Fig. 14 without computing Z_{eff}/Z_d for each value of ψ.

The conditions for a ring with one internal port, and two external ports (as in Fig. 2*d*), have also been computed. The expressions for the input impedance are different for each port, and consequently the conditions for circulation are different. The conditions on x_2 and on

Z_{eff}/Z_d when the input is at port 2 are slightly different from those when the input is at port 1 or 3 but, but adjusting ψ_2 independently of ψ_1 and ψ_3, it may be possible to make the curves more nearly coincident.

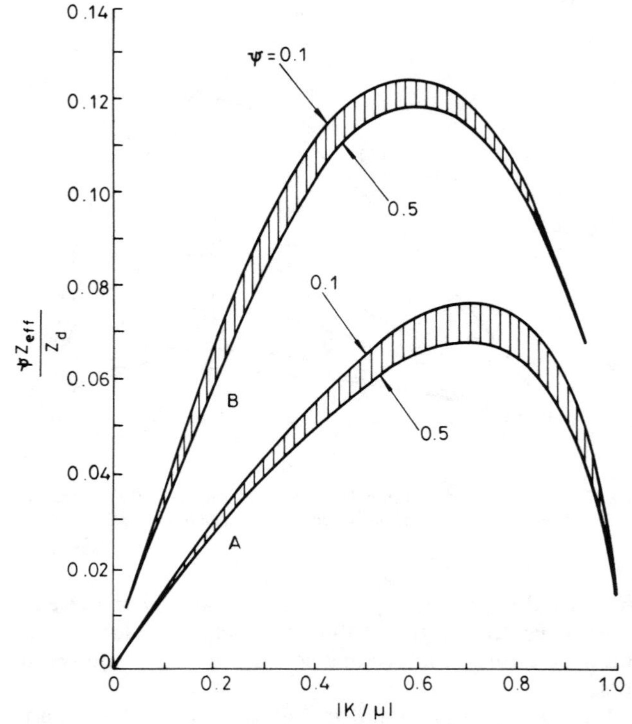

Fig. 14 *Variation of $\psi Z_{eff}/Z_d$ versus κ/μ for coupling angles $0.1 \leqslant \psi \leqslant 0.5$ for symmetrical ring circulator*

a Three external ports
b Three internal ports

5 Conclusions

A multiport ferrite junction with a ring resonator structure has been analysed, and a number of useful features have been derived.

It has been shown that, with a ring structure, circulation can be obtained with a smaller ring diameter and ferrite height than would be required with a disc structure for the same frequency. A further reduction in overall size may be achieved because it may be possible to optimise the ring dimensions to avoid the need for matching transformers.

Depending upon the validity of the magnetic walls assumption, the maximum ratio of inner/outer radii while retaining useful circulation appears to be approximately 0.5. In principle, the central area in the ring circulator could be used for 'inner' ports to which matched loads, diodes, or other components could be connected, thereby permitting increased component density. An inner port could also be used to construct a 4-port circulator with two ports on each side of the ground plane. If a fourth inner or outer port were incorporated a nonreciprocal power/divider/combiner may be realised.

Further theoretical work is in hand to explore some of these possibilities.

6 Acknowledgments

We wish to thank the Ministry of Higher Education of the USSR and the British Council for the grants which enabled V.A. Dmitriyev to visit the UK, and L.E. Davis to visit the USSR, and enabled this work to be carried out. Also, we would like to acknowledge the invaluable assistance provided by the UMIST Computing Service during this work.

7 References

1 BOSMA, H.: 'On the principle of stripline circulation', *Proc. IEE*, 1962, **B109**, (21), pp. 137–146
2 FAY, C.E., and COMSTOCK, R.L.: 'Operation of the ferrite junction circulator', *IEEE Trans.*, 1965, **MTT-13**, pp. 15–27
3 HELSZAJN, J.: 'Principles of microwave ferrite engineering' (Wiley, 1969)
4 WU, Y.S., and ROSENBAUM, F.J.: 'Wide-band operation of microstrip circulator', *IEEE Trans.*, 1974, **MTT-22**, pp. 849–856
5 DE KONING, J.G. *et al.*: 'Full-band low-loss continuous tracking circulation in K-band', *IEEE Trans.*, 1977, **MTT-25**, pp. 152–155
6 AYTER, S., and AYASLI, Y.: 'The frequency behaviour of stripline circulator junctions', *IEEE Trans.*, 1978, **MTT-26**, pp. 197–202
7 HELSZAJN, J.: 'Operation of tracking circulators', *IEEE Trans.*, 1981, **MTT-29**, pp. 700–707
8 HICKSON, M.T., DAVIS, L.E., PAUL, D.K., and SILLARS, S.B.: 'Computer aided design and optimisation of stripline circulators for 18–30 GHz and 18–40 GHz', *IEEE Symposium Digest*, 1991, **MTT-S**, pp. 961–964
9 CASTILLO, J.B., and DAVIS, L.E.: 'Computer-aided design of three-port waveguide junction circulators', *IEEE Trans.*, 1970, **MTT-18**, pp. 25–34
10 CASTILLO, J.B., and DAVIS, L.E.: 'A higher-order approximation for waveguide circulators', *IEEE Trans.*, 1972, **MTT-20**, pp. 410–412
11 KHILLA, A.M.: 'Ring and disk resonator CAD model', *Microwave J.*, 1984
12 HELSZAJN, J. *et al.*: 'Mode charts of gyromagnetic planar ring resonators', *Elect. Lett.*, 1987, **23**, (24), pp. 1290–1291
13 DMITRIYEV, V.A., and KAZANTSEV, V.I.: 'Electrodynamic research of stripline three-port circulators with ring resonator'. Trans. Fifth Int. Conf. on Gyromagnetic Electr., Vol. 1, 6 pp, 1980, Vilnius, USSR
14 DMITRIYEV, V.A.: 'Ferrite circulators with ring resonators', in 'Computer-aided design of microwave devices and systems' (Moscow Inst. of Radio Elect. and Automation (MIREA) 1982), 15 pp
15 ABRAMOV, W.P., DMITRIYEV, V.A., and SHELUCHIN, S.A.: 'Computer-aided design of microstrip ferrite devices' (MHTS Publication, Moscow, USSR, 1986), 54 pp
16 ABRAMOV, W.P., DMITRIYEV, V.A., and SHELUCHIN, S.A.: 'Nonreciprocal microwave junctions using ferrite resonators', in 'Radio and communcation' (Moscow, USSR, 1989), 200 pp
17 NAGAO, T.: 'Double circulation frequency operation of stripline Y-junction circulators', *IEEE Trans.*, 1977, **MTT-25**, (3), pp. 181–189
18 NAGAO, T.: 'Broad-band operation of stripline Y circulators', *IEEE Trans.*, 1977, **MTT-25**, (12), pp. 1042–1047

Chapter 8

Discontinuities in Transmission Structures

CHARACTERIZATION of discontinuities in various kinds of microwave and millimeter-wave transmission structures forms an essential part of circuit design at these frequencies. Before the advent of planar microwave circuits (and the planar transmission structures used in these circuits), considerable effort [8.1] was spent on the characterization of coaxial line and waveguide discontinuities. However, in recent years characterization of discontinuities and junctions in planar microwave transmission structures, such as microstrip lines and coplanar waveguides, has become much more significant because of the applications of these structures in hybrid and monolithic microwave integrated circuits. As postfabrication trimmings and adjustments are neither convenient nor desirable in integrated circuit technology, accurate modeling of discontinuities and junctions (as well as of other circuit components) becomes necessary. Until recently, the absence of accurate discontinuity models was a limiting bottleneck for first-pass success design in monolithic microwave integrated circuits (MMICs). Consequently, substantial research efforts have been made in this area in recent years.

Microstrip discontinuities have been discussed in several recent books [8.2–8.6], because of frequent occurrences of discontinuities in MMIC layouts. Methods for the characterization of microstrip discontinuities can be divided into three groups:

1. Quasi-static methods, in which capacitances and inductances associated with the discontinuities are evaluated
2. Planar waveguide model approach
3. Full-wave analysis, incorporating surface-wave and radiation phenomena

These groups are listed in the order of their increasing accuracy and complexity (and associated computational cost). They are also in their chronological order of development. The quasi-static methods are well documented in literature [8.2–8.6] and are not addressed here.

In this chapter we include six papers, three of which are based on planar waveguide model approach. The rest are based on the full-wave approach. The first (Paper 8.1) is classical in the sense that it introduces a planar waveguide model for the microstrip line. It analyzes a microstrip step-junction and T-junction discontinuity by using an orthogonal series expansion in the various regions of the planar waveguide model. A more detailed description of the planar waveguide model for microstrips is available in [8.7]. Additional results based on this method are also available [8.8–8.10].

The second (Paper 8.2) addresses the following question: "How does one modify the discontinuity geometry such that reactances introduced by the junction are minimized (or compensated for) and hence the resulting discontinuity behavior corresponds to that of an ideal junction?" Compensation of right-angled bends has received more detailed attention [8.11–8.14] than any other configuration.

Paper 8.3 evaluates radiation losses associated with discontinuities and spurious external coupling between discontinuities in microstrip circuits. The method used in this paper is an extension of the planar analysis approach.

As the operating frequency of microwave and millimeter-wave integrated circuits increases, the impact of surface wave and radiation losses becomes more significant. To account for these phenomena and to characterize accurately discontinuities in microstrips and coplanar waveguides, full-wave solutions for Maxwell's equations become essential. The development of full-wave simulation techniques has enabled microwave designers to characterize microstrip discontinuities using rigorous numerical methods with minimum analytical and numerical approximations. Three papers on full-wave analysis are included in this chapter. The first (Paper 8.4) deals with the characterization of microstrip discontinuities; the two other papers discuss the characterization of coplanar waveguide discontinuities.

Paper 8.4 presents a general numerical solver, the analytical foundation of which is based on an integral equation formulated in the spatial domain. The resulting integral equation is then solved using the Galerkin-moment method. On the other hand, Paper 1.4 (in Chapter 1) presents a full-wave spectral-domain analysis for characterizing microstrip discontinuities of arbitrary shapes. It reviews the theoretical background of the full-wave spectral domain approach and discusses its advantages, one of which is the ability to determine independently both radiation and surface-wave losses. As in Paper 8.4, the resulting integral equation is solved by the method of moments. In addition to Papers 8.4 and 1.4, there are a number of papers [8.15–8.20] on the characterization of microstrip discontinuities. These papers are also based on the moment-method solution of the resulting integral equation. Mode-matching technique [8.24] has also been used for analysis of microstrip discontinuities. In this method, conductor thickness effects are included.

Microstrip discontinuities have also been characterized by full-wave methods that are based on solving a differential equation rather than an integral equation. Such methods include the finite-difference approach [8.19–8.20] and the finite-element approach [8.21–8.23].

The remaining two papers (Papers 8.5 and 8.6) in this chapter discuss the application of full-wave analysis methods to coplanar waveguide discontinuities. In Paper 8.5 the authors formulate a space-domain surface-integral equation to characterize shielded coplanar waveguide two-port discontinuities. The integral equation is solved by the method of moments. The use of a three-dimensional finite-difference method to characterize coplanar waveguide discontinuities is demonstrated in Paper 8.6. A three-dimensional method is used to evaluate the equivalent capacitance of various ordinary coplanar waveguide discontinuities. Additional treatment of coplanar waveguide discontinuities can be found in [8.27–8.34].

References

[8.1] Marcuvitz, N., ed. *Waveguide Handbook*. New York: McGraw-Hill, 1951; also London: Peter Peregrinus, 1985.

[8.2] Gupta, K. C. et al. *Microstrip Lines and Slotlines*. Dedham, MA: Artech House, 1979. See Chapters 3 and 4 on "Microstrip Discontinuities—I and II," pp. 107–193.

[8.3] Gupta, K. C. et al. *Computer-Aided Design of Microwave Circuits*. Dedham, MA: Artech House, 1981. See Chapter 6 on "Characterization of Discontinuities—II, Striplines and Microstrip Lines," pp. 179–202.

[8.4] Hoffman, R. K. *Handbook of Microwave Integrated Circuits*. Norwood, MA: Artech House, 1987. See Chapter 10 on "Microstrip Discontinuities," pp. 267–309.

[8.5] Chang, Kai, ed. *Handbook of Microwave and Optical Components*, Volume 1. New York: John Wiley & Sons, See Chapter 2 on "Transmission-Line Discontinuities" by K. C. Gupta, pp. 60–117, 1989.

[8.6] Edwards, Terry. *Foundations for Microstrip Circuit Design*, 2nd ed. Chichester, U.K.: John Wiley & Sons, 1992. See Chapter 5 on "Discontinuities in Microstrip," pp. 127–171.

[8.7] Kompa, G. and R. Mehran. "Planar waveguide model for calculating microstrip components." *Electronic Letters*, Vol. 11, pp. 459–460, September 1975.

[8.8] Menzel, W. and I. Wolff. "A method for calculating the frequency dependent properties of microstrip discontinuities." *IEEE Trans. Microwave Theory Tech.*, Vol. MTT-25, No. 2, pp. 107–112, February 1977.

[8.9] Mehran, R. "Frequency dependent equivalent circuits for microstrip right-angle bends, T-junctions and crossings." *Arch. Elektr. Ubertr.*, Vol. 30, pp. 80–82, 1975.

[8.10] Mehran, R. "Calculation of microstrip bends and Y junctions with arbitrary angle." *IEEE Trans. Microwave Theory Tech.*, Vol. MTT-26, No. 6, pp. 400–405, 1978.

[8.11] Broumas, A. D., H. Ling, and T. Itoh. "Transmission properties of a right-angle microstrip bend with and without a miter." *IEEE Trans. Microwave Theory Tech.*, Vol. MTT-37, No. 5, pp. 925–929, May 1989.

[8.12] Mehran, R. "The frequency dependent scattering matrix of two-fold truncated microstrip bends." *Arch. Elek. Übertragung.*, Vol. 81, pp. 411–415, 1977.

[8.13] Douville, R. J. P. and D. S. James. "Experimental study of symmetric microstrip bends and their compensation." *IEEE Trans. Microwave Theory Tech.*, Vol. MTT-26, No. 3, pp. 175–182, 1978.

[8.14] James, D. S. and R. J. P. Douville. "Compensation of microstrip bends by using square cutouts." *Electronics Letters*, Vol. 12, No. 22, pp. 577–579, 1976.

[8.15] Katehi, P. B. and N. G. Alexopoulos. "Frequency-dependent characteristics of microwave discontinuities in millimeter-wave integrated circuits." *IEEE Trans. Microwave Theory Tech.*, Vol. MTT-33, pp. 1029–1035, October 1985.

[8.16] Jackson, R. W. and D. M. Pozar. "Full-wave analysis of microstrip open-end and gap discontinuities." *IEEE Trans. Microwave Theory Tech.*, Vol. MTT-33, pp. 1036–1042, October 1985.

[8.17] Wu, S. C. et al. "A rigorous dispersive characterization of microstrip cross and T-junctions." *IEEE Trans. Microwave Theory Tech.*, Vol. MTT-38, pp. 1837–1844, December 1990.

[8.18] Horng, T.-S. et al. "A generalized method for distinguishing between radiation and surface wave losses in microstrip discontinuities." *IEEE Trans. Microwave Theory Tech.*, Vol. MTT-38, No. 12, pp. 1800–1807, December 1990.

[8.19] Mosig, J. R. "Arbitrarily shaped microstrip structures and their analysis with a mixed potential integral equation." *IEEE Trans. Microwave Theory Tech.*, Vol. MTT-36, pp. 314–323, February 1988.

[8.20] Mosig, J. R. and F. E. Gardiol. "Integral equation techniques for dynamic analysis of microstrip discontinuities." *Alta Freq.*, Vol. LVII, No. 5, pp. 171–181, June 1988.

[8.21] Feix, N., M. Lalonde, and B. Jecko. "Harmonical characterization of a microstrip bend via the finite difference time domain method." *IEEE Trans. Microwave Theory Tech.*, Vol. MTT-40, No. 5, pp. 955–961, May 1992.

[8.22] Zhang, X. and K. K. Mei. "Time-domain finite difference approach to the calculation of the frequency-dependent characteristics of microstrip discontinuities." *IEEE Trans. Microwave Theory Tech.*, Vol. MTT-36, pp. 1775–1787, December 1988.

[8.23] Sheen, D. M. et al. "Application of the three-dimensional finite-difference time-domain method to the analysis of planar microstrip discontinuities." *IEEE Trans. Microwave Theory Tech.*, Vol. MTT-38, No. 7, pp. 849–857, July 1990.

[8.24] Bögelsack, F. and I. Wolff. "Full-wave analysis of multiport microstrip discontinuities using a superposition principle and mode matching technique," *Int'l. J. Numerical Modelling: Electronic Networks, Devices and Fields*, Vol. 3, pp. 259–268, 1990.

[8.25] Jackson, R. W. "Full-wave finite element analysis of irregular microstrip discontinuities." *IEEE Trans. Microwave Theory Tech.*, Vol. MTT-37, pp. 81–89, January 1989.

[8.26] Khebir, A., A. G. Kouki, and R. Mittra. "Asymptotic boundary conditions for finite element analysis of three-dimensional transmission line discontinuities." *IEEE Trans. Microwave Theory Tech.*, Vol. 38, No. 10, pp. 1427–1432, October 1990.

[8.27] Dib, N. I. et al. "Coplanar waveguide discontinuities for p-i-n diode switches and filter applications." *1990 IEEE MTT-S Int'l. Microwave Symp. Dig.*, pp. 399–402, Dallas, Texas, May 8–10, 1990.

[8.28] Dib, N. I. et al. "A comparative study between shielded and open coplanar waveguide discontinuities." *MIMICAE*, Vol. 2, No. 4, pp. 331–341, October 1992.

[8.29] Simons, R. N. and G. E. Ponchak. "Modelling of some coplanar waveguide discontinuities." *IEEE Trans. Microwave Theory Tech.*, Vol. MTT-36, pp. 1796–1803, December 1988.

[8.30] Jackson, R. W. "Mode conversion at discontinuities in finite-width conductor-backed coplanar waveguide." *IEEE Trans. Microwave Theory Tech.*, Vol. MTT-37, pp. 1582–1589, October 1989.

[8.31] Dib, N. and P. Katehi. "Modeling of shielded CPW discontinuities using the space domain integral equation method (SDIE)." *J Electromag. Waves Appl.*, pp. 503–523, April 1991.

[8.32] Rittweger, M., M. Abdo, and I. Wolff. "Full-wave analysis of coplanar discontinuities considering three-dimensional bond wires." *1991 IEEE MTT-S Int'l. Microwave Symp. Dig.*, pp. 465–468, Boston, Massachusetts, June 10–14, 1991.

[8.33] Harokopus, W. and P. Katehi. "Radiation loss from open CPW discontinuities. *1991 IEEE MTT-S Int'l. Microwave Symp. Dig.*, pp. 743–746, Boston, Massachusetts, June 10–14, 1991.

[8.34] Bröme, R. and R. H. Jansen. "Systematic investigation of coplanar waveguide MIC/MMIC structures using a unified strip/slot 3-d electromagnetic 3D simulator." *1991 IEEE MTT-S Int'l. Microwave Symp. Dig.*, pp. 1081–1084, Boston, Massachusetts, June 10–14, 1991.

CALCULATION METHOD FOR MICROSTRIP DISCONTINUITIES AND T JUNCTIONS

Indexing terms: Stripline components, Waveguide junctions, Modelling

A method for calculating microstrip discontinuities and T junctions is described, and a waveguide model for the microstrip line is defined. With the help of this model and the use of an orthogonal series expansion, a solution of the above problems is found. Numerical results for the scattering matrices of both the discontinuities and the junctions are given.

Calculation methods for stripline discontinuities and junctions have been described by Oliner,[1] Altschuler and Oliner,[2] Franco and Oliner[3] and Campell,[4] using Babinet's principle and the well known solutions for equivalent problems given by Marcuvitz.[5] Leighton and Milnes[6] published a method for calculating T junctions in microstrip techniques, expanding the theory given by Altschuler and Oliner[2] to the case of the microstrip problem.

As is known, there are well tested methods for calculating discontinuities and junctions in the waveguide techniques. These methods use orthogonal-series expansions of the fields in the waveguide (e.g. see References 7–11). A necessary condition for applying orthogonal-series-expansion methods is that a complete set of field solutions for the problem considered must be known. A further condition should be that

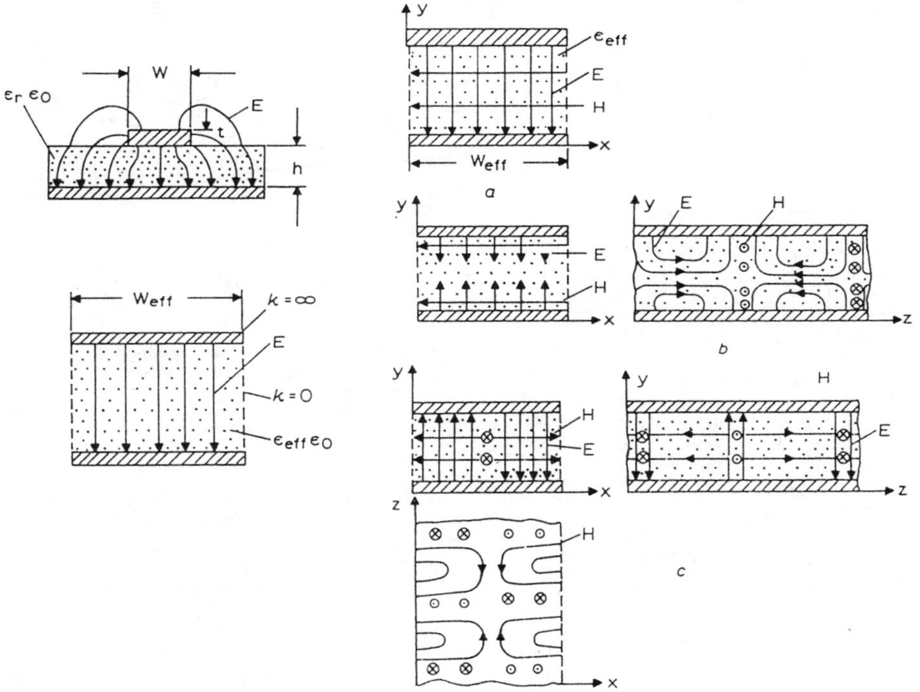

Fig. 1 *Waveguide model for the microstrip line and field distribution of the lowest-order modes of the model*

$W_{eff} = f(w, h, t)$
$\varepsilon_{eff} = g(w, h, t, \varepsilon_r)$
a TEM mode
b E_{10} mode
c H_{10} mode

Reprinted with permission from *Electronics Letters*, vol. 8, no. 7, I. Wolff, G. Kompa, and R. Mehran, "Calculation Method for Microstrip Discontinuities and T-Junctions," pp. 177–179, April 1972, IEE.

the solutions are orthogonal. As is well known, no complete set of solutions for the field problem of the microstrip line has yet been published.

For this reason, we took the waveguide model given by Wheeler[12, 13] for the lowest-order mode, which, to a first approximation, is a TEM mode. Wheeler showed that the behaviour of the TEM mode on the microstrip line can be described by a parallel-plate waveguide of width W_{eff} and relative permittivity ε_{eff}:

$$\left.\begin{aligned} w_{eff} &= \frac{h}{Z_w}\sqrt{\left(\frac{\mu_0}{\varepsilon_{eff}\,\varepsilon_0}\right)} \\ \varepsilon_{eff} &= \left(\frac{\lambda_0}{\lambda_g}\right)^2 \end{aligned}\right\} \quad \ldots\ldots \quad (1)$$

where Z_w is the characteristic impedance defined by Wheeler[12] and λ_g is the wavelength on the microstrip line. Leighton and

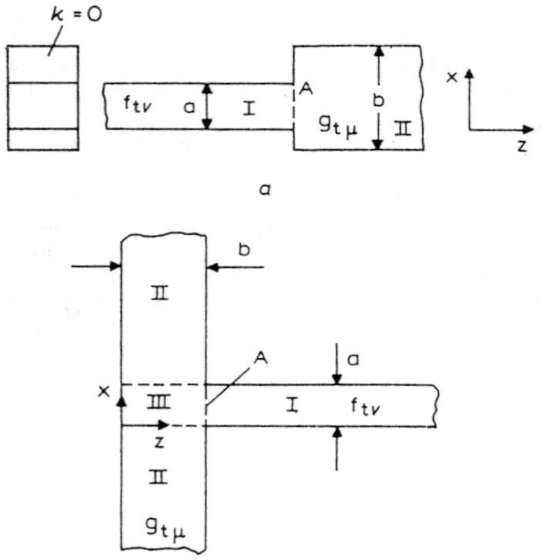

Fig. 2 *Discontinuity and T junction in microstrip technique*
a Discontinuity
b *T* junction

Milnes[6] also use this waveguide model, but they apply Babinet's principle to make use of a solved problem for an equivalent T junction.[5] In contrast to all solutions of the above problem that have been published, we make direct use of the waveguide model of the microstrip line described above to compute the energy stored in the discontinuities and T junctions. This is done by making the assumption that the higher-order modes of the above waveguide model describe, to a first approximation, the physical fields on the microstrip line. This assumption may, at first sight, look very arbitrary, but for solving an eigenvalue problem, an arbitrary, infinite and complete set of solutions, which satisfy the boundary conditions, can be taken. Then, if we take the waveguide model for the microstrip line which is exact only for the TEM mode, and take into account the higher-order modes of this model, we shall obtain better results than those obtained by methods which do not consider any higher-order modes on the microstrip line. Fig. 1 shows the waveguide model and the lowest-order modes on this waveguide.

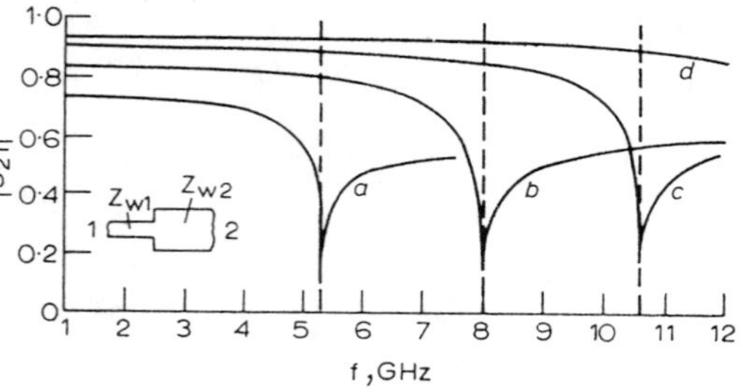

Fig. 3 *Numerical results for the transmission coefficients of a symmetric microstrip discontinuity on a polyguide substrate material ($\varepsilon_r = 2{\cdot}33$, $h = 1{\cdot}5$ mm)*
Change of the characteristic impedance from $Z_{w1} = 50\,\Omega$ to (*a*) $Z_{w2} = 10\,\Omega$, (*b*) $Z_{w2} = 15\,\Omega$, (*c*) $Z_{w2} = 20\,\Omega$ and (*d*) $Z_{w2} = 25\,\Omega$

Fig. 2 shows the discontinuity and the T junction which has been calculated. Using the waveguide model described, we can find an orthogonal and complete set of field solutions f_{tv} in the left-hand guide with effective width a (Fig. 2a) and a complete set $g_{t\mu}$ in the waveguide with effective width b. To compute the scattering matrix of the discontinuity, the transversal magnetic-field strength of the left-hand part of the structure is developed into a series expansion of the functions $g_{t\mu}$, and the transversal electric-field strength of the right-hand part is developed into a series expansion of the functions f_{tv}, as follows:

$$\left.\begin{aligned} H_{tI} &= \sum_{\mu=1}^{\infty} \frac{B_\mu}{z_\mu}(e_z \times g_{t\mu}) \\ E_{tII} &= \sum_{v=1}^{\infty} A_v f_{tv} \end{aligned}\right\} \quad \ldots\ldots\ldots \quad (2)$$

The coefficients A_v and B_μ can be computed so that the boundary conditions in the discontinuity are met. From the amplitudes of the lowest-order TEM mode, the scattering matrix of the discontinuity can be derived. Fig. 3 shows the calculated transmission coefficients for a symmetric discontinuity with the characteristic impedances $Z_{w1} = 50\,\Omega$ and (*a*) $Z_{w2} = 10\,\Omega$, (*b*) $Z_{w2} = 15\Omega$, (*c*) $Z_{w2} = 20\,\Omega$ and (*d*) $Z_{w2} = 25\,\Omega$. The curves are computed for microstrip lines on a polyguide substrate material with a relative permittivity $\varepsilon_r = 2{\cdot}33$ and a height $h = 0{\cdot}625$ mm. As can

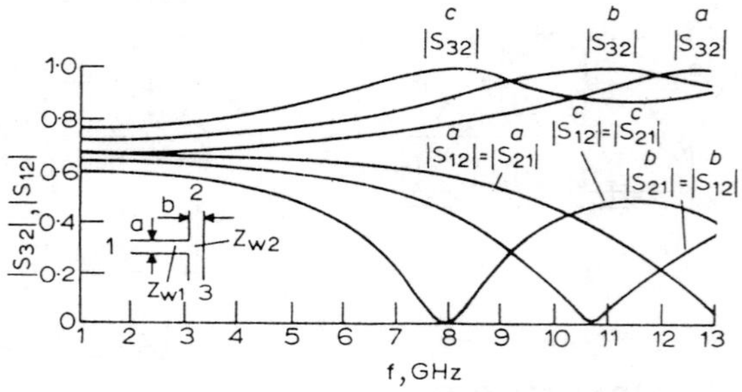

Fig. 4 *Numerical results for transmission coefficients of a microstrip T junction on polyguide substrate material ($\varepsilon_r = 2{\cdot}33$, $h = 1{\cdot}5$ mm)*
$Z_{w1} = 50\,\Omega$
a $Z_{w2} = 50\,\Omega$
b $Z_{w2} = 40\,\Omega$
c $Z_{w2} = 30\,\Omega$

be seen from Fig. 3, the scattering matrix depends strongly on the frequency for those substrate materials, especially if the characteristic impedance Z_{w2} is small. As calculations for aluminium substrate material show, the frequency dependence is negligible at frequencies up to 20 GHz.

If the T junction (Fig. 2*b*) is calculated in the same way, it must be noticed that the field solutions $g_{t\mu}$ of region II do not satisfy the boundary conditions in region III, as has been shown by Lewin[10] for a waveguide problem. Therefore the series expansion which has been used to calculate the discontinuity must be substituted by an integral representation, as shown in References 10 and 11. This means that, for the H_x component of the magnetic-field strength in regions II and III, a relationship

$$H_x = \int_{-\infty}^{+\infty} f(\alpha z)\frac{B}{\beta z_w}\sqrt{\left(\frac{2}{bh}\right)}\sin(\alpha z)\,dz \quad . \quad . \quad (3)$$

must be used. The function $f(\alpha z)$ can be chosen so that the boundary conditions between region I and region III are satisfied and the tangential magnetic-field strength vanishes on the magnetic walls of region II ($z = 0$, $z = b$). From the amplitudes of the TEM mode in both microstrip lines, the scattering matrix of the T junction can be evaluated. Fig. 4 shows the calculated transmission coefficients of the T junction as functions of the frequency. For the T junction also, the frequency dependence is great if substrate material with a small relative permittivity is used (Fig. 4, $\varepsilon_r = 2{\cdot}33$). The frequency dependence once more is small for aluminium substrate material.

Measurements show that the agreement between our theory and experiments is good. For polyguide substrate material, a large influence of radiation losses on the scattering matrix can be measured. For aluminium substrate material, this influence is small.[6] The radiation loss can be calculated by a first approximation (e.g. see Reference 15), and taken into account in our theory. More detailed results of the theory and the experimental work will be published shortly.[14]

I. WOLFF
G. KOMPA
R. MEHRAN

Institut für Hochfrequenztechnik
Technische Hochschule Aachen
W. Germany

10th March 1972

References

1 OLINER, A. A.: 'Equivalent circuits for discontinuities in balanced strip transmission line', *IRE Trans.*, 1955, **MTT-3**, pp. 134–143

2 ALTSCHULER, H. M., and OLINER, A. A.: 'Discontinuities in the center conductor of symmetric strip transmission line', *ibid.*, 1960, **MTT-8**, pp. 328–338

3 FRANCO, A. G., and OLINER, A. A.: 'Symmetric strip transmission line tee junction', *ibid.*, 1962, **MTT-10**, pp. 118–124

4 CAMPBELL, J. J.: 'Application of the solution of certain boundary value problems to the symmetrical four-port junction and specially truncated bends in parallel-plate waveguides and balanced strip-transmission lines', *IEEE Trans.*, 1968, **MTT-16**, pp. 165–176

5 MARCUVITZ, N.: 'Waveguide handbook' (MIT Radiation Laboratory Series, 1950)

6 LEIGHTON, W. M., and MILNES, G.: 'Junction reactance and dimensional tolerance effects on X-Band 3 dB directional couplers', *IEEE Trans.*, 1971, **MTT-19**, pp. 818–824

7 MASTERMAN, P. H., and CLARRICOATS, P. J. B.: 'Computer field-matching solution of waveguide transverse discontinuities', *Proc. IEE*, 1971, **118**, (1), pp. 51–63

8 KNETSCH, H. D.: 'Beitrag zur Theorie spunghafter Querschnitts-veränderungen von Hohlleitern', *Arch. Elek. Übertrag.*, 1968, **22**, pp. 591–600

9 SHARP, E. D.: 'An exact calculation for a T-junction of rectangular waveguide having arbitrary cross-section', *IEEE Trans.*, 1967, **MTT-15**, pp. 109–116

10 LEWIN, L.: 'On the inadequacy of discrete mode-matching techniques in some waveguide discontintuity problems', *ibid.*, 1970, **MTT-18**, pp. 364–369

11 BRÄCKELMANN, W.: 'Hohlleiterverbindungen für Rechteckhohlleiter', *Nachrichtentech. Z.*, 1970, **23**, pp. 2–7

12 WHEELER, H. A.: 'Transmission-line properties of parallel wide strips by a conformal mapping approximation', *IEEE Trans.*, 1964, **MTT-12**, pp. 280–289

13 WHEELER, H. A.: 'Transmission-line properties of parallel wide strips separated by a dielectric sheet', *ibid.*, 1965, **MTT-13**, pp. 172–185

14 WOLFF, I., KOMPA, G., and MEHRAN, R.: 'Streifenleitungsdiskontinuitäten und-Verzweigungen', *Nachrichtentech. Z.* (to be published)

15 LEWIN, L.: 'Radiation from discontinuities in strip-line', *Proc. IEE*, 1960, **107C**, pp. 163–170

Paper 8.2

Compensation of Discontinuities in Planar Transmission Lines

RAKESH CHADHA, MEMBER, IEEE, AND K. C. GUPTA, SENIOR MEMBER, IEEE

Abstract—**Compensation of discontinuity reactances associated with steps, right-angled bends, and T-junctions in planar transmission lines has been carried out by removing appropriate triangular portions from the discontinuity configurations. A two-dimensional analysis using a Green's functions approach has been employed.**

I. INTRODUCTION

PLANAR TRANSMISSION lines (such as stripline, microstrip line, suspended-substrate stripline, inverted microstrip, etc.) are used in the design of microwave and millimeter-wave circuits. Geometrical discontinuities associated with these lines affect the circuit performance. It is thus desirable to reduce the effects of discontinuity reactances. Steps, bends, and T-junctions are some of the commonly occuring discontinuities. Compensation of discontinuity reactances for the steps occuring in stripline transformers and filters has been presented [1]. Douville and James [2] have carried out experimental study of symmetrical microstrip bends and their compensation has been attempted by chamfering the bend. Compensation of T-junctions in striplines and microstrip lines has been reported [3]–[6]. In the compensation proposed by Dydyk [3], strip widths are altered near the junction. The analysis ignores the reactances caused by the steps thus introduced. Also, this method is not very effective over a wide range of frequencies. The other methods [4]–[6] alter the junction to a Y-junction and the branch line is no longer perpendicular to the main line. In some cases, this may be undesirable as it may necessitate introducing bends in layout, etc. In this paper, methods of compensation of reactances in steps, right-angled bends, and T-junctions in planar waveguides are reported. These results can be used for the planar transmission lines for which planar waveguide models are available. Various dimensions considered correspond to those of the planar models. Widths shown in various figures, are effective widths for microstrip configurations.

Two-dimensional (2-D) analysis [7]–[9] is used to analyze accurately the discontinuities without and with the proposed compensations. The 2-D analysis is based upon Green's functions [10], [11] and the segmentation [8], [10] and desegmentation [6], [9] methods. Details of this method of analysis as well as its limitations have been discussed in literature [6], [8], [9], [11] earlier.

Manuscript received March 10, 1982; revised June 28, 1982.

R. Chadha was with the Department of Electrical Engineering, Indian Institute of Technology, Kanpur, India. He is now with the Department of Electrical Engineering, University of Waterloo, Waterloo, Ont., Canada, N2L 3G1.

K. C. Gupta is presently with the Department of Electrical Engineering, University of Kansas, Lawrence, KS 66045, on leave from the Department of Electrical Engineering, Indian Institute of Technology, Kanpur 208016, India.

II. STEPS IN WIDTH

A step discontinuity is present in impedance transformers, half-wave filters, or wherever a change in impedance is involved. Compensation of steps used in stripline transformers and filters has been reported earlier [1]. A one-dimensional analysis is used in [1] and the compensation achieved is not broad band. In this section, compensation of steps with impedance ratios $1:2$ and $1:\sqrt{2}$ has been attempted by chamfering the corners so that the width does not change abruptly.

Fig. 1 (inset) shows a symmetrical step in width (impedance ratio $1:2$) of the line. The discontinuity reactances cause the reflection coefficients on the two sides of the step to be different from their theoretical values of $\pm 1/3$ (at the plane of the step). The magnitude of the reflection coefficient is shown as a function of frequency in Fig. 1 (curve for $\theta = 90°$). The phases of the reflection coefficients are used to obtain the locations of the effective step reference planes, i.e., where S_{11}, S_{22} become real. The normalized electrical lengths Δl_i between the step plane and the effective step reference planes are plotted in Fig. 2. Δl_i is the distance between the step reference plane from the magnetic wall model of the line and the effective step plane computed from S_{ii}. It is taken to be positive if the effective reference plane shifts away from port i and negative if the effective reference plane shifts towards port i. It is desired that the normalized electrical lengths Δl_i either be small or remain constant with frequency and that the magnitude of reflection coefficient be close to $1/3$.

To reduce the effects of the discontinuity reactances, the step is chamfered so that the width varies in a tapered manner as shown in the insert in Fig. 1. The desegmentation method [9] is used to analyze chamfered steps with three values of θ, 30°, 45°, and 60°. Variations of the magnitudes of the reflection coefficients for these three chamfered steps are shown as functions of frequency in Fig. 1. The corresponding normalized electrical lengths for the chamfered steps are shown as functions of frequency in Fig. 2. In these cases, the step plane is considered midway in the taper and the shift to the equivalent step reference

Reprinted from *IEEE Trans. Microwave Theory Tech.*, vol. MTT-30, no. 12, pp. 2151–2156, Dec. 1982.

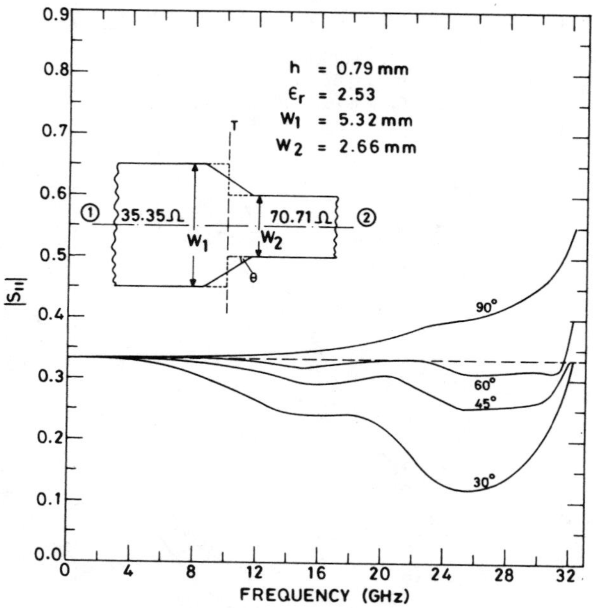

Fig. 1. Reflection coefficients for uncompensated and compensated step discontinuities with a 1:2 impedance ratio.

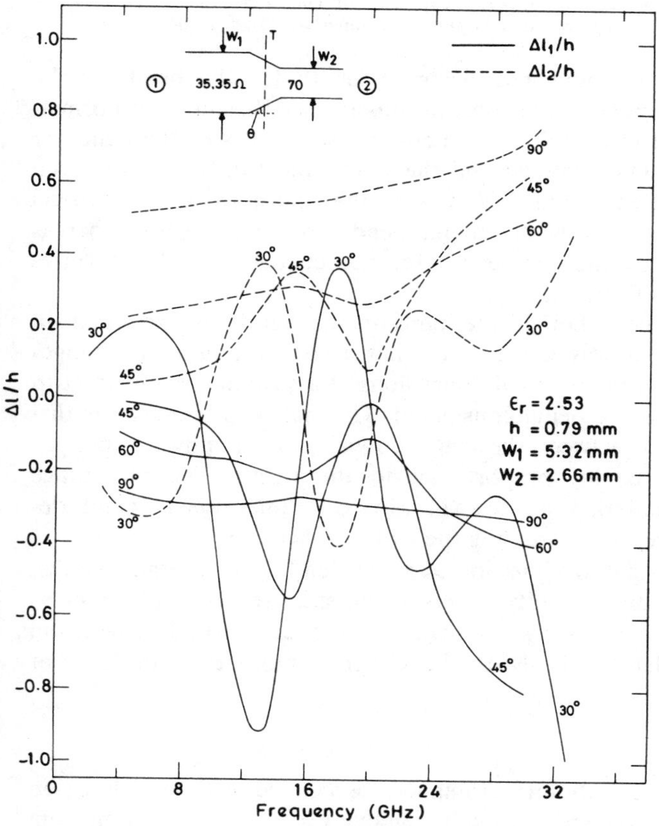

Fig. 2. Normalized shift in the effective step reference planes for the uncompensated and compensated step discontinuities with a 1:2 impedance ratio.

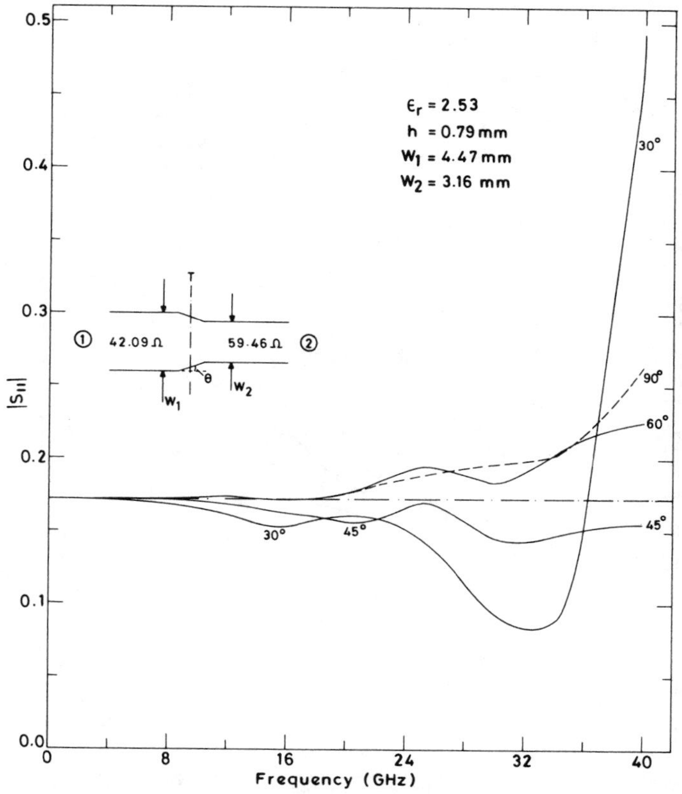

Fig. 3. Reflection coefficients for uncompensated and compensated step discontinuities with a $1:\sqrt{2}$ impedance ratio.

planes is plotted in terms of normalized electrical length. It is seen that the performance of a chamfered step with $\theta = 60°$ is closest to the ideal behavior.

It may be noted that for the case considered in Fig. 1, cutoff frequency for the first higher order (TE_{10}) mode in the wider line is 17.723 GHz. However, for the TEM-mode input, the TE_{10} mode would not be excited because of the symmetry in the step. However, the next higher mode (TE_{20} with a cutoff of 35.445 GHz) would be excited in this case. In Fig. 1, the increase in $|S_{11}|$ seen near 32 GHz is related to the proximity of the cutoff of the TE_{20} mode.

A symmetrical step with an impedance ratio of $1:\sqrt{2}$ has also been considered. Unchamfered ($\theta = 90°$) and chamfered steps (with θ equal to 30°, 45°, and 60°) have been analyzed. The magnitudes of the reflection coefficients in the four cases are plotted as functions of frequency in Fig. 3. The desired magnitude of the reflection coefficient is 0.172. The normalized electrical lengths corresponding to the separation between the step plane (midway in the taper) and the equivalent step reference planes (computed from S_{ii}) are plotted for the four cases in Fig. 4. In this case also, 60° taper may be chosen for a performance closest to the ideal behavior, especially when the shift in the reference plane of the step is considered.

In both of the examples considered, it is seen that for $\theta = 45°$, 60°, or 90°, Δl_1 is negative and Δl_2 is positive. The S_{11} and S_{22} become real at planes which are shifted towards the line of greater width. It may be noted that this shift is from the magnetic wall model and is much smaller than the shift towards the line of smaller width which is involved in obtaining the magnetic wall model from the physical dimensions.

The results for the symmetrical steps discussed above also correspond to their even-mode half-sections (unsym-

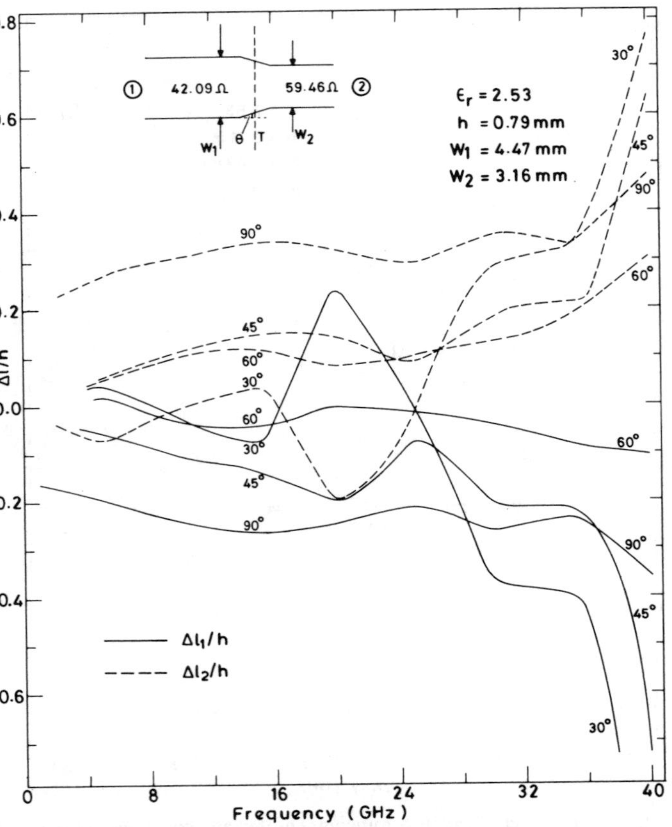

Fig. 4. Normalized shift in the effective step reference planes for the uncompensated and compensated step discontinuities with a $1:\sqrt{2}$ impedance ratio.

metrical step with half the line widths). Chamfering could also be used for steps with other impedance ratios.

III. Right-Angled Bends

Compensation of right-angled microstrip bends has been obtained experimentally by chamfering the corner [2]. However, no analytical results have been reported so far. In this section, theoretical results are presented for a right-angled bend in planar waveguides. It is seen that the percentage of chamfer required remains constant, independent of dielectric constant, width, etc.

Variation in reflection coefficient for an unchamfered right-angled bend is shown in Fig. 5. The high values of $|S_{11}|$ above 25 GHz are related to the next higher mode cutoff at 30.607 GHz. To reduce the effect of the discontinuity reactances, a right-angled triangular portion can be removed from the corner. As the line widths on the two sides of the bend are equal, symmetry suggests that the chamfer be an isosceles triangle. The size of the triangle removed has been optimized, so that the magnitude of the reflection coefficient is minimum. The optimum value of the smaller sides of the triangle equals 0.828 times the width W. This is independent of all other parameters and it makes the distance from the inner corner to the opposite chopped edge also equal to $0.828W$. The variation of reflection coefficient for an optimally chamfered bend is also shown in Fig. 5.

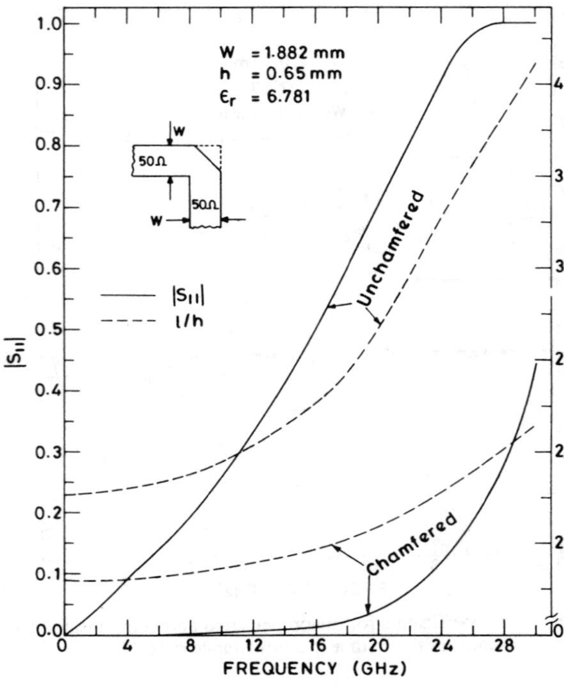

Fig. 5. Reflection coefficients and normalized electrical lengths for uncompensated and optimally compensated right-angled bends.

The equivalent electrical length for the bend can be computed from the transmission coefficient. The normalized electrical length between two planes of the bend for the unchamfered and the chamfered bends has also been plotted in Fig. 5. It is seen that the normalized electrical length for the chamfered bend varies only slightly whereas the normalized length for the unchamfered bend varies significantly.

Even though the percentage chamfering required in planar waveguide model has a constant value, it changes when the physical dimensions of a stripline or a microstrip bend are being considered. This happens because the distance between the magnetic wall and the physical periphery is not proportional to the width. The percentage chamfering needed (in physical dimensions) would decrease on increasing the width-to-height ratio.

If the two lines joined at the bend are of unequal widths, the discontinuity effects would still get reduced by removing a right-angle triangle. In such cases, the line with the wider width should be chopped more than the line of smaller width.

IV. T-Junctions

Discontinuity reactances associated with a T-junction can be compensated by removing a triangular portion from the junction as shown in the Fig. 6 inset. In this section, two symmetric T-junctions with branch line impedances either equal to the main line impedance or $1/\sqrt{2}$ of the main line impedance are considered. These T-junctions are used frequently in stubs, SPDT switches, hybrids, power dividers, couplers, etc.

The magnitudes of the reflection coefficients at the branch line and at the main line for the $1/\sqrt{2}:1:1$ T-junc-

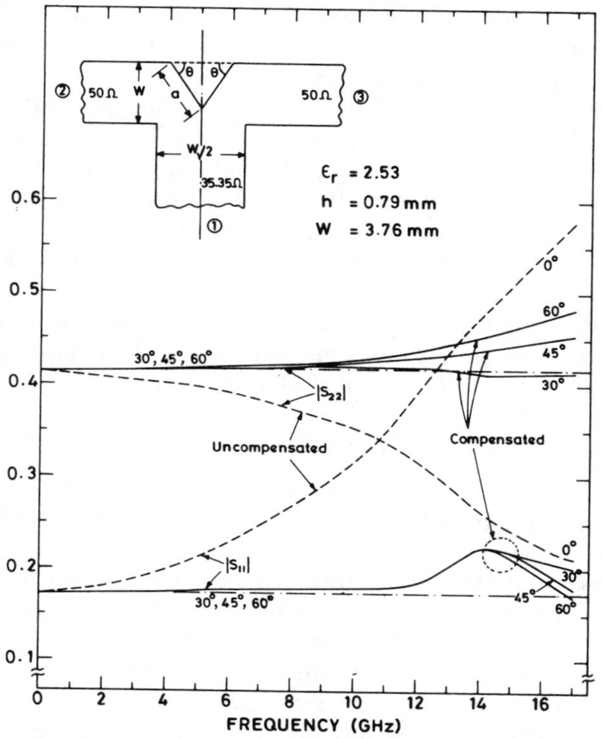

Fig. 6. Main line and branch line reflection coefficients for uncompensated and compensated T-junctions (impedance ratio $1/\sqrt{2}:1:1$).

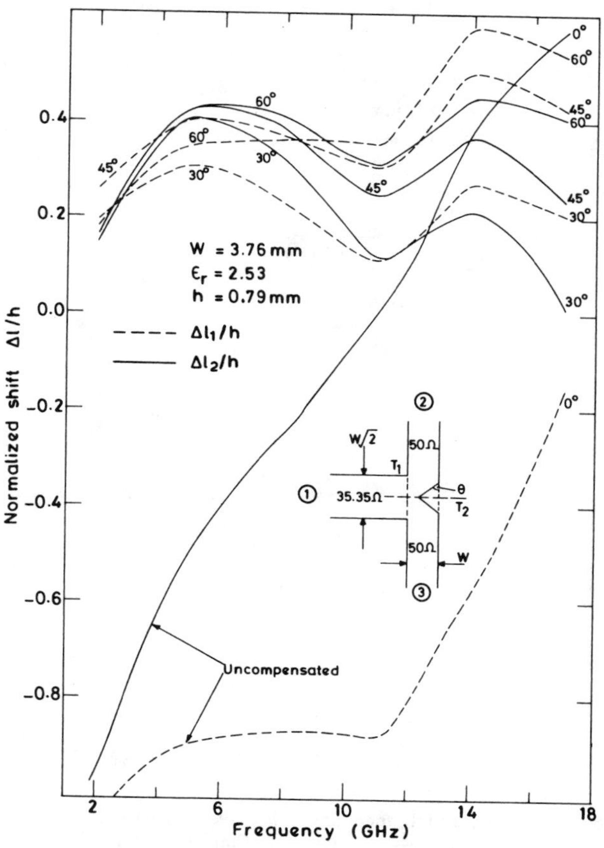

Fig. 7. Normalized shift in the effective junction planes for the uncompensated and compensated T-junctions (impedance ratio $1/\sqrt{2}:1:1$).

tion are shown as functions of frequency in Fig. 6.

As in the case of a step discontinuity, the phases of the reflection coefficients are used to obtain the locations of the effective junction reference planes where S_{11}, S_{22} become real. The shifts Δl_i in the effective junction planes are taken as follows. For the branch port the shift is taken from the plane where the branch line joins the main line (plane T_1 in Fig. 7) and for the ports on the main line the shift is from the plane midway between the two ports on the main line (plane T_2 in Fig. 7). The normalized electrical lengths representing these shifts are plotted in Fig. 7 for ports 1 and 2. The positive and negative signs for Δl_i are taken in the same manner as for the step discontinuity.

For compensation of discontinuity reactances, the removal of isosceles triangles with the angle θ equal to 30°, 45°, and 60° has been considered. The sides "a" of the isosceles triangles removed have been optimized for each of the three values of θ. In optimization, the deviations of the magnitudes of the branch line and the main line reflection coefficients from their ideal values (0.172 and 0.414, respectively) are minimized. The optimum values of "a" are $0.851W$, $0.807W$, and $0.879W$ for θ values equal to 30°, 45°, and 60°, respectively. The variations of the magnitudes of reflection coefficients at the branch line and the main line for the optimized junctions are shown in Fig. 6 for different values of θ. Normalized electrical lengths representing the shifts in the junction planes are plotted in Fig. 7. It is seen that by chopping off an isosceles triangle with $\theta = 30°$ the effects of discontinuity reactances are minimized.

Equi-impedance T-junctions (1 : 1 : 1) have also been considered for compensation. Sides "a" of the isosceles triangles to be removed have again been optimized for the same three values of θ. The deviations of magnitudes of S_{11} and S_{22} from the ideal value of 1/3 are minimized. The optimum values of "a" are $0.601W$, $0.569W$, and $0.666W$ for θ values equal to 30°, 45°, and 60°, respectively. The variations of the magnitudes of the reflection coefficients for the uncompensated and compensated junctions are shown in Fig. 8. Normalized electrical lengths representing the shift in the junction planes are plotted in Fig. 9 for the uncompensated and compensated junctions. In this case, the discontinuity reactances are again minimized by removing an isosceles triangle with $\theta = 30°$.

In optimizing the performance of the junctions, the side of the triangle is chosen such that the magnitudes of the reflection coefficients at ports 1 and 2 are close to their ideal values (giving equal weights to both the deviations). A least-square optimization procedure has been employed. The side of the triangle chopped off would be different for different values of relative weightings given to the deviations in the two reflection coefficients.

The side of the triangle to be removed is found to vary with the frequency at which optimization is carried out. However, this variation is quite small. In the results discussed above, optimization is carried out at two spot frequencies and the mean value is considered for "a". For the case of equi-impedance T-junction, optimum values of a/W (with $\theta = 45°$) at 8 GHz and at 15 GHz are 0.544 and 0.594, respectively. Also, it is observed that for a given Z_0 and height of the substrate, the ratio of optimum side of

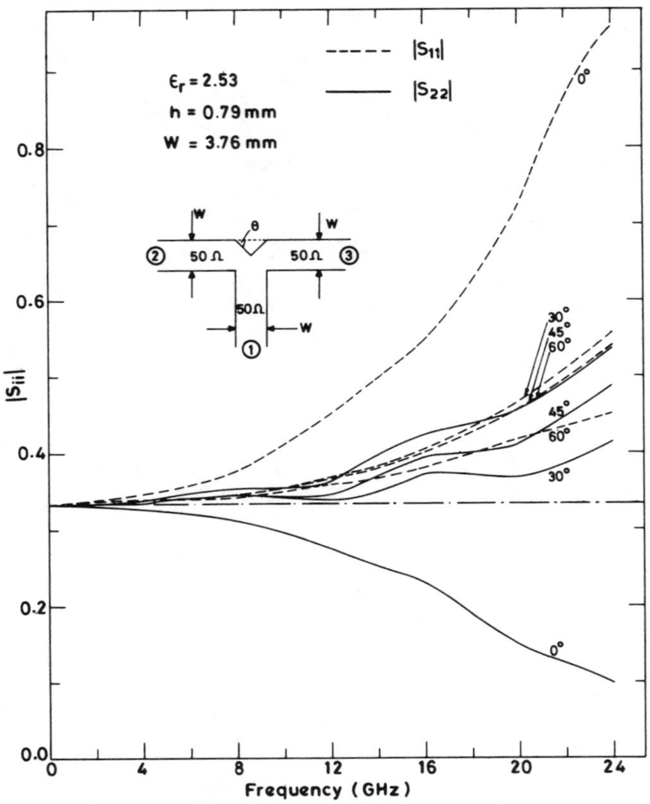

Fig. 8. Main line and branch line reflection coefficients for uncompensated and compensated T-junctions (impedance ratio 1 : 1 : 1).

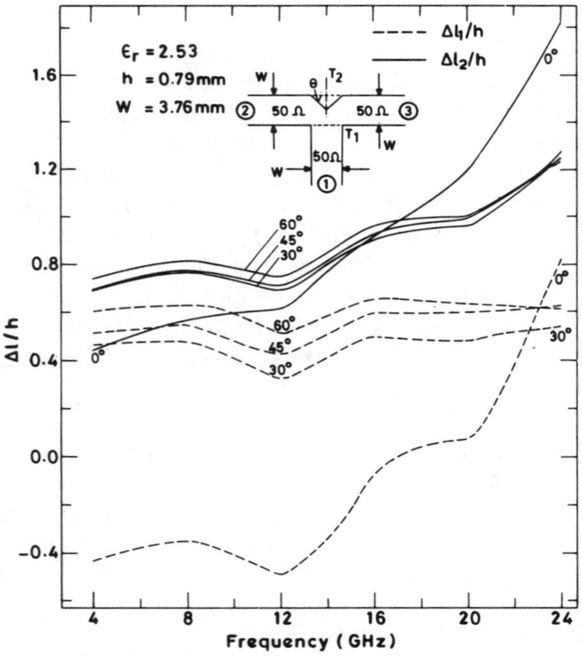

Fig. 9. Normalized shift in the effective junction planes for the uncompensated and compensated T-junctions (impedance ratio 1 : 1 : 1).

the triangle to be removed to the width of the line is independent of the dielectric constant of substrate. Thus if the frequency dependence of the optimum side "a" is ignored, the optimum value of the ratio a/W is independent of all other parameters.

By considering a plane of symmetry located half way through the width of the branch line, the T-junction can be compared with a right-angled bend on one side of this plane. The reflection coefficient at the branch line of the T-junction is equal to that at the corresponding side of the bend. Thus, if only the branch line reflection coefficient is to be optimized, only a right-angled bend need be considered. In such a case, the results discussed in Section III also correspond to a T-junction with branch line impedance equal to one half of the main line impedance.

V. Concluding Remarks

Compensation of discontinuity reactances caused by steps, right-angled bends, and T-junctions in planar waveguides has been discussed. These results can be applied directly to the planar transmission lines for which the outward extension of the periphery to obtain the equivalent magnetic wall model is known. Thus, the results presented should find applications in the design of MIC's using such planar transmission lines.

The compensation has been carried out by removing appropriate triangular portions from the discontinuity configurations. This technique can be used for compensation of other types of discontinuities. The method discussed herein could also be used for compensation of discontinuities in waveguides.

References

[1] K. C. Gupta and R. Chadha, "Design real-world stripline circuits," *Microwaves*, vol. 17, no. 12, pp. 70–80, Dec. 1978.

[2] R. J. P. Douville and D. S. James, "Experimental study of symmetric microstrip bends and their compensation," *IEEE Trans. Microwave Theory Tech.*, vol. MTT-26, pp. 175–182, Mar. 1978.

[3] M. Dydyk, "Master the T-junction and sharpen your MIC designs," *Microwaves*, vol. 16, no. 5, pp. 184–186, May 1977.

[4] W. Menzel, "Frequency dependent transmission properties of microstrip Y-junctions and 120°-bends," *Inst. Elec. Eng. J. Microwaves, Opt., Acoust.*, vol. 2, pp. 55–59, 1978.

[5] R. Mehran, "Compensation of microstrip bends and Y-junctions with arbitrary angle," *IEEE Trans. Microwave Theory Tech.*, vol. MTT-26, pp. 400–405, June 1978.

[6] K. C. Gupta, R. Chadha, and P. C. Sharma, "Two-dimensional analysis for stripline/microstrip circuits," in *1981 IEEE MTT-S Int. Symp. Dig.*, pp. 504–506, June 1981.

[7] T. Okoshi and T. Miyoshi, "The planar circuit—An approach to microwave integrated circuitry," *IEEE Trans. Microwave Theory Tech.*, vol. MTT-20, pp. 245–252, Apr. 1972.

[8] R. Chadha and K. C. Gupta, "Segmentation method using impedance matrices for the analysis of planar microwave circuits," *IEEE Trans. Microwave Theory Tech.*, vol. MTT-29, pp. 71–74, Jan. 1981.

[9] P. C. Sharma and K. C. Gupta, "Desegmentation method for analysis of two-dimensional microwave circuits," *IEEE Trans. Microwave Theory Tech.*, vol. MTT-29, pp. 1094–1098, Oct. 1981.

[10] T. Okoshi and T. Takeuchi, "Analysis of planar circuits by segmentation method," *Electron. Commun. Japan*, vol. 58-B, no. 8, pp. 71–79, Aug. 1975.

[11] R. Chadha and K. C. Gupta, "Green's functions for triangular segments in planar microwave circuits," *IEEE Trans. Microwave Theory Tech.*, vol. MTT-28, pp. 1139–1143, Oct. 1980.

Paper 8.3

MULTIPORT NETWORK MODEL FOR EVALUATING RADIATION LOSS AND SPURIOUS COUPLING BETWEEN DISCONTINUITIES IN MICROSTRIP CIRCUITS*

Albert Sabban and K.C. Gupta

Department of Electrical and Computer Engineering

University of Colorado

Boulder, Colorado 80309-0425

Abstract

This paper presents a convenient method for evaluating radiation from microstrip discontinuities. The multiport network model is used to find voltage distribution around discontinuity edges and an equivalent magnetic current model is used to compute the external fields produced. As an example, the results show that for a 90° bend in 50Ω line on 10 mil thick substrate with ε_r = 2.2, the radiation loss is 0.1 dB at 30 GHz. Electromagnetic coupling between two discontinuities is evaluated by finding the currents induced by the fields of one of the discontinuities at the location of the second discontinuity.

Introduction

Because of the open nature of the microstrip configuration, hybrid and monolithic microwave circuits suffer from radiation originating at various geometrical discontinuities. Two consequences of this radiation phenomenon are: additional signal loss in the circuit and undesired interactions between different parts of the circuit due to external electromagnetic coupling. These phenomena become significant in two different situations: first, when attempts are made to increase circuit density in monolithic microwave circuits, more bends and other discontinuities are introduced and spurious electromagnetic coupling increases considerably. Secondly, in microstrip antenna arrays, relatively thicker substrates are used and the feed network printed on the same substrate can result in substantial spurious radiation.

Estimates of the radiation loss from microstrip discontinuities have been attempted previously [1-6]. Most of these results are based on the Poynting vector method developed by Lewin [1]. In this approach, a line current located at the middle of the microstrip line is taken as a source of radiation. Thus the method is applicable to narrow microstrip lines. It has been applied to a 90° bend, a step discontinuity, an open end, a line terminated in an impedance, and a matched symmetrical T-junction. The approach is not easily extendable to more complicated geometries found in practical microwave circuits.

This paper presents a convenient method for estimating radiation from microstrip discontinuities of generalized shapes. A planar multiport network model of the discontinuity configuration and the segmentation method is used to evaluate voltage distribution around the edges of the discontinuity. This voltage distribution is expressed as an equivalent magnetic current line source distribution which is used to calculate the far-zone field (for radiation loss) or the near-field at the location of the other discontinuity for spurious coupling calculations.

* This work has been sponsored by the NSF Industry/University Cooperative Research Center for Microwave/Millimeter-Wave Computer-Aided Design at the University of Colorado.

This approach is suitable for being included in microwave CAD packages. Numerical results have been obtained for radiation from bends, steps and T-junctions and are found to be in reasonable agreement with the Poynting vector results wherever available.

Multiport Network Modeling of Microstrip Discontinuities

Multiport network modeling of discontinuity configurations [7-10] is based on parallel plate waveguide model [11] of microstrip lines. Similar network modeling approach has been used earlier for analysis of microstrip patch antennas [12] and for calculating mutual coupling between microstrip patches [13].

Planar Network Model for Microstrip Discontinuities

The planar circuit model for microstrip discontinuities is derived from the planar waveguide model for microstrip lines [11]. The planar waveguide model consists of two parallel conductors bounded by magnetic walls in the transverse directions. The electric and magnetic fields inside the planar model are uniform along the thickness of the substrate. The width of the waveguide model $W_e(f)$ is made larger than the physical microstrip width in order to account for the fringing fields at the edges and is given by

$$W_e(f) = \frac{\eta_0 h}{Z_0(f)\sqrt{\varepsilon_{re}(f)}} \tag{1}$$

where $Z_0(f)$ is the characteristic impedance, $\varepsilon_{re}(f)$ is the effective dielectric constant, h is the substrate height and η_0 is the wave impedance in free space.

As the frequency dependences of Z_0 and ε_{re} are incorporated in (1), the dispersion effects get built in the analysis of the discontinuities also. The planar models for discontinuity configuration are obtained by replacing the physical widths of the microstrip lines by effective widths given by (1). The effective dimensions in the discontinuity region are obtained by extrapolating the effective edges for the connecting lines (as shown for a 90° bend in Figure 1a) or having an outward extension equal to that for the adjoining microstrip line (as shown in Figure 1b). The planar modeling approach has been used extensively for characterization and compensation of microstrip discontinuities [7-10].

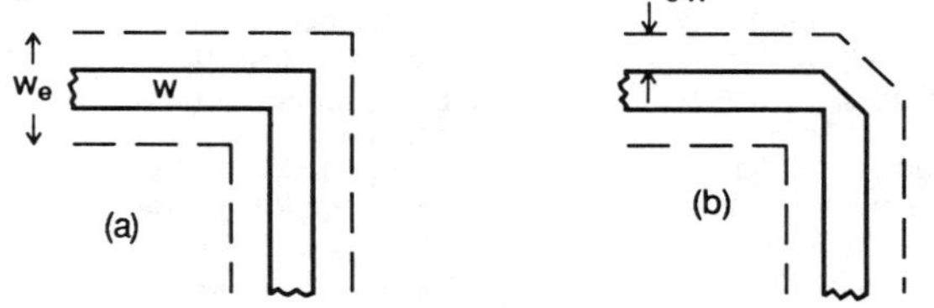

Figure 1. Planar models for microstrip bends.

Reprinted from *IEEE MTT-S Int. Microwave Symp. 1989 Digest*, pp. 707–710, 1989.

Multiport Network Model for Evaluating Edge Voltages

In order to evaluate the external fields produced by a microstrip discontinuity, we first obtain voltages at the edges of a microstrip structure. A multiport network model, similar to that developed for microstrip patch antennas [12,13] is employed. For implementing this method, we add a number of open ports at the edges of the discontinuity structure from which the radiation (or spurious external coupling) is being evaluated. This is shown in Figures 2(a) and 2(b) for a right-angled bend and a compensated right-angled bend, respectively.

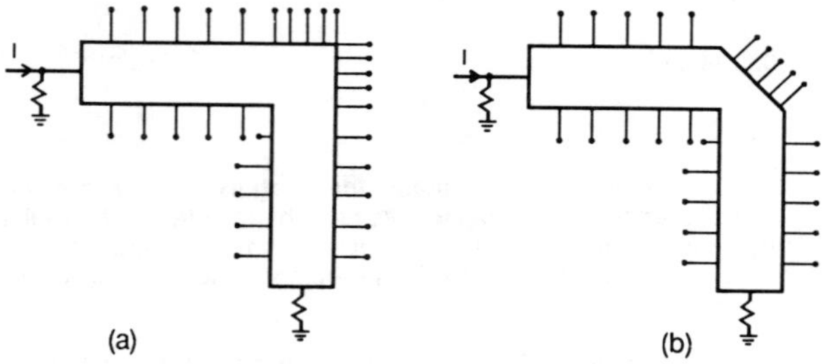

Figure 2. Multiple ports located at the edges of the microstrip discontinuities.

Lengths of transmission lines on two sides of the junction are taken large enough so that the higher order evanescent modes produced by the discontinuities decay out at the locations of external ports 1 and 2. The circuit behavior is simulated by terminating the port 2 in a matched load and adding a matched source to the port 1. Voltages at the N ports at the edges are computed by using the following procedure:

(i) The configuration is broken down in elementary regular segments, connected together at the interfaces by a discrete number of interconnections.

(ii) Z-matrices for each of these elementary segments are evaluated by using the Green's function approach for individual geometries.

(iii) Individual Z-matrices obtained in (ii) are combined together by using the segmentation formula.

(iv) Overall multiport Z-matrix is used for calculating voltages at the N edge ports for a unit current input at the port 1.

As mentioned earlier, a similar procedure has been used for design of microstrip patch antennas [13]. The only distinction in the latter case is the use of edge admittance networks (containing equivalent radiation conductances) which are connected to the edge ports. Because of the non-resonant nature of the microstrip discontinuity structures, the radiated power is small and the edge voltages may be assumed to be unaffected by radiation conductances involved. However, for a more accurate assessment of the radiated power, radiation conductance networks may be added to edge ports and iterative computations may be carried out for evaluating radiation fields.

Evaluation of the Radiated Power

Voltages at the discontinuity edges are represented by equivalent magnetic current sources as shown in Figures 3(a) and 3(b). Each of the magnetic current line sources is divided into small sections over which the field may be assumed to be uniform. The amplitude M of each of the magnetic current elements is twice that of the edge voltage at that location and the phase of M is equal to the phase of the corresponding voltage. The total radiation is computed using the superposition of the far field radiated by each section. Referring to the coordinate system, shown in Figure 4, the far-field

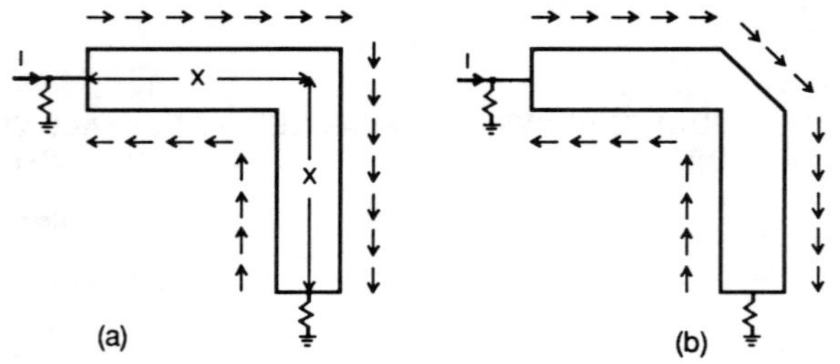

Figure 3. Equivalent magnetic current distribution at discontinuity edges.

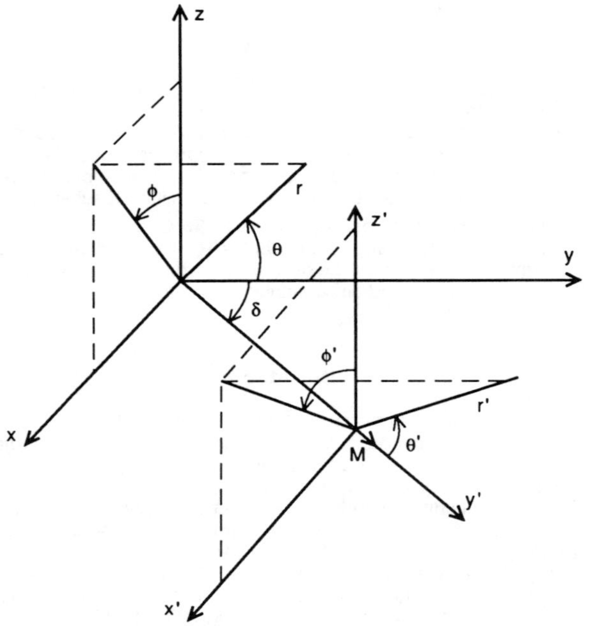

Figure 4. Coordinate system for external field calculations.

pattern may be written in terms of voltages at the various elements. With the voltage at the i-th element as $(V(i)e^{j\alpha(i)})$, we have

$$F(\theta,\phi) = \sum_{i=1}^{N} V(i)\, W(i) \exp\{ (k_0\gamma_0(i) + \alpha(i)\}\, F_i(\theta,\phi) \tag{2}$$

where

$$F_i(\theta,\phi) = \frac{\sin\left(\frac{k_0 W(i)}{2}\cos\theta\right)}{\frac{k_0 W(i)}{2}\cos\theta}\sin\theta$$

$$\gamma_0(i) = X_0(i)\sin\theta\cos\phi + Y_0(i)\cos\theta$$

and N is the number of ports, $X_0(i)$, $Y_0(i)$ specify the location of the i-th magnetic current element, and W(i) is the width of the i-th element.

The radiated power is calculated by the integration of the Poynting vector over the half space and may be written as:

$$P_r=\frac{1}{240\pi}\int_{-\pi/2}^{\pi/2}\int_0^{\pi}\left(|E_\theta|^2+|E_\phi|^2\right)r^2\sin\theta\, d\theta\, d\phi \qquad (3)$$

The fields E_θ and E_ϕ are expressed in terms of $E(\theta,\phi)$ as:

$$E_\theta=\hat{a}_\theta\,(-2j\,F(\theta,\phi)\,F_\theta) \qquad (4)$$

$$E_\phi=\hat{a}_\phi\,(-2j\,F(\theta,\phi)\,F_\phi) \qquad (5)$$

$$F_\phi=\sin\phi'\sin\phi+\cos\delta\cos\phi\cos\phi' \qquad (6)$$

$$F_\theta=-\sin\phi'\cos\theta\cos\phi+\cos\delta\cos\theta\sin\phi+\sin\delta\cos\phi'\sin\theta \qquad (7)$$

$$\cos\theta'=\sin\theta\sin\phi\sin\delta+\cos\theta\cos\delta \qquad (8)$$

$$\cos\phi'=\sin\theta\cos\phi\,/\sqrt{1-\cos^2\theta'} \qquad (9)$$

The radiation loss may be expressed as:

$$\text{Power loss (dB)}=10\log_{10}\left(1-\frac{P_r}{P_i}\right) \qquad (10)$$

where P_i is the input power at port 1.

Evaluation of Spurious Coupling Between Two Discontinuities

The spurious coupling between two discontinuities (due to the external fields) may be incorporated in the Multiport Network Model by connecting an additional multiport network between the two discontinuities as shown in Figure 5. The coupling network MCN is characterized in terms of an admittance matrix $[Y_m]$. Elements of this matrix represent mutual admittances between various sections of the edges of the two discontinuities. The terms of the matrix $[Y_m]$ are obtained by representing the edge fields by small sections of length $d\ell$ of equivalent magnetic current and computing the magnetic field (H_θ, H_r) produced at the j-th subsection of the nearby discontinuity. We have

$$H_\theta=j\frac{k_0 M d\ell\sin\theta}{4\pi\eta_0 r}\left(1+\frac{1}{jk_0 r}-\frac{1}{(k_0 r)^2}\right)e^{-jk_0 r} \qquad (11)$$

$$H_r=\frac{M d\ell\cos\theta}{2\pi\eta_0 r^2}\left(1+\frac{1}{jk_0 r}\right)e^{-jk_0 r} \qquad (12)$$

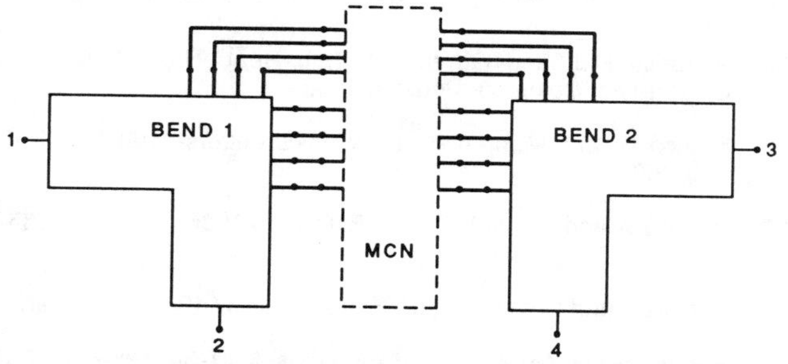

Figure 5. A multiport network (MCN) incorporated for modeling spurious coupling between discontinuities.

The induced current in the j-th element is calculated from the magnetic field as:

$$J_j=(\hat{z}\times\overline{H})\cdot\hat{j} \qquad (13)$$

where $\hat{j}$ is a unit vector normal to the j-th element. The value of Y_{ji} is obtained from the current induced in the j-th subsection as a result of voltage V_i at the i-th subsection.

$$Y_{ji}=J_j\,d\ell_j/V_i \qquad (14)$$

The Z-matrix of the mutual coupling network is the inverse of $[Y_m]$. The segmentation method is used to combine the Z matrix representations of the discontinuities and the coupling network to yield the overall Z matrix. The Z matrix is converted to S-parameters. The effect of the coupling on the circuit performance is obtained from these S-parameters.

It may be noted that a similar procedure has been employed earlier to model the mutual coupling between patches of a microstrip antenna array successfully [13].

Computational Details

Z-Matrix Evaluation

For rectangular planar segments, the impedance matrix elements can be expressed in a single infinite series [14]. The numbers of terms needed for the convergence of the summation was found to be around 100.

The Z matrix elements for sections of transmission line with lengths equal to multiples of half-wavelength become infinitely large. So these line lengths should be avoided. Also for quarter wave sections, some of the Z matrix elements become zero. Thus, for numerical accuracy, it is advisable to avoid quarter wave sections also. The length of each section on either side of the discontinuity should be greater than twice the width of the microstrip line. This allows the higher order evanescent modes to decay out at locations of the external ports (1 and 2 in Figure 2). Also, it is found that to obtain accurate results, the width of the ports at the edges should be less than 0.075 wavelength. Computational results for different widths of interconnected ports were also compared and it is found that the optimum width of interconnected port is about 0.05 wavelength.

In order to identify the regions of the discontinuity configuration that contribute dominantly to the radiated power, several computations were performed by taking different regions around the discontinuity. These computations show that in most of the cases the biggest contribution to the radiated power is due to ports in the region of the discontinuity itself. The results for a right-angled bend, shown in Table 1, illustrate this point.

Table 1: Radiated power as a function of the length of the microstrip line
(ε_r = 12.9, h = 127.1 µm, f = 10 GHz)

Length X (shown in Figure 3)	P_r/P_i *dB*
(0.4 x 0.4) λ	-40.64
(0.35 x 0.35) λ	-40.67
(0.3 x 0.3) λ	-40.77
(0.25 x 0.25) λ	-40.84
(0.2 x 0.2) λ	-41.1
Only ports on bend	-41.1

Radiation from a Right-Angled Microstrip Bend

Results for the power radiated by the 90° bend normalized with respect to the input power is shown in Figure 6 for frequency ranges from 10 GHz to 40 GHz. These results are in good agreement with results based on the complex Poynting vector method [1], which are also plotted in Figure 6. The radiation loss at 40 GHz is a 0.0062 dB for a 50Ω line bend on GaAs (ε_r = 12.9) substrate. The values for radiation loss from a 90° bend in 50Ω line on a substrate with ε_r = 2.2 at 30 GHz and 40 GHz are 0.1 dB and 0.17 dB, respectively.

Radiation from a Microstrip Step Junction

Power radiated from a step junction discontinuity, change in impedance from 50Ω to 10Ω, on ε_r = 2.2 substrate with thickness of 0.79 mm is plotted in Figure 7. A similar computation for a 50Ω to 70.7Ω junction (at 30 GHz, ε_r = 2.2, thickness 0.02 inch), yields the normalized radiated power to be -24.8 dB when the input power is fed from the 50Ω line and -33 dB when the power is fed from the70.7Ω line.

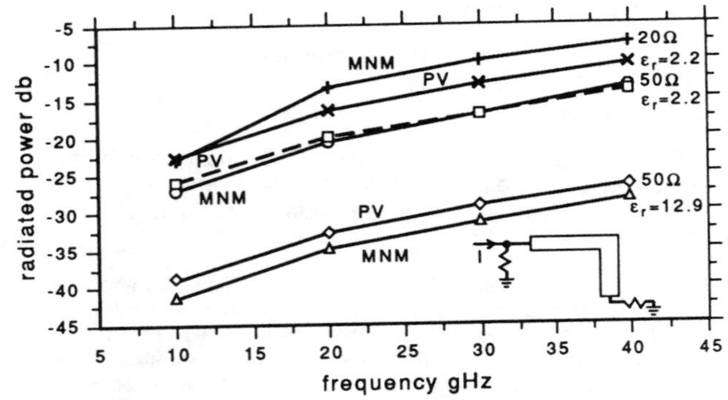

Figure 6. Normalized radiated power from a right angled bend (MNM: Multiport Network Model, and PV: Poynting Vector method [1]).

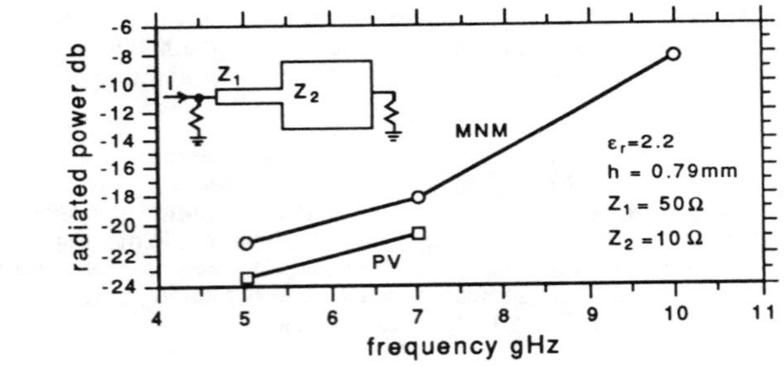

Figure 7. Normalized radiated power from a step (MNM: Multiport Network Model, and PV: Poynting Vector method [1]).

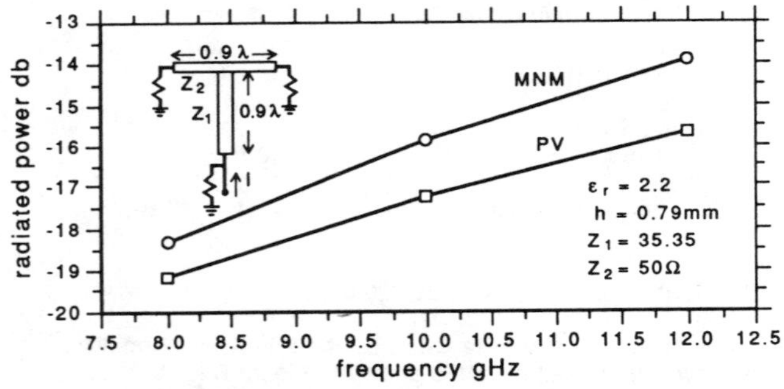

Figure 8: Normalized radiated power from a T-junction (MNM: Multiport network model, and PV: Poynting Vector method [1]).

Radiation from a Tee-Junction Discontinuity

Computed results for power radiated from a T-junction (50W main line with 35.35W branch line on a substrate thickness 1/32" and e_r = 2.2) are shown in Figure 8. The radiation loss at 12 GHz is 0.18 dB. It is found that for such a T-junction that most of the contribution to the radiation losses originates from the region of the junction.

Concluding Remarks

The results reported in this paper show that the multiport network model is a convenient and versatile method for evaluating radiation and spurious coupling associated with discontinuities in microstrip circuits.

Radiation losses from microstrip discontinuities cannot be neglected at higher frequencies. For a 10Ω to 50Ω step change (ε_r = 2.2, h = 0.79 mm) in width, the radiation loss at 10 GHz is 0.7 dB. For a 90° bend in 50Ω line on a 10 mil thick substrate with ε_r = 2.2, the radiation loss at 40 GHz is 0.17 dB. The multiport network model can be used for calculating radiation and coupling from complicated discontinuity configurations such as chamfered bends, compensated T-junctions and cross-junction. More detailed results will be reported at the meeting. Experimental verification is being planned.

Computation results are in good agreement with the Poynting vector method applicable for simple geometries.

This approach has been extended to computation of spurious coupling between two discontinuities in different parts of a microwave circuit. This algorithm is being implemented and numerical results will be presented at the Symposium.

References

[1] Lewin L., Proc. Inst. Elec. Eng.-117(C), 1960, pp.163-170.

[2] Lewin L., IEEE Trans. MTT-26, 1978, pp. 893-894.

[3] Lewin L., Proc. Inst. Elec. Eng.-125, 1978, pp. 633-642.

[4] Abouzahra M. and Lewin L., IEEE Trans. MTT-27, 1979, pp. 722-723.

[5] Abouzahra M.D., IEEE Trans. MTT-29, 1981, pp. 666-668.

[6] Hoffman R.H., chapter 11 in Handbook of Microwave Integrated Circuits, Artech House, 1987, pp. 311-321.

[7] Wolff I. et al., Electronic Letters-8, 1972, pp. 45-77.

[8] Menzel W. and Wolff I., IEEE Trans. MTT-25,1977, pp.107-112.

[9] Chadha R. and Gupta K.C., IEEE Trans. MTT-30, 1982, pp. 2151-2156.

[10] Maramis H.J. and Gupta K.C.,Sci. Rept. No. 96, EM Lab, University of Colorado, Boulder, 1988.

[11] Kompa G. and Mehran R., Electronics Letters-11(9), 1975, pp. 459-460.

[12] Benalla A and Gupta K.C., IEEE Trans. AP-36, 1988, pp. 1337-1342.

[13] Benalla A and Gupta K.C., IEEE Trans. AP-37,1989 (to appear).

[14] Benalla A and Gupta K.C., IEEE Trans. MTT-34, 1986, pp. 733-736.

Accurate Numerical Modeling of Microstrip Junctions and Discontinuities

Doris I. Wu,[1] David C. Chang,[1] and Brad L. Brim[2]

[1]MIMICAD Center, Department of Electrical and Computer Engineering, University of Colorado, Boulder, Colorado 80309

[2]Hewlett Packard Network Measurements Division, Santa Rosa, California 95401

Received April 13, 1990; revised August 22, 1990.

ABSTRACT

A general numerical solver for analyzing microstrip geometries of rectangular shape is presented in this paper. The analytical foundation of this solver is based on an integral equation approach which we formulate in the spatial domain. The unknown surface current on the microstrip is solved by the method of moments using 2D rectangular pulses as the expansion functions. Transmission-line modeling is then used to parameterize a given microstrip junction or discontinuity. Aided by a graphics interface, this solver can analyze complex structures without incurring additional analytical complexity. We illustrate the accuracy and versatility of our solver by applying it to several different microstrip discontinuities ranging from a single-stub to an interdigitated capacitor.

1. INTRODUCTION

Existing computer-aided design (CAD) packages for designing microstrip circuits have several desirable characteristics: they are simple to use and they perform real-time simulations. However, to yield good results it is not only crucial to have accurate models for the different junctions and discontinuities, but effects such as radiations and parasitic couplings must also be accounted for in the simulation process. As pointed out by many in the literature [1,2], the need for more accurate CAD tools has become essential in keeping pace with the latest GaAs MMIC technology.

The purpose of this paper is to describe an integral-equation-based numerical solver, which we developed and implemented in a workstation environment, capable of producing highly accurate design databases for the various bends, junctions, and closely spaced circuit elements. Although the emphasis of this paper will be on the application of our solver, we will first describe briefly the analytical foundation for our solver. As will be illustrated, the strength of our solver lies in its accuracy as well as its generality.

2. ANALYTICAL PROCEDURE

2.1 Mixed Potential Integral Equation

In analyzing microstrip structures, a full two-dimensional (2D) integral equation formulation combined with the method of moments approach generally yields the most accurate results. By solving for the components of the E and H fields normal to the slab surface and imposing the boundary conditions on the surface of the upper conductor [3], we obtain a mixed potential integration equation (MPIE) for microstrip structures. The same MPIE can also be obtained by solving for the vector and scalar potentials associated with the electric field of a microstrip structure [4]. Compared to the more traditional electric-field integral-equation approach, the MPIE lends itself well to spatial domain evaluation. Although the choice of evaluating the integral equation in the spectral or the spatial domain is arbitrary because the two approaches are physically equivalent, they are different in the numerical implementation.

Spectral domain evaluation is most useful for treating microstrips of simple shape, where the current distributions can be expanded using a set of functions having closed-form Fourier-transformed counterparts. Analyses involving semi-infinitely long lines are ideal for this approach since the currents on these lines can be represented by a pair of forward and backward traveling waves, which in turn can be represented by simple Dirac delta functions in the Fourier transform domain. Therefore, in analyzing microstrip discontinuities using this approach, semi-infinitely long lines are often used as standard feed strips [5,6]. However, since the current in and near the discontinuity region is not uniform, a different set of subsectional basis functions is needed to capture the junction effect. This hybrid use of basis functions implies that each set of basis functions must be carefully defined over the structure, and different algorithms are needed to evaluate the moment integrals associated with each type of basis functions. Moreover, to obtain accurate results, this approach also requires the pre-computation of the propagation constant and/or transverse distributions of the current on each distinct feed strip [5–7].

Spatial domain evaluation provides more physical insight since the problem remains in the physical domain. For this approach, the Green's functions are numerically evaluated first and treated as known functions in the integral equation. This implies that the selection of the expansion functions for the current is arbitrary, which renders the approach very versatile. Moreover, subsectional basis functions of simple form can be implemented with straightforward numerical algorithms using the spatial domain approach. Therefore, this approach provides an ideal base for a general solver. The groundwork for a MPIE-based, spatial domain microstrip solver can be attributed to Mosig and Gardiol [4]. Utilizing their basic approach, we present a numerical solver with modified algorithms and expanded generality applicable for most microstrip circuit junctions and discontinuities. As will be shown in this paper, complex structures can be treated using this solver without incurring additional analytical complexity.

The use of potentials is generally preferred in the spatial domain approach because the associated Green's functions are better suited for numerical evaluations. Using an e^{iwt} time convention, the MPIE can be expressed as

$$\vec{E}^s = \frac{-\iota\omega\mu_o}{4\pi}\int_S\left[G_m(\bar{x},\bar{x}')\vec{J}(\bar{x}')\,ds' - \frac{1}{k_o^2}\nabla\nabla'\cdot G_e(\bar{x},\bar{x}')\vec{J}(\bar{x}')\right]dS' = -\vec{E}^{\rm inc}(\bar{x});\ \bar{x}\text{ on }S, \quad (1)$$

where $\vec{J}$ is the unknown current density on the microstrip surface (S), and $\vec{E}^s$ and $\vec{E}^{\rm inc}$ denote the scattered and incident fields, respectively. The scalar Green's functions, G_m and G_e, are associated with the potentials produced by a unit current source on top of the grounded substrate. They are identified as Green's functions of the magnetic and electric type, respectively. By evoking the continuity equation

$$\nabla\cdot\vec{J} + \iota\omega q = 0, \quad (2)$$

it can be shown that G_m is related to the surface current density while G_e is related to the surface charge density, q. Thus, eq. (1) can also be expressed in terms of the surface current and charge densities. For a single-layer grounded dielectric slab in open space, these Green's functions can be expressed in terms of Sommerfeld integrals as [8,9]

$$G_m(R) = \int_0^\infty 2J_0(\xi R)\frac{\xi}{D_{TE}}\,d\xi, \quad (3)$$

$$R = |\bar{x} - \bar{x}'|,$$

$$G_e(R) = \int_0^\infty 2J_0(\xi R)\frac{\xi[u_o + u\tanh(ut)]}{D_{TE}D_{TM}}\,d\xi, \quad (4)$$

where

$$D_{TE} = u_o + u\coth(ut),$$
$$D_{TM} = \epsilon_r u_o + u\tanh(ut), \quad (5)$$

$$u_o = \sqrt{\xi^2 - 1},\ u = \sqrt{\xi^2 - \epsilon_r}. \quad (6)$$

R is the distance between the source and observation points, t is the slab thickness, and k_o is the free space propagation constant. Since these kernels are a function R, they can be precomputed for a given range of R. Once they are known, the integral eq. (1) can be solved readily in the spatial domain.

2.2 Evaluation of the Green's Functions

In evaluating the Sommerfeld integrals of eqs. (3) and (4), it is well known that for large R, the prudent approach is to deform the path of integration to yield at least one residue term and a branch cut integral with exponentially decaying integrand. However, for small distances, the integration along the positive real axis is often more efficient. To achieve efficiency, we subtract out the static term of the Green's function first [4,8] and evaluate the remaining Green's function expression numerically by deforming the contour to wrap around the branch cut [10].

In our solver, the static portion of the Green's function is handled analytically in the MPIE. The Green's functions without the static terms are numerically evaluated at discrete points for a given range of R predetermined by the maximum dimension of the microstrip geometry. Cubic spline approximations are then used to obtain a smooth curve fit for each of the Green's functions.

2.3 Galerkin/Moment Method

To solve the integral eq. (1) for the unknown current and charges, we use the Galerkin/Moment method. For simplicity, we choose to use 2D rectangular pulses as the expansion functions for the current. The expansion functions for the charges are obtained by applying a finite difference approximation to the continuity equation [11,12]. This yields a set of charge cells that are spatially shifted from the current cells. Similar to the current, the distribution in each charge cell also consists of a 2D rectangular pulse. To ensure consistency in our gridding process, we find it is best to divide the microstrip surface into elementary cells for the charges first and generate the current cells from the charge cells. The x- and y-current cells are generated by dividing the charge cells in halves, each along the direction of the current, and combining the adjacent halves systematically to generate new cells for the corresponding currents.

Along the edges of the microstrip, we have an additional boundary condition requiring the normal component of the current to be zero. To enforce this condition, we follow the procedure used in ref. [4] and introduce a narrow strip of cells along the edges normal to the current and force the current to be zero in these cells. For simplicity, this narrow strip is chosen to be half the width of the charge cells along each edge. Figure 1 is an illustration of the x- and y-current grids resulting from a given set of charge cells.

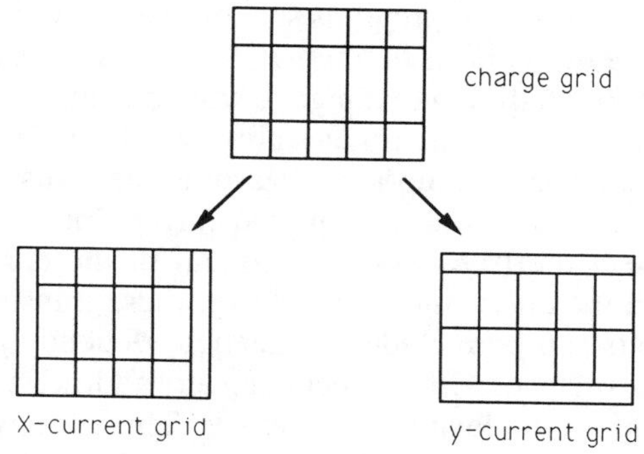

Figure 1. Charge and current grids.

Despite the need for multi-grids to characterize the current and charges, using 2D pulses does offer the advantage that the coupling integrals associated with the static terms, which we subtracted out in the numerical computation of the Green's functions, can be integrated in closed form [13,14]. This ensures a greater accuracy in our numerical computations since the dominant contribution of the moment integrals generally comes from the static terms.

Using 2D rectangular pulses as the weighting functions, we reduce the integral equation to a matrix equation where the unknowns are the expansion coefficients for the surface current. In evaluating the coupling matrix, the nonstatic part of the coupling integral involves a 4D surface integration which is done numerically. Using a voltage-gap source as excitation, the unknown current is solved with standard matrix manipulations such as LU decomposition and back substitutions. Although the use of a voltage-gap source may be physically impractical, it does provide a simple means of exciting the microstrip. As will be shown later, the important characteristics of a microstrip junction can be extracted independent of the source.

3. APPLICATIONS

A numerical solver, which we call PATCH,[1] has been implemented using the analytical procedures

[1]As pointed out by one reviewer, PATCH was the name of the solver originally developed by J. Mosig. Our program should not be confused with the original one. The two solvers contain different numerical algorithms for the evaluation of the Green's functions and the moment integrals.

described in the previous section. Because of the choice of our basis functions, PATCH is most suitable for analyzing rectangularly shaped microstrip structures. To retain generality, we do not make any *a priori* assumptions regarding the transverse distribution of the longitudinal current, nor do we neglect the transverse component of the current from the onset. On the contrary, these transverse effects can be included as needed, depending on the structure, without incurring any additional analytical complexity in our solver. Thus, the types of microstrip junctions and discontinuities we can analyze range from simple structures such as open-end, gap, and L-bend, to more complex elements such as coupled T-junction and interdigitated capacitor. The remaining portion of this paper will focus on the application of our numerical solver. In particular, we will illustrate the versatility and generality of our solver by applying it to different junctions and discontinuities of complex shape.

The usage of our solver is simplified greatly by the development of a graphics interface for drawing and gridding microstrip geometries. This graphics tool, which we call UDRAW [15], provides an input mechanism that is intuitive and easy to use. Complex structures can now be analyzed with the help of this graphics tool. UDRAW simplifies the gridding process by requiring the users to grid only for the charge cells. UDRAW then performs the shifting of charge and current cells internally to create the corresponding sets of cells for the x- and y-currents. Thus, for a given geometry, gridding is done only once. In UDRAW, uniform as well as nonuniform size cells are allowed. Therefore, charge concentration near a discontinuity can be characterized accurately by using smaller-sized cells adaptively.

Since the direct output of PATCH is the current on the microstrip surface, additional processing is needed to extract or de-embed equivalent circuit parameters from the computed current. The procedure we use to de-embed is based on transmission-line modeling and will be described briefly first. As with any electromagnetic solver, PATCH is computationally intensive. To minimize unnecessary computation, we will carry out a convergence study to determine the optimal number of cells to use in our gridding process. To further reduce the number of unknowns, we will also examine a modified one-dimensional method for treating narrow microstrip lines with 90° corners. As will be illustrated, this method is most effective in capturing the junction effect.

Our computing environment consists of a cluster of networked HP 9000/319C workstations. Most of the numerical computations for this paper were done on these HP workstations. However, due to finite capacity, the maximum number of unknowns these workstations can accommodate is approximately 500. Computations involving more than 500 unknowns, such as those encountered in the convergence study, were done using the ETA10 supercomputer resources allocated to us by the John von Neumann Supercomputing Center.

3.1. Extraction of S-Parameters

In characterizing a microstrip junction or discontinuity, the quantities of interest are usually the equivalent circuit parameters, such as the scattering parameters (S-parameters). Since the S-parameters of a junction describe the relationship between waves coming in and going out of a junction, we will use a network description of the discontinuity. Enclosing the reference junction or discontinuity in a black box as illustrated in Figure 2, the S-parameters for a 2-port device can be expressed as

$$b_1 = S_{11}a_1 + S_{12}a_2, \tag{7}$$

$$b_2 = S_{21}a_1 + S_{22}a_2, \tag{8}$$

where the as and the bs are the voltage amplitudes associated with the waves propagating into and out of the network at each port. For a TEM line, these wave variables are related to the voltage and current of the line via the transmission-line equations

$$V(x) = ae^{-\iota\beta x} + be^{\iota\beta x} \tag{9}$$

$$I(x) = \frac{1}{Z_c}(ae^{-\iota\beta x} - be^{\iota\beta x}), \tag{10}$$

where Z_c is the characteristic impedance of the line, and β is the propagation constant. By making the assumption that dominant-mode propagation prevails away from the juntion, we can model the current on a feed strip using eq. (10). Moreover, to ensure dominant-mode propagation, we can ar-

Figure 2. A 2-port device.

tificially extend the length of the feed strip at each port by several wavelengths from the junction reference point. The normalized wave variables, $a' = a/Z_c$ and $b' = b/Z_c$, as well as the propagation constant, can be extracted by "probing" the total current away from the discontinuity. If these normalized wave variables are used directly in eqs. (7) and (8) in place of a and b, we obtain a set of S-parameters that are normalized implicitly to the characteristic impedance of each line. This normalization gives us the advantage that the S-parameters can be obtained without *a priori* knowledge of the characteristic impedance of each line.

To minimize the effect of higher-order modes, the current within a distance of one guided wavelength from both ends of the strip is excluded from the probing process. In general, β can be found very easily by examining the current standing wave pattern. Since the distance between two adjacent peaks is one-half of the guided wavelength (λ_g), β can be found once the guided wavelength is known. To ensure consistent values of a' and b', curve-fitting is used to further smooth out the effect of higher-order modes. Once β is found, we curve fit the real and imaginary parts of the current separately to a sinusoidal distribution of the form $A_{r,m} \sin(\beta x + \phi_{r,m})$, where $A_{r,m}$ and $\phi_{r,m}$ are the optimized amplitude and phase parameters obtained from the curve-fitting routine. The subscripts r and m denote the real and imaginary parts of the current, respectively. The values of a' and b' can be computed from these optimized parameters by expressing the total current in terms of these two curved-fitted functions and equating it to eq. (10). Therefore, they can be computed in a straightforward manner [16]. Unlike the procedure used in ref. [17], our approach for finding the wave variables does not depend on the specific choice of x.

For a 1-port device, the ratio of b' to a' is the desired S_{11} parameter. For a general, nonsymmetrical, N-port device, each port must be excited individually. The current over each extension strip is then fitted separately to the transmission-line equation of eq. (10) to yield the appropriate β, a', and b'. The S-parameters can be computed once the a' and b' are known for each port and each excitation [18].

3.2. Convergence

In using PATCH, the accuracy of the simulation depends in part on the size of the cells, or the number of cells, used in the gridding process. Although the accuracy improves with decreasing cell size, the computation time can increase quite drastically as the total number of cells is increased. To achieve an optimal balance between accuracy and computation time, we examine the convergence of our numerical simulation in order to establish a guideline on the minimum number of cells to use in the gridding process.

Using a four-wavelength-long, open-ended microstrip line as an example, we grid the strip using various numbers of cells per guided wavelength. The transverse component of the current is neglected for now and the transverse distribution of the longitudinal current is assumed to be uniform. The frequency of operation is arbitrarily chosen to be 10 GHz, the permittivity of the slab (ϵ_r) is 9.8, slab thickness is 0.24 mm, and strip width is 0.096 mm. Exciting the strip with a constant voltage-gap source located at far end of the strip, we examine the current as well as the guided wavelength on the open-ended strip for various numbers of cells ranging from 12 to 150 per guided wavelength.

Figures 3 and 4 are plots of the real and imaginary parts of the computed current, respectively, on the strip for the different numbers of cells. As can be seen, the current distributions are sinusoidal. The imaginary part of the current, which is three orders of magnitude smaller than the real part, is directly related to radiation from the open end. Using the $150/\lambda_g$ case as the converged case, Table I shows the percent difference on the amplitudes of the real and imaginary parts of the current for various cell sizes. For example, using

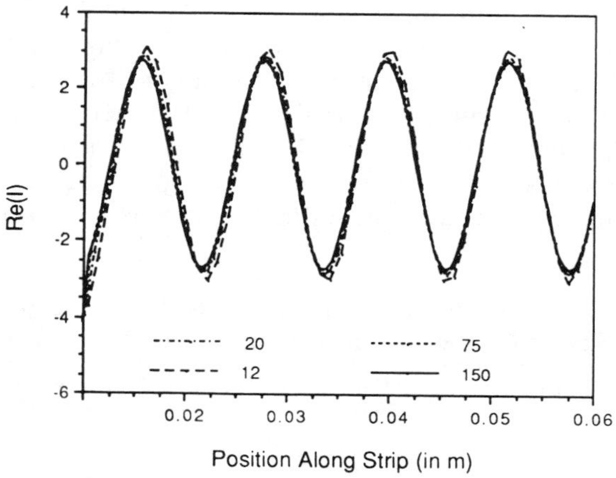

Figure 3. Real part of the current for different numbers of cells.

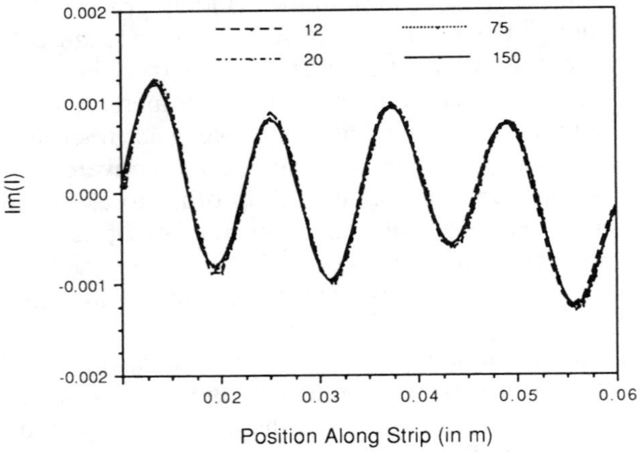

Figure 4. Imaginary part of the current for different numbers of cells.

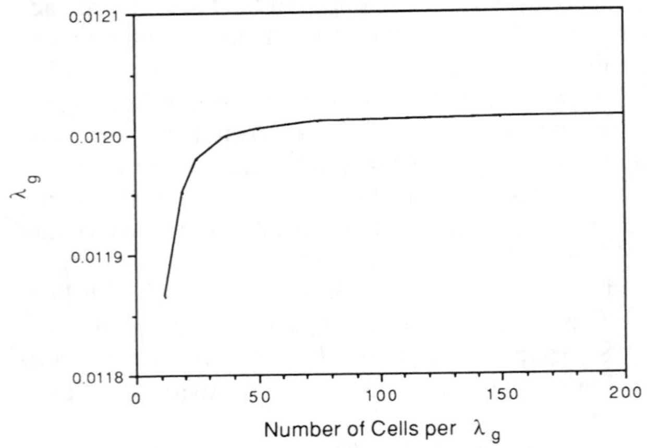

Figure 5. Guided wavelength for different numbers of cells.

a number of 20 cells per guided wavelength can yield current amplitudes that are within 5% of the converged results.

The second quantity we use to evaluate convergence is the guided wavelength. We extract the guided wavelength by curve-fitting the current magnitude over the mid-portion of the strip to a known distribution similar to eq. (10). Figure 5 is a plot of the computed guided wavelength as a function of number of cells. While this plot also indicates a strong convergence, the guided wavelength does not seem to be as sensitive to cell size as the current amplitude. The guided wavelength computed using the most coarse gridding, i.e., 12 per wavelength, is already within 2% of the converged result. This implies that the periodicity of the current is more stable than the magnitude.

The data shown in Table I were obtained using a 1D current approximation. To validate this approximation, we compare it to the full 2D representation. The 2D representation takes into account both the transverse distribution of the longitudinal current as well as the transverse component of the current. For this comparison, we use a section of strip two wavelengths long as an example. The frequency of operation is arbitrarily increased to 60 GHz. The slab thickness is 0.12 mm, ϵ_r is 9.8, and the strip width is 0.11 mm. For the 1D case, we grid the strip using 25 cells per λ_g in the longitudinal direction. For the 2D case, the strip is gridded using 75 cells per λ_g in the longitudinal direction and 5 cells in the transverse direction. This yields a total of 1345 unknowns for both x- and y-current cells.

TABLE I. Percent Difference From the Converged Results for Real and Imaginary Parts of the Current

No. of Cells	% Diff. for Re(I)	% Diff. for Im(I)
12	11.8	5.8
20	4.5	4.6
25	2.6	4.1
75	0.2	0.1

Figures 6 and 7 show the real and imaginary parts of the longitudinal current, respectively, for the 1D and 2D cases. The displayed current for the 2D case is the current averaged over the strip width. The results show that the difference between the two cases is less than 2%. Therefore, for most typical microstrip lines, the 1D current approximation is an adequate representation. Furthermore, a guideline of 20–25 cells per guided wavelength is deemed adequate for our application since it provides a 5% tolerance on the current amplitudes and a 0.5% tolerance on the guided wavelength.

3.3. Modified One-Dimensional Method

In treating simple discontinuities such as an open-end or a gap, a 1D current is generally adequate. For more complex structures involving corners, bends, and change-in-widths, both transverse and longitudinal currents are needed to characterize the junction accurately. However, to include the transverse effects everywhere on a microstrip structure can be costly because the total number of unknowns can become exceedingly large. To alleviate this difficulty, we examine a method which utilizes the simplicity of 1D current for nar-

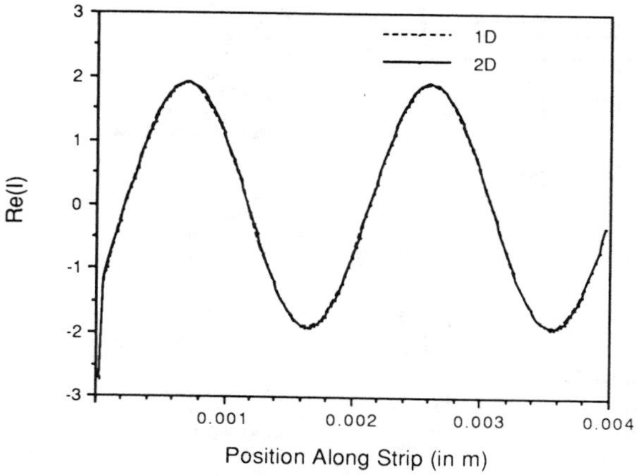

Figure 6. Real part of the current for 1D and 2D representations.

row strip line and at the same time captures the junction effect by using a 2D current in the critical region. This method, which we call a modified 1D method, is intended for treating narrow microstrip lines with 90° bends.

Using an L-bend as an example, we treat the bend as two overlapping single strips in this method and assume that the current in each single strip can be approximated by a 1D current in the longitudinal direction. Each strip is gridded independently. The portion of each strip in the overlapping region is further divided into smaller cells to give us a finer characterization of the junction. Figure 8 shows the composite current representation for the L-bend. It consists of a 2D current expansion in the junction region and a 1D current in each leg of the strip. Although the currents in both legs are treated independently in the gridding process, they are coupled together in the computational process through the charge cells.

To show the adequacy of this modified 1D current, we compare it to the full 2D current representation for the L-bend shown in Figure 8. For the parameters shown in Figure 8, the length of each leg is approximately one guided wavelength. The gridding for the modified 1D representation consists of 6 cells in the junction region and 20 away, yielding a total of 52 cells for both legs. For the 2D approach, we grid the bend using 23 cells in the longitudinal direction and 3 cells in the transverse direction, yielding a total of 302 unknowns for both x- and y-currents. Exciting the structure with a uniform voltage-gap source located at the top end of one leg, we compute the current for both cases.

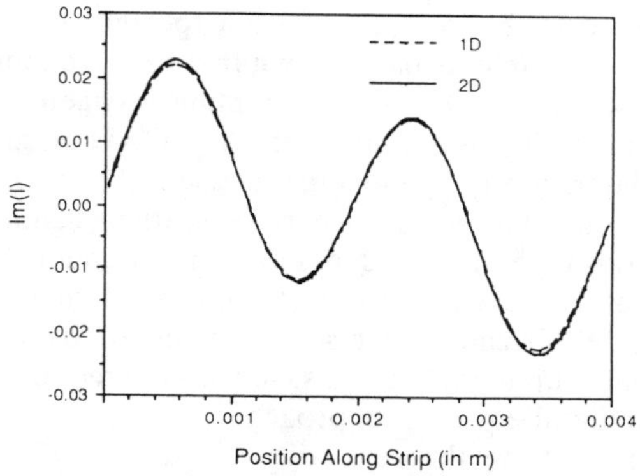

Figure 7. Imaginary part of the current for 1D and 2D representations.

Figures 9 and 10 show the real and imaginary parts of the input admittance, respectively, for the two cases as a function of frequency. As can be seen, the modified 1D approximation is adequate in capturing the junction effect since the plots show a good agreement between the two. Therefore, this modified 1D approach is found to be sufficient for applications such as de-embedding where only the effect of the junction is needed, not the detailed behavior of the junction currents. Additional numerical experimentations have shown that this approximation can provide a 0.5% tolerance on the guided wavelength for $k_o w < 0.1$, where w is the width of the strip.

3.4. Examples

We illustrate the accuracy of our numerical simulations by computing the S-parameters for three different microstrip elements and comparing the results to measured data. The shapes and dimensions of the selected structures are shown in Figure 11. To eliminate redundant gridding, each reference structure is gridded only once using the 25 cells per wavelength guideline for the highest frequency of operation. At each frequency point, we artifically add on a fixed-length, pre-gridded extension strip at each port. The S-parameters are found by probing the currents on both strips using the procedure described in Section 3.1. For all three examples, we assume a perfect dielectric slab and neglect conductor losses.

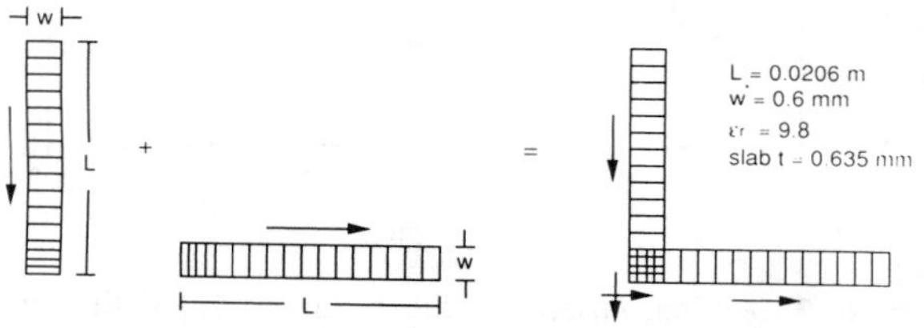

Figure 8. A modified 1D current representation for an L-bend.

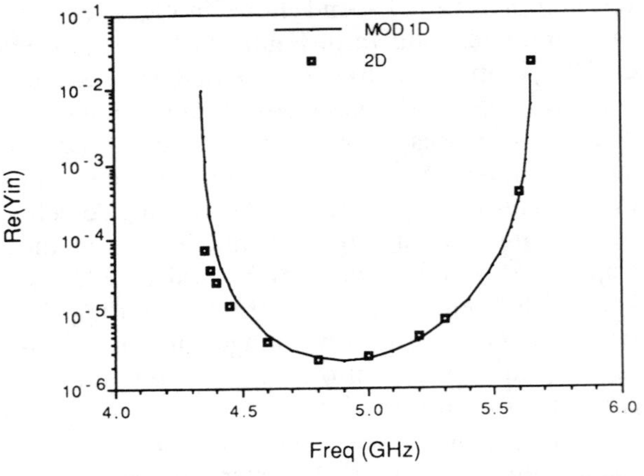

Figure 9. Comparison of the modified 1D with full 2D representation for the real part of the input admittance.

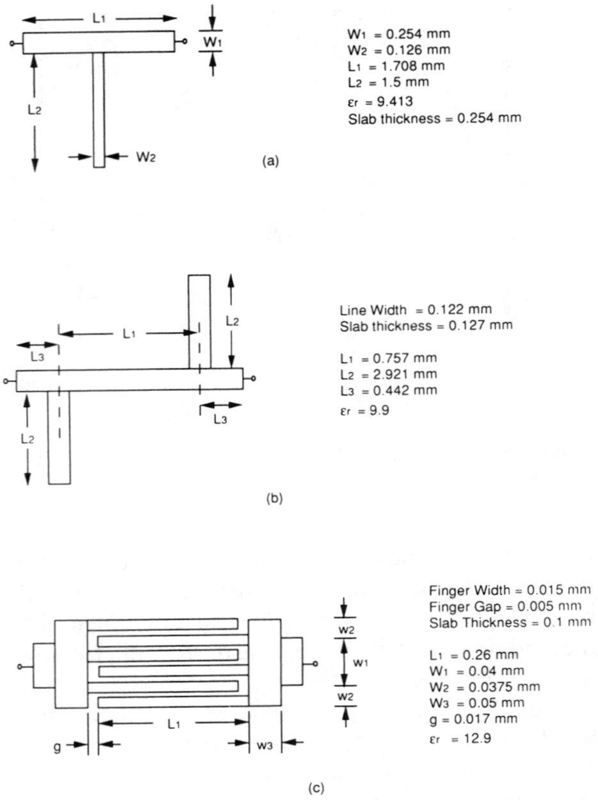

Figure 11. Three simulated structures: (a) single-Tee, (b) double-Tee, and (c) interdigitated capacitor.

The first structure consists of a single open-ended stub connected to a section of microstrip line. A modified 1D current approximation is used for this structure. Including the three-wavelength-long extension strips, the total number of cells for this structure is 194. Figures 12 and 13 are plots of the magnitudes of S_{11} and S_{12}, respectively, for both measured and computed data over the frequency range of 10–40 GHz. As can be seen, our response curves agree quite well with the measured data.

The second structure we tested is a 2-port, coupled-Tee junction. It consists of two open-ended stubs connected in opposite to a section of microstrip line. A modified 1D current approximation is also used for this structure. Using an extension strip length of $4\lambda_g$, the total number of cells for the composite structure is 315. The computed as well as measured data for $|S_{11}|$ and $|S_{12}|$ are shown in Figure 14 and 15, respectively. Except for a slight shift in frequency, the comuted cruves agree well with measured data. The frequency shift in the response curves can be partially attributed to the 2% variability in the dielectric permittivity. Although the phase plots are not shown, the results are similar to the magnitude plots [19].

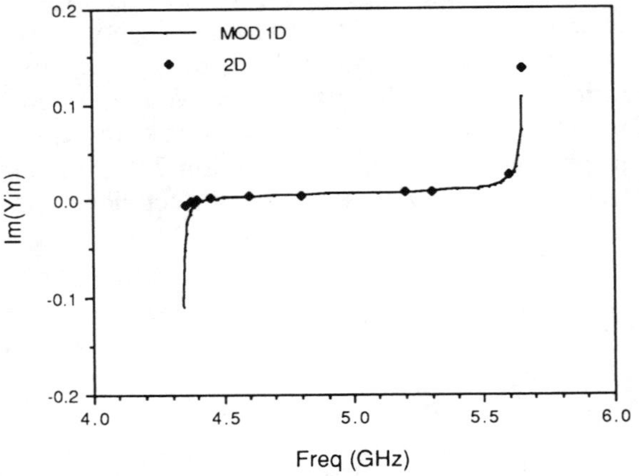

Figure 10. Comparison of the modified 1D with full 2D representation for the imaginary part of the input admittance.

This double-Tee structure is particularly interesting because the double-resonance phenomenon observed in the $|S_{12}|$ plot was unexpected. Laboratory experiments have shown that this phenomenon was caused by mutual coupling between the two Tees [20]. Since most existing CAD circuit simulators are lumped-element based, they failed to predict this occurrence because they cannot take this type of secondary effect into account. In contrast, we are able to predict this phenomenon accurately because effects such as radiations and mutual couplings are intrinsically accounted for in our integral-equation approach.

To illustrate the generality of our solver, we model an interdigitated capacitor as our third example. This commonly used microstrip element is

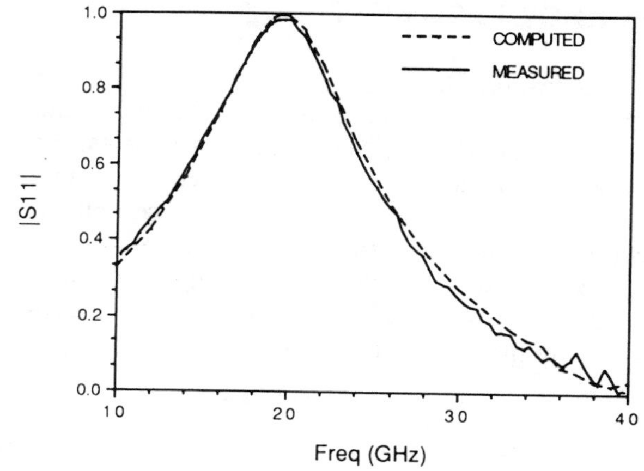

Figure 12. Computed and measured $|S_{11}|$ for the single-Tee.

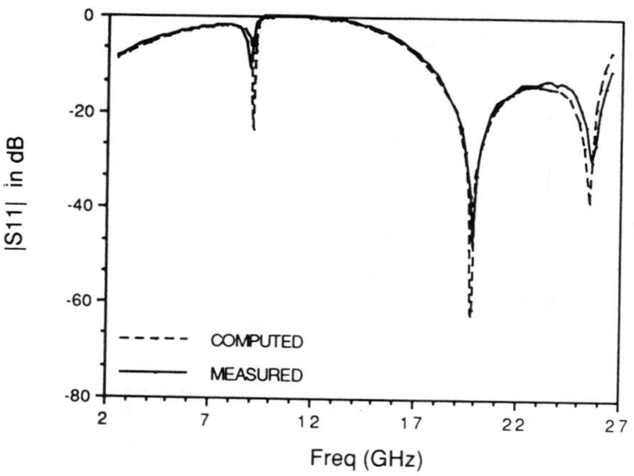

Figure 14. Measured and computed $|S_{11}|$ for the double-Tee.

a complex device because it is composed of many different types of discontinuities, such as open-ends, gaps, T-junctions, and coupled lines. To model this device accurately, the individual discontinuities along with their mutual interactions must be taken into account. Because our solver is comprehensive in nature, it is capable of modeling this device accurately with no more effort than the two devices we simulated previously.

To capture the large strip-width change in the feed strip to capacitor transition, we use a 2D current throughout the reference structure except in the finger region. To reduce the number of unknowns, we also assume that the transverse component of the current in both extension strips is negligible beyond a distance of three substrate thicknesses away from the strip-to-capacitor transition. Using a length of $2.5\lambda_g$ for each extension strip, the total number of unknowns for this structure is 475. Figure 16 shows the comparisons between measured and computed data for both $|S_{11}|$ and $|S_{12}|$. A slight disagreement between the measured and computed values is visible for $|S_{12}|$ and more so for $|S_{11}|$. The largest discrepancy occurs in $|S_{11}|$ at 18 GHz with an absolute difference of 0.08 between the measured and computed values. Although a possible discrepancy between the widths of the feed strips for the measured and simulated structures was discovered later, additional numerical simulations using different strip widths showed no significant changes on the comparison. While this type of accuracy may be acceptable for design comparison purposes, a finer grid in the capacitor region can always be used to further improve the results.

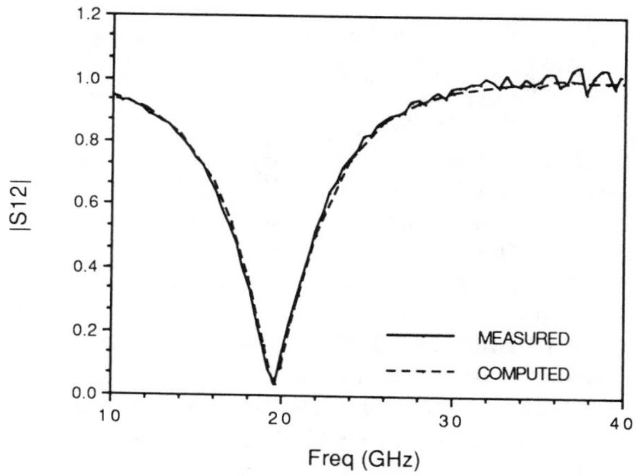

Figure 13. Computed and measured $|S_{12}|$ for the single-Tee.

4. CONCLUSION

In this paper, we outlined the developmental procedure of a moment method-based general solver for microstrip structures. We also illustrated the versatility of our simulation program by applying it to three different structures. Our simulation process began with the drawing and gridding of a geometry using UDRAW. The output files created by UDRAW were used to run the general solver, PATCH. The S-parameters were then extracted from the output of PATCH using a straightforward de-embedding procedure. As we have shown, we can analyze complex structures and yield accurate results. Although the three steps described above are in separate form currently, they can be combined and automated.

While our solver can analyze complex structures with good accuracy, it does require a longer computation time compared to the conventional

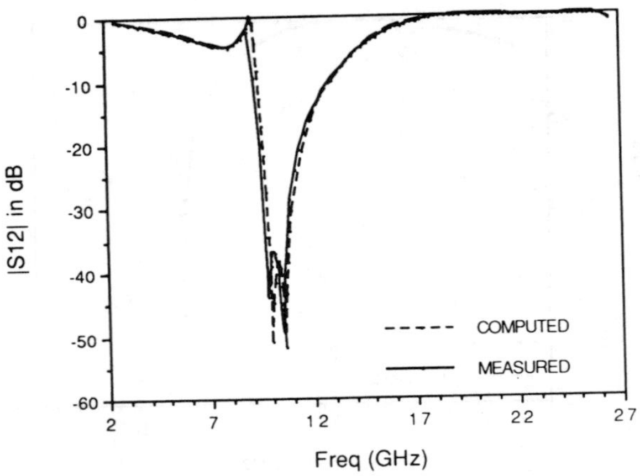

Figure 15. Measured and computed $|S_{12}|$ for the double-Tee.

lumped-element-based circuit simulators. For example, a structure with 200 unknowns would take approximately 30 minutes of CPU time per frequency point on our HP 9000/319C workstations (rated at 2 MIPS). The strength of our current solver is in its ability to model microstrip components accurately. Although our solver in its present form is not suited for real-time simulations, it has not been optimized to the fullest. For example, it does not take advantage of the uniform-sized cells often used in the extension strip sections. Comutation time can be shortened somewhat if we recognize this uniformity and eliminate all repetitive computations of couplings between uniform-sized cells. Moreover, with the availability and the widespread use of workstations, parallel computations can be introduced in a networked environment to further speed up the simulation process.

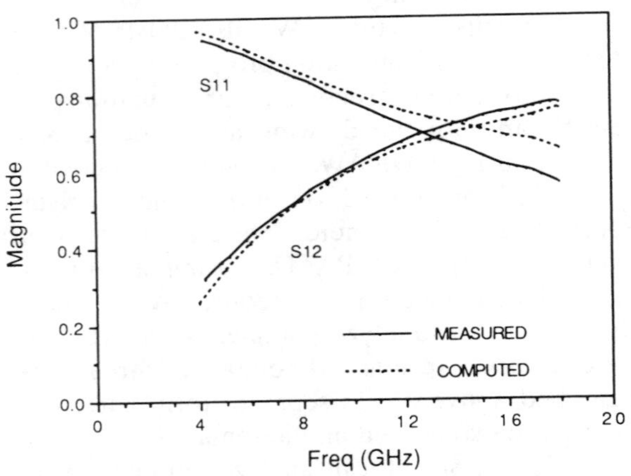

Figure 16. Computed and measured $|S_{11}|$ and $|S_{12}|$ for the interdigitated capacitor.

ACKNOWLEDGMENT

The measured data for the various structures we simulated were provided to us by Hewlett Packard, Hughes Aircraft, and Texas Instruments. We gratefully acknowledge their support. We would also like to acknowledge JvNCC supercomputing center for allocating the requested computer resources. Lastly, all the curve-fitting algorithms used in our de-embedding procedure were provided by the NAG Fortran Library.

REFERENCES

1. R. H. Jansen, R. G. Arnold, and I. G. Eddison, "A comprehensive CAD approach to the design of MMIC's up to mm-wave frequencies," *IEEE Trans. Microwave Theory Tech.*, Vol. 36, 1988, pp. 208–219.
2. R. A. Pucel, "Design considerations for monolithic microwave circuits," *IEEE Trans. Microwave Theory Tech.*, Vol. 29, 1981, pp. 513–534.
3. J. A. Kong, *Theory of Electromagnetics,* Wiley, New York, 1975.
4. J. R. Mosig and F. E. Gardiol, "A dynamic radiation model for microstrip structures," in *Advances in Electronics and Electron Physics,* P. W. Hawkes, ed., Vol. 59, Academic, New York, 1982.
5. R. W. Jackson and D. M. Pozar, "Full-wave analysis of microstrip open-end discontinuities," *IEEE Trans. Microwave Theory Tech.*, Vol. 33, 1985, pp. 1036–1042.
6. R. W. Jackson, "Full-wave, finite element analysis of irregular microstrip discontinuities," *IEEE Trans. Microwave Theory Tech.*, Vol. 37, 1989, pp. 81–89.
7. R. H. Jansen, "The spectral-domain approach for microwave integrated circuits," *IEEE Trans. Microwave Theory Tech.*, Vol. 33, 1985, pp. 1043–1056.
8. J. R. Mosig, "Arbitrarily shaped microstrip structures and their analysis with a mixed potential integral equation," *IEEE Trans. Microwave Theory Tech.*, Vol. 36, 1988, pp. 314–323.
9. D. C. Chang and J. X. Zheng, "Electromagnetic modeling of passive circuit elements in MMIC. Part I: The P-mesh algorithm; Part II: Effect of circuit discontinuities and mutual couplings in microstrip circuits," *IEEE Trans. Microwave Theory Tech.*, to appear.
10. B. L. Brim and D. C. Chang, "Accelerated numerical computation of the spatial domain dyadic Green's function of a grounded dielectric slab," *National Radio Science Meeting Dig.*, January 1987, p. 164.
11. R. F. Harrington, *Field Computations by Moment Methods,* Macmillan, New York, 1968.

12. A. W. Glisson and D. R. Wilton, "Simple and efficient numerical methods for problems of electromagnetic radiation and scattering from surfaces," *IEEE Trans. Antennas Propag.*, Vol. 29, 1980, pp. 593–603.
13. B. Noble, "The numerical solution of the singular integral equation for the charge distribution on a flat rectangular lamina," *PICC Symposium on Differential and Integral Equations,* Rome, September 1960, pp. 530–543.
14. D. R. Wilton, S. M. Rao, A. W. Glisson, D. H. Schaubert, O. M. Al-Bundak, and C. M. Butler, "Potential integrals for uniform and linear source distributions on polygons and polyhedral domains," *IEEE Trans. Antennas Propag.*, Vol. 32, 1984, pp. 276–281.
15. K. J. Russo, "UDRAW User's Manual," prepared for the workshop on *Numerical Modeling Using the BCW Code.* MIMICAD Center, University of Colorado, Boulder, October, 1989.
16. D. I. Wu, "Patch User's Guide," prepared for the workshop on *Numerical Modeling Using the BCW Code.* MIMICAD Center, University of Colorado, Boulder, October, 1989.
17. P. B. Katehi and N. G. Alexopoulos, "Frequency-dependent characteristics of microstrip discontinuities in millimeter-wave integrated circuits," *IEEE Trans. Microwave Theory Tech.*, Vol. 33, 1985, pp. 1029–1035.
18. H. A. Atwater, *Introduction to Microwave Theory,* McGraw-Hill, New York, 1962.
19. D. I. Wu and D. C. Chang, *A highly-accurate numerical modeling of microstrip junctions and discontinuities.* Final Report, MIMICAD Center, University of Colorado, Boulder, April 1990.
20. M. Goldfarb and A. Platzker, "The effect of electromagnetic coupling on MMIC design," *Int. J. Microwave Millimeter-Wave Computer-Aided Eng.*, to appear.

Paper 8.5

Theoretical and Experimental Characterization of Coplanar Waveguide Discontinuities for Filter Applications

Nihad I. Dib, *Student Member, IEEE*, Linda P. B. Katehi, *Senior Member, IEEE*, George E. Ponchak, *Member, IEEE*, and Rainee N. Simons, *Senior Member, IEEE*

Abstract **—A full-wave analysis of shielded coplanar waveguide two-port discontinuities based on the solution of an appropriate surface integral equation in the space domain is presented. Using this method, frequency-dependent scattering parameters for open-end and short-end CPW stubs are computed. The numerically derived results are compared with measurements performed in the frequency range 5–25 GHz and show very good agreement. Equivalent circuit models and closed-form expressions to compute the circuit element values for these discontinuities are also presented.**

I. Introduction

RECENTLY, coplanar waveguide (CPW) technology has attracted a great deal of interest for RF circuit design owing to several advantages over the conventional microstrip line, among them the capability to wafer probe at millimeter-wave frequencies [1]–[6]. Thus, several investigators have undertaken the study of the propagation characteristics of uniform CPW, and extensive data are available in the literature [7]. However, very few models are available on CPW discontinuities, which are useful in the design of circuits such as filters [8]–[12].

Filters are important blocks in microwave circuits and are among the first few circuit elements studied in any new technology. Microstrip or stripline filters have been extensively studied and very accurate design techniques have been presented in the literature [13]. However, CPW filter elements [1], [14] have been investigated only experimentally and lack accurate equivalent circuits. Two such filter elements are the short-end and open-end series stubs, which are shown in Fig. 1(a) and (b) respectively. In this figure *PP′* refers to the reference planes, which are coincident with the input and output ports of the discontinuity. The short-end CPW stub was modeled by Houdart [1] as a series inductor. This model cannot predict the resonant nature of the stub as it approaches $\lambda_g/4$ or the asymmetry of the discontinuity; therefore, it is valid only in the limit as the stub length approaches zero. The model used by Ponchak and Simons [15], an ideal short-end series stub, predicts the resonant nature of the stub but not the asymmetry. The open-end CPW stub was modeled by Houdart [1] as a series capacitor, which is also too simple to predict the resonant nature of the stub or its asymmetry and is valid only for stubs with very small lengths. Williams [14] expanded this model to a capacitive Π network and selected a reference plane which removed the element asymmetry. This improved model is difficult to incorporate into CAD programs since the reference planes are not at the plane of the discontinuity, as shown in Fig. 1. The model by Ponchak and Simons [15], an ideal open-end series stub, again cannot predict the element asymmetry.

This paper attempts, for the first time, to study theoretically and experimentally the two CPW filter elements shown in Fig. 1. The theoretical method used to study these CPW discontinuities is based on a space-domain integral equation (SDIE), which is solved using the method of moments [16]. The main difference between this approach and the one used in [3] and [10]–[12] is that the boundary conditions are applied in the space domain instead of the spectral domain. Thus, in the SDIE method the Fourier transforms of the basis functions, which are used in the method of moments, are not required, which makes it simpler to handle complicated geometries. The SDIE approach has previously been applied to study several CPW discontinuities and has shown very good accuracy, efficiency, and versatility in terms of the geometries it can solve [17], [18]. Using this method, theoretical results for the scattering parameters of the two CPW discontinuities shown in Fig. 1 are computed. Extensive experiments have been performed in the frequency range 5 to 25 GHz to validate the theoretically derived scattering parameters, and a very good agreement has been found. From the scattering parameters, lumped element equivalent circuits have been derived to model the discontinuities. The inductors and capacitors of these models have been represented by closed-form equations, as functions of the stub length, which have potential applications

Manuscript received August 30, 1990; revised January 22, 1991. This work was supported by the National Science Foundation under Contract ECS-8657951.

N. I. Dib and L. P. B. Katehi are with the Department of Electrical Engineering and Computer Science, University of Michigan, Ann Arbor, MI 48109-2122.

G. E. Ponchak and R. N. Simons are with the NASA Lewis Research Center, Cleveland, OH 44135.

IEEE Log Number 9143466.

Reprinted from *IEEE Trans. Microwave Theory Tech.*, vol. 39, no. 5, pp. 873–882, May 1991.

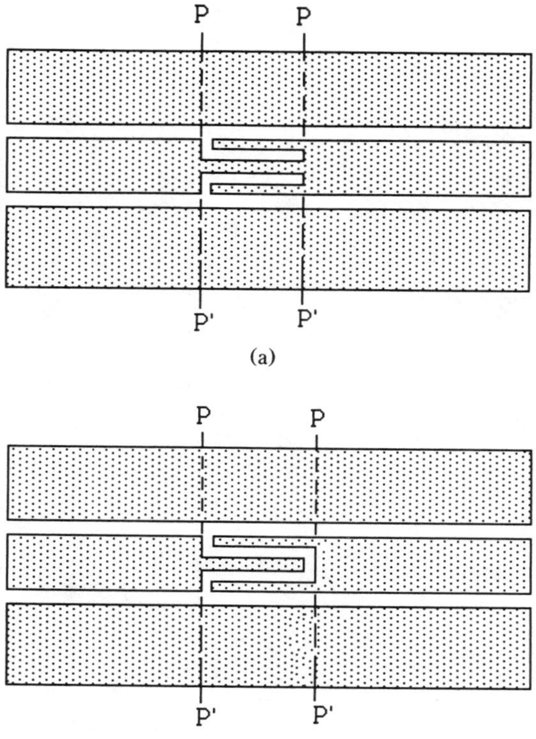

Fig. 1. Coplanar waveguide filter elements. (a) Short-end CPW series stub. (b) Open-end CPW series stub.

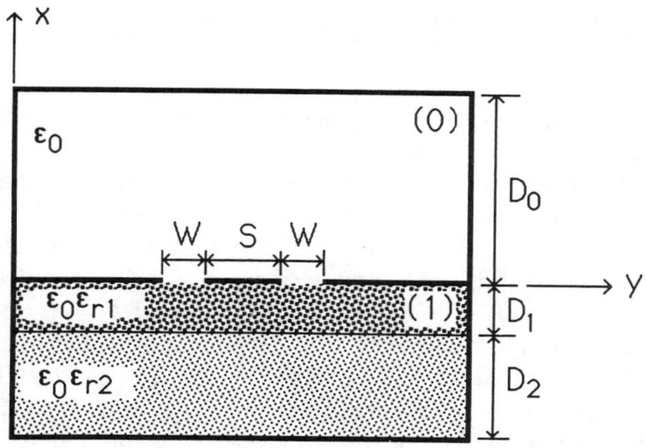

Fig. 2. A cross section of a shielded CPW.

in the design of CPW circuits. In addition, these circuits are capable of modeling the ON and OFF states of CPW p-i-n diode switches [15].

II. Theory

A. Derivation of the Scattering Parameters

Fig. 2 shows a shielded coplanar waveguide with the cavity dimensions chosen such that the CPW fundamental mode is not affected by higher order cavity resonances. The original boundary problem is divided into two simpler ones by introducing an equivalent magnetic current, $\vec{M}_s$, on the slot aperture (see the Appendix). This surface magnetic current radiates an electromagnetic field in the two waveguide regions (above and below the slots) so that the continuity of the electric field on the surface of the slots is satisfied. The remaining boundary condition to be applied is the continuity of the tangential components of the magnetic field on the surface of the slot aperture,

$$\hat{a}_x \times \left(\vec{H}_0 - \vec{H}_1\right) = \vec{J}_s \tag{1}$$

where $\vec{J}_s$ vanishes everywhere on the plane of the slot apertures except at the position of the electric current sources exciting the CPW. $\vec{H}_{0,1}$ are the magnetic fields in the regions directly above and below the slot aperture, respectively (see Fig. 12), and can be expressed in terms of the equivalent magnetic current density, $\vec{M}_s$, as shown below:

$$\vec{H}_0 = \int_{S_{\text{CPW}}}\int \overline{\overline{G}}{}_0^h(\vec{r}/\vec{r}') \cdot \vec{M}_s(\vec{r}')\, ds' \tag{2}$$

$$\vec{H}_1 = -\int_{S_{\text{CPW}}}\int \overline{\overline{G}}{}_1^h(\vec{r}/\vec{r}') \cdot \vec{M}_s(\vec{r}')\, ds'. \tag{3}$$

In (2) and (3), S_{CPW} is the surface of the slot aperture, and $\overline{\overline{G}}{}_{0,1}^h$ is the dyadic Green's function in the two waveguide regions (see the Appendix).

In view of (2) and (3), (1) takes the form

$$\hat{a}_x \times \int_{S_{\text{CPW}}}\int \left[\overline{\overline{G}}{}_0^h + \overline{\overline{G}}{}_1^h\right] \cdot \vec{M}_s(\vec{r}')\, ds' = \vec{J}_s. \tag{4}$$

To obtain the unknown magnetic current distribution, $\vec{M}_s$, (4) is solved by applying the method of moments [16]. First, the slot aperture is subdivided into rectangles. Then, the unknown magnetic current density is expressed as a finite double summation:

$$\vec{M}_s(\vec{r}') = \hat{a}_y \sum_{i=1}^{N_y}\sum_{j=1}^{N_z} V_{y,ij} f_i(y') g_j(z') + \hat{a}_z \sum_{i=1}^{N_y}\sum_{j=1}^{N_z} V_{z,ij} f_j(z') g_i(y') \tag{5}$$

where $\{f_i(y')g_j(z');\ i = 1,\cdots,N_y,\ j = 1,\cdots,N_z\}$ is a family of rooftop functions [19] and $V_{y,ij}$ and $V_{z,ij}$ are the unknown coefficients for the y and z components of the magnetic current density. The subdomain basis functions (rooftop functions) for each current component have piecewise-sinusoidal variation along the longitudinal direction and constant variation along the transverse direction. Using (5), (4) can be written in the form

$$\vec{J}_s + \Delta\vec{J}_s = \hat{a}_x \times \left\{ \sum_{i=1}^{N_y}\sum_{j=1}^{N_z} V_{y,ij} \iint \left[\overline{\overline{G}}{}_0^h + \overline{\overline{G}}{}_1^h\right] \cdot \hat{a}_y f_i(y') g_j(z')\, ds' + \sum_{i=1}^{N_y}\sum_{j=1}^{N_z} V_{z,ij} \iint \left[\overline{\overline{G}}{}_0^h + \overline{\overline{G}}{}_1^h\right] \cdot \hat{a}_z f_j(z') g_i(y')\, ds' \right\} \tag{6}$$

where $\Delta\vec{J}_s$ represents the error introduced from the approximations made in the magnetic current distribution (eq. (5)).

Finally, Galerkin's procedure is used to minimize the error $\Delta\vec{J}_s$ resulting in the following inner products:

$$\iint \left(\hat{a}_x \times \Delta\vec{J}_s\right)\cdot \hat{a}_y f_m(y) g_n(z)\, ds = 0 \tag{7}$$

$$\iint \left(\hat{a}_x \times \Delta\vec{J}_s\right)\cdot \hat{a}_z f_n(z) g_m(y)\, ds = 0 \tag{8}$$

where f_m and g_n are weighting functions identical to the basis functions, $m = 1,\cdots,N_y$ and $n = 1,\cdots,N_z$. In this manner, (6) reduces into a matrix equation of the form

$$\begin{pmatrix} [Y_{yy}] & [Y_{yz}] \\ [Y_{zy}] & [Y_{zz}] \end{pmatrix}\begin{pmatrix} V_y \\ V_z \end{pmatrix} = \begin{pmatrix} I_z \\ I_y \end{pmatrix} \tag{9}$$

where $[Y_{\zeta\xi}]$ $(\zeta,\ \xi = y, z)$ represent blocks of the admittance matrix whose elements are expressed in terms of multiple space integrals, involving trigonometric functions, and are given by

$$Y_{yy} = \iiiint \hat{a}_y f_m(y) g_n(z)\cdot\left[\overline{\overline{G}}{}_0^h + \overline{\overline{G}}{}_1^h\right] \cdot \hat{a}_y f_i(y') g_j(z')\, dy'\, dz'\, dy\, dz \tag{10}$$

$$Y_{yz} = \iiiint \hat{a}_y f_m(y) g_n(z)\cdot\left[\overline{\overline{G}}{}_0^h + \overline{\overline{G}}{}_1^h\right] \cdot \hat{a}_z f_j(z') g_i(y')\, dy'\, dz'\, dy\, dz \tag{11}$$

$$Y_{zy} = \iiiint \hat{a}_z f_n(z) g_m(y)\cdot\left[\overline{\overline{G}}{}_0^h + \overline{\overline{G}}{}_1^h\right] \cdot \hat{a}_y f_i(y') g_j(z')\, dy'\, dz'\, dy\, dz \tag{12}$$

$$Y_{zz} = \iiiint \hat{a}_z f_n(z) g_m(y)\cdot\left[\overline{\overline{G}}{}_0^h + \overline{\overline{G}}{}_1^h\right] \cdot \hat{a}_z f_j(z') g_i(y')\, dy'\, dz'\, dy\, dz \tag{13}$$

where $i, m = 1,\cdots,N_y$ and $j, n = 1,\cdots,N_z$. V_y and V_z are the subvectors of the unknown coefficients for the y and z components of the magnetic current distribution respectively and I_y and I_z are the known excitation subvectors, which are dependent on the impressed feed model.

In order to solve (9), the excitation is modeled by ideal y-directed current sources located at specific node points, as shown in Fig. 3(a), resulting in an excitation vector which has zeros everywhere except at the positions of these current sources (delta gap current generators). Although only a mathematical model, this feeding mechanism has proved to be efficient, accurate, and reliable [17], [20]. In addition, it does not introduce any unwanted numerical complications, as is the case with other excitation techniques [3], [12]. The CPW may be excited in two different ways: with the fields on the two slot apertures in phase (slotline mode) or out of phase (coplanar mode), exhibiting very different characteristics when operating in each mode. The CPW mode is excited by choosing $I'_{g1} = -I_{g1}$ and $I'_{g2} = -I_{g2}$, while the slotline mode is excited by choosing $I'_{g1} = I_{g1}$ and $I'_{g2} = I_{g2}$ (see Fig. 3(a)). However, only the CPW mode will be considered here since it tends to concentrate the fields around the slot aperture

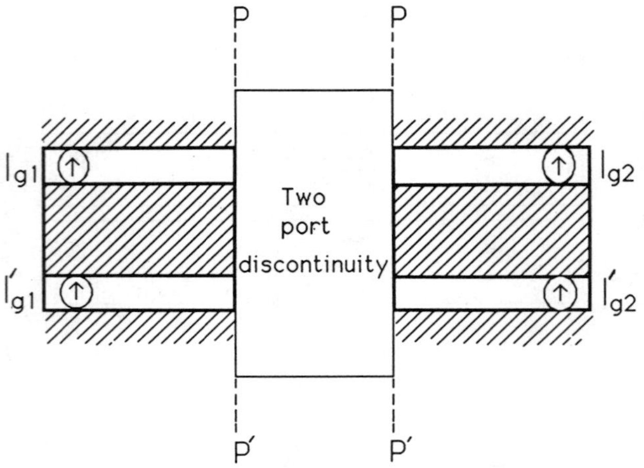

(a)

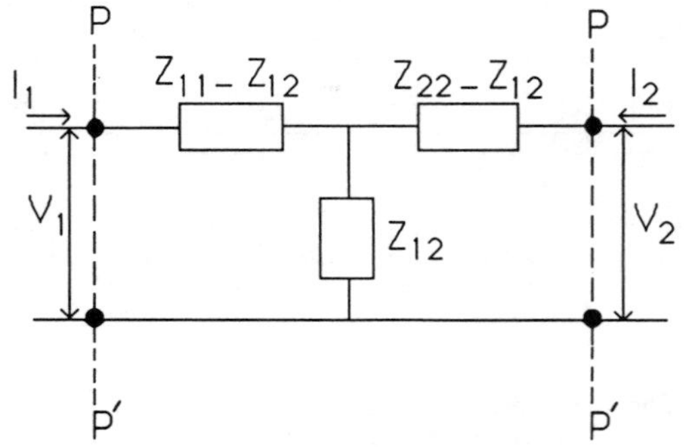

(b)

Fig. 3. (a) A general two-port CPW discontinuity with y-directed electric current sources. (b) Equivalent network representation.

(low radiation losses) and is therefore the commonly used mode in CPW circuits.

The discontinuities behave as an asymmetric two-port network with an equivalent circuit of the type shown in Fig. 3(b). For the evaluation of the network impedance matrix, three different modes of excitation are required ($I_{g1} = I_{g2} = 1$, $I_{g1} = -I_{g2} = 1$, and $I_{g1} = 1$ and $I_{g2} = 0$ were used). It should be noticed that the currents and voltages (I_1, I_2, V_1, V_2) shown in Fig. 3(b) are induced at the discontinuity ports as a consequence of I_{g1} and I_{g2}. The matrix equation (9) is solved to give the magnetic current distribution in the slot aperture for each mode of excitation. Then, the input impedances at ports 1 and 2 are evaluated from the positions of the minima and the maxima of the electric field standing waves in the feeding CPW lines using ideal transmission line theory. Finally, the network impedance matrix and the scattering parameters are derived using the expressions in [21].

B. Convergence Properties of Scattering Parameters

In the expressions for the Green's functions given in the Appendix, the summations over m and n are theoretically infinite. However, for the numerical solution of the matrix equation, these summations are truncated to MSTOP and NSTOP. The values of these two parameters should be chosen such that convergence of the scattering coeffi-

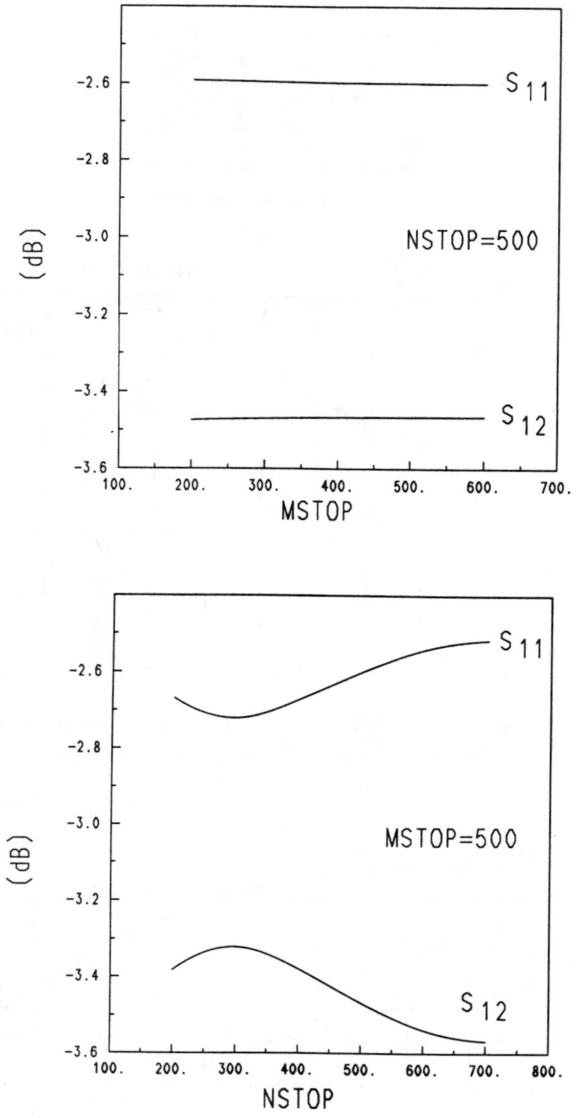

Fig. 4. Typical convergence behavior of the scattering parameters with respect to the number of modes NSTOP and MSTOP of the CPW discontinuities shown in Fig. 1.

cients is ensured. Fig. 4 shows the convergence behavior of the scattering coefficients with respect to each one of the above parameters. In (A10)–(A13), NSTOP and MSTOP correspond to the maximum values of the k_z and k_y eigenvalues considered in the summations. These eigenvalues are given by

$$k_{z\,\max} = \frac{(\text{NSTOP})\pi}{l} \tag{14}$$

$$k_{y\,\max} = \frac{(\text{MSTOP})\pi}{a}. \tag{15}$$

Since the length of the cavity, l, is larger than its width, a, the number of required k_z eigenvalues (NSTOP = 700) is much larger than the number of k_y eigenvalues (MSTOP = 300), as shown in Fig. 4.

Another critical parameter for the convergence of the results is the number of the considered basis functions N_y

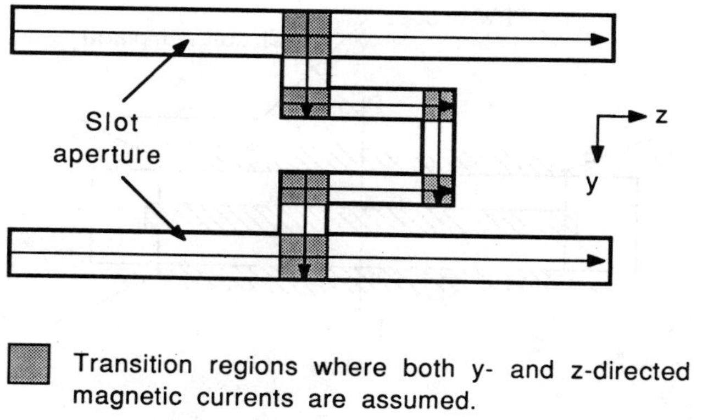

Fig. 5. The assumed magnetic current distribution in the slot aperture.

and N_z. Since the slots are assumed to be fairly thin ($W/\lambda_{\text{CPW}} < 0.1$), only a longitudinal magnetic current in the slot aperture away from the discontinuity is assumed. Furthermore, it has been found through numerical experiments that the transverse current in the feeding line around the discontinuity has a negligible effect on the current distribution and the scattering parameters. Thus, as shown in Fig. 5, both longitudinal and transverse magnetic current components are considered in the transition regions, while only the longitudinal current component is considered in the other regions. In addition, the number of basis functions N_y and N_z is chosen so that convergence of the scattering parameters of the coplanar waveguide discontinuity is achieved. The CPU time required for the evaluation of the scattering parameters depends mainly on the geometry and the electrical size of the structure. Careful consideration of the existing physical symmetries can reduce this computational time substantially.

III. Results and Discussion

Using the space-domain integral equation method, the scattering parameters for the two discontinuities shown in Fig. 1 have been evaluated as a function of the stub length, L, and frequency. The theoretically derived data have been validated through extensive experiments performed in the 5 through 25 GHz frequency range. The circuits were fabricated using conventional MIC techniques on a polished alumina ($\epsilon_r = 9.9$) substrate. The plated gold thickness was 2.8 μm. The RF measurements were performed on an HP 8510 ANA. A probe station with dc–26.5 GHz probes was used for providing the RF connections to the circuits. A two-tier calibration was performed: First, the system was calibrated to the 3.5 mm coaxial cable ends using coaxial open, short, load calibration standards. Then, an LRL calibration was performed to rotate the reference planes to the CPW discontinuities. The LRL calibration standards shown in Fig. 6 were fabricated on the substrate along with the circuits to eliminate errors caused by fabrication nonrepeatability. Air bridges were not needed to connect the ground planes since the discontinuities considered here are symmetric

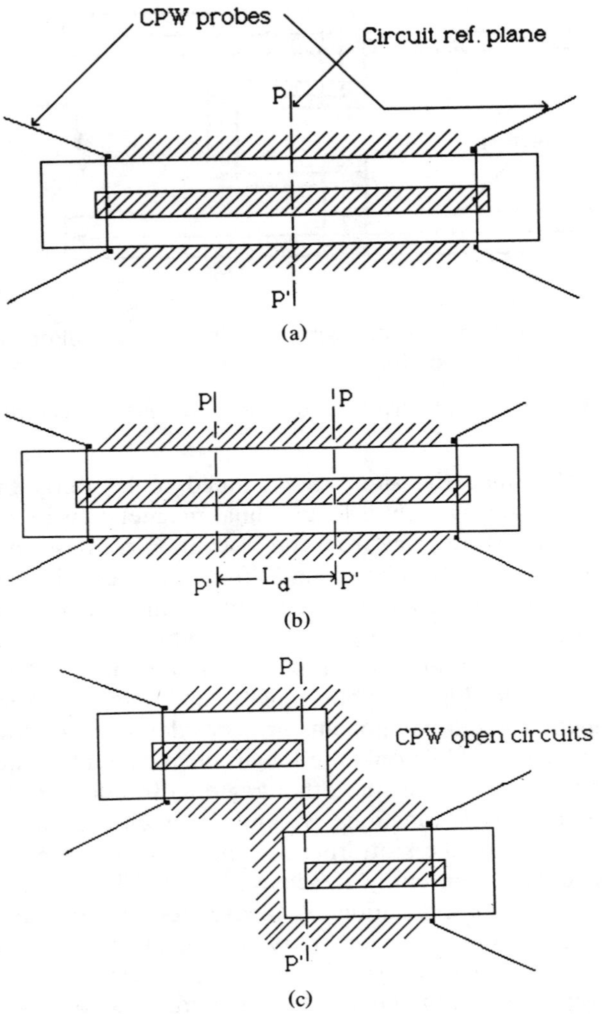

Fig. 6. CPW standards for LRL calibration: (a) Zero length thru line. (b) Delay of length L_d where $0 \leqslant L_d \leqslant \lambda/2$. (c) Open circuit reflection standards.

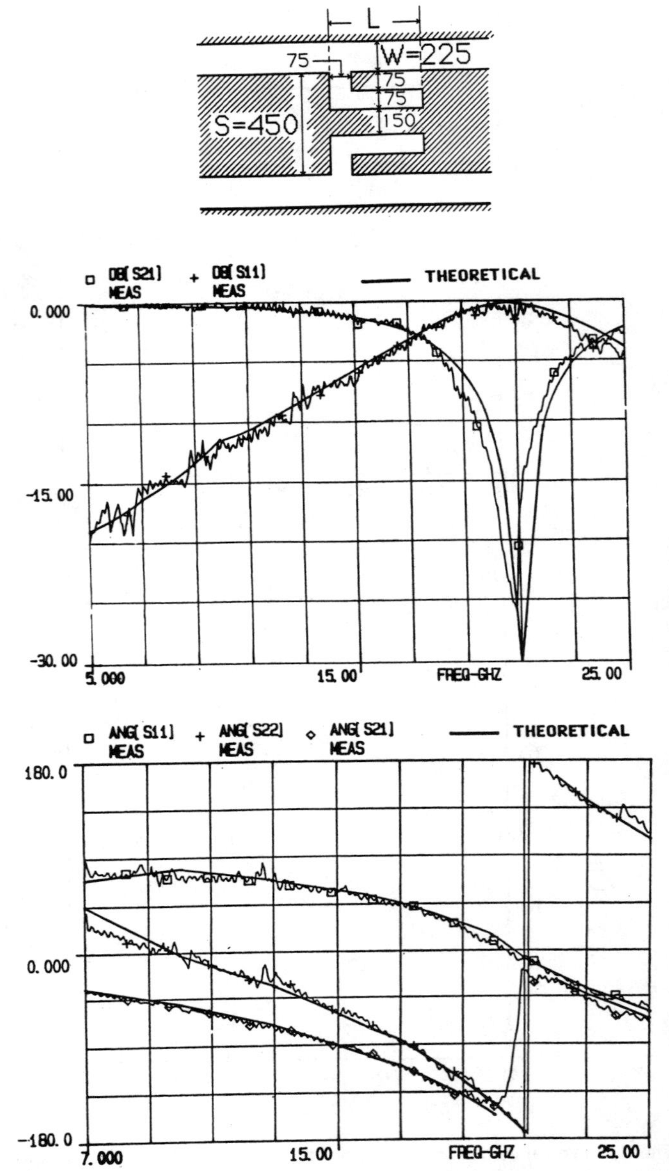

Fig. 7. Scattering parameters for the short-end CPW stub with $L = 1500$ μm ($D_1 = 25$ mil, $D_2 = 125$ mil, $\epsilon_{r1} = 9.9$, $\epsilon_{r2} = 2.2$; other dimensions are in μm).

about the propagation line, and the stubs are printed on the center conductor [22]. The alumina substrate was placed on a 125-mil-thick 5880 RT/Duroid ($\epsilon_r = 2.2$) substrate with copper cladding on one side to serve as the bottom ground plane.

Fig. 7 shows the scattering parameters for the short-end CPW stub of length 1500 μm. It can be seen that the agreement between the theoretical and experimental results is very good. The differences are due to radiation, conductor, and dielectric losses since an unshielded structure was used in the measurements and the theoretical model did not account for losses. From the characteristic behavior of the above stub, it can be concluded that a band-stop filter can be realized by cascading several of these stubs in series.

Fig. 8 shows a comparison between theoretical and experimental values of the resonant frequency as a function of stub length. In addition, superimposed are values of the resonant frequencies for the above CPW stub operating under ideal conditions: no discontinuity effects and zero electromagnetic interactions. It can be noticed that the theoretically computed resonant frequencies for large stub lengths agree very well with the ones predicted for the ideal stub. This indicates that at low resonant frequencies (large stub lengths) the specific stub geometry introduces negligible discontinuity effects and is almost free of parasitic electromagnetic interactions. However, as the frequency increases, the parasitic effects become stronger, resulting in a small difference between the theoretically obtained resonant frequencies and the ones computed for the ideal stub.

Fig. 9 shows the scattering parameters for the open-end CPW stub of length 1500 μm. As in the previous case, the agreement between the theoretical and experimental results is very good. Owing to its performance, such structures can be used to build band-pass filters.

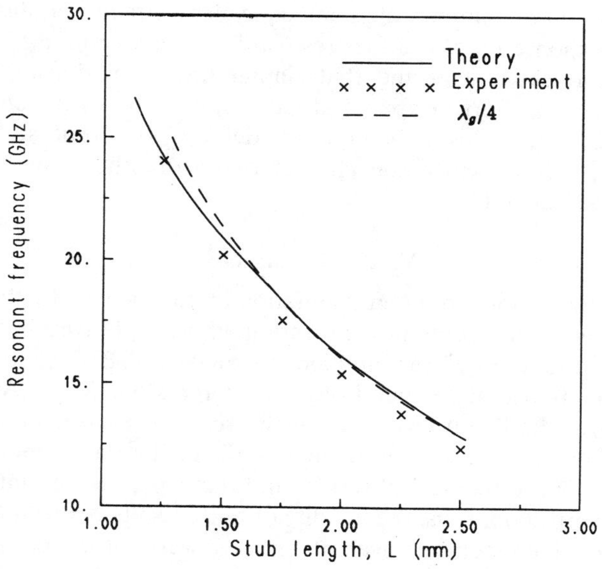

Fig. 8. Resonant frequency of the short-end CPW stub of different lengths ($D_1 = 21.5$ mil, $D_2 = 125$ mil, $\epsilon_{r1} = 9.9$, $\epsilon_{r2} = 2.2$; other dimensions are as in Fig. 10(a)).

In the theoretical analysis, the CPW stubs were assumed to be inside a cavity while for the derivation of experimental data these structures were measured in open environment. The fact that there is very good agreement between theory and experiment suggests that radiation losses were very low. The loss factor of the measured stub discontinuities has been investigated and has shown a maximum value of -10 dB at the stub's resonant frequency. This indicates that it is possible to design CPW stub discontinuities with very low radiation losses.

IV. Equivalent Models

To accurately model the short-end CPW stub over the frequency range from 5 GHz to the first band-stop resonance, the model shown in Fig. 10(b) is proposed. Using the derived scattering parameters (for Fig. 10(a)), the capacitances and inductances are evaluated using commercial optimization software. The following relations have been found, which give the values of the lumped elements in terms of the stub length, L:

$$C_s = 1.32 \times 10^{-4} L + 3.3515 \times 10^{-2} \quad (16)$$

$$C_{f1} = 1.5959 \times 10^{-2} \quad (17)$$

$$C_{f2} = 1.1249 \times 10^{-4} L + 7.522 \times 10^{-3} \quad (18)$$

$$L_1 = 2.6368 \times 10^{-4} L - 6.618 \times 10^{-3} \quad (19)$$

$$L_2 = 1.77 \times 10^{-4} L - 8.35 \times 10^{-4} \quad (20)$$

$$L_3 = 1.8656 \times 10^{-4} L - 8.34 \times 10^{-4} \quad (21)$$

where the stub length, L, is in μm, the inductances in nH, and the capacitances in pF. The above equations, which apply for the configuration shown in Fig. 10(a) only, have been verified for stub lengths, L, through 2500 μm. It should be noted that $C_{f1} = C_{f2}$ and $L_1 = L_3$ when

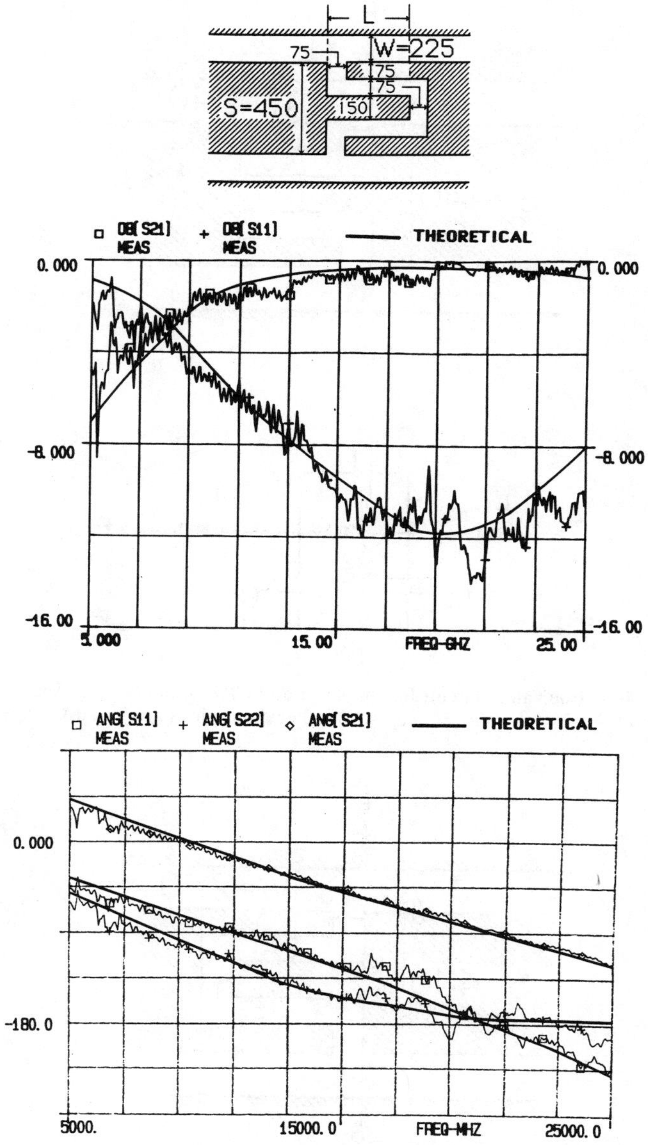

Fig. 9. Scattering parameters for the open-end CPW stub with $L = 1500$ μm ($D_1 = 25$ mil, $D_2 = 125$ mil, $\epsilon_{r1} = 9.9$, $\epsilon_{r2} = 2.2$; other dimensions are in μm).

$L = 75$ μm since a symmetric model is expected for a simple notch in the center conductor of the CPW.

The equivalent circuit shown in Fig. 11(b) is proposed to model the open-end CPW stub over the frequency range from 5 GHz to the first band-pass resonance. The following relations have been found, which give the values of the lumped elements in terms of the stub length, L:

$$C_s = 1.01 \times 10^{-4} L + 1.642 \times 10^{-2} \quad (22)$$

$$C_{f1} = 0.39 \times 10^{-4} L + 1.765 \times 10^{-2} \quad (23)$$

$$C_{f2} = 0.883 \times 10^{-4} L + 1.765 \times 10^{-2} \quad (24)$$

$$L_1 = 1.22 \times 10^{-4} L \quad (25)$$

$$L_2 = 1.43 \times 10^{-4} L \quad (26)$$

$$L_3 = 3.26 \times 10^{-4} L \quad (27)$$

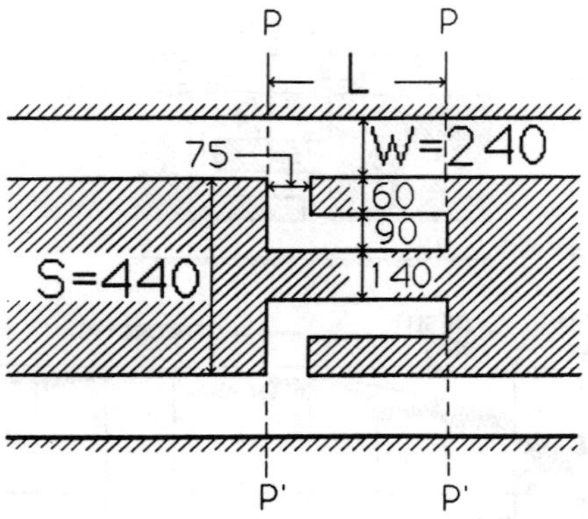

(a)

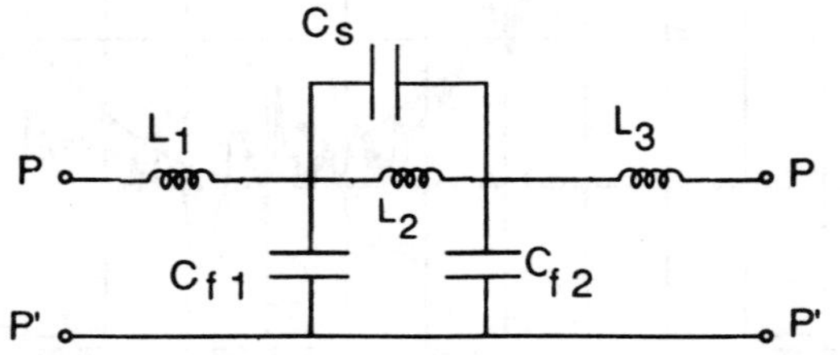

(b)

Fig. 10. Equivalent circuit for the short-end CPW stub ($D_1 = 21.5$ mil, $D_2 = 125$ mil, $\epsilon_{r1} = 9.9$, $\epsilon_{r2} = 2.2$; other dimensions are in μm).

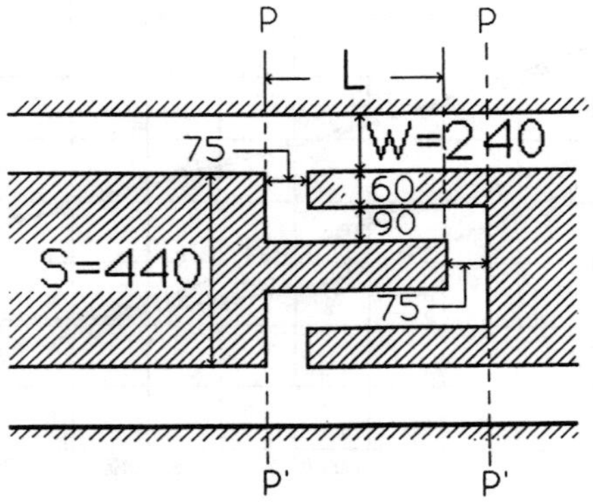

(a)

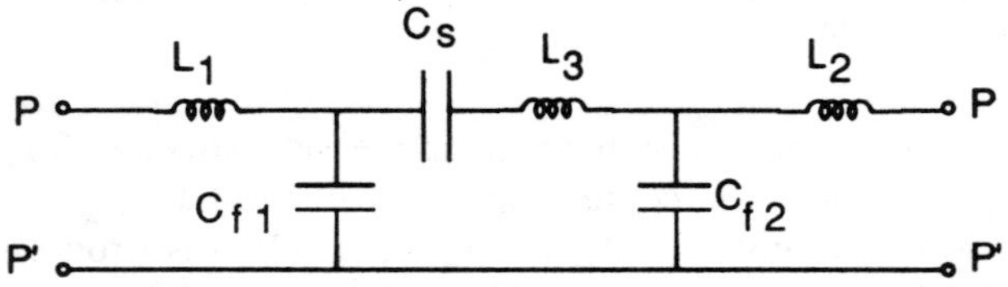

(b)

Fig. 11. Equivalent circuit for the open-end CPW stub ($D_1 = 21.5$ mil, $D_2 = 125$ mil, $\epsilon_{r1} = 9.9$, $\epsilon_{r2} = 2.2$; other dimensions are in μm).

where the stub length, L, is in μm, the inductances in nH, and the capacitances in pF. As in the previous case, the above equations, which apply for the configuration shown in Fig. 11(a) only, have been verified for stub lengths, L, through 2500 μm. In the limit as L approaches zero, the inductances reduce to zero and C_{f1} becomes equal to C_{f2}, resulting in a capacitive Π network which is expected for a series gap.

The above lumped element equivalent circuits predict the response up to the first resonant frequency with a 5% accuracy. It is expected that similar linear relationships apply for a short or open-end CPW stub with any dimensions. Thus, it is enough to model two different stub lengths, from which the characteristics of other lengths can be derived.

V. Conclusions

A space-domain integral equation method solved by the method of moments in conjunction with simple transmission line theory was applied to analyze CPW circuit elements useful for band-pass and band-stop filters. An experimental setup to measure the scattering parameters of those structures has been described. The agreement between the theoretical results and the experimental data was very good; thus, the validity of both results is verified. Lumped element equivalent circuits were proposed to model the above circuit elements, and closed-form expressions to compute the values of the capacitances and inductances were given as functions of stub length.

Appendix

As shown in Fig. 12, a typical CPW discontinuity problem is reduced to deriving the dyadic Green's function in both regions directly above and below the slots. The transmission line theory is used to transform the surrounding layers into impedance boundaries. Using the equivalence principle, the problem is divided into four subproblems (as shown in Fig. 12), where the fields in both regions due to magnetic currents in the y and z directions have to be obtained. After this has been accomplished, the continuity of the tangential fields at the interface is used to arrive at the integral equation. The main steps in the derivation of the fields arising from an infinitesimal z-directed magnetic current inside a cavity (with impedance boundary top side) will be presented here (see Fig. 13).

In the derivations, the following vector potentials for the LSM and LSE modes are assumed:

$$A = \hat{a}_x \psi \qquad F = \hat{a}_x \phi. \tag{A1}$$

By using Maxwell's equations along with

$$H = \frac{1}{\mu} \nabla \times A \tag{A2}$$

$$E = -\frac{1}{\epsilon} \nabla \times F \tag{A3}$$

one can obtain the field components in terms of the magnetic and electric vector potentials.

The differential equations for A and F are solved in view of the pertinent boundary conditions to give the unknown field components. The boundary conditions considered for the solution of this problem are listed below (see Fig. 13, where the magnetic current dipole is raised

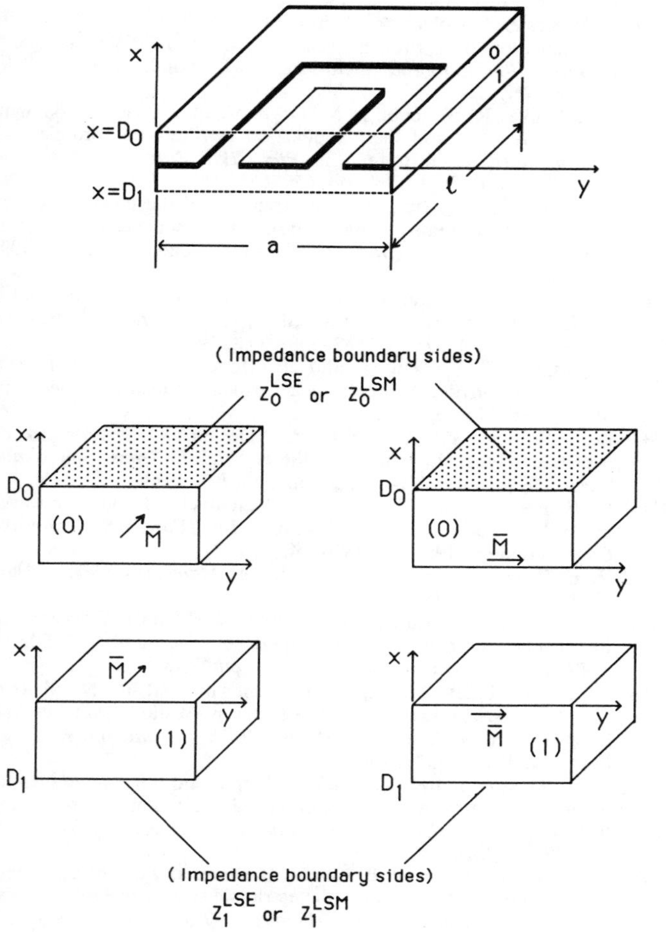

Fig. 12. A typical one-port CPW discontinuity where four subproblems are obtained using the equivalence principle. (All the sides excluding the impedance boundaries are assumed perfectly conducting.)

to simplify the application of the boundary conditions):

$$E_z^{\mathrm{I}} = E_z^{\mathrm{II}} \quad \text{at} \quad x = x' \tag{A4}$$

$$H_y^{\mathrm{I}} = H_y^{\mathrm{II}} \quad \text{at} \quad x = x' \tag{A5}$$

$$H_z^{\mathrm{I}} = H_z^{\mathrm{II}} \quad \text{at} \quad x = x' \tag{A6}$$

$$\left(\frac{E_y^{\mathrm{I}}}{H_z^{\mathrm{I}}}\right)^{\mathrm{LSE}} = Z_0^{\mathrm{LSE}} \quad \text{at} \quad x = D_0 \tag{A7}$$

$$\left(\frac{E_y^{\mathrm{I}}}{H_z^{\mathrm{I}}}\right)^{\mathrm{LSM}} = Z_0^{\mathrm{LSM}} \quad \text{at} \quad x = D_0 \tag{A8}$$

$$E_y^{\mathrm{II}} - E_y^{\mathrm{I}} = \delta(x - x')\delta(y - y')\delta(z - z'). \tag{A9}$$

In (A4)–(A9), D_0 is the thickness of the layer directly above the slot apertures. Z_0^{LSE} and Z_0^{LSM} are the LSE and LSM impedances seen at the interface $x = D_0$. For the structures considered here (Fig. 3), $Z_0^{\mathrm{LSE}} = Z_0^{\mathrm{LSM}} = 0$ since perfect conductors are assumed. Solving (A4)–(A9), the fields in the region directly above the slot aperture caused by a z-directed magnetic dipole are obtained. In the same manner the fields due to a y-directed magnetic dipole can be derived.

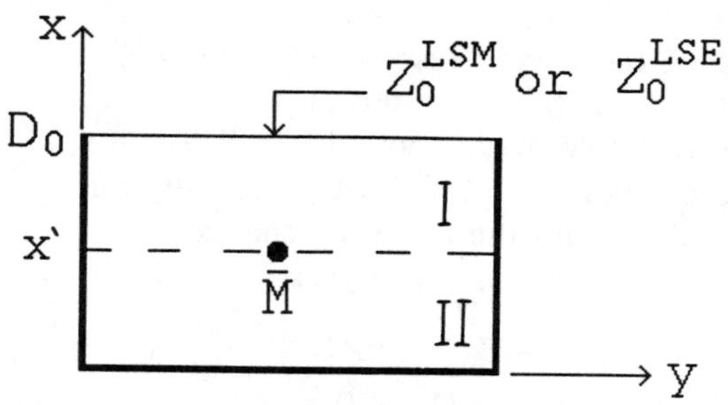

Fig. 13. A z-directed magnetic dipole inside a cavity with an impedance boundary top side.

The components of the dyadic Green's function for the coplanar waveguide problem are given by the following expressions:

$$G_{yy}^{(0)}(\bar r/\bar r') = \sum_{m=0}^{\mathrm{MSTOP}} \sum_{n=0}^{\mathrm{NSTOP}} \frac{2e_n}{al} \frac{1}{k_{x_0}^2 - k_0^2}\left[k_y^2 P_0 + k_z^2 Q_0\right] \cdot \sin(k_y y')\cos(k_z z')\sin(k_y y)\cos(k_z z) \tag{A10}$$

$$G_{zy}^{(0)}(\bar r/\bar r') = \sum_{m=0}^{\mathrm{MSTOP}} \sum_{n=0}^{\mathrm{NSTOP}} \frac{2e_n}{al} \frac{k_y k_z}{k_{x_0}^2 - k_0^2} \cdot [P_0 - Q_0] \cdot \sin(k_y y')\cos(k_z z')\cos(k_y y)\sin(k_z z) \tag{A11}$$

$$G_{yz}^{(0)}(\bar r/\bar r') = \sum_{m=0}^{\mathrm{MSTOP}} \sum_{n=0}^{\mathrm{NSTOP}} \frac{2e_n}{al} \frac{k_y k_z}{k_{x_0}^2 - k_0^2}[P_0 - Q_0] \cdot \cos(k_y y')\sin(k_z z')\sin(k_y y)\cos(k_z z) \tag{A12}$$

$$G_{zz}^{(0)}(\bar r/\bar r') = \sum_{m=0}^{\mathrm{MSTOP}} \sum_{n=0}^{\mathrm{NSTOP}} \frac{2e_m}{al} \frac{1}{k_{x_0}^2 - k_0^2}\left[k_z^2 P_0 + k_y^2 Q_0\right] \cdot \cos(k_y y')\sin(k_z z')\cos(k_y y)\sin(k_z z) \tag{A13}$$

where

$$P_0 = \left(\frac{k_{x_0}}{\omega\mu_0}\right)\frac{\omega\mu_0 + jk_{x_0}Z_0^{\mathrm{LSE}}\tan(k_{x_0}D_0)}{k_{x_0}Z_0^{\mathrm{LSE}} + j\omega\mu_0\tan(k_{x_0}D_0)} \tag{A14}$$

$$Q_0 = \left(\frac{\omega\epsilon_0}{k_{x_0}}\right)\frac{k_{x_0} + j\omega\epsilon_0 Z_0^{\mathrm{LSM}}\tan(k_{x_0}D_0)}{\omega\epsilon_0 Z_0^{\mathrm{LSM}} + jk_{x_0}\tan(k_{x_0}D_0)} \tag{A15}$$

$$e_n = 1, \quad n = 0; \qquad = 2, \quad n \neq 0 \tag{A16}$$

$$e_m = 1, \quad m = 0; \qquad = 2, \quad m \neq 0 \tag{A17}$$

$$k_y = \frac{m\pi}{a} \tag{A18}$$

$$k_z = \frac{n\pi}{l} \tag{A19}$$

$$k_0^2 = \omega^2\mu_0\epsilon_0 \tag{A20}$$

$$k_0^2 = k_{x_0}^2 + k_y^2 + k_z^2. \tag{A21}$$

In the above expressions, a and l are the width and length of the cavity, respectively, and $G_{ij}^{(0)}$ denotes the magnetic field $H_i^{(0)}$ radiated at $x = 0$ by an infinitesimal magnetic dipole M_j located at $x' = 0$ $(i, j = y, z)$.

The components of $\overline{\overline{G}}_1^h$ are essentially the same as in (A10)–(A21) with the following changes:

$$\begin{aligned} Z_0^{\text{LSE}} &\to Z_1^{\text{LSE}} \\ Z_0^{\text{LSM}} &\to Z_1^{\text{LSM}} \\ D_0 &\to D_1 \\ e_n, e_m &\to -e_n, -e_m \\ k_0 &\to k_1 \\ \mu_0, \epsilon_0 &\to \mu_1, \epsilon_1 \end{aligned} \tag{A22}$$

where Z_1^{LSE} and Z_1^{LSM} are the LSE and LSM impedances seen at the interface $x = D_1$.

References

[1] M. Houdart, "Coplanar lines: Application to broadband microwave integrated circuits," in *Proc. 6th European Microwave Conf.* (Rome), 1976, pp. 49–53.

[2] R. A. Pucel, "Design considerations for monolithic microwave circuits," *IEEE Trans. Microwave Theory Tech.*, vol. MTT-29, pp. 513–534, June 1981.

[3] R. W. Jackson, "Considerations in the use of coplanar waveguide for millimeter wave integrated circuits," *IEEE Trans. Microwave Theory Tech.*, vol. MTT-34, pp. 1450–1456, Dec. 1986.

[4] M. Riaziat, E. Par, G. Zdasiuk, S. Bandy, and M. Glenn, "Monolithic millimeter wave CPW circuits," in *1989 IEEE MTT-S Int. Microwave Symp. Dig.* (Long Beach, CA), pp. 525–528.

[5] T. Hirota, Y. Tarusawa, and H. Ogawa, "Uniplanar MMIC hybrids—A Proposed new MMIC structure," *IEEE Trans. Microwave Theory Tech.*, vol. MTT-35, pp. 576–581, June 1987.

[6] M. Riaziat, R. Majidi-Ahi, and I. Feng, "Propagation modes and dispersion characteristics of coplanar waveguides," *IEEE Trans. Microwave Theory Tech.*, vol. 38, pp. 245–251, Mar. 1990.

[7] K. C. Gupta, R. Garge and I. J. Bahl, *Microstrip Lines and Slotlines*. Dedham, MA: Artech House, 1979.

[8] R. N. Simons and G. E. Ponchak, "Modeling of some coplanar waveguide discontinuities," *IEEE Trans. Microwave Theory Tech.*, vol. 36, pp. 1796–1803, Dec. 1988.

[9] N. H. Koster, S. Koblowski, R. Bertenburg, S. Heinen, and I. Wolff, "Investigation of air bridges used for MMICs in CPW technique," in *Proc. 19th European Microwave Conf.* (London), Sept. 1989, pp. 666–671.

[10] G. Kibuuka, R. Bertenburg, M. Naghed and I. Wolff, "Coplanar lumped elements and their application in filters on ceramic and gallium arsenide substrates," in *Proc. 19th European Microwave Conf.* (London), Sept. 1989, pp. 656–661.

[11] C. W. Kuo and T. Itoh, "Characterization of the coplanar waveguide step discontinuity using the transverse resonance method," in *Proc. 19th European Microwave Conf.* (London), Sept. 1989, pp. 662–665.

[12] R. W. Jackson, "Mode conversion at discontinuities in finite-width conductor-backed coplanar waveguide," *IEEE Trans. Microwave Theory Tech.*, vol. 37, pp. 1582–1589, Oct. 1989.

[13] G. Matthaei, L. Young and E. Jones, *Microwave Filters, Impedance-Matching Networks, and Coupling Structures*. Dedham, MA: Artech House, 1980.

[14] D. F. Williams and S. E. Schwarz, "Design and performance of coplanar waveguide band-pass filters," *IEEE Trans. Microwave Theory Tech.*, vol. MTT-31, pp. 558–566, July 1983.

[15] G. E. Ponchak and R. N. Simons, "Channelized coplanar waveguide PIN-diode switches," in *Proc. 19th European Microwave Conf.* (London), 1989, pp. 489–494.

[16] R. F. Harrington, *Field Computation by Moment Methods*. New York: Macmillan, 1968.

[17] N. I. Dib and P. B. Katehi, "Modeling of shielded CPW discontinuities using the space domain integral equation method (SDIE)," *J. Electromagn. Waves and Appl.*, to be published.

[18] N. I. Dib, P. B. Katehi, G. E. Ponchak, and R. N. Simons, "Coplanar waveguide discontinuities for p-i-n diode switches and filter applications," in *1990 IEEE MTT-S Int. Microwave Symp. Dig.*, May 1990, pp. 399–402.

[19] A. W. Glisson and D. R. Wilton, "Simple and efficient numerical methods for problems of electromagnetic radiation and scattering from surfaces," *IEEE Trans. Antennas Propagat.*, vol. AP-28, pp. 593–603, Sept. 1980.

[20] W. P. Harokopus and P. B. Katehi, "Characterization of microstrip discontinuities on multilayer dielectric substrates including radiation losses," *IEEE Trans. Microwave Theory Tech.*, vol. 37, pp. 2058–2065, Dec. 1989.

[21] P. B. Katehi, "A generalized method for the evaluation of mutual coupling in microstrip arrays," *IEEE Trans. Antennas Propagat.*, vol. AP-35, pp. 125–133, Feb. 1987.

[22] R. E. Stegens, "Coplanar waveguide FET amplifiers for satellite communications systems," *Comsat Tech. Rev.*, vol. 9, pp. 255–267, spring 1979.

Equivalent Capacitances of Coplanar Waveguide Discontinuities and Interdigitated Capacitors Using a Three-Dimensional Finite Difference Method

MOHSEN NAGHED AND INGO WOLFF, FELLOW, IEEE

Abstract —Equivalent capacitances of coplanar waveguide discontinuities on multilayered substrates are calculated using a three-dimensional finite difference method. The application of the method is demonstrated for open ends and gaps in microstrip and coplanar waveguides as well as for more complicated structures such as interdigitated capacitors. The effect of conductor metallization thickness is also investigated. The calculated data are in good agreement with experimental results.

I. Introduction

LINE DISCONTINUITIES appear in nearly every microwave circuit and can be approximately described by lumped element equivalent circuits, provided that their dimensions are small compared with the wavelength. For coplanar waveguides, this condition is fulfilled since the structure dimensions are almost independent of substrate thickness and can therefore be chosen very small. Another interesting advantage of coplanar waveguides is the small dispersion of their characteristic parameters due to the fact that the field is mainly concentrated in the spacings between the conductors and does not change much with increasing frequency. This leads to the assumption that the frequency beyond which the static analysis of such structures is no longer valid is higher than in the case of a microstrip line. It is therefore of interest to test the validity of static analysis results for coplanar waveguide discontinuities even at high frequencies.

Several authors have reported on the calculation of equivalent capacitances for various discontinuities [1]–[5]. However these methods are applied to microstrip discontinuities only and cannot be implemented easily in other planar waveguides. Moreover, the methods do not consider the effect of the metallization thickness of conductors, which, especially in the case of coplanar waveguide structures, can become comparable to the geometrical dimensions of the structure. The method presented in this paper is a general approach for finding the static capacitances of discontinuities as an abrupt change in the geometry of conductors in planar waveguides, as shown in Fig. 1. A three-dimensional finite difference method is applied to shielded coplanar structures, and the field distribution inside the shield is calculated using the successive relaxation method. The equivalent capacitances of the structure are then evaluated from the charge distribution on the conductors.

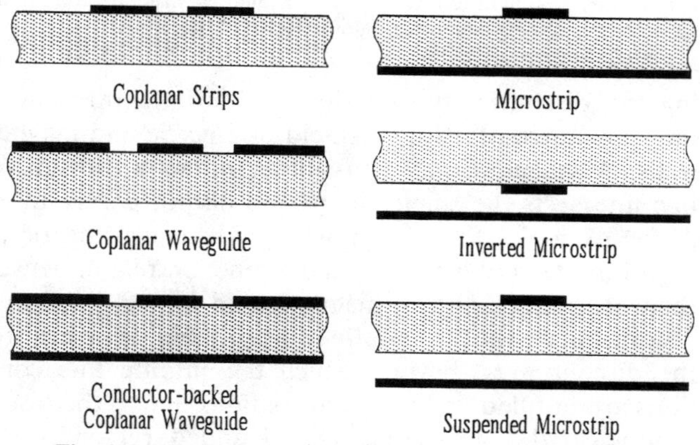

Fig. 1. Planar waveguides supporting quasi-TEM mode.

A program package has been developed which calculates the equivalent circuit capacitances of various ordinary discontinuities. The results for coplanar waveguide open ends and gaps are presented in this paper. Since the equivalent parameters are calculated from the field distribution near the discontinuity, the method can easily be extended to any arbitrary configuration of conductors on substrates with one or two layers. Further applications are demonstrated for coplanar interdigitated capacitors on ceramic and gallium arsenide substrates. The effect of conductor metallization thickness as well as the influence of the shielding is also discussed. Measurements are performed over a broad frequency range to verify the accuracy of the applied method.

II. Numerical Calculation of the Field Distribution

The three-dimensional method presented here is a modified version of the well-known finite difference method described in [6]–[9]. The planar structure contain-

Manuscript received March 23, 1990; revised July 31, 1990.
The authors are with the Department of Electrical Engineering and Sonderforschungsbereich 254, Duisburg University, Bismarckstr. 69, D-4100 Duisburg, West Germany.
IEEE Log Number 9038997.

Reprinted from *IEEE Trans. Microwave Theory Tech.*, vol. 38, no. 12, pp. 1808–1815, Dec. 1990.

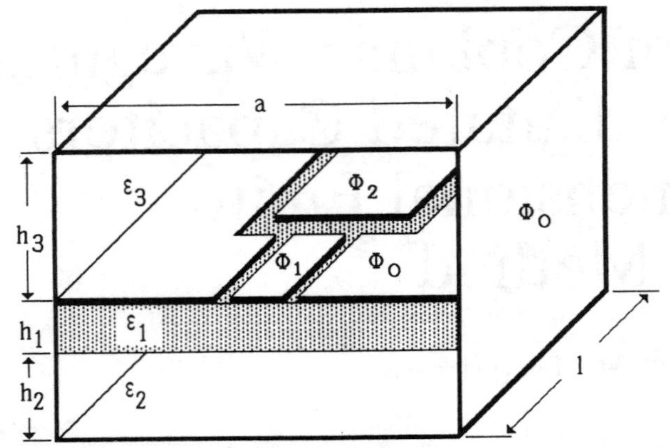

Fig. 2. Three-dimensional multilayered shielded structure containing a planar waveguide discontinuity.

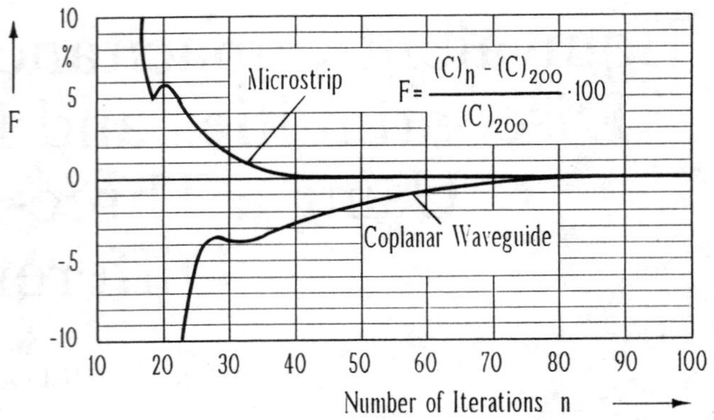

Fig. 3. The relative error of equivalent open-end capacitance as a function of the number of iterations.

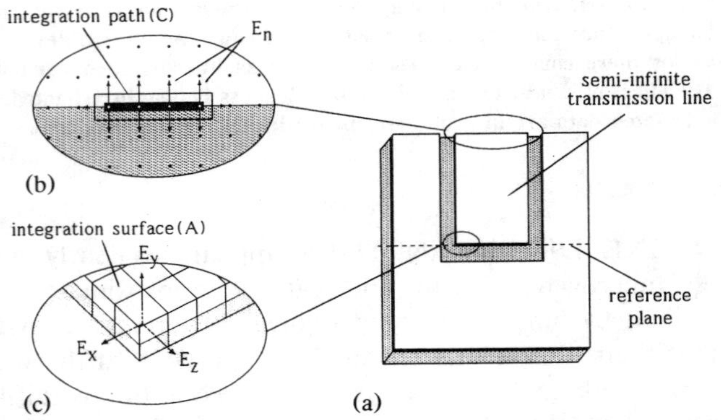

Fig. 4. Calculation of (b) the charge per unit length and (c) the total charge of (a) an open ended coplanar line.

ing the discontinuity and the semi-infinite transmission lines are surrounded by a shield of electric and magnetic walls. A magnetic wall is assumed where a transmission line intersects the shield (input and output ports). In Fig. 2, for example, the planar waveguide consisting of two semi-infinite coplanar lines and a junction region between them is shielded by two magnetic and four electric walls. The magnetic walls are defined at suitable distances from the discontinuity, beyond which the infinite line conditions are fulfilled and the perturbations due to the discontinuity can be neglected. The bounded region is then divided into elementary boxes using a three-dimensional nonequidistant Cartesian grid. The electrical potential can be assumed to be constant inside the elementary boxes and confined at the middle of the box. Since the structure dimensions are small compared with the wavelength, the grid points on the metallization are assumed to be at a constant potential (Fig. 2).

Using Laplace's equation, the electrical potential at any point inside the shield can be written in finite difference form as a linear combination of potentials at neighboring grid points (see the Appendix, eq. (A10)). The "relaxation method" is used for the solution of the resulting equation system. The potential distribution inside the shield can be determined starting with assumed potential values at all the grid points and modifying them as follows:

$$\Phi_{\text{new}} = \Phi_{\text{old}} - k \cdot R \tag{1}$$

where R is the difference between Φ_{old} and the value given by (A10), and k is the relaxation constant, which determines the speed of convergence. The optimal value of k is found here to be 1.8.

The accuracy of the calculated potentials depends on the number of iterations (n). This is demonstrated in Fig. 3. As shown in this figure, 60 iterations are sufficient for an accuracy better than 1% for the calculated equivalent capacitance of an open-ended line. The electrical field at any grid point can then be calculated as

$$\vec{E} = -\vec{\nabla}\Phi. \tag{2}$$

The field distribution obtained in this way is used in the next section to calculate the charge distribution on the conductors.

III. Evaluation of the Equivalent Capacitances

For the following descriptions, let us consider the configuration shown in Fig. 4. The field distribution calculated in (2) is used to find the total charge on the inner conductor (Q_{total}), as well as the charge per unit length on the connected transmission lines (Q'). Q' can be obtained from the relation

$$Q' = \epsilon_0 \epsilon_r \oint_C E_n \cdot ds \tag{3}$$

where C is the contour of the integration region which covers the conductor cross section at the plane of the magnetic wall (Fig. 4(b)). The total charge can be calculated integrating the normal component of the electric field over the inner conductor:

$$Q_{\text{total}} = \epsilon_0 \epsilon_r \oint_A \vec{E} \cdot \vec{n}\, dA. \tag{4}$$

The region of integration now covers the total conductor surface A (Fig. 4(c)). If the potential difference between the inner conductor and the ground planes is V, the capacitance associated with the discontinuity can be calculated from the difference between the total charge,

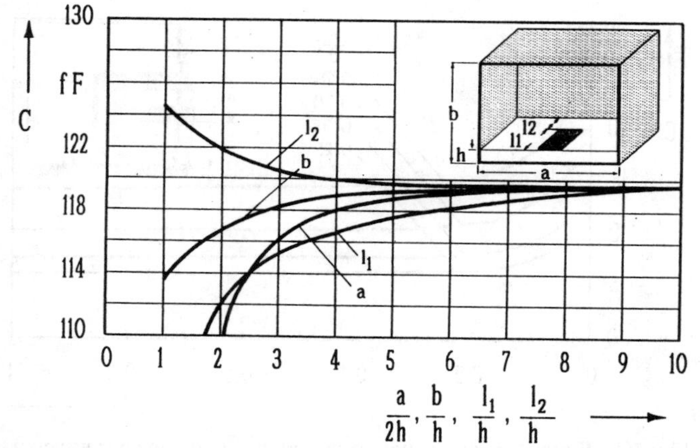

Fig. 5. The effect of shielding on the equivalent capacitance of a microstrip open end.

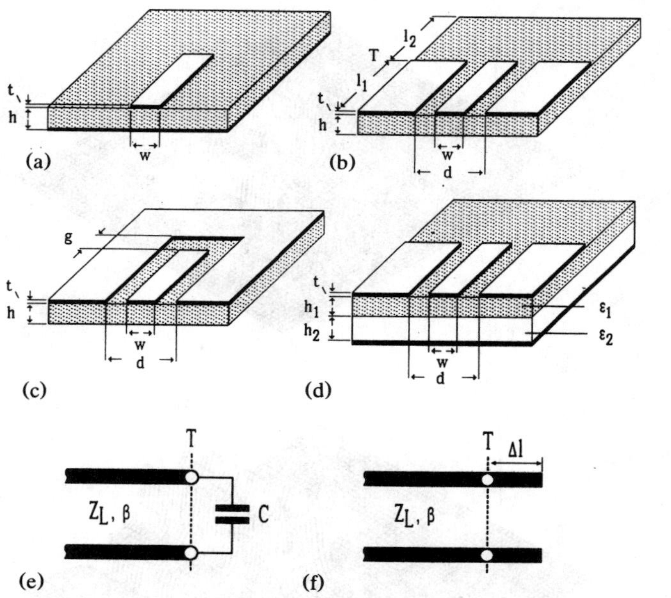

Fig. 6. Different types of open-ended planar waveguides and their equivalent circuit models.

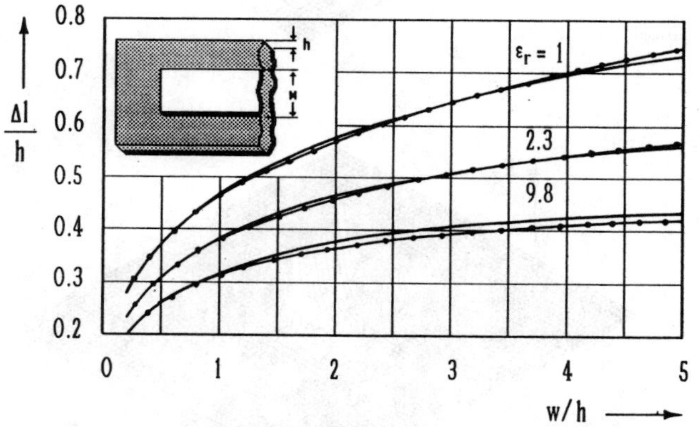

Fig. 7. Equivalent line prolongation of a microstrip open end calculated by the presented method (solid lines) compared with full-wave spectral-domain calculations [10] at 4 GHz (dotted lines).

Q_{total}, and the charge per unit length of the connected transmission line as

$$C_{eq} = (Q_{total} - l \cdot Q')/V \tag{5}$$

where l is defined as the distance between the back-transformed reference plane of the discontinuity (e.g. Fig. 4(a)) and the magnetic wall. The effect of shielding on the equivalent capacitance is studied for a microstrip open end ($\epsilon_r = 9.8$, $h = 0.635$ mm) with the following dimensions:

width-to-height ratio	$w/h = 4$
distance to the lateral electrical walls	$a/2 = 10 \cdot h$
distance to the upper surface	$b = 10 \cdot h$
distance to the front electric wall	$l_2 = 10 \cdot h$
distance to the magnetic wall	$l_1 = 10 \cdot h$.

Each of the last four parameters is changed separately while the others are held constant, and the equivalent capacitance of the open end is computed each time. As shown in Fig. 5, the distance to the magnetic wall l_1 is the critical parameter and must be large enough. The distance to the front electric wall l_2 has a small influence on the calculated results.

IV. Testing Structures

In this section several open ends and gaps in various planar waveguides on ceramic and gallium arsenide substrates are treated. The effect of finite metallization thickness is discussed in more detail in the case of gaps. Moreover, coplanar interdigitated capacitors of various dimensions are investigated and their scattering parameters are computed and compared with measured values.

A. Open Ends

As is well known, the abrupt open end of a transmission line stores electrical energy and can be simulated by an equivalent capacitance C (Fig. 6(e)). An extended length Δl of the transmission line (Fig. 6(f)) can also be used to model the open end. The length Δl is defined as the ratio of the equivalent capacitance to the capacitance per unit length of the line. In this paper both C and Δl are calculated and used for the discussion of the results.

Fig. 6 shows the various open-ended structures which are investigated. To indicate the accuracy of the chosen method, the extended lengths of microstrip open ends of various dielectric constants are plotted in Fig. 7 as a function of width-to-height ratio together with results from the spectral-domain analysis presented in [10]. As shown in the figure, the data calculated with the method presented here are in good agreement with the accurate values based on full-wave calculations.

The next test structure is the coplanar waveguide open end of Fig. 6(b). The field distribution of this structure in the conductor plane (Fig. 8) indicates the scattering of the electric field components of an assumed TEM wave at the end of the line. The calculated data of equivalent capacitances are plotted against the width-to-spacing ratio w/d for different substrate thicknesses (Fig. 9). The equivalent

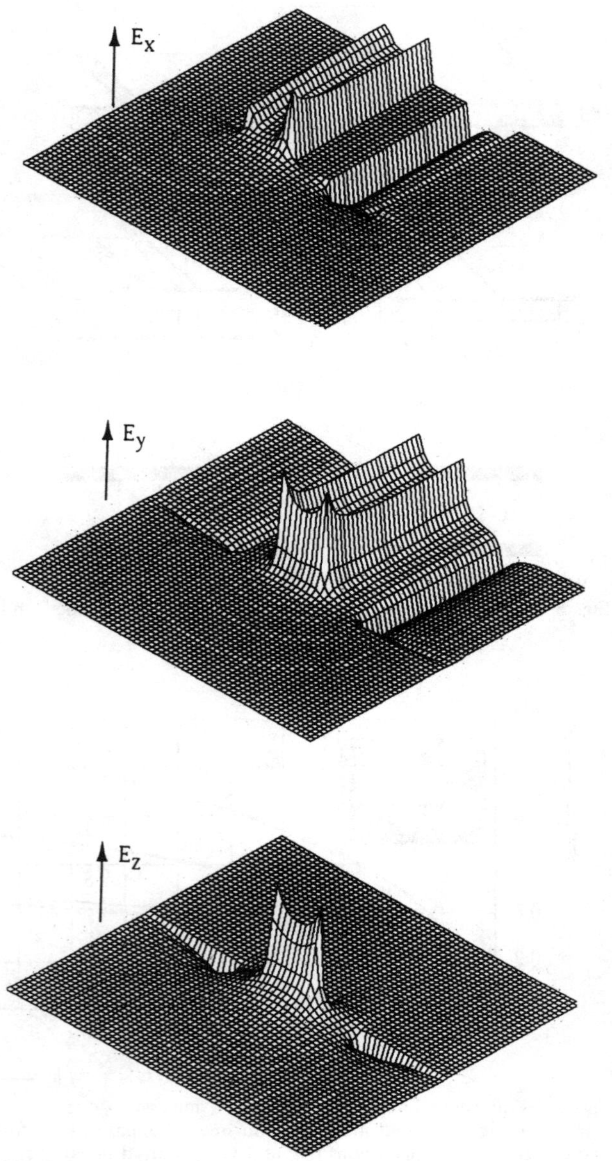

Fig. 8. Electric field components in the conductor plane of an open-ended coplanar waveguide supporting a TEM wave.

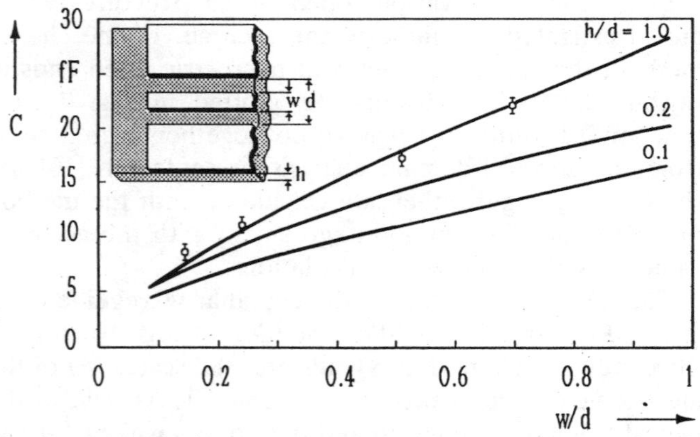

Fig. 9. Calculated equivalent capacitance of an open-ended coplanar waveguide ($\epsilon_r = 9.8$, $h = 0.635$ mm, $t = 0$) together with measured values for $h/d = 1.0$.

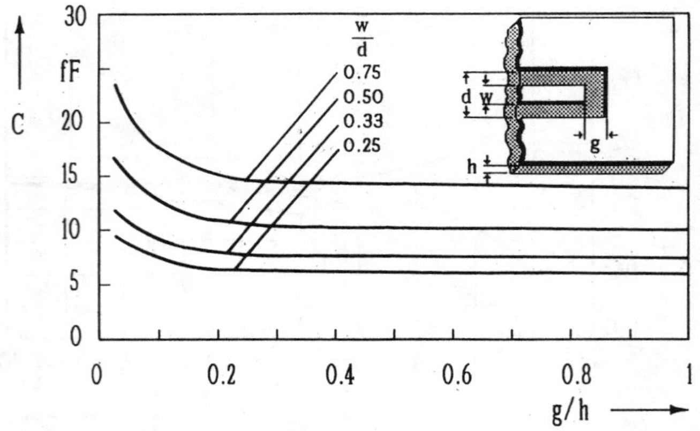

Fig. 10. Equivalent parameters of an open-ended coplanar waveguide with connected ground planes as a function of the spacing g at the end of the line ($\epsilon_r = 12.9$, $h = 0.635$ mm, $d/h = 0.6$, $t = 0$).

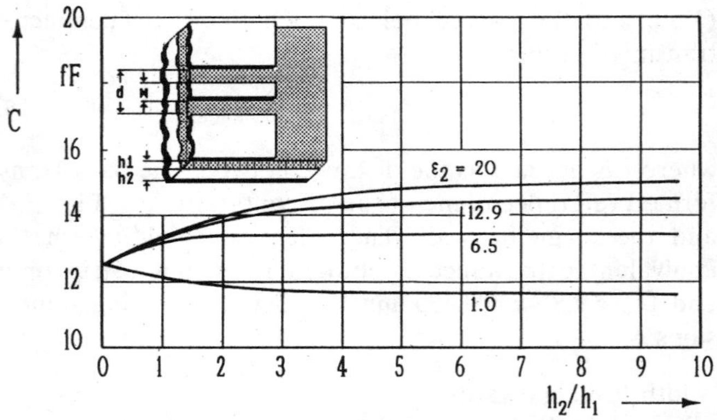

Fig. 11. Equivalent capacitance of an open-ended two-layer conductor-backed coplanar waveguide for various dielectric constants of the bottom layer ($\epsilon_1 = 12.9$, $h_1 = 0.1$ mm, $w/d = 0.5$, $d/h_1 = 4$, $t = 0$).

capacitance increases steadily with the width of the inner conductor. The calculated data are in good agreement with experimental values which are evaluated from the scattering parameter measurements in the frequency range from 45 MHz to 26.5 GHz.

In some cases the ground planes of the coplanar waveguide enclose the open end as shown in Fig. 6(c). The equivalent parameters of such a structure are plotted in Fig. 10 as a function of spacing g for various w/d ratios and constant d/h. It can be observed that for values $g > d$ the variation of the equivalent capacitance is negligible.

Another interesting structure is that of Fig. 6(d), where a two-layer substrate with a back-side metallization serves as a support. In Fig. 11 the calculated capacitances are plotted as a function of bottom layer height, h_2, for various dielectric constants (ϵ_2). The case $\epsilon_1 = \epsilon_2 = 12.9$ represents the conductor-backed coplanar waveguide. The result for $\epsilon_2 = 1.0$ gives additional information on the effect of the electric shield at the bottom.

B. Gaps

Based on the two-port pi network model shown in Fig. 12(e), the equivalent capacitances of gap discontinuities (C_g, C_{p1}, and C_{p2}) are calculated in two steps. First, the

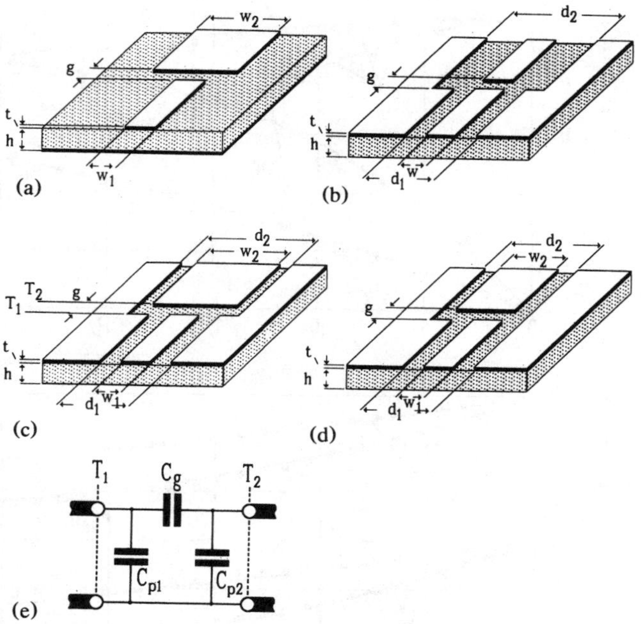

Fig. 12. Gaps in microstrip and coplanar lines and their capacitive pi network model.

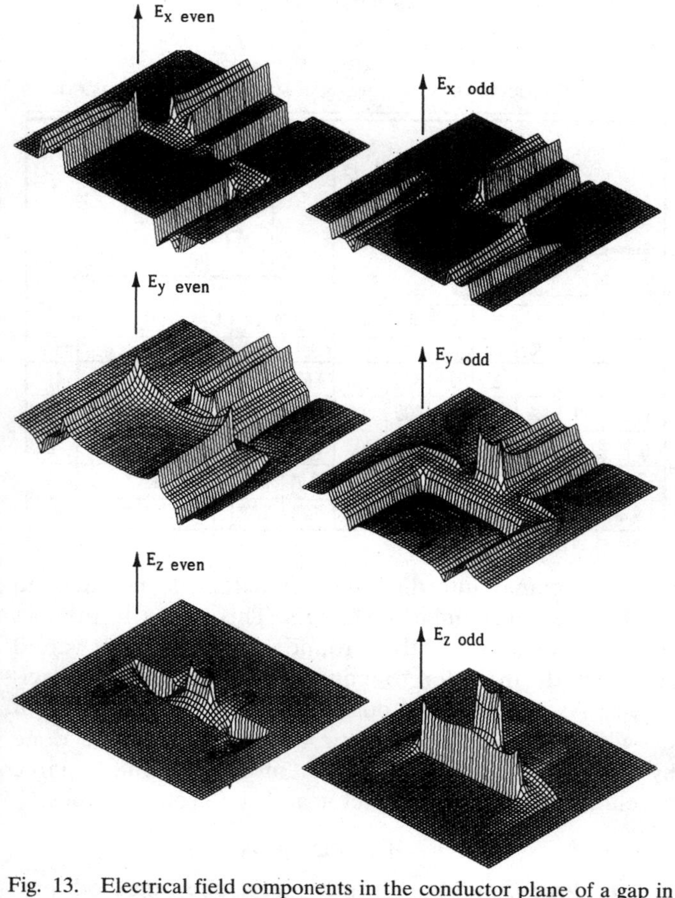

Fig. 13. Electrical field components in the conductor plane of a gap in a coplanar waveguide for the even and odd modes.

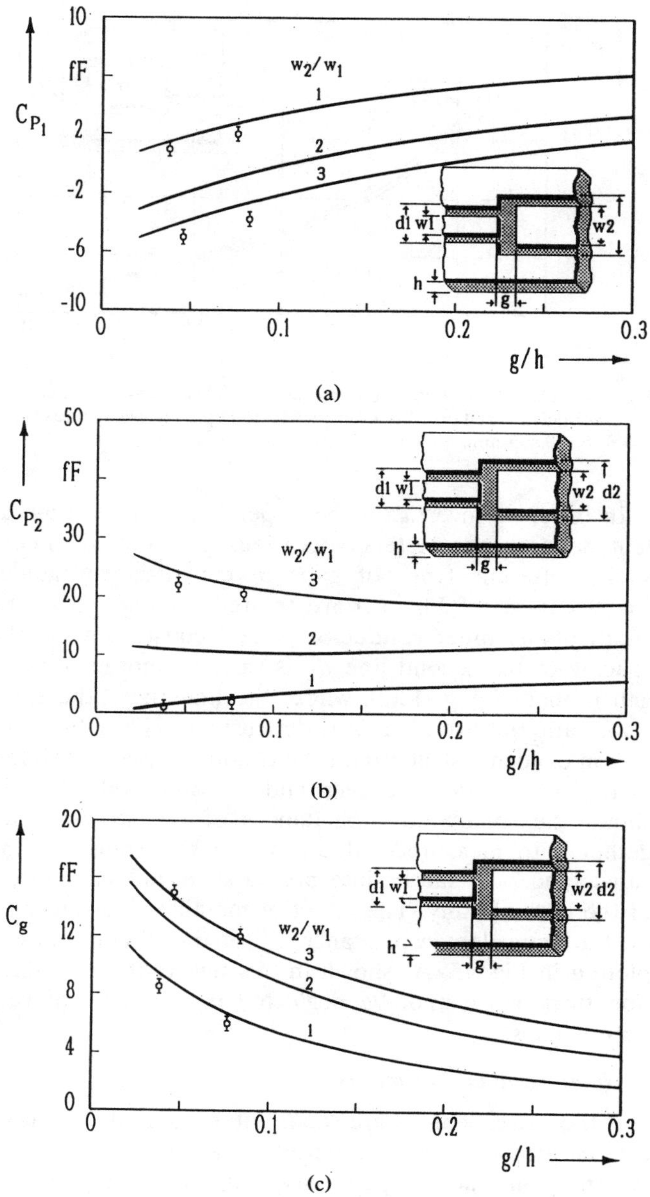

Fig. 14. Equivalent circuit capacitances of a symmetric gap between two 50 Ω coplanar waveguides of different dimensions ($\epsilon_r = 9.8$, $h = 0.635$ mm, $w_1/h = 0.2$, $w_1/d_1 = w_2/d_2 = 0.56$, $t = 0$) together with measured values.

potentials on the inner conductors are assumed to be $\Phi_1 = \Phi_2 = 1$ V (see Fig. 2), and at the ground plane $\Phi_0 = 0$ V (even mode). In this case the capacitances C_{p1} and C_{p2} can be calculated as follows:

$$C_{p1} = \left(Q_{1_{\text{total}}} - l_1 Q_1'\right) \tag{6}$$

$$C_{p2} = \left(Q_{2_{\text{total}}} - l_2 Q_2'\right). \tag{7}$$

In the second step, the conductors are assumed to be at potentials $\Phi_1 = 1$ V, $\Phi_2 = -1$ V, and $\Phi_0 = 0$ V (odd mode). Equations (6) and (7) are used again and the calculated capacitances in this case are called C_{11} and C_{22}. The series capacitance C_g can be determined as

$$C_g = 0.5(C_{11} - C_{p1}) = 0.5(C_{22} - C_{p2}). \tag{8}$$

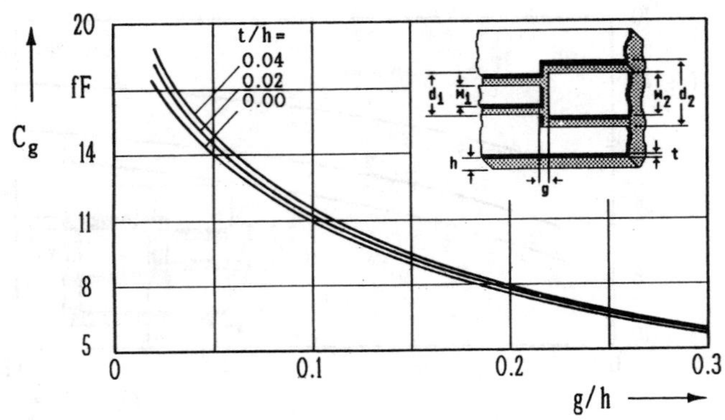

Fig. 15. The effect of conductor metallization thickness on the equivalent circuit capacitance between two 50 Ω coplanar waveguides ($\epsilon_r = 9.8$, $h = 0.635$ mm, $w_1/h = 0.2$, $d_1/h = 0.36$, $w_2/w_1 = 3$).

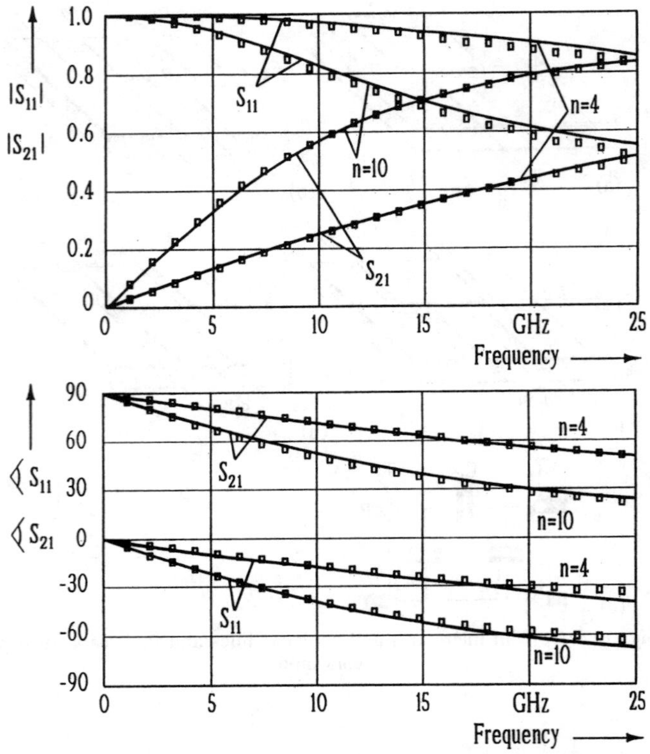

Fig. 16. The calculated (——) and measured (□ □ □) scattering parameters of coplanar interdigitated capacitors ($\epsilon_r = 12.9$, $h = 400$ μm, $l_f = 100$ μm). $n = 4$, $w_f = 17$ μm, $s_f = 3$ μm (see Table I, first line). $n = 10$, $w_f = 12$ μm, $s_f = 2$ μm (see Table I, second line).

In order to investigate the dependence of the equivalent network parameters on various geometrical dimensions, different types of gaps in coplanar waveguides (parts (b)–(d) of Fig. 12) are treated. In Fig. 12(b), the width of the inner conductor is held constant while the spacing of the second line d_2 is varied. Another type of gap is that of Fig. 12(c), where the lines have the same w/d ratio but are of various dimensions. The field distribution of such a structure in the conductor plane is shown in Fig. 13 for both the even and the odd mode. Fig. 14 shows the calculated capacitances of the structure together with measured values. A negative value for the capacitance C_{p1} can be interpreted as an inductive effect of the discontinuity. The effect of metallization thickness on the equivalent capacitance C_g of the discontinuity is plotted in Fig. 15. As shown in this figure, the metallization thickness cannot be neglected in the cases of very narrow gaps.

C. Interdigitated Capacitors

If the finger lengths are small, interdigitated capacitors can in principle be considered as a gap between two transmission lines. In this case the pure capacitive pi network model shown in Fig. 12(e) can be used to describe such a structure. In order to check the validity limits of this model, coplanar interdigitated capacitors with various numbers of fingers and different geometrical sizes on ceramic and gallium arsenide substrates have been fabricated. The capacitances of the equivalent network are computed and the calculated scattering parameters together with results measured using an HP8510B network analyzer (0.045–26.5 GHz) are plotted in Fig. 16. Although a simplified circuit model is used, the agreement is very good up to 25 GHz. The small deviations of the reflection coefficient at the highest frequencies are mainly caused by measurement difficulties at these frequencies due to the calibration.

In Table I, the calculated and experimental data of the capacitors are listed. The experimental values are obtained from measured scattering parameters using optimization routines for the network fitting. In contrast to the microstrip case, the parasitic capacitances C_{p1} and C_{p2} of coplanar interdigitated capacitors do not depend on the size and number of fingers. This is due to the fact that the location of the ground planes can be varied conveniently in order to minimize the parasitic capacitances. As a result, it is possible to have a large value of C_g without increased parasitic effects. This can be done by using an increased number of fingers and a large spacing between the conductor and the ground planes.

TABLE I
EQUIVALENT PARAMETERS OF COPLANAR INTERDIGITATED CAPACITORS

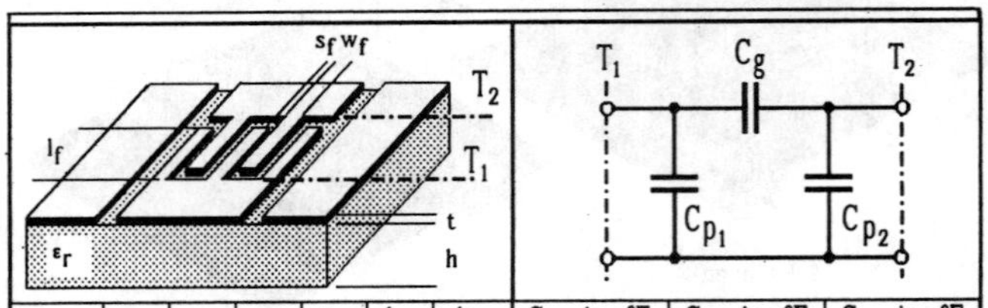

ϵ_r	n_f	w_f µm	s_f µm	t µm	h µm	l_f µm	C_{p1} in fF calc.	C_{p1} in fF meas.	C_{p2} in fF calc.	C_{p2} in fF meas.	C_g in fF calc.	C_g in fF meas.
12.9	4	17	3	3	400	100	9.65	11	9.65	11	40.1	41
12.9	10	12	2	3	400	100	9.9	11	9.9	11	113.3	116
12.9	4	17	3	3	400	200	19.2	20	19.2	20	73	76
9.8	5	38	25	5	635	200	22.94	23	10.78	11	55	56
9.8	7	38	25	5	635	200	21.78	22	11.64	12	85	86

V. CONCLUSIONS

The three-dimensional finite difference method presented in this paper is a simple and accurate procedure

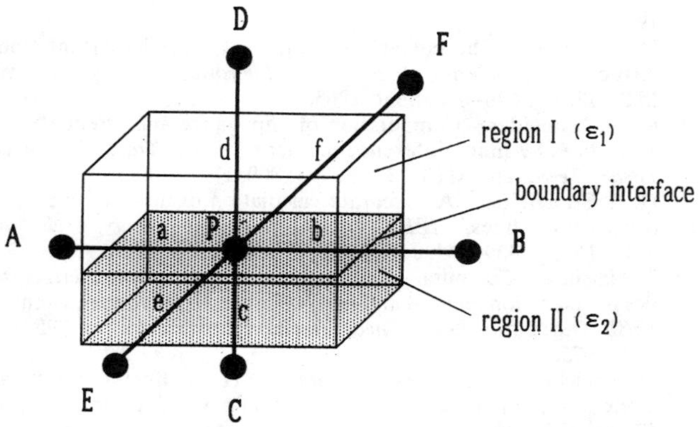

Fig. 17. Configuration of grid points at the boundary interface between two substrate materials.

for computing the static equivalent capacitances of discontinuities in many planar waveguides supporting TEM modes. The main advantage of the method is its flexibility in treating multilayered substrates and different conductor configurations. It can therefore be applied to more complicated structures such as interdigitated capacitors, air bridges, and waveguide transitions. The effect of conductor metallization thickness and shielding walls is also taken into account. The good agreement between the calculated data and measurements up to 25 GHz in the case of coplanar structures indicates the validity of the static analysis even for high frequencies. This is mainly because of the small dispersion of the coplanar line characteristics. However, it must be pointed out that for other dispersive lines such as microstrip lines, the validity of the calculated equivalent parameters at high frequencies is not guaranteed.

The computing time and the memory storage requirement are considerably reduced using a nonequidistant discretization. The computation times for discontinuities discussed in this paper are of the order of seconds (20 to 40 seconds), so this method can be used in the computer-aided design of microwave interdigitated circuits.

Appendix

Considering the configuration shown in Fig. 17 and using Taylor's series expansion, electrical potentials at points (A, D, E) in the immediate vicinity of a point P in region I can be written in finite difference form as

$$\Phi_A = \Phi_P - a \cdot \left.\frac{\partial \Phi}{\partial x}\right|_{\mathrm{I}} + \frac{a^2}{2!} \cdot \left.\frac{\partial^2 \Phi}{\partial x^2}\right|_{\mathrm{I}} - \frac{a^3}{3!} \cdot \left.\frac{\partial^3 \Phi}{\partial x^3}\right|_{\mathrm{I}} + \cdots \quad \text{(A1)}$$

$$\Phi_D = \Phi_P + d \cdot \left.\frac{\partial \Phi}{\partial y}\right|_{\mathrm{I}} + \frac{d^2}{2!} \cdot \left.\frac{\partial^2 \Phi}{\partial y^2}\right|_{\mathrm{I}} + \frac{d^3}{3!} \cdot \left.\frac{\partial^3 \Phi}{\partial y^3}\right|_{\mathrm{I}} + \cdots \quad \text{(A2)}$$

$$\Phi_E = \Phi_P - e \cdot \left.\frac{\partial \Phi}{\partial z}\right|_{\mathrm{I}} + \frac{e^2}{2!} \cdot \left.\frac{\partial^2 \Phi}{\partial z^2}\right|_{\mathrm{I}} - \frac{e^3}{3!} \cdot \left.\frac{\partial^3 \Phi}{\partial z^3}\right|_{\mathrm{I}} + \cdots. \quad \text{(A3)}$$

Ignoring third- and higher order terms and using Laplace's equation $(\Delta\Phi = 0)$, the above relations yield

$$\frac{ed}{a}\Phi_A + \frac{ae}{d}\Phi_D + \frac{ad}{e}\Phi_E - \left(\frac{ed}{a} + \frac{ae}{d} + \frac{ad}{e}\right)\Phi_P + ed\left.\frac{\partial \Phi}{\partial x}\right|_{\mathrm{I}} - ae\left.\frac{\partial \Phi}{\partial y}\right|_{\mathrm{I}} + ad\left.\frac{\partial \Phi}{\partial z}\right|_{\mathrm{I}} = 0. \quad \text{(A4)}$$

Analogously for the grid points (A, D, F), (B, D, E), and (B, D, F), we have

$$\frac{fd}{a}\Phi_A + \frac{af}{d}\Phi_D + \frac{ad}{f}\Phi_F - \left(\frac{fd}{a} + \frac{af}{d} + \frac{ad}{f}\right)\Phi_P + fd\left.\frac{\partial \Phi}{\partial x}\right|_{\mathrm{I}} - af\left.\frac{\partial \Phi}{\partial y}\right|_{\mathrm{I}} - ad\left.\frac{\partial \Phi}{\partial z}\right|_{\mathrm{I}} = 0 \quad \text{(A5)}$$

$$\frac{ed}{b}\Phi_B + \frac{be}{d}\Phi_D + \frac{bd}{e}\Phi_E - \left(\frac{ed}{b} + \frac{be}{d} + \frac{bd}{e}\right)\Phi_P - ed\left.\frac{\partial \Phi}{\partial x}\right|_{\mathrm{I}} - be\left.\frac{\partial \Phi}{\partial y}\right|_{\mathrm{I}} + bd\left.\frac{\partial \Phi}{\partial z}\right|_{\mathrm{I}} = 0 \quad \text{(A6)}$$

$$\frac{fd}{b}\Phi_B + \frac{bf}{d}\Phi_D + \frac{bd}{f}\Phi_F - \left(\frac{fd}{b} + \frac{bf}{d} + \frac{bd}{f}\right)\Phi_P - fd\left.\frac{\partial \Phi}{\partial x}\right|_{\mathrm{I}} - bf\left.\frac{\partial \Phi}{\partial y}\right|_{\mathrm{I}} - bd\left.\frac{\partial \Phi}{\partial z}\right|_{\mathrm{I}} = 0. \quad \text{(A7)}$$

Adding relations (A4) to (A7) yields

$$\left.\frac{\partial \Phi}{\partial y}\right|_{\mathrm{I}} = \frac{d}{a(a+b)}\Phi_A + \frac{d}{b(a+b)}\Phi_B + \frac{1}{d}\Phi_D + \frac{d}{e(e+f)}\Phi_E + \frac{d}{f(e+f)}\Phi_F - \left(\frac{d}{ab} + \frac{1}{d} + \frac{d}{ef}\right)\Phi_P. \quad \text{(A8)}$$

Analogously for region II, the following relation is valid:

$$\left.\frac{\partial \Phi}{\partial y}\right|_{\mathrm{II}} = \frac{-c}{a(a+b)}\Phi_A - \frac{c}{b(a+b)}\Phi_B - \frac{1}{c}\Phi_C - \frac{c}{e(e+f)}\Phi_E - \frac{c}{f(e+f)}\Phi_F + \left(\frac{c}{ab} + \frac{1}{c} + \frac{c}{ef}\right)\Phi_P. \quad \text{(A9)}$$

Using the boundary condition for the y component of the electrical field given by

$$\epsilon_1\left.\frac{\partial \Phi}{\partial y}\right|_{\mathrm{I}} = \epsilon_2\left.\frac{\partial \Phi}{\partial y}\right|_{\mathrm{II}}$$

the potential at point P can be written as follows:

$$\left(\frac{d\epsilon_1 + c\epsilon_2}{ab} + \frac{\epsilon_1}{d} + \frac{\epsilon_2}{c} + \frac{d\epsilon_1 + c\epsilon_2}{ef}\right)\Phi_P = \frac{d\epsilon_1 + c\epsilon_2}{a(a+b)}\Phi_A + \frac{d\epsilon_1 + c\epsilon_2}{b(a+b)}\Phi_B + \frac{\epsilon_2}{c}\Phi_C + \frac{\epsilon_1}{d}\Phi_D + \frac{d\epsilon_1 + c\epsilon_2}{e(e+f)}\Phi_E + \frac{d\epsilon_1 + c\epsilon_2}{f(e+f)}\Phi_F. \quad \text{(A10)}$$

REFERENCES

[1] P. Silvester and P. Benedek, "Equivalent capacitances of microstrip open circuits," *IEEE Trans. Microwave Theory Tech.*, vol. MTT-20, pp. 511–516, Aug. 1972.

[2] P. Benedek and P. Silvester, "Equivalent capacitances of microstrip gaps and steps," *IEEE Trans. Microwave Theory Tech.*, vol. MTT-20, pp. 729–733, Nov. 1972.

[3] M. Maeda, "An analysis of gap in microstrip transmission lines," *IEEE Trans. Microwave Theory Tech.*, vol. MTT-20, pp. 390–396, 1972.

[4] C. Gupta and A. Gopinath, "Equivalent circuit capacitance of microstrip step change in width," *IEEE Trans. Microwave Theory Tech.*, vol. MTT-25, pp. 819–822, Oct. 1977.

[5] P. Silvester and P. Benedek, "Microstrip discontinuity capacitances for right-angle bends, T junctions, and crossings," *IEEE Trans. Microwave Theory Tech.*, vol. MTT-21, pp. 341–346, May 1973.

[6] H. G. Green, "The numerical solution of some important transmission-line problems," *IEEE Trans. Microwave Theory Tech.*, vol. MTT-13, pp. 676–692, Sept. 1965.

[7] M. V. Schneider, "Computation of impedance and attenuation of TEM-lines by finite difference methods," *IEEE Trans. Microwave Theory Tech.*, vol. MTT-13, pp. 793–800, Nov. 1965.

[8] H. E. Stinehelfer, "An accurate calculation of uniform microstrip transmission lines," *IEEE Trans. Microwave Theory Tech.*, vol. MTT-16, pp. 439–444, July 1968.

[9] T. Hatsuda, "Computation of coplanar-type strip-line characteristics by relaxation method and its application to microwave circuits," *IEEE Trans. Microwave Theory Tech.*, vol. MTT-23, pp. 795–802, Oct. 1975.

[10] M. Kirschning, R. H. Jansen, and N. H. L. Koster, "Accurate model for open end effect of microstrip lines," *Electron. Lett.*, vol. 17, pp. 123–124, 1981.

Chapter 9

Filters and Stubs

DURING the past two decades the filtering and matching properties of planar structures with simple (e.g., circular, annular, rectangular, radial) and arbitrary shapes have been investigated extensively. Seven papers on this subject are included in this chapter. The first three reprints (Papers 9.1, 9.2, and 9.3) describe the planar circuit method of analysis and discuss the filtering properties of planar structures having rectangular, circular, and annular shapes. Planar circuit modeling of the radial stub is addressed in Papers 9.4 and 9.5. The last two reprints (Papers 9.6 and 9.7) describe the application of numerical techniques, such as the finite element method and the finite-difference time-domain technique, to the analysis of planar filters and stubs.

The general filtering properties of rectangular structures were first explored in Italy by Bianco, Ridella, and their coworkers [9.1–9.5]. During the same period a Japanese group of researchers led by T. Okoshi developed the concept of planar circuits [9.6] and also explored the filtering properties of circular structures [9.7]. Okoshi's planar circuit approach is extended in Paper 9.1 by D'Inzeo et al., who have incorporated the effective dimensions into the magnetic wall model. The effective magnetic wall model was originally developed by Wolff et al. [9.8] as a useful approach for characterizing microstrip discontinuities.

The method of analysis discussed in Paper 9.1 is limited to two-port networks. It is based on a field expansion in terms of the resonant modes of the planar structure. The effect of the fringing field is taken into account by introducing effective dimensions and effective permittivities for each resonant mode. These effective parameters depend on the field distribution for each mode inside the planar structure. Good agreement between the measured and calculated results was obtained, particularly for the circular microstrip planar structures. The same method of analysis was applied to designing annular filtering structures [9.9] and bias filters for use in active microwave integrated circuits [9.10], as well as for synthesizing low-pass filters of the third order using rectangular microstrip planar segments [9.11].

In Paper 9.2 D'Inzeo and his coworkers develop a new wide-band lumped-element equivalent model for multiport planar circuits of general shape. The elements of this model are almost frequency independent, unlike previously reported models. Recently, however, Giannini et al. [9.12] reported an improved lumped-element equivalent model for rectangular microstrip structures. The new model is reported to provide frequency-independent equivalent elements, broad-band validity, and CAD-oriented implementation. In Paper 9.3 the planar circuit approach is combined with the image parameter method [9.13] to characterize planar structures. Microstrip filters made of rectangular planar components and having wide stopbands with very sharp cutoff characteristics have been investigated.

The radial stub is another planar structure that has been extensively investigated for its filtering [9.14–9.16] and biasing network properties [9.17]. Other applications of this planar structure include resonators [9.18], impedance transformers [9.19], and broad-band matching terminations [9.20, 9.21]. The radial stub structure has been used extensively in monolithic integrated circuits because of its characteristics of providing an accurate, localized zero-impedance point and also for maintaining a low-input impedance value over a wide frequency range [9.22].

Accurate modeling of the radial stub has evolved over the years [9.14, 9.19, 9.23–9.29]. In this chapter two modeling papers (Paper 9.4 and Paper 9.5) are included. Paper 9.4 presents a planar circuit analysis of the microstrip radial stub. The method of analysis is based on expanding the electromagnetic field in terms of the resonant modes of the planar structure. As in the case of rectangular, circular, and annular structures, the fringe field effects are taken into account by using effective values for the permittivity and the radial stub dimensions. Recently, Sorrentino and Roselli [9.29] simplified the slowly convergent infinite series formulations of Paper 9.4 and obtained a highly accurate closed-form expression for the input impedance of radial stub. Paper 9.5 describes a lumped equivalent circuit model for the series and the double-shunt connected radial stub. This new model takes into consideration the conductor, dielectric, as well as the radiation loss of the structure. This paper describes how this model can be implemented into CAD programs such as Touchstone for a quick and accurate analysis of the radial structure.

The application of three-dimensional full-wave numerical techniques to the analysis of planar stubs and filters is illustrated in two reprints included here (Papers 9.6 and 9.7). As discussed in previous chapters, three-dimensional numerical techniques may be divided into two broad categories: one based on the integral equation approach (discussed in Chapter 5) and another based on the differential equation approach (discussed in Chapter 6).

Paper 9.6 proposes a new approach to model planar microstrip circuits with arbitrary shapes. A new model, more accurate and versatile than the one reported earlier in [9.30], is used in a finite-difference time-domain program to analyze a microstrip T-junction, a low-pass microstrip filter, and a two-port circular microstrip resonator.

The last reprint (Paper 9.7) in this chapter is based on the coupled finite-boundary element method. It employs a generalized formulation to analyze planar circuits with complicated geometries and dielectric loads. One of the planar circuits investigated in this paper is a microstrip circular disk filter loaded with a square dielectric cylinder located at the center of the disk. In addition, the filtering properties of a four-post band-pass direct-coupled cavity filter is investigated using this approach.

References

[9.1] Bianco, B. and S. Ridella. "Nonconventional transmission zeros in distributed rectangular structures." *IEEE Trans. Microwave Theory Tech.*, Vol. MTT-20, pp. 297–303, May 1972.

[9.2] Bianco, B. and S. Ridella. "Analysis of a three-layer rectangular structure." In R. Boite, ed., *Network Theory*, pp. 215–227. New York: Gordon & Breach, 1972.

[9.3] Bianco, B., M. Granara, and S. Ridella. "Filtering properties of two dimensional line discontinuities." *Alta Freq.*, Vol. 42, pp. 140E–148E, June 1973.

[9.4] Bianco, B. and P. P. Civalleri. "Basic theory of three-layer N-ports." *Alta Freq.*, Vol. 30, pp. 623–631, August 1969.

[9.5] Civalleri, P. P. and S. Ridella. "Impedance and admittance of distributed three-layer N-ports." *IEEE Trans. Circuit Theory*, Vol. CT-17, pp. 392–398, August 1970.

[9.6] Okoshi, T. and T. Miyoshi. "The planar circuit—An approach to microwave integrated circuitry." *IEEE Trans. Microwave Theory Tech.*, Vol. MTT-20, pp. 245–252, April 1972.

[9.7] Miyoshi, T., and T. Okoshi. "Analysis of microwave planar circuits." *Electronics and Communications in Japan*, Vol. 55-B, No. 8, pp. 24–31, 1972.

[9.8] Wolff, I. and N. Knoppik. "Rectangular and circular disk capacitators and resonators." *IEEE Trans. Microwave Theory Tech.*, Vol. MTT-22, pp. 857–864, October 1974.

[9.9] D'Inzeo, G., F. Giannini, and R. Sorrentino. "Microwave planar networks: the annular structure." *Electronics Letters*, Vol. 14, No. 16, pp. 526–528, August 3, 1978.

[9.10] D'Inzeo, G., F. Giannini, and R. Sorrentino. "Design of circular planar networks for bias filter elements in microwave integrated circuits." *Alta Freq.*, Vol. 48, No. 7, pp. 251E–257E, July 1979.

[9.11] D'Inzeo, G., F. Giannini, and R. Sorrentino. "Novel microwave integrated lowpass filters." *Electronics Letters*, Vol. 15, pp. 258–260, April 1979.

[9.12] Giannini, F., G. Bartolucci, and M. Ruggieri. "Equivalent circuit models for computer aided design of microstrip rectangular structures." *IEEE Trans. Microwave Theory Tech.*, Vol. MTT-40, pp. 378–388, February 1992.

[9.13] Salerno, M., R. Sorrentino, and F. Giannini. "Image parameter design of noncommensurate distributed structures: An application to microstrip low-pass filters." *IEEE Trans. Microwave Theory Tech.*, Vol. MTT-34, pp. 58–65, January 1986.

[9.14] Vinding, J. B. "Radial line stubs as elements in stripline circuits." In *NEREM Records*, pp. 108–109, 1967.

[9.15] De Lima Combra, M. "A new kind of radial stub and some applications." *Proc. 14th European Microwave Conf.*, pp. 516–521, Leige, Belgium, September 10–13, 1984.

[9.16] Giannini, F., M. Salerno, and R. Sorrentino. "Two-octave stopband microstrip lowpass filter design using butterfly stubs." *Proc. 16th European Microwave Conf.*, pp. 292–297, Dublin, Ireland, September 8–12, 1986.

[9.17] Syret, B. A. "A broadband element for microstrip bias or tuning circuits." *IEEE Trans. Microwave Theory Tech.*, Vol. MTT-28, pp. 925–927, August 1980.

[9.18] Chu, A. et al. "GaAs monolithic frequency doubler with series connected varactor diodes." *1984 IEEE MTT-S Int'l. Microwave Symp.*, pp. 51–54.

[9.19] Atwater, H. A. "Microstrip reactive circuit elements." *IEEE Trans. Microwave Theory Tech.*, Vol. MTT-31, pp. 488–491, June 1983.

[9.20] Giannini, F., C. Paoloni, and M. Ruggieri. "A very broadband matched termination utilizing non-grounded radial lines." *Proc. 17th European Microwave Conf.*, pp. 1027–1031, Rome, Italy, September 7–11, 1987.

[9.21] Giannini, F. et al. "Low impedance matching: The radial stub solution." *Microwave and Optical Tech. Letters*, Vol. 2, pp. 291–297, August 1989.

[9.22] Sadhir, V., and I. Bahl. "Radial line structures for broadband microwave circuits applications." *Microwave J.*, pp. 102–123, August 1991.

[9.23] March S. L. "Analyzing lossy radial-line stubs." *IEEE Trans. Microwave Theory Tech.*, Vol. MTT-33, No. 3, pp. 269–271, March 1985.

[9.24] Atwater, H. A. "The design of the radial line stub: A useful microstrip circuit element." *Microwave J.*, Vol. 28, pp. 149–153, August 1985.

[9.25] Giannini, F., M. Ruggieri, and J. Vbra. "Shunt-connected microstrip radial stubs." *IEEE Trans. Microwave Theory Tech.*, Vol. MTT-34, pp. 363–366, March 1986.

[9.26] Giannini, F., C. Paoloni, and J. Vbra. "Losses in microstrip radial stubs." *Proc. 16th European Microwave Conf.*, pp. 523–528, Dublin, Ireland, September 8–12, 1986.

[9.27] Giannini, F. and C. Paoloni. "Broadband lumped equivalent stub." *Electronics Letters*, Vol. 22, No. 9, pp. 485–487, April 24, 1986.

[9.28] Giannini, F. and C. Paoloni. "Modelling of shunt connected single radial stub for CAD applications." *Alta Freq.*, Vol. LVII, No. 5, pp. 227–232, June 1988.

[9.29] Sorrentino, R. and L. Roselli. "A new simple and accurate formula for microstrip radial stub." *IEEE Microwave and Guided Wave Letters*, Vol. 2, No. 12, pp. 480–482, December 1992.

[9.30] Gwarek, W. "Analysis of arbitrarily shaped two-dimensional microwave circuits by finite-difference time-domain method." *IEEE Trans. Microwave Theory Tech.*, Vol. MTT-36, pp. 738–744, April 1988.

Method of Analysis and Filtering Properties of Microwave Planar Networks

GUGLIELMO D'INZEO, FRANCO GIANNINI, CESARE M. SODI, AND ROBERTO SORRENTINO, MEMBER, IEEE

Abstract—**A method of analysis of planar microwave structures, based on a field expansion in term of resonant modes, is presented. A first advantage of the method consists in the possibility of taking into account fringe effects by introducing, for each resonant mode, an equivalent model of the structure. Moreover, the electromagnetic interpretation of the filtering properties of two-port networks, particularly of the transmission zeros, whose nature has been the subject of several discussions, is easily obtained. The existence of two types of transmission zeros, modal and interaction zeros is pointed out. The first ones are due to the structure's resonances, while the second ones are due to the interaction between resonant modes. Several experiments performed on circular and rectangular microstrips in the frequency range 2–18 GHz have shown a good agreement with the theory.**

I. Introduction

AFTER THE STUDY of the transmission properties of microstrip lines, the great diffusion of microwave integrated circuits has led to the analysis of general planar circuits. To this purpose, analytical methods, applied to structures of simple geometry [1]–[3], and numerical methods, apt to the study of more complex geometries [4]–[6], have been developed. In both cases a magnetic wall model has been adopted for the structure because of the formidable boundary value problems. In such a way, however, one not only neglects the dispersion properties of the circuit, which are due to fringe effects, but often obtains erroneous results [7].

To overcome this difficulty, in the case of step discontinuities, i.e., of structures with separable geometry in rectangular coordinates, Menzel and Wolff [8] have recently proposed a method of analysis based on the correction of the magnetic wall model by means of frequency dependent effective parameters. However, it must be observed that effective parameters depend not only on the frequency, but also on the field distribution inside the structure. It is sufficient to instance the disk resonators for which Wolff and Knoppik [9] have shown a frequency dependent equivalent model to exist for each resonant mode, in such a way that a unique equivalent model for the structure cannot be defined. This fact strongly limits the applicability of all the analyses of microstrip structures presented until now. Considerable attention has been devoted to nonuniform lines, i.e., lines with continuously or not continuously varying cross sections. The existence of transmission zeros has been stressed both theoretically and experimentally. In the particular case of a double step discontinuity, the physical nature of such zeros has been discussed for a long time [2], [10]–[13] and they have been ascribed to the excitation of higher order modes of propagation in the line section between the two discontinuities. As will be shown below, such an interpretation, in our opinion, is not correct, also because transmission zeros are present in generic nonuniform lines where the EM field cannot propagate as exp $(-j\beta z)$.

In this paper an analysis of planar circuits based on the theory of resonant cavities is presented. Three important advantages are so obtained. The first consists in the possibility of introducing frequency dependent effective parameters for each resonant mode of the structure in such a way as to obtain an accurate characterization of its frequency behavior. The second is an electromagnetic interpretation of the network's filtering properties, particularly of the transmission zeros, is easily obtained and the above mentioned problems are clarified. Finally, the present method leads to the evaluation of the impedance matrix of the network in the form of a partial fraction expansion with the advantages pointed out by Silvester [6].

The analysis is limited to the important case of two-port networks, since the extension to the general case is

Manuscript received May 16, 1977; revised December 7, 1977. This work has been supported in part by the Consiglio Nazionale delle Ricerche (C. N. R.), Italy.

The authors are with the Istituto di Elettronica, Università di Roma, Rome, Italy.

Reprinted from *IEEE Trans. Microwave Theory Tech.*, vol. MTT-26, no. 7, pp. 462–471, July 1978.

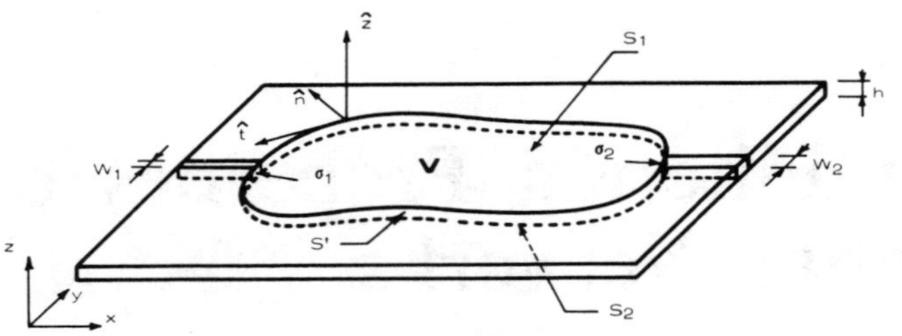

Fig. 1. The planar two-port circuit.

straightforward. The general filtering properties are discussed and criteria for locating transmission zeros are given. Several experimental results for circular and rectangular structures in the frequency range 2–18 GHz show a good agreement with the theoretical ones, obtained using the effective parameters proposed in [9]. Structures with nonseparable geometries could also be studied with the same technique through a numerical method (e.g., a finite element method).

II. Formulation of Field Problem

Fig. 1 shows a microstrip two-port circuit. The main difficulty in the study of such a structure is due to the fact that it is an open one, i.e., the EM field extends to infinity. The central section may be considered as an open resonator; the EM field is mainly concentrated in the cylindrical volume V bounded by the two conducting surfaces S_1 and S_2 and, laterally, by the cylindrical surface S'. It may be expressed as a function of the tangential magnetic field $\boldsymbol{H}_\tau$ on S' in terms of the modes of the cavity V. Following a procedure analogous to that of Kurokawa [14], one obtains

$$\boldsymbol{E}=\Sigma_a e_a \boldsymbol{E}_a+\Sigma_\alpha e_\alpha \boldsymbol{E}_\alpha \qquad (1)$$

$$\boldsymbol{H}=\Sigma_a h_a \boldsymbol{H}_a+\Sigma_\alpha h_\alpha \boldsymbol{H}_\alpha \qquad (2)$$

where $\boldsymbol{E}_a$ and $\boldsymbol{E}_\alpha$ are the orthonormalized eigenvectors of the following eigenvalue problem:

$$\nabla\times\nabla\times\boldsymbol{E}-\nabla\nabla\cdot\boldsymbol{E}-k^2\boldsymbol{E}=0, \quad \text{inside } V \qquad (3a)$$

$$\boldsymbol{n}\times\boldsymbol{E}=0 \quad \nabla\cdot\boldsymbol{E}=0, \quad \text{on } S_1, S_2 \qquad (3b)$$

$$\boldsymbol{n}\cdot\boldsymbol{E}=0 \quad \boldsymbol{n}\times\nabla\times\boldsymbol{E}=0, \quad \text{on } S' \qquad (3c)$$

with the further conditions:

$$\nabla\cdot\boldsymbol{E}_a=0 \quad \nabla\times\boldsymbol{E}_a\neq 0, \quad \text{inside } V \qquad (4a)$$

$$\nabla\times\boldsymbol{E}_\alpha=0, \quad \text{inside } V. \qquad (4b)$$

Similarly, $\boldsymbol{H}_a$ and $\boldsymbol{H}_\alpha$ are the orthonormalized eigenvectors of

$$\nabla\times\nabla\times\boldsymbol{H}-\nabla\nabla\cdot\boldsymbol{H}-k^2\boldsymbol{H}=0, \quad \text{inside } V \qquad (5a)$$

$$\boldsymbol{n}\cdot\boldsymbol{H}=0 \quad \boldsymbol{n}\times\nabla\times\boldsymbol{H}=0, \quad \text{on } S_1, S_2 \qquad (5b)$$

$$\boldsymbol{n}\times\boldsymbol{H}=0 \quad \nabla\cdot\boldsymbol{H}=0, \quad \text{on } S' \qquad (5c)$$

with the conditions

$$\nabla\cdot\boldsymbol{H}_a=0 \quad \nabla\times\boldsymbol{H}_a\neq 0, \quad \text{inside } V \qquad (6a)$$

$$\nabla\times\boldsymbol{H}_\alpha=0, \quad \text{inside } V. \qquad (6b)$$

It is possible to demonstrate that the eigenvalues of (3)–(4) coincide with those of (5)–(6) and that

$$\nabla\times\boldsymbol{H}_a=k_a\boldsymbol{E}_a$$

$$\nabla\times\boldsymbol{E}_a=k_a\boldsymbol{H}_a. \qquad (7)$$

The coefficient of the expansions (1) and (2) may be calculated imposing that the EM field satisfies Maxwell's equations. One obtains

$$e_a=\frac{j\omega\mu}{k_a^2-\omega^2\mu\epsilon}\int_{S'}\boldsymbol{n}\times\boldsymbol{H}_\tau\cdot\boldsymbol{E}_a\,dS$$

$$h_a=\frac{-k_a}{j\omega\mu}e_a$$

$$e_\alpha=\frac{1}{j\omega\mu}\int_{S'}\boldsymbol{n}\times\boldsymbol{H}_\tau\cdot\boldsymbol{E}_\alpha\,dS$$

$$h_\alpha=0. \qquad (8)$$

Once the set of eigenvalue of (3) and (5) is known, the evaluation of the EM field inside V depends on the knowledge of the tangential magnetic field $\boldsymbol{H}_\tau$ on S'. In a first approximation we may assume that $\boldsymbol{H}_\tau$ is different from zero only at the connections σ_i between the cavity and the lines where it has a TEM distribution[1]. Thus it is constant. However, $\boldsymbol{H}_\tau$ is not exactly zero on the remainder of S'; fringe effects can be taken into account by ascribing to the structure effective dimensions and an effective permittivity, according to the widely adopted magnetic wall model of microstrip structures. We shall come back to this point later.

Because of the above simplifying hypotheses, the EM field in the cavity is determined as a function of the magnetic field $\boldsymbol{H}_{\tau 1}=H_1\boldsymbol{t}$ and $\boldsymbol{H}_{\tau 2}=H_2\boldsymbol{t}$ at the outputs, which is independent of z. The structure may, therefore, be considered as a two-dimensional one. It is easily seen that, imposing the condition $\partial/\partial_z=0$ on (1)–(8), the $\boldsymbol{E}_a$'s have only the z component, while the $\boldsymbol{E}_\alpha$'s do not exist, with the exception of only the mode $\boldsymbol{E}_0$ having zero divergence. After simple manipulations, the EM field in the cavity may be expressed as follows:

$$\boldsymbol{E}=\hat{z}\Sigma_a e_a E_a+\hat{z}\,e_0 V^{-1/2} \qquad (9)$$

$$\boldsymbol{H}=\frac{1}{j\omega\mu}\Sigma_a e_a \hat{z}\times\nabla_t E_a \qquad (10)$$

where

$$e_a=\frac{j\omega\mu}{k_a^2-\omega^2\mu\epsilon}\left(\sqrt{\sigma_1}\,P_{a_1}H_1+\sqrt{\sigma_2}\,P_{a_2}H_2\right) \qquad (11a)$$

$$P_{ai}=\sigma_i^{-1/2}\int_{\sigma_i}E_a\,dS, \qquad i=1,2 \qquad (11b)$$

$$e_0=\frac{V^{-1/2}}{j\omega\epsilon}(\sigma_1 H_1+\sigma_2 H_2) \qquad (11c)$$

where $\hat{z}$ is the unit vector of the z axis, V is the volume of

[1]Higher order modes on the uniform lines may be neglected with good approximation if the uniform sections are long enough and their widths are much smaller than the cavity's dimension [1]. In any case, when necessary, higher modes can be taken into account with a rather more complicate algebra.

the cavity, and σ_1 and σ_2 are the surfaces of the outputs of the cavity, i.e., the portions of S' where $\boldsymbol{H}_\tau$ is different from zero. $\boldsymbol{E}_0 = \hat{z} V^{-1/2}$ is the mode having zero curl and zero divergence, belonging to the $\boldsymbol{E}_\alpha$'s. Since (11c) can be obtained from (11a) and (11b) by putting $k_a^2 = 0$ and $E_a = E_0 = V^{-1/2}$, later on this mode will be included among the E_a's.

The eigenfunctions E_a have to satisfy the two-dimensional eigenvalue equation deriving from (3)

$$\nabla_t^2 E + k^2 E = 0 \tag{12}$$

together with the boundary condition

$$\frac{\partial E}{\partial n} = 0 \tag{12'}$$

which derives from the second of (3c); the other boundary conditions are automatically satisfied.

One can note that the a modes are, in this case, TM with respect to the z direction; the o mode, on the contrary, corresponds to the electrostatic field problem.

Once (12) is solved for a particular geometry, the EM field inside the cavity is fully determined through (9)–(11) as a function of the magnetic fields supported by the uniform lines. Nevertheless, a terminal description of the structure as a two-port network is generally preferable. This can be obtained by evaluating the impedance matrix, relative, of course, to the dominant TEM modes of the lines. The amplitude of the electric field E_i on the ith line is obtained by projecting the field (9), calculated at σ_i, on the abstract vector space of the modes of the line and retaining the TEM component [15], i.e.,

$$E_i = \frac{1}{\sigma_i} \int_{\sigma_i} \hat{z} \cdot \boldsymbol{E}\, dS, \qquad i = 1, 2.$$

Through (9) and (11) E_i can be expressed as a function of H_1, H_2

$$\begin{aligned} E_1 &= H_1 j\omega\mu \Sigma_a \frac{P_{a1}^2}{k_a^2 - \omega^2 \mu\epsilon} \\ &\quad + H_2 j\omega\mu \Sigma_a \frac{P_{a1} P_{a2}}{k_a^2 - \omega^2\mu\epsilon} \cdot [\sigma_2/\sigma_1]^{1/2} \\ E_2 &= H_1 j\omega\mu \Sigma_a \frac{P_{a2} P_{a1}}{k_a^2 - \omega^2\mu\epsilon} \cdot [\sigma_1/\sigma_2]^{1/2} \\ &\quad + H_2 j\omega\mu \Sigma_a \frac{P_{a2}^2}{k_a^2 - \omega^2\mu\epsilon}. \end{aligned} \tag{13}$$

If one defines equivalent voltages and currents in such a way as to normalize to unity the characteristic impedances of the lines, i.e.,

$$\begin{aligned} V_i &= E_i \left[\sigma_i \sqrt{\epsilon/\mu} \right]^{1/2} \\ I_i &= H_i \left[\sigma_i \sqrt{\mu/\epsilon} \right]^{1/2}, \qquad i = 1, 2 \end{aligned} \tag{14}$$

from (13) and (14) the following expression of the $[Z]$ matrix is easily obtained

$$[Z] = \Sigma_a [Z_a] \tag{15}$$

with

$$[Z_a] = \frac{j\omega c}{\omega_a^2 - \omega^2} \begin{bmatrix} P_{a1}^2 & P_{a1} P_{a2} \\ P_{a2} P_{a1} & P_{a2}^2 \end{bmatrix} \tag{15'}$$

μ and ϵ are the substrate's permeability and permittivity, respectively, and

$$\omega_a = c k_a$$

are the resonant frequencies of the cavity. If there are ν_a linearly independent eigenfunctions corresponding to the same eigenvalue k_a^2

$$E_a^{(1)}, E_a^{(2)}, \cdots E_a^{(\nu_a)}$$

which, without loss of generality, may be supposed to be ortogonal, (15') should be replaced by

$$[Z_a] = \frac{j\omega c}{\omega_a^2 - \omega^2} \sum_{\nu=1}^{\nu_a} \begin{bmatrix} P_{a1}^{(\nu)^2} & P_{a1}^{(\nu)} P_{a2}^{(\nu)} \\ P_{a1}^{(\nu)} P_{a2}^{(\nu)} & P_{a2}^{(\nu)^2} \end{bmatrix},$$

$$c = 1/\sqrt{\mu\epsilon} \tag{15''}$$

while, in (15), the summation over a should include only distinct ω_a's. $[Z]$ is a purely imaginary matrix since the structure has been supposed without losses. If the network is symmetrical

$$P_{a2} = \epsilon_a P_{a1} \tag{16}$$

where $\epsilon_a = 1$ for even modes and $\epsilon_a = -1$ for odd modes. The impedance parameters may be written

$$\begin{aligned} Z_{11} &= Z_{22} = Z_{\text{ev}} + Z_{\text{od}} \\ Z_{12} &= Z_{21} = Z_{\text{ev}} - Z_{\text{od}} \end{aligned} \tag{17}$$

where

$$\begin{aligned} Z_{\text{ev}} &= j\omega c \Sigma \frac{P_{\text{ev}}^2}{\omega_{\text{ev}}^2 - \omega^2} \\ Z_{\text{od}} &= j\omega c \Sigma \frac{P_{\text{od}}^2}{\omega_{\text{od}}^2 - \omega^2} \end{aligned} \tag{17'}$$

ev being the index of the even modes, od of the odd modes.

The calculation of the $[Z]$ matrix requires the evaluation of the eigenfunctions and eigenvalues E_a, k_a^2 and then of the P_{a1}. This can be done analytically if the structure has a separable geometry; if the geometry is not separable, a numerical method could be adopted.

III. General Filtering Properties

The formulation given in the previous section has led to a complete characterization of the microwave network in terms of its impedance matrix. In order to discuss the filtering properties of the structure, a description in terms of the scattering parameters is preferable since the impedance matrix elements are not quantities easily measurable at microwave frequencies; moreover the scattering matrix provides a more appropriate physical description of the structure behavior. In terms of the impedance parameters the scattering parameters are given by

$$s_{11}=\left[(Z_{11}-1)(Z_{22}+1)-Z_{12}^2\right]/D$$
$$s_{22}=\left[(Z_{11}+1)(Z_{22}-1)-Z_{12}^2\right]/D$$
$$s_{12}=s_{21}=2Z_{12}/D \tag{18}$$
where
$$D=(Z_{11}+1)(Z_{22}+1)-Z_{12}^2. \tag{18$'$}$$
Let us start examining the structure's behavior at the resonant frequency ω_p of one of the modes. It is convenient to write the Z parameters as follows:
$$Z_{11}=j\omega c\frac{Q_{11}}{\omega_p^2-\omega^2}+\hat{Z}_{11}$$
$$Z_{22}=j\omega c\frac{Q_{22}}{\omega_p^2-\omega^2}+\hat{Z}_{22}$$
$$Z_{12}=j\omega c\frac{Q_{12}}{\omega_p^2-\omega^2}+\hat{Z}_{12} \tag{19}$$
where $\hat{Z}_{ij}$ remains finite for $\omega\to\omega_p$. Let us distinguish two cases.

1) $Q_{11}Q_{22}=Q_{12}^2$. This equality is always verified for nondegenerate modes. We further distinguish two subcases.

a) $Q_{11}Q_{22}=Q_{12}^2=0$. Since the case $Q_{11}=Q_{22}=Q_{12}=0$ may be excluded ,[2] the structure has to be nonsymmetrical ($Q_{11}\neq Q_{22}$). From (18) and (19) one immediately obtains
$$s_{12}(\omega_p)=0.$$
According to whether Q_{11} or Q_{22} is different from zero, $s_{11}=1$, or $s_{22}=1$, respectively. The resonant frequency ω_p, therefore, corresponds to a transmission zero. This can be easily explained from an electromagnetic point of view. Suppose $Q_{22}=0$: this means that the p mode is uncoupled to the second port. When an EM field is incident to the first port at the frequency $\omega=\omega_p$, the EM field inside the cavity would become infinite (see (11a)) unless the total (incident plus reflected) magnetic field at the first port is zero; this implies $s_{11}=1$, $s_{12}=0$.

We may, therefore, conclude that in nonsymmetrical structures a transmission zero takes place at the resonant frequency of one mode which is uncoupled to one of the ports. Later on, transmission zeros taking place at resonant frequencies ω_a will be referred to as modal zeros.

b) $Q_{11}Q_{22}=Q_{12}^2\neq 0$. In this case, when $\omega\to\omega_p$ the scattering parameters do not generally assume significant values. Nevertheless, it is worth considering the case of a symmetrical structure ($Q_{11}=Q_{22}=\pm Q_{12}$). For $\omega\to\omega_p$ one obtains from (18) and (19)
$$s_{11}=s_{22}=\frac{\hat{Z}_{11}-\epsilon_p\hat{Z}_{12}}{1+\hat{Z}_{11}-\epsilon_p\hat{Z}_{12}} \qquad s_{12}=s_{21}=\frac{\epsilon_p}{1+\hat{Z}_{11}-\epsilon_p\hat{Z}_{12}}.$$
According to whether p is an even ($\epsilon_p=1$) or an odd ($\epsilon_p=-1$) mode the quantity $\hat{Z}_{11}-\epsilon_p\hat{Z}_{12}$ is equal to $2Z_{\text{od}}$ or to $2Z_{\text{ev}}$ (see (17$'$)), respectively. These quantities are often negligible with respect to unity, so that $s_{11}=s_{22}\cong 0$, $s_{12}\cong\epsilon_p$. In other words, in a symmetrical nondegenerate structure modal transmission zeros do not take place; on the contrary, the ω_a's give generally place to approximate reflection zeros. It is worth specifying that the existence of a reflection zero at or near ω_p depends on the widths of the ports; in some cases, in fact, the reflection zero takes place only if the ports are small enough. Typical examples will be shown below. For the sake of brevity we omit to demonstrate the above statements which, on the other hand, can be easily proved.

2) $Q_{11}Q_{22}\neq Q_{12}^2$. This case can be verified only for a degenerate mode. It is easily seen that at the frequency $\omega=\omega_p$ $s_{11}=s_{22}=1$, $s_{12}=0$. This is another case of modal transmission zero, which is due to a degenerate mode of the cavity, or rather to the superposition of degenerate modes.

Having examined the structure's behavior at the resonant frequencies of the cavity, let us now consider the cases when a transmission zero takes place.

From (18) it follows that for s_{12} to be zero there are only two cases: a) $|D|=\infty$. This condition holds only if $\omega=\omega_a$, and therefore is that of a modal transmission zero. b) $Z_{12}=0$. Since
$$Z_{12}=j\omega c\Sigma_a\frac{P_{a1}P_{a2}}{\omega_a^2-\omega^2} \tag{20}$$
it can be easily inferred that between two consecutive resonant frequencies ω_p, ω_q such that [3]
$$\text{sgn}\,(P_{p1}P_{p2})=\text{sgn}\,(P_{q1}P_{q2}) \tag{21}$$
there is necessarily a frequency $\omega_z\in(\omega_p,\omega_q)$ such that
$$Z_{12}(\omega_z)=0,\ \ s_{12}(\omega_z)=0.$$
In case of mode degeneracy, (21) should be replaced by
$$\text{sgn}\left(\sum_{\nu=1}^{\nu_p}I_{p1}^{(\nu)}I_{p2}^{(\nu)}\right)=\text{sgn}\left(\sum_{\nu=1}^{\nu_q}I_{q1}^{(\nu)}I_{q2}^{(\nu)}\right). \tag{21$'$}$$
In order to find a physical interpretation of this type of transmission zero, suppose the cavity is excited by a field incident to the first port at a frequency located between ω_p and ω_q; if (21) is satisfied, the p and q modes will give place to opposite contributions to the field at the output. In other words, they interact destructively at the second port. At the frequency $\omega=\omega_z$, whose location between ω_p and ω_q depends also on the contribution of all the other modes, there is a totally destructive interaction in such a way that no power can be transferred towards the output. This type of transmission zero will be called interaction zero.

For symmetrical nondegenerate structures, because of (16), (21) becomes

[2]In that case, in fact, the p mode cannot be excited in the structure and therefore can be excluded from any consideration.

[3]If $P_{q1}P_{q2}=0$ (i.e., there is a modal transmission zero at ω_q), in (21) the successive mode must be considered, say, the r mode, such that $P_{r1}P_{r2}\neq 0$.

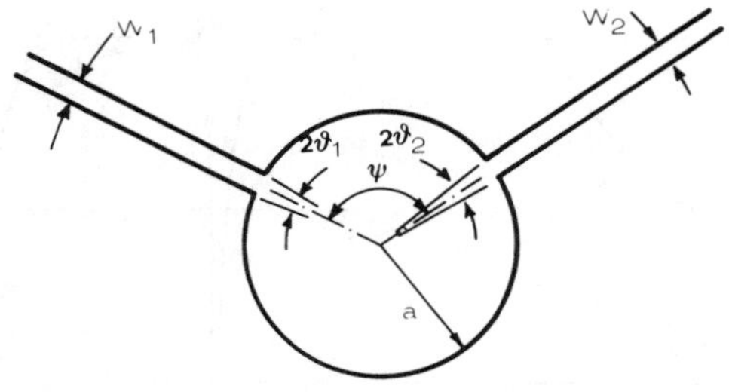

Fig. 2. The circular microstrip.

$$\epsilon_p = \epsilon_q$$

i.e., if two consecutive resonant modes are both even or odd, an interaction transmission zero is located between their resonant frequencies.

IV. The Circular Microstrip

The first case of the two-port network we have considered is the circularly shaped microstrip line shown in Fig. 2. The orthonormalized eigenfunctions of (12) are, in this case,

$$E_{mn} = C_{mn} J_m(k_{mn} r) \begin{Bmatrix} \cos m\phi \\ \sin m\phi \end{Bmatrix}, \qquad \begin{matrix} m = 0, 1, 2 \cdots \\ n = 1, 2, 3 \cdots \end{matrix} \tag{22}$$

where

$$C_{mn} = \frac{1}{\sqrt{V}\, J_m(\xi'_{mn})} \left[\frac{\delta_m}{1 - \left(\frac{m}{\xi'_{mn}} \right)^2} \right]^{1/2}$$

$$V = \pi a^2 h$$

$$k_{mn} = \xi'_{mn}/a = \omega_{mn}/c$$

and

$$\delta_m = \begin{cases} 1, & \text{for } m = 0 \\ 2, & \text{for } m \neq 0 \end{cases}$$

is the Neumann factor; ξ'_{mn} is the nth root of the equation

$$\frac{d}{dx} J_m(x) = 0$$

and h is the substrate's thickness. Besides the set of eigenfunctions (22), the eigenfunction corresponding to $k^2 = 0$ must be considered

$$E_{00} = 1/\sqrt{V}\,.$$

Such an eigenfunction can be obtained from (22) by putting conventionally

$$\left. \frac{m}{\xi'_{mn}} \right|_{\substack{n=0 \\ m=0}} = 0.$$

Equation (22) shows the existence of a pair of degenerate modes for any $m \neq 0$. We shall restrict our attention to the important case of symmetrical structures ($w_1 = w_2 = w$; $\theta_1 = \theta_2 = \theta$). Following the procedure described in the previous section, one obtains for the $[Z]$ matrix elements[4]

$$Z_{11} = Z_{22} = j\omega c \frac{4\theta^2}{\pi w} \sum_{m=0}^{\infty} \sum_{n=0}^{\infty} \frac{A_{mn}^2}{\omega_{mn}^2 - \omega^2}$$

$$Z_{21} = Z_{12} = j\omega c \frac{4\theta^2}{\pi w} \sum_{m=0}^{\infty} \sum_{n=0}^{\infty} \frac{A_{mn}^2}{\omega_{mn}^2 - \omega^2} \cos m\psi \tag{23}$$

where

$$A_{mn}^2 = \left(\frac{\sin m\theta}{m\theta} \right)^2 \frac{\delta_m}{1 - (m/\xi'_{mn})^2}. \tag{23'}$$

Let us consider the structure's behavior at the resonant frequencies of the cavity. The condition of case 2) of the previous section becomes

$$\cos^2 m\psi \neq 1. \tag{24}$$

Therefore, for generic values of ψ and for $m \neq 0$, each resonant frequency corresponds to a modal transmission zero. On the contrary, if $m\psi = s\pi$ ($s = 0, 1, 2, \cdots$) case 2b) is verified, i.e., for θ small enough, a reflection zero takes place near such resonant frequencies. This happens for all the modes of a doubly symmetrical structure ($\psi = \pi$) and, in general, for the $(0, n)$ modes.

Besides modal transmission zeros, interaction zeros take place between consecutive resonant frequencies $\omega_{m_1 n_1}$ and $\omega_{m_2 n_2}$ such that[5]

$$\operatorname{sgn}(\cos m_1\psi) = \operatorname{sgn}(\cos m_2\psi). \tag{25}$$

In the doubly symmetrical case ($\psi = \pi$) (31) becomes[6]

$$(-1)^{m_1} = (-1)^{m_2} \tag{25'}$$

i.e., an interaction zero is located between the resonant frequencies of two consecutive modes having both an even, or an odd, azimutal dependence.

From (25) it follows that transmission zeros can be located between any pair of resonant frequencies by varying the angle ψ between the two uniform lines.

Fig. 3 shows the theoretical behavior of the scattering parameter $|s_{12}|$ of a doubly symmetrical circular microstrip versus the frequency in the range 2–18 GHz. (Expressions (23) have been evaluated taking into account the first 62 modes). This curve has been obtained completely neglecting fringe effects, i.e., ascribing to the EM model the physical dimension of the structure and assuming for ϵ the permittivity of the substrate (alumina, $\epsilon = 10\ \epsilon_0$). The locations of the resonant frequencies ω_{mn} are also indicated in the figure. The structure presents two transmission zeros, which are due to the interaction between the pair of modes (2, 1)–(0, 1) and (1, 2)–(5, 1), accordingly to (25′). (The last resonant frequency is not indicated in the figure, because it is out of scale). Reflection zeros are located near each resonant frequency, with the exception of the modes (4, 1) and (1, 2) which are very close together; for the assumed port widths, corresponding to

[4]As a consequence of the hypothesis that the widths of the ports are much smaller than the cavity's radius, the arcs θ_1 and θ_2 may be confused with the corresponding chord.

[5]If $\cos m_2\psi = 0$, (thus a modal transmission zero takes place at $\omega_{m_2 n_2}$) the successive mode must be considered in (25). See footnote 3.

[6]It may be noted that in this case the structure behaves as a nondegenerate one, since for each k_{mn}^2 only one of the two degenerate modes can be excited, both from the input or from the output.

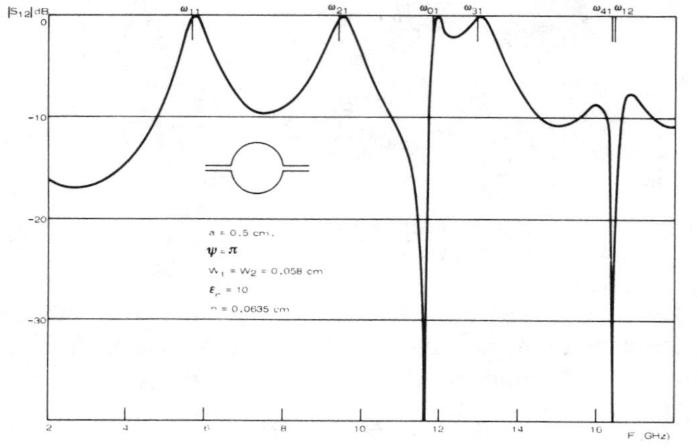

Fig. 3. Transmission coefficient $|s_{12}|$ versus the frequency for a circular microstrip (magnetic wall model without effective parameters).

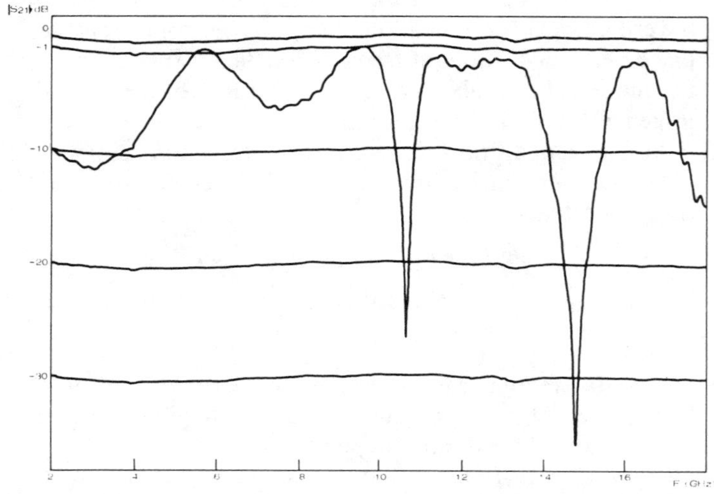

Fig. 4. Experimental behavior of the transmission coefficient $|s_{12}|$ versus the frequency for the same structure as in Fig. 3. Substrate material alumina.

50-Ω lines, reflection zeros do not take place at these frequencies.

Fig. 4 shows the experimental behavior of $|s_{12}|$ for the same structure as in Fig. 3. The comparison between these diagrams shows a good qualitative agreement between theory and experiment up to ~12 GHz. The disagreement consists on the one hand in a slight shifting of the reflection and transmission zero frequencies and, on the other hand, in lower values of the theoretical $|s_{12}|$. The latter fact can be easily explained, since the EM coupling between the uniform lines and the cavity is, in reality, stronger than the theory predicts, because the fringing field of the lines has been neglected. With regard to the frequency shifting of the two diagrams, it must be observed that the experimental resonant frequencies are different from the theoretical ones, as previously pointed out. Over ~12 GHz, the theory yields to inacceptable errors; the two diagrams do not agree even from a qualitative point of view. As has been previously noted [7], the experimental resonant frequencies may differ from the theoretical ones in such a way that the sequence of modes

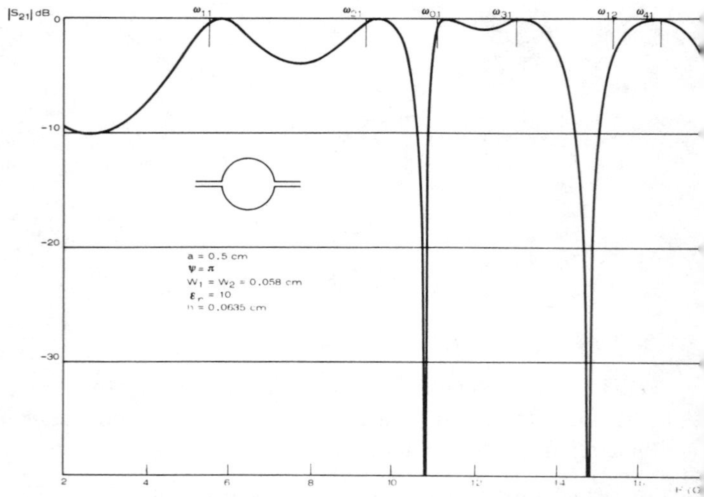

Fig. 5. Transmission coefficient $|s_{12}|$ versus the frequency for the same structure as in Fig. 3 (magnetic wall model according to the present theory).

is different in the two cases. Since interaction zeros depend on such a sequence, their location might be strongly altered.

The above considerations indicate that account must be taken of fringe effects both of the lines and of the cavity. This can be done by ascribing to the lines effective widths and effective permittivities accordingly, for instance, to Wheeler [16] or to Schneider [17], [18]. With regard to the circular resonator, fringe effects depend on the resonant mode and can be taken into account through an equivalent model for each mode, accordingly to Wolff and Knoppik [9]. Expressions (23) should be therefore modified by introducing an effective port width w_{eff}, an effective frequency dependent permittivity of the lines $\epsilon_{\text{eff}}(f)$, an effective cavity radius r_{eff} and, finally, an effective dynamic permittivity of the cavity $\epsilon_{\text{dyn},mn}(f)$, which also depends on the resonant mode.

The results obtained in this way are shown in Fig. 5 and agree very well with the experiments in Fig. 4. In particular, one can note that the resonant frequencies of the modes $(4,1)$ and $(1,2)$ are now interchanged: the second transmission zero is therefore due to the interaction between the $(3,1)$ and $(1,2)$ modes. This phenomenon is analogous to the modal inversion in circular waveguides [19] and could be experimentally verified by means of a field mapping technique [20]. The residual differences between the theoretical and experimental magnitudes of $|s_{12}|$ are essentially due to losses, particularly to radiation losses, which have been completely neglected.

Fig. 6(a) and (b) shows the theoretical behavior of $|s_{12}|$ versus the frequency for a circular microstrip with $\psi = \pi/2$. This structure presents a modal transmission zero at each resonant frequency of one mode having an odd azimutal dependence (see (24)), i.e., in the frequency range considered, of the modes $(1,1)$, $(3,1)$, and $(1,2)$. For $\psi = \pi/2$, (25) shows that interaction zeros take place between the resonant frequencies of modes $(4m_1, n_1) - (4m_2, n_2)$ or $(4m_1+2, n_1) - (4m_2+2, n_2)$; in the present case

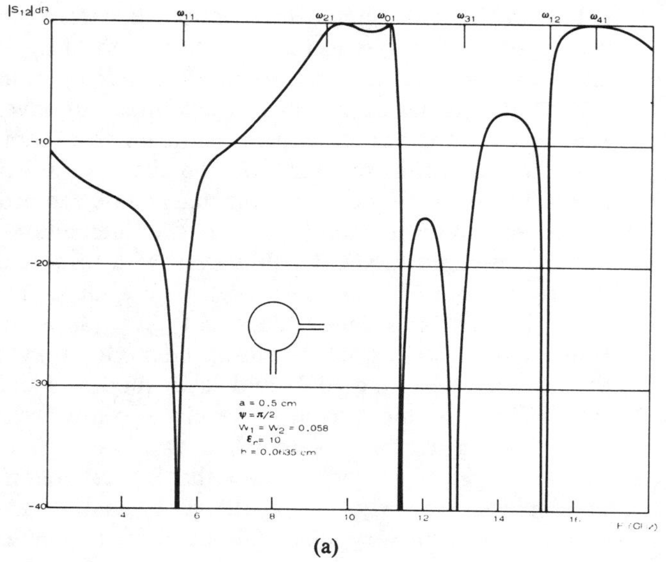

(a)

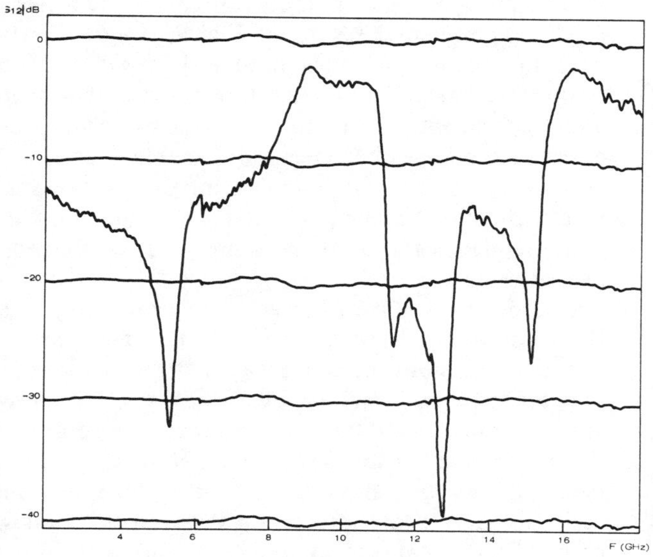

(b)

Fig. 6. Transmission coefficient $|s_{12}|$ versus the frequency, for a circular microstrip. (a) Theory. (b) Experiment.

there is only one interaction zero between the modes (0, 1) and (4, 1).

It is easy to verify that, for $\psi < \pi/2$, an interaction zero takes place between the (0,0) and (1,1) modes. For $\psi \to \pi/2$ the interaction zero tends to the modal zero due to the (1,1) mode. Fig. 7 shows the theoretical frequency location of such transmission zero as a function of the angle ψ for a given cavity's radius. As can be seen, a good agreement with the experiments is obtained. This figure shows the possibility of locating a transmission zero at a given frequency by suitably positioning the output port of the network.

V. The Rectangular Microstrip

Another type of two-port planar network consists of a rectangular central section connected with two uniform lines (Fig. 8). This structure may be considered also as a

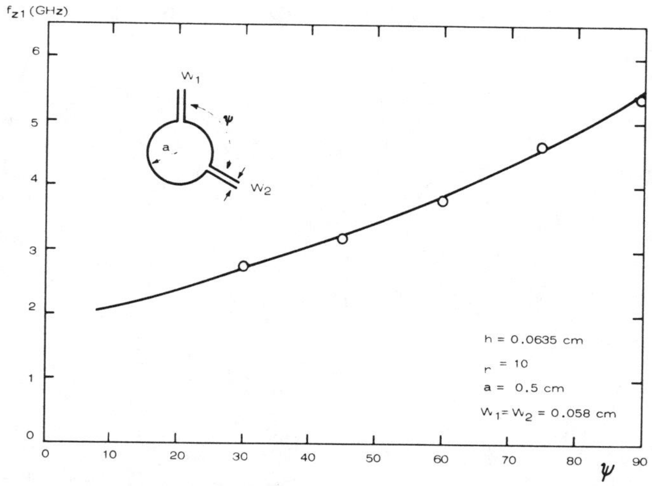

Fig. 7. Frequency location of the first transmission zero presented by a circular microstrip as a function of the angle between the two ports.

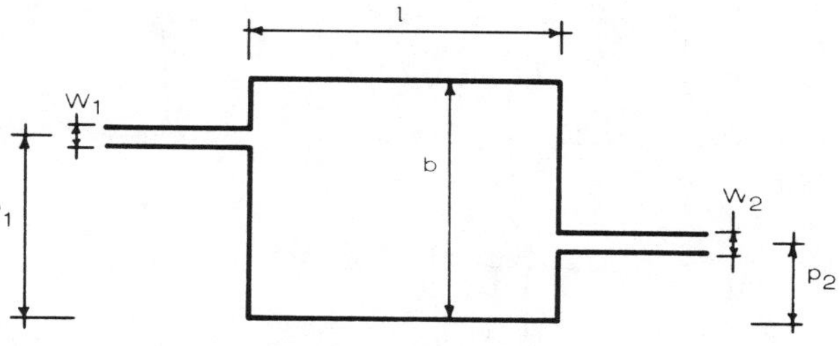

Fig. 8. The rectangular microstrip.

double step discontinuity; on the other hand, for $1 \ll b$ it becomes a stub structure.

Simple calculations yield to the following expressions of the Z parameters of the structure[7]

$$Z_{11} = \frac{j\omega c w_1}{b1} \sum_{m=0}^{\infty} \sum_{n=0}^{\infty} \frac{\delta_m \delta_n f_{n_1}^2}{\omega_{mn}^2 - \omega^2}$$

$$Z_{22} = \frac{j\omega c w_2}{b1} \sum_{m=0}^{\infty} \sum_{n=0}^{\infty} \frac{\delta_m \delta_n f_{n2}^2}{\omega_{mn}^2 - w^2}$$

$$Z_{21} = Z_{12} = \frac{j\omega c \sqrt{w_1 w_2}}{b1} \sum_{m=0}^{\infty} \sum_{n=0}^{\infty} (-1)^m \frac{\delta_m \delta_n f_{n1} f_{n2}}{\omega_{mn}^2 - \omega^2} \qquad (26)$$

where

$$\omega_{mn} = c\pi \sqrt{(m/1)^2 + (n/b)^2}$$

$$f_{ni} = \begin{cases} \cos \dfrac{n\pi p_i}{b} \dfrac{\sin \dfrac{n\pi w_i}{2b}}{\dfrac{n\pi w_i}{2b}}, & \text{for } n \neq 0 \\ 1, & \text{for } n = 0 \end{cases}.$$

[7]The series over m could be evaluated analytically and equivalent expressions to those in [1] would be obtained; nevertheless, they are not suitable for our purpose, since it is necessary to introduce effective parameters for each resonant mode, i.e., for each term of the series (26), in the same way as has been done in the case of the circular microstrip.

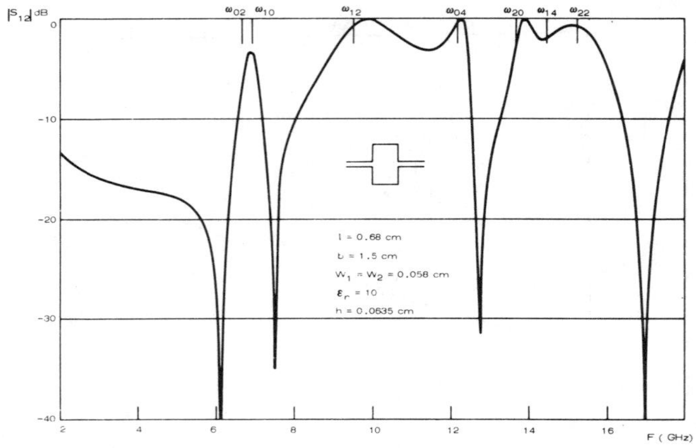

(a)

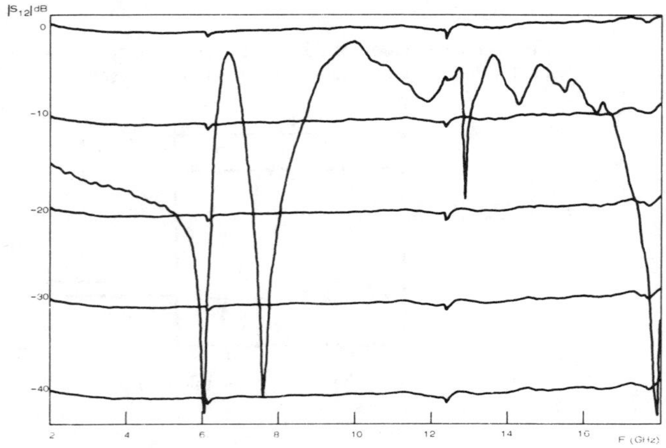

(b)

Fig. 9. Transmission coefficient $|s_{12}|$ versus the frequency for a symmetrical rectangular microstrip. (a) Theory. (b) Experiment.

For symmetrical structures ($w_1 = w_2$, $f_{n1}^2 = f_{n2}^2$), as previously stated, transmission zeros are only of the interaction type; in particular, if $p_1 = p_2$ condition (21) becomes

$$(-1)^{m_1} = (-1)^{m_2}$$

i.e., transmission zeros are located between the resonant frequencies of consecutive modes having both an even or an odd dependence with respect to the x direction.

Nonsymmetrical structures can also present modal transmission zeros, which are located at the resonant frequencies ω_{mn} such that $f_{n1} = 0$ and $f_{n2} \neq 0$, or vice versa.

Fig. 9(a) and (b) shows the theoretical and experimental behaviors of $|s_{12}|$ of a symmetrical rectangular microstrip versus the frequency in the range 2–18 GHz. Theoretical results have been obtained by adopting for the resonator the model suggested by Wolff and Knoppik [9]; the first 400 modes have been retained in (26). The structure presents four transmission zeros due to the interaction of the following pairs of modes: (0,0)–(0,2), (1,0)–(1,2), (0,4)–(2,0), (2,2)–(0,6). It is important to note that the first transmission zero is located close to the resonant frequency of the (0,2) mode. Since this frequency is nothing but the cutoff frequency of the $TE_{20}^{(x)}$ mode of propagation in the wider microstrip section, that transmission zero has been ascribed to the excitation of the $TE_{20}^{(x)}$ mode [10]–[12]. Such an interpretation is, in our opinion, inacceptable. As has been previously pointed out, given the symmetry of the structure, a resonant mode cannot give place to a transmission zero, but on the contrary it can give place to a reflection zero: the (0,2) mode can produce a transmission zero only through the interaction with another resonant mode, which is even with respect to the x direction (i.e., a $(2m,n)$ mode). For a different l/b ratio, in fact, this mode does not give place to any transmission zero, as will be shown later. One may moreover observe that the (0,2) and (1,0) modes, which are rather close together, do not give place to any reflection zero. This is a typical case when two resonant modes interact together in such a way that no reflection zero takes place. However, one could verify that if the port widths would be very small (about 5 μm) two reflection zeros would take place at these resonant frequencies.

The experimental results in Fig. 9(b) agree fairly well with the theoretical ones until ~12.5 GHz; at higher frequencies there is a little shifting between the theoretical and experimental resonant frequencies. Moreover, the effect of the losses becomes appreciable over ~15 GHz. A more accurate and complete model of the structure would therefore require a better evaluation of the resonant frequencies and, at the same time, the introduction of the losses.

In order to confirm what is stated above with regard to the (0,2) mode, another structure has been made with proper dimensions in such a way that this mode is located between two odd modes. Fig. 10(a) and (b) shows the corresponding theoretical and experimental results. As the theory predicts, in this case, the (0,2) mode does not give place to any transmission zero; the three transmission zeros are due to the interaction between the modes (4,0) and (2,2), (3,2) and (5,0), and (4,2) and (6,0). As can be noted the theory agrees very well with the experiment in the whole frequency range 2–18 GHz.

Fig. 11(a) and (b) shows the theoretical and experimental results for a nonsymmetrical rectangular microstrip. In this structure odd modes with respect to the y axis can be excited at the second port, but not at the first one. As a consequence, all the $(m, 2n+1)$ modes give place to modal transmission zeros. It is worth observing that the first zero is located at the resonant frequency of the (0,1) mode, corresponding to the cutoff frequency of the $TE_{10}^{(x)}$ mode of propagation in the central section. This accounts for previous observations [10]–[13] with regard to nonsymmetrical step discontinuities; in this case, the statement that transmission zeros are due to higher order modes is correct, since it is a modal and not an interaction zero. The structure in Fig. 11 presents also three transmission zeros due to the interaction between the modes (2,0)–(0,2), (1,2)–(3,0), and (2,2)–(4,0). In spite of the complexity of the frequency behavior of the structure, the theoretical results may be considered highly satisfactory.

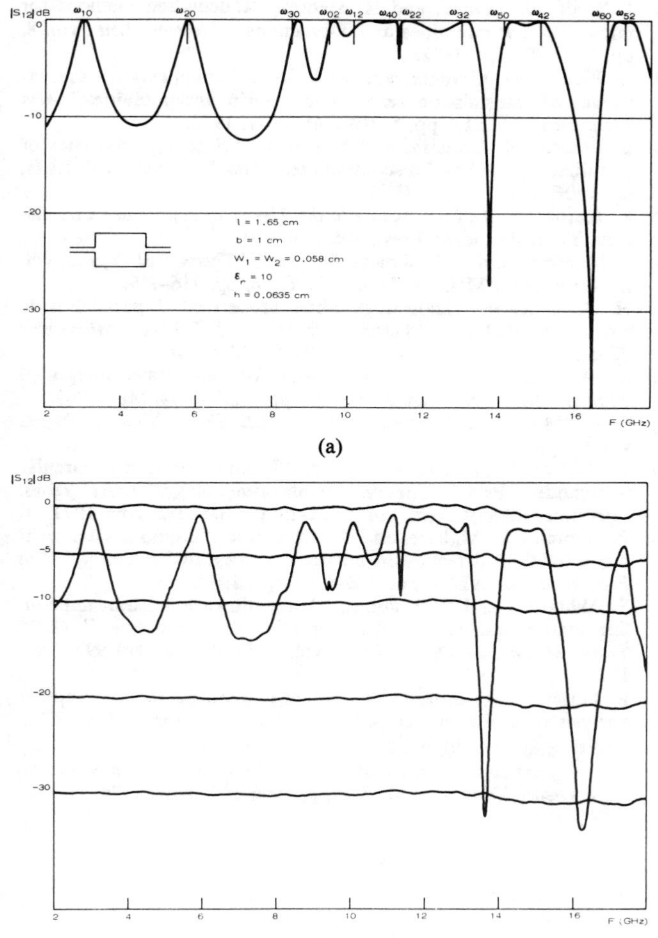

(a)

(b)

Fig. 10. Transmission coefficient $|s_{12}|$ versus the frequency for a symmetrical rectangular microstrip. (a) Theory. (b) Experiment.

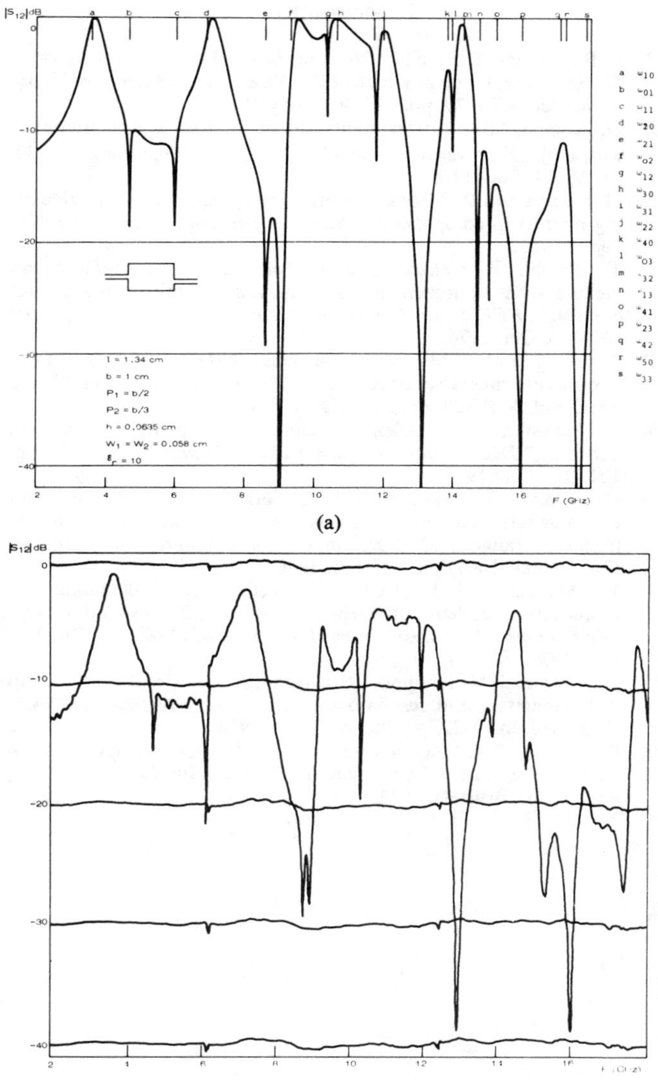

(a)

(b)

Fig. 11. Transmission coefficient $|s_{12}|$ versus the frequency for a nonsymmetrical rectangular microstrip. (a) Theory. (b) Experiment.

VI. Conclusions

A method of analysis of planar microwave structures is presented, which is based on a field expansion in terms of resonant modes. The case of two-port networks is considered, but the extension to N-port circuits is straightforward. The general filtering properties are discussed and, in particular, the physical nature of transmission zeros, which has been the subject of several discussions, is clarified. The existence of two types of transmission zeros, modal and interaction zeros, is pointed out. The first ones are due to the structure's resonances, while the second ones are due to the interaction between resonant modes. The latter are the only ones present in symmetrical nondegenerate structures.

Fringe effects are accounted for in a simple way, namely by introducing in the magnetic wall model effective parameters for each resonant mode, while other methods presented until now are limited by the impossibility of taking into account fringe effects in an adequate way. It is also shown that these effects should not be neglected, since they can produce a modal inversion and, consequently, a strong alteration of the structure's frequency behavior.

Several experimental results performed on circular and rectangular structures are compared with the theoretical ones deduced by assuming the effective parameters suggested in [9]. A good agreement is obtained, particularly for the circular microstrips; for rectangular microstrips a better characterization of the alterations due to fringing fields would be desirable. To that purpose it is necessary to adopt a full-wave analysis for calculating the resonant frequencies and the coupling between the lines and the resonant modes. This can be done on the basis of some methods presented in the literature [21]–[23].

Finally, it is worth pointing out that the method could be also adopted for planar circuits with any geometry, for instance through a finite element analysis.

Acknowledgment

The authors are indebted to the Servizio Microonde of the Elettronica S.p.a. for aid in the measurements; they also wish to thank B. De Santis for drawing the diagrams.

References

[1] B. Bianco and S. Ridella, "Nonconventional transmission zeros in distributed rectangular structures," *IEEE Trans. Microwave Theory Tech.*, vol. MTT-20, pp. 297–303, May 1972.

[2] G. Kompa, "S-Matrix computation of microstrip discontinuities with a planar waveguide model," *Arch. Elek. Übertragung*, vol. 30, pp. 58–64, Feb. 1976.

[3] G. Kompa and R. Mehran, "Planar waveguide model for calculating microstrip components," *Electron. Lett.*, vol. 11, pp. 459–460, Sept. 1975.

[4] T. Okoshi, Y. Uehara, and T. Takeuchi, "The segmentation method—An approach to the analysis of microwave planar circuits," *IEEE Trans. Microwave Theory Tech.*, vol. MTT-24, pp. 662–668, Oct. 1976.

[5] T. Okoshi and T. Miyoshi, "The planar circuit—An approach to microwave integrated circuitry," *IEEE Trans. Microwave Theory Tech.*, vol. MTT-20, pp. 245–252, Apr. 1972.

[6] P. Silvester, "Finite-element analysis of planar microwave networks," *IEEE Trans. Microwave Theory Tech.*, vol. MTT-21, pp. 104–108, Feb. 1973.

[7] G. D'Inzeo, F. Giannini, and R. Sorrentino, "Theoretical and experimental analysis of non-uniform microstrip lines in the frequency range 2–18 GHz," in *Proc. of 6th European Microwave Con.* (Rome, Italy) pp. 627–631, 1976.

[8] W. Menzel and I. Wolff, "A method for calculating the frequency-dependent properties of microstrip discontinuities," *IEEE Trans. Microwave Theory Tech.*, vol. MTT-25, pp. 107–112, Feb. 1977.

[9] I. Wolff and N. Knoppik, "Rectangular and circular microstrip disk capacitors and resonators," *IEEE Trans. Microwave Theory Tech.*, vol. MTT-22, pp. 857–864, Oct. 1974.

[10] G. Kompa, "Excitation and propagation of higher modes in microstrip discontinuities," presented at *Proc. of 3th European Microwave Conf.*, Brussels, 1973.

[11] I. Wolff, G. Kompa, and R. Mehran, "Calculation methods for microstrip discontinuities and T-junctions," *Electron. Lett.*, vol. 8, pp. 177–179, Apr. 1972.

[12] B. Bianco, M. Granara, and S. Ridella, "Comments on the existence of transmission zeros in microstrip discontinuities," *Alta Frequenza*, vol. XLI, pp. 533E–534E, Nov. 1972.

[13] B. Bianco, M. Granara, and S. Ridella, "Filtering properties of two-dimensional lines' discontinuities," *Alta Frequenza*, vol. XLII, pp. 140E–148E, July 1973.

[14] K. Kurokawa, *An Introduction to the Theory of Microwave Circuits.* New York: Academic Press, 1969, ch. 4.

[15] P. M. Morse and H. Feshbach, *Methods of Theoretical-Physics*, vol. I. New York: McGraw-Hill, 1953, ch. 6, pp. 716–719.

[16] H. A. Wheeler, "Transmission-line properties of parallel wide strips separated by a dielectric sheet," *IEEE Trans. Microwave Theory Tech.*, vol. MTT-13, pp. 172–185, Mar. 1965.

[17] M. V. Schneider, "Microstrip lines for microwave integrated circuits," *Bell Syst. Tech. J.*, vol. 48, pp. 1421–1444, May 1969.

[18] ——, "Microstrip dispersion," *Proc. Inst. Elect. Electron. Engrs.*, vol. 60, pp. 144–146, 1972.

[19] G. N. Tsandoulas and W. J. Ince, "Modal inversion in circular waveguides—Part I: Theory and phenomenology," *IEEE Trans. Microwave Theory Tech.*, vol. MTT-19, pp. 386–392, Apr. 1971.

[20] F. Giannini, P. Maltese, and R. Sorrentino, "Liquid crystal technique for field detection in microwave integrated circuitry," *Alta Frequenza*, vol. XLVI, pp. 80E–88E, Apr. 1977.

[21] S. Akhtarzad and P. B. Johns, "Three-dimensional transmission-line matrix computer analysis of microstrip resonators," *IEEE Trans. Microwave Theory Tech.*, vol. MTT-23, pp. 990–997, Dec. 1975.

[22] R. Jansen, "High-order finite element polinomials in the computer analysis of arbitrarily shaped microstrip resonators," *Arch. Elek. Übertragung*, vol. 30, pp. 71–79, Feb. 1976.

[23] T. Itoh, "Analysis of microstrip resonators," *IEEE Trans. Microwave Theory Tech.*, vol. MTT-22, pp. 946–951, Nov. 1974.

Wide-Band Equivalent Circuits of Microwave Planar Networks

GUGLIELMO D'INZEO, FRANCO GIANNINI, AND ROBERTO SORRENTINO, MEMBER IEEE

Abstract—A broad-band equivalent circuit of a generic microwave planar network is derived in terms of lumped constant elements. Contrary to previously proposed equivalent circuits, whose elements are strongly frequency dependent, the elements of the new one show only a smooth dependence on the frequency, because of the dispersion properties of microstrip structures. The equivalent circuit proposed is therefore easy to handle and is shown to be a useful basis for direct synthesis of planar structures. Good agreement with the theory is demonstrated by experiments performed on structures with different geometries up to 12.5 GHz, by using equivalent circuits whose elements are assumed to be constant with the frequency.

I. INTRODUCTION

ONE OF THE major problems in the analysis and design of MIC's is that of determining the parasitics. In a general planar networks parasitics arise essentially from two distinct phenomena: the existence of fringing fields and the excitation of higher order modes at the discontinuities.

While an infinite microstrip line can be characterized by an effective width and an effective permittivity [1], such an approach is no more valid in the case of a planar circuit or, in particular, in the presence of discontinuities. In such cases, in fact, the electromagnetic (EM) field is no more a quasi-TEM one, but results from the contribution of more complicated field distributions to which a variation both of fringe effects and of the EM energy storage have to be ascribed.

A general method of analysis of microwave planar structures, which accounts for fringe field effects, has been recently presented [2]; the approach may be regarded as an extension of the magnetic wall model of microstrip lines [3], [4]. Fringe effects, in fact, are taken into account through effective dimensions and effective permittivities which depend on the field distribution inside the planar structure [5]. On the basis of this method, in the present paper the lumped element equivalent circuit, in which parasitics are automatically taken into account, is derived for a general planar network.

The equivalent circuit approach has been extensively applied to the particular case of a discontinuity between two lines of different characteristic impedances. In such a case the effect of fringing fields consists of an increased capacitance of the structure, while the magnetic energy stored in the higher order evanescent modes excited at the discontinuity can be characterized through a series inductor. Equivalent capacitors and inductors of a single step discontinuity between two semi-infinite microstrip lines have been calculated through a quasi-static approach [6], by Gopinath together with several coworkers [7]–[10]. The resulting equivalent circuit, however, cannot be applied at high frequencies, nor in the case of interacting discontinuities. In fact, if the frequency is not sufficiently low, the contribution of the evanescent higher order modes becomes important as to modify both the fringing field distribution and the EM energy storage at the discontinuity; in addition, when, as in practical cases, more than one discontinuity is present, the interaction between the discontinuities takes place through higher order modes. It follows that this approach is only valid well below cutoff of the first higher mode of the wider microstrip line; in particular, it cannot explain the existence of transmission zeros in planar structures [11], [12].

Following a different theoretical approach, Bianco *et al.* [11] have proposed broad-band equivalent circuits for single and double step discontinuities; fringe effects, however, are taken into account through a rather approximated technique, so that the applicability of these equivalent circuits is strongly limited.

The dynamic approach by Menzel and Wolff [12], which accounts for the frequency variation of the energy stored in the evanescent higher order modes, has been used by Kompa [13] to obtain a lumped-element equivalent circuit of an abrupt impedance step.

However, Kompa's equivalent circuit contains elements which are strongly variable with the frequency, so that it is not easy to handle when broad-band simulation is needed. On the contrary, the elements of the equivalent circuits proposed here are frequency dependent only because of the dispersion properties of microstrip circuits so that, in a first approximation, they may be assumed to be constant with the frequency, also in broad-band simulations. Finally, it should be stressed that, contrary to the equivalent circuits proposed early, the present one can be applied to structures having geometries different from the rectangular one.

Manuscript received March 24, 1980; revised June 12, 1980. This work was partially supported by Consiglio Nazionale delle Ricerche. Part of this work was presented at 2ª Riunione Nazionale di Elettromagnetismo Applicato, Pavia, Italy, 1978.

G. D'Inzeo is with Istituto di Elettronica, Università di Ancona, Via della Montagnola 30, 60100 Ancona, Italy.

F. Giannini and R. Sorrentino are with Istituto di Elettronica, Università di Roma, Via Eudossiana 18, 00184 Roma, Italy.

Reprinted from *IEEE Trans. Microwave Theory Tech.*, vol. MTT-28, no. 10, pp. 1107–1113, Oct. 1980.

II. The General Equivalent Circuit

A generic two-port microwave planar network is sketched in Fig. 1. Following [2], the analysis is carried out firstly by determining in the domain S of the xy plane the orthonormalized set of eigenfunctions of the bidimensional Helmholtz equation

$$\nabla^2\phi_m + k_m^2\phi_m = 0 \quad (1)$$

with homogeneous Neumann boundary conditions. In the hypotheses that losses are negligible, that higher order modes on the connecting lines are evanescent and that line widths are much smaller than the structure's dimensions, the impedance matrix of the network is given by

$$[Z] = \sum_{m=0}^{\infty} [Z_m] \quad (2)$$

with

$$[Z_m] = \frac{j\omega}{\omega_m^2 - \omega^2}\frac{h}{\epsilon_m}\begin{bmatrix} R_{m1}^2 & R_{m1}R_{m2} \\ R_{m2}R_{m1} & R_{m2}^2 \end{bmatrix}. \quad (2a)$$

In (2a) $\omega_m = k_m/\sqrt{\mu\epsilon_m}$ is the mth structure's resonant frequency

$$R_{mi} = \frac{1}{w_{i,\text{eff}}}\int_{l_i}\phi_m \, dl, \qquad i = 1,2 \quad (2b)$$

is the coupling coefficient between the mth resonant mode and the TEM wave traveling on the ith line; l_i is the portion of the contour of the planar structure corresponding to the ith port; k_m^2 and ϕ_m are the mth eigenvalue and eigenfunction of (1), respectively; ϵ_m is the effective permittivity of the mth resonant mode [5]; h is the substrate's thickness; $w_{i,\text{eff}}$ is the effective width of the ith line [1].

The first term of the series (2), for $m=0$, has $\omega_0 = 0$ and thus corresponds to the mode resonating at zero frequency, for which $\phi_0 = 1/\sqrt{S}$, where S is the area of the planar network, and $R_{01} = R_{02} = 1$. From (2)–(2b) it follows that, for a planar network of given geometry, its filtering properties depend on the coupling coefficients, i.e., on the positions and widths of the ports.

According to [5], fringe effects are taken into account in previous formulas by ascribing to the structure effective dimensions, and by introducing a different effective permittivity for each resonant mode. As a consequence, each resonant frequency of the planar structure is shifted with respect to the case of magnetic wall model. A typical example is the modal inversion between the TM_{410} and TM_{120} modes demonstrated in a circular microstrip [14].

As can be easily seen, each term (2a) of (2) can be realized in the form of an antiresonant LC cell (with the exception of the first term which corresponds to a pure capacitance) connected to two ideal transformers whose transformer ratios are given by the coupling coefficients (2b). This leads to the equivalent circuit shown in Fig. 2 as the series connection of an infinite number of such cells. This equivalent circuit is of the type of the one derived in [19]. However, there is a substantial difference between them, since each LC cell of the equivalent circuit of Fig. 2 is derived from an effective model (dimensions and permittivity) of the planar structure, which is in general different for each resonant mode.

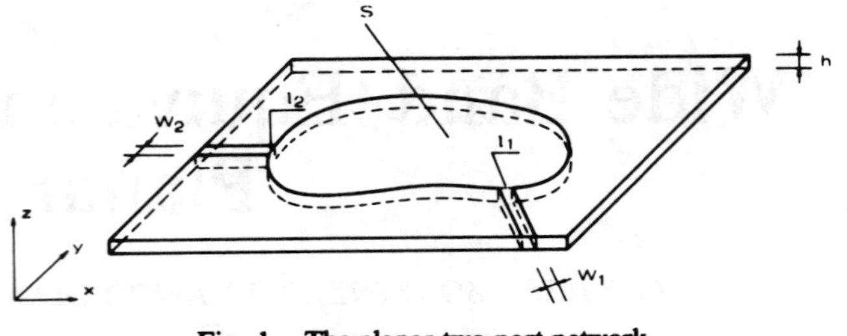

Fig. 1. The planar two-port network.

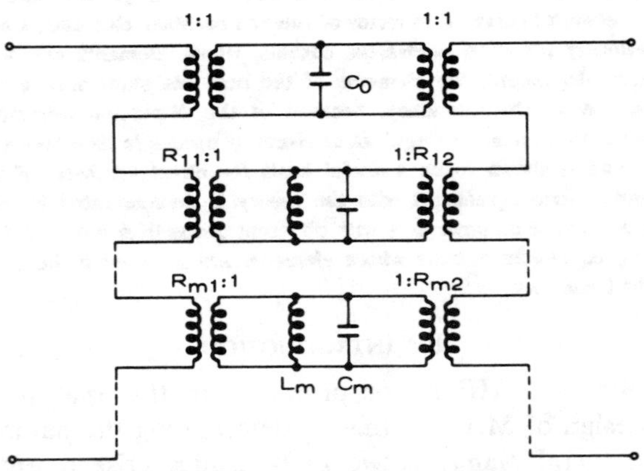

Fig. 2. The general equivalent circuit of a two-port planar network.

For practical application, only a finite number of such resonant cells has to be taken into account, depending on the frequency range of interest and on the approximation to be obtained. It is worth noting that the broad-band equivalent circuits previously proposed [11], [13] contain elements which are so strongly frequency dependent as to become infinite at certain frequencies. On the contrary, the equivalent circuit elements of Fig. 2 are only smoothly frequency dependent because of the well-known dispersion characteristic of the microstrip structure and therefore have always finite values. Furthermore, in most cases, the frequency dependence can be neglected with a fairly good approximation. In all the experiments reported below, in fact, a good agreement is shown up to 12.5 GHz, through the use of equivalent circuits whose elements are assumed to be frequency independent.

Typical features of a microwave planar network can be shown through the equivalent circuit of Fig. 2. For example, suppose the signal frequency coincides with the resonant frequency ω_n of the nth cell; it is immediately found that the transformer radii of such a cell, i.e., the coupling coefficients of the corresponding mode, determine the relationship between the currents I_1, I_2 at the ports of the circuit

$$R_{n1}I_1 + R_{n2}I_2 = 0.$$

If $R_{n1} \neq 0$ and $R_{n2} = 0$, i.e., the nth mode is uncoupled to port 2, then I_1 must be zero, and the circuit presents an open circuit at the port 1 for $\omega = \omega_n$. This has been called a modal transmission zero of the microwave planar network [2], [15]. On the contrary suppose $|R_{n1}|=|R_{n2}|$; if, for $\omega=\omega_n$, the contribution of all the other cells is negligible, as it generally happens, one obtains

$$V_1/I_1=V_2/(-I_2).$$

This means that the impedance at the second port is transferred at the first port, thus $\omega=\omega_n$ is a frequency of null attenuation.

Therefore, depending on the coupling coefficients, the same resonant mode of the microwave structure can give place either to a transmission zero or to a reflection zero.

III. Symmetrical Structures

The equivalent circuit of Fig. 2 is generally of impractical use. A more suitable equivalent circuit can be derived in the case of symmetrical structures. When the planar network is symmetrical, the modal solutions of (1) can be divided into even and odd modes, depending whether ϕ_m has equal or opposite values at the two ports. As a consequence, the coupling coefficients are related by

$$R_{m2}=\alpha_m R_{m1} \tag{3}$$

where $\alpha_m=1$ for even modes, $\alpha_m=-1$ for odd modes. Separating the contributions of the even and odd modes, the impedance parameters (2) of a symmetrical network can be written, because of (3)

$$Z_{11}=Z_{22}=\Sigma_{ev}\frac{j\omega hR_{ev}^2/\epsilon_{ev}}{\omega_{ev}^2-\omega^2}+\Sigma_{od}\frac{j\omega hR_{od}^2/\epsilon_{od}}{\omega_{od}^2-\omega^2}$$
$$=\frac{1}{2}(Z_{ev}+Z_{od})$$
$$Z_{12}=Z_{21}=\Sigma_{ev}\frac{j\omega hR_{ev}^2/\epsilon_{ev}}{\omega_{ev}^2-\omega^2}-\Sigma_{od}\frac{j\omega hR_{od}^2/\epsilon_{od}}{\omega_{od}^2-\omega^2}$$
$$=\frac{1}{2}(Z_{ev}-Z_{od}) \tag{4}$$

where ev and od stand for even and odd, respectively. Expressions (4) lead to the symmetrical lattice realization of Fig. 3, which appears particularly useful since it avoids use of transformers. As is known, such a two-port network presents a pole of attenuation when the signal frequency is such that

$$Z_{ev}(\omega)=Z_{od}(\omega). \tag{5}$$

Now, since Foster's theorem applies to both Z_{ev} and Z_{od}, which are pure reactances, it is evident that (5) is satisfied if Z_{ev} remains finite between two consecutive poles of Z_{od}, or vice versa. In other words, if two consecutive resonant frequencies of the structure are both even (odd), then a pole of attenuation takes place between them. This has been called an interaction transmission zero [2], [15]. On the other hand, in a symmetrical lattice structure, a frequency of null attenuation occurs when

$$Z_{ev}(\omega)Z_{od}(\omega)=\eta^2 \tag{6}$$

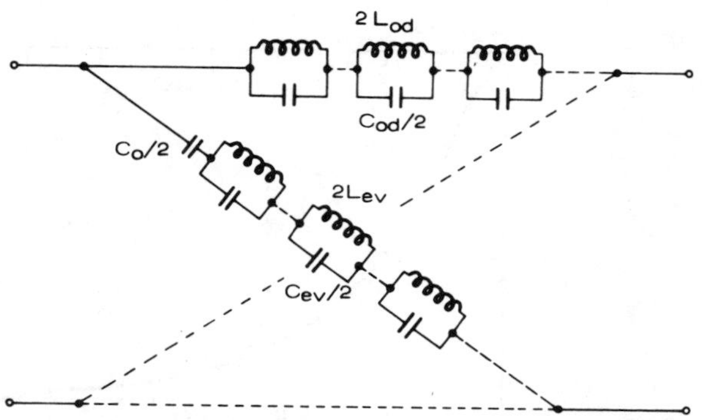

Fig. 3. The equivalent circuit of a two-port symmetrical structure.

where η is the characteristic impedance of the connecting lines. In the limit when η tends to infinite, condition (6) is satisfied when ω is coincident with a pole of Z_{ev} or Z_{od}, i.e., with a resonant frequency of the structure. As can be easily inferred from (2)–(2b), in fact, for $\eta\to\infty$, the impedance parameters of the planar network tend to finite values, except at the resonant frequencies. In practical cases, because of the finite widths of the lines, the frequencies of null attenuation are shifted from the resonant frequencies [16]. In some cases, however, it happens that a resonant mode does not give place to a reflection zero except for η greater than a minimum value. In fact, let us rewrite condition (6) in the form

$$X_{ev}=-\eta^2/X_{od} \tag{6a}$$

where $X_{ev}=-jZ_{ev}$, $X_{od}=-jZ_{od}$. Since, because of Foster's theorem, X_{ev} is an always increasing function of the frequency, in a suitable interval enclosing a pole ω_{ev}, X_{ev} assumes all the values from $-\infty$ to $+\infty$; if $1/X_{od}$ remains finite in this interval, then (6a) is necessarily satisfied in that interval at a frequency $\bar{\omega}$; the higher η the closer $\bar{\omega}$ to ω_{ev}. On the contrary, if $1/X_{od}$ has a pole near ω_{ev}, it can be easily inferred that (6a) is satisfied only for sufficiently high values of η.

An example of application of the equivalent circuit of Fig. 3 for characterizing the frequency behavior of a rectangular planar structure in shown in Fig. 4, where the scattering parameter $|s_{21}|$ of the equivalent circuit is shown as a function of frequency in the range 0–12.5 GHz and is compared with the experimental measurements. As can be seen, good agreement is obtained by taking into account only the first six resonant modes of the structure, five of which fall in the range 0–12.5 GHz. Two poles of attenuation are located between the resonant frequencies $\omega_{00}-\omega_{02}$ (even modes) and $\omega_{10}-\omega_{12}$ (odd modes), according to the rule previously stated. Because of the presence of these poles of attenuation, produced by the interaction between resonant modes involving both an x- and y-field dependence, this structure may not be analyzed through a quasi-static approach, except well below 5 GHz. Let us observe, moreover, that while each one of the resonant frequencies ω_{00}, ω_{12}, ω_{04}, gives place to a reflection zero, ω_{02} and ω_{10} do not. This is due to the fact that the connecting lines impedances are too low (50 Ω) so that

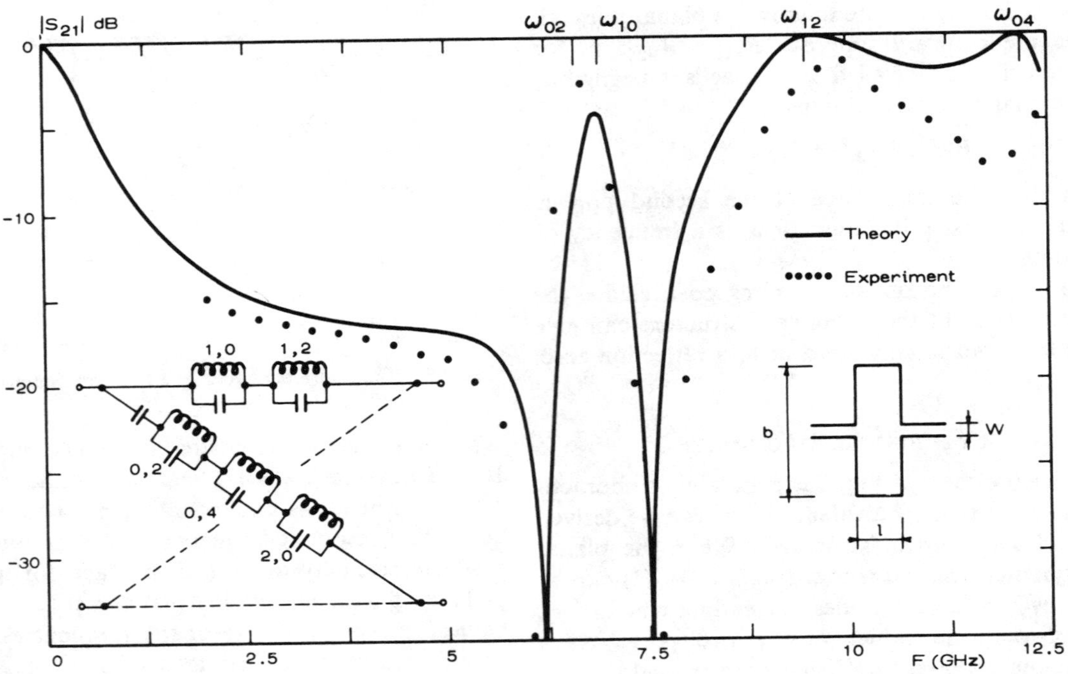

Fig. 4. Comparison between the experimental $|s_{21}|$ of a rectangular structure and the theoretical $|s_{21}|$ of the equivalent circuit indicated. (Alumina substrate, $\epsilon_r=10$, $h=0.0635$ cm; $l=0.68$ cm, $b=1.5$ cm, connecting line impedance $\eta=50\ \Omega$).

condition (6) is not satisfied both for ω_{02} and ω_{10} (see Fig. 5(a)). On the contrary, condition (6) would be satisfied if the connecting lines would have a very high impedance (see Fig. 5(b)).

So far, we have implicitly assumed that only one independent eigensolution of (1) exists for each k_m^2. On the contrary, in typical cases such those of circular, annular or square structures, (1) has two linearly independent solutions corresponding to the same eigenvalue, so that there are two resonant modes with the same resonant frequency. If the two degenerate modes are one even and the other one odd, both the even and the odd branches of the equivalent lattice circuit (Fig. 3) become open circuits at the same resonant frequency, giving place to a transmission zero. This type of zero has been classified among the modal zeros [2] since it occurs at a structure's resonant frequency. Unless the circuit of Fig. 3 is valid also in case of degenerate structures, some simplifications can be made accounting for the mode degeneracy.

As an example, let us consider an annular two-port planar network. According to the analysis of this microwave structure [17], which is based on [2], the elements of the even and odd branches of the equivalent lattice structure can be written as follows:

$$C_{mn}^{(\mathrm{ev})}=\frac{\epsilon_{mn}}{hR_{mn}^2(1+\cos m\psi)};\ C_{mn}^{(\mathrm{od})}=\frac{\epsilon_{mn}}{hR_{mn}^2(1-\cos m\psi)}$$

$$L_{mn}^{(\mathrm{ev})}=\frac{\mu hR_{mn}^2(1+\cos m\psi)}{k_{mn}^2};\ L_{mn}^{(\mathrm{od})}=\frac{\mu hR_{mn}^2(1-\cos m\psi)}{k_{mn}^2} \tag{7}$$

where ψ is the angle between the connecting lines, ϵ_{mn} is the effective permittivity of the degenerate (m,n) mode

$$R_{mn}=1/\sqrt{\pi}\,A_{mn}F_{mn}(k_{mn}r_{\mathrm{out}})\sin m\varphi/m\varphi$$

$$F_{mn}(k_{mn}r)=J_m(k_{mn}r)+KN_m(k_{mn}r)$$

$$A_{mn}=\delta_m\left\{\int_{r_{\mathrm{in}}}^{r_{\mathrm{out}}}r[F_{mn}(k_{mn}r)]^2dr\right\}^{-1}$$

$$K=-J_m'(k_{mn}r_{\mathrm{out}})/N_m'(k_{mn}r_{\mathrm{out}})$$

$$\delta_{\mathrm{in}}=\begin{cases}1, & m=0\\ 2, & m\neq 0\end{cases}.$$

φ is the angle subtended by the ports, and r_{in}, r_{out} are the inner and outer radii of the ring, respectively.

Expressions (7) show that, except when $|\cos m\psi|=1$, each resonant frequency corresponds to a modal zero, since the corresponding degenerate modes contribute to both the even and odd parts of the $[Z]$ matrix. In particular, if $\cos m\psi=0$, the lattice structure can be modified as to realize the corresponding pole of the Z matrix as a private pole of $Z_{11}=Z_{22}$. On the contrary, if $\cos m\psi=1$, or $\cos m\psi=-1$, a resonant cell will appear only at the even, or odd, branch of the lattice equivalent circuit so that, according to the previous discussion, it generally gives place to a reflection zero. Fig. 6 shows the experimental and theoretical behaviors of the scattering parameter $|s_{21}|$ of an annular microstrip with orthogonal connecting lines ($\psi=\pi/2$) versus the frequency. The equivalent

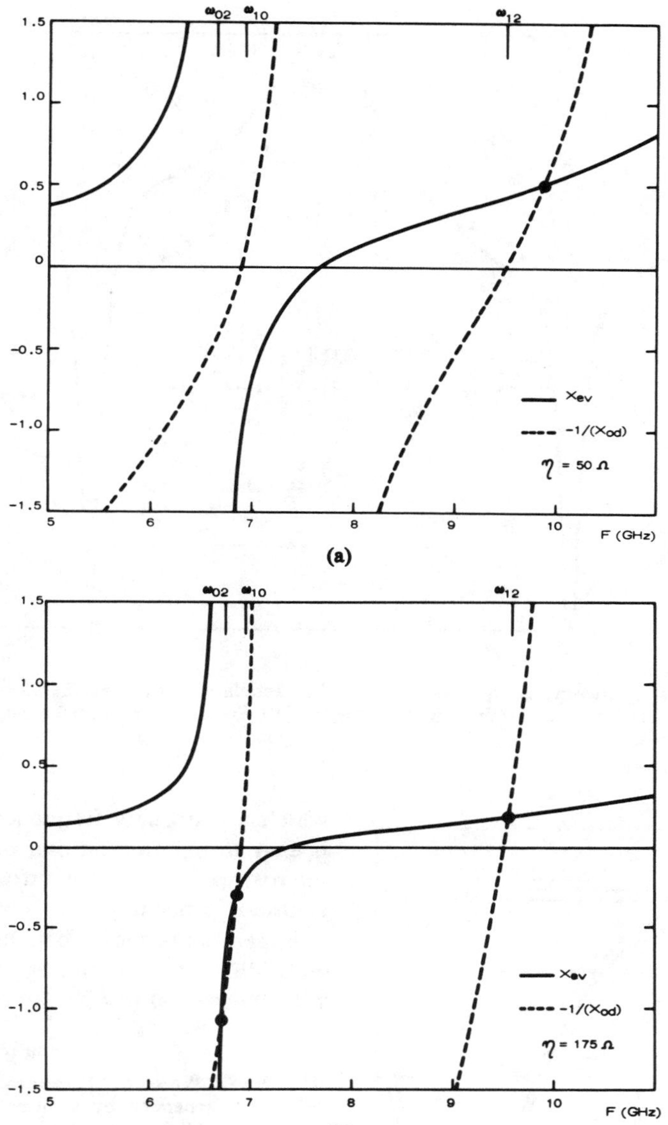

Fig. 5. Plots of X_{ev} and $-1/X_{od}$ of the rectangular structure of Fig. 4 with (a) $\eta = 50\ \Omega$, and (b) $\eta = 175\ \Omega$.

TABLE I

Mode	C(pF)	L(nH)	f_{res}(GHz)
0,0	8.61555	–	0
1,1	3.96161	5.99266	3.26641
2,1	3.27220	1.91647	6.35543
3,1	2.71207	1.10168	9.20741
4,1	2.37537	0.75450	11.88834

circuit used for the theoretical simulation is indicated in the same figure, while the values of the LC elements are quoted in Table I. The resonant frequencies ω_{11} and ω_{31}, for which $\cos m\psi = 0$, give place to two transmission zeros corresponding to two private poles of $Z_{11} = Z_{22}$ (see equivalent circuit) while frequencies of null attenuation are located close to the resonant frequencies ω_{21}, ω_{41} (for which $|\cos m\psi| = 1$).

The availability of lumped-element equivalent circuits of microwave planar network, besides allowing a simple characterization of their behavior, may be also the basis for obtaining synthesis procedures of these structures, starting from the conventional synthesis procedure of lumped element circuits.

An example is illustrated in Fig. 7 where the measured scattering parameter $|s_{21}|$ of a rectangular two-port structure, realizing a Cauer–Chebycheff low-pass filter is plotted versus the frequency.

The rectangular structure has been synthesized according to [18]. For comparison, the same figure shows the behaviors of the CC 31031 prototype and that obtained according to [2].

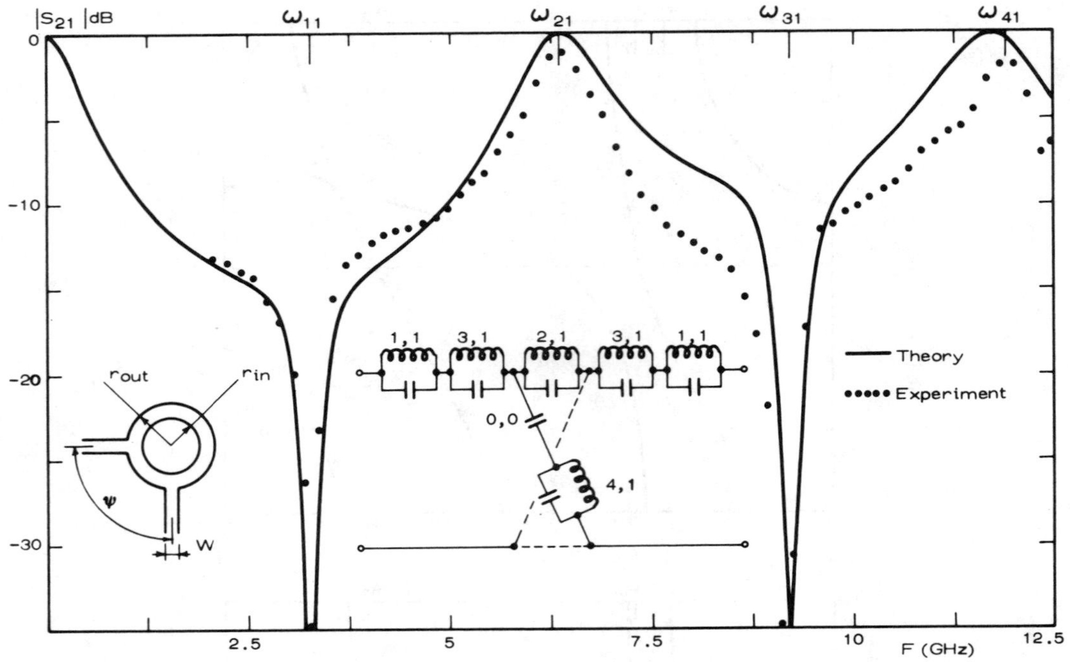

Fig. 6. Comparison between the experimental $|s_{21}|$ of an annular structure and the theoretical $|s_{21}|$ of the equivalent circuit indicated. (Epsilam substrate, $\epsilon_r = 10$, $h = 0.0635$ cm; $r_{in} = 0.4$ cm, $r_{out} = 0.7$ cm, $w = 0.06$ cm, $\psi = 90°$).

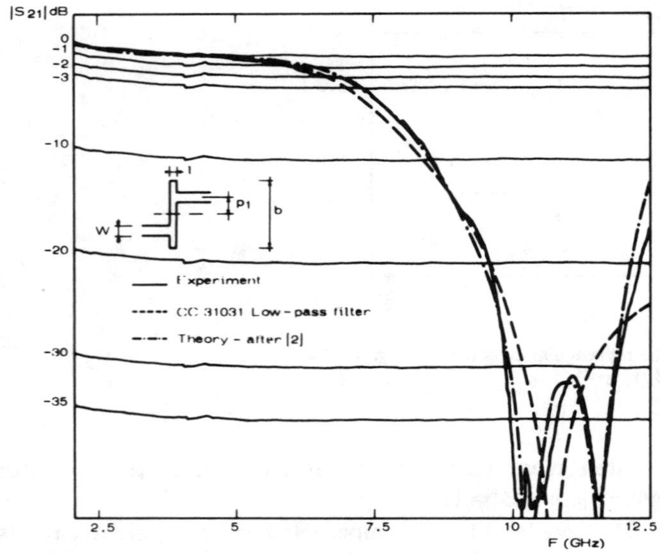

Fig. 7. $|s_{21}|$ of the filtering structure synthesized versus the frequency. (Alumina substrate, $\epsilon_r = 10$, $h = 0.0635$ cm, $l = 0.020$ cm, $b = 0.845$ cm, $w = 0.060$ cm, $p_1 = 0.162$ cm, reference frequency $f_r = 5$ GHz).

IV. Conclusions

A broad-band equivalent circuit of a generic microwave planar network has been derived in terms of lumped-constant elements. These elements are only smoothly frequency dependent, because of the dispersion properties of microstrips, so that they may be considered, with good approximation, to be constant with the frequency, even in broad-band simulations.

Contrary to previously proposed equivalent circuits, which are strongly frequency dependent, the present one is easy to handle and can be a useful basis for designing microstrip planar structures starting from conventional synthesis procedures.

Experiments performed up to 12.5 GHz on structures with different geometries have shown good agreement with theoretical results.

References

[1] H. A. Wheeler, "Transmission-line properties of parallel wide strips separated by a dielectric sheet," *IEEE Trans. Microwave Theory Tech.*, vol. MTT-13, pp. 172–185, Mar. 1965.

[2] G. D'Inzeo, F. Giannini, C. M. Sodi, and R. Sorrentino, "Method of analysis and filtering properties of microwave planar networks," *IEEE Trans. Microwave Theory Tech.*, vol. MTT-26, pp. 462–471, July 1978.

[3] I. Wolff, G. Kompa, and R. Mehran, "Calculation method for microstrip discontinuities and *T*-junctions," *Electron. Lett.*, vol. 8, pp. 177–179, Apr. 1972.

[4] G. Kompa and R. Mehran, "Planar waveguide model for calculating microstrip components," *Electron. Lett.*, vol. 11, pp. 459–460, Sept. 1975.

[5] I. Wolff and N. Knoppik, "Rectangular and circular microstrip disk capacitors and resonators," *IEEE Trans. Microwave Theory Tech.*, vol. MTT-22, pp. 857–864, Oct. 1974.

[6] P. Benedek and P. Silvester, "Equivalent capacitances for microstrip gaps and steps," *IEEE Trans. Microwave Theory Tech.*, vol. MTT-20, pp. 729–733, Nov. 1972.

[7] A. F. Thomson and A. Gopinath, "Calculation of microstrip discontinuity inductances," *IEEE Trans. Microwave Theory Tech.*, vol. MTT-23, pp. 648–655, Aug. 1975.

[8] C. Gupta and A. Gopinath, "Equivalent circuit capacitance of microstrip step change in width," *IEEE Trans. Microwave Theory Tech.*, vol. MTT-25, pp. 819–822, Oct. 1977.

[9] B. Easter, A. Gopinath, and I. M. Stephenson, "Theoretical and experimental methods for evaluating discontinuities in microstrips," *Radio Electron. Eng.*, vol. 48, pp. 73–84, Jan./Feb. 1978.

[10] A. Gopinath and G. Gupta, "Capacitance parameters of discontinuities in microstrip-lines" *IEEE Trans. Microwave Theory Tech.*, vol. MTT-26, pp. 831–836, Oct. 1978.

[11] B. Bianco, M. Granara, and S. Ridella, "Filtering properties of two-dimensional lines' discontinuities," *Alta Freq.*, vol. XLII, pp. 286–294, June 1973.

[12] W. Menzel and I. Wolff, "A method for calculating the frequency-dependent properties of microstrip discontinuities," *IEEE Trans. Microwave Theory Tech.*, vol. MTT-25, pp. 107–112, Feb. 1977.

[13] G. Kompa, "Design of stepped microstrip components," *Radio Electron. Eng.*, vol. 48, pp. 53–63, Jan./Feb. 1978.

[14] G. D'Inzeo, F. Giannini, and R. Sorrentino, "Theoretical and experimental analysis of non-uniform microstrip lines in the frequency range 2-18 GHz," in *Proc. 6th European Microwave Conf.*, (Rome, Italy), 1976, pp. 627–631.

[15] G. D'Inzeo, F. Giannini, P. Maltese, and R. Sorrentino, "On the double nature of transmission zeros in microstrip structures," *Proc. IEEE*, vol. 66, pp. 800–802, July 1978.

[16] T. Okoshi and T. Mihoshi, "The planar circuit—An approach to microwave integrated circuitry," *IEEE Trans. Microwave Theory Tech.*, vol. MTT-20, pp. 245–252, Apr. 1972.

[17] G. D'Inzeo, F. Giannini, R. Sorrentino, and J. Vrba, "Microwave planar networks: the annular structure," *Electron. Lett.*, vol. 14, pp. 526–528, Aug. 1978.

[18] G. D'Inzeo, F. Giannini, and R. Sorrentino, "Novel integrated lowpass filters," *Electron. Lett.*, vol. 15, pp. 258–260, Apr. 1979.

[19] P. P. Civalleri and S. Ridella, "Impedance and admittance matrices of distributed three-layer N-ports," *IEEE Trans. Circuit Theory*, vol. CT-17, pp. 392–398, Aug. 1970.

The Planar-Circuit Image-Parameter Method: a Novel Approach to the Computer-Aided Design of MIC Filters

Guglielmo Forte
Selenia, Industrie Elettroniche Associate S.p.a.
Via Tiburtina, km 12,400 - 00131 Roma -Italy

Mario Salerno, Roberto Sorrentino
Dipartimento di Ingegneria Elettronica,
II Università degli Studi di Roma "Tor Vergata"
Via O. Raimondo, 8 - 00173 Roma - Italy

Abstract. Microstrip low-pass filters with very wide stopbands and very sharp cutoff rates have been designed and realized using a new technique which incorporates two-dimensional effects and related parasitics in the synthesis procedure. This removes a number of limitations inherent to the conventional transmission line approach and makes the present one particularly suited for monolithic microwave circuits.

1. INTRODUCTION

The difficulty in realizing low impedance levels in microstrip configurations constitutes a serious limitation in the design of microstrip circuits, particularly at high frequencies. While the upper limit of realizable characteristic impedance is set by manufacturing tolerances, the lower limit depends on the onset of higher order modes. Typical lower limit is 20-25 ohms [1]. Substantial discrepancies between predicted and actual performances occur when lower values are realized.

Special circuit configurations are adopted in some cases to circumvent the above difficulty. An example is the radial line stubs [2]. The radial line geometry is such that the connection of the stub with the main line can be kept narrow, still providing a low impedance level over a rather wide frequency band.

This, however, does not solve the basic problem, which resides on the inadequacy of the transmission line model to characterize low impedance microstrip circuits. When the transverse dimension of a microstrip element is a non-negligible fraction of the wavelength, it becomes necessary to adopt two-dimensional, or planar, models [3, 4].

In particular, the limit on the low impedance levels affects the realizability of microstrip filters. Filter design is still based on lumped or distributed prototypes that are to be converted into microstrip configurations. Such a procedure is successful provided that the required impedance values are in the range of realizability. When high filter performances are required the conversion into a microstrip structure may be difficult or even impossible. This typically happens for lowpass filters with very wide stopbands, which require very low impedance levels.

To overcome this difficulty, the planar model of microstrip structures should be incorporated into the design procedure. This is not possible in the conventional filter design method, but can be achieved by a suitable extension of the image parameter method (IPM).

Recently, it has been shown that IPM can be applied to distributed filters [5]. Any transformation from a lumped prototype is eliminated; the filter does not need to be commensurate so that more degrees of freedom can be utilized. A further extension to planar structures is demonstrated in this paper. The planar-circuit approach is used to model the microstrip elements which form the filter structure, so that two-dimensional effects are automatically taken into account in the synthesis procedure. Parasitics associated with fringe field effects are included in the model via the effective parameters (permittivity and geometrical dimensions). Furthermore, the manufacturing constraints are incorporated into the design procedure, so that all the filters designed can be actually realized. These advantages render the present method particularly attractive for monolithic microwave integrated circuits, where, owing to the practical impossibility of tuning the circuits realized, the design accuracy is of fundamental importance.

The design is performed interactively on a PC. Elementary planar cells, image matched one to each other, are cascaded together to build up the filter

Reprinted with permission from *Alta Freq.*, vol. LVII, no. 5, pp. 233–239, June 1988.

structure. Each cell is synthesized individually so as to produce a prescribed transmission zero in the stopband. Suitable interpolation formulas are used to speed up the synthesis of the cells. Excellent filter performances have been experimentally demonstrated.

The structure of the microstrip filter is illustrated in section 2. The planar model used to characterize the cells is discussed in section 3. Section 4 describes the design procedure, while the synthesis of the cells is shown in section 5. The results obtained are finally shown in the last section.

2. FILTER STRUCTURE AND CONSTITUTIVE CELLS

Printed circuit lowpass filters have generally a hi-lo type geometry, i.e. they consist of the cascade of alternate high and low impedance line sections [6]. Such a geometry results from the conversion of a lumped LC ladder prototype, where high impedance sections correspond to inductors and low impedance sections to capacitors. These can also be viewed as low impedance double stubs. The wider the upper stop band limit, the lower the stub impedances. When the lumped prototype is designed on an image basis (constant-K filters), m-derived end sections are added to improve the cutoff rate of the filter as well as a to provide a better image matching in the passband. The end sections are realized as high impedance open-ended stubs which provide a transmission zero near the cutoff frequency. It must be noted that, in an extended frequency range, the low impedance sections too can be regarded as open ended stubs.

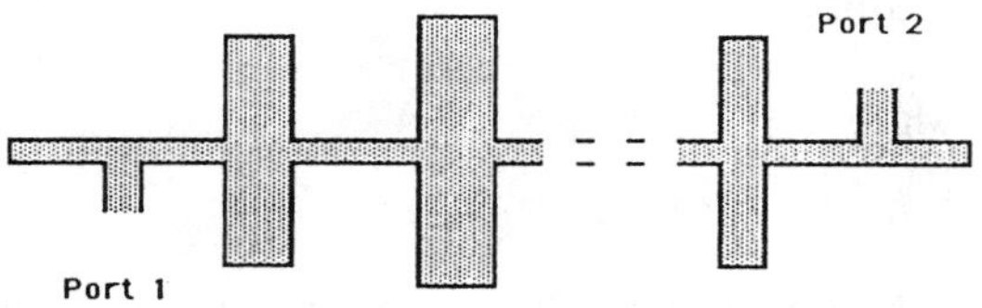

Fig. 1 - The microstrip lowpass filter configuration.

A filter geometry of this type is considered in this paper and is shown in Fig. 1. It consists of an open-ended line section of high characteristic impedance loaded with $m - 2$ low impedance double stubs. The open ends of the main line behave as two additional high impedance stubs. This configuration can exhibit very high performances in terms of high cutoff rates and wide stop bands. It can be designed on an image basis by converting a constant-K with m-derived end sections lumped prototype into a microstrip version using the semi-lumped approximation [7]. Degradations occur because of the approximations involved. A more accurate procedure is based on the direct application of the IPM to a distributed structure [5] but is limited by the occurrence of two-dimensional phenomena in the low impedance stubs, when very wide stop bands are required.

This paper shows the application of the IPM to the design of two-dimensional structures so removing the above limitations.

Filter design by image method consists of cascading a number of elementary cells image-matched in the passband, in such a way as to obtain the prescribed stopband response. The configuration of Fig. 1 results from the cascade of the two types of elementary planar cells shown in Fig. 2. The first type, referred to as the N-cell, is used at the ends of the filter, while m-2 symmetrical cells of the second type, referred to as the T-cell, are used to construct the internal structure of the filter. Both cells are characterized by a planar model, as described in the next section.

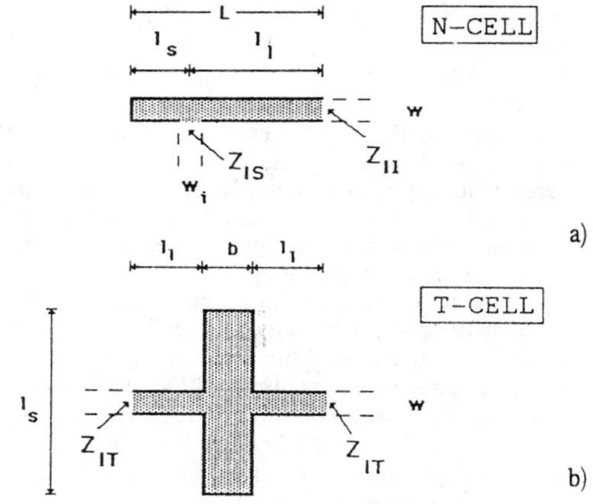

Fig. 2a, b - Planar elementary cells used in the design of Fig. 1: the N-cell and the T-cell.

3. PLANAR MODELS OF THE ELEMENTARY CELLS

The N-cell of Fig. 2a can be considered as a single microstrip rectangular cavity with two ports. The T-cell is a composite cell resulting from a rectangular resonator with two additional line sections. The characterization of the T-cell is reduced to that of the internal rectangle after shifting the reference planes. A planar-circuit formalism can be adopted to characterize these elements as circuit elements [3].

Microstrip lines are modelled using the planar waveguide model, which is defined in terms of an effective width and an effective dielectric constant. The electromagnetic field at the connection of the i-th microstrip line with a planar circuit, can be represented by a modal expansion whose coefficients are voltage and current amplitudes associated with the modes,

$$V_i^{(m)} = \frac{\sqrt{\delta_m}}{w_i} h \int_{w_i} E_z \cos \frac{m \pi l}{w_i} \, dl$$

$$I_i^{(m)} = \sqrt{\delta_m} \int_{w_i} H_\tau \cos \frac{m \pi l}{w_i} \, dl \qquad (1)$$

where $m = 0, 1, 2$ is the order of the mode, w_i is the effective width of the line, thus of i-th the port of the planar circuit, δ_m is the Neumann delta ($\delta_o = 1$, $\delta_m = 2$ for $m \neq 0$); l is the coordinate along the i-th port. E_z is the electric field, directed orthogonal to the substrate, and H_t is the magnetic field, directed in the plane of the substrate.

The EM field inside a planar element is expanded into a series of two-dimensional field configurations E_ν corresponding to the open-circuit resonant modes of the cavity. They are obtained from the following eigenvalue problem

$$\nabla^2 E_\nu + k_\nu^2 E_\nu = 0 \qquad \text{in } S$$
$$\frac{\partial E_\nu}{\partial n} = 0 \qquad \text{on } \partial S \tag{2}$$

S being the surface of the planar circuit and ∂S its contour.

Fringing fileds at the metallization edges are taken into account by proper effective parameters which may depend on the resonant mode [8]. Though approximate and based on phenomenological considerations such an approach to account for parasitics has been shown to provide good agreement with the experiments up to relatively high frequencies (over 12 GHz for microstrip circuits on alumina substrates 0.635 mm thick [9]). The circuit dimensions used throughout this paper are effective quantities. These are related to the actual dimensions of the microstrip structures according to a number of different approximate formulas which can be found, for instance, in [10] and, for the rectangular resonators, in [8].

The generalized impedance matrix formulation relating voltages and currents (1) of the modes of the feeding lines

$$V_i^{(m)} = \sum_{n=0}^{\infty} \sum_{j=1}^{2} Z_{ij}^{(mn)} I_j^{(n)} \tag{3}$$

is then obtained in the form of a series expansion in terms of resonant modes

$$Z_{ij}^{(mn)} = \frac{j\omega\mu h \sqrt{\delta_m \delta_n}}{w_i w_j} \sum_{\nu=0}^{\infty} \frac{g_{\nu i}^{(m)} g_{\nu j}^{(n)}}{k_\nu^2 - k^2} \tag{4}$$

$$g_{\nu i}^{(m)} = \int_{w_i} E_\nu \cos \frac{m\pi l}{w_i} dl \tag{5}$$

where

$$k^2 = \omega^2 \mu \epsilon$$

In the above formulas h is the thickness of the dielectric substrate, $\epsilon = \epsilon_o \epsilon_r$ the electric permittivity and μ the magnetic permeability.

The impedance $Z_{ij}^{(qr)}$ represents the voltage associated with the q-th mode on the (physical) port w_i due to a unit current of the r-th mode at the port w_j. Note that in the generalized matrix representation of the planar network, each physical port is represented by infinite electrical ports, each port being associated with a mode of the corresponding feeding line. In practical computations, of course, only a finite number of ports is considered. The coefficient $g_{\nu i}^{(m)}$ represents the coupling between the guided mode of m-th order at the i-th port with the resonant mode of ν-th order.

For simplicity of notation we have used the only ν-index to label the modes. Actually, the resonant modes depend on two spatial indexes along two directions, so that the series in (4) is a double series. As shown in [3], one of the two series can be evaluated analytically (see also [11]) so as to accelerate numerical computations.

The present design method requires that the planar cells can be modelled as two-port networks. This implies that higher order modes on the physical ports must decay rapidly enough along the feeding lines, so as not to interact with other discontinuities. Electrical ports corresponding to higher order modes are therefore matched. By simple manipulations, the generalized impedance matrix resulting from the planar analysis of each cell is converted into a 2 × 2 impedance matrix involving only the dominant modes ($m = 0$) on the ports. Such a procedure requires the inversion of a matrix with size equal to the number of electrical ports (higher modes) being matched. Inside each cell, on the contrary, the EM field is expanded in terms of two-dimensional eigenfunctions E_ν. Finally, the image impedances Z_{I1}, Z_{I2} and the image propagation factor γ of the cell are computed from the impedance matrix using known formulas [7]:

$$Z_{I1} = \sqrt{\frac{Z_{11}}{Z_{22}}\Delta}\,; \quad Z_{I2} = \sqrt{\frac{Z_{22}}{Z_{11}}\Delta}$$

$$\cosh \gamma = \frac{\sqrt{Z_{11} Z_{22}}}{Z_{21}}$$

where

$$\Delta = \det[Z] = Z_{11}Z_{22} - Z_{21}^2$$

Both N- and T-cells exhibit a lowpass behaviour. The image impedances are real and, correspondingly, the image propagation factor is imaginary up to a first cutoff frequency f_c. For a given substrate (dielectric constant ϵ_r, thickness h), this cutoff frequency depends on the geometrical parameters of the cell. Above f_c a stopband is present, where the image impedances are imaginary and the image propagation factor is real (attenuation). A transmission zero is located in the stopband due to the presence of one stub in the elementary cell. The stopband extends up to a second cutoff frequency f_{c2}. It must be noted that the location of f_{c2} is not related to f_c in an elementary way. This is because of the non-commensurate two-dimensional character of the cell [5].

Because of symmetry, the T-cell has the same image impedance at both ports. The N-cell, which is used at the ends of the filter, has two different image impedances as seen from the external line (Z_{IS}, Fig. 2) or from the adjacent T-cell (Z_{II}). The latter image impedance has the same behaviour as that of the T-cell, as shown in Fig. 3a, b where the typical behaviours of the image impedances of the N- and T-cells respectively are plotted versus the frequency up to the (first) cutoff frequency f_c. In the limit of zero frequency the image impedances tend to a finite value K, while, at f_c, Z_{IS} tends toward infinity and Z_{II} becomes zero.

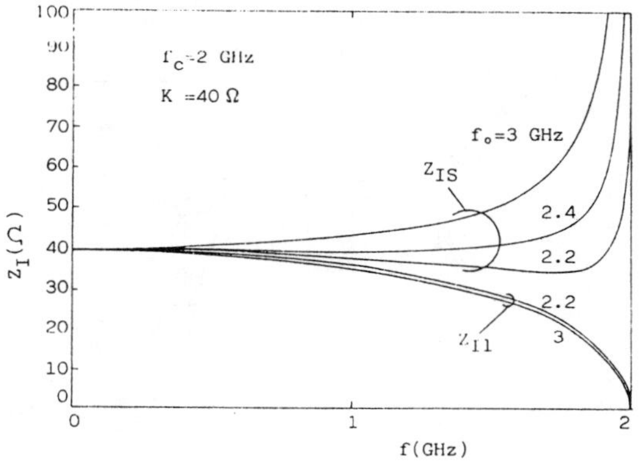

a)

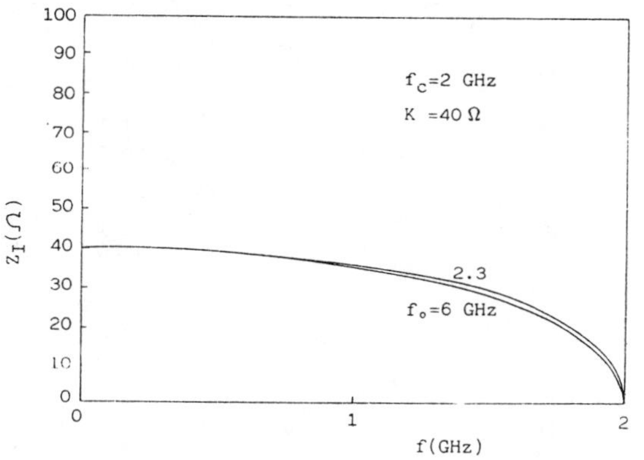

b)

Fig. 3a, b - Image impedances of elementary N- and T-cells as function of frequency and for different values of the transmission zero frequency f_o. The cells have $f_c = 2$ GHz, $K = 40\Omega$, feeding lines with 50Ω characteristic impedance for the N-cell and 90Ω for the T-cell. Substrate thickness $h = 0.635$ mm, dielectric constant $\epsilon_r = 10$.

Curves of Fig. 3 have been plotted for typical values of the electrical parameters of the cells, i.e. $f_c = 2$ GHz, $K = 40\Omega$ and for different values of the transmission zero frequency f_o. The feeding line characteristic impedances are 50Ω for the N-cell and 90Ω for the T-cell. Substrate thickness $h = 0.635$ mm, dielectric constant $\epsilon_r = 10$.

The image impedance behaviours of the N-cell demonstrate that such a cell is useful to provide a good matching between the external terminating resistance and the T-cells. The image impedance on the external side, Z_{IS}, in fact, remains fairly constant with frequency, except near f_c. The N-cell is in fact the planar analog of the m-derived cell for constant-k lumped element filters. As for constant-k with m-derived end cells, planar lowpass filters could successfully be designed using only T-cells, but the use of N-cells at the end of the filter gives rise to a substantial improvement both of the passband response and of the cutoff rate.

4. FILTER DESIGN PROCEDURE

To assure the proper behaviour of the filter in the passband, cascaded elementary cells must be image matched. This requires first that both f_c and K are same for all the cells. In addition, the image impedances of contiguous cells should be as close as possible.

Exactly equal image impedances of different cells are only possible for lumped filters or for their commensurate distributed counterparts. In the classical application of the image method, a perfect image matching is obtained at each internal connection of the filter, while the filter itself is approximately matched to the feeding lines. In the non-commensurate case also the matching condition at the connections between elementary cells is satisfied to a certain approximation [5]. Like in the classical application of the image method, some degradation of the filter performance is expected near the band edges because of the different behaviours near f_c (Fig. 3). This phenomenon, however, is controlled during the design procedure, which is performed interactively on a PC. Possible degradations are detected by observing the computer simulated response of the filter being designed and eventually avoided by a different choice of the design parameters.

The image matching condition guarantees the passband behaviour. To satisfy the stopband requirements of the filter, the elementary cells must be designed in such a way that they produce a proper set of transmission zeros.

Both N- and T-cells are identified by 4 geometrical quantities so that, once f_c and K have been chosen for the filter, only two parameters are still to be given for each planar cell. One additional constraint that arises from the geometrical structure of the filter is that each planar cell has the same line width w (see Fig. 2).

The design starts with the synthesis of the first N-cell, given its transmission zero f_{01} and the input port width w_i. The synthesis of the first N-cell determines w, so that only one quantity is still to be given for each of the remaining m-2 T-cells. These are then synthesized choosing their transmission zeros $f_{0,2} \cdot\cdot f_{0,m-1}$. Finally, the last N-cell is equal to the first one (thus has the same transmission zero f_{01}), as w_i and w must be the same. In conclusion, once f_c and K have been chosen, the m-th order filter is synthesized from a given set of m-1 transmission zeros.

The design is performed interactively on a personal computer and requires a few minutes on a PC AT IBM computer. The procedure is, in principle, similar to that for transmission line filters [5] except that it operates on planar models, thus requires a planar synthesis procedure, which is illustrated in the next section.

5. SYNTHESIS OF PLANAR CELLS

Once the substrate characteristics (thickness h, and permittivity ϵ_r) are given, each planar cell is identified by 4 geometrical quantities. One of these quantities (the port width w_i of N-cells or w of the T-cells) is already determined by the design procedure, so that only three quantities are left to the synthesis, which therefore consists of computing the 3 unknown geometrical parameters of the cell starting from the 3 given electrical parameters f_c (image cutoff frequency), K (impedance level) and f_o (transmission zero).

The synthesis of each cell must be performed on the basis of an iterative application of the analysis routine for computing f_c, K and f_o. As illustrated in section 3, the analysis requires the numerical solution of transcendental equations. It is evident that the synthesis is much more laborious, as it requires a large number of repeated analyses.

It is important to keep the computer time within reasonable limits on a PC. To this end, closed-form empirical formulas can be developed for the approximate synthesis of the cells. The rigorous procedure is applied only for a final adjusting of the cell dimensions.

For the N-cell, closed-form synthesis formulas can be developed easily by modifying the synthesis formulas valid for the transmission line model [5] with the introduction of suitable effective parameters. Planar effects in N-cells, in fact, are generally of moderate importance, due to their high characteristic impedance. T-cells, on the contrary, have very low impedance stubs, so that planar effects produce substantial variations with respect to the one-dimensional transmission line model.

As an example, let us recall the analysis formulas for the N-cell in the one-dimensional case. By simple modifications of the formulas given in [5] one obtains

$$f_c = \frac{c}{4L}$$

$$f_o = \frac{c}{4l_s}$$

$$K = Z_{ol}\sqrt{l_l/L}$$

where $c = 1/\sqrt{\mu\,\epsilon_{eff}}$ is the phase velocity, l_s is the stub length, l_l is the line length, $L = l_s + l_l$ is the total length of the N-cell, Z_{ol} is the characteristic impedance of the cell. For the planar N-cell the above formulas are modified by introducing equivalent geometrical quantities

$$f_c = \frac{c}{4L_{eq}}$$

$$f_o = \frac{c}{4l_{s,eq}}$$

$$K = Z_{ol}\sqrt{l_{l,eq}/L}$$

It is worth stressing that the *equivalent* dimensions introduced above account for planar effects associated with the transverse variation of the electromagnetic field, i.e. with the excitation of higher order modes at the discontinuities. The *effective* dimensions, on the contrary, account for fringing field phenomena at the metallization edges.

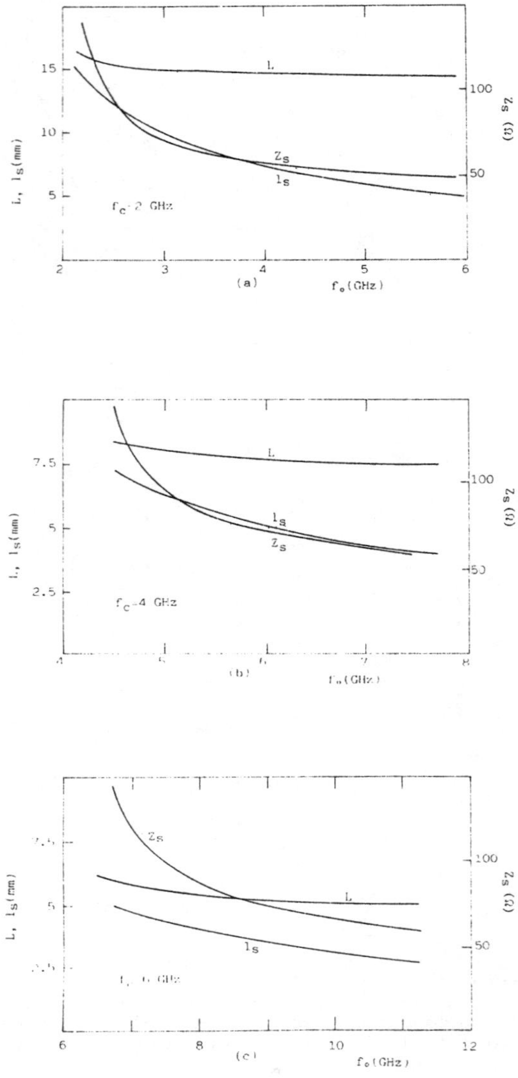

Fig. 4a, b, c - N-cell synthesis curves for three different cutoff frequencies: $f_c = 2$ GHz, 4 GHz and 6 GHz, respectively. Alumina substrate $\epsilon_r = 10$, thickness = 0.635 mm.

Good agreement with the rigorous synthesis procedure was obtained for microstrips on alumina substrate h = 0.635 mm thick using the following empirical values:

$$l_{l,eq} = l_l + \alpha_1 w_i$$

$$l_{s,eq} = l_s - \alpha_2 w_i$$

$$L_{eq} = L - 1.5\alpha_2 w_i$$

with

$$\alpha_1 = 0.166\,(1 - Z_{ol}/Z_o)$$

$$\alpha_2 = 0.022 + 0.054\,Z_{ol}/Z_o$$

Z_o being the reference impedance. This quantity corresponds to the feeding line width w_i.

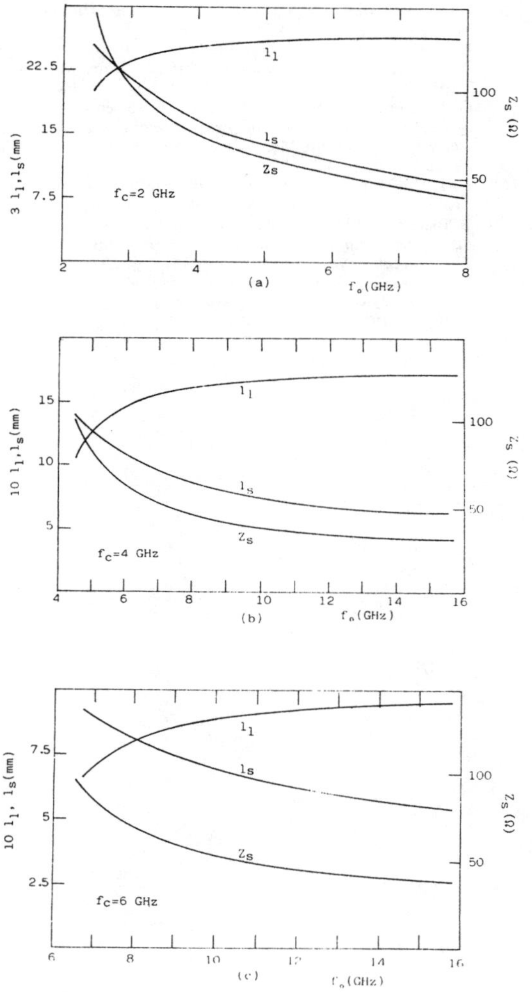

Fig. 5a, b, c – T-cell synthesis curves for three different cutoff frequencies: $f_c = 2$ GHz, 4 GHz and 6 GHz, respectively. Alumina substrate $\epsilon_r = 10$, thickness = 0.635 mm.

The above formulas can be inverted in a straightforward way to compute the geometrical parameters l_s, l_l, from the filter parameters f_c, K and f_o. The accuracy of the above formulas is found to be within a few percent.

Typical synthesis curves are given in Fig. 4 and 5 for N- and T-cells respectively. They refer to microstrip structures on alumina substrate with relative permittivity $\epsilon_r = 10$, 0.635 mm thick. Two geometrical parameters (total length L and stub length l_s for the N-cell, line length l_l and double stub length l_s, for the T-cell) and the stub impedance Z_s are shown versus the transmission zero frequency f_o of the cell for three different cutoff frequencies, $f_c = 2$, 4 and 6 GHz. A typical value of $K = 40\,\Omega$ has been chosen. This value guarantees a good passband behaviour up to near f_c. The forth geometrical parameter of each cell, namely the input port width, is 0.600 mm (corresponding to the 50 Ω line) for the N-cell. For the T-cell this quantity has been chosen as that corresponding to the typical value of 90 Ω. Since N-cells are used to provide a transmission zero near the cutoff frequency, while T-cell are used to widen the stop band, different f_o ranges have been considered.

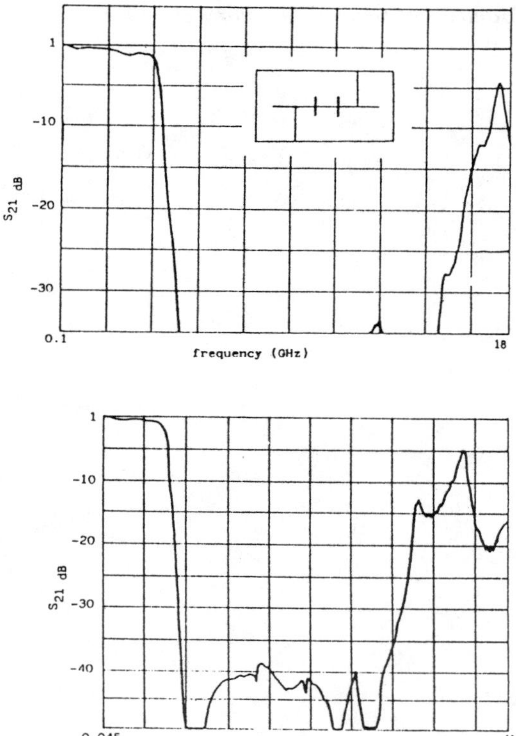

Fig. 6 Experimental responses of two microstrip lowpass filters, designed with $f_c = 4$ and 2 GHz respectively.

RESULTS

Some microstrip filters have been designed according to the present technique, and fabricated on alumina substrate ($\epsilon_r = 10$) $h = 0.635$ mm thick. Fig. 4a, b–shows the experimental responses of two of these filters. Both filters consist of 4 cells (two N– and two T–cells). They have been designed with cutoff frequencies of $f_c = 4$ GHz and 2 GHz, respectively. Very wide stopbands (about 2 octaves) with minimum attenuations of about 35 and 40 dB are demonstrated. These results, which are comparable with those obtained in [12] using radial line stubs, demonstrate the feasibility of the present method.

CONCLUSIONS

The planar circuit approach for characterizing microstrip circuits has been combined with the image parameter method to give a new powerful technique for designing directly microstrip filters with noncommensurate geometries. Microstrip filters with extremely wide stopbands and very sharp cutoff rates have been designed and fabricated.

Manuscript received January 26, 1988.

REFERENCES

[1] T.C. Edwards: *Foundations for Microstrip Circuit Design.* John Wiley, 1981.
[2] J.P. Vinding: *Radial line stubs as elements in strip line circuits.* NEREM Record, 1967, p. 108–109.
[3] R. Sorrentino: *Planar circuits, waveguide models, and segmentation method.* "IEEE Trans. Microwave Theory Tech.", vol. MTT-33, Oct. 1985, p. 1057-1066.
[4] K.C. Gupta et al.: *Computer-Aided Design of Microwave Circuits.* "Artech House", Dedham, MA, 1981.
[5] M. Salerno, R. Sorrentino, F. Giannini: *Image parameter design of noncommensurate distributed structures – An application to microstrip low-pass filters.* "IEEE Trans. Microwave Theory Tech.", vol. MTT-34, Jan. 1986, p. 58–65.
[6] H. Howe: *Stripline Circuit Design.* "Artech House", 1974, Ch. 6.
[7] G.L. Matthaei, L. Young, E.M.T. Jones: *Microwave Filters, Impedance-Matching Networks and Coupling Structures.* McGraw-Hill, New York, 1964.
[8] I. Wolff, N. Knoppik: *Rectangular and circular microstrip disk capacitors and resonators.* "IEEE Trans Microwave Theory Tech.", vol. MTT-22, Oct. 1974, p. 857–864.
[9] G. D'Inzeo, F. Giannini, C.M. Sodi, R. Sorrentino: *Method of analysis and filtering properties of microwave planar networks.* "IEEE Trans. Microwave Theory Tech.", vol. MTT-26, July 1978, p. 462–471.
[10] R.H. Hoffman: *Handbook of Microwave Integrated Circuits.* "Artech House", Dedham, MA, 1987, ch. 3 and 10.
[11] B. Bianco, S. Ridella: *Nonconventional transmission zeros in distributed rectangular structures.* "IEEE Trans. Microwave Theory Tech.", vol. MTT-20, May 1972, p. 297-303.
[12] F. Giannini, M. Salerno, R. Sorrentino: *Two-octave stopband microstrip low-pass filter design using butterfly stubs.* "Proc. 16th Eu. M.C.", 1986, p. 292-297.

Planar Circuit Analysis of Microstrip Radial Stub

FRANCO GIANNINI, MEMBER, IEEE, ROBERT SORRENTINO, MEMBER, IEEE, AND JAN VRBA

Abstract **—Radial-line stubs have been found to work better than low-impedance rectangular stubs when an accurate localization of a zero-point impedance is needed. In this paper, microstrip radial-line stubs are analyzed using a planar circuit technique and characterized for design purposes. Experiments performed on various structures are in excellent agreement with the theory and confirm the suitability of such a structure as an alternative to a conventional straight stub.**

I. Introduction

A NUMBER of microstrip circuits, such as low-pass filters, bias filter elements, mixers, etc., often require the use of shunt stubs with characteristic impedance as low as 10–20 Ω. At high frequencies, however, a microstrip line section with such a low characteristic impedance has a width which is a significant fraction of the wavelength; higher order modes can be easily excited and the structure behavior differs significantly from that predicted on the basis of a one-dimensional transmission-line model. Moreover, the large width of the stub renders its location poorly defined. In order to overcome these problems, the use of radial-line stubs has been suggested [1].

In particular, Vinding [1] has proposed a formula for evaluating the input reactance of microstrip radial stubs. Based on Vinding's formula, Atwater [2] has recently developed a design procedure of radial stubs. Like other formulas for microstrip planar circuits which are based on a mere magnetic wall model, however, Vinding's formula does not yield sufficiently accurate results.

A planar circuit analysis of a microstrip radial stub is presented in this paper. The method is based on the electromagnetic (EM) field expansion in terms of resonant modes of the planar structure [3] and has been successfully applied to other similar problems [4].

II. Electromagnetic Model

The EM-field expansion in terms of resonant modes in a planar circuit has already been applied to rectangular, circular [3], and annular structures [4].

As shown in [3], accurate results can be obtained through a magnetic wall model, provided effective dimensions and effective permittivities properly defined for each resonant

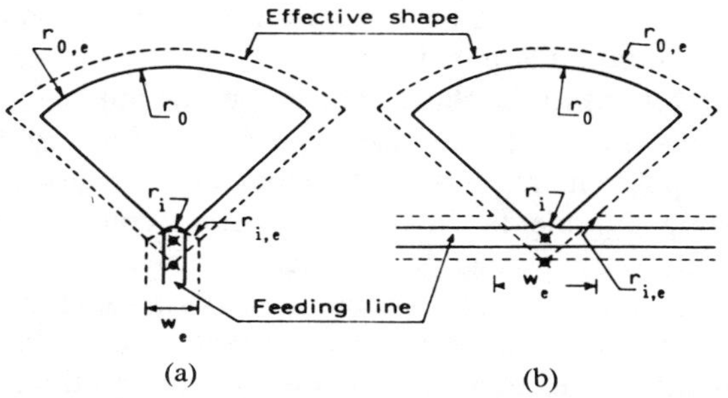

Fig. 1. Geometries of (a) series and (b) shunt connected radial stubs.

mode are used. Moreover, it is worth specifying that different effective parameters should be used depending on the type of connection of the stub to the external circuit. For instance, as sketched in Fig. 1, account has also to be taken of the effective width of the feeding line, so that different effective geometries result when the stub terminates a line section (Fig. 1(a)) or is shunt connected (Fig. 1(b)). In any case, assuming that the effective width of the input port of the planar stub is much smaller than the wavelength, only TM_{on} mode ($n = 0,1,2,\cdots$) resonant modes will be excited. Therefore, the effective dimensions and effective permittivities can be easily obtained following the same procedure as in [5] and [6], and on the basis of the following assumptions.

a) The effective angle is equal to the actual angle α.

b) The end effect at the external edge of the structure is the same as in a circular structure.

c) The effective width w_e of the input port is the same as that of a uniform microstrip line of width w. It can be computed using well-known formulas [7].

One obtains

$$r_{ie} = \frac{w_e}{2\sin(\alpha/2)} \tag{1}$$

$$r_{oe} = r_o\left\{1 + \frac{2h}{\pi r_o}\left[\ln\left(\frac{\pi r_o}{2h}\right) + 1.7726\right]\right\}^{1/2} + \frac{w_e - w}{2}\begin{cases} \dfrac{1}{\sin(\alpha/2)}, & \text{for } \alpha < \pi \\ 1, & \text{for } \pi < \alpha < 3\pi/2 \end{cases} \tag{2}$$

where h is the dielectric substrate thickness, and w and w_e are the actual and effective widths of the feeding line. The other geometrical parameters are defined in Fig. 1, where the actual geometry together with the effective dimensions

Manuscript received May 8, 1984.

F. Giannini is with the University of Rome, Tor Vergata, Department of Electronic Engineering, 00173 Rome, Italy.

R. Sorrentino is with the University of Rome, La Sapienza, Department of Electronics, 00184 Rome, Italy.

J. Vrba is with the Czech Technical University, 16637 Praha 6, Czechoslovakia.

Reprinted from *IEEE Trans. Microwave Theory Tech.*, vol. MTT-32, no. 12, pp. 1652–1655, Dec. 1984.

of the radial stub are shown. For $\alpha \to 0$, the effective geometry of the radial stub approaches that of a usual straight stub.

It is worth specifying that the present model, which assumes the excitation of TM_{on} modes only, has an upper α limit of applicability, which depends on the type of connection of the stub. In the case of the end termination, the model is not valid when α is such that the radii of the sector are very close to the feeding line; because of the proximity of the radial stub sides to the feeding line, modes other than TM_{on} can be excited and a more complicated model should be adopted.

With regard to the shunt connection, the applicability of the model is restricted to lower values of α. This is not only due to physical reasons ($\alpha < \pi$) but also because of the following arguments. First, in the shunt connection, the EM-field propagating on the feeding line is not truly uniform along the width of the input port; secondly, for a given radial stub, the effective width of the port is larger in the shunt connection than in the end connection, as illustrated in Fig. 1.

On the other hand, the use of a very narrow input port can extend the range of applicability of the present simplified theory in the case of the shunt connection.

With regard to the dynamic effective permittivity of the TM_{on} modes, the expressions quoted in [5] have been used.

The normalized input impedance evaluated at the inner radius r_i in terms of TM_{on} resonant modes is given by

$$Z_{\text{in}} = -j\frac{k_g P_{oo}^2}{k^2 \epsilon_{do}} + jk_g \sum_{n=1}^{\infty} \frac{P_{on}^2}{k_{on}^2 - k^2 \epsilon_{dn}} \qquad (3)$$

where k_g is the wavenumber of the feeding line, $k = \omega\sqrt{\epsilon_0 \mu_0}$ is the free-space wavenumber, and ϵ_{dn} and k_{on} are the dynamic effective permittivity [5] and the eigenvalue of the TM_{on} mode [3], [4], respectively. P_{on} is the coupling coefficient between the quasi-TEM mode traveling on the feeding line and the TM_{on} mode excited in the stub

$$P_{on} = \sqrt{\frac{w_e}{\pi}}\,[A_{on} J_0(k_{on} r_{ie}) + B_{on} N_0(k_{on} r_{ie})] \qquad (4)$$

$$A_{on} = \frac{2}{\alpha\pi}\left\{ r_{oe}^2 [J_0(k_{on} r_{oe}) + K_n N_0(k_{on} r_{oe})]^2 - r_{ie}^2 [J_0(k_{on} r_{ie}) + K_n N_0(k_{on} r_{ie})]^2 \right\}^{-1} \qquad (5)$$

$$B_{on} = K_n A_{on};\; K_n = -J_1(k_{on} r_{oe})/N_1(k_{on} r_{oe}) \qquad (6)$$

where J_0 and J_1 are the Bessel functions of the first kind of order 0 and 1; N_0 and N_1 are the Bessel functions of the second kind (Neuman functions) of order 0 and 1.

III. Results

The present theory has been tested on the experiments by Atwater [3], performed on two radial stubs fabricated on a 25-mil alumina substrate, both having an outer radius $r_o = 5.49$ mm [8]. Fig. 2(a) and (b) shows the comparison between the experiments by Atwater and the reactances calculated according to the present theory and by Vinding's formula. Only the first four modes have been used in the expansion. In spite of the low number of modes, the present theory fits the experiments much better than Vinding's formula.

Further experiments were performed on radial stubs of 90° and 120° fabricated on the same substrate as in Atwater's experiments. As shown in Fig. 3(a) and (b), a

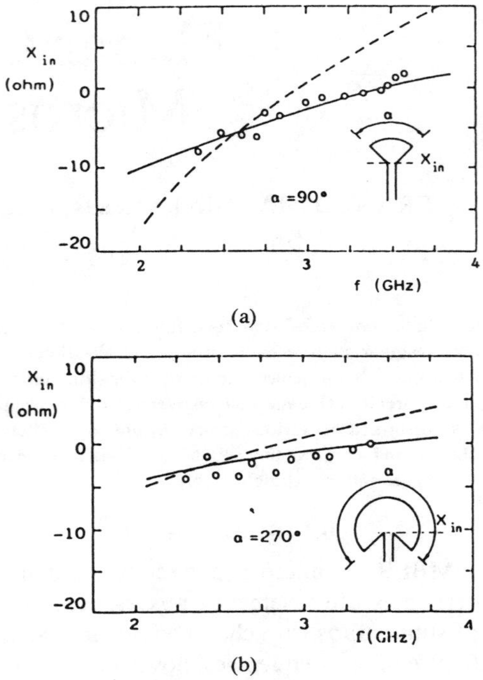

Fig. 2. Input reactance of radial stubs on 0.025-in alumina with (a) sectoral angle $\alpha = 90°$ and (b) sectoral angle $\alpha = 270°$. (——) Present theory, (----) Vinding's formula, (O) experiments after Atwater [2].

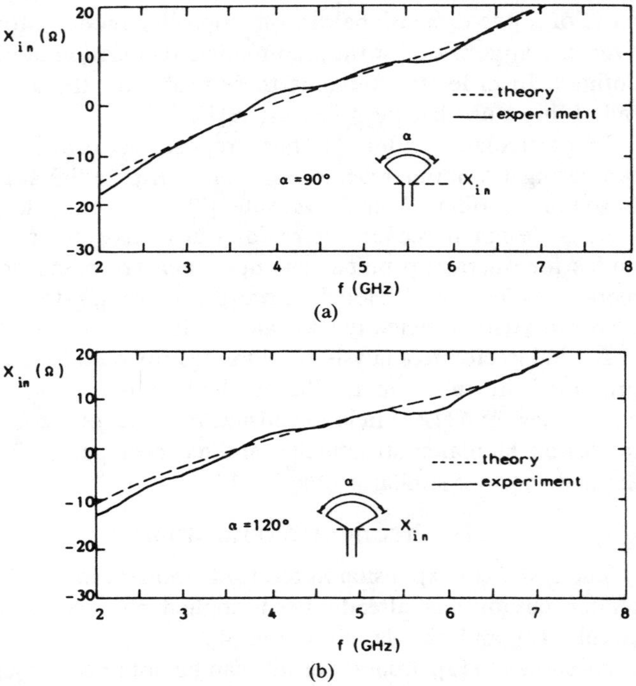

Fig. 3. Frequency behavior of the input reactance of radial stubs with $r_o - r_i = 0.47$ cm; (a) sectoral angle $\alpha = 90°$ and (b) sectoral angle $\alpha = 120°$. Measurements performed on a HP vector analyzer.

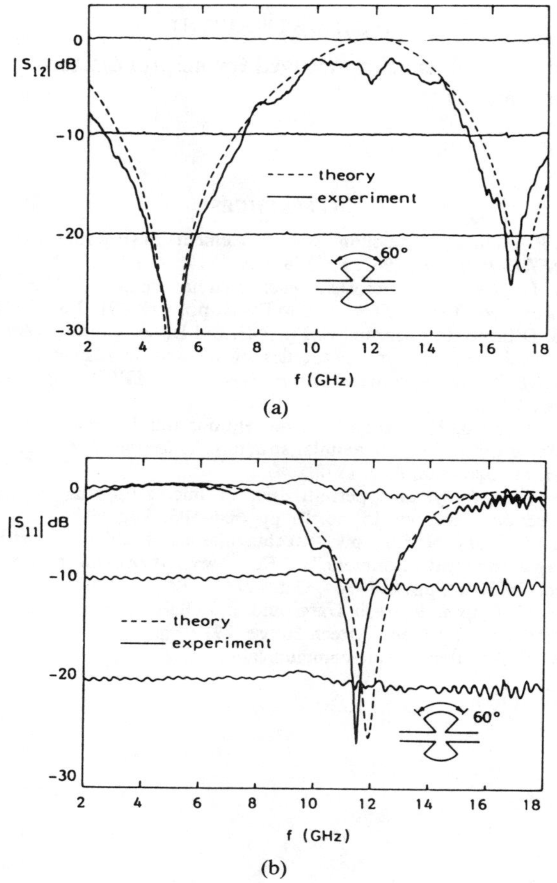

Fig. 4. Frequency behavior of scattering parameters (a) $|S_{12}|$ and (b) $|S_{11}|$ for a "butterfly" stub realized on a 0.0635-cm-thick alumina substrate; $r_o = 0.5$ cm.

very good agreement has been obtained between the theory and the measurements performed on an HP vector network analyzer.

Experiments on shunt stubs were also performed. This is illustrated in Fig. 4, where the scattering parameters $|S_{12}|$, $|S_{11}|$ of a "butterfly" stub, consisting of two 60° shunt stubs with $r_o = 0.5$ cm, are shown in the frequency range 2–18 GHz. The agreement between theory and experiments is highly satisfactory in the whole frequency range.

For design purposes, the radial stub can be usefully characterized in terms of the frequency of zero input reactance f_0 and its equivalent characteristic impedance $Z_0 = (2f_0/\pi)\,dX/df$. The latter quantity is defined as the characteristic impedance of a conventional stub having the same slope parameter at $f = f_0$. Fig. 5(a) and (b) shows the computed behavior of f_0 and Z_0 versus α for microstrip radial stubs with different $r_o - r_i$, series-connected to a 50-Ω feeding line. Experiments performed on radial stubs with $r_o - r_i = 0.47$ cm are in excellent agreement with the theory. Equivalent characteristic impedances as low as 10–15 Ω can be easily obtained with angles α of the order of 120° and $r_o - r_i$ of about 0.5 cm. This figure demonstrates the suitability of a radial-line stub as an alternative to conventional stubs when very low characteristic impedances are required, as suggested by Vinding.

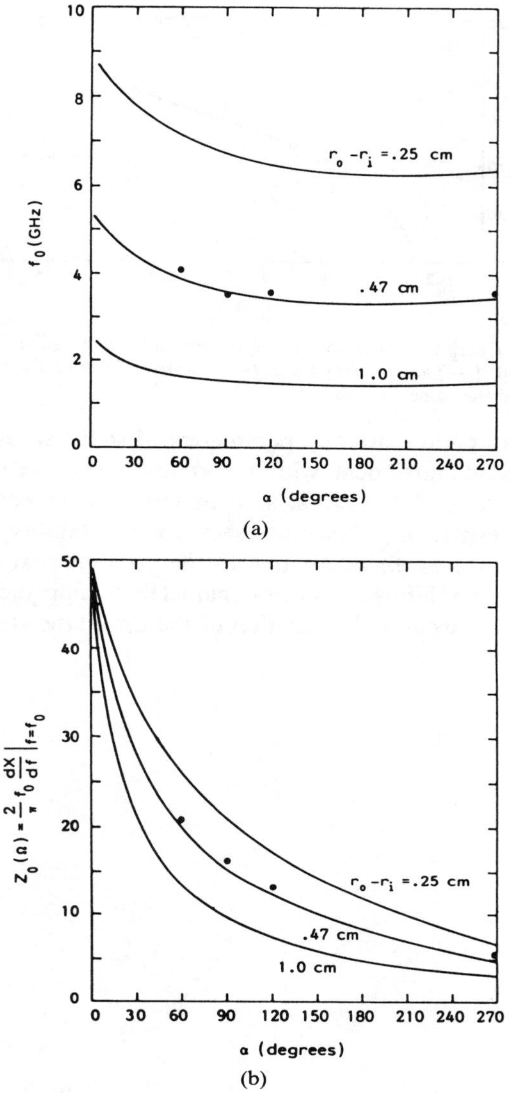

Fig. 5. (a) Frequency of the zero input impedance and (b) equivalent characteristic impedance versus sectoral angle α, for series-connected radial stub.

Finally, the theoretical results of Fig. 6 show that the radial stub can maintain a low-impedance value in a wider frequency range than a conventional straight stub. The input impedance of a 120° radial stub and of a conventional stub with the same slope parameter are plotted versus the normalized frequency f/f_0. The radial stub has an input impedance lower than 20 Ω up to 2 f_0; at this frequency, the conventional stub has an infinite input impedance. Within the approximation used in this short paper, the frequency of infinite reactance of the radial stub is only slightly dependent on the sectoral angle α and close to that of a straight stub $r_o - r_i$ long. Increasing the sectoral angle α, the zero reactance frequency f_0 decreases, thus producing wider low-impedance frequency bands.

II. Conclusions

A planar circuit approach based on the field expansion in terms of resonant modes has been used to characterize a radial-line stub. Measurements of the input impedance as

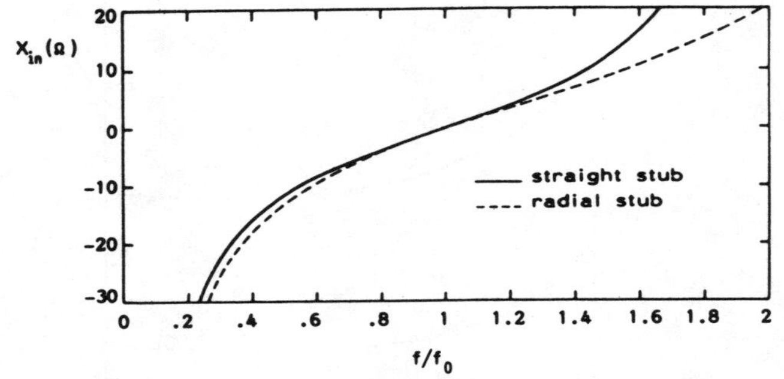

Fig. 6. Computed behaviors of input reactance of a radial stub with $\alpha = 120°$ (----) and a straight stub (——) having the same slope parameter and the same frequency f_0.

well as of the scattering parameters of shunt stubs are in very good agreement with theoretical results. The characterization of the radial stub in terms of an equivalent characteristic impedance has shown the suitability of such a structure as an alternative to the conventional straight stub, as it exhibits a very low characteristic impedance and allows an accurate localization of the impedance reference plane.

Acknowledgment

Prof. T. Itoh is acknowledged for helpful discussions and suggestions.

References

[1] J. P. Vinding, "Radial line stubs as elements in strip line circuits," in *NEREM Record*, 1967, pp. 108–109.

[2] A. H. Atwater, "Microstrip reactive circuit elements," *IEEE Trans. Microwave Theory Tech.*, vol. MTT-31, pp. 488–491, June 1983.

[3] G. D'Inzeo, F. Giannini, C. M. Sodi, and R. Sorrentino, "Method of analysis and filtering properties of microwave planar networks," *IEEE Trans. Microwave Theory Tech.*, vol. MTT-26, pp. 462–471, July 1978.

[4] G. D'Inzeo, F. Giannini, R. Sorrentino, and J. Vrba, "Microwave planar networks: The annular structure," *Electron. Lett.*, vol. 14, no. 16, pp. 526–528, Aug. 1978.

[5] J. Vrba, "Dynamic permittivities of microstrip ring resonators," *Electron. Lett.*, vol. 15, no. 16, pp. 504–505, Aug. 1979.

[6] I. Wolff and N. Knoppik, "Rectangular and circular microstrip disk capacitors and resonators," *IEEE Trans. Microwave Theory Tech.*, vol. MTT-22, pp. 857–864, Oct. 1974.

[7] K. C. Gupta, Ramesh Garg, and I. J. Bahl, *Microstrip Lines and Slotlines*. Dedham: Artech House, 1979, chap. 2.

[8] A. H. Atwater, private communication.

Paper 9.5

CAD-Oriented Lossy Models for Radial Stubs

FRANCO GIANNINI, SENIOR MEMBER, IEEE, CLAUDIO PAOLONI, MEMBER, IEEE, AND MARINA RUGGIERI, MEMBER, IEEE

***Abstract*—A lumped equivalent circuit model for both series and double-shunt (butterfly) connected radial stub has been developed. The model—simple and effective—not only includes conductor and dielectric losses but also radiation ones, which play an important role in microstrip circuit elements. Experiments widely demonstrate its suitability for implementation in available CAD programs. Furthermore, a synthesis procedure for using radial stubs in circuit design is described. An application of the above design procedure and simulation tools in the development of very broad-band nongrounded terminations is also presented.**

I. Introduction

IN MICROWAVE CIRCUITS an alternate approach for the design of biasing, filtering, matching, and virtual grounding structures by resonant straight stubs is the use of radial lines. Broad-band behavior and well-defined low impedance levels are required in both hybrid and monolithic microstrip circuits. Use of radial stubs is a solution for both of these requirements. This structure overcomes the limitation of low-impedance points, where line widths are a significant fraction of a wavelength, particularly at high frequencies. Furthermore, with respect to straight stubs, radial lines have about the same impedance level at frequencies lower or equal to the so-called resonance frequency, while maintaining the lower impedance level over a wider range [1]. Radial stubs are also space-saving and usually present a smaller outer dimension with respect to quarter-wave straight stubs at the same frequency.

Successful circuits adopting these structures have already been demonstrated [2]–[5]. Using a formula which provides the radial line input impedance [6], quite good results have been achieved at low frequencies with low-dielectric-constant substrates [7]. Characterization of radial stubs using effective dimensions and dielectric permittivities instead of actual quantities has extended to higher frequencies and substrates with higher ϵ_r techniques previously developed [8].

A complete characterization of the radial behavior is achieved with the latter method in terms of not only the conventional resonance frequency, but also the equivalent characteristic impedance presented by the radial structure. Furthermore, an accurate characterization of the inner radius has been derived, pointing out the insertion-dependent behavior of the radial stub [1], [8].

Manuscript received April 16, 1987; revised September 28, 1987.
The authors are with the Department of Electronic Engineering, Università di Roma "Tor Vergata," Rome, Italy.
IEEE Log Number 8718677.

Nevertheless, the key point for a more accurate and predictable use of radial lines in microwave circuits is the availability of a complete theoretical model which is easily adapted to available CAD programs, leading to fully automated design procedure. An equivalent lossless lumped circuit model of the radial stub has been previously developed in the case of double-shunt connection [9]. The effectiveness of this model, which is easy to implement in CAD programs, has now been extended to the case of series connection. The new model includes losses for both double-shunt and series insertion.

An extended CAD-oriented circuit is presented together with the supporting theory and an example illustrating its implementation on the Touchstone™ program. Furthermore a synthesis procedure which quickly determines the geometry of a radial stub corresponding to given requirements of a conventional straight one is described. Application of the above design procedure and simulation tools in the development of a very broad-band nongrounded termination is also shown.

II. Theoretical Characterization

The input impedance of a lossless microstrip radial stub has been previously derived by an electromagnetic (e.m.) field expansion in terms of resonant modes, assuming that only TM_{0n} modes be excited [1]. The extention of this theory to the case of lossy microstrip radial stub—in both series (Fig. 1(a)) and double-shunt connection (Fig. 1(b))—is the basis of this characterization, leading to a more accurate description of the radial line behavior.

A previous approach [10] dealt with the lossy nature of this structure, taking into account only losses due to the conductor finite conductivity and, thus, resulting in a less accurate model. In fact, the main attenuation factor is usually related to the radiation of the e.m. energy, due to the open nature of microstrip components.

The different kinds of losses can be taken into account by introducing appropriate quality factors [11] in the above-mentioned formula of the radial input impedance. Therefore, the complex impedance of a lossy radial stub can be expressed as

$$Z_{\text{in}} = \frac{1/Q_{t0} - jK_g P_{00}^2}{K^2 \epsilon_{d,0}} + j \sum_{n=1}^{\infty} \frac{K_g P_{0n}^2}{K_{0n}^2 (1 + j/Q_{t,0n}) - K^2 \epsilon_{d,n}} \tag{1}$$

Reprinted from *IEEE Trans. Microwave Theory Tech.*, vol. 36, no. 2, pp. 305–313, Feb. 1988.

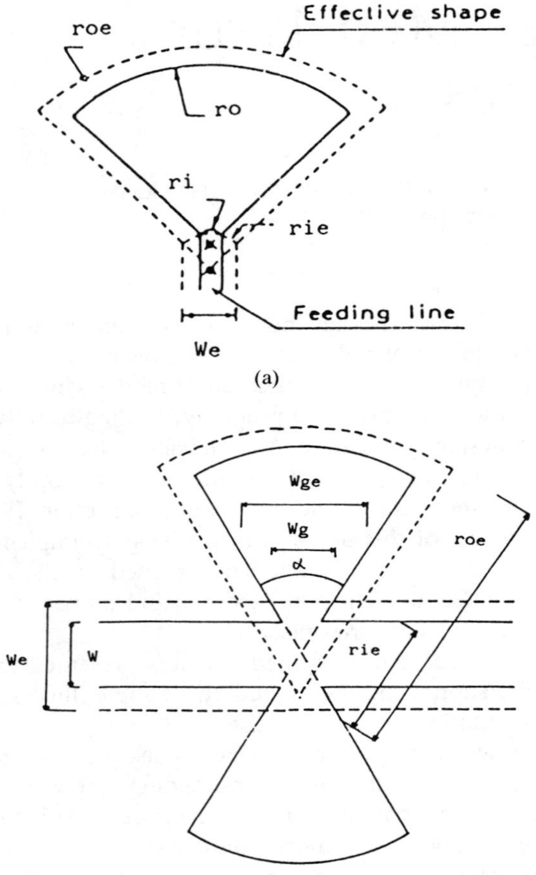

Fig. 1. (a) Series-connected radial stub. (b) Shunt-connected (butterfly) radial stub.

where

$$Q_{t0} = 1/\tan\delta$$

and

K_g	wavenumber of the feeding line,
$K = \omega\sqrt{\mu_0\epsilon_0}$	free-space wavenumber,
$\epsilon_{d,n}$	dynamic effective permettivity of the TM_{0n} mode [12],
K_{0n}	eigenvalue of the TM_{0n} mode (see Appendix),
P_{0n}	coupling coefficient between the quasi-TEM mode traveling on the feeding line and the TM_{0n} mode excited in the stub (see Appendix).

$Q_{t,0n}$ represents the global quality factor and its expression is as follows:

$$Q_{t,0n} = (1/Q_{d,0n} + 1/Q_{c,0n} + 1/Q_r,0n)^{-1} \tag{2}$$

where

$Q_{d,0n}$	quality factor related to losses in the dielectric,
$Q_{c,0n}$	quality factor related to losses in the conductor,
$Q_{r,0n}$	quality factor related to radiation losses.

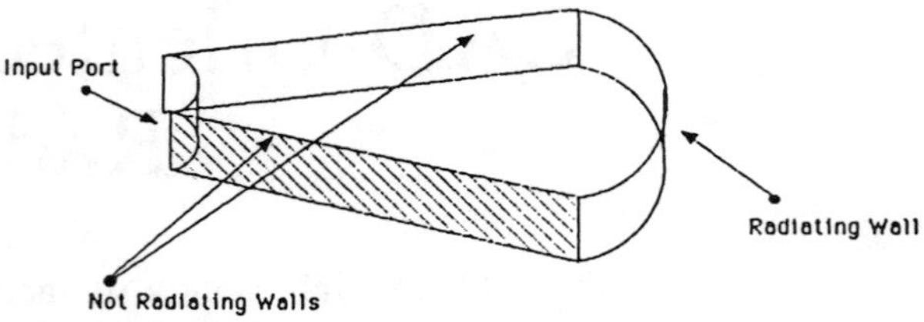

Fig. 2. Sectoral radiating structure.

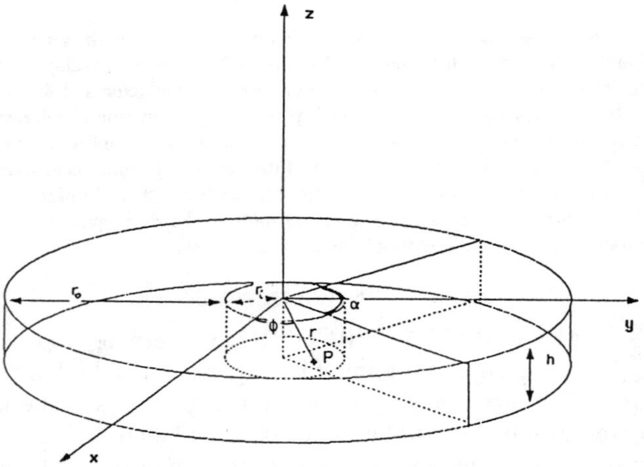

Fig. 3. Annular radiating structure.

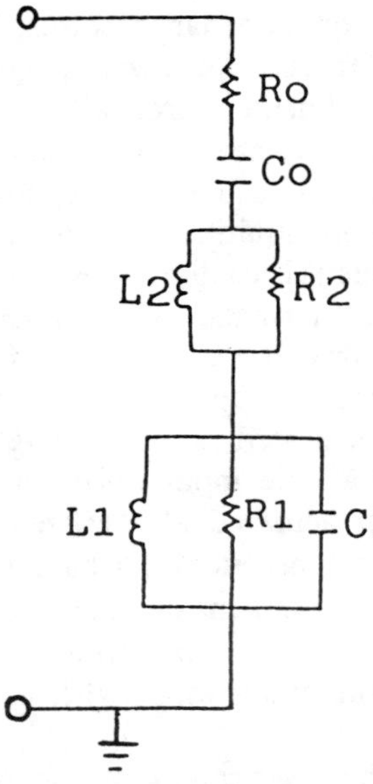

Fig. 4. Lossy lumped equivalent circuit.

The above characterization thus takes into account the three main factors of losses in the radial structure. The Q factors related to ohmic and dielectric losses are the same as those of a circular sector bounded within an angle α. They can thus be easily derived [13] for the generic TM_{0n} mode by means of the following expressions:

$$Q_{c,0n} = \left(\frac{2R_s}{\omega_{0n}\mu_0 h}\right)^{-1} \tag{3}$$

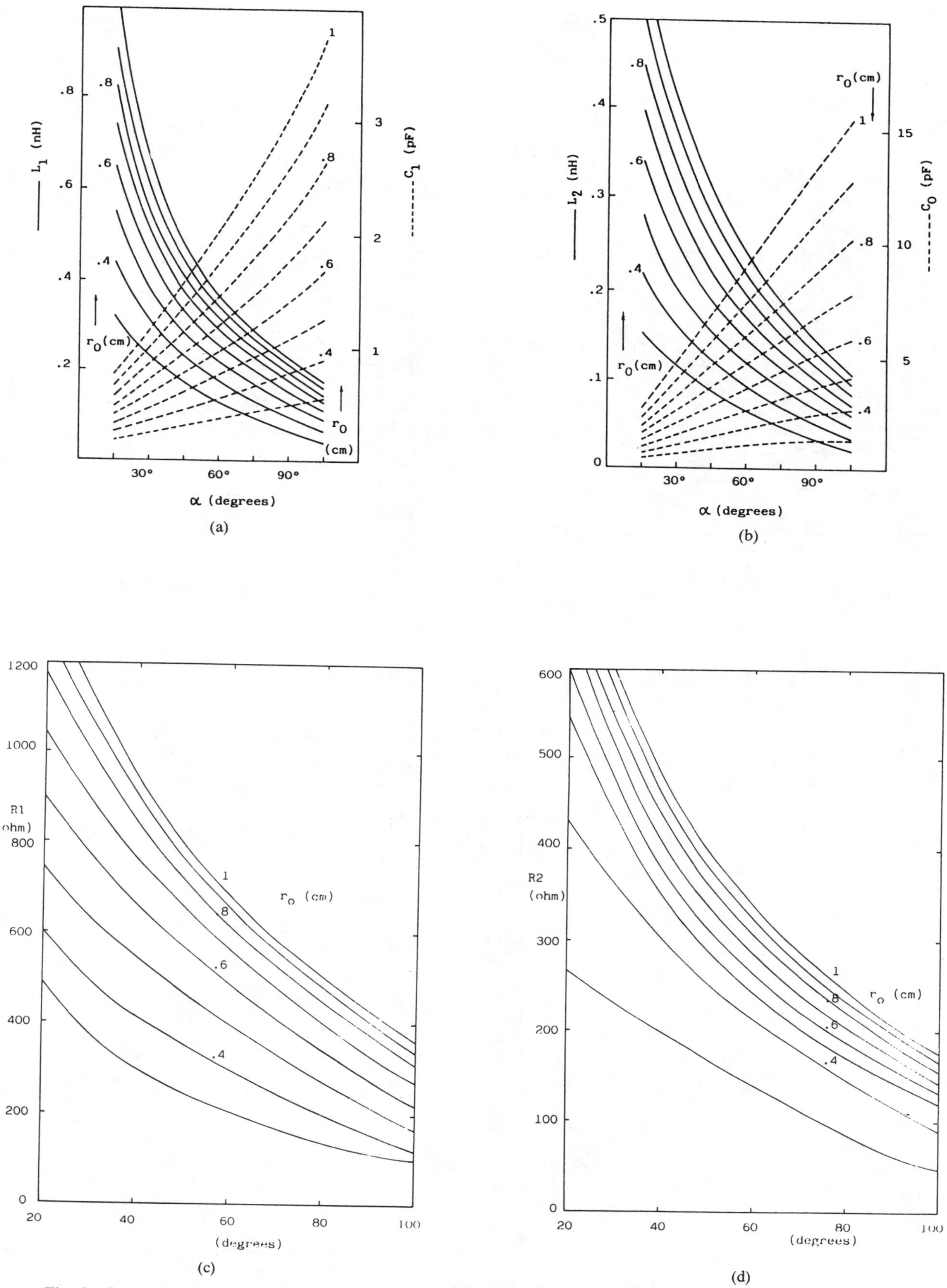

Fig. 5. Lumped equivalent circuits elements versus geometries of a single radial stub ($P = 0.3$ mm) in shunt connection (butterfly). (a) L1, C1. (b) L2, C0. (c) R1. (d) R2. Simulation has been performed on a 0.635-mm-thick substrate with $\epsilon_r = 10$.

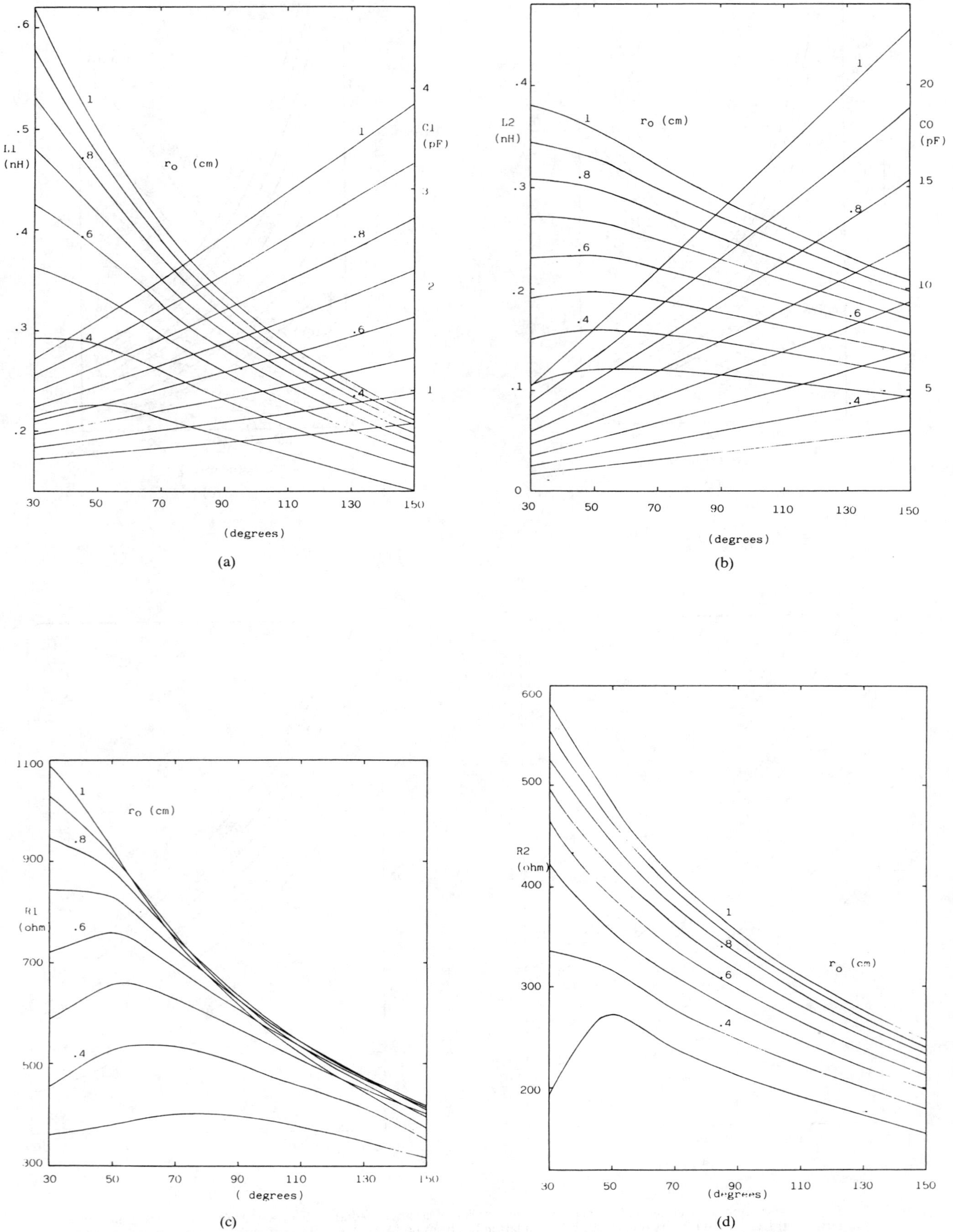

Fig. 6. Lumped equivalent circuit elements versus geometries of a series-connected radial stub (50-Ω feeding line). (a) L1, C1. (b) L2, C0. (c) R1. (d) R2. Simulation has been performed on a 0.635-mm-thick substrate with $\epsilon_r = 10$.

where R_s is the surface resistivity of the conductor, and

$$Q_{d,0n} = 1/\tan\delta \tag{4}$$

where $\tan\delta$ is the loss tangent of the dielectric substrate.

The Q factor related to radiation losses is more difficult to evaluate. A simple approach is to consider the energy related to TM_{0n} modes to be propagating in the radial direction. As a consequence, side walls can be assumed to not radiate; thus radiation losses are due only to the outer surface.

This simplifies the analysis because, for the TM_{0n} mode considered, both the power radiated from the outer surface and the stored energy inside the ring sector of angle α (Fig. 2) are equal to $\alpha/360°$ the power and energy of a ring (Fig. 3) having the same dimensions and supporting the same e.m. field on the outer radiating surface, assumed to be the only radiating one. In fact, even if the spatial e.m. field distribution sustained by the radiating sectoral aperture and the complete ring aperture is different, the starting hypothesis makes it possible to consider the total amount of power radiated by the radial stub as a fraction of the same quantity radiated by the outer surface of a ring structure. Therefore [11], the $Q_{r,0n}$ factor related to radiation losses can be expressed as follows:

$$Q_{r,0n} = \left(\frac{\omega_{0n} h \sqrt{\mu_0 \epsilon_0}}{\epsilon_{d,n}} \frac{I_\theta}{K(r_{ie}, r_{oe})} \right)^{-1} \tag{5}$$

where

$$I_\theta = \int_0^\pi J_1^2(K r_{oe} \sin\theta) \sin\theta \, d\theta$$

$$K(r_{ie}, r_{oe}) = 1 - \frac{r_{ie}^2 \{ J_0(K_{0n} r_{ie}) N_1(K_{0n} r_{ie}) - J_1(K_{0n} r_{ie}) N_0(K_{0n} r_{ie}) \}^2}{r_{oe}^2 \{ J_0(K_{0n} r_{oe}) N_1(K_{0n} r_{ie}) - J_1(K_{0n} r_{ie}) N_0(K_{0n} r_{oe}) \}^2}$$

where $J_{0,1}$ and $N_{0,1}$ are the Bessel and Neumann functions of order 0 and 1, respectively.

III. The Lumped Model

The above theoretical characterization of a lossy radial stub can now be put into the form of a CAD-oriented equivalent circuit, utilizing frequency-independent lumped elements (Fig. 4). The model is valid for both double-shunt and series connection, with proper values assigned to the lumped elements.

In fact, starting with the same physical dimensions, different effective geometries for the shunt and the series connection are obtained, mainly resulting in a variation of the effective inner radius (r_{ie}). This represents a significant improvement over previously developed characterization (e.g. [7]).

The lumped model in Fig. 4 quite completely and accurately characterizes radial stub behavior. In fact it has a predictable range of validity; it adopts lumped elements; and it is substrate and insertion dependent.

The lumped elements of the equivalent circuit can be derived from the above-defined Q_{t0}, $Q_{t,01}$ and $Q_{t,0m}$ ($m >$

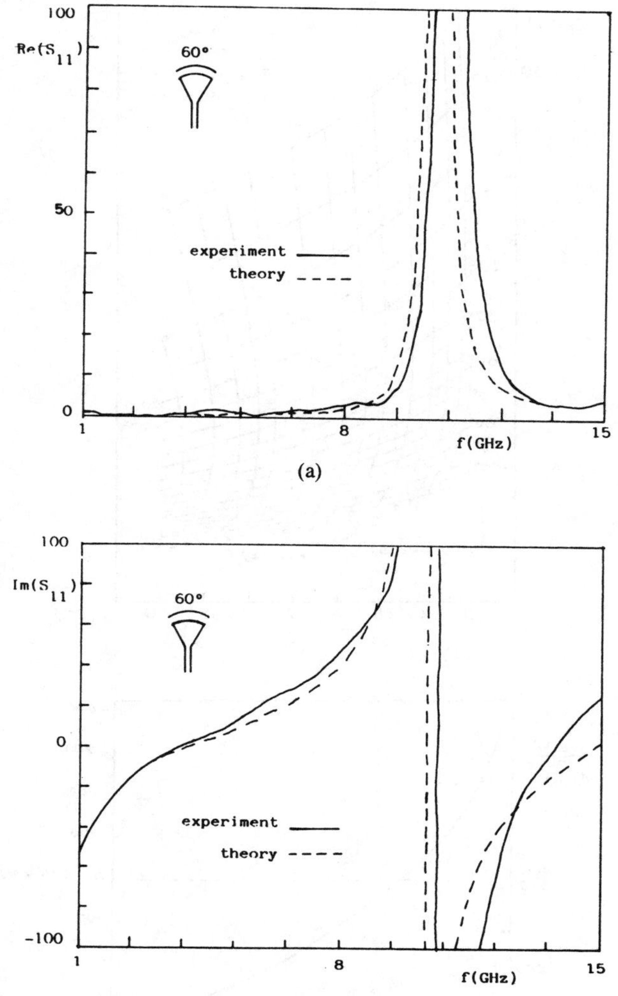

Fig. 7. Experimental and theoretical plots of input impedance for a series-connected radial stub ($r_o = 0.5$ c m, $\alpha = 60°$) realized on a 0.635-mm-thick alumina substrate ($\epsilon_r = 10$). (a) Imaginary part of input impedance. (b) Real part of input impedance.

2), through (1)–(5), as follows:

$$R_0 = \frac{1}{Q_{t_0}\epsilon_{d,0} K^2} \qquad C_0 = \sqrt{\frac{\mu_0}{\epsilon_0 \epsilon_r}} \frac{\epsilon_0 \epsilon_{d,0}}{P_{00}^2} \tag{6}$$

$$R_1 = \frac{K_g P_{01}^2 Q_{t,01}}{K_{01}^2} \qquad L_1 = \frac{\sqrt{\mu_0 \epsilon_0 \epsilon_r} P_{01}^2}{K_{01}^2}$$

$$C_1 = \sqrt{\frac{\mu_0}{\epsilon_0 \epsilon_r}} \frac{\epsilon_0 \epsilon_{d,1}}{P_{01}^2}$$

$$\frac{j\omega L_2 R_2}{j\omega L_2 + R_2} = j \sum_{m=2}^{\infty} \frac{K_g P_{0m}^2}{K_{0m}^2 (1 + j/Q_{tm}) - K^2 \epsilon_{d,m}}.$$

The R_0 value is quite small and can be neglected, while R_1 and R_2 influence the radial stub behavior. Fig. 5(a) and (b) displays the variation of $L_i (i = 1, 2)$, $C_j (j = 0, 1)$ versus the radial stub geometry (angle, radii) for double-shunt connection, also known as the "butterfly" structure. The variations of R_1 and R_2 versus the same parameters are shown in Fig. 5(c) and (d). The same four plots of the

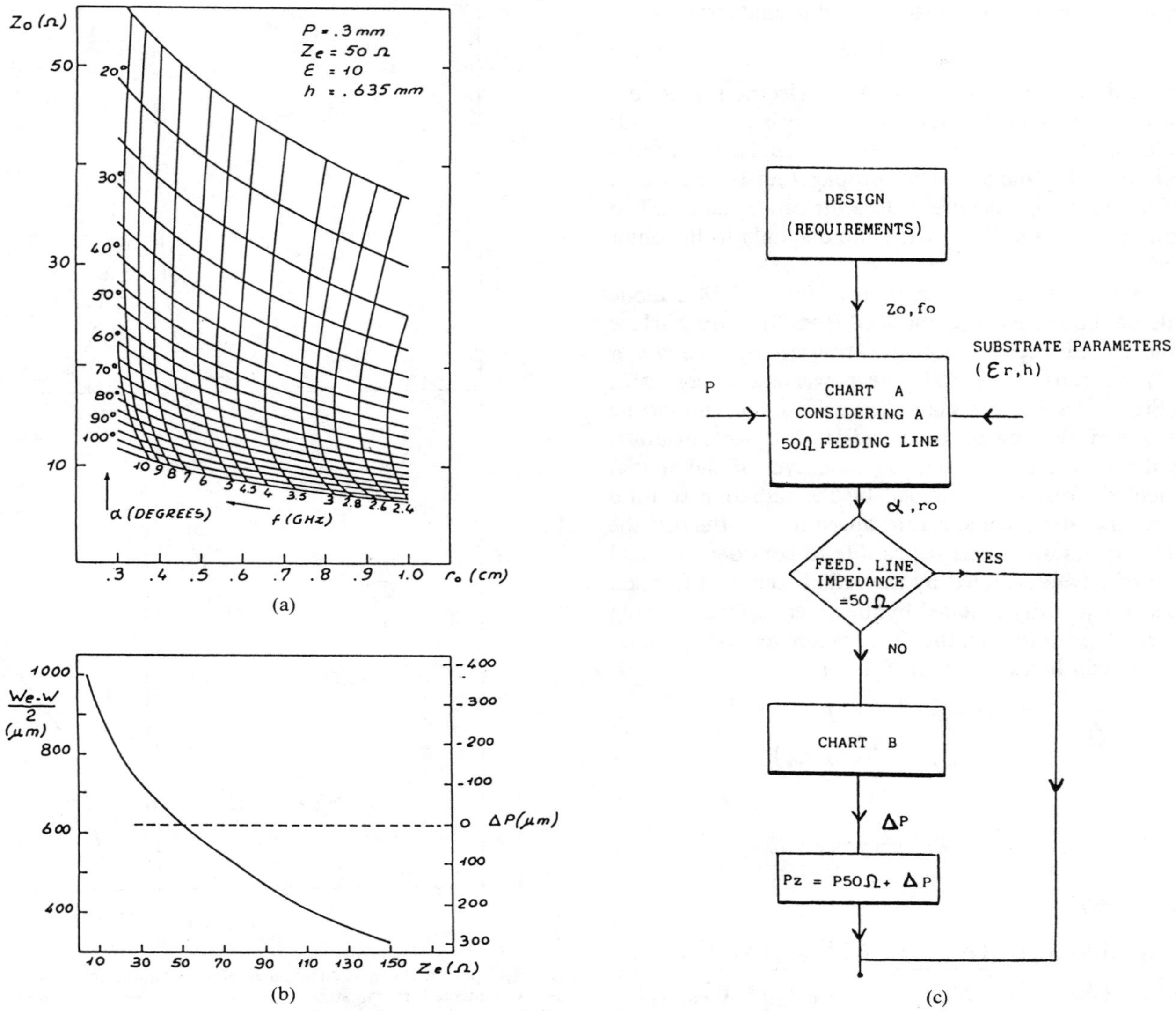

Fig. 8. (a), (b) Design charts for shunt connected radial stub. (c) Flow chart of the synthesis procedure.

series-connected radial stub are shown in Fig. 6(a), (b), (c), and (d). Starting from the physical geometries of a radial stub and going through the above plots, the corresponding lumped elements are quickly derived. Experiments on alumina ($\epsilon_r = 10$) 0.635 mm thick have fully confirmed the effectiveness of the above model. In Fig. 7 the theoretical and experimental values of the real and imaginary parts of the impedance for a series-connected radial stub are compared, displaying them in a restricted range of significant values.

Good agreement between theory and experiment is achieved at least up to a frequency (f_x) corresponding to the frequency of the first resonant mode of the radial structure ($f_x = 1/2\pi\sqrt{L_1 C_1}$), the TM_{01} in this approximation. The agreement can be extended in frequency by adding to the equivalent structure resonant cells corresponding to higher order TM_{0n} modes. Thus, a compromise between frequency effectiveness and model simplicity has to be reached. The lumped model is a complete and accurate way to represent the several dependences of the radial stub behavior. Insertion, substrate parameters, and losses can be taken into account through the lumped elements. The result is a circuit easy to implement in the available CAD programs, leading to quick and accurate analysis of the radial structure.

IV. An Application Example

In parallel to the development of lumped models to implement in CAD programs, a synthesis procedure for radial stubs has also been developed [14]. It consists of deriving the geometrical dimensions (angle α, outer radius r_o), for a fixed insertion depth (P), of a radial stub presenting the same resonance frequency (f_0) and equivalent characteristic impedance (Z_0) as a conventional straight stub. The method, here outlined for a shunt connection, utilizes two sets of curves (shown in Fig. 8(a) and (b)) where $(w_e - w)/2$ is the widening presented by the effective width of one side of the feeding line, with respect to the actual width, and Z_e represents the characteristic impedance of the feeding line. The first curve provides the

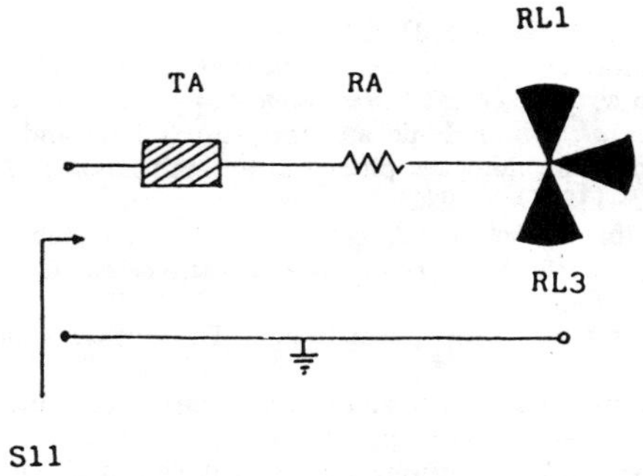

Fig. 9. Schematic of the $3\times60°$ termination.

α and r_o value corresponding to the desired f_0 and Z_0, for a given substrate and P, assuming a 50-Ω feeding line for the shunt radial stub. The second gives the actual insertion depth (P_z) to be provided with a different feeding line width.

The procedure (Fig. 8(c)), which can be fully automated, streamlines designing circuits with radial stubs. It has been successfully adopted in the development of both a two-octave stopband microstrip low-pass filter [14] and a novel termination suitable for very broad-band MIC and MMIC application [15].

The structure (Fig. 9) utilizes three radial lines and no direct grounding. The circuit has been simulated by Touchstone™ adopting the above lumped equivalent model for the radial stubs. Several samples have been developed (Duroid 6010, $\epsilon_r = 10.5$, $h = 25$ mils).

In particular, it is found that the bandwidth over which return loss is better than 20 dB, achievable with a $3\times60°$ termination, is theoretically 20 percent wider than that achieved with a 180° "half-moon" structure. This can be explained observing that—once the resonance frequency and the total angle of the termination are fixed—proper inner and outer radius values can be found in order to improve the return loss figure of the termination itself. Actually the half-moon bandwidth is further narrowed, due to the excitation of circumferential modes. In fact, these may occur if the radial stub transverse dimension is comparable to the longitudinal one, therefore when the angle is large enough (i.e., 180°) and under particular feeding conditions.

A very good agreement between theory and experiment has been obtained for the $3\times60°$ structure (Fig. 10). Also, the improvement of the above termination with respect to the theoretical and experimental performance of a conventional 180° half-moon [16] can be derived by Fig. 11. The bandwidths achieved are also wider than the ones reported for similar structures, adopting either straight [17] or single radial stub [18]. It is worth noting that $3\times60°$ configuration has to adopt different geometries to achieve the same effective ones, thus taking into account the different insertions—butterfly and series—presented by the three radial stubs of the termination. Three radial stubs with the same actual geometries, in fact, result in a configuration with a much narrower bandwidth [15]. The sensitivity of circuit response versus the resistor value has also been checked.

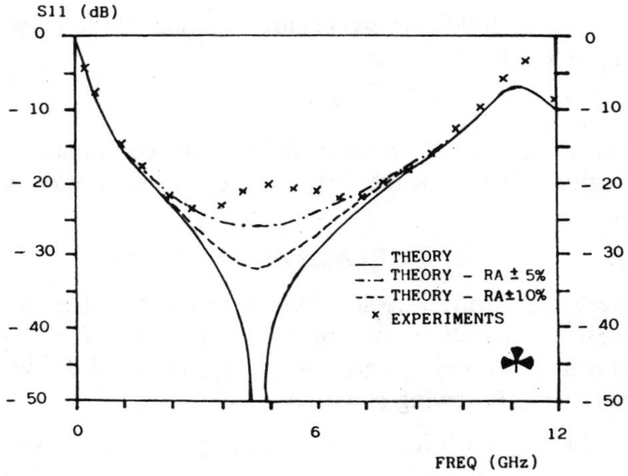

Fig. 10. Theoretical and experimental results of the $3\times60°$ termination.

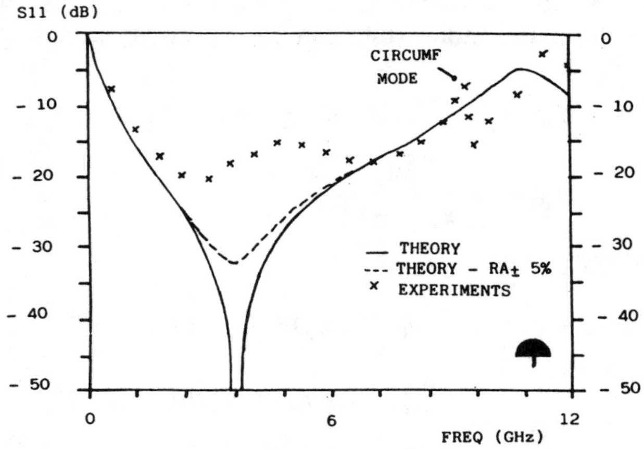

Fig. 11. Theoretical and experimental results of the 180° "half-moon" termination.

The circuits have been analyzed and optimized by Touchstone™, introducing in it the equivalent models.

V. Conclusions

The possibility of using radial stubs as broad-band and low-impedance elements has been enhanced by providing a complete set of lumped circuits modeling their electrical and insertion-dependent lossy behavior. Also, a synthesis method based upon design charts has been developed for an easy implementation of the radial stub in circuit design. The models developed have been adopted in the design of very broadband non-grounded terminations. These designs use the Touchstone™ program, in which the lossy lumped equivalent circuit for both double-shunt and series-connected radial stubs have been developed. The agreement between simulation and results is very satisfactory.

Appendix

In planar circuits, the e.m. field expansion in terms of resonant modes has already been applied to circular [19] and angular [20] structures.

First, the following eigenvalue equation has to be solved for the radial structure:

$$\nabla^2 E_{m,n} + K_{m,n}^2 E_{m,n} = 0 \tag{A1}$$

expressing the ∇ operator in circular coordinates. The solution of (A1) for the TM_{0n} modes can be written in the form

$$E_{0n} = CJ_0(K_{0n}r) + DN_0(K_{0n}r) \tag{A2}$$

where J_0, and N_0 are, respectively, the Bessel and Neumann functions of order 0. Imposing the boundary conditions at $r = r_{ie}$ and $r = r_{oe}$, K_{0n} is found as the nth root of the following equation:

$$J_0'(K_{0n}r_{oe})N_0'(K_{0n}r_{ie}) + J_0'(K_{0n}r_{ie})N_0'(K_{0n}r_{oe}) = 0. \tag{A3}$$

P_{0n}, which is the coupling coefficient between the quasi-TEM mode traveling on the feeding line and the TM_{0n} mode in the radial stub, can be expressed, according to [19], as

$$P_{0n} = \sqrt{w_{ge}}\,[A_{0n}J_0(K_{0n}r_{ie}) + B_{0n}N_0(K_{0n}r_{ie})]$$

$$P_{00} = \sqrt{\frac{2w_{ge}}{\alpha(r_{oe}^2 - r_{ie}^2)}} \tag{A4}$$

where

$$A_{0n} = \sqrt{\frac{2}{\alpha}}\,\{r_{oe}^2[J_0(K_{0n}r_{oe}) + K_nN_0(K_{0n}r_{oe})]^2 - r_{ie}^2[J_0(K_{0n}r_{ie}) + K_nN_0(K_{0n}r_{ie})]^2\}^{-1/2}$$

$$B_{0n} = K_nA_{0n} \qquad K_n = -\frac{J_1(K_{0n}r_{oe})}{N_1(K_{0n}r_{oe})}$$

where J_0, J_1, N_0, and N_1 are the Bessel and Neumann functions of order 0 and 1.

References

[1] F. Giannini, R. Sorrentino, and J. Vrba: "Planar circuit analysis of microstrip radial stub," *IEEE Trans. Microwave Theory Tech.*, vol. MTT-32, pp. 1652–1655, Dec. 1984.

[2] H. A. Atwater, "The design of the radial line stub: A useful microstrip circuit element," *Microwave J.*, pp. 149–153, Nov. 1985.

[3] A. Chu *et al.*, "Monolithic analog phase shifters and frequency multipliers from mm-wave phased array applications," *Microwave J.*, pp. 105–119, Dec. 1986.

[4] M. De Lima Coimbra, "A new kind of radial stub and some applications," in *Proc. 14th European Microwave Conf.* (Liegi), 1984, pp. 516–521.

[5] "Broadband microstrip mixer design: The Butterfly mixer." HP Application note 976.

[6] J. B. Vinding, "Radial line stubs as elements in stripline circuits," in *NEREM Records*, 1967, pp. 108–109.

[7] H. A. Atwater, "Microstrip reactive circuit elements," *IEEE Trans. Microwave Theory Tech.*, vol. MTT-31, pp. 488–491, June 1983.

[8] F. Giannini, M. Ruggieri, and J. Vrba, "Shunt-connected microstrip radial stubs," *IEEE Trans. Microwave Theory Tech.*, vol. MTT-34, pp. 363–366, Mar. 1986.

[9] F. Giannini and C. Paoloni, "Broadband lumped equivalent circuit for shunt-connected radial stub," *Electron. Lett.*, vol. 22, no. 9, pp. 485–487, Apr. 1986.

[10] S. L. March, "Analyzing lossy radial-line stubs," *IEEE Trans. Microwave Theory Tech.*, vol. MTT-33, pp. 269–271, Mar. 1985.

[11] F. Giannini, C. Paoloni, and J. Vrba; "Losses in microstrip radial stubs," in *Proc. 16th European Microwave Conf.* (Dublin), 1986, pp. 523–528.

[12] J. Vrba, "Dynamic permittivities of microstrip ring resonators," *Electron. Lett.*, vol. 15, no. 16, pp. 504–505, Aug. 1979.

[13] I. J. Bahl and P. Barthia, *Microstrip Antennas.* Dedham, MA: Artech House, 1982.

[14] F. Giannini, M. Salerno, and R. Sorrentino: "Two-octave stopband microstrip low-pass filter design," in *Proc. 16th European Microwave Conf.* (Dublin), 1986, pp. 292–297.

[15] F. Giannini and M. Ruggieri, "A broadband termination utilizing multiple radial line structure," in *Proc. MIOP*, vol. 2, no. 6B1, (Wiesbaden, Germany), May 1987.

[16] B. A. Syrett, "A Broadband element for microstrip bias or tuning circuits," *IEEE Trans. Microwave Theory Tech.*, vol. MTT-28, pp. 925–927, Aug. 1980.

[17] I. J. P. Linner and H. B. Lunden, "Theory and design of broadband nongrounded matched loads for planar circuits," *IEEE Trans. Microwave Theory Tech.*, vol. MTT-34, pp. 892–896, Aug. 1986.

[18] A. M. Khilla, "Accurate closed-form expressions for ring and disc-type microstrip resonators," *Microwave J.*, vol. 27 no. 11, pp. 91–105, 1984.

[19] G. D'Inzeo, F. Giannini, C. M. Sodi, and R. Sorrentino, "Method of analysis and filtering properties of microwave planar networks," *IEEE Trans. Microwave Theory Tech.*, vol. MTT-26, pp. 462–471, July 1978.

[20] G. D'Inzeo, F. Giannini, R. Sorrentino, and J. Vrba, "Microwave planar networks: The annular structure," *Electron. Lett.* vol. 14, no. 16, pp. 526–528, Aug. 1978.

An Inhomogeneous Two-Dimensional Model for the Analysis of Microstrip Discontinuities

WOJCIECH K. GWAREK AND CEZARY MROCZKOWSKI

***Abstract*—The paper proposes a new approach to two-dimensional modeling of microstrip circuits of arbitrary shape. A new model was developed to be used for finite-difference time-domain (FDTD) analysis. A characteristic feature of the new model is its inhomogeneity, i.e., the dependence of the parameters at a particular point in the two-dimensional space on the distance of that point from the strip's edge. Examples of FDTD analysis based on the new model are shown.**

I. Introduction

Analysis of microstrip discontinuities has been the focus of researchers for many years and has resulted in an enormous number of publications [1]. Different mathematical models have been used. The choice of the model results from contradicting requirements for high accuracy and small computing effort. Most commercial programs use one-dimensional models consisting of segments of transmission lines and lumped elements. Their use is, however restricted, (see, for example, [2]). The three-dimensional models use Maxwell's equations in three-dimensional space. Although significant progress in applying such models can be noted [3], [4], [12], the analysis of arbitrarily shaped circuits requires very long computing times. In the author's opinion, another generation of computer hardware and software will be required before they become commonly used tools in microwave engineering.

The two-dimensional models use Maxwell's equations in two-dimensional space. They significantly improve the accuracy of analysis when compared with one-dimensional models and require much less computer time than three-dimensional ones. Let us consider the finite-difference time-domain (FDTD) method in two and three dimensions. The change from two to three dimensions not only requires a larger number of mesh points owing to discretization in the third dimension but also increases the number of field components and makes it more difficult to define boundary conditions on curved boundaries and to model lattice terminations. These factors combined usually make the computing time for three-dimensional FDTD larger by a factor of a few hundred if not a few thousand.

Two-dimensional models were employed with different numerical methods by D'Inzeo *et al.* [5], Okoshi [7], and Menzel and Wolff [6]. Recent progress in the development of the FDTD method for arbitrarily shaped two-dimensional circuits opens the way for new classes of 2-D inhomogeneous models. One such model was proposed in [8]. It gave good results for certain types of circuits but was found to be imprecise in other types. The present paper proposes a new model which has been found to be more accurate and versatile then the previous one.

Manuscript received January 30, 1990; revised April 19, 1991. Work done at the Warsaw University of Technology was supported under CPBP 02.02; that done at the University of Duisburg was supported under Sonderforschungsbereich 254.

The authors are with the Institute of Radioelectronics, Warsaw University of Technology, Nowowiejska 15/19, 00-665 Warsaw, Poland.

IEEE Log Number 9101650.

II. Two-Dimensional Inhomogeneous Model

Let us consider a microstrip line of width w build on a substrate of permittivity ϵ_d (Fig. 1) which is characterized by its impedance, $Z(w)$, and effective permittivity, $\epsilon_{\text{eff}}(w)$, or by the unit capacitance, $C(w)$, and unit inductance, $L(w)$:

$$Z(w) = \sqrt{\frac{L(w)}{C(w)}} \tag{1}$$

$$\epsilon_{\text{eff}}(w) = v^2 L(w) C(w). \tag{2}$$

The parameters $Z(w)$ and $\epsilon_{\text{eff}}(w)$ in (1) and (2) can be evaluated by using Hammerstad–Jensen formulas [11].

Let us consider half of the microstrip from one side of the axis of symmetry. If we assume further that an increase in the width of the strip in the center of the line does not change the electromagnetic field near the open edge of the line, then we can assign to each point of the strip two functions representing the "density" of capacitance, $c(r)$, and the "density" of inductance, $l(r)$, where

$$C(w) = \int_0^w c(r)\,dr \tag{3}$$

$$1/L(w) = \int_0^w 1/l(r)\,dr. \tag{4}$$

These densities can be obtained from the known functions $C(w)$ and $L(w)$ using the formulas

$$c(w/2) = \frac{dC(w)}{dw} \tag{5}$$

$$1/l(w/2) = \frac{d(1/L(w))}{dw}. \tag{6}$$

On the basis of (5) and (6) we can define a microstrip model with inhomogeneous medium under the strip:

$$\epsilon_1(r) = c(r)h \tag{7}$$

$$1/\mu_1(r) = \frac{h}{l(r)}. \tag{8}$$

Shapes of the functions $\epsilon_1(r)$ and $\mu_1(r)$ (normalized to ϵ_d and μ_0) for a line on a dielectric substrate of $\epsilon_d = 10$ are shown in Fig. 2. They are plotted for $r > r_1$, where r_1 is assumed to be equal $0.1h$. For values of r approaching 0, direct application of (7) and (8) would lead to infinite growth of $c(w)$ and $l(w)$ owing to the unrealistic assumption that fringing fields are concentrated near the edge of the strip. This would cause numerical problems and would decrease the accuracy. Taking this into

Reprinted from *IEEE Trans. Microwave Theory Tech.*, vol. 39, no. 9, pp. 1655–1658, Sept. 1991.

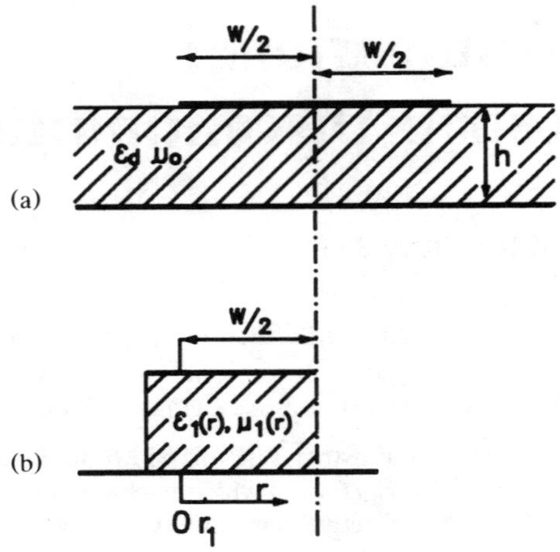

Fig. 1. (a) Cross section of a microstrip line. (b) Model of the line of (a).

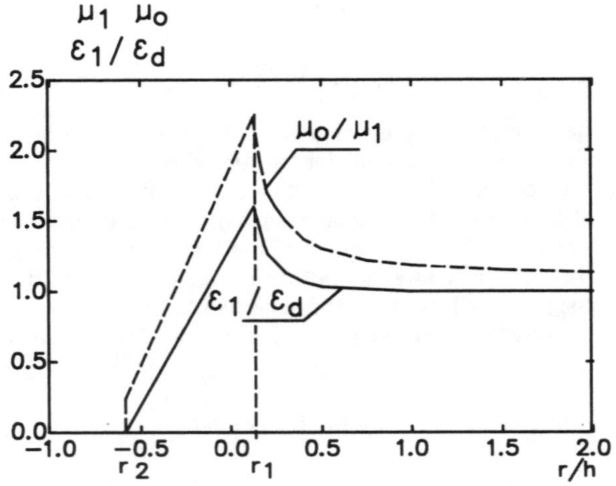

Fig. 2. Examples of the functions ϵ_1/ϵ_d and μ_0/μ_1 for a line on a dielectric substrate of permittivity ϵ_d.

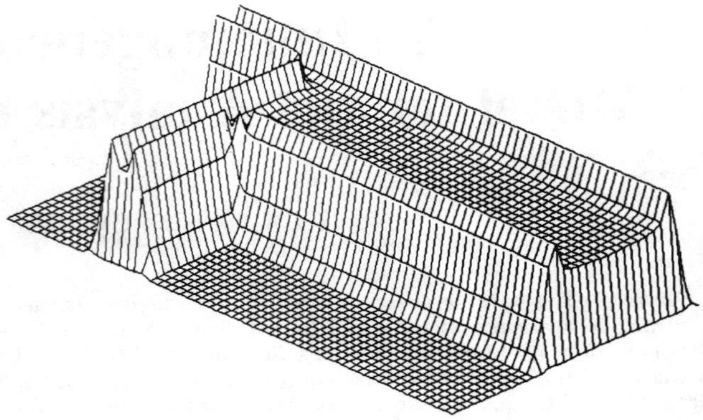

Fig. 3. Inhomogeneous model of the analyzed circuit.

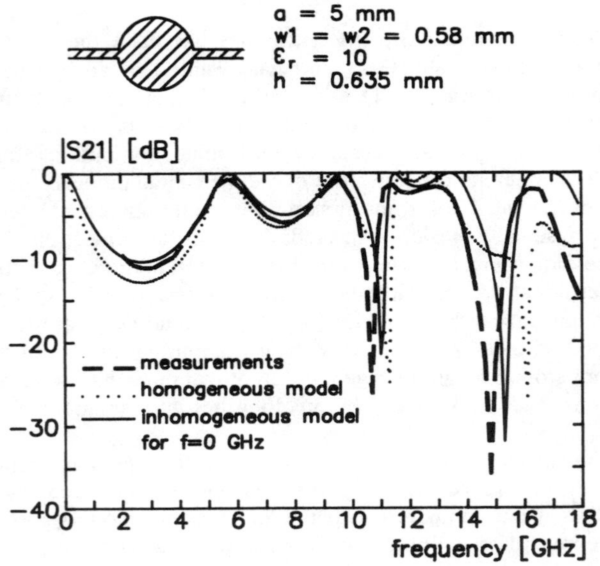

Fig. 4. Comparison of calculated and measured values of $|S_{21}|$ versus frequency for a circular microstrip circuit.

account, we have modified the functions ϵ_1 and μ_1 for $r < r_1$. The modified functions (shown in Fig. 2) are not too far from a physical interpretation and are simple enough for effective numerical application. The unit capacitance, C_0, of the line of width $w = 2r_1$ is assumed to come from a linear distribution of ϵ_1 in the range of r from r_1 to

$$r_2 = r_1 - 2C_0 h / \epsilon_1(r_1). \tag{9}$$

Thus the edge of the model (where the magnetic wall is assumed) is moved with respect to the edge of the real circuit by the distance $|r_2|$. The function of $\mu_1(r)$ is assumed to be of the same shape as in Fig. 2, with the integral of $1/\mu_1$ between r_1 and r_2 corresponding to $1/L(r_1)$.

In the two-dimensional inhomogeneous model we assume that the permittivity and permeability of the medium filling the model at a particular place in the circuit are equal to $\mu(x, y) = \mu_1(r)$ and $\epsilon(x, y) = \epsilon_1(r)$, where r is the distance of the point (x,y) from the nearest edge of the circuit. The two-dimensional distribution of $\epsilon(x, y)$ of the circuit analyzed further in Example 2 is shown in Fig. 3.

The newly developed model was implemented in the Quick-Wave [10] programming package and checked in FDTD calculations of various microstrip circuits. Three of them will be discussed here as examples.

Example 1

We consider a circular microstrip resonator of radius 5 mm connected to a 50 Ω input and output line with the lines making an angle of 180°. The substrate is alumina of $\epsilon = 10$ and $h = 0.635$ mm. This example has been considered in [5], from which the results of measurements are taken. Let us first assume a homogeneous model of the microstrip resonator. The results of FDTD calculations seen in Fig. 4 show large discrepancies with the results of measurement for higher frequencies. As explained in [5], such discrepancies come from the fact that in a homogeneous resonator the resonant frequency of mode 41 (corresponding to mode TM_{410} in a cylindrical resonator with open sidewalls) is below the frequency of mode 12, whereas it has been proved by experiment that in a microstrip resonator the second one appears below the first. In the homogeneous model used in time-domain calculations, it is impossible to produce this shift of frequencies. On the contrary our inhomogeneous 2-D model apparently causes a proper shift of the resonant frequencies, which is visible from the results of calcula-

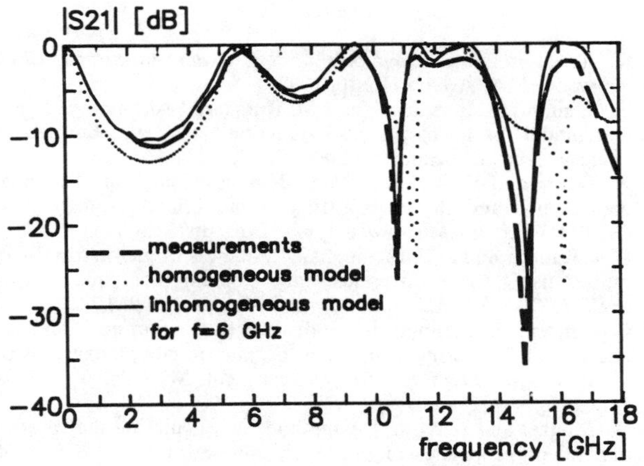

Fig. 5. Modification of the results of the calculation of Fig. 4 under the assumption of an inhomogeneous model for $f = 6$ GHz.

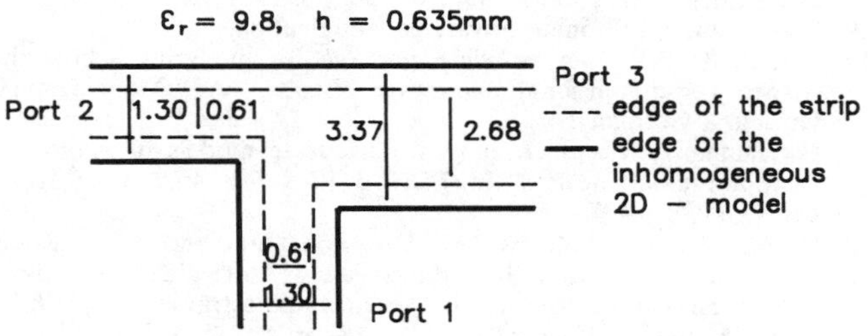

Fig. 6. A microstrip T junction on alumina substrate.

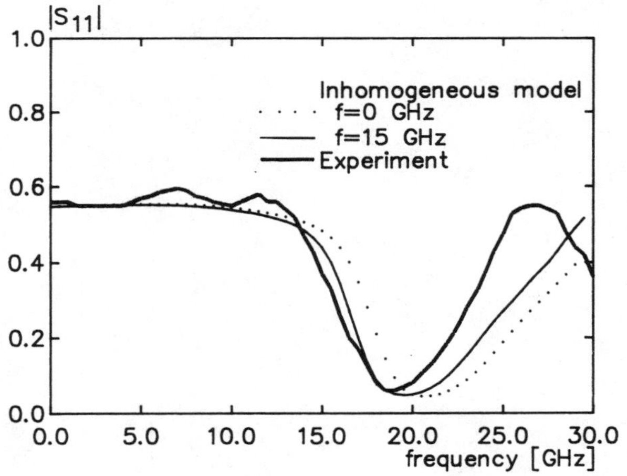

Fig. 7. $|S_{11}|$ of the junction of Fig. 6.

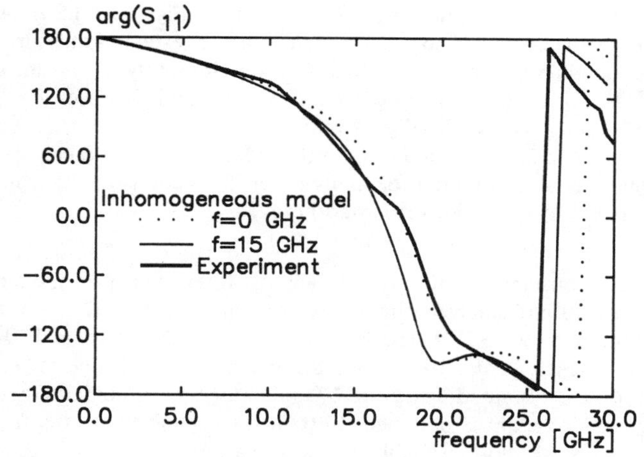

Fig. 8. $\mathrm{Arg}(S_{11})$ of the junction of Fig. 6.

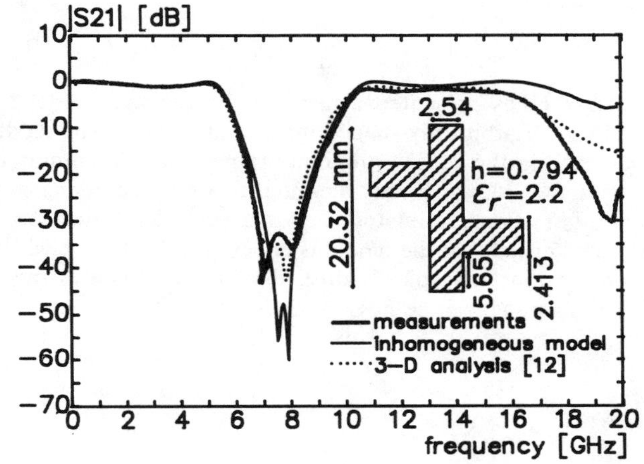

Fig. 9. $|S_{21}|$ of a microstrip low-pass filter.

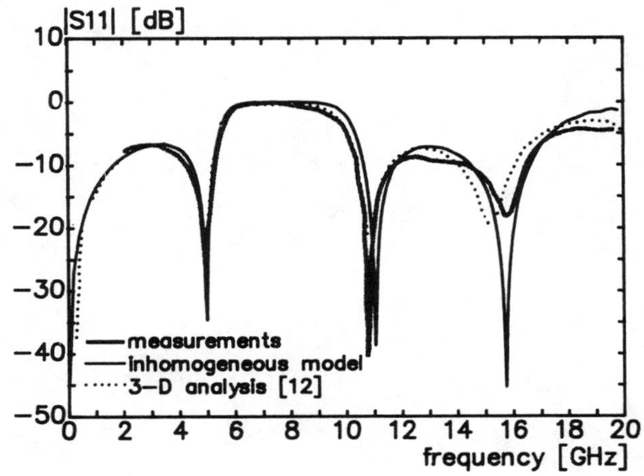

Fig. 10. $|S_{11}|$ of the microstrip low-pass filter of Fig. 9.

tions presented also in Fig. 4. These calculations were obtained using the version of the FDTD method described in [8] and were used in the Quick-Wave program [10]. The unit parameters $L(w)$ and $C(w)$ used for calculations were taken from the closed-form formulas of [1, ch. 3.3.] and [11] for the frequency $f = 0$. It can be seen in Fig. 4 that the calculated higher resonant frequencies are shifted up with respect to those obtained from measurements. The discrepancy is due to microstrip dispersion, which may be taken into account to some extent by assuming the parameters $L(w)$ and $C(w)$ for higher frequencies. In the case considered, after assuming $L(w)$ and $C(w)$ for $f = 6$ GHz, we have obtained results very close to measurements also for higher frequencies (Fig. 5).

Example 2

A microstrip T junction of one 20 Ω line and two 50 Ω lines is considered. Its shape is shown in Fig. 6. The substrate is alumina of $\epsilon = 9.8$ and height $h = 0.635$ mm. Figs. 7 and 8 show the results of calculations compared with the results of measurements taken by Gronau and Wolff at the University of Duisburg [9]. In this example we took values of $L(w)$ and $C(w)$ for $f = 0$ GHz and $f = 15$ GHz (Figs. 7 and 8).

Very good agreement between the measurements and calculations was also obtained for many other T junctions of different shapes, which are not presented here.

Example 3

We consider a microstrip low-pass filter which has been measured and analyzed by a three-dimensional FDTD method by the authors of [12]. The substrate has $\epsilon_r = 2.2$ and $h = 0.794$ mm. In Figs. 9. and 10 we compare the results published in [12] with those obtained using our 2-D inhomogeneous model. Good agreement again is obtained. Certain discrepancies at high frequencies are most probably due to radiation. Our analysis took about 7 min on a PC-386 working under DOS while the 3-D analysis [12] of the same example was reported to take 8 h on a VAX station 3500.

III. Conclusions

The paper has presented a new two-dimensional model for the analysis of arbitrarily shaped microstrip circuits. The model was checked in the FDTD program prepared by the authors to run on a PC. It was found very useful for investigating new designs of junctions, resonators, and patch couplers. However, it must be admitted that the model is effective only in cases where phenomena that are typically three-dimensional, such as radiation and coupling, can be neglected.

References

[1] R. K. Hoffmann, *Handbook of Microwave Integrated Circuits*. Norwood, MA: Artech House, 1987.

[2] F. Giannini, G. Bartolucci, and M. Ruggieri, "An improved equivalent model for microstrip cross-junction," in *Proc. European Microwave Conf*. (Stockholm), 1987.

[3] X. Zhang and K. K. Mei, "Time-domain calculation of microstrip components and the curve-fitting of numerical results," in *1989 IEEE MTT-S Int. Microwave Symp. Dig*., pp. 313–316.

[4] C. J. Railton and J. P. McGeehan, "Analysis of microstrip discontinuities using finite difference time domain technique," in *1989 IEEE MTT-S Int. Microwave Symp. Dig*., pp. 1009–1012.

[5] G.d' Inzeo, F. Giannini, M. Sodi, and R. Sorrentino, "Method of analysis and filtering properties of microwave planar network," *IEEE Trans. Microwave Theory Tech*., vol. MTT-26, pp. 462–471, July 1978.

[6] W. Menzel and I. Wolff, "A method for calculating the frequency-dependent properties of microstrip discontinuities," *IEEE Trans. Microwave Theory Tech*., vol. MTT-25, pp. 107–112, Feb. 1977.

[7] T. Okoshi, *Planar Circuits for Microwaves and Lightwaves*. Berlin and Heidelberg: Springer-Verlag, 1985.

[8] W. Gwarek, "Analysis of arbitrarily-shaped two dimensional microwave circuits by finite-difference time-domain method," *IEEE Trans.Microwave Theory Tech*., vol. 36, pp. 738–744, Apr. 1988.

[9] I. Wolf and G. Gronau, private communication.

[10] "QUICK-WAVE—A software package for analyzing arbitrarily-shaped two-dimensional microwave circuits," ArguMens GmbH, Duisburg, Germany.

[11] E. Hammerstad and O. Jensen, "Accurate models of microstrip computer aided design," in *IEEE MTT-S Int. Microwave Symp. Dig*., 1980, pp. 407–409.

[12] D. M. Sheen, S. M. Ali, M. D. Abouzahra, and J. A. Kong, "Application of the three-dimensional finite-difference time-domain method to the analysis of planar microstrip circuits," *IEEE Trans. Microwave Theory Tech*., vol. 38, pp. 849–857, July 1990.

Characterizing Microwave Planar Circuits Using the Coupled Finite-Boundary Element Method

KE-LI WU, CHEN WU, AND JOHN LITVA

***Abstract*—A general approach is presented for the analysis of microwave planar circuits. The technique is particularly well suited to the analysis of circuits with complicated geometries and dielectric loads. The proposed technique is a hybrid, consisting of an amalgamation of the finite element and the boundary element techniques. The new technique can handle problems with mixed electric and magnetic walls, as well as complicated dielectric loads, such as those composed of ferrite materials. Computed and measured data for various complicated devices are compared, showing excellent agreement.**

I. Introduction

In the development of numerical techniques for analyzing the electromagnetic fields associated with microwave circuits, one has to strike a balance between accuracy and simplicity. Three dimensional full-wave analyses are often impractical because they result in long computational times and prohibitively large memories requirements. In contrast, a simple planar waveguide model can be used in the analysis. This model is particularly useful for solving microwave printed circuit problems. Several methods have been used in the past for the analysis of planar circuits. When the circuit pattern is as simple as a square, rectangle, circle, or annular section, field expansion in terms of resonant models can be used. A general procedure of this technique has been described by Okoshi [1]. When using this method, one must start by calculating the eigenfunctions and eigenvalues for the circuit being analyzed. This can be done analytically if the structure is a separable geometry; if the geometry is not separable, the numerical analysis can be carried out using the contour integral representation of the wave equation.

In practice, it is highly desirable for CAD software to have the capability of analyzing arbitrarily shaped planar circuits. The Finite Element Method (FEM) [2] has been used to handle the discontinuities of arbitrarily shaped planar circuits, even though the computational overhead is large. The boundary integral method or Boundary Element Method (BEM) [3] seems to offer the promise of greatly increased efficiency because it reduces the size of planar circuit problems from two-dimensions to one-dimension. However, when the planar circuits involve complicated or anisotropic dielectric loads, the difficulty of the solution increases quite considerably for one-dimensional algorithms.

In this paper, the coupled finite-boundary element method (CFBM), which was originally developed for waveguide discontinuities [4], is adopted for application to general microwave planar circuit structures. The circuits can include electric walls, magnetic walls, and complex dielectric loads. The CFBM has the merits of both the FEM and the BEM, and can be used to solve complicated problems without requiring excessive computer memory and computation time. Using this method, only the complex media subdomains, which may consist of lossy or anisotropic materials, need to be treated using FEM. Elsewhere, the BEM is used on the boundary to take into account the circuit configuration. Comparing the hybrid technique with the FEM (two-dimensional algorithm) and BEM (one-dimensional algorithm), one concludes that the CFBM can be considered to be a one and one-half dimensional algorithm for planar circuit analysis.

The validity of the application of CFBM to the analysis of planar circuits will be demonstrated with various illustrative examples, which include a cavity filter using metallic posts as inductive shunts, a microstrip disk filter with a dielectric load, and a planar ferrite circulator. The numerical results obtained with CFBM are compared with both measured results and published results. The comparisons are shown to be in good agreement.

Manuscript received January 14, 1992; revised April 23, 1992. This work was supported by the Telecommunication Research Institute of Ontario (TRIO).

The authors are with the Communications Research Laboratory, McMaster University, 1280 Main Street West, Hamilton, ON, Canada L8S 4K1.

IEEE Log Number 9202142.

II. The Planar Circuit Model

There are various techniques that can be used for accurately determining both the equivalent waveguide width $W_{\text{eff}}(f)$ and equivalent filling dielectric constant $\epsilon_{\text{eff}}(f)$. Since they are well known they will not be discussed here for simplicity. When characterizing microstrip or strip line circuits using an equivalent cavity, surrounded by magnetic walls, it is necessary when establishing the equivalent dimensions to take into account the energy stored in the circuit's fringing field. The stored energy can be estimated by assuming that the electric and magnetic fields are constant throughout the increased volume ΔV and are equal to the field at the edge of the circuit. For instance, the increased volume ΔV for a circular disk is taken to be equal to the volume corresponding to the static fringe capacitance ΔC between the radius a and the equivalent radius a_{eq} of the fringe edge, where

$$a_{eq} = a\left\{1 + \frac{d}{\epsilon\pi a^2}\Delta C\right\}^{1/2}, \tag{1}$$

and the static fringe capacitance ΔC can be derived using Kirchhoff 's equation. The equivalent parameters for other geometries can be obtained in a similar way.

It is usually assumed that the equivalent dimensions reflect only the characteristics of the fringing fields when unperturbed by inhomogeneities in the planar circuits, such as a dielectric loads or conductor posts. It can be shown, by comparing the numerical results with the experimental results, that this assumption is fairly reasonable because the equivalent dimensions are determined mostly by the fringing of the outermost fields.

III. General CFBM Formulation for Planar Circuits

The general planar circuit problem shown in Fig. 1 will now be addressed. An M-port device is assumed, where the boundary Γ_Q encloses the inhomogeneous subdomain; boundary Γ_0' is the possible electric wall in the circuit; and the boundary $\Gamma' = \Gamma_0' \cup \Gamma_Q \cup_{m=0}^{M} \Gamma_m$, completely encloses the remaining homogeneous domain.

When using the CFBM technique, complicated subdomains, consisting of dielectric or ferrite posts, can be treated using the

Reprinted from *IEEE Trans. Microwave Theory Tech.*, vol. 40, no. 10, pp. 1963–1966, Oct. 1992.

FEM, while the remaining homogeneous region, surrounded by $\Gamma_Q \cup \Gamma_0' \cup \Gamma'$ can be adequately looked after by using BEM. The detailed formulae and the procedure have been discussed in [4]. Therefore, in the interest of brevity, only the intermediate results are listed here.

By using the Galerkin procedure on the two-dimensional Helmholtz equation and discretizing the complicated subdomain using second-order finite elements, one can obtain the following matrix equation:

$$\begin{bmatrix} [\mathbf{A}]_{Q,\Gamma_Q} & [\mathbf{A}]_{Q,Q} \\ [\mathbf{A}]_{\Gamma_Q,\Gamma_Q} & [\mathbf{A}]_{\Gamma_Q,Q} \end{bmatrix} \begin{Bmatrix} \{E_z\}_{\Gamma_Q} \\ \{E_z\}_Q \end{Bmatrix} = \begin{Bmatrix} \{0\} \\ [\mathbf{D}] \dfrac{\partial E_z}{\partial n}_{\Gamma_Q} \end{Bmatrix} \tag{2}$$

where $\{E_z\}_Q$ and $\{E_z\}_{\Gamma_Q}$ are, respectively, the vectors of the electric field E_z at the nodal points inside domain Q and on the boundary of Q; $[\mathbf{A}]_{Q,Q}$, $[\mathbf{A}]_{Q,\Gamma_Q}$, $[\mathbf{A}]_{\Gamma_Q,Q}$, and $[\mathbf{A}]_{\Gamma_Q,\Gamma_Q}$ are the submatrices associated with $[E_z]_Q$ and $\{E_z\}_{\Gamma_Q}$ in different combinations; and $[D]$ is a square coefficient matrix obtained through the integration of the shape function vector over the boundary Γ_Q. For the second order finite element, $[D]$ can be expressed as

$$[D] = \sum_e{}' \frac{l_e}{15} \begin{bmatrix} 2 & 1 & -0.5 \\ 1 & 8 & 1 \\ -0.5 & 1 & 2 \end{bmatrix} \tag{3}$$

where the l_e is the length of eth boundary element and Σ_e' extends over all the boundary elements. On the other hand, by discretizing the boundary integral equation, which describes the information for the remaining homogeneous domain, using so-called boundary elements, one can subsequently obtain the BEM matrix equation:

$$[H_0, H_0', H_1, H_2, \cdots H_M, H_{\Gamma_Q}] \begin{Bmatrix} \{E_z\} \\ \{E_z\}_{\Gamma_0'} \\ \{E_z\}_{\Gamma_1} \\ \vdots \\ \{E_z\}_{\Gamma_M} \\ \{E_z\}_{\Gamma_Q} \end{Bmatrix} = [G_0, G_0', G_1, G_2, \cdots G_M, G_{\Gamma_Q}] \begin{Bmatrix} \{\partial E_z/\partial n\}_{\Gamma_0} \\ \{\partial E_z/\partial n\}_{\Gamma_0'} \\ \{\partial E_z/\partial n\}_{\Gamma_1} \\ \vdots \\ \{\partial E_z/\partial n\}_{\Gamma_M} \\ \{\partial E_z/\partial n\}_{\Gamma_Q} \end{Bmatrix} \tag{4}$$

The boundary elements used in (4) are similar to finite elements except that their dimensions are usually one less than that of the problem. In the present analysis, second order boundary elements are used for the sake of compatibility with the second order finite elements.

Assuming that the dominant TEM mode is incident from port j, a discretized relation of the electric field and its normal derivative at the planar waveguide ports has been found [3]:

$$\{E_z\}_{\Gamma_i} = 2\delta_{ij}\{f_{i0}\} + [Z]_i \begin{Bmatrix} \dfrac{\partial E_z}{\partial n} \end{Bmatrix}_{\Gamma_i} \tag{5}$$

where

$$[Z]_i = -\sum_{m=0}^{\infty} \frac{1}{j\beta_{im}} \{f_{im}\} \sum_{e^i} \int_{e^i} f_{im}(y_0^{(i)}) \cdot \{N(x^{(i)} = 0, y_0^{(i)})\}\, dy_0^{(i)}$$

$$\cdots, M \tag{6}$$

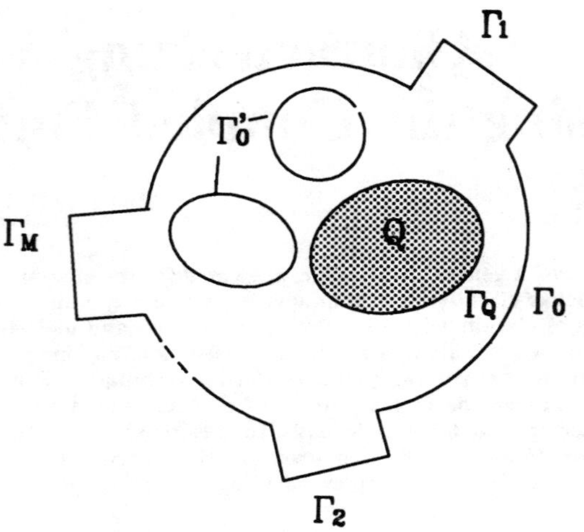

Fig. 1. The geometry of the problem and the computation regions.

and the components of the $\{f_{im}\}$ vector are the values of the mode function $f_{im}(y^{(i)})$ at the nodal points on Γ_i; Σ_{e^i} extends over the elements related to Γ_i; and $\{N\}$ is the second order boundary element shape function.

It should be realized that the boundary conditions that apply to our generalized problem consist of field continuity (compatibility), field normal derivative continuity (equilibrium), as well as the condition that $\partial E_z/\partial n$ equals to zero on the magnetic wall, and E_z vanishes on the conductor surfaces. The complete problem being addressed here can now be described finally in a matrix form, i.e., with the help of (2), (4), (5), and the boundary conditions; it is written in the form:

$$\begin{bmatrix} [H_0, H_1, H_2, \cdots H_M, H_Q, 0] & -[G_0; G_1, \cdots G_M, G_Q] \\ \begin{bmatrix} & 0 & \end{bmatrix} \begin{bmatrix} A \end{bmatrix} & \begin{bmatrix} 0 & 0 & \cdots & 0 & 0 \\ 0 & 0 & \cdots & 0 & D \end{bmatrix} \\ \begin{bmatrix} 0 \end{bmatrix} \begin{bmatrix} 1 & & 0 \\ & 1 & \\ & & \ddots \\ 0 & & 1 \end{bmatrix} \begin{bmatrix} 0. \end{bmatrix} & \begin{bmatrix} -Z_1 & 0 \\ & \ddots \\ 0 & -Z_M \end{bmatrix} \begin{bmatrix} 0 \end{bmatrix} \end{bmatrix} \cdot \begin{Bmatrix} \{E_z\}_{\Gamma_0} \\ \{E_z\}_{\Gamma_1} \\ \vdots \\ \{E_z\}_{\Gamma_M} \\ \{E_z\}_{\Gamma_Q} \\ \{E_z\}_Q \\ \dfrac{\partial E_z}{\partial n}_{\Gamma_0} \\ \dfrac{\partial E_z}{\partial n}_{\Gamma_1} \\ \vdots \\ \dfrac{\partial E_z}{\partial n}_{\Gamma_M} \\ \dfrac{\partial E_z}{\partial n}_{\Gamma_Q} \end{Bmatrix} = \begin{Bmatrix} 0 \\ 0 \\ 0 \\ 2\delta_{ij} f_{i0} \\ 0 \\ \vdots \\ 0 \end{Bmatrix}. \tag{7}$$

In the above equation, [1] is an identity matrix; [0] is an empty matrix, $\{E_z\}_{\Gamma_i}$ and $\{\partial E_z/\partial n\}_{\Gamma_i}$ ($i = 0, 0', 1, 2, \cdots M, Q$) correspond to electric fields and their normal derivatives at the nodal points related to boundary $\Gamma_0, \Gamma_{0'}, \Gamma_1, \Gamma_2, \cdots, \Gamma_M, \Gamma_Q$, respectively; and $\{0\}$ is a null vector.

The solution of the matrix equation determines the scattered electric field distribution across each planar waveguide port. Using the orthogonality of the modes in a planar waveguide, the scattering parameters of the fundamental mode can be easily determined.

IV. Numerical Implementation and Examples

The algorithm described in the preceding section has been implemented as an user-friendly computer package, named PLANAR MICROWAVE CIRCUIT SIMULATOR. This software package has the advantage of requiring only short CPU computer time, as well as having a small memory requirement. This then allows the package to be run on IBM PCs or compatibles. The full-screen graphics interface makes it easy to input the device geometry [5].

Example 1: A Microstrip Line with Conductor Posts (Cavity Filter)

When the substrate in a circuit is homogeneous, the general formula, i.e. (7), degenerates into the BEM-based algorithm. A typical example of this special case is a four-post band-pass filter, using half-wavelength sections as series resonators and shunt posts, as shown in Fig. 2. The filter was originally synthesized using Chebyshev filter theory with a ripple level of 1.0 dB, and constructed using a four-post structure in a Duroid substrate with thickness = 0.49 mm and $\epsilon_r = 2.43$. However, in the present analysis, the filter is considered as a completely integrated system consisting of four circular electric walls and magnetic side walls. The CFBM result agrees well with the results of the multipole expansion method (not shown here) and the measured results are presented in [6].

Example 2: Microstrip Disk Filter with Dielectric Load

A microstrip circular disk filter is considered as an example of a planar circuit with an arbitrary shape. In this study, the effective radius $R_{\text{eff}} = 22.8$ mm is derived from the value of the original radius, i.e. $R = 21.58$ mm. In this example, a square dielectric load with $\epsilon_r = 10.2$ is placed at the center of a disk. The comparison of the predicated and measured results in Fig. 3 indicates that the CFBM is applicable to complex microstrip planar devices. Strictly speaking, fringing effects depend on the resonant mode and can be taken into account more accurately by considering an equivalent model for each mode. This mode dependency is evident in the discrepancy between measurements and simulations at higher frequency. Considering the inevitable radiation loss that will be experienced by this open structure, the small discrepancy can be expected, even though the TRL calibration technique was used for the measurements.

Example 3: A 3-Port Microwave Planar Ferrite Circulator

As an example of a planar circuit of thin substrate with an anisotropic dielectric load, a Y junction with a TT1-109 triangular ferrite post shown in Fig. 4 is considered. The widths of the three planar waveguide ports are the same (i.e. $W_1 = W_2 = W_3 = 8.2$ mm) and the substrate constant is, $\epsilon_r = 2.2$. In the analysis, both magnetic losses and dielectric losses are neglected. The S-parameters obtained using the CFBM are shown in Fig. 4, where it is observed the device acts like a wideband isolator (circulator).

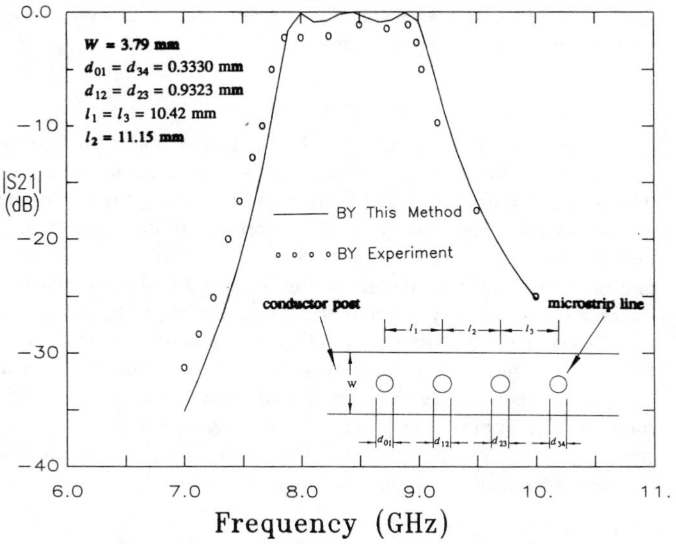

Fig. 2. Computed and measured $|S_{21}|$ for a four-post band-pass direct-coupled cavity filter.

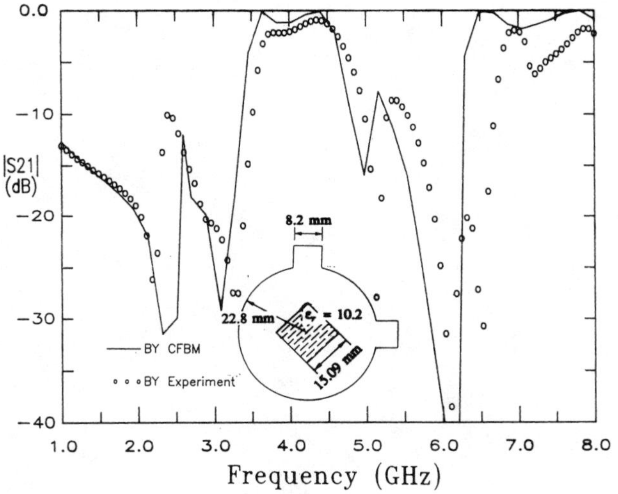

Fig. 3. Computed and measured transmission coefficients for a microstrip circular disk filter with a square dielectric load.

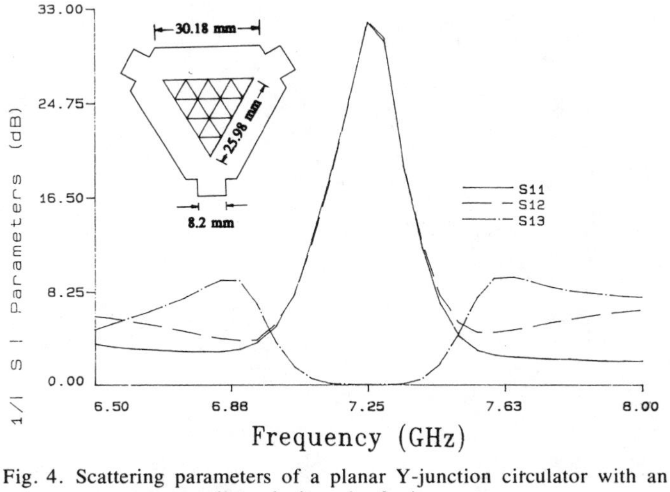

Fig. 4. Scattering parameters of a planar Y-junction circulator with an equilateral triangular ferrite post.

In this example, only 16 triangular second order finite elements are used to obtain a convergent result.

V. Conclusions

A generalized coupled finite and boundary element method has been shown to be applicable to microwave planar circuit problems. The planar waveguide model is used in developing the technique. The technique takes advantage of the strengths of the finite element and boundary element methods. Thus, it can handle complicated and arbitrarily shaped planar circuits with a small computational overhead. The validity of the method was confirmed by comparing the CFBM results, either with published results or with experimental results. The performance of a Y-junction circulator with an equilateral triangular ferrite post was also investigated. For all the numerical examples, the power conservation condition has been found to be satisfied to an accuracy of $\pm 10^{-5}$ to $\pm 10^{-4}$ within the frequency band of the dominant mode.

Acknowledgment

The authors wish to thank Sharon R. Aspden of Rogers Corporation for providing the dielectric material used in this research under the auspices of their university program.

References

[1] T. Okoshi, *Planar Circuits*. Berlin: Springer-Verlag, 1985.

[2] Z. J. Cendes and J. F. Lee, ''The transfinite element method for modeling MMIC devices,'' *IEEE Trans. Microwave Theory Tech.*, vol. 36, pp. 1639–1649, 1988.

[3] K. L. Wu and J. Litva, ''Boundary element method for modelling MIC devices,'' *Electron. Lett.*, vol. 26, no. 8, pp. 518–520, 1990.

[4] K. L. Wu, G. Y. Delisle, D. G. Fang, and M. Lecours, ''Waveguide discontinuity analysis with a coupled finite-boundary element method,'' *IEEE Trans. Microwave Theory Tech.*, vol. 37, pp. 993–998, 1989.

[5] Integrated Antenna Group, *User's Guide of Planar Microwave Circuit Simulator*, Communications Research Laboratory, McMaster University, Hamilton, ON, Canada, 1991.

[6] K. L. Finch and N. G. Alexopoulos, ''Shunt post in microstrip transmission lines,'' *IEEE Trans. Microwave Theory Tech.*, vol. 38, pp. 1585–1594, 1990.

Chapter 10

Hybrids and Power Dividers/Combiners

M. D. Abouzahra
M.I.T. Lincoln Laboratory

K. C. Gupta
University of Colorado at Boulder

10.1 INTRODUCTION

PASSIVE power dividers, hybrids, and power combiners are multiport devices that have found extensive use at microwave and millimeter-wave frequencies. Power dividers, for instance, have been used in antenna-array feed networks to distribute the microwave signal into a number of equal-phase outputs of smaller power [10.1]. Power dividers have also been used to combine several signals and obtain a single output of larger power [10.2]; in this application the device is referred to as a power combiner. The general requirements of a power divider include wide operational frequency bandwidth, good isolation between the output ports, low return loss for input and output ports, and good phase and amplitude balance among the output ports. Power dividers are often of the equal-division type; however, unequal power division is also used. The most commonly used power dividers/combiners are the T-junctions and the Wilkinson-type [10.3–10.5] circuits.

Hybrids, on the other hand, are typically four-port devices that divide an input signal into two equal (or unequal) but out-of-phase signals. If the phase difference between the two equal outputs is 90°, the component is called a 3-dB quadrature hybrid; if the phase difference at the output ports is 180°, the device is called a 180° 3-dB hybrid. Examples of quadrature-type hybrids are branch-line couplers, parallel-line directional couplers, and interdigitated directional couplers (widely referred to as Lange couplers) [10.6]. Rat-race couplers (also referred to as hybrid ring couplers) and magic-T couplers are 180°-type 3-dB hybrids [10.6–10.8].

As demand increases for multiport power dividers/combiners and 3-dB hybrids operating at higher frequencies, the just-mentioned conventional type of hybrids (which are based on the transmission line concept) begin to encounter design and manufacturing difficulties. Issues like parasitic reactances and unwanted line-to-line coupling start to become serious. In view of these difficulties, the possibility of realizing compact two-dimensional hybrids and multiport power dividers has been investigated by many authors over the past two decades. Several new planar-type components have emerged from these investigations [10.9]. Circular microstrip discs, annular microstrip rings, circular microstrip sectors, and rectangular microstrip discs have been used to design 3-dB quadrature hybrids [10.10–10.15], 180° 3-dB hybrids [10.16–10.22], n-way planar equal-power dividers [10.23–10.35], n-way planar unequal-power dividers [10.28, 10.36, 10.37], and n-way radial power dividers/combiners [10.38, 10.39]. In this chapter we shall discuss the analysis and design of these new components and comment on their performance.

10.2 Hybrids

As stated earlier, hybrids are a component group that plays an important role in the design of modern microwave and millimeter-wave integrated circuits. In principle, a hybrid circuit can be represented as a multiport network. In such a network, the port into which the input power is fed is usually called the incident, or input, port. Ports through which the power exits the hybrid circuit are usually called output ports. All other ports are referred to as isolated ports. Hybrids are generally of two types: (1) the 90° type, sometimes called quadrature, and (2) the 180° type. The simplest examples of the 90°- and 180°-type hybrids are, respectively, the branch-line coupler and the rat-race coupler. The problem is that at high frequencies, these conventional designs become harder to implement because of their small dimensions and parasitic reactances at the junctions.

To circumvent these difficulties, alternative hybrid structures have been proposed, some of which are depicted in Fig. 10.1. These new structures are planar in nature and rely on using resonators of regular shapes (i.e., circular, sectorial, and rectangular). In this section representative examples of both 90°- and 180°-type planar hybrid circuits will be reviewed.

Fig. 10.1 Planar hybrid circuits with regular shapes.

10.2.1 Planar-Type Quadrature Hybrids

Three planar geomtries have been shown to exhibit quadrature hybrid characteristics: (1) the optimized ring, (2) the rectangular disc, and (3) the circular disc. Measured and predicted characteristics for each of the three cases will be discussed next.

a. Optimized Ring-Type Quadrature Hybrid [10.10]

A conventional hybrid ring-type quadrature coupler consists of four quarter-wave-long sections of uniform width, as shown in Fig. 10.2. The characteristic impedance of the two parallel lines is $z_o/\sqrt{2}$, and that of the other two lines is z_o. As the operational frequency increases, the width of the through and shunt lines becomes comparable to their lengths. In such a case, the lines in this circuit become two-dimensional; hence the planar circuit approach must be used.

Okoshi and his coworkers [10.10] took advantage of the larger degree of freedom that is inherent in the planar circuit pattern to realize better wide-band hybrid characteristics. A fully computer-oriented synthesis approach was utilized to optimize the circuit pattern and hence obtain wide-band hybrid characteristics. To prevent the possible difficulty of having too many degrees of freedom, the authors assumed the external periphery of the circuit pattern to be circular, as shown in Fig. 10.3. The external diameter, position of external ports, and shape of the internal periphery were adjusted, as shown in Fig. 10.4, to obtain the best wide-band hybrid characteristics.

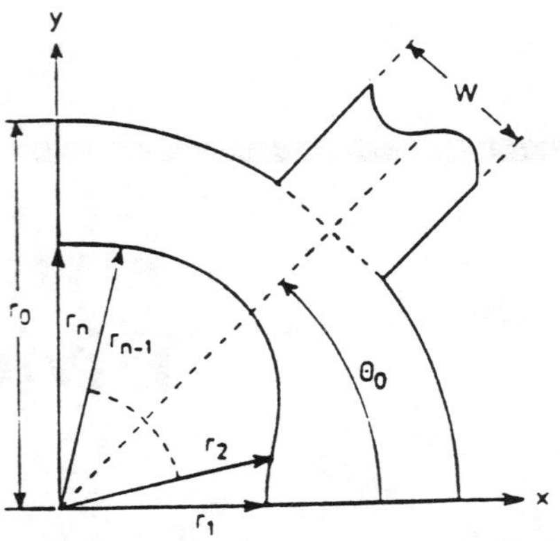

Fig. 10.3 A quarter circuit of the hybrid ring and the pattern variables used in the synthesis.

This computer-oriented synthesis approach was validated by fabricating and measuring the performance of a hybrid circuit. A symmetrical stripline structure was selected. The stripline was 1.45 mm thick, and the dielectric spacer was Rexolite 1442 with a relative dielectric constant of 2.53. External matching stubs and/or quarter-wave impedance transformers were added to obtain wider bandwidth. The measured performance of this circuit is represented in Fig. 10.5 by small circles (S_{11}), squares (S_{12}), triangles (S_{13}), and asterisks (S_{14}). The thin dashed curves represent the computed optimized characteristics. The thicker dashed lines are also computed characteristics, but they are corrected by a circuit loss of 0.58 dB. These loss values were arrived at by the authors upon measuring the insertion loss of a 55-mm-long straight stripline section (with OSM connectors connected on both ends) identical to those used in the hybrids. Figure 10.6 illus-

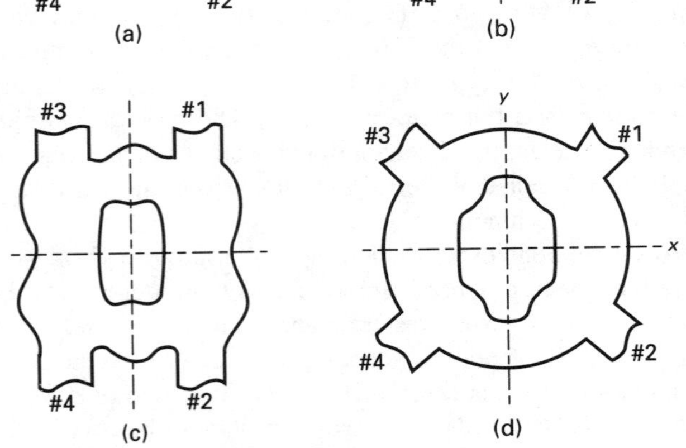

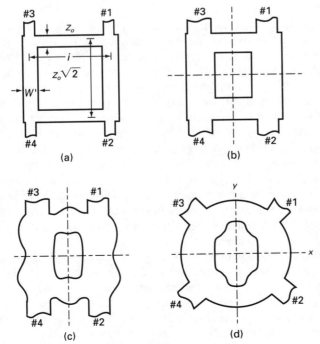

Fig. 10.2 Ring-type quadrature hybrids.

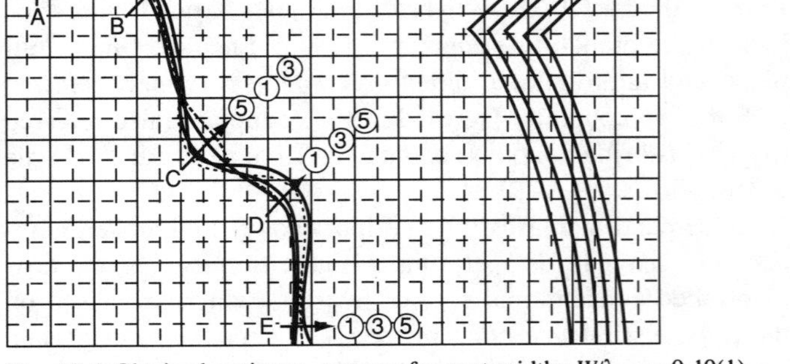

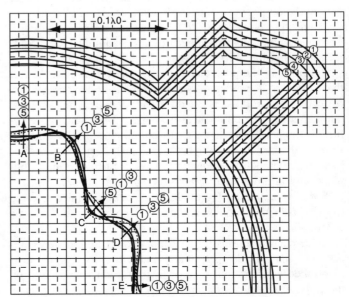

Fig. 10.4 Obtained optimum patterns for port widths W/λ_o = 0.10(1), 0.09 (2), 0.08 (3), 0.07 (4), and 0.06 (5). For the inner periphery, (2, 4) are given as broken and dotted curves, respectively; (1, 3, 5) are all solid curves, and their positions (A, B, C, D, E) to prevent confusion.

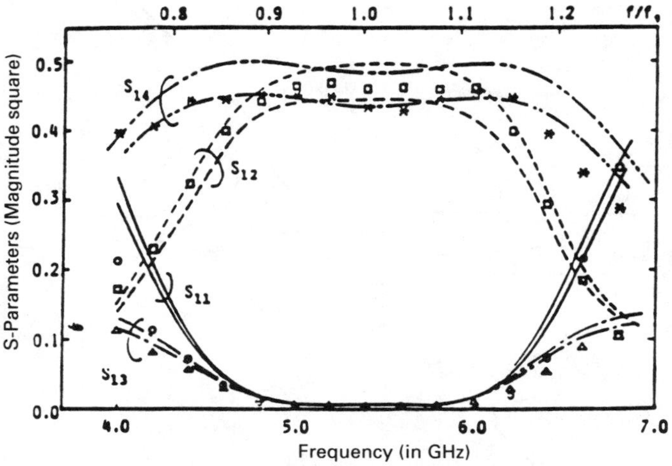

Fig. 10.5 Theoretical and measured amplitude characteristics for an optimized ring hybrid with quarter-wave impedance transformers.

trates the differential phase characteristics given by $\phi = [\arg(S_{12}) - \arg(S_{14})] - 90°$ for these circuits. The difference between the measurements and theoretical predictions is well below 10° over 50 percent bandwidth.

b. Rectangular Disc-Type Quadrature Hybrid [10.11–10.13]

Figure 10.7 illustrates a profile of the rectangular patch quadrature hybrid, first proposed by Burns [10.11] in a 1986 U.S. patent. Burns presented this geometry as an alternative design at 94 GHz to the conventional branch-line coupler. He gave no detailed theoretical treatment apart from two empirical design formulas relating the dimensions of the patch to the wavelength inside the dielectric substrate. The synthesis equations that were used in [10.11] are given by

$$b = 0.46\lambda_g \qquad (10.1)$$

$$a = b + w\sqrt{2} \qquad (10.2)$$

where a and b represent the dimensions of the rectangular patch and w is the width of the feeding ports. Fusco and Stewart [10.12] used Eqs. (10.1) and (10.2) in an attempt to design a 10-GHz quadrature hybrid; they were unsuccessful. As a result, a new design procedure based on the finite-difference method was presented. This method was implemented, and favorable results were obtained.

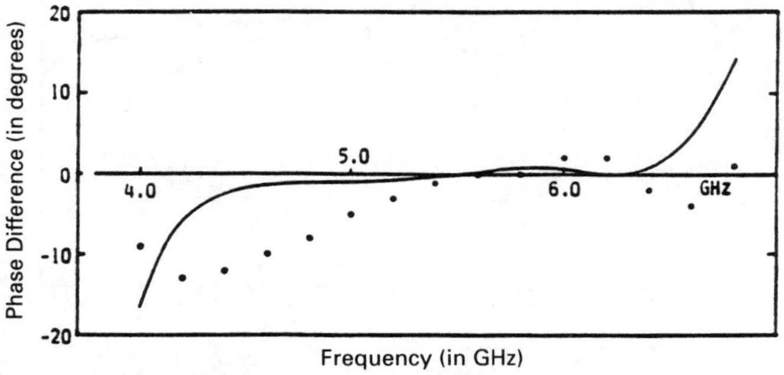

Fig. 10.6 Theoretical and measured phase characteristics of an optimized ring hybrid with quarter-wave impedance transformers.

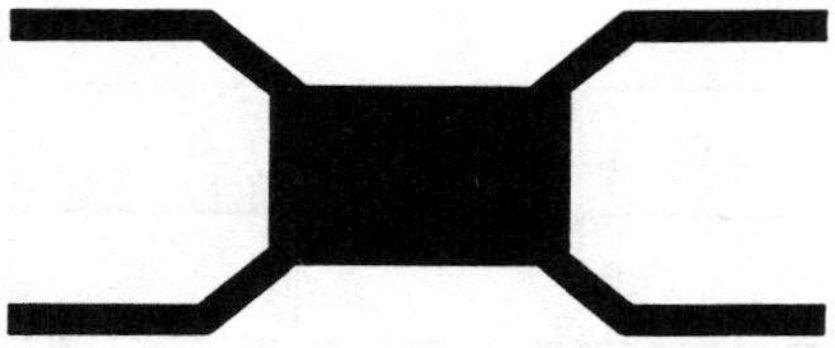

Fig. 10.7 Schematic of a rectangular patch quadrature coupler.

Subsequently, Merugu, Fusco, and Stewart [10.13] presented a three-dimensional, full-wave approach based on the transmission line matrix method for analyzing a square-patch quadrature hybrid. This hybrid was fabricated on a duroid substrate 0.7874 mm thick, with a dielectric constant of 2.20. The side length of the square patch is 8.66 mm (about $52\lambda_g$ at 13.2 GHz). This device was housed in a metallic enclosure, with provisions to vary the height of the top cover. The measured and calculated frequency characteristics for this hybrid, with the height of the top cover located 10 mm above the hybrid, are shown in Figs. 10.8 and 10.9. As shown, this device exhibits an equal power split between the output ports. The power split is -4.75 ± 0.5 dB with respect to the input port over a bandwidth of 7 percent. The port-to-port isolation is better than -16 dB, and the phase variation between the output ports is within $90 \pm 8°$ over the frequency band of interest. All four ports exhibited a voltage standing wave ratio (VSWR) of 2:1 or better.

c. Circular Disc-Type Quadrature Hybrid [10.14, 10.15]

Circular disc microstrip structures have been shown to exhibit attractive characteristics when used as power dividers and combiners [10.17–10.19, 10.23–10.28]. Recently, Page and Judah [10.14] reported that, similar to the rectangular and annular disc structures, planar circular disc structures can also be used as 3-dB quadrature hybrid devices. Figure 10.10 illustrates the symmetrical four-part disc structure that was considered in [10.14]. The planar circuit approach was used to determine the port positions (i.e., the value of Φ) at which the circuit is matched and exhibits equal power division charac-

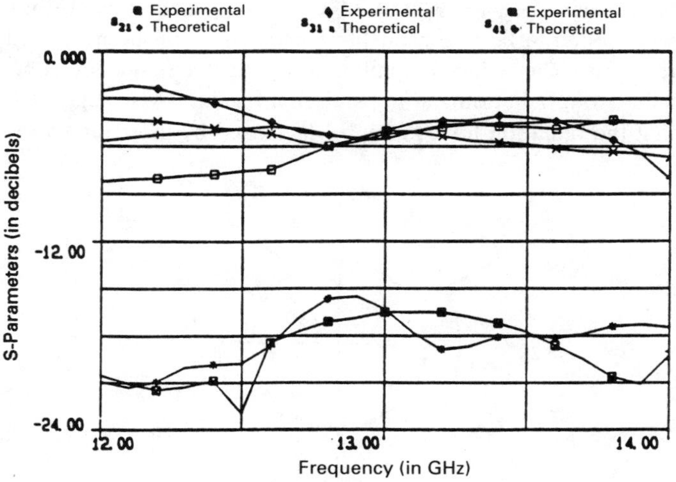

Fig. 10.8 Coupling and isolation characteristics of the square patch coupler. (From [10.13]. Copyright 1989. Reprinted by permission of John Wiley & Sons, Inc.)

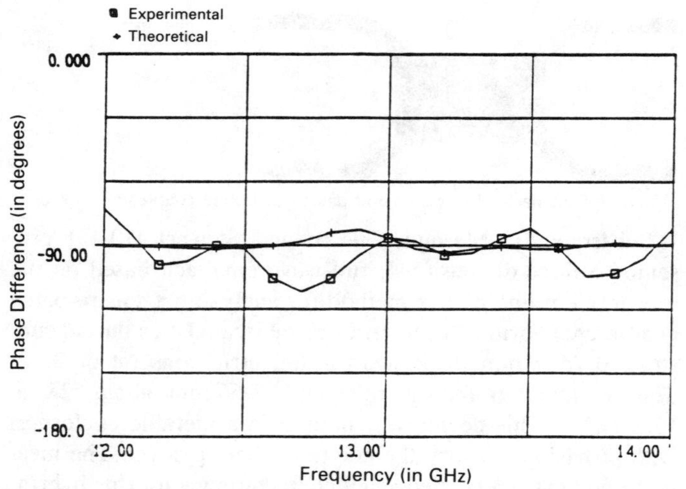

Fig. 10.9 Phase differential between the output ports of the square patch coupler. (From [10.13]. Reprinted by permission of John Wiley & Sons, Inc.)

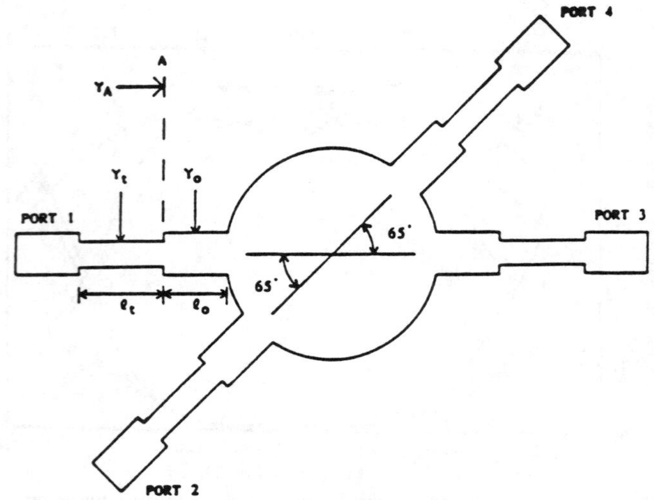

Fig. 10.11 A 3-dB circular disc quadrature hybrid with impedance transformers as matching networks.

teristics. This work is based on the fact that a matched four-port junction, symmetrical about two perpendicular axes, can be represented as a 90° hybrid [10.41].

Later on [10.15], the authors extended their earlier work to provide a flexible design procedure. This procedure was illustrated by considering a four-port disc geometry with a radius of 15 mm on a substrate having a thickness of 1.52 mm and a relative permittivity of 2.5. A contour plot indicating the combinations of frequencies and port separations for which 3-dB quadrature hybrid characteristics can be achieved was presented. It was demonstrated in [10.15] that 3-dB quadrature hybrid characteristics can be achieved over a wide range of frequencies simply by appropriately selecting the port-to-port angular separation Φ and the matching network.

Considering the 65° port separation, it can be seen that a quadrature hybrid operation can be achieved at three different frequencies: 4.34, 5.34, and 7.35 GHz. Figure 10.11 illustrates the geometry of the hybrid. The performance of this device at 7.35 GHz was discussed in [10.14]. Figure 10.12 shows the measured frequency response of this device when operated at a center frequency of 4.34 GHz. The theoretical results were found to be in reasonable agreement with the measured response. The reduced coupling at the center frequency is attributed to radiation and transmission losses. It was suggested that the radiation losses may be decreased by inserting shorting pins along the periphery of the device between the two ports. The slight shift in center frequency is attributed to inadequate modeling of the fringing fields. In designing the device to operate at 4.34 GHz, the parameters of the matching network are l_o = 9.515 mm, Y_o = 0.02 mhos, l_t = 12.1 mm, and Y_t = 0.013 mhos. When operating at 5.36 GHz these parameters become l_o = 24.4 mm, Y_o = 0.02 mhos, l_t = 9.9 mm, and Y_t = 0.012 mhos. In the second case the length of l_o, which is normally a quarter-wave long, has been extended from 5.05 mm to 24.4 mm to avoid the situation where the length of the line is comparable to its width.

No theoretical or measured phase data were presented in [10.15]. Figure 10.13 illustrates the calculated phase difference between the coupling ports as presented in [10.14] for the 7.35-GHz case. Clearly this device offers a very limited bandwidth and hence would need further improvement.

10.2.2 Planar-Type 180° Hybrids

A 180° 3-dB hybrid is a four-port device that produces a 180° phase shift between the two equal outputs while the fourth port remains isolated. The same device can also produce two equal outputs having the same phase. This device is widely referred

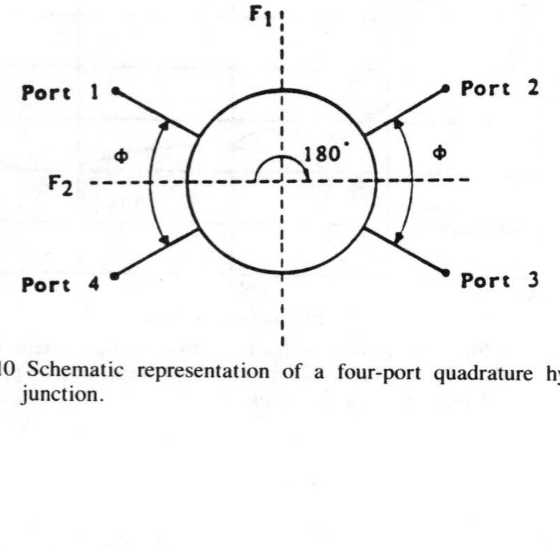

Fig. 10.10 Schematic representation of a four-port quadrature hybrid junction.

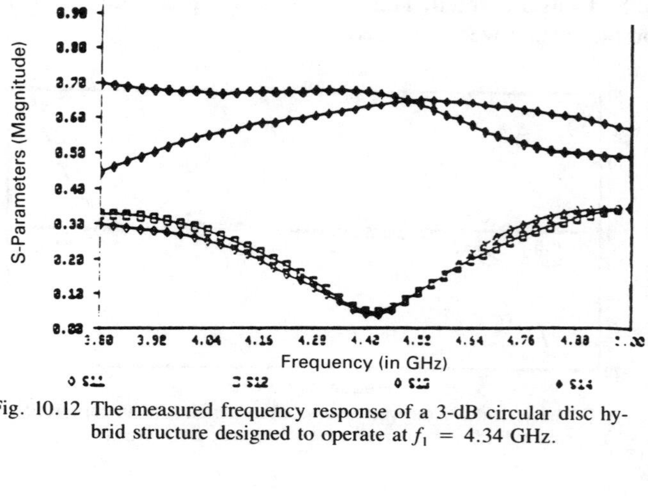

Fig. 10.12 The measured frequency response of a 3-dB circular disc hybrid structure designed to operate at f_1 = 4.34 GHz.

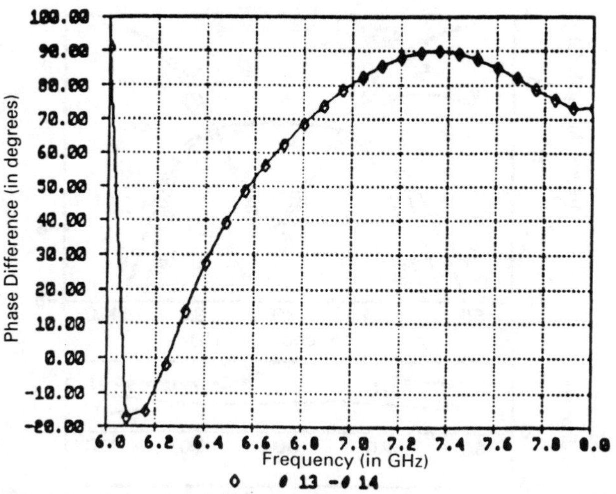

Fig. 10.13 Predicted phase difference between the output ports of the circular disc quadrature hybrid.

to as a magic-T or a rat-race coupler. As the demand for 3-dB hybrids operating at higher frequencies increases, the conventional hybrid circuits (which are based on the transmission line concept) begin to encounter manufacturing and design difficulties. Issues such as parasitic reactances and unwanted line-to-line coupling start to arise. In view of these difficulties, the possibility of realizing compact 180° 3-dB hybrids using simple planar structures has been investigated [10.16–10.22]. In the following sections, circular and rectangular disc-shaped 180° 3-dB hybrid structures are discussed.

a. Circular Disc, 180° 3-dB Hybrid [10.16–10.18]

Okoshi and Takeuchi [10.16] were the first to report on the usefulness of circular disc structures such as 180° 3-dB hybrids. Figure 10.14(a) illustrates the disc-shaped geometry that was proposed in [10.16]. The basic operating principle of this circuit can be understood by considering the dipolar mode excitation of the circular disc resonator depicted in Fig. 10.15. When the dipolar mode is present, the voltage distribution has a null along a diameter located normal to the feeding port (port 1 in this case). Furthermore, the voltage is positive in one semicircular region and negative in the other, as shown in Fig. 10.15; hence no voltage is induced at port 2, and the voltages induced at ports 3 and 4 are negative of that at port 1. In

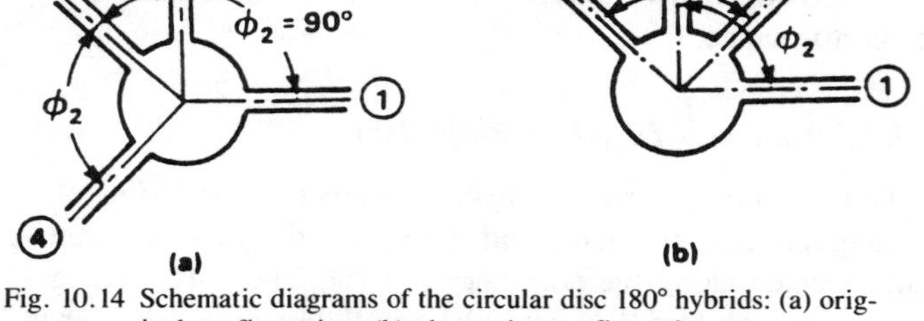

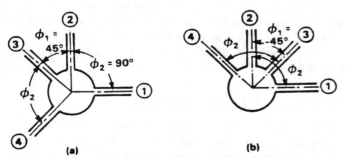

Fig. 10.14 Schematic diagrams of the circular disc 180° hybrids: (a) original configuration; (b) alternative configuration.

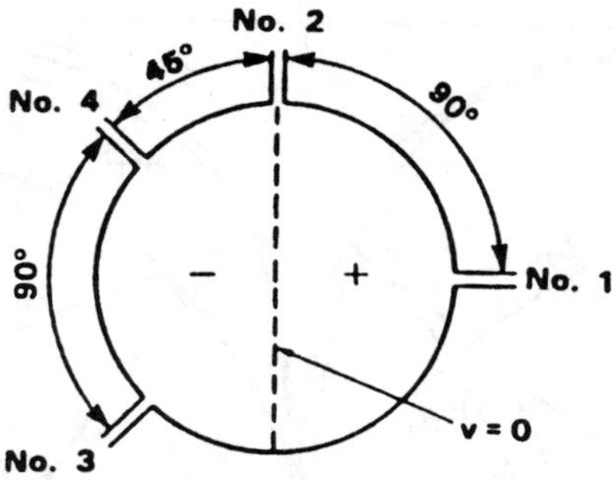

Fig. 10.15 Voltage distribution of the (1, 1) mode of a disc type 3-dB planar hybrid.

other words, when this circuit is fed at port 1, port 2 is isolated, whereas the output powers at ports 3 and 4 are equal in amplitude and phase. On the other hand, if the input power is fed into port 2, then no power will be coupled to port 1, whereas the output powers at ports 3 and 4 are equal in amplitude but 180° out of phase.

Figure 10.14(b) illustrates another geometrical variant of 10.14(a), but with the same hybrid characteristics. The characteristics of these components were computed in [10.16] for a circular stripline disc of radius $R = 7.89$ mm and dielectric substrate with a height of 1.53 mm and $\varepsilon_r = 2.53$. The impedance of the coupling ports was taken to be 50 ohms. The computed characteristics in both cases were found to suffer from unbalanced power division, insufficient coupling, increase in return loss, and separation of optimum frequencies for each performance parameter. The deterioration in the hybrid property of the circuit is believed to be associated with the presence of the higher-order modes, which have not been considered in the analysis. An improvement in the characteristics of this hybrid was obtained [10.17] by rearranging the locations of the four ports around the disc periphery while preserving the symmetry of the circuit. Figure 10.16 shows the computed characteristics for the improved 3-dB disc hybrid whose radius is 7.92 mm and port locations given by $\phi_1 = 53.33°$, $\phi_2 = 101.03°$. These characteristics show a great improvement over those reported in [10.16]. Figure 10.17 demonstrates the measured characteristics of the modified circuit. These measured results are in remarkable agreement with the theoretical results shown in Figure 10.16.

Further improvements to the characteristics of the hybrid geometries illustrated by 10.14(a) and (b) were later obtained [10.18] by (1) adding a capacitive stub at the periphery of the disc, (2) adding a matching network between the resonator and each of the feeding ports, and (3) including the effect of the higher-order modes by rearranging the locations of the four ports. The addition of the stub as shown in Fig. 10.18 makes the doubly degenerate dipole modes nondegenerate. The presence of the stub is modeled by adding a fifth port to the circuit. The fifth port is treated as an open-circuited 50-ohm line of a small length. A proper choice of the length of the stub and the positions of the ports led to a significant improvement in the characteristics of the circuit [10.18]. The bandwidth of the same circuit was broadened by adding high-impedance steps as matching network to all the feeding ports. Figure 10.19

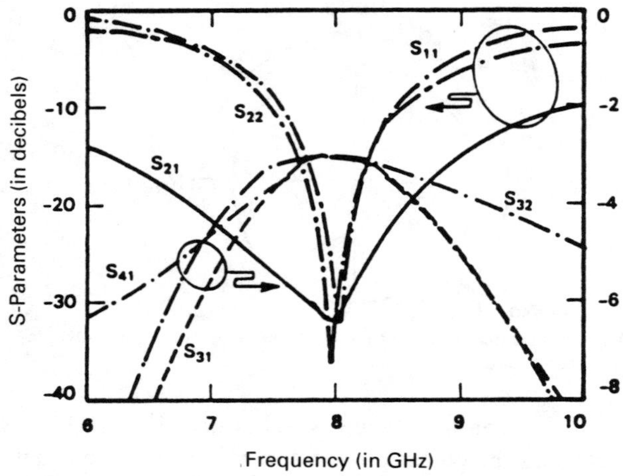

Fig. 10.16 Predicted frequency characteristics of the improved hybrid (from 10.17, reprinted with permission from Institute of Electronics).

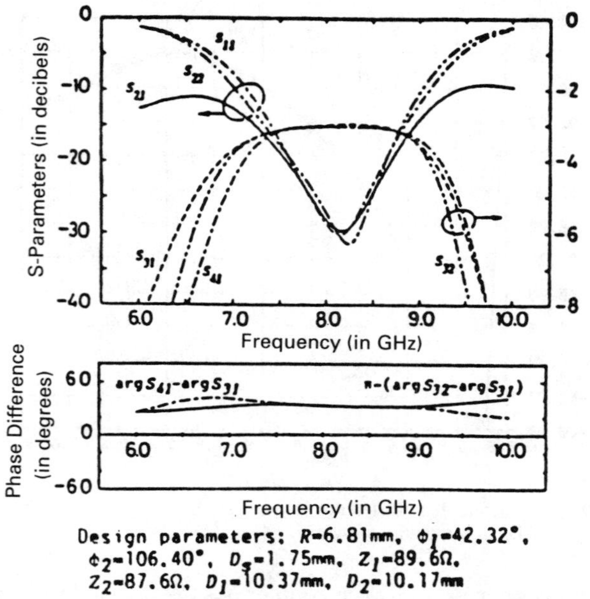

Fig. 10.19 Computed scattering characteristics of the hybrid depicted by Fig. 10.14(a) with matching network, a capacitive stub, and rearranged ports (from 10.18, reprinted with permission from Scripta Technica).

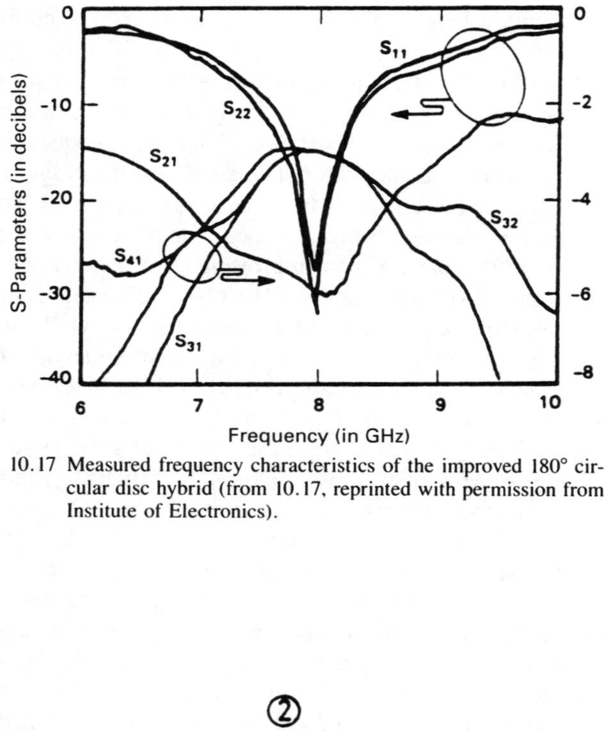

Fig. 10.17 Measured frequency characteristics of the improved 180° circular disc hybrid (from 10.17, reprinted with permission from Institute of Electronics).

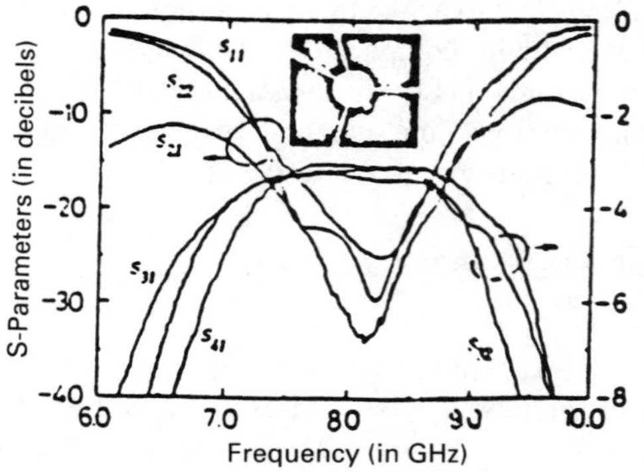

Fig. 10.20 Measured scattering parameters of the hybrid discussed in Fig. 10.19 (from 10.18, reprinted with permission from Scripta Technica).

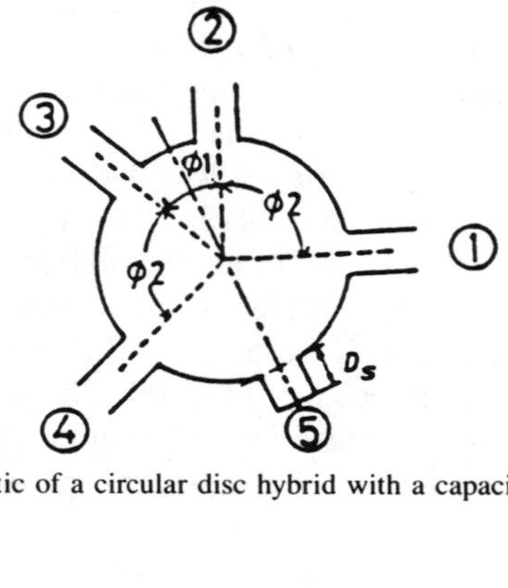

Fig. 10.18 Schematic of a circular disc hybrid with a capacitive stub.

illustrates the calculated characteristics for a hybrid with high-impedance steps, a capacitive stub, and rearranged ports. The measured characteristics for this device are shown in Fig. 10.20. Figures 10.21 and 10.22 display, respectively, the calculated and measured characteristics for a hybrid disc of the type depicted by Fig. 10.14(b) but with high-impedance steps and rearranged ports. Because of the somewhat unacceptable amplitude balance demonstrated in Fig. 10.22, the authors tapered the impedance steps at an angle of 45° and obtained better performance.

b. Rectangular Disc 180° 3-dB Hybrid

This section describes a hybrid geometry consisting of a rectangular disc resonator with two pairs of ports symmetrically located along the periphery [10.19]. The ports arrangements are such that they could be parallel or perpendicular to

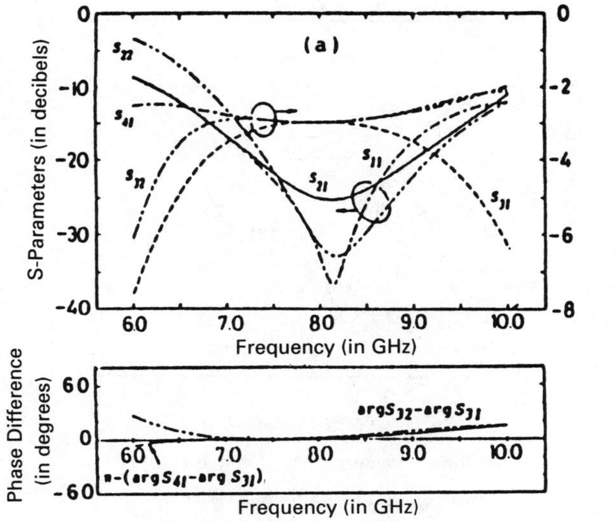

Fig. 10.21 Computer characteristics of the 180° hybrid depicted by Fig. 10.14(b) but with matching networks and rearranged ports (from 10.18, reprinted with permission from Scripta Technica).

one another as shown in Fig. 10.23. As we have seen in other structures, the consisting principle for operating this rectangular structure as a hybrid is the field distribution of the (1, 0) and (0, 1) dipolar modes. Further to what have been reported in [10.19], the voltage anywhere on a square disc excited by a signal fed at any point (x', y') can be expressed in terms of the dipolar modes as follows:

$$V(x, y) = V_o\left[\cos\left(\frac{\pi x'}{a}\right)\cos\left(\frac{\pi x}{a}\right) + \cos\left(\frac{\pi y'}{a}\right)\cos\left(\frac{\pi y}{a}\right)\right] \tag{10.3}$$

where x' and y' represent the location of the source. Taking T_1, T_2, T_3, and T_4 to be equal to T, and assuming that the disc is fed at port 1, the voltage at each port may be written as

$$V_1 = V_o\left[1 + \cos^2\left(\frac{\pi T}{a}\right)\right] \tag{10.4}$$

$$V_2 = 0 \tag{10.5}$$

$$V_3 = -2V_o\cos\left(\frac{\pi T}{a}\right) \tag{10.6}$$

$$V_4 = -V_o\sin^2\left(\frac{\pi T}{a}\right) \tag{10.7}$$

Using Eqs. (10.4–10.7) it can be shown that

$$\frac{|V_3|}{|V_4|} = 2\frac{\cos(\pi T/a)}{\sin^2(\pi T/a)} \tag{10.8}$$

and

$$|V_1|^2 = |V_3|^2 + |V_4|^2 \tag{10.9}$$

Equation (10.4) indicates that port 2 is isolated from port 1. Furthermore, Eq. (10.9) means that a power split equal in phase and amplitude between ports 3 and 4 is possible when $T = 0.364a$.

On the other hand, if the square disc is fed at port 2 instead of 1, then port 1 will be isolated from port 2, and the signals at ports 3 and 4 are equal in amplitude (provided that $T = 0.364a$) but 180° out of phase, as will be shown next. In this case the voltage at each of the ports will be given by

$$V_2 = V_o\left[1 + \cos^2\left(\frac{\pi T}{a}\right)\right] \tag{10.10}$$

$$V_1 = 0 \tag{10.11}$$

$$V_3 = V_o\sin^2\left(\frac{\pi T}{a}\right) \tag{10.12}$$

$$V_4 = -2V_o\cos\left(\frac{\pi T}{a}\right) \tag{10.13}$$

Equations (10.11)–(10.13) show that port 1 is isolated, while ports (3) and (4) are out of phase and have equal amplitudes when $T = 0.346a$. Due to symmetry, one can obtain the same characteristics if the signal were to be fed at ports 3 and 4 instead of 1 and 2.

The scattering parameters for the rectangular disc hybrid can be obtained from the impedance matrix elements that are found by following the Green's function approach discussed in Chapter 2. In [10.19], the scattering parameters for the stripline structure shown in Fig. 10.23 (with a = 13.47 mm and

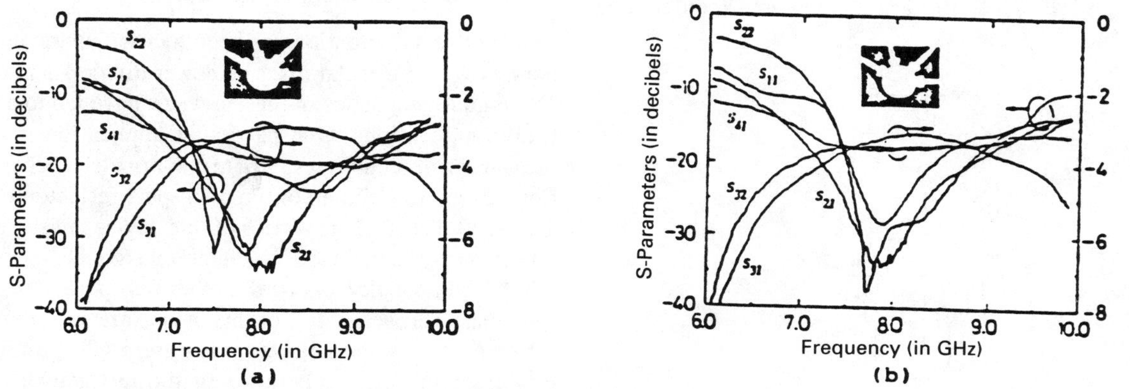

Fig. 10.22 Measured scattering parameters of the hybrid circuit discussed in Fig. 10.21 (from 10.18, reprinted with permission from Scripta Technica)..

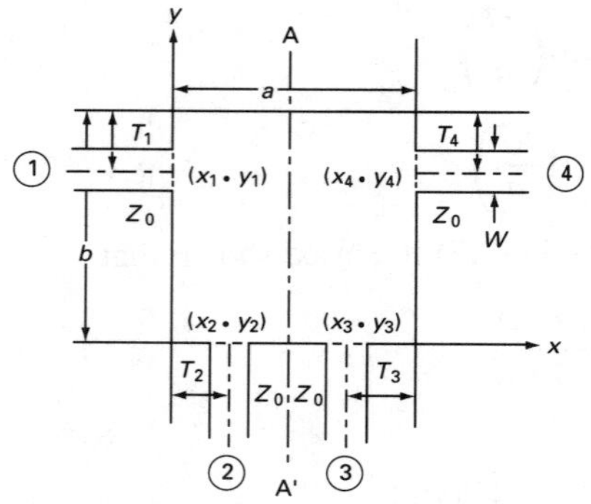

Fig. 10.23 Schematic diagram of a rectangular disc 3-dB 180° hybrid. This circuit becomes fundamental configuration with $a = b$ and $T_1 = T_2 = T_3 = T_4$.

$T = 4.90$ mm) were calculated and found to exhibit unbalanced power division, insufficient coupling, and increased return loss. The dimensions of the square disc were chosen so that the circuit characteristics will be centered at 7 GHz; however, this was not the case, primarily due to the impact of the higher-order modes. As in the case of the circular disc hybrid, the circuit characteristics for the rectangular disc hybrid can be improved significantly by (1) modifying the size of the disc, (2) rearranging the location of the ports, (3) adding a capacitive stub at the circumference of the disc, and (4) adding matching networks in the form of high-impedance steps at each port, as shown in Fig. 10.24. Figures 10.25 and 10.26 show, respectively, the calculated and measured scattering parameters for the improved circuit. Clearly, introducing the earlier-mentioned modifications to the fundamental configuration (Fig. 10.23) has resulted in widening the bandwidth of the device substantially.

10.3 *n*-Way Power Dividers/Combiners

Power dividers and combiner circuits represent a group of microwave components that have found extensive use in communications and radar systems. Two major areas for their applications are (1) distribution of the input signal to, and combining the output from, solid-state amplifiers with limited output power (to obtain higher microwave powers) and (2) distribution of the RF signal to various elements (could be as

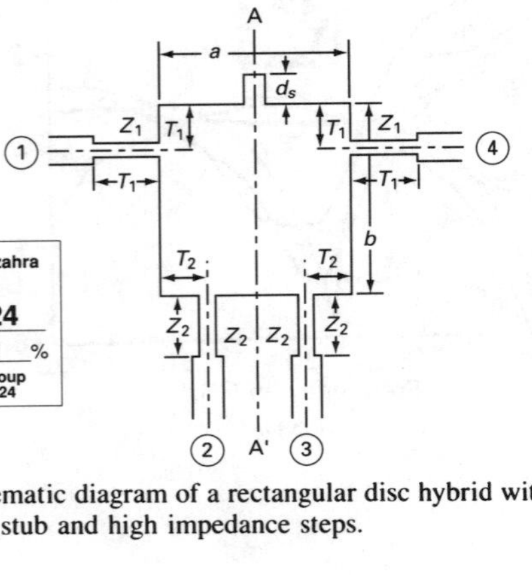

Fig. 10.24 Schematic diagram of a rectangular disc hybrid with a capacitive stub and high impedance steps.

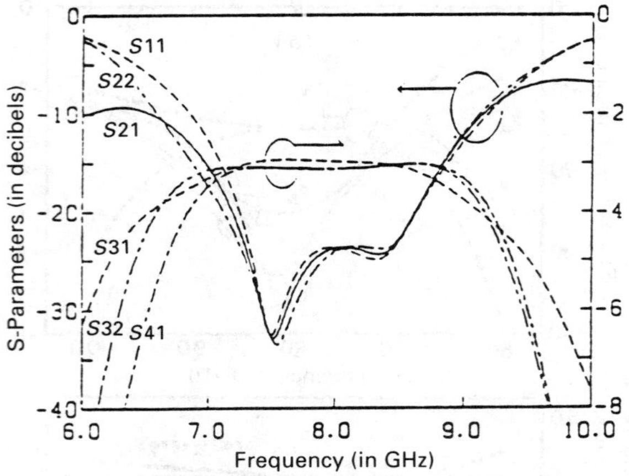

Design parameters: a = 11.78 mm, b = 12.19 mm

T_1 = 1.40 mm, T_2 = 3.58 mm, Z_1 = 90.7 Ω, Z_2 = 91.1 Ω

Z_1 = 9.92 mm, Z_2 = 10.92 mm, ds = 2.83 mm

Fig. 10.25 Predicted characteristics for an improved fundamental hybrid with rearranged ports, modified rectangular disc size, and high impedance steps.

many as thousands) of linear or two-dimensional antenna arrays. In both these applications, we need planar configurations capable of convenient integration with monolithic microwave circuits or with printed microstrip antennas.

Traditionally, microstrip power divider and combiner circuits have been designed based on a configuration proposed by Wilkinson in 1960 [10.3]. An *n*-way Wilkinson circuit consists of *n* quarter-wave lines meeting at a junction. In addition, isolation resistances are added between the quarter-wave lines. The nonplanarity of this configuration, and the parasitic reactances arising at the junction of $n + 1$ lines, makes this type of power divider hard to realize at higher microwave frequencies. In view of these difficulties, several new planar-type power divider-combiner circuits have emerged. These circuits include the edge-fed disc *n*-way power divider [10.23–10.28] (Fig. 10.27(a)), the center-fed disc *n*-way power divider [10.38–10.40] (Fig. 10.27(b)), and the *n*-way sectorial-type power divider [10.29–10.37] (Fig. 10.27(c)). The characteristics of each of these devices are discussed next.

10.3.1 Edge-Fed Discs [10.23–10.28]

In recent years, attention has been focused on the use of multiport, edge-fed circular discs as power dividers and combiners. The unique properties of these devices have led to a number of useful applications. One particular application is in six-port measurement techniques, where a symmetrical five-port junction is used to build an optimal six-port reflectometer [10.23, 10.24, 10.42]. The properties of this device are such that when all the ports are matched, an electrical signal entering one port will be split equally among the other four ports [10.42, 10.43]. The phase characteristics of this device are such that the phase differences between adjacent ports are ±120°, a unique property essential for the operation of six-port network analyzers. Symmetrical six-port microstrip junctions were also investigated and found to yield the power-division characteristics il-

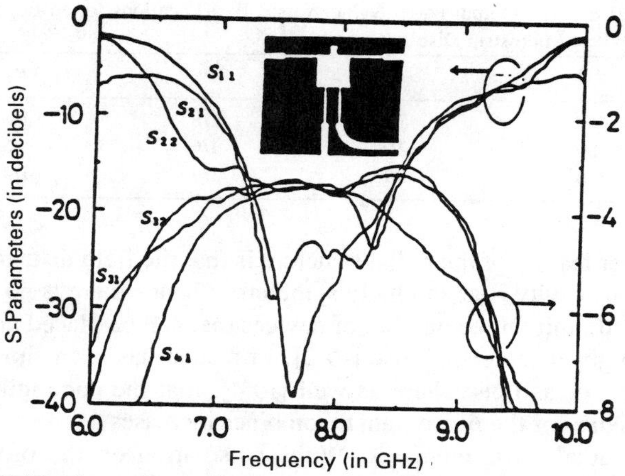

Fig. 10.26 Measured scattering parameters of the hybrid discussed in Fig. 10.33.

lustrated in Fig. 10.28 [10.44]. This device was found to be unsuitable for a six-port measurement system. Similar edge-fed microstrip discs with eight or more rotationally symmetric ports were also reported to have been used as feed networks for lens-fed antenna arrays [10.26].

Figure 10.29 displays another interesting edge-fed microstrip planar power divider [10.28, 10.45]. In this case, the rotational symmetry case has not been preserved; however, some symmetry about the horizontal axis remains. This circuit is comprised of a circular microstrip resonator connected by four coupling ports. The coupling characteristics of this circuit are governed by the locations of the four ports. By positioning port 3 diametrically opposite to port 1 and placing ports 2 and 4 at θ and $-\theta$ relative to port 1, an unequal power divider can be obtained [10.45]. For example, when $\theta = 60°$ and a signal is fed to port 1, half the power will be coupled to port 3, whereas the other half will be split equally between ports 2 and 4. The operating principle of this circuit can be understood by noting the voltage distribution of the (2, 2) mode and the (2, 1) mode resonances. Figure 10.30 shows the measured and calculated characteristics of an experimental model. A good agreement between the measured and calculated results confirms the validity of the planar circuit design approach. The disc radius in the model is 7.60 mm; the substrate thickness and dielectric constant are, respectively, 1/32 inch and 2.2. Figure 10.30 shows that a bandwidth (defined for $|S_{11}| \leq 0.25$ or VSWR ≤ 1.7) of about 41 percent at about 9.5 GHz is obtained. The (2, 2) and (2, 1) resonant modes occurred at 8.4 GHz and 11.2 GHz, respectively.

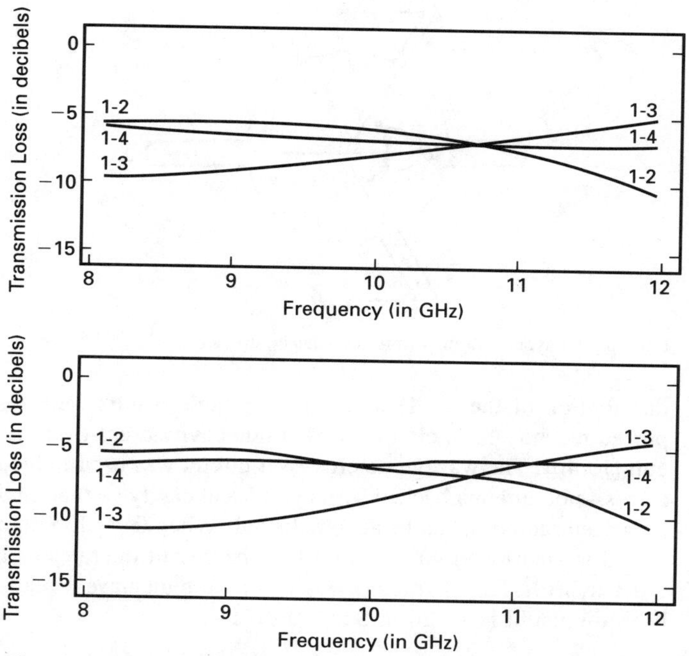

Fig. 10.28 Predicted and measured power division characteristics of a six-way microstrip disc power divider.

Better circuit characteristics can be obtained by changing the characteristic impedance of the various ports to 40 ohms [10.45]. The new $|S_{11}|$ value is 0.0134, compared to the previous one of 0.078 for $Z_o = 50$ ohms. Of course, this design requires the use of external impedance transformation (i.e., matching network) at each of the four ports.

Other circuit configurations leading to different power-division ratios are also possible. Table 10.1 lists the various levels of output powers that can be achieved at ports 2, 3, and 4 for different values of the angle θ. Two such configurations are particularly interesting. The first one is when $\theta = 90°$, making the four coupling ports symmetrically spaced around the disc periphery. In this case, all the input power is coupled to port 3, leaving ports 2 and 4 isolated. At about 15 GHz, the isolation between ports 1 and 2 and ports 1 and 4 is around 20 dB. Figure 10.31 shows the measured and calculated characteristics of this circuit. Once again, the operating characteristics of this circuit can be best understood by noting the voltage

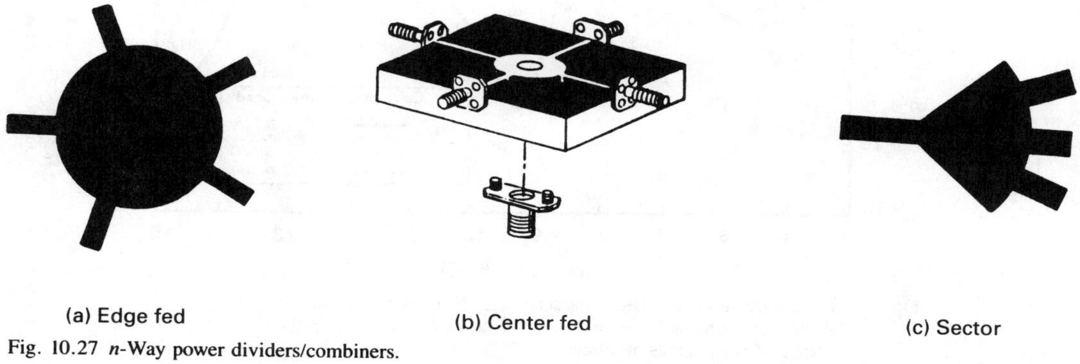

Fig. 10.27 n-Way power dividers/combiners.

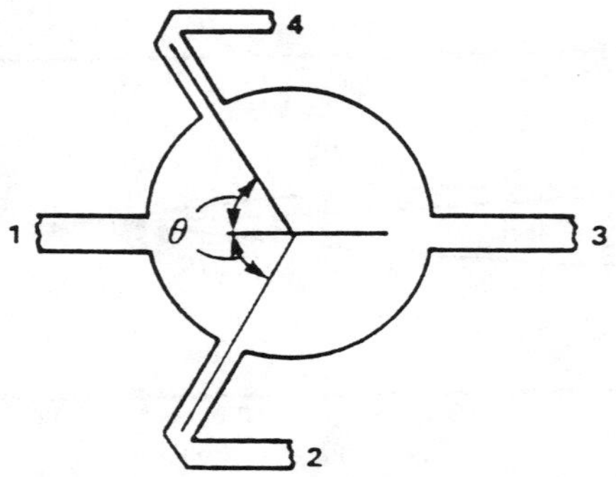

Fig. 10.29 Layout of a three-way disc-shaped divider.

TABLE 10.1 Output Power Values versus Port Locations for an Edge-Fed Four-Port Microstrip Disc

θ	P_3	P_2	P_4
45°	3/4	1/8	1/8
50°	2/3	1/6	1/6
60°	1/2	1/4	1/4

distribution of the (1, 1) mode. Comparable results were reported recently for a circular waveguide cavity cross-over circuit [10.46]. The waveguide crossover circuit was used to feed a cross-slot antenna backed by a cylindrical cavity so that dual polarized radiation can be accomplished. When $\theta = 135°$ and port 3 is positioned 90° from port 1, the circuit operates as a 3-dB hybrid. The characteristics of this circuit have already been discussed in detail in Sec. 10.2.2a.

10.3.2 Center-Fed Discs [10.38–10.40]

Another class of power divider-combiner circuits that have been analyzed by the planar circuit approach is a center-fed disc *n*-way radial power divider-combiner configuration, shown in Fig. 10.32. This structure has many interesting features. First, due to the geometrical symmetry of these circuits, excellent amplitude and phase balance along the radial ports can be obtained. This property has made the *n*-way radial power divider/combiners a very attractive option in many microwave and millimeter-wave applications [10.4, 10.38–10.40, 10.47–10.54]. A second important feature of the *n*-way combining structure is its ability to sum the power of *n*-devices directly in a single step or stage. This feature makes it possible for such structures to have high combining efficiencies. Another feature of the radial structure is that the field distribution in the cavity is azimuthally symmetric; hence there is no electrical limit on the number of devices that can be placed around the given center-fed disc [10.2]. In fact, it has been shown in [10.40], and elsewhere as well [10.4], that the combining efficiency of the *n*-way radial combiner increases as the number of radial ports increases. With an isolation on the order of 20 dB, little interaction takes place between the circumferential ports; hence the phase and amplitude performance of the circuit is not perturbed by random device failures.

Waveguide cavity-type radial combiners/dividers have been used extensively to combine amplifiers (or oscillators) in a single module, thus yielding higher output power capability. Belohoubek and his coworkers [10.38, 10.39] have used a microstrip-type, 30-way center-fed disc power divider/combiner to combine 30 1-watt miniature GaAs FET (Field Effect Transistors) power amplifiers in Ku-band. Amplitude and phase balances of ±0.4 dB and ±3°, respectively, have been achieved over 25–30 percent of bandwidth, with a reported efficiency of 90 percent. In [10.40] we used the planar circuit approach to investigate the characteristics of the 3-, 4-, 5-, 8-, and 10-way center-fed microstrip disc power divider-combiner circuits shown in Fig. 10.33. A disc radius of 15 mm was selected, and 1/32-inch-thick duroid substrate with $\varepsilon_r = 2.2$ was used. All five cases were found to exhibit balanced power-division characteristics; however, those circuits with a smaller number of ports (i.e., 3-way and 4-way) tend to suffer an extra loss. This extra loss was found to decrease as the number of output ports around the disc circumference in-

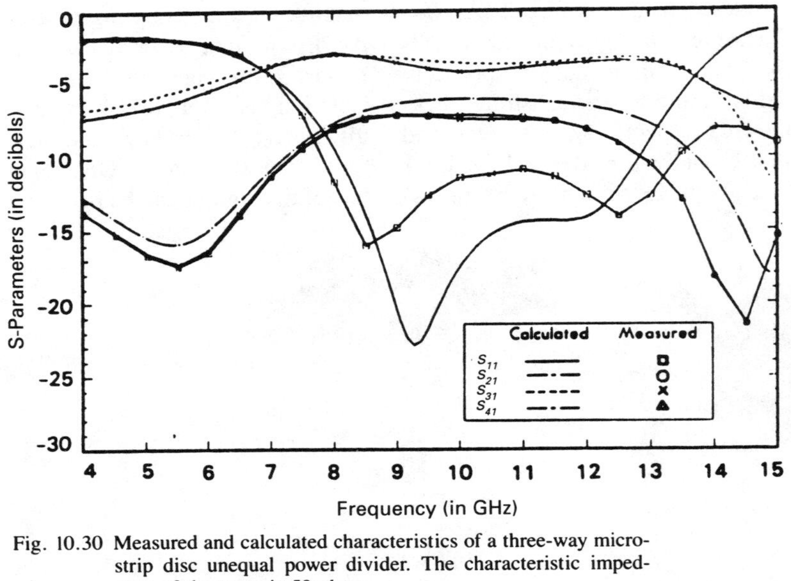

Fig. 10.30 Measured and calculated characteristics of a three-way microstrip disc unequal power divider. The characteristic impedance of the ports is 50 ohms.

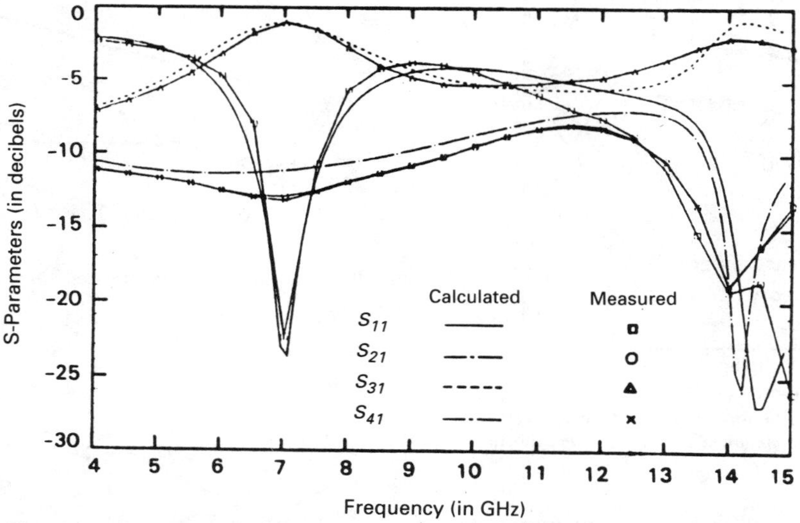

Fig. 10.31 Measured and calculated scattering parameters of a four-port symmetrical microstrip disc.

creases. The term "extra loss" was defined in [10.40] as the difference between the values of the measured and expected (or ideal) $|S_{21}|$. For instance, in the case of a three-way power divider, the expected value of S_{21} is 4.77 dB, whereas the measured value was 10 dB, thus resulting in an extra loss of 5.23 dB. Because the extra loss decreased as the number of output ports increased, it was concluded that the seemingly unexplained extra losses are probably radiation losses. This radiation loss is associated with the fringing fields at the open edge of the disc and is not accounted for in the analysis.

Upon introducing three shorting pins along the disc circumference, such that one shorted pin is located half way between every two adjacent ports, significant reduction in the extra loss has resulted. In fact, the loss was reduced from 5.23 dB to only 0.23 dB; hence an acceptable three-way division performance has been achieved. In the theoretical computation, the shorted pins are modeled as three additional shorted ports. It appears that the inclusion of shorted ports along the exposed or open edges of the disc circumference would be a useful technique for reducing radiation losses, hence improving the performance of multiport circuits, especially those with small number of ports. Figure 10.34 shows a comparison between the measured performance and the characteristics predicted by the planar circuit approach. Good agreement between the two results is apparent. The measured port-to-port isolation and the return loss of the output ports was found to be, respectively, about −5 and −10 dB. Although these levels are not acceptable, better isolation and match can be obtained by introducing (1) matching networks at the output ports and (2) isolation resistors and slots between the output ports, as illustrated in [10.38, 10.39].

The calculated and measured performance of the 10-way center-fed microstrip disc power divider/combiner is shown in Figure 10.35. The measured return loss at any circumferential port and the isolation between any two circumferential ports

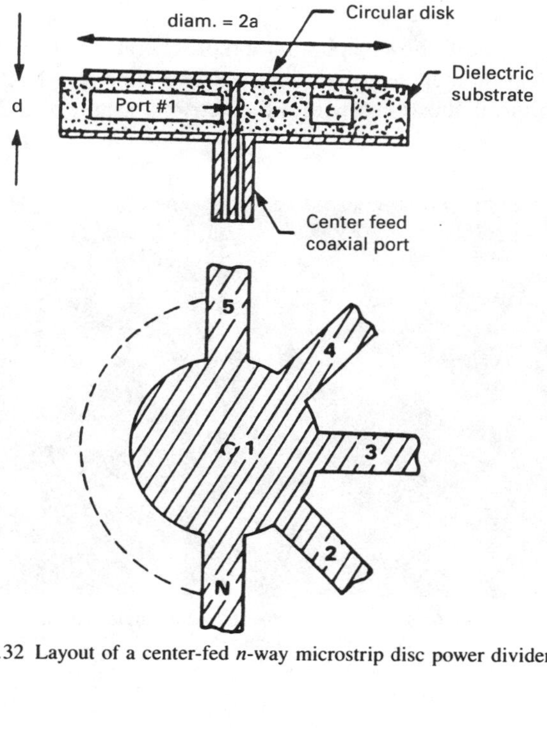

Fig. 10.32 Layout of a center-fed *n*-way microstrip disc power divider.

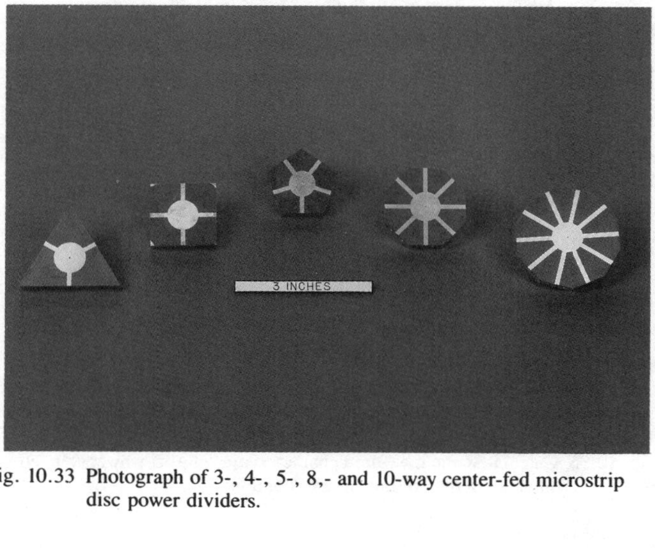

Fig. 10.33 Photograph of 3-, 4-, 5-, 8,- and 10-way center-fed microstrip disc power dividers.

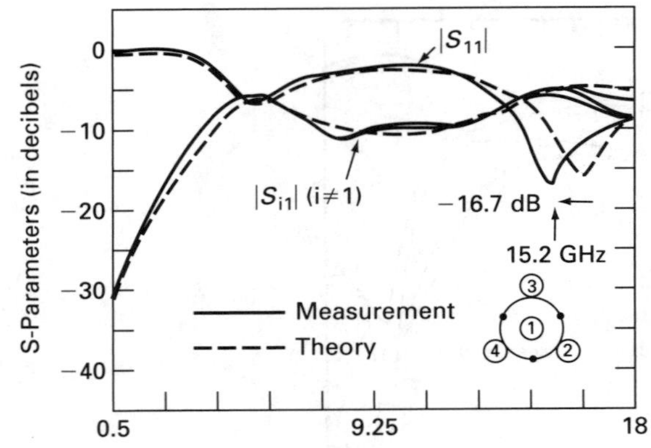

Fig. 10.34 Calculated and measured performance of a three-way center-fed microstrip disc divide-combiner circuit with three shorting pins.

(adjacent, straight across, or otherwise) are no worse than −14 dB (VSWR = 1.5) and −11 dB, respectively, at 17 GHz. Over a frequency range of 15–20 GHz, the corresponding values are better than −12.6 and −9 dB, respectively. This performance was achieved without optimization and without using isolation resistors and matching networks. Isolation values better than 16 dB and wider operational bandwidth can be achieved by adding resistive slots and/or isolation resistors as well as matching networks, as demonstrated in [10.31, 10.32, 10.38, 10.39].

10.3.3 Circular Sectors [10.29–10.37]

This category of planar power divider-combiner circuits was proposed recently [10.29–10.31] as an alternative to the lossy Wilkinson-type corporate feed and edge-fed circular disc–type feed structures. The sector geometry shown in Fig. 10.36 is without vertically oriented ports and has one input port located at the apex and several, almost linearly aligned, output ports located at the curved edge. This type of power divider-combiner circuit could be cascaded in a fan-out or fan-in manner as shown in Fig. 10.37. This truly planar nature makes the sectoral power divider/combiner topologically more suitable

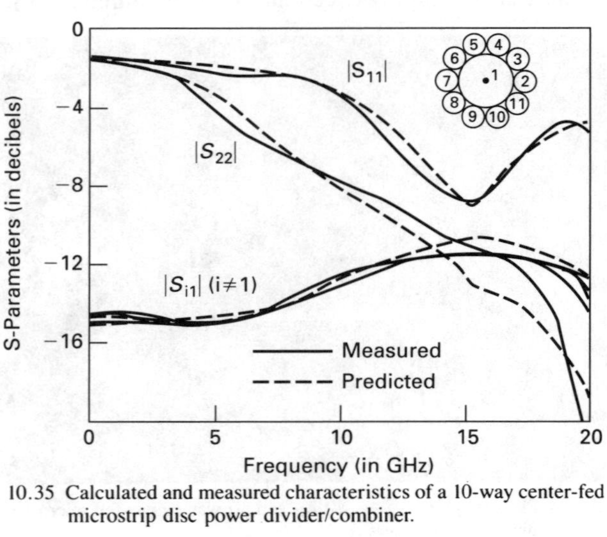

Fig. 10.35 Calculated and measured characteristics of a 10-way center-fed microstrip disc power divider/combiner.

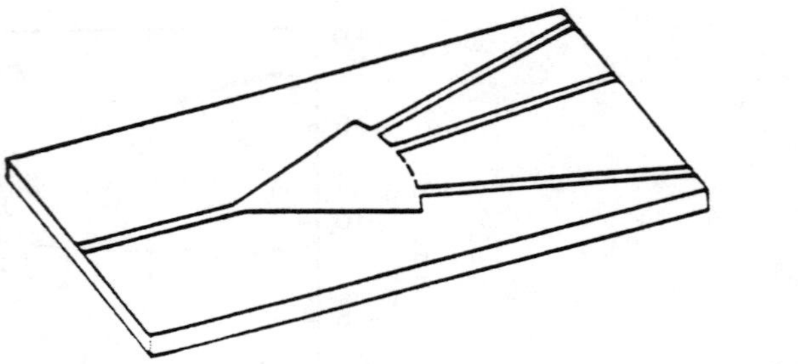

Fig. 10.36 Layout of a microstrip sector *n*-way power divider.

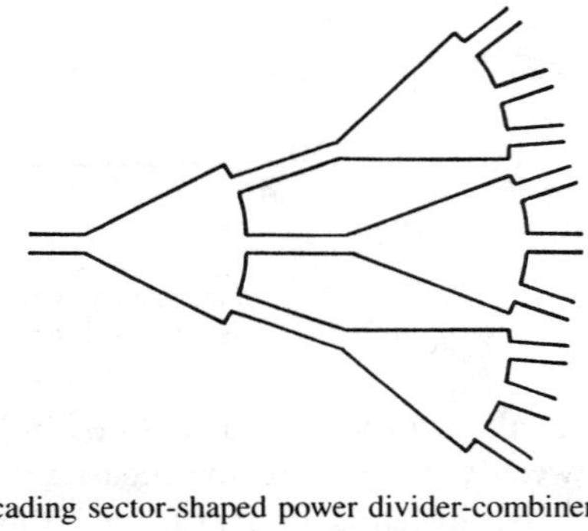

Fig. 10.37 Cascading sector-shaped power divider-combiner circuits.

for monolithic integrated circuit integration and particularly for feeding arrays of antenna elements.

Another promising feature offered by this device is the ability to obtain unequal output power levels at various ports [10.36, 10.37]. This characteristic can be used for the amplitude tapering feature often required in antenna arrays. The power level at the various output ports can be controlled by placing shorting pins along the straight edges of the sector [10.37] or by adjusting the locations and/or widths of the microstrip lines connected to these ports. Altering the widths of the output ports, however, may require the inclusion of quarter-wave impedance transformers so that proper match can be ensured.

Figure 10.38 shows a photograph of several multiway power divider/combiner sectorial circuits fabricated on 1/32-inch-thick duroid substrates with a dielectric constant of 2.2. The

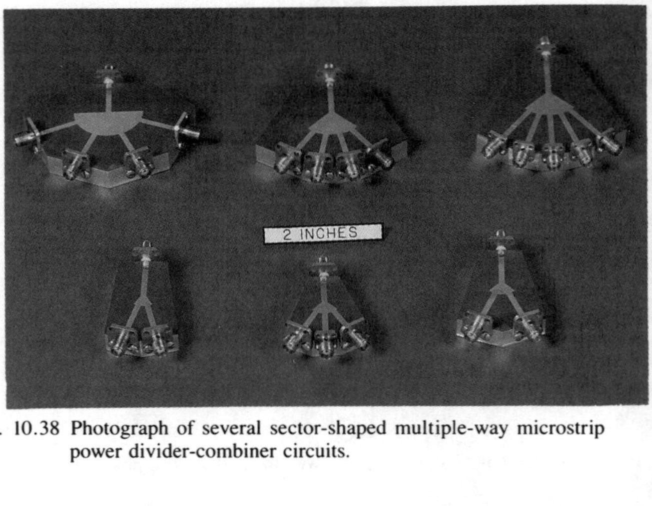

Fig. 10.38 Photograph of several sector-shaped multiple-way microstrip power divider-combiner circuits.

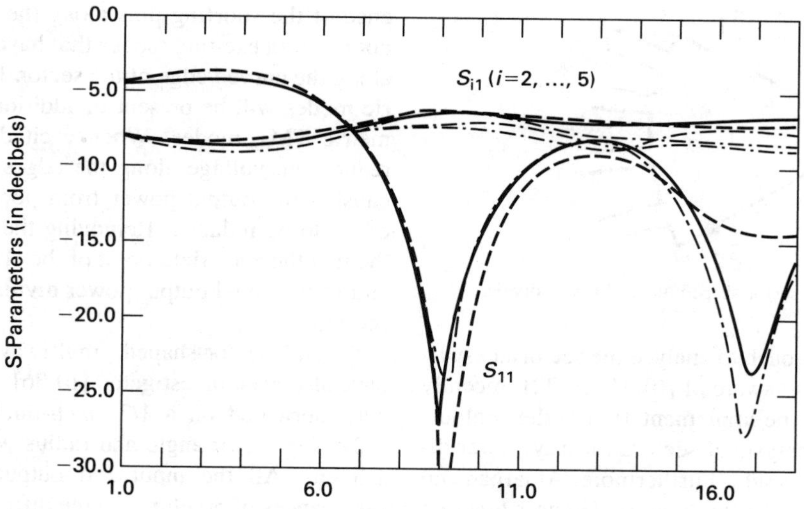

Fig. 10.39 Measured and calculated results for a four-way 90° microstrip sector power divider. Resonant-modes field expansion technique, Green's function approach. Measured.

output ports locations in these devices were determined by dividing the curved edge of the sector into n sections of equal width. One port is then located at the middle of each of these n sections. Such a choice is primarily influenced by the desire to maintain physical symmetry, which helps in obtaining equal power division. These devices were first analyzed by the Green's function approach [10.30], using the double summation impedance matrix expressions outlined in Chapter 2. This analysis is valid for sectors having an angle α equal to submultiple of π. It was found that the convergence of the numerical results was somewhat slow, and at least 30 terms in each summation are needed. Later, Bartolucci et al. showed that computational efficiency can be improved significantly if the resonant mode field expansion technique [10.33] is employed. In [10.33] the authors exploited the symmetry of the device (no angular coordinate dependence for the field at the curved edge) and used the resonant-mode technique to expand the field sustained by a sector-shaped divider in terms of the TM_{0n} resonant-modes only. This approach has resulted in single summation expressions for the elements of the Z-matrix. As a result, computational efficiency has improved substantially, such that more convergent results were obtained using no more than 20 modes, and in some instances as low as 10 modes.

Figure 10.39 [10.33] illustrates a comparison between measured results and calculated results obtained by the Green's function approach and the resonant mode techniques for a four-way 90° sector-shaped power divider with a radius of 15 mm. Equal power division has been achieved at the frequencies that correspond to the resonances of the TM_{0n} azimuthal symmetric modes. The azimuthal symmetric modes yield uniform voltage distribution along the open-circuited circumferential edge. The impedance Green's function results are in agreement with the measured results over the first resonance mode and start to diverge thereafter. On the other hand, the resonant-mode results are in agreement with the measured results over the first and second resonance mode up to about 20 GHz. Clearly, a single summation expression for the Z-matrix elements will lead to a satisfactory numerical convergence and also substantially reduce the computation time. This device offers an excellent amplitude and phase balance over a wide bandwidth. The port-to-port phase balance for the four-way divider has been found to be within ±2° over the frequency band 8–10 GHz. The corresponding theoretical results are ±0.1°. The isolation between the output ports varies from −5 to −25 dB over the lower-frequency band, whereas the corresponding values at the higher-frequency band were between −8 and −14 dB, depending on the port location. These isolation values are obtained without the use of any external resistors. It has been shown in [10.31, 10.32] that for a six-way sector-shaped power divider, at least a 20-dB port-to-port isolation can be realized over a 30 percent bandwidth in X-band. This impressive result was obtained by introducing a resistive slot in the sector and resistive sheets between the adjacent output lines about λ/4 and λ/2 away from the junction, as shown in Fig. 10.40. In [10.31, 10.32] Miyazaki and his coworkers

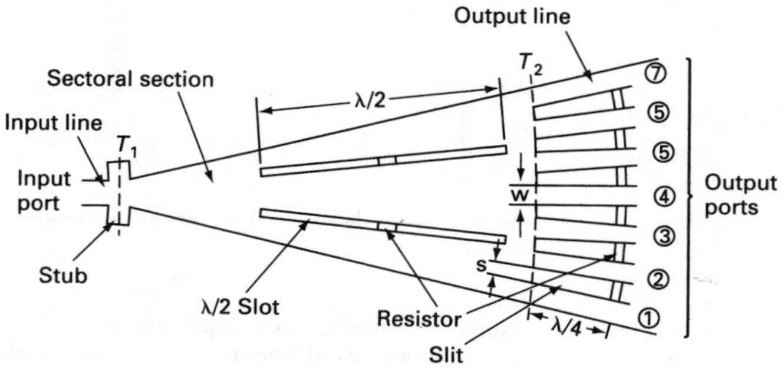

Fig. 10.40 Multiple-way sector divider with slots and resistors for improved isolation characteristics.

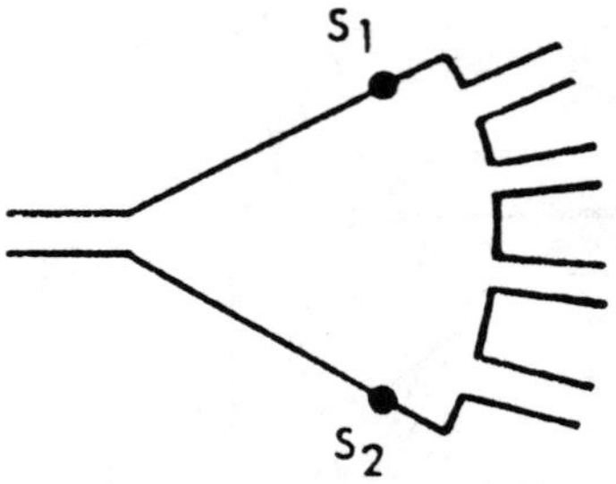

Fig. 10.41 Layout of a four-way sector-shaped unequal power divider.

used a contour integral approach to analyze the sectorial n-way power divider. Apparently unaware of [10.31, 10.32], recently Yeo et al. [10.34] reported the implementation of the contour-integral approach to the analysis of sectorial n-way power dividers with arbitrary sector angle. Furthermore, Alhargan and Judah [10.35] recently derived a single-series Green's function expression valid for sectors with arbitrary angle α. It is claimed in [10.35] that the removal of the sector angle restriction allowed the introduction of an effective sector angle that enabled the authors to account for the fringing fields at the straight edges of the sector.

Sector-shaped power dividers can also be used as unequal power dividers, as shown in [10.36–10.37]. Unequal power division has been achieved by introducing shorting pins (or via holes in the monolithic configuration) on the straight radial edges of the sector, as shown in Fig. 10.41. The shorting pins S_1 and S_2, which are depicted in Fig. 10.41, connect the top conductor to the ground plane. Multiple shorting pins (i.e., more than one) may also be used at the radial edges. The presence of the shorting pins along the radial edges may or may not result in exciting modes that have azimuthal field variation along the curved edge of the sector. If excited, these asymmetric modes will be present in addition to the azimuthally symmetric TM_{0n}-modes. When excited, the asymmetric modes reduce the voltage along the edges of the curved periphery, causing the output power from ports located near the radial edges to be reduced. Removing the shorting pins or locating them at the zero-field point of the (1, 0) mode reduces this circuit to the equal output power divider discussed earlier in this section.

Several sector-shaped, multiway unequal power dividers have also been investigated [10.36]. A four-way power divider was fabricated on a 1/32-inch-thick duroid substrate (ε_r = 2.2). The sector angle and radius were, respectively, 90° and 1.5 cm. All the input and output lines had characteristic impedances of 50 ohms. Three different shorting pin locations were investigated. Circuit characteristics were measured when the shorting pin is located 0.22, 0.33, and 0.44 cm from the corner of the curved periphery. When one shorting pin is located along each straight edge at 0.33 cm away from the curved-edge corner, a better than −14 dB of return loss over the frequency range 5.8–9.0 GHz (a bandwidth of 43.24 percent) has been obtained. For a return loss $S_{11} < -10$ dB the bandwidth is about 77 percent. Over this bandwidth, the output at the outer ports (2 and 5) is 9.5 dB (±0.5 dB); the power output at the inner ports (3 and 4) is 5.5 dB (±0.25 dB). Figure 10.42 depicts a comparison between measured results and theoretical results when the shorting pin is located 0.44 cm away from the corner of the curved periphery. The theoretical

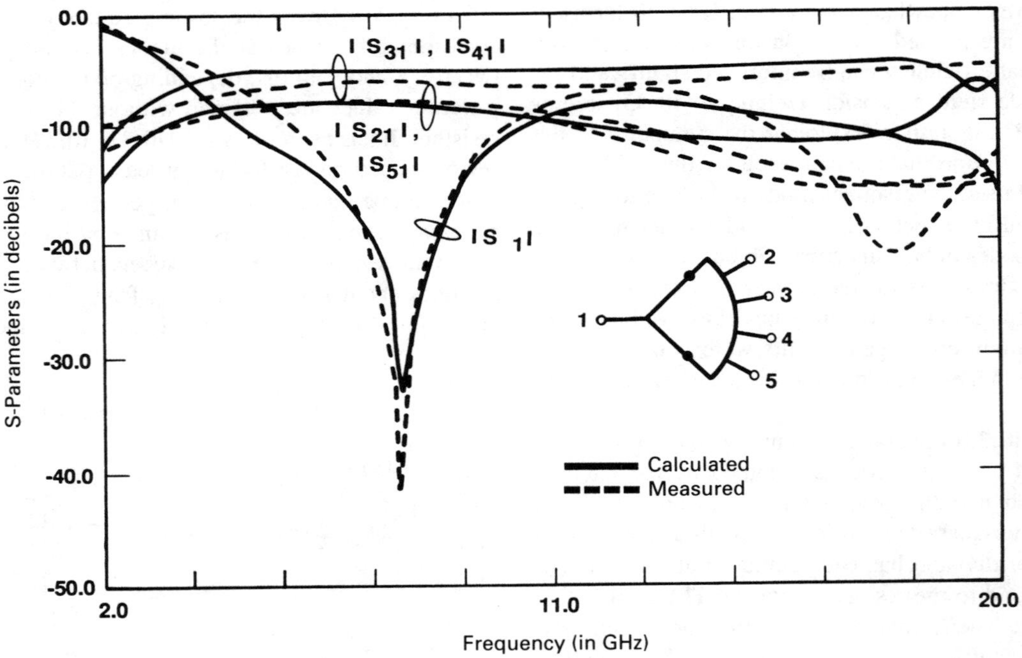

Fig. 10.42 Measured and calculated characteristics of a four-way 90° sector-shaped unequal-power divider with shorting pins located along the straight radial edge 0.44 cm away from the corner of the curved edge.

results were obtained by the Green's function planar circuit approach. Good agreement between measured and calculated results is evident up to about 15 GHz. The lack of agreement above 15 GHz was attributed to poor numerical convergence caused by the limited number of terms ($n = m = 40$) used in the summations. However, it is suspected that, as seen in the case of equal power dividers, less computing time will be used and better numerical convergence can be obtained by following the resonant mode expansion algorithm discussed in [10.33]. When the shorting pins are incorporated in the circuit the same analysis procedure can be used for the equal power divider case, with minor modification. Extra ports will be located at the position of the shorting pins, with the effective width of these ports taken to be equal to the diameter of the shorting pins. The Z-matrix is then computed for the whole circuit with the additional ports open circuited. The Z-matrix is then converted to the corresponding Y-matrix. Shorting of the two ports is equivalent to the removal from the Y-matrix of the rows and columns that correspond to these two ports. The resulting matrix can then be converted to the more familiar S-matrix.

Designs with two shorting pins located on each of the radial edges were also investigated in [10.36]. One interesting result is when the shorting pins are located at a distance of 0.11 and 0.22 cm from the curved edge. The measured return loss in this case was less than -12 dB from 3.8 to 10.5 GHz. The power output levels at the inner ports (3 and 4) and the outer ports (2 and 5) were -6 ± 0.5 dB and -12.5 ± 0.5 dB, respectively. Experiments with a large number of shorting pins (three and five pins) on each radial edge were also conducted but without producing better results.

Fairly wide-band unequal-power dividers may be achievable using this approach. As in the equal-power-division case, the port-to-port isolation is not adequate; however, the inclusion of isolation resistors and slots is expected to yield acceptable isolation characteristics.

10.4 Concluding Remarks

Several new classes of planar power divider-combiner circuits have been discussed in this chapter. The method of analysis and the characteristics of these circuits have been reviewed. Comparison between experimentally measured and theoretically computed results demonstrates that the two-dimensional Green's function approach is adequate for these configurations. The contour-integral approach becomes more appropriate when planar circuits with arbitrarily shaped configurations are considered.

Finally, it must be stated that the characteristics of most of the circuits discussed in this chapter have not been optimized for the best performance; thus one can look forward to a considerably improved performance when computer-aided optimization techniques are used.

References

[10.1] James, J. R., P. S. Hall, and C. Wood. *Microstrip Antenna*, pp. 160–193. New York: Peter Peregrinus, 1981.

[10.2] Russell, K. J. "Microwave power combining techniques." *IEEE Trans. Microwave Theory Tech.*, Vol. MTT-27, pp. 472–478, May 1979.

[10.3] Wilkinson, E. "An n-way hybrid power divider." *IRE Trans. on Microwave Theory and Tech.*, Vol. MTT-8, pp. 116–118, January 1960.

[10.4] Saleh, A. A. M. "Planar electrically symmetric n-way hybrid power dividers/combiners." *IEEE Trans. Microwave Theory Tech.*, Vol. MTT-28, pp. 555–563, June 1984.

[10.5] Pozar, D. M., *Microwave Engineering*, Reading, MA: Chapter 8, pp. 383–449. Addison-Wesley, 1990.

[10.6] Bahl, Inder and Prakash Bhartia. *Microwave Solid State Circuit Design*, Chapter 5, pp. 173–236. New York: John Wiley & Sons, 1988.

[10.7] Mahlerbe, J. A. G., *Microwave Transmission Line Couplers*, pp. 17–82. Norwood, MA: Artech House, 1988.

[10.8] Bahl, I. J. "Filters, hybrids and couplers, power combiners, and matching networks." In *Handbook of Microwave and Optical Components*, Vol. 1, pp.118–190 Chapter 3, New York: John Wiley & Sons, 1989.

[10.9] Gupta, K. C. and M. D. Abouzahra. "Planar power divider and combiner circuits at microwave frequencies," Invited Paper in *1989 SBMO Int'l. Microwave Symp.*, pp. 279–284, São Paulo, Brazil, July 24–27, 1989.

[10.10] Okoshi, T. O., T. Imai, and K. Ito. "Computer-oriented synthesis of optimum circuit pattern of 3 dB hybrid ring by the planar circuit approach." *IEEE Trans. Microwave Theory Tech.*, Vol. MTT-29, pp. 194–202, March 1981.

[10.11] Burns, J. W., "Planar quadrature microwave coupler." United States Patent No. 4,492,939, January 1986.

[10.12] Fusco, V. F. and J. A. C. Stewart. "Design and synthesis of patch microwave couplers." *16th European Microwave Conf.*, pp. 401–406, Dublin, Ireland, September 8–12, 1986.

[10.13] Merugu, L. N., V. F. Fusco and J. A. C. Stewart. "A phase quadrature square patch microwave coupler." *Microwave and Optical Technology Letters*, Vol. 2, No. 12, pp. 424–427, December 1989.

[10.14] Page, M. J. and S. R. Judah. "A microstrip planar disc 3 dB quadrature hybrid." *1989 IEEE Int'l. Microwave Symp.*, pp. 247–250, Long Beach, California, June 13–15, 1989.

[10.15] Page, M. J. and S. R. Judah. "A flexible design procedure for microstrip planar disc 3 dB quadrature hybrids." *IEEE Trans. Microwave Theory Tech.*, Vol. MTT-38, No. 11, pp. 1733–1736, November 1990.

[10.16] Okoshi, T., T. Takeuchi, and J. Hsu. "Planar 3 dB hybrid circuit." *Electronics and Communications in Japan*, Vol. 58-B, No. 8, pp. 80–90, 1975.

[10.17] Ohta, I., T. Yamashita, and I. Hagino. "Optimum ports arrangement for a planar-circuit-type 3 dB hybrid." *Trans. IECE Japan*, Vol. E-67, No. 5, pp. 287–288, May 1984.

[10.18] Ohta, I., I. Hagino, and T. Kaneko. "Improved circular disk 3 dB hybrids." *Electronics and Communications in Japan*, Pt. 2, Vol. 70, No. 12, pp. 66–77, 1987.

[10.19] Ohta, I., H. Kinoshita, T. Kaneko, and K. Fujiwara. "Rectangular disk 3 dB hybrids." *1989 IEEE Int'l. Microwave Symposium*, pp. 235–238, Long Beach, California, June 13–15, 1989.

[10.20] Ohta, I., H. Taniguchi, and T. Taneko. "A rat-race-type directional coupler for loose coupling." *Trans. IECE*, Vol. E-71, No. 4, pp. 304–306, April 1988.

[10.21] Roy, J. S., D. R. Poddar, and S. K. Chowdhury. "Broadband design of ring type microstrip power divider." *Microwave and Optical Technology Letters*, Vol. 3, No. 4, pp. 119–122, April 1990.

[10.22] Kim, D. I. and G. Yang, "Design of new hybrid-ring directional coupler using λ/8 and λ/6 sections." *IEEE Trans. Microwave Theory Tech.*, Vol. MTT-39, No. 10, pp. 1779–1991, October 1991.

[10.23] Bertil Hansson, E. R. and G. P. Riblet. "The matched symmetrical five-port junction as the essential part of an ideal six-port network." *1981 European Microwave Conf.*, pp. 501–506, Amsterdam, The Netherlands, September 7–11, 1981.

[10.24] Riblet, G. P. and E. R. Bertil Hansson. "The use of a matched symmetrical five-port junction to make six-port measurements." *1981*

IEEE Int'l. Microwave Symp., pp. 151–153, Los Angeles, California, June 15–19, 1981.

[10.25] Riblet, G. P. and E. R. Hansson. "Some properties of the matched symmetrical six-port junction." *IEEE Trans. Microwave Theory Tech.*, Vol. MTT-32, No. 2, pp. 165–171, February 1984.

[10.26] Hagelin S., and Börje Carlegrim. "Planar multiport networks with rotational symmetry," In John Clark, ed., *Advanced Electronic Warfare Technology.* Microwave Exhibitions and Publishers, 1984.

[10.27] Abouzahra, M. D. and K. C. Gupta. "Analysis and design of five port circular disc structures for six-port analyzers." *1985 IEEE Int'l. Microwave Symp.*, pp. 449–452, St. Louis, Missouri, June 4–6, 1985.

[10.28] Gupta, K. C. and M. D. Abouzahra. "Analysis and design of four-port and five-port microstrip disc circuits." *IEEE Trans. Microwave Theory Tech.*, Vol. MTT-33, No. 12, pp. 1422–1428, December 1985.

[10.29] Abouzahra, M. D., K. C. Gupta, and A. Dumanian. "Use of circular sector shaped planar circuits for multiport power divider-combiner circuits." *1988 IEEE Int'l. Microwave Symp.* pp. 661–664, New York, May 25–27, 1988.

[10.30] Abouzahra M. D., and K. C. Gupta. "Multiport power divider-combiner circuits using circular-sector-shaped planar components." *IEEE Trans. Microwave Theory Tech.*, Vol. MTT-36, No. 12, pp. 1747–1751, December 1988.

[10.31] Miyazaki, M. et al. "An n-way sectoral planar power divider." *IECE Japan*, Technical Report MW88-27, pp. 23–28 (Japanese), July 1988.

[10.32] Miyazaki, M., O. Ishida, and T. Hashimoto. "A sectoral planar 6-way hybrid power divider with a resistive slot." *19th European Microwave Conf.*, pp. 1–3 London, England, September 4–7, 1989.

[10.33] Bartolucci, G., F. Giannini, and C. Paoloni. "Planar analysis of radial line power dividers." *Int'l. J. Numerical Modelling: Electronic Networks, Devices and Fields*, Vol. 3, pp. 23–31, 1990.

[10.34] Yeo, S. P. et al. "Contour-integral analysis of microstrip sectorial power divider (with arbitrary sector angle)." *Proc. IEE*, Pt. H, Vol. 140, pp. 62–64, February 1993.

[10.35] Alhargan, F. A. and S. R. Judah. "Circular and annular sector planar components of arbitrary angle for N-way power dividers/combiners." *IEEE Trans. Microwave Theory Tech.*, Vol. MTT-42, September 1994.

[10.36] Abouzahra, M. D. and K. C. Gupta. "Multi-way unequal power divider circuits using sector-shaped planar components." *1989 IEEE Int'l. Microwave Symp.*, pp. 321–324. Long Beach, California, June 13–15, 1989.

[10.37] Abouzahra, M. D. and K. C. Gupta. "Multiport power divider-combiner." United States Patent No. 4,947,143, August 1990.

[10.38] Belohoubek, E. et al. "30-way radial power combiner for miniature GaAs FET power amplifiers." *1986 IEEE Int'l. Microwave Symp.*, pp. 515–518, Baltimore, Maryland, June 2–4, 1986.

[10.39] Fathy, A. and D. Kalokitis. "Analysis and design of a 30-way radial combiner for Ku-band applications." *RCA Review*, Vol. 47, pp. 487–508, December 1986.

[10.40] Abouzahra, M. D. and K. C. Gupta, "Multiple-port power divider/combiner circuits using circular microstrip disk configurations." *IEEE Trans. Microwave Theory Tech.*, Vol. MTT-35, No. 12, pp. 1296–1302, December 1987.

[10.41] Riblet, G. P. "An eigenadmittance condition applicable to symmetrical four-port circulators and hybrids." *IEEE Trans. Microwave Theory Tech.*, Vol. MTT-26, pp. 275–279, April 1978.

[10.42] Bertil Hansson, E. R. and G. P. Riblet. "An ideal six-port network consisting of a matched reciprocal lossless five-port and perfect directional coupler." *IEEE Trans. Microwave Theory Tech.*, Vol. MTT-31, No. 3, pp. 284–289, March 1983.

[10.43] Montgomery, C. G., R. H. Dicke, and E. M. Purcell. *Principles of Microwave Circuits.* pp. 455–459. London: Peter Peregrinus, 1987.

[10.44] Riblet, G. P. and E. R. Bertil Hansson. "Some properties of the matched symmetrical six-port junction." *IEEE Trans. Microwave Theory Tech.*, Vol. MTT-32, No. 2, pp. 164–171, February 1984.

[10.45] Gupta, K. C. and M. D. Abouzahra. "Planar circuit analysis." In T. Itoh, ed., *Numerical Techniques for Microwave and Millimeter-Wave Passive Structures*, Chapter 4, pp. 214–333. New York: John Wiley & Sons, 1989.

[10.46] Chang, Kai et al. "Applications of high power four-way power divider and combiner and cross-over circuit." *Microwave J.*, pp. 71–85, November 1991.

[10.47] Schellenberg, J. M. and M. Cohn. "A wideband radial power combiner for F.E.T. amplifiers." *IEEE Int'l. Solid-State Circuits Conf.*, pp. 164–165, San Francisco, California, February 15–17, 1978.

[10.48] Quine, J. P., J. G. McMullen, and H. W. Prather. "MIC power combiners for FET amplifiers," *1979 European Microwave Conf.*, pp. 661–664, September 1979. Brighton, England, September 17–20, 1979.

[10.49] Foti, S. J., R. P. Flam, and W. J. Scharpf, Jr. "60-way radial combiner uses no isolators." *Microwave & RF*, pp. 96–118, July 1984.

[10.50] Hsu, T. and M. Simonutti. "A wideband 60 GHz 16-way power divider/combiner network." *1984 IEEE Int'l. Microwave Symp.*, pp. 175–177, San Francisco, California, May 30–June 1, 1984.

[10.51] Matsumura, H. and H. Mizuno. "Design of microwave power with circular TM_{Om0} mode cavity." *Electronics and Communications in Japan*, Pt. 2, Vol. 70, No. 9, pp. 1–11, 1987.

[10.52] Stone, D. I. "A UHF 16-way power combiner designed by synthesis techniques." *Microwave J.*, pp. 117–120, June 1989.

[10.53] Oz, M. et al. "High efficiency, single-ridged 16-way radial power combiner." *Comsat Technical Review*, Vol. 21, No. 1, pp. 227–251, Spring 1991.

[10.54] Bialkowski, M. E. "Analysis of an n-port consisting of radial cavity and E-coupled rectangular waveguides." *IEEE Trans. Microwave Theory Tech.*, Vol. MTT-40, No. 9, pp. 1840–1843, September 1992.

Chapter 11

Microstrip Patch Antennas

K. C. Gupta
University of Colorado at Boulder

M. D. Abouzahra
M.I.T. Lincoln Laboratory

11.1 INTRODUCTION

AMONG the wide variety of radiating elements configurations available to a microwave engineer, microstrip patch antennas [11.1–11.5] occupy a unique status, largely because of the ease of their integration with microwave feed circuitry.

The basic configuration of a microstrip antenna is a metallic patch printed on a thick, grounded dielectric substrate as shown in Fig. 11.1(a). In most cases, the element is fed with either a coaxial line through the bottom of the substrate or by a coplanar microstrip line. This latter type of excitation allows feed networks and other circuitry to be fabricated on the same substrate as the antenna element, as in the corporate-fed microstrip array shown in Fig. 11.1(b). The microstrip antenna radiates a relatively broad beam broadside to the plane of the substrate. Thus the microstrip antenna has a very low profile and can be fabricated using printed circuit (photolithographic) techniques. This implies that the antenna can be made conformable, and potentially at low cost. Other advantages include the ease of fabrication into linear or planar arrays, as well as the ease of integration with microwave integrated circuits.

To a large extent, the development of microstrip antennas has been driven by systems requirements for antennas with low profile, low weight, low cost, easy integrability into arrays or with microwave integrated circuits, or polarization diversity. Microstrip antennas have found application in both the military and the civil sectors.

Disadvantages of the original microstrip antenna configurations include narrow bandwidth, spurious feed radiation, poor polarization purity, limited power capacity, and tolerance problems. Much of the development work in microstrip antennas has thus gone into trying to overcome these problems.

Microstrip patch antennas and their arrays can be analyzed and designed by using the methods discussed earlier for microstrip circuits. The two-dimensional analysis techniques described in various chapters of this book and the full-wave analysis methods presented in Chapter 5 and 6 have been applied to microstrip antennas extensively. Various approaches for analyzing microstrip antennas are reviewed in this chapter. These approaches include the transmission line model, the multiport network model, and the full-wave numerical simulation.

11.2. TRANSMISSION LINE MODEL FOR MICROSTRIP PATCHES

In this model, a rectangular microstrip antenna patch is viewed as a resonant section of a microstrip transmission line. A detailed description of the transmission line model is given in [11.3, Chapter 10]. The basic concept is shown in Fig. 11.2,

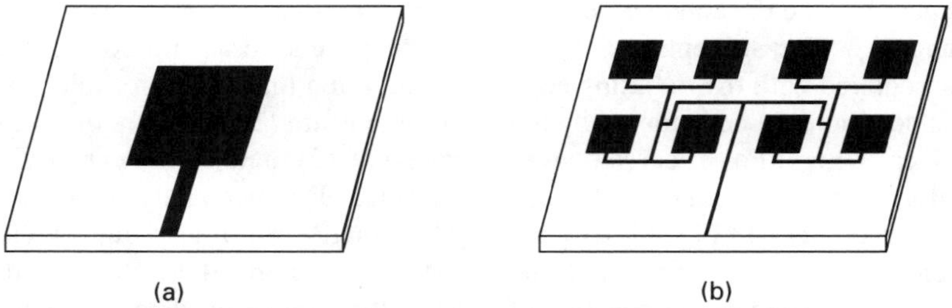

Fig. 11.1 A rectangular microstrip patch and a corporate-fed array of eight patches.

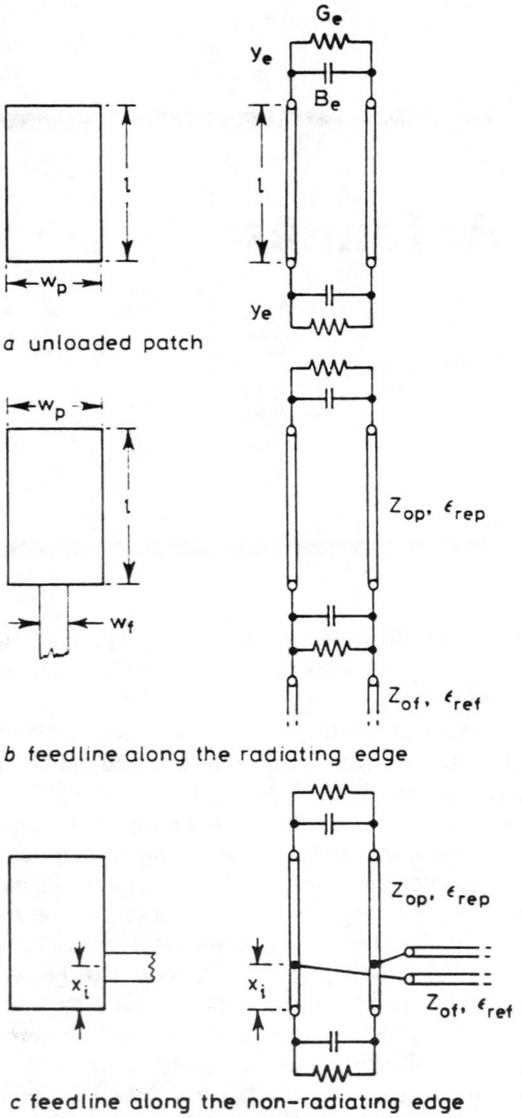

Fig. 11.2 Transmission line models for three rectangular microstrip patch configurations.

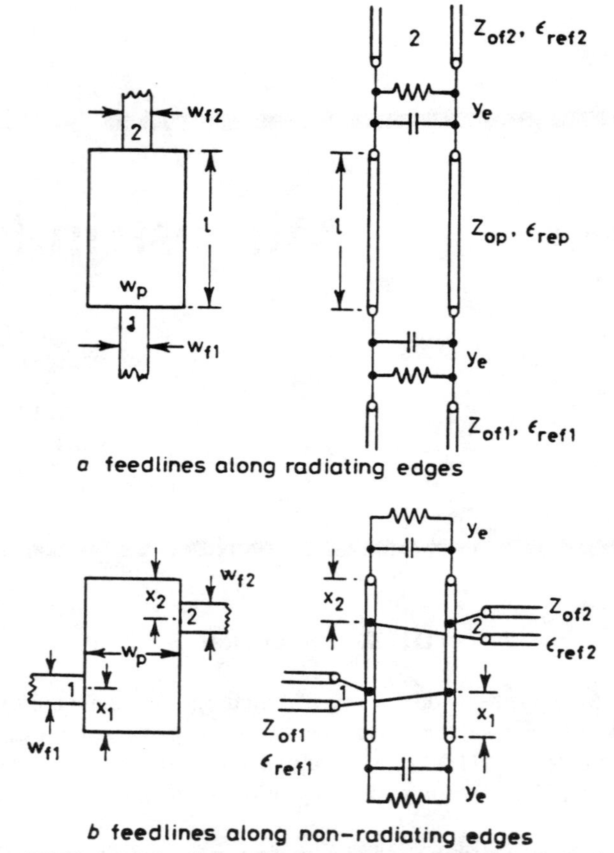

Fig. 11.3 Transmission line modes for two-port rectangular microstrip patch antennas.

which illustrates the transmission line models for an unloaded regular patch (panel a), a rectangular patch with a feedline along the radiating edge (panel b), and a rectangular patch with a feedline along the nonradiating edge (panel c). The term Z_{0p} represents the characteristic impedance of a microstrip line of width W_p, where ε_{rep} is the corresponding effective dielectric constant. B_e and G_e are the capacitive and the conductive components of the edge admittance Y_e. The susceptance B_e accounts for the fringing field associated with the radiating edge of the width W_p, and G_e is the conductance contributed by the radiation field associated with each edge. Power carried away by the surface wave(s) excited along the slab may also be represented by a lumped loss and added to G_e. In Fig. 11.2(b) and 11.2(c), Z_{0f} and ε_{ref} are the characteristic impedance and the effective dielectric constant for the feeding microstrip line of width W_f. In both these cases, the parasitic reactances associated with the junction between the line and the patch have not been taken into account.

Transmission line models may also be developed for two-port rectangular microstrip patches [11.6]. These configurations are used in the design of series-fed linear (or planar) arrays [11.7]. Models for two types of two-port rectangular microstrip patches are shown in Fig. 11.3. Figure 11.3(a) illustrates the equivalent transmission line network when the two ports are located along the radiating edges, and Fig. 11.3(b) shows the transmission line model [11.8] when the two ports are along the nonradiating edges. It has been shown [11.6, 11.8] that, when the two ports are located along the nonradiating edges, transmission from port 1 to port 2 can be controlled by suitable choices of distances x_1 and x_2. Again, the two models shown in Fig. 11.2(a) and 11.2(b) do not incorporate the parasitic reactances associated with the feedline-patch junctions.

There are several limitations inherent in the concept of the transmission line model for microstrip antennas. The basic assumptions are (1) fields are uniform along the width W_p of the patch and (2) there are no currents transverse to the length l of the patch. Detailed analysis of rectangular patches has shown [11.9] that, even at a frequency close to the resonance, field distribution along the radiating edge is not always uniform. Also, the transverse currents are caused by the feeding mechanism and are invariably present. Moreover, the circularly polarized rectangular microstrip antennas (whose operation depends upon the excitation of two orthogonal modes) cannot be represented by the transmission line model just discussed. Clearly, a more accurate method for modeling of microstrip antennas is needed.

11.3 Cavity Model

A planar two-dimensional cavity model for microstrip patch antennas [11.10, 11.11] offers considerable improvement over the one-dimensional transmission line model discussed in the previous section. In this method of modeling, the microstrip patch is considered as a two-dimensional resonator surrounded by a perfect magnetic wall around the periphery. The fields underneath the patch are expanded in terms of the resonant modes of the two-dimensional resonator. This approach is applicable to a variety of patch geometries. These geometries, their corresponding modal variations denoted by ψ_{mn}, and the resonant wave numbers k_{mn} are shown in Table 11.1 (from [11.10]). The electric field E and the magnetic field H are related to the modal functions ψ_{mn} by

$$\mathbf{E}_{mn} = \psi_{mn}\hat{\mathbf{z}} \tag{11.1}$$

$$\mathbf{H}_{mn} = \hat{\mathbf{z}} \times \nabla_t \frac{\psi_{mn}}{j\omega\mu} \tag{11.2}$$

where $\hat{\mathbf{z}}$ is a unit vector normal to the plane of the patch and the time dependence is assumed to be e^{jwt}. The resonant wave numbers k_{mn} are solutions

$$(\nabla_t^2 + k_{mn}^2)\psi_{mn} = 0 \tag{11.3}$$

with

$$\frac{\partial \psi_{mn}}{\partial p} = 0 \tag{11.4}$$

on the magnetic wall (periphery of the patch). The term ∇_t represents the transverse part of the del operator, and p is perpendicular to the magnetic wall.

The fringing fields at the edges are accounted for by extending the patch boundary outward and considering the effective dimensions to be somewhat larger than the physical dimensions of the patch [11.10]. The radiation is accounted for by considering the effective loss tangent of the dielectric to be larger than the actual value. If the radiated power is estimated to be P_r, the effective loss tangent δ_e may be written as

$$\delta_e = \frac{P_r + P_d}{P_d}\delta_d \tag{11.5}$$

where P_d is the power dissipated in the dielectric substrate and δ_d is the loss tangent for the dielectric medium. The effective loss tangent given by Eq. (11.5) can be modified further to incorporate the conductor loss. The modified loss tangent δ_e is given by

$$\delta_e = \frac{P_r + P_d + P_c}{P_d}\delta_d \tag{11.6}$$

The input impedance of the antenna is calculated by finding the power dissipated in the patch for a unit voltage at the feed port and is given by

$$Z_{in} = \frac{|V|^2}{P + 2j\omega(W_E - W_M)} \tag{11.7}$$

where $P = P_d + P_c + P_s + P_r$. The component P_s represents the power carried away by surface waves, W_E is the time-averaged electric stored energy, and W_M is the time-averaged magnetic energy. The voltage V equals $-E_z d$ averaged over the feed-strip width (d is the substrate thickness). The far-zone field and radiated power are computed by replacing the equivalent magnetic-current ribbon on the patch's perimeter by a magnetic line current of magnitude $\mathbf{K}d$ on the ground plane (xy plane). The magnetic-current source is given by

$$\mathbf{K}(x, y) = \mathbf{n} \times \hat{\mathbf{z}}\mathrm{E}(x, y) \tag{11.8}$$

where $\mathbf{n}$ is a unit vector normal to the patch's perimeter and $\hat{\mathbf{z}}\mathrm{E}(x, y)$ is the component of the electric field perpendicular to the ground plane.

A cavity model for microstrip patch antennas may also be formulated by considering a planar two-dimensional resonator with an impedance boundary wall all around the edges of the patch. A direct form of network analog (DFNA) method for the analysis of such a cavity model has been discussed in [11.12].

11.4 Multiport-Network Modeling Approach

11.4.1 Approach

The configuration of an arbitrarily shaped microstrip patch antenna along with the coordinate system employed is shown in Fig. 11.4. As before, the time dependence $e^{j\omega t}$ is assumed for the fields. The substrate material is assumed to be nonmagnetic (but the presence of a magnetic material can be included in the approach). The ground plane and the dielectric substrate are assumed to be infinite in extent.

In the multiport-network modeling (MNM) approach for radiating microstrip patches [11.13, 11.14], the fields underneath the patch, the external fields (radiated, surface wave and fringing fields), and the fields underneath the microstrip feedlines are modeled separately in terms of multiport subnetworks. The fields on either side of an interface between any two subnetworks are matched at discrete number of points by subdividing the common interface into a number of sections. Matching of the fields is achieved by satisfying equivalent Kirchoff's network relations at those interconnecting ports. Equating the voltages at the connected ports is analogous to matching the tangential E-field, and the continuity of currents ensures the continuity of the tangential H-field at the interface. The multiport subnetworks are characterized in terms of either Z-matrices or Y-matrices and are combined together using the segmentation technique discussed in Chapter 4 to obtain the antenna characteristics such as the resonance frequency,

TABLE 11.1 Variation of Modal Fields (ψ_{mn}) and Resonant Wave Numbers (K_{mn}) for Various Patch Geometries Analyzed by the Cavity Method (from [11.10])

Rectangle

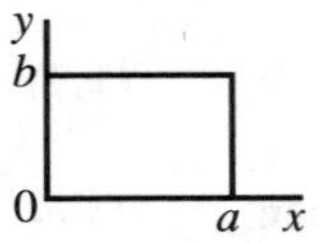

$$\psi_{mn} = \cos\frac{m\pi}{a} \times \cos\frac{n\pi}{b} y$$

$$k_{mn} = \sqrt{\left(\frac{m\pi}{a}\right)^2 + \left(\frac{n\pi}{b}\right)^2}$$

Circle (disc)

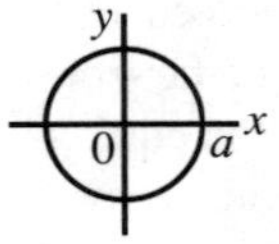

$$\psi_{mn} = J_n(k_{mn}\rho)e^{jn\phi}$$
$$J'_n(k_{mn}a) = 0$$

Circular segment

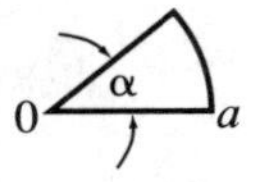

$$\psi_{m\nu} = J_\nu(k_{m\nu}\rho)\cos\nu\phi$$
$$\nu = n\pi/\alpha,\ J'_\nu(k_{m\nu}a) = 0$$

Disk with slot

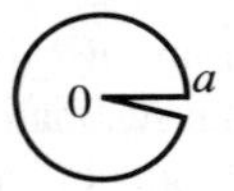

$$\psi_{mn} = J_{n/2}(k_{mn}\rho)\cos n\phi/2$$
$$J_{n/2}(k_{mn}a) = 0$$
$$a \approx 2\pi,\ \nu = n/2$$

Right isosceles

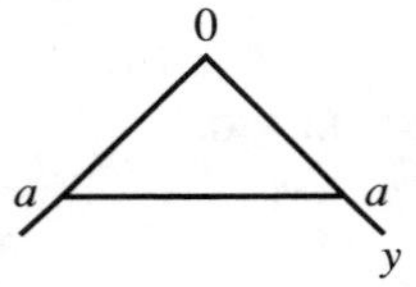

(a) $\psi_m = \cos\frac{m\pi}{a} x - \cos\frac{m\pi}{a} y$

$$k_m = \sqrt{2}\frac{m\pi}{a}$$

(b) $\psi_m = \cos\frac{m\pi}{a} x \cos\frac{m\pi}{a} y$

$$k_m = \sqrt{2}\frac{m\pi}{a}$$

Circular ring

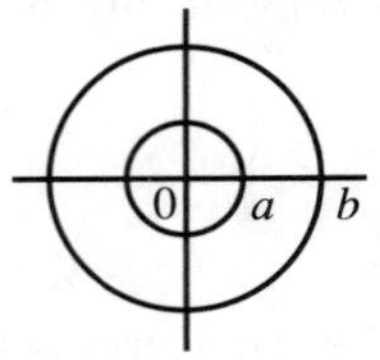

$$\psi_{mn} = [N'_n(k_{mn}a)J_n(k_{mn}\rho) - J'_n(k_{mn}a)N_n(k_{mn}\rho)]e^{in\phi}$$

$$\frac{J'_n(k_{mn}a)}{N'_n(k_{mn}a)} = \frac{J'_n(k_{mn}b)}{N'_n(k_{mn}b)}$$

Circular ring segment

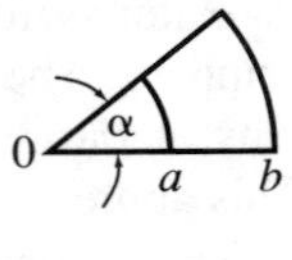

$$\psi_{m\nu} = [N'_\nu(k_{m\nu}a)J_\nu(k_{m\nu}\rho) - J'_\nu(k_{m\nu}a)N_\nu(k_{mn}\rho)]\cos\nu\phi$$

$$\nu = n\pi/\alpha$$

$$\frac{J'_\nu(k_{m\nu}a)}{N'_\nu(k_{m\nu}a)} = \frac{J'_\nu(k_{m\nu}b)}{N'_\nu(k_{m\nu}b)}$$

Ellipse

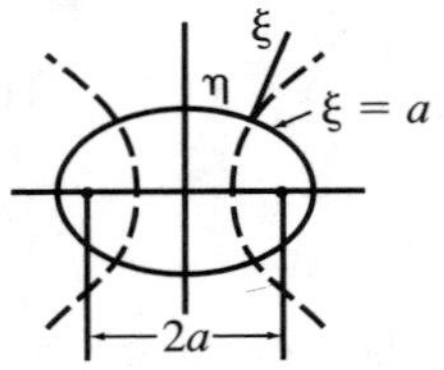

Even modes:
$\psi_{mn} = \mathrm{Re}_m(\zeta,\chi e_n)\, Se_m(\eta,\chi e_n)$
$\mathrm{Re}_m(a, \chi e_n) = 0,\ \chi e_n = kq$
major axis $= 2q \cosh a$
minor axis $= 2q \sinh a$
odd modes:
Replacing e by o in the above

Equilateral triangle

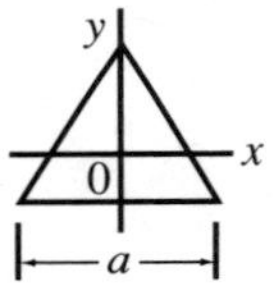

$$\psi_m = \cos\frac{2\pi l}{3b}\left(\frac{u}{2} + b\right) \times \cos\frac{\pi(m+n)(\nu - w)}{9b} + \cos\frac{2\pi m}{3b}\left(\frac{u}{2} + b\right) \times \cos\frac{\pi(n-l)(\nu - w)}{9b} + \cos\frac{2\pi n}{3b}\left(\frac{u}{2} + b\right) \times \cos\frac{\pi(l-m)(\nu - w)}{9b}$$

$$l = -(m+n),\ u = \frac{\sqrt{3}}{2}x + \frac{1}{2}y$$

$$\nu - w = -\frac{\sqrt{3}}{2}x + \frac{3}{2}y$$

$$b = a/2\sqrt{3}$$

$$k_{mn}^2 = \left(\frac{4\pi}{3a}\right)^2(m^2 + n^2 + mn)$$

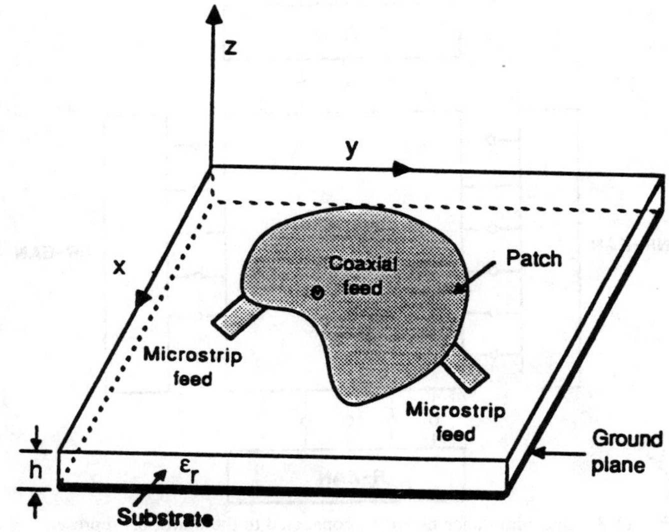

Fig. 11.4 Configuration of an arbitrarily shaped microstrip patch antenna with coordinate system shown.

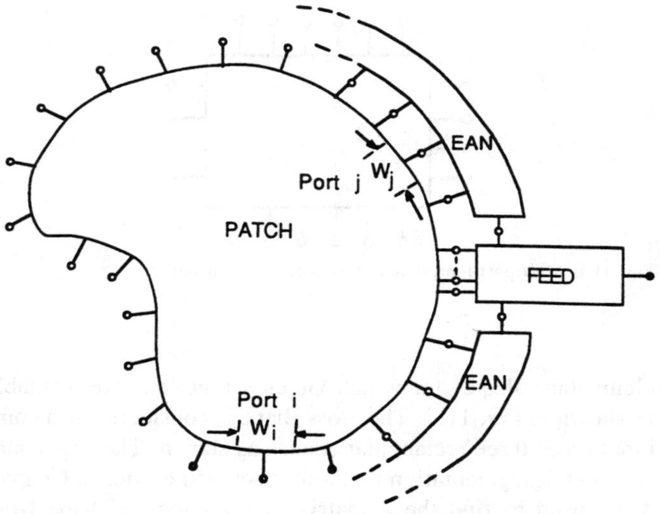

Fig. 11.5 Equivalent multiport-network model of the arbitrarily shaped microstrip patch antenna shown in Fig. 11.4.

the scattering parameters, the bandwidth, and the radiation pattern. The MNM approach can conveniently incorporate the effects of the mutual coupling and the feed-junction reactances.

11.4.2 Modeling of Internal Fields

In most applications, the thickness h of the substrate used for microstrip antennas is small compared to the operating wavelength ($k_0h << 1$). The fields near the edges of the antenna may vary in the direction perpendicular to the patch, but this z-variation decays rapidly as one moves inward and away from the patch edges, leaving only the fields that are uniform along z. So a solution of the electromagnetic fields in the region between the patch and the ground plane can be obtained by considering the patch as a two-dimensional cavity with magnetic wall boundaries. The z-varying fields near the edges (inside and outside the patch) are included in the modeling of the edge fields discussed later.

For patches of regular shapes (rectangles, circles, rings, ellipsoidal discs, sectors of circles and rings, and three types of triangles), the multiport planar component representing the patch can be analyzed by using the two-dimensional impedance Green's functions available for these shapes and discussed in Chapter 2. Elements of the Z-matrix of the patch are derived from the Green's function as

$$Z_{ij} = \frac{1}{W_iW_j} \int_{W_i} \int_{W_j} G(x_i, y_i|x_j, y_j)(ds_i)(ds_j) \tag{11.9}$$

where (x_i, y_i) and (x_j, y_j) denote the locations of the two ports of width W_i and W_j, respectively (as shown in Fig. 11.5). The two line integrals in (11.9) are along the widths of the ports i and j. Ports i and j can be located anywhere inside the patch or at the periphery of the patch. As discussed earlier in Chapter 2, the Green's function G is usually a doubly infinite summation with terms corresponding to the various modes of a planar resonator with magnetic walls. For most of the regular shapes, the Green's function G can be reduced to a single summation by evaluating one of the summation analytically and thus reducing the computation time significantly. The effect of the dielectric losses is incorporated by considering ε_r to be a complex quantity. The conductor losses are also included in an approximate manner by defining (and adding to the dielectric loss tangent) an equivalent loss tangent δ_c as [11.11]

$$\delta_e = \frac{P_c}{P_d}\delta_d \tag{11.10}$$

where P_c is the power dissipation because of the conductor loss in the patch and P_d is the dielectric loss in the substrate with loss tangent δ_d. In the limit of magnetic wall boundary. δ_c is independent of the resonator geometry [11.15] and is equal to $\sqrt{2/(\omega\mu\sigma)}/h$, where σ is the conductivity of the patch metallization.

It may be pointed out that, unlike in the cavity model [11.11], the effective loss tangent used here accounts only for the dielectric and the conductor losses. The radiated power and the exterior losses are incorporated in the edge admittance and the mutual coupling networks discussed later.

The multiport-network representation of the internal fields of a rectangular patch is shown in Fig. 11.6, where a number of ports are located all around the edges. Each port represents a small section (of length W_i) of the edge of the patch. The port width W_i is chosen so small that the fields over the length may be assumed to be uniform. Typically, for a rectangular patch, the number of ports along each radiating edge is taken to be 4, and along each nonradiating edge, the number is taken to be 8. Thus a 24 × 24 matrix is typically adequate for the characterization of the interior fields of a rectangular patch. For patches with composite shapes (such as cross-shaped antenna [11.16, 11.17]), impedance matrix elements Z_{ij} are computed by treating the composite shape as a combination of the

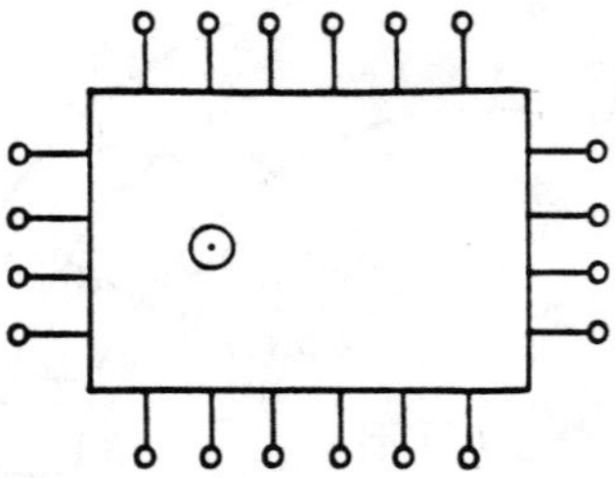

Fig. 11.6 Multiport representation of a rectangular patch.

elementary shapes for which Green's functions are available as shown in Fig. 11.7. The cross shape is considered as a combination of three rectangular shapes as shown. The segmentation and desegmentation methods discussed earlier in Chapter 4 are used to find the Z-matrix of a composite shape from those of elementary segments. If the patch or one of the segments of a composite patch is of an irregular shape for which the Green's function is not available, a contour-integral method as discussed in Chapter 3 can be used to evaluate the elements of the Z-matrix.

11.4.3 Modeling of Edge Fields

In the MNM approach, the edge fields (namely, the fringing fields at the edges, the surface wave fields, and the radiation fields) are modeled by introducing equivalent edge admittance networks (EANs) connected to the edges of the patch (as shown in Figs. 11.5 and 11.8). When a microstrip patch has a polygon shape (rectangular and triangular geometries), the periphery of the patch is divided into edges. Each edge may have a different voltage distribution. For geometries similar to the circular patch, the total periphery is taken as a single edge. The EAN for each edge is a multiport network consisting of combinations of a capacitance C, an inductance L (representing the energy stored in the fringing electric and magnetic fields, respectively), and a conductance G (representing the power carried away by radiation and by surface waves). A portion of a typical EAN is shown in Fig. 11.9. Similar sections are connected to the other ports of the equivalent planar multiport representation of the patch. The elements of the Y-matrix characterizing the EAN networks is computed from the equivalent circuit shown in Fig. 11.9.

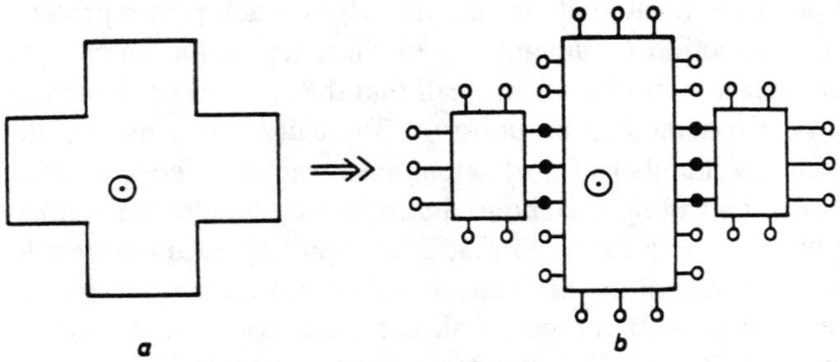

Fig. 11.7 (a) A cross-shaped microstrip patch; (b) multiport-network model of the cross-shaped microstrip patch.

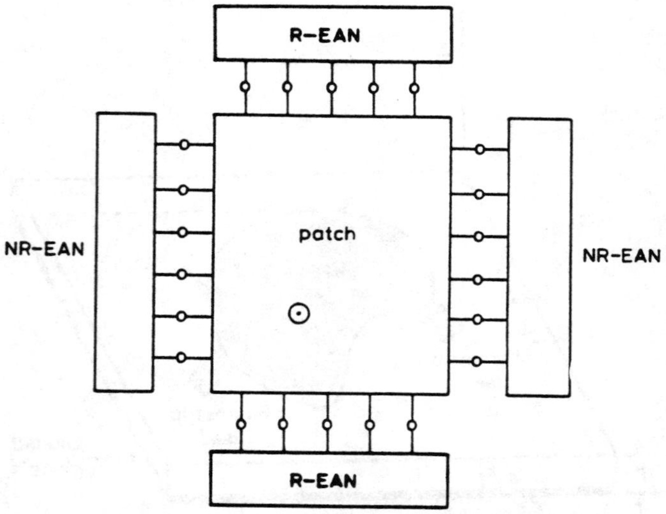

Fig. 11.8 Edge admittance networks connected to the multiport representation of a rectangular patch.

Edge conductance. The edge conductance G, in an EAN, consists of two parts: a radiation conductance G_r and a surface wave conductance G_s. The conductances G_r and G_s associated with an edge of a microstrip patch are defined as equivalent ohmic conductances (distributed or lumped). These conductances, G_r and G_s, when connected to the edge of the patch (continuously or at discrete ports) will dissipate a power equal to that radiated (P_r) and to that launched as surface wave (P_s) by the patch, respectively. When the voltage amplitude distribution along an edge is given by $f(l)$, the conductances G_r and G_s of the edge are obtained as [11.14]

$$G_{r,s} = \frac{2P_{r,s}}{(1/W)\int_0^W f^2(l)dl} \tag{11.11}$$

where l denotes the distance along the edge of the patch and $P_{r,s}$ are computed for a voltage distribution $f(l)$. If we select n uniformly spaced ports (each representing a section of length W/n) along the edge, the conductance G_P connected to each of the ports is taken as $(G_r + G_s)/n$.

The concept of edge conductance can be implemented when $f(l)$ is known a priori. In most of the cases, microstrip antennas

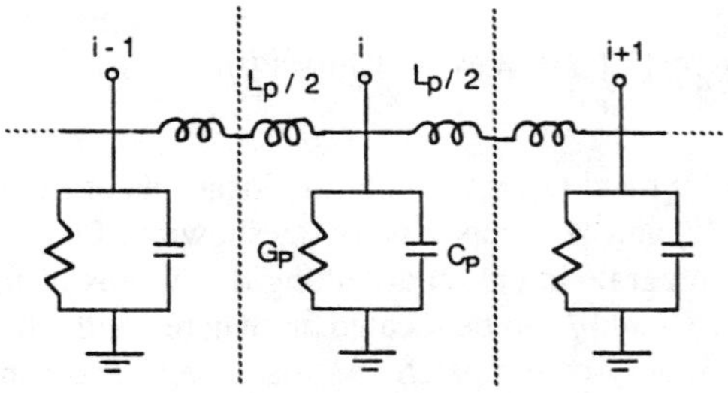

Fig. 11.9 Elements of an edge admittance network.

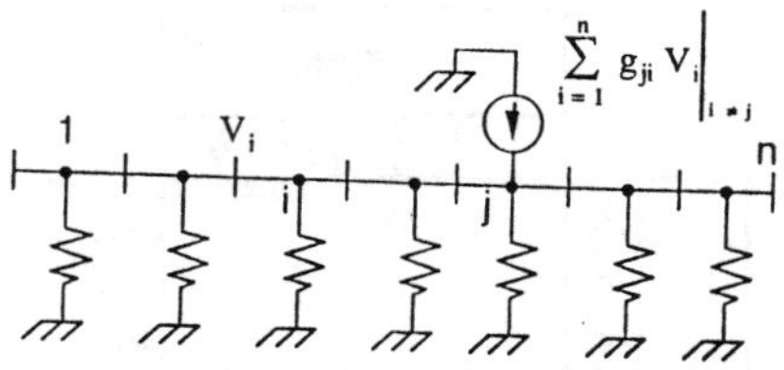

Fig. 11.10 Generalized edge conductance network used when the field distribution along the edge is not known a priori.

are operated near the resonance frequency of the patch and $f(l)$ is known, at least approximately. For more accurate results, iterative computations may be needed. Starting from an approximate $f(l)$, an analysis based on the MNM model is carried out to evaluate the voltages at the n ports on the edge. This computed voltage distribution is then used as a modified $f(l)$, and the computations of G_r and G_s are repeated. A method for computing the radiated power (P_r) from the edges of a microstrip patch (on a thin substrate without cover layer) is discussed in Sec. 4.6. For thin substrates without any cover layer, G_s is much smaller than G_r and may be neglected [11.2, pp. 53–59].

The procedure just described yields an equivalent conductance that is distributed uniformly over sections of the edge. The resulting edge admittance network is a diagonal matrix. The mutual conductance among all sections gets incorporated in the calculation of the power radiated from the edge. This method needs a priori knowledge of the voltage distribution along the edge. A recent modified method [11.18] does not need any such information. In this method, equivalent magnetic-current element sources are placed at the location of each section at the periphery. The elements of the Y-matrix characterizing the edge conductance network are obtained by equating the power flow associated with these magnetic currents (obtained from Poynting vector) to the power dissipated in the equivalent network. The diagonal terms of the Y-matrix are equal to the self-radiation conductances of the corresponding segments. The off-diagonal terms are equal to the real parts of the mutual coupling admittances between the two corresponding segments. It may be noted that in this modified method, the real part of mutual coupling among different sections of the edge is included in the off-diagonal terms of the conductance networks, whereas the reactive part is included (as in the earlier approach also) in the inductance and the capacitance networks.

The generalized edge conductance network, based on this modified approach, looks like that shown in Fig. 11.10. The current source shown in this figure accounts for the mutual coupling among various sections of the edge. Similar current sources are also attached to the other nodes 1 through n. These are not shown in the figure. The transfer conductance g_{ji} is the real part of the complex mutual coupling Y_{ji} between the sections j and i. The mutual admittance Y_{ji} represents the electric current induced in the jth section as a result of a unit magnetic current (or voltage) at the ith section. Computation of Y_{ji} is discussed in a later section. The resulting edge conductance matrix has the following form:

$$[G_e] = \begin{array}{c} 1 \\ \vdots \\ \vdots \\ \vdots \\ n \end{array} \begin{bmatrix} g_{11} & g_{12} & \cdots & g_{1n} \\ g_{21} & \ddots & \ddots & \vdots \\ & & & \\ \vdots & \ddots & \ddots & \vdots \\ g_{n1} & \cdots & \cdots & g_{nn} \end{bmatrix} \tag{11.12}$$

Note that in the edge admittance matrix $[Y_e]$ we need to include the contributions of fringing capacitances and inductances as discussed later in this section. Also, it may be pointed out that actual values of the voltages V_i and V_j are not needed to compute Y_{ji} or g_{ji}. The diagonal terms g_{ii} are self-conductances of individual sections given by

$$g_{ii} = \frac{W_i^2}{90\lambda_0^2} \tag{11.13}$$

where W_i is the width of the ith section and λ_0 is the free-space wavelength at the operating frequency.

Various formulas for edge conductance are summarized in [11.3, Chap. 11, Sec. 11.4.1].

Edge Capacitance. The edge capacitance C accounts for the energy stored in the fringing electric field (inside and outside) at the edge of the patch. The fringing capacitance is defined as the excess of the total capacitance of the patch over that which would exist if the patch were considered as a two-dimensional capacitor with magnetic walls at the open edges. As in the cases of the edge conductance, the edge capacitance is also distributed uniformly over the n ports ($C_P = C/n$). Formulas for the edge capacitance are available for simple shapes, namely, rectangular [11.19, 11.20] and circular [11.21] geometries. The evaluation of the edge capacitance of an arbitrarily shaped microstrip patch is described in [11.22].

Edge Inductance. The edge inductance L accounts for the energy stored in the fringing magnetic field at the edge of the patch. As in the cases of G and C, the edge inductance is also distributed uniformly over the n ports ($L_P = L/n$). For the nonradiating edges of a rectangular patch operating in the dominant mode, the fringing inductance per unit length L_e and the fringing capacitance per unit length C_e (for $\varepsilon_r = 1$) are related by $L_e = \mu_0\varepsilon_0/C_e(\varepsilon_r = 1)$.

When the voltage distribution along an edge is uniform, the voltages at two adjacent ports on the edge are equal, and hence, no current flows through the edge inductance. In this case, the edge inductance need not be included, and the EAN network simplifies to parallel pairs of capacitance and conductance only. This situation occurs at the radiating edges of a rectangular patch operating in the dominant mode.

The common practice for evaluating the resonance frequency of two-dimensional resonators (incorporating the fringing fields) is to use effective dimensions larger than the physical dimensions. Calculation of these effective dimensions is commonly carried out by finding the total electrostatic capacitance and equating it to the parallel-plate capacitance of effective dimensions. However, it needs to be pointed out here that the use of effective dimensions modifies the magnetic field distribution also, in addition to adding the fringing capacitance at the edges. The use of effective dimensions for the analysis of microstrip patches, using the cavity model with magnetic wall boundary, is consistent with the MNM approach that includes both inductance and capacitance. The fringing edge inductance L_e per unit length and the fringing capacitance per unit length are also related by $L_e = \mu_0\varepsilon_0/C_e(\varepsilon_r = 1)$.

11.4.4 Modeling of Mutual Coupling

The flexibility of the multiport-network model leads to several advantages when compared with the conventional cavity model. For example, the multiport-network model allows us to incorporate the effect of mutual coupling among different edges of a patch or among patches of an array [11.23] by defining a mutual coupling network (MCN). The edge admittance terms associated with various ports at the edge constitute the diagonal terms of the admittance matrix for the MCN. The nondiagonal terms of this matrix are obtained from the "reaction" between the equivalent magnetic current sources at the two corresponding sections of the edges.

When the periphery of a patch is considered as one edge (circular geometries), the mutual coupling among different sections of that edge is included in the EAN parameters (G, C, and L), which are computed using electromagnetic analysis. Hence, in this case, no MCN network is needed. On the other hand, when the periphery is divided into edges (polygon shapes), an MCN network must be included among all edges of the patch. A detailed analysis of the effect of mutual coupling on rectangular microstrip patches has been reported in [11.23] and is summarized in Appendix A to this chapter.

11.4.5 Modeling of Feed-Junction Reactances

For a microstrip-fed patch, the parasitic reactances at the junction between the microstrip feedline and the patch can be incorporated in the MNM by modeling a small section of the feedline as a rectangular planar segment (as shown in Figs. 11.5 and 11.11) connected to the patch at a finite number of (typically five) ports. The exact number of ports depends upon the variation of the fields along the common interface. For example, a larger number of ports is needed when the feedline is along the nonradiating edge of a rectangular patch than when it is along the radiating edge. One port on each side of the FEED segment (near the feed-junction plane as shown in Figs. 11.5 and 11.11) is used to connect EAN network to the FEED segment. This connection assures the continuity of the current flow tangential to the periphery of the patch.

The effect of fringing fields along the transmission line length is accounted for by using an equivalent effective width and dielectric constant for the transmission line. The length of the feedline segment should be long enough ($\geq \lambda_0/8$) so that the higher-order evanescent modes (excited by the junction) decay and only the dominant quasi-TEM mode is present at the input of the feedline. Thus only one port is needed at the input end of the line section. The elements of the Z-matrix characterizing the planar FEED segment are computed using Eq. (11.9). The effect of the feed junction is to add an equivalent inductive reactance at the feed point.

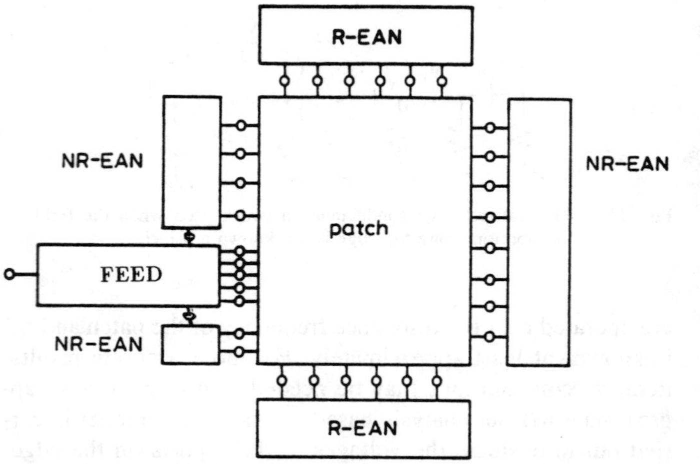

Fig. 11.11 Incorporation of feed-junction reactance in multiport-network model of a rectangular patch.

11.4.6 Analysis of Multiport-Network Model

The network analysis methods discussed in Chapter 4 are appropriate tools for analyzing the MNM model of microstrip antennas. First, Z-matrices for the different subnetworks modeling the microstrip patch antenna are computed and the segmentation technique is then used to combine these Z-matrices and to obtain the overall Z-matrix for the antenna. The segmentation method is employed by combining two segments at a time. This process is computationally more efficient than the procedure in which all the segments are combined at the same time. This is because combining two segments at a time requires the inversion of smaller matrices whose sizes depend upon the number of interconnected ports of the two segments. Use of the segmentation method to yield the antenna characteristics is also discussed in this section.

Computing the Z-matrix for the antenna. The impedance matrices for planar components (patch and the transmission line feeds) are obtained by using the Green's function approach described in Chapter 2. The Z-matrix for the EAN network is obtained by inverting the Y-matrix characterization of the edge fields described in Sec. 11.4.3. The segmentation method (Chapter 4) is then used to combine these three Z-matrices, two at a time, to yield one Z-matrix characterizing the patch. For example, the applications of segmentation method to a single-port antenna yield the input impedance. For a multiple-port patch, the segmentation procedure yields a Z-matrix referenced to these multiple ports. Elements of the S-matrix (reflection and transmission coefficients) are obtained from the Z-matrix using standard conversion formulas.

The resonance frequency of a one-port patch is obtained by finding the frequency where the reflection coefficient at the input port is zero (or minimum). Also, the impedance bandwidth for a specified VSWR is computed from the variation of the input impedance with frequency.

Evaluation of the radiation characteristics. The application of the segmentation method also yields the voltages at the radiating edges (at the interconnected ports between the PATCH and the EAN segments). These voltage distributions are expressed as equivalent magnetic-current distributions along the edges and are used for the evaluation of the radiation characteristics.

The far-field electric radiation vector L for a microstrip patch (on thin substrate without any cover layer) is related to the port voltages V_j at the patch periphery by [11.2]

$$\bar{L} = -2 \sum_{j}^{N} \bar{a}_t W_j V_j \, e^{ik_0\rho_j \sin\theta \cos(\phi - \phi_j)} \tag{11.14}$$

where the factor 2 arises because of the image of the magnetic-current element with respect to the ground plane. The other parameters used in (11.14) are shown in Fig. 11.12. The summation in (11.14) is with respect to all the n ports at the radiating edges of the patch. V_j is the average voltage at the jth port of width W_j. The radiation fields are related to the radiation vector L by [11.2]:

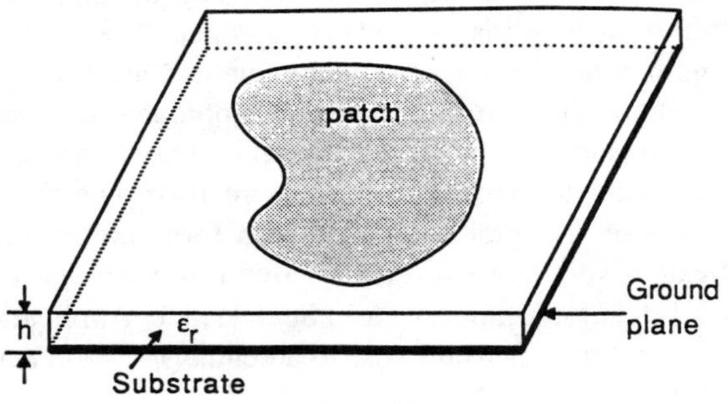

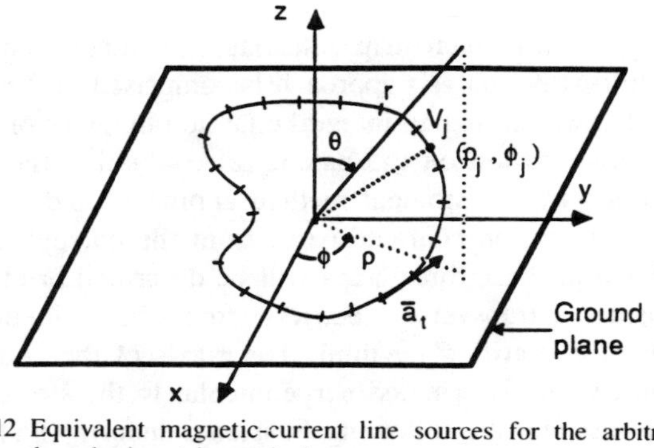

Fig. 11.12 Equivalent magnetic-current line sources for the arbitrary shaped microstrip patch shown.

$$E_\theta = -i \frac{\bar{L} \cdot \bar{a}_\phi}{2\lambda_0 r} \exp(-ik_0 r) \tag{11.15}$$

$$E_\phi = i \frac{\bar{L} \cdot \bar{a}_\theta}{2\lambda_0 r} \exp(-ik_0 r) \tag{11.16}$$

and $H_\theta = -E_\phi/\eta_0$, $H_\phi = E_\theta/\eta_0$, λ_0 is the free-space wavelength, and η_0 is the free-space impedance ($\cong 377\ \Omega$). The total power P_r radiated by the patch is given by

$$P_r = \int_{\theta=0}^{\pi/2} \int_{\phi=0}^{2\pi} \frac{|\bar{L} \times \bar{a}_r|^2}{8\lambda^2 \eta_0} \sin\theta \, d\theta \, d\phi \tag{11.17}$$

where ($\mathbf{a}_\phi$, $\mathbf{a}_\theta$, $\mathbf{a}_r$) are unit vectors in spherical coordinates. The radiated power P_r is used for the computation of the radiation conductance G_r (given by eq. 11.11).

11.4.7 Sensitivity Analysis and Optimization

The sensitivity analysis is an extension of the usual antenna analysis discussed in Sec. 11.4.6. It describes the variations in the performance (input impedance, radiation pattern, bandwidth, etc.) of the microstrip patch antenna that result from small changes in the various designable parameters (such as dimensions, etc.) and nondesignable parameters (such as substrate dielectric constant, height of substrate, etc.). Two major applications of the sensitivity analysis are (1) tolerance analysis (i.e., evaluation of performance limits resulting from statistical tolerances in values of various parameters) and (2) optimization of the patch configuration (to yield the optimum performance) using gradient optimization techniques that are more efficient than the direct search methods (not requiring the sensitivity analysis).

Sensitivity analysis of multiport networks has been dealt with extensively in the literature [11.24, 11.25]. Two distinct approaches are available. The first approach is the incremental or finite-difference method, which consists of changing one of the parameters slightly and going through the network analysis again. Thus, if sensitivity calculations are needed with respect to n parameters, we need to run the analysis program at least $n + 1$ times. This can become computationally very expensive. The second approach uses the concept of "adjoint network," which is derived from the original network under consideration. Sensitivity calculations using the adjoint network approach need only two analyses to be carried out, one for the original network and the other for the adjoint network. For this reason, the adjoint network approach is used extensively for the CAD of electronic networks.

The key advantage of using the multiport-network modeling approach for the CAD of microstrip patch antenna is that the adjoint network method can now be used for the sensitivity analysis of microstrip antennas also. Furthermore, as the multiport-network model of a microstrip patch is a passive network, the adjoint network is identical to the original network itself. Thus only one analysis is needed to get the sensitivity information, hence leading to a remarkable im-

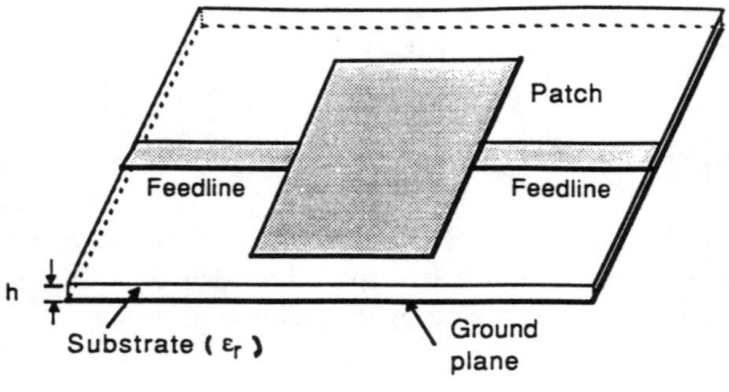

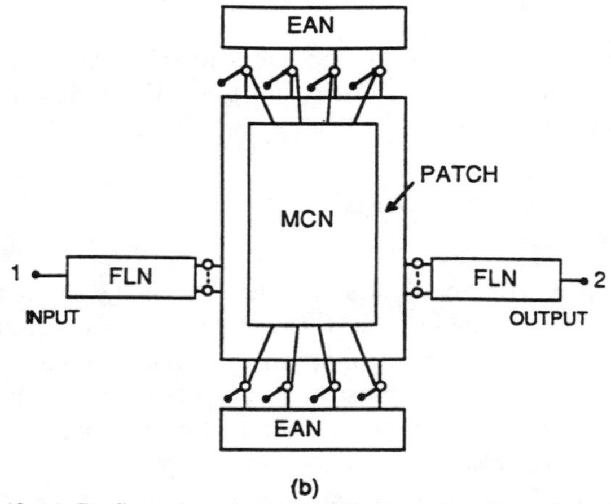

Fig. 11.13 (a) Configuration of a two-port rectangular patch; (b) equivalent multiport network model of the two-port rectangular patch.

provement in the computational speed. It may be noted that if the full-wave moment-method solution discussed later in this chapter is used for microstrip patches, the adjoint network approach is no longer applicable. In this case, the full-wave analysis will need to be carried out $n + 1$ times to calculate the antenna performance sensitivities with respect to the n parameters. This consideration may make the CAD implementation using the full-wave moment method solution impractical.

The implementation of the sensitivity analysis for microstrip patches using the MNM approach may be applied to the case of a two-port rectangular patch (Fig. 11.13) whose MNM is shown in Fig. 11.13(b). We have six subnetworks in this MNM representation. For calculating the antenna performance, we need Z-matrix (or Y-matrix) characterization for each of these subnetworks. For the sensitivity analysis, we need also the differential Z-matrix for the six subnetworks. This is equivalent to calculating the sensitivities for each of the six components. This can be carried out analytically for the three rectangular planar segments and numerically (or from empirical closed form expressions) for the MCN and EAN segments. Knowing the sensitivities of these components, the sensitivity of the overall network (in this case, the MNM of the patch antenna) may be computed using the adjoint network approach. For sensitivity calculations, the external ports of interest are the input and output ports and four ports on each of the radiating edges. Sensitivity computations in terms of Z-matrices may be expressed as [11.25, p. 598]

$$I_p^t \Delta Z \, I_p = I_b^t \, \Delta Z_b \, I_b \tag{11.18}$$

where ΔZ_b represents the differential Z-matrix of the component whose parameter Φ is being changed by $\Delta\Phi$ and I_b represents a current vector (with currents at various ports of this component). The superscript t denotes the transpose of a vector and I_p is a current excitation vector at the external ports. The term ΔZ represents the change in the overall (10×10 in this case) Z-matrix and yields the sensitivity values. Typical values for I_p are $[1, 0, 0, \ldots, 0]^t$ for an excitation at port 1 or $[0, 1, 0, \ldots, 0]^t$ when the effect of the reflected signal on the antenna performance is being computed. The elements of I_b are obtained by analyzing the MNM for the patch.

Values of $\Delta Z/\Delta\Phi$ obtained from relation (11.18) may be used directly as gradients in optimization algorithms requiring gradient evaluation.

The multiport network modeling approach for the computer-aided design of radiating microstrip patches has been employed for a variety of microstrip patch configurations. Some of these configurations are summarized later, in Sec. 11.6.

11.5 Full-Wave Analysis of Microstrip Antennas

In Secs. 11.2, 11.3, and 11.4, three different approaches for designing microstrip antennas have been reviewed. Techniques discussed in Secs. 11.3 and 11.4 are based on two-dimensional formulation, and hence they are approximate. The full-wave analysis methods [11.3, Chap. 5, 8, and 12] utilize rigorous electromagnetic technique and are less approximate. Both the integral equation approach discussed in Chapter 5 and the difference equation–based methods discussed in Chapter 6 are applicable to microstrip patch antenna analysis also. Integral equation based formulation has been used extensively. An overview of this full-wave analysis approach [11.26] is summarized in Sec. 11.5.1. Applications of FD and FE analysis to microstrip antennas [11.48–11.52] are summarized in Sec. 11.5.2.

11.5.1 Integral Equation–Based Full-Wave Analysis

When applied to microstrip antennas, the integral equation–based full-wave analysis approach is comprised of three basic steps: (1) formulating an integral equation in terms of electric current distribution on the patch, (2) evaluating the current distribution by the moment method approach, and (3) evaluating the radiation characteristics from the current distribution. Each of these three steps will be described briefly.

Consider a transverse electric current sheet situated in a horizontally stratified medium. The z-axis of the chosen coordinate system is oriented perpendicular to the stratification boundaries, and the following Fourier transform (FT) pair is applied to Maxwell equations

$$F\{\mathbf{A}(\rho,z)\} = \tilde{\mathbf{A}}(\mathbf{k}_t,z) = \iint_{-\infty}^{\infty} \mathbf{A}(\rho,z)e^{-jk_t\cdot\rho}d\rho \tag{11.19a}$$

$$F^{-1}\{\tilde{\mathbf{A}}(\mathbf{k}_t,z)\} = \mathbf{A}(\rho,z) = (1/4\pi^2)\iint_{-\infty}^{\infty} \tilde{\mathbf{A}}(\mathbf{k}_t,z)e^{-jk_t\cdot\rho}d\mathbf{k}_t \tag{11.19b}$$

where

k_t = transverse vector wave number
$= k_x\hat{\mathbf{x}} + k_y\hat{\mathbf{y}} = k_t(\cos\alpha\hat{\mathbf{x}} + \sin\alpha\hat{\mathbf{y}}) = k_t\hat{\mathbf{k}}_t;\ k_t = |\mathbf{k}_t|$

The transforms $\tilde{\mathbf{E}}$ and $\tilde{\mathbf{H}}$ of the electric and the magnetic fields, respectively, are also composed in the following manner:

$$\tilde{\mathbf{E}} = \tilde{E}_z\hat{\mathbf{z}} + \hat{\mathbf{E}}_t \qquad \tilde{\mathbf{H}} = \tilde{H}_z + \tilde{\mathbf{H}}_t \tag{11.20a}$$

$$\tilde{\mathbf{E}}_t = V'\hat{\mathbf{k}}_t + V''(\hat{\mathbf{k}}_t + \hat{\mathbf{z}}) \tag{11.20b}$$

$$\tilde{\mathbf{H}}_t = I'(\hat{\mathbf{z}} \times \hat{\mathbf{k}}_t) + I''\hat{\mathbf{k}}_t \tag{11.20c}$$

This decomposition is completely analogous to the $\mathbf{E}$ (or TM_z) and H (or TE_z) modal representations of the field in waveguides. Each plane wave in the Fourier spectrum of the field can be considered as a mode with the corresponding wave number k_t, V' and I' play the role of the $\mathbf{E}$ (TM_z) modal coefficients, and V'' and I'' play the role of the $\mathbf{H}$(TE_z) coefficients. Substitution of (11.20) into the transformed Maxwell equations can be shown [11.27] to reduce the latter to two sets of independent scalar equations having the following form:

$$\frac{d}{dz}V(z,z') + jk_zZI(z,z') = 0 \tag{11.21a}$$

$$\frac{d}{dz}I(z,z') + jk_zYV(z,z') = i \tag{11.21b}$$

Equations (11.21) are valid for both E- and H-modal coefficients, and the quantities appearing therein are defined as follows:

$$\begin{aligned} z &= \text{observation point} \\ z' &= \text{source point} \\ i &= \begin{cases}\delta(z-z'), & z, z' \text{ are in the same layer} \\ 0, & z, z' \text{ are not in the same later}\end{cases} \\ \delta(\cdot) &= \text{Dirac delta function} \\ Z &= 1/Y = \begin{cases} Z' = k_2/\omega\varepsilon, & E\cdot\text{modes} \\ Z'' = \omega\mu/k_z, & H\cdot\text{modes}\end{cases} \\ k_z &= \sqrt{k^2 - k_t^2};\ k^2 = \omega^2\varepsilon\mu \text{ (the branch of the square root is chosen such as to satisfy the radiation condition)} \end{aligned} \tag{11.21c}$$

The original vector problem is thus reduced to a scalar one, governed by standard transmission line Eqs. (11.21). In this context, the individual dielectric layers are viewed as sections of transmission lines having characteristic impedance Z and propagation constant k_z. Continuity conditions at interfaces and other conditions at boundary and source can be applied to the scalar quantities V and I.

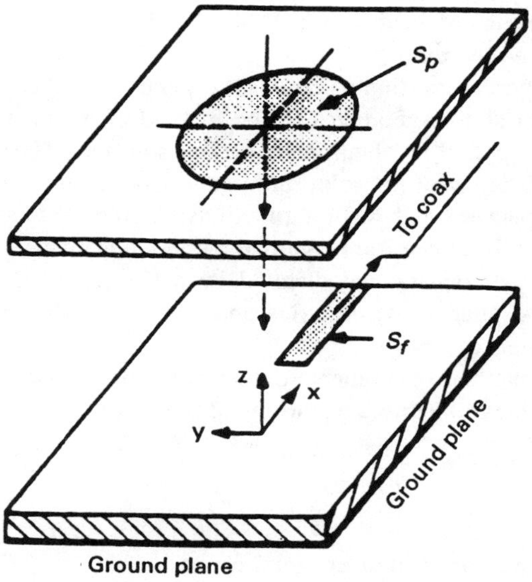

Fig. 11.14 Patch antenna electromagnetically excited by a microstrip transmission line.

With these definitions, the FT of the electric-field dyadic Green's function for a transverse electric current sheet takes on the following form:

$$\tilde{\overline{\overline{\mathbf{G}}}}(\mathbf{k}_t;\, z,z') = -V'(z,z')\hat{\mathbf{k}}_t\hat{\mathbf{k}}_t - V''\cdot(z,z')(\hat{\mathbf{k}}_t \times \hat{\mathbf{z}})(\hat{\mathbf{k}}_t \times \hat{\mathbf{z}}) + (k_t/\omega\varepsilon)\, I'(z,z')\hat{\mathbf{z}}\hat{\mathbf{k}}_t \tag{11.22}$$

Derivation of the Green's function for a $\hat{\mathbf{z}}$-directed magnetic current is completely analogous to that shown [11.27].

In physical space (x, y, z), the Green's function can be used to cast the microstrip antenna problem in the form of a vector integral equation. Figure 11.14 depicts a microstrip structure consisting of a patch S_p and a microstrip feed S_f, which can be located on the same plane as the patch or at a different level. Enforcement of the boundary condition, requiring the total electric field to vanish on the microstrip, yields the desired integral equation,

$$\mathbf{z} \times [\mathbf{E}^i(\mathbf{r}) + \mathbf{E}(\mathbf{r})] = 0, \qquad \mathbf{r} \in [S_p + S_f] \tag{11.23}$$

where $\mathbf{E}^i(\mathbf{r})$ is the electric field at $\mathbf{r}$ due to the impressed source, and

$$\mathbf{E}(\mathbf{r}) = \iint_{S_p} \overline{\overline{\mathbf{G}}}(\mathbf{r},\mathbf{r}')\cdot\mathbf{J}_p(\mathbf{r}')\,dr' + \iint_{S_f} \overline{\overline{\mathbf{G}}}(\mathbf{r},\mathbf{r}')\cdot\mathbf{J}_f(\mathbf{r}')\,d\mathbf{r}' \tag{11.24a}$$

$$= F^{-1}\{\tilde{\overline{\overline{\mathbf{G}}}}(\mathbf{k}_t;\, z, z')\cdot\tilde{\mathbf{J}}_p(\mathbf{k}_t)\} + F^{-1}\{\tilde{\mathbf{G}}(\mathbf{k}_t;\, z, z')\cdot\tilde{\mathbf{J}}_f(\mathbf{k}_t)\} \tag{11.24b}$$

where $\mathbf{J}_p$ and $\mathbf{J}_f$ are the currents on the patch and the feed, respectively. The microstrip is assumed to be made of infinitesimally thin conductors; hence $\mathbf{J}_p$ and $\mathbf{J}_f$ are actually sums of currents flowing on its top and bottom surfaces. Equation

(11.24b) is obtained from (11.24a) by application by the convolution theorem.

Analytical solution of Eqs. (11.23) and (11.24) is precluded even for simple geometries. The method of moments [11.28] can be applied to obtain a numerical solution. The accuracy and efficiency of the solution largely depend on the choice of the expansion and testing functions. In the method of moments, this choice can be guided by previous investigations of similar geometries and dictated by the inherent geometrical features, such as symmetries and edges and also numerical convenience.

The method of moments reduces the integral equation to a set of linear algebraic equations of the form

$$[Z][I] = [V] \tag{11.25}$$

where

$[I]$ = vector of undetermined expansion function coefficients
$[V]$ = excitation vector
$[Z]$ = moment method matrix

The elements of $[Z]$, denoted by Z_{mn}, are given by

$$Z_{mn} = \iint_{D(J_m)} \mathbf{J}_m(\mathbf{r}) \cdot \left\{ \iint_{D(\mathbf{J}_n)} \overline{\overline{\mathbf{G}}}(\mathbf{r}, \mathbf{r}') \cdot \mathbf{J}_n(\mathbf{r}')\, dt' \right\} dr \tag{11.26a}$$

$$= \frac{1}{4\pi^2} \iint_{-\infty}^{\infty} \tilde{\mathbf{J}}_m^*(\mathbf{k}_t) \cdot \tilde{\overline{\overline{\mathbf{G}}}}(\mathbf{k}_t; z, z') \cdot \tilde{\mathbf{J}}_n(\mathbf{k}_t)\, d\mathbf{k}_t \tag{11.26b}$$

where

$\mathbf{J}_n(\mathbf{r}')$ = expansion basis function
$\mathbf{J}_m(\mathbf{r}')$ = testing function
$D(\cdot)$ = domain of the function
* = complex conjugation

The elements of $[V]$ can be written as

$$V_m = -\iint_{D(\mathbf{J}_m)} \mathbf{E}_{\tan}(\mathbf{r}) \cdot \mathbf{J}_m(\mathbf{r})\, d\mathbf{r} \tag{11.27a}$$

$$= \frac{-1}{4\pi^2} \int_{-\infty}^{\infty}\int \tilde{\mathbf{E}}^i_{\tan}(\mathbf{k}_t) \cdot \tilde{\mathbf{J}}_m^*(\mathbf{k}_t)\, d\mathbf{k}_t \tag{11.27b}$$

where $\mathbf{E}^i_{\tan}$ $(\mathbf{r})$ is the component of the impressed field transverse to the $\hat{z}$-axis. Equations (11.26b) and (11.27b) are obtained from their respective predecessors (11.26a) and (11.27a) by application of convolution and Parseval theorems. The former are more advantageous from the computational standpoint, as they require fewer numerical integrations to be performed. Often, the expansion and testing functions are drawn from the same set. The moment-method solution in that case is known as the Galerkin technique.

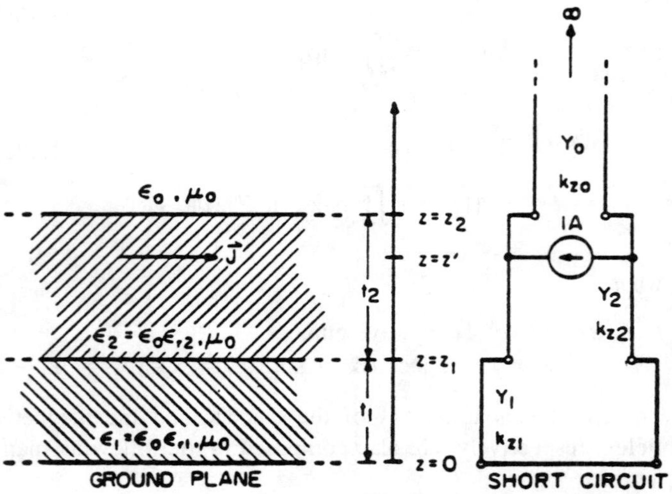

Fig. 11.15 Two-layered ground dielectric slab and its equivalent transmission line circuit representation.

As an example illustrating the solution of Eq. (11.21), consider the case of a two-layered slab backed by a ground plane. Figure 11.15 shows the physical configuration alongside its equivalent transmission line model. The dielectric layers are characterized by their permittivities ε_1 and ε_2 and respective thicknesses t_1 and t_2. The source point $z'$is located in the second layer (i.e., $z_1 \leq z' \leq z_2$). The modal Green's function for this example is given by

$$V(z, z') = \begin{cases} V_2(z_2, z')e^{jk_{z0}(z-z_2)} & z_2 \leq z \quad (11.28a) \\ V_2(z, z') & z_1 \leq z \leq z_2 \quad (11.28b) \\ V_2(z_1, z')\dfrac{\sin k_{z1}z}{\sin k_{z1}z_1} & 0 \leq z \leq z_1 \quad (11.28c) \end{cases}$$

$$I(z, z') = -\frac{1}{jk_z Z}\frac{d}{dz}V(z, z') \tag{11.28d}$$

where

$$V_2(z, z') = \frac{1}{\overleftrightarrow{Y}(z_1)}[\cos k_{z2}(z_< - z_1) + j\overleftarrow{Y}'(z_1)\sin k_{z2}(z_< - z_1)] \times [\cos k_{z2}(z_> - z_1) - j\overrightarrow{Y}'(z_1)\sin k_{z2}(z_> - z_1)] \tag{11.29}$$

$$\overrightarrow{Y}'(z_1) = \frac{Y_2 - jY_0 \cot k_{z2}t_2}{Y_0 - jY_2 \cot k_{z2}t_2} \qquad \overrightarrow{Y}(z_1) = Y_2\overrightarrow{Y}'(z_1)$$

$$\overleftarrow{Y}'(z_1) = -j\frac{Y_1}{Y_2}\cot k_{z1}t_1 \qquad \overleftarrow{Y}(z_1) = Y_2\overleftarrow{Y}'(z_1)$$

$$\overleftrightarrow{Y}(z_1) = \overleftarrow{Y}(z_1) + \overrightarrow{Y}(z_1)$$

$$k_{zq} = (\varepsilon_q k_0^2 - k_t^2)^{1/2} \qquad q = 1 \text{ or } 2$$

and

$$z_> = \max(z, z')$$

$$z_< = \min(z, z')$$

Note that Eqs. (11.28) and (11.29) are equally valid for both E– and H-modal coefficients. $V'(z, z')$ and $V''(z, z')$ are ob-

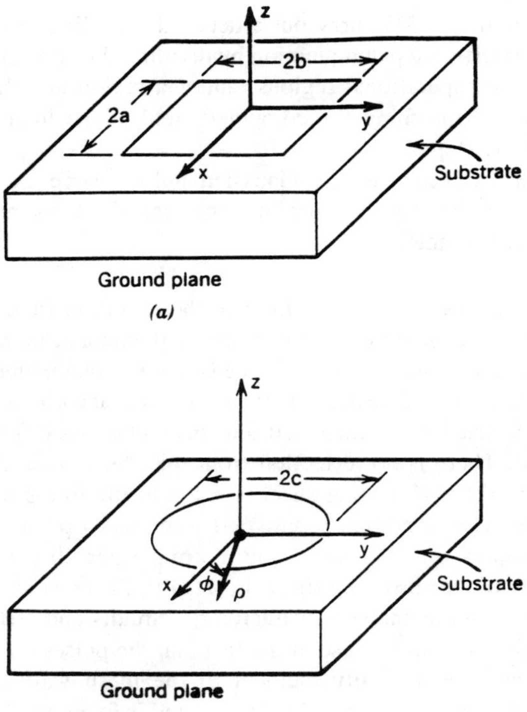

Fig. 11.16 (a) Rectangular patch geometry; (b) circular patch geometry.

tained, therefore, by substitution of the respective modal admittances Y_q, as defined in Eq. (11.21c), for the region with ε_{rq}, $q = 0, 1, 2$.

The basic functions employed in the Green's function moment-method solutions of the microstrip antenna problem, as indeed in all moment-method solutions, are of two varieties: entire domain or subsectional. As implied by the nomenclature, in the first case the unknowns are expanded in a complete set of basis functions, having the entire structure (e.g., entire patch or feedline) as the domain. In the second case, the currents are approximated piecewise by simple functions whose domains are considerably smaller than the structure itself and the wavelength.

Examples of the entire-domain basis functions are as follows.

a. Rectangular Patch (Fig. 11.16(a))

$$\text{(i) } J_x = \frac{\sin r\pi(x/a)}{\sqrt{a^2 - x^2}} \frac{\sin\left(s - \frac{1}{2}\right)\pi(y/b)}{\sqrt{b^2 - y^2}}, \quad r, s = 1, 2, \ldots \quad (11.30)$$

$$J_y = \frac{\cos(r - 1)\pi(x/a)}{\sqrt{a^2 - x^2}} \frac{\cos\left(s - \frac{1}{2}\right)\pi(y/b)}{\sqrt{b^2 - y^2}}, \quad r, s = 1, 2, \ldots$$

$$\text{(ii) } J_x = \sqrt{a^2 - x^2}\, U_m\left(\frac{x}{a}\right) \frac{T_n(y/b)}{\sqrt{b^2 - y^2}}, \quad m, n = 0, 1, 2, \ldots \quad (11.31)$$

$$J_y = \frac{T_m(x/a)}{\sqrt{a^2 - x^2}} \sqrt{b^2 - y^2}\, U_n(y/b), \quad m, n = 0, 1, 2, \ldots$$

where T_m and U_n are the Chebyshev polynomials of the first and second kind, respectively.

b. Circular patch (Fig. 11.16(b))

$$J_\rho = \sqrt{1 - (\rho/c)^2}\, U_\nu(\rho/c) \cos m\phi, \quad m, \nu = 0, 1, 2, \ldots; m + \nu = \text{odd} \quad (11.32)$$

$$J_\phi = \frac{1}{\sqrt{1 - (\rho/c)^2}} T_\zeta(\rho/c) \sin n\phi, \quad n, \zeta = 0, 1, 2, \ldots; n + \zeta = \text{odd}$$

In (11.30) through (11.32) the polynomials under the square root are included to fulfill the edge condition requirements.

An example of the subsectional basis function, which is widely used microstrip modeling work, is the piecewise sinusoidal (PWS) function

$$J_x = \frac{\sin k(\Delta - |x - x'|)}{\sin k\Delta} \mathbf{P}(y - y', w), \quad (11.33)$$
$$x' - \Delta < x < x' + \Delta$$

where

$$P(y - y', w) = \begin{cases} \frac{1}{2w}, & y' - w < y < y' + w \\ 0 & \text{elsewhere} \end{cases} \quad (11.34)$$

and k is an empirically determined constant. As $k \to 0$, the PWS mode approaches the so-called "roof-top" function

$$J_x = \left(1 - \frac{|x - x'|}{\Delta}\right) \mathbf{P}(y - y', w) \quad (11.34)$$

It should be noted that a few judiciously chosen entire-domain expansion functions usually suffice to represent the currents on the microstrip antenna accurately, in which case the MM matrix is of relatively low order. Use of subsectional bases leads to larger matrices. However, they are more versatile in their applicability to a much wider class of geometries.

The chief drawback of this method is the expense associated with computation of the improper integrals, which give the moment-method matrix elements. Careful analysis to reveal the possible sources of computational difficulty, followed by application of appropriate numerical methods, can increase significantly the speed and accuracy of numerical integration. A common first step is to analyze the Green's function in the complex plane, in order to characterize its singularities [11.27]. Of particular importance for microstrip antenna calculations are the pole singularities of the Green's function, corresponding to the surface-wave modes. In most microstrip problems, those are located on or close to the patch of integration (real k_t-axis). Straightforward numerical integration of functions possessing such singularities is time and cost intensive and, at times, unreliable. However, provided that the pole locations and residues are calculated, the well-known singularity subtraction technique can be applied to effectively remove the pole [11.29].

Another apparent difficulty in the evaluation of (11.26b) and (11.27b) is the infinite range of integration. Numerical treatment of the improper integrals usually involves truncation of the integration limits. The finite limits sufficient to achieve a certain degree of relative convergence greatly depend on the behavior of the integrands at infinity. To a certain extent, this depends on the choice of the basis and testing functions, whose transforms are involved in the integrands. Ultimately, other criteria for their selection may outweigh the numerical considerations. Again, simple numerical methods can be utilized to speed up convergence of the integrals. One standard technique is analogous to the singularity subtraction method. It consists of subtraction from the original integrand a function, which is asymptotically equivalent to the original integrand for large values of the arguments. The integration limits of the original integrals are thereby reduced, since the new integrand will quickly approach zero. The asymptotic function, which usually possesses a relatively simple form, can in many cases be integrated in closed form and subsequently added to the final result. This method has been applied successfully to accelerate infinite series as well as integrals [11.30, 11.31].

11.5.2 Microstrip Antenna Analysis Using Finite-Difference and Finite-Element Methods

In addition to the integral equation formulation for full-wave analysis of microstrip patches, finite-difference time-domain [11.48–11.51] and finite-element boundary-integral [11.52] methods have also been used for these antennas.

Finite-difference time-domain method. Finite-difference time-domain method is discussed in Papers 6.1 and 6.2, reprinted in this volume. An example of the application of the FD-TD method to a rectangular microstrip patch antenna is included in Paper 6.2.

Basic formulation of the FD-TD method (as a central difference discretization of Maxwell's curl equations in both time and space) is well known and is identical for circuit and antenna problems. However, details of the implementation reported by various authors differ in respect of excitation treatment, boundary conditions, and postprocessing of results to obtain frequency parameters of interest.

Excitation treatment. Reineix and Jecko [11.48] use a sine-modulated Gaussian voltage source with

$$V(t) = e^{-a^2(t-t_0)^2} \sin 2\pi f_0(t - t_0)$$

Parameters a, f_0, and t_0 are selected such that the signal spectrum covers the antenna response, and $V(t)$ is zero at $t = 0$. A localized electric field $E(t)$, generated by the excitation voltage, is given by

$$E(t) = \text{grad}\,[V(t) - R_g i(t)]$$

where R_g is the generator internal impedance and $i(t)$ is the current through the source. FD-TD simulation shows that after a very short transient, a constant oscillation period, corresponding to the antenna resonance, is established.

Sheen et al. [11.49] consider a microstrip line-fed rectangular patch. A Gaussian (in time) pulse of vertically oriented E-field under the strip, at the line-patch junction, is taken as the excitation source. After the Gaussian pulse has been fully launched, the absorbing boundary condition is switched on at the source wall.

Wu et al. [11.51] carry out a detailed modeling of coaxial feed to microstrip patch junction by dividing the configuration into two computational regions (antenna region and the feed region) and matching the computed field at the interface at every time step.

Other variations of the Gaussian pulse source excitation may as well be employed for implementing FD-TD simulation of microstrip antennas.

Boundary conditions. To simulate the radiation fields in an infinite space surrounding the microstrip antenna, an absorbing boundary condition (ABC) needs to be implemented carefully. Reineix and Jecko [11.48] have used absorbing sheets (with electric loss σ and a fictitious magnetic loss σ^*) for this purpose. There is no reflection from this sheet when the impedance of the absorbing sheet is equal to the free-space impedance. This condition is satisfied when $\mu_0 = \mu'$, $\varepsilon_0 = \varepsilon'$, and $\sigma^*/\mu_0 = \sigma/\varepsilon_0$, where μ' and ε' are permeability and permittivity of the absorbing sheet. In [11.49], Sheen et al. placed emphasis on the analysis of microstrip circuits and ABC employed by them makes use of the fact that the pulses on microstrip lines are normally incident to the mesh walls. Mur's [11.53] first approximate ABC is used and is found to allow for accurate simulations of the various microstrip circuits considered. Wu et al. [11.51] point out that because of the variations of microstrip phase velocity with frequency, one needs a dispersive boundary condition which can absorb fields in a wide frequency band. They use an extension of a technique originally proposed by Higdon [11.54] for wide-angle absorption of incident waves. The condition for absorbing plane waves traveling with velocity v_1 and v_2 is given by

$$\left(\frac{\partial}{\partial z} - \frac{1}{v_1}\frac{\partial}{\partial t}\right)\left(\frac{\partial}{\partial z} - \frac{1}{v_2}\frac{\partial}{\partial t}\right)E = 0$$

By concatenating several terms similar to the two on the left-hand side, the number of velocities at which absorption is optimized can be increased.

Frequency parameters of interest. Postprocessing for FD-TD simulation results is necessary to obtain frequency-domain parameters of interest for microstrip antennas. These parameters are input impedance for a single-feed antenna, S-parameters for antennas with more than one port, and far-field radiation pattern. Evaluation of network parameters from FD-TD simulation is discussed in Paper 6.2, included in this book.

Reineix and Jecko [11.48] use FD-TD simulation to obtain the electric current density $i(t)$ at each point on the patch as a function of time. Fourier transform yields current distribution as a function of frequency, which is used to calculate Hertzian potential. Far-zone radiated field is calculated from the Hertzian potential. A different procedure is used by Wu et al. [11.51]. They evaluate electric field $E(f)$ at the top surface of the substrate by transformation of time-domain E-field data. Equivalent surface magnetic-current distribution $M_s(f)$ is obtained from $E(f)$, and the free-space Green's function for a magnetic current source is used to obtain the radiation pattern.

Finite-element method applied to antennas. Papers 6.5 and 6.6 reprinted in this volume discuss applications of finite-element method to analysis of microstrip circuits. As in the case of FD-TD method, FEM can also be extended to antenna analysis by incorporating suitable absorbing boundary conditions for simulation of the infinite-external region.

A combination of the finite-element and boundary-integral methods has been used [11.55] for the electromagnetic characterization of the scattering by a three-dimensional cavity-backed aperture in an infinite ground plane. A system of equations is formulated for the solution of the aperture fields and those inside the cavity. Specifically, the finite element method is employed to formulate the fields in the cavity region, and the boundary-integral approach is used in conjunction with the equivalence principle to represent the fields above the ground plane. This general approach has been extended [11.52] to the analysis of microstrip patch antennas and arrays residing in a cavity recessed in a ground plane. Results for a single patch and also for a 3 × 3 array of patches are presented.

One can look forward to increased applications of finite-difference, finite-element, and boundary-element methods to some specific microstrip patch antenna configurations.

11.6 Examples of Microstrip Patch Antennas and Arrays

11.6.1 Two-Port Rectangular Patches

Two-port rectangular microstrip patches constitute elements of series-fed arrays [11.7, 11.32]. For the accurate design of the aperture phase and the amplitude distribution of the array, accurate computations of the two-port transmission characteristics of a single patch are needed. The usual design requirements for these two-port patches are a match at the input port ($S_{11} = 0$) and a prespecified value of S_{21} to ensure the required amplitude taper in the array. A multiport-network modeling approach for this configuration has been described in detail [11.14]. Figure 11.13 depicts an equivalent multiport network of the two-port patch. In this MNM model, EANs are added at the radiating edges only. Fringing fields at the non-radiating edges (where the input and output lines are located) are accounted for by using an effective width and effective dielectric constant for the patch. This modification simplifies the MNM model for a rectangular patch operating in the dominant mode.

The design of a two-port patch with given input match and given transmission coefficient also requires the optimization of the design parameters. This can be carried out by MNM approach discussed in Sec. 11.4. A very good starting point in the optimization process is the use of results based on the dominant-mode analysis [11.33, 11.7]. In this approximate analysis, only the dominant mode of the cavity is included in the analysis. The effects of fringing fields and the radiated power are included by using an effective length for the patch and an effective loss tangent for the substrate dielectric medium.

Sensitivity analysis using the incremental (finite difference) method has also been carried out for this configuration [11.34]. Rectangular patches fabricated on 1/64-inch-thick Duroid® substrates ($\varepsilon_r = 2.2$), with values of $|S_{21}|$ in the range 0 to 0.98, can be conveniently designed when the patch widths are in the range of 1 mm to 6 mm (at 7.5 GHz).

11.6.2 Two-Port Circular Microstrip Patches

Two-port circular microstrip patch (shown in Figure 11.17(a)) can also be used as an element of a series-fed arrays. An equivalent multiport network for this configuration is shown in Figure 11.17(b). In the MNM approach, the fields underneath the patch are computed by considering the patch as a two-dimensional circular resonator (denoted by "patch" in the figure) surrounded by a magnetic wall. The elements of the Z-matrix for this segment are computed by inserting the proper Green's function into Eq. (11.9). The Green's function for a circular planar segment has been reduced to a single summation [11.33, 11.35].

The effects of microstrip feed-junction reactances are included by modeling each of the feedlines as a rectangular planar segment (with effective width and effective dielectric constant) and connected to the "patch" segment at five ports. Furthermore, the radiated power from the patch and the effects of fringing fields at the open edges are accounted for by connecting an edge admittance network EAN to the open periphery. The periphery in this case is considered as one edge. The EAN consists of inductances, capacitances, and conductances. The fringing inductances are needed here because of the cosine distribution of the voltage along the circumference.

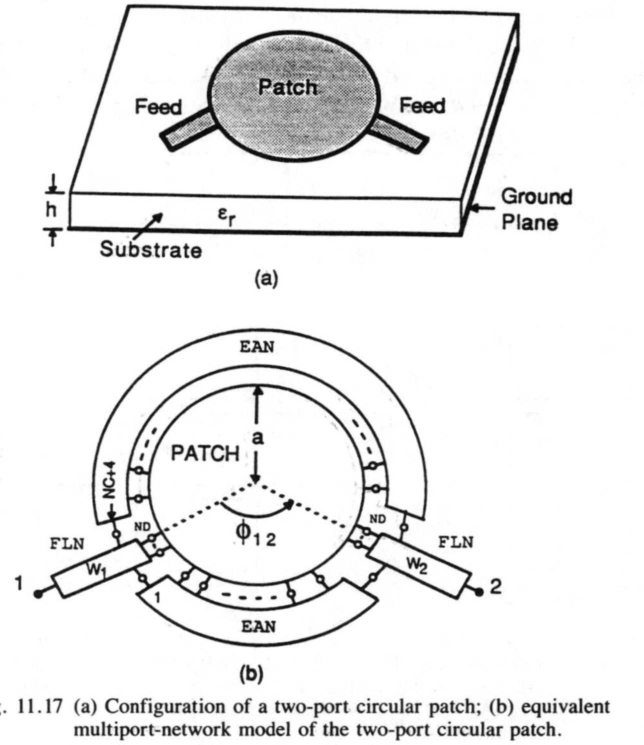

Fig. 11.17 (a) Configuration of a two-port circular patch; (b) equivalent multiport-network model of the two-port circular patch.

The number of ports NC (at the patch periphery where EAN is connected to the "patch") is chosen such that the angular width of each port is ≤ 8°.

It is possible to control the transmission coefficient by varying the angular separation between the input and output ports. However, to achieve a match at the input port, the characteristic impedances of the input and output lines have to be selected appropriately for every angular separation.

11.6.3 Wide-Band Multiple Resonator Microstrip Antennas

Microstrip patch radiators are well known to be narrowband antennas. One of the approaches to increase the bandwidth available from these antennas is the use of multiple resonators in the same plane (see Fig. 11.18). In [11.37], two configurations yielding broad bandwidth are proposed, namely, two nonradiating edges gap-coupled microstrip antennas (NEGCOMA) and four edges gap-coupled microstrip antennas (FEGCOMA). The bandwidth of microstrip patch antennas can be increased by using two radiating edges gap-coupled microstrip antennas (REGCOMA) [11.38]. The resonators consist of resonators of slightly different resonant lengths.

The multiport network modeling approach was used to analyze and to optimize the design of these configurations. The coupling gaps between the resonators are modeled as capacitive π-networks. Since the two edges are very close to each other, the capacitive coupling dominates, and it is not necessary to use an MCN. The series capacitance of the π-network is shunted by a conductance to account for the radiation from the gap. To account for the radiated power from the other edges of the patches, an edge admittance network is connected to these edges. The length of the parasitic elements, the gap width between the resonators, and the feed-point location have been optimized to yield wider impedance bandwidth (VSWR ≤2.0). The impedance bandwidths obtained using NEGCOMA

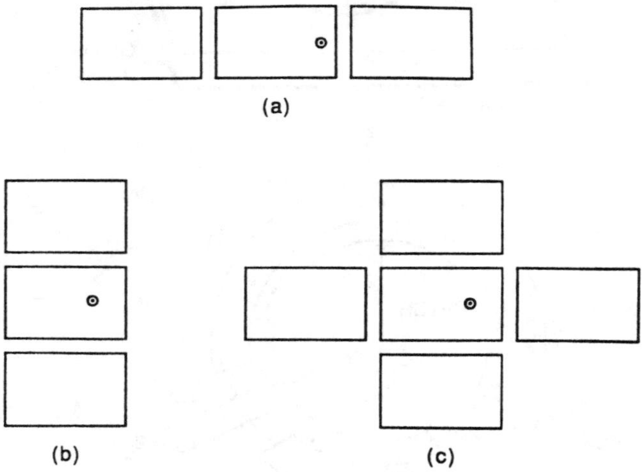

Fig. 11.18 (a) Configuration of a radiating edges gap-coupled microstrip patch antenna (REGCOMA), (b) configuration of a nonradiating edges gap-coupled microstrip patch antenna (NEGCOMA), (c) a four-edges gap-coupled microstrip patch antenna (FEGCOMA).

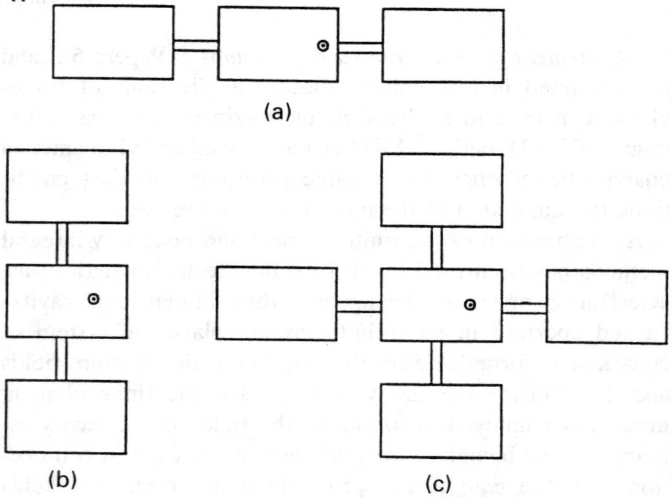

Fig. 11.19 (a) Configuration of a radiating edges directly coupled microstrip patch antenna (REDCOMA), (b) configuration of a nonradiating edges directly coupled microstrip patch antenna (NEDCOMA), (c) a four-edges directly coupled microstrip antenna (FEDCOMA).

and FEGCOMA are as large as four times and seven times that of a single rectangular patch, respectively. The bandwidth obtained from REGCOMA is as large as five times that of a single rectangular patch. The radiation patterns of these configurations vary over the bandwidth.

An alternative to the gap coupling is the direct coupling between resonators, using short sections of microstrip line [11.39] as shown in Fig. 11.19. These configurations have also been analyzed by using the multiport-network modeling approach. The short sections of microstrip line are modeled as rectangular planar segments. In this case also, the length of these segments is slightly larger than twice the substrate thickness to minimize the capacitive coupling through the gaps. The length of the additional resonators, the dimensions of the connecting strips, and the feed-point location have been optimized to obtain wider bandwidth. The MNM approach can be used also to optimize the antenna's configurations with respect to both the impedance bandwidth and the radiation characteristics simultaneously. Very good agreement between theory and experiment has been reported for the three configurations [11.37–11.39].

11.6.4 Circularly Polarized Antenna Elements

Rectangular and circular microstrip patches excited in a single dominant mode provide linearly polarized radiation. However, there are several different approaches that can be used for obtaining circular polarization (CP). A two-dimensional analysis for a class of CP configurations that use modifications of single-feed patches is discussed in [11.40]. The configurations used are nearly square, square with truncated opposite corners, and square with diagonal slot microstrip patches. These geometries are shown in Fig. 11.20. The multiport-network modeling approach has been used to analyze and optimize the design of these configurations. The desegmentation method has also been used to find the Z-matrix of the segment that

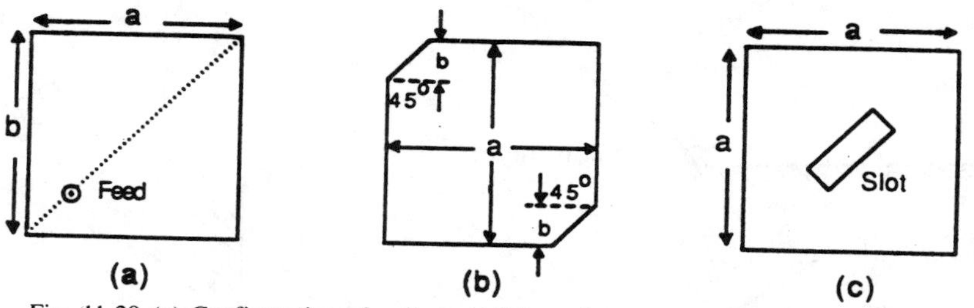

Fig. 11.20 (a) Configuration of a diagonal-fed nearly square patch, (b) configuration of a truncated-corners square patch, (c) configuration of a square patch with a diagonal slot.

models the internal fields of these patches. The fringing fields along the edges are included by extending the physical periphery of the patch outward to obtain a planar model with magnetic wall boundary. The radiated power is taken into account by connecting an edge admittance network that consists of conductance only. The MNM analysis yields the input impedance and the radiation characteristics.

The parameters of the antenna have been optimized to obtain the best axial ratio. It has been observed that an external matching network would be required to achieve a perfect match to a 50Ω feedline. Among the three types of antennas, the square patch antenna with a diagonal slot has the largest axial ratio bandwidth. The patch with truncated opposite corners has the best axial ratio but the least axial ratio bandwidth. The theoretical and the experimental results are found to be in reasonable agreement. Two other types of circularly polarized antennas, a square ring and a cross-shaped patch, have also been analyzed by the segmentation method [11.41].

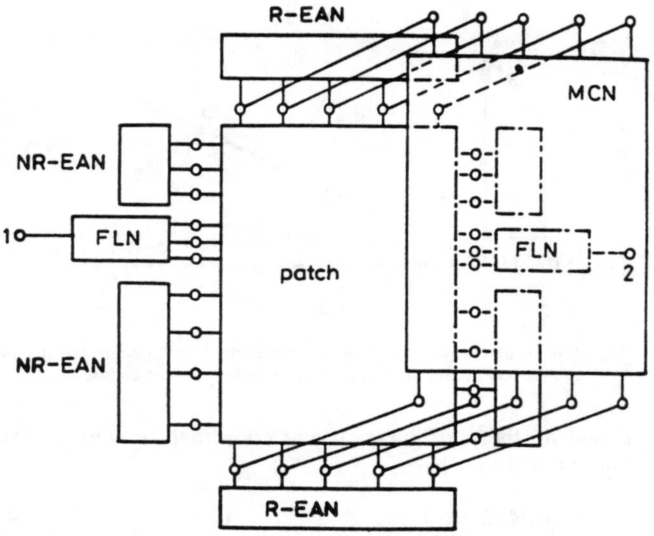

Fig. 11.A.1 Incorporation of mutual coupling in multiport-network model of a rectangular patch.

Appendix 11.A
Mutual Coupling Networks for Microstrip Patches

As discussed in Sec. 11.4.4, the mutual interaction between the fringing fields associated with any two edges of a microstrip patch can be expressed in terms of a network model. The basic idea of using an admittance element to represent mutual coupling between two radiating edges was initially presented in [11.42] in conjunction with the transmission line model of microstrip antennas. The multiport mutual-coupling network shown in Figure 11.A.1 is an extension of this concept and is suitable to be incorporated with the multiport-network model of microstrip patch antennas discussed in this chapter.

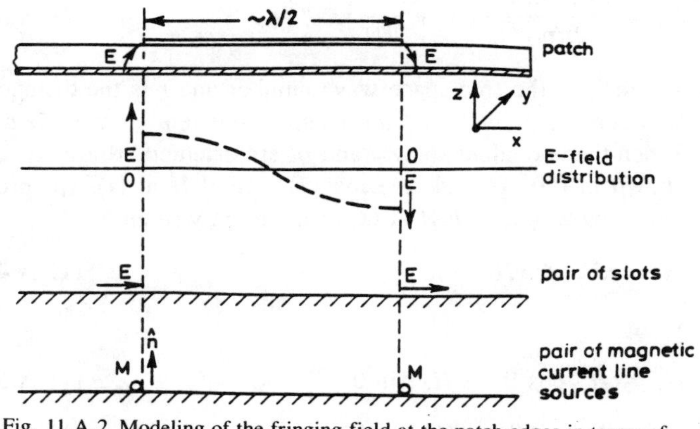

Fig. 11.A.2 Modeling of the fringing field at the patch edges in terms of magnetic-current line sources.

Evaluation of mutual-coupling networks. To compute the external mutual coupling between various edges of a microstrip patch antenna located on a thin substrate, the field at the edge is modeled by equivalent line sources of magnetic current. The equivalent sources are placed directly on the ground plane at the location of the edges. This is illustrated in Figure 11.A.2. Magnetic currents M are given by

$$\boldsymbol{M} = -\hat{\mathbf{n}} \times \mathbf{E}d \tag{11.A.1}$$

where d is the height of the substrate, the product $(-\mathbf{E}d)$ is the voltage $V(x, y)$ defined as $-\mathbf{E}_z(x, y)d$, and $\hat{\mathbf{n}}$ denotes a unit vector normal to the ground plane. Due to the indicated direction of the E-field, both magnetic-current sources $\boldsymbol{M}$ are along the $-y$-direction, that is, directed out of the plane of the paper.

The coupling between two magnetic-current line sources is evaluated by dividing each of the line sources into small sections, each of length dl. The magnetic field produced by each of these sections (on the line source 1) at the locations of the various sections on the line source 2 can be written by using fields of a magnetic-current dipole in free space [11.43].

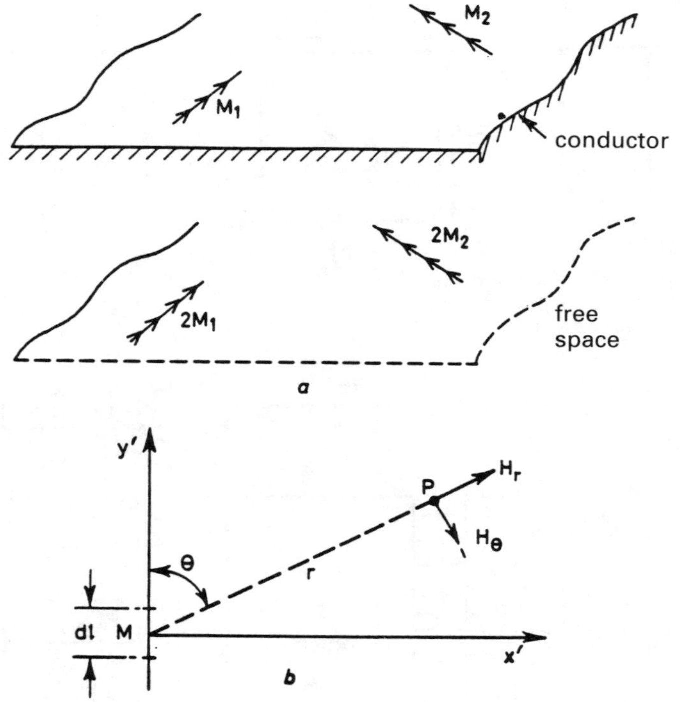

Fig. 11.A.3 (a) Two arbitrarily spaced magnetic current elements; (b) the coordinate system used for computation of fields.

Based on the configuration and coordinate system shown in Fig. 11.A.3, we have

$$H_\theta = j\frac{k_0 M\, dl \sin\theta}{4\pi\eta_0 r}\left[1 + \frac{1}{jk_0 r} - \frac{1}{(k_0 r)^2}\right]e^{-jk_0 r} \tag{11.A.2}$$

$$H_r = \frac{M\, dl \cos\theta}{2\pi\eta_0 r^2}\left(1 + \frac{1}{jk_0 r}\right)e^{-jk_0 r} \tag{11.A.3}$$

where k_0 is the free-space wave number and r is the distance between the point P and the magnetic-current element $M\ dl$. When the two edges, (say, i and j), are oriented arbitrarily, as shown in Fig. 11.A.4, the magnetic field **H** at (x_j, y_j), produced by a source $dl_i M$ at (x_i, y_i), can be written as

$$\mathbf{H} = \mathbf{x}H_x + \mathbf{y}H_y \tag{11.A.4}$$

with

$$H_y = H_{y'} \cos\theta_i - H_{x'} \sin\theta_i \tag{11.A.5}$$

$$H_x = H_{y'} \sin\theta_i + H_{x'} \cos\theta_i \tag{11.A.6}$$

where

$$H_{x'} = H_\theta \cos\theta + H_r \sin\theta \tag{11.A.7}$$

$$H_{y'} = -H_\theta \sin\theta + H_r \cos\theta \tag{11.A.8}$$

The (x, y) and (x', y') coordinate systems are illustrated in Fig. 11.A.4(b). The mutual admittance between sections j and i can be written in terms of the electric current density $\boldsymbol{J}_j$, (induced in the upper surface of the edge segment j) as follows:

$$j_j = \mathbf{n} \times \boldsymbol{H} = (-H_x \cos\delta_j - H_y \sin\delta_j)\mathbf{j} \tag{11.A.9}$$

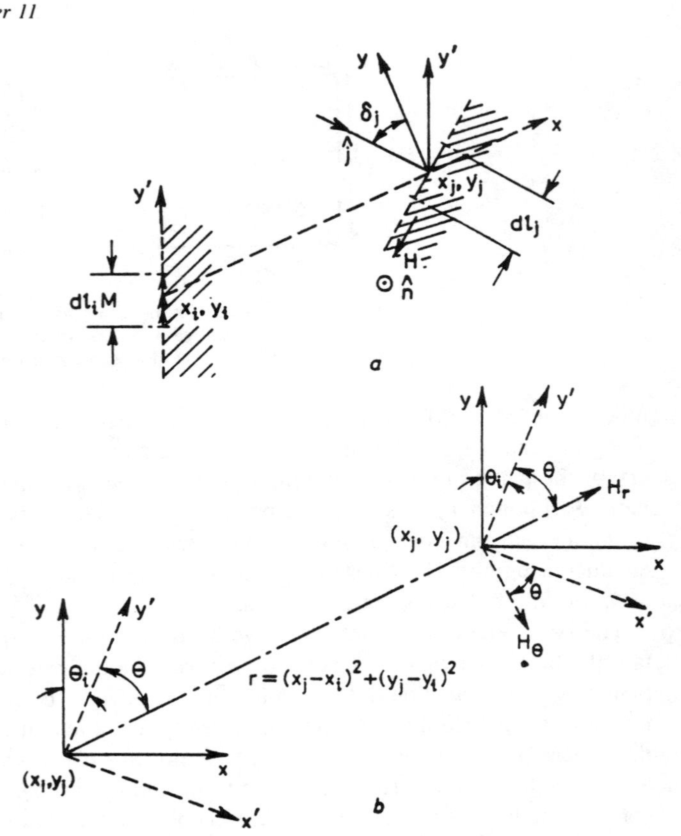

Fig. 11.A.4 (a) Configuration showing sections i and j of two patch edges; (b) two different coordinate systems used for mutual-coupling calculations.

The current density induced on the surface of the edge section j, underneath the patch, is $-\boldsymbol{J}_j$. The mutual admittance between sections i and j is given by the negative of the current flow into section j (underneath the patch) divided by the voltage at section i; that is,

$$Y_{ji} = -\left(-J_j \frac{dl_j}{M_i}\right) \tag{11.A.10}$$

The second minus sign in Eq. (11.A.10) accounts for the fact that the current used to define the admittance matrix is directed into the network as shown in Fig. 11.A.5. The two edges shown in Fig. 11.A.4(a) may be the edges of the same radiating patch or those of two different patches in an array. When coupling between two adjacent patches in an array environment is being computed, several individual edges of the two patches contribute to the mutual-coupling network. An MCN configuration taking the four radiating edges into account is shown in Fig. 11.A.6. Here, the MCN is connected to four ports along each radiating edge. In practice, the number of sections considered on each edge for mutual-coupling calculations is usually larger (typically, 12). However, while using MCN for antenna analysis (by segmentation), a small number of ports along radiating edges (typically, 4) is sufficient. Thus the original mutual admittance matrix (48 × 48 for 12 ports along each edge) is reduced to a smaller-sized matrix (16 × 16 as shown) by paralleling the ports in subgroups of three each. The contributions of nonradiating edges can also be incorporated in the MCN. Detailed computations

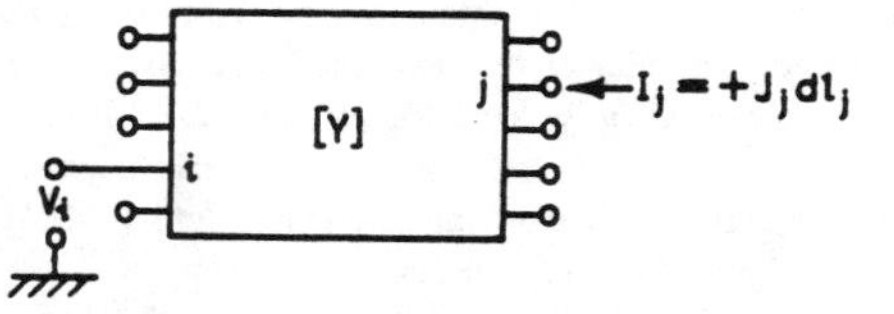

Fig. 11.A.5 Representation of the mutual coupling by an admittance matrix.

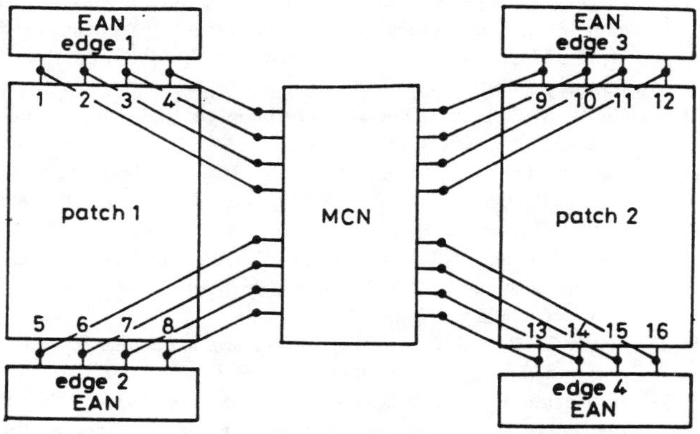

Fig. 11.A.6 A mutual-coupling network representing the coupling between two adjacent patches in an array.

[11.23], however, point out that the mutual-coupling contribution by nonradiating edges is usually small and, hence, may be ignored as a first-order approximation.

Mutual-coupling computations based on the foregoing formulation have been verified [11.23] by comparison with the available experimental results [11.44]. Some of these results are shown in Figure 11.A.7 for E-plane coupling and in Fig. 11.A.8 for H-plane coupling between two probe-fed rectangular patches. A very good agreement between the computation (solid line) and experimental results (diamond points) is evident. Also shown in these figures are results based on transmission line theory and obtained by Van Lil et al. [11.42].

It may be noted that the preceding method of evaluating mutual coupling (based on the equivalent magnetic current model

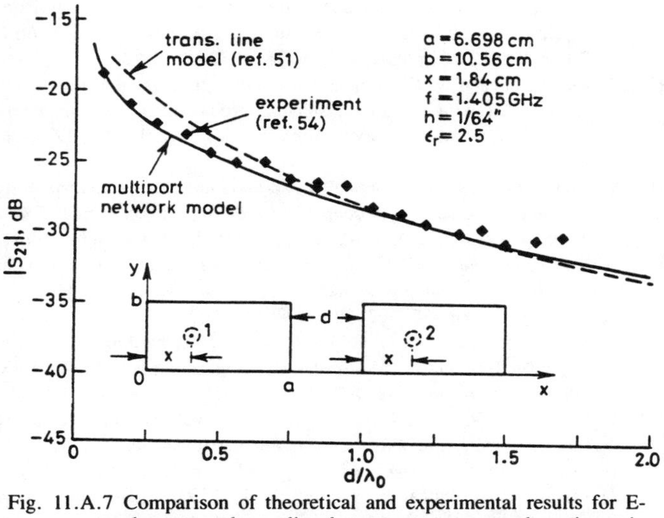

Fig. 11.A.7 Comparison of theoretical and experimental results for E-plane mutual coupling between two rectangular microstrip patches (from [11.53]).

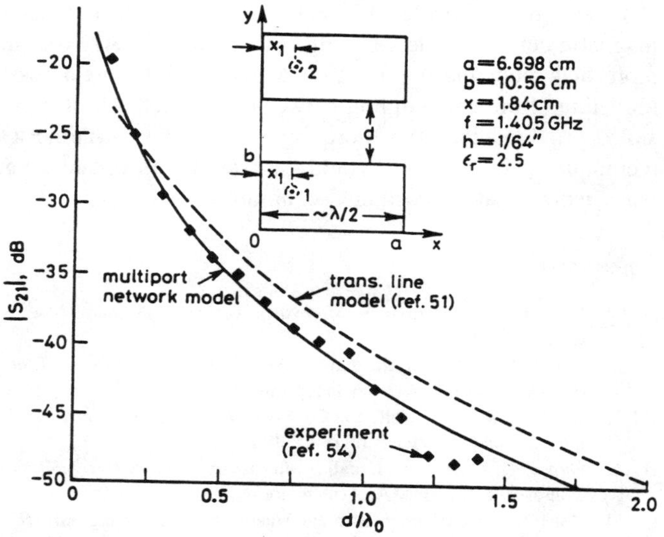

Fig. 11.A.8 Comparison of theoretical and experimental results for H-plane mutual coupling between two rectangular microstrip patches (from [11.53]).

shown in Fig. 11.A.2) is valid only for electrically thin substrates where the effect of surface waves along the substrate is negligible. This modeling approach has been recently extended to microstrip patches on thin substrates that have a relatively thick dielectric cover layer [11.45]. The equivalent magnetic current model used in this case is shown in Fig. 11.A.9. The basic approach is similar to that for the case without a cover layer, which was discussed earlier. Equations (11.A.2) and (11.A.3) are replaced by H_θ and H_r in the presence of the cover layer. These field components are now dominated by the effect of surface waves in the thicker cover layer.

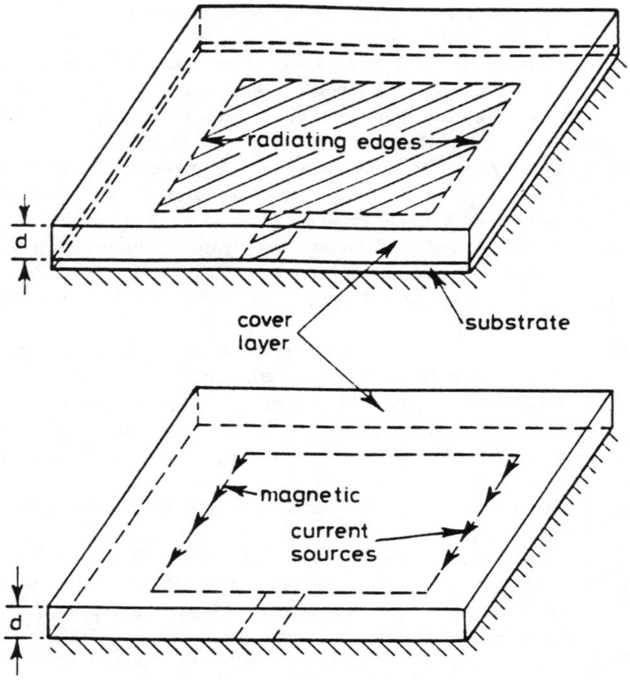

Fig. 11.A.9 Magnetic-current source model for computation of mutual coupling in case of microstrip patches on a thin substrate but covered with a thick dielectric layer.

When the substrate thickness is increased, equivalent magnetic-current models for Fig. 11.A.2 and 11.A.9 become more and more inaccurate. Conceptually, multiport-network modeling of mutual coupling between two patches is still possible, provided that the more rigorous analytical-numerical techniques [11.46, 11.47] could be extended to obtain a network representation of mutual coupling.

References

[11.1] Bahl, I. J. and P. Bhartia. *Microstrip Antennas.* Dedham, MA: Artech House, 1980.

[11.2] James, J. R., P. S. Hall, and C. Wood. *Microstrip Antenna Theory and Design.* London: Peter Peregrinus, 1981.

[11.3] James, J. R., P. S. Hall, and C. Wood. *Handbook of Microstrip Antennas.* London: Peter Peregrinus, 1989.

[11.4] Gupta, K. C. and A. Benalla. *Microstrip Antenna Design* Reprint Volume. Norwood, MA: Artech House, 1988.

[11.5] "Special issue on microstrip antennas." In D. C. Chang, ed., *IEEE Trans. Antennas Propagat.*, Vol. AP-29, No. 1, January 1981 (182 pages).

[11.6] Gupta, K. C. "Two-port transmission characteristics of rectangular microstrip patch radiators." *1985 IEEE AP-S Int'l. Antennas Propagat. Symp. Dig.*, pp. 71–74.

[11.7] Metzler, T. "Microstrip Series Arrays," *IEEE Trans. Antennas Propagat.*, Vol. AP-29, No. 1, pp. 174–178, January 1981.

[11.8] Benalla, A. and K. C. Gupta. "Transmission line model for 2-port rectangular microstrip patches with ports at the non-radiating edges. *Electronics, Letters* Vol. 23, pp. 882–884, August 1987.

[11.9] Benalla, A. and K. C. Gupta. "Two-dimensional analysis of one-port and two-port microstrip antennas," Electromagnetics Laboratory, Scientific Rept. 85. p. 48, University of Colorado (Boulder), May 1986.

[11.10] Lo, Y. T. et al. "Theory and experiment on microstrip antennas." *IEEE Trans. Antenna Propagat.*, Vol. AP-27, pp. 137–145, March 1979.

[11.11] Richards, W. F. et al. "An improved theory for microstrip antennas and applications." *IEEE Trans. Antenna Propagat.*, Vol. AP-29, pp. 38–46, January 1981.

[11.12] Coffey, E. L. and T. H. Lehman. "A new analysis technique for calculating the self and mutual impedance of microstrip antennas," Proc. Workshop Printed Circuit Antenna Technology, New Mexico State University (Albuquerque), pp. 33.1–33.21, 1979.

[11.13] Gupta, K. C. "Multiport-network modelling approach for computer-aided design of microstrip patches and arrays." *1987 IEEE AP-S Int'l. Symp. Antennas Propagat.*, Blacksburg, Virginia, June 1987.

[11.14] Benalla, A. and K. C. Gupta. "Multiport-network model and transmission characteristics of two-port rectangular microstrip antennas." *IEEE Trans. Antennas Propagat.*, Vol. AP-36, pp. 1337–1342, October 1988.

[11.15] Okoshi, T. *Planar Circuits for Microwaves and Lightwaves*, Appendix A, p. A2.1.2. Berlin: Springer-Verlag, 1985.

[11.16] Palanisamy, V. and R. Garg. "Analysis of arbitrary shaped microstrip patch antennas using segmentation technique and cavity model." *IEEE Trans. Antennas Propagat.*, Vol. AP-34, pp. 1208–1213, 1986.

[11.17] Palanisamy, V., and R. Garg. "Analysis of circularly polarized square ring and crossed-strip microstrip antennas." *IEEE Trans. Antenna Propagat.*, Vol. AP-34, pp. 1340–1346, 1986.

[11.18] Benalla, A. and K. C. Gupta. "Generalized edge-admittance network for modeling of radiation from microstrip patch antennas." National Radio Science Meeting, Boulder, Colorado, January 7–10, 1992.

[11.19] Kuester, E. F. et al. "The thin-substrate approximation for reflection from the end of a slab-loaded parallel plate waveguide with application to microstrip patch antenna." *IEEE Trans. Antennas Propagat*, Vol. AP-30, pp. 910–917, September 1982.

[11.20] Gogoi, A. and K. C. Gupta. "Wiener-Hopf computation of edge admittances for microstrip patch radiators, *Archiv für Elektronik und Übertragungstechnik*, Vol. 36, pp. 247–251, 1982.

[11.21] Chew, W. C. and J. A. Kong. "Microstrip capacitance for circular disk through matched asymptotic expansions." *SIAM J. Appl. Math.*, Vol. 42, pp. 302–317, 1982.

[11.22] Kuester, E. F. "Explicit approximation for the static capacitance of a microstrip patch of arbitrary shape." *J. Electromagnetic Waves Appl.*, Vol. 2, No. 1, pp. 103–135, 1987.

[11.23] Benalla, A. and K. C. Gupta. "Multiport network approach for modeling the mutual coupling effects in microstrip patch antennas and arrays." *IEEE Trans. Antennas Propagat.*, Vol. AP-37, pp. 148–152, February 1989.

[11.24] Gupta, K. C. et al., *Computer-Aided Design of Microwave Circuits.* Dedham, MA: Artech House, 1981.

[11.25] Chua, L. O. and P. M. Lin. *Computer-Aided Analysis of Electronic Circuits, Algorithms and Computational Techniques.* Englewood Cliffs, NJ: Prentice Hall, 1975.

[11.26] Lo, Y. T. et al., "Microstrip antenna." In Kai Chang, ed., *Handbook of Microwave and Optical Components*, Vol. 1, Chapter 13. New York: John Wiley & Sons, 1989.

[11.27] Felsen, L. B. and N. Marcuvitz. *Radiation and Scattering of Waves.* Englewood Cliffs, NJ: Prentice Hall, 1973.

[11.28] Harrington, R. F. *Field Computation by Moment Methods.* New York: Macmillan, 1968; reprinted: Melbourne, FL: R. E. Krieger, 1982.

[11.29] Davies, P. J. and P. Rabinowitz. *Method of Numerical Integration.* New York: Academic Press, 1975.

[11.30] Pozar, D. M. "Improved computational efficiency for the moment method solution of printed dipoles and patches." *Electromagnetics*, Vol. 3, Nos. 3–4, pp. 299–309, July–December 1983.

[11.31] Jackson, D. R. and N. G. Alexopoules. "An asymptotic extraction technique for evaluating sommerfeld-type integrals." *IEEE Trans. Antennas Propagat.*, Vol. AP-34, No. 12, pp. 1467–1470, December 1986.

[11.32] Wong, W. R. and D. L. Sengupta. "A class of broadband patch microstrip travelling wave antennas." *IEEE Trans. Antennas Propagat.*, Vol. AP-32, pp. 98–100, January 1984.

[11.33] Benalla, A. and K. C. Gupta. "Design procedure for linear series-fed arrays of microstrip patches covered with a thick dielectric layer." University of Colorado, Sci. Rept. No. 100, September 1989 (292 p.).

[11.34] Benalla, A. and K. C. Gupta. "A method for sensitivity analysis of series-fed arrays of rectangular microstrip patches." National Radio Science Meeting, p. 65, Boulder, Colorado, January 12–15, 1987.

[11.35] Benalla, A. and K. C. Gupta. "Analysis of two-port circular microstrip patch antennas using multiport network model." *Proc. 1989 Int'l. Symp. Antennas Propagat*, Tokyo, Japan, August 1989.

[11.36] Gupta, K. C. and A. Benalla. "Two-port transmission characteristics of circular microstrip patch antennas." *Digest IEEE AP-S Int'l. Symp. 1986* (Philadelphia, PA), pp. 821–824.

[11.37] Kumar, G. and K. C. Gupta. "Non-radiating edges and four edges gap-coupled with multiple resonator, broadband microstrip antennas." *IEEE Trans. Antennas Propagat.*, Vol. AP-33, pp. 173–178, 1985.

[11.38] Kumar, G. and K. C. Gupta. "Broad-band microstrip antennas using additional resonators gap-coupled to radiating edges." *IEEE Trans. Antennas Propagat.*, Vol. AP-32, pp. 1375–1379, December 1984.

[11.39] Kumar, G. and K. C. Gupta. "Directly coupled multiple resonator wide-band microstrip antennas." *IEEE Trans. Antennas Propagat.*, Vol. AP-33, pp. 588–593, June 1985.

[11.40] Sharma, P. C. and K. C. Gupta. "Analysis and optimized design of single feed circularly polarized microstrip antennas." *IEEE Trans. Antennas Propagat.*, Vol. AP-31, pp. 949–955, November 1983.

[11.41] Palanisamy, V. and R. Garg. "Analysis of circularly polarized square ring and square ring and cross-strip microstrip patches." *IEEE Trans. Antennas Propagat.*, Vol. AP-34, pp. 1340–1346, November 1986.

[11.42] Van Lil, E. H. and A. R. Van de Capelle. "Transmission line model for mutual coupling between microstrip antennas." *IEEE Trans. Antennas Propagat.*, Vol. AP-32, pp. 816–821, August 1984.

[11.43] Balanis, C. A. *Antenna Theory Analysis and Design*, p. 169. New York, Harper & Row, 1982.

[11.44] Jedlicka, R. P. et al. ''Measured mutual coupling between microstrip antennas.'' *IEEE Trans. Antennas Propagat.*, Vol. AP-29, pp. 147–149, January 1981.

[11.45] Tu, Y., K. C. Gupta, and D. C. Chang. ''Mutual coupling computations for rectangular microstrip patch antennas with a dielectric cover layer.'' National Radio Science Meeting, Boulder, Colorado, January 12–15, 1987.

[11.46] Pozar, D. M. ''Input impedance and mutual coupling of rectangular microstrip antennas.'' *IEEE Trans. Antennas Propagat.*, Vol. AP-30, pp. 1191–1196, November 1982.

[11.47] Jackson, D. R. et al. ''An exact coupling theory for microstrip patches.'' *1987 IEEE AP-S Int'l. Symp. Antennas Propagat. Dig.*, Vol. 2, pp. 790–793, 1987.

[11.48] Reineix, R. and B. Jecko. ''Analysis of microstrip patch antennas using finite difference time-domain method.'' *IEEE Trans. Antennas Propagat.*, Vol. 37, pp. 1361–1368, November 1989.

[11.49] Sheen, D. M. et al., ''Application of three-dimensional finite-difference time domain method to the analysis of planar microwave circuits.'' *IEEE Trans. Microwave Theory Tech.*, Vol. 38, pp. 849–857, July 1990.

[11.50] Wu, C. et al. ''Modeling of coaxial-fed microstrip patch antennas using finite-difference time-domain method.'' *Electronics Letters*, Vol. 27, No. 19, pp. 1691–1692, 1991.

[11.51] Wu, C. et al. ''Accurate characterization of planar printed antennas using finite-difference time-domain method.'' *IEEE Trans. Antennas Propagat.*, Vol. 40, No. 5, pp. 526–534, May 1992.

[11.52] Jin, J.-M. and J. L. Volakis. ''A hybrid finite element method for scattering and radiation by microstrip patch antennas and arrays residing in a cavity.'' *IEEE Trans. Antennas Propagat.*, Vol. 39, No. 11, pp. 1598–1604, November 1991.

[11.53] Mur, G. ''Absorbing boundary conditions for the finite difference approximation of the time domain electromagnetic field equations.'' *IEEE Trans. Electromagn. Compat.*, Vol. EMC-23, pp. 377–382, November 1981.

[11.54] Higdon, R. L. ''Numerical absorbing boundary conditions for the wave equation.'' *Math. Comput.*, Vol. 49, pp. 65–91, 1987.

[11.55] Jin, J.-M. and J. L. Volakis. ''A finite-element boundary-intragral formulation for scattering by three-dimensional cavity-backed apertures.'' *IEEE Trans. Antennas Propagat.*, Vol. 39, No. 1, pp. 97–104, January 1991.

Author Index

Subject Index

N

O

P

W

Y

Z

Editors' Biographies

Kuldip C. Gupta received the B.E. and M.E. degrees in Electrical Communication Engineering from the Indian Institute of Science, Bangalore, India, in 1961 and 1962, respectively, and the Ph.D. degree from Birla Institute of Technology and Science, Pilani, India, in 1969.

Dr. Gupta has been at the University of Colorado since 1983, initially as a Visiting Professor and later as a Professor. Presently, he is also the Research Coordinator for NSF IUCR Center for Microwave and Millimeter-Wave Computer-Aided Design at the University of Colorado. Earlier, he had a long stay (since 1969) at the Indian Institute of Technology, Kanpur, where he had been a Professor in Electrical Engineering since 1975. On leave from IITK, he has been a Visiting Professor at the University of Waterloo, Canada; at the Ecole Polytechnique Federale de Lausanne, Switzerland; at the Technical University of Denmark (Lyngby); at the Eidgenossische Technische Hochschule, Zurich; and at the University of Kansas, Lawrence. From 1971 to 1979 he was the Coordinator for the Phased Array Radar Group of the Advanced Center for Electronics Systems at the Indian Institute of Technology.

Dr. Gupta's current research interests are in the area of computer-aided design techniques for microwave and millimeter-wave integrated circuits and integrated antennas. He is author or coauthor of five books: *Microwave Integrated Circuits* (Wiley Eastern, 1974; Halsted Press of John Wiley, 1974), *Microstrip Lines and Slotlines* (Artech House, 1979), *Microwaves* (Wiley Eastern, 1979; Halsted Press of John Wiley, 1980; Editorial Limusa Mexico, 1983), *CAD of Microwave Circuits* (Artech House, 1981; Chinese Scientific Press, 1986; Radio i Syvaz, 1987), and *Microstrip Antenna Design* (Artech House, 1988). Also, he has contributed chapters to the *Handbook of Microstrip Antennas* (Peter Peregrinus, 1989), the *Handbook of Microwave and Optical Components,* Volume 1 (John Wiley, 1989), *Microwave Solid State Circuit Design* (John Wiley, 1988), and *Numerical Techniques for Microwave and Millimeter Wave Passive Structures* (John Wiley, 1989). Dr. Gupta has published over 130 research papers and holds two patents in the microwave area.

Dr. Gupta is a Fellow of IEEE and a fellow of the Institution of Electronics and Telecommunication Engineers (India). He is chairman of the MTT-S Technical Committee on CAD (MTT-1), and a member of the Committee on Microwave Field Theory (MTT-15), and serves on the Technical Program Committee for MTT-S International Symposia. He is the founding editor of *International Journal of Microwave and Millimeter-Wave Computer-Aided Engineering,* published by John Wiley. He is on the editorial boards of *IEEE MTT-S Transactions,* of *Microwave &*

Optical Technology Letters (John Wiley), and of three journals of IETE (India). He is listed in *Who's Who in America, Who's Who in the World,* and *Who's Who in American Education.*

Mohamed D. Abouzahra was born in Beirut, Lebanon on June 15, 1953. He received the B.S. Degree (with distinction) in electronics and communications from Cairo University, Cairo, Egypt, in 1976 and the M.S. and Ph.D. degrees in electrical engineering from the University of Colorado, Boulder, in 1978 and 1984, respectively.

From 1979 to 1984 he worked at the University of Colorado at Boulder as a research and teaching assistant. In 1984 Dr. Abouzahra joined the technical staff of the Massachusetts Institute of Technology, Lincoln Laboratory. Since 1991 he has been an assistant leader at the millimeter-wave radar station, which is located in the Kwajalein Atoll of the Marshall Islands. The millimeter-wave radar station is funded by the U.S. Army Kwajalein Atoll (USAKA) and is operated under the scientific direction of the M.I.T. Lincoln Laboratory.

Dr. Abouzahra's professional interests include radar systems, radar scattering, computer-aided microwave and millimeter-wave measurements, planar circuits analysis, and applied mathematics. Dr. Abouzahra holds one patent in the microwave area and has authored or coauthored over 60 technical publications. He has also coauthored a contributing chapter in *Numerical Techniques for Microwave and Millimeter Wave Passive Structures* (John Wiley, 1989) and three chapters in *Properties of Polylogarithms* (American Mathematical Society, 1991).

Dr. Abouzahra is a senior member of IEEE, and a member in the Institute for Systems and Components of the Electromagnetic Academy, and serves as a member of the editorial board of the *IEEE Transactions on Microwave Theory and Techniques,* and the *International Journal on Microwave and Millimeter-Wave Computer-Aided Engineering.*